T0133609

Electromagnetics

Third Edition

Electromagnetics
Third Edition

Edward J. Rothwell

Michael J. Cloud

CRC Press
Taylor & Francis Group
Boca Raton London New York

CRC Press is an imprint of the
Taylor & Francis Group, an **informa** business

CRC Press
Taylor & Francis Group
6000 Broken Sound Parkway NW, Suite 300
Boca Raton, FL 33487-2742

First issued in paperback 2022

© 2018 by Taylor & Francis Group, LLC
CRC Press is an imprint of Taylor & Francis Group, an Informa business

No claim to original U.S. Government works

ISBN-13: 978-1-498-79656-9 (hbk)
ISBN-13: 978-1-03-233917-7 (pbk)
DOI: 10.1201/9781351057554

This book contains information obtained from authentic and highly regarded sources. Reasonable efforts have been made to publish reliable data and information, but the author and publisher cannot assume responsibility for the validity of all materials or the consequences of their use. The authors and publishers have attempted to trace the copyright holders of all material reproduced in this publication and apologize to copyright holders if permission to publish in this form has not been obtained. If any copyright material has not been acknowledged please write and let us know so we may rectify in any future reprint.

Except as permitted under U.S. Copyright Law, no part of this book may be reprinted, reproduced, transmitted, or utilized in any form by any electronic, mechanical, or other means, now known or hereafter invented, including photocopying, microfilming, and recording, or in any information storage or retrieval system, without written permission from the publishers.

For permission to photocopy or use material electronically from this work, please access www.copyright.com (http://www.copyright.com/) or contact the Copyright Clearance Center, Inc. (CCC), 222 Rosewood Drive, Danvers, MA 01923, 978-750-8400. CCC is a not-for-profit organization that provides licenses and registration for a variety of users. For organizations that have been granted a photocopy license by the CCC, a separate system of payment has been arranged.

Trademark Notice: Product or corporate names may be trademarks or registered trademarks, and are used only for identification and explanation without intent to infringe.

Publisher's Note

The publisher has gone to great lengths to ensure the quality of this reprint but points out that some imperfections in the original copies may be apparent.

Visit the eResources: https://www.crcpress.com/9781498796569

Visit the Taylor & Francis Web site at
http://www.taylorandfrancis.com

and the CRC Press Web site at
http://www.crcpress.com

To Professor Kun-Mu Chen: Thank you for three decades of guidance, support, encouragement, and inspiration.

Contents

Preface

This is the third edition of our book *Electromagnetics*. It is intended as a text for use by engineering graduate students in a first-year sequence where the basic concepts learned as undergraduates are solidified and a conceptual foundation is established for future work in research. It should also prove useful for practicing engineers who wish to improve their understanding of the principles of electromagnetics, or brush up on those fundamentals that have become cloudy with the passage of time.

The assumed background of the reader is limited to standard undergraduate topics in physics and mathematics. These include complex arithmetic, vector analysis, ordinary differential equations, and certain topics normally covered in a "signals and systems" course (e.g., convolution and the Fourier transform). Further analytical tools, such as contour integration, dyadic analysis, and separation of variables, are covered in a self-contained mathematical appendix.

The organization of the book, as with the second edition, is in seven chapters. In Chapter 1 we present essential background on the field concept, as well as information related specifically to the electromagnetic field and its sources. Chapter 2 is concerned with a presentation of Maxwell's theory of electromagnetism. Here attention is given to several useful forms of Maxwell's equations, the nature of the four field quantities and of the postulate in general, some fundamental theorems, and the wave nature of the time-varying field. The electrostatic and magnetostatic cases are treated in Chapter 3, and an introduction to quasistatics is provided. In Chapter 4 we cover the representation of the field in the frequency domains: both temporal and spatial. The behavior of common engineering materials is also given some attention. The use of potential functions is discussed in Chapter 5, along with other field decompositions including the solenoidal–lamellar, transverse–longitudinal, and TE–TM types. In Chapter 6 we present the powerful integral solution to Maxwell's equations by the method of Stratton and Chu. Finally, in Chapter 7 we provide introductory coverage of integral equations and discuss how they may be used to solve a variety of problems in electromagnetics, including several classic problems in radiation and scattering. A main mathematical appendix near the end of the book contains brief but sufficient treatments of Fourier analysis, vector transport theorems, complex-plane integration, dyadic analysis, and boundary value problems. Several subsidiary appendices provide useful tables of identities, transforms, and so on.

The third edition of *Electromagnetics* includes a large amount of new material. Most significantly, each chapter (except the first) now has a culminating section titled "Application." The material introduced in each chapter is applied to an area of some practical interest, to solidify concepts and motivate their study. In Chapter 2 we examine particle motion in static electric and magnetic fields, with applications to electron guns and cathode ray tubes. Also included is a bit of material on relativistic motion to demonstrate that knowledge of relativity can be important in engineering as well as physics. We take a detailed look at using structures to shield static and quasistatic electric and magnetic fields in Chapter 3, and compute shielding effectiveness for several canonical problems. In Chapter 4 we examine the important problem of material characterization, concentrating on planar structures. We show how measured data can be used to determine ϵ and μ, and describe several well-known techniques for extracting the electromagnetic parameters of planar samples using reflection and transmission measurements. In Chapter 5 we look at waveguides and transmission lines from an electromagnetic perspective. We treat many classic structures such as rectangular, circular, and triangular waveguides, coaxial cables, fiber-optic cables, slab waveguides, strip transmission lines, horn waveguides, and conical transmission lines. We also examine waveguides filled with ferrite and partially filled with dielectrics. In Chapter 6 we investigate the properties of antennas, concentrating on characteristics such as gain, pattern, bandwidth, and radiation resistance. We treat both wire antennas, such as dipoles and loops, and aperture antennas, such as dish antennas. Finally, in Chapter 7 we revisit shielding and study the penetration of electromagnetic waves through ground plane apertures by solving integral equations. The material on the impedance properties of antennas, covered earlier in Chapter 7, is augmented by computing the radiation properties of slot antennas using the method of moments.

We have added many new examples, and now set them off in a smaller typeface, indented and enclosed by solid arrows ▶ ◀. This makes them easier to locate. A large fraction of the new examples are numerical, describing practical situations the reader may expect to encounter. Many include graphs or tables exploring solution behavior as relevant parameters are varied, thereby providing significant visual feedback. There are 247 numbered examples in the third edition, including several in the mathematical appendix.

Several other new topics are covered in the third edition. A detailed section on electro- and magneto-quasistatics has been added to Chapter 3. This allows the concepts of capacitance and inductance to be introduced within the context of time-changing fields without having to worry about solving Maxwell's equations. The diffusion equation is derived, and skin depth and internal impedance are examined — quantities crucial for understanding shielding at low frequencies. New material on the direct solution to Gauss' law and Ampere's law is also added to Chapter 3. The section on plane wave propagation in layered media, covered in Chapter 4, is considerably enhanced. The wave-matrix method for obtaining the reflected and transmitted plane-wave fields in multi-layered media is described, and many new canonical problems are examined that are used in the section on material characterization. In Chapter 5, the material on TE-TM decomposition of the electromagnetic fields is extended to anisotropic materials. The decomposition is applied later in the chapter to analyze the fields in a waveguide filled with a magnetized ferrite. Also new in Chapter 5 is a description of the solenoidal-lamellar decomposition of solutions to the vector wave equation and vector spherical wave functions. The guided waves application section of Chapter 5 includes new material on cascaded waveguide systems, including using mode matching to determine the reflection and transmission of higher-order modes at junctions between differing waveguides. Finally, new material on the application of Fourier transforms to partial differential equations has been added to the mathematical appendix. Solutions to Laplace's equation are examined, and the Fourier transform representation of the three-dimensional Green's function for the Helmholtz equation is derived.

We have also made some minor notational changes. We represented complex permittivity and permeability in the previous editions as $\tilde{\epsilon} = \tilde{\epsilon}' + j\tilde{\epsilon}''$ and $\tilde{\mu} = \tilde{\mu}' + j\tilde{\mu}''$, with $\tilde{\epsilon}'' \leq 0$ and $\tilde{\mu}'' \leq 0$ for passive materials. Some readers found this objectionable, since it is more common to write, for instance, $\tilde{\epsilon} = \tilde{\epsilon}' - j\tilde{\epsilon}''$ with $\tilde{\epsilon}'' \geq 0$ for passive materials. To avoid additional confusion we now write $\tilde{\epsilon} = \operatorname{Re}\tilde{\epsilon} + j\operatorname{Im}\tilde{\epsilon}$, where $\operatorname{Im}\tilde{\epsilon} \leq 0$ for passive materials. We find this notation unambiguous, although perhaps a bit cumbersome. There were instances in the first two editions where we wrote a wavenumber as $k = k' + jk''$. We now use the unambiguous $k = \operatorname{Re}k + j\operatorname{Im}k$. We have transposed the term *Kronig–Kramers*, used in previous editions, to *Kramers–Kronig*, which is in more common use. Finally, we have changed the terms *Lorentz condition* and *Lorentz gauge* to *Lorenz condition* and *Lorenz gauge*; although Hendrik Antoon Lorentz has long been associated with the equation $\nabla \cdot \mathbf{A} = -\mu\epsilon\partial\phi/\partial t$, it is now recognized that this equation originated with Ludvig Valentin Lorenz [93, 204].

We would like to express our deep gratitude to colleagues who contributed to the development of the book. The reciprocity-based derivation of the Stratton–Chu formula was provided by Dennis P. Nyquist, as was the material on wave reflection from multiple layers. The groundwork for our discussion of the Kramers–Kronig relations was laid by Michael Havrilla. Material on the time-domain reflection coefficient was developed by Jungwook Suk. Korede Oladimeji contributed to the work on field penetration through slots, while Raenita Fenner provided material on error analysis for material characterization. We owe thanks to Leo Kempel, David Infante, and Ahmet Kizilay for carefully reading large portions of the first edition manuscript, and to Christopher Coleman for helping to prepare the figures. Benjamin Crowgey assisted us by reading the new application section on material characterization, and by providing material on the TE–TM decomposition of fields in anisotropic media. We are indebted to John E. Ross for kindly permitting us to employ one of his computer programs for scattering from a sphere and another for numerical Fourier transformation. Michael I. Mishchenko and Christophe Caloz offered valuable feedback and encouragement based on their experiences with the second edition. Finally, we would like to thank Kyra Lindholm, Michele Dimont, and Nora Konopka of Taylor & Francis for their guidance and support throughout the publication process, and John Gandour for designing the cover.

Authors

Edward J. Rothwell received the BS degree from Michigan Technological University, the MS and EE degrees from Stanford University, and the PhD from Michigan State University, all in electrical engineering. He has been a faculty member in the Department of Electrical and Computer Engineering at Michigan State University since 1985, and currently holds the Dennis P. Nyquist Professorship in Electromagnetics. Before coming to Michigan State he worked at Raytheon and Lincoln Laboratory. Dr. Rothwell has published numerous articles in professional journals involving his research in electromagnetics and related subjects. He is a member of Phi Kappa Phi, Sigma Xi, URSI Commission B, and is a Fellow of the IEEE.

Michael J. Cloud received the BS, MS, and PhD degrees from Michigan State University, all in electrical engineering. He has been a faculty member in the Department of Electrical and Computer Engineering at Lawrence Technological University since 1987, and currently holds the rank of associate professor. Dr. Cloud has coauthored twelve other books. He is a senior member of the IEEE.

1

Introductory concepts

1.1 Notation, conventions, and symbology

Any book that covers a broad range of topics will likely harbor some problems with notation and symbology. This results from having the same symbol used in different areas to represent different quantities, and also from having too many quantities to represent. Rather than invent new symbols, we choose to stay close to the standards and warn the reader about any symbol used to represent more than one distinct quantity.

The basic nature of a physical quantity is indicated by typeface or by the use of a diacritical mark. Scalars are shown in ordinary typeface: q, Φ, for example. Vectors are shown in boldface: $\mathbf{E}, \mathbf{\Pi}$. Dyadics are shown in boldface with an overbar: $\bar{\epsilon}, \bar{\mathbf{A}}$. Frequency-dependent quantities are indicated by a tilde, whereas time-dependent quantities are written without additional indication; thus we write $\tilde{\mathbf{E}}(\mathbf{r}, \omega)$ and $\mathbf{E}(\mathbf{r}, t)$. (Some quantities, such as impedance, are used in the frequency domain to interrelate Fourier spectra; although these quantities are frequency dependent they are seldom written in the time domain, and hence we do not attach tildes to their symbols.) We often combine diacritical marks: for example, $\tilde{\bar{\epsilon}}$ denotes a frequency domain dyadic. We distinguish carefully between phasor and frequency domain quantities. The variable ω is used for the frequency variable of the Fourier spectrum, while $\check{\omega}$ is used to indicate the constant frequency of a time harmonic signal. We thus further separate the notion of a phasor field from a frequency domain field by using a *check* to indicate a phasor field: $\check{\mathbf{E}}(\mathbf{r})$. However, there is often a simple relationship between the two, such as $\check{\mathbf{E}} = \tilde{\mathbf{E}}(\check{\omega})$.

We designate the field and source point position vectors by $\mathbf{r}$ and $\mathbf{r}'$, respectively, and the corresponding relative displacement or distance vector by $\mathbf{R}$:

$$\mathbf{R} = \mathbf{r} - \mathbf{r}'.$$

A hat designates a vector as a unit vector (e.g., $\hat{\mathbf{x}}$). The sets of coordinate variables in rectangular, cylindrical, and spherical coordinates are denoted by

$$(x, y, z),$$
$$(\rho, \phi, z),$$
$$(r, \theta, \phi),$$

respectively. (In the spherical system ϕ is the azimuthal angle and θ is the polar angle.) We freely use the "del" operator notation ∇ for gradient, curl, divergence, Laplacian, and so on.

The SI (MKS) system of units is employed throughout the book.

1.2   The field concept of electromagnetics

Introductory treatments of electromagnetics often stress the role of the field in force transmission: the individual fields **E** and **B** are defined via the mechanical force on a small test charge. This is certainly acceptable, but does not tell the whole story. We might, for example, be left with the impression that the EM field always arises from an interaction between charged objects. Often coupled with this is the notion that the field concept is meant merely as an aid to the calculation of force, a kind of notational convenience not placed on the same physical footing as force itself. In fact, fields are more than useful — they are fundamental. Before discussing electromagnetic fields in more detail, let us attempt to gain a better perspective on the field concept and its role in modern physical theory. Fields play a central role in any attempt to describe physical reality. They are as real as the physical substances we ascribe to everyday experience. In the words of Einstein [55],

> "It seems impossible to give an obvious qualitative criterion for distinguishing between matter and field or charge and field."

We must therefore put fields and particles of matter on the same footing: both carry energy and momentum, and both interact with the observable world.

1.2.1   Historical perspective

Early nineteenth century physical thought was dominated by the *action at a distance* concept, formulated by Newton more than 100 years earlier in his immensely successful theory of gravitation. In this view the influence of individual bodies extends across space, instantaneously affects other bodies, and remains completely unaffected by the presence of an intervening medium. Such an idea was revolutionary; until then *action by contact*, in which objects are thought to affect each other through physical contact or by contact with the intervening medium, seemed the obvious and only means for mechanical interaction. Priestly's experiments in 1766 and Coulomb's torsion-bar experiments in 1785 seemed to indicate that the force between two electrically charged objects behaves in strict analogy with gravitation: both forces obey inverse square laws and act along a line joining the objects. Oersted, Ampere, Biot, and Savart soon showed that the magnetic force on segments of current-carrying wires also obeys an inverse square law.

The experiments of Faraday in the 1830s placed doubt on whether action at a distance really describes electric and magnetic phenomena. When a material (such as a dielectric) is placed between two charged objects, the force of interaction decreases; thus, the intervening medium does play a role in conveying the force from one object to the other. To explain this, Faraday visualized "lines of force" extending from one charged object to another. The manner in which these lines were thought to interact with materials they intercepted along their path was crucial in understanding the forces on the objects. This also held for magnetic effects. Of particular importance was the number of lines passing through a certain area (the *flux*), which was thought to determine the amplitude of the effect observed in Faraday's experiments on electromagnetic induction.

Faraday's ideas presented a new world view: electromagnetic phenomena occur in the region surrounding charged bodies, and can be described in terms of the laws governing the "field" of his lines of force. Analogies were made to the stresses and strains in material objects, and it appeared that Faraday's force lines created equivalent electromagnetic stresses and strains in media surrounding charged objects. His law of induction was formulated not in terms of positions of bodies, but in terms of lines of magnetic force. Inspired by Faraday's ideas, Gauss restated Coulomb's law in terms of flux lines, and Maxwell extended the idea to time-changing fields through his concept of displacement current.

In the 1860s Maxwell created what Einstein called "the most important invention since Newton's time" — a set of equations describing an entirely field-based theory of electromagnetism. These equations do not model the forces acting between bodies, as do Newton's law of gravitation and Coulomb's law, but rather describe only the dynamic, time-evolving structure of the electromagnetic field. Thus, bodies are not seen to interact with each other, but rather with the (very real) electromagnetic field they create, an interaction described by a supplementary equation (the Lorentz force law). To better understand the interactions in terms of mechanical concepts, Maxwell also assigned properties of stress and energy to the field.

Using constructs that we now call the electric and magnetic fields and potentials, Maxwell synthesized all known electromagnetic laws and presented them as a system of differential and algebraic equations. By the end of the nineteenth century, Hertz had devised equations involving only the electric and magnetic fields, and had derived the laws of circuit theory (Ohm's law and Kirchhoff's laws) from the field expressions. His experiments with high-frequency fields verified Maxwell's predictions of the existence of electromagnetic waves propagating at finite velocity, and helped solidify the link between electromagnetism and optics. But one problem remained: if the electromagnetic fields propagated by stresses and strains on a medium, how could they propagate through a vacuum? A substance called the *luminiferous aether*, long thought to support the transverse waves of light, was put to the task of carrying the vibrations of the electromagnetic field as well. However, the pivotal experiments of Michelson and Morely showed that the aether was fictitious, and the physical existence of the field was firmly established.

The essence of the field concept can be conveyed through a simple thought experiment. Consider two stationary charged particles in free space. Since the charges are stationary, we know that (1) another force is present to balance the Coulomb force between the charges, and (2) the momentum and kinetic energy of the system are zero. Now suppose one charge is quickly moved and returned to rest at its original position. Action at a distance would require the second charge to react immediately (Newton's third law), but by Hertz's experiments it does not. There appears to be no change in energy of the system: both particles are again at rest in their original positions. However, after a time (given by the distance between the charges divided by the speed of light) we find that the second charge does experience a change in electrical force and begins to move away from its state of equilibrium. But by doing so it has gained net kinetic energy and momentum, and the energy and momentum of the system seem larger than at the start. This can only be reconciled through field theory. If we regard the field as a physical entity, then the nonzero work required to initiate the motion of the first charge and return it to its initial state can be seen as increasing the energy of the field. A disturbance propagates at finite speed and, upon reaching the second charge, transfers energy into kinetic energy of the charge. Upon its acceleration this charge also sends out a wave of field disturbance, carrying energy with it, eventually reaching the first charge and creating a second reaction. At any given time, the net energy and momentum of the system, composed of both the bodies and the field, remain constant. We thus come to regard the electromagnetic field as a true physical entity: an entity capable of carrying energy and momentum.

1.2.2   Formalization of field theory

Before we can invoke physical laws, we must find a way to describe the *state* of the system we intend to study. We generally begin by identifying a set of *state variables* that can depict the physical nature of the system. In a mechanical theory such as Newton's law of gravitation, the state of a system of point masses is expressed in terms of the instantaneous positions and momenta of the individual particles. Hence $6N$ state variables are needed to describe the state of a system of N particles, each particle having three position coordinates and three momentum components. The time evolution of the system state is determined by a supplementary force function (e.g., gravitational attraction), the initial state (initial conditions), and Newton's second law $\mathbf{F} = d\mathbf{P}/dt$.

Descriptions using finite sets of state variables are appropriate for action-at-a-distance interpretations of physical laws such as Newton's law of gravitation or the interaction of charged particles. If Coulomb's law were taken as the force law in a mechanical description of electromagnetics, the state of a system of particles could be described completely in terms of their positions, momenta, and charges. Of course, charged particle interaction is not this simple. An attempt to augment Coulomb's force law with Ampere's force law would not account for kinetic energy loss via radiation. Hence we abandon* the mechanical viewpoint in favor of the field viewpoint, selecting a different set of state variables. The essence of field theory is to regard electromagnetic phenomena as affecting all of space. We shall find that we can describe the field in terms of the four vector quantities $\mathbf{E}$, $\mathbf{D}$, $\mathbf{B}$, and $\mathbf{H}$. Because these fields exist by definition at each point in space and each time t, a finite set of state variables cannot describe the system.

Here then is an important distinction between field theories and mechanical theories: the state of a field at any instant can only be described by an infinite number of state variables. Mathematically we describe fields in terms of functions of continuous variables; however, we must be careful not to confuse all quantities described as "fields" with those fields innate to a scientific field theory. For instance, we may refer to a temperature "field" in the sense that we can describe temperature as a function of space and time. However, we do *not* mean by this that temperature obeys a set of physical laws analogous to those obeyed by the electromagnetic field.

What special character, then, can we ascribe to the electromagnetic field that has meaning beyond that given by its mathematical implications? In this book, $\mathbf{E}$, $\mathbf{D}$, $\mathbf{B}$, and $\mathbf{H}$ are integral parts of a *field-theory description* of electromagnetics. In any field theory we need two types of fields: a *mediating field* generated by a source, and a field describing the source itself. In free-space electromagnetics the mediating field consists of $\mathbf{E}$ and $\mathbf{B}$, while the source field is the distribution of charge or current. An important consideration is that the source field must be independent of the mediating field that it "sources." Additionally, fields are generally regarded as unobservable: they can only be measured indirectly through interactions with observable quantities. We need a link to mechanics to observe $\mathbf{E}$ and $\mathbf{B}$: we might measure the change in kinetic energy of a particle as it interacts with the field through the Lorentz force. The Lorentz force becomes the force function in the mechanical interaction that uniquely determines the (observable) mechanical state of the particle.

A field is associated with a set of *field equations* and a set of *constitutive relations*. The field equations describe, through partial derivative operations, both the spatial distribution and temporal evolution of the field. The constitutive relations describe the effect

---

*Attempts have been made to formulate electromagnetic theory purely in action-at-a-distance terms, but this viewpoint has not been generally adopted [59].

of the supporting medium on the fields and are dependent upon the physical state of the medium. The state may include macroscopic effects, such as mechanical stress and thermodynamic temperature, as well as the microscopic, quantum-mechanical properties of matter.

The value of the field at any position and time in a bounded region V is then determined uniquely by specifying the sources within V, the initial state of the fields within V, and the value of the field or finitely many of its derivatives on the surface bounding V. If the boundary surface also defines a surface of discontinuity between adjacent regions of differing physical characteristics, or across discontinuous sources, then *jump conditions* may be used to relate the fields on either side of the surface.

The variety of forms of field equations is restricted by many physical principles, including reference-frame invariance, conservation, causality, symmetry, and simplicity. Causality prevents the field at time $t = 0$ from being influenced by events occurring at subsequent times $t > 0$. Of course, we prefer that a field equation be mathematically robust and well-posed to permit solutions that are unique and stable.

Many of these ideas are well illustrated by a consideration of electrostatics. We can describe the electrostatic field through a mediating scalar field $\Phi(x, y, z)$ known as the electrostatic potential. The spatial distribution of the field is governed by Poisson's equation

$$\frac{\partial^2 \Phi}{\partial x^2} + \frac{\partial^2 \Phi}{\partial y^2} + \frac{\partial^2 \Phi}{\partial z^2} = -\frac{\rho}{\epsilon_0},$$

where $\rho = \rho(x, y, z)$ is the source charge density. No temporal derivatives appear, and the spatial derivatives determine the spatial behavior of the field. The function ρ represents the spatially averaged distribution of charge that acts as the source term for the field Φ. Note that ρ incorporates no information about Φ. To uniquely specify the field at any point, we must still specify its behavior over a boundary surface. We could, for instance, specify Φ on five of the six faces of a cube and the normal derivative $\partial \Phi / \partial n$ on the remaining face. Finally, we cannot directly observe the static potential field, but we can observe its interaction with a particle. We relate the static potential field theory to the realm of mechanics via the electrostatic force $\mathbf{F} = q\mathbf{E}$ acting on a particle of charge q.

In future chapters we shall present a classical field theory for macroscopic electromagnetics. In that case the mediating field quantities are $\mathbf{E}$, $\mathbf{D}$, $\mathbf{B}$, and $\mathbf{H}$, and the source field is the current density $\mathbf{J}$.

1.3 The sources of the electromagnetic field

Electric charge is an intriguing natural entity. Human awareness of charge and its effects dates back to at least 600 BC, when the Greek philosopher Thales of Miletus observed that rubbing a piece of amber could enable the amber to attract bits of straw. Although *charging by friction* is probably still the most common and familiar manifestation of electric charge, systematic experimentation has revealed much more about the behavior of charge and its role in the physical universe. There are two kinds of charge, to which Benjamin Franklin assigned the respective names *positive* and *negative*. Franklin observed that charges of opposite kind attract and charges of the same kind repel. He also found that an increase in one kind of charge is accompanied by an increase in the other, and so first described the principle of *charge conservation*. Twentieth-century physics has added dramatically to the understanding of charge:

1. Electric charge is a fundamental property of matter, as is mass or dimension.

2. Charge is *quantized*: there exists a smallest quantity (*quantum*) of charge that can be associated with matter. No smaller amount has been observed, and larger amounts always occur in integral multiples of this quantity.

3. The charge quantum is associated with the smallest subatomic particles, and these particles interact through electrical forces. In fact, matter is organized and arranged through electrical interactions; for example, our perception of physical contact is merely the macroscopic manifestation of countless charges in our fingertips pushing against charges in the things we touch.

4. Electric charge is an *invariant*: the value of charge on a particle does not depend on the speed of the particle. In contrast, the mass of a particle increases with speed.

5. Charge acts as the source of an electromagnetic field; the field is an entity that can carry energy and momentum away from the charge via propagating waves.

We begin our investigation of the properties of the electromagnetic field with a detailed examination of its source.

1.3.1   Macroscopic electromagnetics

We are interested primarily in those electromagnetic effects that can be predicted by classical techniques using continuous sources (charge and current densities). Although macroscopic electromagnetics is limited in scope, it is useful in many situations encountered by engineers. These include, for example, the determination of currents and voltages in lumped circuits, torques exerted by electrical machines, and fields radiated by antennas. Macroscopic predictions can fall short in cases where quantum effects are important: e.g., with devices such as tunnel diodes. Even so, quantum mechanics can often be coupled with classical electromagnetics to determine the macroscopic electromagnetic properties of important materials.

Electric charge is not of a continuous nature. The quantization of atomic charge — $\pm e$ for electrons and protons, $\pm e/3$ and $\pm 2e/3$ for quarks — is one of the most precisely established principles in physics (verified to 1 part in 10^{21}). The value of e itself is known to great accuracy. The 2014 recommendation of the Committee on Data for Science and Technology (CODATA) is

$$e = 1.6021766208 \times 10^{-19} \text{ Coulombs (C)},$$

with an uncertainty of $0.0000000098 \times 10^{-19}$ C. However, the discrete nature of charge is not easily incorporated into everyday engineering concerns. The strange world of the individual charge — characterized by particle spin, molecular moments, and thermal vibrations — is well described only by quantum theory. There is little hope that we can learn to describe electrical machines using such concepts. Must we therefore retreat to the macroscopic idea and ignore the discretization of charge completely? A viable alternative is to use atomic theories of matter to estimate the useful scope of macroscopic electromagnetics.

Remember, we are completely free to postulate a theory of nature whose scope may be limited. Like continuum mechanics, which treats distributions of matter as if they were continuous, macroscopic electromagnetics is regarded as valid because it is verified by experiment over a certain range of conditions. This applicability range generally corresponds to dimensions on a laboratory scale, implying a very wide range of validity for engineers.

1.3.1.1 Macroscopic effects as averaged microscopic effects

Macroscopic electromagnetics can hold in a world of discrete charges because applications usually occur over physical scales that include vast numbers of charges. Common devices, generally much larger than individual particles, "average" the rapidly varying fields that exist in the spaces between charges, and this allows us to view a source as a continuous "smear" of charge. To determine the range of scales over which the macroscopic viewpoint is valid, we must compare averaged values of microscopic fields to the macroscopic fields we measure in the lab. But if the effects of the individual charges are describable only in terms of quantum notions, this task will be daunting at best. A simple compromise, which produces useful results, is to extend the macroscopic theory right down to the microscopic level and regard discrete charges as "point" entities that produce electromagnetic fields according to Maxwell's equations. Then, in terms of scales much larger than the classical radius of an electron ($\approx 10^{-14}$ m), the expected rapid fluctuations of the fields in the spaces between charges is predicted. Finally, we ask: over what spatial scale must we average the effects of the fields and the sources in order to obtain agreement with the macroscopic equations?

In the spatial averaging approach, a convenient weighting function $f(\mathbf{r})$ is chosen, and is normalized so that $\int f(\mathbf{r})\,dV = 1$. An example is the Gaussian distribution

$$f(\mathbf{r}) = (\pi a^2)^{-3/2} e^{-r^2/a^2},$$

where a is the approximate radial extent of averaging. The spatial average of a microscopic quantity $F(\mathbf{r}, t)$ is given by

$$\langle F(\mathbf{r}, t) \rangle = \int F(\mathbf{r} - \mathbf{r}', t) f(\mathbf{r}')\,dV'. \tag{1.1}$$

The scale of validity of the macroscopic model can be found by determining the averaging radius a that produces good agreement between the averaged microscopic fields and the macroscopic fields.

1.3.1.2 The macroscopic volume charge density

At this point we do not distinguish between the "free" charge that is unattached to a molecular structure and the charge found near the surface of a conductor. Nor do we consider the dipole nature of polarizable materials or the microscopic motion associated with molecular magnetic moment or the magnetic moment of free charge. For the consideration of free-space electromagnetics, we assume charge exhibits either three degrees of freedom (*volume charge*), two degrees of freedom (*surface charge*), or one degree of freedom (*line charge*).

In typical matter, the microscopic fields vary spatially over dimensions of 10^{-10} m or less, and temporally over periods (determined by atomic motion) of 10^{-13} s or less. At the surface of a material such as a good conductor where charge often concentrates, averaging with a radius on the order of 10^{-10} m may be required to resolve the rapid variation in the distribution of individual charged particles. However, within a solid or liquid material, or within a free-charge distribution characteristic of a dense gas or an electron beam, a radius of 10^{-8} m proves useful, containing typically 10^6 particles. A diffuse gas, on the other hand, may have a particle density so low that the averaging radius takes on laboratory dimensions, and in such a case the microscopic theory must be employed even at macroscopic dimensions.

Once the averaging radius has been determined, the value of the charge density may be found via (1.1). The volume density of charge for an assortment of point sources can be written in terms of the three-dimensional Dirac delta as

$$\rho^o(\mathbf{r}, t) = \sum_i q_i \delta(\mathbf{r} - \mathbf{r}_i(t)),$$

where $\mathbf{r}_i(t)$ is the position of the charge q_i at time t. Substitution into (1.1) gives

$$\rho(\mathbf{r}, t) = \langle \rho^o(\mathbf{r}, t) \rangle = \sum_i q_i f(\mathbf{r} - \mathbf{r}_i(t)) \tag{1.2}$$

as the averaged charge density appropriate for use in a macroscopic field theory. Because the oscillations of the atomic particles are statistically uncorrelated over the distances used in spatial averaging, the time variations of microscopic fields are not present in the macroscopic fields and temporal averaging is unnecessary. In (1.2) the time dependence of the spatially averaged charge density is due entirely to bulk motion of the charge aggregate (macroscopic charge motion).

With the definition of macroscopic charge density given by (1.2), we can determine the total charge $Q(t)$ in any macroscopic volume region V using

$$Q(t) = \int_V \rho(\mathbf{r}, t) \, dV. \tag{1.3}$$

We have

$$Q(t) = \sum_i q_i \int_V f(\mathbf{r} - \mathbf{r}_i(t)) \, dV = \sum_{\mathbf{r}_i(t) \in V} q_i.$$

Here we ignore the small discrepancy produced by charges lying within distance a of the boundary of V. It is common to employ a box B having volume ΔV:

$$f(\mathbf{r}) = \begin{cases} 1/\Delta V, & \mathbf{r} \in B, \\ 0, & \mathbf{r} \notin B. \end{cases}$$

In this case

$$\rho(\mathbf{r}, t) = \frac{1}{\Delta V} \sum_{\mathbf{r} - \mathbf{r}_i(t) \in B} q_i. \tag{1.4}$$

The size of B is chosen with the same considerations as to atomic scale as was the averaging radius a. Discontinuities at the box edges introduce some difficulties concerning charges that move in and out of the box because of molecular motion.

▶ **Example 1.1:** Volume charge density for a uniform arrangement of point charges

Point charges of value Q_0 are located on a regular three-dimensional grid with separation a. Assume the charges are located at $x = \pm \ell a$, $y = \pm m a$, and $z = \pm n a$ for $\ell, m, n = 0, 1, 2, \ldots$. Calculate the volume charge density at the origin using a cube centered at the origin as the averaging function. Assume the cube has side length: (a) a; (b) $3a$; (c) $(2k+1)a$. What is the limiting value of the charge density as the box becomes large?

Solution: We need only find the total charge within each box and use (1.4).

(a) With side length a we count one charge within the box and have

$$\rho(0, t) = \frac{Q_0}{a^3}.$$

(b) A box of side length $3a$ contains 27 charges and

$$\rho(0, t) = \frac{27Q_0}{(3a)^3} = \frac{Q_0}{a^3}.$$

(c) A box of side length $(2k + 1)a$ contains $(2k + 1)^3$ charges and

$$\rho(0, t) = \frac{(2k + 1)^3 Q_0}{[(2k + 1)a]^3} = \frac{Q_0}{a^3}.$$

Thus, the averaged density is independent of the box size if the side length is an odd multiple of a. ◀

As an example with a distribution of non-identical charges, we consider the following.

▶ **Example 1.2:** Volume charge density for an arrangement of nonuniform point charges

Point charges of value $Q_n = n^2 Q_0$ reside on the x-axis at the points $x = \pm na$ for $n = 1, 2, 3, \ldots$, on the y-axis at the points $y = \pm na$ for $n = 1, 2, 3, \ldots$, and on the z-axis at the points $z = \pm na$ for $n = 1, 2, 3, \ldots$. Calculate the volume charge density at the origin using a cube centered at the origin as the averaging function. Assume the cube has side length: (a) a; (b) $3a$; (c) $5a$; (d) $7a$; (e) $(2k + 1)a$. What is the limiting value of the charge density as the box becomes large?

Solution:

(a) With a side length a we find no charges within the box and thus $\rho(0, t) = \rho(\mathbf{r} = 0, t) = 0$. In this case the box is too small to provide a meaningful definition of charge density.

(b) With side length $3a$ we have a total charge in the box of $Q = 6(1)^2 Q_0$ and

$$\rho(0, t) = \frac{6Q_0}{(3a)^3} = 0.222222 \frac{Q_0}{a^3}.$$

(c) A box of side length $5a$ contains the total charge $Q = 6(1)^2 Q_0 + 6(2)^2 Q_0 = 30Q_0$ and

$$\rho(0, t) = \frac{30Q_0}{(5a)^3} = 0.24 \frac{Q_0}{a^3}.$$

(d) A box of side length $7a$ contains the total charge $Q = 6(1)^2 Q_0 + 6(2)^2 Q_0 + 6(3)^2 Q_0 = 84Q_0$ and

$$\rho(0, t) = \frac{84Q_0}{(7a)^3} = 0.244898 \frac{Q_0}{a^3}.$$

(e) A box of side length $(2k + 1)a$ contains the total charge

$$Q = 6Q_0 \sum_{i=1}^{k} i^2 = k(k + 1)(2k + 1)Q_0$$

and

$$\rho(0, t) = \frac{k(k + 1)(2k + 1)}{[(2k + 1)a]^3} Q_0 = \frac{k(k + 1)}{(2k + 1)^2} \frac{Q_0}{a^3}.$$

As the box gets large, the charge density approaches

$$\rho(0, t) = \lim_{k \to \infty} \frac{k(k+1)}{(2k+1)^2} \frac{Q_0}{a^3} = \frac{1}{4} \frac{Q_0}{a^3} = 0.25 \frac{Q_0}{a^3}.$$

Thus, we can define a meaningful charge density at the origin by choosing a sufficiently large box. ◄

The next example, involving a nonuniform distribution of identical charges, is presented for comparison.

▶ **Example 1.3:** Volume charge density for a nonuniform arrangement of point charges

Charges of value Q_0 Coulombs are placed at the vertices of concentric cubes centered on the origin. The sides of the cubes, having length $2na$ ($n = 1, 2, 3, \ldots$), are parallel to the coordinate axes. Find the volume charge density at the origin using a cube as the averaging function. Assume the cube has side length: (a) a; (b) $3a$; (c) $5a$ (d) $(2k+1)a$.

Solution:

(a) With side length a the box contains no charges and $\rho(0, t) = 0$. Here the box is too small to provide a meaningful definition of charge density.

(b) A box of side $3a$ contains eight charges:

$$\rho(0, t) = \frac{8Q_0}{(3a)^3} = 0.296296 \frac{Q_0}{a^3}.$$

(c) A box of side $5a$ contains sixteen charges:

$$\rho(0, t) = \frac{16Q_0}{(5a)^3} = 0.128 \frac{Q_0}{a^3}.$$

(d) A box of side $(2k+1)a$ contains $8k$ charges:

$$\rho(0, t) = \frac{8kQ_0}{[(2k+1)a]^3} = \frac{8k}{(2k+1)^3} \frac{Q_0}{a^3}.$$

Because the charge density decreases with increasing box size, it is difficult to assign a meaningful volume charge density. ◄

1.3.1.3 The macroscopic volume current density

Electric charge in motion is referred to as *electric current*. Charge motion can be associated with external forces and with microscopic fluctuations in position. Assuming charge q_i has velocity $\mathbf{v}_i(t) = d\mathbf{r}_i(t)/dt$, the charge aggregate has volume current density

$$\mathbf{J}^o(\mathbf{r}, t) = \sum_i q_i \mathbf{v}_i(t) \, \delta(\mathbf{r} - \mathbf{r}_i(t)).$$

Spatial averaging gives the macroscopic volume current density

$$\mathbf{J}(\mathbf{r}, t) = \langle \mathbf{J}^o(\mathbf{r}, t) \rangle = \sum_i q_i \mathbf{v}_i(t) f(\mathbf{r} - \mathbf{r}_i(t)). \tag{1.5}$$

Spatial averaging at time t eliminates currents associated with microscopic motions that are uncorrelated at the scale of the averaging radius (again, we do not consider the

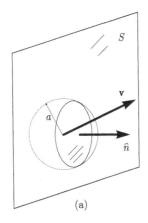 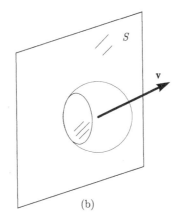

(a)   (b)

FIGURE 1.1
Intersection of the averaging function of a point charge with a surface S, as the charge crosses S with velocity $\mathbf{v}$: (a) at some time $t = t_1$, and (b) at $t = t_2 > t_1$. The averaging function is represented by a sphere of radius a.

magnetic moments of particles). The assumption of a sufficiently large averaging radius leads to

$$\mathbf{J}(\mathbf{r}, t) = \rho(\mathbf{r}, t)\, \mathbf{v}(\mathbf{r}, t).$$

The total flux $I(t)$ of current through a surface S is given by

$$I(t) = \int_S \mathbf{J}(\mathbf{r}, t) \cdot \hat{\mathbf{n}}\, dS$$

where $\hat{\mathbf{n}}$ is the unit normal to S. Hence, using (1.5), we have

$$I(t) = \sum_i q_i \frac{d}{dt}(\mathbf{r}_i(t) \cdot \hat{\mathbf{n}}) \int_S f(\mathbf{r} - \mathbf{r}_i(t))\, dS$$

if $\hat{\mathbf{n}}$ stays approximately constant over the extent of the averaging function and S is not in motion. We see that the integral effectively intersects S with the averaging function surrounding each moving point charge (Figure 1.1). The time derivative of $\mathbf{r}_i \cdot \hat{\mathbf{n}}$ represents the velocity at which the averaging function is "carried across" the surface.

Electric current takes a variety of forms, each described by the relation $\mathbf{J} = \rho\mathbf{v}$. Isolated charged particles (positive and negative) and charged insulated bodies moving through space compose *convection currents*. Negatively charged electrons moving through the positive background lattice within a conductor compose a *conduction current*. Empirical evidence suggests that conduction currents are also described by the relation $\mathbf{J} = \sigma\mathbf{E}$ known as *Ohm's law*. A third type of current, called *electrolytic current*, results from the flow of positive or negative ions through a fluid.

1.3.2   Impressed vs. secondary sources

In addition to the simple classification given above, we may classify currents as *primary* or *secondary*, depending on the action that sets the charge in motion.

It is helpful to separate primary or "impressed" sources, which are independent of the fields they source, from secondary sources, which result from interactions between the sourced fields and the medium in which the fields exist. Most familiar is the conduction current set up in a conducting medium by an externally applied electric field. The impressed source concept is particularly important in circuit theory, where independent voltage sources are modeled as providing primary voltage excitations that are independent of applied load. In this way they differ from the secondary or "dependent" sources that react to the effect produced by the application of primary sources.

In applied electromagnetics, the primary source may be so distant that return effects resulting from local interaction of its impressed fields can be ignored. Other examples of primary sources include the applied voltage at the input of an antenna, the current on a probe inserted into a waveguide, and the currents producing a power-line field in which a biological body is immersed.

1.3.3 Surface and line source densities

Because they are spatially averaged effects, macroscopic sources and the fields they source cannot have true spatial discontinuities. However, it is often convenient to work with sources in one or two dimensions. Surface and line source densities are idealizations of actual, continuous macroscopic densities.

The entity we describe as a *surface charge* is a continuous volume charge distributed in a thin layer across some surface S. If the thickness of the layer is small compared to laboratory dimensions, it is useful to assign to each point $\mathbf{r}$ on the surface a quantity describing the amount of charge contained within a cylinder oriented normal to the surface and having infinitesimal cross-section dS. We call this quantity the *surface charge density* $\rho_s(\mathbf{r}, t)$, and write the volume charge density as

$$\rho(\mathbf{r}, w, t) = \rho_s(\mathbf{r}, t) f(w, \Delta),$$

where w is distance from S in the normal direction and Δ in some way parameterizes the "thickness" of the charge layer at $\mathbf{r}$. The continuous density function $f(x, \Delta)$ satisfies

$$\int_{-\infty}^{\infty} f(x, \Delta) \, dx = 1$$

and

$$\lim_{\Delta \to 0} f(x, \Delta) = \delta(x).$$

For instance, we might have

$$f(x, \Delta) = \frac{e^{-x^2/\Delta^2}}{\Delta\sqrt{\pi}}. \tag{1.6}$$

With this definition, the total charge contained in a cylinder normal to the surface at $\mathbf{r}$ and having cross-sectional area dS is

$$dQ(t) = \int_{-\infty}^{\infty} [\rho_s(\mathbf{r}, t) \, dS] \, f(w, \Delta) \, dw = \rho_s(\mathbf{r}, t) \, dS,$$

and the total charge contained within any cylinder oriented normal to S is

$$Q(t) = \int_S \rho_s(\mathbf{r}, t) \, dS.$$

We may describe a line charge as a thin "tube" of volume charge distributed along some contour Γ. The amount of charge contained between two planes normal to the contour and separated by a distance dl is described by the *line charge density* $\rho_l(\mathbf{r}, t)$. The volume charge density associated with the contour is then

$$\rho(\mathbf{r}, \rho, t) = \rho_l(\mathbf{r}, t) f_s(\rho, \Delta),$$

where ρ is the radial distance from the contour in the plane normal to Γ and $f_s(\rho, \Delta)$ is a density function with the properties

$$\int_0^\infty f_s(\rho, \Delta) 2\pi \rho \, d\rho = 1$$

and

$$\lim_{\Delta \to 0} f_s(\rho, \Delta) = \frac{\delta(\rho)}{2\pi\rho}.$$

For example, we might have

$$f_s(\rho, \Delta) = \frac{e^{-\rho^2/\Delta^2}}{\pi\Delta^2}. \tag{1.7}$$

Then the total charge contained between planes separated by a distance dl is

$$dQ(t) = \int_0^\infty [\rho_l(\mathbf{r}, t) \, dl] \, f_s(\rho, \Delta) 2\pi \rho \, d\rho = \rho_l(\mathbf{r}, t) \, dl$$

and the total charge contained between planes placed at the ends of a contour Γ is

$$Q(t) = \int_\Gamma \rho_l(\mathbf{r}, t) \, dl.$$

We may define surface and line currents similarly. A surface current is merely a volume current confined to the vicinity of a surface S. The volume current density may be represented using a surface current density function $\mathbf{J}_s(\mathbf{r}, t)$, defined at each point $\mathbf{r}$ on the surface so that

$$\mathbf{J}(\mathbf{r}, w, t) = \mathbf{J}_s(\mathbf{r}, t) f(w, \Delta).$$

Here $f(w, \Delta)$ is some appropriate density function such as (1.6), and the surface current vector obeys $\hat{\mathbf{n}} \cdot \mathbf{J}_s = 0$ where $\hat{\mathbf{n}}$ is normal to S. The total current flowing through a strip of width dl arranged perpendicular to S at $\mathbf{r}$ is

$$dI(t) = \int_{-\infty}^\infty [\mathbf{J}_s(\mathbf{r}, t) \cdot \hat{\mathbf{n}}_l(\mathbf{r}) \, dl] \, f(w, \Delta) \, dw = \mathbf{J}_s(\mathbf{r}, t) \cdot \hat{\mathbf{n}}_l(\mathbf{r}) \, dl$$

where $\hat{\mathbf{n}}_l$ is normal to the strip at $\mathbf{r}$ (and thus also tangential to S at $\mathbf{r}$). The total current passing through a strip intersecting with S along a contour Γ is thus

$$I(t) = \int_\Gamma \mathbf{J}_s(\mathbf{r}, t) \cdot \hat{\mathbf{n}}_l(\mathbf{r}) \, dl.$$

We may describe a line current as a thin "tube" of volume current distributed about some contour Γ and flowing parallel to it. The amount of current passing through a plane normal to the contour is described by the *line current density* $J_l(\mathbf{r}, t)$. The volume current density associated with the contour may be written as

$$\mathbf{J}(\mathbf{r}, \rho, t) = \hat{\mathbf{u}}(\mathbf{r}) J_l(\mathbf{r}, t) f_s(\rho, \Delta),$$

where $\hat{\mathbf{u}}$ is a unit vector along Γ, ρ is the radial distance from the contour in the plane normal to Γ, and $f_s(\rho, \Delta)$ is a density function such as (1.7). The total current passing through any plane normal to Γ at $\mathbf{r}$ is

$$I(t) = \int_0^\infty \left[J_l(\mathbf{r}, t) \hat{\mathbf{u}}(\mathbf{r}) \cdot \hat{\mathbf{u}}(\mathbf{r}) \right] f_s(\rho, \Delta) 2\pi \rho \, d\rho$$

$$= J_l(\mathbf{r}, t).$$

It is often convenient to employ *singular* models for continuous source densities. For instance, it is mathematically simpler to regard a surface charge as residing only in the surface S than to regard it as being distributed about the surface. Of course, the source is then discontinuous since it is zero everywhere outside the surface. We may obtain a representation of such a charge distribution by letting the thickness parameter Δ in the density functions recede to zero, thus concentrating the source into a plane or a line. We describe the limit of the density function in terms of the δ-function. For instance, the volume charge distribution for a surface charge located about the xy-plane is

$$\rho(x, y, z, t) = \rho_s(x, y, t) f(z, \Delta).$$

As $\Delta \to 0$ we have

$$\rho(x, y, z, t) = \rho_s(x, y, t) \lim_{\Delta \to 0} f(z, \Delta)$$

$$= \rho_s(x, y, t) \delta(z).$$

It is a simple matter to represent singular source densities in this way as long as the surface or line is easily parameterized in terms of constant values of coordinate variables. However, care must be taken to represent the δ-function properly.

▶ **Example 1.4:** Charge density on the surface of a cone

Charge is located on the surface of a cone in spherical coordinates. If the surface charge density is $\rho_s(r, \phi, t)$, determine the volume charge density.

Solution: The density of charge on the surface of a cone at $\theta = \theta_0$ may be described using the distance along the normal to this surface, which is given by $r\theta - r\theta_0$:

$$\rho(r, \theta, \phi, t) = \rho_s(r, \phi, t) \, \delta(r[\theta - \theta_0]).$$

Using the property $\delta(ax) = \delta(x)/a$, we can also write this as

$$\rho(r, \theta, \phi, t) = \rho_s(r, \phi, t) \frac{\delta(\theta - \theta_0)}{r}. \quad \blacktriangleleft$$

▶ **Example 1.5:** Total charge on the surface of a sphere

Charge on the surface of a sphere is described by the volume charge density

$$\rho(\mathbf{r}, t) = 2e^{-3t} r^2 \cos^2 \theta \cos^2 \phi \, \delta(r - 3) \, \mu\text{C/m}^3 \qquad (0 \le \theta \le \pi, \ 0 \le \phi \le 2\pi).$$

Compute the total charge $Q(t)$ on the surface.

Solution: Using (1.3) we have

$$Q(t) = \int \rho \, dV = \int_0^{2\pi} \int_0^{\pi} \int_0^{\infty} 2e^{-3t} r^2 \cos^2 \theta \cos^2 \phi \, \delta(r-3) r^2 \sin \theta \, dr \, d\theta \, d\phi$$

$$= 2e^{-3t} \int_0^{\infty} r^4 \delta(r-3) \, dr \int_0^{\pi} \cos^2 \theta \sin \theta \, d\theta \int_0^{2\pi} \cos^2 \phi \, d\phi$$

$$= 2e^{-3t}(81)(\tfrac{2}{3})(\pi) = 108\pi e^{-3t} \ \mu\text{C}. \ \blacktriangleleft$$

▶ **Example 1.6:** Total charge on a segment

The density of charge on a line segment parallel to the x-axis is given by the volume charge density

$$\rho(\mathbf{r}) = 10y^2 \sin x \, \delta(y-4)\delta(z) \ \text{nC/m}^3 \qquad (0 \le x \le \pi).$$

Compute the total charge Q on the line.

Solution: Using (1.3) we have

$$Q = \int_{-\infty}^{\infty} \int_{-\infty}^{\infty} \int_0^{\pi} 10y^2 \sin x \, \delta(y-4)\delta(z) \, dx \, dy \, dz$$

$$= 10 \int_0^{\pi} \sin x \, dx \int_{-\infty}^{\infty} y^2 \delta(y-4) \, dy \int_{-\infty}^{\infty} \delta(z) \, dz = 320 \ \text{nC}. \ \blacktriangleleft$$

1.3.4 Charge conservation

There are four fundamental conservation laws in physics: conservation of energy, momentum, angular momentum, and charge. These laws are said to be *absolute*; they have never been observed to fail. In that sense they are true empirical laws of physics.

However, in modern physics the fundamental conservation laws have come to represent more than just observed facts. Each law is now associated with a fundamental symmetry of the universe; conversely, each known symmetry is associated with a conservation principle. For example, energy conservation can be shown to arise from the observation that the universe is symmetric with respect to time; the laws of physics do not depend on choice of time origin $t = 0$. Similarly, momentum conservation arises from the observation that the laws of physics are invariant under translation, while angular momentum conservation arises from invariance under rotation.

The law of conservation of charge also arises from a symmetry principle. But instead of being spatial or temporal in character, it is related to the invariance of electrostatic potential. Experiments show that there is no absolute potential, only potential difference. The laws of nature are invariant with respect to what we choose as the "reference" potential. This in turn is related to the invariance of Maxwell's equations under gauge transforms; the values of the electric and magnetic fields do not depend on which gauge transformation we use to relate the scalar potential Φ to the vector potential $\mathbf{A}$.

We may state the conservation of charge as follows:

The net charge in any closed system remains constant with time.

This does not mean that individual charges cannot be created or destroyed, only that the total charge in any isolated system must remain constant. Thus it is possible for a positron with charge e to annihilate an electron with charge $-e$ without changing the net charge of the system. Only if a system is not closed can its net charge be altered; since moving charge constitutes current, we can say that the total charge within a system depends on the current passing through the surface enclosing the system. This is the essence of the continuity equation. To derive this important result we consider a closed system within which the charge remains constant, and apply the Reynolds transport theorem (see § A.3).

1.3.4.1   The continuity equation

Consider a region of space occupied by a distribution of charge whose velocity is given by the vector field $\mathbf{v}$. We surround a portion of charge by a surface S and let S deform as necessary to "follow" the charge as it moves. Since S always contains precisely the same charged particles, we have an isolated system for which the time rate of change of total charge must vanish. An expression for the time rate of change is given by the Reynolds transport theorem (A.68); we have[†]

$$\frac{DQ}{Dt} = \frac{D}{Dt} \int_{V(t)} \rho \, dV = \int_{V(t)} \frac{\partial \rho}{\partial t} \, dV + \oint_{S(t)} \rho \mathbf{v} \cdot d\mathbf{S} = 0.$$

The "D/Dt" notation indicates that the volume region $V(t)$ moves with its enclosed particles. Since $\rho \mathbf{v}$ represents current density, we can write

$$\int_{V(t)} \frac{\partial \rho(\mathbf{r}, t)}{\partial t} \, dV + \oint_{S(t)} \mathbf{J}(\mathbf{r}, t) \cdot d\mathbf{S} = 0. \tag{1.8}$$

In this *large-scale form* of the continuity equation, the partial derivative term describes the time rate of change of the charge density for a fixed spatial position $\mathbf{r}$. At any time t, the time rate of change of charge density integrated over a volume is exactly compensated by the total current exiting through the surrounding surface.

We can obtain the continuity equation in *point form* by applying the divergence theorem to the second term of (1.8) to get

$$\int_{V(t)} \left[\frac{\partial \rho(\mathbf{r}, t)}{\partial t} + \nabla \cdot \mathbf{J}(\mathbf{r}, t) \right] dV = 0.$$

Since $V(t)$ is arbitrary, we can set the integrand to zero to obtain

$$\frac{\partial \rho(\mathbf{r}, t)}{\partial t} + \nabla \cdot \mathbf{J}(\mathbf{r}, t) = 0. \tag{1.9}$$

This expression involves the time derivative of ρ with $\mathbf{r}$ fixed. We can also find an expression in terms of the material derivative by using the transport equation (A.69). Enforcing conservation of charge by setting that expression to zero, we have

$$\frac{D\rho(\mathbf{r}, t)}{Dt} + \rho(\mathbf{r}, t) \nabla \cdot \mathbf{v}(\mathbf{r}, t) = 0.$$

Here $D\rho/Dt$ is the time rate of change of the charge density experienced by an observer moving with the current.

[†]Note that in Appendix A we use the symbol $\mathbf{u}$ to represent the velocity of a material and $\mathbf{v}$ to represent the velocity of an artificial surface.

We can state the large-scale form of the continuity equation in terms of a stationary volume. Integrating (1.9) over a stationary volume region V and using the divergence theorem, we find that

$$\int_V \frac{\partial \rho(\mathbf{r}, t)}{\partial t} \, dV = - \oint_S \mathbf{J}(\mathbf{r}, t) \cdot d\mathbf{S}.$$

Since V is not changing with time, we have

$$\frac{dQ(t)}{dt} = \frac{d}{dt} \int_V \rho(\mathbf{r}, t) \, dV = - \oint_S \mathbf{J}(\mathbf{r}, t) \cdot d\mathbf{S}. \tag{1.10}$$

Hence any increase of total charge within V must be produced by current entering V through S.

▶ **Example 1.7:** Use of the continuity equation

The volume density of charge in a bounded region of space is given by

$$\rho(\mathbf{r}, t) = \rho_0 r e^{-\beta t}.$$

Find $\mathbf{J}$ and $\mathbf{v}$, and verify both versions of the continuity equation in point form.

Solution: The spherical symmetry of ρ requires that $\mathbf{J} = \hat{\mathbf{r}} J_r$. Application of (1.10) over a sphere of radius a gives

$$4\pi \frac{d}{dt} \int_0^a \rho_0 r e^{-\beta t} r^2 \, dr = -4\pi J_r(a) a^2.$$

Hence

$$\mathbf{J} = \hat{\mathbf{r}} \beta \rho_0 \frac{r^2}{4} e^{-\beta t}$$

and therefore

$$\nabla \cdot \mathbf{J} = \frac{1}{r^2} \frac{\partial}{\partial r} (r^2 J_r) = \beta \rho_0 r e^{-\beta t}.$$

The velocity is

$$\mathbf{v} = \frac{\mathbf{J}}{\rho} = \hat{\mathbf{r}} \beta \frac{r}{4},$$

and we have $\nabla \cdot \mathbf{v} = 3\beta/4$. To verify the continuity equations, we compute the time derivatives

$$\frac{\partial \rho}{\partial t} = -\beta \rho_0 r e^{-\beta t},$$

$$\frac{D\rho}{Dt} = \frac{\partial \rho}{\partial t} + \mathbf{v} \cdot \nabla \rho = -\beta \rho_0 r e^{-\beta t} + \left(\hat{\mathbf{r}} \beta \frac{r}{4} \right) \cdot \left(\hat{\mathbf{r}} \rho_0 e^{-\beta t} \right) = -\tfrac{3}{4} \beta \rho_0 r e^{-\beta t}.$$

Note that the charge density decreases with time less rapidly for a moving observer than for a stationary one (3/4 as fast): the moving observer is following the charge outward, and $\rho \propto r$. Now we can check the continuity equations. First we see

$$\frac{D\rho}{Dt} + \rho \nabla \cdot \mathbf{v} = -\tfrac{3}{4} \beta \rho_0 r e^{-\beta t} + (\rho_0 r e^{-\beta t}) \left(\tfrac{3}{4} \beta \right) = 0,$$

as required for a moving observer; second we see

$$\frac{\partial \rho}{\partial t} + \nabla \cdot \mathbf{J} = -\beta \rho_0 r e^{-\beta t} + \beta \rho_0 e^{-\beta t} = 0,$$

as required for a stationary observer. ◀

▶ **Example 1.8:** Use of the continuity equation

The charge density in a region of free space is given by

$$\rho(\mathbf{r}, t) = \rho_0 \sin(\omega t - \beta z).$$

An observer moves with constant velocity $\mathbf{v} = v\hat{\mathbf{z}}$. At what speed v does the observer measure zero time rate of change of charge?

Solution: The total derivative is given by (A.60) as

$$\frac{d\rho}{dt} = \frac{\partial \rho}{\partial t} + \mathbf{v} \cdot \nabla \rho.$$

Computing

$$\frac{\partial \rho}{\partial t} = \omega \rho_0 \cos(\omega t - \beta z) \quad \text{and} \quad \nabla \rho = -\hat{\mathbf{z}} \beta \rho_0 \cos(\omega t - \beta z)$$

we get

$$\frac{d\rho}{dt} = \omega \rho_0 \cos(\omega t - \beta z) - v\beta \rho_0 \cos(\omega t - \beta z) = (\omega - v\beta)\rho_0 \cos(\omega t - \beta z).$$

Thus, the observer will measure no time rate of change of the charge while moving with speed $v = \omega / \beta$. ◀

1.3.4.2   The continuity equation in fewer dimensions

The continuity equation can also be used to relate current and charge on a surface or along a line. By conservation of charge we can write

$$\frac{d}{dt} \int_S \rho_s(\mathbf{r}, t) \, dS = - \oint_\Gamma \mathbf{J}_s(\mathbf{r}, t) \cdot \hat{\mathbf{m}} \, dl$$

where $\hat{\mathbf{m}}$ is the vector normal to the curve Γ and tangential to the surface S. By the surface divergence theorem (B.26), the corresponding point form is

$$\frac{\partial \rho_s(\mathbf{r}, t)}{\partial t} + \nabla_s \cdot \mathbf{J}_s(\mathbf{r}, t) = 0.$$

Here $\nabla_s \cdot \mathbf{J}_s$ is the surface divergence of the vector field $\mathbf{J}_s$. For instance, in rectangular coordinates in the $z = 0$ plane, we have

$$\nabla_s \cdot \mathbf{J}_s = \frac{\partial J_{sx}}{\partial x} + \frac{\partial J_{sy}}{\partial y}.$$

In cylindrical coordinates on the cylinder $\rho = a$, we would have

$$\nabla_s \cdot \mathbf{J}_s = \frac{1}{a} \frac{\partial J_{s\phi}}{\partial \phi} + \frac{\partial J_{sz}}{\partial z}.$$

A detailed description of vector operations on a surface may be found in Tai [188], while many identities may be found in Van Bladel [203].

The equation of continuity for a line is easily established by reference to Figure 1.2. Here the net charge exiting the surface during time Δt is given by

$$\Delta t [I(u_2, t) - I(u_1, t)].$$

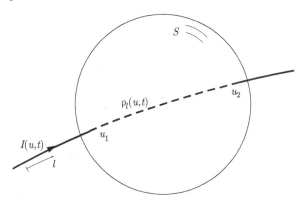

FIGURE 1.2
Linear form of the continuity equation.

Thus, the rate of net increase of charge within the system is

$$\frac{dQ(t)}{dt} = \frac{d}{dt} \int \rho_l(\mathbf{r}, t)\, dl = -[I(u_2, t) - I(u_1, t)].$$

The corresponding point form is found by letting the length of the curve approach zero:

$$\frac{\partial I(l, t)}{\partial l} + \frac{\partial \rho_l(l, t)}{\partial t} = 0,$$

where l is arc length along the curve.

▶ **Example 1.9:** Using the 1-D continuity equation with a line source

The line current on a circular loop antenna is approximately

$$I(\phi, t) = I_0 \cos\left(\frac{\omega a}{c}\phi\right) \cos \omega t,$$

where a is the radius of the loop, ω is the frequency of operation, and c is the speed of light. Find the line charge density on the loop.

Solution: Since $l = a\phi$, we can write

$$I(l, t) = I_0 \cos\left(\frac{\omega l}{c}\right) \cos \omega t.$$

Thus

$$\frac{\partial I(l, t)}{\partial l} = -I_0 \frac{\omega}{c} \sin\left(\frac{\omega l}{c}\right) \cos \omega t = -\frac{\partial \rho_l(l, t)}{\partial t}.$$

Integrating with respect to time and ignoring any constant (static) charge, we have

$$\rho_l(l, t) = \frac{I_0}{c} \sin\left(\frac{\omega l}{c}\right) \sin \omega t$$

or

$$\rho_l(\phi, t) = \frac{I_0}{c} \sin\left(\frac{\omega a}{c}\phi\right) \sin \omega t.$$

Note that we could have used the chain rule

$$\frac{\partial I(\phi,t)}{\partial l} = \frac{\partial I(\phi,t)}{\partial \phi}\frac{\partial \phi}{\partial l} \quad \text{and} \quad \frac{\partial \phi}{\partial l} = \left[\frac{\partial l}{\partial \phi}\right]^{-1} = \frac{1}{a}$$

to calculate the spatial derivative. ◄

We can apply the volume density continuity equation (1.9) directly to surface and line distributions written in singular notation.

▶ **Example 1.10:** Using the 3-D continuity equation with a line source

Consider the loop in the previous example. Write the volume current using singular notation and use the volume continuity equation to find the line charge density.

Solution: We write the volume current density corresponding to the line current as

$$\mathbf{J}(\mathbf{r},t) = \hat{\phi}\,\delta(\rho - a)\delta(z)I(\phi,t).$$

Substitution into (1.9) then gives

$$\nabla \cdot [\hat{\phi}\delta(\rho - a)\delta(z)I(\phi,t)] = -\frac{\partial \rho(\mathbf{r},t)}{\partial t}.$$

The divergence formula for cylindrical coordinates gives

$$\delta(\rho - a)\delta(z)\frac{\partial I(\phi,t)}{\rho\,\partial \phi} = -\frac{\partial \rho(\mathbf{r},t)}{\partial t}.$$

Next we substitute for $I(\phi,t)$ to get

$$-\frac{I_0}{\rho}\frac{\omega a}{c}\sin\left(\frac{\omega a}{c}\phi\right)\delta(\rho - a)\delta(z)\cos\omega t = -\frac{\partial \rho(\mathbf{r},t)}{\partial t}.$$

Finally, integrating with respect to time and ignoring any constant term, we have

$$\rho(\mathbf{r},t) = \frac{I_0}{c}\delta(\rho - a)\delta(z)\sin\left(\frac{\omega a}{c}\phi\right)\sin\omega t,$$

where we have set $\rho = a$ because of the presence of the factor $\delta(\rho - a)$. From this we see immediately that

$$\rho_l(\phi,t) = \frac{I_0}{c}\sin\left(\frac{\omega a}{c}\phi\right)\sin\omega t,$$

which is identical to the line charge density found in Example 1.9. ◄

1.3.5 Magnetic charge

We take for granted that electric fields are produced by electric charges, whether stationary or in motion. The smallest element of electric charge is the electric *monopole*: a single discretely charged particle from which the electric field diverges. In contrast, experiments show that magnetic fields are created only by currents or by time changing electric fields; hence, magnetic fields have moving electric charge as their source. The elemental source of magnetic field is the magnetic *dipole*, representing a tiny loop of electric current (or a spinning electric particle). The observation made in 1269 by Pierre de Maricourt, that even the smallest magnet has two poles, still holds today.

In a world filled with symmetry at the fundamental level, we find it hard to understand why there should not be a source from which the magnetic field diverges. We would call such a source *magnetic charge*, and the most fundamental quantity of magnetic charge would be exhibited by a *magnetic monopole*. In 1931 Paul Dirac invigorated the search for magnetic monopoles by making the first strong theoretical argument for their existence. Dirac showed that the existence of magnetic monopoles would imply the quantization of electric charge, and would thus provide an explanation for one of the great puzzles of science. Since that time, magnetic monopoles have become important players in the "Grand Unified Theories" of modern physics, and in cosmological theories of the origin of the universe.

If magnetic monopoles are ever found to exist, there will be both positive and negative charged particles whose motions will constitute currents. We can define a macroscopic magnetic charge density ρ_m and current density $\mathbf{J}_m$ exactly as we did with electric charge, and use conservation of magnetic charge to provide a continuity equation:

$$\nabla \cdot \mathbf{J}_m(\mathbf{r}, t) + \frac{\partial \rho_m(\mathbf{r}, t)}{\partial t} = 0.$$

With these new sources, Maxwell's equations become appealingly symmetric. Despite uncertainties about the existence and physical nature of magnetic monopoles, magnetic charge and current have become an integral part of electromagnetic theory. We often use the concept of *fictitious* magnetic sources to make Maxwell's equations symmetric, and then derive various equivalence theorems for use in the solution of important problems. Thus we can put the idea of magnetic sources to use regardless of whether these sources actually exist.

1.4 Problems

1.1 Write the volume charge density for a singular surface charge located on the sphere $r = r_0$, entirely in terms of spherical coordinates. Find the total charge on the sphere.

1.2 Repeat Problem 1.1 for a charged half plane $\phi = \phi_0$.

1.3 Write the volume charge density for a singular surface charge located on the cylinder $\rho = \rho_0$, entirely in terms of cylindrical coordinates. Find the total charge on the cylinder.

1.4 Repeat Problem 1.3 for a charged half plane $\phi = \phi_0$.

1.5 A current flows radially outward from the z-axis of cylindrical coordinates. The volume charge density associated with the current is given by

$$\rho(\mathbf{r}, t) = \rho_0 \rho^2 e^{-\beta t}$$

where ρ_0 and β are constants. (a) Calculate the current density. (b) Show that the following two forms of the continuity equation are satisfied:

$$\frac{\partial \rho}{\partial t} + \nabla \cdot \mathbf{J} = 0, \qquad \frac{D\rho}{Dt} + \rho \nabla \cdot \mathbf{v} = 0.$$

1.6 Compute the total charge associated with the following volume charge densities:

(a) $\rho = 4r^2 \cos^2 \theta \, \delta(\theta - \pi/4)$, $0 \leq \phi \leq 2\pi$, $0 \leq r \leq 2$.

(b) $\rho = 4\cos^2 \phi \, \delta(\rho - 2)$, $0 \leq z \leq 2$, $0 \leq \phi \leq 2\pi$.

(c) $\rho = 4z^3 \, \delta(x)\delta(y)$, $0 \leq z \leq 2$.

(d) $\rho = 4\rho z\delta(\rho - 3)$, $0 \leq z \leq 2$, $0 \leq \phi \leq 2\pi$.

(e) $\rho = 5(z + 2)\delta(z)\delta(\rho - 3)$, $0 \leq \phi \leq \pi$.

1.7 A charge distribution has density $\rho(\mathbf{r}, t) = 4x^2 e^{-4t}$ C/m^3. Calculate the time rate of change of ρ measured by an observer moving with velocity $\mathbf{v} = \hat{\mathbf{x}} A x$. Find the value of A such that the observer measures the same value of ρ at all times t.

1.8 The charge density in a certain region of free space is given by $\rho(\mathbf{r}, t) = 4r^2 e^{-\beta t}$ C/m^3. Find the time rate of change of the charge density measured by an observer moving with velocity $\mathbf{v} = \hat{\mathbf{r}} 2r$ m/s. For what value of β is the measured time rate of change identically zero?

2

Maxwell's theory of electromagnetism

2.1 The postulate

In 1864, James Clerk Maxwell proposed one of the most successful theories in the history of science. In a famous memoir to the Royal Society [127], he presented nine equations summarizing all known laws on electricity and magnetism. This was more than a mere cataloging of the laws of nature. By postulating the need for an additional term to make the set of equations self-consistent, Maxwell was able to put forth what is still considered a complete theory of macroscopic electromagnetism. The beauty of Maxwell's equations led Boltzmann to ask, "Was it a god who wrote these lines ...?" [180].

Since that time, authors have struggled to find the best way to present Maxwell's theory. Although it is possible to study electromagnetics from an "empirical–inductive" viewpoint (roughly following the historical order of development, beginning with static fields), it is only by postulating the complete theory that we can do justice to Maxwell's vision. His concept of the existence of an electromagnetic "field" (as introduced by Faraday) is fundamental to this theory, and has become one of the most significant principles of modern science.

We find controversy even over the best way to present Maxwell's equations. Maxwell worked at a time before vector notation was completely in place, and thus chose to use scalar variables and equations to represent the fields. Certainly the true beauty of Maxwell's equations emerges when they are written in vector form, and the use of tensors reduces the equations to their underlying physical simplicity. We shall use vector notation in this book because of its wide acceptance by engineers, but we still must decide whether it is more appropriate to present the vector equations in integral or point form.

On one side of this debate, the brilliant mathematician David Hilbert felt that the fundamental natural laws should be posited as axioms, each best described in terms of integral equations [158]. This idea has been championed by Truesdell and Toupin [199]. On the other side, we may quote from the great physicist Arnold Sommerfeld: "The general development of Maxwell's theory must proceed from its differential form; for special problems the integral form may, however, be more advantageous" ([180], p. 23). Special relativity flows naturally from the point forms, with fields easily converted between moving reference frames. For stationary media, it seems to us that the only difference between the two approaches arises in how we handle discontinuities in sources and materials. If we choose to use the point forms of Maxwell's equations, then we must also postulate the boundary conditions at surfaces of discontinuity. This is pointed out clearly by Tai [189], who also notes that if the integral forms are used, then their validity across regions of discontinuity should be stated as part of the postulate.

We have decided to use the point form in this text. In doing so we follow a long history begun by Hertz in 1890 [86] when he wrote down Maxwell's differential equations as a set of axioms, recognizing the equations as the launching point for the theory of electromagnetism. Also, by postulating Maxwell's equations in point form, we can take full advantage of modern developments in the theory of partial differential equations; in particular, the idea of a "well-posed" theory determines what sort of information must be specified to make the postulate useful.

We must also decide which form of Maxwell's differential equations to use as the basis of our postulate. There are several competing forms, each differing as to the manner in which materials are considered. The oldest and most widely used form was suggested by Minkowski in 1908 [132]. In the Minkowski form, the differential equations contain no mention of the materials supporting the fields; all information about material media is relegated to the constitutive relationships. This places simplicity of the differential equations above intuitive understanding of the behavior of fields in materials. We choose the Maxwell–Minkowski form as the basis of our postulate, primarily for ease of manipulation. But we also recognize the value of other versions of Maxwell's equations. We shall present the basic ideas behind the Boffi form, which places some information about materials into the differential equations (although constitutive relationships are still required). Missing, however, is any information regarding the velocity of a moving medium. By using the polarization and magnetization vectors **P** and **M** rather than the fields **D** and **H**, it is sometimes easier to visualize the meaning of the field vectors and to understand (or predict) the nature of the constitutive relations.

The Chu and Amperian forms of Maxwell's equations have been promoted as useful alternatives to the Minkowski and Boffi forms. These include explicit information about the velocity of a moving material, and differ somewhat from the Boffi form in the physical interpretation of the electric and magnetic properties of matter. Although each of these models matter in terms of charged particles immersed in free space, magnetization in the Boffi and Amperian forms arises from electric current loops, while the Chu form employs magnetic dipoles. In all three forms, polarization is modeled using electric dipoles. For a detailed discussion of the Chu and Amperian forms, the reader should consult the work of Kong [108], Tai [190], Penfield and Haus [149], or Fano, Chu, and Adler [61].

Importantly, all of these various forms of Maxwell's equations produce the same values of the physical fields (at least external to the material where the fields are measurable).

We must include several other constituents, besides the field equations, to make the postulate complete. To form a complete field theory, we need a source field, a mediating field, and a set of field differential equations. This allows us to mathematically describe the relationship between effect (the mediating field) and cause (the source field). In a well-posed postulate we must also include a set of constitutive relationships and a specification of some field relationship over a bounding surface and at an initial time. If the electromagnetic field is to have physical meaning, we must link it to some observable quantity such as force. Finally, to allow the solution of problems involving mathematical discontinuities, we must specify certain boundary, or "jump," conditions.

2.1.1   The Maxwell–Minkowski equations

In Maxwell's macroscopic theory of electromagnetics, the source field consists of the vector field $\mathbf{J}(\mathbf{r},t)$ (the current density) and the scalar field $\rho(\mathbf{r},t)$ (the charge density). In Minkowski's form of Maxwell's equations, the mediating field is the *electromagnetic field* consisting of the set of four vector fields $\mathbf{E}(\mathbf{r},t)$, $\mathbf{D}(\mathbf{r},t)$, $\mathbf{B}(\mathbf{r},t)$, and $\mathbf{H}(\mathbf{r},t)$. The field equations are the four partial differential equations referred to as the *Maxwell*

$$\nabla \times \mathbf{E}(\mathbf{r}, t) = -\frac{\partial}{\partial t} \mathbf{B}(\mathbf{r}, t), \tag{2.1}$$

$$\nabla \times \mathbf{H}(\mathbf{r}, t) = \mathbf{J}(\mathbf{r}, t) + \frac{\partial}{\partial t} \mathbf{D}(\mathbf{r}, t), \tag{2.2}$$

$$\nabla \cdot \mathbf{D}(\mathbf{r}, t) = \rho(\mathbf{r}, t), \tag{2.3}$$

$$\nabla \cdot \mathbf{B}(\mathbf{r}, t) = 0, \tag{2.4}$$

along with the continuity equation

$$\nabla \cdot \mathbf{J}(\mathbf{r}, t) = -\frac{\partial}{\partial t} \rho(\mathbf{r}, t). \tag{2.5}$$

Here (2.1) is called *Faraday's law*, (2.2) is called *Ampere's law*, (2.3) is called *Gauss's law*, and (2.4) is called the *magnetic Gauss's law*. For brevity we shall often leave the dependence on $\mathbf{r}$ and t implicit, and refer to the Maxwell–Minkowski equations as simply the "Maxwell equations," or "Maxwell's equations."

Equations (2.1)–(2.5), the point forms of the field equations, describe the relationships between the fields and their sources at each point in space where the fields are continuously differentiable (i.e., the derivatives exist and are continuous). Such points are called *ordinary points*. We shall not attempt to define the fields at other points, but instead seek conditions relating the fields across surfaces containing these points. Normally this is necessary on surfaces across which either sources or material parameters are discontinuous.

The electromagnetic fields carry SI units as follows: $\mathbf{E}$ is measured in Volts per meter (V/m), $\mathbf{B}$ is measured in Teslas (T), $\mathbf{H}$ is measured in Amperes per meter (A/m), and $\mathbf{D}$ is measured in Coulombs per square meter (C/m^2). In older texts we find the units of $\mathbf{B}$ given as Webers per square meter (Wb/m^2) to reflect the role of $\mathbf{B}$ as a flux vector; in that case the Weber (Wb = T·m^2) is regarded as a unit of magnetic flux.

2.1.1.1 The interdependence of Maxwell's equations

It is often claimed that the divergence equations (2.3) and (2.4) may be derived from the curl equations (2.1) and (2.2). While this is true, it is *not* proper to say that only the two curl equations are required to describe Maxwell's theory. This is because an additional physical assumption, not present in the two curl equations, is required to complete the derivation. Either the divergence equations must be specified, or the values of certain constants that fix the initial conditions on the fields must be specified. It is customary to specify the divergence equations and include them with the curl equations to form the complete set we now call "Maxwell's equations."

To identify the interdependence, we take the divergence of (2.1) to get

$$\nabla \cdot (\nabla \times \mathbf{E}) = \nabla \cdot \left(-\frac{\partial \mathbf{B}}{\partial t} \right),$$

hence

$$\frac{\partial}{\partial t} (\nabla \cdot \mathbf{B}) = 0$$

by (B.55). This requires that $\nabla \cdot \mathbf{B}$ be constant with time, say $\nabla \cdot \mathbf{B}(\mathbf{r}, t) = C_B(\mathbf{r})$.

The constant C_B must be specified as part of the postulate of Maxwell's theory, and the choice we make is subject to experimental validation. We postulate that $C_B(\mathbf{r}) = 0$, which leads us to (2.4). Note that if we can identify a time prior to which $\mathbf{B}(\mathbf{r}, t) \equiv 0$, then $C_B(\mathbf{r})$ must vanish. For this reason, $C_B(\mathbf{r}) = 0$ and (2.4) are often called the "initial conditions" for Faraday's law [160].

Next we take the divergence of (2.2) to find that

$$\nabla \cdot (\nabla \times \mathbf{H}) = \nabla \cdot \mathbf{J} + \frac{\partial}{\partial t}(\nabla \cdot \mathbf{D}).$$

Using (2.5) and (B.55), we obtain

$$\frac{\partial}{\partial t}(\rho - \nabla \cdot \mathbf{D}) = 0$$

and thus $\rho - \nabla \cdot \mathbf{D}$ must be some temporal constant $C_D(\mathbf{r})$. Again, we must postulate the value of C_D as part of the Maxwell theory. We choose $C_D(\mathbf{r}) = 0$ and thus obtain Gauss's law (2.3). If we can identify a time prior to which both $\mathbf{D}$ and ρ are everywhere equal to zero, then $C_D(\mathbf{r})$ must vanish. Hence $C_D(\mathbf{r}) = 0$ and (2.3) may be regarded as "initial conditions" for Ampere's law. Combining the two sets of initial conditions, we find that the curl equations imply the divergence equations, as long as we can find a time prior to which all of the fields $\mathbf{E}, \mathbf{D}, \mathbf{B}, \mathbf{H}$ and the sources $\mathbf{J}$ and ρ are equal to zero (since all the fields are related through the curl equations, and the charge and current are related through the continuity equation). Conversely, the empirical evidence supporting the two divergence equations implies that such a time should exist.

Throughout this book we shall refer to the two curl equations as the "fundamental" Maxwell equations, and to the two divergence equations as the "auxiliary" equations. The fundamental equations describe the relationships between the fields, while, as we have seen, the auxiliary equations provide a sort of initial condition. This does not imply that the auxiliary equations are of lesser importance; indeed, they are required to establish uniqueness of the fields, to derive the wave equations for the fields, and to properly describe static fields.

2.1.1.2   Field vector terminology

Various terms are used for the field vectors, sometimes harkening back to the descriptions used by Maxwell himself, and often based on the physical nature of the fields. We are attracted to Sommerfeld's separation of the fields into *entities of intensity* ($\mathbf{E}, \mathbf{B}$) and *entities of quantity* ($\mathbf{D}, \mathbf{H}$). In this system $\mathbf{E}$ is called the *electric field strength*, $\mathbf{B}$ the *magnetic field strength*, $\mathbf{D}$ the *electric excitation*, and $\mathbf{H}$ the *magnetic excitation* [180]. Maxwell separated the fields into a set ($\mathbf{E}, \mathbf{H}$) of vectors that appear within line integrals to give work-related quantities, and a set ($\mathbf{B}, \mathbf{D}$) of vectors that appear within surface integrals to give flux-related quantities; we shall see this clearly when considering the integral forms of Maxwell's equations. By this system, authors such as Jones [98] and Ramo, Whinnery, and Van Duzer [156] call $\mathbf{E}$ the *electric intensity*, $\mathbf{H}$ the *magnetic intensity*, $\mathbf{B}$ the *magnetic flux density*, and $\mathbf{D}$ the *electric flux density*.

Maxwell himself designated names for each of the vector quantities. In his classic paper "A Dynamical Theory of the Electromagnetic Field," [177] Maxwell referred to the quantity we now designate $\mathbf{E}$ as the *electromotive force*, the quantity $\mathbf{D}$ as the *electric displacement* (with a time rate of change given by his now famous "displacement current"), the quantity $\mathbf{H}$ as the *magnetic force*, and the quantity $\mathbf{B}$ as the *magnetic induction* (although he described $\mathbf{B}$ as a density of lines of magnetic force). Maxwell also included a quantity designated *electromagnetic momentum* as an integral part of his theory. We now know this as the vector potential **A**, which is not generally included as a part of the electromagnetics postulate.

Many authors follow the original terminology of Maxwell, with some slight modifications. For instance, Stratton [183] calls **E** the *electric field intensity*, **H** the *magnetic field intensity*, **D** the *electric displacement*, and **B** the *magnetic induction*. Jackson [92] calls **E** the *electric field*, **H** the *magnetic field*, **D** the *displacement*, and **B** the *magnetic induction*.

Other authors choose freely among combinations of these terms. For instance, Kong [108] calls **E** the *electric field strength*, **H** the *magnetic field strength*, **B** the *magnetic flux density*, and **D** the *electric displacement*. We do not wish to inject further confusion into the issue of nomenclature; still, we find it helpful to use as simple a naming system as possible. We shall refer to **E** as the *electric field*, **H** as the *magnetic field*, **D** as the *electric flux density*, and **B** as the *magnetic flux density*. When we use the term *electromagnetic field*, we imply the entire set of field vectors (**E**, **D**, **B**, **H**) used in Maxwell's theory.

2.1.1.3 Invariance of Maxwell's equations

Maxwell's differential equations are valid for any system in uniform relative motion with respect to the laboratory frame of reference in which we normally do our measurements. The field equations describe the relationships between the source and mediating fields *within that frame of reference*. This property was first proposed for moving material media by Minkowski in 1908 (using the term *covariance*) [132]. For this reason, Maxwell's equations expressed in the form (2.1)–(2.2) are referred to as the *Minkowski form*.

2.1.2 Connection to mechanics

Our postulate must include a connection between the abstract quantities of charge and field and a measurable physical quantity. A convenient means of linking electromagnetics to other classical theories is through mechanics. We postulate that charges experience mechanical forces given by the *Lorentz force equation*. If a small volume element dV contains a total charge $\rho\,dV$, then the force experienced by that charge when moving at velocity **v** in an electromagnetic field is

$$d\mathbf{F} = \rho\,dV\,\mathbf{E} + \rho\mathbf{v}\,dV \times \mathbf{B}.$$

As with any postulate, we verify this equation through experiment. Note that we write the Lorentz force in terms of charge $\rho\,dV$, rather than charge density ρ, since charge is an invariant quantity under a Lorentz transformation.

The important links between the electromagnetic fields and energy and momentum must also be postulated. We postulate that the quantity

$$\mathbf{S}_{em} = \mathbf{E} \times \mathbf{H}$$

represents the transport density of electromagnetic power, and that the quantity

$$\mathbf{g}_{em} = \mathbf{D} \times \mathbf{B}$$

represents the transport density of electromagnetic momentum.

2.2   The well-posed nature of the postulate

It is important to investigate whether Maxwell's equations, along with the point form of the continuity equation, suffice as a useful theory of electromagnetics. Certainly we must agree that a theory is "useful" as long as it is defined as such by the scientists and engineers who employ it. In practice a theory is considered useful if it accurately predicts the behavior of nature under given circumstances, and even a theory that often fails may be useful if it is the best available. We choose here to take a more narrow view and investigate whether the theory is "well-posed."

A mathematical model for a physical problem is said to be *well-posed*, or *correctly set*, if three conditions hold:

1. the model has at least one solution (*existence*);

2. the model has at most one solution (*uniqueness*);

3. the solution is continuously dependent on the data supplied.

The importance of the first condition is obvious: if the electromagnetic model has no solution, it will be of little use to scientists and engineers. The importance of the second condition is equally obvious: if we apply two different solution methods to the same model and get two different answers, the model will not be very helpful in analysis or design work. The third point is more subtle; it is often extended in a practical sense to the following statement:

3'. Small changes in the data supplied produce equally small changes in the solution.

That is, the solution is not sensitive to errors in the data. To make sense of this we must decide which quantity is specified (the independent quantity) and which remains to be calculated (the dependent quantity). Commonly the source field (charge) is taken as the independent quantity, and the mediating (electromagnetic) field is computed from it; in such cases it can be shown that Maxwell's equations are well-posed. Taking the electromagnetic field to be the independent quantity, we can produce situations in which the computed quantity (charge or current) changes wildly with small changes in the specified fields. These situations (called *inverse problems*) are of great importance in remote sensing, where the field is measured and the properties of the object probed are thereby deduced.

At this point we shall concentrate on the "forward" problem of specifying the source field (charge) and computing the mediating field (the electromagnetic field). In this case we may question whether the first of the three conditions (existence) holds. We have twelve unknown quantities (the scalar components of the four vector fields), but only eight equations to describe them (from the scalar components of the two fundamental Maxwell equations and the two scalar auxiliary equations). With fewer equations than unknowns, we cannot be sure that a solution exists, and we refer to Maxwell's equations as being *indefinite*. To overcome this problem, we must specify more information in the form of constitutive relations among the field quantities $\mathbf{E}$, $\mathbf{B}$, $\mathbf{D}$, $\mathbf{H}$, and $\mathbf{J}$. When these are properly formulated, the number of unknowns and the number of equations are equal and Maxwell's equations are in *definite form*. If we provide more equations than unknowns, the solution may be nonunique. When we model the electromagnetic properties of materials, we must supply precisely the right amount of information in the constitutive relations, or our postulate will not be well-posed.

Once Maxwell's equations are in definite form, standard methods for partial differential equations can be used to determine whether the electromagnetic model is well-posed. In a nutshell, the system (2.1)–(2.2) of hyperbolic differential equations is well-posed if and only if we specify $\mathbf{E}$ and $\mathbf{H}$ throughout a volume region V at some time instant, and also specify, at all subsequent times,

1. the tangential component of $\mathbf{E}$ over all of the boundary surface S, or

2. the tangential component of $\mathbf{H}$ over all of S, or

3. the tangential component of $\mathbf{E}$ over part of S, and the tangential component of $\mathbf{H}$ over the remainder of S.

Proof of all three of the conditions of well-posedness is quite tedious, but a simplified uniqueness proof is often given in textbooks on electromagnetics. The procedure used by Stratton [183] is reproduced below. The interested reader should refer to Hansen and Yaghjian [79] for a discussion of the existence of solutions to Maxwell's equations.

2.2.1   Uniqueness of solutions to Maxwell's equations

Consider a simply connected region of space V bounded by a surface S, where both V and S contain only ordinary points. The fields within V are associated with a current distribution $\mathbf{J}$, which may be internal to V (entirely or in part). By the initial conditions that imply the auxiliary Maxwell's equations, we know there is a time, say $t = 0$, prior to which the current is zero for all time, and thus by causality the fields throughout V are identically zero for all times $t < 0$. We next assume that the fields are specified throughout V at some time $t_0 > 0$, and seek conditions under which they are determined uniquely for all $t > t_0$.

Let the field set $(\mathbf{E}_1, \mathbf{D}_1, \mathbf{B}_1, \mathbf{H}_1)$ be a solution to Maxwell's equations (2.1)–(2.2) associated with the current $\mathbf{J}$ (along with an appropriate set of constitutive relations), and let $(\mathbf{E}_2, \mathbf{D}_2, \mathbf{B}_2, \mathbf{H}_2)$ be a second solution associated with $\mathbf{J}$. To determine the conditions for uniqueness of the fields, we look for a situation that results in $\mathbf{E}_1 = \mathbf{E}_2$, $\mathbf{B}_1 = \mathbf{B}_2$, and so on. The electromagnetic fields must obey

$$\nabla \times \mathbf{E}_1 = -\frac{\partial \mathbf{B}_1}{\partial t}, \qquad\qquad \nabla \times \mathbf{H}_1 = \mathbf{J} + \frac{\partial \mathbf{D}_1}{\partial t},$$

$$\nabla \times \mathbf{E}_2 = -\frac{\partial \mathbf{B}_2}{\partial t}, \qquad\qquad \nabla \times \mathbf{H}_2 = \mathbf{J} + \frac{\partial \mathbf{D}_2}{\partial t}.$$

Subtracting, we have

$$\nabla \times (\mathbf{E}_1 - \mathbf{E}_2) = -\frac{\partial (\mathbf{B}_1 - \mathbf{B}_2)}{\partial t}, \tag{2.6}$$

$$\nabla \times (\mathbf{H}_1 - \mathbf{H}_2) = \frac{\partial (\mathbf{D}_1 - \mathbf{D}_2)}{\partial t}, \tag{2.7}$$

hence defining $\mathbf{E}_0 = \mathbf{E}_1 - \mathbf{E}_2$, $\mathbf{B}_0 = \mathbf{B}_1 - \mathbf{B}_2$, and so on, we have

$$\mathbf{H}_0 \cdot (\nabla \times \mathbf{E}_0) = -\mathbf{H}_0 \cdot \frac{\partial \mathbf{B}_0}{\partial t},$$

$$\mathbf{E}_0 \cdot (\nabla \times \mathbf{H}_0) = \mathbf{E}_0 \cdot \frac{\partial \mathbf{D}_0}{\partial t}.$$

Subtracting again, we have

$$\mathbf{E}_0 \cdot (\nabla \times \mathbf{H}_0) - \mathbf{H}_0 \cdot (\nabla \times \mathbf{E}_0) = \mathbf{E}_0 \cdot \frac{\partial \mathbf{D}_0}{\partial t} + \mathbf{H}_0 \cdot \frac{\partial \mathbf{B}_0}{\partial t},$$

hence

$$-\nabla \cdot (\mathbf{E}_0 \times \mathbf{H}_0) = \mathbf{E}_0 \cdot \frac{\partial \mathbf{D}_0}{\partial t} + \mathbf{H}_0 \cdot \frac{\partial \mathbf{B}_0}{\partial t}$$

by (B.50). Integrating both sides throughout V and using the divergence theorem on the left-hand side, we get

$$-\oint_S (\mathbf{E}_0 \times \mathbf{H}_0) \cdot \mathbf{dS} = \int_V \left(\mathbf{E}_0 \cdot \frac{\partial \mathbf{D}_0}{\partial t} + \mathbf{H}_0 \cdot \frac{\partial \mathbf{B}_0}{\partial t} \right) dV.$$

Breaking S into two arbitrary portions and using (B.6), we obtain

$$\int_{S_1} \mathbf{E}_0 \cdot (\hat{\mathbf{n}} \times \mathbf{H}_0) \, dS - \int_{S_2} \mathbf{H}_0 \cdot (\hat{\mathbf{n}} \times \mathbf{E}_0) \, dS = \int_V \left(\mathbf{E}_0 \cdot \frac{\partial \mathbf{D}_0}{\partial t} + \mathbf{H}_0 \cdot \frac{\partial \mathbf{B}_0}{\partial t} \right) dV.$$

Now if $\hat{\mathbf{n}} \times \mathbf{E}_0 = 0$ or $\hat{\mathbf{n}} \times \mathbf{H}_0 = 0$ over all of S, or some combination of these conditions holds over all of S, then

$$\int_V \left(\mathbf{E}_0 \cdot \frac{\partial \mathbf{D}_0}{\partial t} + \mathbf{H}_0 \cdot \frac{\partial \mathbf{B}_0}{\partial t} \right) dV = 0. \qquad (2.8)$$

This expression implies a relationship between $\mathbf{E}_0$, $\mathbf{D}_0$, $\mathbf{B}_0$, and $\mathbf{H}_0$. Since V is arbitrary, we see that one possibility is simply to have $\mathbf{D}_0$ and $\mathbf{B}_0$ constant with time. However, since the fields are identically zero for $t < 0$, if they are constant for all time then those constant values must be zero. Another possibility is to have one of each pair $(\mathbf{E}_0, \mathbf{D}_0)$ and $(\mathbf{H}_0, \mathbf{B}_0)$ equal to zero. Then, by (2.6) and (2.7), $\mathbf{E}_0 = 0$ implies $\mathbf{B}_0 = 0$, and $\mathbf{D}_0 = 0$ implies $\mathbf{H}_0 = 0$. Thus $\mathbf{E}_1 = \mathbf{E}_2$, $\mathbf{B}_1 = \mathbf{B}_2$, and so on, and the solution is unique throughout V. However, we cannot in general rule out more complicated relationships. The number of possibilities depends on the additional constraints on the relationship between $\mathbf{E}_0$, $\mathbf{D}_0$, $\mathbf{B}_0$, and $\mathbf{H}_0$ that we must supply to describe the material supporting the field — i.e., the constitutive relationships. For a simple medium described by the time-constant permittivity ϵ and permeability μ, (2.8) becomes

$$\int_V \left(\mathbf{E}_0 \cdot \epsilon \frac{\partial \mathbf{E}_0}{\partial t} + \mathbf{H}_0 \cdot \mu \frac{\partial \mathbf{H}_0}{\partial t} \right) dV = 0,$$

or

$$\frac{1}{2} \frac{\partial}{\partial t} \int_V (\epsilon \mathbf{E}_0 \cdot \mathbf{E}_0 + \mu \mathbf{H}_0 \cdot \mathbf{H}_0) \, dV = 0.$$

Since the integrand is always positive or zero (and not constant with time, as mentioned above), the only possible conclusion is that $\mathbf{E}_0$ and $\mathbf{H}_0$ must both be zero, and thus the fields are unique.

When establishing more complicated constitutive relations, we must be careful to ensure that they lead to a unique solution, and that the condition for uniqueness is understood. In the case above, the assumption $\hat{\mathbf{n}} \times \mathbf{E}_0|_S = 0$ implies that the tangential components of $\mathbf{E}_1$ and $\mathbf{E}_2$ are identical over S — that is, we must give specific values of these quantities on S to ensure uniqueness. A similar statement holds for the condition $\hat{\mathbf{n}} \times \mathbf{H}_0|_S = 0$. Requiring that constitutive relations lead to a unique solution is known as *just setting*, and is one of several factors that must be considered, as discussed in the next section.

Uniqueness implies that the electromagnetic state of an isolated region of space may be determined without the knowledge of conditions outside the region. If we wish to solve Maxwell's equations for that region, we need know only the source density within the region and the values of the tangential fields over the bounding surface. The effects of a complicated external world are thus reduced to the specification of surface fields. This concept has numerous applications to problems in antennas, diffraction, and guided waves.

2.2.2   Constitutive relations

We must supply a set of constitutive relations to complete the conditions for well-posedness. We generally split these relations into two sets. The first describes the relationships between the electromagnetic field quantities, and the second describes mechanical interaction between the fields and resulting secondary sources. All of these relations depend on the properties of the medium supporting the electromagnetic field. Material phenomena are quite diverse, and it is remarkable that the Maxwell–Minkowski equations hold for all phenomena yet discovered. All material effects, from nonlinearity to chirality to temporal dispersion, are described by the constitutive relations.

The specification of constitutive relationships is required in many areas of physical science to describe the behavior of "ideal materials": mathematical models of actual materials encountered in nature. For instance, in continuum mechanics the constitutive equations describe the relationship between material motions and stress tensors [210]. Truesdell and Toupin [199] give an interesting set of "guiding principles" for the concerned scientist to use when constructing constitutive relations. These include consideration of *consistency* (with the basic conservation laws of nature), *coordinate invariance* (independence of coordinate system), *isotropy and aeolotropy* (dependence on, or independence of, orientation), *just setting* (constitutive parameters should lead to a unique solution), *dimensional invariance* (similarity), *material indifference* (nondependence on the observer), and *equipresence* (inclusion of *all* relevant physical phenomena in *all* of the constitutive relations across disciplines).

The constitutive relations generally involve a set of constitutive parameters and a set of constitutive operators. The constitutive parameters may be as simple as constants of proportionality between the fields or they may be components in a dyadic relationship. The constitutive operators may be linear and integro-differential in nature, or may imply some nonlinear operation on the fields. If the constitutive parameters are spatially constant within a certain region, we term the medium *homogeneous* within that region. If the constitutive parameters vary spatially, the medium is *inhomogeneous*. If the constitutive parameters are constants with time, we term the medium *stationary*; if they are time-changing, the medium is *nonstationary*. If the constitutive operators involve time derivatives or integrals, the medium is said to be *temporally dispersive*; if space derivatives or integrals are involved, the medium is *spatially dispersive*. Examples of all these effects can be found in common materials. It is important to note that the constitutive parameters may depend on other physical properties of the material, such as temperature, mechanical stress, and isomeric state, just as the mechanical constitutive parameters of a material may depend on the electromagnetic properties (principle of equipresence).

Many effects produced by linear constitutive operators, such as those associated with temporal dispersion, have been studied primarily in the frequency domain. In this case temporal derivative and integral operations produce complex constitutive parameters. It is becoming equally important to characterize these effects directly in the time domain for use with direct time-domain field solving techniques such as the finite-difference time-domain (FDTD) method. We shall cover the very basic properties of dispersive media in this section. A detailed description of frequency-domain fields (and a discussion of complex constitutive parameters) is deferred until later in this book.

It is difficult to find a simple and consistent means for classifying materials by their electromagnetic effects. One way is to separate linear and nonlinear materials, then categorize linear materials by the way in which the fields are coupled through the constitutive relations:

1. *Isotropic* materials are those in which $\mathbf{D}$ is related to $\mathbf{E}$, $\mathbf{B}$ is related to $\mathbf{H}$, and the secondary source current $\mathbf{J}$ is related to $\mathbf{E}$, with the field direction in each pair aligned.

2. In *anisotropic* materials the pairings are the same, but the fields in each pair are generally not aligned.

3. In *biisotropic* materials (such as chiral media) the fields $\mathbf{D}$ and $\mathbf{B}$ depend on both $\mathbf{E}$ and $\mathbf{H}$, but with no realignment of $\mathbf{E}$ or $\mathbf{H}$; for instance, $\mathbf{D}$ is given by the addition of a scalar times $\mathbf{E}$ plus a second scalar times $\mathbf{H}$. Thus the contributions to $\mathbf{D}$ involve no changes to the directions of $\mathbf{E}$ and $\mathbf{H}$.

4. *Bianisotropic* materials exhibit the most general behavior: $\mathbf{D}$ and $\mathbf{H}$ depend on both $\mathbf{E}$ and $\mathbf{B}$, with an arbitrary realignment of either or both of these fields.

In 1888, Roentgen showed experimentally that a material isotropic in its own stationary reference frame exhibits bianisotropic properties when observed from a moving frame. Only recently have materials bianisotropic in their own rest frame been discovered. In 1894 Curie predicted that in a stationary material, based on symmetry, an electric field might produce magnetic effects and a magnetic field might produce electric effects. These effects, coined *magnetoelectric* by Landau and Lifshitz in 1957, were sought unsuccessfully by many experimentalists during the first half of the twentieth century. In 1959 the Soviet scientist I.E. Dzyaloshinskii predicted that, theoretically, the antiferromagnetic material chromium oxide (Cr_2O_3) should display magnetoelectric effects. The magneto-electric effect was finally observed soon after by D.N. Astrov in a single crystal of Cr_2O_3 using a 10 kHz electric field. Since then the effect has been observed in many different materials. Recently, highly exotic materials with useful electromagnetic properties have been proposed and studied in depth, including chiroplasmas and chiroferrites [211]. As the technology of materials synthesis advances, a host of new and intriguing media will certainly be created.

The most general forms of the constitutive relations between the fields may be written in symbolic form as

$$\mathbf{D} = \mathbf{D}[\mathbf{E}, \mathbf{B}],$$
$$\mathbf{H} = \mathbf{H}[\mathbf{E}, \mathbf{B}].$$

That is, $\mathbf{D}$ and $\mathbf{H}$ have some mathematically descriptive relationship to $\mathbf{E}$ and $\mathbf{B}$. The specific forms of the relationships may be written in terms of dyadics as [109]

$$c\mathbf{D} = \bar{\mathbf{P}} \cdot \mathbf{E} + \bar{\mathbf{L}} \cdot (c\mathbf{B}), \tag{2.9}$$
$$\mathbf{H} = \bar{\mathbf{M}} \cdot \mathbf{E} + \bar{\mathbf{Q}} \cdot (c\mathbf{B}), \tag{2.10}$$

where each of the quantities $\bar{\mathbf{P}}, \bar{\mathbf{L}}, \bar{\mathbf{M}}, \bar{\mathbf{Q}}$ may be dyadics in the usual sense, or dyadic operators containing space or time derivatives or integrals, or some nonlinear operations on the fields. We may write these expressions as a single matrix equation

$$\begin{bmatrix} c\mathbf{D} \\ \mathbf{H} \end{bmatrix} = [\bar{\mathbf{C}}] \begin{bmatrix} \mathbf{E} \\ c\mathbf{B} \end{bmatrix} \tag{2.11}$$

where the 6×6 matrix

$$[\bar{\mathbf{C}}] = \begin{bmatrix} \bar{\mathbf{P}} & \bar{\mathbf{L}} \\ \bar{\mathbf{M}} & \bar{\mathbf{Q}} \end{bmatrix}.$$

This most general relationship between fields is the property of a bianisotropic material.

We may wonder why $\mathbf{D}$ is not related to $(\mathbf{E}, \mathbf{B}, \mathbf{H})$, $\mathbf{E}$ to $(\mathbf{D}, \mathbf{B})$, etc. The reason is that since the field pairs $(\mathbf{E}, \mathbf{B})$ and $(\mathbf{D}, \mathbf{H})$ convert identically under a Lorentz transformation, a constitutive relation that maps fields as in (2.11) is form invariant, as are the Maxwell–Minkowski equations. That is, although the constitutive parameters may vary numerically between observers moving at different velocities, the form of the relationship given by (2.11) is maintained.

Many authors choose to relate $(\mathbf{D}, \mathbf{B})$ to $(\mathbf{E}, \mathbf{H})$, often because the expressions are simpler and can be more easily applied to specific problems. For instance, in a linear, isotropic material (as shown below), $\mathbf{D}$ is directly proportional to $\mathbf{E}$ and $\mathbf{B}$ is directly proportional to $\mathbf{H}$. To provide the appropriate expression for the constitutive relations, we need only remap (2.11). This gives

$$\mathbf{D} = \bar{\epsilon} \cdot \mathbf{E} + \bar{\xi} \cdot \mathbf{H}, \tag{2.12}$$

$$\mathbf{B} = \bar{\zeta} \cdot \mathbf{E} + \bar{\mu} \cdot \mathbf{H}, \tag{2.13}$$

or

$$\begin{bmatrix} \mathbf{D} \\ \mathbf{B} \end{bmatrix} = [\bar{\mathbf{C}}_{EH}] \begin{bmatrix} \mathbf{E} \\ \mathbf{H} \end{bmatrix}, \tag{2.14}$$

where the new constitutive parameters $\bar{\epsilon}, \bar{\xi}, \bar{\zeta}, \bar{\mu}$ can be easily found from the original constitutive parameters $\bar{\mathbf{P}}, \bar{\mathbf{L}}, \bar{\mathbf{M}}, \bar{\mathbf{Q}}$. We do note, however, that in the form (2.12)–(2.13) the Lorentz invariance of the constitutive equations is not obvious.

In the following sections we shall characterize some of the most common materials according to these classifications. With this approach, effects such as temporal or spatial dispersion are not part of the classification process, but arise from the nature of the constitutive parameters. Hence we shall not dwell on the particulars of the constitutive parameters, but shall concentrate on the form of the constitutive relations.

2.2.2.1 Constitutive relations for fields in free space

In a vacuum, the fields are related by the simple constitutive equations

$$\mathbf{D} = \epsilon_0 \mathbf{E}, \tag{2.15}$$

$$\mathbf{H} = \mu_0^{-1} \mathbf{B}. \tag{2.16}$$

The quantities μ_0 and ϵ_0 are, respectively, the *free-space permeability* and *permittivity constants*. It is convenient to use three numerical quantities to describe the electromagnetic properties of free space — μ_0, ϵ_0, and the speed of light c — and interrelate them through the equation

$$c = 1/(\mu_0 \epsilon_0)^{1/2}.$$

Historically it has been the practice to define μ_0, measure c, and compute ϵ_0. The speed of light is now a defined constant in the SI system of units; the meter is based on this constant. Thus, each of the quantities c, ϵ_0, and μ_0 is exact:

$$c = 299,792,458 \text{ m/s},$$

$$\mu_0 = 4\pi \times 10^{-7} \text{ H/m},$$

$$\epsilon_0 = 8.854187817620\ldots \times 10^{-12} \text{ F/m}.$$

With the two constitutive equations, we have enough information to put Maxwell's equations into definite form. Traditionally (2.15) and (2.16) are substituted into (2.1)–(2.2) to give

$$\nabla \times \mathbf{E} = -\frac{\partial \mathbf{B}}{\partial t},$$

$$\nabla \times \mathbf{B} = \mu_0 \mathbf{J} + \mu_0 \epsilon_0 \frac{\partial \mathbf{E}}{\partial t}.$$

These are two vector equations in two vector unknowns (equivalently, six scalar equations in six scalar unknowns).

In terms of the general constitutive relation (2.11), we find that free space is isotropic with

$$\bar{\mathbf{P}} = \bar{\mathbf{Q}} = \frac{1}{\eta_0}\bar{\mathbf{I}}, \qquad \bar{\mathbf{L}} = \bar{\mathbf{M}} = 0,$$

where $\eta_0 = (\mu_0/\epsilon_0)^{1/2}$ is called the *intrinsic impedance of free space*. This emphasizes the fact that free space has, along with c, only a single empirical constant associated with it (i.e., ϵ_0 or η_0). Since no derivative or integral operators appear in the constitutive relations, free space is nondispersive.

2.2.2.2   Constitutive relations in a linear isotropic material

In a linear isotropic material there is proportionality between $\mathbf{D}$ and $\mathbf{E}$ and between $\mathbf{B}$ and $\mathbf{H}$. The constants of proportionality are the *permittivity* ϵ and the *permeability* μ. If the material is nondispersive, the constitutive relations take the form

$$\mathbf{D} = \epsilon\mathbf{E}, \qquad \mathbf{B} = \mu\mathbf{H},$$

where ϵ and μ may depend on position for inhomogeneous materials. Often the permittivity and permeability are referenced to the permittivity and permeability of free space according to

$$\epsilon = \epsilon_r \epsilon_0, \qquad \mu = \mu_r \mu_0.$$

Here the dimensionless quantities ϵ_r and μ_r are called, respectively, the *relative permittivity* and *relative permeability*.

When dealing with the Maxwell–Boffi equations (§ 2.4), the difference between the material and free space values of $\mathbf{D}$ and $\mathbf{H}$ becomes important. Thus, for linear isotropic materials, we often write the constitutive relations as

$$\mathbf{D} = \epsilon_0 \mathbf{E} + \epsilon_0 \chi_e \mathbf{E}, \tag{2.17}$$

$$\mathbf{B} = \mu_0 \mathbf{H} + \mu_0 \chi_m \mathbf{H}, \tag{2.18}$$

where the dimensionless quantities $\chi_e = \epsilon_r - 1$ and $\chi_m = \mu_r - 1$ are called, respectively, the *electric* and *magnetic susceptibilities* of the material. In terms of (2.11) we have

$$\bar{\mathbf{P}} = \frac{\epsilon_r}{\eta_0}\bar{\mathbf{I}}, \qquad \bar{\mathbf{Q}} = \frac{1}{\eta_0 \mu_r}\bar{\mathbf{I}}, \qquad \bar{\mathbf{L}} = \bar{\mathbf{M}} = 0.$$

Generally, a material will have either its electric or magnetic properties dominant. If $\mu_r = 1$ and $\epsilon_r \neq 1$, then the material is generally called a *perfect dielectric* or a *perfect insulator*, and is said to be an electric material. If $\epsilon_r = 1$ and $\mu_r \neq 1$, the material is said to be a magnetic material.

A linear isotropic material may also have conduction properties. In a *conducting material*, a constitutive relation is generally used to describe the mechanical interaction of field and charge by relating the electric field to a secondary electric current. For a nondispersive isotropic material, the current is aligned with, and proportional to, the electric field; there are no temporal operators in the constitutive relation, which is simply

$$\mathbf{J} = \sigma \mathbf{E}. \tag{2.19}$$

This is known as *Ohm's law*. Here σ is the *conductivity* of the material.

If $\mu_r \approx 1$ and σ is very small, the material is generally called a *good dielectric*. If σ is very large, the material is generally called a *good conductor*. The conditions by which we say the conductivity is "small" or "large" are usually established using the frequency response of the material. Materials that are good dielectrics over broad ranges of frequency include various glasses and plastics such as fused quartz, polyethylene, and teflon. Materials that are good conductors over broad ranges of frequency include common metals such as gold, silver, and copper.

For dispersive linear isotropic materials, the constitutive parameters become nonstationary (time dependent), and the constitutive relations involve time operators. (Note that the name *dispersive* describes the tendency for pulsed electromagnetic waves to spread out, or disperse, in materials of this type.) If we assume that the relationships given by (2.17), (2.18), and (2.19) retain their product form in the frequency domain, then by the convolution theorem we have in the time domain the constitutive relations

$$\mathbf{D}(\mathbf{r}, t) = \epsilon_0 \left[\mathbf{E}(\mathbf{r}, t) + \int_{-\infty}^{t} \chi_e(\mathbf{r}, t - t') \mathbf{E}(\mathbf{r}, t') \, dt' \right], \tag{2.20}$$

$$\mathbf{B}(\mathbf{r}, t) = \mu_0 \left[\mathbf{H}(\mathbf{r}, t) + \int_{-\infty}^{t} \chi_m(\mathbf{r}, t - t') \mathbf{H}(\mathbf{r}, t') \, dt' \right], \tag{2.21}$$

$$\mathbf{J}(\mathbf{r}, t) = \int_{-\infty}^{t} \sigma(\mathbf{r}, t - t') \mathbf{E}(\mathbf{r}, t') \, dt'. \tag{2.22}$$

These expressions were first introduced by Volterra in 1912 [199]. We see that for a linear dispersive material of this type the constitutive operators are time integrals, and that the behavior of $\mathbf{D}(t)$ depends not only on the value of $\mathbf{E}$ at time t, but on its values at all past times. Thus, in dispersive materials there is a "time lag" between the effect of the applied field and the polarization or magnetization that results. In the frequency domain, temporal dispersion is associated with complex values of the constitutive parameters, which, to describe a causal relationship, cannot be constant with frequency. The nonzero imaginary component is identified with the dissipation of electromagnetic energy as heat. Causality is implied by the upper limit being t in the convolution integrals, which indicates that $\mathbf{D}(t)$ cannot depend on future values of $\mathbf{E}(t)$. This assumption leads to a relationship between the real and imaginary parts of the frequency domain constitutive parameters as described through the Kramers–Kronig equations.

▶ **Example 2.1:** Fields in an inhomogeneous dielectric medium

A source-free region is filled with an inhomogeneous dielectric having permeability $\mu(\mathbf{r}) = \mu_0$

and permittivity $\epsilon(\mathbf{r}) = \epsilon_0 \epsilon_r e^{Kz}$. Assume the electric and magnetic fields take the form

$$\mathbf{E}(\mathbf{r}, t) = \hat{\mathbf{y}} E_y(z, t), \qquad \mathbf{H}(\mathbf{r}, t) = \hat{\mathbf{x}} H_x(z, t).$$

If

$$H_x(z, t) = H_0 e^{\frac{K}{2} z} J_1 \left(\frac{2k}{K} e^{\frac{K}{2} z} \right) \cos \omega t,$$

find $E_y(z, t)$ using Ampere's law. Then show that $H_x(z, t)$ and $E_y(z, t)$ obey Faraday's law. Here $k = \omega \sqrt{\mu_0 \epsilon_0 \epsilon_r}$, and $J_1(x)$ is the ordinary Bessel function of the first kind.

Solution: By Ampere's law we have

$$\nabla \times \mathbf{H} = \hat{\mathbf{y}} \frac{\partial H_x}{\partial z} = \hat{\mathbf{y}} \epsilon(z) \frac{\partial E_y}{\partial t}.$$

We compute

$$\frac{\partial H_x}{\partial z} = H_0 e^{\frac{K}{2} z} J_1' \left(\frac{2k}{K} e^{\frac{K}{2} z} \right) k e^{\frac{K}{2} z} \cos \omega t + H_0 \frac{K}{2} e^{\frac{K}{2} z} \left(\frac{2k}{K} e^{\frac{K}{2} z} \right) \cos \omega t.$$

Using $J_1'(x) = J_0(x) - J_1(x)/x$ we find that the J_1 terms cancel, leaving

$$\frac{\partial H_x}{\partial z} = H_0 k e^{Kz} J_0 \left(\frac{2k}{K} e^{\frac{K}{2} z} \right) \cos \omega t. \tag{2.23}$$

Integrating with respect to time we obtain

$$E_y(z, t) = H_0 \frac{k}{\omega \epsilon(z)} e^{Kz} J_0 \left(\frac{2k}{K} e^{\frac{K}{2} z} \right) \sin \omega t = H_0 \eta J_0 \left(\frac{2k}{K} e^{\frac{K}{2} z} \right) \sin \omega t,$$

where $\eta = \sqrt{\mu_0 / (\epsilon_0 \epsilon_r)}$ is a wave impedance with units of ohms.

Next we wish to verify Faraday's law

$$\nabla \times \mathbf{E} = \hat{\mathbf{x}} \frac{\partial E_y}{\partial z} = \hat{\mathbf{x}} \mu_0 \frac{\partial H_x}{\partial t}.$$

The derivative of E_y is

$$\frac{\partial E_y}{\partial z} = -H_0 \eta k e^{\frac{K}{2} z} J_1 \left(\frac{2k}{K} e^{\frac{K}{2} z} \right) \sin \omega t$$

since $J_0'(x) = -J_1(x)$. The derivative of H_x is

$$\frac{\partial H_x}{\partial t} = -H_0 \omega e^{\frac{K}{2} z} J_1 \left(\frac{2k}{K} e^{\frac{K}{2} z} \right) \sin \omega t$$

so that

$$\mu_0 \frac{\partial H_x}{\partial t} = -H_0 \eta k e^{\frac{K}{2} z} J_1 \left(\frac{2k}{K} e^{\frac{K}{2} z} \right) \sin \omega t$$

and Faraday's law is satisfied. ◄

2.2.2.3 Constitutive relations for fields in perfect conductors

In a perfect electric conductor (PEC) or a perfect magnetic conductor (PMC), the fields are exactly specified as the null field:

$$\mathbf{E} = \mathbf{D} = \mathbf{B} = \mathbf{H} = 0.$$

By Ampere's and Faraday's laws we must also have $\mathbf{J} = \mathbf{J}_m = 0$; hence, by the continuity equation, $\rho = \rho_m = 0$.

In addition to the null field, we have the condition that the tangential electric field on the surface of a PEC must be zero. Similarly, the tangential magnetic field on the surface of a PMC must be zero. This implies (§ 2.8.3) that an electric surface current may exist on the surface of a PEC but not on the surface of a PMC, while a magnetic surface current may exist on the surface of a PMC but not on the surface of a PEC.

A PEC may be regarded as the limit of a conducting material as $\sigma \to \infty$. In many practical cases, good conductors such as gold and copper can be assumed to be perfect electric conductors, which greatly simplifies the application of boundary conditions. No physical material is known to behave as a PMC, but the concept is mathematically useful for applying symmetry conditions (in which a PMC is sometimes referred to as a "magnetic wall") and for use in developing equivalence theorems.

2.2.2.4 Constitutive relations in a linear anisotropic material

In a linear anisotropic material there are relationships between $\mathbf{B}$ and $\mathbf{H}$ and between $\mathbf{D}$ and $\mathbf{E}$, but the field vectors are not aligned as in the isotropic case. We can thus write

$$\mathbf{D} = \bar{\epsilon} \cdot \mathbf{E}, \qquad \mathbf{B} = \bar{\mu} \cdot \mathbf{H}, \qquad \mathbf{J} = \bar{\sigma} \cdot \mathbf{E},$$

where $\bar{\epsilon}$ is called the *permittivity dyadic*, $\bar{\mu}$ is the *permeability dyadic*, and $\bar{\sigma}$ is the *conductivity dyadic*. In terms of the general constitutive relation (2.11) we have

$$\bar{\mathbf{P}} = c\bar{\epsilon}, \qquad \bar{\mathbf{Q}} = c^{-1}\bar{\mu}^{-1}, \qquad \bar{\mathbf{L}} = \bar{\mathbf{M}} = 0.$$

Many different types of materials demonstrate anisotropic behavior, including optical crystals, magnetized plasmas, and ferrites. Plasmas and ferrites are examples of *gyrotropic* media. With the proper choice of coordinate system, the frequency-domain permittivity or permeability can be written in matrix form as

$$[\tilde{\bar{\epsilon}}] = \begin{bmatrix} \epsilon_{11} & \epsilon_{12} & 0 \\ -\epsilon_{12} & \epsilon_{11} & 0 \\ 0 & 0 & \epsilon_{33} \end{bmatrix}, \qquad [\tilde{\bar{\mu}}] = \begin{bmatrix} \mu_{11} & \mu_{12} & 0 \\ -\mu_{12} & \mu_{11} & 0 \\ 0 & 0 & \mu_{33} \end{bmatrix}. \tag{2.24}$$

Each of the matrix entries may be complex. For the special case of a *lossless* gyrotropic material, the matrices become *hermitian*:

$$[\tilde{\bar{\epsilon}}] = \begin{bmatrix} \epsilon & -j\delta & 0 \\ j\delta & \epsilon & 0 \\ 0 & 0 & \epsilon_3 \end{bmatrix}, \qquad [\tilde{\bar{\mu}}] = \begin{bmatrix} \mu & -j\kappa & 0 \\ j\kappa & \mu & 0 \\ 0 & 0 & \mu_3 \end{bmatrix}, \tag{2.25}$$

where ϵ, ϵ_3, δ, μ, μ_3, and κ are real numbers.

Crystals have received particular attention because of their birefringent properties. A birefringent crystal can be characterized by a symmetric permittivity dyadic that has real permittivity parameters in the frequency domain; equivalently, the constitutive relations do not involve constitutive operators. A coordinate system called the *principal system*, with axes called the *principal axes*, can always be found so that the permittivity dyadic in that system is diagonal:

$$[\tilde{\bar{\epsilon}}] = \begin{bmatrix} \epsilon_x & 0 & 0 \\ 0 & \epsilon_y & 0 \\ 0 & 0 & \epsilon_z \end{bmatrix}.$$

The geometrical structure of a crystal determines the relationship between ϵ_x, ϵ_y, and ϵ_z. If $\epsilon_x = \epsilon_y < \epsilon_z$, then the crystal is *positive uniaxial* (e.g., quartz). If $\epsilon_x = \epsilon_y > \epsilon_z$, the crystal is *negative uniaxial* (e.g., calcite). If $\epsilon_x \neq \epsilon_y \neq \epsilon_z$, the crystal is *biaxial* (e.g., mica). In uniaxial crystals the z-axis is called the *optical axis*.

If the anisotropic material is dispersive, we can generalize the convolutional form of the isotropic dispersive media to obtain the constitutive relations

$$\mathbf{D}(\mathbf{r},t) = \epsilon_0 \left[\mathbf{E}(\mathbf{r},t) + \int_{-\infty}^{t} \bar{\chi}_e(\mathbf{r},t-t') \cdot \mathbf{E}(\mathbf{r},t')\,dt' \right],$$

$$\mathbf{B}(\mathbf{r},t) = \mu_0 \left[\mathbf{H}(\mathbf{r},t) + \int_{-\infty}^{t} \bar{\chi}_m(\mathbf{r},t-t') \cdot \mathbf{H}(\mathbf{r},t')\,dt' \right],$$

$$\mathbf{J}(\mathbf{r},t) = \int_{-\infty}^{t} \bar{\sigma}(\mathbf{r},t-t') \cdot \mathbf{E}(\mathbf{r},t')\,dt'.$$

2.2.2.5 Constitutive relations for biisotropic materials

A biisotropic material is an isotropic magnetoelectric material. Here we have $\mathbf{D}$ related to $\mathbf{E}$ and $\mathbf{B}$, and $\mathbf{H}$ related to $\mathbf{E}$ and $\mathbf{B}$, but with no realignment of the fields as in anisotropic (or bianisotropic) materials. Perhaps the simplest example is the *Tellegen medium* devised by B.D.H. Tellegen in 1948 [193], having

$$\mathbf{D} = \epsilon\mathbf{E} + \xi\mathbf{H}, \tag{2.26}$$

$$\mathbf{B} = \xi\mathbf{E} + \mu\mathbf{H}. \tag{2.27}$$

Note that these relations may be rearranged in a number of ways, including

$$\mathbf{D} = \left(\epsilon - \frac{\xi^2}{\mu} \right)\mathbf{E} + \frac{\xi}{\mu}\mathbf{B}, \tag{2.28}$$

$$\mathbf{H} = -\frac{\xi}{\mu}\mathbf{E} + \frac{1}{\mu}\mathbf{B}. \tag{2.29}$$

Tellegen proposed that his hypothetical material be composed of small (but macroscopic) ferromagnetic particles suspended in a liquid. This is an example of a *synthetic* material, constructed from ordinary materials to have an exotic electromagnetic behavior. Other examples include artificial dielectrics made from metallic particles imbedded in lightweight foams [57], and *chiral materials* made from small metallic helices suspended in resins [119].

▶ **Example 2.2:** Fields in a Tellegen medium

A source-free region is filled with a homogeneous Tellegen medium described by the constitutive relations (2.26)–(2.27). Assume the electric field has the form of a wave,

$$\mathbf{E}(\mathbf{r},t) = \hat{\mathbf{x}}E_0 f\left(t - \frac{z}{v} \right).$$

Here $f(t)$ is an arbitrary time function and

$$v = \frac{1}{\sqrt{\mu\epsilon - \xi^2}}$$

is the wave velocity with $\xi^2 < \mu\epsilon$ [27]. Use Maxwell's equations to find $\mathbf{B}$, $\mathbf{D}$, and $\mathbf{H}$. Verify that the fields satisfy both curl equations. Also find the angles between $\mathbf{B}$ and $\mathbf{E}$, between

B and **D**, and between **E** and **D**.

Solution: To find **B** we use Faraday's law

$$\nabla \times \mathbf{E} = -\frac{\partial \mathbf{B}}{\partial t}.$$

The curl of **E** is

$$\nabla \times \mathbf{E} = \hat{\mathbf{y}}\frac{\partial E_x}{\partial z} = -\hat{\mathbf{y}}\frac{E_0}{v}f'\left(t - \frac{z}{v}\right),$$

where $f'(u) = df(u)/du$. Time integration yields **B**:

$$\mathbf{B} = \hat{\mathbf{y}}\frac{E_0}{v}f\left(t - \frac{z}{v}\right).$$

Hence **E** and **B** are orthogonal. We can find **D** using (2.28). Since

$$\epsilon - \frac{\xi^2}{\mu} = \epsilon - \frac{\mu\epsilon - \frac{1}{v^2}}{\mu} = \frac{1}{v^2\mu},$$

we have

$$\mathbf{D} = \left(\epsilon - \frac{\xi^2}{\mu}\right)\mathbf{E} + \frac{\xi}{\mu}\mathbf{B} = \frac{E_0}{\mu v}\left[\hat{\mathbf{x}}\frac{1}{v} + \hat{\mathbf{y}}\xi\right]f\left(t - \frac{z}{v}\right).$$

Similarly, **H** is given by (2.29):

$$\mathbf{H} = -\frac{\xi}{\mu}\mathbf{E} + \frac{1}{\mu}\mathbf{B} = \frac{E_0}{\mu}\left[\hat{\mathbf{y}}\frac{1}{v} - \hat{\mathbf{x}}\xi\right]f\left(t - \frac{z}{v}\right).$$

We wish to verify the field expressions by substituting into Ampere's law

$$\nabla \times \mathbf{H} = \frac{\partial \mathbf{D}}{\partial t}.$$

The curl of **H** is

$$\nabla \times \mathbf{H} = -\hat{\mathbf{x}}\frac{\partial H_y}{\partial z} + \hat{\mathbf{y}}\frac{\partial H_x}{\partial z} = \frac{E_0}{\mu v}\left[\hat{\mathbf{x}}\frac{1}{v} + \hat{\mathbf{y}}\xi\right]f'\left(t - \frac{z}{v}\right).$$

But we also have

$$\frac{\partial \mathbf{D}}{\partial t} = \frac{E_0}{\mu v}\left[\hat{\mathbf{x}}\frac{1}{v} + \hat{\mathbf{y}}\xi\right]f'\left(t - \frac{z}{v}\right),$$

and thus Ampere's law is satisfied.

To find the angle between **B** and **D**, we note that a unit vector along **D** is

$$\hat{\mathbf{u}}_D = \frac{\hat{\mathbf{x}}\frac{1}{v} + \hat{\mathbf{y}}\xi}{\sqrt{(1/v)^2 + \xi^2}} = \frac{\hat{\mathbf{x}}\frac{1}{v} + \hat{\mathbf{y}}\xi}{\sqrt{\mu\epsilon - \xi^2 + \xi^2}} = \frac{\hat{\mathbf{x}}\frac{1}{v} + \hat{\mathbf{x}}\xi}{\sqrt{\mu\epsilon}}.$$

Hence

$$\hat{\mathbf{u}}_B \cdot \hat{\mathbf{u}}_D = \cos\theta_{BD} = \hat{\mathbf{y}} \cdot \frac{\hat{\mathbf{x}}\frac{1}{v} + \hat{\mathbf{y}}\xi}{\sqrt{\mu\epsilon}} = \frac{\xi}{\sqrt{\mu\epsilon}}$$

gives the angle θ_{BD} between **B** and **D**. So **B** and **E** are orthogonal but **B** and **D** are typically not. In like manner the angle θ_{ED} between **E** and **D** is given by

$$\hat{\mathbf{u}}_E \cdot \hat{\mathbf{u}}_D = \cos\theta_{ED} = \hat{\mathbf{x}} \cdot \frac{\hat{\mathbf{x}}\frac{1}{v} + \hat{\mathbf{y}}\xi}{\sqrt{\mu\epsilon}} = \frac{1}{v\sqrt{\mu\epsilon}} = \sqrt{1 - \frac{\xi^2}{\mu\epsilon}},$$

and thus, unlike in a simple dielectric, **E** and **D** are not collinear. ◀

Chiral materials are also biisotropic, and have the constitutive relations

$$\mathbf{D} = \epsilon\mathbf{E} - \chi\frac{\partial\mathbf{H}}{\partial t}, \tag{2.30}$$

$$\mathbf{B} = \mu\mathbf{H} + \chi\frac{\partial\mathbf{E}}{\partial t}, \tag{2.31}$$

where the constitutive parameter χ is called the *chirality parameter*. Note the presence of temporal derivative operators. Alternatively,

$$\mathbf{D} = \epsilon(\mathbf{E} + \beta\nabla \times \mathbf{E}),$$
$$\mathbf{B} = \mu(\mathbf{H} + \beta\nabla \times \mathbf{H}),$$

by Faraday's and Ampere's laws. Chirality is a natural state of symmetry; many natural substances are chiral materials, including DNA and many sugars. The time derivatives in (2.30)–(2.31) produce rotation of the polarization of time harmonic electromagnetic waves propagating in chiral media.

2.2.2.6 Constitutive relations in nonlinear media

Nonlinear electromagnetic effects have been studied by scientists and engineers since the beginning of the era of electrical technology. Familiar examples include saturation and hysteresis in ferromagnetic materials and the behavior of p–n junctions in solid-state rectifiers. The invention of the laser extended interest in nonlinear effects to the realm of optics, where phenomena such as parametric amplification and oscillation, harmonic generation, and magneto-optic interactions have found applications in modern devices [174].

Provided that the external field applied to a nonlinear electric material is small compared to the internal molecular fields, the relationship between $\mathbf{E}$ and $\mathbf{D}$ can be expanded in a Taylor series of the electric field. For an anisotropic material exhibiting no hysteresis effects, the constitutive relation is [134]

$$D_i(\mathbf{r}, t) = \epsilon_0 E_i(\mathbf{r}, t) + \sum_{j=1}^{3} \chi_{ij}^{(1)} E_j(\mathbf{r}, t) + \sum_{j,k=1}^{3} \chi_{ijk}^{(2)} E_j(\mathbf{r}, t) E_k(\mathbf{r}, t)$$

$$+ \sum_{j,k,l=1}^{3} \chi_{ijkl}^{(3)} E_j(\mathbf{r}, t) E_k(\mathbf{r}, t) E_l(\mathbf{r}, t) + \cdots \tag{2.32}$$

where the index $i = 1, 2, 3$ refers to the three components of the fields $\mathbf{D}$ and $\mathbf{E}$. The first sum in (2.32) is identical to the constitutive relation for linear anisotropic materials. Thus, $\chi_{ij}^{(1)}$ is identical to the susceptibility dyadic of a linear anisotropic medium considered earlier. The quantity $\chi_{ijk}^{(2)}$ is called the *second-order susceptibility*, and is a three-dimensional matrix (or third rank tensor) describing the nonlinear electric effects quadratic in $\mathbf{E}$. Similarly $\chi_{ijkl}^{(3)}$ is called the *third-order susceptibility*, and is a four-dimensional matrix (or fourth-rank tensor) describing the nonlinear electric effects cubic in $\mathbf{E}$. Numerical values of $\chi_{ijk}^{(2)}$ and $\chi_{ijkl}^{(3)}$ are given in Shen [174] for a variety of crystals.

When the material shows hysteresis effects, $\mathbf{D}$ at any point $\mathbf{r}$ and time t is due not only to the value of $\mathbf{E}$ at that point and at that time, but to the values of $\mathbf{E}$ at all points and times. That is, the material displays both temporal and spatial dispersion.

2.3 Maxwell's equations in moving frames

The essence of special relativity is that the mathematical forms of Maxwell's equations are identical in all *inertial reference frames*: frames moving with uniform velocities relative to the *laboratory frame of reference* in which we perform our measurements. This *form invariance* of Maxwell's equations is a specific example of the general physical *principle of covariance*. In the laboratory frame we write the differential equations of Maxwell's theory as

$$\nabla \times \mathbf{E}(\mathbf{r}, t) = -\frac{\partial \mathbf{B}(\mathbf{r}, t)}{\partial t},$$

$$\nabla \times \mathbf{H}(\mathbf{r}, t) = \mathbf{J}(\mathbf{r}, t) + \frac{\partial \mathbf{D}(\mathbf{r}, t)}{\partial t},$$

$$\nabla \cdot \mathbf{D}(\mathbf{r}, t) = \rho(\mathbf{r}, t),$$

$$\nabla \cdot \mathbf{B}(\mathbf{r}, t) = 0,$$

$$\nabla \cdot \mathbf{J}(\mathbf{r}, t) = -\frac{\partial \rho(\mathbf{r}, t)}{\partial t}.$$

Similarly, in an inertial frame having four-dimensional coordinates $(\mathbf{r}', t')$ we have

$$\nabla' \times \mathbf{E}'(\mathbf{r}', t') = -\frac{\partial \mathbf{B}'(\mathbf{r}', t')}{\partial t'},$$

$$\nabla' \times \mathbf{H}'(\mathbf{r}', t') = \mathbf{J}'(\mathbf{r}', t') + \frac{\partial \mathbf{D}'(\mathbf{r}', t')}{\partial t'},$$

$$\nabla' \cdot \mathbf{D}'(\mathbf{r}', t') = \rho'(\mathbf{r}', t'),$$

$$\nabla' \cdot \mathbf{B}'(\mathbf{r}', t') = 0,$$

$$\nabla' \cdot \mathbf{J}'(\mathbf{r}', t') = -\frac{\partial \rho'(\mathbf{r}', t')}{\partial t'}.$$

The primed fields measured in the moving system do *not* have the same numerical values as the unprimed fields measured in the laboratory. To convert between $\mathbf{E}$ and $\mathbf{E}'$, $\mathbf{B}$ and $\mathbf{B}'$, and so on, we must find a way to convert between the coordinates $(\mathbf{r}, t)$ and $(\mathbf{r}', t')$.

2.3.1 Field conversions under Galilean transformation

We shall assume that the primed coordinate system moves with constant velocity $\mathbf{v}$ relative to the laboratory frame (Figure 2.1). Prior to the early part of the twentieth century, converting between the primed and unprimed coordinate variables was intuitive and obvious: it was thought that time must be measured identically in each coordinate system, and that the relationship between the space variables can be determined simply by the displacement of the moving system at time $t = t'$. Under these assumptions, and under the further assumption that the two systems coincide at time $t = 0$, we can write

$$t' = t, \qquad x' = x - v_x t, \qquad y' = y - v_y t, \qquad z' = z - v_z t,$$

or simply

$$t' = t, \qquad \mathbf{r}' = \mathbf{r} - \mathbf{v} t.$$

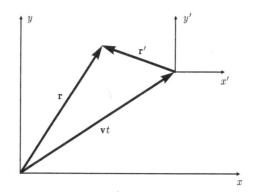

FIGURE 2.1
Primed coordinate system moving with velocity **v** relative to laboratory (unprimed)
coordinate system.

This is called a *Galilean transformation*. We can use the chain rule to describe the
manner in which differential operations transform, i.e., to relate derivatives with respect
to the laboratory coordinates to derivatives with respect to the inertial coordinates. We
have, for instance,

$$\frac{\partial}{\partial t} = \frac{\partial t'}{\partial t} \frac{\partial}{\partial t'} + \frac{\partial x'}{\partial t} \frac{\partial}{\partial x'} + \frac{\partial y'}{\partial t} \frac{\partial}{\partial y'} + \frac{\partial z'}{\partial t} \frac{\partial}{\partial z'}$$

$$= \frac{\partial}{\partial t'} - v_x \frac{\partial}{\partial x'} - v_y \frac{\partial}{\partial y'} - v_z \frac{\partial}{\partial z'}$$

$$= \frac{\partial}{\partial t'} - (\mathbf{v} \cdot \nabla'). \tag{2.33}$$

Similarly

$$\frac{\partial}{\partial x} = \frac{\partial}{\partial x'}, \qquad \frac{\partial}{\partial y} = \frac{\partial}{\partial y'}, \qquad \frac{\partial}{\partial z} = \frac{\partial}{\partial z'},$$

from which

$$\nabla \times \mathbf{A}(\mathbf{r}, t) = \nabla' \times \mathbf{A}(\mathbf{r}, t), \qquad \nabla \cdot \mathbf{A}(\mathbf{r}, t) = \nabla' \cdot \mathbf{A}(\mathbf{r}, t), \tag{2.34}$$

for each vector field **A**.

Newton was aware that the laws of mechanics are invariant with respect to Galilean
transformations. Do Maxwell's equations also behave in this way? Let us use the Galilean
transformation to determine which relationship between the primed and unprimed fields
results in form invariance of Maxwell's equations. We first examine $\nabla' \times \mathbf{E}$, the spatial
rate of change of the laboratory field with respect to the inertial frame spatial coordinates:

$$\nabla' \times \mathbf{E} = \nabla \times \mathbf{E} = -\frac{\partial \mathbf{B}}{\partial t} = -\frac{\partial \mathbf{B}}{\partial t'} + (\mathbf{v} \cdot \nabla')\mathbf{B}$$

by (2.34) and (2.33). Rewriting the last term by (B.51) we have

$$(\mathbf{v} \cdot \nabla')\mathbf{B} = -\nabla' \times (\mathbf{v} \times \mathbf{B})$$

since **v** is constant and $\nabla' \cdot \mathbf{B} = \nabla \cdot \mathbf{B} = 0$, hence

$$\nabla' \times (\mathbf{E} + \mathbf{v} \times \mathbf{B}) = -\frac{\partial \mathbf{B}}{\partial t'}. \tag{2.35}$$

Similarly

$$\nabla' \times \mathbf{H} = \nabla \times \mathbf{H} = \mathbf{J} + \frac{\partial \mathbf{D}}{\partial t} = \mathbf{J} + \frac{\partial \mathbf{D}}{\partial t'} + \nabla' \times (\mathbf{v} \times \mathbf{D}) - \mathbf{v}(\nabla' \cdot \mathbf{D})$$

where $\nabla' \cdot \mathbf{D} = \nabla \cdot \mathbf{D} = \rho$ so that

$$\nabla' \times (\mathbf{H} - \mathbf{v} \times \mathbf{D}) = \frac{\partial \mathbf{D}}{\partial t'} - \rho\mathbf{v} + \mathbf{J}. \tag{2.36}$$

Also

$$\nabla' \cdot \mathbf{J} = \nabla \cdot \mathbf{J} = -\frac{\partial \rho}{\partial t} = -\frac{\partial \rho}{\partial t'} + (\mathbf{v} \cdot \nabla')\rho$$

and we may use (B.48) to write

$$(\mathbf{v} \cdot \nabla')\rho = \mathbf{v} \cdot (\nabla'\rho) = \nabla' \cdot (\rho\mathbf{v}),$$

obtaining

$$\nabla' \cdot (\mathbf{J} - \rho\mathbf{v}) = -\frac{\partial \rho}{\partial t'}. \tag{2.37}$$

Equations (2.35), (2.36), and (2.37) show that the forms of Maxwell's equations in the inertial and laboratory frames are identical, provided that

$$\mathbf{E}' = \mathbf{E} + \mathbf{v} \times \mathbf{B}, \tag{2.38}$$
$$\mathbf{D}' = \mathbf{D}, \tag{2.39}$$
$$\mathbf{H}' = \mathbf{H} - \mathbf{v} \times \mathbf{D}, \tag{2.40}$$
$$\mathbf{B}' = \mathbf{B}, \tag{2.41}$$
$$\mathbf{J}' = \mathbf{J} - \rho\mathbf{v}, \tag{2.42}$$
$$\rho' = \rho. \tag{2.43}$$

That is, (2.38)–(2.43) result in form invariance of Faraday's law, Ampere's law, and the continuity equation under a Galilean transformation. These equations express the fields measured by a moving observer in terms of those measured in the laboratory frame. To convert the opposite way, we need only use the principle of relativity. Neither observer can tell whether he or she is stationary — only that the other observer is moving relative to him or her. To obtain the fields in the laboratory frame, we simply change the sign on $\mathbf{v}$ and swap primed with unprimed fields in (2.38)–(2.43):

$$\mathbf{E} = \mathbf{E}' - \mathbf{v} \times \mathbf{B}',$$
$$\mathbf{D} = \mathbf{D}',$$
$$\mathbf{H} = \mathbf{H}' + \mathbf{v} \times \mathbf{D}',$$
$$\mathbf{B} = \mathbf{B}',$$
$$\mathbf{J} = \mathbf{J}' + \rho'\mathbf{v},$$
$$\rho = \rho'.$$

According to (2.42), a moving observer interprets charge stationary in the laboratory frame as an additional current moving opposite the direction of his or her motion. This seems reasonable. However, while $\mathbf{E}$ depends on both $\mathbf{E}'$ and $\mathbf{B}'$, the field $\mathbf{B}$ is unchanged under the transformation. Why should $\mathbf{B}$ have this special status? In fact, we may uncover an inconsistency among the transformations by considering free space where (2.15) and (2.16) hold: in this case (2.38) gives

$$\mathbf{D}'/\epsilon_0 = \mathbf{D}/\epsilon_0 + \mathbf{v} \times \mu_0 \mathbf{H}$$

or

$$\mathbf{D}' = \mathbf{D} + \mathbf{v} \times \mathbf{H}/c^2$$

rather than (2.39). Similarly, from (2.40) we get

$$\mathbf{B}' = \mathbf{B} - \mathbf{v} \times \mathbf{E}/c^2$$

instead of (2.41). Using these, the set of transformations becomes

$$\mathbf{E}' = \mathbf{E} + \mathbf{v} \times \mathbf{B}, \tag{2.44}$$
$$\mathbf{D}' = \mathbf{D} + \mathbf{v} \times \mathbf{H}/c^2, \tag{2.45}$$
$$\mathbf{H}' = \mathbf{H} - \mathbf{v} \times \mathbf{D},$$
$$\mathbf{B}' = \mathbf{B} - \mathbf{v} \times \mathbf{E}/c^2, \tag{2.46}$$
$$\mathbf{J}' = \mathbf{J} - \rho\mathbf{v},$$
$$\rho' = \rho. \tag{2.47}$$

These can also be written using dyadic notation as

$$\mathbf{E}' = \bar{\mathbf{I}} \cdot \mathbf{E} + \bar{\boldsymbol{\beta}} \cdot (c\mathbf{B}),$$
$$c\mathbf{B}' = -\bar{\boldsymbol{\beta}} \cdot \mathbf{E} + \bar{\mathbf{I}} \cdot (c\mathbf{B}),$$

and

$$c\mathbf{D}' = \bar{\mathbf{I}} \cdot (c\mathbf{D}) + \bar{\boldsymbol{\beta}} \cdot \mathbf{H},$$
$$\mathbf{H}' = -\bar{\boldsymbol{\beta}} \cdot (c\mathbf{D}) + \bar{\mathbf{I}} \cdot \mathbf{H},$$

where

$$[\bar{\boldsymbol{\beta}}] = \begin{bmatrix} 0 & -\beta_z & \beta_y \\ \beta_z & 0 & -\beta_x \\ -\beta_y & \beta_x & 0 \end{bmatrix}$$

with $\boldsymbol{\beta} = \mathbf{v}/c$. This set of equations is self-consistent among Maxwell's equations. However, the equations are not consistent with the assumption of a Galilean transformation of the coordinates, and thus Maxwell's equations are not covariant under a Galilean transformation. Maxwell's equations are only covariant under a Lorentz transformation as described in the next section. Expressions (2.44)–(2.46) turn out to be accurate to order v/c, hence are the results of a *first-order Lorentz transformation*. Only when v is an appreciable fraction of c do the field conversions resulting from the first-order Lorentz transformation differ markedly from those resulting from a Galilean transformation; those resulting from the true Lorentz transformation require even higher velocities to differ markedly from the first-order expressions. Engineering accuracy is often accomplished using the Galilean transformation. This pragmatic observation leads to quite a bit of confusion when considering the large-scale forms of Maxwell's equations, as we shall soon see.

2.3.2   Field conversions under Lorentz transformation

To find the proper transformation under which Maxwell's equations are covariant, we must discard our notion that time progresses the same in the primed and the unprimed frames. The proper transformation of coordinates that guarantees covariance of Maxwell's equations is the *Lorentz transformation*

$$ct' = \gamma ct - \gamma \boldsymbol{\beta} \cdot \mathbf{r}, \tag{2.48}$$

$$\mathbf{r}' = \bar{\boldsymbol{\alpha}} \cdot \mathbf{r} - \gamma \boldsymbol{\beta} ct, \tag{2.49}$$

where

$$\bar{\boldsymbol{\alpha}} = \bar{\mathbf{I}} + (\gamma - 1)\frac{\boldsymbol{\beta}\boldsymbol{\beta}}{\beta^2}, \qquad \beta = |\boldsymbol{\beta}|.$$

Here

$$\gamma = \frac{1}{\sqrt{1 - \beta^2}}$$

is called the *Lorentz factor*. The Lorentz transformation is obviously more complicated than the Galilean transformation; only as $\boldsymbol{\beta} \to 0$ are the Lorentz and Galilean transformations equivalent.

Not surprisingly, field conversions between inertial reference frames are more complicated with the Lorentz transformation than with the Galilean transformation. For simplicity we assume that the velocity of the moving frame has only an x-component: $\mathbf{v} = \hat{\mathbf{x}}v$. Later we can generalize this to any direction. Equations (2.48) and (2.49) become

$$x' = x + (\gamma - 1)x - \gamma vt, \tag{2.50}$$

$$y' = y, \tag{2.51}$$

$$z' = z, \tag{2.52}$$

$$ct' = \gamma ct - \gamma \frac{v}{c}x, \tag{2.53}$$

and the chain rule gives

$$\frac{\partial}{\partial x} = \gamma \frac{\partial}{\partial x'} - \gamma \frac{v}{c^2} \frac{\partial}{\partial t'}, \tag{2.54}$$

$$\frac{\partial}{\partial y} = \frac{\partial}{\partial y'}, \tag{2.55}$$

$$\frac{\partial}{\partial z} = \frac{\partial}{\partial z'}, \tag{2.56}$$

$$\frac{\partial}{\partial t} = -\gamma v \frac{\partial}{\partial x'} + \gamma \frac{\partial}{\partial t'}. \tag{2.57}$$

We begin by examining Faraday's law in the laboratory frame. In component form we have

$$\frac{\partial E_z}{\partial y} - \frac{\partial E_y}{\partial z} = -\frac{\partial B_x}{\partial t}, \tag{2.58}$$

$$\frac{\partial E_x}{\partial z} - \frac{\partial E_z}{\partial x} = -\frac{\partial B_y}{\partial t}, \tag{2.59}$$

$$\frac{\partial E_y}{\partial x} - \frac{\partial E_x}{\partial y} = -\frac{\partial B_z}{\partial t}. \tag{2.60}$$

These become

$$\frac{\partial E_z}{\partial y'} - \frac{\partial E_y}{\partial z'} = \gamma v \frac{\partial B_x}{\partial x'} - \gamma \frac{\partial B_x}{\partial t'}, \tag{2.61}$$

$$\frac{\partial E_x}{\partial z'} - \gamma \frac{\partial E_z}{\partial x'} + \gamma \frac{v}{c^2} \frac{\partial E_z}{\partial t'} = \gamma v \frac{\partial B_y}{\partial x'} - \gamma \frac{\partial B_y}{\partial t'}, \tag{2.62}$$

$$\gamma \frac{\partial E_y}{\partial x'} - \gamma \frac{v}{c^2} \frac{\partial E_y}{\partial t'} - \frac{\partial E_x}{\partial y'} = \gamma v \frac{\partial B_z}{\partial x'} - \gamma \frac{\partial B_z}{\partial t'}, \tag{2.63}$$

after we use (2.54)–(2.57) to convert the derivatives in the laboratory frame to derivatives with respect to the moving frame coordinates. To simplify (2.61) we consider

$$\nabla \cdot \mathbf{B} = \frac{\partial B_x}{\partial x} + \frac{\partial B_y}{\partial y} + \frac{\partial B_z}{\partial z} = 0.$$

Converting the laboratory frame coordinates to the moving frame coordinates, we have

$$\gamma \frac{\partial B_x}{\partial x'} - \gamma \frac{v}{c^2} \frac{\partial B_x}{\partial t'} + \frac{\partial B_y}{\partial y'} + \frac{\partial B_z}{\partial z'} = 0$$

or

$$-\gamma v \frac{\partial B_x}{\partial x'} = -\gamma \frac{v^2}{c^2} \frac{\partial B_x}{\partial t'} + v \frac{\partial B_y}{\partial y'} + v \frac{\partial B_z}{\partial z'}.$$

Substituting this into (2.61) and rearranging (2.62) and (2.63), we obtain

$$\frac{\partial}{\partial y'} \gamma (E_z + v B_y) - \frac{\partial}{\partial z'} \gamma (E_y - v B_z) = -\frac{\partial B_x}{\partial t'},$$

$$\frac{\partial E_x}{\partial z'} - \frac{\partial}{\partial x'} \gamma (E_z + v B_y) = -\frac{\partial}{\partial t'} \gamma \left(B_y + \frac{v}{c^2} E_z \right),$$

$$\frac{\partial}{\partial x'} \gamma (E_y - v B_z) - \frac{\partial E_x}{\partial y'} = -\frac{\partial}{\partial t'} \gamma \left(B_z - \frac{v}{c^2} E_y \right).$$

Comparison with (2.58)–(2.60) shows that form invariance of Faraday's law under the Lorentz transformation requires

$$E_x' = E_x, \qquad E_y' = \gamma (E_y - v B_z), \qquad E_z' = \gamma (E_z + v B_y),$$

and

$$B_x' = B_x, \qquad B_y' = \gamma \left(B_y + \frac{v}{c^2} E_z \right), \qquad B_z' = \gamma \left(B_z - \frac{v}{c^2} E_y \right).$$

To generalize $\mathbf{v}$ to any direction, we simply note that the components of the fields parallel to the velocity direction are identical in the moving and laboratory frames, while the components perpendicular to the velocity direction convert according to a simple cross-product rule. After similar analyses with Ampere's and Gauss's laws (see Example 2.3), we find that

$$\mathbf{E}_\parallel' = \mathbf{E}_\parallel, \qquad \mathbf{B}_\parallel' = \mathbf{B}_\parallel, \qquad \mathbf{D}_\parallel' = \mathbf{D}_\parallel, \qquad \mathbf{H}_\parallel' = \mathbf{H}_\parallel,$$

$$\mathbf{E}_\perp' = \gamma (\mathbf{E}_\perp + \boldsymbol{\beta} \times c\mathbf{B}_\perp), \tag{2.64}$$

$$c\mathbf{B}_\perp' = \gamma (c\mathbf{B}_\perp - \boldsymbol{\beta} \times \mathbf{E}_\perp), \tag{2.65}$$

$$c\mathbf{D}_\perp' = \gamma (c\mathbf{D}_\perp + \boldsymbol{\beta} \times \mathbf{H}_\perp), \tag{2.66}$$

$$\mathbf{H}_\perp' = \gamma (\mathbf{H}_\perp - \boldsymbol{\beta} \times c\mathbf{D}_\perp), \tag{2.67}$$

and

$$\mathbf{J}'_\parallel = \gamma(\mathbf{J}_\parallel - \rho\mathbf{v}), \tag{2.68}$$

$$\mathbf{J}'_\perp = \mathbf{J}_\perp, \tag{2.69}$$

$$c\rho' = \gamma(c\rho - \boldsymbol{\beta} \cdot \mathbf{J}), \tag{2.70}$$

where the symbols $\parallel$ and $\perp$ designate the components of the field parallel and perpendicular to $\mathbf{v}$, respectively.

▶ **Example 2.3:** Reference frame transformation

Consider Ampere's law and Gauss's law, written in terms of rectangular components in the laboratory frame of reference. Assume that an inertial frame moves with velocity $\mathbf{v} = \hat{x}v$ with respect to the laboratory frame. Using the Lorentz transformation given by (2.50)–(2.53), show that

$$c\mathbf{D}'_\perp = \gamma(c\mathbf{D}_\perp + \boldsymbol{\beta} \times \mathbf{H}_\perp), \qquad \mathbf{J}'_\parallel = \gamma(\mathbf{J}_\parallel - \rho\mathbf{v}),$$
$$\mathbf{H}'_\perp = \gamma(\mathbf{H}_\perp - \boldsymbol{\beta} \times c\mathbf{D}_\perp), \qquad \mathbf{J}'_\perp = \mathbf{J}_\perp,$$

where "$\perp$" means perpendicular to the direction of the velocity and "$\parallel$" means parallel to the direction of the velocity.

Solution: Examine Ampere's law in rectangular coordinates:

$$\frac{\partial H_z}{\partial y} - \frac{\partial H_y}{\partial z} = J_x + \frac{\partial D_x}{\partial t}, \tag{2.71}$$

$$\frac{\partial H_x}{\partial z} - \frac{\partial H_z}{\partial x} = J_y + \frac{\partial D_y}{\partial t}, \tag{2.72}$$

$$\frac{\partial H_y}{\partial x} - \frac{\partial H_x}{\partial y} = J_z + \frac{\partial D_z}{\partial t}. \tag{2.73}$$

Assume $\mathbf{v} = \hat{x}v$ and use the derivative transforms (2.54)–(2.57):

$$\frac{\partial}{\partial x} = \gamma\frac{\partial}{\partial x'} - \gamma\frac{v}{c^2}\frac{\partial}{\partial t'}, \qquad \frac{\partial}{\partial y} = \frac{\partial}{\partial y'}, \qquad \frac{\partial}{\partial t} = -\gamma v\frac{\partial}{\partial x'} + \gamma\frac{\partial}{\partial t'}, \qquad \frac{\partial}{\partial z} = \frac{\partial}{\partial z'}.$$

With these, (2.71)–(2.73) become

$$\frac{\partial H_z}{\partial y'} - \frac{\partial H_y}{\partial z'} = J_x - \gamma v\frac{\partial D_x}{\partial x'} + \gamma\frac{\partial D_x}{\partial t'}, \tag{2.74}$$

$$\frac{\partial H_x}{\partial z'} - \gamma\frac{\partial H_z}{\partial x'} + \gamma\frac{v}{c^2}\frac{\partial H_z}{\partial t'} = J_y - \gamma v\frac{\partial D_y}{\partial x'} + \gamma\frac{\partial D_y}{\partial t'}, \tag{2.75}$$

$$\gamma\frac{\partial H_y}{\partial x'} - \gamma\frac{v}{c^2}\frac{\partial H_y}{\partial t'} - \frac{\partial H_x}{\partial y'} = J_z - \gamma v\frac{\partial D_z}{\partial x'} + \gamma\frac{\partial D_z}{\partial t'}. \tag{2.76}$$

In addition, note that Gauss's law $\nabla \cdot \mathbf{D} = \rho$ becomes

$$\gamma\frac{\partial D_x}{\partial x'} - \gamma\frac{v}{c^2}\frac{\partial D_x}{\partial t'} + \frac{\partial D_y}{\partial y'} + \frac{\partial D_z}{\partial z'} = \rho. \tag{2.77}$$

So we can substitute

$$\gamma v\frac{\partial D_x}{\partial x'} = \rho v + \gamma\frac{v^2}{c^2}\frac{\partial D_x}{\partial t'} - v\frac{\partial D_y}{\partial y'} - v\frac{\partial D_z}{\partial z'}$$

into (2.74) and obtain

$$\frac{\partial}{\partial y'}(H_z - vD_y) - \frac{\partial}{\partial z'}(H_y + vD_z) = J_x - \rho v + \gamma\left(1 - \frac{v^2}{c^2}\right)\frac{\partial D_x}{\partial t'}.$$

But

$$1 - \frac{v^2}{c^2} = 1 - \beta^2 = \frac{1}{\gamma^2},$$

so

$$\frac{\partial}{\partial y'}\gamma(H_z - vD_y) - \frac{\partial}{\partial z'}\gamma(H_y + vD_z) = \gamma J_x - \gamma\rho v + \frac{\partial D_x}{\partial t'}. \tag{2.78}$$

Next rearrange (2.75) and (2.76) to get

$$\frac{\partial H_x}{\partial z'} - \frac{\partial}{\partial x'}\gamma(H_z - vD_y) = J_y + \frac{\partial}{\partial t'}\gamma\left(D_y - \frac{v}{c^2}H_z\right), \tag{2.79}$$

$$\frac{\partial}{\partial x'}\gamma(H_y + vD_z) - \frac{\partial H_x}{\partial y'} = J_z + \frac{\partial}{\partial t'}\gamma\left(D_z + \frac{v}{c^2}H_y\right). \tag{2.80}$$

Comparing (2.78)–(2.80) to Ampere's law in the moving frame, we see

$$H_z' = \gamma(H_z - vD_y), \qquad H_x' = H_x, \qquad D_y' = \gamma\left(D_y - \frac{v}{c^2}H_z\right),$$

$$H_y' = \gamma(H_y + vD_z), \qquad D_x' = D_x, \qquad D_z' = \gamma\left(D_z + \frac{v}{c^2}H_y\right),$$

$$J_x' = \gamma(J_x - \rho v), \qquad J_y' = J_y, \qquad J_z' = J_z.$$

We can generalize this result to an arbitrary direction of **v** by letting ‖ represent the part of a vector in the direction of **v** and ⊥ the part perpendicular to **v**. Since the perpendicular part obeys a cross-product rule, we have

$$c\mathbf{D}_\perp' = \gamma(c\mathbf{D}_\perp + \boldsymbol{\beta} \times \mathbf{H}_\perp), \qquad\qquad \mathbf{J}_\parallel' = \gamma(\mathbf{J}_\parallel - \rho\mathbf{v}),$$

$$\mathbf{H}_\perp' = \gamma(\mathbf{H}_\perp - \boldsymbol{\beta} \times c\mathbf{D}_\perp), \qquad\qquad \mathbf{J}_\perp' = \mathbf{J}_\perp,$$

where $\boldsymbol{\beta} = \mathbf{v}/c$. It is easy to verify that these results yield the special case treated above when

$$\mathbf{v} = v\hat{\mathbf{x}}, \qquad \mathbf{H}_\perp' = \hat{\mathbf{y}}H_y' + \hat{\mathbf{z}}H_z', \qquad \mathbf{H}_\perp = \hat{\mathbf{y}}H_y + \hat{\mathbf{z}}H_z, \qquad \mathbf{D}_\perp = \hat{\mathbf{y}}D_y + \hat{\mathbf{z}}D_z,$$

for instance.

Demonstration that $c\rho' = \gamma(c\rho - \boldsymbol{\beta}\cdot\mathbf{J})$ is left for Problem 2.2. ◄

These conversions are self-consistent, and the Lorentz transformation is the transformation under which Maxwell's equations are covariant. If $v^2 \ll c^2$, then $\gamma \approx 1$ and to first order (2.64)–(2.70) reduce to (2.44)–(2.47). If $v/c \ll 1$, then the first-order fields reduce to the Galilean fields (2.38)–(2.43).

To convert in the opposite direction, we can swap primed and unprimed fields and change the sign on **v**:

$$\mathbf{E}_\perp = \gamma(\mathbf{E}_\perp' - \boldsymbol{\beta} \times c\mathbf{B}_\perp'),$$

$$c\mathbf{B}_\perp = \gamma(c\mathbf{B}_\perp' + \boldsymbol{\beta} \times \mathbf{E}_\perp'),$$

$$c\mathbf{D}_\perp = \gamma(c\mathbf{D}_\perp' - \boldsymbol{\beta} \times \mathbf{H}_\perp'),$$

$$\mathbf{H}_\perp = \gamma(\mathbf{H}_\perp' + \boldsymbol{\beta} \times c\mathbf{D}_\perp'),$$

and

$$\mathbf{J}_\parallel = \gamma(\mathbf{J}'_\parallel + \rho'\mathbf{v}),$$

$$\mathbf{J}_\perp = \mathbf{J}'_\perp,$$

$$c\rho = \gamma(c\rho' + \boldsymbol{\beta} \cdot \mathbf{J}').$$

The conversion formulas can be written much more succinctly in dyadic notation:

$$\mathbf{E}' = \gamma\bar{\boldsymbol{\alpha}}^{-1} \cdot \mathbf{E} + \gamma\bar{\boldsymbol{\beta}} \cdot (c\mathbf{B}), \tag{2.81}$$

$$c\mathbf{B}' = -\gamma\bar{\boldsymbol{\beta}} \cdot \mathbf{E} + \gamma\bar{\boldsymbol{\alpha}}^{-1} \cdot (c\mathbf{B}), \tag{2.82}$$

$$c\mathbf{D}' = \gamma\bar{\boldsymbol{\alpha}}^{-1} \cdot (c\mathbf{D}) + \gamma\bar{\boldsymbol{\beta}} \cdot \mathbf{H}, \tag{2.83}$$

$$\mathbf{H}' = -\gamma\bar{\boldsymbol{\beta}} \cdot (c\mathbf{D}) + \gamma\bar{\boldsymbol{\alpha}}^{-1} \cdot \mathbf{H}, \tag{2.84}$$

and

$$c\rho' = \gamma(c\rho - \boldsymbol{\beta} \cdot \mathbf{J}), \tag{2.85}$$

$$\mathbf{J}' = \bar{\boldsymbol{\alpha}} \cdot \mathbf{J} - \gamma\boldsymbol{\beta}c\rho, \tag{2.86}$$

where $\bar{\boldsymbol{\alpha}}^{-1} \cdot \bar{\boldsymbol{\alpha}} = \bar{\mathbf{I}}$, and thus

$$\bar{\boldsymbol{\alpha}}^{-1} = \bar{\boldsymbol{\alpha}} - \gamma\boldsymbol{\beta}\boldsymbol{\beta}.$$

Maxwell's equations are covariant under a Lorentz transformation but not under a Galilean transformation; the laws of mechanics are invariant under a Galilean transformation but not under a Lorentz transformation. How then should we analyze interactions between electromagnetic fields and particles or materials? Einstein realized that the laws of mechanics needed revision to make them Lorentz covariant: in fact, under his theory of special relativity all physical laws should demonstrate Lorentz covariance. Interestingly, charge is then Lorentz invariant, whereas mass is not (recall that invariance refers to a quantity, whereas covariance refers to the form of a natural law). We shall not attempt to describe all the ramifications of special relativity, but instead refer the reader to any of the excellent and readable texts on the subject, including those by Bohm [19], Einstein [54], and Born [21], and to the nice historical account by Miller [132]. However, we shall examine the importance of Lorentz invariants in electromagnetic theory.

2.3.2.1 Lorentz invariants

Although the electromagnetic fields are not Lorentz invariant (e.g., the numerical value of $\mathbf{E}$ measured by one observer differs from that measured by another observer in uniform relative motion), several quantities do give identical values regardless of the velocity of motion. Most fundamental are the speed of light and the quantity of electric charge which, unlike mass, is the same in all frames of reference. Other important *Lorentz invariants* include $\mathbf{E} \cdot \mathbf{B}$, $\mathbf{H} \cdot \mathbf{D}$, and the quantities

$$\mathbf{B} \cdot \mathbf{B} - \mathbf{E} \cdot \mathbf{E}/c^2,$$

$$\mathbf{B} \cdot \mathbf{H} - \mathbf{E} \cdot \mathbf{D},$$

$$\mathbf{H} \cdot \mathbf{H} - c^2\mathbf{D} \cdot \mathbf{D},$$

$$c\mathbf{B} \cdot \mathbf{D} + \mathbf{E} \cdot \mathbf{H}/c.$$

▶ **Example 2.4:** Lorentz invariants

Show that the quantity $\mathbf{E} \cdot \mathbf{B}$ is a Lorentz invariant.

Solution: Let $\mathbf{E} = \mathbf{E}_\parallel + \mathbf{E}_\perp$ and $\mathbf{B} = \mathbf{B}_\parallel + \mathbf{B}_\perp$. Then

$$\mathbf{E} \cdot \mathbf{B} = \mathbf{E}_\parallel \cdot \mathbf{B}_\parallel + \mathbf{E}_\parallel \cdot \mathbf{B}_\perp + \mathbf{E}_\perp \cdot \mathbf{B}_\parallel + \mathbf{E}_\perp \cdot \mathbf{B}_\perp = \mathbf{E}_\parallel \cdot \mathbf{B}_\parallel + \mathbf{E}_\perp \cdot \mathbf{B}_\perp.$$

Now examine

$$\mathbf{E}' \cdot \mathbf{B}' = \mathbf{E}'_\parallel \cdot \mathbf{B}'_\parallel + \mathbf{E}'_\perp \cdot \mathbf{B}'_\perp.$$

We have $\mathbf{E}' = \mathbf{E}'_\parallel + \mathbf{E}'_\perp$ and $\mathbf{B}' = \mathbf{B}'_\parallel + \mathbf{B}'_\perp$, where

$$\mathbf{E}'_\perp = \gamma(\mathbf{E}_\perp + \boldsymbol{\beta} \times c\mathbf{B}_\perp), \qquad\qquad \mathbf{E}'_\parallel = \mathbf{E}_\parallel,$$
$$\mathbf{B}'_\perp = \frac{\gamma}{c}(c\mathbf{B}_\perp - \boldsymbol{\beta} \times \mathbf{E}_\perp), \qquad\qquad \mathbf{B}'_\parallel = \mathbf{B}_\parallel.$$

So

$$\mathbf{E}' \cdot \mathbf{B}' = \mathbf{E}_\parallel \cdot \mathbf{B}_\parallel + \frac{\gamma^2}{c}(\mathbf{E}_\perp + \boldsymbol{\beta} \times c\mathbf{B}_\perp) \cdot (c\mathbf{B}_\perp - \boldsymbol{\beta} \times \mathbf{E}_\perp)$$

$$= \mathbf{E}_\parallel \cdot \mathbf{B}_\parallel + \gamma^2 \mathbf{E}_\perp \cdot \mathbf{B}_\perp + \frac{\gamma^2}{c}\mathbf{E}_\perp \cdot (\boldsymbol{\beta} \times \mathbf{E}_\perp) - \gamma^2(\boldsymbol{\beta} \times c\mathbf{B}_\perp) \cdot \mathbf{B}_\perp$$

$$- \gamma^2(\boldsymbol{\beta} \times \mathbf{B}_\perp) \cdot (\boldsymbol{\beta} \times \mathbf{E}_\perp).$$

Use of (B.6) gives

$$\mathbf{E}_\perp \cdot (\boldsymbol{\beta} \times \mathbf{E}_\perp) = \boldsymbol{\beta} \cdot (\mathbf{E}_\perp \times \mathbf{E}_\perp) = 0,$$
$$\mathbf{B}_\perp \cdot (\boldsymbol{\beta} \times \mathbf{B}_\perp) = \boldsymbol{\beta} \cdot (\mathbf{B}_\perp \times \mathbf{B}_\perp) = 0,$$

and then use of (B.8) gives

$$(\boldsymbol{\beta} \times \mathbf{B}_\perp) \cdot (\boldsymbol{\beta} \times \mathbf{E}_\perp) = (\mathbf{B}_\perp \cdot \mathbf{E}_\perp)(\boldsymbol{\beta} \cdot \boldsymbol{\beta}) - (\mathbf{B}_\perp \cdot \boldsymbol{\beta})(\mathbf{E}_\perp \cdot \boldsymbol{\beta}) = \beta^2 \mathbf{B}_\perp \cdot \mathbf{E}_\perp.$$

Finally,

$$\mathbf{E}' \cdot \mathbf{B}' = \mathbf{E}_\parallel \cdot \mathbf{B}_\parallel + \gamma^2 \mathbf{E}_\perp \cdot \mathbf{B}_\perp (1 - \beta^2) = \mathbf{E}_\parallel \cdot \mathbf{B}_\parallel + \mathbf{E}_\perp \cdot \mathbf{B}_\perp = \mathbf{E} \cdot \mathbf{B}.$$

Consideration of the remaining Lorentz invariants is Problem 2.3. ◀

To see the importance of the Lorentz invariants, consider the special case of fields in empty space. If $\mathbf{E} \cdot \mathbf{B} = 0$ in one reference frame, then it is zero in all reference frames. Then if $\mathbf{B} \cdot \mathbf{B} - \mathbf{E} \cdot \mathbf{E}/c^2 = 0$ in any reference frame, the ratio of E to B is always c^2 regardless of the reference frame in which the fields are measured. This is the characteristic of a plane wave in free space.

If $\mathbf{E} \cdot \mathbf{B} = 0$ and $c^2 B^2 > E^2$, then we can find a reference frame using the conversion formulas (2.81)–(2.86) (see Problem 2.5) in which the electric field is zero but the magnetic field is nonzero. In this case we call the fields *purely magnetic* in any reference frame, even if both $\mathbf{E}$ and $\mathbf{B}$ are nonzero. Similarly, if $\mathbf{E} \cdot \mathbf{B} = 0$ and $c^2 B^2 < E^2$ then we can find a reference frame in which the magnetic field is zero but the electric field is nonzero. We call fields of this type *purely electric*.

The Lorentz force is not Lorentz invariant. Consider a point charge at rest in the laboratory frame. While we measure only an electric field in the laboratory frame, an inertial observer measures both electric and magnetic fields. A test charge Q in the laboratory frame experiences the Lorentz force $\mathbf{F} = Q\mathbf{E}$; in an inertial frame the same charge experiences $\mathbf{F}' = Q\mathbf{E}' + Q\mathbf{v} \times \mathbf{B}'$ (see Problem 2.6). The conversion formulas show that $\mathbf{F}$ and $\mathbf{F}'$ are not identical.

We see that both $\mathbf{E}$ and $\mathbf{B}$ are integral components of the electromagnetic field: the separation of the field into electric and magnetic components depends on the motion of the reference frame in which measurements are made. This has obvious implications when considering static electric and magnetic fields.

2.3.2.2 Derivation of Maxwell's equations from Coulomb's law

Consider a point charge at rest in the laboratory frame. If the magnetic component of force on this charge arises naturally through motion of an inertial reference frame, and if this force can be expressed in terms of Coulomb's law in the laboratory frame, then perhaps the magnetic field can be derived directly from Coulomb's law and the Lorentz transformation. Perhaps it is possible to derive all of Maxwell's theory with Coulomb's law and Lorentz invariance as the only postulates.

Several authors, notably Purcell [155] and Elliott [56], have used this approach. However, Jackson [92] has pointed out that many additional assumptions are required to deduce Maxwell's equations beginning with Coulomb's law. Feynman [63] is critical of the approach, pointing out that we must introduce a vector potential that adds to the scalar potential from electrostatics in order to produce an entity that transforms according to the laws of special relativity. In addition, the assumption of Lorentz invariance seems to involve circular reasoning, since the Lorentz transformation was originally introduced to make Maxwell's equations covariant. But Lucas and Hodgson [123] point out that the Lorentz transformation can be deduced from other fundamental principles (such as causality and the isotropy of space), and that the postulate of a vector potential is reasonable. Schwartz [172] gives a detailed derivation of Maxwell's equations from Coulomb's law, outlining the necessary assumptions.

2.3.2.3 Transformation of constitutive relations

Minkowski's interest in the covariance of Maxwell's equations was aimed not merely at the relationship between fields in different moving frames of reference, but at an understanding of the electrodynamics of moving media. He wished to ascertain the effect of a moving material body on the electromagnetic fields in some region of space. By proposing the covariance of Maxwell's equations in materials as well as in free space, he extended Maxwell's theory to moving material bodies.

We have seen in (2.81)–(2.84) that $(\mathbf{E}, c\mathbf{B})$ and $(c\mathbf{D}, \mathbf{H})$ convert identically under a Lorentz transformation. Since the most general form of the constitutive relations relate $c\mathbf{D}$ and $\mathbf{H}$ to the field pair $(\mathbf{E}, c\mathbf{B})$ (see § 2.2.2) as

$$\begin{bmatrix} c\mathbf{D} \\ \mathbf{H} \end{bmatrix} = [\bar{\mathbf{C}}] \begin{bmatrix} \mathbf{E} \\ c\mathbf{B} \end{bmatrix},$$

this form of the constitutive relations must be Lorentz covariant. That is, in the reference frame of a moving material we have

$$\begin{bmatrix} c\mathbf{D}' \\ \mathbf{H}' \end{bmatrix} = [\bar{\mathbf{C}}'] \begin{bmatrix} \mathbf{E}' \\ c\mathbf{B}' \end{bmatrix},$$

and should be able to convert $[\bar{\mathbf{C}}']$ to $[\bar{\mathbf{C}}]$. We should be able to find the constitutive matrix describing the relationships among the fields observed in the laboratory frame.

It is somewhat laborious to obtain the constitutive matrix $[\bar{\mathbf{C}}]$ for an arbitrary moving medium. Detailed expressions for isotropic, bianisotropic, gyrotropic, and uniaxial media are given by Kong [108]. The rather complicated expressions can be written in a more compact form if we consider the expressions for $\mathbf{B}$ and $\mathbf{D}$ in terms of the pair $(\mathbf{E}, \mathbf{H})$. For a linear isotropic material such that $\mathbf{D}' = \epsilon'\mathbf{E}'$ and $\mathbf{B}' = \mu'\mathbf{H}'$ in the moving frame, the relationships in the laboratory frame are [108]

$$\mathbf{B} = \mu'\bar{\mathbf{A}} \cdot \mathbf{H} - \boldsymbol{\Omega} \times \mathbf{E}, \tag{2.87}$$

$$\mathbf{D} = \epsilon'\bar{\mathbf{A}} \cdot \mathbf{E} + \boldsymbol{\Omega} \times \mathbf{H}, \tag{2.88}$$

where

$$\bar{\mathbf{A}} = \frac{1 - \beta^2}{1 - n^2\beta^2} \left[\bar{\mathbf{I}} - \frac{n^2 - 1}{1 - \beta^2} \boldsymbol{\beta}\boldsymbol{\beta} \right], \tag{2.89}$$

$$\boldsymbol{\Omega} = \frac{n^2 - 1}{1 - n^2\beta^2} \frac{\boldsymbol{\beta}}{c}, \tag{2.90}$$

and where $n = c(\mu'\epsilon')^{1/2}$ is the optical index of the medium. A moving material that is isotropic in its own moving reference frame is bianisotropic in the laboratory frame. If, for instance, we tried to measure the relationship between the fields of a moving isotropic fluid, but used instruments that were stationary in our laboratory (e.g., attached to our measurement bench) we would find that $\mathbf{D}$ depends not only on $\mathbf{E}$ but also on $\mathbf{H}$, and that $\mathbf{D}$ aligns with neither $\mathbf{E}$ nor $\mathbf{H}$. That a moving material isotropic in its own frame of reference is bianisotropic in the laboratory frame was known long ago. Roentgen showed experimentally in 1888 that a dielectric moving through an electric field becomes magnetically polarized, while H.A. Wilson showed in 1905 that a dielectric moving through a magnetic field becomes electrically polarized [141].

If $v^2/c^2 \ll 1$, we can consider the form of the constitutive equations for a first-order Lorentz transformation. Ignoring terms to order v^2/c^2 in (2.89) and (2.90), we obtain $\bar{\mathbf{A}} = \bar{\mathbf{I}}$ and $\boldsymbol{\Omega} = \mathbf{v}(n^2 - 1)/c^2$. Then, by (2.87) and (2.88),

$$\mathbf{B} = \mu'\mathbf{H} - (n^2 - 1)\frac{\mathbf{v} \times \mathbf{E}}{c^2}, \tag{2.91}$$

$$\mathbf{D} = \epsilon'\mathbf{E} + (n^2 - 1)\frac{\mathbf{v} \times \mathbf{H}}{c^2}. \tag{2.92}$$

We can also derive these from the first-order field conversion equations (2.44)–(2.46). From (2.44) and (2.45) we have

$$\mathbf{D}' = \mathbf{D} + \mathbf{v} \times \mathbf{H}/c^2 = \epsilon'\mathbf{E}' = \epsilon'(\mathbf{E} + \mathbf{v} \times \mathbf{B}).$$

Eliminating $\mathbf{B}$ via (2.46), we have

$$\mathbf{D} + \mathbf{v} \times \mathbf{H}/c^2 = \epsilon'\mathbf{E} + \epsilon'\mathbf{v} \times (\mathbf{v} \times \mathbf{E}/c^2) + \epsilon'\mathbf{v} \times \mathbf{B}' = \epsilon'\mathbf{E} + \epsilon'\mathbf{v} \times \mathbf{B}'$$

where we have neglected terms of order v^2/c^2. Since $\mathbf{B}' = \mu'\mathbf{H}' = \mu'(\mathbf{H} - \mathbf{v} \times \mathbf{D})$, we have

$$\mathbf{D} + \mathbf{v} \times \mathbf{H}/c^2 = \epsilon'\mathbf{E} + \epsilon'\mu'\mathbf{v} \times \mathbf{H} - \epsilon'\mu'\mathbf{v} \times \mathbf{v} \times \mathbf{D}.$$

Using $n^2 = c^2\mu'\epsilon'$ and neglecting the last term since it is of order v^2/c^2, we obtain

$$\mathbf{D} = \epsilon'\mathbf{E} + (n^2 - 1)\frac{\mathbf{v} \times \mathbf{H}}{c^2},$$

which is identical to the expression (2.92) obtained by approximating the exact result to first order. Similar steps produce (2.91). In a Galilean frame where $v/c \ll 1$, the expressions reduce to $\mathbf{D} = \epsilon'\mathbf{E}$ and $\mathbf{B} = \mu'\mathbf{H}$, and the isotropy of the fields is preserved.

For a conducting medium having

$$\mathbf{J}' = \sigma'\mathbf{E}'$$

in a moving reference frame, Cullwick [43] shows that in the laboratory frame

$$\mathbf{J} = \sigma'\gamma(\bar{\mathbf{I}} - \boldsymbol{\beta}\boldsymbol{\beta}) \cdot \mathbf{E} + \sigma'\gamma c\boldsymbol{\beta} \times \mathbf{B}.$$

For $v \ll c$ we can set $\gamma \approx 1$ and see that $\mathbf{J} = \sigma'(\mathbf{E} + \mathbf{v} \times \mathbf{B})$ to first order.

2.3.2.4 Constitutive relations in deforming or rotating media

The transformations discussed in the previous sections hold for media in uniform relative motion. When a material body undergoes deformation or rotation, the concepts of special relativity are not directly applicable. However, authors such as Pauli [148] and Sommerfeld [180] have maintained that Minkowski's theory is *approximately* valid for deforming or rotating media if $\mathbf{v}$ is taken to be the *instantaneous* velocity at each point within the body. The reasoning is that at any instant in time each point within the body has a velocity $\mathbf{v}$ that may be associated with some inertial reference frame (generally different for each point). Thus the constitutive relations for the material at that point, within some small time interval taken about the observation time, may be assumed to be those of a stationary material, and the relations measured by an observer within the laboratory frame may be computed using the inertial frame for that point. This *instantaneous rest-frame* theory is most accurate at small accelerations $d\mathbf{v}/dt$. Van Bladel [202] outlines its shortcomings. See also Anderson and Ryon [3] and Mo [136] for detailed discussions of the electromagnetic properties of material media in accelerating frames of reference.

2.4 The Maxwell–Boffi equations

In any version of Maxwell's theory, the mediating field is the electromagnetic field described by four field vectors. In Minkowski's form of Maxwell's equations we use $\mathbf{E}$, $\mathbf{D}$, $\mathbf{B}$, and $\mathbf{H}$. As an alternative, consider the electromagnetic field as represented by the vector fields $\mathbf{E}$, $\mathbf{B}$, $\mathbf{P}$, and $\mathbf{M}$, and described by

$$\nabla \times \mathbf{E} = -\frac{\partial \mathbf{B}}{\partial t}, \tag{2.93}$$

$$\nabla \times (\mathbf{B}/\mu_0 - \mathbf{M}) = \mathbf{J} + \frac{\partial}{\partial t}(\epsilon_0 \mathbf{E} + \mathbf{P}), \tag{2.94}$$

$$\nabla \cdot (\epsilon_0 \mathbf{E} + \mathbf{P}) = \rho, \tag{2.95}$$

$$\nabla \cdot \mathbf{B} = 0. \tag{2.96}$$

These *Maxwell–Boffi equations* are named after L. Boffi, who formalized them for moving media [18]. The quantity $\mathbf{P}$ is the *polarization vector*, and $\mathbf{M}$ is the *magnetization vector*. The use of $\mathbf{P}$ and $\mathbf{M}$ in place of $\mathbf{D}$ and $\mathbf{H}$ is sometimes called an application of the *principle of Ampere and Lorentz* [199].

Let us examine the ramification of using (2.93)–(2.96) as the basis for a postulate of electromagnetics. These equations are similar to the Maxwell–Minkowski equations used earlier; must we rebuild all the underpinning of a new postulate, or can we use our original arguments based on the Minkowski form? For instance, how do we invoke uniqueness if we no longer have the field $\mathbf{H}$? What represents the flux of energy, formerly found using $\mathbf{E} \times \mathbf{H}$? And, importantly, are (2.93)–(2.94) form invariant under a Lorentz transformation?

It turns out that the set of vector fields $(\mathbf{E}, \mathbf{B}, \mathbf{P}, \mathbf{M})$ is merely a linear mapping of the set $(\mathbf{E}, \mathbf{D}, \mathbf{B}, \mathbf{H})$. As pointed out by Tai [190], any linear mapping of the four field vectors from Minkowski's form onto any other set of four field vectors will preserve the covariance of Maxwell's equations. Boffi chose to keep $\mathbf{E}$ and $\mathbf{B}$ intact and to introduce only two new fields; he could have kept $\mathbf{H}$ and $\mathbf{D}$ instead, or used a mapping that introduced four completely new fields (as did Chu). Many authors retain $\mathbf{E}$ and $\mathbf{H}$. This is somewhat more cumbersome since these vectors do not convert as a pair under a Lorentz transformation. A discussion of the idea of field vector "pairing" appears in § 2.6.

The usefulness of the Boffi form lies in the specific mapping chosen. Comparison of (2.93)–(2.96) to (2.1)–(2.4) quickly reveals that

$$\mathbf{P} = \mathbf{D} - \epsilon_0 \mathbf{E}, \tag{2.97}$$

$$\mathbf{M} = \mathbf{B}/\mu_0 - \mathbf{H}. \tag{2.98}$$

We see that $\mathbf{P}$ is the difference between $\mathbf{D}$ in a material and $\mathbf{D}$ in free space, while $\mathbf{M}$ is the difference between $\mathbf{H}$ in free space and $\mathbf{H}$ in a material. In free space, $\mathbf{P} = \mathbf{M} = 0$.

▶ **Example 2.5:** Manipulation of dyadic constitutive parameters

Find the 6×6 matrix $\bar{\mathbf{U}}$ that allows $\mathbf{P}$ and $\mathbf{M}$ to be written in terms of $\mathbf{E}$ and $\mathbf{B}$ as

$$\begin{bmatrix} \mathbf{P} \\ \mathbf{M} \end{bmatrix} = [\bar{\mathbf{U}}] \begin{bmatrix} \mathbf{E} \\ \mathbf{B} \end{bmatrix}.$$

Solution: Start with (2.9):

$$c\mathbf{D} = \bar{\mathbf{P}} \cdot \mathbf{E} + \bar{\mathbf{L}} \cdot (c\mathbf{B}).$$

Use $\mathbf{D} = \epsilon_0 \mathbf{E} + \mathbf{P}$ to get

$$\epsilon_0 \mathbf{E} + \mathbf{P} = \bar{\mathbf{P}} \cdot \left(\frac{1}{c} \mathbf{E} \right) + \bar{\mathbf{L}} \cdot \mathbf{B}.$$

Rearrangement gives

$$\mathbf{P} = \bar{\mathbf{P}} \cdot \left(\frac{1}{c} \mathbf{E} \right) - \epsilon_0 \mathbf{E} + \bar{\mathbf{L}} \cdot \mathbf{B},$$

or

$$\mathbf{P} = \left(\frac{1}{c} \bar{\mathbf{P}} - \epsilon_0 \bar{\mathbf{I}} \right) \cdot \mathbf{E} + \bar{\mathbf{L}} \cdot \mathbf{B}. \tag{2.99}$$

Next start with (2.10):

$$\mathbf{H} = \bar{\mathbf{M}} \cdot \mathbf{E} + \bar{\mathbf{Q}} \cdot (c\mathbf{B}).$$

Use $\mathbf{H} = \mathbf{B}/\mu_0 - \mathbf{M}$ to get

$$\frac{\mathbf{B}}{\mu_0} - \mathbf{M} = \bar{\mathbf{M}} \cdot \mathbf{E} + \bar{\mathbf{Q}} \cdot (c\mathbf{B}).$$

Rearrangement gives

$$\mathbf{M} = -\bar{\mathbf{M}} \cdot \mathbf{E} - \bar{\mathbf{Q}} \cdot (c\mathbf{B}) + \frac{\mathbf{B}}{\mu_0},$$

or

$$\mathbf{M} = -\bar{\mathbf{M}} \cdot \mathbf{E} - \left(c\bar{\mathbf{Q}} - \frac{1}{\mu_0}\bar{\mathbf{I}} \right) \cdot \mathbf{B}. \tag{2.100}$$

Thus, $\bar{\mathbf{U}}$ is

$$[\bar{\mathbf{U}}] = \begin{bmatrix} \left(\frac{1}{c}\bar{\mathbf{P}} - \epsilon_0\bar{\mathbf{I}}\right) & \bar{\mathbf{L}} \\ -\bar{\mathbf{M}} & -\left(c\bar{\mathbf{Q}} - \frac{1}{\mu_0}\bar{\mathbf{I}}\right) \end{bmatrix} \cdot \blacktriangleleft$$

2.4.1 Equivalent polarization and magnetization sources

The Boffi formulation provides a new way to regard $\mathbf{E}$ and $\mathbf{B}$. Maxwell grouped $(\mathbf{E}, \mathbf{H})$ as a pair of "force vectors" to be associated with line integrals (or curl operations in the point forms of his equations), and $(\mathbf{D}, \mathbf{B})$ as a pair of "flux vectors" associated with surface integrals (or divergence operations). That is, $\mathbf{E}$ is interpreted as belonging to the computation of "emf" as a line integral, while $\mathbf{B}$ is interpreted as a density of magnetic "flux" passing through a surface. Similarly, $\mathbf{H}$ yields the "mmf" about some closed path and $\mathbf{D}$ the electric flux through a surface. The introduction of $\mathbf{P}$ and $\mathbf{M}$ allows us to also regard $\mathbf{E}$ as a flux vector and $\mathbf{B}$ as a force vector — in essence, allowing the two fields $\mathbf{E}$ and $\mathbf{B}$ to take on the duties that required four fields in Minkowski's form. To see this, we rewrite the Maxwell–Boffi equations as

$$\nabla \times \mathbf{E} = -\frac{\partial \mathbf{B}}{\partial t},$$

$$\nabla \times \frac{\mathbf{B}}{\mu_0} = \left(\mathbf{J} + \nabla \times \mathbf{M} + \frac{\partial \mathbf{P}}{\partial t} \right) + \frac{\partial \epsilon_0 \mathbf{E}}{\partial t},$$

$$\nabla \cdot (\epsilon_0 \mathbf{E}) = (\rho - \nabla \cdot \mathbf{P}),$$

$$\nabla \cdot \mathbf{B} = 0,$$

and compare them to the Maxwell–Minkowski equations for sources in free space:

$$\nabla \times \mathbf{E} = -\frac{\partial \mathbf{B}}{\partial t},$$

$$\nabla \times \frac{\mathbf{B}}{\mu_0} = \mathbf{J} + \frac{\partial \epsilon_0 \mathbf{E}}{\partial t},$$

$$\nabla \cdot (\epsilon_0 \mathbf{E}) = \rho,$$

$$\nabla \cdot \mathbf{B} = 0.$$

The forms are preserved if we identify $\partial \mathbf{P}/\partial t$ and $\nabla \times \mathbf{M}$ as new types of current density, and $\nabla \cdot \mathbf{P}$ as a new type of charge density. We define

$$\mathbf{J}_P = \frac{\partial \mathbf{P}}{\partial t} \tag{2.101}$$

as an *equivalent polarization current* density, and

$$\mathbf{J}_M = \nabla \times \mathbf{M}$$

as an *equivalent magnetization current* density (sometimes called the *equivalent Amperian currents of magnetized matter* [199]). We define

$$\rho_P = -\nabla \cdot \mathbf{P}$$

as an *equivalent polarization charge* density (sometimes called the *Poisson–Kelvin* equivalent charge distribution [199]). Then the Maxwell–Boffi equations become simply

$$\nabla \times \mathbf{E} = -\frac{\partial \mathbf{B}}{\partial t}, \tag{2.102}$$

$$\nabla \times \frac{\mathbf{B}}{\mu_0} = (\mathbf{J} + \mathbf{J}_M + \mathbf{J}_P) + \frac{\partial \epsilon_0 \mathbf{E}}{\partial t}, \tag{2.103}$$

$$\nabla \cdot (\epsilon_0 \mathbf{E}) = (\rho + \rho_P), \tag{2.104}$$

$$\nabla \cdot \mathbf{B} = 0. \tag{2.105}$$

Here is the new view. A material can be viewed as composed of charged particles of matter immersed in free space. When these charges are properly considered as "equivalent" polarization and magnetization charges, all field effects (describable through flux and force vectors) can be handled by the two fields $\mathbf{E}$ and $\mathbf{B}$. Whereas in Minkowski's form $\mathbf{D}$ diverges from ρ, in Boffi's form $\mathbf{E}$ diverges from a *total* charge density consisting of ρ and ρ_P. Whereas in the Minkowski form $\mathbf{H}$ curls around $\mathbf{J}$, in the Boffi form $\mathbf{B}$ curls around the total current density consisting of $\mathbf{J}$, $\mathbf{J}_M$, and $\mathbf{J}_P$.

This view was pioneered by Lorentz, who by 1892 considered matter as consisting of bulk molecules in a vacuum that would respond to an applied electromagnetic field [132]. The resulting motion of the charged particles of matter then became another source term for the "fundamental" fields $\mathbf{E}$ and $\mathbf{B}$. Using this reasoning, he was able to reduce the fundamental Maxwell equations to two equations in two unknowns, demonstrating a simplicity appealing to many (including Einstein). Of course, to apply this concept we must be able to describe how the charged particles respond to an applied field. Simple microscopic models of the constituents of matter are generally used: some combination of electric and magnetic dipoles, or of loops of electric and magnetic current.

The Boffi equations are mathematically appealing since they now specify both the curl and divergence of the two field quantities $\mathbf{E}$ and $\mathbf{B}$. By the Helmholtz theorem we know that a field vector is uniquely specified when both its curl and divergence are given. But this assumes that the equivalent sources produced by $\mathbf{P}$ and $\mathbf{M}$ are true source fields in the same sense as $\mathbf{J}$. We have precluded this by insisting in Chapter 1 that the source field must be independent of the mediating field it sources. If we view $\mathbf{P}$ and $\mathbf{M}$ as merely a mapping from the original vector fields of Minkowski's form, we still have four vector fields with which to contend. And with these must also be a mapping of the constitutive relationships, which now link the fields $\mathbf{E}$, $\mathbf{B}$, $\mathbf{P}$, and $\mathbf{M}$. Rather than argue the actual physical existence of the equivalent sources, we note that a real benefit of the new view is that under certain circumstances the equivalent source quantities can be determined through physical reasoning. Hence we can create physical models of $\mathbf{P}$ and $\mathbf{M}$ and deduce their links to $\mathbf{E}$ and $\mathbf{B}$. We may then find it easier to understand and deduce the constitutive relationships. However, we do not in general consider $\mathbf{E}$ and $\mathbf{B}$ to be in any way more "fundamental" than $\mathbf{D}$ and $\mathbf{H}$.

2.4.2 Covariance of the Boffi form

Because of the linear relationships (2.97) and (2.98), covariance of the Maxwell–Minkowski equations carries over to the Maxwell–Boffi equations. However, the conversion between fields in different moving reference frames will now involve $\mathbf{P}$ and $\mathbf{M}$. Since Faraday's law is unchanged in the Boffi form, we still have

$$\mathbf{E}'_\| = \mathbf{E}_\|,$$
$$\mathbf{B}'_\| = \mathbf{B}_\|,$$
$$\mathbf{E}'_\perp = \gamma(\mathbf{E}_\perp + \boldsymbol{\beta} \times c\mathbf{B}_\perp),$$
$$c\mathbf{B}'_\perp = \gamma(c\mathbf{B}_\perp - \boldsymbol{\beta} \times \mathbf{E}_\perp).$$

To see how $\mathbf{P}$ and $\mathbf{M}$ convert, we note that in the laboratory frame $\mathbf{D} = \epsilon_0\mathbf{E} + \mathbf{P}$ and $\mathbf{H} = \mathbf{B}/\mu_0 - \mathbf{M}$, while in the moving frame $\mathbf{D}' = \epsilon_0\mathbf{E}' + \mathbf{P}'$ and $\mathbf{H}' = \mathbf{B}'/\mu_0 - \mathbf{M}'$. Thus

$$\mathbf{P}'_\| = \mathbf{D}'_\| - \epsilon_0\mathbf{E}'_\| = \mathbf{D}_\| - \epsilon_0\mathbf{E}_\| = \mathbf{P}_\|$$

and

$$\mathbf{M}'_\| = \mathbf{B}'_\|/\mu_0 - \mathbf{H}'_\| = \mathbf{B}_\|/\mu_0 - \mathbf{H}_\| = \mathbf{M}_\|.$$

For the perpendicular components

$$\mathbf{D}'_\perp = \gamma(\mathbf{D}_\perp + \boldsymbol{\beta} \times \mathbf{H}_\perp/c) = \epsilon_0\mathbf{E}'_\perp + \mathbf{P}'_\perp = \epsilon_0\left[\gamma(\mathbf{E}_\perp + \boldsymbol{\beta} \times c\mathbf{B}_\perp)\right] + \mathbf{P}'_\perp;$$

substitution of $\mathbf{H}_\perp = \mathbf{B}_\perp/\mu_0 - \mathbf{M}_\perp$ then gives

$$\mathbf{P}'_\perp = \gamma(\mathbf{D}_\perp - \epsilon_0\mathbf{E}_\perp) - \gamma\epsilon_0\boldsymbol{\beta} \times c\mathbf{B}_\perp + \gamma\boldsymbol{\beta} \times \mathbf{B}_\perp/(c\mu_0) - \gamma\boldsymbol{\beta} \times \mathbf{M}_\perp/c$$

or

$$c\mathbf{P}'_\perp = \gamma(c\mathbf{P}_\perp - \boldsymbol{\beta} \times \mathbf{M}_\perp).$$

Similarly,

$$\mathbf{M}'_\perp = \gamma(\mathbf{M}_\perp + \boldsymbol{\beta} \times c\mathbf{P}_\perp).$$

Hence

$$\mathbf{E}'_\| = \mathbf{E}_\|, \quad \mathbf{B}'_\| = \mathbf{B}_\|, \quad \mathbf{P}'_\| = \mathbf{P}_\|, \quad \mathbf{M}'_\| = \mathbf{M}_\|, \quad \mathbf{J}'_\perp = \mathbf{J}_\perp, \tag{2.106}$$

and

$$\mathbf{E}'_\perp = \gamma(\mathbf{E}_\perp + \boldsymbol{\beta} \times c\mathbf{B}_\perp), \tag{2.107}$$
$$c\mathbf{B}'_\perp = \gamma(c\mathbf{B}_\perp - \boldsymbol{\beta} \times \mathbf{E}_\perp), \tag{2.108}$$
$$c\mathbf{P}'_\perp = \gamma(c\mathbf{P}_\perp - \boldsymbol{\beta} \times \mathbf{M}_\perp), \tag{2.109}$$
$$\mathbf{M}'_\perp = \gamma(\mathbf{M}_\perp + \boldsymbol{\beta} \times c\mathbf{P}_\perp), \tag{2.110}$$
$$\mathbf{J}'_\| = \gamma(\mathbf{J}_\| - \rho\mathbf{v}). \tag{2.111}$$

In the case of the first-order Lorentz transformation, we can set $\gamma \approx 1$ to obtain

$$\mathbf{E}' = \mathbf{E} + \mathbf{v} \times \mathbf{B}, \tag{2.112}$$

$$\mathbf{B}' = \mathbf{B} - \frac{\mathbf{v} \times \mathbf{E}}{c^2}, \tag{2.113}$$

$$\mathbf{P}' = \mathbf{P} - \frac{\mathbf{v} \times \mathbf{M}}{c^2}, \tag{2.114}$$

$$\mathbf{M}' = \mathbf{M} + \mathbf{v} \times \mathbf{P}, \tag{2.115}$$

$$\mathbf{J}' = \mathbf{J} - \rho\mathbf{v}. \tag{2.116}$$

To convert from the moving frame to the laboratory frame we simply swap primed with unprimed fields and let $\mathbf{v} \to -\mathbf{v}$.

▶ **Example 2.6:** Covariance of the Maxwell-Boffi equations

A linear isotropic medium has

$$\mathbf{D}' = \epsilon_0 \epsilon_r' \mathbf{E}', \qquad \mathbf{B}' = \mu_0 \mu_r' \mathbf{H}',$$

in a moving reference frame. Find $\mathbf{P}$ and $\mathbf{M}$ in terms of the laboratory frame electromagnetic fields $\mathbf{E}$ and $\mathbf{B}$. For simplicity, consider only the first-order equations.

Solution: From (2.97) we have

$$\mathbf{P}' = \epsilon_0 \epsilon_r' \mathbf{E}' - \epsilon_0 \mathbf{E}' = \epsilon_0 \chi_e' \mathbf{E}'$$

where $\chi_e' = \epsilon_r' - 1$ is the electric susceptibility of the moving material. Similarly (2.98) yields

$$\mathbf{M}' = \frac{\mathbf{B}'}{\mu_0} - \frac{\mathbf{B}'}{\mu_0 \mu_r'} = \frac{\mathbf{B}' \chi_m'}{\mu_0 \mu_r'}$$

where $\chi_m' = \mu_r' - 1$ is the magnetic susceptibility of the moving material. To find how $\mathbf{P}$ and $\mathbf{M}$ are related to $\mathbf{E}$ and $\mathbf{B}$ in the laboratory frame, we consider the first-order expressions. From (2.114) we have

$$\mathbf{P} = \mathbf{P}' + \frac{\mathbf{v} \times \mathbf{M}'}{c^2} = \epsilon_0 \chi_e' \mathbf{E}' + \frac{\mathbf{v} \times \mathbf{B}' \chi_m'}{\mu_0 \mu_r' c^2}.$$

Substituting for $\mathbf{E}'$ and $\mathbf{B}'$ from (2.112) and (2.113), and using $\mu_0 c^2 = 1/\epsilon_0$, we have

$$\mathbf{P} = \epsilon_0 \chi_e' (\mathbf{E} + \mathbf{v} \times \mathbf{B}) + \epsilon_0 \frac{\chi_m'}{\mu_r'} \mathbf{v} \times \left(\mathbf{B} - \frac{\mathbf{v} \times \mathbf{E}}{c^2} \right).$$

Neglecting the last term since it varies as v^2/c^2, we get

$$\mathbf{P} = \epsilon_0 \chi_e' \mathbf{E} + \epsilon_0 \left(\chi_e' + \frac{\chi_m'}{\mu_r'} \right) \mathbf{v} \times \mathbf{B}. \tag{2.117}$$

Similarly,

$$\mathbf{M} = \frac{\chi_m'}{\mu_0 \mu_r'} \mathbf{B} - \epsilon_0 \left(\chi_e' + \frac{\chi_m'}{\mu_r'} \right) \mathbf{v} \times \mathbf{E}. \blacktriangleleft \tag{2.118}$$

2.5 Large-scale form of Maxwell's equations

We can write Maxwell's equations in a form that incorporates the spatial variation of the field in a certain region of space. To do this, we integrate the point form of Maxwell's equations over a region of space, then perform some succession of manipulations until we arrive at a form that provides us some benefit in our work with electromagnetic fields. The results are particularly useful for understanding the properties of electric and magnetic circuits, and for predicting the behavior of electrical machinery.

We shall consider two important situations: a mathematical surface that moves with constant velocity $\mathbf{v}$ and with constant shape, and a surface that moves and deforms arbitrarily.

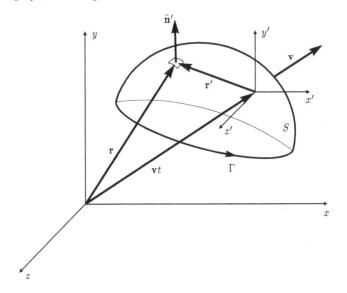

FIGURE 2.2
Open surface having velocity **v** relative to laboratory (unprimed) coordinate system.
Surface is non-deforming.

2.5.1 Surface moving with constant velocity

Consider an open surface S moving with constant velocity **v** relative to the laboratory
frame (Figure 2.2). Assume every point on the surface is an ordinary point. At any
instant t we can express the relationship between the fields at points on S in either
frame. In the laboratory frame we have

$$\nabla \times \mathbf{E} = -\frac{\partial \mathbf{B}}{\partial t}, \qquad \nabla \times \mathbf{H} = \frac{\partial \mathbf{D}}{\partial t} + \mathbf{J},$$

while in the moving frame

$$\nabla' \times \mathbf{E}' = -\frac{\partial \mathbf{B}'}{\partial t'}, \qquad \nabla' \times \mathbf{H}' = \frac{\partial \mathbf{D}'}{\partial t'} + \mathbf{J}'.$$

If we integrate over S and use Stokes's theorem, we get for the laboratory frame

$$\oint_{\Gamma} \mathbf{E} \cdot \mathbf{dl} = -\int_{S} \frac{\partial \mathbf{B}}{\partial t} \cdot \mathbf{dS}, \tag{2.119}$$

$$\oint_{\Gamma} \mathbf{H} \cdot \mathbf{dl} = \int_{S} \frac{\partial \mathbf{D}}{\partial t} \cdot \mathbf{dS} + \int_{S} \mathbf{J} \cdot \mathbf{dS}, \tag{2.120}$$

and for the moving frame

$$\oint_{\Gamma'} \mathbf{E}' \cdot \mathbf{dl}' = -\int_{S'} \frac{\partial \mathbf{B}'}{\partial t'} \cdot \mathbf{dS}', \tag{2.121}$$

$$\oint_{\Gamma'} \mathbf{H}' \cdot \mathbf{dl}' = \int_{S'} \frac{\partial \mathbf{D}'}{\partial t'} \cdot \mathbf{dS}' + \int_{S'} \mathbf{J}' \cdot \mathbf{dS}'. \tag{2.122}$$

Here boundary contour Γ has sense determined by the right-hand rule. We use the
notation Γ', S', etc., to indicate that all integrations for the moving frame are computed

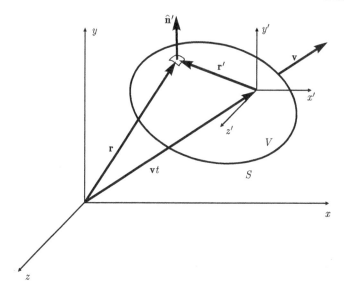

FIGURE 2.3
Nondeforming volume region having velocity **v** relative to laboratory (unprimed) coordinate system.

using space and time variables in that frame. Equation (2.119) is the *integral form of Faraday's law*, while (2.120) is the *integral form of Ampere's law*.

Faraday's law states that the net circulation of **E** about a contour Γ (sometimes called the *electromotive force* or *emf*) is determined by the flux of the time-rate of change of the flux vector **B** passing through the surface bounded by Γ. Ampere's law states that the circulation of **H** (sometimes called the *magnetomotive force* or *mmf*) is determined by the flux of the current **J** plus the flux of the time-rate of change of the flux vector **D**. It is the term containing $\partial \mathbf{D}/\partial t$ that Maxwell recognized as necessary to make his equations consistent; since it has units of current, it is often referred to as the *displacement current* term.

Equations (2.119)–(2.120) are the large-scale or integral forms of Maxwell's equations. They are the integral-form equivalents of the point forms, and are form invariant under Lorentz transformation. If we express the fields in terms of the moving reference frame, we can write

$$\oint_{\Gamma'} \mathbf{E}' \cdot \mathbf{dl}' = -\frac{d}{dt'} \int_{S'} \mathbf{B}' \cdot \mathbf{dS}', \tag{2.123}$$

$$\oint_{\Gamma'} \mathbf{H}' \cdot \mathbf{dl}' = \frac{d}{dt'} \int_{S'} \mathbf{D}' \cdot \mathbf{dS}' + \int_{S'} \mathbf{J}' \cdot \mathbf{dS}'. \tag{2.124}$$

These hold for a stationary surface, since the surface would be stationary to an observer who moves with it. We are therefore justified in removing the partial derivative from the integral. Although the surfaces and contours considered here are purely mathematical, they often coincide with actual physical boundaries. The surface may surround a moving material medium, for instance, or the contour may conform to a wire moving in an electrical machine.

We can also convert the auxiliary equations to large-scale form. Consider a volume region V surrounded by a surface S that moves with velocity **v** relative to the laboratory frame (Figure 2.3). Integrating the point form of Gauss's law over V we have

$$\int_V \nabla \cdot \mathbf{D} \, dV = \int_V \rho \, dV.$$

Using the divergence theorem and recognizing that the integral of charge density is total charge, we obtain

$$\oint_S \mathbf{D} \cdot d\mathbf{S} = \int_V \rho \, dV = Q(t) \tag{2.125}$$

where $Q(t)$ is the total charge contained within V at time t. This large-scale form of Gauss's law states that the total flux of $\mathbf{D}$ passing through a closed surface is identical to the electric charge Q contained within. Similarly,

$$\oint_S \mathbf{B} \cdot d\mathbf{S} = 0 \tag{2.126}$$

is the large-scale magnetic field Gauss's law. It states that the total flux of $\mathbf{B}$ passing through a closed surface is zero, since there are no magnetic charges contained within (i.e., magnetic charge does not exist).

Since charge is an invariant quantity, the large-scale forms of the auxiliary equations take the same form in a moving reference frame:

$$\oint_{S'} \mathbf{D}' \cdot d\mathbf{S}' = \int_{V'} \rho' \, dV' = Q(t)$$

and

$$\oint_{S'} \mathbf{B}' \cdot d\mathbf{S}' = 0.$$

The large-scale forms of the auxiliary equations may be derived from the large-scale forms of Faraday's and Ampere's laws. To obtain Gauss's law, we let the open surface in Ampere's law become a closed surface. Then $\oint \mathbf{H} \cdot d\mathbf{l}$ vanishes, and application of the large-scale form of the continuity equation (1.8) produces (2.125). The magnetic Gauss's law (2.126) is found from Faraday's law (2.119) by a similar transition from an open surface to a closed surface.

The values obtained from the expressions (2.119)–(2.120) will *not* match those obtained from (2.121)–(2.122), and we can use the Lorentz transformation field conversions to study how they differ. That is, we can write either side of the laboratory equations in terms of the moving reference frame fields, or vice versa. For most engineering applications where $v/c \ll 1$, this is not done via the Lorentz transformation field relations, but rather via the Galilean approximations to these relations (see Tai [191] for details on using the Lorentz transformation field relations). We consider the most common situation in the next section.

2.5.1.1 Kinematic form of the large-scale Maxwell equations

Confusion can result from the fact that the large-scale forms of Maxwell's equations can be written in a number of ways. A popular formulation of Faraday's law, the *emf formulation*, revolves around the concept of electromotive force. Unfortunately, various authors offer different definitions of emf in a moving circuit.

Consider a nondeforming contour in space, moving with constant velocity $\mathbf{v}$ relative to the laboratory frame (Figure 2.4). In terms of the laboratory fields we have the large-scale form of Faraday's law (2.119). The flux term on the right-hand side of this equation

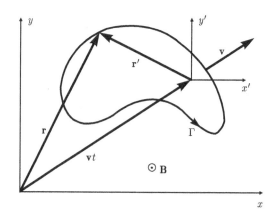

FIGURE 2.4
Nondeforming closed contour moving with velocity **v** through a magnetic field **B** given in the laboratory (unprimed) coordinate system.

can be written differently by employing the Helmholtz transport theorem (A.65). If a nondeforming surface S moves with uniform velocity **v** relative to the laboratory frame, and a vector field $\mathbf{A}(\mathbf{r}, t)$ is expressed in the stationary frame, then the time derivative of the flux of **A** through S is

$$\frac{d}{dt} \int_S \mathbf{A} \cdot d\mathbf{S} = \int_S \left[\frac{\partial \mathbf{A}}{\partial t} + \mathbf{v}(\nabla \cdot \mathbf{A}) - \nabla \times (\mathbf{v} \times \mathbf{A}) \right] \cdot d\mathbf{S}. \qquad (2.127)$$

Using this with (2.119) we have

$$\oint_\Gamma \mathbf{E} \cdot d\mathbf{l} = -\frac{d}{dt} \int_S \mathbf{B} \cdot d\mathbf{S} + \int_S \mathbf{v}(\nabla \cdot \mathbf{B}) \cdot d\mathbf{S} - \int_S \nabla \times (\mathbf{v} \times \mathbf{B}) \cdot d\mathbf{S}.$$

Remembering that $\nabla \cdot \mathbf{B} = 0$ and using Stokes's theorem on the last term, we obtain

$$\oint_\Gamma (\mathbf{E} + \mathbf{v} \times \mathbf{B}) \cdot d\mathbf{l} = -\frac{d}{dt} \int_S \mathbf{B} \cdot d\mathbf{S} = -\frac{d\Psi(t)}{dt}$$

where the *magnetic flux*

$$\int_S \mathbf{B} \cdot d\mathbf{S} = \Psi(t)$$

represents the flux of **B** through S. Following Sommerfeld [180], we may set

$$\mathbf{E}^* = \mathbf{E} + \mathbf{v} \times \mathbf{B}$$

to obtain the *kinematic form of Faraday's law*

$$\oint_\Gamma \mathbf{E}^* \cdot d\mathbf{l} = -\frac{d}{dt} \int_S \mathbf{B} \cdot d\mathbf{S} = -\frac{d\Psi(t)}{dt}. \qquad (2.128)$$

(The asterisk should not be confused with the notation for complex conjugate.)

Much confusion arises from the similarity between (2.128) and (2.123). In fact, these expressions are different and give different results. This is because $\mathbf{B}'$ in (2.123) is measured *in the frame of the moving circuit*, while **B** in (2.128) is measured in the frame of the laboratory. Further confusion arises from various definitions of emf. Many authors (e.g., Hermann Weyl [213]) define emf to be the circulation of $\mathbf{E}^*$. In that case the emf is equal to the negative time rate of change of the flux of the *laboratory frame* magnetic field $\mathbf{B}$ through S. Since the Lorentz force experienced by a charge q moving *with the contour* is given by $q\mathbf{E}^* = q(\mathbf{E} + \mathbf{v} \times \mathbf{B})$, this emf is the circulation of Lorentz force per unit charge along the contour. If the contour is aligned with a conducting circuit, then in some cases this emf can be given physical interpretation as the work required to move a charge around the entire circuit through the conductor against the Lorentz force. Unfortunately the usefulness of this definition of emf is lost if the time or space rate of change of the fields is so large that no true loop current can be established (hence Kirchoff's law cannot be employed). Such a problem must be treated as an electromagnetic "scattering" problem with consideration given to retardation effects. Detailed discussions of the physical interpretation of $\mathbf{E}^*$ in the definition of emf are given by Cullwick [43] and Scanlon et al. [168].

Other authors choose to define emf as the circulation of the electric field *in the frame of the moving contour*. In this case the circulation of $\mathbf{E}'$ in (2.123) is the emf, and is related to the flux of the magnetic field *in the frame of the moving circuit*. As pointed out above, the result differs from that based on the Lorentz force. If we wish, we can also write this emf in terms of the fields expressed in the laboratory frame. To do this we must convert $\partial \mathbf{B}'/\partial t'$ to the laboratory fields using the rules for a Lorentz transformation. The result, given by Tai [191], is quite complicated and involves both the magnetic *and* electric laboratory-frame fields.

The moving-frame emf as computed from the Lorentz transformation is rarely used as a working definition of emf, mostly because circuits moving at relativistic velocities are seldom used by engineers. Unfortunately, more confusion arises for the case $v \ll c$, since for a Galilean frame the Lorentz-force and moving-frame emfs become identical. This is apparent if we use (2.41) to replace $\mathbf{B}'$ with the laboratory frame field $\mathbf{B}$, and (2.38) to replace $\mathbf{E}'$ with the combination of laboratory frame fields $\mathbf{E} + \mathbf{v} \times \mathbf{B}$. Then (2.123) becomes

$$\oint_\Gamma \mathbf{E}' \cdot d\mathbf{l} = \oint_\Gamma (\mathbf{E} + \mathbf{v} \times \mathbf{B}) \cdot d\mathbf{l} = -\frac{d}{dt} \int_S \mathbf{B} \cdot d\mathbf{S},$$

which is identical to (2.128). For circuits moving with low velocity then, the circulation of $\mathbf{E}'$ can be interpreted as work per unit charge. As an added bit of confusion, the term

$$\oint_\Gamma (\mathbf{v} \times \mathbf{B}) \cdot d\mathbf{l} = \int_S \nabla \times (\mathbf{v} \times \mathbf{B}) \cdot d\mathbf{S}$$

is sometimes called *motional emf*, since it is the component of the circulation of $\mathbf{E}^*$ that is directly attributable to the motion of the circuit.

Although less commonly done, we can also rewrite Ampere's law (2.120) using (2.127). This gives

$$\oint_\Gamma \mathbf{H} \cdot d\mathbf{l} = \int_S \mathbf{J} \cdot d\mathbf{S} + \frac{d}{dt} \int_S \mathbf{D} \cdot d\mathbf{S} - \int_S (\mathbf{v}\nabla \cdot \mathbf{D}) \cdot d\mathbf{S} + \int_S \nabla \times (\mathbf{v} \times \mathbf{D}) \cdot d\mathbf{S}.$$

Using $\nabla \cdot \mathbf{D} = \rho$ and using Stokes's theorem on the last term, we obtain

$$\oint_\Gamma (\mathbf{H} - \mathbf{v} \times \mathbf{D}) \cdot d\mathbf{l} = \frac{d}{dt} \int_S \mathbf{D} \cdot d\mathbf{S} + \int_S (\mathbf{J} - \rho\mathbf{v}) \cdot d\mathbf{S}.$$

Finally, letting $\mathbf{H}^* = \mathbf{H} - \mathbf{v} \times \mathbf{D}$ and $\mathbf{J}^* = \mathbf{J} - \rho\mathbf{v}$ we can write the *kinematic form of Ampere's law*:

$$\oint_\Gamma \mathbf{H}^* \cdot d\mathbf{l} = \frac{d}{dt} \int_S \mathbf{D} \cdot d\mathbf{S} + \int_S \mathbf{J}^* \cdot d\mathbf{S}. \tag{2.129}$$

In a Galilean frame where we use (2.38)–(2.43), we see that (2.129) is identical to

$$\oint_{\Gamma} \mathbf{H}' \cdot \mathbf{dl} = \frac{d}{dt'} \int_{S} \mathbf{D}' \cdot \mathbf{dS} + \int_{S} \mathbf{J}' \cdot \mathbf{dS}$$

where the primed fields are measured in the frame of the moving contour. This equivalence does *not* hold when the Lorentz transformation is used to represent the primed fields.

2.5.1.2   Alternative form of the large-scale Maxwell equations

We can write Maxwell's equations in an alternative large-scale form involving only surface and volume integrals. This will be useful later for establishing the field jump conditions across a material or source discontinuity. Again we begin with Maxwell's equations in point form, but instead of integrating them over an open surface we integrate over a volume region V moving with velocity $\mathbf{v}$ (Figure 2.3). In the laboratory frame this gives

$$\int_{V} (\nabla \times \mathbf{E}) \, dV = -\int_{V} \frac{\partial \mathbf{B}}{\partial t} \, dV,$$

$$\int_{V} (\nabla \times \mathbf{H}) \, dV = \int_{V} \left(\frac{\partial \mathbf{D}}{\partial t} + \mathbf{J} \right) dV.$$

An application of curl theorem (B.30) then gives

$$\oint_{S} (\hat{\mathbf{n}} \times \mathbf{E}) \, dS = -\int_{V} \frac{\partial \mathbf{B}}{\partial t} \, dV, \tag{2.130}$$

$$\oint_{S} (\hat{\mathbf{n}} \times \mathbf{H}) \, dS = \int_{V} \left(\frac{\partial \mathbf{D}}{\partial t} + \mathbf{J} \right) dV. \tag{2.131}$$

Similar results are obtained for the fields in the moving frame:

$$\oint_{S'} (\hat{\mathbf{n}}' \times \mathbf{E}') \, dS' = -\int_{V'} \frac{\partial \mathbf{B}'}{\partial t'} \, dV',$$

$$\oint_{S'} (\hat{\mathbf{n}}' \times \mathbf{H}') \, dS' = \int_{V'} \left(\frac{\partial \mathbf{D}'}{\partial t'} + \mathbf{J}' \right) dV'.$$

These large-scale forms are an alternative to (2.119)–(2.122). They are also form-invariant under a Lorentz transformation.

An alternative to the kinematic formulation of (2.128) and (2.129) can be achieved by applying a kinematic identity for a moving volume region. If V is surrounded by a surface S that moves with velocity $\mathbf{v}$ relative to the laboratory frame, and if a vector field $\mathbf{A}$ is measured in the laboratory frame, then the vector form of the general transport theorem (A.70) states that

$$\frac{d}{dt} \int_{V} \mathbf{A} \, dV = \int_{V} \frac{\partial \mathbf{A}}{\partial t} \, dV + \oint_{S} \mathbf{A}(\mathbf{v} \cdot \hat{\mathbf{n}}) \, dS. \tag{2.132}$$

Applying this to (2.130) and (2.131) we have

$$\oint_{S} [\hat{\mathbf{n}} \times \mathbf{E} - (\mathbf{v} \cdot \hat{\mathbf{n}})\mathbf{B}] \, dS = -\frac{d}{dt} \int_{V} \mathbf{B} \, dV,$$

$$\oint_{S} [\hat{\mathbf{n}} \times \mathbf{H} + (\mathbf{v} \cdot \hat{\mathbf{n}})\mathbf{D}] \, dS = \int_{V} \mathbf{J} \, dV + \frac{d}{dt} \int_{V} \mathbf{D} \, dV.$$

We can also apply (2.132) to the large-scale form of the continuity equation (1.8) and obtain the expression for a volume region moving with velocity $\mathbf{v}$:

$$\oint_S (\mathbf{J} - \rho\mathbf{v}) \cdot d\mathbf{S} = -\frac{d}{dt} \int_V \rho \, dV.$$

2.5.2 Moving, deforming surfaces

Because (2.127) holds for arbitrarily moving surfaces, the kinematic versions (2.128) and (2.129) hold when $\mathbf{v}$ is interpreted as an instantaneous velocity. However, if the surface and contour lie within a material body that moves relative to the laboratory frame, the constitutive equations relating $\mathbf{E}$, $\mathbf{D}$, $\mathbf{B}$, $\mathbf{H}$, and $\mathbf{J}$ in the laboratory frame differ from those relating the fields in the stationary frame of the body (if the body is not accelerating), and thus the concepts of § 2.3.2.3 must be employed. This is important when boundary conditions at a moving surface are needed. Particular care must be taken when the body accelerates, since the constitutive relations are then only approximate.

The representation (2.123)–(2.124) is also generally valid, provided we *define* the primed fields as those converted from laboratory fields using the Lorentz transformation with instantaneous velocity $\mathbf{v}$. Here we should use a different inertial frame for each point in the integration, and align the frame with the velocity vector $\mathbf{v}$ at the instant t. We certainly may do this since we can choose to integrate any function we wish. However, this representation may not find wide application.

We thus choose the following expressions, valid for arbitrarily moving surfaces containing only regular points, as our general forms of the large-scale Maxwell equations:

$$\oint_{\Gamma(t)} \mathbf{E}^* \cdot d\mathbf{l} = -\frac{d}{dt} \int_{S(t)} \mathbf{B} \cdot d\mathbf{S} = -\frac{d\Psi(t)}{dt},$$

$$\oint_{\Gamma(t)} \mathbf{H}^* \cdot d\mathbf{l} = \frac{d}{dt} \int_{S(t)} \mathbf{D} \cdot d\mathbf{S} + \int_{S(t)} \mathbf{J}^* \cdot d\mathbf{S},$$

where

$$\mathbf{E}^* = \mathbf{E} + \mathbf{v} \times \mathbf{B}, \qquad \mathbf{H}^* = \mathbf{H} - \mathbf{v} \times \mathbf{D}, \qquad \mathbf{J}^* = \mathbf{J} - \rho\mathbf{v},$$

and where all fields are taken to be measured in the laboratory frame with $\mathbf{v}$ the instantaneous velocity of points on the surface and contour relative to that frame. The constitutive parameters must be considered carefully if the contours and surfaces lie in a moving material medium.

Kinematic identity (2.132) is also valid for arbitrarily moving surfaces. Thus we have the following, valid for arbitrarily moving surfaces and volumes containing only regular points:

$$\oint_{S(t)} [\hat{\mathbf{n}} \times \mathbf{E} - (\mathbf{v} \cdot \hat{\mathbf{n}})\mathbf{B}] \, dS = -\frac{d}{dt} \int_{V(t)} \mathbf{B} \, dV,$$

$$\oint_{S(t)} [\hat{\mathbf{n}} \times \mathbf{H} + (\mathbf{v} \cdot \hat{\mathbf{n}})\mathbf{D}] \, dS = \int_{V(t)} \mathbf{J} \, dV + \frac{d}{dt} \int_{V(t)} \mathbf{D} \, dV.$$

We also find that the two Gauss's law expressions,

$$\oint_{S(t)} \mathbf{D} \cdot d\mathbf{S} = \int_{V(t)} \rho \, dV, \qquad \oint_{S(t)} \mathbf{B} \cdot d\mathbf{S} = 0,$$

remain valid.

2.5.3 Large-scale form of the Boffi equations

The Maxwell–Boffi equations can be written in large-scale form using the same approach as with the Maxwell–Minkowski equations. Integrating (2.102) and (2.103) over an open surface S and applying Stokes's theorem, we have

$$\oint_\Gamma \mathbf{E} \cdot \mathbf{dl} = -\int_S \frac{\partial \mathbf{B}}{\partial t} \cdot \mathbf{dS}, \tag{2.133}$$

$$\oint_\Gamma \mathbf{B} \cdot \mathbf{dl} = \mu_0 \int_S \left(\mathbf{J} + \mathbf{J}_M + \mathbf{J}_P + \frac{\partial \epsilon_0 \mathbf{E}}{\partial t} \right) \cdot \mathbf{dS}, \tag{2.134}$$

for fields in the laboratory frame, and

$$\oint_{\Gamma'} \mathbf{E}' \cdot \mathbf{dl}' = -\int_{S'} \frac{\partial \mathbf{B}'}{\partial t'} \cdot \mathbf{dS}',$$

$$\oint_{\Gamma'} \mathbf{B}' \cdot \mathbf{dl}' = \mu_0 \int_{S'} \left(\mathbf{J}' + \mathbf{J}'_M + \mathbf{J}'_P + \frac{\partial \epsilon_0 \mathbf{E}'}{\partial t'} \right) \cdot \mathbf{dS}',$$

for fields in a moving frame. We see that Faraday's law is unmodified by the introduction of polarization and magnetization. Hence our prior discussion of emf for moving contours remains valid. However, Ampere's law must be interpreted somewhat differently. The flux vector $\mathbf{B}$ also acts as a force vector, and its circulation is proportional to the out-flux of total current, consisting of $\mathbf{J}$ plus the equivalent magnetization and polarization currents plus the displacement current *in free space*, through the surface bounded by the circulation contour.

The large-scale forms of the auxiliary equations can be found by integrating (2.104) and (2.105) over a volume region and applying the divergence theorem. This gives

$$\oint_S \mathbf{E} \cdot \mathbf{dS} = \frac{1}{\epsilon_0} \int_V (\rho + \rho_P)\, dV, \qquad \oint_S \mathbf{B} \cdot \mathbf{dS} = 0,$$

for the laboratory frame fields, and

$$\oint_{S'} \mathbf{E}' \cdot \mathbf{dS}' = \frac{1}{\epsilon_0} \int_{V'} (\rho' + \rho'_P)\, dV', \qquad \oint_{S'} \mathbf{B}' \cdot \mathbf{dS}' = 0,$$

for the moving frame fields. Here we find the force vector $\mathbf{E}$ also acting as a flux vector, with the outflux of $\mathbf{E}$ over a closed surface proportional to the sum of the electric and polarization charges enclosed by the surface.

To provide the alternative representation, we integrate the point forms over V and use the curl theorem to obtain

$$\oint_S (\hat{\mathbf{n}} \times \mathbf{E})\, dS = -\int_V \frac{\partial \mathbf{B}}{\partial t}\, dV, \tag{2.135}$$

$$\oint_S (\hat{\mathbf{n}} \times \mathbf{B})\, dS = \mu_0 \int_V \left(\mathbf{J} + \mathbf{J}_M + \mathbf{J}_P + \frac{\partial \epsilon_0 \mathbf{E}}{\partial t} \right) dV, \tag{2.136}$$

for the laboratory frame fields, and

$$\oint_{S'} (\hat{\mathbf{n}}' \times \mathbf{E}')\, dS' = -\int_{V'} \frac{\partial \mathbf{B}'}{\partial t'}\, dV',$$

$$\oint_{S'} (\hat{\mathbf{n}}' \times \mathbf{B}')\, dS' = \mu_0 \int_{V'} \left(\mathbf{J}' + \mathbf{J}'_M + \mathbf{J}'_P + \frac{\partial \epsilon_0 \mathbf{E}'}{\partial t'} \right) dV',$$

for the moving frame fields.

The large-scale forms of the Boffi equations can also be put into kinematic form using either (2.127) or (2.132). Using (2.127) on (2.133) and (2.134) we have

$$\oint_{\Gamma(t)} \mathbf{E}^* \cdot \mathbf{dl} = -\frac{d}{dt} \int_{S(t)} \mathbf{B} \cdot \mathbf{dS}, \tag{2.137}$$

$$\oint_{\Gamma(t)} \mathbf{B}^\dagger \cdot \mathbf{dl} = \int_{S(t)} \mu_0 \mathbf{J}^\dagger \cdot \mathbf{dS} + \frac{1}{c^2} \frac{d}{dt} \int_{S(t)} \mathbf{E} \cdot \mathbf{dS}, \tag{2.138}$$

where

$$\mathbf{E}^* = \mathbf{E} + \mathbf{v} \times \mathbf{B},$$

$$\mathbf{B}^\dagger = \mathbf{B} - \frac{1}{c^2} \mathbf{v} \times \mathbf{E},$$

$$\mathbf{J}^\dagger = \mathbf{J} + \mathbf{J}_M + \mathbf{J}_P - (\rho + \rho_P)\mathbf{v}.$$

Here $\mathbf{B}^\dagger$ is equivalent to the first-order Lorentz transformation representation of the field in the moving frame (2.46). (The dagger $\dagger$ should not be confused with the symbol for the hermitian operation.) Using (2.132) on (2.135) and (2.136) we have

$$\oint_{S(t)} [\hat{\mathbf{n}} \times \mathbf{E} - (\mathbf{v} \cdot \hat{\mathbf{n}})\mathbf{B}] \, dS = -\frac{d}{dt} \int_{V(t)} \mathbf{B} \, dV, \tag{2.139}$$

and

$$\oint_{S(t)} \left[\hat{\mathbf{n}} \times \mathbf{B} + \frac{1}{c^2} (\mathbf{v} \cdot \hat{\mathbf{n}})\mathbf{E} \right] dS = \mu_0 \int_{V(t)} (\mathbf{J} + \mathbf{J}_M + \mathbf{J}_P) \, dV + \frac{1}{c^2} \frac{d}{dt} \int_{V(t)} \mathbf{E} \, dV. \tag{2.140}$$

In each case the fields are measured in the laboratory frame, and $\mathbf{v}$ is measured with respect to the laboratory frame and may vary arbitrarily over the surface or contour.

2.6 The nature of the four field quantities

Since the very inception of Maxwell's theory, its students have been distressed by the fact that while there are four electromagnetic fields $(\mathbf{E}, \mathbf{D}, \mathbf{B}, \mathbf{H})$, there are only two fundamental equations (the curl equations) to describe their interrelationship. The relegation of additional required information to constitutive equations that vary widely between classes of materials seems to lessen the elegance of the theory. While some may find elegant the separation of equations into a set expressing the basic wave nature of electromagnetism and a set describing how the fields interact with materials, the history of the discipline is one of categorizing and pairing fields as "fundamental" and "supplemental" in hopes of reducing the model to two equations in two unknowns.

Lorentz led the way in this area. With his electrical theory of matter, all material effects could be interpreted in terms of atomic charge and current immersed in free space. We have seen how the Maxwell–Boffi equations seem to eliminate the need for $\mathbf{D}$ and $\mathbf{H}$, and indeed for simple media where there is a linear relation between the remaining "fundamental" fields and the induced polarization and magnetization, it appears that only

E and **B** are required. However, for more complicated materials that display nonlinear and bianisotropic effects, we are only able to supplant **D** and **H** with two other fields **P** and **M**, along with (possibly complicated) constitutive relations relating them to **E** and **B**.

Even those authors who do not wish to eliminate two of the fields tend to categorize the fields into pairs based on physical arguments, implying that one or the other pair is in some way "more fundamental." Maxwell himself separated the fields into the pair (**E**, **H**) that appears within line integrals to give work and the pair (**B**, **D**) that appears within surface integrals to give flux. In what other ways might we pair the four vectors?

Most prevalent is the splitting of the fields into electric and magnetic pairs: (**E**, **D**) and (**B**, **H**). In Poynting's theorem **E** · **D** describes one component of stored energy (called "electric energy") and **B** · **H** describes another component (called "magnetic energy"). These pairs also occur in Maxwell's stress tensor. In statics, the fields decouple into electric and magnetic sets. But biisotropic and bianisotropic materials demonstrate how separation into electric and magnetic effects can become problematic.

In the study of electromagnetic waves, the ratio of E to H appears to be an important quantity, called the "intrinsic impedance." The pair (**E**, **H**) also determines the Poynting flux of power, and is required to establish the uniqueness of the electromagnetic field. In addition, constitutive relations for simple materials usually express (**D**, **B**) in terms of (**E**, **H**). Models for these materials are often conceived by viewing the fields (**E**, **H**) as interacting with the atomic structure in such a way as to produce secondary effects describable by (**D**, **B**). These considerations, along with Maxwell's categorization into a pair of work vectors and a pair of flux vectors, lead many authors to formulate electromagnetics with **E** and **H** as the "fundamental" quantities. But the pair (**B**, **D**) gives rise to electromagnetic momentum and is also perpendicular to the direction of wave propagation in an anisotropic material; in these senses, we might argue that these fields must be equally "fundamental."

Perhaps the best motivation for grouping fields comes from relativistic considerations. We have found that (**E**, **B**) transform together under a Lorentz transformation, as do (**D**, **H**). In each of these pairs we have one polar vector (**E** or **D**) and one axial vector (**B** or **H**). A polar vector retains its meaning under a change in handedness of the coordinate system, while an axial vector does not. The Lorentz force involves one polar vector (**E**) and one axial vector (**B**) that we also call "electric" and "magnetic." If we follow the lead of some authors and *choose* to define **E** and **B** through measurements of the Lorentz force, then we recognize that **B** must be axial since it is not measured directly, but as part of the cross product **v** × **B** that changes its meaning if we switch from a right-hand to a left-hand coordinate system. The other polar vector (**D**) and axial vector (**H**) arise through the "secondary" constitutive relations. Following this reasoning we might claim that **E** and **B** are "fundamental."

Sommerfeld also associates **E** with **B** and **D** with **H**. The vectors **E** and **B** are called *entities of intensity*, describing "how strong," while **D** and **H** are called *entities of quantity*, describing "how much." This is in direct analogy with stress (intensity) and strain (quantity) in materials. We might also say that the entities of intensity describe a "cause" while the entities of quantity describe an "effect." In this view **E** "induces" (causes) a polarization **P**, and the field $\mathbf{D} = \epsilon_0 \mathbf{E} + \mathbf{P}$ is the result. Similarly, **B** creates **M**, and $\mathbf{H} = \mathbf{B}/\mu_0 - \mathbf{M}$ is the result. Interestingly, each of the terms describing energy and momentum in the electromagnetic field (**D** · **E**, **B** · **H**, **E** × **H**, **D** × **B**) involves the interaction of an entity of intensity with an entity of quantity.

Although there is a natural tendency to group things together based on conceptual similarity, there appears to be little reason to believe that any of the four field vectors are more "fundamental" than the rest. Perhaps we are fortunate that we can apply Maxwell's theory without worrying too much about such questions of underlying philosophy.

2.7 Maxwell's equations with magnetic sources

Researchers have yet to discover the "magnetic monopole": a magnetic source from which magnetic field would diverge. This has not stopped speculation on the form that Maxwell's equations might take if such a discovery were made. Arguments based on fundamental principles of physics (such as symmetry and conservation laws) indicate that in the presence of magnetic sources, Maxwell's equations would assume the forms

$$\nabla \times \mathbf{E} = -\mathbf{J}_m - \frac{\partial \mathbf{B}}{\partial t}, \tag{2.141}$$

$$\nabla \times \mathbf{H} = \mathbf{J} + \frac{\partial \mathbf{D}}{\partial t}, \tag{2.142}$$

$$\nabla \cdot \mathbf{B} = \rho_m, \tag{2.143}$$

$$\nabla \cdot \mathbf{D} = \rho, \tag{2.144}$$

where $\mathbf{J}_m$ is a volume magnetic current density describing the flow of magnetic charge in exactly the same manner as $\mathbf{J}$ describes the flow of electric charge. The density of this magnetic charge is given by ρ_m and should, by analogy with electric charge density, obey a conservation law

$$\nabla \cdot \mathbf{J}_m + \frac{\partial \rho_m}{\partial t} = 0.$$

This is the magnetic source continuity equation.

It is interesting to inquire as to the units of $\mathbf{J}_m$ and ρ_m. From (2.141) we see that if $\mathbf{B}$ has units of Wb/m^2, then $\mathbf{J}_m$ has units of (Wb/s)/m^2. Similarly, (2.143) shows that ρ_m must have units of Wb/m^3. Hence magnetic charge is measured in Wb, magnetic current in Wb/s. This gives a nice symmetry with electric sources where charge is measured in C and current in C/s.* The physical symmetry is equally appealing: magnetic flux lines diverge from magnetic charge, and the total flux passing through a surface is given by the total magnetic charge contained within the surface. This is best seen by considering the large-scale forms of Maxwell's equations for stationary surfaces. We need only modify (2.123) to include the magnetic current term; this gives

$$\oint_\Gamma \mathbf{E} \cdot \mathbf{dl} = -\int_S \mathbf{J}_m \cdot \mathbf{dS} - \frac{d}{dt} \int_S \mathbf{B} \cdot \mathbf{dS},$$

$$\oint_\Gamma \mathbf{H} \cdot \mathbf{dl} = \int_S \mathbf{J} \cdot \mathbf{dS} + \frac{d}{dt} \int_S \mathbf{D} \cdot \mathbf{dS}.$$

---

*We note that if the modern unit of T is used to describe $\mathbf{B}$, then ρ_m is described using the more cumbersome units of T/m, while $\mathbf{J}_m$ is given in terms of T/s. Thus, magnetic charge is measured in Tm^2 and magnetic current in (Tm^2)/s.

If we modify (2.126) to include magnetic charge, we get the auxiliary equations

$$\oint_S \mathbf{D} \cdot \mathbf{dS} = \int_V \rho \, dV,$$

$$\oint_S \mathbf{B} \cdot \mathbf{dS} = \int_V \rho_m \, dV.$$

Any of the large-scale forms of Maxwell's equations can be similarly modified to include magnetic current and charge. For arbitrarily moving surfaces, we have

$$\oint_{\Gamma(t)} \mathbf{E}^* \cdot \mathbf{dl} = -\frac{d}{dt} \int_{S(t)} \mathbf{B} \cdot \mathbf{dS} - \int_{S(t)} \mathbf{J}_m^* \cdot \mathbf{dS},$$

$$\oint_{\Gamma(t)} \mathbf{H}^* \cdot \mathbf{dl} = \frac{d}{dt} \int_{S(t)} \mathbf{D} \cdot \mathbf{dS} + \int_{S(t)} \mathbf{J}^* \cdot \mathbf{dS},$$

where

$$\mathbf{E}^* = \mathbf{E} + \mathbf{v} \times \mathbf{B}, \qquad\qquad \mathbf{J}^* = \mathbf{J} - \rho \mathbf{v},$$

$$\mathbf{H}^* = \mathbf{H} - \mathbf{v} \times \mathbf{D}, \qquad\qquad \mathbf{J}_m^* = \mathbf{J}_m - \rho_m \mathbf{v},$$

and all fields are taken to be measured in the laboratory frame with $\mathbf{v}$ the instantaneous velocity of points on the surface and contour relative to the laboratory frame. We also have the alternative forms

$$\oint_S (\hat{\mathbf{n}} \times \mathbf{E}) \, dS = \int_V \left(-\frac{\partial \mathbf{B}}{\partial t} - \mathbf{J}_m \right) dV, \tag{2.145}$$

$$\oint_S (\hat{\mathbf{n}} \times \mathbf{H}) \, dS = \int_V \left(\frac{\partial \mathbf{D}}{\partial t} + \mathbf{J} \right) dV, \tag{2.146}$$

and

$$\oint_{S(t)} [\hat{\mathbf{n}} \times \mathbf{E} - (\mathbf{v} \cdot \hat{\mathbf{n}})\mathbf{B}] \, dS = -\int_{V(t)} \mathbf{J}_m \, dV - \frac{d}{dt} \int_{V(t)} \mathbf{B} \, dV, \tag{2.147}$$

$$\oint_{S(t)} [\hat{\mathbf{n}} \times \mathbf{H} + (\mathbf{v} \cdot \hat{\mathbf{n}})\mathbf{D}] \, dS = \int_{V(t)} \mathbf{J} \, dV + \frac{d}{dt} \int_{V(t)} \mathbf{D} \, dV, \tag{2.148}$$

and the two Gauss's law expressions

$$\oint_{S(t)} \mathbf{D} \cdot \hat{\mathbf{n}} \, dS = \int_{V(t)} \rho \, dV,$$

$$\oint_{S(t)} \mathbf{B} \cdot \hat{\mathbf{n}} \, dS = \int_{V(t)} \rho_m \, dV.$$

Magnetic sources also allow us to develop equivalence theorems in which difficult problems involving boundaries are replaced by simpler problems involving magnetic sources. Although these sources may not physically exist, the mathematical solutions are completely valid.

2.8 Boundary (jump) conditions

If we restrict ourselves to regions of space without spatial (jump) discontinuities in either the sources or the constitutive relations, we can find meaningful solutions to the Maxwell differential equations. We also know that for given sources, if the fields are specified on a closed boundary and at an initial time, the solutions are unique. The standard approach to treating regions that do contain spatial discontinuities is to isolate the discontinuities on surfaces. That is, we introduce surfaces that serve to separate space into regions in which the differential equations are solvable and the fields are well defined. To make the solutions in adjoining regions unique, we must specify the tangential fields on each side of the adjoining surface. If we can relate the fields across the boundary, we can propagate the solution from one region to the next; in this way, information about the source in one region is effectively passed on to the solution in an adjacent region. For uniqueness, only relations between the tangential components need be specified.

We shall determine the appropriate boundary conditions via two distinct approaches. We first model a thin source layer and consider a discontinuous surface source layer as a limiting case of the continuous thin layer. With no true discontinuity, Maxwell's differential equations hold everywhere. We then consider a true spatial discontinuity between material surfaces (with possible surface sources lying along the discontinuity). We must then isolate the region containing the discontinuity and *postulate* a field relationship that is both physically meaningful and experimentally verifiable.

We shall also consider both stationary and moving boundary surfaces, and surfaces containing magnetic as well as electric sources.

2.8.1 Boundary conditions across a stationary, thin source layer

In § 1.3.3 we discussed how in the macroscopic sense a surface source is actually a volume distribution concentrated near a surface S. We write the charge and current in terms of the point $\mathbf{r}$ on the surface and the normal distance x from the surface at $\mathbf{r}$ as

$$\rho(\mathbf{r}, x, t) = \rho_s(\mathbf{r}, t) f(x, \Delta),$$

$$\mathbf{J}(\mathbf{r}, x, t) = \mathbf{J}_s(\mathbf{r}, t) f(x, \Delta),$$

where $f(x, \Delta)$ is the source density function obeying

$$\int_{-\infty}^{\infty} f(x, \Delta) \, dx = 1. \tag{2.149}$$

The parameter Δ describes the "width" of the source layer normal to the reference surface.

We use (2.130)–(2.131) to study field behavior across the source layer. Consider a volume region V that intersects the source layer as shown in Figure 2.5. Let the top and bottom surfaces be parallel to the reference surface, and label the fields on the top and bottom surfaces with subscripts 1 and 2, respectively. Since points on and within V are all regular, (2.131) yields

$$\int_{S_1} \hat{\mathbf{n}}_1 \times \mathbf{H}_1 \, dS + \int_{S_2} \hat{\mathbf{n}}_2 \times \mathbf{H}_2 \, dS + \int_{S_3} \hat{\mathbf{n}}_3 \times \mathbf{H} \, dS = \int_V \left(\mathbf{J} + \frac{\partial \mathbf{D}}{\partial t} \right) dV.$$

We now choose $\delta = k\Delta$ ($k > 1$) so that most of the source lies within V. As $\Delta \to 0$ the thin source layer recedes to a surface layer, and the volume integral of displacement

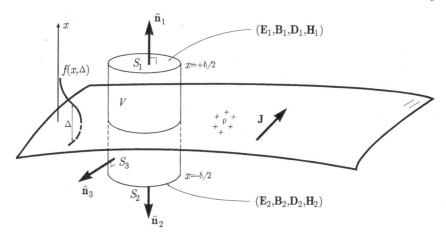

FIGURE 2.5
Derivation of the electromagnetic boundary conditions across a thin continuous source layer.

current and the integral of tangential $\mathbf{H}$ over S_3 both approach zero by continuity of the fields. By symmetry, $S_1 = S_2$ and $\hat{\mathbf{n}}_1 = -\hat{\mathbf{n}}_2 = \hat{\mathbf{n}}_{12}$, where $\hat{\mathbf{n}}_{12}$ is the surface normal directed into region 1 from region 2. Thus

$$\int_{S_1} \hat{\mathbf{n}}_{12} \times (\mathbf{H}_1 - \mathbf{H}_2)\, dS = \int_V \mathbf{J}\, dV.$$

Note that

$$\int_V \mathbf{J}\, dV = \int_{S_1} \int_{-\delta/2}^{\delta/2} \mathbf{J}\, dS\, dx = \int_{-\delta/2}^{\delta/2} f(x, \Delta)\, dx \int_{S_1} \mathbf{J}_s(\mathbf{r}, t)\, dS.$$

Since we assume that the majority of the source current lies within V, the integral can be evaluated using (2.149) to give

$$\int_{S_1} [\hat{\mathbf{n}}_{12} \times (\mathbf{H}_1 - \mathbf{H}_2) - \mathbf{J}_s]\, dS = 0,$$

hence

$$\hat{\mathbf{n}}_{12} \times (\mathbf{H}_1 - \mathbf{H}_2) = \mathbf{J}_s.$$

The tangential magnetic field across a thin source distribution is discontinuous by an amount equal to the surface current density.

Similar steps with Faraday's law give

$$\hat{\mathbf{n}}_{12} \times (\mathbf{E}_1 - \mathbf{E}_2) = 0.$$

The tangential electric field is continuous across a thin source.

We can also derive conditions on the normal components of the fields, although these are not required for uniqueness. Gauss's law (2.125) applied to the volume V in Figure 2.5 gives

$$\int_{S_1} \mathbf{D}_1 \cdot \hat{\mathbf{n}}_1\, dS + \int_{S_2} \mathbf{D}_2 \cdot \hat{\mathbf{n}}_2\, dS + \int_{S_3} \mathbf{D} \cdot \hat{\mathbf{n}}_3\, dS = \int_V \rho\, dV.$$

As $\Delta \to 0$, the thin source layer recedes to a surface layer. The integral of normal $\mathbf{D}$ over S_3 tends to zero by continuity of the fields. By symmetry, $S_1 = S_2$ and $\hat{\mathbf{n}}_1 = -\hat{\mathbf{n}}_2 = \hat{\mathbf{n}}_{12}$. Thus

$$\int_{S_1} (\mathbf{D}_1 - \mathbf{D}_2) \cdot \hat{\mathbf{n}}_{12} \, dS = \int_V \rho \, dV.$$

The volume integral is

$$\int_V \rho \, dV = \int_{S_1} \int_{-\delta/2}^{\delta/2} \rho \, dS \, dx = \int_{-\delta/2}^{\delta/2} f(x, \Delta) \, dx \int_{S_1} \rho_s(\mathbf{r}, t) \, dS.$$

Since $\delta = k\Delta$ has been chosen so that most of the source charge lies within V, (2.149) gives

$$\int_{S_1} [(\mathbf{D}_1 - \mathbf{D}_2) \cdot \hat{\mathbf{n}}_{12} - \rho_s] \, dS = 0,$$

hence

$$(\mathbf{D}_1 - \mathbf{D}_2) \cdot \hat{\mathbf{n}}_{12} = \rho_s.$$

The normal component of $\mathbf{D}$ is discontinuous across a thin source distribution by an amount equal to the surface charge density. Similar steps with the magnetic Gauss's law yield

$$(\mathbf{B}_1 - \mathbf{B}_2) \cdot \hat{\mathbf{n}}_{12} = 0.$$

The normal component of $\mathbf{B}$ is continuous across a thin source layer.

We can follow similar steps when a thin magnetic source layer is present. When evaluating Faraday's law we must include magnetic surface current and when evaluating the magnetic Gauss's law we must include magnetic charge. However, since such sources are not physical, we postpone their consideration until the next section, where appropriate boundary conditions are postulated rather than derived.

2.8.2 Boundary conditions holding across a stationary layer of field discontinuity

Provided that we model a surface source as a limiting case of a very thin but continuous volume source, we can derive boundary conditions across a surface layer. We might ask whether we can extend this idea to surfaces of materials where the constitutive parameters change from one region to another. Indeed, if we take Lorentz' viewpoint and visualize a material as a conglomerate of atomic charge, we should be able to apply this same idea. After all, a material should demonstrate a continuous transition (in the macroscopic sense) across its boundary, and we can employ the Maxwell–Boffi equations to describe the relationship between the "equivalent" sources and the electromagnetic fields.

We should note, however, that the limiting concept is not without its critics. Stokes suggested as early as 1848 that jump conditions should never be derived from smooth solutions [199]. Let us therefore pursue the boundary conditions for a surface of true field discontinuity. This will also allow us to treat a material modeled as having a true discontinuity in its material parameters (which we can always take as a mathematical model of a more gradual transition) before we have studied in a deeper sense the physical properties of materials. This approach, taken by many textbooks, must be done carefully.

There is a logical difficulty with this approach, lying in the application of the large-scale forms of Maxwell's equations. Many authors postulate Maxwell's equations in point form, integrate to obtain the large-scale forms, then apply the large-scale forms to regions of discontinuity. Unfortunately, the large-scale forms thus obtained are only valid in the same regions where their point form antecedents were valid — discontinuities must be excluded. Schelkunoff [169] has criticized this approach, calling it a "swindle" rather than a proof, and has suggested that the proper way to handle true discontinuities is to postulate the large-scale forms of Maxwell's equations, *and* to include as part of the postulate the assumption that the large-scale forms are valid at points of field discontinuity. Does this mean we must reject our postulate of the point-form Maxwell equations and reformulate everything in terms of the large-scale forms? Fortunately, no. Tai [189] has pointed out that it is still possible to postulate the point forms, as long as we also postulate appropriate boundary conditions that make the large-scale forms, as derived from the point forms, valid at surfaces of discontinuity. In essence, both approaches require an additional postulate for surfaces of discontinuity: the large-scale forms require a postulate of applicability to discontinuous surfaces, and from there the boundary conditions can be derived; the point forms require a postulate of the boundary conditions that result in the large-scale forms being valid on surfaces of discontinuity. Let us examine how the latter approach works.

Consider a surface across which the constitutive relations are discontinuous, containing electric and magnetic surface currents and charges $\mathbf{J}_s$, ρ_s, $\mathbf{J}_{ms}$, and ρ_{ms} (Figure 2.6). We locate a volume region V_1 above the surface of discontinuity; this volume is bounded by a surface S_1 and another surface S_{10}, which is parallel to, and a small distance $\delta/2$ above, the surface of discontinuity. A second volume region V_2 is similarly situated below the surface of discontinuity. Because these regions exclude the surface of discontinuity, we can use (2.146) to get

$$\int_{S_1} \hat{\mathbf{n}} \times \mathbf{H}\, dS + \int_{S_{10}} \hat{\mathbf{n}} \times \mathbf{H}\, dS = \int_{V_1} \left(\mathbf{J} + \frac{\partial \mathbf{D}}{\partial t} \right) dV,$$

$$\int_{S_2} \hat{\mathbf{n}} \times \mathbf{H}\, dS + \int_{S_{20}} \hat{\mathbf{n}} \times \mathbf{H}\, dS = \int_{V_2} \left(\mathbf{J} + \frac{\partial \mathbf{D}}{\partial t} \right) dV.$$

Adding these we obtain

$$\int_{S_1+S_2} \hat{\mathbf{n}} \times \mathbf{H}\, dS - \int_{V_1+V_2} \left(\mathbf{J} + \frac{\partial \mathbf{D}}{\partial t} \right) dV$$

$$- \int_{S_{10}} \hat{\mathbf{n}}_{10} \times \mathbf{H}_1\, dS - \int_{S_{20}} \hat{\mathbf{n}}_{20} \times \mathbf{H}_2\, dS = 0, \qquad (2.150)$$

where we have used subscripts to delineate the fields on each side of the discontinuity surface.

If δ is very small (but nonzero), then $\hat{\mathbf{n}}_{10} = -\hat{\mathbf{n}}_{20} = \hat{\mathbf{n}}_{12}$ and $S_{10} = S_{20}$. Letting $S_1 + S_2 = S$ and $V_1 + V_2 = V$, we can write (2.150) as

$$\int_S (\hat{\mathbf{n}} \times \mathbf{H})\, dS - \int_V \left(\mathbf{J} + \frac{\partial \mathbf{D}}{\partial t} \right) dV = \int_{S_{10}} \hat{\mathbf{n}}_{12} \times (\mathbf{H}_1 - \mathbf{H}_2)\, dS. \qquad (2.151)$$

Now suppose we use the same volume region V, but let it intersect the surface of discontinuity (Figure 2.6), and suppose that the large-scale form of Ampere's law holds even if V contains points of field discontinuity. We must include the surface current in the computation. Since $\int_V \mathbf{J}\, dV$ becomes $\int_S \mathbf{J}_s\, dS$ on the surface, we have

$$\int_S (\hat{\mathbf{n}} \times \mathbf{H})\, dS - \int_V \left(\mathbf{J} + \frac{\partial \mathbf{D}}{\partial t} \right) dV = \int_{S_{10}} \mathbf{J}_s\, dS.$$

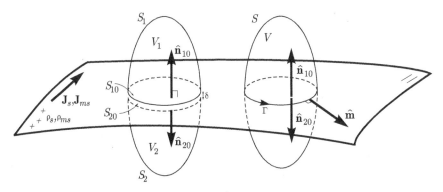

FIGURE 2.6
Derivation of the electromagnetic boundary conditions across a discontinuous source layer.

We wish to have this give the same value for the integrals over V and S as (2.151), which included in its derivation no points of discontinuity. This is true provided that

$$\hat{\mathbf{n}}_{12} \times (\mathbf{H}_1 - \mathbf{H}_2) = \mathbf{J}_s. \tag{2.152}$$

Thus, under the condition (2.152) we may interpret the large-scale form of Ampere's law (as derived from the point form) as being valid for regions containing discontinuities. Note that this condition is not "derived," but must be regarded as a postulate that results in the large-scale form holding for surfaces of discontinuous field.

Similar reasoning can be used to determine the appropriate boundary condition on tangential $\mathbf{E}$ from Faraday's law. Corresponding to (2.151) we obtain

$$\int_S (\hat{\mathbf{n}} \times \mathbf{E})\, dS - \int_V \left(-\mathbf{J}_m - \frac{\partial \mathbf{B}}{\partial t} \right) dV = \int_{S_{10}} \hat{\mathbf{n}}_{12} \times (\mathbf{E}_1 - \mathbf{E}_2)\, dS. \tag{2.153}$$

Employing (2.145) over the region containing the field discontinuity surface, we get

$$\int_S (\hat{\mathbf{n}} \times \mathbf{E})\, dS - \int_V \left(-\mathbf{J}_m - \frac{\partial \mathbf{B}}{\partial t} \right) dV = - \int_{S_{10}} \mathbf{J}_{ms}\, dS. \tag{2.154}$$

To have (2.153) and (2.154) produce identical results, we postulate

$$\hat{\mathbf{n}}_{12} \times (\mathbf{E}_1 - \mathbf{E}_2) = -\mathbf{J}_{ms}$$

as the boundary condition appropriate to a surface of field discontinuity containing a magnetic surface current.

We can also postulate boundary conditions on the normal fields to make Gauss's laws valid for surfaces of discontinuous fields. Integrating (2.125) over the regions V_1 and V_2 and adding, we obtain

$$\int_{S_1+S_2} \mathbf{D} \cdot \hat{\mathbf{n}}\, dS - \int_{S_{10}} \mathbf{D}_1 \cdot \hat{\mathbf{n}}_{10}\, dS - \int_{S_{20}} \mathbf{D}_2 \cdot \hat{\mathbf{n}}_{20}\, dS = \int_{V_1+V_2} \rho\, dV.$$

As $\delta \to 0$ this becomes

$$\int_S \mathbf{D} \cdot \hat{\mathbf{n}}\, dS - \int_V \rho\, dV = \int_{S_{10}} (\mathbf{D}_1 - \mathbf{D}_2) \cdot \hat{\mathbf{n}}_{12}\, dS. \tag{2.155}$$

If we integrate Gauss's law over the entire region V, including the surface of discontinuity, we get

$$\oint_S \mathbf{D} \cdot \hat{\mathbf{n}} \, dS = \int_V \rho \, dV + \int_{S_{10}} \rho_s \, dS. \tag{2.156}$$

In order to get identical answers from (2.155) and (2.156), we must have

$$(\mathbf{D}_1 - \mathbf{D}_2) \cdot \hat{\mathbf{n}}_{12} = \rho_s$$

as the boundary condition appropriate to a surface of field discontinuity containing an electric surface charge. Similarly, we must postulate

$$(\mathbf{B}_1 - \mathbf{B}_2) \cdot \hat{\mathbf{n}}_{12} = \rho_{ms}$$

as the condition appropriate to a surface of field discontinuity containing a magnetic surface charge.

We can determine an appropriate boundary condition on current by using the large-scale form of the continuity equation. Applying (1.8) over each of the volume regions of Figure 2.6 and adding the results, we have

$$\int_{S_1+S_2} \mathbf{J} \cdot \hat{\mathbf{n}} \, dS - \int_{S_{10}} \mathbf{J}_1 \cdot \hat{\mathbf{n}}_{10} \, dS - \int_{S_{20}} \mathbf{J}_2 \cdot \hat{\mathbf{n}}_{20} \, dS = -\int_{V_1+V_2} \frac{\partial \rho}{\partial t} \, dV.$$

As $\delta \to 0$ we have

$$\int_S \mathbf{J} \cdot \hat{\mathbf{n}} \, dS - \int_{S_{10}} (\mathbf{J}_1 - \mathbf{J}_2) \cdot \hat{\mathbf{n}}_{12} \, dS = -\int_V \frac{\partial \rho}{\partial t} \, dV. \tag{2.157}$$

Applying the continuity equation over the entire region V and allowing it to intersect the discontinuity surface, we get

$$\int_S \mathbf{J} \cdot \hat{\mathbf{n}} \, dS + \int_\Gamma \mathbf{J}_s \cdot \hat{\mathbf{m}} \, dl = -\int_V \frac{\partial \rho}{\partial t} \, dV - \int_{S_{10}} \frac{\partial \rho_s}{\partial t} \, dS.$$

By the two-dimensional divergence theorem (B.26) we can write this as

$$\int_S \mathbf{J} \cdot \hat{\mathbf{n}} \, dS + \int_{S_{10}} \nabla_s \cdot \mathbf{J}_s \, dS = -\int_V \frac{\partial \rho}{\partial t} \, dV - \int_{S_{10}} \frac{\partial \rho_s}{\partial t} \, dS.$$

In order for this expression to produce the same values of the integrals over S and V as in (2.157), we require

$$\nabla_s \cdot \mathbf{J}_s = -\hat{\mathbf{n}}_{12} \cdot (\mathbf{J}_1 - \mathbf{J}_2) - \frac{\partial \rho_s}{\partial t},$$

which we take as our postulate of the boundary condition on current across a surface containing discontinuities. A similar set of steps carried out using the continuity equation for magnetic sources yields

$$\nabla_s \cdot \mathbf{J}_{ms} = -\hat{\mathbf{n}}_{12} \cdot (\mathbf{J}_{m1} - \mathbf{J}_{m2}) - \frac{\partial \rho_{ms}}{\partial t}.$$

In summary, we have the following boundary conditions for fields across a surface containing discontinuities:

$$\hat{\mathbf{n}}_{12} \times (\mathbf{H}_1 - \mathbf{H}_2) = \mathbf{J}_s, \tag{2.158}$$

$$\hat{\mathbf{n}}_{12} \times (\mathbf{E}_1 - \mathbf{E}_2) = -\mathbf{J}_{ms}, \tag{2.159}$$

$$\hat{\mathbf{n}}_{12} \cdot (\mathbf{D}_1 - \mathbf{D}_2) = \rho_s, \tag{2.160}$$

$$\hat{\mathbf{n}}_{12} \cdot (\mathbf{B}_1 - \mathbf{B}_2) = \rho_{ms}, \tag{2.161}$$

and

$$\hat{\mathbf{n}}_{12} \cdot (\mathbf{J}_1 - \mathbf{J}_2) = -\nabla_s \cdot \mathbf{J}_s - \frac{\partial \rho_s}{\partial t}, \tag{2.162}$$

$$\hat{\mathbf{n}}_{12} \cdot (\mathbf{J}_{m1} - \mathbf{J}_{m2}) = -\nabla_s \cdot \mathbf{J}_{ms} - \frac{\partial \rho_{ms}}{\partial t}, \tag{2.163}$$

where $\hat{\mathbf{n}}_{12}$ points into region 1 from region 2.

2.8.3 Boundary conditions at the surface of a perfect conductor

We can easily specialize the results of the previous section to the case of perfect electric or magnetic conductors. In § 2.2.2.3 we saw that the constitutive relations for perfect conductors requires the null field within the material. In addition, a PEC requires zero tangential electric field, while a PMC requires zero tangential magnetic field. Using (2.158)–(2.163), we find that the boundary conditions for a perfect electric conductor are

$$\hat{\mathbf{n}} \times \mathbf{H} = \mathbf{J}_s, \tag{2.164}$$
$$\hat{\mathbf{n}} \times \mathbf{E} = 0, \tag{2.165}$$
$$\hat{\mathbf{n}} \cdot \mathbf{D} = \rho_s, \tag{2.166}$$
$$\hat{\mathbf{n}} \cdot \mathbf{B} = 0, \tag{2.167}$$

and

$$\hat{\mathbf{n}} \cdot \mathbf{J} = -\nabla_s \cdot \mathbf{J}_s - \frac{\partial \rho_s}{\partial t}, \qquad \hat{\mathbf{n}} \cdot \mathbf{J}_m = 0.$$

For a PMC the conditions are

$$\hat{\mathbf{n}} \times \mathbf{H} = 0,$$
$$\hat{\mathbf{n}} \times \mathbf{E} = -\mathbf{J}_{ms},$$
$$\hat{\mathbf{n}} \cdot \mathbf{D} = 0,$$
$$\hat{\mathbf{n}} \cdot \mathbf{B} = \rho_{ms},$$

and

$$\hat{\mathbf{n}} \cdot \mathbf{J}_m = -\nabla_s \cdot \mathbf{J}_{ms} - \frac{\partial \rho_{ms}}{\partial t}, \qquad \hat{\mathbf{n}} \cdot \mathbf{J} = 0.$$

We note that the normal vector $\hat{\mathbf{n}}$ points out of the conductor and into the adjacent region of nonzero fields.

2.8.4 Boundary conditions across a stationary layer of field discontinuity using equivalent sources

So far we have avoided using the physical interpretation of the equivalent sources in the Maxwell–Boffi equations so that we might investigate the behavior of fields across true discontinuities. Now that we have the appropriate boundary conditions, it is interesting to interpret them in terms of the equivalent sources.

If we put $\mathbf{H} = \mathbf{B}/\mu_0 - \mathbf{M}$ into (2.158) and rearrange, we get

$$\hat{\mathbf{n}}_{12} \times (\mathbf{B}_1 - \mathbf{B}_2) = \mu_0 (\mathbf{J}_s + \hat{\mathbf{n}}_{12} \times \mathbf{M}_1 - \hat{\mathbf{n}}_{12} \times \mathbf{M}_2). \tag{2.168}$$

The terms on the right involving $\hat{\mathbf{n}}_{12} \times \mathbf{M}$ have the units of surface current and are called *equivalent magnetization surface currents*. Defining

$$\mathbf{J}_{Ms} = -\hat{\mathbf{n}} \times \mathbf{M}$$

where $\hat{\mathbf{n}}$ is directed normally outward from the material region of interest, we can rewrite (2.168) as

$$\hat{\mathbf{n}}_{12} \times (\mathbf{B}_1 - \mathbf{B}_2) = \mu_0(\mathbf{J}_s + \mathbf{J}_{Ms1} + \mathbf{J}_{Ms2}).$$

We note that J_{Ms} replaces atomic charge moving along the surface of a material with an equivalent surface current in free space.

If we substitute $\mathbf{D} = \epsilon_0 \mathbf{E} + \mathbf{P}$ into (2.160) and rearrange, we get

$$\hat{\mathbf{n}}_{12} \cdot (\mathbf{E}_1 - \mathbf{E}_2) = \frac{1}{\epsilon_0}(\rho_s - \hat{\mathbf{n}}_{12} \cdot \mathbf{P}_1 + \hat{\mathbf{n}}_{12} \cdot \mathbf{P}_2). \tag{2.169}$$

The terms on the right involving $\hat{\mathbf{n}}_{12} \cdot \mathbf{P}$ have the units of surface charge and are called *equivalent polarization surface charges*. Defining

$$\rho_{Ps} = \hat{\mathbf{n}} \cdot \mathbf{P},$$

we can rewrite (2.169) as

$$\hat{\mathbf{n}}_{12} \cdot (\mathbf{E}_1 - \mathbf{E}_2) = \frac{1}{\epsilon_0}(\rho_s + \rho_{Ps1} + \rho_{Ps2}).$$

We note that ρ_{Ps} replaces atomic charge adjacent to a surface of a material with an equivalent surface charge in free space.

In summary, the boundary conditions at a stationary surface of discontinuity written in terms of equivalent sources are

$$\hat{\mathbf{n}}_{12} \times (\mathbf{B}_1 - \mathbf{B}_2) = \mu_0(\mathbf{J}_s + \mathbf{J}_{Ms1} + \mathbf{J}_{Ms2}),$$

$$\hat{\mathbf{n}}_{12} \times (\mathbf{E}_1 - \mathbf{E}_2) = -\mathbf{J}_{ms},$$

$$\hat{\mathbf{n}}_{12} \cdot (\mathbf{E}_1 - \mathbf{E}_2) = \frac{1}{\epsilon_0}(\rho_s + \rho_{Ps1} + \rho_{Ps2}),$$

$$\hat{\mathbf{n}}_{12} \cdot (\mathbf{B}_1 - \mathbf{B}_2) = \rho_{ms}.$$

2.8.5 Boundary conditions across a moving layer of field discontinuity

With a moving material body it is often necessary to apply boundary conditions describing the behavior of the fields across the surface of the body. If a surface of discontinuity moves with constant velocity v, the boundary conditions (2.158)–(2.163) hold as long as all fields are expressed *in the frame of the moving surface*. We can also derive boundary conditions for a deforming surface moving with arbitrary velocity by using equations (2.147)–(2.148). In this case all fields are expressed in the laboratory frame. Proceeding through the same set of steps that gave us (2.158)–(2.161), we find

$$\hat{\mathbf{n}}_{12} \times (\mathbf{H}_1 - \mathbf{H}_2) + (\hat{\mathbf{n}}_{12} \cdot \mathbf{v})(\mathbf{D}_1 - \mathbf{D}_2) = \mathbf{J}_s, \tag{2.170}$$

$$\hat{\mathbf{n}}_{12} \times (\mathbf{E}_1 - \mathbf{E}_2) - (\hat{\mathbf{n}}_{12} \cdot \mathbf{v})(\mathbf{B}_1 - \mathbf{B}_2) = -\mathbf{J}_{ms}, \tag{2.171}$$

$$\hat{\mathbf{n}}_{12} \cdot (\mathbf{D}_1 - \mathbf{D}_2) = \rho_s, \tag{2.172}$$

$$\hat{\mathbf{n}}_{12} \cdot (\mathbf{B}_1 - \mathbf{B}_2) = \rho_{ms}. \tag{2.173}$$

Note that when $\hat{\mathbf{n}}_{12} \cdot \mathbf{v} = 0$ these boundary conditions reduce to those for a stationary surface. This occurs not only when $\mathbf{v} = 0$ but also when the velocity is parallel to the surface.

The reader must be wary when employing (2.170)–(2.173). Since the fields are measured in the laboratory frame, if the constitutive relations are substituted into the boundary conditions they must also be represented in the laboratory frame. It is probable that the material parameters would be known in the rest frame of the material, in which case a conversion to the laboratory frame would be necessary.

2.9 Fundamental theorems

In this section we shall consider some of the important theorems of electromagnetics that pertain directly to Maxwell's equations. They may be derived without reference to the solutions of Maxwell's equations, and are not connected with any specialization of the equations or any specific application or geometrical configuration. In this sense these theorems are fundamental to the study of electromagnetics.

2.9.1 Linearity

Recall that a mathematical operator L is *linear* if

$$L(\alpha_1 f_1 + \alpha_2 f_2) = \alpha_1 L(f_1) + \alpha_2 L(f_2)$$

holds for any two functions $f_{1,2}$ in the domain of L and any two scalar constants $\alpha_{1,2}$. A standard observation regarding the equation

$$L(f) = s, \tag{2.174}$$

where L is a linear operator and s is a given forcing function, is that if f_1 and f_2 are solutions to

$$L(f_1) = s_1, \qquad L(f_2) = s_2, \tag{2.175}$$

respectively, and

$$s = s_1 + s_2, \tag{2.176}$$

then

$$f = f_1 + f_2 \tag{2.177}$$

is a solution to (2.174). This is the *principle of superposition*; if convenient, we can decompose s in Equation (2.174) as a sum (2.176) and solve the two resulting equations (2.175) independently. The solution to (2.174) is then (2.177), "by superposition." Of course, we are free to split the right side of (2.174) into more than two terms — the method extends directly to any finite number of terms.

Because the operators $\nabla\cdot$, $\nabla\times$, and $\partial/\partial t$ are all linear, Maxwell's equations can be treated by this method. If, for instance,

$$\nabla \times \mathbf{E}_1 = -\frac{\partial \mathbf{B}_1}{\partial t}, \qquad \nabla \times \mathbf{E}_2 = -\frac{\partial \mathbf{B}_2}{\partial t},$$

then

$$\nabla \times \mathbf{E} = -\frac{\partial \mathbf{B}}{\partial t}$$

where $\mathbf{E} = \mathbf{E}_1 + \mathbf{E}_2$ and $\mathbf{B} = \mathbf{B}_1 + \mathbf{B}_2$. The motivation for decomposing terms in a particular way is often based on physical considerations; we give one example here and defer others to later sections of the book. We saw earlier that Maxwell's equations can be written in terms of both electric and (fictitious) magnetic sources as in equations (2.141)–(2.144). Let $\mathbf{E} = \mathbf{E}_e + \mathbf{E}_m$ where $\mathbf{E}_e$ is produced by electric-type sources and $\mathbf{E}_m$ is produced by magnetic-type sources, and decompose the other fields similarly. Then

$$\nabla \times \mathbf{E}_e = -\frac{\partial \mathbf{B}_e}{\partial t}, \qquad \nabla \times \mathbf{H}_e = \mathbf{J} + \frac{\partial \mathbf{D}_e}{\partial t}, \qquad \nabla \cdot \mathbf{D}_e = \rho, \qquad \nabla \cdot \mathbf{B}_e = 0,$$

with a similar equation set for the magnetic sources. We may, if desired, solve these two equation sets independently for $\mathbf{E}_e, \mathbf{D}_e, \mathbf{B}_e, \mathbf{H}_e$ and $\mathbf{E}_m, \mathbf{D}_m, \mathbf{E}_m, \mathbf{H}_m$, and then use superposition to obtain the total fields $\mathbf{E}, \mathbf{D}, \mathbf{B}, \mathbf{H}$.

2.9.2 Duality

The intriguing symmetry of Maxwell's equations leads us to an observation that can reduce the effort required to compute solutions. Consider a closed surface S enclosing a region of space that includes an electric source current $\mathbf{J}$ and a magnetic source current $\mathbf{J}_m$. The fields $(\mathbf{E}_1, \mathbf{D}_1, \mathbf{B}_1, \mathbf{H}_1)$ within the region (which may also contain arbitrary media) are described by

$$\nabla \times \mathbf{E}_1 = -\mathbf{J}_m - \frac{\partial \mathbf{B}_1}{\partial t}, \tag{2.178}$$

$$\nabla \times \mathbf{H}_1 = \mathbf{J} + \frac{\partial \mathbf{D}_1}{\partial t}, \tag{2.179}$$

$$\nabla \cdot \mathbf{D}_1 = \rho, \tag{2.180}$$

$$\nabla \cdot \mathbf{B}_1 = \rho_m. \tag{2.181}$$

Suppose we have been given a mathematical description of the sources $(\mathbf{J}, \mathbf{J}_m)$ and have solved for the field vectors $(\mathbf{E}_1, \mathbf{D}_1, \mathbf{B}_1, \mathbf{H}_1)$. Of course, we must also have been supplied with a set of boundary values and constitutive relations in order to make the solution unique. We note that if we replace the formula for $\mathbf{J}$ with the formula for $\mathbf{J}_m$ in (2.179) (and ρ with ρ_m in (2.180)) and also replace $\mathbf{J}_m$ with $-\mathbf{J}$ in (2.178) (and ρ_m with $-\rho$ in (2.181)) we get a new problem to solve, with a different solution. However, the symmetry of the equations allows us to specify the solution immediately. The new set of curl equations requires

$$\nabla \times \mathbf{E}_2 = \mathbf{J} - \frac{\partial \mathbf{B}_2}{\partial t}, \tag{2.182}$$

$$\nabla \times \mathbf{H}_2 = \mathbf{J}_m + \frac{\partial \mathbf{D}_2}{\partial t}. \tag{2.183}$$

As long as we can resolve the question of how the constitutive parameters must be altered to reflect these replacements, we can conclude by comparing (2.182) with (2.179) and (2.183) with (2.178) that the solution to these equations is merely

$$\mathbf{E}_2 = \mathbf{H}_1,$$
$$\mathbf{B}_2 = -\mathbf{D}_1,$$
$$\mathbf{D}_2 = \mathbf{B}_1,$$
$$\mathbf{H}_2 = -\mathbf{E}_1.$$

That is, if we have solved the original problem, we can use those solutions to find the new ones. This is an application of the general *principle of duality*.

Unfortunately, this approach is a little awkward since the units of the sources and fields in the two problems are different. We can make the procedure more convenient by multiplying Ampere's law by $\eta_0 = (\mu_0/\epsilon_0)^{1/2}$. Then we have

$$\nabla \times \mathbf{E} = -\mathbf{J}_m - \frac{\partial \mathbf{B}}{\partial t},$$

$$\nabla \times (\eta_0 \mathbf{H}) = (\eta_0 \mathbf{J}) + \frac{\partial(\eta_0 \mathbf{D})}{\partial t}.$$

Thus, if the original problem has solution $(\mathbf{E}_1, \eta_0 \mathbf{D}_1, \mathbf{B}_1, \eta_0 \mathbf{H}_1)$, then the dual problem with $\mathbf{J}$ replaced by $\mathbf{J}_m/\eta_0$ and $\mathbf{J}_m$ replaced by $-\eta_0 \mathbf{J}$ has solution

$$\mathbf{E}_2 = \eta_0 \mathbf{H}_1, \tag{2.184}$$

$$\mathbf{B}_2 = -\eta_0 \mathbf{D}_1, \tag{2.185}$$

$$\eta_0 \mathbf{D}_2 = \mathbf{B}_1, \tag{2.186}$$

$$\eta_0 \mathbf{H}_2 = -\mathbf{E}_1. \tag{2.187}$$

The units on the quantities in the two problems are now identical.

Of course, the constitutive parameters for the dual problem must be altered from those of the original problem to reflect the change in field quantities. From (2.12) and (2.13) we know that the most general forms of the constitutive relations (those for linear, bianisotropic media) are

$$\mathbf{D}_1 = \bar{\xi}_1 \cdot \mathbf{H}_1 + \bar{\epsilon}_1 \cdot \mathbf{E}_1, \tag{2.188}$$

$$\mathbf{B}_1 = \bar{\mu}_1 \cdot \mathbf{H}_1 + \bar{\zeta}_1 \cdot \mathbf{E}_1, \tag{2.189}$$

for the original problem, and

$$\mathbf{D}_2 = \bar{\xi}_2 \cdot \mathbf{H}_2 + \bar{\epsilon}_2 \cdot \mathbf{E}_2, \tag{2.190}$$

$$\mathbf{B}_2 = \bar{\mu}_2 \cdot \mathbf{H}_2 + \bar{\zeta}_2 \cdot \mathbf{E}_2, \tag{2.191}$$

for the dual problem. Substitution of (2.184)–(2.187) into (2.188) and (2.189) gives

$$\mathbf{D}_2 = (-\bar{\zeta}_1) \cdot \mathbf{H}_2 + \left(\frac{\bar{\mu}_1}{\eta_0^2}\right) \cdot \mathbf{E}_2, \tag{2.192}$$

$$\mathbf{B}_2 = \left(\eta_0^2 \bar{\epsilon}_1\right) \cdot \mathbf{H}_2 + (-\bar{\xi}_1) \cdot \mathbf{E}_2. \tag{2.193}$$

Comparing (2.192) with (2.190) and (2.193) with (2.191), we conclude that

$$\bar{\zeta}_2 = -\bar{\xi}_1, \qquad \bar{\xi}_2 = -\bar{\zeta}_1, \qquad \bar{\mu}_2 = \eta_0^2 \bar{\epsilon}_1, \qquad \bar{\epsilon}_2 = \bar{\mu}_1/\eta_0^2.$$

As an important special case, we see that for a linear, isotropic medium specified by a permittivity ϵ and permeability μ, the dual problem is obtained by replacing ϵ_r with μ_r and μ_r with ϵ_r. The solution to the dual problem is then given by

$$\mathbf{E}_2 = \eta_0 \mathbf{H}_1$$

$$\eta_0 \mathbf{H}_2 = -\mathbf{E}_1,$$

as before. We thus see that the medium in the dual problem must have electric properties numerically equal to the magnetic properties of the medium in the original problem, and magnetic properties numerically equal to the electric properties of the medium in the original problem. This is rather inconvenient for most applications. Alternatively, we may divide Ampere's law by $\eta = (\mu/\epsilon)^{1/2}$ instead of η_0. Then the dual problem has the replacements

$$\mathbf{J} \mapsto \mathbf{J}_m/\eta, \qquad \mathbf{J}_m \mapsto -\eta\mathbf{J},$$

and the solution to the dual problem is given by

$$\mathbf{E}_2 = \eta\mathbf{H}_1,$$
$$\eta\mathbf{H}_2 = -\mathbf{E}_1.$$

In this case there is no need to swap ϵ_r and μ_r, since information about these parameters is incorporated into the replacement sources.

We must also remember that to obtain a unique solution we need to specify the boundary values of the fields. In a true dual problem, the boundary values of the fields used in the original problem are used on the swapped fields in the dual problem. A typical example of this is when the condition of zero tangential electric field on a perfect electric conductor is replaced by the condition of zero tangential magnetic field on the surface of a perfect magnetic conductor. However, duality can also be used to obtain the mathematical form of the field expressions, often in a homogeneous (source-free) situation, and boundary values can be applied later to specify the solution appropriate to the problem geometry. This approach is often used to compute waveguide modal fields and the electromagnetic fields scattered from objects. In these cases a TE/TM field decomposition is employed, and duality is used to find one part of the decomposition once the other is known.

2.9.2.1   Duality of electric and magnetic point source fields

By duality, we can sometimes use the known solution to one problem to solve a related problem by merely substituting different variables into the known mathematical expression. An example of this is the case in which we have solved for the fields produced by a certain distribution of electric sources and wish to determine the fields when the same distribution is used to describe magnetic sources.

Let us consider the case when the source distribution is that of a point current, or *Hertzian dipole*, immersed in free space. As we shall see in Chapter 5, the fields for a general source may be found by using the fields produced by these point sources. We begin by finding the fields produced by an electric dipole source at the origin aligned along the z-axis, $\mathbf{J} = \hat{\mathbf{z}}I_0\delta(\mathbf{r})$, then use duality to find the fields produced by a magnetic current source $\mathbf{J}_m = \hat{\mathbf{z}}I_{m0}\delta(\mathbf{r})$.

The fields produced by the electric source must obey

$$\nabla \times \mathbf{E}_e = -\frac{\partial}{\partial t}\mu_0\mathbf{H}_e, \tag{2.194}$$

$$\nabla \times \mathbf{H}_e = \hat{\mathbf{z}}I_0\delta(\mathbf{r}) + \frac{\partial}{\partial t}\epsilon_0\mathbf{E}_e, \tag{2.195}$$

$$\nabla \cdot \epsilon_0\mathbf{E}_e = \rho, \tag{2.196}$$

$$\nabla \cdot \mathbf{H}_e = 0, \tag{2.197}$$

while those produced by the magnetic source must obey

$$\nabla \times \mathbf{E}_m = -\hat{\mathbf{z}} I_{m0} \delta(\mathbf{r}) - \frac{\partial}{\partial t} \mu_0 \mathbf{H}_m, \tag{2.198}$$

$$\nabla \times \mathbf{H}_m = \frac{\partial}{\partial t} \epsilon_0 \mathbf{E}_m, \tag{2.199}$$

$$\nabla \cdot \mathbf{E}_m = 0, \tag{2.200}$$

$$\nabla \cdot \mu_0 \mathbf{H}_m = \rho_m. \tag{2.201}$$

We see immediately that the second set of equations is the dual of the first, as long as we scale the sources appropriately. Multiplying (2.198) by $-I_0/I_{m0}$ and (2.199) by $I_0 \eta_0^2 / I_{m0}$, we have the curl equations

$$\nabla \times \left(-\frac{I_0}{I_{m0}} \mathbf{E}_m \right) = \hat{\mathbf{z}} I_0 \delta(\mathbf{r}) + \frac{\partial}{\partial t} \left(\mu_0 \frac{I_0}{I_{m0}} \mathbf{H}_m \right), \tag{2.202}$$

$$\nabla \times \left(\frac{I_0 \eta_0^2}{I_{m0}} \mathbf{H}_m \right) = -\frac{\partial}{\partial t} \left(-\epsilon_0 \frac{I_0 \eta_0^2}{I_{m0}} \mathbf{E}_m \right). \tag{2.203}$$

Comparing (2.203) with (2.194) and (2.202) with (2.195) we see that

$$\mathbf{E}_m = -\frac{I_{m0}}{I_0} \mathbf{H}_e, \qquad \mathbf{H}_m = \frac{I_{m0}}{I_0} \frac{\mathbf{E}_e}{\eta_0^2}.$$

We note that it is impossible to have a point current source without accompanying point charge sources terminating each end of the dipole current. The point charges are required to satisfy the continuity equation, and vary in time as the moving charge that establishes the current accumulates at the ends of the dipole. From (2.195) we see that the magnetic field curls around the combination of the electric field and electric current source, while from (2.194) the electric field curls around the magnetic field, and from (2.196) diverges from the charges located at the ends of the dipole. From (2.198) we see that the electric field must curl around the combination of the magnetic field and magnetic current source, while (2.199) and (2.201) show that the magnetic field curls around the electric field and diverges from the magnetic charge.

2.9.2.2 Duality in a source-free region

Consider a closed surface S enclosing a source-free region of space. For simplicity, assume that the medium within S is linear, isotropic, and homogeneous. The fields within S are described by Maxwell's equations

$$\nabla \times \mathbf{E}_1 = -\frac{\partial}{\partial t} \mu \mathbf{H}_1,$$

$$\nabla \times \eta \mathbf{H}_1 = \frac{\partial}{\partial t} \epsilon \eta \mathbf{E}_1,$$

$$\nabla \cdot \epsilon \mathbf{E}_1 = 0,$$

$$\nabla \cdot \mu \mathbf{H}_1 = 0.$$

Under these conditions the concept of duality takes on a different face. The symmetry of the equations is such that the mathematical form of the solution for $\mathbf{E}$ is the same as that for $\eta\mathbf{H}$. That is, the fields

$$\mathbf{E}_2 = \eta\mathbf{H}_1,$$

$$\mathbf{H}_2 = -\mathbf{E}_1/\eta,$$

are also a solution to Maxwell's equations, and thus the dual problem merely involves replacing $\mathbf{E}$ by $\eta\mathbf{H}$ and $\mathbf{H}$ by $-\mathbf{E}/\eta$. However, the final forms of $\mathbf{E}$ and $\mathbf{H}$ will *not* be identical after appropriate boundary values are imposed.

This form of duality is very important for the solution of fields within waveguides or the fields scattered by objects where the sources are located outside the region where the fields are evaluated.

2.9.3   Reciprocity

The reciprocity theorem, also called the *Lorentz reciprocity theorem*, describes a specific and often useful relationship between sources and the electromagnetic fields they produce. Under certain special circumstances, we find that an interaction between independent source and mediating fields, called "reaction," is a spatially symmetric quantity. The reciprocity theorem is used in the study of guided waves to establish the orthogonality of guided wave modes, in microwave network theory to obtain relationships between terminal characteristics, and in antenna theory to demonstrate the equivalence of transmission and reception patterns.

Consider a closed surface S enclosing a volume V. Assume that the fields within and on S are produced by two independent source fields. The source $(\mathbf{J}_a, \mathbf{J}_{ma})$ produces the field $(\mathbf{E}_a, \mathbf{D}_a, \mathbf{B}_a, \mathbf{H}_a)$ as described by Maxwell's equations

$$\nabla \times \mathbf{E}_a = -\mathbf{J}_{ma} - \frac{\partial \mathbf{B}_a}{\partial t},$$

$$\nabla \times \mathbf{H}_a = \mathbf{J}_a + \frac{\partial \mathbf{D}_a}{\partial t},$$

while the source field $(\mathbf{J}_b, \mathbf{J}_{mb})$ produces the field $(\mathbf{E}_b, \mathbf{D}_b, \mathbf{B}_b, \mathbf{H}_b)$ as described by

$$\nabla \times \mathbf{E}_b = -\mathbf{J}_{mb} - \frac{\partial \mathbf{B}_b}{\partial t},$$

$$\nabla \times \mathbf{H}_b = \mathbf{J}_b + \frac{\partial \mathbf{D}_b}{\partial t}.$$

The sources may be distributed in any way relative to S: they may lie completely inside, completely outside, or partially inside and partially outside. Material media may lie within S, and their properties may depend on position.

Let us examine the quantity

$$R \equiv \nabla \cdot (\mathbf{E}_a \times \mathbf{H}_b - \mathbf{E}_b \times \mathbf{H}_a).$$

By (B.50) we have

$$R = \mathbf{H}_b \cdot \nabla \times \mathbf{E}_a - \mathbf{E}_a \cdot \nabla \times \mathbf{H}_b - \mathbf{H}_a \cdot \nabla \times \mathbf{E}_b + \mathbf{E}_b \cdot \nabla \times \mathbf{H}_a$$

so that by Maxwell's curl equations

$$R = \left[\mathbf{H}_a \cdot \frac{\partial \mathbf{B}_b}{\partial t} - \mathbf{H}_b \cdot \frac{\partial \mathbf{B}_a}{\partial t} \right] - \left[\mathbf{E}_a \cdot \frac{\partial \mathbf{D}_b}{\partial t} - \mathbf{E}_b \cdot \frac{\partial \mathbf{D}_a}{\partial t} \right]$$

$$+ \left[\mathbf{J}_a \cdot \mathbf{E}_b - \mathbf{J}_b \cdot \mathbf{E}_a - \mathbf{J}_{ma} \cdot \mathbf{H}_b + \mathbf{J}_{mb} \cdot \mathbf{H}_a \right].$$

The useful relationships we seek occur when the first two bracketed quantities on the right-hand side of the above expression are zero. Whether this is true depends not only on the behavior of the fields, but on the properties of the medium at the point in question. Though we have assumed that the sources of the field sets are independent, it is apparent that they must share a similar time dependence in order for the terms within each of the bracketed quantities to cancel. Of special interest is the case where the two sources are both sinusoidal in time with identical frequencies, but with differing spatial distributions. We shall consider this case in detail in § 4.10.2 after we have discussed the properties of the time-harmonic field. Importantly, we will find that only certain characteristics of the constitutive parameters allow cancellation of the bracketed terms; materials with these characteristics are called *reciprocal*, and the fields they support are said to display the property of *reciprocity*. To see what this property entails, we set the bracketed terms to zero and integrate over a volume V to obtain

$$\oint_S (\mathbf{E}_a \times \mathbf{H}_b - \mathbf{E}_b \times \mathbf{H}_a) \cdot d\mathbf{S} = \int_V (\mathbf{J}_a \cdot \mathbf{E}_b - \mathbf{J}_b \cdot \mathbf{E}_a - \mathbf{J}_{ma} \cdot \mathbf{H}_b + \mathbf{J}_{mb} \cdot \mathbf{H}_a) \, dV,$$

which is the time-domain version of the *Lorentz reciprocity theorem*.

Two special cases of this theorem are important to us. If all sources lie outside S, we have *Lorentz's lemma*

$$\oint_S (\mathbf{E}_a \times \mathbf{H}_b - \mathbf{E}_b \times \mathbf{H}_a) \cdot d\mathbf{S} = 0.$$

This remarkable expression shows that a relationship exists between the fields produced by completely independent sources, and is useful for establishing waveguide mode orthogonality for time-harmonic fields. If sources reside within S but the surface integral is equal to zero, we have

$$\int_V (\mathbf{J}_a \cdot \mathbf{E}_b - \mathbf{J}_b \cdot \mathbf{E}_a - \mathbf{J}_{ma} \cdot \mathbf{H}_b + \mathbf{J}_{mb} \cdot \mathbf{H}_a) \, dV = 0.$$

This occurs when the surface is bounded by a special material (such as an impedance sheet or a perfect conductor), or when the surface recedes to infinity; the expression is useful for establishing the reciprocity conditions for networks and antennas. We shall interpret it for time harmonic fields in § 4.10.2.

2.9.4 Similitude

A common approach in physical science involves the introduction of normalized variables to provide for scaling of problems along with a chance to identify certain physically significant parameters. Similarity as a general principle can be traced back to the earliest attempts to describe physical effects with mathematical equations, with serious study undertaken by Galileo. Helmholtz introduced the first systematic investigation in 1873, and the concept was rigorized by Reynolds ten years later [218]. Similitude is now considered a fundamental guiding principle in the modeling of materials [199].

The process often begins with a consideration of the fundamental differential equations. In electromagnetics we may introduce a set of dimensionless field and source variables

$$\underline{\mathbf{E}}, \quad \underline{\mathbf{D}}, \quad \underline{\mathbf{B}}, \quad \underline{\mathbf{H}}, \quad \underline{\mathbf{J}}, \quad \underline{\rho}, \tag{2.204}$$

by setting

$$\mathbf{E} = \underline{\mathbf{E}} k_E, \quad \mathbf{B} = \underline{\mathbf{B}} k_B, \quad \mathbf{D} = \underline{\mathbf{D}} k_D,$$

$$\mathbf{H} = \underline{\mathbf{H}} k_H, \quad \mathbf{J} = \underline{\mathbf{J}} k_J, \quad \rho = \underline{\rho} k_\rho. \tag{2.205}$$

Here we regard the quantities $k_E, k_B, \ldots$ as base units for the discussion, while the dimensionless quantities (2.204) serve to express the actual fields $\mathbf{E}, \mathbf{B}, \ldots$ in terms of these base units. Of course, the time and space variables can also be scaled: we can write

$$t = \underline{t}k_t, \qquad l = \underline{l}k_l, \tag{2.206}$$

if l is any length of interest. Again, the quantities $\underline{t}$ and $\underline{l}$ are dimensionless measure numbers used to express the actual quantities t and l relative to the chosen base amounts k_t and k_l. With (2.205) and (2.206), Maxwell's curl equations become

$$\underline{\nabla} \times \underline{\mathbf{E}} = -\frac{k_B}{k_E}\frac{k_l}{k_t}\frac{\partial \underline{\mathbf{B}}}{\partial \underline{t}}, \qquad \underline{\nabla} \times \underline{\mathbf{H}} = \frac{k_J k_l}{k_H}\underline{\mathbf{J}} + \frac{k_D}{k_H}\frac{k_l}{k_t}\frac{\partial \underline{\mathbf{D}}}{\partial \underline{t}} \tag{2.207}$$

while the continuity equation becomes

$$\underline{\nabla} \cdot \underline{\mathbf{J}} = -\frac{k_\rho}{k_J}\frac{k_l}{k_t}\frac{\partial \underline{\rho}}{\partial \underline{t}}, \tag{2.208}$$

where $\underline{\nabla}$ has been normalized by k_l. These are examples of field equations cast into dimensionless form — it is easily verified that the *similarity parameters*

$$\frac{k_B}{k_E}\frac{k_l}{k_t}, \qquad \frac{k_J k_l}{k_H}, \qquad \frac{k_D}{k_H}\frac{k_l}{k_t}, \qquad \frac{k_\rho}{k_J}\frac{k_l}{k_t}, \tag{2.209}$$

are dimensionless. The idea behind electromagnetic similitude is that a given set of normalized values $\underline{\mathbf{E}}, \underline{\mathbf{B}}, \ldots$ can satisfy equations (2.207) and (2.208) for many different physical situations, provided that the numerical values of the coefficients (2.209) are all fixed across those situations. Indeed, the differential equations would be identical.

To make this discussion a bit more concrete, let us assume a conducting linear medium where

$$\mathbf{D} = \epsilon\mathbf{E}, \qquad \mathbf{B} = \mu\mathbf{H}, \qquad \mathbf{J} = \sigma\mathbf{E},$$

and use

$$\epsilon = \underline{\epsilon}k_\epsilon, \qquad \mu = \underline{\mu}k_\mu, \qquad \sigma = \underline{\sigma}k_\sigma,$$

to express the material parameters in terms of dimensionless values $\underline{\epsilon}$, $\underline{\mu}$, and $\underline{\sigma}$. Then

$$\underline{\mathbf{D}} = \frac{k_\epsilon k_E}{k_D}\underline{\epsilon}\underline{\mathbf{E}}, \qquad \underline{\mathbf{B}} = \frac{k_\mu k_H}{k_B}\underline{\mu}\underline{\mathbf{H}}, \qquad \underline{\mathbf{J}} = \frac{k_\sigma k_E}{k_J}\underline{\sigma}\underline{\mathbf{E}},$$

and equations (2.207) become

$$\underline{\nabla} \times \underline{\mathbf{E}} = -\left(\frac{k_\mu k_l}{k_t}\frac{k_H}{k_E}\right)\underline{\mu}\frac{\partial \underline{\mathbf{H}}}{\partial \underline{t}},$$

$$\underline{\nabla} \times \underline{\mathbf{H}} = \left(k_\sigma k_l\frac{k_E}{k_H}\right)\underline{\sigma}\underline{\mathbf{E}} + \left(\frac{k_\epsilon k_l}{k_t}\frac{k_E}{k_H}\right)\underline{\epsilon}\frac{\partial \underline{\mathbf{E}}}{\partial \underline{t}}.$$

Defining

$$\alpha = \frac{k_\mu k_l}{k_t}\frac{k_H}{k_E}, \qquad \gamma = k_\sigma k_l\frac{k_E}{k_H}, \qquad \beta = \frac{k_\epsilon k_l}{k_t}\frac{k_E}{k_H},$$

we see that under the current assumptions similarity holds between two electromagnetics problems only if $\alpha\underline{\mu}$, $\gamma\underline{\sigma}$, and $\beta\underline{\epsilon}$ are numerically the same in both problems. A necessary condition for similitude, then, is that the products

$$(\alpha\underline{\mu})(\beta\underline{\epsilon}) = k_\mu k_\epsilon\left(\frac{k_l}{k_t}\right)^2\underline{\mu}\underline{\epsilon}, \qquad (\alpha\underline{\mu})(\gamma\underline{\sigma}) = k_\mu k_\sigma\frac{k_l^2}{k_t}\underline{\mu}\underline{\sigma},$$

(which do not involve k_E or k_H) stay constant between problems. We see, for example, that we may compensate for a halving of the length scale k_l by (a) a quadrupling of the permeability $\underline{\mu}$, or (b) a simultaneous halving of the time scale k_t and doubling of the conductivity $\underline{\sigma}$. A much less subtle special case is that for which $\underline{\sigma} = 0$, $k_\epsilon = \epsilon_0$, $k_\mu = \mu_0$, and $\underline{\epsilon} = \underline{\mu} = 1$; we then have free space and must simply maintain

$$k_l / k_t = \text{constant}$$

so that the time and length scales stay proportional. In the sinusoidal steady state, for instance, the frequency would be made to vary inversely with the length scale.

2.9.5 Conservation theorems

The misconception that Poynting's theorem can be "derived" from Maxwell's equations is widespread and ingrained. We must, in fact, *postulate* the idea that the electromagnetic field can be associated with an energy flux propagating at the speed of light. Since the form of the postulate is patterned after the well-understood laws of mechanics, we begin by developing the basic equations of momentum and energy balance in mechanical systems. Then we shall see whether it is sensible to ascribe these principles to the electromagnetic field.

Maxwell's theory allows us to describe, using Maxwell's equations, the behavior of the electromagnetic fields within a (possibly) finite region V of space. The presence of any sources or material objects outside V are made known through the specification of tangential fields over the boundary of V, as required for uniqueness. Thus, the influence of external effects can always be viewed as being transported across the boundary. This is true of mechanical as well as electromagnetic effects. A charged material body can be acted on by physical contact with another body, by gravitational forces, and by the Lorentz force, each effect resulting in momentum exchange across the boundary of the object. These effects must all be taken into consideration if we are to invoke momentum conservation, resulting in a very complicated situation. This suggests that we try to decompose the problem into simpler "systems" based on physical effects.

2.9.5.1 The system concept in the physical sciences

The idea of decomposing a complicated system into simpler, self-contained systems is quite common in the physical sciences. Penfield and Haus [149] invoke this concept by introducing an *electromagnetic system* where the effects of the Lorentz force equation are considered to accompany a *mechanical system* where effects of pressure, stress, and strain are considered, and a *thermodynamic system* where the effects of heat exchange are considered. These systems can all be interrelated in a variety of ways. For instance, as a material heats up, it can expand, and the resulting mechanical forces can alter the electrical properties of the material. We will follow Penfield and Haus by considering separate electromagnetic and mechanical subsystems; other systems may be added analogously.

If we separate the various systems by physical effect, we will need to know how to "reassemble the information." Two conservation theorems are very helpful in this regard: conservation of energy, and conservation of momentum. Engineers often employ these theorems to make tacit use of the system idea. For instance, when studying electromagnetic waves propagating in a waveguide, it is common practice to compute wave attenuation by calculating the Poynting flux of power into the walls of the guide. The power lost from the wave is said to "heat up the waveguide walls," which indeed it does.

This is an admission that the electromagnetic system is not "closed": it requires the inclusion of a thermodynamic system in order that energy be conserved. Of course, the detailed workings of the thermodynamic system are often ignored, indicating that any thermodynamic "feedback" mechanism is weak. In the waveguide example, for instance, the heating of the metallic walls does not alter their electromagnetic properties enough to couple back into an effect on the fields in the walls or in the guide. If such effects were important, they would have to be included in the conservation theorem via the boundary fields; it is therefore reasonable to associate with these fields a "flow" of energy or momentum into V. Thus, we wish to develop conservation laws that include not only the Lorentz force effects within V, but a flow of external effects into V through its boundary surface.

To understand how external influences may affect the electromagnetic subsystem, we look to the behavior of the mechanical subsystem as an analogue. In the electromagnetic system, effects are felt both internally to a region (because of the Lorentz force effect) and through the system boundary (by the dependence of the internal fields on the boundary fields). In the mechanical and thermodynamic systems, a region of mass is affected both internally (through transfer of heat and gravitational forces) and through interactions occurring across its surface (through transfers of energy and momentum, by pressure and stress). One beauty of electromagnetic theory is that we can find a mathematical symmetry between electromagnetic and mechanical effects which parallels the above conceptual symmetry. This makes applying conservation of energy and momentum to the total system (electromagnetic, thermodynamic, and mechanical) very convenient.

2.9.5.2   Conservation of momentum and energy in mechanical systems

We begin by reviewing the interactions of material bodies in a mechanical system. For simplicity we concentrate on fluids (analogous to charge in space); the extension of these concepts to solid bodies is straightforward.

Consider a fluid with mass density ρ_m. The momentum of a small subvolume of the fluid is given by $\rho_m \mathbf{v}\, dV$, where $\mathbf{v}$ is the velocity of the subvolume. So the momentum density is $\rho_m \mathbf{v}$. Newton's second law states that a force acting throughout the subvolume results in a change in its momentum, given by

$$\frac{D}{Dt}(\rho_m \mathbf{v}\, dV) = \mathbf{f}\, dV, \tag{2.210}$$

where $\mathbf{f}$ is the volume force density and the D/Dt notation shows that we are interested in the rate of change of the momentum as observed by the moving fluid element (see § A.3). Here $\mathbf{f}$ could be the weight force, for instance. Addition of the results for all elements of the fluid body gives

$$\frac{D}{Dt}\int_V \rho_m \mathbf{v}\, dV = \int_V \mathbf{f}\, dV \tag{2.211}$$

as the change in momentum for the entire body. If on the other hand the force exerted on the body is through contact with its surface, the change in momentum is

$$\frac{D}{Dt}\int_V \rho_m \mathbf{v}\, dV = \oint_S \mathbf{t}\, dS \tag{2.212}$$

where $\mathbf{t}$ is the "surface traction."

We can write the time-rate of change of momentum in a more useful form by applying the Reynolds transport theorem (A.68):

$$\frac{D}{Dt} \int_V \rho_m \mathbf{v} \, dV = \int_V \frac{\partial}{\partial t} (\rho_m \mathbf{v}) \, dV + \oint_S (\rho_m \mathbf{v}) \mathbf{v} \cdot d\mathbf{S}. \tag{2.213}$$

Superposing (2.211) and (2.212) and substituting into (2.213) we have

$$\int_V \frac{\partial}{\partial t} (\rho_m \mathbf{v}) \, dV + \oint_S (\rho_m \mathbf{v}) \mathbf{v} \cdot d\mathbf{S} = \int_V \mathbf{f} \, dV + \oint_S \mathbf{t} \, dS. \tag{2.214}$$

If we define the dyadic quantity

$$\bar{\mathbf{T}}_k = \rho_m \mathbf{v} \mathbf{v},$$

then (2.214) can be written as

$$\int_V \frac{\partial}{\partial t} (\rho_m \mathbf{v}) \, dV + \oint_S \hat{\mathbf{n}} \cdot \bar{\mathbf{T}}_k \, dS = \int_V \mathbf{f} \, dV + \oint_S \mathbf{t} \, dS. \tag{2.215}$$

This *principle of linear momentum* [214] can be interpreted as a large-scale form of conservation of kinetic linear momentum. Here $\hat{\mathbf{n}} \cdot \bar{\mathbf{T}}_k$ represents the flow of kinetic momentum across S, and the sum of this momentum transfer and the change of momentum within V stands equal to the forces acting internal to V and upon S.

The surface traction may be related to the surface normal $\hat{\mathbf{n}}$ through a dyadic quantity $\bar{\mathbf{T}}_m$ called the mechanical *stress tensor*:

$$\mathbf{t} = \hat{\mathbf{n}} \cdot \bar{\mathbf{T}}_m.$$

With this we may write (2.215) as

$$\int_V \frac{\partial}{\partial t} (\rho_m \mathbf{v}) \, dV + \oint_S \hat{\mathbf{n}} \cdot \bar{\mathbf{T}}_k \, dS = \int_V \mathbf{f} \, dV + \oint_S \hat{\mathbf{n}} \cdot \bar{\mathbf{T}}_m \, dS$$

and apply the dyadic form of the divergence theorem (B.25) to get

$$\int_V \frac{\partial}{\partial t} (\rho_m \mathbf{v}) \, dV + \int_V \nabla \cdot (\rho_m \mathbf{v} \mathbf{v}) \, dV = \int_V \mathbf{f} \, dV + \int_V \nabla \cdot \bar{\mathbf{T}}_m \, dV.$$

Combining the volume integrals and setting the integrand to zero, we have

$$\frac{\partial}{\partial t} (\rho_m \mathbf{v}) + \nabla \cdot (\rho_m \mathbf{v} \mathbf{v}) = \mathbf{f} + \nabla \cdot \bar{\mathbf{T}}_m,$$

which is the point-form equivalent of (2.215). Note that the second term on the right-hand side is nonzero only for points residing on the surface of the body. Finally, letting $\mathbf{g}$ denote momentum density, we obtain the simple expression

$$\nabla \cdot \bar{\mathbf{T}}_k + \frac{\partial \mathbf{g}_k}{\partial t} = \mathbf{f}_k, \tag{2.216}$$

where

$$\mathbf{g}_k = \rho_m \mathbf{v}$$

is the density of kinetic momentum, and

$$\mathbf{f}_k = \mathbf{f} + \nabla \cdot \bar{\mathbf{T}}_m \tag{2.217}$$

is the total force density.

Equation (2.216) is somewhat analogous to the electric charge continuity equation (1.9). For each point of the body, the total outflux of kinetic momentum plus the time rate of change of kinetic momentum equals the total force. The resemblance to (1.9) is strong, except for the nonzero term on the right-hand side. The charge continuity equation represents a closed system: charge cannot spontaneously appear and add an extra term to the right-hand side of (1.9). On the other hand, the change in total momentum at a point can exceed that given by the momentum flowing out of the point if there is another "source" (e.g., gravity for an internal point, or pressure on a boundary point).

To obtain a momentum conservation expression that resembles the continuity equation, we must consider a "subsystem" with terms that exactly counterbalance the extra expressions on the right-hand side of (2.216). For a fluid acted on only by external pressure, the sole effect enters through the traction term, and [149]

$$\nabla \cdot \bar{\mathbf{T}}_m = -\nabla p \tag{2.218}$$

where p is the pressure exerted on the fluid body. Now, using (B.69), we can write

$$-\nabla p = -\nabla \cdot \bar{\mathbf{T}}_p \tag{2.219}$$

where

$$\bar{\mathbf{T}}_p = p\bar{\mathbf{I}}$$

and $\bar{\mathbf{I}}$ is the unit dyad. Finally, using (2.219), (2.218), and (2.217) in (2.216), we obtain

$$\nabla \cdot (\bar{\mathbf{T}}_k + \bar{\mathbf{T}}_p) + \frac{\partial}{\partial t}\mathbf{g}_k = 0$$

and we have an expression for a closed system including all possible effects. Now, note that we can form the above expression as

$$\left(\nabla \cdot \bar{\mathbf{T}}_k + \frac{\partial}{\partial t}\mathbf{g}_k\right) + \left(\nabla \cdot \bar{\mathbf{T}}_p + \frac{\partial}{\partial t}\mathbf{g}_p\right) = 0 \tag{2.220}$$

where $\mathbf{g}_p = 0$ since there are no volume effects associated with pressure. This can be viewed as the sum of two closed subsystems

$$\nabla \cdot \bar{\mathbf{T}}_k + \frac{\partial}{\partial t}\mathbf{g}_k = 0, \tag{2.221}$$

$$\nabla \cdot \bar{\mathbf{T}}_p + \frac{\partial}{\partial t}\mathbf{g}_p = 0.$$

We now have the desired viewpoint. The conservation formula for the complete closed system can be viewed as a sum of formulas for open subsystems, each having the form of a conservation law for a closed system. In case we must include the effects of gravity, for instance, we need only determine $\bar{\mathbf{T}}_g$ and $\mathbf{g}_g$ such that

$$\nabla \cdot \bar{\mathbf{T}}_g + \frac{\partial}{\partial t}\mathbf{g}_g = 0$$

and add this new conservation equation to (2.220). If we can find a conservation expression of form similar to (2.221) for an "electromagnetic subsystem," we can include its effects along with the mechanical effects by merely adding together the conservation laws. We shall find just such an expression later in this section.

We stated in § 1.3.4 that there are four fundamental conservation principles. We have now discussed linear momentum; the principle of angular momentum follows similarly. Our next goal is to find an expression similar to (2.220) for conservation of energy. We may expect the conservation of energy expression to obey a similar law of superposition. We begin with the fundamental definition of work: for a particle moving with velocity $\mathbf{v}$ under the influence of a force $\mathbf{f}_k$, the work is given by $\mathbf{f}_k \cdot \mathbf{v}$. Dot multiplying (2.210) by $\mathbf{v}$ and replacing $\mathbf{f}$ by $\mathbf{f}_k$ (to represent both volume and surface forces), we get

$$\mathbf{v} \cdot \frac{D}{Dt}(\rho_m \mathbf{v}) \, dV = \mathbf{v} \cdot \mathbf{f}_k \, dV$$

or equivalently

$$\frac{D}{Dt}(\tfrac{1}{2}\rho_m \mathbf{v} \cdot \mathbf{v}) \, dV = \mathbf{v} \cdot \mathbf{f}_k \, dV.$$

Integration over a volume and application of the Reynolds transport theorem (A.68) then gives

$$\int_V \frac{\partial}{\partial t}(\tfrac{1}{2}\rho_m v^2) \, dV + \oint_S \hat{\mathbf{n}} \cdot (\mathbf{v}\tfrac{1}{2}\rho_m v^2) \, dS = \int_V \mathbf{f}_k \cdot \mathbf{v} \, dV.$$

Hence the sum of the time rate of change in energy internal to the body and the flow of kinetic energy across the boundary must equal the work done by internal and surface forces acting on the body. In point form,

$$\nabla \cdot \mathbf{S}_k + \frac{\partial}{\partial t} W_k = \mathbf{f}_k \cdot \mathbf{v} \qquad (2.222)$$

where

$$\mathbf{S}_k = \mathbf{v}\tfrac{1}{2}\rho_m v^2$$

is the density of the flow of kinetic energy, and

$$W_k = \tfrac{1}{2}\rho_m v^2$$

is the kinetic energy density. Again, the system is not closed (the right-hand side of (2.222) is not zero) because the balancing forces are not included. As was done with the momentum equation, the effect of the work done by the pressure forces can be described in a closed-system-type equation:

$$\nabla \cdot \mathbf{S}_p + \frac{\partial}{\partial t} W_p = 0. \qquad (2.223)$$

Combining (2.222) and (2.223), we have

$$\nabla \cdot (\mathbf{S}_k + \mathbf{S}_p) + \frac{\partial}{\partial t}(W_k + W_p) = 0,$$

the energy conservation equation for the closed system.

2.9.5.3 Conservation in the electromagnetic subsystem

We would now like to achieve closed-system conservation theorems for the electromagnetic subsystem so that we can add in the effects of electromagnetism. For the momentum equation, we can proceed exactly as we did with the mechanical system. We begin with

$$\mathbf{f}_{em} = \rho \mathbf{E} + \mathbf{J} \times \mathbf{B}.$$

This force term should appear on one side of the point form of the momentum conservation equation. The term on the other side must involve the electromagnetic fields, since they are the mechanism for exerting force on the charge distribution. Substituting for $\mathbf{J}$ from (2.2) and for ρ from (2.3) we have

$$\mathbf{f}_{em} = \mathbf{E}(\nabla \cdot \mathbf{D}) - \mathbf{B} \times (\nabla \times \mathbf{H}) + \mathbf{B} \times \frac{\partial \mathbf{D}}{\partial t}.$$

Using

$$\mathbf{B} \times \frac{\partial \mathbf{D}}{\partial t} = -\frac{\partial}{\partial t}(\mathbf{D} \times \mathbf{B}) + \mathbf{D} \times \frac{\partial \mathbf{B}}{\partial t}$$

and substituting from Faraday's law for $\partial \mathbf{B}/\partial t$ we have

$$-[\mathbf{E}(\nabla \cdot \mathbf{D}) - \mathbf{D} \times (\nabla \times \mathbf{E}) + \mathbf{H}(\nabla \cdot \mathbf{B}) - \mathbf{B} \times (\nabla \times \mathbf{H})] + \frac{\partial}{\partial t}(\mathbf{D} \times \mathbf{B}) = -\mathbf{f}_{em}. \quad (2.224)$$

Here we have also added the null term $\mathbf{H}(\nabla \cdot \mathbf{B})$.

The forms of (2.224) and (2.216) would be identical if the bracketed term could be written as the divergence of a dyadic function $\bar{\mathbf{T}}_{em}$. This is indeed possible for linear, homogeneous, bianisotropic media, provided that the constitutive matrix $[\bar{\mathbf{C}}_{EH}]$ in (2.14) is symmetric [108]. In that case

$$\bar{\mathbf{T}}_{em} = \tfrac{1}{2}(\mathbf{D} \cdot \mathbf{E} + \mathbf{B} \cdot \mathbf{H})\bar{\mathbf{I}} - \mathbf{DE} - \mathbf{BH}, \quad (2.225)$$

which is called the *Maxwell stress tensor*. Let us demonstrate this equivalence for a linear, isotropic, homogeneous material. Putting $\mathbf{D} = \epsilon\mathbf{E}$ and $\mathbf{H} = \mathbf{B}/\mu$ into (2.224) we obtain

$$\nabla \cdot \mathbf{T}_{em} = -\epsilon\mathbf{E}(\nabla \cdot \mathbf{E}) + \frac{1}{\mu}\mathbf{B} \times (\nabla \times \mathbf{B}) + \epsilon\mathbf{E} \times (\nabla \times \mathbf{E}) - \frac{1}{\mu}\mathbf{B}(\nabla \cdot \mathbf{B}). \quad (2.226)$$

Now (B.52) gives

$$\nabla(\mathbf{A} \cdot \mathbf{A}) = 2\mathbf{A} \times (\nabla \times \mathbf{A}) + 2(\mathbf{A} \cdot \nabla)\mathbf{A}$$

so that

$$\mathbf{E}(\nabla \cdot \mathbf{E}) - \mathbf{E} \times (\nabla \times \mathbf{E}) = \mathbf{E}(\nabla \cdot \mathbf{E}) + (\mathbf{E} \cdot \nabla)\mathbf{E} - \tfrac{1}{2}\nabla(E^2).$$

Finally, (B.61) and (B.69) give

$$\mathbf{E}(\nabla \cdot \mathbf{E}) - \mathbf{E} \times (\nabla \times \mathbf{E}) = \nabla \cdot \left(\mathbf{EE} - \tfrac{1}{2}\bar{\mathbf{I}}\mathbf{E} \cdot \mathbf{E}\right).$$

Substituting this expression and a similar one for $\mathbf{B}$ into (2.226) we have

$$\nabla \cdot \bar{\mathbf{T}}_{em} = \nabla \cdot [\tfrac{1}{2}(\mathbf{D} \cdot \mathbf{E} + \mathbf{B} \cdot \mathbf{H})\bar{\mathbf{I}} - \mathbf{DE} - \mathbf{BH}],$$

which matches (2.225).

Replacing the term in brackets in (2.224) by $\nabla \cdot \bar{\mathbf{T}}_{em}$, we get

$$\nabla \cdot \bar{\mathbf{T}}_{em} + \frac{\partial \mathbf{g}_{em}}{\partial t} = -\mathbf{f}_{em} \quad (2.227)$$

where

$$\mathbf{g}_{em} = \mathbf{D} \times \mathbf{B}.$$

Equation (2.227) is the point form of the electromagnetic conservation of momentum theorem. It is mathematically identical in form to the mechanical theorem (2.216). Integration over a volume gives the large-scale form

$$\oint_S \bar{\mathbf{T}}_{em} \cdot d\mathbf{S} + \int_V \frac{\partial \mathbf{g}_{em}}{\partial t} \, dV = -\int_V \mathbf{f}_{em} \, dV.$$

If we interpret this as we interpreted the conservation theorems from mechanics, the first term on the left-hand side represents the flow of electromagnetic momentum across the boundary of V, while the second term represents the change in momentum within V. The sum of these two quantities is exactly compensated by the total Lorentz force acting on the charges within V. Thus we identify $\mathbf{g}_{em}$ as the transport density of electromagnetic momentum.

Because (2.227) is not zero on the right-hand side, it does not represent a closed system. If the Lorentz force is the only force acting on the charges within V, then the mechanical reaction to the Lorentz force should be described by Newton's third law. Thus we have the kinematic momentum conservation formula

$$\nabla \cdot \bar{\mathbf{T}}_k + \frac{\partial \mathbf{g}_k}{\partial t} = \mathbf{f}_k = -\mathbf{f}_{em}.$$

Subtracting this expression from (2.227) we obtain

$$\nabla \cdot (\bar{\mathbf{T}}_{em} - \bar{\mathbf{T}}_k) + \frac{\partial}{\partial t}(\mathbf{g}_{em} - \mathbf{g}_k) = 0,$$

which describes momentum conservation for the closed system.

It is also possible to derive a conservation theorem for electromagnetic energy that resembles the corresponding theorem for mechanical energy. Earlier we noted that $\mathbf{v} \cdot \mathbf{f}$ represents the volume density of work produced by moving an object at velocity $\mathbf{v}$ under the action of a force $\mathbf{f}$. For the electromagnetic subsystem the work is produced by charges moving against the Lorentz force. So the volume density of work delivered to the currents is

$$w_{em} = \mathbf{v} \cdot \mathbf{f}_{em} = \mathbf{v} \cdot (\rho \mathbf{E} + \mathbf{J} \times \mathbf{B}) = (\rho \mathbf{v}) \cdot \mathbf{E} + \rho \mathbf{v} \cdot (\mathbf{v} \times \mathbf{B}). \tag{2.228}$$

Using (B.6) on the second term in (2.228) we get

$$w_{em} = (\rho \mathbf{v}) \cdot \mathbf{E} + \rho \mathbf{B} \cdot (\mathbf{v} \times \mathbf{v}).$$

The second term vanishes by definition of the cross product. This is the familiar property that the magnetic field does no work on moving charge. Hence

$$w_{em} = \mathbf{J} \cdot \mathbf{E}. \tag{2.229}$$

This important relation says that charge moving in an electric field experiences a force which results in energy transfer to (or from) the charge. We wish to write this energy transfer in terms of an energy flux vector, as we did with the mechanical subsystem.

As with our derivation of the conservation of electromagnetic momentum, we wish to relate the energy transfer to the electromagnetic fields. Substitution of $\mathbf{J}$ from (2.2) into (2.229) gives

$$w_{em} = (\nabla \times \mathbf{H}) \cdot \mathbf{E} - \frac{\partial \mathbf{D}}{\partial t} \cdot \mathbf{E},$$

hence

$$w_{em} = -\nabla \cdot (\mathbf{E} \times \mathbf{H}) + \mathbf{H} \cdot (\nabla \times \mathbf{E}) - \frac{\partial \mathbf{D}}{\partial t} \cdot \mathbf{E}$$

by (B.50). Substituting for $\nabla \times \mathbf{E}$ from (2.1) we have

$$w_{em} = -\nabla \cdot (\mathbf{E} \times \mathbf{H}) - \left(\mathbf{E} \cdot \frac{\partial \mathbf{D}}{\partial t} + \mathbf{H} \cdot \frac{\partial \mathbf{B}}{\partial t} \right).$$

This is not quite of the form (2.222) since a single term representing the time rate of change of energy density is not present. However, for a linear isotropic medium in which ϵ and μ do not depend on time (i.e., a nondispersive medium) we have

$$\mathbf{E} \cdot \frac{\partial \mathbf{D}}{\partial t} = \epsilon \mathbf{E} \cdot \frac{\partial \mathbf{E}}{\partial t} = \frac{1}{2} \epsilon \frac{\partial}{\partial t} (\mathbf{E} \cdot \mathbf{E}) = \frac{1}{2} \frac{\partial}{\partial t} (\mathbf{D} \cdot \mathbf{E}), \qquad (2.230)$$

$$\mathbf{H} \cdot \frac{\partial \mathbf{B}}{\partial t} = \mu \mathbf{H} \cdot \frac{\partial \mathbf{H}}{\partial t} = \frac{1}{2} \mu \frac{\partial}{\partial t} (\mathbf{H} \cdot \mathbf{H}) = \frac{1}{2} \frac{\partial}{\partial t} (\mathbf{H} \cdot \mathbf{B}). \qquad (2.231)$$

Using this we obtain

$$\nabla \cdot \mathbf{S}_{em} + \frac{\partial}{\partial t} W_{em} = -\mathbf{f}_{em} \cdot \mathbf{v} = -\mathbf{J} \cdot \mathbf{E} \qquad (2.232)$$

where $W_{em} = \frac{1}{2}(\mathbf{D} \cdot \mathbf{E} + \mathbf{B} \cdot \mathbf{H})$ and

$$\mathbf{S}_{em} = \mathbf{E} \times \mathbf{H}. \qquad (2.233)$$

Equation (2.232) is the point form of the energy conservation theorem, also called *Poynting's theorem* after J.H. Poynting who first proposed it. The quantity $\mathbf{S}_{em}$ given in (2.233) is known as the *Poynting vector*. Integrating (2.232) over a volume and using the divergence theorem, we obtain the large-scale form

$$-\int_V \mathbf{J} \cdot \mathbf{E} \, dV = \int_V \frac{1}{2} \frac{\partial}{\partial t} (\mathbf{D} \cdot \mathbf{E} + \mathbf{B} \cdot \mathbf{H}) \, dV + \oint_S (\mathbf{E} \times \mathbf{H}) \cdot d\mathbf{S}. \qquad (2.234)$$

This also holds for a nondispersive, linear, bianisotropic medium with a symmetric constitutive matrix [108, 180].

We see that the electromagnetic energy conservation theorem (2.232) is identical in form to the mechanical energy conservation theorem (2.222). Thus, if the system is composed of just the kinetic and electromagnetic subsystems, the mechanical force exactly balances the Lorentz force, and (2.232) and (2.222) add to give

$$\nabla \cdot (\mathbf{S}_{em} + \mathbf{S}_k) + \frac{\partial}{\partial t} (W_{em} + W_k) = 0, \qquad (2.235)$$

showing that energy is conserved for the entire system.

As in the mechanical system, we identify W_{em} as the volume electromagnetic energy density in V, and $\mathbf{S}_{em}$ as the density of electromagnetic energy flowing across the boundary of V. This interpretation is somewhat controversial, as discussed below.

2.9.5.4 Interpretation of the energy and momentum conservation theorems

There has been some controversy regarding Poynting's theorem (and, equally, the momentum conservation theorem). While there is no question that Poynting's theorem is mathematically correct, we may wonder whether we are justified in associating W_{em} with

W_k and $\mathbf{S}_{em}$ with $\mathbf{S}_k$ merely because of the similarities in their mathematical expressions. Certainly there is some justification for associating W_k, the kinetic energy of particles, with W_{em}, since we shall show that for static fields the term $\frac{1}{2}(\mathbf{D} \cdot \mathbf{E} + \mathbf{B} \cdot \mathbf{H})$ represents the energy required to assemble the charges and currents into a certain configuration. However, the term $\mathbf{S}_{em}$ is more problematic. In a mechanical system, $\mathbf{S}_k$ represents the flow of kinetic energy associated with moving particles — does that imply that $\mathbf{S}_{em}$ represents the flow of electromagnetic energy? That is the position generally taken, and it is widely supported by experimental evidence. However, the interpretation is not clear-cut.

If we associate $\mathbf{S}_{em}$ with the flow of electromagnetic energy at a point in space, then we must define what a flow of electromagnetic energy is. We naturally associate the flow of kinetic energy with moving particles; with what do we associate the flow of electromagnetic energy? Maxwell felt that electromagnetic energy must flow through space as a result of the mechanical stresses and strains associated with an unobserved substance called the "aether." A more modern interpretation is that the electromagnetic fields propagate as a wave through space at finite velocity; when those fields encounter a charged particle a force is exerted, work is done, and energy is "transferred" from the field to the particle. Hence the energy flow is associated with the "flow" of the electromagnetic wave.

Unfortunately, it is uncertain whether $\mathbf{E} \times \mathbf{H}$ is the appropriate quantity to associate with this flow, since only its divergence appears in Poynting's theorem. We could add any other term $\mathbf{S}'$ that satisfies $\nabla \cdot \mathbf{S}' = 0$ to $\mathbf{S}_{em}$ in (2.232), and the conservation theorem would be unchanged. (Equivalently, we could add to (2.234) any term that integrates to zero over S.) There is no such ambiguity in the mechanical case because kinetic energy is rigorously defined. We are left, then, to postulate that $\mathbf{E} \times \mathbf{H}$ represents the density of energy flow associated with an electromagnetic wave (based on the symmetry with mechanics), and to look to experimental evidence as justification. In fact, experimental evidence does point to the correctness of this hypothesis, and the quantity $\mathbf{E} \times \mathbf{H}$ is widely and accurately used to compute the energy radiated by antennas, carried by waveguides, etc.

Confusion also arises regarding the interpretation of W_{em}. Since this term is so conveniently paired with the mechanical volume kinetic energy density in (2.235) it would seem that we should interpret it as an electromagnetic energy density. As such, we can think of this energy as "localized" in certain regions of space. This viewpoint has been criticized [183, 149, 59] since the large-scale form of energy conservation for a space region only requires that the total energy in the region be specified, and the integrand (energy density) giving this energy is not unique. It is also felt that energy should be associated with a "configuration" of objects (such as charged particles) and not with an arbitrary point in space. However, we retain the concept of localized energy because it is convenient and produces results consistent with experiment.

The validity of extending the static field interpretation of $\frac{1}{2}(\mathbf{D} \cdot \mathbf{E} + \mathbf{B} \cdot \mathbf{H})$ as the energy "stored" by a charge and a current arrangement to the time-varying case has also been questioned. If we do extend this view to the time-varying case, Poynting's theorem suggests that every point in space somehow has an energy density associated with it, and the flow of energy from that point (via $\mathbf{S}_{em}$) must be accompanied by a change in the stored energy at that point. This again gives a very useful and intuitively satisfying point of view. Since we can associate the flow of energy with the propagation of the electromagnetic fields, we can view the fields in any region of space as having the potential to do work on charged particles in that region. If there are charged particles in that region then work is done, accompanied by a transfer of energy to the particles and a reduction in the amplitudes of the fields.

We must also remember that the association of stored electromagnetic energy density W_{em} with the mechanical energy density W_k is only possible if the medium is nondispersive. If we cannot make the assumptions that justify (2.230) and (2.231), then Poynting's theorem must take the form

$$-\int_V \mathbf{J} \cdot \mathbf{E} \, dV = \int_V \left(\mathbf{E} \cdot \frac{\partial \mathbf{D}}{\partial t} + \mathbf{H} \cdot \frac{\partial \mathbf{B}}{\partial t} \right) dV + \oint_S (\mathbf{E} \times \mathbf{H}) \cdot d\mathbf{S}. \qquad (2.236)$$

For dispersive media, the volume term on the right-hand side describes not only the stored electromagnetic energy, but also the energy dissipated within the material produced by a time lag between the field applied to the medium and the resulting polarization or magnetization of the atoms. This is clearly seen in (2.20), which shows that $\mathbf{D}(t)$ depends on the value of $\mathbf{E}$ at time t and at all past times. The stored energy and dissipative terms are hard to separate, but we can see that there must always be a stored energy term by substituting $\mathbf{D} = \epsilon_0 \mathbf{E} + \mathbf{P}$ and $\mathbf{H} = \mathbf{B}/\mu_0 - \mathbf{M}$ into (2.236) to obtain

$$-\int_V [(\mathbf{J} + \mathbf{J}_P) \cdot \mathbf{E} + \mathbf{J}_H \cdot \mathbf{H}] \, dV$$

$$= \frac{1}{2} \frac{\partial}{\partial t} \int_V (\epsilon_0 \mathbf{E} \cdot \mathbf{E} + \mu_0 \mathbf{H} \cdot \mathbf{H}) \, dV + \oint_S (\mathbf{E} \times \mathbf{H}) \cdot d\mathbf{S}. \qquad (2.237)$$

Here J_P is the equivalent polarization current (2.101) and J_H is an analogous *magnetic polarization current* given by

$$\mathbf{J}_H = \mu_0 \frac{\partial \mathbf{M}}{\partial t}.$$

In this form we easily identify the quantity $\frac{1}{2}(\epsilon_0 \mathbf{E} \cdot \mathbf{E} + \mu_0 \mathbf{H} \cdot \mathbf{H})$ as the electromagnetic energy density for the fields $\mathbf{E}$ and $\mathbf{H}$ *in free space*. Any dissipation produced by polarization and magnetization lag is now handled by the interaction between the fields and equivalent current, just as $\mathbf{J} \cdot \mathbf{E}$ describes the interaction of the electric current (source and secondary) with the electric field. Unfortunately, the equivalent current interaction terms also include the additional stored energy that results from polarizing and magnetizing the material atoms, and again the effects are hard to separate.

Finally, let us consider the case of static fields. Setting the time derivative to zero in (2.234), we have

$$-\int_V \mathbf{J} \cdot \mathbf{E} \, dV = \oint_S (\mathbf{E} \times \mathbf{H}) \cdot d\mathbf{S}.$$

This shows that energy flux is required to maintain steady current flow. For instance, we need both an electromagnetic and a thermodynamic subsystem to account for energy conservation in the case of steady current flow through a resistor. The Poynting flux describes the electromagnetic energy entering the resistor and the thermodynamic flux describes the heat dissipation. For the sum of the two subsystems, conservation of energy requires

$$\nabla \cdot (\mathbf{S}_{em} + \mathbf{S}_{th}) = -\mathbf{J} \cdot \mathbf{E} + P_{th} = 0.$$

To compute the heat dissipation, we can use

$$P_{th} = \mathbf{J} \cdot \mathbf{E} = -\nabla \cdot \mathbf{S}_{em}$$

and thus either use the boundary fields or the fields and current internal to the resistor to find the dissipated heat.

2.9.5.5 Boundary conditions on the Poynting vector

The large-scale form of Poynting's theorem may be used to determine the behavior of the Poynting vector on either side of a boundary surface. We proceed exactly as in § 2.8.2. Consider a surface S across which the electromagnetic sources and constitutive parameters are discontinuous (Figure 2.6). As before, let $\hat{\mathbf{n}}_{12}$ be the unit normal directed into region 1. We now simplify the notation and write $\mathbf{S}$ instead of $\mathbf{S}_{em}$. If we apply Poynting's theorem

$$\int_V \left(\mathbf{J} \cdot \mathbf{E} + \mathbf{E} \cdot \frac{\partial \mathbf{D}}{\partial t} + \mathbf{H} \cdot \frac{\partial \mathbf{B}}{\partial t} \right) dV + \oint_S \mathbf{S} \cdot \mathbf{n} \, dS = 0$$

to the two separate surfaces shown in Figure 2.6, we obtain

$$\int_V \left(\mathbf{J} \cdot \mathbf{E} + \mathbf{E} \cdot \frac{\partial \mathbf{D}}{\partial t} + \mathbf{H} \cdot \frac{\partial \mathbf{B}}{\partial t} \right) dV + \int_S \mathbf{S} \cdot \mathbf{n} \, dS = \int_{S_{10}} \hat{\mathbf{n}}_{12} \cdot (\mathbf{S}_1 - \mathbf{S}_2) \, dS. \quad (2.238)$$

If on the other hand we apply Poynting's theorem to the entire volume region including the surface of discontinuity and include the contribution produced by surface current, we get

$$\int_V \left(\mathbf{J} \cdot \mathbf{E} + \mathbf{E} \cdot \frac{\partial \mathbf{D}}{\partial t} + \mathbf{H} \cdot \frac{\partial \mathbf{B}}{\partial t} \right) dV + \int_S \mathbf{S} \cdot \mathbf{n} \, dS = -\int_{S_{10}} \mathbf{J}_s \cdot \mathbf{E} \, dS. \quad (2.239)$$

Since we are uncertain whether to use $\mathbf{E}_1$ or $\mathbf{E}_2$ in the surface term on the right-hand side, if we wish to have the integrals over V and S in (2.238) and (2.239) produce identical results, we must postulate the two conditions

$$\hat{\mathbf{n}}_{12} \times (\mathbf{E}_1 - \mathbf{E}_2) = 0$$

and

$$\hat{\mathbf{n}}_{12} \cdot (\mathbf{S}_1 - \mathbf{S}_2) = -\mathbf{J}_s \cdot \mathbf{E}. \quad (2.240)$$

The first condition is merely the continuity of tangential electric field as originally postulated in § 2.8.2; it allows us to be nonspecific as to which value of $\mathbf{E}$ we use in the second condition, which is the desired boundary condition on $\mathbf{S}$.

It is interesting to note that (2.240) may also be derived directly from the two postulated boundary conditions on tangential $\mathbf{E}$ and $\mathbf{H}$. Here we write with the help of (B.6)

$$\hat{\mathbf{n}}_{12} \cdot (\mathbf{S}_1 - \mathbf{S}_2) = \hat{\mathbf{n}}_{12} \cdot (\mathbf{E}_1 \times \mathbf{H}_1 - \mathbf{E}_2 \times \mathbf{H}_2) = \mathbf{H}_1 \cdot (\hat{\mathbf{n}}_{12} \times \mathbf{E}_1) - \mathbf{H}_2 \cdot (\hat{\mathbf{n}}_{12} \times \mathbf{E}_2).$$

Since $\hat{\mathbf{n}}_{12} \times \mathbf{E}_1 = \hat{\mathbf{n}}_{12} \times \mathbf{E}_2 = \hat{\mathbf{n}}_{12} \times \mathbf{E}$, we have

$$\hat{\mathbf{n}}_{12} \cdot (\mathbf{S}_1 - \mathbf{S}_2) = (\mathbf{H}_1 - \mathbf{H}_2) \cdot (\hat{\mathbf{n}}_{12} \times \mathbf{E}) = [-\hat{\mathbf{n}}_{12} \times (\mathbf{H}_1 - \mathbf{H}_2)] \cdot \mathbf{E}.$$

Finally, using $\hat{\mathbf{n}}_{12} \times (\mathbf{H}_1 - \mathbf{H}_2) = \mathbf{J}_s$ we arrive at (2.240).

The arguments above suggest an interesting way to look at the boundary conditions. Once we identify $\mathbf{S}$ with the flow of electromagnetic energy, we may consider the condition on normal $\mathbf{S}$ as a fundamental statement of the conservation of energy. This statement implies continuity of tangential $\mathbf{E}$ in order to have an unambiguous interpretation for the meaning of the term $\mathbf{J}_s \cdot \mathbf{E}$. Then, with continuity of tangential $\mathbf{E}$ established, we can derive the condition on tangential $\mathbf{H}$ directly.

2.9.5.6   An alternative formulation of the conservation theorems

As we saw in the sections above, our derivation of the conservation theorems lacks strong motivation. We manipulated Maxwell's equations until we found expressions that resembled those for mechanical momentum and energy, but in the process found that the validity of the expressions is somewhat limiting. For instance, we needed to assume a linear, homogeneous, bianisotropic medium in order to identify the Maxwell stress tensor (2.225) and the energy densities in Poynting's theorem (2.234). In the end, we were reduced to postulating the meaning of the individual terms in the conservation theorems in order for the whole to have meaning.

An alternative approach is popular in physics. It involves postulating a single Lagrangian density function for the electromagnetic field, and then applying the stationary property of the action integral. The results are precisely the same conservation expressions for linear momentum and energy as obtained from manipulating Maxwell's equations (plus the equation for conservation of angular momentum), obtained with fewer restrictions regarding the constitutive relations. This process also separates the stored energy, Maxwell stress tensor, momentum density, and Poynting vector as natural components of a tensor equation, allowing a better motivated interpretation of the meaning of these components. Since this approach is also a powerful tool in mechanics, its application is more strongly motivated than merely manipulating Maxwell's equations. Of course, some knowledge of the structure of the electromagnetic field is required to provide an appropriate postulate of the Lagrangian density. Interested readers should consult Kong [108], Jackson [92], Doughty [50], or Tolstoy [196].

2.10   The wave nature of the electromagnetic field

Throughout this chapter our goal has been a fundamental understanding of Maxwell's theory of electromagnetics. We have concentrated on developing and understanding the equations relating the field quantities, but have done little to understand the nature of the field itself. We would now like to investigate, in a very general way, the behavior of the field. We shall not attempt to solve a vast array of esoteric problems, but shall instead concentrate on a few illuminating examples.

The electromagnetic field can take on a wide variety of characteristics. Static fields differ qualitatively from those that undergo rapid time variations. Time-varying fields exhibit wave behavior and carry energy away from their sources. In the case of slow time variation, this wave nature may often be neglected in favor of the nearby coupling of sources we know as the inductance effect, hence circuit theory may suffice to describe the field-source interaction. In the case of extremely rapid oscillations, particle concepts may be needed to describe the field.

The dynamic coupling between the various field vectors in Maxwell's equations provides a means of characterizing the field. Static fields are characterized by decoupling of the electric and magnetic fields. Quasistatic fields exhibit some coupling, but the wave characteristic of the field is ignored. Tightly coupled fields are dominated by the wave effect, but may still show a static-like spatial distribution near the source. Any such "near-zone" effects are generally ignored for fields at light-wave frequencies, and the particle nature of light must often be considered.

2.10.1 Electromagnetic waves

An early result of Maxwell's theory was the prediction and later verification by Heinrich Hertz of the existence of electromagnetic waves. We now know that nearly any time-varying source produces waves, and that these waves have certain important properties. An electromagnetic wave is a propagating electromagnetic field that travels with finite velocity as a disturbance through a medium. The field itself is the disturbance, rather than merely representing a physical displacement or other effect on the medium. This fact is fundamental for understanding how electromagnetic waves can travel through a true vacuum. Many specific characteristics of the wave, such as velocity and polarization, depend on the properties of the medium through which it propagates. The evolution of the disturbance also depends on these properties: we say that a material exhibits "dispersion" if the disturbance undergoes a change in its temporal behavior as the wave progresses. As waves travel they carry energy and momentum away from their source. This energy may be later returned to the source or delivered to some distant location. Waves are also capable of transferring energy to, or withdrawing energy from, the medium through which they propagate. When energy is carried outward from the source never to return, we refer to the process as "electromagnetic radiation." The effects of radiated fields can be far-reaching; indeed, radio astronomers observe waves that originated at the very edges of the universe.

Light is an electromagnetic phenomenon, and many of the familiar characteristics of light that we recognize from our everyday experience may be applied to all electromagnetic waves. For instance, radio waves bend (or "refract") in the ionosphere much as light waves bend while passing through a prism. Microwaves reflect from conducting surfaces in the same way that light waves reflect from a mirror; detecting these reflections forms the basis of radar. Electromagnetic waves may also be "confined" by reflecting boundaries to form waves standing in one or more directions. With this concept we can use waveguides or transmission lines to guide electromagnetic energy from spot to spot, or to concentrate it in the cavity of a microwave oven.

The manifestations of electromagnetic waves are so diverse that no one book can possibly describe the entire range of phenomena or applications. In this section we shall merely introduce the reader to some of the most fundamental concepts of electromagnetic wave behavior. In the process we shall also introduce the three most often studied types of traveling electromagnetic waves: plane waves, spherical waves, and cylindrical waves. In later sections we shall study some of the complicated interactions of these waves with objects and boundaries, in the form of guided waves and scattering problems.

Mathematically, electromagnetic waves arise as a subset of solutions to Maxwell's equations. These solutions obey the electromagnetic "wave equation," which may be derived from Maxwell's equations under certain circumstances. Not all electromagnetic fields satisfy the wave equation. Obviously, time-invariant fields cannot represent evolving wave disturbances, and must obey the static field equations. Time-varying fields in certain metals may obey the diffusion equation rather than the wave equation, and must thereby exhibit different behavior. In the study of quasistatic fields we often ignore the displacement current term in Maxwell's equations, producing solutions that are most important near the sources of the fields and having little associated radiation. When the displacement term is significant we produce solutions with the properties of waves.

In deriving electromagnetic wave equations we transform the first-order coupled partial differential equations we know as Maxwell's equations into uncoupled second-order equations. That is, we perform a set of operations (and make appropriate assumptions) to reduce the set of four differential equations in the four unknown fields, **E**, **D**, **B**, and

H, into a set of differential equations, each involving a single unknown (usually **E** or **H**).

▶ **Example 2.7:** Wave equation for fields in an inhomogeneous dielectric medium

A source-free region is filled with an inhomogeneous dielectric material having permeability $\mu(\mathbf{r}) = \mu_0$ and permittivity $\epsilon(\mathbf{r}) = \epsilon(z)$. Assume the electric and magnetic fields take the form

$$\mathbf{E}(\mathbf{r}, t) = \hat{\mathbf{y}} E_y(z, t), \qquad \mathbf{H}(\mathbf{r}, t) = \hat{\mathbf{x}} H_x(z, t).$$

Derive the wave equations for E_y and H_x.

Solution: By Example 2.1 we have from Ampere's law and Faraday's law

$$\frac{\partial H_x}{\partial z} = \epsilon(z) \frac{\partial E_y}{\partial t}, \tag{2.241}$$

$$\frac{\partial E_y}{\partial z} = \mu_0 \frac{\partial H_x}{\partial t}. \tag{2.242}$$

We differentiate (2.241) with respect to z,

$$\frac{\partial^2 H_x}{\partial z^2} = \epsilon(z) \frac{\partial^2 E_y}{\partial z \partial t} + \frac{\partial \epsilon(z)}{\partial z} \frac{\partial E_y}{\partial t}, \tag{2.243}$$

and (2.242) with respect to t,

$$\frac{\partial^2 E_y}{\partial z \partial t} = \mu_0 \frac{\partial^2 H_x}{\partial t^2}. \tag{2.244}$$

Substituting (2.244) and (2.241) into (2.243), we obtain the wave equation for H_x:

$$\frac{\partial^2 H_x}{\partial z^2} - \frac{1}{\epsilon(z)} \frac{\partial \epsilon(z)}{\partial z} \frac{\partial H_x}{\partial z} - \epsilon(z) \mu_0 \frac{\partial^2 H_x}{\partial t^2} = 0. \tag{2.245}$$

To find the wave equation for E_y we differentiate (2.242) with respect to z,

$$\frac{\partial^2 E_y}{\partial z^2} = \mu_0 \frac{\partial^2 H_x}{\partial z \partial t}, \tag{2.246}$$

and (2.241) with respect to t,

$$\frac{\partial^2 H_x}{\partial z \partial t} = \epsilon(z) \frac{\partial^2 E_y}{\partial t^2}. \tag{2.247}$$

Substitution of (2.247) into (2.246) yields

$$\frac{\partial^2 E_y}{\partial z^2} - \epsilon(z) \mu_0 \frac{\partial^2 E_y}{\partial t^2} = 0. \blacktriangleleft \tag{2.248}$$

▶ **Example 2.8:** Solution to the wave equation for fields in an inhomogeneous dielectric

A source-free region of space is filled with an inhomogeneous dielectric material having permeability $\mu(\mathbf{r}) = \mu_0$ and permittivity $\epsilon(\mathbf{r}) = \epsilon_0 \epsilon_r e^{Kz}$. Show by substitution that the fields given in Example 2.1 satisfy the wave equations (2.245) and (2.248).

Solution: From Example 2.1 we have the magnetic field

$$H_x(z, t) = H_0 e^{\frac{K}{2} z} J_1 \left(\frac{2k}{K} e^{\frac{K}{2} z} \right) \cos \omega t.$$

The derivative of H_x with respect to z is given by (2.23):

$$\frac{\partial H_x}{\partial z} = H_0 k e^{Kz} J_0\left(\frac{2k}{K} e^{\frac{K}{2}z}\right)\cos\omega t.$$

The second derivative is

$$\frac{\partial^2 H_x}{\partial z^2} = \left[H_0 kK e^{Kz} J_0\left(\frac{2k}{K} e^{\frac{K}{2}z}\right) - H_0 k^2 e^{Kz} e^{\frac{K}{2}z} J_1\left(\frac{2k}{K} e^{\frac{K}{2}z}\right)\right]\cos\omega t.$$

We also have the second derivative of H_x with respect to t,

$$\frac{\partial^2 H_x}{\partial t^2} = -H_0 \omega^2 e^{\frac{K}{2}z} J_1\left(\frac{2k}{K} e^{\frac{K}{2}z}\right)\cos\omega t,$$

and the term

$$\frac{1}{\epsilon(z)}\frac{\partial \epsilon(z)}{\partial z} = K.$$

Substituting these into (2.245) we get

$$\frac{\partial^2 H_x}{\partial z^2} - \frac{1}{\epsilon(z)}\frac{\partial \epsilon(z)}{\partial z}\frac{\partial H_x}{\partial z} - \epsilon(z)\mu_0\frac{\partial^2 H_x}{\partial t^2}$$

$$= \left[H_0 kK e^{Kz} J_0\left(\frac{2k}{K} e^{\frac{K}{2}z}\right) - H_0 k^2 e^{Kz} e^{\frac{K}{2}z} J_1\left(\frac{2k}{K} e^{\frac{K}{2}z}\right)\right.$$

$$\left. - H_0 kK e^{Kz} J_0\left(\frac{2k}{K} e^{\frac{K}{2}z}\right) + \mu_0\epsilon_0\epsilon_r e^{Kz} H_0 \omega^2 e^{\frac{K}{2}z} J_1\left(\frac{2k}{K} e^{\frac{K}{2}z}\right)\right]\cos\omega t.$$

Since $k^2 = \omega^2\mu_0\epsilon_0\epsilon_r$, the terms in the brackets cancel and the right-hand side is zero. The wave equation for H_x is satisfied.

To verify the equation for E_y we begin with the electric field from Example 2.1,

$$E_y = H_0\eta J_0\left(\frac{2k}{K} e^{\frac{K}{2}z}\right)\sin\omega t,$$

which has the first derivative

$$\frac{\partial E_y}{\partial z} = -H_0\eta k e^{\frac{K}{2}z} J_1\left(\frac{2k}{K} e^{\frac{K}{2}z}\right)\sin\omega t.$$

The second derivative is

$$\frac{\partial^2 E_y}{\partial z^2} = -H_0\eta k^2 e^{Kz} J_0\left(\frac{2k}{K} e^{\frac{K}{2}z}\right)\sin\omega t$$

since $J_1'(x) = J_0(x) - J_1(x)/x$. Next, two time derivatives of E_y yield

$$\frac{\partial^2 E_y}{\partial t^2} = -H_0\omega^2\eta J_0\left(\frac{2k}{K} e^{\frac{K}{2}z}\right)\sin\omega t.$$

Substitution into (2.248) gives

$$\frac{\partial^2 E_y}{\partial z^2} - \epsilon(z)\mu_0\frac{\partial^2 E_y}{\partial t^2} = -H_0\eta e^{Kz} J_0\left(\frac{2k}{K} e^{\frac{K}{2}z}\right)\left[k^2 - \omega^2\mu_0\epsilon_0\epsilon_r\right]\sin\omega t.$$

The term in brackets is zero and the wave equation is satisfied. ◄

2.10.2   Wave equation for bianisotropic materials

It is possible to derive wave equations for $\mathbf{E}$ and $\mathbf{H}$ even for the most general cases of inhomogeneous, bianisotropic media, as long as the constitutive parameters $\bar{\mu}$ and $\bar{\xi}$ are constant with time. Substituting the constitutive relations (2.12)–(2.13) into the Maxwell–Minkowski curl equations (2.141)–(2.142) we get

$$\nabla \times \mathbf{E} = -\frac{\partial}{\partial t}(\bar{\zeta} \cdot \mathbf{E} + \bar{\mu} \cdot \mathbf{H}) - \mathbf{J}_m, \tag{2.249}$$

$$\nabla \times \mathbf{H} = \frac{\partial}{\partial t}(\bar{\epsilon} \cdot \mathbf{E} + \bar{\xi} \cdot \mathbf{H}) + \mathbf{J}. \tag{2.250}$$

Separate equations for $\mathbf{E}$ and $\mathbf{H}$ are facilitated by introducing a new dyadic operator $\bar{\nabla}$, which when dotted with a vector field $\mathbf{V}$ gives the curl:

$$\bar{\nabla} \cdot \mathbf{V} = \nabla \times \mathbf{V}. \tag{2.251}$$

It is easy to verify that in rectangular coordinates $\bar{\nabla}$ is

$$[\bar{\nabla}] = \begin{bmatrix} 0 & -\partial/\partial z & \partial/\partial y \\ \partial/\partial z & 0 & -\partial/\partial x \\ -\partial/\partial y & \partial/\partial x & 0 \end{bmatrix}.$$

With this notation, Maxwell's curl equations (2.249)–(2.250) become simply

$$\left(\bar{\nabla} + \frac{\partial}{\partial t}\bar{\zeta}\right) \cdot \mathbf{E} = -\frac{\partial}{\partial t}\bar{\mu} \cdot \mathbf{H} - \mathbf{J}_m, \tag{2.252}$$

$$\left(\bar{\nabla} - \frac{\partial}{\partial t}\bar{\xi}\right) \cdot \mathbf{H} = \frac{\partial}{\partial t}\bar{\epsilon} \cdot \mathbf{E} + \mathbf{J}. \tag{2.253}$$

Obtaining separate equations for $\mathbf{E}$ and $\mathbf{H}$ is straightforward. Defining the inverse dyadic $\bar{\mu}^{-1}$ through

$$\bar{\mu} \cdot \bar{\mu}^{-1} = \bar{\mu}^{-1} \cdot \bar{\mu} = \bar{\mathbf{I}},$$

we can write (2.252) as

$$\frac{\partial}{\partial t}\mathbf{H} = -\bar{\mu}^{-1} \cdot \left(\bar{\nabla} + \frac{\partial}{\partial t}\bar{\zeta}\right) \cdot \mathbf{E} - \bar{\mu}^{-1} \cdot \mathbf{J}_m \tag{2.254}$$

where we have assumed that $\bar{\mu}$ is independent of time. Assuming that $\bar{\xi}$ is also independent of time, we can differentiate (2.253) with respect to time to obtain

$$\left(\bar{\nabla} - \frac{\partial}{\partial t}\bar{\xi}\right) \cdot \frac{\partial \mathbf{H}}{\partial t} = \frac{\partial^2}{\partial t^2}(\bar{\epsilon} \cdot \mathbf{E}) + \frac{\partial \mathbf{J}}{\partial t}.$$

Substituting $\partial \mathbf{H}/\partial t$ from (2.254) and rearranging, we get

$$\left[\left(\bar{\nabla} - \frac{\partial}{\partial t}\bar{\xi}\right) \cdot \bar{\mu}^{-1} \cdot \left(\bar{\nabla} + \frac{\partial}{\partial t}\bar{\zeta}\right) + \frac{\partial^2}{\partial t^2}\bar{\epsilon}\right] \cdot \mathbf{E} = -\left(\bar{\nabla} - \frac{\partial}{\partial t}\bar{\xi}\right) \cdot \bar{\mu}^{-1} \cdot \mathbf{J}_m - \frac{\partial \mathbf{J}}{\partial t}. \tag{2.255}$$

This is the general wave equation for $\mathbf{E}$. Using an analogous set of steps, and assuming $\bar{\epsilon}$ and $\bar{\zeta}$ are independent of time, we can find

$$\left[\left(\bar{\nabla} + \frac{\partial}{\partial t}\bar{\zeta}\right) \cdot \bar{\epsilon}^{-1} \cdot \left(\bar{\nabla} - \frac{\partial}{\partial t}\bar{\xi}\right) + \frac{\partial^2}{\partial t^2}\bar{\mu}\right] \cdot \mathbf{H} = \left(\bar{\nabla} + \frac{\partial}{\partial t}\bar{\zeta}\right) \cdot \bar{\epsilon}^{-1} \cdot \mathbf{J} - \frac{\partial \mathbf{J}_m}{\partial t}. \tag{2.256}$$

This is the wave equation for **H**. The case in which the constitutive parameters are time-dependent will be handled using frequency domain techniques in later chapters.

▶ **Example 2.9:** Wave equation for fields in a Tellegen medium

A source-free region of space contains a homogeneous Tellegen medium with constitutive parameters given by (2.26)–(2.27) or equivalently (2.28)–(2.29). Derive a wave equation for $\mathbf{E}(\mathbf{r}, t)$. Show that the field

$$\mathbf{E}(\mathbf{r}, t) = \hat{\mathbf{x}} E_0 f\left(t - \frac{z}{v}\right)$$

of Example 2.2 satisfies this wave equation. Here $\xi^2 < \mu\epsilon$, and

$$v = \frac{1}{\sqrt{\mu\epsilon - \xi^2}}.$$

Solution: Substituting (2.28) and (2.29) into Ampere's law, we have

$$\nabla \times \left[-\frac{\xi}{\mu}\mathbf{E} + \frac{1}{\mu}\mathbf{B}\right] = \frac{\partial}{\partial t}\left[\left(\epsilon - \frac{\xi^2}{\mu}\right)\mathbf{E} + \frac{\xi}{\mu}\mathbf{B}\right].$$

By Faraday's law we transform this to

$$\frac{\xi}{\mu}\frac{\partial \mathbf{B}}{\partial t} + \frac{1}{\mu}\nabla \times \mathbf{B} = \frac{\partial}{\partial t}\left[\left(\epsilon - \frac{\xi^2}{\mu}\right)\mathbf{E} + \frac{\xi}{\mu}\mathbf{B}\right]$$

and simplify:

$$\nabla \times \mathbf{B} = (\mu\epsilon - \xi^2)\frac{\partial \mathbf{E}}{\partial t} = \frac{1}{v^2}\frac{\partial \mathbf{E}}{\partial t}. \tag{2.257}$$

Taking the curl of Faraday's law and substituting from (2.257) we obtain

$$\nabla \times \nabla \times \mathbf{E} + \frac{1}{v^2}\frac{\partial^2 \mathbf{E}}{\partial t^2} = 0. \tag{2.258}$$

Expansion of $\nabla \times \nabla \times \mathbf{E}$ then gives

$$\nabla(\nabla \cdot \mathbf{E}) - \nabla^2\mathbf{E} + \frac{1}{v^2}\frac{\partial^2 \mathbf{E}}{\partial t^2} = 0.$$

Finally, note that by (2.28) we have

$$\nabla \cdot \mathbf{E} = \left(\epsilon - \frac{\xi^2}{\mu}\right)^{-1}\nabla \cdot \left[\mathbf{D} - \frac{\xi}{\mu}\mathbf{B}\right] = 0$$

since both $\nabla \cdot \mathbf{B} = 0$ and $\nabla \cdot \mathbf{D} = 0$. Hence we have the wave equation

$$\nabla^2\mathbf{E} - \frac{1}{v^2}\frac{\partial^2 \mathbf{E}}{\partial t^2} = 0.$$

To show that

$$\mathbf{E}(\mathbf{r}, t) = \hat{\mathbf{x}} E_0 f\left(t - \frac{z}{v}\right)$$

satisfies the wave equation, we note that

$$\nabla^2\mathbf{E} = \hat{\mathbf{x}}\frac{\partial^2 E_x}{\partial z^2}$$

$$= \hat{\mathbf{x}}\frac{1}{v^2}E_0 f''\left(t - \frac{z}{v}\right)$$

and

$$\frac{1}{v^2}\frac{\partial^2 \mathbf{E}}{\partial t^2} = \hat{\mathbf{x}}\frac{1}{v^2}E_0 f''\left(t - \frac{z}{v}\right).$$

These terms are identical and the wave equation is satisfied. ◄

Note that we can also obtain the wave equation for a Tellegen medium by specializing (2.255), as in the next example.

▶ **Example 2.10:** Specialization of general wave equation for a Tellegen medium

A source-free region contains a homogeneous Tellegen medium with constitutive parameters given by (2.26)–(2.27) or equivalently (2.28)–(2.29). Derive the wave equation (2.258) for $\mathbf{E}(\mathbf{r}, t)$ by specializing (2.255).

Solution: Multiplying out the terms in (2.255), we have

$$\bar{\nabla}\cdot\left[\bar{\boldsymbol{\mu}}^{-1}\cdot\bar{\nabla}\cdot\mathbf{E}\right] - \frac{\partial}{\partial t}\left[\bar{\boldsymbol{\xi}}\cdot\bar{\boldsymbol{\mu}}^{-1}\cdot\frac{\partial}{\partial t}\left(\bar{\boldsymbol{\zeta}}\cdot\mathbf{E}\right)\right] - \frac{\partial}{\partial t}\left[\bar{\boldsymbol{\xi}}\cdot\bar{\boldsymbol{\mu}}^{-1}\cdot\bar{\nabla}\cdot\mathbf{E}\right]$$

$$+ \bar{\nabla}\cdot\left[\bar{\boldsymbol{\mu}}^{-1}\cdot\frac{\partial}{\partial t}\left(\bar{\boldsymbol{\zeta}}\cdot\mathbf{E}\right)\right] + \frac{\partial^2}{\partial t^2}\left[\bar{\boldsymbol{\epsilon}}\cdot\mathbf{E}\right] = 0.$$

Now, use

$$\bar{\boldsymbol{\mu}}^{-1} = \mu^{-1}\bar{\mathbf{I}}, \qquad \bar{\boldsymbol{\xi}} = \bar{\boldsymbol{\zeta}} = \xi\bar{\mathbf{I}}, \qquad \bar{\boldsymbol{\epsilon}} = \epsilon\bar{\mathbf{I}}$$

and recall that ϵ, μ, and ξ are constant with both space and time. This gives

$$\frac{1}{\mu}\bar{\nabla}\cdot\bar{\nabla}\cdot\mathbf{E} - \frac{\xi^2}{\mu}\frac{\partial^2\mathbf{E}}{\partial t^2} - \frac{\xi}{\mu}\frac{\partial}{\partial t}\bar{\nabla}\cdot\mathbf{E} + \frac{\xi}{\mu}\frac{\partial}{\partial t}\bar{\nabla}\cdot\mathbf{E} + \epsilon\frac{\partial^2\mathbf{E}}{\partial t^2} = 0.$$

Simplification gives

$$\bar{\nabla}\cdot\bar{\nabla}\cdot\mathbf{E} + (\mu\epsilon - \xi^2)\frac{\partial^2\mathbf{E}}{\partial t^2} = 0.$$

Finally, using $\bar{\nabla}\cdot\bar{\nabla}\cdot\mathbf{E} = \nabla\times\nabla\times\mathbf{E}$ and remembering that $v = 1/\sqrt{\mu\epsilon - \xi^2}$, we have

$$\nabla\times\nabla\times\mathbf{E} + \frac{1}{v^2}\frac{\partial^2\mathbf{E}}{\partial t^2} = 0,$$

which is identical to (2.258). It is clear that (2.256) will produce the identical wave equation for $\mathbf{H}$. ◄

Wave equations for anisotropic, isotropic, and homogeneous media are easily obtained from (2.255) and (2.256) as special cases. For example, the wave equations for a homogeneous, isotropic medium can be found by setting $\bar{\boldsymbol{\zeta}} = \bar{\boldsymbol{\xi}} = 0$, $\bar{\boldsymbol{\mu}} = \mu\bar{\mathbf{I}}$, and $\bar{\boldsymbol{\epsilon}} = \epsilon\bar{\mathbf{I}}$:

$$\frac{1}{\mu}\bar{\nabla}\cdot(\bar{\nabla}\cdot\mathbf{E}) + \epsilon\frac{\partial^2\mathbf{E}}{\partial t^2} = -\frac{1}{\mu}\bar{\nabla}\cdot\mathbf{J}_m - \frac{\partial\mathbf{J}}{\partial t},$$

$$\frac{1}{\epsilon}\bar{\nabla}\cdot(\bar{\nabla}\cdot\mathbf{H}) + \mu\frac{\partial^2\mathbf{H}}{\partial t^2} = \frac{1}{\epsilon}\bar{\nabla}\cdot\mathbf{J} - \frac{\partial\mathbf{J}_m}{\partial t}.$$

Returning to standard curl notation, we find that these become

$$\nabla\times(\nabla\times\mathbf{E}) + \mu\epsilon\frac{\partial^2\mathbf{E}}{\partial t^2} = -\nabla\times\mathbf{J}_m - \mu\frac{\partial\mathbf{J}}{\partial t}, \tag{2.259}$$

$$\nabla\times(\nabla\times\mathbf{H}) + \mu\epsilon\frac{\partial^2\mathbf{H}}{\partial t^2} = \nabla\times\mathbf{J} - \epsilon\frac{\partial\mathbf{J}_m}{\partial t}. \tag{2.260}$$

In each of the wave equations it appears that operations on the electromagnetic fields have been separated from operations on the source terms. However, we have not yet invoked any coupling between the fields and sources associated with secondary interactions. That is, we need to separate the impressed sources, which are independent of the fields they source, with secondary sources resulting from interactions between the sourced fields and the medium in which the fields exist. The simple case of an isotropic conducting medium will be discussed below.

2.10.3 Wave equation using equivalent sources

An alternative approach for studying wave behavior in general media is to use the Maxwell–Boffi form of the field equations

$$\nabla \times \mathbf{E} = -\frac{\partial \mathbf{B}}{\partial t}, \tag{2.261}$$

$$\nabla \times \frac{\mathbf{B}}{\mu_0} = (\mathbf{J} + \mathbf{J}_M + \mathbf{J}_P) + \frac{\partial \epsilon_0 \mathbf{E}}{\partial t}, \tag{2.262}$$

$$\nabla \cdot (\epsilon_0 \mathbf{E}) = (\rho + \rho_P), \tag{2.263}$$

$$\nabla \cdot \mathbf{B} = 0. \tag{2.264}$$

Taking the curl of (2.261) we have

$$\nabla \times (\nabla \times \mathbf{E}) = -\frac{\partial}{\partial t} \nabla \times \mathbf{B}.$$

Substituting for $\nabla \times \mathbf{B}$ from (2.262) we then obtain

$$\nabla \times (\nabla \times \mathbf{E}) + \mu_0 \epsilon_0 \frac{\partial^2 \mathbf{E}}{\partial t^2} = -\mu_0 \frac{\partial}{\partial t} (\mathbf{J} + \mathbf{J}_M + \mathbf{J}_P),$$

which is the wave equation for $\mathbf{E}$. Taking the curl of (2.262) and substituting from (2.261) we obtain the wave equation

$$\nabla \times (\nabla \times \mathbf{B}) + \mu_0 \epsilon_0 \frac{\partial^2 \mathbf{B}}{\partial t^2} = \mu_0 \nabla \times (\mathbf{J} + \mathbf{J}_M + \mathbf{J}_P)$$

for $\mathbf{B}$. Solution of the wave equations is often facilitated by writing the curl–curl operation in terms of the vector Laplacian. Using (B.53), and substituting for the divergence from (2.263) and (2.264), we can write the wave equations as

$$\nabla^2 \mathbf{E} - \mu_0 \epsilon_0 \frac{\partial^2 \mathbf{E}}{\partial t^2} = \frac{1}{\epsilon_0} \nabla (\rho + \rho_P) + \mu_0 \frac{\partial}{\partial t} (\mathbf{J} + \mathbf{J}_M + \mathbf{J}_P), \tag{2.265}$$

$$\nabla^2 \mathbf{B} - \mu_0 \epsilon_0 \frac{\partial^2 \mathbf{B}}{\partial t^2} = -\mu_0 \nabla \times (\mathbf{J} + \mathbf{J}_M + \mathbf{J}_P). \tag{2.266}$$

The simplicity of these equations relative to (2.255) and (2.256) is misleading. We have not considered the constitutive equations relating the polarization $\mathbf{P}$ and magnetization $\mathbf{M}$ to the fields, nor have we considered interactions leading to secondary sources.

2.10.4 Wave equation in a conducting medium

As an example of the type of wave equation that arises when secondary sources are included, consider a homogeneous isotropic conducting medium described by permittivity

ϵ, permeability μ, and conductivity σ. In a conducting medium we must separate the source field into a causative impressed term $\mathbf{J}^i$ that is independent of the fields it sources, and a secondary term $\mathbf{J}^s$ that is an effect of the sourced fields. In an isotropic conducting medium the effect is described by Ohm's law: $\mathbf{J}^s = \sigma\mathbf{E}$. Writing the total current as $\mathbf{J} = \mathbf{J}^i + \mathbf{J}^s$, and assuming that $\mathbf{J}_m = 0$, we write the wave equation (2.259) as

$$\nabla \times (\nabla \times \mathbf{E}) + \mu\epsilon\frac{\partial^2\mathbf{E}}{\partial t^2} = -\mu\frac{\partial(\mathbf{J}^i + \sigma\mathbf{E})}{\partial t}.$$

Using (B.53) and substituting $\nabla \cdot \mathbf{E} = \rho/\epsilon$, we can write the wave equation for $\mathbf{E}$ as

$$\nabla^2\mathbf{E} - \mu\sigma\frac{\partial\mathbf{E}}{\partial t} - \mu\epsilon\frac{\partial^2\mathbf{E}}{\partial t^2} = \mu\frac{\partial\mathbf{J}^i}{\partial t} + \frac{1}{\epsilon}\nabla\rho. \tag{2.267}$$

Substituting $\mathbf{J} = \mathbf{J}^i + \sigma\mathbf{E}$ into (2.260) and using (B.53), we obtain

$$\nabla(\nabla \cdot \mathbf{H}) - \nabla^2\mathbf{H} + \mu\epsilon\frac{\partial^2\mathbf{H}}{\partial t^2} = \nabla \times \mathbf{J}^i + \sigma\nabla \times \mathbf{E}.$$

Since $\nabla \times \mathbf{E} = -\partial\mathbf{B}/\partial t$ and $\nabla \cdot \mathbf{H} = \nabla \cdot \mathbf{B}/\mu = 0$, we have

$$\nabla^2\dot{\mathbf{H}} - \mu\sigma\frac{\partial\mathbf{H}}{\partial t} - \mu\epsilon\frac{\partial^2\mathbf{H}}{\partial t^2} = -\nabla \times \mathbf{J}^i. \tag{2.268}$$

This is the wave equation for $\mathbf{H}$.

2.10.4.1 Scalar wave equation for a conducting medium

In many applications, particularly those involving planar boundary surfaces, it is convenient to decompose the vector wave equation into cartesian components. Using $\nabla^2\mathbf{V} = \hat{\mathbf{x}}\nabla^2 V_x + \hat{\mathbf{y}}\nabla^2 V_y + \hat{\mathbf{z}}\nabla^2 V_z$ in (2.267) and in (2.268), we find that the rectangular components of $\mathbf{E}$ and $\mathbf{H}$ must obey the scalar wave equation

$$\nabla^2\psi(\mathbf{r}, t) - \mu\sigma\frac{\partial\psi(\mathbf{r}, t)}{\partial t} - \mu\epsilon\frac{\partial^2\psi(\mathbf{r}, t)}{\partial t^2} = s(\mathbf{r}, t). \tag{2.269}$$

For the electric field wave equation, we have

$$\psi = E_\alpha, \qquad s = \mu\frac{\partial J_\alpha^i}{\partial t} + \frac{1}{\epsilon}\hat{\boldsymbol{\alpha}} \cdot \nabla\rho,$$

where $\alpha = x, y, z$. For the magnetic field wave equation, we have

$$\psi = H_\alpha, \qquad s = \hat{\boldsymbol{\alpha}} \cdot (-\nabla \times \mathbf{J}^i).$$

2.10.5 Fields determined by Maxwell's equations vs. fields determined by the wave equation

Although we derive the wave equations directly from Maxwell's equations, we may wonder whether the solutions to second-order differential equations such as (2.259)–(2.260) are necessarily the same as the solutions to the first-order Maxwell equations. Hansen and Yaghjian [79] show that if all information about the fields is supplied by the sources $\mathbf{J}(\mathbf{r}, t)$ and $\rho(\mathbf{r}, t)$, rather than by specification of field values on boundaries, the solutions to Maxwell's equations and the wave equations are equivalent as long as the second derivatives of the quantities

$$\nabla \cdot \mathbf{E}(\mathbf{r}, t) - \rho(\mathbf{r}, t)/\epsilon, \qquad \nabla \cdot \mathbf{H}(\mathbf{r}, t),$$

are continuous functions of $\mathbf{r}$ and t. If boundary values are supplied in an attempt to guarantee uniqueness, then solutions to the wave equation and to Maxwell's equations may differ. This is particularly important when comparing numerical solutions obtained directly from Maxwell's equations (using the FDTD method, say) to solutions obtained from the wave equation. "Spurious" solutions having no physical significance are a continual plague for engineers who employ numerical techniques. The interested reader should see Jiang et al. [96].

We note that these conclusions do not hold for static fields. The conditions for equivalence of the first-order and second-order static field equations are considered in § 3.3.5.

2.10.6 Transient uniform plane waves in a conducting medium

We can learn a great deal about the wave nature of the electromagnetic field by solving the wave equation (2.267) under simple circumstances. In Chapter 5 we shall solve for the field produced by an arbitrary distribution of impressed sources, but here we seek a simple solution to the homogeneous form of the equation. This allows us to study the phenomenology of wave propagation without worrying about the consequences of specific source functions. We shall also assume a high degree of symmetry so that we are not bogged down in details about the vector directions of the field components.

We seek a solution of the wave equation in which the fields are invariant over a chosen planar surface. The resulting fields are said to compose a *uniform plane wave*. Although we can envision a uniform plane wave as being created by a uniform surface source of doubly infinite extent, plane waves are also useful as models for spherical waves over localized regions of the wavefront.

We choose the plane of field invariance to be the xy-plane and later generalize the resulting solution to any planar surface by a simple rotation of the coordinate axes. Since the fields vary with z only we choose to write the wave equation (2.267) in rectangular coordinates, giving for a source-free region of space[†]

$$\hat{\mathbf{x}}\frac{\partial^2 E_x(z,t)}{\partial z^2} + \hat{\mathbf{y}}\frac{\partial^2 E_y(z,t)}{\partial z^2} + \hat{\mathbf{z}}\frac{\partial^2 E_z(z,t)}{\partial z^2} - \mu\sigma\frac{\partial \mathbf{E}(z,t)}{\partial t} - \mu\epsilon\frac{\partial^2 \mathbf{E}(z,t)}{\partial t^2} = 0. \quad (2.270)$$

If we return to Maxwell's equations, we soon find that not all components of $\mathbf{E}$ are present in the plane-wave solution. Faraday's law states that

$$\nabla \times \mathbf{E}(z,t) = -\hat{\mathbf{x}}\frac{\partial E_y(z,t)}{\partial z} + \hat{\mathbf{y}}\frac{\partial E_x(z,t)}{\partial z} = \hat{\mathbf{z}} \times \frac{\partial \mathbf{E}(z,t)}{\partial z} = -\mu\frac{\partial \mathbf{H}(z,t)}{\partial t}. \quad (2.271)$$

We see that $\partial H_z/\partial t = 0$, hence H_z must be constant with respect to time. Because a nonzero constant field component does not exhibit wave-like behavior, we can only have $H_z = 0$ in our wave solution. Similarly, Ampere's law in a homogeneous conducting region free from impressed sources states that

$$\nabla \times \mathbf{H}(z,t) = \mathbf{J} + \frac{\partial \mathbf{D}(z,t)}{\partial t} = \sigma\mathbf{E}(z,t) + \epsilon\frac{\partial \mathbf{E}(z,t)}{\partial t}$$

[†]The term "source free" applied to a conducting region implies that the region is devoid of impressed sources, *and*, because of the relaxation effect, has no free charge. See the discussion in Jones [98].

or

$$-\hat{\mathbf{x}}\frac{\partial H_y(z,t)}{\partial z} + \hat{\mathbf{y}}\frac{\partial H_x(z,t)}{\partial z} = \hat{\mathbf{z}} \times \frac{\partial \mathbf{H}(z,t)}{\partial z} = \sigma\mathbf{E}(z,t) + \epsilon\frac{\partial \mathbf{E}(z,t)}{\partial t}. \qquad (2.272)$$

This implies that

$$\sigma E_z(z,t) + \epsilon\frac{\partial E_z(z,t)}{\partial t} = 0,$$

which is a differential equation for E_z with solution

$$E_z(z,t) = E_0(z)\,e^{-\frac{\sigma}{\epsilon}t}.$$

Since we are interested only in wave-type solutions, we choose $E_z = 0$.

Hence $E_z = H_z = 0$, and thus both $\mathbf{E}$ and $\mathbf{H}$ are perpendicular to the z-direction. Using (2.271) and (2.272), we also see that

$$\frac{\partial}{\partial t}(\mathbf{E} \cdot \mathbf{H}) = \mathbf{E} \cdot \frac{\partial \mathbf{H}}{\partial t} + \mathbf{H} \cdot \frac{\partial \mathbf{E}}{\partial t}$$

$$= -\frac{1}{\mu}\mathbf{E} \cdot \left(\hat{\mathbf{z}} \times \frac{\partial \mathbf{E}}{\partial z}\right) - \mathbf{H} \cdot \left(\frac{\sigma}{\epsilon}\mathbf{E}\right) + \frac{1}{\epsilon}\mathbf{H} \cdot \left(\hat{\mathbf{z}} \times \frac{\partial \mathbf{H}}{\partial z}\right)$$

or

$$\left(\frac{\partial}{\partial t} + \frac{\sigma}{\epsilon}\right)(\mathbf{E} \cdot \mathbf{H}) = \frac{1}{\mu}\hat{\mathbf{z}} \cdot \left(\mathbf{E} \times \frac{\partial \mathbf{E}}{\partial z}\right) - \frac{1}{\epsilon}\hat{\mathbf{z}} \cdot \left(\mathbf{H} \times \frac{\partial \mathbf{H}}{\partial z}\right).$$

We seek solutions of the type $\mathbf{E}(z,t) = \hat{\mathbf{p}}E(z,t)$ and $\mathbf{H}(z,t) = \hat{\mathbf{q}}H(z,t)$, where $\hat{\mathbf{p}}$ and $\hat{\mathbf{q}}$ are constant unit vectors. Under this condition we have $\mathbf{E} \times \partial\mathbf{E}/\partial z = 0$ and $\mathbf{H} \times \partial\mathbf{H}/\partial z = 0$, giving

$$\left(\frac{\partial}{\partial t} + \frac{\sigma}{\epsilon}\right)(\mathbf{E} \cdot \mathbf{H}) = 0.$$

Thus we also have $\mathbf{E} \cdot \mathbf{H} = 0$, and find that $\mathbf{E}$ must be perpendicular to $\mathbf{H}$. So $\mathbf{E}$, $\mathbf{H}$, and $\hat{\mathbf{z}}$ compose a mutually orthogonal triplet of vectors. A wave having this property is said to be *TEM to the z-direction* or simply TEM$_z$. Here "TEM" stands for *transverse electromagnetic*, indicating the orthogonal relationship between the field vectors and the z-direction. Note that

$$\hat{\mathbf{p}} \times \hat{\mathbf{q}} = \pm\hat{\mathbf{z}}.$$

The constant direction described by $\hat{\mathbf{p}}$ is called the *polarization* of the plane wave.

We are now ready to solve the source-free wave equation (2.270). If we dot both sides of the homogeneous expression by $\hat{\mathbf{p}}$ we obtain

$$\hat{\mathbf{p}} \cdot \hat{\mathbf{x}}\frac{\partial^2 E_x}{\partial z^2} + \hat{\mathbf{p}} \cdot \hat{\mathbf{y}}\frac{\partial^2 E_y}{\partial z^2} - \mu\sigma\frac{\partial(\hat{\mathbf{p}} \cdot \mathbf{E})}{\partial t} - \mu\epsilon\frac{\partial^2(\hat{\mathbf{p}} \cdot \mathbf{E})}{\partial t^2} = 0.$$

Noting that

$$\hat{\mathbf{p}} \cdot \hat{\mathbf{x}}\frac{\partial^2 E_x}{\partial z^2} + \hat{\mathbf{p}} \cdot \hat{\mathbf{y}}\frac{\partial^2 E_y}{\partial z^2} = \frac{\partial^2}{\partial z^2}(\hat{\mathbf{p}} \cdot \hat{\mathbf{x}}E_x + \hat{\mathbf{p}} \cdot \hat{\mathbf{y}}E_y) = \frac{\partial^2}{\partial z^2}(\hat{\mathbf{p}} \cdot \mathbf{E}),$$

we have the wave equation

$$\frac{\partial^2 E(z,t)}{\partial z^2} - \mu\sigma\frac{\partial E(z,t)}{\partial t} - \mu\epsilon\frac{\partial^2 E(z,t)}{\partial t^2} = 0. \qquad (2.273)$$

Similarly, dotting both sides of (2.268) with $\hat{\mathbf{q}}$ and setting $\mathbf{J}^i = 0$ we obtain

$$\frac{\partial^2 H(z,t)}{\partial z^2} - \mu\sigma\frac{\partial H(z,t)}{\partial t} - \mu\epsilon\frac{\partial^2 H(z,t)}{\partial t^2} = 0. \qquad (2.274)$$

In a source-free homogeneous conducting region, $\mathbf{E}$ and $\mathbf{H}$ satisfy identical wave equations.

Solutions are considered in § A.2.5. There we solve for the total field for all z, t given the value of the field and its derivative over the $z = 0$ plane. This solution can be directly applied to find the total field of a plane wave reflected by a perfect conductor. Let us begin by considering the lossless case where $\sigma = 0$, and assuming the region $z < 0$ contains a perfect electric conductor. The conditions on the field in the $z = 0$ plane are determined by the required boundary condition on a perfect conductor: the tangential electric field must vanish. From (2.272) we see that since $\mathbf{E} \perp \hat{\mathbf{z}}$, requiring

$$\left. \frac{\partial H(z,t)}{\partial z} \right|_{z=0} = 0 \tag{2.275}$$

gives $\mathbf{E}(0, t) = 0$ and thus satisfies the boundary condition. Writing

$$H(0,t) = H_0 f(t), \quad \left. \frac{\partial H(z,t)}{\partial z} \right|_{z=0} = H_0 g(t) = 0, \tag{2.276}$$

and setting $\Omega = 0$ in (A.42) we obtain the solution to (2.274):

$$H(z,t) = \frac{H_0}{2} f\left(t - \frac{z}{v}\right) + \frac{H_0}{2} f\left(t + \frac{z}{v}\right), \tag{2.277}$$

where $v = 1/(\mu\epsilon)^{1/2}$. Since we designate the vector direction of $\mathbf{H}$ as $\hat{\mathbf{q}}$, the vector field is

$$\mathbf{H}(z,t) = \hat{\mathbf{q}} \frac{H_0}{2} f\left(t - \frac{z}{v}\right) + \hat{\mathbf{q}} \frac{H_0}{2} f\left(t + \frac{z}{v}\right). \tag{2.278}$$

From (2.271) we also have the solution for $\mathbf{E}(z, t)$:

$$\mathbf{E}(z,t) = \hat{\mathbf{p}} \frac{v\mu H_0}{2} f\left(t - \frac{z}{v}\right) - \hat{\mathbf{p}} \frac{v\mu H_0}{2} f\left(t + \frac{z}{v}\right), \tag{2.279}$$

where $\hat{\mathbf{p}} \times \hat{\mathbf{q}} = \hat{\mathbf{z}}$. The boundary conditions $E(0, t) = 0$ and $H(0, t) = H_0 f(t)$ are easily verified by substitution.

This solution displays the quintessential behavior of electromagnetic waves. We may interpret the term $f(t + z/v)$ as a wave-field disturbance, propagating at velocity v in the $-z$-direction, incident from $z > 0$ upon the conductor. The term $f(t - z/v)$ represents a wave-field disturbance propagating in the $+z$-direction with velocity v, reflected from the conductor. By "propagating" we mean that if we increment time, the disturbance will occupy a spatial position determined by incrementing z by vt. For free space where $v = 1/(\mu_0\epsilon_0)^{1/2}$, the velocity of propagation is the speed of light c.

▶ **Example 2.11:** Propagation of a transient plane wave in a lossless medium

Fresh water, with the constitutive parameters $\mu = \mu_0$ and $\epsilon = 81\epsilon_0$, fills a source-free region of space. Using the rectangular pulse waveform

$$f(t) = \text{rect}(t/\tau) \tag{2.280}$$

with $\tau = 1 \ \mu s$, plot (2.277) as a function of position for fixed values of time and interpret the result.

Solution: Figure 2.7 shows the spatial distribution of the magnetic field for various times. We see that the disturbance is spatially distributed as a rectangular pulse of extent $L = 2v\tau = 66.6$ m, where $v = c/\sqrt{81} = 3.33 \times 10^7$ m/s is the wave velocity, and where 2τ is the temporal duration of the pulse. At $t = -8$ μs the leading edge of the pulse is at $z = 233$ m, while at -4 μs the pulse has traveled a distance $z = vt = (3.33 \times 10^7) \times (4 \times 10^{-6}) = 133$ m in the $-z$-direction, and the leading edge is thus at 100 m. At $t = -1$ μs the leading edge strikes the conductor and begins to induce a current in the conductor surface. This current sets up the reflected wave, which begins to travel in the opposite $(+z)$ direction. At $t = -0.5$ μs a portion of the wave begins to travel in the $+z$-direction while the trailing portion of the disturbance continues to travel in the $-z$-direction. At $t = 1$ μs the wave is completely reflected from the surface, and thus consists only of the component traveling in the $+z$-direction. Note that if we plot the total field in the $z = 0$ plane, the sum of the forward and backward traveling waves produces the pulse waveform (2.280) as expected.

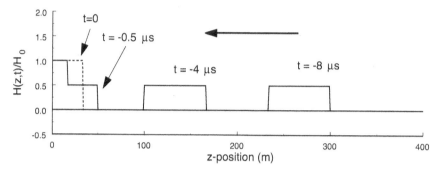

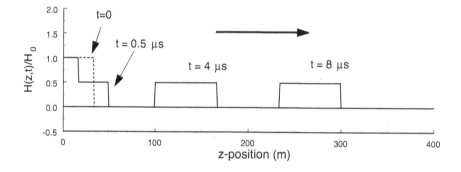

FIGURE 2.7

Propagation of a transient plane wave in a lossless medium. ◄

Using the expressions for **E** and **H** we can determine many interesting characteristics of the wave. We see that the terms $f(t \pm z/v)$ represent the components of the waves traveling in the $\mp z$-directions, respectively. If we were to isolate these waves from each other (by, for instance, measuring them as functions of time at a position where they do not overlap) we would find from (2.278) and (2.279) that the ratio of E to H for a wave traveling in either direction is

$$\left| \frac{E(z,t)}{H(z,t)} \right| = v\mu = (\mu/\epsilon)^{1/2},$$

independent of the time and position of the measurement. This ratio, denoted by η and carrying units of ohms, is called the *intrinsic impedance* of the medium through which the wave propagates. Thus, if we let $E_0 = \eta H_0$ we can write

$$\mathbf{E}(z,t) = \hat{\mathbf{p}}\frac{E_0}{2}f\left(t - \frac{z}{v}\right) - \hat{\mathbf{p}}\frac{E_0}{2}f\left(t + \frac{z}{v}\right).$$

We can easily determine the current induced in the conductor by applying the boundary condition (2.164):

$$\mathbf{J}_s = \hat{\mathbf{n}} \times \mathbf{H}|_{z=0} = \hat{\mathbf{z}} \times [H_0 \hat{\mathbf{q}} f(t)] = -\hat{\mathbf{p}} H_0 f(t). \tag{2.281}$$

We can also determine the pressure exerted on the conductor due to the Lorentz force interaction between the fields and the induced current. The total force on the conductor can be computed by integrating the Maxwell stress tensor (2.225) over the xy-plane:

$$\mathbf{F}_{em} = -\int_S \bar{\mathbf{T}}_{em} \cdot \mathbf{dS}.$$

The surface traction is

$$\mathbf{t} = \bar{\mathbf{T}}_{em} \cdot \hat{\mathbf{n}} = \left[\tfrac{1}{2}(\mathbf{D} \cdot \mathbf{E} + \mathbf{B} \cdot \mathbf{H})\bar{\mathbf{I}} - \mathbf{DE} - \mathbf{BH}\right] \cdot \hat{\mathbf{z}}.$$

Since $\mathbf{E}$ and $\mathbf{H}$ are both normal to $\hat{\mathbf{z}}$, the last two terms in this expression are zero. Also, the boundary condition on $\mathbf{E}$ implies that it vanishes in the xy-plane. Thus

$$\mathbf{t} = \tfrac{1}{2}(\mathbf{B} \cdot \mathbf{H})\hat{\mathbf{z}} = \hat{\mathbf{z}}\tfrac{1}{2}\mu H^2(t).$$

With $H_0 = E_0/\eta$ we have

$$\mathbf{t} = \hat{\mathbf{z}}\frac{E_0^2}{2\eta^2}\mu f^2(t). \tag{2.282}$$

▶ **Example 2.12:** Nuclear electromagnetic pulse

Consider a high-altitude nuclear electromagnetic pulse (HEMP) generated by the explosion of a large nuclear weapon in the upper atmosphere. Such an explosion could generate a transient electromagnetic wave of short (submicrosecond) duration with an electric field amplitude of 50,000 V/m in air [201]. Find the peak pressure exerted on a perfect conductor if the wave impinges at normal incidence.

Solution: Using (2.282), we find that the wave would exert a peak pressure of $P = |\mathbf{t}| = .011$ Pa $= 1.6 \times 10^{-6}$ lb/in^2 if reflected from a perfect conductor at normal incidence. Obviously, even for this extreme field level the pressure produced by a transient electromagnetic wave is quite small. However, from (2.281) we find that the current induced in the conductor would have a peak value of 133 A/m. Even a small portion of this current could destroy a sensitive electronic circuit if it were to leak through an opening in the conductor. This is an important concern for engineers designing circuitry to be used in high-field environments, and demonstrates why the concepts of current and voltage can often supersede the concept of force in terms of importance. ◀

Finally, let us see how the terms in the Poynting power balance theorem relate. Consider a cubic region V bounded by the planes $z = z_1$ and $z = z_2$, $z_2 > z_1$. We choose the field waveform $f(t)$ and locate the planes so that we can isolate either the forward or backward traveling wave. Since there is no current in V, Poynting's theorem (2.234) becomes

$$\frac{1}{2}\frac{\partial}{\partial t}\int_V (\epsilon \mathbf{E} \cdot \mathbf{E} + \mu \mathbf{H} \cdot \mathbf{H})\, dV = -\oint_S (\mathbf{E} \times \mathbf{H}) \cdot \mathbf{dS}.$$

Consider the wave traveling in the $-z$-direction. Substitution from (2.278) and (2.279) gives the time-rate of change of stored energy as

$$S_{\text{cube}}(t) = \frac{1}{2}\frac{\partial}{\partial t}\int_V \left[\epsilon E^2(z,t) + \mu H^2(z,t)\right] dV$$

$$= \frac{1}{2}\frac{\partial}{\partial t}\int_x\int_y dx\,dy \int_{z_1}^{z_2}\left[\epsilon\frac{(v\mu)^2 H_0^2}{4}f^2\left(t+\frac{z}{v}\right) + \mu\frac{H_0^2}{4}f^2\left(t+\frac{z}{v}\right)\right] dz$$

$$= \frac{1}{2}\frac{\partial}{\partial t}\mu\frac{H_0^2}{2}\int_x\int_y dx\,dy\int_{z_1}^{z_2} f^2\left(t+\frac{z}{v}\right) dz.$$

Integration over x and y gives the area A of the cube face. Putting $u = t + z/v$ we see that

$$S = A\mu\frac{H_0^2}{4}\frac{\partial}{\partial t}\int_{t+z_1/v}^{t+z_2/v} f^2(u)v\,du.$$

Leibnitz' rule for differentiation (A.31) then gives

$$S_{\text{cube}}(t) = A\frac{\mu v H_0^2}{4}\left[f^2\left(t+\frac{z_2}{v}\right) - f^2\left(t+\frac{z_1}{v}\right)\right]. \tag{2.283}$$

Again substituting for $E(t+z/v)$ and $H(t+z/v)$ we can write

$$S_{\text{cube}}(t) = -\oint_S (\mathbf{E}\times\mathbf{H})\cdot d\mathbf{S}$$

$$= -\int_x\int_y \frac{v\mu H_0^2}{4}f^2\left(t+\frac{z_1}{v}\right)(-\hat{\mathbf{p}}\times\hat{\mathbf{q}})\cdot(-\hat{\mathbf{z}})\,dx\,dy$$

$$- \int_x\int_y \frac{v\mu H_0^2}{4}f^2\left(t+\frac{z_2}{v}\right)(-\hat{\mathbf{p}}\times\hat{\mathbf{q}})\cdot(\hat{\mathbf{z}})\,dx\,dy.$$

The second term represents the energy change in V produced by the backward traveling wave entering the cube by passing through the plane at $z = z_2$, while the first term represents the energy change in V produced by the wave exiting the cube by passing through the plane $z = z_1$. Contributions from the sides, top, and bottom are zero since $\mathbf{E}\times\mathbf{H}$ is perpendicular to $\hat{\mathbf{n}}$ over those surfaces. Since $\hat{\mathbf{p}}\times\hat{\mathbf{q}} = \hat{\mathbf{z}}$, we get

$$S_{\text{cube}}(t) = A\frac{\mu v H_0^2}{4}\left[f^2\left(t+\frac{z_2}{v}\right) - f^2\left(t+\frac{z_1}{v}\right)\right],$$

which matches (2.283) and thus verifies Poynting's theorem. We may interpret this result as follows. The propagating electromagnetic disturbance carries energy through space. The energy within any region is associated with the field in that region, and can change with time as the propagating wave carries a flux of energy across the boundary of the region. The energy continues to propagate even if the source is changed or is extinguished altogether. That is, the behavior of the leading edge of the disturbance is determined by causality — it is affected by obstacles it encounters, but not by changes in the source that occur after the leading edge has been established.

When propagating through a dissipative region, a plane wave takes on a somewhat different character. Again applying the conditions (2.275) and (2.276), we obtain from

(A.42) the solution to the wave equation (2.274):

$$H(z,t) = \frac{H_0}{2}e^{-\frac{\Omega}{v}z}f\left(t - \frac{z}{v}\right) + \frac{H_0}{2}e^{\frac{\Omega}{v}z}f\left(t + \frac{z}{v}\right)$$

$$- \frac{z\Omega^2 H_0}{2v}e^{-\Omega t}\int_{t-\frac{z}{v}}^{t+\frac{z}{v}} f(u)e^{\Omega u}\frac{J_1\left(\frac{\Omega}{v}\sqrt{z^2 - (t-u)^2v^2}\right)}{\frac{\Omega}{v}\sqrt{z^2 - (t-u)^2v^2}}\,du \qquad (2.284)$$

where $\Omega = \sigma/2\epsilon$. The first two terms resemble those for the lossless case, modified by an exponential damping factor. This accounts for the loss in amplitude that must accompany the transfer of energy from the propagating wave to joule loss (heat) within the conducting medium. The remaining term appears only when the medium is lossy, and results in an extension of the disturbance through the medium because of the currents induced by the passing wavefront. This "wake" follows the leading edge of the disturbance.

▶ **Example 2.13:** Propagation of a transient plane wave in a dissipative medium

In Example 2.11 we examined the propagation of a pulse through fresh water, assuming no loss. Again consider fresh water with the constitutive parameters $\mu = \mu_0$ and $\epsilon = 81\epsilon_0$, but also assume that it is lossy with a conductivity $\sigma = 2\times10^{-4}$ S/m. As in Example 2.11, using the rectangular pulse waveform $f(t) = \text{rect}(t/\tau)$ with $\tau = 1\ \mu$s, plot (2.284) as a function of position for fixed values of time and interpret the result.

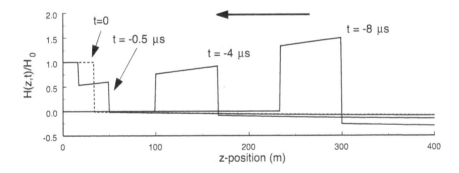

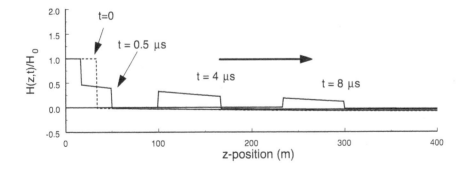

FIGURE 2.8
Propagation of a transient plane wave in a dissipative medium.

Solution: Figure 2.8 shows the spatial distribution of the magnetic field for various times. Comparing to the lossless results of Figure 2.7, it is clear that the addition of loss leads to the wake effect discussed above. As the wave travels to the left it attenuates and leaves a trailing remnant behind. Upon reaching the conductor it reflects much as in the lossless case, resulting in a time dependence at $z = 0$ given by the finite-duration rectangular pulse $f(t)$. In order for the pulse to be of finite duration, the wake left by the reflected pulse must exactly cancel the wake associated with the incident pulse that continues to arrive after the reflection. As the reflected pulse sweeps forward, the wake is obliterated everywhere behind. ◄

If we were to verify the Poynting theorem for a dissipative medium (which we shall not attempt because of the complexity of the computation), we would need to include the $\mathbf{E} \cdot \mathbf{J}$ term. Here $\mathbf{J}$ is the induced conduction current and the integral of $\mathbf{E} \cdot \mathbf{J}$ accounts for the joule loss within a region V balanced by the difference in Poynting energy flux carried into and out of V.

Once we have the fields for a wave propagating along the z-direction, it is a simple matter to generalize these results to any propagation direction. Assume that $\hat{\mathbf{u}}$ is normal to the surface of a plane over which the fields are invariant. Then $u = \hat{\mathbf{u}} \cdot \mathbf{r}$ describes the distance from the origin along the direction $\hat{\mathbf{u}}$. We need only replace z by $\hat{\mathbf{u}} \cdot \mathbf{r}$ in any of the expressions obtained above to determine the fields of a plane wave propagating in the u-direction. We must also replace the orthogonality condition $\hat{\mathbf{p}} \times \hat{\mathbf{q}} = \hat{\mathbf{z}}$ with $\hat{\mathbf{p}} \times \hat{\mathbf{q}} = \hat{\mathbf{u}}$. For instance, the fields associated with a wave propagating through a lossless medium in the positive u-direction are, from (2.278)–(2.279),

$$\mathbf{H}(\mathbf{r}, t) = \hat{\mathbf{q}} \frac{H_0}{2} f\left(t - \frac{\hat{\mathbf{u}} \cdot \mathbf{r}}{v}\right), \qquad \mathbf{E}(\mathbf{r}, t) = \hat{\mathbf{p}} \frac{v\mu H_0}{2} f\left(t - \frac{\hat{\mathbf{u}} \cdot \mathbf{r}}{v}\right).$$

2.10.7 Propagation of cylindrical waves in a lossless medium

Much as we envisioned a uniform plane wave arising from a uniform planar source, we can imagine a uniform cylindrical wave arising from a uniform line source. Although this line source must be infinite in extent, uniform cylindrical waves (unlike plane waves) display the physical behavior of diverging from their source while carrying energy outwards to infinity.

A *uniform cylindrical wave* has fields that are invariant over a cylindrical surface: $\mathbf{E}(\mathbf{r}, t) = \mathbf{E}(\rho, t)$, $\mathbf{H}(\mathbf{r}, t) = \mathbf{H}(\rho, t)$. For simplicity, we shall assume that waves propagate in a homogeneous, isotropic, linear, and lossless medium described by permittivity ϵ and permeability μ. From Maxwell's equations we find that requiring the fields to be independent of ϕ and z puts restrictions on the remaining vector components. Faraday's law states

$$\nabla \times \mathbf{E}(\rho, t) = -\hat{\phi} \frac{\partial E_z(\rho, t)}{\partial \rho} + \hat{\mathbf{z}} \frac{1}{\rho} \frac{\partial}{\partial \rho}[\rho E_\phi(\rho, t)] = -\mu \frac{\partial \mathbf{H}(\rho, t)}{\partial t}. \tag{2.285}$$

Equating components, we see that $\partial H_\rho / \partial t = 0$, and because our interest lies in wave solutions we take $H_\rho = 0$. Ampere's law in a homogeneous lossless region free from impressed sources states in a similar manner

$$\nabla \times \mathbf{H}(\rho, t) = -\hat{\phi} \frac{\partial H_z(\rho, t)}{\partial \rho} + \hat{\mathbf{z}} \frac{1}{\rho} \frac{\partial}{\partial \rho}[\rho H_\phi(\rho, t)] = \epsilon \frac{\partial \mathbf{E}(\rho, t)}{\partial t}. \tag{2.286}$$

Equating components, we find that $E_\rho = 0$. Since $E_\rho = H_\rho = 0$, both $\mathbf{E}$ and $\mathbf{H}$ are perpendicular to the ρ-direction. Note that if there is only a z-component of $\mathbf{E}$ then there is only a ϕ-component of **H**. This case, termed *electric polarization*, results in

$$\frac{\partial E_z(\rho,t)}{\partial \rho} = \mu\frac{\partial H_\phi(\rho,t)}{\partial t}.$$

Similarly, if there is only a z-component of **H** then there is only a ϕ-component of **E**. This case, termed *magnetic polarization*, results in

$$-\frac{\partial H_z(\rho,t)}{\partial \rho} = \epsilon\frac{\partial E_\phi(\rho,t)}{\partial t}.$$

Since $\mathbf{E} = \hat{\phi}E_\phi + \hat{z}E_z$ and $\mathbf{H} = \hat{\phi}H_\phi + \hat{z}H_z$, we can always decompose a cylindrical electromagnetic wave into cases of electric and magnetic polarization. In each case the resulting field is TEM$_\rho$ since the vectors $\mathbf{E}, \mathbf{H}$, and $\hat{\rho}$ are mutually orthogonal.

Wave equations for E_z in the electric polarization case and for H_z in the magnetic polarization case can be found in the usual manner. Taking the curl of (2.285) and substituting from (2.286) we find

$$\nabla \times (\nabla \times \mathbf{E}) = -\hat{z}\frac{1}{\rho}\frac{\partial}{\partial \rho}\left(\rho\frac{\partial E_z}{\partial \rho}\right) - \hat{\phi}\frac{\partial}{\partial \rho}\left(\frac{1}{\rho}\frac{\partial}{\partial \rho}[\rho E_\phi]\right)$$

$$= -\frac{1}{v^2}\frac{\partial^2 \mathbf{E}}{\partial t^2} = -\frac{1}{v^2}\left(\hat{z}\frac{\partial^2 E_z}{\partial t^2} + \hat{\phi}\frac{\partial^2 E_\phi}{\partial t^2}\right)$$

where $v = 1/(\mu\epsilon)^{1/2}$. Noting that $E_\phi = 0$ for the electric polarization case, we obtain the wave equation for E_z. A similar set of steps beginning with the curl of (2.286) gives an identical equation for H_z. Thus

$$\frac{1}{\rho}\frac{\partial}{\partial \rho}\left(\rho\frac{\partial}{\partial \rho}\begin{bmatrix}E_z\\H_z\end{bmatrix}\right) - \frac{1}{v^2}\frac{\partial^2}{\partial t^2}\begin{bmatrix}E_z\\H_z\end{bmatrix} = 0. \qquad (2.287)$$

We can obtain a solution for (2.287) in much the same way as we do for the wave equations in § A.2.4. We begin by substituting for $E_z(\rho,t)$ in terms of its temporal Fourier representation

$$E_z(\rho,t) = \frac{1}{2\pi}\int_{-\infty}^{\infty}\tilde{E}_z(\rho,\omega)e^{j\omega t}\,d\omega$$

to obtain

$$\frac{1}{2\pi}\int_{-\infty}^{\infty}\left[\frac{1}{\rho}\frac{\partial}{\partial \rho}\left(\rho\frac{\partial}{\partial \rho}\tilde{E}_z(\rho,\omega)\right) + \frac{\omega^2}{v^2}\tilde{E}_z(\rho,\omega)\right]e^{j\omega t}\,d\omega = 0.$$

The Fourier integral theorem implies that the integrand is zero. Then, expanding out the ρ derivatives, we find that $\tilde{E}_z(\rho,\omega)$ obeys the ordinary differential equation

$$\frac{d^2\tilde{E}_z}{d\rho^2} + \frac{1}{\rho}\frac{d\tilde{E}_z}{d\rho} + k^2\tilde{E}_z = 0$$

where $k = \omega/v$. This is merely Bessel's differential equation (A.126). It is a second-order equation with two independent solutions chosen from the list

$$J_0(k\rho), \quad Y_0(k\rho), \quad H_0^{(1)}(k\rho), \quad H_0^{(2)}(k\rho).$$

We find that $J_0(k\rho)$ and $Y_0(k\rho)$ are useful for describing standing waves between boundaries while $H_0^{(1)}(k\rho)$ and $H_0^{(2)}(k\rho)$ are useful for describing waves propagating in the

ρ-direction. Of these, $H_0^{(1)}(k\rho)$ represents waves traveling inward while $H_0^{(2)}(k\rho)$ represents waves traveling outward. Concentrating on the outward traveling wave, we find that

$$\tilde{E}_z(\rho, \omega) = \tilde{A}(\omega) \left[-j\frac{\pi}{2} H_0^{(2)}(k\rho) \right] = \tilde{A}(\omega)\tilde{g}(\rho, \omega).$$

Here $A(t) \leftrightarrow \tilde{A}(\omega)$ is the disturbance waveform, assumed to be a real, causal function. To make $E_z(\rho, t)$ real we require that the inverse transform of $\tilde{g}(\rho, \omega)$ be real. This requires the inclusion of the $-j\pi/2$ factor in $\tilde{g}(\rho, \omega)$. Inverting we have

$$E_z(\rho, t) = A(t) * g(\rho, t) \tag{2.288}$$

where $g(\rho, t) \leftrightarrow (-j\pi/2)H_0^{(2)}(k\rho)$.

The inverse transform needed to obtain $g(\rho, t)$ may be found in [28]:

$$g(\rho, t) = \mathcal{F}^{-1}\left\{ -j\frac{\pi}{2}H_0^{(2)}\left(\omega\frac{\rho}{v}\right) \right\} = \frac{U\left(t - \frac{\rho}{v}\right)}{\sqrt{t^2 - \frac{\rho^2}{v^2}}},$$

where $U(t)$ is the unit step function defined in (A.6). Substituting this into (2.288) and writing the convolution in integral form, we have

$$E_z(\rho, t) = \int_{-\infty}^{\infty} A(t - t') \frac{U(t' - \rho/v)}{\sqrt{t'^2 - \rho^2/v^2}}\, dt'.$$

The change of variable $x = t' - \rho/v$ then gives

$$E_z(\rho, t) = \int_0^{\infty} \frac{A(t - x - \rho/v)}{\sqrt{x^2 + 2x\rho/v}}\, dx. \tag{2.289}$$

Those interested in the details of the inverse transform should see Chew [35].

▶ **Example 2.14:** Propagation of a transient cylindrical wave in a lossless medium

Fresh water, with the constitutive parameters $\mu = \mu_0$ and $\epsilon = 81\epsilon_0$, fills a source-free region of space. Using the rectangular pulse waveform

$$A(t) = E_0[U(t) - U(t - \tau)]$$

with $\tau = 2$ μs, plot (2.289) for fixed values of time and interpret the result.

Solution: This situation is the same as that in the plane-wave case of Example 2.11 above, except that the pulse waveform begins at $t = 0$. Substituting for $A(t)$ into (2.289) and using

$$\int \frac{dx}{\sqrt{x}\sqrt{x + a}} = 2\ln\left[\sqrt{x} + \sqrt{x + a}\right]$$

we can write the electric field in closed form as

$$E_z(\rho, t) = 2E_0 \ln\left[\frac{\sqrt{x_2} + \sqrt{x_2 + 2\rho/v}}{\sqrt{x_1} + \sqrt{x_1 + 2\rho/v}} \right], \tag{2.290}$$

where $x_2 = \max[0, t - \rho/v]$ and $x_1 = \max[0, t - \rho/v - \tau]$. The field is plotted in Figure 2.9 for various values of time. Note that the leading edge of the disturbance propagates outward at a velocity v and a wake trails behind the disturbance. This wake is similar to that for a plane wave in a dissipative medium, but it exists in this case even though the medium is lossless. We can think of the wave as being created by a line source of infinite extent, pulsed by the disturbance waveform. Although current changes simultaneously everywhere along the line, it takes the disturbance longer to propagate to an observation point in the $z = 0$ plane from source points $z \neq 0$ than from the source point at $z = 0$. Thus, the field at an arbitrary observation point ρ arrives from different source points at different times. If we look at Figure 2.9 we note that there is always a nonzero field near $\rho = 0$ (or any value of $\rho < vt$) regardless of the time, since at any given t the disturbance is arriving from some point along the line source.

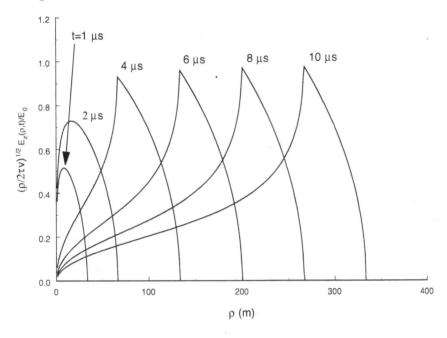

FIGURE 2.9
Propagation of a transient cylindrical wave in a lossless medium.

We also see in Figure 2.9 that as ρ becomes large the peak value of the propagating disturbance approaches a certain value. This value occurs at $t_m = \rho/v + \tau$ or, equivalently, $\rho_m = v(t - \tau)$. If we substitute this value into (2.290) we find that

$$E_z(\rho, t_m) = 2E_0 \ln \left[\sqrt{\frac{\tau}{2\rho/v}} + \sqrt{1 + \frac{\tau}{2\rho/v}} \right].$$

For large values of ρ/v,

$$E_z(\rho, t_m) \approx 2E_0 \ln \left[1 + \sqrt{\frac{\tau}{2\rho/v}} \right] \approx E_0 \sqrt{\frac{2\tau v}{\rho}}$$

since $\ln(1 + x) \approx x$ for $x \ll 1$. Thus, as $\rho \to \infty$ we have $\mathbf{E} \times \mathbf{H} \sim 1/\rho$ and the flux of energy passing through a cylindrical surface of area $\rho \, d\phi \, dz$ is independent of ρ. This result is similar to that seen for spherical waves where $\mathbf{E} \times \mathbf{H} \sim 1/r^2$. ◀

2.10.8 Propagation of spherical waves in a lossless medium

In the previous section we found solutions that describe uniform cylindrical waves dependent only on the radial variable ρ. It turns out that similar solutions are not possible in spherical coordinates; fields that only depend on r cannot satisfy Maxwell's equations since, as shown in § 2.10.10, a source having the appropriate symmetry for the production of uniform spherical waves in fact produces no field at all external to the region it occupies. As we shall see in Chapter 5, the fields produced by localized sources are in general quite complex. However, certain solutions that are only slightly nonuniform may be found, and these allow us to investigate the most important properties of spherical waves. We shall find that spherical waves diverge from a localized point source and expand outward with finite velocity, carrying energy away from the source.

Consider a homogeneous, lossless, source-free region of space characterized by permittivity ϵ and permeability μ. We seek solutions to the wave equation that are TEM$_r$ in spherical coordinates ($H_r = E_r = 0$), and independent of the azimuthal angle ϕ. Thus we may write

$$\mathbf{E}(\mathbf{r},t) = \hat{\boldsymbol{\theta}} E_\theta(r,\theta,t) + \hat{\boldsymbol{\phi}} E_\phi(r,\theta,t),$$

$$\mathbf{H}(\mathbf{r},t) = \hat{\boldsymbol{\theta}} H_\theta(r,\theta,t) + \hat{\boldsymbol{\phi}} H_\phi(r,\theta,t).$$

Maxwell's equations show that not all of these vector components are required. Faraday's law states that

$$\nabla \times \mathbf{E}(r,\theta,t) = \hat{\mathbf{r}} \frac{1}{r\sin\theta} \frac{\partial}{\partial\theta}[\sin\theta E_\phi(r,\theta,t)] - \hat{\boldsymbol{\theta}} \frac{1}{r}\frac{\partial}{\partial r}[rE_\phi(r,\theta,t)] + \hat{\boldsymbol{\phi}}\frac{1}{r}\frac{\partial}{\partial r}[rE_\theta(r,\theta,t)]$$

$$= -\mu\frac{\partial \mathbf{H}(r,\theta,t)}{\partial t}. \tag{2.291}$$

Since we require $H_r = 0$, we must have

$$\frac{\partial}{\partial\theta}[\sin\theta E_\phi(r,\theta,t)] = 0.$$

This implies that either $E_\phi \sim 1/\sin\theta$ or $E_\phi = 0$. We shall choose $E_\phi = 0$ and investigate whether the resulting fields satisfy the remaining Maxwell equations.

In a source-free region of space, we have $\nabla\cdot\mathbf{D} = \epsilon\nabla\cdot\mathbf{E} = 0$. Since we now have only a θ-component of the electric field, this requires

$$\frac{1}{r}\frac{\partial}{\partial\theta}E_\theta(r,\theta,t) + \frac{\cot\theta}{r}E_\theta(r,\theta,t) = 0.$$

From this we see that when $E_\phi = 0$ the component E_θ must obey

$$E_\theta(r,\theta,t) = \frac{f_E(r,t)}{\sin\theta}.$$

By (2.291) there is only a ϕ-component of magnetic field, and it must obey $H_\phi(r,\theta,t) = f_H(r,t)/\sin\theta$ where

$$-\mu\frac{\partial}{\partial t}f_H(r,t) = \frac{1}{r}\frac{\partial}{\partial r}[rf_E(r,t)]. \tag{2.292}$$

Thus the spherical wave has the property $\mathbf{E} \perp \mathbf{H} \perp \mathbf{r}$, and is TEM to the r-direction.

We can obtain a wave equation for E_θ by taking the curl of (2.291) and substituting from Ampere's law:

$$\nabla\times(\nabla\times\mathbf{E}) = -\hat{\boldsymbol{\theta}}\frac{1}{r}\frac{\partial^2}{\partial r^2}[rE_\theta] = \nabla\times\left(-\mu\frac{\partial}{\partial t}\mathbf{H}\right) = -\mu\frac{\partial}{\partial t}\left(\sigma\mathbf{E} + \epsilon\frac{\partial}{\partial t}\mathbf{E}\right).$$

This gives

$$\frac{\partial^2}{\partial r^2}[rf_E(r,t)] - \mu\sigma\frac{\partial}{\partial t}[rf_E(r,t)] - \mu\epsilon\frac{\partial^2}{\partial t^2}[rf_E(r,t)] = 0,$$

which is the desired wave equation for $\mathbf{E}$. Proceeding similarly we find that H_ϕ obeys

$$\frac{\partial^2}{\partial r^2}[rf_H(r,t)] - \mu\sigma\frac{\partial}{\partial t}[rf_H(r,t)] - \mu\epsilon\frac{\partial^2}{\partial t^2}[rf_H(r,t)] = 0.$$

We see that the wave equation for rf_E is identical to that for the plane-wave field E_z (2.273). Thus, we can use the solution obtained in § A.2.5, as we did with the plane wave, with a few subtle differences. First, we cannot have $r < 0$. Second, we do not anticipate a solution representing a wave traveling in the $-r$-direction — i.e., a wave converging toward the origin. (In other situations we might need such a solution in order to form a standing wave between two spherical boundary surfaces, but here we are only interested in the basic propagating behavior of spherical waves.) Thus, we choose as our solution the term (A.46) and find for a lossless medium where $\Omega = 0$

$$E_\theta(r,\theta,t) = \frac{1}{r\sin\theta}A\left(t - \frac{r}{v}\right). \tag{2.293}$$

From (2.292) we see that

$$H_\phi = \frac{1}{\mu v}\frac{1}{r\sin\theta}A\left(t - \frac{r}{v}\right). \tag{2.294}$$

Since $\mu v = (\mu/\epsilon)^{1/2} = \eta$, we can also write this as

$$\mathbf{H} = \frac{\hat{\mathbf{r}} \times \mathbf{E}}{\eta}.$$

We note that our solution is not appropriate for unbounded space since the fields have a singularity at $\theta = 0$. Thus we must exclude the z-axis. This can be accomplished by using PEC cones of angles θ_1 and θ_2, $\theta_2 > \theta_1$. Because the electric field $\mathbf{E} = \hat{\boldsymbol{\theta}}E_\theta$ is normal to these cones, the boundary condition that tangential $\mathbf{E}$ vanishes is satisfied.

It is informative to see how the terms in the Poynting power balance theorem relate for a spherical wave. Consider the region between the spherical surfaces $r = r_1$ and $r = r_2$, $r_2 > r_1$. Since there is no current within the volume region, Poynting's theorem (2.234) becomes

$$\frac{1}{2}\frac{\partial}{\partial t}\int_V (\epsilon\mathbf{E}\cdot\mathbf{E} + \mu\mathbf{H}\cdot\mathbf{H})\,dV = -\oint_S (\mathbf{E}\times\mathbf{H})\cdot d\mathbf{S}.$$

From (2.293) and (2.294), the time-rate of change of stored energy is

$$P_{\text{sphere}}(t) = \frac{1}{2}\frac{\partial}{\partial t}\int_V [\epsilon E^2(r,\theta,t) + \mu H^2(r,\theta,t)]\,dV$$

$$= \frac{1}{2}\frac{\partial}{\partial t}\int_0^{2\pi} d\phi \int_{\theta_1}^{\theta_2} \frac{d\theta}{\sin\theta} \int_{r_1}^{r_2} \left[\epsilon\frac{1}{r^2}A^2\left(t-\frac{r}{v}\right) + \mu\frac{1}{r^2}\frac{1}{(v\mu)^2}A^2\left(t-\frac{r}{v}\right)\right] r^2\,dr$$

$$= 2\pi\epsilon F\frac{\partial}{\partial t}\int_{r_1}^{r_2} A^2\left(t-\frac{r}{v}\right)dr$$

where

$$F = \ln\left[\frac{\tan(\theta_2/2)}{\tan(\theta_1/2)}\right].$$

Putting $u = t - r/v$ we see that

$$P_{\text{sphere}}(t) = -2\pi\epsilon F \frac{\partial}{\partial t} \int_{t-r_1/v}^{t-r_2/v} A^2(u) v \, du.$$

An application of Leibnitz' rule for differentiation (A.31) gives

$$P_{\text{sphere}}(t) = -\frac{2\pi}{\eta} F \left[A^2 \left(t - \frac{r_2}{v} \right) - A^2 \left(t - \frac{r_1}{v} \right) \right]. \tag{2.295}$$

Next we find the Poynting flux term:

$$
\begin{aligned}
P_{\text{sphere}}(t) &= -\oint_S (\mathbf{E} \times \mathbf{H}) \cdot d\mathbf{S} \\
&= -\int_0^{2\pi} d\phi \int_{\theta_1}^{\theta_2} \left[\frac{1}{r_1} A \left(t - \frac{r_1}{v} \right) \hat{\boldsymbol{\theta}} \right] \times \left[\frac{1}{r_1} \frac{1}{\mu v} A \left(t - \frac{r_1}{v} \right) \hat{\boldsymbol{\phi}} \right] \cdot (-\hat{\mathbf{r}}) r_1^2 \frac{d\theta}{\sin\theta} \\
&\quad - \int_0^{2\pi} d\phi \int_{\theta_1}^{\theta_2} \left[\frac{1}{r_2} A \left(t - \frac{r_2}{v} \right) \hat{\boldsymbol{\theta}} \right] \times \left[\frac{1}{r_2} \frac{1}{\mu v} A \left(t - \frac{r_2}{v} \right) \hat{\boldsymbol{\phi}} \right] \cdot \hat{\mathbf{r}} r_2^2 \frac{d\theta}{\sin\theta}.
\end{aligned}
$$

The first term represents the power carried by the traveling wave into the volume region by passing through the spherical surface at $r = r_1$, while the second term represents the power carried by the wave out of the region by passing through the surface $r = r_2$. Integration gives

$$P_{\text{sphere}}(t) = -\frac{2\pi}{\eta} F \left[A^2 \left(t - \frac{r_2}{v} \right) - A^2 \left(t - \frac{r_1}{v} \right) \right], \tag{2.296}$$

which matches (2.295), thus verifying Poynting's theorem.

It is also interesting to compute the total energy passing through a surface of radius r_0. From (2.296) we see that the flux of energy (power density) passing outward through the surface $r = r_0$ is

$$P_{\text{sphere}}(t) = \frac{2\pi}{\eta} F A^2 \left(t - \frac{r_0}{v} \right).$$

The total energy associated with this flux can be computed by integrating over all time: we have

$$E = \frac{2\pi}{\eta} F \int_{-\infty}^{\infty} A^2 \left(t - \frac{r_0}{v} \right) dt = \frac{2\pi}{\eta} F \int_{-\infty}^{\infty} A^2(u) \, du$$

after making the substitution $u = t - r_0/v$. The total energy passing through a spherical surface is independent of the radius of the sphere. This is an important property of spherical waves. The $1/r$ dependence of the electric and magnetic fields produces a power density that decays with distance in precisely the right proportion to compensate for the r^2-type increase in the surface area through which the power flux passes.

2.10.9 Energy radiated by sources

We may use Poynting's theorem to show that the energy radiated by a bounded impressed source $\mathbf{J}^i$ in a simple medium is nonnegative, and thus the following inequality holds [217]:

$$-\int_{-\infty}^{t} \int_V \mathbf{J}^i \cdot \mathbf{E} \, dV \, dt \geq \int_{-\infty}^{t} \int_V \sigma |\mathbf{E}|^2 \, dV \, dt + \frac{1}{2} \int_V \left(\epsilon |\mathbf{E}|^2 + \mu |\mathbf{H}|^2 \right) dV.$$

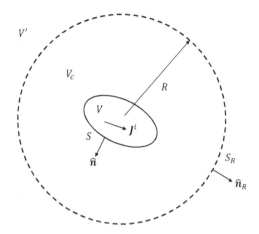

FIGURE 2.10

Geometry for establishing radiated energy inequality. Note that because V is bounded, it is contained entirely inside a sphere of radius R.

The equality is true for *nonradiating sources*, which are discussed in the next section.

To establish the inequality, consider the bounded region V shown in Figure 2.10. Applied to V, Poynting's theorem reads

$$-\oint_S (\mathbf{E} \times \mathbf{H}) \cdot \hat{\mathbf{n}} \, dS = \int_V \mathbf{J}^i \cdot \mathbf{E} \, dV + \int_V \sigma |\mathbf{E}|^2 \, dV + \frac{1}{2} \frac{\partial}{\partial t} \int_V (\epsilon |\mathbf{E}|^2 + \mu |\mathbf{H}|^2) \, dV. \quad (2.297)$$

Now assume that V is completely contained within a sphere of radius R such that the region internal to the sphere but external to V is V_c. Assuming the impressed source currents are all inside V as shown, application of Poynting's theorem to V_c gives

$$\oint_S (\mathbf{E} \times \mathbf{H}) \cdot \hat{\mathbf{n}} \, dS - \oint_{S_R} (\mathbf{E} \times \mathbf{H}) \cdot \hat{\mathbf{n}}_R \, dS = \int_{V_c} \sigma |\mathbf{E}|^2 \, dV + \frac{1}{2} \frac{\partial}{\partial t} \int_{V_c} (\epsilon |\mathbf{E}|^2 + \mu |\mathbf{H}|^2) \, dV.$$

In the limit as $R \to \infty$, the second term vanishes by the radiation conditions for $\mathbf{E}$ and $\mathbf{H}$ on S_R:

$$\oint_S (\mathbf{E} \times \mathbf{H}) \cdot \hat{\mathbf{n}} \, dS = \int_{V'} \sigma |\mathbf{E}|^2 \, dV + \frac{1}{2} \frac{\partial}{\partial t} \int_{V'} (\epsilon |\mathbf{E}|^2 + \mu |\mathbf{H}|^2) \, dV,$$

where V' is the region external to the sphere. The left-hand side is the electromagnetic power flux from V into V' at time t, which we will denote by $P(t)$. Its time integral is the total electromagnetic energy that has left V by time t. We have

$$\int_{-\infty}^{t} P(t) \, dt = \int_{-\infty}^{t} \oint_S (\mathbf{E} \times \mathbf{H}) \cdot \hat{\mathbf{n}} \, dS \, dt$$

$$= \int_{-\infty}^{t} \left[\int_{V'} \sigma |\mathbf{E}|^2 \, dV + \frac{1}{2} \frac{\partial}{\partial t} \int_{V'} (\epsilon |\mathbf{E}|^2 + \mu |\mathbf{H}|^2) \, dV \right] dt$$

$$= \int_{-\infty}^{t} \int_{V'} \sigma |\mathbf{E}|^2 \, dV \, dt + \frac{1}{2} \int_{V'} (\epsilon |\mathbf{E}|^2 + \mu |\mathbf{H}|^2) \, dV$$

$$\geq 0,$$

with equality if and only if $\mathbf{E} \equiv 0$ and $\mathbf{H} \equiv 0$ in V'. Hence by (2.297)

$$-\int_{-\infty}^{t} \left[\int_V \mathbf{J}^i \cdot \mathbf{E} \, dV + \int_V \sigma|\mathbf{E}|^2 \, dV + \frac{1}{2} \frac{\partial}{\partial t} \int_V (\epsilon|\mathbf{E}|^2 + \mu|\mathbf{H}|^2) \, dV \right] dt \geq 0$$

which gives

$$-\int_{-\infty}^{t} \int_V \mathbf{J}^i \cdot \mathbf{E} \, dV \, dt \geq \int_{-\infty}^{t} \int_V \sigma|\mathbf{E}|^2 \, dV \, dt + \frac{1}{2} \int_V (\epsilon|\mathbf{E}|^2 + \mu|\mathbf{H}|^2) \, dV.$$

We see that the energy introduced by the impressed currents is greater than or equal to the energy dissipated as Joule heat in V plus the energy stored in V. We can expect a positive net radiation of energy from nearly any source. The conditions under which a source does not radiate are given in the next section.

Note that if $\mathbf{J}^i \equiv 0$ in V, the last inequality becomes

$$0 \geq \int_{-\infty}^{t} \int_V \sigma|\mathbf{E}|^2 \, dV \, dt + \frac{1}{2} \int_V (\epsilon|\mathbf{E}|^2 + \mu|\mathbf{H}|^2) \, dV,$$

which implies that $\mathbf{E} \equiv \mathbf{0} \equiv \mathbf{H}$ in V.

2.10.10   Nonradiating sources

Not all time-dependent sources produce electromagnetic waves. In fact, certain localized source distributions produce no fields external to the region containing the sources. Such distributions are said to be *nonradiating*, and the fields they produce (within their source regions) lack wave characteristics.

Let us consider a specific example involving two concentric spheres. The inner sphere, carrying a uniformly distributed total charge $-Q$, is rigid and has a fixed radius a; the outer sphere, carrying uniform charge $+Q$, is a flexible balloon that can be stretched to any radius $b = b(t)$. The two surfaces are initially stationary, some external force being required to hold them in place. Now suppose we apply a time-varying force that results in $b(t)$ changing from $b(t_1) = b_1$ to $b(t_2) = b_2 > b_1$. This creates a radially directed time-varying current $\hat{\mathbf{r}} J_r(\mathbf{r}, t)$. By symmetry, J_r depends only on r and produces a field $\mathbf{E}$ that depends only on r and is directed radially. An application of Gauss's law over a sphere of radius $r_0 > b_2$, which contains zero total charge, gives

$$4\pi r_0^2 E_r(r_0, t) = 0,$$

hence $\mathbf{E}(\mathbf{r}, t) = 0$ for $r > r_0$ and all time t. So $\mathbf{E} = 0$ external to the current distribution and no outward traveling wave is produced. Gauss's law also shows that $\mathbf{E} = 0$ inside the rigid sphere, while between the spheres

$$\mathbf{E}(\mathbf{r}, t) = -\hat{\mathbf{r}} \frac{Q}{4\pi\epsilon_0 r^2}.$$

Now work is certainly required to stretch the balloon and overcome the Lorentz force between the two charged surfaces. But an application of Poynting's theorem over a surface enclosing both spheres shows that no energy is carried away by an electromagnetic wave. Where does the expended energy go? The presence of only two nonzero terms in Poynting's theorem clearly indicates that the power term $\int_V \mathbf{E} \cdot \mathbf{J} \, dV$ corresponding to the external work must be balanced exactly by a change in stored energy. As the radius of the balloon increases, so does the region of nonzero field as well as the stored energy.

In free space any current source expressible in the form

$$\mathbf{J}(\mathbf{r},t) = \nabla\left(\frac{\partial\psi(\mathbf{r},t)}{\partial t}\right) \tag{2.298}$$

and localized to a volume region V, such as the current in the example above, is nonradiating. Indeed, Ampere's law states that

$$\nabla\times\mathbf{H} = \epsilon_0\frac{\partial\mathbf{E}}{\partial t} + \nabla\left(\frac{\partial\psi(\mathbf{r},t)}{\partial t}\right) \tag{2.299}$$

for $\mathbf{r}\in V$; taking the curl, we have

$$\nabla\times(\nabla\times\mathbf{H}) = \epsilon_0\frac{\partial\nabla\times\mathbf{E}}{\partial t} + \nabla\times\nabla\left(\frac{\partial\psi(\mathbf{r},t)}{\partial t}\right).$$

But the second term on the right is zero, so

$$\nabla\times(\nabla\times\mathbf{H}) = \epsilon_0\frac{\partial\nabla\times\mathbf{E}}{\partial t}$$

and this equation holds for all $\mathbf{r}$. By Faraday's law we can rewrite it as

$$\left((\nabla\times\nabla\times) + \frac{1}{c^2}\frac{\partial^2}{\partial t^2}\right)\mathbf{H}(\mathbf{r},t) = 0.$$

So $\mathbf{H}$ obeys the homogeneous wave equation everywhere, and $\mathbf{H}=0$ follows from causality. The laws of Ampere and Faraday may also be combined with (2.298) to show that

$$\left((\nabla\times\nabla\times) + \frac{1}{c^2}\frac{\partial^2}{\partial t^2}\right)\left[\mathbf{E}(\mathbf{r},t) + \frac{1}{\epsilon_0}\nabla\psi(\mathbf{r},t)\right] = 0$$

for all $\mathbf{r}$. By causality,

$$\mathbf{E}(\mathbf{r},t) = -\frac{1}{\epsilon_0}\nabla\psi(\mathbf{r},t) \tag{2.300}$$

everywhere. But since $\psi(\mathbf{r},t)=0$ external to V, we must also have $\mathbf{E}=0$ there. Note that $\mathbf{E}=-\nabla\psi/\epsilon_0$ is consistent with Ampere's law (2.299) provided that $\mathbf{H}=0$ everywhere.

We see that sources having spherical symmetry such that

$$\mathbf{J}(\mathbf{r},t) = \hat{\mathbf{r}}J_r(r,t) = \nabla\left(\frac{\partial\psi(r,t)}{\partial t}\right) = \hat{\mathbf{r}}\frac{\partial^2\psi(r,t)}{\partial r\partial t}$$

obey (2.298) and are therefore nonradiating. Hence the fields associated with any outward traveling spherical wave must possess some angular variation. This holds, for example, for the fields far removed from a time-varying source of finite extent.

As pointed out by Lindell [120], nonradiating sources are not merely hypothetical. The outflowing currents produced by a highly symmetric nuclear explosion in outer space or in a homogeneous atmosphere would produce no electromagnetic field outside the source region. The large electromagnetic-pulse effects discussed in Example 2.12 are due to inhomogeneities in the earth's atmosphere. We also note that the fields produced by a radiating source $\mathbf{J}^r(\mathbf{r},t)$ do not change external to the source if we superpose a nonradiating component $\mathbf{J}^{nr}(\mathbf{r},t)$ to create a new source $\mathbf{J}=\mathbf{J}^{nr}+\mathbf{J}^r$. We say that the two sources $\mathbf{J}$ and $\mathbf{J}^r$ are *equivalent* for the region V external to the sources. This presents difficulties in remote sensing where investigators are often interested in reconstructing an unknown source by probing the fields external to (and usually far away from) the source region. Unique reconstruction is possible only if the fields within the source region are also measured.

For the time harmonic case, Devaney and Wolf [48] provide the most general possible form for a nonradiating source. See § 4.11.10 for details.

2.11 Application: single charged particle motion in static electric and magnetic fields

Moving charge is of course current, so understanding the dynamics of moving charge is essential to understanding electromagnetics. The challenge is that particle motion can be extremely complicated depending on the environment where the motion takes place. Particles in a material lattice behave differently from those in a vacuum. Particles accelerated by a time-varying field obey much different physics than particles moving through a static field. Particles in a dense plasma, where interactions are strong, behave differently from particles in a highly diffuse gas where collisions are rare and particle-to-particle forces are weak. Many technologies are based on the specifics of these complex motions, including semiconductor devices, particle accelerators, microwave amplifiers, plasma displays, and spacecraft thrusters. Accurate descriptions of particle dynamics (in either deterministic or statistical fashion) are therefore essential. Even so, much insight can be gained by examining the simplest cases of individual charges moving through static fields, and several applications are relevant, such as cathode ray tubes, traveling-wave tube amplifiers, and particle separators.

2.11.1 Fundamental equations of motion

The trajectory of a particle of rest mass m_0 traveling with velocity $\mathbf{v}$ through an electric field $\mathbf{E}(\mathbf{r}, t)$ and a magnetic field $\mathbf{B}(\mathbf{r}, t)$ (measured in the laboratory frame) is determined from the Lorentz force [92]

$$\frac{d\mathbf{p}}{dt} = q(\mathbf{E} + \mathbf{v} \times \mathbf{B}), \tag{2.301}$$

where $\mathbf{p}$ is the particle momentum. Here we neglect all other forces acting on the particle (such as gravity). For particles moving at relativistic speeds, the momentum is given by $\mathbf{p} = \gamma m_0 \mathbf{v}$, where γ is the Lorentz factor

$$\gamma = \frac{1}{\sqrt{1 - \frac{v^2}{c^2}}}.$$

Relativistic effects may be neglected in many applications. However, electron speeds in gyrotrons typically exceed $c/2$ and relativistic effects cannot be ignored [39]. In free-electron lasers, or ubitrons [151], electron speed may be as high as $0.999998c$ [8, 112]. In large particle accelerators the electrons may move even faster — see Example 2.16 below.

▶ **Example 2.15:** Electron acceleration

Assume we should start considering relativistic effects when $\gamma = 1.1$. How fast must an electron travel to achieve this value of γ? Through what voltage must the electron be accelerated to reach this speed?

Solution: When $\gamma = 1.1$ we have

$$\frac{v}{c} = \sqrt{\frac{\gamma^2 - 1}{\gamma^2}} = 0.4166,$$

which indicates an electron traveling nearly half the speed of light. When accelerated through a voltage of V volts, an electron gains an energy of eV Joules, where e is the *unsigned* charge on the electron. If $V = 1$ V, we say that the electron has an energy of one electron volt, or 1 eV. Since this energy must be equal to the kinetic energy of the electron, we may find the required voltage by equating eV with this kinetic energy. If we include relativistic effects, the kinetic energy of the electron is given by [114]

$$E_k = mc^2 - m_0 c^2,$$

where m_0 is the rest mass of the electron and $m = \gamma m_0$. Thus

$$eV = m_0(\gamma - 1)c^2 \qquad (2.302)$$

and so

$$V = \frac{m_0}{e}(\gamma - 1)c^2.$$

Setting $m_0 = 9.109384 \times 10^{-31}$ kg, $e = 1.6021766 \times 10^{-19}$ C, and $\gamma = 1.1$, we find $V = 51.1$ kV. Hence we might expect to experience relativistic particle speeds only in technologies that employ high acceleration voltages. ◀

▶ **Example 2.16:** Relativistic velocity

A free electron laser uses an electron beam with an electron energy of 250 MeV. How fast do the electrons travel? The Stanford Linear Accelerator Center can impart an energy of 50 GeV to electrons. How fast do these electrons travel?

Solution: The relationship between electron energy and electron speed is given by (2.302). Solving this equation for v/c gives

$$\frac{v}{c} = \frac{\sqrt{1 + 2\frac{m_0 c^2}{eV}}}{1 + \frac{m_0 c^2}{eV}}.$$

Setting $V = 250 \times 10^6$ V gives $v/c = 0.999998$ and $\gamma = 490$. Setting $V = 50 \times 10^9$ V gives $v/c = 0.99999999995$ and $\gamma = 97849$. Obviously, these technologies produce electron speeds near the speed of light. ◀

Another form of (2.301) offers some insight into the relativistic behavior of the moving particle. By the product rule,

$$\frac{d\mathbf{p}}{dt} = \frac{d}{dt}(m_0 \gamma \mathbf{v}) = m_0 \mathbf{v}\frac{d\gamma}{dt} + m_0 \gamma \frac{d\mathbf{v}}{dt} = q(\mathbf{E} + \mathbf{v} \times \mathbf{B}).$$

We see that if the particle speed changes with time, then so does γ, and the first term contributes to the change in momentum. Note that

$$\mathbf{v} \cdot \frac{d\mathbf{p}}{dt} = m_0 v^2 \frac{d\gamma}{dt} + m_0 \gamma \mathbf{v} \cdot \frac{d\mathbf{v}}{dt} = q\mathbf{v} \cdot \mathbf{E}.$$

But

$$\frac{d}{dt}(\mathbf{v} \cdot \mathbf{v}) = \mathbf{v} \cdot \frac{d\mathbf{v}}{dt} + \frac{d\mathbf{v}}{dt} \cdot \mathbf{v} = 2\mathbf{v} \cdot \frac{d\mathbf{v}}{dt},$$

so

$$v^2 \frac{d\gamma}{dt} + \gamma \frac{1}{2} \frac{dv^2}{dt} = \frac{q}{m_0} \mathbf{v} \cdot \mathbf{E}.$$

Then, using

$$\frac{d\gamma}{dt} = \frac{d}{dt} \frac{1}{\sqrt{1 - \frac{v^2}{c^2}}} = \frac{1}{2c^2} \gamma^3 \frac{dv^2}{dt}$$

we find that

$$v^2 \frac{d\gamma}{dt} + \gamma \frac{1}{2} \frac{dv^2}{dt} = \frac{d\gamma}{dt} \left(v^2 + \frac{c^2}{\gamma^2} \right) = c^2 \frac{d\gamma}{dt} = \frac{q}{m_0} \mathbf{v} \cdot \mathbf{E},$$

and thus

$$\frac{d\gamma}{dt} = \frac{q}{m_0 c^2} \mathbf{v} \cdot \mathbf{E}.$$

Hence, changes in particle speed (and thus in γ) are due only to the electric field and not the magnetic field (i.e., the magnetic field can cause acceleration only by changing the particle direction). Then, substituting this into (2.301), we get an alternative form of the equation of motion:

$$\gamma m_0 \frac{d\mathbf{v}}{dt} + \frac{q\mathbf{v}}{c^2} \mathbf{v} \cdot \mathbf{E} = q(\mathbf{E} + \mathbf{v} \times \mathbf{B}). \tag{2.303}$$

For most of the cases considered below, we ignore relativistic effects by assuming $v^2/c^2 \ll 1$ so that $\gamma \approx 1$. Then we have simply

$$\frac{d\mathbf{v}}{dt} = \frac{q}{m}(\mathbf{E} + \mathbf{v} \times \mathbf{B}), \tag{2.304}$$

since $m = m_0$.

2.11.2 Nonrelativistic particle motion in a uniform, static electric field

Neglecting relativistic effects, the trajectory of a particle in a uniform static electric field is determined from (2.304) using

$$\frac{d\mathbf{v}}{dt} = \frac{q}{m}\mathbf{E}, \qquad \mathbf{v} = \frac{d\mathbf{r}}{dt}. \tag{2.305}$$

Integration gives the velocity vector at time t as

$$\mathbf{v}(t) = \mathbf{v}_0 + \frac{q}{m}\mathbf{E}t, \tag{2.306}$$

where $\mathbf{v}_0$ is the initial velocity. Hence, the particle gains speed linearly with time, with the direction of motion determined by the sign of q. The particle's kinetic energy is

$$E_k(t) = \tfrac{1}{2}m|\mathbf{v}|^2 = \tfrac{1}{2}m \left| \mathbf{v}_0 + \frac{q}{m}\mathbf{E}t \right|^2,$$

which increases quadratically with time. Since velocity is the time derivative of position, we can also write (2.306) as

$$\frac{d\mathbf{r}}{dt} = \mathbf{v}_0 + \frac{q}{m}\mathbf{E}t.$$

Integration gives the particle position:

$$\mathbf{r}(t) = \mathbf{r}_0 + \mathbf{v}_0 t + \frac{1}{2}\frac{q}{m}\mathbf{E}t^2. \qquad (2.307)$$

▶ **Example 2.17:** Longitudinal electron acceleration: the electron gun

An electron gun creates a collimated stream of electrons for use in, e.g., a cathode ray tube, a microwave source or amplifier, or an electron microscope. A heated cathode emits electrons with low initial velocity $\mathbf{v}_0$, which are then accelerated in a static electric field. The electric field is created by applying a voltage between the cathode and an anode, separated by a distance d so that $\mathbf{E}_0 = -\hat{\mathbf{x}}V_0/d$ (Figure 2.11). Consider an electron emitted at the origin at time $t = 0$. Determine the speed and kinetic energy of the electron when it passes through the anode at $x = d$.

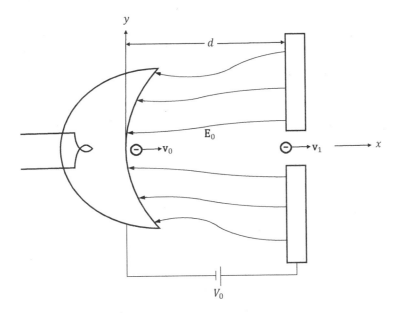

FIGURE 2.11
Electron gun.

Solution: The electron velocity is given by (2.306):

$$\mathbf{v}(t) = \mathbf{v}_0 + \frac{q}{m}\mathbf{E}t = \left(v_0 + \frac{e}{m}E_0 t\right)\hat{\mathbf{x}}.$$

Its kinetic energy is

$$E_k(t) = \tfrac{1}{2}m\left(v_0 + \frac{e}{m}E_0 t\right)^2,$$

and its position is, by (2.307),

$$\mathbf{r}(t) = \mathbf{r}_0 + \mathbf{v}_0 t + \frac{1}{2}\frac{e}{m}E_0 t^2 \hat{\mathbf{x}} = \left(v_0 t + \frac{1}{2}\frac{e}{m}E_0 t^2 \right) \hat{\mathbf{x}}.$$

Assuming the initial velocity can be neglected, the electron will reach the anode when

$$\frac{1}{2}\frac{e}{m}E_0 t^2 = d$$

or

$$t = \sqrt{\frac{2d^2 m}{eV_0}}.$$

At that time the speed is

$$v = \frac{e}{m}E_0 t = \sqrt{\frac{2q_e V_0}{m}}$$

and the kinetic energy is

$$E_k = \tfrac{1}{2}m \left(\sqrt{\frac{2eV_0}{m}} \right)^2 = eV_0.$$

This result is expected from the electrostatic relationship between energy and voltage, as discussed in the next chapter. ◄

► **Example 2.18:** Lateral electron acceleration: deflection in a CRT

A classic application of electrostatic particle deflection is the cathode ray tube, used as a display device in some oscilloscopes. An electron gun launches electrons into an electric field; the field deflects the electrons so that they impact a fluorescent screen at a defined point, and the screen in turn emits visible photons. Alterations in the voltage cause a pattern of light to be traced on the screen.

Consider Figure 2.12. An electron gun accelerates an electron through a voltage V_0 so that its initial velocity at the origin is $\mathbf{v}_0 = v_0 \hat{\mathbf{x}}$. At that point it enters the region between two deflector plates; the plates are held at a potential difference V_d so that the electric field is (neglecting fringing) $\mathbf{E}_0 = -\hat{\mathbf{y}}V_d/d$. If the deflector plates have length L, find the position y_2 at which the electron impacts a screen located a distance D beyond the plates. Also find the kinetic energy gained by the electron in passing through the deflector plates.

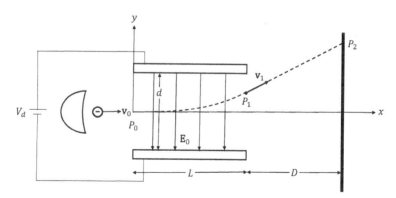

FIGURE 2.12
Electron deflection in a cathode ray tube.

Solution: The initial kinetic energy of the electron is

$$E_k = \tfrac{1}{2}mv_0^2 = eV_0,$$

so the initial velocity is

$$\mathbf{v}_0 = \hat{\mathbf{x}}\sqrt{\frac{2eV_0}{m}}.$$

The velocity at any time while the electron is between the plates is given by (2.306):

$$\mathbf{v}(t) = v_0\hat{\mathbf{x}} + \frac{e}{m}E_0 t\hat{\mathbf{y}}.$$

Hence the kinetic energy is

$$E_k(t) = eV_0 + \frac{1}{2}\frac{e^2}{m}E_0^2 t^2.$$

The electron position is given by (2.307):

$$\mathbf{r}(t) = v_0 t\hat{\mathbf{x}} + \frac{1}{2}\frac{e}{m}E_0 t^2\hat{\mathbf{y}}.$$

The particle will reach the end of the deflector plates when $t = t_1 = L/v_0$. At this time the velocity is

$$\mathbf{v}_1 = v_0\hat{\mathbf{x}} + \frac{e}{m}E_0 t_1\hat{\mathbf{y}} = v_0\hat{\mathbf{x}} + \frac{e}{m}E_0\frac{L}{v_0}\hat{\mathbf{y}},$$

the position is

$$\mathbf{r}_1 = v_0 t_1\hat{\mathbf{x}} + \frac{1}{2}\frac{e}{m}E_0 t_1^2\hat{\mathbf{y}} = L\hat{\mathbf{x}} + \frac{1}{4}\frac{V_d}{V_0}\frac{L^2}{d}\hat{\mathbf{y}},$$

and the kinetic energy is

$$E_k(t_1) = eV_0 + \frac{1}{2}\frac{e^2}{m}E_0^2 t_1^2 = eV_0 + eV_d\left(\frac{1}{4}\frac{V_d}{V_0}\frac{L^2}{d^2}\right).$$

After passing through the deflector plates, the electron will no longer gain kinetic energy; therefore the gain in kinetic energy is

$$\Delta E_k = eV_d\left(\frac{1}{4}\frac{V_d}{V_0}\frac{L^2}{d^2}\right),$$

and the gain relative to the initial kinetic energy is

$$\frac{\Delta E_k}{E_{k0}} = \frac{eV_d\left(\frac{1}{4}\frac{V_d}{V_0}\frac{L^2}{d^2}\right)}{eV_0} = \left(\frac{1}{2}\frac{V_d}{V_0}\frac{L}{d}\right)^2.$$

After the electron passes through the deflector plates it continues unaccelerated, with position given by

$$\mathbf{r}(t) = \mathbf{r}_1 + \mathbf{v}_1(t - t_1),$$

reaching the screen at time $t - t_1 = D/v_0$. At this time we have

$$x_2 = L + v_0\frac{D}{v_0} = L + D$$

and

$$y_2 = \frac{1}{4}\frac{V_d}{V_0}\frac{L^2}{d} + \frac{e}{m}E_0\frac{LD}{v_0^2} = \frac{1}{2}\frac{V_d}{V_0}\frac{L}{d}\left(\frac{L}{2} + D\right).$$

Large deflections may be achieved with large deflection voltages relative to the initial acceleration voltage, with wide deflector plates, or with a long travel distance to the screen. ◀

2.11.3   Nonrelativistic particle motion in a nonuniform, static electric field; electron optics

Nonuniform static electric fields are used as *electrostatic lenses* to direct and focus electron beams. Applications include particle extractors, electron microprobes, and electron microscopes [106]. Regions of nonuniform fields are created using closely spaced electrodes, apertures in screens, and gaps in cylinders, and particles passing through these regions undergo alteration of their trajectories to produce focused beams.

It is found that particle trajectories in nonuniform fields follow rules similar to those followed by light rays in geometrical optics. Hence the term *electron optics* is often assigned to the study of these trajectories. When light rays cross an interface between differing optical media, they undergo refraction and the direction of the ray path changes. When an electron crosses an interface between differing regions of electrostatic potential, the trajectory of the particle changes in a manner analogous to light refraction. A light ray bends toward the interface normal when entering a region of greater optical index. In electron optics, the electron trajectory bends toward the normal as the electron enters a region of higher potential. However, while a light ray may experience a discontinuous alteration of its direction when crossing an interface between differing materials, the potential in vacuum is never truly discontinuous (although the gradient can be large), and thus the path of an electron always changes continuously. We can see this effect in Figure 2.12. The equipotential surfaces in the region between the deflection plates are parallel to the plates, and as the electron crosses each equipotential surface into a region of higher potential, its path changes slightly, bending toward the normal to the surfaces. The result is the curved parabolic trajectory described in the equations of Example 2.18. In optics, the path of a light ray through a continuously changing optical medium is described by Fermat's principle of least time:

$$\delta \int_{\Gamma, A}^{B} n(\mathbf{r}) \, dl = 0,$$

where $n(\mathbf{r})$ is the optical index of the nonuniform medium. That is, the path Γ taken by the light ray between the points A and B minimizes the optical path length, and the ray takes the least amount of time to get from A to B. Similarly, the curved path of an electron in a nonuniform electric field may be described by a version of Fermat's principle,

$$\delta \int_{\Gamma, A}^{B} v(\mathbf{r}) \, dl = 0,$$

where $v(\mathbf{r})$ is the particle speed in the nonuniform field [131]. This implies that the path Γ is such that the particle takes the least time traveling from A to B, just as in optics.

Closed-form expressions for paths of charged particles in nonuniform electric fields may be difficult to obtain. Hence numerical solutions to the kinematic equations are often sought. A simple numerical solution, often sufficient for illustrative purposes, can be obtained using a finite-difference, or *Euler* method. The algorithm is as follows:

1. choose a time step δt and a number of iterations N;
2. set $t_0 = 0$; set the initial particle position $\mathbf{r}_0$ and velocity $\mathbf{v}_0$;
3. update the velocity using (2.305): $\mathbf{v}_{i+1} = \mathbf{v}_i + (q/m)\mathbf{E}\,\delta t$;
4. update the position: $\mathbf{r}_{i+1} = \mathbf{r}_i + \mathbf{v}_{i+1}\,\delta t$;
5. update the time: $t_{i+1} = t_i + \delta t$;
6. return to step 3 and repeat N times.

This algorithm is used to calculate electron trajectories in the following example.

▶ **Example 2.19**: Electrostatic lens

A very simple electrostatic lens may be created using two point charges (or equivalently two uniformly charged spheres). Assume identical charges $Q = +10$ nC are fixed at the points $(0, 0.5)$ m and $(0, -0.5)$ m in the xy-plane. Electrons are injected at various positions y_0 at $x = -4$ m in the xy-plane, with initial speed $v_0 = 6.2 \times 10^6$ m/s, and initial trajectories passing through the origin. Use the finite-difference approach to plot the electron trajectories and determine their speeds over time.

Solution: With $\delta t = 2$ ns, the finite-difference technique outlined above gives the electron trajectories shown in Figure 2.13. As an electron travels it intersects equipotential surfaces at a slight angle, and its trajectory bends slightly toward the normal. This effect accumulates until the particle passes between the charges along a path nearly parallel to the x-axis. The point charges thus act as a diverging lens, transforming the converging electron paths to a collimated beam parallel to the x-axis. Of course, in this simple example collimation only occurs for electrons traveling in the xy-plane. To collimate particles converging from a cathode, for instance, a more complicated three-dimensional arrangement of electrodes is required.

Note that the trajectories in Figure 2.13 are for individual particles, not for a beam of electrons, and thus there is no interaction between particles. In an actual electron beam, mutual repulsion between electrons will spread the beam, even if it is at some point completely collimated. Hence long electron beams must be continually refocused, using either nonuniform static electric fields or nonuniform static magnetic fields. For instance, in many linear-beam microwave devices, such as traveling-wave tubes, the beam is collimated over a long distance using alternating-pole permanent magnets [152]. Electron trajectories in the presence of magnetic fields is considered in later sections.

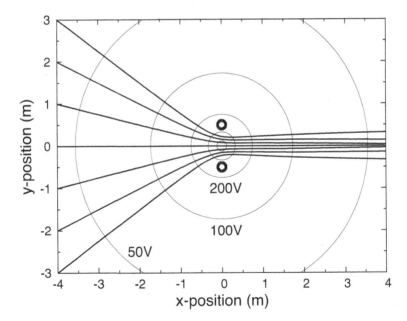

FIGURE 2.13

Trajectory of electrons deflected by a pair of positive point charges. Equipotential lines are for 50, 100, 200, 300, and 350 V.

Figure 2.14 shows the speed of electrons injected at $y_0 = 0$ and at $y_0 = 3m$. Electrons accelerate as they are drawn toward the positive charges by the outward-pointing electric field. They reach a maximum speed as they pass nearest the charges and then start to decelerate. Note that the electron velocity is maximum at the point of maximum potential along the trajectory, so the kinetic energy $mv^2/2$ is greatest there. However, the potential energy $-eV$ is lowest at this point, and thus the sum of the kinetic and potential energies is conserved.

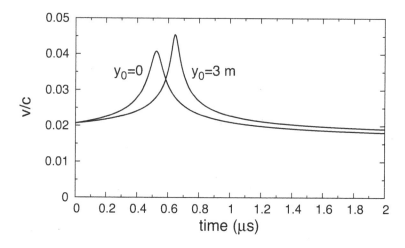

FIGURE 2.14

Speed of electrons deflected by a pair of positive point charges. ◄

2.11.4   Nonrelativistic particle motion in a uniform, static magnetic field

In the absence of relativistic effects, the trajectory of a particle in a uniform static magnetic field is determined from (2.304) using

$$\frac{d\mathbf{v}}{dt} = \frac{q}{m}\mathbf{v} \times \mathbf{B}, \qquad \mathbf{v} = \frac{d\mathbf{r}}{dt}. \tag{2.308}$$

As mentioned earlier, the magnetic field may accelerate a particle by changing its direction, but it cannot change the particle speed. It is also clear that a particle following a trajectory parallel to $\mathbf{B}$ will not be affected by the magnetic field, since $\mathbf{v} \times \mathbf{B}$ is zero. Hence it is convenient to decompose the velocity into components parallel and perpendicular to the magnetic field,

$$\mathbf{v} = \mathbf{v}_\parallel + \mathbf{v}_\perp, \tag{2.309}$$

such that

$$\mathbf{v}_\parallel \times \mathbf{B} = 0, \qquad \mathbf{v}_\perp \times \mathbf{B} \neq 0.$$

Substituting (2.309) into (2.308) we have

$$\frac{d\mathbf{v}_\parallel}{dt} = 0, \tag{2.310}$$

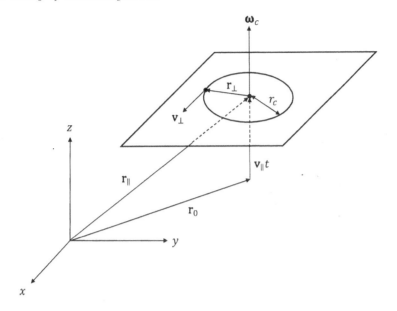

FIGURE 2.15

Transverse-plane motion of a charged particle in a uniform static magnetic field.

which says that the magnetic field does not affect the velocity in the direction of the magnetic field, and

$$\frac{d\mathbf{v}_\perp}{dt} = \frac{q}{m}\mathbf{v}_\perp \times \mathbf{B} = \boldsymbol{\omega}_c \times \mathbf{v}_\perp, \tag{2.311}$$

which says that the particle will accelerate in the plane perpendicular to the magnetic field. Here

$$\boldsymbol{\omega}_c = -\frac{q}{m}\mathbf{B}$$

is a vector describing the particle acceleration. Its magnitude

$$\omega_c = \frac{|q|}{m}|\mathbf{B}| = \frac{|q|}{m}B$$

is called the *cyclotron frequency*.

The particle trajectory may be found by decomposing the position vector as

$$\mathbf{r} = \mathbf{r}_\parallel + \mathbf{r}_\perp$$

so that

$$\mathbf{v}_\parallel = \frac{d\mathbf{r}_\parallel}{dt}, \qquad \mathbf{v}_\perp = \frac{d\mathbf{r}_\perp}{dt}. \tag{2.312}$$

From (2.310) we find that $\mathbf{v}_\parallel$ is constant with time and thus

$$\mathbf{r}_\parallel = \mathbf{r}_0 + \mathbf{v}_\parallel t.$$

From (2.311) we have

$$\frac{d\mathbf{v}_\perp}{dt} = \boldsymbol{\omega}_c \times \frac{d\mathbf{r}_\perp}{dt},$$

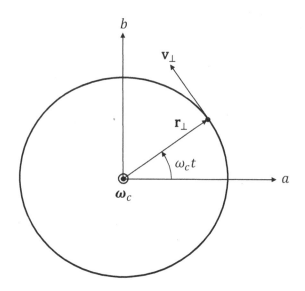

FIGURE 2.16
Motion of a charged particle in a uniform static magnetic field about its guiding center.

and thus

$$\mathbf{v}_\perp = \boldsymbol{\omega}_c \times \mathbf{r}_\perp \tag{2.313}$$

since $\boldsymbol{\omega}_c$ is constant with time. Now, since the particle speed is constant, so are both $|\mathbf{v}_\perp|$ and $|\mathbf{r}_\perp| = r_c$. So $\mathbf{v}_\perp$ is tangential to a circle of radius r_c, and represents the velocity of a particle moving in circular motion in the plane perpendicular to $\mathbf{B}$. The center of this circle, or *guiding center*, is located by the position vector $\mathbf{r}_\parallel = \mathbf{r}_0 + \mathbf{v}_\parallel t$ as shown in Figure 2.15.

We can write an explicit formula for the particle trajectory by considering Figure 2.16. Define the unit vector $\hat{\boldsymbol{\omega}}_c = \boldsymbol{\omega}_c / \omega_c$, and pick a time $t = 0$ such that

$$\hat{\mathbf{a}} = \frac{\mathbf{r}_\perp(t=0)}{r_c}.$$

Then $\hat{\mathbf{b}} = \hat{\boldsymbol{\omega}}_c \times \hat{\mathbf{a}}$, and we can write the position and velocity vectors as

$$\mathbf{v} = \mathbf{v}_\parallel + v_a \hat{\mathbf{a}} + v_b \hat{\mathbf{b}}, \qquad \mathbf{r} = \mathbf{r}_0 + r_a \hat{\mathbf{a}} + r_b \hat{\mathbf{b}}.$$

With this, we have from (2.312) and (2.313)

$$\frac{d}{dt}(\hat{\mathbf{a}} r_a + \hat{\mathbf{b}} r_b) = \omega_c \hat{\boldsymbol{\omega}}_c \times (\hat{\mathbf{a}} r_a + \hat{\mathbf{b}} r_b).$$

But $\hat{\boldsymbol{\omega}}_c \times \hat{\mathbf{b}} = \hat{\boldsymbol{\omega}}_c \times (\hat{\boldsymbol{\omega}}_c \times \hat{\mathbf{a}}) = -\hat{\mathbf{a}}$, so we have the two equations

$$\frac{dr_a}{dt} = -\omega_c r_b, \qquad \frac{dr_b}{dt} = \omega_c r_a.$$

Substituting for r_b, we obtain

$$\frac{d^2 r_a}{dt^2} + \omega_c^2 r_a = 0,$$

which has the solution

$$r_a = r_c \cos \omega_c t$$

by our choice of time reference. Then

$$r_b = -\frac{1}{\omega_c}\frac{dr_a}{dt} = r_c \sin \omega_c t.$$

Thus,

$$\mathbf{r}_\perp = \hat{\mathbf{a}} r_c \cos \omega_c t + \hat{\mathbf{b}} r_c \sin \omega_c t.$$

Finally, the particle trajectory is given by

$$\mathbf{r} = \mathbf{r}_\parallel + \mathbf{r}_\perp = \mathbf{v}_\parallel t + r_c(\hat{\mathbf{a}} \cos \omega_c t + \hat{\boldsymbol{\omega}}_c \times \hat{\mathbf{a}} \sin \omega_c t).$$

The result of superposing the linear and circular motions is a helical trajectory. The helix can be either left-handed or right-handed, depending on the sign of q and the direction of $\mathbf{v}_\parallel$ relative to the direction of $\mathbf{B}$. The particle will travel once around the helix in time $T_c = 2\pi/\omega_c$ (known as the *Larmor period* [16]), and in that time will also progress a distance

$$p = \frac{2\pi}{\omega_c} v_\parallel$$

along the helix axis. This distance p is called the *pitch* of the helix. The *pitch angle* α is the angle made between the direction of motion and the *ab*-plane, or

$$\tan \alpha = \frac{p}{2\pi r_c} = \frac{v_\parallel}{\omega_c r_c} = \frac{v_\parallel}{v_\perp}.$$

▶ **Example 2.20**: Spiraling particles in the ionosphere

Particles impinging on the earth from the solar wind are guided along the earth's magnetic field lines toward the poles. Protons and electrons typically travel at 3×10^5 m/s when reaching the upper ionosphere at altitude 250 km where the earth's magnetic field is around 50×10^{-6} T [157]. Assuming the particles spiral with a pitch angle of $45°$, find the radius and pitch of a helical trajectory. Plot the spiraling trajectory of protons for a typical case of geometrical parameters.

Solution: For $\alpha = 45°$, we have $v_\parallel = v_\perp = v/\sqrt{2} = 212$ km/s. Using

$$\omega_c = \frac{|q|}{m} B$$

we have $\omega_c = 4.8$ kr/s for protons and $\omega_c = 8.8$ Mr/s for electrons. Finally, using $v_\perp = \omega_c r_c$ we find that $r_c = 44$ m for protons and $r_c = 2.4$ cm for electrons. So electrons spiral much more tightly than do protons with the same speed. This is because of the much smaller mass of the electron. For protons the helical pitch is $p = 2\pi v_\parallel/\omega_c = 278$ m, so each proton spirals about four times during each kilometer of axial distance traveled.

To plot the trajectory we must choose the geometrical properties of the path. For illustrative purposes let

$$\hat{\boldsymbol{\omega}}_c = 0.2\hat{\mathbf{x}} + 0.2\hat{\mathbf{y}} + \sqrt{0.92}\hat{\mathbf{z}},$$

$$\hat{\mathbf{a}} = \sqrt{0.5}\hat{\mathbf{x}} - \sqrt{0.5}\hat{\mathbf{y}},$$

$$\hat{\mathbf{b}} = \sqrt{0.46}\hat{\mathbf{x}} + \sqrt{0.46}\hat{\mathbf{y}} - \sqrt{0.08}\hat{\mathbf{z}},$$

$$\mathbf{r}_0 = 50\hat{\mathbf{x}} + 50\hat{\mathbf{y}} \text{ m}.$$

Note that $\hat{\mathbf{a}} \cdot \hat{\boldsymbol{\omega}} = 0$, as required, and $\hat{\mathbf{b}} = \hat{\boldsymbol{\omega}}_c \times \hat{\mathbf{a}}$. Figure 2.17 shows the trajectory plotted for a proton over about seven cycles of the helix (from $z = 0$ to $z = 2$ km). The dotted line shows the trajectory of the guiding center as determined by the formula $\mathbf{r} = \mathbf{r}_0 + \mathbf{v}_{\parallel} t$, which starts at the point $(50, 50)$ m in the $z = 0$ plane. As expected, the proton spirals upward about the trajectory of the guiding center, which makes an angle of $\cos^{-1}(0.92) = 23°$ with the z-axis.

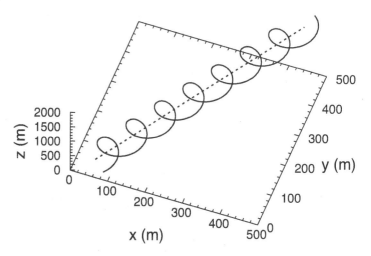

FIGURE 2.17

Helical trajectory of a proton in the ionosphere. Dotted line shows the trajectory of the guiding center. ◄

▶ **Example 2.21:** Electron deflection in a static magnetic field

Example 2.18 showed how an electron may be deflected using a static electric field, forming the basis for the cathode ray tube. Until the advent of flat screen technology, CRTs were also used in television sets and computer monitors. These types of CRTs use magnetic fields to deflect an electron beam, thereby forming an image on a phosphorescent screen.

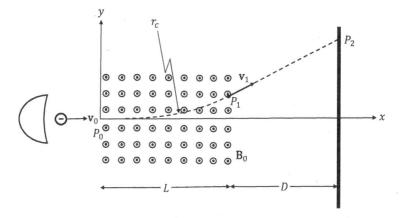

FIGURE 2.18

Electron deflection by a uniform static magnetic field in a cathode ray tube.

Consider Figure 2.18. An electron gun accelerates an electron through a voltage V_0 so that its initial velocity at the origin is $\mathbf{v}_0 = v_0\hat{\mathbf{x}}$. It then enters a region of uniform static magnetic field $\mathbf{B} = B_0\hat{\mathbf{z}}$, created by a permanent magnet or by an electromagnet. If the magnetic field region has length L, find the position y_2 at which the electron strikes a screen located a distance D beyond the plates.

Solution: When the electron enters the magnetic field it is deflected by the Lorentz force and adopts a circular trajectory. So the entry point is on the tangent to the circular trajectory experienced by the electron, and we have the initial guiding center of gyration as $\mathbf{r}_0 = r_c\hat{\mathbf{y}}$. Because the initial trajectory is perpendicular to the magnetic field, and because the magnetic field cannot change the electron speed, the parallel component of the velocity is zero, while the perpendicular component has a magnitude v_0. So $v_\parallel = 0$ and $v_\perp = v_0$. The cyclotron frequency vector and radius of curvature of the electron trajectory are given by

$$\boldsymbol{\omega}_c = \hat{\mathbf{z}}\frac{e}{m}B_0, \qquad r_c = \frac{v_\perp}{\omega_c} = \frac{v_0 m}{e B_0}.$$

Assuming the electron enters the magnetic field at $t = 0$, we must have $\hat{\mathbf{a}} = -\hat{\mathbf{y}}$ and $\hat{\mathbf{b}} = \hat{\boldsymbol{\omega}}_c \times \hat{\mathbf{a}} = \hat{\mathbf{x}}$. Also $\mathbf{r}_\parallel = 0$ because $v_\parallel = 0$, and the guiding center of gyration of the electron remains stationary at $\mathbf{r}_0 = r_c\hat{\mathbf{y}}$. Hence we can describe the trajectory of the electron while it is in the magnetic field as

$$\mathbf{r} = \mathbf{r}_0 + \mathbf{r}_\perp = r_c\hat{\mathbf{y}} + r_c(-\hat{\mathbf{y}}\cos\omega_c t + \hat{\mathbf{x}}\sin\omega_c t)$$
$$= \hat{\mathbf{x}}r_c\sin\omega_c t + \hat{\mathbf{y}}r_c(1 - \cos\omega_c t).$$

The electron reaches the edge of the deflection region when $x(t) = L$, or

$$r_c\sin\omega_c t = L.$$

This occurs at time

$$t = \frac{1}{\omega_c}\sin^{-1}\left\{\frac{L}{r_c}\right\}.$$

At that time the electron has traveled to point P_1 where the y coordinate is

$$y = r_c\left[1 - \cos\left(\sin^{-1}\left\{\frac{L}{r_c}\right\}\right)\right] = r_c\left[1 - \frac{\sqrt{r_c^2 - L^2}}{r_c}\right] = r_c - \sqrt{r_c^2 - L^2}.$$

Once the electron leaves the magnetic field region, it travels with uniform velocity until impacting the screen at $x = L + D$. At the exit point P_1 we have

$$\mathbf{r}_\perp = L\hat{\mathbf{x}} - \sqrt{r_c^2 - L^2}\,\hat{\mathbf{y}},$$

and thus the electron has exit velocity

$$\mathbf{v}_\perp = \boldsymbol{\omega}_c \times \mathbf{r}_\perp = v_0\frac{L}{r_c}\hat{\mathbf{y}} + v_0\frac{\sqrt{r_c^2 - L^2}}{r_c}\hat{\mathbf{x}}.$$

The electron will therefore travel from the exit point to the screen in time

$$t = \frac{D r_c}{v_0\sqrt{r_c^2 - L^2}}.$$

Finally, at this time the y-coordinate of the electron position is

$$y_2 = v_0\frac{L}{r_c}\left[\frac{D r_c}{v_0\sqrt{r_c^2 - L^2}}\right] + r_c - \sqrt{r_c^2 - L^2},$$

which can be written as

$$y_2 = r_c \left[1 - \sqrt{1 - \frac{L^2}{r_c^2}} + \frac{LD}{r_c^2} \frac{1}{\sqrt{1 - \frac{L^2}{r_c^2}}} \right].$$

For small deflection ($L \ll r_c$), we may approximate the square root function with the first two terms of the binomial series and find that

$$y_2 \approx \frac{L^2}{r_c} \left(\frac{D}{L} + \frac{1}{2} \right).$$

So larger deflections are produced by smaller values of r_c (which requires larger values of B_0) and by larger values of D and L. ◀

2.11.5   Nonrelativistic particle motion in uniform, static electric and magnetic fields: $\mathbf{E} \times \mathbf{B}$ drift

In the simultaneous presence of both electric and magnetic fields, a particle trajectory is determined from

$$\frac{d\mathbf{v}}{dt} = \frac{q}{m} \left(\mathbf{E} + \mathbf{v} \times \mathbf{B} \right), \qquad \mathbf{v} = \frac{d\mathbf{r}}{dt}. \tag{2.314}$$

We may solve these by decomposing the electric field, the velocity vector, and the position vector into components parallel and perpendicular to $\mathbf{B}$:

$$\mathbf{E} = \mathbf{E}_\| + \mathbf{E}_\perp, \qquad \mathbf{v} = \mathbf{v}_\| + \mathbf{v}_\perp, \qquad \mathbf{r} = \mathbf{r}_\| + \mathbf{r}_\perp.$$

Substituting into (2.314) and equating parallel and perpendicular vectors, we obtain

$$\frac{dv_\|}{dt} = \frac{q}{m} E_\|, \qquad\qquad v_\| = \frac{dr_\|}{dt}, \tag{2.315}$$

$$\frac{d\mathbf{v}_\perp}{dt} = \frac{q}{m} \left(\mathbf{E}_\perp + \mathbf{v}_\perp \times \mathbf{B} \right), \qquad\qquad \mathbf{v}_\perp = \frac{d\mathbf{r}_\perp}{dt}. \tag{2.316}$$

The first equation is identical to that for a particle moving in a uniform, static electric field, and can be integrated to give

$$v_\|(t) = v_{0\|} + \frac{q}{m} E_\| t, \qquad r_\|(t) = r_{0\|} + v_{0\|} t + \frac{1}{2} \frac{q}{m} E_\| t^2.$$

So the particle accelerates in the direction of the magnetic field under the force of the parallel component of the electric field. Motion perpendicular to $\mathbf{B}$ is more complicated; we expect some combination of magnetic and electric forces, with the magnetic field causing a gyrating motion and the electric field normal to $\mathbf{B}$ causing acceleration along the perpendicular direction. The actual effect is more intriguing.

Equation (2.316) may be solved using a clever decomposition of the perpendicular velocity. Let

$$\mathbf{v}_\perp = \mathbf{u} + \mathbf{w}$$

where $\mathbf{w}$ is time-independent. Such a decomposition is always possible, as we can always have $\mathbf{u} = \mathbf{v}_\perp - \mathbf{w}$ so that

$$\mathbf{v}_\perp = (\mathbf{v}_\perp - \mathbf{w}) + \mathbf{w}.$$

Hence $\mathbf{w}$ may be chosen as any time-independent vector. Substitution into (2.316) gives

$$\frac{d\mathbf{u}}{dt} = \frac{q}{m} \left[\mathbf{E}_\perp + (\mathbf{u} + \mathbf{w}) \times \mathbf{B} \right] = \frac{q}{m} \mathbf{u} \times \mathbf{B} + \frac{q}{m} \left(\mathbf{E}_\perp + \mathbf{w} \times \mathbf{B} \right). \tag{2.317}$$

Since $\mathbf{w}$ is arbitrary, let us choose it so that

$$\frac{q}{m}\left(\mathbf{E}_\perp + \mathbf{w} \times \mathbf{B}\right) = 0$$

or

$$\mathbf{w} \times \mathbf{B} = -\mathbf{E}_\perp.$$

Crossing this equation with $\mathbf{B}$ we get

$$\mathbf{B} \times (\mathbf{w} \times \mathbf{B}) = -\mathbf{B} \times \mathbf{E}_\perp = -\mathbf{B} \times (\mathbf{E} - \mathbf{E}_\parallel) = \mathbf{E} \times \mathbf{B},$$

since $\mathbf{B} \times \mathbf{E}_\parallel = 0$. Expansion of the cross product gives

$$\mathbf{B} \times (\mathbf{w} \times \mathbf{B}) = \mathbf{w}(\mathbf{B} \cdot \mathbf{B}) - \mathbf{B}(\mathbf{w} \cdot \mathbf{B}) = B^2 \mathbf{w},$$

since $\mathbf{w}$ is normal to $\mathbf{B}$. Thus

$$\mathbf{w} = \frac{\mathbf{E} \times \mathbf{B}}{|\mathbf{B}|^2}. \tag{2.318}$$

We also have from (2.317) the equation for $\mathbf{u}$:

$$\frac{d\mathbf{u}}{dt} = \frac{q}{m}\mathbf{u} \times \mathbf{B}.$$

This is identical to (2.311), which describes the gyrating motion of a particle in a uniform magnetic field. Therefore $\mathbf{v}_\perp$ is a superposition of two velocities. The first gives rise to the gyrating motion that would be executed by the particle if there were a magnetic field but no electric field. The second is a *constant* velocity in the direction of $\mathbf{E} \times \mathbf{B}$. So the gyrating particle *drifts* along this direction. The effect of the electric field in the direction transverse to $\mathbf{B}$ is called, appropriately, the $\mathbf{E} \times \mathbf{B}$ drift.

Several observations are warranted. First note that any constant force may substitute for the electric force $q\mathbf{E}$, or augment this force. For instance, if a gravitational field produces a force $\mathbf{F}_g$, then the equation of motion is

$$\frac{d\mathbf{v}}{dt} = \frac{q}{m}\left(\mathbf{E} + \frac{\mathbf{F}_g}{q} + \mathbf{v} \times \mathbf{B}\right).$$

Decomposing $\mathbf{F}_g$ into parallel and perpendicular components, we see that the velocity in the parallel direction is

$$v_\parallel(t) = v_{0\parallel} + \frac{q}{m}\left(E_\parallel + \frac{F_{g\parallel}}{q}\right)t,$$

while that in the perpendicular direction represents gyration superimposed with a drift velocity

$$\mathbf{w} = \frac{1}{|\mathbf{B}|^2}\left(\mathbf{E} + \frac{\mathbf{F}_g}{q}\right) \times \mathbf{B}.$$

Second, note that the drift velocity (2.318) is independent of q. Third, if the particle motion is observed in a Galilean reference frame moving with uniform velocity $\mathbf{w}$, the particle will seem to gyrate under the influence of $\mathbf{B}$ but the effect from the electric field $\mathbf{E}_\perp$ will not be observed. (Acceleration along the parallel direction will be observed, however.) These last two observations suggest that it may be insightful to examine a particle moving at relativistic speeds in certain Lorentzian reference frames. This is done in § 2.11.7.

► **Example 2.22:** $\mathbf{E} \times \mathbf{B}$ drift for normal fields

Study the trajectory of a particle immersed in perpendicular fields $\mathbf{B} = B_0\hat{\mathbf{z}}$ and $\mathbf{E} = E_0\hat{\mathbf{y}}$.

Solution: Since $\mathbf{E}$ is perpendicular to $\mathbf{B}$ there is no parallel component of $\mathbf{E}$, hence no z-directed acceleration. The equation of motion (2.316) is

$$\frac{d\mathbf{v}_\perp}{dt} = \frac{q}{m}[\hat{\mathbf{x}}v_y B_0 + \hat{\mathbf{y}}(E_0 - v_x B_0)],$$

or in component form

$$\frac{dv_x}{dt} = \frac{q}{m}v_y B_0, \tag{2.319}$$

$$\frac{dv_y}{dt} = \frac{q}{m}(E_0 - v_x B_0). \tag{2.320}$$

Differentiating (2.320) and substituting from (2.319), we have

$$\frac{d^2 v_y}{dt^2} = -\frac{q}{m}B_0\frac{dv_x}{dt} = -\omega_c^2 v_y,$$

or

$$\frac{d^2 v_y}{dt^2} + \omega_c^2 v_y = 0.$$

This has the solution

$$v_y(t) = v_\perp \cos(\omega_c t + \phi),$$

where ϕ is determined by the $t = 0$ reference. From (2.320) we then have

$$v_x = \frac{E_0}{B_0} - \left(\pm\frac{1}{\omega_c}\frac{dv_y}{dt}\right) = \frac{E_0}{B_0} \pm v_\perp \sin\omega_c t.$$

Here the upper and lower signs apply to the cases of positively and negatively charged particles, respectively. The first term in this expression is exactly the $\mathbf{E} \times \mathbf{B}$ drift velocity,

$$\mathbf{w} = \frac{\mathbf{E} \times \mathbf{B}}{|\mathbf{B}|^2} = \hat{\mathbf{x}}\frac{E_0}{B_0}.$$

Integrating the velocity vector, we find the trajectory:

$$x(t) = x_0 + \frac{E_0}{B_0}t \mp r_c \cos(\omega_c t + \phi), \tag{2.321}$$

$$y(t) = y_0 + r_c \sin(\omega_c t + \phi), \tag{2.322}$$

where $r_c = v_\perp/\omega_c$. ◄

► **Example 2.23:** $\mathbf{E} \times \mathbf{B}$ cycloid motion

A proton travels through the uniform static fields

$$\mathbf{B} = 50 \times 10^{-6}\hat{\mathbf{z}} \text{ T}, \qquad \mathbf{E} = E_0\hat{\mathbf{y}} \text{ V/m},$$

with speed 3×10^5 m/s. Plot its trajectory.

Solution: Take $x_0 = 0$, $y_0 = r_c$, and a time reference such that $\phi = -\pi/2$. Then by (2.321) and (2.322)

$$x(t) = \frac{E_0}{B_0}t - r_c \sin \omega_c t, \qquad y(t) = r_c(1 - \cos \omega_c t),$$

which we can also write as

$$x(t) = r_c \left[\frac{E_0}{B_0 v_\perp}(\omega_c t) - \sin \omega_c t \right], \qquad y(t) = r_c(1 - \cos \omega_c t).$$

This has the form of a *trochoid* shifted along the y-direction,

$$x(t) = r_c(A\theta - \sin \theta), \qquad y(t) = r_c(1 - A) + r_c(A - \cos \theta),$$

where $\theta = \omega_c t$ and $A = E_0/(B_0 v_\perp)$. When $A = 1$ so that $E_0/B_0 = v_\perp$, the trajectory is a *cycloid*:

$$x(t) = r_c(\theta - \sin \theta), \qquad y(t) = r_c(1 - \cos \theta).$$

When $A > 1$ the trajectory is a y-shifted *curate cycloid*. When $A < 1$ the trajectory is a y-shifted *prolate cycloid*.

Figure 2.19 shows the trajectories where $E_0/B_0 = v_\perp$ (cycloid), $E_0/B_0 = 0.5v_\perp$ (prolate cycloid), and $E_0/B_0 = 2v_\perp$ (curate cycloid). Note that the prolate cycloid has been shifted an additional $4r_c$ in the y-direction, and the curate cycloid $-4r_c$ in the y-direction, to permit easier visualization. The $\mathbf{E} \times \mathbf{B}$ drift along the x-direction is clearly visible.

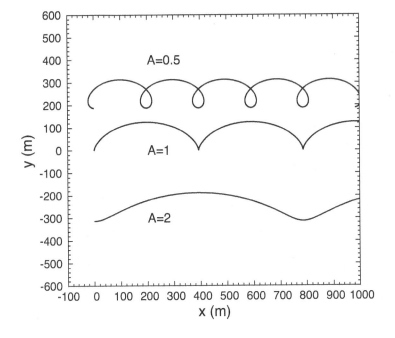

FIGURE 2.19
Trajectory of a proton in perpendicular $\mathbf{B}$ and $\mathbf{E}$ fields. $A = E_0/(B_0 v_\perp)$. ◄

2.11.6 Nonrelativistic particle motion in a nonuniform, static magnetic field

As with nonuniform electric fields, the equations describing particle motion in nonuniform static magnetic fields can be difficult to solve in closed form. However, if the radius of gyration is small compared to the scale describing the spatial variation of the magnetic field, the effect of field nonuniformity may be described in an approximate manner [66]. It is found that a gradient in $|\mathbf{B}|$ in the direction perpendicular to $\mathbf{B}$ produces a drift in the gyrating particle trajectory. The drift velocity is given by

$$\mathbf{v}_\nabla = \mp \frac{v_\perp^2}{2\omega_c} \frac{\mathbf{B} \times \nabla|\mathbf{B}|}{|\mathbf{B}|^2},$$

and is thus analogous with the $\mathbf{E} \times \mathbf{B}$ drift of the previous section. Here the sign is the sign of the charge on the particle. This effect is called *magnetic gradient drift*, or ∇B *drift*.

A second drift effect experienced by particles in a nonuniform static magnetic field is due to the curvature of the field lines. The velocity associated with this *curvature drift* may be expressed as

$$\mathbf{v}_\kappa = \mp \frac{v_\parallel^2}{\omega_c} \frac{\mathbf{R}_c \times \mathbf{B}}{|\mathbf{R}_c|^2|\mathbf{B}|}.$$

Here $\mathbf{R}_c$ is a vector describing the curvature of the magnetic field with $|\mathbf{R}_c|$ the radius of curvature, and the sign is that of the charge on the particle. For detailed derivations the reader should consult [66].

2.11.7 Relativistic particle motion in a uniform, static magnetic field

From (2.303) we have the equation of motion for particles moving at relativistic speeds in a uniform, static magnetic field:

$$\frac{d\mathbf{v}}{dt} = \frac{q}{\gamma m_0} \mathbf{v} \times \mathbf{B}.$$

Comparison with (2.308) shows that the only difference between a particle moving relativistically and nonrelativistically is an increase in the mass with increasing speed. Hence the trajectories described in § 2.11.4 are valid in the relativistic case provided we replace

$$\omega_c \mapsto \hat{\omega}_c = \frac{|q|}{m_0\gamma}|\mathbf{B}|.$$

The particles follow helical trajectories, but with gyro frequencies that decrease as particle speed increases.

2.11.8 Relativistic particle motion in uniform, static electric and magnetic fields

So far we have described the motion of charged particles traveling through electric and magnetic fields with reference to a stationary laboratory frame. How does the motion appear to an observer traveling with some uniform velocity relative to the laboratory frame? To answer this we can employ the transformation equations of § 2.3.

We first consider particles moving so slowly that Lorentzian relativistic effects can be neglected. The equations of motion are (2.315) and (2.316). In particular, the perpendicular velocity satisfies

$$\frac{d\mathbf{v}_\perp}{dt} = \frac{q}{m}(\mathbf{E}_\perp + \mathbf{v}_\perp \times \mathbf{B}).$$

Let us take a frame moving with a uniform velocity $\mathbf{u}_\perp$ that is perpendicular to the magnetic field. In this frame the equation of motion is

$$\frac{d\mathbf{v}'_\perp}{dt} = \frac{q}{m}(\mathbf{E}'_\perp + \mathbf{v}'_\perp \times \mathbf{B}'), \tag{2.323}$$

where $\mathbf{v}'_\perp$, $\mathbf{E}'$, and $\mathbf{B}'$ are the fields observed in the moving frame. The Galilean transformation given in (2.38) and (2.41) relates the fields observed in the moving frame to those observed in the laboratory frame:

$$\mathbf{E}' = \mathbf{E} + \mathbf{u}_\perp \times \mathbf{B},$$
$$\mathbf{B}' = \mathbf{B}.$$

To use these, define the unit vector $\hat{\mathbf{B}} = \mathbf{B}/|\mathbf{B}|$ such that $\mathbf{E}_\| = \hat{\mathbf{B}}(\hat{\mathbf{B}} \cdot \mathbf{E})$. Then

$$\mathbf{E}_\perp = \mathbf{E} - \mathbf{E}_\| = \mathbf{E} - \hat{\mathbf{B}}(\hat{\mathbf{B}} \cdot \mathbf{E}).$$

Thus,

$$\begin{aligned}
\mathbf{E}'_\perp &= \mathbf{E}' - \hat{\mathbf{B}}(\hat{\mathbf{B}} \cdot \mathbf{E}') \\
&= \mathbf{E}' - \hat{\mathbf{B}}[\hat{\mathbf{B}} \cdot (\mathbf{E} + \mathbf{u}_\perp \times \mathbf{B})] \\
&= \mathbf{E}' - \hat{\mathbf{B}}(\hat{\mathbf{B}} \cdot \mathbf{E}) - \hat{\mathbf{B}}[\hat{\mathbf{B}} \cdot (\mathbf{u}_\perp \times \mathbf{B})].
\end{aligned}$$

But $\hat{\mathbf{B}} \cdot (\mathbf{u}_\perp \times \mathbf{B}) = \mathbf{u}_\perp \cdot (\mathbf{B} \times \hat{\mathbf{B}}) = 0$, so

$$\begin{aligned}
\mathbf{E}'_\perp &= \mathbf{E}' - \hat{\mathbf{B}}(\hat{\mathbf{B}} \cdot \mathbf{E}) \\
&= \mathbf{E} + \mathbf{u}_\perp \times \mathbf{B} - \hat{\mathbf{B}}(\hat{\mathbf{B}} \cdot \mathbf{E}) \\
&= \mathbf{E}_\perp + \mathbf{u}_\perp \times \mathbf{B}.
\end{aligned}$$

Substitution into (2.323) gives the equation of motion for the moving observer:

$$\frac{d\mathbf{v}'_\perp}{dt} = \frac{q}{m}\left(\mathbf{E}_\perp + \mathbf{u}_\perp \times \mathbf{B} + \mathbf{v}'_\perp \times \mathbf{B}\right).$$

Now let us determine the velocity $\mathbf{u}_\perp$ needed to make

$$\mathbf{E}_\perp + \mathbf{u}_\perp \times \mathbf{B} = 0.$$

Crossing both sides with $\mathbf{B}$, we get

$$\mathbf{B} \times \mathbf{E}_\perp + \mathbf{B} \times (\mathbf{u}_\perp \times \mathbf{B}) = 0.$$

Expanding the vector triple product and noting that $\mathbf{u}_\perp \cdot \mathbf{B} = 0$, we have

$$\mathbf{u}_\perp = \frac{\mathbf{E} \times \mathbf{B}}{|\mathbf{B}|^2}.$$

This is exactly the drift velocity of the particle identified in (2.318). So if the observer moves with this velocity, the equation of motion becomes

$$\frac{d\mathbf{v}'_\perp}{dt} = \frac{q}{m}\mathbf{v}'_\perp \times \mathbf{B}.$$

This equation for $\mathbf{v}'_\perp$ is identical to the equation for $\mathbf{v}_\perp$ for an observer in the laboratory frame when $\mathbf{E}_\perp = 0$. To the observer drifting along with the particle, only the gyrating motion is apparent; the drift is not observed. This is intuitively sensible. Note that the moving observer *will* still see the acceleration of the charge due to the presence of a parallel electric field $\mathbf{E}_\parallel$.

If we cannot ignore the relativistic speed of the particle, the equation of motion becomes

$$\frac{d\mathbf{p}}{dt} = q(\mathbf{E} + \mathbf{v} \times \mathbf{B})$$

where $\mathbf{p} = \gamma m_0 \mathbf{v}$. Now suppose an observer moves at uniform velocity $\mathbf{u}_p$ perpendicular to the magnetic field. In the observer's frame, the equation of motion is

$$\frac{d\mathbf{p}'}{dt} = q(\mathbf{E}' + \mathbf{v}' \times \mathbf{B}').$$

We decompose the fields into components parallel and perpendicular to $\mathbf{B}$. The perpendicular component obeys

$$\frac{d\mathbf{p}'_p}{dt} = q(\mathbf{E}'_p + \mathbf{v}'_p \times \mathbf{B}'). \tag{2.324}$$

We use the subscript p in this case to avoid confusion with the Lorentz field transformations. From § 2.3.2 we have the transformations from the laboratory frame to the moving frame:

$$\mathbf{E}'_\parallel = \mathbf{E}_\parallel, \tag{2.325}$$

$$\mathbf{B}'_\parallel = \mathbf{B}_\parallel, \tag{2.326}$$

$$\mathbf{E}'_\perp = \gamma(\mathbf{E}_\perp + \mathbf{u}_p \times \mathbf{B}_\perp), \tag{2.327}$$

$$\mathbf{B}'_\perp = \gamma\mathbf{B}_\perp - \frac{\gamma}{c^2}\mathbf{u}_p \times \mathbf{E}_\perp. \tag{2.328}$$

Here the subscripts $\parallel$ and $\perp$ indicate vectors parallel and perpendicular to the $\mathbf{u}_p$ direction, respectively.

By (2.327) we note that the condition

$$\mathbf{E}_\perp + \mathbf{u}_p \times \mathbf{B}_\perp = 0$$

results in $\mathbf{E}'_\perp = 0$. Consequently, crossing both sides of the equation with $\mathbf{B}$, we find that

$$\mathbf{u}_p = \frac{\mathbf{E} \times \mathbf{B}}{|\mathbf{B}|^2} \tag{2.329}$$

as seen above. Crossing both sides of the equation with $\mathbf{u}_p$, we obtain

$$\mathbf{B} = \frac{\mathbf{u}_p \times \mathbf{E}_\perp}{u_p^2}.$$

Substitution into (2.328) gives

$$\mathbf{B}'_\perp = \gamma\mathbf{B} - \gamma\frac{u_p^2}{c^2}\mathbf{B} = \frac{\mathbf{B}}{\gamma} \tag{2.330}$$

since $\mathbf{B}_\perp = \mathbf{B}$ and $\gamma = 1/\sqrt{1 - u_p^2/c^2}$. Now write $\hat{\mathbf{B}} = \mathbf{B}/|\mathbf{B}|$ and examine

$$\mathbf{E}'_p = \mathbf{E}' - \hat{\mathbf{B}}(\hat{\mathbf{B}} \cdot \mathbf{E}') = \mathbf{E}'_\perp + \mathbf{E}'_\parallel - \hat{\mathbf{B}}(\hat{\mathbf{B}} \cdot [\mathbf{E}'_\perp + \mathbf{E}'_\parallel]).$$

Since $\mathbf{E}'_\perp = 0$ by our choice of $\mathbf{u}_p$, we have

$$\mathbf{E}'_p = \mathbf{E}_\| - \hat{\mathbf{B}}(\hat{\mathbf{B}} \cdot \mathbf{E}_\|),$$

where we have also used (2.325). But by (2.329) we see that $\mathbf{u}_p$ is perpendicular to both $\mathbf{E}$ and $\mathbf{B}$, hence $\mathbf{E}_\| = 0$. Thus $\mathbf{E}'_p = 0$. Substitution of this into (2.324) gives

$$\frac{d\mathbf{p}'_p}{dt} = q\mathbf{v}'_p \times \mathbf{B}' = q\mathbf{v}'_p \times (\mathbf{B}'_\perp + \mathbf{B}_\|),$$

since $\mathbf{B}'_\| = \mathbf{B}_\|$ from (2.326). But $\mathbf{u}_p$ is perpendicular to $\mathbf{B}$ so $\mathbf{B}_\| = 0$. Thus by (2.330) we finally have

$$\frac{d\mathbf{p}'_p}{dt} = q\mathbf{v}'_p \times \frac{\mathbf{B}}{\gamma}.$$

As with the Galilean transformation at nonrelativistic speeds, the Lorentzian transformation produces an equation for $\mathbf{v}'_\perp$ that is identical to the equation for $\mathbf{v}_\perp$ for an observer in the laboratory frame when $\mathbf{E}_\perp = 0$ (except for the presence of γ.) To the observer drifting along with the particle, only the gyrating motion is apparent.

Note that the condition (2.329) has restrictions, since u_p cannot exceed the speed of light. This requires

$$|\mathbf{E}|/|\mathbf{B}| < c.$$

For instance, in the ionosphere where a typical value of $|\mathbf{B}|$ is 50×10^{-6} T, for an electric field exceeding 15 kV/m it is not possible to find a moving frame in which $\mathbf{E}'_\perp = 0$. Interestingly, when $|\mathbf{E}|/|\mathbf{B}| > c$ it is possible to find a reference frame where $\mathbf{B}'_\perp = 0$, and thus the moving observer sees no gyrating motion.

2.12   Problems

2.1 Show that the constitutive equations relating $\mathbf{E}$, $\mathbf{H}$, $\mathbf{P}$, and $\mathbf{M}$ are

$$\left[\frac{1}{c}\bar{\mathbf{P}} - \epsilon_0 \bar{\mathbf{I}}\right] \cdot \mathbf{E} + [\mu_0 \bar{\mathbf{L}}] \cdot \mathbf{H} = \mathbf{P} - [\mu_0 \bar{\mathbf{L}}] \cdot \mathbf{M}$$

and

$$\bar{\mathbf{M}} \cdot \mathbf{E} + (\mu_0 c \bar{\mathbf{Q}} - \bar{\mathbf{I}}) \cdot \mathbf{H} = -\mathbf{M} - (\mu_0 c \bar{\mathbf{Q}} - \bar{\mathbf{I}}) \cdot \mathbf{M}.$$

Hint: start with Equations (2.99) and (2.100) from Example 2.5.

2.2 Consider Ampere's law and Gauss's law written in terms of rectangular components in the laboratory frame of reference. Assume that an inertial frame moves with velocity $\mathbf{v} = \hat{\mathbf{x}}v$ with respect to the laboratory frame. Using the Lorentz transformation given by (2.50)–(2.53), show that

$$c\rho' = \gamma(c\rho - \boldsymbol{\beta} \cdot \mathbf{J}).$$

Hint: Substitute (2.74) into Gauss' law (2.77).

2.3 Show that the following quantities are invariant under Lorentz transformation:

(a) $\mathbf{H} \cdot \mathbf{D}$,

(b) $\mathbf{B} \cdot \mathbf{B} - \mathbf{E} \cdot \mathbf{E}/c^2$,

(c) $\mathbf{H} \cdot \mathbf{H} - c^2 \mathbf{D} \cdot \mathbf{D}$,

(d) $\mathbf{B} \cdot \mathbf{H} - \mathbf{E} \cdot \mathbf{D}$,

(e) $c\mathbf{B} \cdot \mathbf{D} + \mathbf{E} \cdot \mathbf{H}/c$.

2.4 Show that if $c^2 B^2 > E^2$ holds in one reference frame, then it holds in all other reference frames. Repeat for the inequality $c^2 B^2 < E^2$.

2.5 Show that if $\mathbf{E} \cdot \mathbf{B} = 0$ and $c^2 B^2 > E^2$ holds in one reference frame, then a reference frame may be found such that $\mathbf{E} = 0$. Show that if $\mathbf{E} \cdot \mathbf{B} = 0$ and $c^2 B^2 < E^2$ holds in one reference frame, then a reference frame may be found such that $\mathbf{B} = 0$.

2.6 A test charge Q at rest in the laboratory frame experiences a force $\mathbf{F} = Q\mathbf{E}$ as measured by an observer in the laboratory frame. An observer in an inertial frame measures a force on the charge given by $\mathbf{F}' = Q\mathbf{E}' + Q\mathbf{v} \times \mathbf{B}'$. Show that $\mathbf{F} \neq \mathbf{F}'$ and find the formula for converting between $\mathbf{F}$ and $\mathbf{F}'$.

2.7 Consider a material moving with velocity $\mathbf{v}$ with respect to the laboratory frame of reference. When the fields are measured in the moving frame, the material is found to be isotropic with $\mathbf{D}' = \epsilon' \mathbf{E}'$ and $\mathbf{B}' = \mu' \mathbf{H}'$. Show that the fields measured in the laboratory frame are given by (2.87) and (2.88), indicating that the material is bianisotropic when measured in the laboratory frame.

2.8 Show that by assuming $v^2/c^2 \ll 1$ in (2.44)–(2.46) we may obtain (2.91).

2.9 Derive the following expressions that allow us to convert the value of the magnetization measured in the laboratory frame of reference to the value measured in a moving frame:
$$\mathbf{M}'_\perp = \gamma(\mathbf{M}_\perp + \boldsymbol{\beta} \times c\mathbf{P}_\perp), \qquad \mathbf{M}'_\parallel = \mathbf{M}_\parallel.$$

2.10 Beginning with the expressions (2.44)–(2.46) for the field conversions under a first-order Lorentz transformation, show that
$$\mathbf{P}' = \mathbf{P} - \frac{\mathbf{v} \times \mathbf{M}}{c^2}, \qquad \mathbf{M}' = \mathbf{M} + \mathbf{v} \times \mathbf{P}.$$

2.11 Consider a simple isotropic material moving through space with velocity $\mathbf{v}$ relative to the laboratory frame. The relative permittivity and permeability of the material measured in the moving frame are ϵ'_r and μ'_r, respectively. Show that the magnetization as measured in the laboratory frame is related to the laboratory frame electric field and magnetic flux density as
$$\mathbf{M} = \frac{\chi'_m}{\mu_0 \mu'_r} \mathbf{B} - \epsilon_0 \left(\chi'_e + \frac{\chi'_m}{\mu'_r} \right) \mathbf{v} \times \mathbf{E}$$

when a first-order Lorentz transformation is used. Here $\chi'_e = \epsilon'_r - 1$ and $\chi'_m = \mu'_r - 1$.

2.12 Consider a simple isotropic material moving through space with velocity $\mathbf{v}$ relative to the laboratory frame. The relative permittivity and permeability of the material measured in the moving frame are ϵ'_r and μ'_r, respectively. Derive the formulas for the magnetization and polarization in the laboratory frame in terms of $\mathbf{E}$ and $\mathbf{B}$ measured in the laboratory frame by using the Lorentz transformations (2.106) and (2.107)–(2.110). Show that these expressions reduce to (2.117) and (2.118) under the assumption of a first-order Lorentz transformation ($v^2/c^2 \ll 1$).

2.13 Derive the kinematic form of the large-scale Maxwell–Boffi equations (2.137) and (2.138). Derive the alternative form of the large-scale Maxwell–Boffi equations (2.139) and (2.140).

2.14 Modify the kinematic form of the Maxwell–Boffi equations (2.137)–(2.138) to account for the presence of magnetic sources. Repeat for the alternative forms (2.139)–(2.140).

2.15 Consider a thin magnetic source distribution concentrated near a surface S. The magnetic charge and current densities are given by

$$\rho_m(\mathbf{r}, x, t) = \rho_{ms}(\mathbf{r}, t) f(x, \Delta), \qquad \mathbf{J}_m(\mathbf{r}, x, t) = \mathbf{J}_{ms}(\mathbf{r}, t) f(x, \Delta),$$

where $f(x, \Delta)$ satisfies

$$\int_{-\infty}^{\infty} f(x, \Delta) \, dx = 1.$$

Let $\Delta \to 0$ and derive the boundary conditions on $(\mathbf{E}, \mathbf{D}, \mathbf{B}, \mathbf{H})$ across S.

2.16 Beginning with the kinematic forms of Maxwell's equations (2.147)–(2.148), derive the boundary conditions for a moving surface

$$\hat{\mathbf{n}}_{12} \times (\mathbf{H}_1 - \mathbf{H}_2) + (\hat{\mathbf{n}}_{12} \cdot \mathbf{v})(\mathbf{D}_1 - \mathbf{D}_2) = \mathbf{J}_s,$$
$$\hat{\mathbf{n}}_{12} \times (\mathbf{E}_1 - \mathbf{E}_2) - (\hat{\mathbf{n}}_{12} \cdot \mathbf{v})(\mathbf{B}_1 - \mathbf{B}_2) = -\mathbf{J}_{ms}.$$

2.17 Beginning with Maxwell's equations and the constitutive relationships for a bianisotropic medium (2.12)–(2.13), derive the wave equation for $\mathbf{H}$ (2.256). Specialize the result for the case of an anisotropic medium.

2.18 Consider an isotropic but inhomogeneous material, so that

$$\mathbf{D}(\mathbf{r}, t) = \epsilon(\mathbf{r})\mathbf{E}(\mathbf{r}, t), \qquad \mathbf{B}(\mathbf{r}, t) = \mu(\mathbf{r})\mathbf{H}(\mathbf{r}, t).$$

Show that the wave equations for the fields within this material may be written as

$$\nabla^2 \mathbf{E} - \mu\epsilon \frac{\partial^2 \mathbf{E}}{\partial t^2} + \nabla\left[\mathbf{E} \cdot \left(\frac{\nabla\epsilon}{\epsilon}\right)\right] - (\nabla \times \mathbf{E}) \times \left(\frac{\nabla\mu}{\mu}\right) = \mu\frac{\partial \mathbf{J}}{\partial t} + \nabla\left(\frac{\rho}{\epsilon}\right),$$

$$\nabla^2 \mathbf{H} - \mu\epsilon \frac{\partial^2 \mathbf{H}}{\partial t^2} + \nabla\left[\mathbf{H} \cdot \left(\frac{\nabla\mu}{\mu}\right)\right] - (\nabla \times \mathbf{H}) \times \left(\frac{\nabla\epsilon}{\epsilon}\right) = -\nabla \times \mathbf{J} - \mathbf{J} \times \left(\frac{\nabla\epsilon}{\epsilon}\right).$$

2.19 Repeat Example 2.7 for

$$\epsilon = \epsilon_0, \qquad \mu = \mu(z).$$

2.20 Specialize the wave equations in Problem 2.18 to the case of a source-free inhomogeneous dielectric with $\mu(\mathbf{r}) = \mu_0$ and $\epsilon(\mathbf{r}) = \epsilon(z)$. If

$$\mathbf{E}(\mathbf{r}, t) = \hat{\mathbf{y}} E_y(z, t), \qquad \mathbf{H}(\mathbf{r}, t) = \hat{\mathbf{x}} H_x(z, t),$$

show that the wave equations reduce to those found in Example 2.7.

2.21 Consider a homogeneous, isotropic material in which $\mathbf{D} = \epsilon \mathbf{E}$ and $\mathbf{B} = \mu \mathbf{H}$. Using the definitions of the equivalent sources, show that the wave equations (2.265)–(2.266) are equivalent to (2.259)–(2.260). Assume $\mathbf{J}_m = 0$.

2.22 When a material is only slightly conducting, and thus Ω is very small, we often neglect the third term in the plane wave solution (2.284). Reproduce the plot of Figure 2.8 with this term omitted and compare. Discuss how the omitted term affects the shape of the propagating waveform.

2.23 A total charge Q is evenly distributed over a spherical surface. The surface expands outward at constant velocity so that the radius of the surface is $b = vt$ at time t. (a) Use Gauss's law to find $\mathbf{E}$ everywhere as a function of time. (b) Show that $\mathbf{E}$ may be found from a potential function

$$\psi(\mathbf{r}, t) = \frac{Q}{4\pi r} \frac{vt - r}{vt} U(r - vt)$$

according to (2.300). Here $U(t)$ is the unit step function. (c) Write down the form of $\mathbf{J}$ for the expanding sphere and show that since it may be found from (2.298) it is a nonradiating source.

2.24 The instantaneous electric field inside a source-free, homogeneous, isotropic region is given by (assume all integration constants are zero)

$$\mathbf{E}(\mathbf{r}, t) = [\hat{\mathbf{x}} A(x + y) + \hat{\mathbf{y}} B(x - y)] \cos \omega t.$$

Find (a) the relationship between A and B, and (b) the instantaneous magnetic field $\mathbf{H}$.

2.25 The instantaneous electric field inside an empty (source free) rectangular waveguide is given by (assume all integration constants are zero)

$$\mathbf{E}(\mathbf{r}, t) = \hat{\mathbf{y}} E_0 \sin\left(\frac{\pi}{a} x\right) \cos(\omega t - \beta z).$$

Find (a) the instantaneous magnetic field $\mathbf{H}$, and (b) β.

2.26 The electric field in a certain region of space is given by

$$\mathbf{E}(\mathbf{r}, t) = \hat{\mathbf{y}} 10 x^2 e^{-\alpha t} U(t)$$

where $U(t)$ is the unit step function. Find $\mathbf{D}(\mathbf{r}, t)$ if

(a) $\epsilon(\mathbf{r}, t) = 3\epsilon_0 y^2$;

(b) $\chi_e(\mathbf{r}, t) = 3\epsilon_0 y^2 U(t)$;

(c) $\bar{\epsilon}(\mathbf{r}, t) = 3\epsilon_0 \hat{\mathbf{x}}\hat{\mathbf{x}} + 2\epsilon_0 \hat{\mathbf{x}}\hat{\mathbf{y}} + 4\epsilon_0 \hat{\mathbf{z}}\hat{\mathbf{y}}$;

(d) the region is a chiral material with $\epsilon = 7\epsilon_0$ and $\beta = 10$.

2.27 Consider a uniform volume charge density ρ_0 in free space. A spherical surface expands outward with radius $R(t) = v_0 t$. Compute $\oint_S \mathbf{J}^* \cdot \mathbf{dS}$ and show that it is equal to $-\frac{d}{dt} \int_V \rho \, dV$.

2.28 A spherical dielectric shell occupies the region $a \leq r \leq b$ in free space, where $a = 2$ and $b = 4$ cm. The polarization and electric fields within the shell are given by

$$\mathbf{P} = \hat{\mathbf{r}} \frac{31.87 \times 10^{-12}}{r^2} \quad \text{C/m}^2, \qquad \mathbf{E} = \hat{\mathbf{r}} \frac{0.45}{r^2} \quad \text{V/m}.$$

Find (a) the equivalent polarization surface charge density on each surface, (b) the total polarization surface charge, (c) the equivalent polarization volume charge density, (d) the total polarization volume charge, and (e) the relative permittivity of the material.

2.29 Consider a homogeneous Tellegen medium with constitutive relations $\mathbf{D} = \epsilon\mathbf{E} + \xi\mathbf{H}$ and $\mathbf{B} = \xi\mathbf{E} + \mu\mathbf{H}$. Maxwell's equations are solved for an electric source current density $\mathbf{J}$ and a magnetic source current density $\mathbf{J}_m$, and the fields are found to be $(\mathbf{E}_1, \mathbf{D}_1, \mathbf{B}_1, \mathbf{H}_1)$. The problem is then solved again with $\mathbf{J}$ replaced by $\mathbf{J}_m/\eta_0$ and $\mathbf{J}_m$ replaced by $-\eta_0\mathbf{J}$, and the fields are found to be $(\mathbf{E}_2, \mathbf{D}_2, \mathbf{B}_2, \mathbf{H}_2)$. Determine the constitutive parameters for the dual problem.

2.30 Consider the power balance equation

$$\mathbf{J} \cdot \mathbf{E} = -\nabla \cdot (\mathbf{E} \times \mathbf{H}) - \left(\mathbf{E} \cdot \frac{\partial \mathbf{D}}{\partial t} + \mathbf{H} \cdot \frac{\partial \mathbf{B}}{\partial t} \right).$$

Show that for a Tellegen medium we can write this as

$$\nabla \cdot \mathbf{S}_{em} + \frac{\partial}{\partial t} U_{em} = -\mathbf{J} \cdot \mathbf{E}$$

where U_{em} is a term related to the stored energy.

2.31 Show by substitution that

$$\psi(z, t) = \frac{1}{2} \left[f\left(t - \frac{z}{v}\right) + f\left(t + \frac{z}{v}\right) \right] + \frac{v}{2} \int_{t - \frac{z}{v}}^{t + \frac{z}{v}} g(\tau) \, d\tau$$

is a solution to the scalar wave equation

$$\frac{\partial^2 \psi}{\partial z^2} - \frac{1}{v^2} \frac{\partial^2 \psi}{\partial t^2} = 0.$$

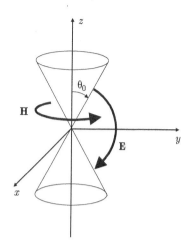

FIGURE 2.20
Conical transmission line.

2.32 The electric field in free space is given (in spherical coordinates) by

$$\mathbf{E}(\mathbf{r}, t) = \hat{\boldsymbol{\theta}} \frac{E_0}{r} \cos(\omega t - \beta r).$$

(a) Find $\mathbf{H}(\mathbf{r}, t)$. Assume all integration constants are zero. (b) Calculate the Poynting flux passing through a sphere of radius r.

2.33 The instantaneous magnetic field in a source-free region of free space is given by

$$\mathbf{H}(\mathbf{r}, t) = \hat{\mathbf{z}} H_0 \cos(\omega t - \beta x).$$

(a) Find the instantaneous electric field (set integration constants to zero). (b) Find β.

2.34 The fields in a region of space are given by

$$\mathbf{E} = \hat{\boldsymbol{\rho}} \frac{1}{\rho} \cos(\omega t - \beta z), \qquad \mathbf{H} = \hat{\boldsymbol{\phi}} \frac{1}{\eta \rho} \cos(\omega t - \beta z),$$

where $\beta = \omega \sqrt{\mu_0 \epsilon_0}$ and $\eta = \sqrt{\mu_0 / \epsilon_0}$. Consider a rectangular loop in the xz-plane, with height b along the z-direction and width a along the x-direction. The loop moves along the x-axis with velocity $\mathbf{v} = \hat{\mathbf{x}} v_0$ such that its left side is at $x(t) = v_0 t$ at time t.

(a) Compute $\oint \mathbf{E} \cdot \mathbf{dl}$ and show that it is equal to $-\int \frac{\partial \mathbf{B}}{\partial t} \cdot \mathbf{dS}$.

(b) Compute $\oint (\mathbf{E} + \mathbf{v} \times \mathbf{B}) \cdot \mathbf{dl}$ and show that it is equal to $-\frac{d}{dt} \int \mathbf{B} \cdot \mathbf{dS}$.

2.35 A conical transmission line (Figure 2.20) consists of a perfectly conducting cone aligned along the $+z$-axis at $\theta = \theta_0$ and an identical cone aligned along the $-z$-axis at $\theta = \pi - \theta_0$. Compute (a) the voltage $V(r, t)$ found by integrating the electric field; (b) the current $I(r, t)$ found by integrating the surface current density; (c) the characteristic resistance of the line, $R_c = V(r, t) / I(r, t)$.

3

The static and quasistatic electromagnetic fields

3.1 Statics and quasistatics

Perhaps the most carefully studied area of electromagnetics is that in which the fields are time-invariant. This area, known generally as *statics*, offers the most direct opportunities for solution of the governing equations and the clearest physical pictures of the electromagnetic field. We therefore devote the majority of the present chapter to a treatment of statics. We also endeavor to describe the behavior of fields that are slowly changing with time. These *quasistatic* fields obey many of the same relations as static fields, but also give rise to phenomena such as diffusion that are important in many applications of electromagnetics.

We begin to seek and examine specific solutions to the field equations; however, our selection of examples is shaped by a search for insight into the behavior of the field itself, rather than by a desire to catalog the solutions of numerous statics problems. We note at the outset that a static field is physically sensible only as a limiting case of a time-varying field as the latter approaches a time-invariant equilibrium, and then only in local regions. The static field equations we shall study thus represent an idealized model of the physical fields. And while the quasistatic field equations are more broadly applicable, the underlying assumptions that make their solutions tractable also limit their validity. To observe the most general behavior of the electromagnetic field, we must seek solutions to the full system of Maxwell's equations. This is the subject of subsequent chapters.

3.2 Static fields and steady currents

If we examine the Maxwell–Minkowski equations (2.1)–(2.4) and set the time derivatives to zero, we obtain the *static field Maxwell equations*

$$\nabla \times \mathbf{E}(\mathbf{r}) = 0,$$
$$\nabla \cdot \mathbf{D}(\mathbf{r}) = \rho(\mathbf{r}),$$
$$\nabla \times \mathbf{H}(\mathbf{r}) = \mathbf{J}(\mathbf{r}),$$
$$\nabla \cdot \mathbf{B}(\mathbf{r}) = 0.$$

We note that if the fields are to be everywhere time-invariant, then the sources $\mathbf{J}$ and ρ must also be everywhere time-invariant. Under this condition the dynamic coupling between the fields described by Maxwell's equations disappears; any connection between $\mathbf{E}$, $\mathbf{D}$, $\mathbf{B}$, and $\mathbf{H}$ imposed by the time-varying nature of the field is gone. For static fields we also require that any dynamic coupling between fields in the constitutive relations vanish. In this *static field limit* we cannot derive the divergence equations from the curl equations, since we can no longer use the initial condition argument that the fields were identically zero prior to some time.

The static field equations are useful for approximating many physical situations in which the fields rapidly settle to a local, macroscopically-static state. This may occur so rapidly and so completely that, in a practical sense, the static equations describe the fields within our ability to measure and to compute. Such is the case when a capacitor is rapidly charged using a battery in series with a resistor; for example, a 1 pF capacitor charging through a 1 Ω resistor reaches 99.99% of its total charge static limit within 10 ps.

3.2.1   Decoupling of the electric and magnetic fields

For the rest of this chapter we assume there is no coupling between $\mathbf{E}$ and $\mathbf{H}$ or between $\mathbf{D}$ and $\mathbf{B}$ in the constitutive relations. Then the static equations decouple into two independent sets of equations in terms of two independent sets of fields. The static electric field set ($\mathbf{E},\mathbf{D}$) is described by

$$\nabla \times \mathbf{E}(\mathbf{r}) = 0, \tag{3.1}$$

$$\nabla \cdot \mathbf{D}(\mathbf{r}) = \rho(\mathbf{r}). \tag{3.2}$$

Integrating these over a stationary contour and surface, respectively, we have the large-scale forms

$$\oint_{\Gamma} \mathbf{E} \cdot \mathbf{dl} = 0,$$

$$\oint_{S} \mathbf{D} \cdot \mathbf{dS} = \int_{V} \rho \, dV.$$

The static magnetic field set ($\mathbf{B},\mathbf{H}$) is described by

$$\nabla \times \mathbf{H}(\mathbf{r}) = \mathbf{J}(\mathbf{r}),$$

$$\nabla \cdot \mathbf{B}(\mathbf{r}) = 0,$$

or, in large-scale form,

$$\oint_{\Gamma} \mathbf{H} \cdot \mathbf{dl} = \int_{S} \mathbf{J} \cdot \mathbf{dS},$$

$$\oint_{S} \mathbf{B} \cdot \mathbf{dS} = 0.$$

We can also specialize the Maxwell–Boffi equations to static form. Assuming the fields, sources, and equivalent sources are time-invariant, the electrostatic field $\mathbf{E}(\mathbf{r})$ is described by the point-form equations

$$\nabla \times \mathbf{E} = 0,$$

$$\nabla \cdot \mathbf{E} = \frac{1}{\epsilon_0} \left(\rho - \nabla \cdot \mathbf{P} \right),$$

or the equivalent large-scale equations

$$\oint_\Gamma \mathbf{E} \cdot d\mathbf{l} = 0,$$

$$\oint_S \mathbf{E} \cdot d\mathbf{S} = \frac{1}{\epsilon_0} \int_V (\rho - \nabla \cdot \mathbf{P}) \, dV.$$

Similarly, the magnetostatic field $\mathbf{B}$ is described by

$$\nabla \times \mathbf{B} = \mu_0 \left(\mathbf{J} + \nabla \times \mathbf{M} \right),$$

$$\nabla \cdot \mathbf{B} = 0,$$

or

$$\oint_\Gamma \mathbf{B} \cdot d\mathbf{l} = \mu_0 \int_S (\mathbf{J} + \nabla \times \mathbf{M}) \cdot d\mathbf{S},$$

$$\oint_S \mathbf{B} \cdot d\mathbf{S} = 0.$$

It is important to note that any separation of the electromagnetic field into independent static electric and magnetic portions is illusory. As we mentioned in § 2.3.2, the electric and magnetic components of the EM field depend on the motion of the observer. An observer stationary with respect to a single charge measures only a static electric field, while an observer in uniform motion with respect to the charge measures both electric and magnetic fields.

3.2.2 Static field equilibrium and conductors

Suppose we could arrange a group of electric charges into a static configuration in free space. The charges would produce an electric field, resulting in a force on the distribution via the Lorentz force law, and hence would begin to move. Regardless of how we arrange the charges, they cannot maintain their original static configuration without the help of some mechanical force to counterbalance the electrical force. This is a statement of Earnshaw's theorem, discussed in detail in § 3.5.2.

The situation is similar for charges within and on electric conductors. A conductor is a material having many charges free to move under external influences, both electric and non-electric. In a metallic conductor, electrons move against a background lattice of positive charges. An *uncharged conductor* is neutral: the amount of negative charge carried by the electrons is equal to the positive charge in the background lattice. The distribution of charges in an uncharged conductor is such that the macroscopic electric field is zero inside and outside the conductor. When the conductor is exposed to an additional electric field, the electrons move under the influence of the Lorentz force, creating a *conduction current*. Rather than accelerating indefinitely, conduction electrons experience collisions with the lattice, thereby giving up their kinetic energy. Macroscopically, the charge motion can be described in terms of a time-average velocity, hence a macroscopic current density can be assigned to the density of moving charge. The relationship between the applied, or "impressed," field and the resulting current density is given by *Ohm's law*; in a linear, isotropic, nondispersive material, this is

$$\mathbf{J}(\mathbf{r}, t) = \sigma(\mathbf{r})\mathbf{E}(\mathbf{r}, t).$$

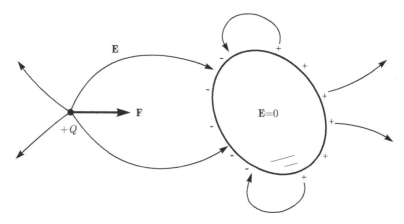

FIGURE 3.1

Positive point charge in the vicinity of an insulated, uncharged conductor.

The conductivity σ describes the impediment to charge motion through the lattice: the higher the conductivity, the farther an electron may move on average before undergoing a collision.

Let us examine how a state of equilibrium is established in a conductor. We shall consider several important situations. First, suppose we bring a positively charged particle into the vicinity of a neutral, insulated conductor (we say that a conductor is "insulated" if no means exists for depositing excess charge onto the conductor). The Lorentz force on the free electrons in the conductor results in their motion toward the particle (Figure 3.1). A reaction force $\mathbf{F}$ attracts the particle to the conductor. If the particle and the conductor are both held rigidly in space by an external mechanical force, the electrons within the conductor continue to move toward the surface. In a metal, when these electrons reach the surface and try to continue further, they experience a rapid reversal in the direction of the Lorentz force, drawing them back toward the surface. A sufficiently large force (described by the *work function* of the metal) will be able to draw these charges from the surface, but anything less will permit the establishment of a stable equilibrium at the surface. If σ is large then equilibrium is established quickly, and a nonuniform static charge distribution appears on the conductor surface. The electric field within the conductor must settle to zero at equilibrium, since a nonzero field would be associated with a current $\mathbf{J} = \sigma\mathbf{E}$. In addition, the component of the field tangential to the surface must be zero or the charge would be forced to move along the surface. *At equilibrium, the field within and tangential to a conductor must be zero.* Note also that equilibrium cannot be established without external forces to hold the conductor and particle in place.

Next, suppose we bring a positively charged particle into the vicinity of a *grounded* (rather than insulated) conductor as in Figure 3.2. Use of the term "grounded" means that the conductor is attached via a filamentary conductor to a remote reservoir of charge known as *ground*; in practical applications the earth acts as this charge reservoir. Charges are drawn from or returned to the reservoir, without requiring any work, in response to the Lorentz force on the charge within the conducting body. As the particle approaches, negative charge is drawn to the body and then along the surface until a static equilibrium is re-established. Unlike the insulated body, the grounded conductor in equilibrium has excess negative charge, the amount of which depends on the proximity of the particle. Again, both particle and conductor must be held in place by external mechanical forces,

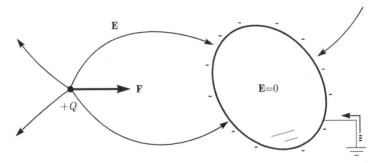

FIGURE 3.2
Positive point charge near a grounded conductor.

and the total field produced by both the static charge on the conductor and the particle must be zero at points interior to the conductor.

Finally, consider the process whereby excess charge placed inside a conducting body redistributes as equilibrium is established. We assume an isotropic, homogeneous conducting body with permittivity ϵ and conductivity σ. An initially static charge with density $\rho_0(\mathbf{r})$ is introduced at time $t = 0$. The charge density must obey the continuity equation

$$\nabla \cdot \mathbf{J}(\mathbf{r}, t) = -\frac{\partial \rho(\mathbf{r}, t)}{\partial t};$$

since $\mathbf{J} = \sigma \mathbf{E}$, we have

$$\sigma \nabla \cdot \mathbf{E}(\mathbf{r}, t) = -\frac{\partial \rho(\mathbf{r}, t)}{\partial t}.$$

By Gauss's law, $\nabla \cdot \mathbf{E}$ can be eliminated:

$$\frac{\sigma}{\epsilon} \rho(\mathbf{r}, t) = -\frac{\partial \rho(\mathbf{r}, t)}{\partial t}.$$

Solving this differential equation for the unknown $\rho(\mathbf{r}, t)$ we have

$$\rho(\mathbf{r}, t) = \rho_0(\mathbf{r}) e^{-\sigma t/\epsilon}.$$

The charge density within a homogeneous, isotropic conducting body decreases exponentially with time, regardless of the original charge distribution and shape of the body. Of course, the total charge must be constant, and thus charge within the body travels to the surface where it distributes itself in such a way that the field internal to the body approaches zero at equilibrium. The rate at which the volume charge dissipates is determined by the *relaxation time* ϵ/σ; for copper (a good conductor) this is an astonishingly small 10^{-19} s. Even distilled water, a relatively poor conductor, has $\epsilon/\sigma = 10^{-6}$ s. Thus we see how rapidly static equilibrium can be approached.

3.2.3   Steady current

Since time-invariant fields must arise from time-invariant sources, we have from the continuity equation

$$\nabla \cdot \mathbf{J}(\mathbf{r}) = 0. \tag{3.3}$$

In large-scale form this is

$$\oint_S \mathbf{J} \cdot d\mathbf{S} = 0. \tag{3.4}$$

A current with the property (3.3) is said to be a *steady current*. By (3.4), a steady current must be completely lineal (and infinite in extent) or must form closed loops. However, if a current forms loops then the individual moving charges must undergo acceleration (from the change in direction of velocity). Since a single accelerating particle radiates energy in the form of an electromagnetic wave, we might expect a large steady loop current to produce a great deal of radiation. In fact, if we superpose the fields produced by the many particles composing a steady current, we find that a steady current produces no radiation [92]. Remarkably, to obtain this result we must consider the exact relativistic fields, and thus our finding is precise within the limits of our macroscopic assumptions.

If we try to create a steady current in free space, the flowing charges will tend to disperse because of the Lorentz force from the field set up by the charges, and the resulting current will not form closed loops. A beam of electrons or ions will produce both an electric field (because of the nonzero net charge of the beam) and a magnetic field (because of the current). At nonrelativistic particle speeds, the electric field produces an outward force on the charges that is much greater than the inward (or *pinch*) force produced by the magnetic field. Application of an additional, external force will allow the creation of a *collimated beam* of charge, as occurs in an electron tube where a series of permanent magnets can be used to create a beam of steady current.

More typically, steady currents are created using wire conductors to guide the moving charge. When an external force, such as the electric field created by a battery, is applied to an uncharged conductor, the free electrons will begin to move through the positive lattice, forming a current. Each electron moves only a short distance before colliding with the positive lattice, and if the wire is bent into a loop the resulting macroscopic current will be steady in the sense that the temporally and spatially averaged microscopic current will obey $\nabla \cdot \mathbf{J} = 0$. We note from the examples above that any charges attempting to leave the surface of the wire are drawn back by the electrostatic force produced by the resulting imbalance in electrical charge. For conductors, the "drift" velocity associated with the moving electrons is proportional to the applied field:

$$\mathbf{u}_d = -\mu_e \mathbf{E}$$

where μ_e is the *electron mobility*. The mobility of copper (3.2×10^{-3} m^2/V · s) is such that an applied field of 1 V/m results in a drift velocity of only a third of a centimeter per second.

Integral properties of a steady current

Steady currents obey several useful integral properties. To develop these properties we need an integral identity. Let $f(\mathbf{r})$ and $g(\mathbf{r})$ be scalar functions, continuous and with continuous derivatives in a volume region V. Let $\mathbf{J}$ represent a steady current field of finite extent, completely contained within V. We begin by using (B.48) to expand

$$\nabla \cdot (fg\mathbf{J}) = fg(\nabla \cdot \mathbf{J}) + \mathbf{J} \cdot \nabla(fg).$$

Noting that $\nabla \cdot \mathbf{J} = 0$ and using (B.47), we get

$$\nabla \cdot (fg\mathbf{J}) = (f\mathbf{J}) \cdot \nabla g + (g\mathbf{J}) \cdot \nabla f.$$

Now let us integrate over V and employ the divergence theorem:

$$\oint_S (fg)\mathbf{J} \cdot d\mathbf{S} = \int_V [(f\mathbf{J}) \cdot \nabla g + (g\mathbf{J}) \cdot \nabla f]\, dV.$$

Since $\mathbf{J}$ is contained entirely within S, we must have $\hat{\mathbf{n}} \cdot \mathbf{J} = 0$ everywhere on S. Hence

$$\int_V [(f\mathbf{J}) \cdot \nabla g + (g\mathbf{J}) \cdot \nabla f]\, dV = 0. \tag{3.5}$$

We can obtain a useful relation by letting $f = 1$ and $g = x_i$ in (3.5), where $(x, y, z) = (x_1, x_2, x_3)$. This gives

$$\int_V J_i(\mathbf{r})\, dV = 0, \tag{3.6}$$

where $J_1 = J_x$ and so on. Hence the volume integral of any rectangular component of $\mathbf{J}$ is zero. Similarly, letting $f = g = x_i$ we find that

$$\int_V x_i J_i(\mathbf{r})\, dV = 0. \tag{3.7}$$

With $f = x_i$ and $g = x_j$ we obtain

$$\int_V [x_i \mathbf{J}_j(\mathbf{r}) + x_j \mathbf{J}_i(\mathbf{r})]\, dV = 0. \tag{3.8}$$

▶ **Example 3.1:** Azimuthal current

The current density in free space is given by

$$\mathbf{J}(\mathbf{r}) = \hat{\phi} J_0 \frac{z}{a}\frac{\rho}{b} \qquad (a \leq \rho \leq b,\ 0 \leq z \leq h)$$

where J_0 is a constant in $\mathrm{A/m}^2$. Show that $\mathbf{J}$ is a steady current for which (3.7) holds.

Solution: We have

$$\nabla \cdot \mathbf{J} = \frac{1}{\rho}\frac{\partial}{\partial \phi}\left(J_0 \frac{z}{a}\frac{\rho}{b}\right) = 0.$$

To verify (3.7) we note that

$$\int_V z J_z(\mathbf{r})\, dV = 0$$

since $J_z = 0$, and that

$$\int_V x J_x(\mathbf{r})\, dV = \int_0^h \int_0^{2\pi} \int_a^b (\rho\cos\phi)\left(-\sin\phi\, J_0 \frac{z}{a}\frac{\rho}{b}\right)\rho\, d\rho\, d\phi\, dz = 0,$$

$$\int_V y J_y(\mathbf{r})\, dV = \int_0^h \int_0^{2\pi} \int_a^b (\rho\sin\phi)\left(\cos\phi\, J_0 \frac{z}{a}\frac{\rho}{b}\right)\rho\, d\rho\, d\phi\, dz = 0,$$

since the integrations over ϕ produce zeros. ◀

3.3   Electrostatics

3.3.1   Direct solutions to Gauss's law

When the spatial distribution of charge is highly symmetric, Gauss's law (in either integral or point form) may be solved directly for the electric field. Any of the following symmetry conditions is appropriate:

1. The charge depends only on the spherical coordinate r so that $\mathbf{D}(\mathbf{r}) = D_r(r)\hat{\mathbf{r}}$.

2. The charge depends only on the cylindrical coordinate ρ so that $\mathbf{D}(\mathbf{r}) = D_\rho(\rho)\hat{\boldsymbol{\rho}}$.

3. The charge depends on a single rectangular coordinate, e.g., x, so that $\mathbf{D}(\mathbf{r}) = D_x(x)\hat{\mathbf{x}}$.

When employing the integral form of Gauss's law, the procedure is to choose a flux surface (called a *Gaussian surface*) over which the electric field is either constant in magnitude and parallel to the surface normal, or perpendicular to the normal (or over which some combination of these conditions holds). Then the electric field may be removed from the integral and determined. The point form is employed by separating the field volume into regions in which the partial differential equation may be reduced to an ordinary differential equation solvable by direct integration. The solutions are then connected across the adjoining surfaces using boundary conditions.

▶ **Example 3.2:** Solution to the integral form of Gauss's law for a line charge

The z-axis carries a line charge of uniform density ρ_l in a homogeneous medium with permittivity ϵ. Find $\mathbf{E}$ for $\rho > 0$.

Solution: The line charge can be described by the volume charge density

$$\rho(\mathbf{r}) = \rho_l \frac{\delta(\rho)}{2\pi\rho}.$$

Since $\rho(\mathbf{r})$ depends only on the radial coordinate ρ, we have $\mathbf{D}(\mathbf{r}) = D_\rho(\rho)\hat{\boldsymbol{\rho}}$. Choose as a flux surface a cylinder of radius ρ_0 and height h centered on the z-axis. Gauss's law requires

$$\int_{\text{bottom}} \mathbf{D} \cdot \hat{\mathbf{n}}\, dS + \int_{\text{top}} \mathbf{D} \cdot \hat{\mathbf{n}}\, dS + \int_{\text{side}} \mathbf{D} \cdot \hat{\mathbf{n}}\, dS = \int_V \rho\, dV,$$

where the net outflux integral has been split into contributions from the bottom, top, and side surfaces. The integrals over the top and bottom surfaces are zero, since the surface normal is perpendicular to the field there. However,

$$\int_{\text{side}} \mathbf{D} \cdot \hat{\mathbf{n}}\, dS = \int_0^h \int_0^{2\pi} D_\rho(\rho_0)\hat{\boldsymbol{\rho}} \cdot \hat{\boldsymbol{\rho}}\rho_0\, d\phi\, dz$$

$$= \rho_0 D_\rho(\rho_0) \int_0^h dz \int_0^{2\pi} d\phi = 2\pi h \rho_0 D_\rho(\rho_0).$$

The charge enclosed by the cylinder is

$$Q_{\text{enc}} = \int_0^h \int_0^{2\pi} \int_0^{\rho_0} \rho_l \frac{\delta(\rho)}{2\pi\rho} \rho\, d\rho\, d\phi\, dz = \frac{\rho_l}{2\pi} \int_0^h dz \int_0^{2\pi} d\phi \int_0^{\rho_0} \delta(\rho)\, d\rho = \rho_l h.$$

Hence, by Gauss's law, $2\pi h \rho_0 D_\rho(\rho_0) = \rho_l h$, and we obtain

$$D_\rho(\rho_0) = \frac{\rho_l}{2\pi\rho_0}$$

where the subscript "0" can be dropped as ρ_0 is arbitrary. Therefore

$$\mathbf{E}(\rho) = \hat{\rho}\,\frac{\rho_l}{2\pi\epsilon\rho}. \quad \blacktriangleleft$$

▶ **Example 3.3:** Solution to the integral form of Gauss's law for a volume charge

A ball of charge having radius a resides at the coordinate origin in a homogeneous medium with permittivity ϵ. Taking the volume charge density as a constant value ρ_v for $r \leq a$, find $\mathbf{E}$ everywhere.

Solution: Since the charge depends only on the radial variable r of spherical coordinates, we have $\mathbf{D}(\mathbf{r}) = D_r(r)\hat{\mathbf{r}}$. Choose as a flux surface the sphere $r = r_0$. Gauss's law requires

$$\int_S \mathbf{D} \cdot \hat{\mathbf{n}}\,dS = Q_{\text{enc}}$$

where

$$\int_S \mathbf{D} \cdot \hat{\mathbf{n}}\,dS = \int_0^{2\pi}\int_0^{\pi} D_r(r_0)\hat{\mathbf{r}} \cdot \hat{\mathbf{r}} r_0^2 \sin\theta\,d\theta\,d\phi$$

$$= r_0^2 D_r(r_0)\int_0^{2\pi} d\phi \int_0^{\pi}\sin\theta\,d\theta = 4\pi r_0^2 D_r(r_0)$$

and

$$Q_{\text{enc}} = \begin{cases} \int_0^{2\pi}\int_0^{\pi}\int_0^{r_0}\rho_v r^2 \sin\theta\,dr\,d\theta\,d\phi = \frac{4}{3}\pi r_0^3 \rho_v, & r_0 < a, \\ \int_0^{2\pi}\int_0^{\pi}\int_0^{a}\rho_v r^2 \sin\theta\,dr\,d\theta\,d\phi = \frac{4}{3}\pi a^3 \rho_v, & r_0 > a, \end{cases}$$

is the charge enclosed by the sphere $r = r_0$. We obtain

$$\mathbf{E}(r) = \begin{cases} \hat{\mathbf{r}}\,\dfrac{\rho_v r}{3\epsilon}, & r < a, \\[2mm] \hat{\mathbf{r}}\,\dfrac{\rho_v a^3}{3\epsilon r^2}, & r > a. \end{cases} \quad \blacktriangleleft$$

▶ **Example 3.4:** Solution to the point form of Gauss's law for a volume charge

An infinite cylinder of volume charge is aligned with the z-axis in a homogeneous medium of permittivity ϵ. The radius of the cylinder is a, and the charge density is a constant ρ_v within the cylinder. Use the point form of Gauss's law to find $\mathbf{E}$.

Solution: The cylindrical symmetry of the charge distribution implies that $\mathbf{D}(\mathbf{r}) = D_\rho(\rho)\hat{\rho}$. Hence

$$\nabla \cdot \mathbf{D} = \frac{1}{\rho}\frac{\partial}{\partial\rho}\left(\rho D_\rho\right).$$

For $\rho \leq a$, Gauss's law gives

$$\frac{1}{\rho}\frac{\partial}{\partial\rho}(\rho D_\rho) = \rho_v$$

so that by integration

$$D_\rho = \frac{\rho_v}{2}\rho + \frac{C_1}{\rho}.$$

Boundedness of the field on the z-axis requires $C_1 = 0$. For $\rho > a$ we have

$$\frac{1}{\rho}\frac{\partial}{\partial\rho}(\rho D_\rho) = 0$$

so that $D_\rho = C_2/\rho$. Continuity of the normal component of $\mathbf{D}$ at $\rho = a$ (cf., § 3.3.3) requires

$$\frac{\rho_v}{2}a = \frac{C_2}{a},$$

giving $C_2 = \rho_v a^2/2$. Therefore

$$\mathbf{E}(\rho) = \begin{cases} \hat{\rho}\,\dfrac{\rho_v\rho}{2\epsilon}, & \rho \leq a, \\[2ex] \hat{\rho}\,\dfrac{\rho_v a^2}{2\epsilon\rho}, & \rho > a. \end{cases} \qquad \blacktriangleleft$$

3.3.2   The electrostatic potential and work

The equation

$$\oint_\Gamma \mathbf{E}\cdot\mathbf{dl} = 0 \tag{3.9}$$

satisfied by the electrostatic field $\mathbf{E}(\mathbf{r})$ is particularly interesting. A field with zero circulation is said to be *conservative*. To see why, let us examine the work required to move a particle of charge Q around a closed path in the presence of $\mathbf{E}(\mathbf{r})$. Since work is the line integral of force and $\mathbf{B} = 0$, the work expended by the external system moving the charge against the Lorentz force is

$$W = -\oint_\Gamma (Q\mathbf{E} + Q\mathbf{v}\times\mathbf{B})\cdot\mathbf{dl} = -Q\oint_\Gamma \mathbf{E}\cdot\mathbf{dl} = 0.$$

This property is analogous to the conservation property for a classical gravitational field: any potential energy gained by raising a point mass is lost when the mass is lowered.

Direct experimental verification of the electrostatic conservative property is difficult, aside from the fact that the motion of Q may alter $\mathbf{E}$ by interacting with the sources of $\mathbf{E}$. By moving Q with nonuniform velocity (i.e., with acceleration at the beginning of the loop, direction changes in transit, and deceleration at the end) we observe a radiative loss of energy, and this energy cannot be regained by the mechanical system providing the motion. To avoid this problem we may assume that the charge is moved so slowly, or in such small increments, that it does not radiate. We shall use this concept later to determine the "assembly energy" in a charge distribution.

3.3.2.1   The electrostatic potential

By the point form of (3.9),

$$\nabla\times\mathbf{E}(\mathbf{r}) = 0,$$

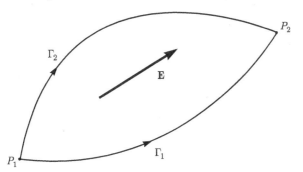

FIGURE 3.3
Demonstration of path independence of the electric field line integral.

we can introduce a scalar field $\Phi = \Phi(\mathbf{r})$ such that

$$\mathbf{E}(\mathbf{r}) = -\nabla\Phi(\mathbf{r}). \tag{3.10}$$

The function Φ carries units of volts and is known as the *electrostatic potential*. Let us consider the work expended by an external agent in moving a charge between points P_1 at $\mathbf{r}_1$ and P_2 at $\mathbf{r}_2$:

$$W_{21} = -Q\int_{P_1}^{P_2} -\nabla\Phi(\mathbf{r})\cdot d\mathbf{l} = Q\int_{P_1}^{P_2} d\Phi(\mathbf{r}) = Q\left[\Phi(\mathbf{r}_2) - \Phi(\mathbf{r}_1)\right].$$

The work W_{21} is clearly independent of the path taken between P_1 and P_2; the quantity

$$V_{21} = \frac{W_{21}}{Q} = \Phi(\mathbf{r}_2) - \Phi(\mathbf{r}_1) = -\int_{P_1}^{P_2} \mathbf{E}\cdot d\mathbf{l},$$

called the *potential difference*, has an obvious physical meaning as work per unit charge required to move a particle against an electric field between two points.

Of course, the large-scale form (3.9) also implies the path-independence of work in the electrostatic field. Indeed, we may pass an arbitrary closed contour Γ through P_1 and P_2 and then split it into two pieces Γ_1 and Γ_2 as shown in Figure 3.3. Since

$$-Q\oint_{\Gamma_1-\Gamma_2} \mathbf{E}\cdot d\mathbf{l} = -Q\int_{\Gamma_1} \mathbf{E}\cdot d\mathbf{l} + Q\int_{\Gamma_2} \mathbf{E}\cdot d\mathbf{l} = 0,$$

we have

$$-Q\int_{\Gamma_1} \mathbf{E}\cdot d\mathbf{l} = -Q\int_{\Gamma_2} \mathbf{E}\cdot d\mathbf{l}$$

as desired.

We sometimes refer to $\Phi(\mathbf{r})$ as the *absolute electrostatic potential*. Choosing a suitable reference point P_0 at location $\mathbf{r}_0$ and writing the potential difference as

$$V_{21} = [\Phi(\mathbf{r}_2) - \Phi(\mathbf{r}_0)] - [\Phi(\mathbf{r}_1) - \Phi(\mathbf{r}_0)],$$

we can justify calling $\Phi(\mathbf{r})$ the *absolute potential referred to* P_0. Thus, we write

$$\Phi(\mathbf{r}) = -\int_{P_0}^{P} \mathbf{E}\cdot d\mathbf{l}, \tag{3.11}$$

where P is located at $\mathbf{r}$. Note that P_0 might describe a locus of points, rather than a single point, since many points can be at the same potential. Although we can choose any reference point without changing the resulting value of $\mathbf{E}$ found from (3.10), for simplicity we often choose $\mathbf{r}_0$ such that $\Phi(\mathbf{r}_0) = 0$.

▶ **Example 3.5:** Absolute potential of a volume source

Find Φ for the charge distribution of Example 3.3.

Solution: Recall that

$$\mathbf{E}(\mathbf{r}) = \begin{cases} \hat{\mathbf{r}} \dfrac{\rho_v r}{3\epsilon}, & r < a, \\[2mm] \hat{\mathbf{r}} \dfrac{\rho_v a^3}{3\epsilon r^2}, & r > a. \end{cases}$$

We take a reference point at radius $r_0 > a$; this will permit us to let $r_0 \to \infty$. When $r > a$, the potential is merely

$$\Phi(r) = -\int_{r_0}^r E_r(r')\,dr'$$

$$= -\frac{\rho_v a^3}{3\epsilon} \int_{r_0}^r \frac{dr'}{r'^2}$$

$$= \frac{\rho_v a^3}{3\epsilon} \left(\frac{1}{r} - \frac{1}{r_0} \right).$$

Otherwise

$$\Phi(r) = -\int_{r_0}^a E_r(r')\,dr' - \int_a^r E_r(r')\,dr'$$

$$= -\frac{\rho_v a^3}{3\epsilon} \int_{r_0}^a \frac{dr'}{r^2} - \frac{\rho_v}{3\epsilon} \int_a^r r'\,dr'$$

$$= \frac{\rho_v a^2}{3\epsilon r_0}(r_0 - a) + \frac{\rho_v}{6\epsilon}(a^2 - r^2).$$

As $r_0 \to \infty$, we obtain the *absolute potential referred to infinity*:

$$\Phi(r) = \begin{cases} \dfrac{\rho_v a^3}{3\epsilon r}, & r > a, \\[2mm] \dfrac{\rho_v}{6\epsilon}(3a^2 - r^2), & r < a. \end{cases} \quad \blacktriangleleft$$

Several properties of the electrostatic potential make it convenient for describing static electric fields. We know that, at equilibrium, the electrostatic field within a conducting body must vanish. By (3.10) the potential at all points within the body must therefore have the same constant value. It follows that the surface of a conductor is an *equipotential surface*: a surface for which $\Phi(\mathbf{r})$ is constant.

As an infinite reservoir of charge that can be tapped through a filamentary conductor, the entity we call "ground" must also be an equipotential object. If we connect a conductor to ground, we have seen that charge may flow freely onto the conductor. Since no work is expended, "grounding" a conductor obviously places the conductor at the same absolute potential as ground. For this reason, ground is often assigned the role as the potential reference with an absolute potential of zero volts. Later we shall see that for sources of finite extent, ground must be located at infinity.

3.3.3 Boundary conditions

3.3.3.1 Boundary conditions for the electrostatic field

The boundary conditions found for the dynamic electric field remain valid in the electrostatic case. Thus

$$\hat{\mathbf{n}}_{12} \times (\mathbf{E}_1 - \mathbf{E}_2) = 0 \tag{3.12}$$

and

$$\hat{\mathbf{n}}_{12} \cdot (\mathbf{D}_1 - \mathbf{D}_2) = \rho_s. \tag{3.13}$$

Here $\hat{\mathbf{n}}_{12}$ points into region 1 from region 2. Because the static curl and divergence equations are independent, so are the boundary conditions (3.12) and (3.13).

For a linear and isotropic dielectric where $\mathbf{D} = \epsilon \mathbf{E}$, Equation (3.13) becomes

$$\hat{\mathbf{n}}_{12} \cdot (\epsilon_1 \mathbf{E}_1 - \epsilon_2 \mathbf{E}_2) = \rho_s. \tag{3.14}$$

Alternatively, using $\mathbf{D} = \epsilon_0 \mathbf{E} + \mathbf{P}$, we can write (3.13) as

$$\hat{\mathbf{n}}_{12} \cdot (\mathbf{E}_1 - \mathbf{E}_2) = \frac{1}{\epsilon_0} (\rho_s + \rho_{Ps1} + \rho_{Ps2}) \tag{3.15}$$

where

$$\rho_{Ps} = \hat{\mathbf{n}} \cdot \mathbf{P}$$

is the polarization surface charge with $\hat{\mathbf{n}}$ pointing outward from the material body.

We can also write the boundary conditions in terms of the electrostatic potential. With $\mathbf{E} = -\nabla \Phi$, Equation (3.12) becomes

$$\Phi_1(\mathbf{r}) = \Phi_2(\mathbf{r}) \tag{3.16}$$

for all points $\mathbf{r}$ on the surface. Actually Φ_1 and Φ_2 may differ by a constant; because this constant is eliminated when the gradient is taken to find $\mathbf{E}$, it is generally ignored. We can write (3.15) as

$$\epsilon_0 \left(\frac{\partial \Phi_1}{\partial n} - \frac{\partial \Phi_2}{\partial n} \right) = -\rho_s - \rho_{Ps1} - \rho_{Ps2}$$

where the normal derivative is taken in the $\hat{\mathbf{n}}_{12}$ direction. For a linear, isotropic dielectric, (3.13) becomes

$$\epsilon_1 \frac{\partial \Phi_1}{\partial n} - \epsilon_2 \frac{\partial \Phi_2}{\partial n} = -\rho_s. \tag{3.17}$$

Again, we note that (3.16) and (3.17) are independent.

3.3.3.2 Boundary conditions for steady electric current

The boundary condition on the normal component of current found in § 2.8.2 remains valid in the steady current case. Assume that the boundary exists between two linear, isotropic conducting regions having constitutive parameters (ϵ_1, σ_1) and (ϵ_2, σ_2), respectively. By (2.162) we have

$$\hat{\mathbf{n}}_{12} \cdot (\mathbf{J}_1 - \mathbf{J}_2) = -\nabla_s \cdot \mathbf{J}_s \tag{3.18}$$

where $\hat{\mathbf{n}}_{12}$ points into region 1 from region 2. A surface current will not appear on the boundary between two regions having finite conductivity, although a surface charge may accumulate there during the transient period when the currents are established [34]. If

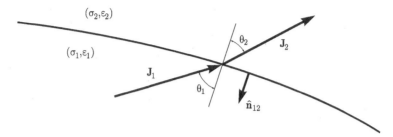

FIGURE 3.4
Refraction of steady current at a material interface.

charge is influenced to move from the surface, it will move into the adjacent regions, rather than along the surface, and a new charge will replace it, supplied by the current. Thus, for finite conducting regions (3.18) becomes

$$\hat{\mathbf{n}}_{12} \cdot (\mathbf{J}_1 - \mathbf{J}_2) = 0. \tag{3.19}$$

A boundary condition on the tangential component of current can also be found. Substituting $\mathbf{E} = \mathbf{J}/\sigma$ into (3.12) we have

$$\hat{\mathbf{n}}_{12} \times \left(\frac{\mathbf{J}_1}{\sigma_1} - \frac{\mathbf{J}_2}{\sigma_2} \right) = 0.$$

We can also write this as

$$\frac{\mathbf{J}_{1t}}{\sigma_1} = \frac{\mathbf{J}_{2t}}{\sigma_2} \tag{3.20}$$

where

$$\mathbf{J}_{1t} = \hat{\mathbf{n}}_{12} \times \mathbf{J}_1, \qquad \mathbf{J}_{2t} = \hat{\mathbf{n}}_{12} \times \mathbf{J}_2.$$

We may combine the boundary conditions for the normal components of current and electric field to better understand the behavior of current at a material boundary. Substituting $\mathbf{E} = \mathbf{J}/\sigma$ into (3.14) we have

$$\frac{\epsilon_1}{\sigma_1} J_{1n} - \frac{\epsilon_2}{\sigma_2} J_{2n} = \rho_s \tag{3.21}$$

where $J_{1n} = \hat{\mathbf{n}}_{12} \cdot \mathbf{J}_1$ and $J_{2n} = \hat{\mathbf{n}}_{12} \cdot \mathbf{J}_2$. Combining (3.21) with (3.19), we have

$$\rho_s = J_{1n} \left(\frac{\epsilon_1}{\sigma_1} - \frac{\epsilon_2}{\sigma_2} \right) = E_{1n} \left(\epsilon_1 - \frac{\sigma_1}{\sigma_2} \epsilon_2 \right) = J_{2n} \left(\frac{\epsilon_1}{\sigma_1} - \frac{\epsilon_2}{\sigma_2} \right) = E_{2n} \left(\epsilon_1 \frac{\sigma_2}{\sigma_1} - \epsilon_2 \right)$$

where

$$E_{1n} = \hat{\mathbf{n}}_{12} \cdot \mathbf{E}_1, \qquad E_{2n} = \hat{\mathbf{n}}_{12} \cdot \mathbf{E}_2.$$

Unless $\epsilon_1 \sigma_2 - \sigma_1 \epsilon_2 = 0$, a surface charge will exist on the interface between dissimilar current-carrying conductors.

We may also combine the vector components of current on each side of the boundary to determine the effects of the boundary on current direction (Figure 3.4). Let $\theta_{1,2}$ denote the angle between $\mathbf{J}_{1,2}$ and $\hat{\mathbf{n}}_{12}$ so that

$$J_{1n} = J_1 \cos \theta_1, \qquad J_{1t} = J_1 \sin \theta_1,$$
$$J_{2n} = J_2 \cos \theta_2, \qquad J_{2t} = J_2 \sin \theta_2.$$

Then $J_1 \cos \theta_1 = J_2 \cos \theta_2$ by (3.19), while $\sigma_2 J_1 \sin \theta_1 = \sigma_1 J_2 \sin \theta_2$ by (3.20). Hence

$$\sigma_2 \tan \theta_1 = \sigma_1 \tan \theta_2. \tag{3.22}$$

It is interesting to consider the case of current incident from a conducting material onto an insulating material. If region 2 is an insulator, then $J_{2n} = J_{2t} = 0$; by (3.19) we have $J_{1n} = 0$. But (3.20) does not require $J_{1t} = 0$; with $\sigma_2 = 0$ the right-hand side of (3.20) is indeterminate and thus J_{1t} may be nonzero. In other words, when current moving through a conductor approaches an insulating surface, it bends and flows tangential to the surface. This concept is useful in explaining how wires guide current.

Interestingly, (3.22) shows that when $\sigma_2 \ll \sigma_1$ we have $\theta_2 \to 0$; current passing from a conducting region into a slightly conducting region does so normally.

3.3.4 Uniqueness of the electrostatic field

In § 2.2.1 we found that the electromagnetic field is unique within a region V when the tangential component of $\mathbf{E}$ is specified over the surrounding surface. Unfortunately, this condition is not appropriate in the electrostatic case. We should remember that an additional requirement for uniqueness of solution to Maxwell's equations is that the field be specified throughout V at some time t_0. For a static field this would completely determine $\mathbf{E}$ without need for the surface field!

Let us determine conditions for uniqueness beginning with the static field equations. Consider a region V surrounded by a surface S. Static charge may be located entirely or partially within V, or entirely outside V, and produces a field within V. The region may also contain any arrangement of conductors or other materials. Suppose $(\mathbf{D}_1, \mathbf{E}_1)$ and $(\mathbf{D}_2, \mathbf{E}_2)$ represent solutions to the static field equations within V with source $\rho(\mathbf{r})$. We wish to find conditions that guarantee both $\mathbf{E}_1 = \mathbf{E}_2$ and $\mathbf{D}_1 = \mathbf{D}_2$.

Since $\nabla \cdot \mathbf{D}_1 = \rho$ and $\nabla \cdot \mathbf{D}_2 = \rho$, the difference field $\mathbf{D}_0 = \mathbf{D}_2 - \mathbf{D}_1$ obeys the homogeneous equation

$$\nabla \cdot \mathbf{D}_0 = 0. \tag{3.23}$$

Consider the quantity

$$\nabla \cdot (\mathbf{D}_0 \Phi_0) = \Phi_0 (\nabla \cdot \mathbf{D}_0) + \mathbf{D}_0 \cdot (\nabla \Phi_0)$$

where $\mathbf{E}_0 = \mathbf{E}_2 - \mathbf{E}_1 = -\nabla \Phi_0 = -\nabla(\Phi_2 - \Phi_1)$. We integrate over V and use the divergence theorem and (3.23) to obtain

$$\oint_S \Phi_0 \left(\mathbf{D}_0 \cdot \hat{\mathbf{n}} \right) dS = \int_V \mathbf{D}_0 \cdot (\nabla \Phi_0) \, dV = - \int_V \mathbf{D}_0 \cdot \mathbf{E}_0 \, dV. \tag{3.24}$$

Now suppose that $\Phi_0 = 0$ everywhere on S, or that $\hat{\mathbf{n}} \cdot \mathbf{D}_0 = 0$ everywhere on S, or that $\Phi_0 = 0$ over part of S and $\hat{\mathbf{n}} \cdot \mathbf{D}_0 = 0$ elsewhere on S. Then

$$\int_V \mathbf{D}_0 \cdot \mathbf{E}_0 \, dV = 0. \tag{3.25}$$

Since V is arbitrary, either $\mathbf{D}_0 = 0$ or $\mathbf{E}_0 = 0$. Assuming $\mathbf{E}$ and $\mathbf{D}$ are linked by the constitutive relations, we have $\mathbf{E}_1 = \mathbf{E}_2$ and $\mathbf{D}_1 = \mathbf{D}_2$.

Hence the fields within V are unique provided that either Φ, the normal component of $\mathbf{D}$, or some combination of the two, is specified over S. We often use a multiply connected surface to exclude conductors. By (3.13) we see that specification of the normal component of $\mathbf{D}$ on a conductor is equivalent to specification of the surface charge density. Thus we must specify the potential or surface charge density over all conducting surfaces.

One other condition results in zero on the left-hand side of (3.24). If S recedes to infinity and Φ_0 and $\mathbf{D}_0$ decrease sufficiently fast, then (3.25) still holds and uniqueness is guaranteed. If $\mathbf{D}, \mathbf{E} \sim 1/r^2$ as $r \to \infty$, then $\Phi \sim 1/r$ and the surface integral in (3.24) tends to zero since the area of an expanding sphere increases only as r^2. We shall find later in this section that for sources of finite extent, the fields do indeed vary inversely with distance squared from the source. Hence we may allow S to expand and encompass all space.

For the case in which conducting bodies are immersed in an infinite homogeneous medium and the static fields must be determined throughout all space, a multiply connected surface is used with one part receding to infinity and the remaining parts surrounding the conductors. Here uniqueness is guaranteed by specifying the potentials or charges on the surfaces of the conducting bodies.

3.3.5 Poisson's and Laplace's equations

For computational purposes it is often convenient to deal with the differential versions

$$\nabla \times \mathbf{E}(\mathbf{r}) = 0, \tag{3.26}$$

$$\nabla \cdot \mathbf{D}(\mathbf{r}) = \rho(\mathbf{r}), \tag{3.27}$$

of the electrostatic field equations. We must supplement these with constitutive relations between $\mathbf{E}$ and $\mathbf{D}$; at this point we focus our attention on linear, isotropic materials for which

$$\mathbf{D}(\mathbf{r}) = \epsilon(\mathbf{r})\mathbf{E}(\mathbf{r}).$$

Using this in (3.27) along with $\mathbf{E} = -\nabla\Phi$ (justified by (3.26)), we can write

$$\nabla \cdot [\epsilon(\mathbf{r})\nabla\Phi(\mathbf{r})] = -\rho(\mathbf{r}). \tag{3.28}$$

This is *Poisson's equation*. The corresponding homogeneous equation

$$\nabla \cdot [\epsilon(\mathbf{r})\nabla\Phi(\mathbf{r})] = 0, \tag{3.29}$$

holding at points $\mathbf{r}$ where $\rho(\mathbf{r}) = 0$, is *Laplace's equation*. Equations (3.28) and (3.29) are valid for inhomogeneous media. By (B.48) we can write

$$\nabla\Phi(\mathbf{r}) \cdot \nabla\epsilon(\mathbf{r}) + \epsilon(\mathbf{r})\nabla \cdot [\nabla\Phi(\mathbf{r})] = -\rho(\mathbf{r}).$$

For a homogeneous medium, $\nabla\epsilon = 0$; since $\nabla \cdot (\nabla\Phi) \equiv \nabla^2\Phi$, we have

$$\nabla^2\Phi(\mathbf{r}) = -\rho(\mathbf{r})/\epsilon \tag{3.30}$$

in such a medium. Correspondingly,

$$\nabla^2\Phi(\mathbf{r}) = 0$$

at points where $\rho(\mathbf{r}) = 0$.

Poisson's and Laplace's equations can be solved by separation of variables, Fourier transformation, conformal mapping, and numerical techniques such as the finite difference and moment methods. In Appendix A we consider the separation of variables solution to

Laplace's equation in three major coordinate systems for a variety of problems. For an introduction to numerical techniques, the reader is referred to the books by Sadiku [167], Harrington [81], and Peterson [150]. Solution to Poisson's equation is often undertaken using the method of Green's functions, which we shall address later in this section. We shall also consider the solution to Laplace's equation for bodies immersed in an applied, or "impressed," field.

3.3.5.1 Uniqueness of solution to Poisson's equation

Before attempting any solutions, we must ask two very important questions. How do we know that solving the second-order differential equation produces the same values for $\mathbf{E} = -\nabla\Phi$ as solving the first-order equations directly for $\mathbf{E}$? And, if these solutions are the same, what are the conditions for uniqueness of solution to Poisson's and Laplace's equations? To answer the first question, a sufficient condition is to have Φ twice differentiable. We shall not attempt to prove this, but shall instead show that the condition for uniqueness of the second-order equations is the same as that for the first-order equations.

Consider a region of space V surrounded by a surface S. Static charge may be located entirely or partially within V, or entirely outside V, and produces a field within V. This region may also contain any arrangement of conductors or other materials. Now, assume that Φ_1 and Φ_2 represent solutions to the static field equations within V with source $\rho(\mathbf{r})$. We wish to find conditions under which $\Phi_1 = \Phi_2$.

Since we have

$$\nabla \cdot [\epsilon(\mathbf{r})\nabla\Phi_1(\mathbf{r})] = -\rho(\mathbf{r}), \qquad \nabla \cdot [\epsilon(\mathbf{r})\nabla\Phi_2(\mathbf{r})] = -\rho(\mathbf{r}),$$

the difference field $\Phi_0 = \Phi_2 - \Phi_1$ obeys

$$\nabla \cdot [\epsilon(\mathbf{r})\nabla\Phi_0(\mathbf{r})] = 0. \tag{3.31}$$

That is, Φ_0 obeys Laplace's equation. Now consider the quantity

$$\nabla \cdot (\epsilon\Phi_0\nabla\Phi_0) = \epsilon|\nabla\Phi_0|^2 + \Phi_0\nabla \cdot (\epsilon\nabla\Phi_0).$$

Integration over V and use of the divergence theorem and (3.31) gives

$$\oint_S \Phi_0(\mathbf{r})[\epsilon(\mathbf{r})\nabla\Phi_0(\mathbf{r})] \cdot d\mathbf{S} = \int_V \epsilon(\mathbf{r})|\nabla\Phi_0(\mathbf{r})|^2 \, dV.$$

As with the first order equations, we see that specifying either $\Phi(\mathbf{r})$ or $\epsilon(\mathbf{r})\nabla\Phi(\mathbf{r}) \cdot \hat{\mathbf{n}}$ over S results in $\Phi_0(\mathbf{r}) = 0$ throughout V, hence $\Phi_1 = \Phi_2$. As before, specifying $\epsilon(\mathbf{r})\nabla\Phi(\mathbf{r}) \cdot \hat{\mathbf{n}}$ for a conducting surface is equivalent to specifying the surface charge on S.

3.3.5.2 Integral solution to Poisson's equation: the static Green's function

The method of Green's functions is one of the most useful techniques for solving Poisson's equation. We seek a solution for a single point source, then use Green's second identity to write the solution for an arbitrary charge distribution in terms of a superposition integral.

We seek the solution to Poisson's equation for a region of space V as shown in Figure 3.5. The region is assumed homogeneous with permittivity ϵ, and its surface is multiply-connected, consisting of a bounding surface S_B and any number of closed surfaces internal to V. We denote by S the composite surface consisting of S_B and the N internal surfaces S_n, $n = 1, \ldots, N$. The internal surfaces are used to exclude material bodies, such as the

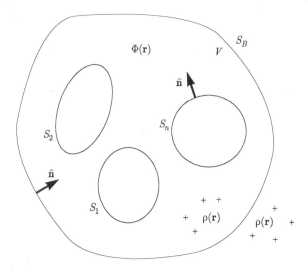

FIGURE 3.5
Computation of potential from known sources and values on bounding surfaces.

plates of a capacitor, which may be charged and on which the potential is assumed to be known. To solve for $\Phi(\mathbf{r})$ within V we must know the potential produced by a point source. This potential, called the *Green's function*, is denoted $G(\mathbf{r}|\mathbf{r}')$; it has two arguments because it satisfies Poisson's equation at $\mathbf{r}$ when the source is located at $\mathbf{r}'$:

$$\nabla^2 G(\mathbf{r}|\mathbf{r}') = -\delta(\mathbf{r} - \mathbf{r}'). \tag{3.32}$$

Later we shall demonstrate that in all cases of interest to us the Green's function is symmetric in its arguments:

$$G(\mathbf{r}'|\mathbf{r}) = G(\mathbf{r}|\mathbf{r}'). \tag{3.33}$$

This property of G is known as *reciprocity*.

Our development rests on the mathematical result (B.36) known as *Green's second identity*. We can derive this by subtracting the identities

$$\nabla \cdot (\phi \nabla \psi) = \phi \nabla \cdot (\nabla \psi) + (\nabla \phi) \cdot (\nabla \psi),$$
$$\nabla \cdot (\psi \nabla \phi) = \psi \nabla \cdot (\nabla \phi) + (\nabla \psi) \cdot (\nabla \phi),$$

to obtain

$$\nabla \cdot (\phi \nabla \psi - \psi \nabla \phi) = \phi \nabla^2 \psi - \psi \nabla^2 \phi.$$

Integrating this over a volume region V with respect to the dummy variable $\mathbf{r}'$ and using the divergence theorem, we obtain

$$\int_V [\phi(\mathbf{r}')\nabla'^2\psi(\mathbf{r}') - \psi(\mathbf{r}')\nabla'^2\phi(\mathbf{r}')]\, dV' = -\oint_S [\phi(\mathbf{r}')\nabla'\psi(\mathbf{r}') - \psi(\mathbf{r}')\nabla'\phi(\mathbf{r}')] \cdot d\mathbf{S}'.$$

The negative sign on the right-hand side occurs because $\hat{\mathbf{n}}$ is an *inward* normal to V. Finally, since $\partial\psi(\mathbf{r}')/\partial n' = \hat{\mathbf{n}}' \cdot \nabla'\psi(\mathbf{r}')$, we have

$$\int_V [\phi(\mathbf{r}')\nabla'^2\psi(\mathbf{r}') - \psi(\mathbf{r}')\nabla'^2\phi(\mathbf{r}')]\, dV' = -\oint_S \left[\phi(\mathbf{r}')\frac{\partial\psi(\mathbf{r}')}{\partial n'} - \psi(\mathbf{r}')\frac{\partial\phi(\mathbf{r}')}{\partial n'}\right] dS'$$

as desired.

To solve for Φ in V we shall make some seemingly unmotivated substitutions into this identity. First note that by (3.32) and (3.33) we can write

$$\nabla'^2 G(\mathbf{r}|\mathbf{r}') = -\delta(\mathbf{r}' - \mathbf{r}). \tag{3.34}$$

We now set $\phi(\mathbf{r}') = \Phi(\mathbf{r}')$ and $\psi(\mathbf{r}') = G(\mathbf{r}|\mathbf{r}')$ to obtain

$$\int_V [\Phi(\mathbf{r}')\nabla'^2 G(\mathbf{r}|\mathbf{r}') - G(\mathbf{r}|\mathbf{r}')\nabla'^2\Phi(\mathbf{r}')]\,dV' =$$
$$-\oint_S \left[\Phi(\mathbf{r}')\frac{\partial G(\mathbf{r}|\mathbf{r}')}{\partial n'} - G(\mathbf{r}|\mathbf{r}')\frac{\partial\Phi(\mathbf{r}')}{\partial n'}\right]\,dS', \tag{3.35}$$

hence

$$\int_V \left[\Phi(\mathbf{r}')\delta(\mathbf{r}' - \mathbf{r}) - G(\mathbf{r}|\mathbf{r}')\frac{\rho(\mathbf{r}')}{\epsilon}\right]\,dV' = \oint_S \left[\Phi(\mathbf{r}')\frac{\partial G(\mathbf{r}|\mathbf{r}')}{\partial n'} - G(\mathbf{r}|\mathbf{r}')\frac{\partial\Phi(\mathbf{r}')}{\partial n'}\right]\,dS'.$$

By the sifting property of the Dirac delta

$$\Phi(\mathbf{r}) = \int_V G(\mathbf{r}|\mathbf{r}')\frac{\rho(\mathbf{r}')}{\epsilon}\,dV' + \oint_{S_B} \left[\Phi(\mathbf{r}')\frac{\partial G(\mathbf{r}|\mathbf{r}')}{\partial n'} - G(\mathbf{r}|\mathbf{r}')\frac{\partial\Phi(\mathbf{r}')}{\partial n'}\right]\,dS'$$
$$+ \sum_{n=1}^N \oint_{S_n} \left[\Phi(\mathbf{r}')\frac{\partial G(\mathbf{r}|\mathbf{r}')}{\partial n'} - G(\mathbf{r}|\mathbf{r}')\frac{\partial\Phi(\mathbf{r}')}{\partial n'}\right]\,dS'. \tag{3.36}$$

With this we may compute the potential anywhere within V in terms of the charge density within V and the values of the potential and its normal derivative over S. We must simply determine $G(\mathbf{r}|\mathbf{r}')$ first.

Let us take a moment to specialize (3.36) to the case of unbounded space. Provided that the sources are of finite extent, as $S_B \to \infty$ we shall find that

$$\Phi(\mathbf{r}) = \int_V G(\mathbf{r}|\mathbf{r}')\frac{\rho(\mathbf{r}')}{\epsilon}\,dV' + \sum_{n=1}^N \oint_{S_n} \left[\Phi(\mathbf{r}')\frac{\partial G(\mathbf{r}|\mathbf{r}')}{\partial n'} - G(\mathbf{r}|\mathbf{r}')\frac{\partial\Phi(\mathbf{r}')}{\partial n'}\right]\,dS'.$$

3.3.5.3 Useful derivative identities

Many differential operations on the displacement vector $\mathbf{R} = \mathbf{r} - \mathbf{r}'$ occur in the study of electromagnetics. The identities

$$\nabla R = -\nabla' R = \hat{\mathbf{R}}, \qquad \nabla\left(\frac{1}{R}\right) = -\nabla'\left(\frac{1}{R}\right) = -\frac{\hat{\mathbf{R}}}{R^2}, \tag{3.37}$$

follow from direct differentiation of the rectangular coordinate representation

$$\mathbf{R} = \hat{\mathbf{x}}(x - x') + \hat{\mathbf{y}}(y - y') + \hat{\mathbf{z}}(z - z').$$

Other identities are more difficult to establish.

▶ **Example 3.6:** Laplacian of $1/R$

Establish the identity

$$\nabla^2\left(\frac{1}{R}\right) = -4\pi\delta(\mathbf{r} - \mathbf{r}'), \tag{3.38}$$

which is crucial to potential theory.

Solution: We shall prove the equivalent version

$$\nabla'^2 \left(\frac{1}{R}\right) = -4\pi\delta(\mathbf{r}' - \mathbf{r})$$

by showing that

$$\int_V f(\mathbf{r}')\nabla'^2 \left(\frac{1}{R}\right) dV' = \begin{cases} -4\pi f(\mathbf{r}), & \mathbf{r} \in V, \\ 0, & \mathbf{r} \notin V, \end{cases} \tag{3.39}$$

holds for any continuous function $f(\mathbf{r})$. By direct differentiation we have

$$\nabla'^2 \left(\frac{1}{R}\right) = 0 \text{ for } \mathbf{r}' \neq \mathbf{r},$$

hence the second part of (3.39) is established. This also shows that if $\mathbf{r} \in V$ then the domain of integration in (3.39) can be restricted to a sphere of arbitrarily small radius ε centered at $\mathbf{r}$ (Figure 3.6). The result we seek is found in the limit as $\varepsilon \to 0$. Thus we are interested in computing

$$\int_V f(\mathbf{r}')\nabla'^2 \left(\frac{1}{R}\right) dV' = \lim_{\varepsilon \to 0} \int_{V_\varepsilon} f(\mathbf{r}')\nabla'^2 \left(\frac{1}{R}\right) dV'.$$

Since f is continuous at $\mathbf{r}' = \mathbf{r}$, we have by the mean value theorem

$$\int_V f(\mathbf{r}')\nabla'^2 \left(\frac{1}{R}\right) dV' = f(\mathbf{r}) \lim_{\varepsilon \to 0} \int_{V_\varepsilon} \nabla'^2 \left(\frac{1}{R}\right) dV'.$$

The integral over V_ε can be computed using $\nabla'^2(1/R) = \nabla' \cdot \nabla'(1/R)$ and the divergence theorem:

$$\int_{V_\varepsilon} \nabla'^2 \left(\frac{1}{R}\right) dV' = \int_{S_\varepsilon} \hat{\mathbf{n}}' \cdot \nabla' \left(\frac{1}{R}\right) dS',$$

where S_ε bounds V_ε. Noting that $\hat{\mathbf{n}}' = -\hat{\mathbf{R}}$, using (3.37), and writing the integral in spherical coordinates $(\varepsilon, \theta, \phi)$ centered at the point $\mathbf{r}$, we have

$$\int_V f(\mathbf{r}')\nabla'^2 \left(\frac{1}{R}\right) dV' = f(\mathbf{r}) \lim_{\varepsilon \to 0} \int_0^{2\pi} \int_0^\pi -\hat{\mathbf{R}} \cdot \left(\frac{\hat{\mathbf{R}}}{\varepsilon^2}\right) \varepsilon^2 \sin\theta \, d\theta \, d\phi$$

$$= -4\pi f(\mathbf{r}).$$

Hence the first part of (3.39) is also established.

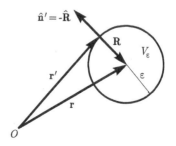

FIGURE 3.6

Geometry for establishing the singular property of $\nabla^2(1/R)$. ◄

3.3.5.4 The Green's function for unbounded space

In view of (3.38), one solution to (3.32) is

$$G(\mathbf{r}|\mathbf{r}') = \frac{1}{4\pi|\mathbf{r} - \mathbf{r}'|}. \tag{3.40}$$

This simple Green's function is generally used to find the potential produced by charge in unbounded space. Here $N = 0$ (no internal surfaces) and $S_B \to \infty$. Thus

$$\Phi(\mathbf{r}) = \int_V G(\mathbf{r}|\mathbf{r}')\frac{\rho(\mathbf{r}')}{\epsilon}\,dV' + \lim_{S_B \to \infty} \oint_{S_B} \left[\Phi(\mathbf{r}')\frac{\partial G(\mathbf{r}|\mathbf{r}')}{\partial n'} - G(\mathbf{r}|\mathbf{r}')\frac{\partial \Phi(\mathbf{r}')}{\partial n'}\right]dS'.$$

We have seen that the Green's function varies inversely with distance from the source, and thus expect that, as a superposition of point-source potentials, $\Phi(\mathbf{r})$ will also vary inversely with distance from a source of finite extent as that distance becomes large with respect to the size of the source. The normal derivatives then vary inversely with distance squared. Thus, each term in the surface integrand will vary inversely with distance cubed, while the surface area itself varies with distance squared. The result is that the surface integral vanishes as the surface recedes to infinity, giving

$$\Phi(\mathbf{r}) = \int_V G(\mathbf{r}|\mathbf{r}')\frac{\rho(\mathbf{r}')}{\epsilon}\,dV'.$$

By (3.40) we then have

$$\Phi(\mathbf{r}) = \frac{1}{4\pi\epsilon}\int_V \frac{\rho(\mathbf{r}')}{|\mathbf{r} - \mathbf{r}'|}\,dV' \tag{3.41}$$

where the integration is performed over all of space. Since $\Phi(\mathbf{r}) \to 0$ as $|\mathbf{r}| \to \infty$, points at infinity are a convenient reference for the absolute potential.

Later we shall need to know the amount of work required to move a charge Q from infinity to a point P located at $\mathbf{r}$. If a potential field is produced by charge located in unbounded space, moving an additional charge into position requires the work

$$W_{21} = -Q\int_\infty^P \mathbf{E} \cdot d\mathbf{l} = Q[\Phi(\mathbf{r}) - \Phi(\infty)] = Q\Phi(\mathbf{r}). \tag{3.42}$$

3.3.5.5 Coulomb's law

We can obtain $\mathbf{E}$ from (3.41) by direct differentiation. We have

$$\mathbf{E}(\mathbf{r}) = -\frac{1}{4\pi\epsilon}\nabla\int_V \frac{\rho(\mathbf{r}')}{|\mathbf{r} - \mathbf{r}'|}\,dV'$$

$$= -\frac{1}{4\pi\epsilon}\int_V \rho(\mathbf{r}')\nabla\left(\frac{1}{|\mathbf{r} - \mathbf{r}'|}\right)dV',$$

hence

$$\mathbf{E}(\mathbf{r}) = \frac{1}{4\pi\epsilon}\int_V \rho(\mathbf{r}')\frac{\mathbf{r} - \mathbf{r}'}{|\mathbf{r} - \mathbf{r}'|^3}\,dV' \tag{3.43}$$

by (3.37). So Coulomb's law follows from the two fundamental postulates of electrostatics (3.1) and (3.2).

▶ **Example 3.7:** Electric field of an infinite uniform line charge

A line charge of uniform density ρ_l extends along the z-axis in a uniform medium of permittivity ϵ. Use Coulomb's law to find $\mathbf{E}$.

Solution: By (3.43) we have

$$\mathbf{E}(\mathbf{r}) = \frac{1}{4\pi\epsilon} \int_\Gamma \rho_l(z') \frac{\mathbf{r} - \mathbf{r}'}{|\mathbf{r} - \mathbf{r}'|^3} \, dl'.$$

Then, since $\mathbf{r} = \hat{\mathbf{z}}z + \hat{\boldsymbol{\rho}}\rho$, $\mathbf{r}' = \hat{\mathbf{z}}z'$, and $dl' = dz'$, we have

$$\mathbf{E}(\boldsymbol{\rho}) = \frac{\rho_l}{4\pi\epsilon} \int_{-\infty}^{\infty} \frac{\hat{\boldsymbol{\rho}}\rho + \hat{\mathbf{z}}(z - z')}{[\rho^2 + (z - z')^2]^{3/2}} \, dz'.$$

Using a change of variables, the z-component of the field is found to be

$$E_z(\boldsymbol{\rho}) = \frac{\rho_l}{4\pi\epsilon} \int_{-\infty}^{\infty} \frac{u \, du}{[\rho^2 + u^2]^{3/2}} = 0$$

since the integrand is odd in u. Thus $\mathbf{E}$ has only a ρ-component, which varies only with ρ. The remaining integral gives

$$\mathbf{E}(\boldsymbol{\rho}) = \hat{\boldsymbol{\rho}} \frac{\rho_l}{2\pi\epsilon\rho}, \tag{3.44}$$

which was obtained from Gauss's law in Example 3.2. ◀

3.3.5.6 Green's function for unbounded space: two dimensions

We define the two-dimensional Green's function as the potential at a point $\mathbf{r} = \boldsymbol{\rho} + \hat{\mathbf{z}}z$ produced by a z-directed line source of constant density located at $\mathbf{r}' = \boldsymbol{\rho}'$. Perhaps the easiest way to find this is to start with a line charge on the z-axis, the field for which was found in Example 3.7. The absolute potential referred to a radius ρ_0 can be found by computing the line integral of $\mathbf{E}$ from ρ to ρ_0. Using the field (3.44) for a line charge gives

$$\Phi(\boldsymbol{\rho}) = -\frac{\rho_l}{2\pi\epsilon} \int_{\rho_0}^{\rho} \frac{d\rho'}{\rho'} = \frac{\rho_l}{2\pi\epsilon} \ln\left(\frac{\rho_0}{\rho}\right).$$

We may choose any reference point ρ_0 except $\rho_0 = 0$ or $\rho_0 = \infty$. This choice is equivalent to the addition of an arbitrary constant, hence we can also write

$$\Phi(\boldsymbol{\rho}) = \frac{\rho_l}{2\pi\epsilon} \ln\left(\frac{1}{\rho}\right) + C.$$

The potential for a general two-dimensional charge distribution in unbounded space is by superposition

$$\Phi(\boldsymbol{\rho}) = \int_{S_T} \frac{\rho_T(\boldsymbol{\rho}')}{\epsilon} G(\boldsymbol{\rho}|\boldsymbol{\rho}') \, dS', \tag{3.45}$$

where the Green's function is the potential of a unit line source located at $\boldsymbol{\rho}'$:

$$G(\boldsymbol{\rho}|\boldsymbol{\rho}') = \frac{1}{2\pi} \ln\left(\frac{\rho_0}{|\boldsymbol{\rho} - \boldsymbol{\rho}'|}\right). \tag{3.46}$$

Here S_T denotes the transverse (xy) plane, and ρ_T denotes the two-dimensional charge distribution (C/m^2) within that plane. In cylindrical coordinates we have

$$|\boldsymbol{\rho} - \boldsymbol{\rho}'| = \sqrt{\rho^2 + \rho'^2 - 2\rho\rho' \cos(\phi - \phi')}$$

and (3.46) becomes

$$G(\boldsymbol{\rho}|\boldsymbol{\rho}') = \frac{1}{2\pi} \ln\left(\frac{\rho_0}{\rho}\right) - \frac{1}{4\pi} \ln\left[1 + \left(\frac{\rho'}{\rho}\right)^2 - 2\left(\frac{\rho'}{\rho}\right)\cos(\phi - \phi')\right] \tag{3.47}$$

or

$$G(\boldsymbol{\rho}|\boldsymbol{\rho}') = \frac{1}{2\pi} \ln\left(\frac{\rho_0}{\rho'}\right) - \frac{1}{4\pi} \ln\left[1 + \left(\frac{\rho}{\rho'}\right)^2 - 2\left(\frac{\rho}{\rho'}\right)\cos(\phi - \phi')\right].$$

We note that the potential field (3.45) of a two-dimensional source decreases logarithmically with distance. Only the potential produced by a source of finite extent decreases inversely with distance.

▶ **Example 3.8:** Potential of an infinite cylinder of charge

An infinite cylinder of uniform charge density ρ_v and radius a is centered on the z-axis in a uniform medium of permittivity ϵ. Compute Φ at a radius ρ referred to a radius ρ_0, where $\rho > \rho_0 > a$.

Solution: We use (3.47) in (3.45) to obtain

$$\Phi(\rho) = \frac{\rho_v}{2\pi\epsilon} \int_0^{2\pi} \int_0^a \ln\left(\frac{\rho_0}{\rho}\right) \rho' \, d\rho' \, d\phi'$$

$$- \frac{\rho_v}{4\pi\epsilon} \int_0^{2\pi} \int_0^a \ln\left[1 + \left(\frac{\rho'}{\rho}\right)^2 - 2\left(\frac{\rho'}{\rho}\right)\cos(\phi - \phi')\right] \rho' \, d\rho' \, d\phi'.$$

Integrating over ϕ' and using the handbook integral [74]

$$\int_0^{2\pi} \ln\left(1 - 2a\cos x + a^2\right) dx = \begin{cases} 0, & a^2 < 1, \\ 2\pi \ln a^2, & a^2 > 1, \end{cases}$$

we get

$$\Phi(\rho) = \frac{\rho_v}{\epsilon} \ln\left(\frac{\rho_0}{\rho}\right) \int_0^a \rho' \, d\rho' = \frac{\rho_v a^2}{2\epsilon} \ln\left(\frac{\rho_0}{\rho}\right).$$

We can also find the potential using the line integral of **E** from (3.11). We use the electric field for $\rho > a$ found in Example 3.4,

$$\mathbf{E}(\rho) = \hat{\boldsymbol{\rho}} \frac{\rho_v a^2}{2\epsilon\rho},$$

to confirm that

$$\Phi(\rho) = -\int_{\rho_0}^{\rho} \frac{\rho_v a^2}{2\epsilon\rho} \, d\rho = \frac{\rho_v a^2}{2\epsilon} \ln\left(\frac{\rho_0}{\rho}\right). \blacktriangleleft$$

3.3.5.7 Dirichlet and Neumann Green's functions

The unbounded space Green's function may be inconvenient for expressing the potential in a region having internal surfaces. In fact, (3.36) shows that to use this function we would be forced to specify both Φ and its normal derivative over all surfaces. This, of course, would exceed the actual requirements for uniqueness.

Many functions can satisfy (3.32). For instance,

$$G(\mathbf{r}|\mathbf{r}') = \frac{A}{|\mathbf{r} - \mathbf{r}'|} + \frac{B}{|\mathbf{r} - \mathbf{r}_i|} \tag{3.48}$$

satisfies (3.32) if $\mathbf{r}_i \notin V$. Evaluation of (3.35) with the Green's function (3.48) reproduces the general formulation (3.36) since the Laplacian of the second term in (3.48) is identically zero in V. In fact, we can add any function to the free-space Green's function, provided that the additional term obeys Laplace's equation within V:

$$G(\mathbf{r}|\mathbf{r}') = \frac{A}{|\mathbf{r} - \mathbf{r}'|} + F(\mathbf{r}|\mathbf{r}'), \qquad \nabla'^2 F(\mathbf{r}|\mathbf{r}') = 0. \tag{3.49}$$

A good choice for $G(\mathbf{r}|\mathbf{r}')$ will minimize the effort required to evaluate $\Phi(\mathbf{r})$. Examining (3.36) we notice two possibilities. If we demand that

$$G(\mathbf{r}|\mathbf{r}') = 0 \text{ for all } \mathbf{r}' \in S \tag{3.50}$$

then the surface integral terms in (3.36) involving $\partial\Phi/\partial n'$ will vanish. The Green's function satisfying (3.50) is known as the *Dirichlet Green's function*. Let us designate it by G_D and use reciprocity to write (3.50) as

$$G_D(\mathbf{r}|\mathbf{r}') = 0 \text{ for all } \mathbf{r} \in S.$$

The resulting specialization of (3.36),

$$\Phi(\mathbf{r}) = \int_V G_D(\mathbf{r}|\mathbf{r}')\frac{\rho(\mathbf{r}')}{\epsilon}\,dV' + \oint_{S_B} \Phi(\mathbf{r}')\frac{\partial G_D(\mathbf{r}|\mathbf{r}')}{\partial n'}\,dS'$$
$$+ \sum_{n=1}^{N} \oint_{S_n} \Phi(\mathbf{r}')\frac{\partial G_D(\mathbf{r}|\mathbf{r}')}{\partial n'}\,dS', \tag{3.51}$$

requires the specification of Φ (but not its normal derivative) over the boundary surfaces. In case S_B and S_n surround and are adjacent to perfect conductors, the Dirichlet boundary condition has an important physical meaning. The corresponding Green's function is the potential at point $\mathbf{r}$ produced by a point source at $\mathbf{r}'$ in the presence of the conductors when the conductors are grounded — i.e., held at zero potential. Then we must specify the actual constant potentials on the conductors to determine Φ everywhere within V using (3.51). The additional term $F(\mathbf{r}|\mathbf{r}')$ in (3.49) accounts for the potential produced by surface charges on the grounded conductors.

By analogy with (3.50) it is tempting to try to define another electrostatic Green's function according to

$$\frac{\partial G(\mathbf{r}|\mathbf{r}')}{\partial n'} = 0 \text{ for all } \mathbf{r}' \in S. \tag{3.52}$$

But this choice is not permissible if V is a finite-sized region. Let us integrate (3.34) over V and employ the divergence theorem and the sifting property to get

$$\oint_S \frac{\partial G(\mathbf{r}|\mathbf{r}')}{\partial n'}\,dS' = -1; \tag{3.53}$$

in conjunction with this, Equation (3.52) would imply the false statement $0 = -1$. Suppose instead that we introduce a Green's function according to

$$\frac{\partial G(\mathbf{r}|\mathbf{r}')}{\partial n'} = -\frac{1}{A} \text{ for all } \mathbf{r}' \in S \tag{3.54}$$

where A is the total area of S. This choice avoids a contradiction in (3.53); it does not nullify any terms in (3.36), but does reduce the surface integral terms involving Φ to constants. Taken together, these terms all compose a single additive constant on the right-hand side; although the corresponding potential $\Phi(\mathbf{r})$ is thereby determined only to within this additive constant, the value of $\mathbf{E}(\mathbf{r}) = -\nabla\Phi(\mathbf{r})$ will be unaffected. By reciprocity we can rewrite (3.54) as

$$\frac{\partial G_N(\mathbf{r}|\mathbf{r}')}{\partial n} = -\frac{1}{A} \text{ for all } \mathbf{r} \in S.$$

The Green's function G_N so defined is known as the *Neumann Green's function*. Observe that if V is not finite-sized then $A \to \infty$ and according to (3.54) the choice (3.52) becomes allowable.

Finding the Green's function that obeys one of the boundary conditions for a given geometry is often a difficult task. Nevertheless, certain canonical geometries make the Green's function approach straightforward and simple. Such is the case in image theory, when a charge is located near a simple conducting body such as a ground screen or a sphere. In these cases the function $F(\mathbf{r}|\mathbf{r}')$ consists of a single correction term as in (3.48). We shall consider these simple cases in examples to follow.

3.3.5.8 Reciprocity of the static Green's function

It remains to show that

$$G(\mathbf{r}|\mathbf{r}') = G(\mathbf{r}'|\mathbf{r})$$

for any of the Green's functions introduced above. The unbounded-space Green's function is reciprocal by inspection; $|\mathbf{r} - \mathbf{r}'|$ is unaffected by interchanging $\mathbf{r}$ and $\mathbf{r}'$. However, we can give a more general treatment covering this case as well as the Dirichlet and Neumann cases. We begin with

$$\nabla^2 G(\mathbf{r}|\mathbf{r}') = -\delta(\mathbf{r} - \mathbf{r}').$$

In Green's second identity we let $\phi(\mathbf{r}) = G(\mathbf{r}|\mathbf{r}_a)$ and $\psi(\mathbf{r}) = G(\mathbf{r}|\mathbf{r}_b)$ where $\mathbf{r}_a$ and $\mathbf{r}_b$ are arbitrary points, and integrate over the unprimed coordinates:

$$\int_V [G(\mathbf{r}|\mathbf{r}_a)\nabla^2 G(\mathbf{r}|\mathbf{r}_b) - G(\mathbf{r}|\mathbf{r}_b)\nabla^2 G(\mathbf{r}|\mathbf{r}_a)]\,dV =$$
$$-\oint_S \left[G(\mathbf{r}|\mathbf{r}_a)\frac{\partial G(\mathbf{r}|\mathbf{r}_b)}{\partial n} - G(\mathbf{r}|\mathbf{r}_b)\frac{\partial G(\mathbf{r}|\mathbf{r}_a)}{\partial n} \right] dS.$$

If G is the unbounded-space Green's function, the surface integral must vanish since $S_B \to \infty$. It must also vanish under Dirichlet or Neumann boundary conditions. Since

$$\nabla^2 G(\mathbf{r}|\mathbf{r}_a) = -\delta(\mathbf{r} - \mathbf{r}_a), \qquad \nabla^2 G(\mathbf{r}|\mathbf{r}_b) = -\delta(\mathbf{r} - \mathbf{r}_b),$$

we have

$$\int_V [G(\mathbf{r}|\mathbf{r}_a)\delta(\mathbf{r} - \mathbf{r}_b) - G(\mathbf{r}|\mathbf{r}_b)\delta(\mathbf{r} - \mathbf{r}_a)]\,dV = 0,$$

hence

$$G(\mathbf{r}_b|\mathbf{r}_a) = G(\mathbf{r}_a|\mathbf{r}_b)$$

by the sifting property. By the arbitrariness of $\mathbf{r}_a$ and $\mathbf{r}_b$, reciprocity is established.

3.3.5.9 Image interpretation for solutions to Poisson's equation

For simple geometries, Poisson's equation may be solved as part of a *boundary value problem* (§ A.5). Occasionally such a solution has an appealing interpretation as the superposition of potentials produced by the physical charge and its "images." We shall consider here the case of planar media and subsequently use the results to predict the potential produced by charge near a conducting sphere.

Consider a layered dielectric medium where various regions of space are separated by planes at constant values of z. Material region i occupies volume region V_i and has permittivity ϵ_i; it may or may not contain source charge. The solution to Poisson's equation is given by (3.36). The contribution

$$\Phi^p(\mathbf{r}) = \int_V G(\mathbf{r}|\mathbf{r}')\frac{\rho(\mathbf{r}')}{\epsilon}\,dV'$$

produced by sources within V is known as the *primary potential*. The term

$$\Phi^s(\mathbf{r}) = \oint_S \left[\Phi(\mathbf{r}')\frac{\partial G(\mathbf{r}|\mathbf{r}')}{\partial n'} - G(\mathbf{r}|\mathbf{r}')\frac{\partial \Phi(\mathbf{r}')}{\partial n'}\right] dS',$$

on the other hand, involves an integral over the surface fields and is known as the *secondary potential*. This term is linked to effects outside V. Since the "sources" of Φ^s (i.e., the surface fields) lie on the boundary of V, Φ^s satisfies Laplace's equation within V. We may therefore use other, more convenient, representations of Φ^s provided they satisfy Laplace's equation. However, as solutions to a homogeneous equation they are of indefinite form until linked to appropriate boundary values.

Since the geometry is invariant in the x and y directions, we represent each potential function in terms of a 2-D Fourier transform over these variables. We leave the z dependence intact so that we may apply boundary conditions directly in the spatial domain. The transform representations of the Green's functions for the primary and secondary potentials are derived in Appendix A. From (A.56) we see that the primary potential within region V_i can be written as

$$\Phi_i^p(\mathbf{r}) = \int_{V_i} G^p(\mathbf{r}|\mathbf{r}')\frac{\rho(\mathbf{r}')}{\epsilon_i}\,dV'$$

where

$$G^p(\mathbf{r}|\mathbf{r}') = \frac{1}{4\pi|\mathbf{r} - \mathbf{r}'|} = \frac{1}{(2\pi)^2}\int_{-\infty}^{\infty} \frac{e^{-k_\rho|z-z'|}}{2k_\rho}e^{j\mathbf{k}_\rho\cdot(\mathbf{r}-\mathbf{r}')}\,d^2k_\rho \qquad (3.55)$$

is the primary Green's function with $\mathbf{k}_\rho = \hat{\mathbf{x}}k_x + \hat{\mathbf{y}}k_y$, $k_\rho = |\mathbf{k}_\rho|$, and $d^2k_\rho = dk_x\,dk_y$. We also find in (A.57) that a solution of Laplace's equation can be written as

$$\Phi^s(\mathbf{r}) = \frac{1}{(2\pi)^2}\int_{-\infty}^{\infty} \left[A(\mathbf{k}_\rho)e^{k_\rho z} + B(\mathbf{k}_\rho)e^{-k_\rho z}\right] e^{j\mathbf{k}_\rho\cdot\mathbf{r}}\,d^2k_\rho$$

where $A(\mathbf{k}_\rho)$ and $B(\mathbf{k}_\rho)$ must be found by the application of appropriate boundary conditions.

▶ **Example 3.9:** Charge distribution above a conducting plane

Find Φ for a charge distribution $\rho(\mathbf{r})$ in free space above the grounded conducting plane $z = 0$.

Solution: We will find the potential in the region $z > 0$ using a Fourier transform representation. The total potential is a sum of primary and secondary terms:

$$\Phi(x,y,z) = \int_V \left[\frac{1}{(2\pi)^2} \int_{-\infty}^{\infty} \frac{e^{-k_\rho|z-z'|}}{2k_\rho} e^{j\mathbf{k}_\rho \cdot (\mathbf{r}-\mathbf{r}')} d^2 k_\rho \right] \frac{\rho(\mathbf{r}')}{\epsilon_0} dV'$$
$$+ \frac{1}{(2\pi)^2} \int_{-\infty}^{\infty} \left[B(\mathbf{k}_\rho) e^{-k_\rho z} \right] e^{j\mathbf{k}_\rho \cdot \mathbf{r}} d^2 k_\rho,$$

where the integral is over the region $z > 0$. Here we have set $A(\mathbf{k}_\rho) = 0$ because $e^{k_\rho z}$ grows with increasing z. Since the plane is grounded we must have $\Phi(x,y,0) = 0$. Because $z < z'$ when we apply this condition, we have $|z - z'| = z' - z$ and thus

$$\Phi(x,y,0) = \frac{1}{(2\pi)^2} \int_{-\infty}^{\infty} \left[\int_V \frac{\rho(\mathbf{r}')}{\epsilon_0} \frac{e^{-k_\rho z'}}{2k_\rho} e^{-j\mathbf{k}_\rho \cdot \mathbf{r}'} dV' + B(\mathbf{k}_\rho) \right] e^{j\mathbf{k}_\rho \cdot \mathbf{r}} d^2 k_\rho = 0.$$

By the Fourier integral theorem

$$B(\mathbf{k}_\rho) = -\int_V \frac{\rho(\mathbf{r}')}{\epsilon_0} \frac{e^{-k_\rho z'}}{2k_\rho} e^{-j\mathbf{k}_\rho \cdot \mathbf{r}'} dV',$$

hence the total potential is

$$\Phi(x,y,z) = \int_V \left[\frac{1}{(2\pi)^2} \int_{-\infty}^{\infty} \frac{e^{-k_\rho|z-z'|} - e^{-k_\rho(z+z')}}{2k_\rho} e^{j\mathbf{k}_\rho \cdot (\mathbf{r}-\mathbf{r}')} d^2 k_\rho \right] \frac{\rho(\mathbf{r}')}{\epsilon_0} dV'$$
$$= \int_V G(\mathbf{r}|\mathbf{r}') \frac{\rho(\mathbf{r}')}{\epsilon_0} dV'$$

where $G(\mathbf{r}|\mathbf{r}')$ is the Green's function for the region above a grounded planar conductor.

We can interpret this Green's function as a sum of the primary Green's function (3.55) and a secondary Green's function

$$G^s(\mathbf{r}|\mathbf{r}') = -\frac{1}{(2\pi)^2} \int_{-\infty}^{\infty} \frac{e^{-k_\rho(z+z')}}{2k_\rho} e^{j\mathbf{k}_\rho \cdot (\mathbf{r}-\mathbf{r}')} d^2 k_\rho. \qquad (3.56)$$

For $z > 0$ the term $z + z'$ can be replaced by $|z + z'|$. Then, comparing (3.56) with (3.55), we see that

$$G^s(\mathbf{r}|x',y',z') = -G^p(\mathbf{r}|x',y',-z') = -\frac{1}{4\pi|\mathbf{r} - \mathbf{r}_i'|}$$

where $\mathbf{r}_i' = \hat{\mathbf{x}}x' + \hat{\mathbf{y}}y' - \hat{\mathbf{z}}z'$. Because the Green's function is the potential of a point charge, we may interpret the secondary Green's function as produced by a negative unit charge placed in a position $-z'$ immediately beneath the positive unit charge that produces G^p (Figure 3.7). This secondary charge is the "image" of the primary charge. That two such charges would produce a null potential on the ground plane is easily verified.

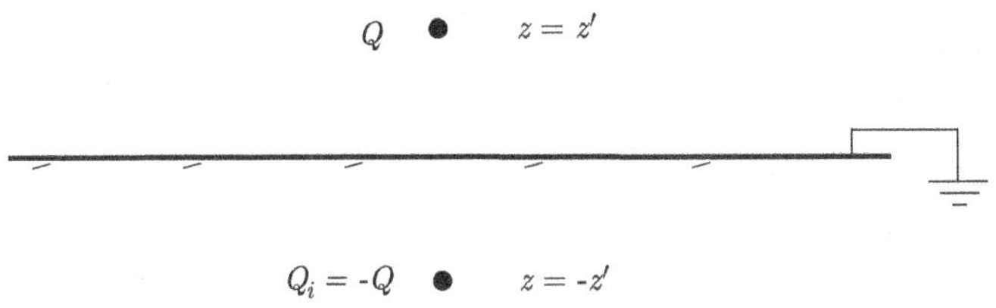

FIGURE 3.7
Construction of electrostatic Green's function for a ground plane. ◀

▶ **Example 3.10:** Charge distribution above an interface between dielectric regions

Find the potential for a charge distribution $\rho(\mathbf{r})$ above a planar interface separating two homogeneous dielectric media.

Solution: Let region 1 occupy $z > 0$ with permittivity ϵ_1, and let region 2 occupy $z < 0$ with permittivity ϵ_2. In region 1 we can write the total potential as a sum of primary and secondary components, discarding the term that grows with z:

$$
\Phi_1(x, y, z) = \int_V \left[\frac{1}{(2\pi)^2} \int_{-\infty}^{\infty} \frac{e^{-k_\rho |z-z'|}}{2k_\rho} e^{j\mathbf{k}_\rho \cdot (\mathbf{r}-\mathbf{r}')} \, d^2 k_\rho \right] \frac{\rho(\mathbf{r}')}{\epsilon_1} \, dV'
$$
$$
+ \frac{1}{(2\pi)^2} \int_{-\infty}^{\infty} \left[B(\mathbf{k}_\rho) e^{-k_\rho z} \right] e^{j\mathbf{k}_\rho \cdot \mathbf{r}} \, d^2 k_\rho. \tag{3.57}
$$

With no source in region 2, the potential there must obey Laplace's equation and therefore consists of only a secondary component:

$$
\Phi_2(\mathbf{r}) = \frac{1}{(2\pi)^2} \int_{-\infty}^{\infty} \left[A(\mathbf{k}_\rho) e^{k_\rho z} \right] e^{j\mathbf{k}_\rho \cdot \mathbf{r}} \, d^2 k_\rho. \tag{3.58}
$$

To determine A and B we impose (3.16) and (3.17). By (3.16) we have

$$
\frac{1}{(2\pi)^2} \int_{-\infty}^{\infty} \left[\int_V \frac{\rho(\mathbf{r}')}{\epsilon_1} \frac{e^{-k_\rho z'}}{2k_\rho} e^{-j\mathbf{k}_\rho \cdot \mathbf{r}'} \, dV' + B(\mathbf{k}_\rho) - A(\mathbf{k}_\rho) \right] e^{j\mathbf{k}_\rho \cdot \mathbf{r}} \, d^2 k_\rho = 0,
$$

hence

$$
\int_V \frac{\rho(\mathbf{r}')}{\epsilon_1} \frac{e^{-k_\rho z'}}{2k_\rho} e^{-j\mathbf{k}_\rho \cdot \mathbf{r}'} \, dV' + B(\mathbf{k}_\rho) - A(\mathbf{k}_\rho) = 0
$$

by the Fourier integral theorem. Applying (3.17) at $z = 0$ with $\hat{\mathbf{n}}_{12} = \hat{\mathbf{z}}$, and noting that there is no excess surface charge, we find

$$
\int_V \rho(\mathbf{r}') \frac{e^{-k_\rho z'}}{2k_\rho} e^{-j\mathbf{k}_\rho \cdot \mathbf{r}'} \, dV' - \epsilon_1 B(\mathbf{k}_\rho) - \epsilon_2 A(\mathbf{k}_\rho) = 0.
$$

The solutions

$$
A(\mathbf{k}_\rho) = \frac{2\epsilon_1}{\epsilon_1 + \epsilon_2} \int_V \frac{\rho(\mathbf{r}')}{\epsilon_1} \frac{e^{-k_\rho z'}}{2k_\rho} e^{-j\mathbf{k}_\rho \cdot \mathbf{r}'} \, dV',
$$

$$
B(\mathbf{k}_\rho) = \frac{\epsilon_1 - \epsilon_2}{\epsilon_1 + \epsilon_2} \int_V \frac{\rho(\mathbf{r}')}{\epsilon_1} \frac{e^{-k_\rho z'}}{2k_\rho} e^{-j\mathbf{k}_\rho \cdot \mathbf{r}'} \, dV',
$$

are then substituted into (3.57) and (3.58) to give

$$
\Phi_1(\mathbf{r}) = \int_V \left[\frac{1}{(2\pi)^2} \int_{-\infty}^{\infty} \frac{e^{-k_\rho |z-z'|} + \frac{\epsilon_1 - \epsilon_2}{\epsilon_1 + \epsilon_2} e^{-k_\rho (z+z')}}{2k_\rho} e^{j\mathbf{k}_\rho \cdot (\mathbf{r}-\mathbf{r}')} \, d^2 k_\rho \right] \frac{\rho(\mathbf{r}')}{\epsilon_1} \, dV'
$$

$$
= \int_V G_1(\mathbf{r}|\mathbf{r}') \frac{\rho(\mathbf{r}')}{\epsilon_1} \, dV',
$$

$$
\Phi_2(\mathbf{r}) = \int_V \left[\frac{1}{(2\pi)^2} \int_{-\infty}^{\infty} \frac{2\epsilon_2}{\epsilon_1 + \epsilon_2} \frac{e^{-k_\rho (z'-z)}}{2k_\rho} e^{j\mathbf{k}_\rho \cdot (\mathbf{r}-\mathbf{r}')} \, d^2 k_\rho \right] \frac{\rho(\mathbf{r}')}{\epsilon_2} \, dV'
$$

$$
= \int_V G_2(\mathbf{r}|\mathbf{r}') \frac{\rho(\mathbf{r}')}{\epsilon_2} \, dV'.
$$

Since $z' > z$ for all points in region 2, we can replace $z' - z$ by $|z - z'|$ in the formula for Φ_2.

As with Example 3.9, let us compare the result to the form of the primary Green's function (3.55). We see that

$$G_1(\mathbf{r}|\mathbf{r}') = \frac{1}{4\pi|\mathbf{r} - \mathbf{r}'|} + \frac{\epsilon_1 - \epsilon_2}{\epsilon_1 + \epsilon_2}\frac{1}{4\pi|\mathbf{r} - \mathbf{r}_1'|},$$

$$G_2(\mathbf{r}|\mathbf{r}') = \frac{2\epsilon_2}{\epsilon_1 + \epsilon_2}\frac{1}{4\pi|\mathbf{r} - \mathbf{r}_2'|},$$

where $\mathbf{r}_1' = \hat{\mathbf{x}}x' + \hat{\mathbf{y}}y' - \hat{\mathbf{z}}z'$ and $\mathbf{r}_2' = \hat{\mathbf{x}}x' + \hat{\mathbf{y}}y' + \hat{\mathbf{z}}z'$. So we can also write

$$\Phi_1(\mathbf{r}) = \frac{1}{4\pi}\int_V \left[\frac{1}{|\mathbf{r} - \mathbf{r}'|} + \frac{\epsilon_1 - \epsilon_2}{\epsilon_1 + \epsilon_2}\frac{1}{|\mathbf{r} - \mathbf{r}_1'|}\right]\frac{\rho(\mathbf{r}')}{\epsilon_1}\,dV',$$

$$\Phi_2(\mathbf{r}) = \frac{1}{4\pi}\int_V \left[\frac{2\epsilon_2}{\epsilon_1 + \epsilon_2}\frac{1}{|\mathbf{r} - \mathbf{r}_2'|}\right]\frac{\rho(\mathbf{r}')}{\epsilon_2}\,dV'.$$

Note that $\Phi_2 \to \Phi_1$ as $\epsilon_2 \to \epsilon_1$.

There is an image interpretation for the secondary Green's functions. The secondary Green's function for region 1 appears as a potential produced by an image of the primary charge located at $-z'$ in an infinite medium of permittivity ϵ_1, and with an amplitude of $(\epsilon_1 - \epsilon_2)/(\epsilon_1 + \epsilon_2)$ times the primary charge. The Green's function in region 2 is produced by an image charge located at z' (i.e., at the location of the primary charge) in an infinite medium of permittivity ϵ_2 with an amplitude of $2\epsilon_2/(\epsilon_1 + \epsilon_2)$ times the primary charge. ◄

▶ **Example 3.11:** Charge distribution external to a grounded conducting sphere

Find the potential for a charge distribution $\rho(\mathbf{r})$ external to a grounded conducting sphere.

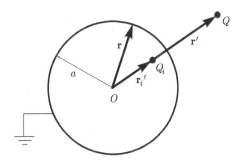

FIGURE 3.8

Green's function for a grounded conducting sphere.

Solution: Consider a point charge Q placed near a grounded conducting sphere in free space as shown in Figure 3.8. Based on our experience with planar layered media, we hypothesize that the secondary potential will be produced by an image charge; hence we try the simple Green's function

$$G^s(\mathbf{r}|\mathbf{r}') = \frac{A(\mathbf{r}')}{4\pi|\mathbf{r} - \mathbf{r}_i'|}$$

where the amplitude A and location $\mathbf{r}_i'$ of the image are to be determined. We further assume, based on our experience with planar problems, that the image charge will reside inside the sphere along a line joining the origin to the primary charge. Since $\mathbf{r} = a\hat{\mathbf{r}}$ for all points on the sphere, the total Green's function must obey the Dirichlet condition

$$G(\mathbf{r}|\mathbf{r}')|_{r=a} = \left.\frac{1}{4\pi|\mathbf{r}-\mathbf{r}'|}\right|_{r=a} + \left.\frac{A(\mathbf{r}')}{4\pi|\mathbf{r}-\mathbf{r}_i'|}\right|_{r=a}$$

$$= \frac{1}{4\pi|a\hat{\mathbf{r}} - r'\hat{\mathbf{r}}'|} + \frac{A(\mathbf{r}')}{4\pi|a\hat{\mathbf{r}} - r_i'\hat{\mathbf{r}}'|} = 0$$

in order to have the potential, given by (3.36), vanish on the sphere surface. Factoring a from the first denominator and r_i' from the second, we obtain

$$\frac{1}{4\pi a|\hat{\mathbf{r}} - \frac{r'}{a}\hat{\mathbf{r}}'|} + \frac{A(\mathbf{r}')}{4\pi r_i'|\frac{a}{r_i'}\hat{\mathbf{r}} - \hat{\mathbf{r}}'|} = 0.$$

Now $|k\hat{\mathbf{r}} - k'\hat{\mathbf{r}}'| = (k^2 + k'^2 - 2kk'\cos\gamma)^{1/2}$ where γ is the angle between $\hat{\mathbf{r}}$ and $\hat{\mathbf{r}}'$ and k, k' are constants; this means that $|k\hat{\mathbf{r}} - \hat{\mathbf{r}}'| = |\hat{\mathbf{r}} - k\hat{\mathbf{r}}'|$. Hence as long as

$$\frac{r'}{a} = \frac{a}{r_i'}, \qquad \frac{A}{r_i'} = -\frac{1}{a},$$

the total Green's function vanishes everywhere on the surface of the sphere. The image charge is therefore located within the sphere at $\mathbf{r}_i' = a^2\mathbf{r}'/r'^2$ and has amplitude $A = -a/r'$. (Note that both the location and amplitude of the image depend on the location of the primary charge.) With this Green's function and (3.51), the potential of an arbitrary source placed near a grounded conducting sphere is

$$\Phi(\mathbf{r}) = \int_V \frac{\rho(\mathbf{r}')}{\epsilon}\frac{1}{4\pi}\left[\frac{1}{|\mathbf{r}-\mathbf{r}'|} - \frac{a/r'}{|\mathbf{r} - \frac{a^2}{r'^2}\mathbf{r}'|}\right] dV'. \quad \blacktriangleleft$$

The Green's function derived in Example 3.11 may be used to compute the surface charge density induced on a sphere by a unit point charge: it is merely necessary to find the normal component of electric field from the gradient of $\Phi(\mathbf{r})$. We leave this as an exercise for the reader, who may then integrate the surface charge and thereby show that the total charge induced on the sphere is equal to the image charge. So the total charge induced on a grounded sphere by a point charge q at a point $r = r'$ is $Q = -qa/r'$.

It is possible to find the total charge induced on the sphere without finding the image charge first. This is an application of Green's reciprocation theorem (§ 3.5.4). According to (3.164), if we can find the potential V_P at a point $\mathbf{r}$ produced by the sphere when it is isolated and carrying a total charge Q_0, then the total charge Q induced on the grounded sphere in the vicinity of a point charge q placed at $\mathbf{r}$ is given by

$$Q = -q\frac{V_P}{V_1}$$

where V_1 is the potential of the isolated sphere. We can apply this formula by noting that an isolated sphere carrying charge Q_0 produces a field

$$\mathbf{E}(\mathbf{r}) = \hat{\mathbf{r}}\frac{Q_0}{4\pi\epsilon r^2}.$$

Integration from a radius r to infinity gives the potential referred to infinity: $\Phi(\mathbf{r}) = Q_0/4\pi\epsilon r$. So the potential of the isolated sphere is $V_1 = Q_0/4\pi\epsilon a$, while the potential at radius r' is $V_P = Q_0/4\pi\epsilon r'$. Substitution gives $Q = -qa/r'$ as before.

3.3.6 Force and energy

3.3.6.1 Maxwell's stress tensor

The electrostatic version of Maxwell's stress tensor can be obtained from (2.225) by setting $\mathbf{B} = \mathbf{H} = 0$:

$$\bar{\mathbf{T}}_e = \tfrac{1}{2}(\mathbf{D} \cdot \mathbf{E})\bar{\mathbf{I}} - \mathbf{D}\mathbf{E}.$$

The total electric force on the charges in a region V bounded by the surface S is given by the relation

$$\mathbf{F}_e = -\oint_S \bar{\mathbf{T}}_e \cdot d\mathbf{S} = \int_V \mathbf{f}_e \, dV$$

where $\mathbf{f}_e = \rho \mathbf{E}$ is the electric force volume density.

In particular, suppose that S is adjacent to a solid conducting body embedded in a dielectric having permittivity $\epsilon(\mathbf{r})$. Since all the charge is at the surface of the conductor, the force within V acts directly on the surface. Thus, $-\bar{\mathbf{T}}_e \cdot \hat{\mathbf{n}}$ is the surface force density (*traction*) $\mathbf{t}$. Using $\mathbf{D} = \epsilon \mathbf{E}$, and remembering that the fields are normal to the conductor, we find that

$$\bar{\mathbf{T}}_e \cdot \hat{\mathbf{n}} = \tfrac{1}{2}\epsilon E_n^2 \hat{\mathbf{n}} - \epsilon \mathbf{E}\mathbf{E} \cdot \hat{\mathbf{n}} = -\tfrac{1}{2}\epsilon E_n^2 \hat{\mathbf{n}} = -\tfrac{1}{2}\rho_s \mathbf{E}.$$

The surface force density is perpendicular to the surface.

▶ **Example 3.12:** Force on a charged sphere

Find the force acting on a rigid conducting sphere of radius a carrying total charge Q in a homogeneous medium.

Solution: At equilibrium the charge is distributed uniformly with surface density $\rho_s = Q/4\pi a^2$, producing a field $\mathbf{E} = \hat{\mathbf{r}}Q/4\pi\epsilon r^2$ external to the sphere. Hence a force density

$$\mathbf{t} = \frac{1}{2}\hat{\mathbf{r}}\frac{Q^2}{\epsilon(4\pi a^2)^2}$$

acts at each point on the surface. This would cause the sphere to expand outward if the structural integrity of the material were to fail. Integration over the entire sphere yields

$$\mathbf{F} = \frac{1}{2}\frac{Q^2}{\epsilon(4\pi a^2)^2}\int_S \hat{\mathbf{r}} \, dS = 0.$$

However, integration of $\mathbf{t}$ over the upper hemisphere yields

$$\mathbf{F} = \frac{1}{2}\frac{Q^2}{\epsilon(4\pi a^2)^2}\int_0^{2\pi}\int_0^{\pi/2} \hat{\mathbf{r}}a^2 \sin\theta \, d\theta \, d\phi.$$

Substitution of $\hat{\mathbf{r}} = \hat{\mathbf{x}}\sin\theta\cos\phi + \hat{\mathbf{y}}\sin\theta\sin\phi + \hat{\mathbf{z}}\cos\theta$ leads immediately to $F_x = F_y = 0$, but the z-component is

$$F_z = \frac{1}{2}\frac{Q^2}{\epsilon(4\pi a^2)^2}\int_0^{2\pi}\int_0^{\pi/2} a^2 \cos\theta\sin\theta \, d\theta \, d\phi = \frac{Q^2}{32\epsilon\pi a^2}.$$

This result can also be obtained by integrating $-\bar{\mathbf{T}}_e \cdot \hat{\mathbf{n}}$ over the entire xy-plane with $\hat{\mathbf{n}} = -\hat{\mathbf{z}}$. Since $-\bar{\mathbf{T}}_e \cdot (-\hat{\mathbf{z}}) = \hat{\mathbf{z}}\epsilon \mathbf{E} \cdot \mathbf{E}/2$, we have

$$\mathbf{F} = \hat{\mathbf{z}}\frac{1}{2}\frac{Q^2\epsilon}{(4\pi\epsilon)^2}\int_0^{2\pi}\int_a^{\infty}\frac{r \, dr \, d\phi}{r^4} = \hat{\mathbf{z}}\frac{Q^2}{32\epsilon\pi a^2}. \quad ◀$$

▶ **Example 3.13:** Force on adjacent line charges

Consider two identical line charges parallel to the z-axis and located at $x = \pm d/2$, $y = 0$ in free space. Find the force on one line charge due to the other by integrating Maxwell's stress tensor over the yz-plane.

Solution: From (3.44) we find that the total electric field on the yz-plane is

$$\mathbf{E}(y,z) = \frac{y}{y^2 + (d/2)^2} \frac{\rho_l}{\pi \epsilon_0} \, \hat{\mathbf{y}}$$

where ρ_l is the line charge density. The force density for either line charge is $-\bar{\mathbf{T}}_e \cdot \hat{\mathbf{n}}$, where we use $\hat{\mathbf{n}} = \pm \hat{\mathbf{x}}$ to obtain the force on the charge at $x = \mp d/2$. The force density for the charge at $x = -d/2$ is

$$\bar{\mathbf{T}}_e \cdot \hat{\mathbf{n}} = \frac{1}{2}(\mathbf{D} \cdot \mathbf{E})\bar{\mathbf{I}} \cdot \hat{\mathbf{x}} - \mathbf{D}\mathbf{E} \cdot \hat{\mathbf{x}} = \frac{\epsilon_0}{2} \left[\frac{y}{y^2 + (d/2)^2} \frac{\rho_l}{\pi \epsilon_0} \right]^2 \hat{\mathbf{x}}$$

and the total force is

$$\mathbf{F}_- = -\int_{-\infty}^{\infty} \int_{-\infty}^{\infty} \frac{\rho_l^2}{2\pi^2 \epsilon_0} \frac{y^2}{[y^2 + (d/2)^2]^2} \, \hat{\mathbf{x}} \, dy \, dz.$$

On a per-unit-length basis, the force is

$$\frac{\mathbf{F}_-}{l} = -\hat{\mathbf{x}} \frac{\rho_l^2}{2\pi^2 \epsilon_0} \int_{-\infty}^{\infty} \frac{y^2}{[y^2 + (d/2)^2]^2} \, dy = -\hat{\mathbf{x}} \frac{\rho_l^2}{2\pi d \epsilon_0}.$$

Note that the force is repulsive as expected. ◀

3.3.6.2 Electrostatic stored energy

In § 2.9.5 we considered the energy relations for the electromagnetic field. Those relations remain valid in the static case. Since our interpretation of the dynamic relations was guided in part by our knowledge of the energy stored in a static field, we must, for completeness, carry out a study of that effect here.

The energy of a static configuration is taken to be the work required to assemble the configuration from a chosen starting point. For a configuration of static charges, the *stored electric energy* is the energy required to assemble the configuration, starting with all charges removed to infinite distance (the assumed zero potential reference). If the assembled charges are not held in place by an external mechanical force, they will move, thereby converting stored electric energy into other forms of energy (e.g., kinetic energy and radiation).

By (3.42), the work required to move a point charge q from a reservoir at infinity to a point P at $\mathbf{r}$ in a potential field Φ is

$$W = q\Phi(\mathbf{r}).$$

If instead we have a continuous charge density ρ present, and wish to increase this to $\rho + \delta\rho$ by bringing in a small quantity of charge $\delta\rho$, a total work

$$\delta W = \int_{V_\infty} \delta\rho(\mathbf{r})\Phi(\mathbf{r}) \, dV \tag{3.59}$$

is required, and the potential field is increased to $\Phi + \delta\Phi$. Here V_∞ denotes all of space.

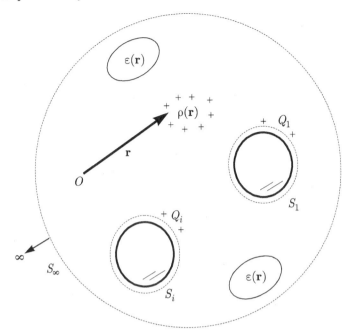

FIGURE 3.9
Computation of electrostatic stored energy via the assembly energy of a charge distribution.

(We could restrict the integral to the region containing the charge, but we shall find it helpful to extend the domain of integration to all of space.)

Now consider the situation shown in Figure 3.9. Here we have charge in the form of both volume densities and surface densities on conducting bodies. Also present may be linear material bodies. We can think of assembling the charge in two distinctly different ways. We could, for instance, bring small portions of charge (or point charges) together to form the distribution ρ. Or, we could slowly build up ρ by adding infinitesimal, but spatially identical, distributions. That is, we can create the distribution ρ from a zero initial state by repeatedly adding a charge distribution

$$\delta\rho(\mathbf{r}) = \frac{\rho(\mathbf{r})}{N},$$

where N is a large number. Whenever we add $\delta\rho$ we must perform the work given by (3.59), but we also increase the potential proportionately (remembering that all materials are assumed linear). At each step, more work is required. The total work is

$$W = \sum_{n=1}^{N} \int_{V_\infty} \delta\rho(\mathbf{r})[(n-1)\delta\Phi(\mathbf{r})]\,dV = \left[\sum_{n=1}^{N}(n-1)\right] \int_{V_\infty} \frac{\rho(\mathbf{r})}{N} \frac{\Phi(\mathbf{r})}{N}\,dV. \qquad (3.60)$$

We must use an infinite number of steps so that no energy is lost to radiation at any step (since the charge we add each time is infinitesimally small). Using

$$\sum_{n=1}^{N}(n-1) = \frac{N(N-1)}{2},$$

(3.60) becomes

$$W = \frac{1}{2} \int_{V_\infty} \rho(\mathbf{r}) \Phi(\mathbf{r}) \, dV \tag{3.61}$$

as $N \to \infty$. Finally, since some assembled charge will be in the form of a volume density and some in the form of the surface density on conductors, we can generalize (3.61) to

$$W = \frac{1}{2} \int_{V'} \rho(\mathbf{r}) \Phi(\mathbf{r}) \, dV + \frac{1}{2} \sum_{i=1}^{I} Q_i V_i. \tag{3.62}$$

Here V' is the region outside the conductors, Q_i is the total charge on the ith conductor $(i = 1, \ldots, I)$, and V_i is the absolute potential (referred to infinity) of the ith conductor.

▶ **Example 3.14:** Energy stored in a ball of volume charge, using potential

Use (3.62) to find the stored energy for the charge distribution of Example 3.5.

Solution: Substituting

$$\Phi(\mathbf{r}) = \frac{\rho_v}{6\epsilon}(3a^2 - r^2)$$

into (3.62), we obtain

$$W = \frac{1}{2} \int_0^{2\pi} \int_0^{\pi} \int_0^a \frac{\rho_v^2}{6\epsilon}(3a^2 - r^2)r^2 \sin\theta \, dr \, d\theta \, d\phi = \frac{4\pi\rho_v^2 a^5}{15\epsilon}. \quad ◀$$

An intriguing property of electrostatic energy is that the charges on the conductors will arrange themselves, while seeking static equilibrium, into a minimum-energy configuration (Thomson's theorem). See § 3.5.3.

In keeping with our field-centered view of electromagnetics, we now wish to write the energy (3.61) entirely in terms of the field vectors $\mathbf{E}$ and $\mathbf{D}$. Since $\rho = \nabla \cdot \mathbf{D}$ we have

$$W = \frac{1}{2} \int_{V_\infty} [\nabla \cdot \mathbf{D}(\mathbf{r})] \Phi(\mathbf{r}) \, dV.$$

Then, by (B.48),

$$W = \frac{1}{2} \int_{V_\infty} \nabla \cdot [\Phi(\mathbf{r})\mathbf{D}(\mathbf{r})] \, dV - \frac{1}{2} \int_{V_\infty} \mathbf{D}(\mathbf{r}) \cdot [\nabla\Phi(\mathbf{r})] \, dV.$$

The divergence theorem and (3.10) imply

$$W = \frac{1}{2} \oint_{S_\infty} \Phi(\mathbf{r})\mathbf{D}(\mathbf{r}) \cdot d\mathbf{S} + \frac{1}{2} \int_{V_\infty} \mathbf{D}(\mathbf{r}) \cdot \mathbf{E}(\mathbf{r}) \, dV$$

where S_∞ is the bounding surface that recedes toward infinity to encompass all of space. Because $\Phi \sim 1/r$ and $D \sim 1/r^2$ as $r \to \infty$, the integral over S_∞ tends to zero and

$$W = \frac{1}{2} \int_{V_\infty} \mathbf{D}(\mathbf{r}) \cdot \mathbf{E}(\mathbf{r}) \, dV. \tag{3.63}$$

Hence we may compute the assembly energy in terms of the fields supported by the charge ρ.

▶ **Example 3.15:** Energy stored in a ball of volume charge, using the electric field

Use (3.63) to find the stored energy for the charge distribution of Example 3.3.

Solution: Substituting

$$
\mathbf{E}(r) = \begin{cases}
\hat{\mathbf{r}} \dfrac{\rho_v r}{3\epsilon}, & r < a, \\[2mm]
\hat{\mathbf{r}} \dfrac{\rho_v a^3}{3\epsilon r^2}, & r > a,
\end{cases}
$$

into (3.63), we obtain

$$
\begin{aligned}
W &= \frac{1}{2} \int_0^{2\pi} \int_0^{\pi} \int_0^a \left(\frac{\rho_v r}{3} \hat{\mathbf{r}} \right) \cdot \left(\frac{\rho_v r}{3\epsilon} \hat{\mathbf{r}} \right) r^2 \sin\theta \, dr \, d\theta \, d\phi \\
&\quad + \frac{1}{2} \int_0^{2\pi} \int_0^{\pi} \int_a^{\infty} \left(\frac{\rho_v a^3}{3r^2} \hat{\mathbf{r}} \right) \cdot \left(\frac{\rho_v a^3}{3\epsilon r^2} \hat{\mathbf{r}} \right) r^2 \sin\theta \, dr \, d\theta \, d\phi \\
&= \frac{2\pi \rho_v^2}{9\epsilon} \left(\int_0^a r^4 \, dr + a^6 \int_a^{\infty} \frac{dr}{r^2} \right) = \frac{4\pi \rho_v^2 a^5}{15\epsilon}
\end{aligned}
$$

as in Example 3.14. ◀

It is significant that the assembly energy W is identical to the term within the time derivative in Poynting's theorem (2.234). Hence our earlier interpretation, that this term represents the time-rate of change of energy "stored" in the electric field, has a firm basis. Of course, the assembly energy is a static concept, and our generalization to dynamic fields is purely intuitive. We face similar questions regarding the meaning of energy density, and whether energy can be "localized" in space. The discussions in § 2.9.5 still apply.

3.3.7 Multipole expansion

Consider an arbitrary but spatially localized charge distribution of total charge Q in an unbounded homogeneous medium (Figure 3.10). We have already obtained the potential (3.41) of the source; as we move the observation point away, Φ should decrease in a manner roughly proportional to $1/r$. The actual variation depends on the nature of the charge distribution and can be complicated. Often this dependence is dominated by a specific inverse power of distance for observation points far from the source, and we can investigate it by expanding the potential in powers of $1/r$. Although such *multipole expansions* of the potential are rarely used to perform actual computations, they can provide insight into both the behavior of static fields and the physical meaning of the polarization vector $\mathbf{P}$.

Let us place our origin of coordinates somewhere within the charge distribution, as shown in Figure 3.10, and expand the Green's function spatial dependence in a three-dimensional Taylor series about the origin:

$$
\frac{1}{R} = \sum_{n=0}^{\infty} \frac{1}{n!} (\mathbf{r}' \cdot \nabla')^n \left. \frac{1}{R} \right|_{\mathbf{r}'=0} = \frac{1}{r} + (\mathbf{r}' \cdot \nabla') \left. \frac{1}{R} \right|_{\mathbf{r}'=0} + \frac{1}{2} (\mathbf{r}' \cdot \nabla')^2 \left. \frac{1}{R} \right|_{\mathbf{r}'=0} + \cdots, \quad (3.64)
$$

where $R = |\mathbf{r} - \mathbf{r}'|$. Convergence occurs if $|\mathbf{r}| > |\mathbf{r}'|$. In the notation $(\mathbf{r}' \cdot \nabla')^n$ we interpret a power on a derivative operator as the order of the derivative. Substituting (3.64) into

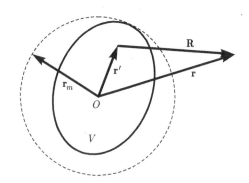

FIGURE 3.10
Multipole expansion.

(3.41) we obtain

$$\Phi(\mathbf{r}) = \frac{1}{4\pi\epsilon} \int_V \rho(\mathbf{r}') \left[\frac{1}{R} \bigg|_{\mathbf{r}'=0} + (\mathbf{r}' \cdot \nabla') \frac{1}{R} \bigg|_{\mathbf{r}'=0} + \frac{1}{2} (\mathbf{r}' \cdot \nabla')^2 \frac{1}{R} \bigg|_{\mathbf{r}'=0} + \cdots \right] dV'. \quad (3.65)$$

For the second term we can use (3.37) to write

$$(\mathbf{r}' \cdot \nabla') \frac{1}{R} \bigg|_{\mathbf{r}'=0} = \mathbf{r}' \cdot \left(\nabla' \frac{1}{R} \right) \bigg|_{\mathbf{r}'=0} = \mathbf{r}' \cdot \left(\frac{\hat{\mathbf{R}}}{R^2} \right) \bigg|_{\mathbf{r}'=0} = \mathbf{r}' \cdot \frac{\hat{\mathbf{r}}}{r^2}. \quad (3.66)$$

The third term is complicated. Let us denote (x, y, z) by (x_1, x_2, x_3) and perform an expansion in rectangular coordinates:

$$(\mathbf{r}' \cdot \nabla')^2 \frac{1}{R} \bigg|_{\mathbf{r}'=0} = \sum_{i=1}^{3} \sum_{j=1}^{3} x_i' x_j' \frac{\partial^2}{\partial x_i' \partial x_j'} \frac{1}{R} \bigg|_{\mathbf{r}'=0}.$$

It turns out [173] that this can be written as

$$(\mathbf{r}' \cdot \nabla')^2 \frac{1}{R} \bigg|_{\mathbf{r}'=0} = \frac{1}{r^3} \hat{\mathbf{r}} \cdot (3\mathbf{r}'\mathbf{r}' - r'^2 \bar{\mathbf{I}}) \cdot \hat{\mathbf{r}}.$$

Substitution into (3.65) gives

$$\Phi(r) = \frac{Q}{4\pi\epsilon r} + \frac{\hat{\mathbf{r}} \cdot \mathbf{p}}{4\pi\epsilon r^2} + \frac{1}{2} \frac{\hat{\mathbf{r}} \cdot \bar{\mathbf{Q}} \cdot \hat{\mathbf{r}}}{4\pi\epsilon r^3} + \cdots, \quad (3.67)$$

which is the multipole expansion for $\Phi(r)$. It converges for all $r > r_m$ where r_m is the radius of the smallest sphere completely containing the charge centered at $\mathbf{r}' = 0$ (Figure 3.10). In (3.67) the terms Q, $\mathbf{p}$, $\bar{\mathbf{Q}}$, and so on are called the *multipole moments* of $\rho(\mathbf{r})$. The first moment is merely the total charge

$$Q = \int_V \rho(\mathbf{r}') \, dV'.$$

The second moment is the *electric dipole moment vector*

$$\mathbf{p} = \int_V \mathbf{r}' \rho(\mathbf{r}') \, dV'.$$

The third moment is the *electric quadrupole moment dyadic*

$$\bar{\mathbf{Q}} = \int_V (3\mathbf{r}'\mathbf{r}' - r'^2\bar{\mathbf{I}})\rho(\mathbf{r}')\,dV'.$$

The expansion (3.67) allows us to identify the dominant power of r for $r \gg r_m$. The first nonzero term in (3.67) dominates the potential at points far from the source. Interestingly, the first nonvanishing moment is independent of the location of the origin of $\mathbf{r}'$, while all subsequent higher moments depend on the location of the origin [92].

▶ **Example 3.16:** Multipole moments of a point charge

Compute the multipole moments of a single point charge q located at $\mathbf{r}_0$.

Solution: Write $\rho(\mathbf{r}) = q\delta(\mathbf{r} - \mathbf{r}_0)$. The first moment of ρ is

$$Q = \int_V q\delta(\mathbf{r}' - \mathbf{r}_0)\,dV' = q.$$

Note that this is independent of $\mathbf{r}_0$. The second moment

$$\mathbf{p} = \int_V \mathbf{r}'q\delta(\mathbf{r}' - \mathbf{r}_0)\,dV' = q\mathbf{r}_0$$

depends on $\mathbf{r}_0$, as does the third moment

$$\bar{\mathbf{Q}} = \int_V (3\mathbf{r}'\mathbf{r}' - r'^2\bar{\mathbf{I}})q\delta(\mathbf{r}' - \mathbf{r}_0)\,dV' = q(3\mathbf{r}_0\mathbf{r}_0 - r_0^2\bar{\mathbf{I}}).$$

If $\mathbf{r}_0 = 0$ then only the first moment is nonzero, as is obvious from (3.41). ◀

▶ **Example 3.17:** Multipole moments of a dipole

Compute the multipole moments of the dipole shown in Figure 3.11.

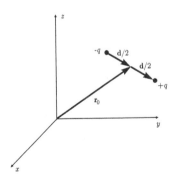

FIGURE 3.11
A dipole distribution.

Solution: Write $\rho(\mathbf{r}) = -q\delta(\mathbf{r} - \mathbf{r}_0 + \mathbf{d}/2) + q\delta(\mathbf{r} - \mathbf{r}_0 - \mathbf{d}/2)$. The first moment is

$$Q = \int_V [-q\delta(\mathbf{r} - \mathbf{r}_0 + \mathbf{d}/2) + q\delta(\mathbf{r} - \mathbf{r}_0 - \mathbf{d}/2)]\,dV' = -q + q = 0.$$

The second moment is

$$\mathbf{p} = \int_V \mathbf{r}'[-q\delta(\mathbf{r} - \mathbf{r}_0 + \mathbf{d}/2) + q\delta(\mathbf{r} - \mathbf{r}_0 - \mathbf{d}/2)]\, dV'$$
$$= -q(\mathbf{r}_0 - \mathbf{d}/2) + q(\mathbf{r}_0 + \mathbf{d}/2) = q\mathbf{d}.$$

The third moment is

$$\bar{\mathbf{Q}} = \int_V (3\mathbf{r}'\mathbf{r}' - r'^2\bar{\mathbf{I}})[-q\delta(\mathbf{r} - \mathbf{r}_0 + \mathbf{d}/2) + q\delta(\mathbf{r} - \mathbf{r}_0 - \mathbf{d}/2)]\, dV'$$
$$= -q[3(\mathbf{r}_0 - \mathbf{d}/2)(\mathbf{r}_0 - \mathbf{d}/2) - |\mathbf{r}_0 - \mathbf{d}/2|^2\bar{\mathbf{I}}]$$
$$+ q[3(\mathbf{r}_0 + \mathbf{d}/2)(\mathbf{r}_0 + \mathbf{d}/2) - |\mathbf{r}_0 + \mathbf{d}/2|^2\bar{\mathbf{I}}].$$

Use of $|\mathbf{r}_0 \pm \mathbf{d}/2|^2 = (\mathbf{r}_0 \pm \mathbf{d}/2) \cdot (\mathbf{r}_0 \pm \mathbf{d}/2) = r_0^2 + d^2/4 \pm \mathbf{r}_0 \cdot \mathbf{d}$ gives

$$\bar{\mathbf{Q}} = q[3(\mathbf{r}_0\mathbf{d} + \mathbf{d}\mathbf{r}_0) - 2(\mathbf{r}_0 \cdot \mathbf{d})\bar{\mathbf{I}}].$$

Only the first nonzero moment, in this case $\mathbf{p}$, is independent of $\mathbf{r}_0$. For $\mathbf{r}_0 = 0$ the only nonzero multipole moment would be the dipole moment $\mathbf{p}$. If the dipole is aligned along the z-axis with $\mathbf{d} = d\hat{\mathbf{z}}$ and $\mathbf{r}_0 = 0$, then the exact potential is

$$\Phi(\mathbf{r}) = \frac{1}{4\pi\epsilon} \frac{p\cos\theta}{r^2}.$$

By (3.10) we have

$$\mathbf{E}(\mathbf{r}) = \frac{1}{4\pi\epsilon} \frac{p}{r^3}(\hat{\mathbf{r}}2\cos\theta + \hat{\boldsymbol{\theta}}\sin\theta), \tag{3.68}$$

which is the classic result for the electric field of a dipole. ◄

▶ **Example 3.18:** Multipole moments of a quadrupole

Compute the multipole moments of the quadrupole shown in Figure 3.12.

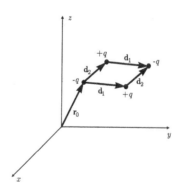

FIGURE 3.12
A quadrupole distribution.

Solution: Write the charge density as

$$\rho(\mathbf{r}) = -q\delta(\mathbf{r} - \mathbf{r}_0) + q\delta(\mathbf{r} - \mathbf{r}_0 - \mathbf{d}_1) + q\delta(\mathbf{r} - \mathbf{r}_0 - \mathbf{d}_2) - q\delta(\mathbf{r} - \mathbf{r}_0 - \mathbf{d}_1 - \mathbf{d}_2).$$

The first moment is

$$Q = \int_V [-q\delta(\mathbf{r} - \mathbf{r}_0) + q\delta(\mathbf{r} - \mathbf{r}_0 - \mathbf{d}_1)$$
$$+ q\delta(\mathbf{r} - \mathbf{r}_0 - \mathbf{d}_2) - q\delta(\mathbf{r} - \mathbf{r}_0 - \mathbf{d}_1 - \mathbf{d}_2)] \, dV' = -q + q + q - q = 0.$$

The second moment is

$$\mathbf{p} = \int_V \mathbf{r}'[-q\delta(\mathbf{r} - \mathbf{r}_0) + q\delta(\mathbf{r} - \mathbf{r}_0 - \mathbf{d}_1)$$
$$+ q\delta(\mathbf{r} - \mathbf{r}_0 - \mathbf{d}_2) - q\delta(\mathbf{r} - \mathbf{r}_0 - \mathbf{d}_1 - \mathbf{d}_2)] \, dV'$$
$$= -q\mathbf{r}_0 + q(\mathbf{r}_0 + \mathbf{d}_1) + q(\mathbf{r}_0 + \mathbf{d}_2) - q(\mathbf{r}_0 + \mathbf{d}_1 + \mathbf{d}_2) = 0.$$

The third moment is

$$\bar{\mathbf{Q}} = \int_V (3\mathbf{r}'\mathbf{r}' - r'^2\bar{\mathbf{I}})[-q\delta(\mathbf{r} - \mathbf{r}_0) + q\delta(\mathbf{r} - \mathbf{r}_0 - \mathbf{d}_1)$$
$$+ q\delta(\mathbf{r} - \mathbf{r}_0 - \mathbf{d}_2) - q\delta(\mathbf{r} - \mathbf{r}_0 - \mathbf{d}_1 - \mathbf{d}_2)] \, dV'$$
$$= -q(3\mathbf{r}_0\mathbf{r}_0 - r_0^2\bar{\mathbf{I}})$$
$$+ q[3(\mathbf{r}_0 + \mathbf{d}_1)(\mathbf{r}_0 + \mathbf{d}_1) - (\mathbf{r}_0 + \mathbf{d}_1) \cdot (\mathbf{r}_0 + \mathbf{d}_1)\bar{\mathbf{I}}]$$
$$+ q[3(\mathbf{r}_0 + \mathbf{d}_2)(\mathbf{r}_0 + \mathbf{d}_2) - (\mathbf{r}_0 + \mathbf{d}_2) \cdot (\mathbf{r}_0 + \mathbf{d}_2)\bar{\mathbf{I}}]$$
$$- q[3(\mathbf{r}_0 + \mathbf{d}_1 + \mathbf{d}_2)(\mathbf{r}_0 + \mathbf{d}_1 + \mathbf{d}_2) - (\mathbf{r}_0 + \mathbf{d}_1 + \mathbf{d}_2) \cdot (\mathbf{r}_0 + \mathbf{d}_1 + \mathbf{d}_2)\bar{\mathbf{I}}].$$

Simplification yields
$$\bar{\mathbf{Q}} = q[-3(\mathbf{d}_1\mathbf{d}_2 + \mathbf{d}_2\mathbf{d}_1) + 2(\mathbf{d}_1 \cdot \mathbf{d}_2)\bar{\mathbf{I}}].$$

As expected, the first two moments of ρ vanish, while the third is independent of $\mathbf{r}_0$. ◄

It is tedious to carry (3.67) beyond the quadrupole term using the Taylor expansion. Another approach is to expand $1/R$ in spherical harmonics. Referring to Appendix E.3 we find that

$$\frac{1}{|\mathbf{r} - \mathbf{r}'|} = 4\pi \sum_{n=0}^{\infty} \sum_{m=-n}^{n} \frac{1}{2n+1} \frac{r'^n}{r^{n+1}} Y_{nm}^*(\theta', \phi') Y_{nm}(\theta, \phi)$$

(see Jackson [92] or Arfken [6] for a detailed derivation). This expansion converges for $|\mathbf{r}| > |\mathbf{r}_m|$. Substitution into (3.41) gives

$$\Phi(\mathbf{r}) = \frac{1}{\epsilon} \sum_{n=0}^{\infty} \frac{1}{r^{n+1}} \left[\frac{1}{2n+1} \sum_{m=-n}^{n} q_{nm} Y_{nm}(\theta, \phi) \right] \tag{3.69}$$

where

$$q_{nm} = \int_V \rho(\mathbf{r}')r'^n Y_{nm}^*(\theta', \phi') \, dV'.$$

We can now identify any inverse power of r in the multipole expansion, but at the price of dealing with a double summation. For a charge distribution with axial symmetry (no ϕ-variation), only the coefficient q_{n0} is nonzero. The relation

$$Y_{n0}(\theta, \phi) = \sqrt{\frac{2n+1}{4\pi}} P_n(\cos\theta)$$

allows us to simplify (3.69) and obtain

$$\Phi(\mathbf{r}) = \frac{1}{4\pi\epsilon} \sum_{n=0}^{\infty} \frac{1}{r^{n+1}} q_n P_n(\cos\theta) \qquad (3.70)$$

where

$$q_n = 2\pi \int_{r'} \int_{\theta'} \rho(r', \theta') r'^n P_n(\cos\theta') r'^2 \sin\theta' \, d\theta' \, dr'.$$

▶ **Example 3.19:** Multipole expansion coefficients for a spherical charge

Compute the multipole expansion coefficients for a spherical distribution of charge given by

$$\rho(\mathbf{r}) = \frac{3Q}{\pi a^3} \cos\theta \qquad (r \le a).$$

Solution: This can be viewed as two adjacent hemispheres carrying total charges $\pm Q$. Since $\cos\theta = P_1(\cos\theta)$, we compute

$$q_n = 2\pi \int_0^a \int_0^\pi \frac{3Q}{\pi a^3} P_1(\cos\theta') r'^n P_n(\cos\theta') r'^2 \sin\theta' \, d\theta' \, dr'$$

$$= 2\pi \frac{3Q}{\pi a^3} \frac{a^{n+3}}{n+3} \int_0^\pi P_1(\cos\theta) P_n(\cos\theta') \sin\theta' \, d\theta'.$$

Using the orthogonality relation (E.125) we find

$$q_n = 2\pi \frac{3Q}{\pi a^3} \frac{a^{n+3}}{n+3} \delta_{1n} \frac{2}{2n+1}.$$

Hence the only nonzero coefficient is $q_1 = Qa$ and

$$\Phi(\mathbf{r}) = \frac{1}{4\pi\epsilon} \frac{1}{r^2} Qa P_1(\cos\theta) = \frac{Qa}{4\pi\epsilon r^2} \cos\theta.$$

Since a dipole with moment $\mathbf{p} = \hat{\mathbf{z}} Qa$ produces this same potential when $r \gg a$, we could replace the sphere with point charges $\mp Q$ at $z = \mp a/2$ without changing the field far from the source. ◀

3.3.7.1 Physical interpretation of the polarization vector in a dielectric

We have used the Maxwell–Minkowski equations to determine the electrostatic potential of a charge distribution in the presence of a dielectric medium. Alternatively, we can use the Maxwell–Boffi equations

$$\nabla \times \mathbf{E} = 0, \qquad (3.71)$$

$$\nabla \cdot \mathbf{E} = \frac{1}{\epsilon_0}(\rho - \nabla \cdot \mathbf{P}). \qquad (3.72)$$

Equation (3.71) allows us to define a scalar potential through (3.10). Substitution into (3.72) gives

$$\nabla^2 \Phi(\mathbf{r}) = -\frac{1}{\epsilon_0}[\rho(\mathbf{r}) + \rho_P(\mathbf{r})] \qquad (3.73)$$

where $\rho_P = -\nabla \cdot \mathbf{P}$. This has the form of Poisson's equation (3.30), but with charge density term $\rho(\mathbf{r}) + \rho_P(\mathbf{r})$. Hence the solution is

$$\Phi(\mathbf{r}) = \frac{1}{4\pi\epsilon_0} \int_V \frac{\rho(\mathbf{r}') - \nabla' \cdot \mathbf{P}(\mathbf{r}')}{|\mathbf{r} - \mathbf{r}'|} \, dV'.$$

To this we must add any potential produced by surface sources such as ρ_s. If there is a discontinuity in the dielectric region, there is also a surface polarization source $\rho_{Ps} = \hat{\mathbf{n}} \cdot \mathbf{P}$ according to (3.15). Separating the volume into regions with bounding surfaces S_i across which the permittivity is discontinuous, we may write

$$\Phi(\mathbf{r}) = \frac{1}{4\pi\epsilon_0} \int_V \frac{\rho(\mathbf{r}')}{|\mathbf{r} - \mathbf{r}'|} \, dV' + \frac{1}{4\pi\epsilon_0} \int_S \frac{\rho_s(\mathbf{r}')}{|\mathbf{r} - \mathbf{r}'|} \, dS'$$
$$+ \sum_i \left[\frac{1}{4\pi\epsilon_0} \int_{V_i} \frac{-\nabla' \cdot \mathbf{P}(\mathbf{r}')}{|\mathbf{r} - \mathbf{r}'|} \, dV' + \frac{1}{4\pi\epsilon_0} \oint_{S_i} \frac{\hat{\mathbf{n}}' \cdot \mathbf{P}(\mathbf{r}')}{|\mathbf{r} - \mathbf{r}'|} \, dS' \right], \qquad (3.74)$$

where $\hat{\mathbf{n}}$ points outward from region i. Using the divergence theorem on the fourth term and employing (B.48), we obtain

$$\Phi(\mathbf{r}) = \frac{1}{4\pi\epsilon_0} \int_V \frac{\rho(\mathbf{r}')}{|\mathbf{r} - \mathbf{r}'|} \, dV' + \frac{1}{4\pi\epsilon_0} \int_S \frac{\rho_s(\mathbf{r}')}{|\mathbf{r} - \mathbf{r}'|} \, dS'$$
$$+ \sum_i \left[\frac{1}{4\pi\epsilon_0} \int_{V_i} \mathbf{P}(\mathbf{r}') \cdot \nabla' \left(\frac{1}{|\mathbf{r} - \mathbf{r}'|} \right) dV' \right].$$

Since $\nabla'(1/R) = \hat{\mathbf{R}}/R^2$, the third term is a sum of integrals of the form

$$\frac{1}{4\pi\epsilon} \int_{V_i} \mathbf{P}(\mathbf{r}') \cdot \frac{\hat{\mathbf{R}}}{R^2} \, dV.$$

Comparing this to the second term of (3.67), we see that this integral represents a volume superposition of dipole terms where $\mathbf{P}$ is a volume density of dipole moments.

Thus, a dielectric with permittivity ϵ is equivalent to a volume distribution of dipoles in free space. No higher-order moments are required, and no zero-order moments are needed, since any net charge is included in ρ. Note that we have arrived at this conclusion based only on Maxwell's equations and the assumption of a linear, isotropic relationship between $\mathbf{D}$ and $\mathbf{E}$. Assuming our macroscopic theory is correct, we are tempted to make assumptions about the behavior of matter on a microscopic level (e.g., atoms exposed to fields are polarized and their electron clouds are displaced from their positively charged nuclei), but this area of science is better studied from the viewpoints of particle physics and quantum mechanics.

3.3.7.2 Potential of an azimuthally symmetric charged spherical surface

In several of our example problems we shall be interested in evaluating the potential of a charged spherical surface. When the charge is azimuthally symmetric, the potential is particularly simple.

We will need the value of the integral

$$F(\mathbf{r}) = \frac{1}{4\pi} \int_S \frac{f(\theta')}{|\mathbf{r} - \mathbf{r}'|} \, dS' \qquad (3.75)$$

where $\mathbf{r} = r\hat{\mathbf{r}}$ describes an arbitrary observation point and $\mathbf{r}' = a\hat{\mathbf{r}}'$ identifies the source point on the surface of the sphere of radius a. The integral is most easily done using the expansion (E.204) for $|\mathbf{r} - \mathbf{r}'|^{-1}$ in spherical harmonics. We have

$$F(\mathbf{r}) = a^2 \sum_{n=0}^{\infty} \sum_{m=-n}^{n} \frac{Y_{nm}(\theta,\phi)}{2n+1} \frac{r_<^n}{r_>^{n+1}} \int_{-\pi}^{\pi} \int_0^{\pi} f(\theta') Y_{nm}^*(\theta',\phi') \sin\theta' \, d\theta' \, d\phi'$$

where $r_< = \min\{r,a\}$ and $r_> = \max\{r,a\}$. Using orthogonality of the exponentials, we find that only the $m = 0$ terms contribute:

$$F(\mathbf{r}) = 2\pi a^2 \sum_{n=0}^{\infty} \frac{Y_{n0}(\theta,\phi)}{2n+1} \frac{r_<^n}{r_>^{n+1}} \int_0^{\pi} f(\theta') Y_{n0}^*(\theta',\phi') \sin\theta' \, d\theta'.$$

Finally, since

$$Y_{n0} = \sqrt{\frac{2n+1}{4\pi}} P_n(\cos\theta)$$

we have

$$F(\mathbf{r}) = \frac{1}{2} a^2 \sum_{n=0}^{\infty} P_n(\cos\theta) \frac{r_<^n}{r_>^{n+1}} \int_0^{\pi} f(\theta') P_n(\cos\theta') \sin\theta' \, d\theta'. \qquad (3.76)$$

▶ **Example 3.20:** Potential integral for $f(\theta) = \cos\theta$

Compute the integral $F(\mathbf{r})$ in (3.75) for $f(\theta) = \cos\theta$.

Solution: Note that $f(\theta) = \cos\theta = P_1(\cos\theta)$. Then by (3.76)

$$F(\mathbf{r}) = \frac{1}{2} a^2 \sum_{n=0}^{\infty} P_n(\cos\theta) \frac{r_<^n}{r_>^{n+1}} \int_0^{\pi} P_1(\cos\theta') P_n(\cos\theta') \sin\theta' \, d\theta'.$$

The orthogonality of the Legendre polynomials can be used to show that

$$\int_0^{\pi} P_1(\cos\theta') P_n(\cos\theta') \sin\theta' \, d\theta' = \tfrac{2}{3}\delta_{1n},$$

hence

$$F(\mathbf{r}) = \frac{a^2}{3} \cos\theta \, \frac{r_<}{r_>^2}. \quad ◀ \qquad (3.77)$$

3.3.8 Field produced by a permanently polarized body

Certain materials, called *electrets*, exhibit polarization in the absence of an external electric field. A permanently polarized material produces an electric field both internal and external to the material, hence there must be a charge distribution to support the fields. We can interpret this charge as being caused by the permanent separation of atomic charge within the material, but if we are only interested in the macroscopic field then we need not worry about the microscopic implications of such materials. Instead, we can use the Maxwell–Boffi equations and find the potential produced by the material by using (3.74). Thus, the field of an electret with known polarization $\mathbf{P}$ occupying volume region V in free space is dipolar in nature and is given by

$$\Phi(\mathbf{r}) = \frac{1}{4\pi\epsilon_0} \int_V \frac{-\nabla' \cdot \mathbf{P}(\mathbf{r}')}{|\mathbf{r} - \mathbf{r}'|} \, dV' + \frac{1}{4\pi\epsilon_0} \oint_S \frac{\hat{n}' \cdot \mathbf{P}(\mathbf{r}')}{|\mathbf{r} - \mathbf{r}'|} \, dS'$$

where $\hat{\mathbf{n}}$ points out of the volume region V.

▶ **Example 3.21:** Potential of a permanently polarized sphere

A material sphere of radius a, is permanently polarized along its axis with uniform polarization $\mathbf{P}(\mathbf{r}) = \hat{\mathbf{z}}P_0$. Compute Φ everywhere.

Solution: We have the equivalent source densities

$$\rho_p = -\nabla \cdot \mathbf{P} = 0, \qquad \rho_{Ps} = \hat{\mathbf{n}} \cdot \mathbf{P} = \hat{\mathbf{r}} \cdot \hat{\mathbf{z}}P_0 = P_0 \cos\theta.$$

Then

$$\Phi(\mathbf{r}) = \frac{1}{4\pi\epsilon_0} \oint_S \frac{\rho_{Ps}(\mathbf{r}')}{|\mathbf{r} - \mathbf{r}'|} \, dS' = \frac{1}{4\pi\epsilon_0} \oint_S \frac{P_0 \cos\theta'}{|\mathbf{r} - \mathbf{r}'|} \, dS'.$$

The integral takes the form (3.75), hence by (3.77) the solution is

$$\Phi(\mathbf{r}) = P_0 \frac{a^2}{3\epsilon_0} \cos\theta \frac{r_<}{r_>^2}. \quad \blacktriangleleft \tag{3.78}$$

▶ **Example 3.22:** Multipole expansion of the potential of a permanently polarized sphere

Consider the potential of the permanently polarized material sphere found in Example 3.21. Perform a multipole expansion of the potential external to the sphere, and show that it is a dipole field.

Solution: For $r > a$ we can use the multipole expansion (3.70) to obtain

$$\Phi(\mathbf{r}) = \frac{1}{4\pi\epsilon_0} \sum_{n=0}^{\infty} \frac{1}{r^{n+1}} q_n P_n(\cos\theta) \qquad (r > a)$$

where

$$q_n = 2\pi \int_0^\pi \rho_{Ps}(\theta') a^n P_n(\cos\theta') a^2 \sin\theta' \, d\theta'.$$

Substituting for ρ_{Ps} and remembering that $\cos\theta = P_1(\cos\theta)$, we have

$$q_n = 2\pi a^{n+2} P_0 \int_0^\pi P_1(\cos\theta') P_n(\cos\theta') \sin\theta' \, d\theta'.$$

Using the orthogonality relation (E.125) we find

$$q_n = 2\pi a^{n+2} P_0 \delta_{1n} \frac{2}{2n + 1}.$$

Therefore, the only nonzero coefficient is $q_1 = \frac{4}{3}\pi a^3 P_0$, and

$$\Phi(\mathbf{r}) = \frac{1}{4\pi\epsilon_0} \frac{1}{r^2} \frac{4\pi a^3 P_0}{3} P_1(\cos\theta) = \frac{P_0 a^3}{3\epsilon_0 r^2} \cos\theta \qquad (r > a).$$

This is a dipole field, and matches (3.78) as expected. ◀

3.3.9   Potential of a dipole layer

Surface charge layers sometimes occur in bipolar form, such as in the membrane surrounding an animal cell. These can be modeled as a dipole layer consisting of parallel

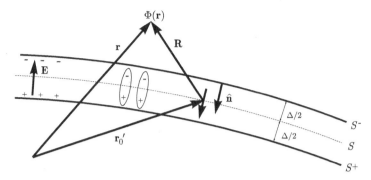

FIGURE 3.13
A dipole layer.

surface charges of opposite sign.

Consider a surface S located in free space as shown in Figure 3.13. Parallel to this surface, and a distance $\Delta/2$ below, is located a surface charge layer of density $\rho_s(\mathbf{r}) = P_s(\mathbf{r})$. Also parallel to S, but a distance $\Delta/2$ above, is a surface charge layer of density $\rho_s(\mathbf{r}) = -P_s(\mathbf{r})$. We define the *surface dipole moment density* D_s as

$$D_s(\mathbf{r}) = \Delta\, P_s(\mathbf{r}).$$

Letting the position vector $\mathbf{r}_0'$ point to the surface S we can write the potential (3.41) produced by the two charge layers as

$$\Phi(\mathbf{r}) = \frac{1}{4\pi\epsilon_0}\int_{S+} P_s(\mathbf{r}')\frac{1}{|\mathbf{r}-\mathbf{r}_0'-\hat{\mathbf{n}}'\frac{\Delta}{2}|}\, dS' - \frac{1}{4\pi\epsilon_0}\int_{S-} P_s(\mathbf{r}')\frac{1}{|\mathbf{r}-\mathbf{r}_0'+\hat{\mathbf{n}}'\frac{\Delta}{2}|}\, dS'.$$

We are interested in the case in which the two charge layers collapse onto the surface S, and wish to compute the potential produced by a given dipole moment density. When $\Delta \to 0$ we have $\mathbf{r}_0' \to \mathbf{r}'$ and may write

$$\Phi(\mathbf{r}) = \lim_{\Delta\to 0}\frac{1}{4\pi\epsilon_0}\int_S \frac{D_s(\mathbf{r}')}{\Delta}\left(\frac{1}{|\mathbf{R}-\hat{\mathbf{n}}'\frac{\Delta}{2}|} - \frac{1}{|\mathbf{R}+\hat{\mathbf{n}}'\frac{\Delta}{2}|}\right) dS',$$

where $\mathbf{R} = \mathbf{r} - \mathbf{r}'$. By the binomial theorem, the limit of the parenthetical term can be written as

$$\lim_{\Delta\to 0}\left\{\left[R^2 + \left(\frac{\Delta}{2}\right)^2 - 2\mathbf{R}\cdot\hat{\mathbf{n}}'\frac{\Delta}{2}\right]^{-\frac{1}{2}} - \left[R^2 + \left(\frac{\Delta}{2}\right)^2 + 2\mathbf{R}\cdot\hat{\mathbf{n}}'\frac{\Delta}{2}\right]^{-\frac{1}{2}}\right\}$$

$$= \lim_{\Delta\to 0}\left\{R^{-1}\left[1 + \frac{\hat{\mathbf{R}}\cdot\hat{\mathbf{n}}'}{R}\frac{\Delta}{2}\right] - R^{-1}\left[1 - \frac{\hat{\mathbf{R}}\cdot\hat{\mathbf{n}}'}{R}\frac{\Delta}{2}\right]\right\} = \Delta\hat{\mathbf{n}}'\cdot\frac{\mathbf{R}}{R^3}.$$

Thus

$$\Phi(\mathbf{r}) = \frac{1}{4\pi\epsilon_0}\int_S \mathbf{D}_s(\mathbf{r}')\cdot\frac{\mathbf{R}}{R^3}\, dS' \tag{3.79}$$

where $\mathbf{D}_s = \hat{\mathbf{n}}D_s$ is the surface vector dipole moment density. The potential of a dipole layer decreases more rapidly ($\sim 1/r^2$) than that of a unipolar charge layer. We saw

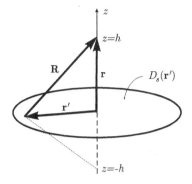

FIGURE 3.14
Auxiliary disk for studying the potential distribution across a dipole layer.

similar behavior in the dipole term of the multipole expansion (3.67) for a general charge distribution.

We can use (3.79) to study the behavior of the potential across a dipole layer. As we approach the layer from above, the greatest contribution to Φ comes from the charge region immediately beneath the observation point. Assuming that the surface dipole moment density is continuous beneath the point, we can compute the difference in the fields across the layer at point $\mathbf{r}$ by replacing the arbitrary surface layer by a disk of constant surface dipole moment density $\mathbf{D}_0 = \mathbf{D}_s(\mathbf{r})$. For simplicity we center the disk at $z = 0$ in the xy-plane as shown in Figure 3.14 and compute the potential difference ΔV across the layer; i.e., $\Delta V = \Phi(h) - \Phi(-h)$ on the disk axis as $h \to 0$. Using (3.79) along with $\mathbf{r}' = \pm h\hat{\mathbf{z}} - \rho'\hat{\boldsymbol{\rho}}'$, we obtain

$$\Delta V = \lim_{h\to 0} \left[\frac{1}{4\pi\epsilon_0} \int_0^{2\pi} \int_0^a (\hat{\mathbf{z}} D_0) \cdot \frac{\hat{\mathbf{z}} h - \hat{\boldsymbol{\rho}}' \rho'}{(h^2 + \rho'^2)^{3/2}} \rho' \, d\rho' \, d\phi' \right.$$
$$\left. - \frac{1}{4\pi\epsilon_0} \int_0^{2\pi} \int_0^a (\hat{\mathbf{z}} D_0) \cdot \frac{-\hat{\mathbf{z}} h - \hat{\boldsymbol{\rho}}' \rho'}{(h^2 + \rho'^2)^{3/2}} \rho' \, d\rho' \, d\phi' \right]$$

where a is the disk radius. Integration yields

$$\Delta V = \frac{D_0}{2\epsilon_0} \lim_{h\to 0} \left[\frac{-2}{\sqrt{1 + (a/h)^2}} + 2 \right] = \frac{D_0}{\epsilon_0},$$

independent of a. Generalizing this to an arbitrary surface dipole moment density, we find that the boundary condition on the potential is given by

$$\Phi_2(\mathbf{r}) - \Phi_1(\mathbf{r}) = \frac{D_s(\mathbf{r})}{\epsilon_0} \tag{3.80}$$

where "1" denotes the positive side of the dipole moments and "2" the negative side. Physically, the potential difference in (3.80) is produced by the line integral of $\mathbf{E}$ "internal" to the dipole layer. Since there is no field internal to a unipolar surface layer, V is continuous across a surface containing charge ρ_s but having $D_s = 0$.

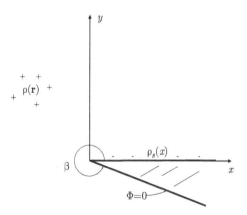

FIGURE 3.15
A conducting edge.

3.3.10 Behavior of electric charge density near a conducting edge

Sharp corners are often encountered in the application of electrostatics to practical geometries. The behavior of the charge distribution near these corners must be understood in order to develop numerical techniques for solving more complicated problems. We can use a simple model of a corner if we restrict our interest to the region near the edge. Consider the intersection of two planes as shown in Figure 3.15. The region near the intersection represents the corner we wish to study. We assume that the planes are held at zero potential and that the charge on the surface is induced by a two-dimensional charge distribution $\rho(\mathbf{r})$, or by a potential difference between the edge and another conductor far removed from the edge.

We can find the potential in the region near the edge by solving Laplace's equation in cylindrical coordinates. This problem is studied in Appendix A where the separation of variables solution is found to be either (A.128) or (A.129). Using (A.129) and enforcing $\Phi = 0$ at both $\phi = 0$ and $\phi = \beta$, we obtain the null solution. Hence the solution must take the form (A.128):

$$\Phi(\rho, \phi) = [A_\phi \sin(k_\phi \phi) + B_\phi \cos(k_\phi \phi)](a_\rho \rho^{-k_\phi} + b_\rho \rho^{k_\phi}).$$

Since the origin is included, we cannot have negative powers of ρ and must put $a_\rho = 0$. The boundary condition $\Phi(\rho, 0) = 0$ requires $B_\phi = 0$. The condition $\Phi(\rho, \beta) = 0$ then requires $\sin(k_\phi \beta) = 0$, which holds only if $k_\phi = n\pi/\beta$, $n = 1, 2, \ldots$. The general solution for the potential near the edge is therefore

$$\Phi(\rho, \phi) = \sum_{n=1}^{N} A_n \sin\left(\frac{n\pi}{\beta}\phi\right) \rho^{n\pi/\beta}$$

where the constants A_n depend on the excitation source or system of conductors. (Note that if the corner is held at potential $V_0 \neq 0$, we must merely add V_0 to the solution.) The charge on the conducting surfaces can be computed from the boundary condition on normal $\mathbf{D}$. Using (3.10) we have

$$E_\phi = -\frac{1}{\rho}\frac{\partial}{\partial \phi}\sum_{n=1}^{N} A_n \sin\left(\frac{n\pi}{\beta}\phi\right)\rho^{n\pi/\beta} = -\sum_{n=1}^{N} A_n \frac{n\pi}{\beta}\cos\left(\frac{n\pi}{\beta}\phi\right)\rho^{(n\pi/\beta)-1},$$

hence

$$\rho_s(x) = -\epsilon \sum_{n=1}^{N} A_n \frac{n\pi}{\beta} x^{(n\pi/\beta)-1}$$

on the surface at $\phi = 0$. Near the edge, at small values of x, the variation of ρ_s is dominated by the lowest power of x. (Here we ignore those special excitation arrangements that produce $A_1 = 0$.) Thus

$$\rho_s(x) \sim x^{(\pi/\beta)-1}. \tag{3.81}$$

The behavior of the charge clearly depends on the wedge angle β. For a sharp edge (half plane) we put $\beta = 2\pi$ and find that the field varies as $x^{-1/2}$. This *square-root edge singularity* is very common on thin plates, fins, etc., and means that charge tends to accumulate near the edge of a flat conducting surface. For a right-angle corner where $\beta = 3\pi/2$, there is the somewhat weaker singularity $x^{-1/3}$. When $\beta = \pi$, the two surfaces fold out into an infinite plane and the charge, not surprisingly, is invariant with x to lowest order near the folding line. When $\beta < \pi$ the corner becomes interior and we find that the charge density varies with a positive power of distance from the edge. For very sharp interior angles the power is large, meaning that little charge accumulates on the inner surfaces near an interior corner.

3.3.11 Solution to Laplace's equation for bodies immersed in an impressed field

An important class of problems is based on the idea of placing a body into an existing electric field, assuming that the field arises from sources so remote that the introduction of the body does not alter the original field. The pre-existing field is often referred to as the *applied* or *impressed field*, and the solution external to the body is usually formulated as the sum of the applied field and a *secondary* or *scattered field* that satisfies Laplace's equation. This total field differs from the applied field, and must satisfy the appropriate boundary condition on the body. If the body is a conductor then the total potential must be constant everywhere on the boundary surface. If the body is a solid homogeneous dielectric then the total potential field must be continuous across the boundary.

▶ **Example 3.23:** Potential for a dielectric sphere in a uniform electric field

A dielectric sphere of permittivity ϵ and radius a, is centered at the origin in free space and immersed in a uniform electric field $\mathbf{E}_0(\mathbf{r}) = E_0 \hat{\mathbf{z}}$. Find Φ everywhere.

Solution: By (3.10) the applied potential field is

$$\Phi_0(\mathbf{r}) = -E_0 z = -E_0 r \cos\theta$$

(to within a constant). Outside the sphere $(r > a)$ we write the total potential field as

$$\Phi_2(\mathbf{r}) = \Phi_0(\mathbf{r}) + \Phi^s(\mathbf{r}),$$

where $\Phi^s(\mathbf{r})$ is the secondary or scattered potential. Since Φ^s must satisfy Laplace's equation, we can write it as a separation of variables solution (§ A.5.3). By azimuthal symmetry, the potential has an r-dependence as in (A.146), and a θ-dependence as in (A.142) with $B_\theta = 0$ and $m = 0$. Thus Φ^s has a representation identical to (A.147), except that we cannot use terms that are unbounded as $r \to \infty$. We therefore use

$$\Phi^s(r, \theta) = \sum_{n=0}^{\infty} B_n r^{-(n+1)} P_n(\cos \theta). \tag{3.82}$$

The potential inside the sphere also obeys Laplace's equation, so we can use the same form (A.147) while discarding terms unbounded at the origin. Thus

$$\Phi_1(r, \theta) = \sum_{n=0}^{\infty} A_n r^n P_n(\cos \theta) \tag{3.83}$$

for $r < a$. To find the constants A_n and B_n, we apply (3.16) and (3.17) to the total field. Application of (3.16) at $r = a$ gives

$$-E_0 a \cos \theta + \sum_{n=0}^{\infty} B_n a^{-(n+1)} P_n(\cos \theta) = \sum_{n=0}^{\infty} A_n a^n P_n(\cos \theta).$$

Multiplying through by $P_m(\cos \theta) \sin \theta$, integrating from $\theta = 0$ to $\theta = \pi$, and using the orthogonality relationship (E.125), we obtain

$$-E_0 a + a^{-2} B_1 = A_1 a, \tag{3.84}$$

$$B_n a^{-(n+1)} = A_n a^n \quad (n \neq 1), \tag{3.85}$$

where we have used $P_1(\cos \theta) = \cos \theta$. Next, since $\rho_s = 0$, Equation (3.17) requires

$$\epsilon_1 \frac{\partial \Phi_1(\mathbf{r})}{\partial r} = \epsilon_2 \frac{\partial \Phi_2(\mathbf{r})}{\partial r}$$

at $r = a$. This gives

$$-\epsilon_0 E_0 \cos \theta + \epsilon_0 \sum_{n=0}^{\infty} [-(n+1) B_n] a^{-n-2} P_n(\cos \theta) = \epsilon \sum_{n=0}^{\infty} [n A_n] a^{n-1} P_n(\cos \theta).$$

By orthogonality of the Legendre functions, we have

$$-\epsilon_0 E_0 - 2\epsilon_0 B_1 a^{-3} = \epsilon A_1, \tag{3.86}$$

$$-\epsilon_0(n+1) B_n a^{-n-2} = \epsilon n A_n a^{n-1} \quad (n \neq 1). \tag{3.87}$$

Equations (3.85) and (3.87) cannot both hold unless $A_n = B_n = 0$ for $n \neq 1$. Solving (3.84) and (3.86), we have

$$A_1 = -E_0 \frac{3\epsilon_0}{\epsilon + 2\epsilon_0}, \qquad B_1 = E_0 a^3 \frac{\epsilon - \epsilon_0}{\epsilon + 2\epsilon_0}.$$

Hence

$$\Phi_1(\mathbf{r}) = -E_0 \frac{3\epsilon_0}{\epsilon + 2\epsilon_0} r \cos \theta = -E_0 z \frac{3\epsilon_0}{\epsilon + 2\epsilon_0}, \tag{3.88}$$

$$\Phi_2(\mathbf{r}) = -E_0 r \cos \theta + E_0 \frac{a^3}{r^2} \frac{\epsilon - \epsilon_0}{\epsilon + 2\epsilon_0} \cos \theta. \; \blacktriangleleft \tag{3.89}$$

▶ **Example 3.24:** Electric field for a dielectric sphere in a uniform electric field

Consider the dielectric sphere immersed in a uniform electric field described in Example 3.23. Find **E** inside the sphere and show that it is uniform and aligned with the applied field. Is the internal field larger or smaller than the applied field? Explain.

Solution: The electric field inside the sphere is given by

$$\mathbf{E}_1(\mathbf{r}) = -\nabla \Phi_1(\mathbf{r}) = \hat{\mathbf{z}} \frac{\partial}{\partial z} \left(-E_0 z \frac{3\epsilon_0}{\epsilon + 2\epsilon_0} \right) = \hat{\mathbf{z}} E_0 \frac{3\epsilon_0}{\epsilon + 2\epsilon_0}.$$

We see that $\mathbf{E}_1$ is constant with position and is aligned with the applied external field $\mathbf{E}_0 = E_0 \hat{\mathbf{z}}$. However, $\mathbf{E}_1$ is weaker than $\mathbf{E}_0$, since $\epsilon > \epsilon_0$. To explain this, we compute the polarization charge within and on the sphere. Using $\mathbf{D} = \epsilon \mathbf{E} = \epsilon_0 \mathbf{E} + \mathbf{P}$, we have

$$\mathbf{P}_1 = \hat{\mathbf{z}}(\epsilon - \epsilon_0) E_0 \frac{3\epsilon_0}{\epsilon + 2\epsilon_0}. \tag{3.90}$$

The volume polarization charge density $-\nabla \cdot \mathbf{P}$ is zero, while the polarization surface charge density is

$$\rho_{Ps} = \hat{\mathbf{r}} \cdot \mathbf{P} = (\epsilon - \epsilon_0) E_0 \frac{3\epsilon_0}{\epsilon + 2\epsilon_0} \cos\theta.$$

Hence the secondary electric field can be attributed to an induced surface polarization charge, and is in a direction opposing the applied field. According to the Maxwell–Boffi viewpoint, we should be able to replace the sphere by the surface polarization charge in free space and use (3.41) to reproduce (3.88)–(3.89). This is left as an exercise for the reader. ◄

3.4 Magnetostatics

The large-scale forms of the magnetostatic field equations are

$$\oint_\Gamma \mathbf{H} \cdot d\mathbf{l} = \int_S \mathbf{J} \cdot d\mathbf{S}, \tag{3.91}$$

$$\oint_S \mathbf{B} \cdot d\mathbf{S} = 0,$$

while the point forms are

$$\nabla \times \mathbf{H}(\mathbf{r}) = \mathbf{J}(\mathbf{r}), \tag{3.92}$$

$$\nabla \cdot \mathbf{B}(\mathbf{r}) = 0.$$

Note the interesting dichotomy between the electrostatic field equations and the magnetostatic field equations. Whereas the electrostatic field exhibits zero curl and a divergence proportional to the source (charge), the magnetostatic field has zero divergence and a curl proportional to the source (current). Because the vector relationship between the magnetostatic field and its source is of a more complicated nature than the scalar relationship between the electrostatic field and its source, more effort is required to develop a strong understanding of magnetic phenomena. Also, it must always be remembered that although the equations describing the electrostatic and magnetostatic field sets decouple, the phenomena themselves remain linked. Since current is moving charge, electrical phenomena are associated with the establishment of the current that supports a magnetostatic field. We know, for example, that in order to have current in a wire, an electric field must be present to drive electrons through the wire.

3.4.1 Direct solutions to Ampere's law

In circumstances where current is highly symmetric, Ampere's law (in either integral or point form) may be solved directly for the magnetic field. Any of these symmetry conditions is appropriate:

1. The current is z-directed and depends only on the cylindrical coordinate ρ. Then $\mathbf{H}(\mathbf{r}) = H_\phi(\rho)\hat{\boldsymbol{\phi}}$.

2. The current is ϕ-directed and depends only on the cylindrical coordinate ρ. Then $\mathbf{H}(\mathbf{r}) = H_z(\rho)\hat{\mathbf{z}}$.

3. The current is x-directed and depends only on the variable z in rectangular coordinates. Then $\mathbf{H}(\mathbf{r}) = H_y(z)\hat{\mathbf{y}}$. Similarly for y or z-directed currents.

To employ the integral form of Ampere's law, we choose an integration path (*Amperian path*) on which the magnetic field is either constant in magnitude and tangential to the path, or perpendicular to the path (or on which some combination of these conditions holds). Then the magnetic field may be removed from the integral and determined. For employment of the law in point form, the volume is separated into regions in each of which the partial differential equation reduces to an ordinary differential equation, solvable by direct integration. The solutions are then connected across the adjoining surfaces using boundary conditions.

▶ **Example 3.25:** Solution to the integral form of Ampere's law for a line current

Find $\mathbf{H}$ produced by a filamentary wire carrying current I along the z-axis.

Solution: We start with the volume current density

$$\mathbf{J}(\mathbf{r}) = \hat{\mathbf{z}}\, I \frac{\delta(\rho)}{2\pi\rho}.$$

Since the current is z-directed and depends only on the cylindrical radius variable ρ, we have $\mathbf{H}(\mathbf{r}) = H_\phi(\rho)\hat{\boldsymbol{\phi}}$. Choose a circular path Γ of radius ρ_0 in the $z = 0$ plane. Then Ampere's law requires

$$\oint_\Gamma \mathbf{H} \cdot d\mathbf{l} = \int_S \mathbf{J} \cdot d\mathbf{S},$$

where

$$\oint_\Gamma \mathbf{H} \cdot d\mathbf{l} = \int_0^{2\pi} H_\phi(\rho_0)\hat{\boldsymbol{\phi}} \cdot \hat{\boldsymbol{\phi}}\rho_0\, d\phi = 2\pi\rho_0 H_\phi(\rho_0)$$

and the current linked by Γ is

$$\int_S \mathbf{J} \cdot d\mathbf{S} = \int_0^{2\pi} \int_0^{\rho_0} I\frac{\delta(\rho)}{2\pi\rho}\hat{\mathbf{z}} \cdot \hat{\mathbf{z}}\rho\, d\rho\, d\phi = I.$$

Hence, $2\pi\rho_0 H_\phi(\rho_0) = I$, and we obtain

$$H_\phi(\rho_0) = \frac{I}{2\pi\rho_0}.$$

Noting that ρ_0 is arbitrary, we finally write

$$\mathbf{H}(\rho) = \hat{\boldsymbol{\phi}}\,\frac{I}{2\pi\rho}. \quad \blacktriangleleft \tag{3.93}$$

▶ **Example 3.26:** Solution to the integral form of Ampere's law for a volume current

A uniform cylindrical beam of charge having radius a carries current I in free space, and is aligned on the z-axis. Find $\mathbf{H}$ everywhere.

Solution: The volume current density of the beam is

$$\mathbf{J}(\mathbf{r}) = \begin{cases} \hat{\mathbf{z}}\dfrac{I}{\pi a^2}, & \rho \le a, \\ 0, & \rho > a. \end{cases}$$

Since $\mathbf{J}(\mathbf{r})$ depends only on ρ, we have

$$\mathbf{H}(\mathbf{r}) = H_\phi(\rho)\hat{\boldsymbol{\phi}}.$$

Take Γ as a circle of radius ρ_0 centered in the $z = 0$ plane. Then

$$\oint_\Gamma \mathbf{H} \cdot \mathbf{dl} = \int_0^{2\pi} H_\phi(\rho_0)\hat{\boldsymbol{\phi}} \cdot \hat{\boldsymbol{\phi}}\rho_0\, d\phi = 2\pi\rho_0 H_\phi(\rho_0)$$

and

$$\int_S \mathbf{J} \cdot \mathbf{dS} = \begin{cases} \displaystyle\int_0^{2\pi}\int_0^{\rho_0} \dfrac{I}{\pi a^2}\rho\, d\rho\, d\phi = I\dfrac{\rho_0^2}{a^2}, & \rho_0 \le a, \\ \displaystyle\int_0^{2\pi}\int_0^{a} \dfrac{I}{\pi a^2}\rho\, d\rho\, d\phi = I, & \rho_0 > a. \end{cases}$$

In view of the arbitrariness of ρ_0, we finally write

$$\mathbf{H}(\rho) = \begin{cases} \hat{\boldsymbol{\phi}}\dfrac{I\rho}{2\pi a^2}, & \rho \le a, \\ \hat{\boldsymbol{\phi}}\dfrac{I}{2\pi\rho}, & \rho > a. \end{cases} \tag{3.94}$$

At points outside the beam, the form of $\mathbf{H}$ matches that for a filamentary wire carrying current I on the z-axis (Example 3.25). ◀

▶ **Example 3.27:** Solution to the point form of Ampere's law for a volume current

Use the point form of Ampere's law to solve the problem of Example 3.26.

Solution: With $\mathbf{H}(\mathbf{r}) = H_\phi(\rho)\hat{\boldsymbol{\phi}}$ as in Example 3.26, we have

$$\nabla \times \mathbf{H} = \hat{\mathbf{z}}\frac{1}{\rho}\frac{\partial}{\partial\rho}(\rho H_\phi).$$

For $\rho \le a$, Ampere's law yields

$$\frac{1}{\rho}\frac{\partial}{\partial\rho}(\rho H_\phi) = \frac{I}{\pi a^2}$$

and hence by integration

$$H_\phi = \frac{I}{2\pi a^2}\rho + \frac{C_1}{\rho}.$$

Finiteness of H_ϕ as $\rho \to 0$ requires $C_1 = 0$. For $\rho > a$ we have

$$\frac{1}{\rho}\frac{\partial}{\partial\rho}(\rho H_\phi) = 0,$$

hence $H_\phi = C_2/\rho$. Continuity of tangential $\mathbf{H}$ at $\rho = a$ (§ 3.4.5) requires

$$\frac{I}{2\pi a} = \frac{C_2}{a}$$

and yields $C_2 = I/2\pi$. The answer is once again (3.94). ◄

3.4.2   The magnetic scalar potential

Under certain conditions, the equations of magnetostatics have the same form as those of electrostatics. If $\mathbf{J} = 0$ in a region V, the magnetostatic equations are

$$\nabla \times \mathbf{H}(\mathbf{r}) = 0, \tag{3.95}$$

$$\nabla \cdot \mathbf{B}(\mathbf{r}) = 0; \tag{3.96}$$

compare with (3.1)–(3.2) when $\rho = 0$. Using (3.95) we can define a *magnetic scalar potential* Φ_m:

$$\mathbf{H} = -\nabla\Phi_m. \tag{3.97}$$

The negative sign is chosen for consistency with (3.10). We can then define a magnetic potential difference between two points as

$$V_{m21} = -\int_{P_1}^{P_2} \mathbf{H} \cdot d\mathbf{l} = -\int_{P_1}^{P_2} -\nabla\Phi_m(\mathbf{r}) \cdot d\mathbf{l} = \int_{P_1}^{P_2} d\Phi_m(\mathbf{r}) = \Phi_m(\mathbf{r}_2) - \Phi_m(\mathbf{r}_1).$$

Unlike the electrostatic potential difference, V_{m21} is not unique. Consider Figure 3.16, which shows a plane passing through the cross-section of a wire carrying total current I. Although there is no current within the region V (external to the wire), Equation (3.91) still gives

$$\int_{\Gamma_2} \mathbf{H} \cdot d\mathbf{l} - \int_{\Gamma_3} \mathbf{H} \cdot d\mathbf{l} = I.$$

Thus

$$\int_{\Gamma_2} \mathbf{H} \cdot d\mathbf{l} = \int_{\Gamma_3} \mathbf{H} \cdot d\mathbf{l} + I,$$

and the integral $\int_\Gamma \mathbf{H} \cdot d\mathbf{l}$ is not path-independent. However,

$$\int_{\Gamma_1} \mathbf{H} \cdot d\mathbf{l} = \int_{\Gamma_2} \mathbf{H} \cdot d\mathbf{l}$$

since no current passes through the surface bounded by $\Gamma_1 - \Gamma_2$. So we can artificially impose uniqueness by demanding that no path cross a cut such as that indicated by the line L in the figure.

Because V_{m21} is not unique, the field $\mathbf{H}$ is nonconservative. In point form this is shown by the fact that $\nabla \times \mathbf{H}$ is not identically zero. We are not too concerned about energy-related implications of the nonconservative nature of $\mathbf{H}$; the electric point charge has no magnetic analogue that might fail to conserve potential energy if moved around in a magnetic field.

Assuming a linear, isotropic region where $\mathbf{B}(\mathbf{r}) = \mu(\mathbf{r})\mathbf{H}(\mathbf{r})$, we can substitute (3.97) into (3.96) and expand to obtain

$$\nabla\mu(\mathbf{r}) \cdot \nabla\Phi_m(\mathbf{r}) + \mu(\mathbf{r})\nabla^2\Phi_m(\mathbf{r}) = 0.$$

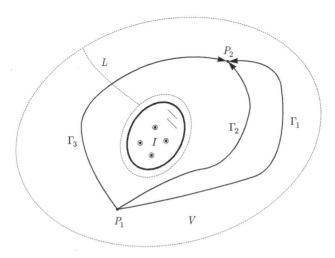

FIGURE 3.16
Magnetic potential.

For a homogeneous medium, this reduces to Laplace's equation

$$\nabla^2 \Phi_m = 0.$$

We can also obtain an analogue to Poisson's equation of electrostatics if we use

$$\mathbf{B} = \mu_0(\mathbf{H} + \mathbf{M}) = -\mu_0 \nabla \Phi_m + \mu_0 \mathbf{M}$$

in (3.96); we have

$$\nabla^2 \Phi_m = -\rho_M$$

where

$$\rho_M = -\nabla \cdot \mathbf{M}$$

is called the *equivalent magnetization charge density*. This form can be used to describe fields of permanent magnets in the absence of $\mathbf{J}$. Comparison with (3.73) shows that ρ_M is analogous to the polarization charge ρ_P.

Since Φ_m obeys Poisson's equation, the details regarding uniqueness and the construction of solutions follow from those of the electrostatic case. If we include the possibility of a surface density of magnetization charge, then the integral solution for Φ_m in unbounded space is

$$\Phi_m(\mathbf{r}) = \frac{1}{4\pi} \int_V \frac{\rho_M(\mathbf{r}')}{|\mathbf{r} - \mathbf{r}'|} dV' + \frac{1}{4\pi} \int_S \frac{\rho_{Ms}(\mathbf{r}')}{|\mathbf{r} - \mathbf{r}'|} dS'. \tag{3.98}$$

Here ρ_{Ms}, the surface density of magnetization charge, is identified as $\hat{\mathbf{n}} \cdot \mathbf{M}$ in the boundary condition (3.121).

3.4.3 The magnetic vector potential

Although the magnetic scalar potential is useful for describing fields of permanent magnets and for solving certain boundary value problems, it does not include the effects of source current. A second type of potential function, called the *magnetic vector potential*, can be used with complete generality to describe the magnetostatic field. Because $\nabla \cdot \mathbf{B} = 0$, we can write by (B.55)

$$\mathbf{B}(\mathbf{r}) = \nabla \times \mathbf{A}(\mathbf{r}) \tag{3.99}$$

where $\mathbf{A}$ is the vector potential. Now $\mathbf{A}$ is not determined by (3.99) alone, since the gradient of any scalar field can be added to $\mathbf{A}$ without changing the value of $\nabla \times \mathbf{A}$. Such "gauge transformations" are discussed in Chapter 5, where we find that $\nabla \cdot \mathbf{A}$ must also be specified for uniqueness of $\mathbf{A}$.

The vector potential can be used to develop a simple formula for the magnetic flux passing through an open surface S:

$$\Psi_m = \int_S \mathbf{B} \cdot d\mathbf{S} = \int_S (\nabla \times \mathbf{A}) \cdot d\mathbf{S} = \oint_\Gamma \mathbf{A} \cdot d\mathbf{l}, \tag{3.100}$$

where Γ is the contour bounding S.

In the linear isotropic case where $\mathbf{B} = \mu\mathbf{H}$, we can find a partial differential equation for $\mathbf{A}$ by substituting (3.99) into (3.92). Using (B.49) we have

$$\nabla \times \left[\frac{1}{\mu(\mathbf{r})} \nabla \times \mathbf{A}(\mathbf{r}) \right] = \mathbf{J}(\mathbf{r}),$$

hence

$$\frac{1}{\mu(\mathbf{r})} \nabla \times [\nabla \times \mathbf{A}(\mathbf{r})] - [\nabla \times \mathbf{A}(\mathbf{r})] \times \nabla \left(\frac{1}{\mu(\mathbf{r})} \right) = \mathbf{J}(\mathbf{r}).$$

In a homogeneous region, we have

$$\nabla \times (\nabla \times \mathbf{A}) = \mu \mathbf{J} \tag{3.101}$$

or

$$\nabla(\nabla \cdot \mathbf{A}) - \nabla^2 \mathbf{A} = \mu \mathbf{J} \tag{3.102}$$

by (B.53). As mentioned above, we must eventually specify $\nabla \cdot \mathbf{A}$. Although the choice is arbitrary, certain selections make the computation of $\mathbf{A}$ both mathematically tractable and physically meaningful. The "Coulomb gauge condition" $\nabla \cdot \mathbf{A} = 0$ reduces (3.102) to

$$\nabla^2 \mathbf{A} = -\mu \mathbf{J}. \tag{3.103}$$

The vector potential concept can also be applied to the Maxwell–Boffi magnetostatic equations

$$\nabla \times \mathbf{B} = \mu_0 (\mathbf{J} + \nabla \times \mathbf{M}), \tag{3.104}$$

$$\nabla \cdot \mathbf{B} = 0. \tag{3.105}$$

By (3.105) we may still define $\mathbf{A}$ through (3.99). Substituting this into (3.104), we have, under the Coulomb gauge,

$$\nabla^2 \mathbf{A} = -\mu_0 (\mathbf{J} + \mathbf{J}_M) \tag{3.106}$$

where $\mathbf{J}_M = \nabla \times \mathbf{M}$ is the magnetization current density.

3.4.3.1 Integral solution for the vector potential

The differential equations (3.103) and (3.106) are vector versions of Poisson's equation, and may be solved quite easily for unbounded space by decomposing the vector source into rectangular components. For instance, dotting (3.103) with $\hat{\mathbf{x}}$ we find that

$$\nabla^2 A_x = -\mu J_x.$$

This scalar version of Poisson's equation has solution

$$A_x(\mathbf{r}) = \frac{\mu}{4\pi} \int_V \frac{J_x(\mathbf{r}')}{|\mathbf{r} - \mathbf{r}'|} \, dV'$$

in unbounded space. Repeating this for each component and assembling the results, we obtain the solution for the vector potential in an unbounded homogeneous medium:

$$\mathbf{A}(\mathbf{r}) = \frac{\mu}{4\pi} \int_V \frac{\mathbf{J}(\mathbf{r}')}{|\mathbf{r} - \mathbf{r}'|} \, dV'. \tag{3.107}$$

Any surface sources can be easily included through a surface integral:

$$\mathbf{A}(\mathbf{r}) = \frac{\mu}{4\pi} \int_V \frac{\mathbf{J}(\mathbf{r}')}{|\mathbf{r} - \mathbf{r}'|} \, dV' + \frac{\mu}{4\pi} \int_S \frac{\mathbf{J}_s(\mathbf{r}')}{|\mathbf{r} - \mathbf{r}'|} \, dS'.$$

In unbounded free space containing materials represented by $\mathbf{M}$, we have

$$\mathbf{A}(\mathbf{r}) = \frac{\mu_0}{4\pi} \int_V \frac{\mathbf{J}(\mathbf{r}') + \mathbf{J}_M(\mathbf{r}')}{|\mathbf{r} - \mathbf{r}'|} \, dV' + \frac{\mu_0}{4\pi} \int_S \frac{\mathbf{J}_s(\mathbf{r}') + \mathbf{J}_{Ms}(\mathbf{r}')}{|\mathbf{r} - \mathbf{r}'|} \, dV' \tag{3.108}$$

where $\mathbf{J}_{Ms} = -\hat{\mathbf{n}} \times \mathbf{M}$ is the surface density of magnetization current as described in (3.122). It may be verified directly from (3.108) that $\nabla \cdot \mathbf{A} = 0$.

▶ **Example 3.28:** Magnetic field of a line current segment

Find $\mathbf{B}$ produced by a filamentary wire carrying a current I, extending from $-L$ to L along the z-axis within a material of permeability μ.

Solution: The volume current density is

$$\mathbf{J}(\mathbf{r}) = \hat{\mathbf{z}} \, I \frac{\delta(\rho)}{2\pi\rho}.$$

In cylindrical coordinates, we have $\mathbf{r} = z\hat{\mathbf{z}} + \rho\hat{\boldsymbol{\rho}}$ and $\mathbf{r}' = z'\hat{\mathbf{z}} + \rho'\hat{\boldsymbol{\rho}}'$, so that

$$|\mathbf{r} - \mathbf{r}'| = \sqrt{(z - z')^2 + \rho^2 + \rho'^2 - 2\rho\rho' \cos(\phi - \phi')}.$$

Substitution into (3.107) gives

$$\mathbf{A}(\mathbf{r}) = \hat{\mathbf{z}} \frac{\mu I}{4\pi} \int_{-L}^{L} \int_0^{2\pi} \int_0^{\infty} \frac{\delta(\rho')}{2\pi\rho'} \frac{\rho' \, d\rho' \, d\phi' \, dz'}{\sqrt{(z - z')^2 + \rho^2 + \rho'^2 - 2\rho\rho' \cos(\phi - \phi')}}$$

$$= \hat{\mathbf{z}} \frac{\mu I}{4\pi} \int_{-L}^{L} \left[\frac{dz'}{\sqrt{(z - z')^2 + \rho^2}} \right] \int_0^{2\pi} \frac{d\phi'}{2\pi} = \hat{\mathbf{z}} \frac{\mu I}{4\pi} \int_{z-L}^{z+L} \frac{du}{\sqrt{u^2 + \rho^2}}.$$

Rather than computing the integral and then taking the curl, we will differentiate under the integral sign:

$$\mathbf{B}(\mathbf{r}) = \nabla \times \mathbf{A}(\mathbf{r}) = -\hat{\phi}\,\frac{\partial A_z(\rho)}{\partial \rho} = -\hat{\phi}\,\frac{\mu I}{4\pi}\int_{z-L}^{z+L}\frac{d}{d\rho}\left[\frac{du}{\sqrt{u^2+\rho^2}}\right]$$

$$= \hat{\phi}\,\frac{\mu I}{4\pi}\int_{z-L}^{z+L}\frac{\rho}{(\rho^2+u^2)^{3/2}}\,du.$$

Integration gives

$$\mathbf{B}(\rho,z) = \hat{\phi}\,\frac{\mu I}{4\pi\rho}\left[\frac{z+L}{\sqrt{\rho^2+(z+L)^2}} - \frac{z-L}{\sqrt{\rho^2+(z-L)^2}}\right]. \quad \blacktriangleleft \qquad (3.109)$$

▶ **Example 3.29:** Magnetic field of an infinite line current

Find **H** produced by an infinitely long filamentary wire carrying a current I along the z-axis through a material of permeability μ.

Solution: We simply let $L \to \infty$ in (3.109):

$$\mathbf{B}(\mathbf{r}) = \lim_{L\to\infty}\hat{\phi}\,\frac{\mu I}{4\pi\rho}\left[\frac{\frac{z}{L}+1}{\sqrt{\left(\frac{\rho}{L}\right)^2+(\frac{z}{L}+1)^2}} - \frac{\frac{z}{L}-1}{\sqrt{\left(\frac{\rho}{L}\right)^2+(\frac{z}{L}-1)^2}}\right] = \hat{\phi}\,\frac{\mu I}{2\pi\rho}.$$

Using $\mathbf{B} = \mu\mathbf{H}$ we have (3.93) of Example 3.25. ◀

▶ **Example 3.30:** Vector potential of a circular loop current

Find **A** produced by a circular wire loop of radius a carrying a current I, centered in the $z = 0$ plane in a material of permeability μ (Figure 3.17).

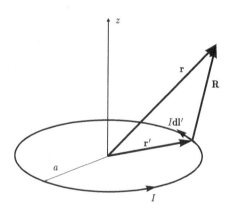

FIGURE 3.17
Circular loop of wire.

Solution: The volume current density is

$$\mathbf{J}(\mathbf{r}') = I\hat{\phi}'\delta(z')\delta(\rho'-a).$$

In cylindrical coordinates, we have

$$\mathbf{r} = z\hat{\mathbf{z}} + \rho\hat{\boldsymbol{\rho}}, \qquad \mathbf{r}' = z'\hat{\mathbf{z}} + \rho'\hat{\boldsymbol{\rho}}',$$

so that

$$|\mathbf{r} - \mathbf{r}'| = \sqrt{(z - z')^2 + \rho^2 + \rho'^2 - 2\rho\rho'\cos(\phi - \phi')}.$$

Substitution into (3.107) gives

$$\mathbf{A}(\mathbf{r}) = \frac{\mu I}{4\pi} \int_{-\infty}^{\infty} \int_{0}^{2\pi} \int_{0}^{\infty} \hat{\boldsymbol{\phi}}' \frac{\delta(z')\delta(\rho' - a)\rho'\,d\rho'\,d\phi'\,dz'}{\sqrt{(z - z')^2 + \rho^2 + \rho'^2 - 2\rho\rho'\cos(\phi - \phi')}}$$

$$= \frac{\mu I a}{4\pi} \int_{0}^{2\pi} \hat{\boldsymbol{\phi}}' \frac{d\phi'}{\sqrt{z^2 + \rho^2 + a^2 - 2\rho a\cos(\phi - \phi')}}.$$

Since $\hat{\boldsymbol{\rho}} \cdot \hat{\boldsymbol{\phi}}' = \sin(\phi - \phi')$, the integrand for A_ρ is odd in $\phi - \phi'$ and thus $A_\rho = 0$. Using $\hat{\boldsymbol{\phi}} \cdot \hat{\boldsymbol{\phi}}' = \cos(\phi - \phi')$ and the change of variable $u = \phi - \phi'$ gives the remaining component as

$$A_\phi(\mathbf{r}) = \frac{\mu I a}{4\pi} \int_{0}^{2\pi} \frac{\cos u\,du}{\sqrt{z^2 + \rho^2 + a^2 - 2\rho a\cos u}}.$$

We put the integral into standard form by setting $u = \pi - 2x$:

$$\mathbf{A}(\mathbf{r}) = -\frac{\mu I a}{4\pi} \hat{\boldsymbol{\phi}} \int_{-\pi/2}^{\pi/2} \frac{1 - 2\sin^2 x}{[\rho^2 + a^2 + z^2 + 2a\rho(1 - 2\sin^2 x)]^{1/2}}\, 2\,dx.$$

Letting

$$k^2 = \frac{4a\rho}{(a + \rho)^2 + z^2}, \qquad F^2 = (a + \rho)^2 + z^2,$$

we have

$$\mathbf{A}(\mathbf{r}) = -\frac{\mu I a}{4\pi} \hat{\boldsymbol{\phi}} \frac{4}{F} \int_{0}^{\pi/2} \frac{1 - 2\sin^2 x}{[1 - k^2\sin^2 x]^{1/2}}\, dx.$$

Then, since

$$\frac{1 - 2\sin^2 x}{[1 - k^2\sin^2 x]^{1/2}} = \frac{k^2 - 2}{k^2}[1 - k^2\sin^2 x]^{-1/2} + \frac{2}{k^2}[1 - k^2\sin^2 x]^{1/2},$$

we have

$$\mathbf{A}(\mathbf{r}) = \hat{\boldsymbol{\phi}} \frac{\mu I}{\pi k} \sqrt{\frac{a}{\rho}} \left[\left(1 - \frac{1}{2}k^2 \right) K(k^2) - E(k^2) \right]. \qquad (3.110)$$

Here

$$K(k^2) = \int_{0}^{\pi/2} \frac{du}{[1 - k^2\sin^2 u]^{1/2}}, \qquad E(k^2) = \int_{0}^{\pi/2} [1 - k^2\sin^2 u]^{1/2}\,du,$$

are complete elliptic integrals of the first and second kinds, respectively. ◄

▶ **Example 3.31:** Magnetic field of a circular loop current

Find $\mathbf{B}$ produced by a circular wire loop of radius a carrying a current I, centered in the $z = 0$ plane in a material of permeability μ.

Solution: We have $\mathbf{B} = \nabla \times \mathbf{A}$ where $\mathbf{A}$ is given by (3.110). From

$$\nabla \times \mathbf{A} = -\hat{\boldsymbol{\rho}}\frac{\partial A_\phi}{\partial z} + \hat{\mathbf{z}}\frac{1}{\rho}\frac{\partial}{\partial \rho}(\rho A_\phi)$$

we find $B_\phi = 0$. The remaining components are found by computing the indicated derivatives of A_ϕ. This process is somewhat tedious and is left to the reader as Problem 3.17. The results,

$$B_z = \frac{\mu I}{2\pi F} \left[\frac{a^2 - \rho^2 - z^2}{(a - \rho)^2 + z^2} E(k^2) + K(k^2) \right],$$

$$B_\rho = \frac{\mu I}{2\pi} \left(\frac{z}{\rho} \right) \frac{1}{F} \left[\frac{a^2 + \rho^2 + z^2}{(a - \rho)^2 + z^2} E(k^2) - K(k^2) \right],$$

should be compared with those of Example 3.34. ◄

3.4.3.2 Magnetic field of a small circular current loop

If in Example 3.31 the distance to the observation point is large compared to the size of the loop, then $r^2 = \rho^2 + z^2 \gg a^2$ and hence $k^2 \ll 1$. Using the expansions [41]

$$K(k^2) = \frac{\pi}{2} \left[1 + \frac{1}{4} k^2 + \frac{9}{64} k^4 + \cdots \right], \qquad E(k^2) = \frac{\pi}{2} \left[1 - \frac{1}{4} k^2 - \frac{3}{64} k^4 - \cdots \right],$$

in (3.110) and keeping the first nonzero term, we find that

$$\mathbf{A}(\mathbf{r}) \approx \hat{\phi} \frac{\mu I}{4\pi r^2} (\pi a^2) \sin \theta. \tag{3.111}$$

Defining the *magnetic dipole moment* of the loop as

$$\mathbf{m} = \hat{z} I \pi a^2,$$

we can write (3.111) as

$$\mathbf{A}(\mathbf{r}) = \frac{\mu}{4\pi} \frac{\mathbf{m} \times \hat{\mathbf{r}}}{r^2}.$$

Generalization to an arbitrarily oriented circular loop with center located at $\mathbf{r}_0$ is accomplished by writing $\mathbf{m} = \hat{n} I A$, where A is the loop area and $\hat{n}$ is normal to the loop in the right-hand sense. Then

$$\mathbf{A}(\mathbf{r}) = \frac{\mu}{4\pi} \mathbf{m} \times \frac{\mathbf{r} - \mathbf{r}_0}{|\mathbf{r} - \mathbf{r}_0|^3}.$$

We shall find, upon investigating the general multipole expansion of $\mathbf{A}$ below, that this holds for any planar loop.

The magnetic field of the loop can be found by direct application of (3.99). For the case $r^2 \gg a^2$, we take the curl of (3.111) and find that

$$\mathbf{B}(\mathbf{r}) = \frac{\mu}{4\pi} \frac{m}{r^3} (\hat{\mathbf{r}} \, 2 \cos \theta + \hat{\boldsymbol{\theta}} \, \sin \theta). \tag{3.112}$$

Comparison with (3.68) shows why we often refer to a small loop as a *magnetic dipole*. But (3.112) is approximate, and since there are no magnetic monopoles we cannot construct an exact magnetic analogue to the electric dipole. On the other hand, we shall find below that the multipole expansion of a finite-extent steady current begins with the dipole term (since the current must form closed loops). We may regard small loops as the elemental units of steady current from which all other currents may be constructed.

3.4.4 Multipole expansion

It is possible to derive a general multipole expansion for $\mathbf{A}$ analogous to (3.69). But the vector nature of $\mathbf{A}$ requires that we use vector spherical harmonics, hence the result is far more complicated than (3.69). A simpler approach yields the first few terms and requires only the Taylor expansion of $1/R$. Consider a steady current localized near the origin and contained within a sphere of radius r_m. We substitute the expansion (3.64) into (3.107) to obtain

$$\mathbf{A}(\mathbf{r}) = \frac{\mu}{4\pi} \int_V \mathbf{J}(\mathbf{r}') \left[\frac{1}{R}\bigg|_{r'=0} + (\mathbf{r}' \cdot \nabla') \frac{1}{R}\bigg|_{r'=0} + \frac{1}{2}(\mathbf{r}' \cdot \nabla')^2 \frac{1}{R}\bigg|_{r'=0} + \cdots \right] dV', \quad (3.113)$$

which we view as

$$\mathbf{A}(\mathbf{r}) = \mathbf{A}^{(0)}(\mathbf{r}) + \mathbf{A}^{(1)}(\mathbf{r}) + \mathbf{A}^{(2)}(\mathbf{r}) + \cdots .$$

The first term is merely

$$\mathbf{A}^{(0)}(\mathbf{r}) = \frac{\mu}{4\pi r} \int_V \mathbf{J}(\mathbf{r}') \, dV' = \frac{\mu}{4\pi r} \sum_{i=1}^{3} \hat{\mathbf{x}}_i \int_V J_i(\mathbf{r}') \, dV'$$

where $(x, y, z) = (x_1, x_2, x_3)$. However, by (3.6) each of the integrals is zero and we have

$$\mathbf{A}^{(0)}(\mathbf{r}) = 0;$$

the leading term in the multipole expansion of $\mathbf{A}$ for a general steady current distribution vanishes.

Using (3.66), we can write the second term as

$$\mathbf{A}^{(1)}(\mathbf{r}) = \frac{\mu}{4\pi r^3} \int_V \mathbf{J}(\mathbf{r}') \sum_{i=1}^{3} x_i x_i' \, dV' = \frac{\mu}{4\pi r^3} \sum_{j=1}^{3} \hat{\mathbf{x}}_j \sum_{i=1}^{3} x_i \int_V x_i' J_j(\mathbf{r}') \, dV'. \quad (3.114)$$

By adding the null relation (3.8) we can write

$$\int_V x_i' J_j \, dV' = \int_V x_i' J_j \, dV' + \int_V [x_i' J_j + x_j' J_i] \, dV' = 2 \int_V x_i' J_j \, dV' + \int_V x_j' J_i \, dV'$$

or

$$\int_V x_i' J_j \, dV' = \frac{1}{2} \int_V [x_i' J_j - x_j' J_i] \, dV'. \quad (3.115)$$

By this and (3.114), the second term in the multipole expansion is

$$\mathbf{A}^{(1)}(\mathbf{r}) = \frac{\mu}{4\pi r^3} \frac{1}{2} \int_V \sum_{j=1}^{3} \hat{\mathbf{x}}_j \sum_{i=1}^{3} x_i [x_i' J_j - x_j' J_i] \, dV' = -\frac{\mu}{4\pi r^3} \frac{1}{2} \int_V \mathbf{r} \times [\mathbf{r}' \times \mathbf{J}(\mathbf{r}')] \, dV'.$$

Defining the *dipole moment vector*

$$\mathbf{m} = \frac{1}{2} \int_V \mathbf{r} \times \mathbf{J}(\mathbf{r}) \, dV \quad (3.116)$$

we have

$$\mathbf{A}^{(1)}(\mathbf{r}) = \frac{\mu}{4\pi} \mathbf{m} \times \left(\frac{\hat{\mathbf{r}}}{r^2} \right) = -\frac{\mu}{4\pi} \mathbf{m} \times \nabla \frac{1}{r}. \quad (3.117)$$

This is the *dipole moment potential* for the steady current $\mathbf{J}$. Since steady currents of finite extent consist of loops, the dipole component is generally the first nonzero term in the expansion of $\mathbf{A}$. Higher-order components may be calculated, but extension of (3.113) beyond the dipole term is quite tedious and will not be attempted.

▶ **Example 3.32:** Dipole moment of a planar loop

Compute the dipole moment of the planar but otherwise arbitrary loop shown in Figure 3.18.

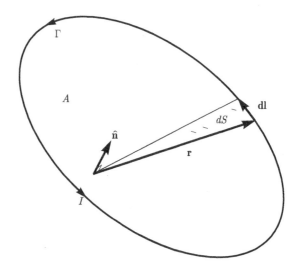

FIGURE 3.18
A planar wire loop.

Solution: Specializing (3.116) for a line current, we have

$$\mathbf{m} = \frac{I}{2} \oint_\Gamma \mathbf{r} \times d\mathbf{l}.$$

Examining Figure 3.18, we see that $\frac{1}{2}\mathbf{r} \times d\mathbf{l} = \hat{\mathbf{n}}\, dS$, where dS is the area of the sector swept out by $\mathbf{r}$ as it moves along $d\mathbf{l}$, and $\hat{\mathbf{n}}$ is the normal to the loop in the right-hand sense. Thus

$$\mathbf{m} = \hat{\mathbf{n}} I A \tag{3.118}$$

where A is the area of the loop. ◀

3.4.4.1 Physical interpretation of M in a magnetic material

In (3.108) we presented an expression for the vector potential produced by a magnetized material in terms of equivalent magnetization surface and volume currents. Suppose a magnetized medium is separated into volume regions with bounding surfaces across which the permeability is discontinuous. With $\mathbf{J}_M = \nabla \times \mathbf{M}$ and $\mathbf{J}_{Ms} = -\hat{\mathbf{n}} \times \mathbf{M}$, we obtain

$$\mathbf{A}(\mathbf{r}) = \frac{\mu_0}{4\pi} \int_V \frac{\mathbf{J}(\mathbf{r}')}{|\mathbf{r} - \mathbf{r}'|}\, dV' + \frac{\mu_0}{4\pi} \int_S \frac{\mathbf{J}_s(\mathbf{r}')}{|\mathbf{r} - \mathbf{r}'|}\, dS'$$

$$+ \sum_i \frac{\mu_0}{4\pi} \left[\int_{V_i} \frac{\nabla' \times \mathbf{M}(\mathbf{r}')}{|\mathbf{r} - \mathbf{r}'|}\, dV' + \int_{S_i} \frac{-\hat{\mathbf{n}}' \times \mathbf{M}(\mathbf{r}')}{|\mathbf{r} - \mathbf{r}'|}\, dS' \right].$$

Here $\hat{\mathbf{n}}$ points outward from region V_i. Using the curl theorem on the fourth term and employing (B.49), we have

$$\mathbf{A}(\mathbf{r}) = \frac{\mu_0}{4\pi} \int_V \frac{\mathbf{J}(\mathbf{r}')}{|\mathbf{r} - \mathbf{r}'|} \, dV' + \frac{\mu_0}{4\pi} \int_S \frac{\mathbf{J}_s(\mathbf{r}')}{|\mathbf{r} - \mathbf{r}'|} \, dS'$$
$$+ \sum_i \left[\frac{\mu_0}{4\pi} \int_{V_i} \mathbf{M}(\mathbf{r}') \times \nabla' \left(\frac{1}{|\mathbf{r} - \mathbf{r}'|} \right) dV' \right].$$

But $\nabla'(1/R) = \hat{\mathbf{R}}/R^2$, hence the third term is a sum of integrals of the form

$$\frac{\mu_0}{4\pi} \int_{V_i} \mathbf{M}(\mathbf{r}') \times \frac{\hat{\mathbf{R}}}{R^2} \, dV'.$$

Comparison with (3.117) shows that this integral represents a volume superposition of dipole moments where $\mathbf{M}$ is a volume density of magnetic dipole moments. Hence a magnetic material with permeability μ is equivalent to a volume distribution of magnetic dipoles in free space. As with our interpretation of the polarization vector in a dielectric, we base this conclusion only on Maxwell's equations and the assumption of a linear, isotropic relationship between $\mathbf{B}$ and $\mathbf{H}$.

3.4.5 Boundary conditions for the magnetostatic field

The boundary conditions found for the dynamic magnetic field remain valid in the magnetostatic case. Hence

$$\hat{\mathbf{n}}_{12} \times (\mathbf{H}_1 - \mathbf{H}_2) = \mathbf{J}_s \tag{3.119}$$

and

$$\hat{\mathbf{n}}_{12} \cdot (\mathbf{B}_1 - \mathbf{B}_2) = 0, \tag{3.120}$$

where $\hat{\mathbf{n}}_{12}$ points into region 1 from region 2. Since the magnetostatic curl and divergence equations are independent, so are the boundary conditions (3.119) and (3.120). We can also write (3.120) in terms of equivalent sources by (2.98):

$$\hat{\mathbf{n}}_{12} \cdot (\mathbf{H}_1 - \mathbf{H}_2) = \rho_{Ms1} + \rho_{Ms2}, \tag{3.121}$$

where $\rho_{Ms} = \hat{\mathbf{n}} \cdot \mathbf{M}$ is called the *equivalent magnetization surface charge density*. Here $\hat{\mathbf{n}}$ points outward from the material body.

For a linear, isotropic material described by $\mathbf{B} = \mu\mathbf{H}$, Equation (3.119) becomes

$$\hat{\mathbf{n}}_{12} \times \left(\frac{\mathbf{B}_1}{\mu_1} - \frac{\mathbf{B}_2}{\mu_2} \right) = \mathbf{J}_s.$$

With (2.98) we can also write (3.119) as

$$\hat{\mathbf{n}}_{12} \times (\mathbf{B}_1 - \mathbf{B}_2) = \mu_0 \left(\mathbf{J}_s + \mathbf{J}_{Ms1} + \mathbf{J}_{Ms2} \right) \tag{3.122}$$

where $\mathbf{J}_{Ms} = -\hat{\mathbf{n}} \times \mathbf{M}$ is the equivalent magnetization surface current density.

We may also write the boundary conditions in terms of the scalar or vector potential. Using $\mathbf{H} = -\nabla\Phi_m$, we can write (3.119) as

$$\Phi_{m1}(\mathbf{r}) = \Phi_{m2}(\mathbf{r}) \tag{3.123}$$

provided that the surface current $\mathbf{J}_s = 0$. As was the case with (3.16), the possibility of an additive constant here is generally ignored. To write (3.120) in terms of Φ_m we first note that $\mathbf{B}/\mu_0 - \mathbf{M} = -\nabla\Phi_m$; substitution into (3.120) gives

$$\frac{\partial\Phi_{m1}}{\partial n} - \frac{\partial\Phi_{m2}}{\partial n} = -\rho_{Ms1} - \rho_{Ms2} \tag{3.124}$$

where the normal derivative is taken in the direction of $\hat{\mathbf{n}}_{12}$. For a linear isotropic material where $\mathbf{B} = \mu\mathbf{H}$, we have

$$\mu_1\frac{\partial\Phi_{m1}}{\partial n} = \mu_2\frac{\partial\Phi_{m2}}{\partial n}. \tag{3.125}$$

Note that (3.123) and (3.125) are independent.

Boundary conditions on $\mathbf{A}$ may be derived using the approach of § 2.8.2. Consider Figure 2.6. Here the surface may carry either an electric surface current $\mathbf{J}_s$ or an equivalent magnetization current $\mathbf{J}_{Ms}$, and thus may be a surface of discontinuity between differing magnetic media. If we integrate $\nabla \times \mathbf{A}$ over the volume regions V_1 and V_2 and add the results, we find that

$$\int_{V_1} \nabla \times \mathbf{A}\, dV + \int_{V_2} \nabla \times \mathbf{A}\, dV = \int_{V_1+V_2} \mathbf{B}\, dV.$$

By the curl theorem,

$$\int_{S_1+S_2} \hat{\mathbf{n}} \times \mathbf{A}\, dS + \int_{S_{10}} -\hat{\mathbf{n}}_{10} \times \mathbf{A}_1\, dS + \int_{S_{20}} -\hat{\mathbf{n}}_{20} \times \mathbf{A}_2\, dS = \int_{V_1+V_2} \mathbf{B}\, dV,$$

where $\mathbf{A}_1$ is the field on the surface S_{10} and $\mathbf{A}_2$ is the field on S_{20}. As $\delta \to 0$, the surfaces S_1 and S_2 combine to give S. Also S_{10} and S_{20} coincide, as do the normals $\hat{\mathbf{n}}_{10} = -\hat{\mathbf{n}}_{20} = \hat{\mathbf{n}}_{12}$. Thus

$$\int_S (\hat{\mathbf{n}} \times \mathbf{A})\, dS - \int_V \mathbf{B}\, dV = \int_{S_{10}} \hat{\mathbf{n}}_{12} \times (\mathbf{A}_1 - \mathbf{A}_2)\, dS. \tag{3.126}$$

Now let us integrate over the entire volume region V including the surface of discontinuity. This gives

$$\int_S (\hat{\mathbf{n}} \times \mathbf{A})\, dS - \int_V \mathbf{B}\, dV = 0,$$

and for agreement with (3.126) we must have

$$\hat{\mathbf{n}}_{12} \times (\mathbf{A}_1 - \mathbf{A}_2) = 0.$$

A similar development shows that

$$\hat{\mathbf{n}}_{12} \cdot (\mathbf{A}_1 - \mathbf{A}_2) = 0.$$

Therefore $\mathbf{A}$ is continuous across a surface carrying electric or magnetization current.

3.4.6 Uniqueness of the magnetostatic field

Because the uniqueness conditions established for the dynamic field do not apply to magnetostatics, we begin with the magnetostatic field equations. Consider a region of space V bounded by a surface S. There may be source currents and magnetic materials both inside and outside V. Assume $(\mathbf{B}_1, \mathbf{H}_1)$ and $(\mathbf{B}_2, \mathbf{H}_2)$ are solutions to the magnetostatic field equations with source $\mathbf{J}$. We seek conditions under which $\mathbf{B}_1 = \mathbf{B}_2$ and $\mathbf{H}_1 = \mathbf{H}_2$.

The difference field $\mathbf{H}_0 = \mathbf{H}_2 - \mathbf{H}_1$ obeys $\nabla \times \mathbf{H}_0 = 0$. Using (B.50) we examine the quantity

$$\nabla \cdot (\mathbf{A}_0 \times \mathbf{H}_0) = \mathbf{H}_0 \cdot (\nabla \times \mathbf{A}_0) - \mathbf{A}_0 \cdot (\nabla \times \mathbf{H}_0) = \mathbf{H}_0 \cdot (\nabla \times \mathbf{A}_0)$$

where $\mathbf{A}_0$ is defined by $\mathbf{B}_0 = \mathbf{B}_2 - \mathbf{B}_1 = \nabla \times \mathbf{A}_0 = \nabla \times (\mathbf{A}_2 - \mathbf{A}_1)$. Integrating over V we obtain

$$\oint_S (\mathbf{A}_0 \times \mathbf{H}_0) \cdot d\mathbf{S} = \int_V \mathbf{H}_0 \cdot (\nabla \times \mathbf{A}_0)\, dV = \int_V \mathbf{H}_0 \cdot \mathbf{B}_0\, dV.$$

Then, since $(\mathbf{A}_0 \times \mathbf{H}_0) \cdot \hat{\mathbf{n}} = -\mathbf{A}_0 \cdot (\hat{\mathbf{n}} \times \mathbf{H}_0)$, we have

$$-\oint_S \mathbf{A}_0 \cdot (\hat{\mathbf{n}} \times \mathbf{H}_0) dS = \int_V \mathbf{H}_0 \cdot \mathbf{B}_0\, dV. \tag{3.127}$$

If $\mathbf{A}_0 = 0$ or $\hat{\mathbf{n}} \times \mathbf{H}_0 = 0$ everywhere on S, or $\mathbf{A}_0 = 0$ on part of S and $\hat{\mathbf{n}} \times \mathbf{H}_0 = 0$ on the remainder, then

$$\int_V \mathbf{H}_0 \cdot \mathbf{B}_0\, dS = 0. \tag{3.128}$$

So $\mathbf{H}_0 = 0$ or $\mathbf{B}_0 = 0$ by arbitrariness of V. Assuming $\mathbf{H}$ and $\mathbf{B}$ are linked by the constitutive relations, we have $\mathbf{H}_1 = \mathbf{H}_2$ and $\mathbf{B}_1 = \mathbf{B}_2$. The fields within V are unique provided that $\mathbf{A}$, the tangential component of $\mathbf{H}$, or some combination of the two, is specified over the bounding surface S.

One other condition will cause the left-hand side of (3.127) to vanish. If S recedes to infinity, then, provided that the potential functions vanish sufficiently fast, the condition (3.128) still holds and uniqueness is guaranteed. Equation (3.107) shows that $\mathbf{A} \sim 1/r$ as $r \to \infty$, hence $\mathbf{B}, \mathbf{H} \sim 1/r^2$. So uniqueness is ensured by the specification of $\mathbf{J}$ in unbounded space.

3.4.6.1 Integral solution for the vector potential

We have used the scalar Green's theorem to find a solution for the electrostatic potential within a region V in terms of the source charge in V and the values of the potential and its normal derivative on the boundary surface S. Analogously, we may find $\mathbf{A}$ within V in terms of the source current in V and the values of $\mathbf{A}$ and its derivatives on S. The vector relationship between $\mathbf{B}$ and $\mathbf{A}$ complicates the derivation somewhat, requiring Green's second identity for vector fields.

Let $\mathbf{P}$ and $\mathbf{Q}$ be continuous with continuous first and second derivatives throughout V and on S. The divergence theorem shows that

$$\int_V \nabla \cdot [\mathbf{P} \times (\nabla \times \mathbf{Q})]\, dV = \int_S [\mathbf{P} \times (\nabla \times \mathbf{Q})] \cdot d\mathbf{S}.$$

By virtue of (B.50) we have

$$\int_V [(\nabla \times \mathbf{Q}) \cdot (\nabla \times \mathbf{P}) - \mathbf{P} \cdot (\nabla \times \{\nabla \times \mathbf{Q}\})]\, dV = \int_S [\mathbf{P} \times (\nabla \times \mathbf{Q})] \cdot d\mathbf{S}.$$

We now interchange $\mathbf{P}$ and $\mathbf{Q}$ and subtract the result from the above, obtaining

$$\int_V [\mathbf{Q} \cdot (\nabla \times \{\nabla \times \mathbf{P}\}) - \mathbf{P} \cdot (\nabla \times \{\nabla \times \mathbf{Q}\})]\, dV$$

$$= \int_S [\mathbf{P} \times (\nabla \times \mathbf{Q}) - \mathbf{Q} \times (\nabla \times \mathbf{P})] \cdot d\mathbf{S}. \tag{3.129}$$

Note that $\hat{\mathbf{n}}$ points outward from V. This is *Green's second identity for vector fields.*

Now assume that V contains a magnetic material of uniform permeability μ and set

$$\mathbf{P} = \mathbf{A}(\mathbf{r}'), \qquad \mathbf{Q} = \mathbf{c}/R,$$

in (3.129) written in terms of primed coordinates. Here $\mathbf{c}$ is a constant vector, nonzero but otherwise arbitrary. We first examine the volume integral terms. Note that

$$\nabla' \times (\nabla' \times \mathbf{Q}) = \nabla' \times \left(\nabla' \times \frac{\mathbf{c}}{R}\right) = -\nabla'^2\left(\frac{\mathbf{c}}{R}\right) + \nabla'\left[\nabla' \cdot \left(\frac{\mathbf{c}}{R}\right)\right].$$

By (B.59) and (3.38) we have

$$\nabla'^2\left(\frac{\mathbf{c}}{R}\right) = \frac{1}{R}\nabla'^2\mathbf{c} + \mathbf{c}\nabla'^2\left(\frac{1}{R}\right) + 2\left(\nabla'\frac{1}{R} \cdot \nabla'\right)\mathbf{c} = \mathbf{c}\nabla'^2\left(\frac{1}{R}\right) = -\mathbf{c}4\pi\delta(\mathbf{r} - \mathbf{r}'),$$

hence

$$\mathbf{P} \cdot [\nabla' \times (\nabla' \times \mathbf{Q})] = 4\pi\mathbf{c} \cdot \mathbf{A}\delta(\mathbf{r} - \mathbf{r}') + \mathbf{A} \cdot \nabla'\left[\nabla' \cdot \left(\frac{\mathbf{c}}{R}\right)\right].$$

Since $\nabla \cdot \mathbf{A} = 0$, the second term on the right-hand side can be rewritten using (B.48):

$$\nabla' \cdot (\psi\mathbf{A}) = \mathbf{A} \cdot (\nabla'\psi) + \psi\nabla' \cdot \mathbf{A} = \mathbf{A} \cdot (\nabla'\psi).$$

Thus

$$\mathbf{P} \cdot [\nabla' \times (\nabla' \times \mathbf{Q})] = 4\pi\mathbf{c} \cdot \mathbf{A}\delta(\mathbf{r} - \mathbf{r}') + \nabla' \cdot \left[\mathbf{A}\left\{\mathbf{c} \cdot \nabla'\left(\frac{1}{R}\right)\right\}\right],$$

where we have again used (B.48). The other volume integral term can be found by substituting from (3.101):

$$\mathbf{Q} \cdot [\nabla' \times (\nabla' \times \mathbf{P})] = \mu\frac{1}{R}\mathbf{c} \cdot \mathbf{J}(\mathbf{r}').$$

Next we investigate the surface integral terms. Consider

$$\hat{\mathbf{n}}' \cdot [\mathbf{P} \times (\nabla' \times \mathbf{Q})] = \hat{\mathbf{n}}' \cdot \left\{\mathbf{A} \times \left[\nabla' \times \left(\frac{\mathbf{c}}{R}\right)\right]\right\}$$

$$= \hat{\mathbf{n}}' \cdot \left\{\mathbf{A} \times \left[\frac{1}{R}\nabla' \times \mathbf{c} - \mathbf{c} \times \nabla'\left(\frac{1}{R}\right)\right]\right\}$$

$$= -\hat{\mathbf{n}}' \cdot \left\{\mathbf{A} \times \left[\mathbf{c} \times \nabla'\left(\frac{1}{R}\right)\right]\right\}.$$

This can be put in slightly different form via (B.8). Note that

$$(\mathbf{A} \times \mathbf{B}) \cdot (\mathbf{C} \times \mathbf{D}) = \mathbf{A} \cdot [\mathbf{B} \times (\mathbf{C} \times \mathbf{D})] = (\mathbf{C} \times \mathbf{D}) \cdot (\mathbf{A} \times \mathbf{B}) = \mathbf{C} \cdot [\mathbf{D} \times (\mathbf{A} \times \mathbf{B})],$$

hence

$$\hat{\mathbf{n}}' \cdot [\mathbf{P} \times (\nabla' \times \mathbf{Q})] = -\mathbf{c} \cdot \left[\nabla'\left(\frac{1}{R}\right) \times (\hat{\mathbf{n}}' \times \mathbf{A})\right].$$

The other surface term is given by

$$\hat{\mathbf{n}}' \cdot [\mathbf{Q} \times (\nabla' \times \mathbf{P})] = \hat{\mathbf{n}}' \cdot \left[\frac{\mathbf{c}}{R} \times (\nabla' \times \mathbf{A})\right] = \hat{\mathbf{n}}' \cdot \left(\frac{\mathbf{c}}{R} \times \mathbf{B}\right) = -\frac{\mathbf{c}}{R} \cdot (\hat{\mathbf{n}}' \times \mathbf{B}).$$

We can substitute each of the terms into (3.129) and obtain

$$\mu \mathbf{c} \cdot \int_V \frac{\mathbf{J}(\mathbf{r}')}{R} \, dV' - 4\pi \mathbf{c} \cdot \int_V \mathbf{A}(\mathbf{r}') \delta(\mathbf{r} - \mathbf{r}') \, dV' - \mathbf{c} \cdot \oint_S [\hat{\mathbf{n}}' \cdot \mathbf{A}(\mathbf{r}')] \nabla' \left(\frac{1}{R}\right) dS'$$

$$= -\mathbf{c} \cdot \oint_S \nabla' \left(\frac{1}{R}\right) \times [\hat{\mathbf{n}}' \times \mathbf{A}(\mathbf{r}')] \, dS' + \mathbf{c} \cdot \oint_S \frac{1}{R} \hat{\mathbf{n}}' \times \mathbf{B}(\mathbf{r}') \, dS'.$$

Since $\mathbf{c}$ is arbitrary, we can remove the dot products to obtain a vector equation. Then

$$\mathbf{A}(\mathbf{r}) = \frac{\mu}{4\pi} \int_V \frac{\mathbf{J}(\mathbf{r}')}{R} \, dV' - \frac{1}{4\pi} \oint_S \left\{ [\hat{\mathbf{n}}' \times \mathbf{A}(\mathbf{r}')] \times \nabla' \left(\frac{1}{R}\right) \right.$$

$$\left. + \frac{1}{R} \hat{\mathbf{n}}' \times \mathbf{B}(\mathbf{r}') + [\hat{\mathbf{n}}' \cdot \mathbf{A}(\mathbf{r}')] \nabla' \left(\frac{1}{R}\right) \right\} dS'. \tag{3.130}$$

We have expressed $\mathbf{A}$ in a closed region in terms of the sources within the region and the values of $\mathbf{A}$ and $\mathbf{B}$ on the surface. While uniqueness requires specification of *either* $\mathbf{A}$ or $\hat{\mathbf{n}} \times \mathbf{B}$ on S, the expression (3.130) includes *both* quantities. This is similar to (3.36) for electrostatic fields, which required both the scalar potential and its normal derivative.

The reader may be troubled by the fact that we require $\mathbf{P}$ and $\mathbf{Q}$ to be somewhat well behaved, then proceed to involve the singular function $\mathbf{c}/R$ and integrate over the singularity. We choose this approach to simplify the presentation; a more rigorous approach which excludes the singular point with a small sphere also gives (3.130). This approach was used in § 3.3.5.3 to establish (3.38). The interested reader should see Stratton [183] for details on the application of this technique to obtain (3.130).

It is interesting to note that as $S \to \infty$, the surface integral vanishes, since $\mathbf{A} \sim 1/r$ and $\mathbf{B} \sim 1/r^2$, and we recover (3.107). Moreover, (3.130) returns the null result when evaluated at points outside S (see Stratton [183]). We shall see this again when studying the integral solutions for electrodynamic fields in § 6.1.3.

Finally, with

$$\mathbf{Q} = \nabla' \left(\frac{1}{R}\right) \times \mathbf{c}$$

we can find an integral expression for $\mathbf{B}$ within an enclosed region, representing a generalization of the Biot–Savart law. However, this case will be covered in the more general development of § 6.1.1.

3.4.6.2 The Biot–Savart law

We can obtain an expression for $\mathbf{B}$ in unbounded space by performing the curl operation directly on the vector potential:

$$\mathbf{B}(\mathbf{r}) = \nabla \times \frac{\mu}{4\pi} \int_V \frac{\mathbf{J}(\mathbf{r}')}{|\mathbf{r} - \mathbf{r}'|} \, dV' = \frac{\mu}{4\pi} \int_V \nabla \times \frac{\mathbf{J}(\mathbf{r}')}{|\mathbf{r} - \mathbf{r}'|} \, dV'.$$

Using (B.49) and $\nabla \times \mathbf{J}(\mathbf{r}') = 0$, we have

$$\mathbf{B}(\mathbf{r}) = -\frac{\mu}{4\pi} \int_V \mathbf{J}(\mathbf{r}') \times \nabla \frac{1}{|\mathbf{r} - \mathbf{r}'|} \, dV'.$$

The *Biot–Savart law*

$$\mathbf{B}(\mathbf{r}) = \frac{\mu}{4\pi} \int_V \mathbf{J}(\mathbf{r}') \times \frac{\hat{\mathbf{R}}}{R^2} \, dV' \tag{3.131}$$

follows from (3.37).

▶ **Example 3.33:** Magnetic field of an infinite line current from the Biot-Savart law

Find the magnetic field of an infinitely long filamentary wire carrying a current I along the z-axis through a material of permeability μ.

Solution: The volume current density is

$$\mathbf{J}(\mathbf{r}) = \hat{\mathbf{z}}\, I \frac{\delta(\rho)}{2\pi\rho}.$$

In cylindrical coordinates, $\mathbf{r} = z\hat{\mathbf{z}} + \rho\hat{\boldsymbol{\rho}}$ and $\mathbf{r}' = z'\hat{\mathbf{z}} + \rho'\hat{\boldsymbol{\rho}}'$, so that

$$\mathbf{R} = \mathbf{r} - \mathbf{r}' = (z - z')\hat{\mathbf{z}} + (\rho\cos\phi - \rho'\cos\phi')\hat{\mathbf{x}} + (\rho\sin\phi - \rho'\sin\phi')\hat{\mathbf{y}},$$

and

$$R = |\mathbf{r} - \mathbf{r}'| = \sqrt{(z - z')^2 + \rho^2 + \rho'^2 - 2\rho\rho'\cos(\phi - \phi')}.$$

Substitution into (3.131) gives

$$\mathbf{B}(\mathbf{r}) = \frac{\mu I}{4\pi} \int_{-\infty}^{\infty} \int_{0}^{2\pi} \int_{0}^{\infty} \frac{\delta(\rho')}{2\pi\rho'} \cdot$$

$$\cdot \frac{(\rho\cos\phi - \rho'\cos\phi')\hat{\mathbf{y}} - (\rho\sin\phi - \rho'\sin\phi')\hat{\mathbf{x}}}{[(z-z')^2 + \rho^2 + \rho'^2 - 2\rho\rho'\cos(\phi - \phi')]^{3/2}} \rho'\, d\rho'\, d\phi'\, dz'$$

$$= \frac{\mu I}{4\pi} \int_{-\infty}^{\infty} \int_{0}^{2\pi} \frac{\rho(\hat{\mathbf{y}}\cos\phi - \hat{\mathbf{x}}\sin\phi)}{[(z-z')^2 + \rho^2]^{3/2}} \frac{d\phi'}{2\pi}\, dz'$$

$$= \hat{\boldsymbol{\phi}} \frac{\mu I \rho}{4\pi} \int_{-\infty}^{\infty} \frac{du}{[u^2 + \rho^2]^{3/2}} \int_{0}^{2\pi} \frac{d\phi'}{2\pi}$$

$$= \hat{\boldsymbol{\phi}} \frac{\mu I}{2\pi\rho},$$

which is identical to the results of Examples 3.25 and 3.29. ◀

We use delta-functions in the example above to represent the line current. Alternatively, we can replace $\mathbf{J}\, dV'$ by $I\, d\mathbf{l}'$ and obtain

$$\mathbf{B}(\mathbf{r}) = I \frac{\mu}{4\pi} \int_{\Gamma} d\mathbf{l}' \times \frac{\hat{\mathbf{R}}}{R^2}. \tag{3.132}$$

For an infinitely long line current on the z-axis, we then have the more direct result

$$\mathbf{B}(\mathbf{r}) = I \frac{\mu}{4\pi} \int_{-\infty}^{\infty} \hat{\mathbf{z}} \times \frac{\hat{\mathbf{z}}(z - z') + \hat{\boldsymbol{\rho}}\rho}{[(z-z')^2 + \rho^2]^{3/2}}\, dz'$$

$$= \hat{\boldsymbol{\phi}} \frac{\mu I}{2\pi\rho}. \tag{3.133}$$

▶ **Example 3.34:** Magnetic field of a circular loop current from the Biot–Savart law

Find $\mathbf{B}$ produced by a circular wire loop of radius a carrying a current I, centered in the

$z = 0$ plane within a material of permeability μ (Figure 3.17).

Solution: In cylindrical coordinates, $\mathbf{r} = z\hat{\mathbf{z}} + \rho\hat{\boldsymbol{\rho}}$ and $\mathbf{r}' = a\hat{\boldsymbol{\rho}}'$ so that

$$\mathbf{R} = \mathbf{r} - \mathbf{r}' = z\hat{\mathbf{z}} + (\rho\cos\phi - a\cos\phi')\hat{\mathbf{x}} + (\rho\sin\phi - a\sin\phi')\hat{\mathbf{y}},$$

and

$$R = |\mathbf{r} - \mathbf{r}'| = \sqrt{z^2 + \rho^2 + a^2 - 2a\rho\cos(\phi - \phi')}.$$

Since $\mathbf{dl}' = a\,d\phi'\hat{\boldsymbol{\phi}}' = a\,d\phi'(-\hat{\mathbf{x}}\sin\phi' + \hat{\mathbf{y}}\cos\phi')$, we also have

$$\begin{aligned}
\mathbf{dl}' \times \mathbf{R} &= a\,d\phi'\left(-\hat{\mathbf{x}}\sin\phi' + \hat{\mathbf{y}}\cos\phi'\right) \times \\
&\quad \left[z\hat{\mathbf{z}} + (\rho\cos\phi - a\cos\phi')\hat{\mathbf{x}} + (\rho\sin\phi - a\sin\phi')\hat{\mathbf{y}}\right] \\
&= \hat{\boldsymbol{\rho}}a\,d\phi'z\cos(\phi - \phi') + \hat{\boldsymbol{\phi}}a\,d\phi'z\sin(\phi - \phi') + \hat{\mathbf{z}}a\,d\phi'\left[a - \rho\cos(\phi - \phi')\right].
\end{aligned}$$

Substitution into (3.132) yields

$$B_\phi = \frac{\mu I}{4\pi}az\int_0^2 \pi\frac{\sin u\,du}{[z^2 + \rho^2 + a^2 - 2a\rho\cos u]^{3/2}} = 0.$$

We also have

$$B_z = \frac{\mu I}{4\pi}a\int_0^2 \pi\frac{a - \rho\cos u\,du}{[z^2 + \rho^2 + a^2 - 2a\rho\cos u]^{3/2}},$$

$$B_\rho = \frac{\mu I}{4\pi}az\int_0^2 \pi\frac{\cos u\,du}{[z^2 + \rho^2 + a^2 - 2a\rho\cos u]^{3/2}}.$$

These last two integrals may be converted to standard form via the substitution $u = \pi - 2x$:

$$B_z = \frac{\mu I}{2\pi F}\left[\frac{a^2 - \rho^2 - z^2}{(a - \rho)^2 + z^2}E(k^2) + K(k^2)\right],$$

$$B_\rho = \frac{\mu I}{2\pi}\left(\frac{z}{\rho}\right)\frac{1}{F}\left[\frac{a^2 + \rho^2 + z^2}{(a - \rho)^2 + z^2}E(k^2) - K(k^2)\right],$$

exactly as in Example 3.31. The details are left as Problem 3.17. Here the quantities k and F are defined in Example 3.30, as are the elliptic integrals $E(k^2)$ and $K(k^2)$. ◄

▶ **Example 3.35:** Lower bound on the magnetic field of a filamentary loop current using the Biot–Savart law

A steady current I flows around a planar filamentary loop C. Assuming the loop is *star-shaped* with respect to a given point P inside C (i.e., that every ray leaving P cuts C in a single point), show that the magnitude of the magnetic field at P satisfies

$$H_P = |\mathbf{H}(\mathbf{r} = \mathbf{r}_P)| \geq \frac{I}{2}\sqrt{\frac{\pi}{A}}$$

where A is the loop area.

Solution: Without loss of generality we can place the loop in the x-y plane with the origin of coordinates at the point P (Figure 3.19). Then $\mathbf{R} = \mathbf{r} - \mathbf{r}' = -\mathbf{r}'$, and the Biot–Savart law states

$$\mathbf{H} = \frac{1}{4\pi}\oint_C \frac{I\mathbf{dl}' \times \mathbf{R}}{|\mathbf{R}|^3}.$$

We note that $\mathbf{dl'} \times \mathbf{R} = \hat{\mathbf{z}} R \, dl' \sin \alpha$ and so

$$H_P = \frac{I}{4\pi} \oint_C \frac{\sin \alpha \, dl'}{R^2}.$$

But since from Figure 3.19 we see that $dl' \sin \alpha = R \, d\phi'$,

$$H_P = \frac{I}{4\pi} \int_0^{2\pi} \frac{d\phi'}{R}.$$

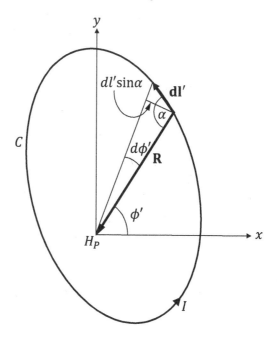

FIGURE 3.19
A star-shaped filamentary current loop.

To establish the bound on H_P we need *Hölder's inequality for integrals*, which states that

$$\int_a^b f(x) g(x) \, dx \le \left(\int_a^b f^p(x) \, dx \right)^{1/p} \left(\int_a^b g^q(x) \, dx \right)^{1/q},$$

where $p > 1$ and q is its *conjugate exponent* satisfying $p^{-1} + q^{-1} = 1$. Setting $p = 3/2$ and $q = 3$, we have

$$2\pi = \int_0^{2\pi} d\phi' = \int_0^{2\pi} \left(\frac{1}{R} \right)^{2/3} (R^2)^{1/3} \, d\phi'$$

$$\le \left(\int_0^{2\pi} \frac{d\phi'}{R} \right)^{2/3} \left(\int_0^{2\pi} R^2 \, d\phi' \right)^{1/3} \le \left(\frac{4\pi H_P}{I} \right)^{2/3} (2A)^{1/3},$$

which yields the result. By invoking the condition for equality in Hölder's inequality, we can show that equality holds in the present situation if and only if R is constant: i.e., C is a circle. ◄

3.4.7 Force and energy

3.4.7.1 Ampere force on a system of currents

If a steady current $\mathbf{J}(\mathbf{r})$ occupying a region V is exposed to a magnetic field, the force on the moving charge is given by the Lorentz force law

$$d\mathbf{F}(\mathbf{r}) = \mathbf{J}(\mathbf{r}) \times \mathbf{B}(\mathbf{r}). \qquad (3.134)$$

This can be integrated to give the total force on the current distribution:

$$\mathbf{F} = \int_V \mathbf{J}(\mathbf{r}) \times \mathbf{B}(\mathbf{r}) \, dV.$$

It is apparent that the charge flow composing a steady current must be constrained in some way, or the Lorentz force will accelerate the charge and destroy the steady nature of the current. This constraint is often provided by a conducting wire.

▶ **Example 3.36:** Magnetic force on a cylindrical wire

An infinitely long wire of circular cross-section is centered on the z-axis in free space and carries a current I. Find the force density within the wire and integrate to find the total force on the wire.

Solution: If current I is uniformly distributed over the cross-section, then the current density is

$$\mathbf{J}(\mathbf{r}) = \begin{cases} \hat{\mathbf{z}} \dfrac{I}{\pi a^2}, & \rho \leq a, \\ 0, & \rho > a, \end{cases}$$

where a is the wire radius. Thus, the magnetic field produced by the wire is identical to the field produced by the beam of charge in Example 3.26:

$$\mathbf{B}(\rho) = \begin{cases} \hat{\boldsymbol{\phi}} \dfrac{\mu_0 I \rho}{2\pi a^2}, & \rho \leq a, \\ \hat{\boldsymbol{\phi}} \dfrac{\mu_0 I}{2\pi \rho}, & \rho > a. \end{cases} \qquad (3.135)$$

The force density within the wire,

$$d\mathbf{F} = \mathbf{J} \times \mathbf{B} = -\hat{\boldsymbol{\rho}} \frac{\mu_0 I^2 \rho}{2\pi^2 a^4},$$

is directed inward and tends to compress the wire. Volume integration gives zero net force because $\int_0^{2\pi} \hat{\boldsymbol{\rho}} \, d\phi = 0$. ◀

▶ **Example 3.37:** Magnetic force on a split cylindrical wire

The cylindrical wire of Example 3.36 is split in half axially. Find the force on the top half, described by $0 \leq \phi \leq \pi$.

Solution: Using $\hat{\boldsymbol{\rho}} = \hat{\mathbf{x}} \cos\phi + \hat{\mathbf{y}} \sin\phi$, we obtain

$$F_x = -\frac{\mu_0 I^2}{2\pi^2 a^4} \int dz \int_0^a \rho^2 \, d\rho \int_0^\pi \cos\phi \, d\phi = 0$$

and

$$F_y = -\frac{\mu_0 I^2}{2\pi^2 a^4} \int dz \int_0^a \rho^2 \, d\rho \int_0^\pi \sin\phi \, d\phi$$

$$= -\frac{\mu_0 I^2}{3\pi^2 a} \int dz.$$

The force per unit length

$$\frac{\mathbf{F}}{l} = -\hat{\mathbf{y}} \frac{\mu_0 I^2}{3\pi^2 a} \tag{3.136}$$

is directed toward the bottom half as expected. The force on the bottom half is directed oppositely by Newton's third law. ◄

If the wire takes the form of a loop carrying current I, then (3.134) becomes

$$d\mathbf{F}(\mathbf{r}) = I d\mathbf{l}(\mathbf{r}) \times \mathbf{B}(\mathbf{r}) \tag{3.137}$$

and the total force acting is

$$\mathbf{F} = I \oint_\Gamma d\mathbf{l}(\mathbf{r}) \times \mathbf{B}(\mathbf{r}).$$

We can write the force on $\mathbf{J}$ in terms of the current producing $\mathbf{B}$. Assuming this latter current $\mathbf{J}'$ occupies region V', the Biot–Savart law (3.131) yields

$$\mathbf{F} = \frac{\mu}{4\pi} \int_V \mathbf{J}(\mathbf{r}) \times \int_{V'} \mathbf{J}(\mathbf{r}') \times \frac{\mathbf{r} - \mathbf{r}'}{|\mathbf{r} - \mathbf{r}'|^3} \, dV' \, dV.$$

This can be specialized to describe the force between line currents. Assume current 1, following a path Γ_1 along the direction $d\mathbf{l}$, carries current I_1, while current 2, following path Γ_2 along the direction $d\mathbf{l}'$, carries current I_2. Then the force on current 1 is

$$\mathbf{F}_1 = I_1 I_2 \frac{\mu}{4\pi} \oint_{\Gamma_1} \oint_{\Gamma_2} d\mathbf{l} \times \left(d\mathbf{l}' \times \frac{\mathbf{r} - \mathbf{r}'}{|\mathbf{r} - \mathbf{r}'|^3} \right).$$

This equation, known as *Ampere's force law*, can be cast in a better form for computational purposes. We use (B.7) and $\nabla(1/R)$ from (3.37):

$$\mathbf{F}_1 = I_1 I_2 \frac{\mu}{4\pi} \oint_{\Gamma_2} d\mathbf{l}' \oint_{\Gamma_1} d\mathbf{l} \cdot \nabla' \left(\frac{1}{|\mathbf{r} - \mathbf{r}'|} \right) - I_1 I_2 \frac{\mu}{4\pi} \oint_{\Gamma_1} \oint_{\Gamma_2} (d\mathbf{l} \cdot d\mathbf{l}') \frac{\mathbf{r} - \mathbf{r}'}{|\mathbf{r} - \mathbf{r}'|^3}.$$

The first term involves an integral of a perfect differential about a closed path, producing a null result. Thus

$$\mathbf{F}_1 = -I_1 I_2 \frac{\mu}{4\pi} \oint_{\Gamma_1} \oint_{\Gamma_2} (d\mathbf{l} \cdot d\mathbf{l}') \frac{\mathbf{r} - \mathbf{r}'}{|\mathbf{r} - \mathbf{r}'|^3}.$$

▶ **Example 3.38:** Force on parallel wires

Find the force between two parallel wires separated by a distance d and carrying currents I_1 and I_2 in a medium with permeability μ (Figure 3.20).

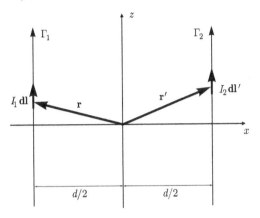

FIGURE 3.20
Parallel current-carrying wires.

Solution: To find the force on wire 1 we set

$$\mathbf{r} = -\tfrac{1}{2}d\,\hat{\mathbf{z}} + z\,\hat{\mathbf{z}}, \qquad \mathbf{r}' = \tfrac{1}{2}d\,\hat{\mathbf{z}} + z'\,\hat{\mathbf{z}}, \qquad \mathbf{dl} = \hat{\mathbf{z}}\,dz, \qquad \mathbf{dl}' = \hat{\mathbf{z}}\,dz'.$$

Thus,

$$\mathbf{r} - \mathbf{r}' = -d\hat{\mathbf{x}} + (z - z')\hat{\mathbf{z}}, \qquad |\mathbf{r} - \mathbf{r}'| = \sqrt{d^2 + (z - z')^2}$$

and

$$\mathbf{F}_1 = -I_1 I_2 \frac{\mu}{4\pi} \int \left[\int_{-\infty}^{\infty} \frac{-d\hat{\mathbf{x}} + (z - z')\hat{\mathbf{z}}}{[d^2 + (z - z')^2]^{3/2}} \, dz' \right] dz.$$

The z-component of the force is

$$F_{1z} = -I_1 I_2 \frac{\mu}{4\pi} \int \left[\int_{-\infty}^{\infty} \frac{u}{[d^2 + u^2]^{3/2}} \, dz' \right] dz = 0$$

because the integrand is odd. The x-component is

$$F_{1x} = I_1 I_2 \frac{\mu d}{4\pi} \int \left[\int_{-\infty}^{\infty} \frac{1}{[d^2 + u^2]^{3/2}} \, du \right] dz$$

$$= I_1 I_2 \frac{\mu d}{4\pi} \left[\frac{u}{d^2\sqrt{d^2 + u^2}} \bigg|_{-\infty}^{\infty} \right] \int dz = I_1 I_2 \frac{\mu}{2\pi d} \hat{\mathbf{x}} \int dz,$$

so the force per unit length is

$$\frac{\mathbf{F}_1}{l} = \hat{\mathbf{x}} \, I_1 I_2 \frac{\mu}{2\pi d}. \tag{3.138}$$

The force is attractive if $I_1 I_2 \geq 0$ (i.e., if the currents flow in the same direction). ◄

3.4.7.2 Maxwell's stress tensor

The magnetostatic version of the stress tensor can be obtained from (2.225) by setting $\mathbf{E} = \mathbf{D} = 0$:

$$\bar{\mathbf{T}}_m = \tfrac{1}{2}(\mathbf{B} \cdot \mathbf{H})\bar{\mathbf{I}} - \mathbf{BH}.$$

The total magnetic force on the current in a region V surrounded by surface S is given by

$$\mathbf{F}_m = -\oint_S \bar{\mathbf{T}}_m \cdot \mathbf{dS} = \int_V \mathbf{f}_m \, dV$$

where $\mathbf{f}_m = \mathbf{J} \times \mathbf{B}$ is the magnetic force volume density.

▶ **Example 3.39:** Force on parallel wires from Maxwell's stress tensor

Find the force between two parallel wires separated by a distance d and carrying identical currents I in a medium with permeability μ. This is the case shown in Figure 3.20 with $I_1 = I_2 = I$.

Solution: The force on the wire at $x = -d/2$ can be computed by integrating $\bar{\mathbf{T}}_m \cdot \hat{\mathbf{n}}$ over the yz-plane with $\hat{\mathbf{n}} = \hat{\mathbf{x}}$. Using (3.133) we see that in this plane the total magnetic field is

$$\mathbf{B} = \frac{\mu I}{2\pi} \frac{-\hat{\mathbf{x}} \sin \xi + \hat{\mathbf{y}} \cos \xi}{\sqrt{y^2 + \left(\frac{d}{2}\right)^2}} + \frac{\mu I}{2\pi} \frac{-\hat{\mathbf{x}} \sin \xi - \hat{\mathbf{y}} \cos \xi}{\sqrt{y^2 + \left(\frac{d}{2}\right)^2}} = -\hat{\mathbf{x}} \frac{\mu I}{\pi} \frac{\sin \xi}{\sqrt{y^2 + \left(\frac{d}{2}\right)^2}}$$

where

$$\sin \xi = \frac{y}{\sqrt{y^2 + \left(\frac{d}{2}\right)^2}}.$$

Thus,

$$\mathbf{B} = -\hat{\mathbf{x}} \mu \frac{I}{\pi} \frac{y}{y^2 + d^2/4},$$

and

$$\bar{\mathbf{T}}_m \cdot \hat{\mathbf{n}} = \frac{1}{2} B_x \frac{B_x}{\mu} \hat{\mathbf{x}} - \hat{\mathbf{x}} B_x \frac{B_x}{\mu} = -\frac{1}{2} \frac{B_x^2}{\mu} \hat{\mathbf{x}} = -\mu_0 \frac{I^2}{2\pi^2} \frac{y^2}{[y^2 + d^2/4]^2} \hat{\mathbf{x}}.$$

Upon integration, we obtain

$$\mathbf{F}_1 = \mu \frac{I^2}{2\pi^2} \hat{\mathbf{x}} \int dz \int_{-\infty}^{\infty} \frac{y^2}{[y^2 + d^2/4]^2} \, dy$$

$$= \mu \frac{I^2}{2\pi^2} \hat{\mathbf{x}} \int dz \left[\frac{1}{d} \tan^{-1}\left(\frac{2y}{d}\right) - \frac{y}{2\left[y^2 + \left(\frac{d}{2}\right)^2\right]} \right]\Bigg|_{-\infty}^{\infty} = I^2 \frac{\mu}{2\pi d} \hat{\mathbf{x}} \int dz.$$

The resulting force per unit length agrees with (3.138) when $I_1 = I_2 = I$. ◀

3.4.7.3   Torque in a magnetostatic field

The torque exerted on a current-carrying conductor immersed in a magnetic field plays an important role in many engineering applications. If a rigid body is exposed to a force field of volume density $d\mathbf{F}(\mathbf{r})$, the torque on that body about a certain origin is given by

$$\mathbf{T} = \int_V \mathbf{r} \times d\mathbf{F} \, dV$$

where integration is performed over the body and $\mathbf{r}$ extends from the origin of torque. If the force arises from the interaction of a current with a magnetostatic field, then $d\mathbf{F} = \mathbf{J} \times \mathbf{B}$ and

$$\mathbf{T} = \int_V \mathbf{r} \times (\mathbf{J} \times \mathbf{B}) \, dV.$$

For a line current, we can replace $\mathbf{J} \, dV$ with $I\mathbf{dl}$ to obtain

$$\mathbf{T} = I \int_\Gamma \mathbf{r} \times (\mathbf{dl} \times \mathbf{B}). \tag{3.139}$$

▶ **Example 3.40:** Torque on a circular loop in a nonuniform magnetic field

A circular loop of radius a is centered in the $z = 0$ plane in free space, and immersed in a magnetic field given by $\mathbf{B} = \hat{\mathbf{y}} B_0 (x/a)^2$ in the region of the loop. Find the torque acting on the loop.

Solution: We have $\mathbf{r} = a\boldsymbol{\rho}$ and $\mathbf{dl} = a\,d\phi\,\hat{\boldsymbol{\phi}}$ so that by (B.7),

$$\mathbf{r} \times (\mathbf{dl} \times \mathbf{B}) = \mathbf{dl}(\mathbf{r} \cdot \mathbf{B}) - \mathbf{B}(\mathbf{r} \cdot \mathbf{dl}) = \mathbf{dl}(\mathbf{r} \cdot \mathbf{B}),$$

since $\mathbf{r} \cdot \mathbf{dl} = 0$. Next, note that $\mathbf{r} \cdot \mathbf{B} = a\hat{\boldsymbol{\rho}} \cdot \hat{\mathbf{y}} B_0 (x/a)^2 = B_0 a \sin\phi \cos^2\phi$. Hence,

$$\mathbf{r} \times (\mathbf{dl} \times \mathbf{B}) = \hat{\boldsymbol{\phi}} B_0 a^2\, d\phi \sin\phi \cos^2\phi.$$

Substitution into (3.139) gives

$$T_x = -I B_0 a^2 \int_0^{2\pi} \sin^2\phi \cos^2\phi\, d\phi = -\frac{I}{4} B_0 \pi a^2, \tag{3.140}$$

$$T_y = I B_0 a^2 \int_0^{2\pi} \sin\phi \cos^3\phi\, d\phi = 0.$$

The torque attempts to rotate the loop about the x-axis. ◀

If $\mathbf{B}$ is uniform, then by (B.7) we have

$$\mathbf{T} = \int_V [\mathbf{J}(\mathbf{r} \cdot \mathbf{B}) - \mathbf{B}(\mathbf{r} \cdot \mathbf{J})]\, dV.$$

The second term can be written as

$$\int_V \mathbf{B}(\mathbf{r} \cdot \mathbf{J})\, dV = \mathbf{B} \sum_{i=1}^3 \int_V x_i J_i\, dV = 0$$

where $(x_1, x_2, x_3) = (x, y, z)$, and where we have employed (3.7). Thus

$$\mathbf{T} = \int_V \mathbf{J}(\mathbf{r} \cdot \mathbf{B})\, dV = \sum_{j=1}^3 \hat{\mathbf{x}}_j \int_V J_j \sum_{i=1}^3 x_i B_i\, dV = \sum_{i=1}^3 B_i \sum_{j=1}^3 \hat{\mathbf{x}}_j \int_V J_j x_i\, dV.$$

We can replace the integral using (3.115) to get

$$\mathbf{T} = \frac{1}{2} \int_V \sum_{j=1}^3 \hat{\mathbf{x}}_j \sum_{i=1}^3 B_i [x_i J_j - x_j J_i]\, dV$$

$$= -\frac{1}{2} \int_V \mathbf{B} \times (\mathbf{r} \times \mathbf{J})\, dV.$$

Since $\mathbf{B}$ is uniform, we have, by (3.116),

$$\mathbf{T} = \mathbf{m} \times \mathbf{B} \tag{3.141}$$

where $\mathbf{m}$ is the dipole moment. For a planar loop, (3.118) yields $\mathbf{T} = I A \hat{\mathbf{n}} \times \mathbf{B}$.

▶ **Example 3.41:** Torque on a circular loop in a uniform magnetic field

Modify Example 3.40 for a loop in a uniform magnetic field $\mathbf{B} = \hat{\mathbf{y}}B_0$.

Solution: We have $\hat{\mathbf{n}} = \hat{\mathbf{z}}$, so $\mathbf{T} = IA\hat{\mathbf{n}} \times \mathbf{B} = I(\pi a^2)\hat{\mathbf{z}} \times \hat{\mathbf{y}}B_0 = -\hat{\mathbf{x}}IB_0\pi a^2$. Comparison with (3.140) shows that the field nonuniformity reduces the torque by a factor of four. ◀

3.4.7.4   Joule's law

In § 2.9.5 we showed that when a moving charge interacts with an electric field in a volume region V, energy is transferred between the field and the charge. If the source of that energy is outside V, the energy is carried into V as an energy flux over the boundary surface S. The energy balance described by Poynting's theorem (2.234) also holds for static fields supported by steady currents: we must simply recognize that we have no time-rate of change of stored energy. Thus

$$-\int_V \mathbf{J} \cdot \mathbf{E} \, dV = \oint_S (\mathbf{E} \times \mathbf{H}) \cdot \mathbf{dS}. \tag{3.142}$$

The term

$$P = -\int_V \mathbf{J} \cdot \mathbf{E} \, dV \tag{3.143}$$

describes the rate at which energy is supplied to the fields by the currents within V; we have $P > 0$ if there are sources within V that result in energy transferred to the fields, and $P < 0$ if there is energy transferred to the currents. The latter case occurs when there are conducting materials in V. Within these conductors,

$$P = -\int_V \sigma \mathbf{E} \cdot \mathbf{E} \, dV. \tag{3.144}$$

Here $P < 0$; energy is transferred from the fields to the currents, and from the currents into heat (i.e., into lattice vibrations via collisions). Equation (3.144) is called *Joule's law*, and the transfer of energy from the fields into heat is *Joule heating*. Joule's law is the power relationship for a conducting material.

▶ **Example 3.42:** Power balance for a wire segment

A straight section of conducting wire having circular cross-section of radius a carries a total current I uniformly distributed over its cross-section. The wire is centered on the z-axis and extends between the planes $z = 0, L$. Assume a potential difference V between the ends, and verify satisfaction of the power balance relation (3.142).

Solution: Using (3.135) we see that at the surface of the wire

$$\mathbf{H} = \hat{\boldsymbol{\phi}} \frac{I}{2\pi a}, \qquad \mathbf{E} = \hat{\mathbf{z}} \frac{V}{L}.$$

The corresponding Poynting flux is

$$\mathbf{E} \times \mathbf{H} = -\hat{\boldsymbol{\rho}} \frac{IV}{2\pi aL},$$

which, since it is $-\hat{\boldsymbol{\rho}}$ directed, implies that energy flows into wire volume through the curved side surface. The total power flowing into the wire is just

$$\oint_S (\mathbf{E} \times \mathbf{H}) \cdot d\mathbf{S} = \int_0^{2\pi} \int_0^L \left(-\hat{\rho} \frac{IV}{2\pi a L} \right) \cdot \hat{\rho} a \, d\phi \, dz = -IV,$$

as expected from circuit theory. The density of the power supplied to the fields within the wire by the current is

$$\mathbf{J} \cdot \mathbf{E} = \frac{IV}{\pi a^2 L},$$

and thus the total power supplied to the fields is

$$-\int_V \mathbf{J} \cdot \mathbf{E} \, dV = -\int_0^L \int_0^{2\pi} \int_0^a \frac{IV}{\pi a^2 L} \rho \, d\rho \, d\phi \, dz = -IV,$$

again as expected from circuit theory. Since the power supplied to the fields equals the flow of power across the boundary of the wire, the power balance relation (3.142) is verified. ◄

3.4.7.5 Stored magnetic energy

We have shown that the energy stored in a static charge distribution may be regarded as the "assembly energy" required to bring charges from infinity against the Coulomb force. By proceeding very slowly with this assembly, we are able to avoid any complications resulting from the motion of the charges.

Similarly, we may equate the energy stored in a steady current distribution to the energy required for its assembly from current filaments* brought in from infinity. However, the calculation of assembly energy is more complicated in this case: moving a current filament into the vicinity of existing filaments changes the total magnetic flux passing through the existing loops, regardless of how slowly we assemble the filaments. As described by Faraday's law, this change in flux must be associated with an induced emf, which will tend to change the current flowing in the filament (and any existing filaments) unless energy is expended to keep the current constant (by the application of a battery emf in the opposite direction). We therefore regard the assembly energy as consisting of two parts: (1) the energy required to bring a filament with *constant* current from infinity against the Ampere force, and (2) the energy required to keep the current in this filament, and any existing filaments, constant. We ignore the energy required to keep the steady current flowing through an isolated loop (i.e., the energy needed to overcome Joule losses).

We begin by computing the amount of energy required to bring a filament with current I from infinity to a given position within an applied magnetostatic field $\mathbf{B}(\mathbf{r})$. In this first step we assume that the field is supported by localized sources, hence vanishes at infinity, and that it will not be altered by the motion of the filament. The force on each small segment of the filament is given by Ampere's force law (3.137), and the total force is found by integration. Suppose an external agent displaces the filament incrementally from a starting position 1 to an ending position 2 along a vector $\delta\mathbf{r}$, as shown in Figure 3.21. The work required is

$$\delta W = -(I d\mathbf{l} \times \mathbf{B}) \cdot \delta\mathbf{r} = (I d\mathbf{l} \times \delta\mathbf{r}) \cdot \mathbf{B}$$

for each segment of the wire. Figure 3.21 shows that $d\mathbf{l} \times \delta\mathbf{r}$ describes a small patch of surface area between the starting and ending positions of the filament, hence $-(d\mathbf{l} \times \delta\mathbf{r}) \cdot \mathbf{B}$

---

*Recall that a flux tube of a vector field is bounded by streamlines of the field. A current filament is a flux tube of current having vanishingly small, but nonzero, cross-section.

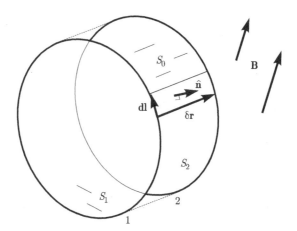

FIGURE 3.21

Calculation of work to move a filamentary loop in an applied magnetic field.

is the *outward* flux of **B** through the patch. Integrating over all segments composing the filament, we obtain

$$\Delta W = I \oint_{\Gamma} (\mathbf{dl} \times \delta\mathbf{r}) \cdot \mathbf{B} = -I \int_{S_0} \mathbf{B} \cdot \mathbf{dS}$$

for the total work required to displace the entire filament through $\delta\mathbf{r}$; here the surface S_0 is described by the superposition of all patches. If S_1 and S_2 are the surfaces bounded by the filament in its initial and final positions, respectively, then S_1, S_2, and S_0 taken together form a closed surface. The outward flux of **B** through this surface is

$$\oint_{S_0+S_1+S_2} \mathbf{B} \cdot \mathbf{dS} = 0$$

so that

$$\Delta W = -I \int_{S_0} \mathbf{B} \cdot \mathbf{dS} = I \int_{S_1+S_2} \mathbf{B} \cdot \mathbf{dS}$$

where $\hat{\mathbf{n}}$ is outward from the closed surface. Finally, let $\Psi_{1,2}$ be the flux of **B** through $S_{1,2}$ in the direction determined by **dl** and the right-hand rule. Then

$$\Delta W = -I(\Psi_2 - \Psi_1) = -I\Delta\Psi. \tag{3.145}$$

Now suppose that the initial position of the filament is at infinity. We bring the filament into a final position within the field **B** through a succession of small displacements, each requiring work (3.145). By superposition over all displacements, the total work is $W = -I(\Psi - \Psi_{\infty})$, where Ψ_{∞} and Ψ are the fluxes through the filament in its initial and final positions, respectively. However, since the source of the field is localized, we know that **B** is zero at infinity. Therefore $\Psi_{\infty} = 0$ and

$$W = -I\Psi = -I \int_S \mathbf{B} \cdot \hat{\mathbf{n}} \, dS \tag{3.146}$$

where $\hat{\mathbf{n}}$ is determined from **dl** in the right-hand sense.

Now let us find the work required to position two current filaments in a field-free region of space, starting with both filaments at infinity. Assume filament 1 carries current I_1 and filament 2 carries current I_2, and that we hold these currents constant as we move the filaments into position. We can think of assembling these filaments in two ways: by placing filament 1 first, or by placing filament 2 first. In either case, placing the first filament requires no work since (3.146) is zero. The work required to place the second filament is $W_1 = -I_1\Psi_1$ if filament 2 is placed first, where Ψ_1 is the flux passing through filament 1 in its final position, caused by the presence of filament 2. If filament 1 is placed first, the work required is $W_2 = -I_2\Psi_2$. Since the work cannot depend on which loop is placed first, we have $W_1 = W_2 = W$, where we can use either $W = -I_1\Psi_1$ or $W = -I_2\Psi_2$. It is even more convenient, as we shall see, to average these values and use

$$W = -\tfrac{1}{2}\left(I_1\Psi_1 + I_2\Psi_2\right). \tag{3.147}$$

We must determine the energy required to keep the currents constant as we move the filaments into position. When moving the first filament into place there is no induced emf, since no applied field is yet present. However, when moving the second filament into place we will change the flux linked by *both* the first and second loops. This change of flux will induce an emf in each of the loops, and this will change the current. To keep the current constant we must supply an opposing emf. Let dW_{emf}/dt be the rate of work required to keep the current constant. Then by (2.128) and (3.143) we have

$$\frac{dW_{\mathrm{emf}}}{dt} = -\int_V \mathbf{J}\cdot\mathbf{E}\,dV = -I\int \mathbf{E}\cdot\mathbf{dl} = -I\frac{d\Psi}{dt}.$$

Integrating, we find the total work ΔW required to keep the current constant in either loop as the flux through the loop is changed by an amount $\Delta\Psi$:

$$\Delta W_{\mathrm{emf}} = I\Delta\Psi.$$

So the total work required to keep I_1 constant as the loops are moved from infinity (where the flux is zero) to their final positions is $I_1\Psi_1$. Similarly, a total work $I_2\Psi_2$ is required to keep I_2 constant during the same process. Adding these to (3.147), the work required to position the loops, we obtain the complete assembly energy

$$W = \tfrac{1}{2}\left(I_1\Psi_1 + I_2\Psi_2\right)$$

for two filaments. The extension to N filaments is

$$W_m = \frac{1}{2}\sum_{n=1}^{N} I_n\Psi_n. \tag{3.148}$$

Consequently, the energy of a single current filament is

$$W_m = \tfrac{1}{2}I\Psi.$$

We may interpret this as the "assembly energy" required to bring the single loop into existence by bringing vanishingly small loops (magnetic dipoles) in from infinity. We may also interpret it as the energy required to establish the current in this single filament against the back emf. That is, if we establish I by slowly increasing the current from zero in N small steps $\Delta I = I/N$, an energy $\Psi_n\Delta I$ will be required at each step. Since Ψ_n increases proportionally to I, we have

$$W_m = \sum_{n=1}^{N} \frac{I}{N}\left[(n-1)\frac{\Psi}{N}\right]$$

where Ψ is the flux when the current is fully established. Since $\sum_{n=1}^{N}(n-1) = N(N-1)/2$ we obtain

$$W_m = \tfrac{1}{2}I\Psi$$

as $N \to \infty$.

A volume current $\mathbf{J}$ can be treated as though it were composed of N current filaments. Equations (3.100) and (3.148) give

$$W_m = \frac{1}{2}\sum_{n=1}^{N} I_n \oint_{\Gamma_n} \mathbf{A} \cdot \mathbf{dl}.$$

Since the total current is

$$I = \int_{CS} \mathbf{J} \cdot \mathbf{dS} = \sum_{n=1}^{N} I_n$$

where CS denotes the cross-section of the steady current, we have as $N \to \infty$

$$W_m = \frac{1}{2}\int_V \mathbf{A} \cdot \mathbf{J}\, dV. \tag{3.149}$$

Alternatively, using (3.107), we may write

$$W_m = \frac{1}{2}\int_V \int_V \frac{\mathbf{J}(\mathbf{r}) \cdot \mathbf{J}(\mathbf{r}')}{|\mathbf{r} - \mathbf{r}'|}\, dV\, dV'.$$

Note the similarity between (3.149) and (3.61). We now manipulate (3.149) into a form involving only the electromagnetic fields. By Ampere's law

$$W_m = \frac{1}{2}\int_V \mathbf{A} \cdot (\nabla \times \mathbf{H})\, dV.$$

Using (B.50) and the divergence theorem, we can write

$$W_m = \frac{1}{2}\oint_S (\mathbf{H} \times \mathbf{A}) \cdot \mathbf{dS} + \frac{1}{2}\int_V \mathbf{H} \cdot (\nabla \times \mathbf{A})\, dV.$$

We now let S expand to infinity. This does not change the value of W_m since we do not enclose any more current; however, since $\mathbf{A} \sim 1/r$ and $\mathbf{H} \sim 1/r^2$, the surface integral vanishes. Thus, remembering that $\nabla \times \mathbf{A} = \mathbf{B}$, we have

$$W_m = \frac{1}{2}\int_{V_\infty} \mathbf{H} \cdot \mathbf{B}\, dV \tag{3.150}$$

where V_∞ denotes all of space.

Although we do not provide a derivation, (3.150) is also valid within linear materials. For nonlinear materials, the total energy required to build up a magnetic field from $\mathbf{B}_1$ to $\mathbf{B}_2$ is

$$W_m = \frac{1}{2}\int_{V_\infty} \left[\int_{\mathbf{B}_1}^{\mathbf{B}_2} \mathbf{H} \cdot d\mathbf{B} \right] dV. \tag{3.151}$$

This accounts for the work required to drive a ferromagnetic material through its hysteresis loop. Readers interested in a complete derivation of (3.151) should consult Stratton [183].

▶ **Example 3.43:** Magnetic energy stored between coaxial cylinders

Two thin-walled, coaxial, current-carrying cylinders have radii a, b $(b > a)$. The intervening region is a linear magnetic material having permeability μ. Assume that the inner and outer conductors carry total currents I in the $\pm z$ directions, respectively. Compute the stored magnetic energy per unit length.

Solution: From the large-scale form of Ampere's law we find that

$$\mathbf{H} = \begin{cases} 0, & \rho \leq a, \\ \hat{\phi} \, I/2\pi\rho, & a \leq \rho \leq b, \\ 0, & \rho > b, \end{cases} \tag{3.152}$$

hence by (3.150)

$$W_m = \frac{1}{2} \int dz \int_0^{2\pi} \int_a^b \frac{\mu I^2}{(2\pi\rho)^2} \rho \, d\rho \, d\phi.$$

Integration gives the energy value per unit length:

$$\frac{W_m}{l} = \mu \frac{I^2}{4\pi} \ln(b/a). \quad \blacktriangleleft \tag{3.153}$$

▶ **Example 3.44:** Magnetic energy stored between and within coaxial cylinders

Consider the coaxial cylinders described in Example 3.43, except that the inner cylinder is solid and current is spread uniformly throughout. Compute the stored magnetic energy per unit length.

Solution: The field between the cylinders is still given by (3.152) but within the inner conductor we have

$$\mathbf{H} = \hat{\phi} \, \frac{I\rho}{2\pi a^2}$$

by (3.135). Thus, to (3.153) we must add the energy

$$\frac{W_{m,\text{inside}}}{l} = \frac{1}{2} \int_0^{2\pi} \int_0^a \frac{\mu_0 I^2 \rho^2}{(2\pi a^2)^2} \rho \, d\rho \, d\phi$$

$$= \mu_0 \frac{I^2}{16\pi}$$

stored within the solid wire. The result is

$$\frac{W_m}{l} = \mu_0 \frac{I^2}{4\pi} [\mu_r \ln(b/a) + \tfrac{1}{4}]. \quad \blacktriangleleft$$

3.4.8 Magnetic field of a permanently magnetized body

We now have the tools necessary to compute the magnetic field produced by a permanent magnet (a body with permanent magnetization $\mathbf{M}$). As an example, we shall find the field due to a uniformly magnetized ball in three different ways: by computing the vector potential integral and taking the curl, by computing the scalar potential integral and taking the gradient, and by finding the scalar potential using separation of variables and applying the boundary condition across the surface of the sphere.

▶ **Example 3.45:** Magnetic field of uniformly magnetized ball by curl of vector potential

Consider a ball of radius a, residing in free space and having permanent magnetization $\mathbf{M}(\mathbf{r}) = M_0\hat{\mathbf{z}}$. Find the magnetic field by computing the curl of the vector potential.

Solution: The equivalent magnetization current and charge densities are

$$\mathbf{J}_M = \nabla \times \mathbf{M} = 0,$$

$$\mathbf{J}_{Ms} = -\hat{\mathbf{n}} \times \mathbf{M} = -\hat{\mathbf{r}} \times M_0\hat{\mathbf{z}} = M_0\hat{\boldsymbol{\phi}}\sin\theta,$$

$$\rho_M = -\nabla \cdot \mathbf{M} = 0,$$

$$\rho_{Ms} = \hat{\mathbf{n}} \cdot \mathbf{M} = \hat{\mathbf{r}} \cdot M_0\hat{\mathbf{z}} = M_0\cos\theta. \tag{3.154}$$

The vector potential is produced by the equivalent magnetization surface current. Using (3.108) we find that

$$\mathbf{A}(\mathbf{r}) = \frac{\mu_0}{4\pi}\int_S \frac{\mathbf{J}_{Ms}}{|\mathbf{r}-\mathbf{r}'|}\,dS' = \frac{\mu_0}{4\pi}\int_{-\pi}^{\pi}\int_0^{\pi}\frac{M_0\hat{\boldsymbol{\phi}}'\sin\theta'}{|\mathbf{r}-\mathbf{r}'|}\sin\theta'\,d\theta'\,d\phi'.$$

Since $\hat{\boldsymbol{\phi}}' = -\hat{\mathbf{x}}\sin\phi' + \hat{\mathbf{y}}\cos\phi'$, the rectangular components of $\mathbf{A}$ are

$$\begin{Bmatrix} -A_x \\ A_y \end{Bmatrix} = \frac{\mu_0}{4\pi}\int_{-\pi}^{\pi}\int_0^{\pi}\frac{M_0 \begin{matrix} \sin\phi' \\ \cos\phi' \end{matrix}\sin\theta'}{|\mathbf{r}-\mathbf{r}'|}a^2\sin\theta'\,d\theta'\,d\phi'.$$

The integrals are most easily computed via the spherical harmonic expansion (E.204) for the inverse distance $|\mathbf{r}-\mathbf{r}'|^{-1}$:

$$\begin{Bmatrix} -A_x \\ A_y \end{Bmatrix} = \mu_0 M_0 a^2 \sum_{n=0}^{\infty}\sum_{m=-n}^{n}\frac{Y_{nm}(\theta,\phi)}{2n+1}\frac{r_<^n}{r_>^{n+1}}\int_{-\pi}^{\pi}\int_0^{\pi}\begin{matrix} \sin\phi' \\ \cos\phi' \end{matrix}\sin^2\theta' Y_{nm}^*(\theta',\phi')\,d\theta'\,d\phi'.$$

All the integrals vanish except when $n = 1$, $m = \pm 1$. Since

$$Y_{1,-1}(\theta,\phi) = \sqrt{\frac{3}{8\pi}}\sin\theta e^{-j\phi},$$

$$Y_{1,1}(\theta,\phi) = -\sqrt{\frac{3}{8\pi}}\sin\theta e^{j\phi},$$

we have

$$\begin{Bmatrix} -A_x \\ A_y \end{Bmatrix} = \mu_0 M_0 \frac{a^2}{3}\frac{r_<}{r_>^2}\frac{3}{8\pi}\sin\theta\int_0^{\pi}\sin^3\theta'\,d\theta'$$

$$\cdot\left[e^{-j\phi}\int_{-\pi}^{\pi}\begin{matrix} \sin\phi' \\ \cos\phi' \end{matrix}e^{j\phi'}\,d\phi' + e^{j\phi}\int_{-\pi}^{\pi}\begin{matrix} \sin\phi' \\ \cos\phi' \end{matrix}e^{-j\phi'}\,d\phi'\right].$$

Carrying out the integrals, we find that

$$\begin{Bmatrix} -A_x \\ A_y \end{Bmatrix} = \mu_0 M_0 \frac{a^2}{3}\frac{r_<}{r_>^2}\sin\theta\begin{Bmatrix} \sin\phi \\ \cos\phi \end{Bmatrix}$$

or

$$\mathbf{A} = \mu_0 M_0 \frac{a^2}{3}\frac{r_<}{r_>^2}\sin\theta\,\hat{\boldsymbol{\phi}}.$$

Finally, $\mathbf{B} = \nabla \times \mathbf{A}$ gives

$$\mathbf{B} = \begin{cases} \dfrac{2\mu_0 M_0}{3}\hat{\mathbf{z}}, & r < a, \\[3mm] \dfrac{\mu_0 M_0 a^3}{3r^3}(\hat{\mathbf{r}}\, 2\cos\theta + \hat{\boldsymbol{\theta}}\sin\theta), & r > a. \end{cases} \tag{3.155}$$

Within the sphere, $\mathbf{B}$ is uniform and in the same direction as $\mathbf{M}$, while outside it has the form of the magnetic dipole field with moment $m = \frac{4}{3}\pi a^3 M_0$. ◄

▶ **Example 3.46:** Magnetic field of uniformly magnetized ball by gradient of scalar potential found from potential integral

Consider a ball of radius a, residing in free space and having permanent magnetization $\mathbf{M}(\mathbf{r}) = M_0\hat{\mathbf{z}}$. Find the magnetic field everywhere by computing the gradient of the scalar potential found using the potential integral (3.98).

Solution: The equivalent magnetization surface charge density ρ_{Ms} is given by (3.154). Substituting into (3.98), we have

$$\Phi_m(\mathbf{r}) = \frac{1}{4\pi}\int_S \frac{\rho_{Ms}(\mathbf{r}')}{|\mathbf{r}-\mathbf{r}'|}\, dS' = \frac{1}{4\pi}\int_{-\pi}^{\pi}\int_0^{\pi} \frac{M_0\cos\theta'}{|\mathbf{r}-\mathbf{r}'|}\sin\theta'\, d\theta'\, d\phi'.$$

This integral has the form of (3.75) with $f(\theta) = M_0\cos\theta$. Thus, from (3.77),

$$\Phi_m(\mathbf{r}) = M_0\frac{a^2}{3}\cos\theta\frac{r_<}{r_>^2}. \tag{3.156}$$

The magnetic field $\mathbf{H}$ is then

$$\mathbf{H} = -\nabla\Phi_m = \begin{cases} -\dfrac{M_0}{3}\hat{\mathbf{z}}, & r < a, \\[3mm] \dfrac{M_0 a^3}{3r^3}(\hat{\mathbf{r}}\, 2\cos\theta + \hat{\boldsymbol{\theta}}\sin\theta), & r > a. \end{cases}$$

Inside the sphere $\mathbf{B}$ is given by $\mathbf{B} = \mu_0(\mathbf{H}+\mathbf{M})$, while outside it is merely $\mathbf{B} = \mu_0\mathbf{H}$. These observations lead again to (3.155). ◄

▶ **Example 3.47:** Magnetic field of uniformly magnetized ball by gradient of scalar potential found from solution to Laplace's equation

Consider a ball of radius a, residing in free space and having permanent magnetization $\mathbf{M}(\mathbf{r}) = M_0\hat{\mathbf{z}}$. Find the magnetic field everywhere by computing the gradient of the scalar potential found by solving Laplace's equation.

Solution: Since the scalar potential obeys Laplace's equation both inside and outside the sphere, Φ_m may be found as the separation of variables solution discussed in § A.5.3. We can repeat our earlier arguments for the dielectric sphere in an impressed electric field (§ 3.3.11). Copying Equations (3.82) and (3.83), we can write for $r \le a$

$$\Phi_{m1}(r,\theta) = \sum_{n=0}^{\infty} A_n r^n P_n(\cos\theta), \tag{3.157}$$

and for $r \geq a$

$$\Phi_{m2}(r, \theta) = \sum_{n=0}^{\infty} B_n r^{-(n+1)} P_n(\cos \theta). \qquad (3.158)$$

The boundary condition (3.123) at $r = a$ requires that

$$\sum_{n=0}^{\infty} A_n a^n P_n(\cos \theta) = \sum_{n=0}^{\infty} B_n a^{-(n+1)} P_n(\cos \theta);$$

upon application of the orthogonality of the Legendre functions, this becomes

$$A_n a^n = B_n a^{-(n+1)}. \qquad (3.159)$$

We can write (3.124) as

$$-\frac{\partial \Phi_{m1}}{\partial r} + \frac{\partial \Phi_{m2}}{\partial r} = -\rho_{Ms}$$

so that at $r = a$

$$-\sum_{n=0}^{\infty} A_n n a^{n-1} P_n(\cos \theta) - \sum_{n=0}^{\infty} B_n (n+1) a^{-(n+2)} P_n(\cos \theta) = -M_0 \cos \theta.$$

After application of orthogonality, this becomes

$$A_1 + 2B_1 a^{-3} = M_0, \qquad (3.160)$$

$$na^{n-1} A_n = -(n+1) B_n a^{-(n+2)} \qquad (n \neq 1). \qquad (3.161)$$

Solving (3.159) and (3.160) simultaneously for $n = 1$, we find

$$A_1 = \frac{M_0}{3}, \qquad B_1 = \frac{M_0}{3} a^3.$$

We also see that (3.159) and (3.161) are inconsistent unless $A_n = B_n = 0$, $n \neq 1$. Substituting these results into (3.157) and (3.158), we have

$$\Phi_m = \begin{cases} \dfrac{M_0}{3} r \cos \theta, & r \leq a, \\[2mm] \dfrac{M_0}{3} \dfrac{a^3}{r^2} \cos \theta, & r \geq a, \end{cases}$$

which is (3.156). The gradient operation reproduces (3.155). ◄

3.5   Static field theorems

3.5.1   Mean value theorem of electrostatics

The average value of the electrostatic potential over a sphere is equal to the potential at the center of the sphere, provided that the sphere encloses no electric charge. To see this, write

$$\Phi(\mathbf{r}) = \frac{1}{4\pi\epsilon} \int_V \frac{\rho(\mathbf{r}')}{R} \, dV' + \frac{1}{4\pi} \oint_S \left[-\Phi(\mathbf{r}') \frac{\hat{\mathbf{R}}}{R^2} + \frac{\nabla' \Phi(\mathbf{r}')}{R} \right] \cdot d\mathbf{S}';$$

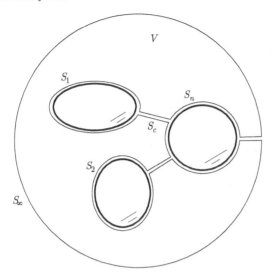

FIGURE 3.22
System of conductors used to derive Thomson's theorem.

put $\rho \equiv 0$ in V, and use the obvious facts that if S is a sphere centered at point $\mathbf{r}$ then (1) R is constant on S and (2) $\hat{\mathbf{n}}' = -\hat{\mathbf{R}}$:

$$\Phi(\mathbf{r}) = \frac{1}{4\pi R^2} \oint_S \Phi(\mathbf{r}')\, dS' - \frac{1}{4\pi R} \oint_S \mathbf{E}(\mathbf{r}') \cdot \mathbf{dS}'.$$

The last term vanishes by Gauss's law, giving the desired result.

3.5.2 Earnshaw's theorem

It is impossible for a charge to rest in stable equilibrium under the influence of electrostatic forces alone. This is an easy consequence of the mean value theorem of electrostatics, which precludes the existence of a point where Φ can assume a maximum or a minimum.

3.5.3 Thomson's theorem

Static charge on a system of perfect conductors distributes itself so that the electric stored energy is a minimum. Figure 3.22 shows a system of n conducting bodies held at potentials $\Phi_1, \ldots, \Phi_n$. Suppose the potential field associated with the actual distribution of charge on these bodies is Φ, giving

$$W_e = \frac{\epsilon}{2} \int_V \mathbf{E} \cdot \mathbf{E}\, dV = \frac{\epsilon}{2} \int_V \nabla\Phi \cdot \nabla\Phi\, dV$$

for the actual stored energy. Now assume a slightly different charge distribution, resulting in a new potential $\Phi' = \Phi + \delta\Phi$ that satisfies the same boundary conditions (i.e., assume $\delta\Phi = 0$ on each conducting body). The stored energy associated with this hypothetical situation is

$$W_e' = W_e + \delta W_e = \frac{\epsilon}{2} \int_V \nabla(\Phi + \delta\Phi) \cdot \nabla(\Phi + \delta\Phi)\, dV$$

so that

$$\delta W_e = \epsilon \int_V \nabla \Phi \cdot \nabla(\delta \Phi) \, dV + \frac{\epsilon}{2} \int_V |\nabla(\delta \Phi)|^2 \, dV;$$

Thomson's theorem will be proved if we can show that

$$\int_V \nabla \Phi \cdot \nabla(\delta \Phi) \, dV = 0, \qquad (3.162)$$

because then we shall have

$$\delta W_e = \frac{\epsilon}{2} \int_V |\nabla(\delta \Phi)|^2 \, dV \geq 0.$$

To establish (3.162), we use Green's first identity

$$\int_V (\nabla u \cdot \nabla v + u \nabla^2 v) \, dV = \oint_S u \nabla v \cdot \mathbf{dS}$$

with $u = \delta \Phi$ and $v = \Phi$:

$$\int_V \nabla \Phi \cdot \nabla(\delta \Phi) \, dV = \oint_S \delta \Phi \, \nabla \Phi \cdot \mathbf{dS}.$$

Here S is composed of (1) the exterior surfaces S_k ($k = 1, \ldots, n$) of the n bodies, (2) the surfaces S_c of the "cuts" that are introduced in order to keep V a simply-connected region (a condition for the validity of Green's identity), and (3) the sphere S_∞ of very large radius r. Thus

$$\int_V \nabla \Phi \cdot \nabla(\delta \Phi) \, dV = \sum_{k=1}^n \int_{S_k} \delta \Phi \, \nabla \Phi \cdot \mathbf{dS} + \int_{S_c} \delta \Phi \, \nabla \Phi \cdot \mathbf{dS} + \int_{S_\infty} \delta \Phi \, \nabla \Phi \cdot \mathbf{dS}.$$

The first term on the right vanishes because $\delta \Phi = 0$ on each S_k. The second term vanishes because the contributions from opposite sides of each cut cancel (note that $\hat{\mathbf{n}}$ occurs in pairs that are oppositely directed). The third term vanishes because $\Phi \sim 1/r$, $\nabla \Phi \sim 1/r^2$, and $dS \sim r^2$ where $r \to \infty$ for points on S_∞.

Thomson's theorem is used to find the voltage across series capacitors in Example 3.49.

3.5.4 Green's reciprocation theorem

Consider a system of n conducting bodies as in Figure 3.23. An associated mathematical surface S_t consists of the exterior surfaces $S_1, \ldots, S_n$ of the n bodies, taken together with a surface S that enclosed all of the bodies. Suppose Φ and Φ' are electrostatic potentials produced by two distinct distributions of stationary charge over the set of conductors. Then $\nabla^2 \Phi = 0 = \nabla^2 \Phi'$, and Green's second identity gives

$$\oint_{S_t} \left(\Phi \frac{\partial \Phi'}{\partial n} - \Phi' \frac{\partial \Phi}{\partial n} \right) dS = 0$$

or

$$\sum_{k=1}^n \int_{S_k} \Phi \frac{\partial \Phi'}{\partial n} \, dS + \int_S \Phi \frac{\partial \Phi'}{\partial n} \, dS = \sum_{k=1}^n \int_{S_k} \Phi' \frac{\partial \Phi}{\partial n} \, dS + \int_S \Phi' \frac{\partial \Phi}{\partial n} \, dS.$$

Now let S be a sphere of very large radius R so that at points on S we have

$$\Phi, \Phi' \sim \frac{1}{R}, \qquad \frac{\partial \Phi}{\partial n}, \frac{\partial \Phi'}{\partial n} \sim \frac{1}{R^2}, \qquad dS \sim R^2;$$

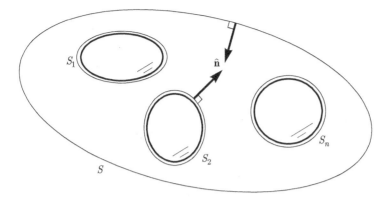

FIGURE 3.23
System of conductors used to derive Green's reciprocation theorem.

then, as $R \to \infty$,

$$\sum_{k=1}^{n} \int_{S_k} \Phi \frac{\partial \Phi'}{\partial n} \, dS = \sum_{k=1}^{n} \int_{S_k} \Phi' \frac{\partial \Phi}{\partial n} \, dS.$$

Furthermore, the conductors are equipotentials so that

$$\sum_{k=1}^{n} \Phi_k \int_{S_k} \frac{\partial \Phi'}{\partial n} \, dS = \sum_{k=1}^{n} \Phi'_k \int_{S_k} \frac{\partial \Phi}{\partial n} \, dS$$

and we therefore have

$$\sum_{k=1}^{n} q'_k \Phi_k = \sum_{k=1}^{n} q_k \Phi'_k \tag{3.163}$$

where the kth conductor $(k = 1, \ldots, n)$ has potential Φ_k when it carries charge q_k, and has potential Φ'_k when it carries charge q'_k. This is *Green's reciprocation theorem*.

▶ **Example 3.48:** Charge induced on a grounded conductor

A total charge q is placed on a conductor, as shown in Figure 3.24, and as a result the potential of the conductor is V_1. Next, the conductor is grounded and a point charge q'_p is placed at the point P. Find the total charge q' induced on the grounded conductor.

Solution: Consider the two situations shown in Figure 3.24. In the first situation, compute the potential V_p at point P with the conductor ungrounded. In the second situation, model the point charge q_P near the grounded conductor as a very small conducting body designated as body 2 and located at point P in space. For the first situation, set $q_1 = q$, $q_2 = 0$, $\Phi_1 = V_1$, $\Phi_2 = V_P$, while for the second, set $q'_1 = q'$, $q'_2 = q'_P$, $\Phi'_1 = 0$, $\Phi'_2 = V'_P$. Substituting these into Green's reciprocation theorem (3.163) specialized to two bodies,

$$q'_1 \Phi_1 + q'_2 \Phi_2 = q_1 \Phi'_1 + q_2 \Phi'_2,$$

gives $q'V_1 + q'_P V_P = 0$ so that

$$q' = -q'_P V_P / V_1. \tag{3.164}$$

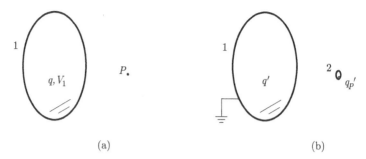

(a)   (b)

FIGURE 3.24
Application of Green's reciprocation theorem. (a) The "unprimed" situation. (b) The "primed" situation. ◄

3.6   Quasistatics

Subsequent chapters are concerned with the solutions to Maxwell's equations for generally time-varying fields (including the time harmonic case). Good approximations to time-varying fields can sometimes be achieved without resorting to complete solutions of Maxwell's equations. Important situations arise when the temporal variation of the field is slow compared to the transit time of the electromagnetic disturbance across an object. A classic example is AC circuit theory, where elements are said to be *lumped* when they are of sizes small compared to the wavelength of the AC signal. A rigorous, electromagnetics-based description of circuit theory is provided in the book by King [103], which also describes the circumstances under which the radiation properties of circuit elements may be neglected. The most extensive use of quasistatics is in circuit theory, but it also finds wide application in other areas, such as induction heating and eddy current braking.

In this and the following section we provide a short introduction to the topic of quasistatics and its application to shielding. For more extensive coverage of quasistatics, the interested reader is referred to the book by Haus and Melcher [83]. The two specific cases of *electro-quasistatics* and *magneto-quasistatics* are considered in turn.

3.6.1   Electro-quasistatics

An electromagnetic system is said to be *electro-quasistatic* (EQS) when it can be sufficiently treated using the point form equations

$$\nabla \times \mathbf{E}(\mathbf{r}, t) = 0, \tag{3.165}$$

$$\nabla \times \mathbf{H}(\mathbf{r}, t) = \mathbf{J}(\mathbf{r}, t) + \frac{\partial \mathbf{D}(\mathbf{r}, t)}{\partial t}, \tag{3.166}$$

$$\nabla \cdot \mathbf{D}(\mathbf{r}, t) = \rho(\mathbf{r}, t), \tag{3.167}$$

$$\nabla \cdot \mathbf{B}(\mathbf{r}, t) = 0, \tag{3.168}$$

$$\nabla \cdot \mathbf{J}(\mathbf{r}, t) + \frac{\partial \rho(\mathbf{r}, t)}{\partial t} = 0, \tag{3.169}$$

or the large-scale equations

$$\oint_\Gamma \mathbf{E}(\mathbf{r}, t) \cdot \mathbf{dl} = 0, \tag{3.170}$$

$$\oint_\Gamma \mathbf{H}(\mathbf{r}, t) \cdot \mathbf{dl} = \int_S \mathbf{J}(\mathbf{r}, t) \cdot \mathbf{dS} + \frac{d}{dt} \int_S \mathbf{D}(\mathbf{r}, t) \cdot \mathbf{dS}, \tag{3.171}$$

$$\oint_S \mathbf{D}(\mathbf{r}, t) \cdot \mathbf{dS} = \int_V \rho(\mathbf{r}, t) \, dV, \tag{3.172}$$

$$\oint_S \mathbf{B}(\mathbf{r}, t) \cdot \mathbf{dS} = 0, \tag{3.173}$$

$$\oint_S \mathbf{J}(\mathbf{r}, t) \cdot \mathbf{dS} = -\frac{d}{dt} \int_V \rho(\mathbf{r}, t) \, dV. \tag{3.174}$$

Of course, the notion of "sufficiently" is subjective and situation dependent. In addition to these equations we also have the Lorentz force equation

$$\mathbf{f}_{em}(\mathbf{r}, t) = \rho \mathbf{E}(\mathbf{r}, t) + \mathbf{J} \times \mathbf{B}(\mathbf{r}, t),$$

and, for linear, isotropic, nondispersive materials, the constitutive relations

$$\mathbf{D}(\mathbf{r}, t) = \epsilon(\mathbf{r}) \mathbf{E}(\mathbf{r}, t),$$

$$\mathbf{J}(\mathbf{r}, t) = \sigma(\mathbf{r}) \mathbf{E}(\mathbf{r}, t).$$

We can also express the equations for EQS systems in the frequency or phasor domains. (See Chapter 4 for detailed discussion of these domains.) In the frequency domain, (3.165)–(3.169) become

$$\nabla \times \tilde{\mathbf{E}}(\mathbf{r}, \omega) = 0,$$

$$\nabla \times \tilde{\mathbf{H}}(\mathbf{r}, \omega) = \tilde{\mathbf{J}}(\mathbf{r}, \omega) + j\omega \tilde{\mathbf{D}}(\mathbf{r}, \omega),$$

$$\nabla \cdot \tilde{\mathbf{D}}(\mathbf{r}, \omega) = \tilde{\rho}(\mathbf{r}, \omega),$$

$$\nabla \cdot \tilde{\mathbf{B}}(\mathbf{r}, \omega) = 0,$$

$$\nabla \cdot \tilde{\mathbf{J}}(\mathbf{r}, \omega) + j\omega \tilde{\rho}(\mathbf{r}, \omega) = 0,$$

while (3.170)–(3.174) become

$$\oint_\Gamma \tilde{\mathbf{E}}(\mathbf{r}, \omega) \cdot \mathbf{dl} = 0,$$

$$\oint_\Gamma \tilde{\mathbf{H}}(\mathbf{r}, \omega) \cdot \mathbf{dl} = \int_S \tilde{\mathbf{J}}(\mathbf{r}, \omega) \cdot \mathbf{dS} + j\omega \int_S \tilde{\mathbf{D}}(\mathbf{r}, \omega) \cdot \mathbf{dS},$$

$$\oint_S \tilde{\mathbf{D}}(\mathbf{r}, \omega) \cdot \mathbf{dS} = \int_V \tilde{\rho}(\mathbf{r}, \omega) \, dV,$$

$$\oint_S \tilde{\mathbf{B}}(\mathbf{r}, \omega) \cdot \mathbf{dS} = 0,$$

$$\oint_S \tilde{\mathbf{J}}(\mathbf{r}, \omega) \cdot \mathbf{dS} = -j\omega \int_V \tilde{\rho}(\mathbf{r}, \omega) \, dV.$$

3.6.1.1 Characteristics of an EQS system

The primary characteristic of an EQS system is that the electric field is the predominant descriptor of system behavior. The magnetic field, while secondary, is important to understand energy transfer. Several other characteristics of an EQS system may be identified:

1. field interactions leading to wave phenomena are assumed insignificant;

2. a unique potential difference is defined between any two points in space;

3. capacitive effects dominate inductive effects, and the latter may be ignored;

4. stored electric energy is much greater than stored magnetic energy, and the latter may be ignored.

The first characteristic is evident from the absence of the magnetic flux term in Faraday's law. Without the link between the curl of $\mathbf{E}$ and magnetic flux, wave phenomena are not possible. This allows the separation of space and time dependence of the dominant electric field:

$$\mathbf{E}(\mathbf{r}, t) = f_E(t)\mathbf{E}(\mathbf{r}). \tag{3.175}$$

In the frequency domain, the analogous separation is with space and frequency:

$$\tilde{\mathbf{E}}(\mathbf{r}, \omega) = \tilde{f}_E(\omega)\mathbf{E}(\mathbf{r}).$$

The magnetic field is secondary and dependent on the electric field through (3.166). As such, its time dependence is dictated both by the electric field and the current density. For example, in a conductor where $\mathbf{J}(\mathbf{r}, t) = \sigma\mathbf{E}(\mathbf{r}, t)$, the magnetic field will have terms dependent both on the electric field waveform (through the current) and on its derivative (through the displacement current).

The second characteristic follows from (3.165), which says that $\mathbf{E}(\mathbf{r}, t)$ may be completely specified using a potential function:

$$\mathbf{E}(\mathbf{r}, t) = -\nabla\Phi(\mathbf{r}, t).$$

Since $\mathbf{E}$ and Φ are related through a spatial derivative, we can also write $\Phi(\mathbf{r}, t) = f_E(t)\Phi(\mathbf{r})$ with

$$\mathbf{E}(\mathbf{r}) = -\nabla\Phi(\mathbf{r}). \tag{3.176}$$

This allows the specification of a unique potential difference between two points as

$$V(t) = -\int_a^b \mathbf{E}(\mathbf{r}, t) \cdot \mathbf{dl} = f_E(t)\left[-\int_a^b \mathbf{E}(\mathbf{r}) \cdot \mathbf{dl}\right].$$

Since $\nabla \cdot \mathbf{D}(\mathbf{r}, t) = \rho(\mathbf{r}, t)$ by (3.167), $\rho(\mathbf{r}, t) = f_E(t)\rho(\mathbf{r})$, and taking the divergence of (3.176) we find that for a homogeneous medium $\Phi(\mathbf{r})$ satisfies Poisson's equation

$$\nabla^2\Phi(\mathbf{r}) = -\frac{\rho(\mathbf{r})}{\epsilon},$$

or, in a source-free region, Laplace's equation

$$\nabla^2\Phi(\mathbf{r}) = 0.$$

We also find that a perfect electric conductor must present an equipotential surface, and that the boundary conditions on $\mathbf{E}$ and $\mathbf{D}$ are those for the electrostatic field. Since the electric field and potential obey the same spatial relationships as in electrostatics, many of the calculations under the EQS assumptions are no more difficult than in electrostatics.

This also allows the introduction of capacitance and resistance for EQS systems, as shown below.

The third characteristic is evident from the absence of the magnetic flux term $\int_S \mathbf{B} \cdot d\mathbf{S}$ in Faraday's law, which precludes the description of inductive effects. In contrast, the presence of displacement current in Ampere's law allows for the description of capacitive effects. Capacitance is discussed in detail below.

The fourth characteristic is evident in Poynting's theorem. To establish Poynting's theorem for EQS, begin with the formula for electromagnetic energy density (2.229)

$$w_{em} = \mathbf{J} \cdot \mathbf{E}.$$

Substitution of $\mathbf{J}$ from (3.166) gives $w_{em} = (\nabla \times \mathbf{H}) \cdot \mathbf{E} - \frac{\partial \mathbf{D}}{\partial t} \cdot \mathbf{E}$ and thus

$$w_{em} = -\nabla \cdot (\mathbf{E} \times \mathbf{H}) + \mathbf{H} \cdot (\nabla \times \mathbf{E}) - \frac{\partial \mathbf{D}}{\partial t} \cdot \mathbf{E}$$

by (B.50). Substituting $\nabla \times \mathbf{E} = 0$ from (3.165), we have

$$w_{em} = -\nabla \cdot (\mathbf{E} \times \mathbf{H}) - \mathbf{E} \cdot \frac{\partial \mathbf{D}}{\partial t}.$$

For a linear, isotropic, nondispersive medium, we have

$$\mathbf{E} \cdot \frac{\partial \mathbf{D}}{\partial t} = \epsilon \mathbf{E} \cdot \frac{\partial \mathbf{E}}{\partial t} = \frac{1}{2} \epsilon \frac{\partial}{\partial t} (\mathbf{E} \cdot \mathbf{E}) = \frac{1}{2} \frac{\partial}{\partial t} (\mathbf{D} \cdot \mathbf{E}),$$

giving

$$\nabla \cdot \mathbf{S}_{em} + \frac{\partial}{\partial t} W_{em} = -\mathbf{J} \cdot \mathbf{E}$$

where

$$W_{em} = \tfrac{1}{2} \mathbf{D} \cdot \mathbf{E}$$

and

$$\mathbf{S}_{em} = \mathbf{E} \times \mathbf{H}.$$

Integration over a volume and use of the divergence theorem gives the large-scale form of Poynting's theorem for EQS systems:

$$-\int_V \mathbf{J} \cdot \mathbf{E} \, dV = \int_V \frac{1}{2} \frac{\partial}{\partial t} (\mathbf{D} \cdot \mathbf{E}) \, dV + \oint_S (\mathbf{E} \times \mathbf{H}) \cdot d\mathbf{S}. \tag{3.177}$$

Note that compared to the general form of Poynting's theorem (2.234), the stored magnetic energy term is missing. However, the interpretation of Poynting's vector $\mathbf{S}_{em}$ as the density of electromagnetic energy flow remains.

3.6.1.2 Capacitance and resistance

Two important concepts in electro-quasistatics are *capacitance* and *resistance*. Consider the two perfectly conducting bodies shown in Figure 3.25. They are immersed in a uniform, isotropic, nondispersive medium of permittivity ϵ and conductivity σ. A voltage $V_0(t)$ is applied between the bodies such that conductor A receives a positive charge $Q(t)$ and conductor B receives a negative charge $-Q(t)$. Under the EQS assumption, the conductors are equipotentials, and the voltage $V_0(t)$ is given by the line integral of the electric field between any point on conductor A and any point on conductor B:

$$V_0(t) = -\int_B^A \mathbf{E}(\mathbf{r}, t) \cdot d\mathbf{l}.$$

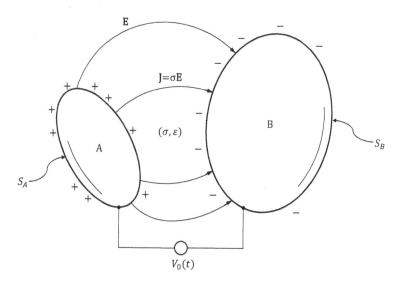

FIGURE 3.25
A two-body system.

The charge induced on A may be found by using the boundary condition (2.166). The surface charge density is

$$\rho_s(\mathbf{r}, t) = \mathbf{n} \cdot \mathbf{D}(\mathbf{r}, t) = \epsilon \mathbf{n} \cdot \mathbf{E}(\mathbf{r}, t),$$

hence the total charge is

$$Q(t) = \int_{S_A} \epsilon \mathbf{n} \cdot \mathbf{E}(\mathbf{r}, t) \, dS.$$

The capacitance is then defined as

$$C = \frac{Q(t)}{V_0(t)} = \frac{\int_{S_A} \epsilon \mathbf{n} \cdot \mathbf{E}(\mathbf{r}, t) \, dS}{-\int_B^A \mathbf{E}(\mathbf{r}, t) \cdot \mathbf{dl}} = \frac{\int_{S_A} \epsilon \mathbf{n} \cdot \mathbf{E}(\mathbf{r}) \, dS}{-\int_B^A \mathbf{E}(\mathbf{r}) \cdot \mathbf{dl}}. \tag{3.178}$$

Since the electric field may be written as the product shown in (3.175), the capacitance is time-independent. We can easily find the stored electric energy in terms of capacitance using (3.62). Substituting $Q(t) = CV_0(t)$, we have

$$W(t) = \frac{1}{2}Q(t)V_0(t) = \frac{1}{2}CV_0^2(t). \tag{3.179}$$

▶ **Example 3.49:** Voltage distribution across capacitors connected in series

A dc voltage source of value V_s is connected in series with two initially uncharged capacitors C_1 and C_2 as shown in Figure 3.26. Use the minimum property of electrostatic energy (Thomson's theorem) to determine how the source voltage divides across the individual capacitors. Also find the individual stored energy and charge values.

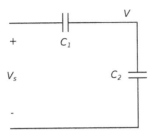

FIGURE 3.26

Two capacitors in series with a voltage source.

Solution: Using (3.179), the electric stored energy is given by

$$W(V) = W_1(V) + W_2(V) = \tfrac{1}{2}C_1(V_s - V)^2 + \tfrac{1}{2}C_2 V^2.$$

The necessary condition for an extremum, $W'(V) = 0$, yields the value

$$V = \mathcal{V} = \frac{C_1}{C_1 + C_2} V_s.$$

This value gives

$$W_1(\mathcal{V}) = \frac{1}{2}\frac{C_1 C_2^2}{(C_1 + C_2)^2} V_s^2, \qquad W_2(\mathcal{V}) = \frac{1}{2}\frac{C_1^2 C_2}{(C_1 + C_2)^2} V_s^2,$$

and

$$W(\mathcal{V}) = W_1(\mathcal{V}) + W_2(\mathcal{V}) = \frac{1}{2}\frac{C_1 C_2}{C_1 + C_2} V_s^2 = \tfrac{1}{2}C_e V_s^2$$

where

$$C_e = \frac{C_1 C_2}{C_1 + C_2}.$$

Use of the relation $Q = CV$ gives $Q_1 = Q_2 = C_e V_s = Q_e$. To see that $\mathcal{V}$ minimizes $W(V)$, we can write

$$W(\mathcal{V} + \delta) = \tfrac{1}{2}C_1 \left[V_s - \left(\frac{C_1}{C_1 + C_2} V_s + \delta \right) \right]^2 + \tfrac{1}{2}C_2 \left[\frac{C_1}{C_1 + C_2} V_s + \delta \right]^2$$

$$= \tfrac{1}{2}C_e V_s^2 + \tfrac{1}{2}(C_1 + C_2)\delta^2$$

and see that

$$W(\mathcal{V} + \delta) - W(\mathcal{V}) = \tfrac{1}{2}(C_1 + C_2)\delta^2 \geq 0$$

for any real value of δ. Additional problems may be found in [71]. ◄

The capacitance concept may be introduced using electrostatic fields and defined using (3.178), but its usefulness is not evident until electro-quasistatics is considered. Since we retain the displacement current term in Ampere's law (3.166) we can compute the total displacement current flowing from conductor A to conductor B as

$$I_d(t) = \int_{S_A} \frac{\partial \mathbf{D}(\mathbf{r}, t)}{\partial t} \cdot \mathbf{n} \, dS.$$

Assuming S_A is stationary,

$$I_d(t) = \frac{d}{dt} \int_{S_A} \mathbf{D} \cdot \mathbf{n} \, dS = \frac{dQ(t)}{dt} = C\frac{dV_0(t)}{dt}.$$

This expression may be used to describe the behavior of circuits containing bulk capacitive elements (capacitors). A system of conductors is considered in Problem 3.22.

To determine the resistance of the EQS system, the current passing from conductor A to conductor B is computed:

$$I(t) = \int_{S_A} \sigma \mathbf{E}(\mathbf{r}, t) \cdot \mathbf{n} \, dS.$$

Resistance is then defined as

$$R = \frac{V_0(t)}{I(t)} = \frac{-\int_B^A \mathbf{E}(\mathbf{r}, t) \cdot \mathbf{dl}}{\int_{S_A} \sigma \mathbf{E}(\mathbf{r}, t) \cdot \mathbf{n} \, dS} = \frac{-\int_B^A \mathbf{E}(\mathbf{r}) \cdot \mathbf{dl}}{\int_{S_A} \sigma \mathbf{E}(\mathbf{r}) \cdot \mathbf{n} \, dS}. \tag{3.180}$$

Since the electric field may be written as the product shown in (3.175), the resistance is independent of time. Note that when (3.180) is multiplied by (3.178) we get the interesting relationship

$$RC = \epsilon/\sigma. \tag{3.181}$$

► **Example 3.50:** Electro-quasistatic analysis of a resistive capacitor

A resistive capacitor consists of a cylinder of material with uniform permittivity ϵ and uniform conductivity σ. Circular perfectly conducting plates are attached to the top and bottom of the cylinder (Figure 3.27). A voltage $V_0(t)$ is applied across the plates. Find the electric and magnetic fields in the resistor using EQS analysis. Neglect fringing. Investigate the system power balance and identify the capacitance and resistance of the system.

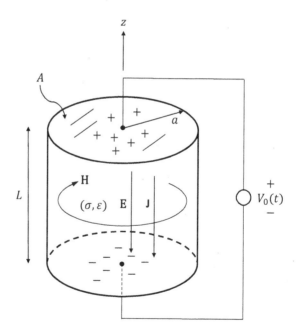

FIGURE 3.27
A resistive capacitor.

Solution: Here we treat the device as a *lumped element* by employing electro-quasistatics. Using symmetry and neglecting fringing, we expect that

$$\mathbf{E}(\mathbf{r}, t) = \hat{\mathbf{z}} \, E_z(\rho, z, t).$$

However, since there is no free charge, Gauss's law implies $\partial E_z / \partial z = 0$ and thus E_z is independent of z. Also, since the conductors are equipotential surfaces,

$$V_0(t) = - \int_0^L E_z(\rho, t) \, dz = -E_z(\rho, t) L,$$

and thus E_z must be independent of ρ, and

$$E_z(t) = -\frac{V_0(t)}{L}.$$

To find $\mathbf{H}$, examine Ampere's law (3.166). Since $\mathbf{J}$ and $\mathbf{D}$ are z-directed and spatially constant within the resistor, we must have $\mathbf{H}(\mathbf{r}, t) = \hat{\boldsymbol{\phi}} H_\phi(\rho, t)$ and

$$\nabla \times \mathbf{H} = \hat{\mathbf{z}} \frac{1}{\rho} \frac{\partial (\rho H_\phi)}{\partial \rho} = \hat{\mathbf{z}} \sigma E_z + \hat{\mathbf{z}} \epsilon \frac{\partial E_z}{\partial t} = -\hat{\mathbf{z}} \frac{\sigma}{L} V_0(t) - \hat{\mathbf{z}} \frac{\epsilon}{L} \frac{dV_0(t)}{dt}.$$

Integration over ρ gives

$$H_\phi(\rho, t) = -\frac{\rho}{2} \frac{\sigma}{L} V_0(t) - \frac{\rho}{2} \frac{\epsilon}{L} \frac{dV_0(t)}{dt}.$$

Note the presence of terms with time variation proportional to both the electric field and to its derivative.

The charge density on the top (positively charged) plate is, by the boundary condition (2.166),

$$\rho_s(t) = -\hat{\mathbf{z}} \cdot \left(-\hat{\mathbf{z}} \epsilon \frac{V_0(t)}{L} \right) = \epsilon \frac{V_0(t)}{L},$$

hence the total charge is

$$Q(t) = \int_S \rho_s(t) \, dS = \epsilon A \frac{V_0(t)}{L}$$

where $A = \pi a^2$ is the plate area. With this, the capacitance is

$$C = \frac{Q(t)}{V_0(t)} = \epsilon \frac{A}{L}.$$

The current density within the material is $\mathbf{J}(t) = \sigma \mathbf{E}(t)$ so that the total current flowing in the resistor is

$$I(t) = \int_S \mathbf{J}(t) \cdot \hat{\mathbf{n}} \, dS = A \sigma \frac{V_0(t)}{L}.$$

The resistance is then

$$R = \frac{V_0(t)}{I(t)} = \frac{L}{\sigma A}. \tag{3.182}$$

Note that $RC = \epsilon / \sigma$ as in (3.181).

Finally, we examine the terms in Poynting's theorem (3.177). Write the power balance as

$$-\oint_S (\mathbf{E} \times \mathbf{H}) \cdot d\mathbf{S} = \int_V \mathbf{J} \cdot \mathbf{E} \, dV + \int_V \frac{1}{2} \frac{\partial}{\partial t} (\mathbf{D} \cdot \mathbf{E}) \, dV$$

or

$$P_{\text{in}} = P_R + P_W,$$

which states that the power entering the system is transferred to power dissipated in the resistor plus power in the time-rate of change of stored electric energy. The instantaneous power dissipated in the resistor is

$$P_R(t) = \int_V \mathbf{J}(t) \cdot \mathbf{E}(t)\, dV = \int_V \sigma |E_z(t)|^2\, dV$$

$$= \sigma L A \left(\frac{V_0(t)}{L} \right)^2 = \frac{V_0^2(t)}{\left(\frac{L}{\sigma A} \right)} = \frac{V_0^2(t)}{R}.$$

The time-rate of change of stored energy is

$$P_W(t) = \int_V \frac{1}{2} \frac{\partial}{\partial t} (\mathbf{D} \cdot \mathbf{E})\, dV = \int_V \frac{1}{2} \frac{\partial}{\partial t} \epsilon |E_z(t)|^2\, dV = \frac{1}{2} \int_V \frac{d}{dt} \epsilon \left(\frac{V_0(t)}{L} \right)^2 dV$$

$$= \epsilon A \frac{V_0(t)}{L} \frac{dV_0(t)}{dt} = V_0(t) \left[C \frac{dV_0(t)}{dt} \right].$$

The sum of these powers is

$$P_R(t) + P_W(t) = \frac{V_0^2(t)}{R} + V_0(t) \left[C \frac{dV_0(t)}{dt} \right]. \tag{3.183}$$

Finally, the Poynting vector at the outer surface of the resistor is given by

$$\mathbf{E}(t) \times \mathbf{H}(\rho,t)\big|_{\rho=a} = -\frac{V_0(t)}{L} \left[-\frac{a}{2} \frac{\sigma}{L} V_0(t) - \frac{a}{2} \frac{\epsilon}{L} \frac{dV_0(t)}{dt} \right] (\hat{\mathbf{z}} \times \hat{\boldsymbol{\phi}})$$

$$= -\left[\frac{\sigma a}{2L^2} V_0^2(t) + \frac{\epsilon a}{2L} \frac{V_0(t)}{L} \frac{dV_0(t)}{dt} \right] \hat{\boldsymbol{\rho}},$$

and thus the power enters the resistor radially from the outside. The power entering the system is therefore

$$P_{\text{in}}(t) = \oint_S (\mathbf{E} \times \mathbf{H}) \cdot d\mathbf{S} = -2\pi a L \left[\frac{\sigma a}{2L^2} V_0^2(t) + \frac{\epsilon a}{2L} \frac{V_0(t)}{L} \frac{dV_0(t)}{dt} \right] \hat{\boldsymbol{\rho}} \cdot (-\hat{\boldsymbol{\rho}})$$

$$= \frac{V_0^2(t)}{R} + V_0(t) \left[C \frac{dV_0(t)}{dt} \right],$$

which balances with (3.183). ◄

3.6.2 Magneto-quasistatics

An electromagnetic system is said to be *magneto-quasistatic* (MQS) when it can be analyzed to good accuracy using the point forms

$$\nabla \times \mathbf{E}(\mathbf{r},t) = -\frac{\partial \mathbf{B}(\mathbf{r},t)}{\partial t}, \tag{3.184}$$

$$\nabla \times \mathbf{H}(\mathbf{r},t) = \mathbf{J}(\mathbf{r},t), \tag{3.185}$$

$$\nabla \cdot \mathbf{D}(\mathbf{r},t) = \rho(\mathbf{r},t), \tag{3.186}$$

$$\nabla \cdot \mathbf{B}(\mathbf{r},t) = 0, \tag{3.187}$$

$$\nabla \cdot \mathbf{J}(\mathbf{r},t) = 0, \tag{3.188}$$

or the integral forms

$$\oint_\Gamma \mathbf{E}(\mathbf{r}, t) \cdot \mathbf{dl} = -\frac{d}{dt} \int_S \mathbf{B}(\mathbf{r}, t) \cdot \mathbf{dS}, \tag{3.189}$$

$$\oint_\Gamma \mathbf{H}(\mathbf{r}, t) \cdot \mathbf{dl} = \int_S \mathbf{J}(\mathbf{r}, t) \cdot \mathbf{dS}, \tag{3.190}$$

$$\oint_S \mathbf{D}(\mathbf{r}, t) \cdot \mathbf{dS} = \int_V \rho(\mathbf{r}, t) \, dV, \tag{3.191}$$

$$\oint_S \mathbf{B}(\mathbf{r}, t) \cdot \mathbf{dS} = 0, \tag{3.192}$$

$$\oint_S \mathbf{J}(\mathbf{r}, t) \cdot \mathbf{dS} = 0. \tag{3.193}$$

In addition, we have the Lorentz force equation

$$\mathbf{f}_{em}(\mathbf{r}, t) = \rho \mathbf{E}(\mathbf{r}, t) + \mathbf{J} \times \mathbf{B}(\mathbf{r}, t),$$

and, for linear, isotropic, nondispersive materials, the constitutive relations

$$\mathbf{D}(\mathbf{r}, t) = \epsilon(\mathbf{r}) \mathbf{E}(\mathbf{r}, t),$$

$$\mathbf{J}(\mathbf{r}, t) = \sigma(\mathbf{r}) \mathbf{E}(\mathbf{r}, t).$$

The frequency domain counterparts to (3.184)–(3.188) are

$$\nabla \times \tilde{\mathbf{E}}(\mathbf{r}, \omega) = -j\omega \tilde{\mathbf{B}}(\mathbf{r}, \omega), \tag{3.194}$$

$$\nabla \times \tilde{\mathbf{H}}(\mathbf{r}, \omega) = \tilde{\mathbf{J}}(\mathbf{r}, \omega), \tag{3.195}$$

$$\nabla \cdot \tilde{\mathbf{D}}(\mathbf{r}, \omega) = \tilde{\rho}(\mathbf{r}, \omega), \tag{3.196}$$

$$\nabla \cdot \tilde{\mathbf{B}}(\mathbf{r}, \omega) = 0, \tag{3.197}$$

$$\nabla \cdot \tilde{\mathbf{J}}(\mathbf{r}, \omega) = 0. \tag{3.198}$$

and the counterparts to (3.189)–(3.193) are

$$\oint_\Gamma \tilde{\mathbf{E}}(\mathbf{r}, \omega) \cdot \mathbf{dl} = -j\omega \int_S \tilde{\mathbf{B}}(\mathbf{r}, \omega) \cdot \mathbf{dS},$$

$$\oint_\Gamma \tilde{\mathbf{H}}(\mathbf{r}, \omega) \cdot \mathbf{dl} = \int_S \tilde{\mathbf{J}}(\mathbf{r}, \omega) \cdot \mathbf{dS},$$

$$\oint_S \tilde{\mathbf{D}}(\mathbf{r}, \omega) \cdot \mathbf{dS} = \int_V \tilde{\rho}(\mathbf{r}, \omega) \, dV,$$

$$\oint_S \tilde{\mathbf{B}}(\mathbf{r}, \omega) \cdot \mathbf{dS} = 0,$$

$$\oint_S \tilde{\mathbf{J}}(\mathbf{r}, \omega) \cdot \mathbf{dS} = 0.$$

Some qualitative characteristics of MQS systems are as follows:

1. there is no time-rate of change of charge density and thus all currents are steady;

2. inductive effects dominate over capacitive effects, and the latter may be ignored;

3. stored magnetic energy is much greater than stored electric energy, and the latter may be ignored.

The first of these characteristics follows from (3.188). In a homogeneous conducting material where $\mathbf{J}(\mathbf{r}, t) = \sigma \mathbf{E}(\mathbf{r}, t)$ and $\mathbf{D}(\mathbf{r}, t) = \epsilon \mathbf{E}(\mathbf{r}, t)$, the condition of a steady current requires

$$\nabla \cdot \mathbf{J}(\mathbf{r}, t) = 0 = \sigma \nabla \cdot \mathbf{E}(\mathbf{r}, t) = \frac{\sigma}{\epsilon} \nabla \cdot \mathbf{D}(\mathbf{r}, t),$$

and thus $\rho(\mathbf{r}, t) = 0$. The second characteristic is evident by the absence of the displacement current term in (3.185). In contrast, the presence of the magnetic flux term in Faraday's law (3.184) allows for inductive effects. The third characteristic is evident in Poynting's theorem. To establish Poynting's theorem for MQS, begin with the formula for electromagnetic energy density (2.229)

$$w_{em} = \mathbf{J} \cdot \mathbf{E}.$$

Substitution of $\mathbf{J}$ from (3.185) gives $w_{em} = (\nabla \times \mathbf{H}) \cdot \mathbf{E}$ and thus

$$w_{em} = -\nabla \cdot (\mathbf{E} \times \mathbf{H}) + \mathbf{H} \cdot (\nabla \times \mathbf{E})$$

by (B.50). Substituting $\nabla \times \mathbf{E} = -\partial \mathbf{B}/\partial t$ from (3.184), we have

$$w_{em} = -\nabla \cdot (\mathbf{E} \times \mathbf{H}) - \left(\mathbf{H} \cdot \frac{\partial \mathbf{B}}{\partial t} \right).$$

For a linear, isotropic, nondispersive medium, we have

$$\mathbf{H} \cdot \frac{\partial \mathbf{B}}{\partial t} = \mu \mathbf{H} \cdot \frac{\partial \mathbf{H}}{\partial t} = \frac{1}{2} \mu \frac{\partial}{\partial t} (\mathbf{H} \cdot \mathbf{H}) = \frac{1}{2} \frac{\partial}{\partial t} (\mathbf{B} \cdot \mathbf{H}),$$

giving

$$\nabla \cdot \mathbf{S}_{em} + \frac{\partial}{\partial t} W_{em} = -\mathbf{J} \cdot \mathbf{E}$$

where

$$W_{em} = \tfrac{1}{2} \mathbf{B} \cdot \mathbf{H}$$

and

$$\mathbf{S}_{em} = \mathbf{E} \times \mathbf{H}.$$

Integration over a volume and use of the divergence theorem gives the large-scale form of Poynting's theorem for MQS systems:

$$-\int_V \mathbf{J} \cdot \mathbf{E} \, dV = \int_V \frac{1}{2} \frac{\partial}{\partial t} (\mathbf{B} \cdot \mathbf{H}) \, dV + \oint_S (\mathbf{E} \times \mathbf{H}) \cdot \mathbf{dS}.$$

Note that in comparison with the general form of Poynting's theorem (2.234), the stored electric energy term is missing. However, the interpretation of the Poynting vector $\mathbf{S}_{em}$ as the density of electromagnetic energy flow remains.

3.6.2.1 Magnetic potentials for MQS systems

By the magnetic source law (3.187) we can write the MQS field in terms of a vector potential $\mathbf{A}(\mathbf{r}, t)$ as

$$\mathbf{B}(\mathbf{r}, t) = \nabla \times \mathbf{A}(\mathbf{r}, t).$$

Easily calculated is the magnetic flux passing through an open surface:

$$\Psi_m(t) = \int_S \mathbf{B}(\mathbf{r}, t) \cdot d\mathbf{S} = \int_S [\nabla \times \mathbf{A}(\mathbf{r}, t)] \cdot d\mathbf{S} = \oint_\Gamma \mathbf{A}(\mathbf{r}, t) \cdot d\mathbf{l}. \tag{3.199}$$

In a homogeneous isotropic material where $\mathbf{B}(\mathbf{r}, t) = \mu \mathbf{H}(\mathbf{r}, t)$, Ampere's law (3.185) requires

$$\nabla \times [\nabla \times \mathbf{A}(\mathbf{r}, t)] = \mu \mathbf{J}(\mathbf{r}, t).$$

Expanding the curl–curl operator and assuming the Coulomb gauge condition $\nabla \cdot \mathbf{A}(\mathbf{r}, t) = 0$, we get a differential equation for $\mathbf{A}$:

$$\nabla^2 \mathbf{A}(\mathbf{r}, t) = -\mu \mathbf{J}(\mathbf{r}, t).$$

In a source-free region of space, we have by Ampere's law (3.185)

$$\nabla \times \mathbf{H}(\mathbf{r}, t) = 0,$$

and thus we may write the magnetic field in terms of a scalar potential function as

$$\mathbf{H}(\mathbf{r}, t) = -\nabla \Phi_m(\mathbf{r}, t).$$

In a homogeneous isotropic region where $\mathbf{B}(\mathbf{r}, t) = \mu \mathbf{H}(\mathbf{r}, t)$, the magnetic source law (3.187) requires

$$\nabla \cdot \mathbf{B}(\mathbf{r}, t) = -\mu \nabla \cdot [\nabla \Phi_m(\mathbf{r}, t)] = 0$$

and so Φ_m obeys Laplace's equation:

$$\nabla^2 \Phi_m(\mathbf{r}, t) = 0.$$

3.6.2.2 Fields in nonconducting media and inductance

For currents in free space, or in nonconducting materials, the interpretation of MQS is similar to EQS. The spatial and temporal dependences of the magnetic field separate and the spatial dependence obeys the magnetostatic field equations. The situation is much different in conductors, where the presence of a strong time-changing magnetic flux density ($\partial \mathbf{B}/\partial t$) induces *eddy currents*, which are associated with Joule heating (particularly in transformer cores). These currents reside near the surface of the conductor in what is termed the *skin effect region*. This is discussed further below.

In nonconducting media the time and spatial dependencies of the current and magnetic field are multiplicative:

$$\mathbf{H}(\mathbf{r}, t) = f_H(t)\mathbf{H}(\mathbf{r}), \qquad \mathbf{J}(\mathbf{r}, t) = f_H(t)\mathbf{J}(\mathbf{r}). \tag{3.200}$$

Thus, Ampere's law relates the spatial dependence of the fields through

$$\nabla \times \mathbf{H}(\mathbf{r}) = \mathbf{J}(\mathbf{r}).$$

In a nondispersive medium where $\mathbf{B}(\mathbf{r}, t) = \mu \mathbf{H}(\mathbf{r}, t)$, the time and spatial dependences of the magnetic flux density also separate:

$$\mathbf{B}(\mathbf{r}, t) = f_H(t)\mu \mathbf{H}(\mathbf{r}) = f_H(t)\mathbf{B}(\mathbf{r}).$$

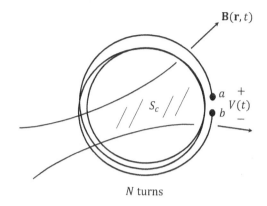

FIGURE 3.28

A wire coil with a gap.

The magnetic vector and scalar potentials separate similarly as

$$\mathbf{A}(\mathbf{r}, t) = f_H(t)\mathbf{A}(\mathbf{r})$$

and

$$\Phi_m(\mathbf{r}, t) = f_H(t)\Phi_m(\mathbf{r}).$$

This means the spatial dependencies of the fields are related to those of the potentials as

$$\mathbf{B}(\mathbf{r}) = \nabla \times \mathbf{A}(\mathbf{r}), \qquad \mathbf{H}(\mathbf{r}) = -\nabla\Phi_m(\mathbf{r}),$$

and therefore obey the same relationships as the magnetostatic fields (e.g., the Biot-Savart law).

Through Faraday's law (3.189) we have the electromotive force (circulation of $\mathbf{E}$) as

$$\text{emf}(t) = \oint_\Gamma \mathbf{E}(\mathbf{r}, t) \cdot d\mathbf{l} = -\frac{d\Psi_m(t)}{dt}$$

where

$$\Psi_m(t) = \int_S \mathbf{B}(\mathbf{r}, t) \cdot d\mathbf{S} = f_H(t) \int_S \mathbf{B}(\mathbf{r}) \cdot d\mathbf{S}$$

is the magnetic flux. Of particular importance is when the circulation is taken around a coil with a gap, as shown in Figure 3.28. The coil consists of N closely wound turns of perfectly conducting wire. Since the tangential component of $\mathbf{E}(\mathbf{r}, t)$ vanishes everywhere along the integration path except in the gap, the entire contribution is given by this "gap voltage" as

$$V(t) = -\int_b^a \mathbf{E}(\mathbf{r}, t) \cdot d\mathbf{l} = -\frac{d}{dt} \int_{S_C} \mathbf{B}(\mathbf{r}, t) \cdot d\mathbf{S},$$

where S_C is the surface bounded by the coil windings. Care is required to compute the flux. Since the surface S_C is folded back on itself N times, the total flux will be N times that through one turn of the coil, which is assumed to bound a shared surface S. It is

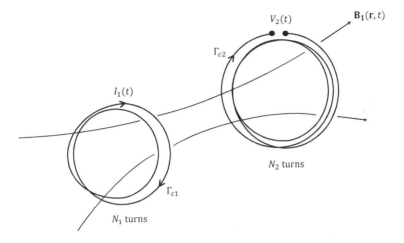

FIGURE 3.29
Coupled coils.

therefore helpful to define the *flux linkage* Λ as the total flux "linked" by the current as it traverses its path. Then

$$\Lambda(t) = N \int_S \mathbf{B}(\mathbf{r}, t) \cdot \mathbf{dS} = N f_H(t) \int_S \mathbf{B}(\mathbf{r}) \cdot \mathbf{dS},$$

and

$$V(t) = -\frac{d\Lambda(t)}{dt}. \tag{3.201}$$

Although flux linkage is easily defined in the case of current-carrying wire coils in free space, the concept is a bit trickier when the current is volume distributed (e.g., within a thick wire). We shall consider this when treating the internal inductance of a wire below.

The gap voltage (3.201) can be related to the current creating the magnetic flux density through a quantity called *inductance*. Consider two tightly wound coils of wire immersed in a nonconducting medium (Figure 3.29). The first coil has N_1 turns and carries steady current $I_1(t) = I f_H(t)$, while the second has N_2 turns with a gap. The first coil creates a magnetic flux density $\mathbf{B}_1(\mathbf{r}, t) = \mu \mathbf{H}_1(\mathbf{r}, t)$ by Ampere's law. Since the time dependence separates as in (3.200), the flux linked by the second coil is given by (3.199):

$$\Lambda_2(t) = f_H(t) \oint_{\Gamma_{c2}} \mathbf{A}_1(\mathbf{r}) \cdot \mathbf{dl}_2 = N_2 f_H(t) \oint_{\Gamma_2} \mathbf{A}_1(\mathbf{r}) \cdot \mathbf{dl}_2,$$

where Γ_{c2} is the contour following all N_2 turns of coil 2 and Γ_2 follows one of the closely spaced turns of the coil. Furthermore, since the spatial dependence of the vector potential produced by coil 1 satisfies the magnetostatic field relations, it can be computed using (3.107) as

$$\mathbf{A}_1(\mathbf{r}) = \frac{\mu}{4\pi} \oint_{\Gamma_{C1}} \frac{I_1 \mathbf{dl}_1'}{|\mathbf{r} - \mathbf{r}_1'|}.$$

Here Γ_{C_1} is the contour following all N_1 turns of coil 1. Since the turns of the coil are closely spaced, the vector potential can also be written as

$$\mathbf{A}_1(\mathbf{r}) = N_1 \frac{\mu}{4\pi} \oint_{\Gamma_1} \frac{I_1 \mathbf{dl}_1'}{|\mathbf{r} - \mathbf{r}_1'|},$$

where Γ_1 bounds one of the closely-spaced turns. It is thus helpful to define the flux linked by coil 2 to include the effects of both N_1 and N_2 as

$$\Lambda_2(t) = N_1 N_2 f_H(t) \frac{\mu}{4\pi} \oint_{\Gamma_2} \oint_{\Gamma_1} \frac{I_1}{|\mathbf{r}_2 - \mathbf{r}_1'|} \, \mathbf{dl}_1' \cdot \mathbf{dl}_2. \tag{3.202}$$

Finally, the *mutual inductance* between the coils is defined as

$$L_{21} = \frac{\Lambda_2(t)}{I_1(t)}, \tag{3.203}$$

such that the voltage appearing at the gap of coil 2 is

$$V_2(t) = L_{21} \frac{dI_1(t)}{dt}.$$

Substitution of (3.202) into (3.203) gives

$$L_{21} = N_1 N_2 \frac{\mu}{4\pi} \oint_{\Gamma_2} \oint_{\Gamma_1} \frac{\mathbf{dl}_1' \cdot \mathbf{dl}_2}{|\mathbf{r}_2 - \mathbf{r}_1'|}. \tag{3.204}$$

This expression is called the *Neumann formula* for the mutual inductance. We may also define the *self inductance* as the ratio of flux linked by any coil due to the current in that same coil. Specializing (3.204) by letting $\mathbf{r}_2 = \mathbf{r}$, $\mathbf{r}_1' = \mathbf{r}'$, $N_1 = N_2 = N$, and $\Gamma_1 = \Gamma_2 = \Gamma$, we have Neumann's formula for the self inductance:

$$L = N^2 \frac{\mu}{4\pi} \oint_{\Gamma} \oint_{\Gamma} \frac{\mathbf{dl}' \cdot \mathbf{dl}}{|\mathbf{r} - \mathbf{r}'|}.$$

Note that because of the product relationship (3.200), the self and mutual inductances are time-independent. The inductance of a system of loops is considered in Problem 3.25.

▶ **Example 3.51:** Mutual inductance of coaxial circular loops

Two single-turn coaxial wire loops in free space are separated axially by distance d. Each wire has circular cross-section with radius a, and the mean radii of the loops are b_1 and b_2. See Figure 3.30. Find the mutual inductance between the loops.

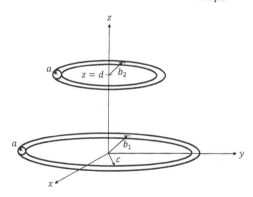

FIGURE 3.30
Coaxial loops.

Solution: Assume the current is a line current residing on the inner edge of the loop periphery, at radius $c = b_1 - a$.

Although the current may be placed at the center of the wire, an offset by the loop radius will be useful in the next example for determining self inductance. For the mutual inductance we use (3.204) with $\mathbf{r}_1' = c\hat{\boldsymbol{\rho}}'$, $\mathbf{r}_2 = d\hat{\mathbf{z}} + b_2\hat{\boldsymbol{\rho}}$, $\mathbf{dl}_1' = cd\phi'\hat{\boldsymbol{\phi}}'$, and $\mathbf{dl}_2 = b_2 d\phi\hat{\boldsymbol{\phi}}$. Substitution gives

$$L_{21} = \frac{\mu_0}{4\pi} cb_2 \int_0^{2\pi} \int_0^{2\pi} \frac{\cos(\phi' - \phi)\, d\phi'\, d\phi}{\sqrt{d^2 + c^2 + b_2^2 - 2cb_2\cos(\phi' - \phi)}}.$$

The change of variables $u = \phi' - \phi$ renders one integral trivial, leaving

$$L_{21} = \frac{\mu_0}{2} cb_2 \int_0^{2\pi} \frac{\cos u\, du}{\sqrt{d^2 + c^2 + b_2^2 - 2cb_2\cos u}}.$$

With the change of variables $u = \pi - 2\beta$ (so that $\cos u = 2\sin^2\beta - 1$) we have

$$L_{21} = \mu_0 2cb_2 \int_0^{\frac{\pi}{2}} \frac{(2\sin^2\beta - 1)\, d\beta}{\sqrt{d^2 + c^2 + b_2^2 - 2cb_2(2\sin^2\beta - 1)}}.$$

The term under the radical can be written as

$$d^2 + c^2 + b_2^2 - 2cb_2(2\sin^2\beta - 1) = \frac{4cb_2}{k^2}(1 - k^2\sin^2\beta)$$

where

$$k^2 = \frac{4cb_2}{d^2 + (c + b_2)^2}.$$

Substituting and rearranging, we get

$$L_{21} = \mu_0 \sqrt{cb_2}\left[\left(\frac{2}{k} - k\right) F(k, \tfrac{\pi}{2}) - \frac{2}{k} E(k, \tfrac{\pi}{2})\right]. \tag{3.205}$$

Here

$$F(k, \psi) = \int_0^{\psi} \frac{d\beta}{\sqrt{1 - k^2\sin^2\beta}}$$

is the *incomplete elliptic integral of the first kind*, and

$$E(k, \psi) = \int_0^{\psi} \sqrt{1 - k^2\sin^2\beta}\, d\beta$$

is the *incomplete elliptic integral of the second kind*. ◄

▶ **Example 3.52:** External self inductance of a circular loop

Find the self inductance of a circular loop.

Solution: We can set $b_2 = b_1$ and $d = 0$ in (3.205):

$$L = \mu_0(2b_1 - a)\left[\left(1 - \frac{k^2}{2}\right) F(k, \tfrac{\pi}{2}) - E(k, \tfrac{\pi}{2})\right],$$

where

$$k^2 = \frac{4b_1(b_1 - a)}{(2b_1 - a)^2}.$$

If the wire is thin ($a \ll b_1$), then $k \to 1^-$ and we can use the limiting approximations [1]

$$F(1^-, \tfrac{\pi}{2}) \approx \ln\left(\frac{4}{\sqrt{1-k^2}}\right), \qquad E(1^-, \tfrac{\pi}{2}) \approx 1,$$

to obtain

$$L \approx \mu_0 b_1 \left[\ln\left(\frac{8b_1}{a}\right) - 2\right]$$

for $a/b_1 \ll 1$. ◄

In a conducting material with $\mathbf{J} = \sigma \mathbf{E}$, Ampere's law (3.185) becomes

$$\nabla \times \mathbf{H}(\mathbf{r}, t) = \sigma \mathbf{E}(\mathbf{r}, t).$$

But Faraday's law (3.184) relates the electric field to the magnetic field through the magnetic flux density and the relation $\mathbf{B} = \mu \mathbf{H}$. This complicated interaction requires that inductance be defined somewhat differently within conductors. Only when the link back to the magnetic flux density is ignored can the usual definition of inductance as magnetic flux per unit current be used, as in the following example.

▶ **Example 3.53:** Internal inductance of a wire — uniform current assumption

A long cylindrical metallic wire of radius a, conductivity σ, and permeability μ lies along the z-axis. Compute its internal inductance.

Solution: We take the current density to be z-directed and uniform over the wire cross-section. By (3.188), $\mathbf{J}$ is also independent of z. Thus

$$\mathbf{J}(\mathbf{r}, t) = \hat{\mathbf{z}} \frac{I(t)}{\pi a^2},$$

where $I(t)$ is the current carried by the wire. Since $\mathbf{E} = \mathbf{J}/\sigma$ is also uniform, $\nabla \times \mathbf{E} = 0$ and thus by Faraday's law (3.184) we have $\partial \mathbf{B}/\partial t = 0$. This will only hold for fields varying slowly with time, such as low-frequency time-harmonic fields. Hence the internal inductance obtained from the constant current assumption applies only to currents having slow time variation. We explore this further in Examples 3.55 and 3.56.

To find the magnetic field, examine Ampere's law (3.185). Since $\mathbf{J}$ is z-directed and uniform within the wire, we have $\mathbf{H}(\mathbf{r}, t) = \hat{\boldsymbol{\phi}} H_\phi(\rho, t)$ and

$$\nabla \times \mathbf{H} = \hat{\mathbf{z}} \frac{1}{\rho} \frac{\partial(\rho H_\phi)}{\partial \rho} = \hat{\mathbf{z}} J_z = \hat{\mathbf{z}} \frac{I}{\pi a^2} \qquad (\rho < a).$$

Integration over ρ gives

$$H_\phi(\rho, t) = \frac{\rho I(t)}{2\pi a^2}.$$

The flux linkage for a section of wire of length h must be considered carefully. The differential flux passing through a rectangle of height dz extending from ρ to $\rho + d\rho$ is given by

$$d\Phi_m(\rho, t) = \hat{\boldsymbol{\phi}} B_\phi(\rho, t) \cdot \hat{\boldsymbol{\phi}} \, d\rho \, dz.$$

However, not all of the current is linked by this flux; only the current out to radius ρ will link, giving the total linking current as $I(t)\rho^2/a^2$. Hence the differential flux linkage is

$$d\Lambda(\rho, t) = \mu \frac{\rho \left[\frac{\rho^2}{a^2} I(t)\right]}{2\pi a^2} \, d\rho \, dz,$$

and the total flux linkage is

$$\Lambda(t) = \frac{\mu I(t)}{2\pi a^4} \int_{z=0}^{h} \int_{\rho=0}^{a} \rho^3 d\rho = h\frac{\mu I(t)}{8\pi}.$$

The inductance per unit length is therefore

$$\frac{L}{h} = \frac{\Lambda(t)}{I(t)} = \frac{\mu}{8\pi}. \tag{3.206}$$

Because this was computed using only the magnetic flux internal to the wire, it is called the *internal inductance* of the wire. Note that it is independent of a. ◀

When the link between magnetic flux density and electric field is considered, it is found that the current within a conductor cannot be uniform, but must obey a specific relationship called the diffusion equation, as discussed next.

3.6.2.3   MQS and conductors: diffusion, eddy currents, and skin depth

Consider a homogeneous conductor with permeability μ and conductivity σ. By Ampere's law (3.185),

$$\nabla \times \mathbf{H}(\mathbf{r}, t) = \mathbf{J}(\mathbf{r}, t) = \sigma \mathbf{E}(\mathbf{r}, t).$$

Taking the curl and substituting from Faraday's law (3.184), we have

$$\nabla \times \nabla \times \mathbf{H}(\mathbf{r}, t) = \sigma \nabla \times \mathbf{E}(\mathbf{r}, t) = -\mu\sigma \frac{\partial \mathbf{H}(\mathbf{r}, t)}{\partial t}.$$

Finally, expanding $\nabla \times \nabla \times \mathbf{H} = \nabla(\nabla \cdot \mathbf{H}) - \nabla^2 \mathbf{H}$ and using $\nabla \cdot \mathbf{H} = \mu \nabla \cdot \mathbf{B} = 0$ from (3.187), we get

$$\nabla^2 \mathbf{H}(\mathbf{r}, t) = \mu\sigma \frac{\partial \mathbf{H}(\mathbf{r}, t)}{\partial t}.$$

So, in a conductor, $\mathbf{H}$ satisfies a *diffusion equation*. For frequency domain fields this takes the form

$$\nabla^2 \tilde{\mathbf{H}}(\mathbf{r}, \omega) = j\omega\mu\sigma \tilde{\mathbf{H}}(\mathbf{r}, \omega) = \gamma^2 \tilde{\mathbf{H}}(\mathbf{r}, \omega), \tag{3.207}$$

where

$$\gamma = \sqrt{j\omega\mu\sigma} = \frac{1+j}{\delta}.$$

Here

$$\delta = \frac{1}{\sqrt{\pi f \mu \sigma}}$$

is called the *skin depth* for reasons discussed in Example 3.54. The fact that $\tilde{\mathbf{H}}$ obeys a diffusion equation suggests that the magnetic field *diffuses* into the conductor from the outer surface. The following examples will confirm this.

It is interesting to compare the diffusion equation to the Helmholtz (wave) equation that incorporates the effects of displacement current. In Chapter 4 we show that the magnetic field in a simple, homogeneous, conducting medium obeys the Helmholtz equation

$$\nabla^2 \tilde{\mathbf{H}}(\mathbf{r}, \omega) + k^2 \tilde{\mathbf{H}}(\mathbf{r}, \omega) = 0,$$

where

$$k = \omega\sqrt{\mu\epsilon}\sqrt{1 - j\frac{\sigma}{\omega\epsilon}}$$

is the wave number. At low frequencies where $\sigma/(\omega\epsilon) \gg 1$, we have $k^2 \approx -j\omega\mu\sigma = -\gamma^2$ and the Helmholtz equation reduces to the diffusion equation (3.207). Again, MQS analysis is appropriate for systems with slow time variation. The condition $\sigma/(\omega\epsilon) \gg 1$ states that the conduction current $\sigma\tilde{E}$ dominates the displacement current $j\omega\epsilon\tilde{E}$.

The electric field and current density also obey the diffusion equation. To show this, take the curl of Faraday's law (3.184) to get

$$\nabla \times \nabla \times \mathbf{E}(\mathbf{r}, t) = -\mu\frac{\partial}{\partial t}\nabla \times \mathbf{H}(\mathbf{r}, t)$$

and substitute from Ampere's law (3.185):

$$\nabla \times \nabla \times \mathbf{E}(\mathbf{r}, t) = -\mu\frac{\partial \mathbf{J}(\mathbf{r}, t)}{\partial t} = -\mu\sigma\frac{\partial \mathbf{E}(\mathbf{r}, t)}{\partial t}.$$

Then expand the curl and use $\nabla \cdot \mathbf{E} = \sigma\nabla \cdot \mathbf{J} = 0$ from the continuity equation (3.188):

$$\nabla^2 \mathbf{E}(\mathbf{r}, t) = \mu\sigma\frac{\partial \mathbf{E}(\mathbf{r}, t)}{\partial t},$$

or

$$\nabla^2 \tilde{\mathbf{E}}(\mathbf{r}, \omega) = \gamma^2 \tilde{\mathbf{E}}(\mathbf{r}, \omega)$$

in the frequency domain. Finally, since $\mathbf{E} = \mathbf{J}/\sigma$,

$$\nabla^2 \mathbf{J}(\mathbf{r}, t) = \mu\sigma\frac{\partial \mathbf{J}(\mathbf{r}, t)}{\partial t}$$

and

$$\nabla^2 \tilde{\mathbf{J}}(\mathbf{r}, \omega) = \gamma^2 \tilde{\mathbf{J}}(\mathbf{r}, \omega). \tag{3.208}$$

Finally, we show that the frequency-domain vector potential also obeys a diffusion equation. Substituting $\tilde{\mathbf{B}} = \mu\tilde{\mathbf{H}} = \nabla \times \tilde{\mathbf{A}}$ into Faraday's law (3.194), we obtain

$$\nabla \times (\tilde{\mathbf{E}} - j\omega\tilde{\mathbf{A}}) = 0$$

and thus

$$\tilde{\mathbf{E}} = -j\omega\tilde{\mathbf{A}} - \nabla\tilde{\phi}. \tag{3.209}$$

Substitution into Ampere's law (3.195) gives

$$\nabla \times \nabla \times \tilde{\mathbf{A}} = \mu\tilde{\mathbf{J}} = \mu\sigma\tilde{\mathbf{E}} = -j\omega\mu\sigma\tilde{\mathbf{A}} - \mu\sigma\nabla\tilde{\phi}.$$

By the identity $\nabla \times \nabla \times \tilde{\mathbf{A}} = \nabla(\nabla \cdot \tilde{\mathbf{A}}) - \nabla^2\tilde{\mathbf{A}}$ we have

$$\nabla(\nabla \cdot \tilde{\mathbf{A}} + \mu\sigma\tilde{\phi}) - \nabla^2\tilde{\mathbf{A}} = -j\omega\mu\sigma\tilde{\mathbf{A}}.$$

Finally, under the gauge condition

$$\nabla \cdot \tilde{\mathbf{A}} = -\mu\sigma\tilde{\phi} \tag{3.210}$$

we obtain

$$\nabla^2 \tilde{\mathbf{A}}(\mathbf{r}, \omega) = \gamma^2 \tilde{\mathbf{A}}(\mathbf{r}, \omega). \tag{3.211}$$

To see how diffusion affects the spatial dependence of the current, consider the following simple example.

▶ **Example 3.54:** Diffusion into a planar conductor

A conductor with uniform permeability μ and conductivity σ occupies the half space $x > 0$. The region $x \leq 0$ consists of free space. An electric field $\tilde{\mathbf{E}}(\mathbf{r}, \omega) = \hat{\mathbf{z}} \tilde{E}_0(\omega)$ exists in the free space region at the interface. Find the spatial distribution of current in the conductor.

Solution: By symmetry, the conduction current depends only on x: $\tilde{\mathbf{J}}(\mathbf{r}, \omega) = \hat{\mathbf{z}} \tilde{J}_z(x, \omega)$. The diffusion equation (3.208), i.e.,

$$\frac{d^2 \tilde{J}_z(x, \omega)}{dx^2} - \gamma^2 \tilde{J}_z(x, \omega) = 0,$$

has solution

$$\tilde{J}_z(x, \omega) = C_1(\omega) e^{-\gamma x} + C_2(\omega) e^{+\gamma x}$$

where C_1 and C_2 are constants in x and

$$\gamma(\omega) = \sqrt{j\omega\mu\sigma} = \frac{1+j}{\delta}. \tag{3.212}$$

We reject the growing exponential as nonphysical, leaving

$$\tilde{J}_z(x, \omega) = \sigma \tilde{E}_0(\omega) e^{-\gamma x} = \sigma \tilde{E}_0(\omega) e^{-\frac{x}{\delta}} e^{-j\frac{x}{\delta}} \tag{3.213}$$

where we have required continuity of $\tilde{\mathbf{E}}$ across the interface.

The quantity δ in (3.212), given by

$$\delta(\omega) = \frac{1}{\sqrt{\pi f \mu \sigma}},$$

is the conductor *skin depth*. It describes the depth of diffusion, or penetration, of the current into the conductor. Within one skin depth of the surface, the current density is reduced by a factor of $1/e$ to 37% of its value at the surface. The region $0 \leq x \leq \delta$ is referred to as the *skin* of the conductor since the majority of the current resides there.

The total current per unit width flowing in the conductor may be computed as

$$\frac{\tilde{I}(\omega)}{w} = \int_0^\infty \tilde{\mathbf{J}}(x, \omega) \cdot \hat{\mathbf{z}} \, dx = \int_0^\infty \sigma \tilde{E}_0(\omega) e^{-\frac{(1+j)}{\delta} x} \, dx = \sigma \tilde{E}_0(\omega) \frac{\delta}{1+j}. \tag{3.214}$$

Using this, we can define a *surface impedance* as

$$Z_s(\omega) = \frac{\tilde{E}_0(\omega)}{\tilde{I}(\omega)/w} = \frac{1+j}{\sigma\delta},$$

which has units of ohms. Often the units are given as "Ohms per square" ($\Omega/\square$) since each square section of the surface determines the impedance of the conductor beneath, regardless of the size of the square.

The magnetic field may be found from Ampere's law. By symmetry we have $\tilde{\mathbf{H}}(\mathbf{r}, \omega) = \hat{\mathbf{y}} \tilde{H}_y(x, \omega)$ and

$$\nabla \times \tilde{\mathbf{H}}(\mathbf{r}, \omega) = \hat{\mathbf{z}} \frac{d\tilde{H}_y(x, \omega)}{dx} = \hat{\mathbf{z}} \tilde{J}_z(x, \omega) = \hat{\mathbf{z}} \sigma \tilde{E}_0(\omega) e^{-\frac{1+j}{\delta} x}.$$

Integration yields

$$\tilde{\mathbf{H}}(\mathbf{r}, \omega) = -\hat{\mathbf{y}} \frac{\delta}{1+j} \sigma \tilde{E}_0(\omega) e^{-\frac{1+j}{\delta} x}.$$

If we apply the boundary condition on the *internal* magnetic field at the interface

$$\hat{\mathbf{n}} \times \tilde{\mathbf{H}}(\mathbf{r}, \omega) = -\hat{\mathbf{x}} \times \left[-\hat{\mathbf{y}} \frac{\delta}{1+j} \sigma \tilde{E}_0(\omega) e^{-\frac{1+j}{\delta}x} \right]\Bigg|_{x=0} = \sigma \tilde{E}_0(\omega) \frac{\delta}{1+j}, \qquad (3.215)$$

we get the total current per unit width (3.214) found earlier. ◄

The simple geometry used in this example leads to a conduction current that is linearly directed. In practical situations, the current is induced by the presence of a localized magnetic flux density imposed from the air region, possibly by a coil of wire. The resulting current will produce an additional magnetic field within the conductor that will also leak back into the air region and oppose the impressed magnetic flux. Since all currents under the assumption of MQS are steady, the currents induced in the conductor will form loops, or *eddies*. These eddy currents are significant sources of loss in transformers. However, they can also reveal structural defects within conductors in a process known as *nondestructive evaluation* [85].

The surface impedance found in the example has both real and imaginary components. The real component,

$$R_s = \frac{1}{\sigma\delta}, \qquad (3.216)$$

is called the *surface resistance*. The imaginary part shows that the surface impedance is inductively reactive and allows us to define a *surface inductance* L_s via

$$\omega L_s = \frac{1}{\sigma\delta}. \qquad (3.217)$$

Produced by the linkage of magnetic flux within the conductor, this parameter is also called the *internal inductance*. This effect was examined for a circular wire in Example 3.53. There, however, we assumed the current was uniform. We now know that current may concentrate near the surface of a planar conductor, and anticipate a similar effect in a wire. Let us revisit Example 3.53 without the assumption of a uniform current.

▶ **Example 3.55:** Internal impedance of a wire — diffusion and skin depth

A long cylindrical metallic wire of radius a, conductivity σ, and permeability μ lies along the z-axis. Compute the internal impedance by solving the diffusion equation. Plot the current density in the wire as a function of position for various values of a/δ, where δ is the skin depth.

Solution: As in Example 3.54, assume an external electric field $\tilde{\mathbf{E}}(\mathbf{r}, \omega) = \hat{\mathbf{z}}\tilde{E}_0(\omega)$ at the wire boundary in the air region. By symmetry, the current within the wire is $\tilde{\mathbf{J}}(\mathbf{r}, \omega) = \hat{\mathbf{z}}\tilde{J}_z(\rho, \omega)$. The diffusion equation (3.208) requires

$$\nabla^2 \tilde{\mathbf{J}}(\mathbf{r}, \omega) = \hat{\mathbf{z}}\nabla^2 \tilde{J}_z(\rho, \omega) = \hat{\mathbf{z}}\frac{1}{\rho}\frac{\partial}{\partial \rho}\left(\rho\frac{\partial \tilde{J}_z(\rho, \omega)}{\partial \rho} \right) = \hat{\mathbf{z}}\,\gamma^2 \tilde{J}_z(\rho, \omega).$$

Rearrangement gives the second-order ordinary differential equation

$$\frac{d^2 \tilde{J}_z}{d\rho^2} + \frac{1}{\rho}\frac{d\tilde{J}_z}{d\rho} - \gamma^2 \tilde{J}_z = 0.$$

This is the modified Bessel equation of order zero; its solution is

$$\tilde{J}_z(\rho, \omega) = C(\omega) I_0(\gamma\rho), \qquad (3.218)$$

where $I_0(x)$ is the modified Bessel function of the first kind and order zero and $C(\omega)$ is independent of ρ (we rejected the nonphysical solution $K_0(\gamma\rho)$). To find $C(\omega)$ we use $\tilde{\mathbf{E}} = \tilde{\mathbf{J}}/\sigma$ and require continuity of $\tilde{\mathbf{E}}$ at the boundary:

$$\tilde{E}_0(\omega) = C(\omega)\frac{I_0(\gamma a)}{\sigma}.$$

Thus $C(\omega) = \sigma\tilde{E}_0(\omega)/I_0(\gamma a)$ and

$$\tilde{J}_z(\rho,\omega) = \sigma\tilde{E}_0(\omega)\frac{I_0(\gamma\rho)}{I_0(\gamma a)}.$$

Recall that $\gamma = (1+j)/\delta$ where $\delta = 1/\sqrt{\pi f\mu\sigma}$ is the skin depth. Thus, the arguments of the Bessel functions are

$$\gamma\rho = (1+j)\frac{\rho}{\delta}, \qquad \gamma a = (1+j)\frac{a}{\delta},$$

and we are prompted to normalize all distances to the skin depth. Figure 3.31 shows the magnitude of the current density for various values of wire radius. When the radius is a skin depth or less, the current is nearly uniform. However, as the radius increases, the current begins to concentrate near the edge of the wire; for large a/δ it lies primarily within one skin depth of the surface (cf., Example 3.54).

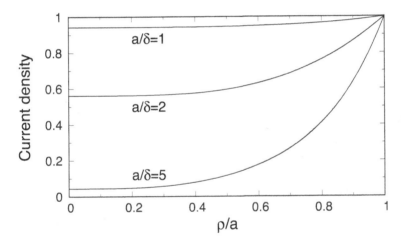

FIGURE 3.31
Normalized current density $|\tilde{J}_z(\rho)/\tilde{J}_z(a)|$ in a circular wire.

The total current carried by the wire is

$$\tilde{I}(\omega) = \int_S \hat{\mathbf{z}}\cdot\hat{\mathbf{z}}\tilde{J}_z(\rho,\omega)\,dS = \int_0^{2\pi}\int_0^a \sigma\tilde{E}_0(\omega)\frac{I_0(\gamma\rho)}{I_0(\gamma a)}\rho\,d\rho\,d\phi.$$

Integration over ϕ and the change of variables $x = \gamma\rho$ give

$$\tilde{I}(\omega) = \frac{2\pi\sigma\tilde{E}_0(\omega)}{\gamma^2 I_0(\gamma a)}\int_0^{\gamma a} xI_0(x)\,dx.$$

The integral is $\int xI_0(x)\,dx = xI_1(x)$. Evaluation at the limits gives

$$\tilde{I}(\omega) = \frac{2\pi a\sigma\tilde{E}_0(\omega)}{\gamma}\frac{I_1(\gamma a)}{I_0(\gamma a)}.$$

Finally, the internal impedance per unit length is just

$$\frac{Z_i}{h} = \frac{\tilde{E}_0(\omega)}{\tilde{I}(\omega)} = \frac{\gamma}{2\pi a \sigma} \frac{I_0(\gamma a)}{I_1(\gamma a)} = \frac{1}{2\pi a \sigma \delta} F(\gamma a), \tag{3.219}$$

where

$$F(\gamma a) = (1 + j) \frac{I_0(\gamma a)}{I_1(\gamma a)} = F_r(\gamma a) + j F_i(\gamma a). \tag{3.220}$$

To determine the internal inductance, note that Z_i is complex with $\mathrm{Im}\, Z_i > 0$. Thus

$$\frac{Z_i}{h} = \frac{R_i}{h} + j\omega \frac{L_i}{h}.$$

With this, we can write the internal inductance per unit length as

$$\frac{L_i}{h} = \frac{1}{2\pi a \delta \sigma \omega} F_i(\gamma a) = \left(\frac{2\delta}{a}\right) \frac{\mu}{8\pi} F_i(\gamma a) \tag{3.221}$$

and the resistance per unit length as

$$\frac{R_i}{h} = \frac{1}{2\pi a \delta \sigma} F_r(\gamma a) = \left(\frac{a}{2\delta}\right) \frac{1}{\pi a^2 \sigma} F_r(\gamma a). \ \blacktriangleleft \tag{3.222}$$

Because the Bessel function arguments in the internal impedance are complex, it is difficult to gain physical understanding of the dependence of Z_i on frequency. It is therefore helpful to examine low- and high-frequency limits of the expression.

▶ **Example 3.56:** Internal impedance of a wire at low frequency

Specialize (3.219) to the low-frequency case.

Solution: The condition $|\gamma a| \ll 1$ implies $\sqrt{\omega \mu \sigma}\, a \ll 1$ or $a \ll \delta/\sqrt{2}$, so the wire radius is much less than a skin depth. Hence we may use two-term approximations for the Bessel functions:

$$I_0(z) \approx 1 + \frac{z^2}{4},$$

$$I_1(z) \approx \frac{z}{2} + \frac{z^3}{16}.$$

Now (3.218) shows that the current density in the wire is nearly constant with ρ. So the low-frequency assumption is consistent with the uniform current assumption of Example 3.53.

Substituting the Bessel function approximations into (3.220), we get

$$F(\gamma a) \approx (1 + j) \frac{1 + \frac{\gamma^2 a^2}{4}}{\frac{\gamma a}{2}\left(1 + \frac{\gamma^2 a^2}{8}\right)}$$

$$\approx (1 + j)\frac{2}{\gamma a}\left(1 + \frac{\gamma^2 a^2}{4}\right)\left(1 - \frac{\gamma^2 a^2}{8}\right),$$

where we have used $1/(1 + x) \approx 1 - x$. Retaining terms to order γ^2, we have

$$F(\gamma a) \approx (1 + j)\frac{2}{\gamma a}\left(1 + \frac{\gamma^2 a^2}{8}\right).$$

Finally, using $\gamma = (1 + j)/\delta$, we have

$$F(\gamma a) \approx \frac{2\delta}{a} + j\frac{a}{2\delta}.$$

Hence the low-frequency resistance per unit length is

$$\frac{R_{\text{low}}}{h} \approx \frac{1}{\pi a^2 \sigma}, \tag{3.223}$$

while the low-frequency inductance per unit length is

$$\frac{L_{\text{low}}}{h} \approx \frac{\mu}{8\pi}. \tag{3.224}$$

The resistance per unit length matches the value (3.182) computed in Example 3.50 under EQS. The internal inductance per unit length matches the value (3.206) computed using the constant current approximation in Example 3.53. So the general expression reduces to that found earlier for low frequency. ◄

▶ **Example 3.57:** Internal impedance of a wire at high frequency

Specialize (3.219) to the high frequency case.

Solution: At high frequencies we have $|\gamma a| \gg 1$, hence $\sqrt{\omega\mu\sigma}a \gg 1$ so that $a \gg \delta$. Using the large argument approximations [1]

$$I_0(z) \approx \frac{e^z}{\sqrt{2\pi z}}\left(1 + \frac{1}{8z}\right), \qquad I_1(z) \approx \frac{e^z}{\sqrt{2\pi z}}\left(1 - \frac{3}{8z}\right),$$

in (3.220), we obtain

$$F(\gamma a) \approx (1 + j)\left(1 + \frac{1}{8\gamma a}\right)\left(1 + \frac{3}{8\gamma a}\right).$$

Keeping terms to order $1/(\gamma a)$ and remembering that $\gamma = (1 + j)/\delta$, we have $F(\gamma a) \approx 1 + j + \delta/2a$. Thus we have the high-frequency resistance per unit length

$$\frac{R_{\text{high}}}{h} \approx \left(\frac{a}{2\delta} + \frac{1}{4}\right)\frac{R_{\text{low}}}{h}, \tag{3.225}$$

and the high-frequency inductance per unit length

$$\frac{L_{\text{high}}}{h} \approx \left(\frac{2\delta}{a}\right)\frac{L_{\text{low}}}{h} = \frac{\mu}{4\pi}\left(\frac{\delta}{a}\right). \tag{3.226}$$

It is interesting to compare the high-frequency resistance given by (3.225) to the low-frequency resistance for a wire where the current is concentrated within one skin depth of the surface, but uniformly distributed, as shown in Figure 3.32. Use of (3.223) with the annular area $A = \pi a^2 - \pi(a - \delta)^2$ gives the resistance per unit length

$$\frac{R_{\text{ann}}}{h} = \frac{1}{\sigma\left[\pi a^2 - \pi(a - \delta)^2\right]} = \frac{1}{\sigma\pi a\delta\left[2 - \frac{\delta}{a}\right]}.$$

Thus we have

$$\frac{R_{\text{ann}}}{h} = \left(\frac{a}{2\delta}\frac{1}{1 - \frac{\delta}{2a}}\right)\frac{R_{\text{low}}}{h} \approx \left(\frac{a}{2\delta} + \frac{1}{4}\right)\frac{R_{\text{low}}}{h},$$

where we have used $1/(1-x) \approx 1+x$. This expression is identical to the high-frequency approximation for the resistance per unit length (3.225). Hence the annular model of Figure 3.32 is often used to compute high-frequency resistance.

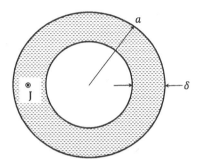

FIGURE 3.32

Current in a wire at high frequency uniformly distributed in an annulus within one skin depth of the surface. ◄

▶ **Example 3.58:** Comparison of wire internal impedance formulas

Compare the exact formula for the internal impedance of a wire to the low and high-frequency approximations by plotting values as a function of a/δ.

Solution: It is most efficient to plot the inductance and resistance relative to their low-frequency values. Figure 3.33 shows the inductance L_i found using (3.221) normalized to the low-frequency value from (3.224). (Note that this ratio is independent of wire length.) Also plotted is the high-frequency result (3.226) normalized to (3.224). For $0 < a/\delta < 1$ the low-frequency formula gives very good results. For $a/\delta > 3$, the high-frequency formula is very accurate. Similar results are shown in Figure 3.34 for the internal resistance. Here R_i is plotted using (3.222) normalized to (3.223), along with the high-frequency result (3.225) normalized to (3.223). Again, for $0 < a/\delta < 1$ the low-frequency formula gives very good results. The high-frequency result is quite accurate for $a/\delta > 2$.

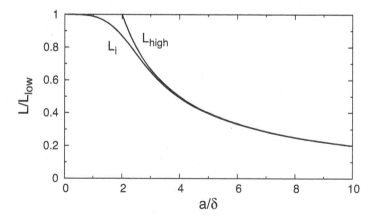

FIGURE 3.33

Internal inductance of a wire normalized to its low-frequency value.

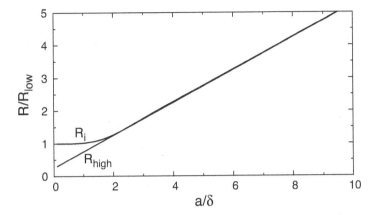

FIGURE 3.34
Internal resistance of a wire normalized to its low-frequency value. ◀

3.7 Application: electromagnetic shielding

Shielding is an essential concept in the area of *electromagnetic compatibility* [146]. There are many instances in which an undesired electromagnetic field is either created by an electrical system or interacts and interferes with a device or system. A simple solution is to enclose the system in a metal container, called a *shield*, thereby disrupting the flow of electromagnetic energy or lines of flux into or out of the system. Quite simply, the presence of the shield reduces the strength of the field from one side of the enclosure to the other. The physical basis for this reduction depends on the properties of the fields. In the case of static electric fields, the charges induced on the surface of a metal enclosure produce an oppositely directed field that cancels the impressed field. Alternatively, the enclosure may be constructed from a material with a large dielectric constant, in which case the polarization charge creates a canceling field. In the case of magnetostatic fields, a highly permeable material may be used to reduce the field by inducing magnetization currents in the shield. Fields associated with an electromagnetic wave may be reduced by a combination of reflection and absorption of wave energy. Dynamic fields that may be better described using quasistatic concepts may also be reduced by shielding, in which case canceling effects are often produced by the induction of eddy currents within the shield, or by ducting the field through the shield. Each of these situations is considered below. In every case, the amount by which the field is reduced may be described using the *shielding effectiveness.*

3.7.1 Shielding effectiveness

Shielding effectiveness, designated SE and described in dB, quantifies the amount that the field strength is reduced due to the presence of the shield [29]. For the case of electrostatics, or quasistatics with the electric field dominant, the shielding effectiveness is taken as the ratio of the electric field magnitudes on either side of the shield:

$$SE = 20 \log_{10} \frac{|E_i|}{|E_t|}.$$

Here E_i is some measure of the field on the side of the shield where the source of the field resides, and E_t is some measure of the field on the other side. For instance, E_i could be some component of the impressed (or applied) field, while E_t could be the magnitude of the total field. Since the purpose of the shield is to reduce the field strength, SE is normally a positive number in dB.

For the case of magnetostatics, or quasistatics with the magnetic field dominant, the shielding effectiveness is defined as

$$SE = 20 \log_{10} \frac{|H_i|}{|H_t|}.$$

When the field on the source side is associated with an impinging electromagnetic wave, generally the electric field ratio is used, since $\mathbf{E}$ can be more easily measured, and the ratio of $\mathbf{E}$ to $\mathbf{H}$ is often prescribed by the physics of the wave. In this case E_i is usually the amplitude of the field incident on the shield, and E_t is the amplitude of the transmitted field.

Shielding effectiveness is not a straightforward concept. For complicated structures it depends on geometry and polarization, and for electromagnetic shielding it depends on the arrival angle of the wave. However, several simple canonical problems are available to illustrate the underlying properties of shielding enclosures.

3.7.2 Electrostatic shielding

Electrostatic shielding is usually accomplished with properly grounded conductors. If the conductors are perfect, the shield can completely eliminate the presence of the electrostatic field. This can be shown quite elegantly for shields of arbitrary shape. Partial shielding may be accomplished using material with a high dielectric constant.

3.7.2.1 Shielding using perfectly conducting enclosures

An effective way to provide electrostatic shielding is to use a metallic enclosure. If the metal is assumed to be perfectly conducting, then it is easy to show that no electric field penetrates the shield.

Consider a closed, grounded, perfectly conducting shell with charge outside but not inside (Figure 3.35). By (3.51) the potential at points inside the shell may be written in terms of the Dirichlet Green's function as

$$\Phi(\mathbf{r}) = \oint_{S_B} \Phi(\mathbf{r}') \frac{\partial G_D(\mathbf{r}|\mathbf{r}')}{\partial n'} \, dS',$$

where S_B is tangential to the inner surface of the shell and we have used $\rho = 0$ within the shell. Because $\Phi(\mathbf{r}') = 0$ for all $\mathbf{r}'$ on S_B, we have $\Phi(\mathbf{r}) = 0$ everywhere in the region enclosed by the shell. This result is independent of the charge outside the shell, and the interior region is "shielded" from the effects of that charge.

Conversely, consider a grounded perfectly conducting shell with charge contained inside. If we surround the outside of the shell by a surface S_1 and let S_B recede to infinity, then (3.51) becomes

$$\Phi(\mathbf{r}) = \lim_{S_B \to \infty} \oint_{S_B} \Phi(\mathbf{r}') \frac{\partial G_D(\mathbf{r}|\mathbf{r}')}{\partial n'} \, dS' + \oint_{S_1} \Phi(\mathbf{r}') \frac{\partial G_D(\mathbf{r}|\mathbf{r}')}{\partial n'} \, dS'.$$

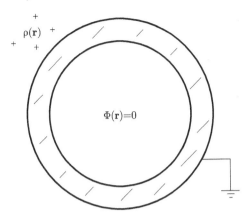

FIGURE 3.35
Electrostatic shielding by a perfectly conducting shell.

Again there is no charge in V (since the charge lies completely inside the shell). The contribution from S_B vanishes. Since S_1 lies adjacent to the outer surface of the shell, $\Phi(\mathbf{r}') \equiv 0$ on S_1. Thus $\Phi(\mathbf{r}) = 0$ for all points outside the conducting shell.

3.7.2.2 Perfectly conducting enclosures with apertures

A metallic enclosure may have apertures to allow passage of cables or to provide ventilation. The presence of apertures degrades the shielding effectiveness, and it is worthwhile to explore the penetration of the electrostatic field by considering a simple canonical problem.

Figure 3.36 shows an infinite conducting screen, grounded at zero potential, with a circular aperture. Above the screen is a uniform impressed field $\mathbf{E}_0 = -E_0\hat{\mathbf{z}}$, associated with potential $\Phi(\mathbf{r}) = E_0 z$. The field induces charge on the conductor with azimuthal symmetry. This in turn produces a scattered potential $\Phi^s(\rho, z)$ that is azimuthally symmetric and even in z, and an electric field $\mathbf{E}^s(\rho, z)$ that is azimuthally symmetric and odd in z. The total potential is

$$\Phi(\rho, z) = \begin{cases} -E_0 z + \Phi^s(\rho, z), & z > 0, \\ \Phi^s(\rho, z), & z < 0, \end{cases}$$

while the z-component of the total electric field is

$$E_z(\rho, z) = \begin{cases} -E_0 + E_z^s(\rho, z), & z > 0, \\ E_z^s(\rho, z), & z < 0. \end{cases}$$

Within the aperture, the normal component of the electric field is continuous:

$$-E_0 + E_z^s(\rho, 0^+) = E_z^s(\rho, 0^-) \qquad (0 \leq \rho < a).$$

Hence, because $E_z^s(\rho, z)$ is odd,

$$E_z^s(\rho, 0^+) = -E_z^s(\rho, 0^-) = \frac{E_0}{2} \qquad (0 \leq \rho < a)$$

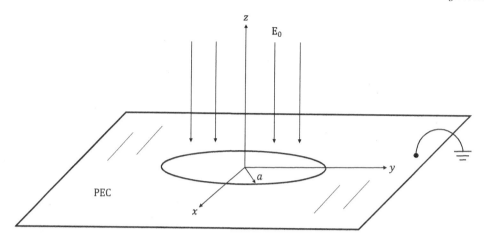

FIGURE 3.36
Conducting screen with circular aperture.

or

$$\left.\frac{\partial \Phi^s(\rho, z)}{\partial z}\right|_{z=0^+} = -\frac{E_0}{2} \qquad (0 \le \rho < a). \tag{3.227}$$

Since the potential is azimuthally symmetric, it can be represented using the inverse transform formula (A.58):

$$\Phi^s(\rho, z) = \int_0^\infty B(k_\rho)e^{-k_\rho z} J_0(k_\rho \rho)\, dk_\rho \qquad (z > 0).$$

Substituting this into (3.227) and noting that the scattered potential must vanish on the conducting plane, we are led to a set of *dual integral equations* for the amplitude function $B(k_\rho)$:

$$\int_0^\infty B(k_\rho)J_0(k_\rho \rho)k_\rho\, dk_\rho = \frac{E_0}{2} \qquad (0 \le \rho < a),$$

$$\int_0^\infty B(k_\rho)J_0(k_\rho \rho)\, dk_\rho = 0 \qquad (\rho \ge a).$$

The solution to these is [92]

$$B(k_\rho) = \frac{E_0 a^2}{\pi} j_1(k_\rho a), \tag{3.228}$$

where $j_1(x)$ is the ordinary spherical Bessel function of the first kind. Finally, with (3.228) we have the z-component of the field penetrating the slot:

$$E_z^s(\rho, z) = -\int_0^\infty \frac{E_0 a^2}{\pi} j_1(k_\rho a)e^{-k_\rho |z|} J_0(k_\rho \rho)\, k_\rho\, dk_\rho \qquad (z < 0).$$

This may be written in normalized form through the use of the formula

$$j_1(x) = \frac{\sin x}{x^2} - \frac{\cos x}{x}$$

and the substitution $u = k_\rho a$:

$$\frac{E_z^s(\rho, z)}{E_0} = -\frac{1}{\pi} \int_0^\infty e^{-u\frac{|z|}{a}} \left(\frac{\sin u}{u} - \cos u \right) J_0 \left(u\frac{\rho}{a} \right) du \qquad (z < 0). \qquad (3.229)$$

This canonical problem permits us to explore the charge induced on the underside of the screen as the rim of the hole is approached. The conductor boundary condition $\rho_s(\rho) = \hat{n} \cdot \mathbf{D}$ yields $\rho_s(\rho) = -\epsilon_0 E_z^s(\rho, 0)$, and thus

$$\frac{\rho_s(\rho)}{\epsilon_0 E_0} = \frac{1}{\pi} \int_0^\infty \left(\frac{\sin u}{u} - \cos u \right) J_0 \left(u\frac{\rho}{a} \right) du.$$

Using the handbook integrals [74]

$$\int_0^\infty J_0(\alpha x) \cos(x)\, dx = \frac{1}{\sqrt{\alpha^2 - 1}} \qquad (\alpha > 1),$$

$$\int_0^\infty J_0(\alpha x) \frac{\sin x}{x}\, dx = \csc^{-1}(\alpha) \qquad (\alpha > 1),$$

we have

$$\frac{\rho_s(\rho)}{\epsilon_0 E_0} = \frac{1}{\pi} \left[\csc^{-1} \left(\frac{\rho}{a} \right) - \frac{1}{\sqrt{\left(\frac{\rho}{a} \right)^2 - 1}} \right] \qquad (\rho > a). \qquad (3.230)$$

Thus, as $\rho \to a^+$,

$$\frac{\rho_s(\rho)}{\epsilon_0 E_0} \approx \frac{1}{2} - \frac{1}{\pi\sqrt{2}} \frac{1}{\sqrt{\frac{\rho}{a} - 1}}. \qquad (3.231)$$

We see that the charge near the sharp edge of the hole is singular, varying as an inverse square root of distance from the edge, as predicted by (3.81).

To determine the shielding effectiveness, the field below the conducting screen must be computed. The field is quite large immediately below the edge of the aperture, due to the charge singularity discussed above. However, the field rapidly decays away from the edge, and it is helpful to examine the field along the z-axis to determine how the shielding effectiveness depends on distance beneath the screen. Since there is only a z-component of the scattered field on the z-axis, we may set $\rho = 0$ in (3.229) to obtain

$$\frac{E_z^s(0, z)}{E_0} = -\frac{1}{\pi} \int_0^\infty e^{-u\frac{|z|}{a}} \left(\frac{\sin u}{u} - \cos u \right) du \qquad (z < 0).$$

This can be evaluated in closed form using the integrals [74]

$$\int_0^\infty e^{-\beta x} \cos x\, dx = \frac{\beta}{\beta^2 + 1}, \qquad \int_0^\infty e^{-\beta x} \frac{\sin x}{x}\, dx = \tan^{-1} \frac{1}{\beta},$$

which give

$$\frac{E_z^s(0, z)}{E_0} = -\frac{1}{\pi} \left(\tan^{-1} \frac{a}{|z|} - \frac{1}{\frac{|z|}{a} + \frac{a}{|z|}} \right) \qquad (z < 0). \qquad (3.232)$$

As expected, we have $E_z^s(0, z)/E_0 \to -1/2$ as $z \to 0^-$ (see Problem 3.48).

► **Example 3.59:** Charge density near a circular aperture in a conducting screen

A circular aperture of radius a is cut into a conducting screen as shown in Figure 3.36. Plot the density of the charge induced on the underside of the screen as a function of the distance from the aperture edge. Compare to the expected square-root edge singularity.

Solution: The charge density, given by (3.230), is plotted in Figure 3.37. Also shown is the approximate charge density (3.231). As $\rho \to a^+$, the charge density displays the square-root edge singularity. The approximation is quite good for distances up to about $0.1a$ from the aperture. Beyond that, the simple approximation rapidly loses accuracy.

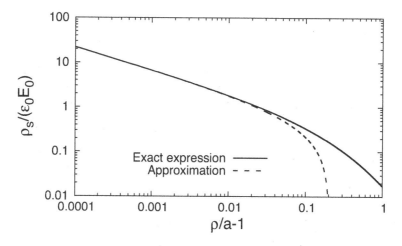

FIGURE 3.37

Charge density on a conducting screen with a circular aperture. ◄

► **Example 3.60:** Electrostatic shielding by a PEC screen with a hole

A circular aperture of radius a is cut into a conducting screen as shown in Figure 3.36. Plot the shielding effectiveness as a function of distance from the hole along the z-axis. Determine an approximation for the shielding effectiveness, valid when the observation point is far from the hole, and compare to the exact result.

Solution: Figure 3.38 shows a plot of the shielding effectiveness

$$SE = 20 \log_{10} \frac{|E_0|}{|E_z^s|}$$

as a function of the distance from the hole along the z-axis, found using (3.232). At $z = 0$, at the center of the hole, we have $SE = 20 \log_{10}(1/0.50) = 6$ dB. The shielding effectiveness increases with increasing depth. An expression for SE at large depth may be obtained by using the small argument approximation $\tan^{-1}(x) \approx x - \frac{1}{3}x^3$ in (3.232):

$$\frac{E_z^s(0, z)}{E_0} \approx \frac{2}{3\pi} \left| \frac{a}{z} \right|^3 .$$

With this,

$$SE \approx 13.46 \text{ dB} + 60 \log_{10} |z/a| \qquad (|z/a| \gg 1). \qquad (3.233)$$

Thus, when the distance from the hole is large, the shielding effectiveness increases by 60 dB for every factor of 10 increase in $|z|$. Equation (3.233) is plotted as the dashed line in Figure 3.38; it is clear that the approximation is quite good (within a difference of less than a dB) beyond about $|z/a| = 3$.

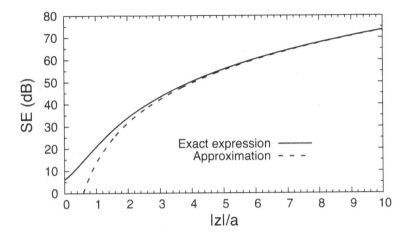

FIGURE 3.38
Electrostatic shielding effectiveness of a PEC screen with a circular hole. ◄

► **Example 3.61:** Shielding effectiveness for a PEC screen with a hole

At what distance beneath a PEC screen will a 40 dB shielding effectiveness be achieved when the screen has a hole of radius a? Repeat for $SE = 80$ dB.

Solution: Restricting ourselves to the axis of the hole, we can use (3.232) or obtain the shielding effectiveness from Figure 3.38. We see that $|z/a|$ must be about 2.62 for $SE = 40$ dB. Thus, the field drops from $0.5E_0$ at the center of the hole to $0.01E_0$ at a distance a bit more than one hole diameter. If we use the approximation (3.233) we obtain $|z/a| \approx 2.77$, which is reasonably close even though $|z/a|$ is not significantly larger than unity.

We might try to use Figure 3.38 for $SE = 80$ dB, but the curve is not plotted far enough to find this. Instead, since $|z/a|$ must be larger than 10, we note that $|z/a| \gg 1$ and use the approximation (3.233). Setting

$$13.46 \text{ dB} + 60 \log_{10} |z/a| = 80 \text{ dB}$$

we find $|z/a| = 12.85$. ◄

3.7.2.3 Shielding with high-permittivity dielectric materials

A cylindrical shell is often used as a canonical problem to study shielding effectiveness. Its two-dimensional nature simplifies analysis, and limiting cases such as a very thin shell or a high permittivity shell may be easily examined.

Consider a cylindrical shell of dielectric material in free space, aligned along the z-axis and immersed in a uniform applied field $\mathbf{E}_0 = E_0\hat{\mathbf{x}}$ (Figure 3.39). The inner radius of the shell is a, the outer radius is b, and the permittivity of the shell is ϵ. To determine the shielding effectiveness, we must compute the internal field. Assume the impressed field

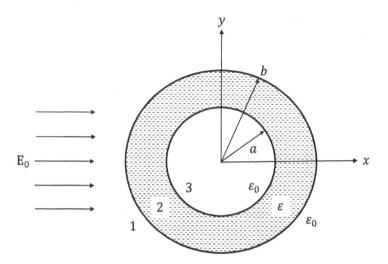

FIGURE 3.39

A cylindrical dielectric shell illuminated by a uniform electric field.

is created by sources far removed from the cylinder. Then we can express the potentials as solutions to Laplace's equation in cylindrical coordinates.

Let us begin by examining the separation of variables solution (A.128). Since $\mathbf{E}_0 = -\nabla\Phi_0$ where $\Phi_0 = -E_0 x$, only the term with $k_\phi = 1$ is needed to represent the applied field. By continuity of the potentials, this will also be true for the scattered potential in each region. In region 1 we have both the scattered and applied potentials, whereas in regions 2 and 3 we have only scattered potentials:

$$\Phi_1(\mathbf{r}) = Ab^2 E_0 \rho^{-1} \cos\phi - E_0\rho\cos\phi = Ab^2 E_0 \frac{x}{x^2 + y^2} - E_0 x,$$

$$\Phi_2(\mathbf{r}) = Bb^2 E_0 \rho^{-1} \cos\phi + CE_0\rho\cos\phi = Bb^2 E_0 \frac{x}{x^2 + y^2} + CE_0 x,$$

$$\Phi_3(\mathbf{r}) = DE_0\rho\cos\phi = DE_0 x.$$

Here we have chosen the decaying scattered potential in region 1 so that the total potential remains finite as $\rho \to \infty$, and the growing potential in region 3 so that the potential is finite at $\rho = 0$; we must use both in region 2. The coefficients A, B, C, D are found by applying boundary conditions at $\rho = a$ and $\rho = b$. By continuity of the scalar potential across each boundary, we have

$$A - 1 = B + C, \qquad \frac{b^2}{a^2}B + C = D.$$

By (3.17), the quantity $\epsilon \partial\Phi/\partial\rho$ is also continuous at $\rho = a$ and $\rho = b$; this gives two more equations:

$$-A - 1 = -\epsilon_r B + \epsilon_r C, \qquad -\frac{b^2}{a^2}B + C = \frac{1}{\epsilon_r}D.$$

The solutions are

$$A = \left(1 - \frac{a^2}{b^2}\right)\frac{\epsilon_r^2 - 1}{K}, \qquad B = -2\frac{a^2}{b^2}\frac{\epsilon_r - 1}{K}, \qquad C = -2\frac{\epsilon_r + 1}{K}, \qquad D = -4\frac{\epsilon_r}{K},$$

where

$$K = (\epsilon_r + 1)^2 - (\epsilon_r - 1)^2 \frac{a^2}{b^2}.$$

The electric field in each region may be computed using $\mathbf{E} = -\nabla\Phi$. Taking the gradient in rectangular coordinates, we get

$$\frac{\mathbf{E}_1}{E_0} = \hat{\mathbf{x}}\left[\frac{X^2 - Y^2}{(X^2 + Y^2)^2}A + 1\right] + \hat{\mathbf{y}}\left[\frac{2XY}{(X^2 + Y^2)^2}A\right], \qquad (3.234)$$

$$\frac{\mathbf{E}_2}{E_0} = \hat{\mathbf{x}}\left[\frac{X^2 - Y^2}{(X^2 + Y^2)^2}B - C\right] + \hat{\mathbf{y}}\left[\frac{2XY}{(X^2 + Y^2)^2}B\right], \qquad (3.235)$$

$$\frac{\mathbf{E}_3}{E_0} = -\hat{\mathbf{x}}D. \qquad (3.236)$$

Here $X = x/b$, $Y = y/b$.

▶ **Example 3.62:** Shielding by a dielectric shell

Consider a dielectric shell with $a/b = 0.9$ and $\epsilon_r = 10$. Plot the electric field streamlines for a shell immersed in a uniform electric field $\mathbf{E} = E_0\,\hat{\mathbf{x}}$.

Solution: The streamlines may be plotted using the total electric field given by (3.234)–(3.236). Since the streamlines follow the vector field, they may be traced by projecting along the direction of the field. The result is shown in Figure 3.40.

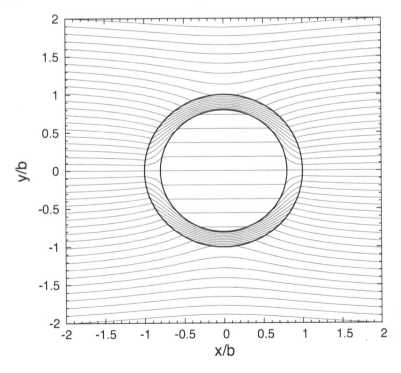

FIGURE 3.40

Electric field streamlines for a dielectric shell immersed in a uniform static electric field.

Note that the electric field lines are concentrated within the wall of the dielectric shell and are thus ducted away from the interior. This effect causes the field strength within the interior to be lower than the applied field strength E_0. Ducting of the electric flux is enhanced as the dielectric constant is increased, but since materials with very high dielectric constants are difficult to create, significant levels of shielding effectiveness are usually not attainable by this method. In contrast, the availability of high-permeability materials allows the technique to be used to provide considerable levels of magnetic shielding effectiveness. This is explored in examples below. ◀

From (3.236) we can write the field interior to the dielectric shell as

$$\frac{\mathbf{E}_t}{E_0} = \kappa\hat{\mathbf{x}},$$

where $\kappa = 4\epsilon_r/K$. This field is uniform, and, since $\kappa < 1$ for $\epsilon_r > 1$, it is weaker than the applied field. So a simple definition of shielding effectiveness for this canonical problem is

$$SE = 20\log_{10}\frac{|E_0|}{|E_t|} = 20\log_{10}\frac{K}{4\epsilon_r}. \tag{3.237}$$

To identify conditions under which the shield will be effective, write

$$\frac{K}{4\epsilon_r} = \frac{b^2 + a^2}{2b^2}\left(\frac{\epsilon_r^2 + 1}{2\epsilon_r}\Delta\frac{b+a}{b^2+a^2} + 1\right),$$

where $\Delta = b - a$ is the shield thickness. A large shielding effectiveness requires the first term in brackets to be much larger than unity, which is satisfied when

$$\frac{\epsilon_r^2 + 1}{\epsilon_r} \gg 4\frac{b}{\Delta}.$$

Since $\Delta < b$, we can only have a good shielding effectiveness when $\epsilon_r \gg 1$, in which case

$$\frac{K}{4\epsilon_r} \approx \left(1 + \frac{a}{b}\right)\frac{\epsilon_r}{4}\frac{\Delta}{b}.$$

Finally, if the shield is also thin so that $a \approx b$, then

$$SE = 20\log_{10}\left(\frac{\epsilon_r\Delta}{2b}\right). \tag{3.238}$$

▶ **Example 3.63:** Electrostatic shielding using a medium-permittivity dielectric shell

Compute the shielding effectiveness for the dielectric shell of Example 3.62.

Solution: Using $a/b = 0.9$ and $\epsilon_r = 10$, we compute $K = 55.39$. Substituting into (3.237) we get $SE = 2.8\,\text{dB}$. So the interior field is reduced only slightly below the external field. Significant reductions require very large dielectric constants, as demonstrated next. ◀

▶ **Example 3.64:** Electrostatic shielding using a high-permittivity dielectric shell

A dielectric must have a very large permittivity to provide useful shielding. For a shield with $a/b = 0.99$, what value of ϵ_r is needed to attain 40 dB shielding effectiveness?

Solution: Since the dielectric constant is large and $a \approx b$, we can use (3.238):

$$\frac{\epsilon_r \Delta}{2b} = 10^{40/20} = 100.$$

Using $\Delta = b - a$, we obtain

$$\frac{\epsilon_r}{2} \left(1 - \frac{a}{b}\right) = 100$$

and hence $\epsilon_r = 20000$. It is unlikely that a dielectric constant will be this large. ◀

3.7.3   Magnetostatic shielding

The canonical problem of a cylindrical shell of magnetic material can be easily adapted from the electrostatic canonical problem considered above. Assume the shell is immersed in an applied magnetic field $\mathbf{H}_0 = H_0\hat{\mathbf{x}}$. Write $\mathbf{H}_0 = -\nabla\Phi_0$ where $\Phi_0 = H_0 x$ is the impressed magnetic scalar potential, and proceed as in the electrostatic case. We find that we need only change ϵ_r to μ_r to obtain the shielding effectiveness. For a thin shield $(a \approx b)$ we have

$$SE = 20\log_{10}\left(\frac{\mu_r \Delta}{2b}\right). \tag{3.239}$$

▶ **Example 3.65:** Magnetostatic shielding using a high-permeability cylindrical shell

Repeat Example 3.64 for the case of magnetostatic shielding. Again let $a/b = 0.99$, and assume the permeability of the shell is large. What value of μ_r is needed to achieve 40 dB shielding effectiveness?

Solution: Since the permeability is large and $a \approx b$, we can use (3.239):

$$\frac{\mu_r \Delta}{2b} = 10^{40/20} = 100.$$

With $\Delta = b - a$ we get

$$\frac{\mu_r}{2} \left(1 - \frac{a}{b}\right) = 100$$

and hence $\mu_r = 20000$, a value that is not unreasonable. ◀

A canonical problem often used to predict shielding effectiveness is the spherical shell. Although for thin shells of high permeability, the SE predicted using the spherical shell is similar to that predicted by the cylindrical shell, it is worth examining the spherical shell as a three-dimensional alternative.

Consider a spherical shell of highly permeable material (Figure 3.41); assume it is immersed in a uniform impressed field $\mathbf{H}_0 = H_0\hat{\mathbf{z}}$. We wish to determine the internal field and the factor by which it is reduced from the external applied field. Because there are no sources (the applied field is assumed to be created by sources far removed), we may use magnetic scalar potentials to represent the fields everywhere. We may represent the scalar potentials using a separation of variables solution to Laplace's equation, with a contribution only from the $n = 1$ term in the series. In region 1 we have both scattered and applied potentials, where the applied potential is just $\Phi_0 = -H_0 z = -H_0 r\cos\theta$,

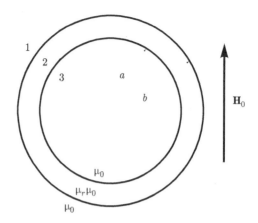

FIGURE 3.41
Spherical shell of magnetic material.

since $\mathbf{H}_0 = -\nabla \Phi_0 = H_0 \hat{\mathbf{z}}$. We have

$$\Phi_1(\mathbf{r}) = A H_0 b^3 r^{-2} \cos\theta - H_0 r \cos\theta, \tag{3.240}$$
$$\Phi_2(\mathbf{r}) = (B H_0 b^3 r^{-2} + C H_0 r) \cos\theta, \tag{3.241}$$
$$\Phi_3(\mathbf{r}) = D H_0 r \cos\theta. \tag{3.242}$$

We choose (3.82) for the scattered potential in region 1 so that it decays as $\mathbf{r} \to \infty$, and (3.83) for the scattered potential in region 3 so that it remains finite at $r = 0$. In region 2 we have no restrictions and therefore include both contributions. The coefficients A, B, C, D are found by applying the appropriate boundary conditions at $r = a$ and $r = b$. By continuity of the scalar potential across each boundary, we have

$$A - 1 = B + C, \qquad B\frac{b^3}{a^3} + C = D.$$

By (3.125), the quantity $\mu \partial \Phi / \partial r$ is also continuous at $r = a$ and $r = b$; this gives two more equations:

$$-2A - 1 = -2B\mu_r + C\mu_r, \qquad -2B\frac{b^3}{a^3} + C = \frac{D}{\mu_r}.$$

Simultaneous solution yields

$$D = -\frac{9\mu_r}{K}$$

where

$$K = (2 + \mu_r)(1 + 2\mu_r) - 2(a/b)^3(\mu_r - 1)^2.$$

Substituting this into (3.242) and using $\mathbf{H} = -\nabla \Phi_m$, we find that

$$\mathbf{H} = \kappa H_0 \hat{\mathbf{z}}$$

within the enclosure, where $\kappa = 9\mu_r/K$. This field is uniform, and, since $\kappa < 1$ for $\mu_r > 1$, it is weaker than the applied field.

For $\mu_r \gg 1$ we have $K \approx 2\mu_r^2[1 - (a/b)^3]$. Denoting the shell thickness by $\Delta = b - a$, we find that $K \approx 6\mu_r^2\Delta/a$ when $\Delta/a \ll 1$. Thus, for a shell with $\mu_r \gg 1$ and $\Delta/a \ll 1$, the shielding effectiveness is

$$SE = 20 \log_{10}\left(\frac{2\mu_r\Delta}{3b}\right). \tag{3.243}$$

Compare this formula to (3.239) for the case of a cylindrical shell.

▶ **Example 3.66:** Magnetostatic shielding using a high-permeability spherical shell

Repeat Example 3.65 for the case of a spherical shell. Again let $a/b = 0.99$, and assume the permeability of the shell is large. What value of μ_r is required to achieve 40 dB shielding effectiveness?

Solution: Since the permeability is large and $a \approx b$, we can use (3.243):

$$\frac{2\mu_r\Delta}{3b} = 10^{40/20} = 100.$$

Using $\Delta = b - a$, we find

$$\frac{2\mu_r}{3}\left(1 - \frac{a}{b}\right) = 100,$$

hence $\mu_r = 15000$. This is smaller than the permeability found for a cylindrical shield in Example 3.65. ◀

3.7.4 Quasistatic shielding

Metallic shields are often used to prevent interference from AC magnetic fields. Let's repeat the canonical problem of a cylindrical shell of magnetic material immersed in a magnetic field considered in § 3.7.3, except now we allow the shell to be conducting and we assume that the impressed field varies sinusoidally with time. See Figure 3.42. The shield (region 2) is homogeneous with permeability $\mu = \mu_r\mu_0$ and conductivity σ. Both regions 1 and 3 are free space. We further assume that the frequency is low enough that the system is MQS and wave effects may be ignored.

The impressed magnetic field is taken to be x-directed with $\tilde{\mathbf{H}}_0(\mathbf{r}, \omega) = \hat{\mathbf{x}}\tilde{H}_0(\omega)$. This can be written as the curl of a z-directed vector potential:

$$\tilde{\mathbf{A}}_0(\mathbf{r}, \omega) = \hat{\mathbf{z}}\,\mu_0 y\tilde{H}_0(\omega) = \hat{\mathbf{z}}\,\mu_0\rho\sin\phi\tilde{H}_0(\omega).$$

By symmetry, we expect the scattered vector potential to be z-directed. Thus, in regions 1 and 3, which are source-free and nonconducting, the scattered vector potential obeys Laplace's equation

$$\nabla^2\tilde{A}_z(\mathbf{r}, \omega) = 0,$$

while in region 2, which is conducting, $\tilde{A}_z$ obeys the diffusion equation (3.211)

$$\nabla^2\tilde{A}_z(\mathbf{r}, \omega) - \gamma^2\tilde{A}_z(\mathbf{r}, \omega) = 0. \tag{3.244}$$

To find the vector potential in regions 1 and 3, examine the separation of variables solution to Laplace's equation (A.128). Since the impressed vector potential varies as $\sin\phi$ we only need this term in the expansion of $\tilde{A}_z$. In region 1 we have both the impressed and scattered potentials, while in region 3 we have only the scattered potential.

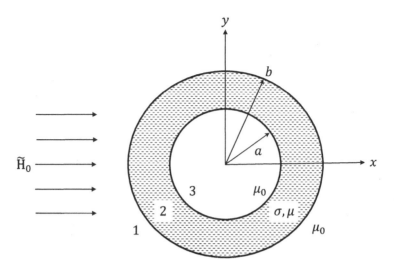

FIGURE 3.42
Cylindrical shell of conducting magnetic material illuminated by an AC magnetic field.

Furthermore, we must have the potential decay as $\rho \to \infty$ in region 1, and remain finite in region 3. Thus we may write

$$\tilde{A}_{z1} = A\rho^{-1}\sin\phi + \mu_0\tilde{H}_0\rho\sin\phi, \qquad \tilde{A}_{z3} = D\rho\sin\phi.$$

In region 2 we must use the solution to the diffusion equation (3.244), which in cylindrical coordinates is written as

$$\frac{1}{\rho}\frac{\partial}{\partial\rho}\left(\rho\frac{\partial\tilde{A}_z}{\partial\rho}\right) - \gamma^2\tilde{A}_z = 0. \tag{3.245}$$

Taking a separation of variables approach, we write $\tilde{A}_z$ as a product solution

$$\tilde{A}_z = P(\rho)\Phi(\phi)$$

and substitute into (3.245). We find that $P(\rho)$ and $\Phi(\phi)$ satisfy the separated ordinary differential equations

$$\frac{d^2\Phi}{d\phi^2} + n^2\Phi = 0$$

and

$$\frac{d^2P}{d\rho^2} + \frac{1}{\rho}\frac{dP}{d\rho} - \left(\gamma^2 + \frac{n^2}{\rho^2}\right)P = 0. \tag{3.246}$$

So the ϕ-dependence of $\tilde{A}_z$ is trigonometric. Equation (3.246) is Bessel's modified differential equation with solutions $I_n(\gamma\rho)$ and $K_n(\gamma\rho)$. Thus, with $n = 1$ to match the impressed vector potential, the vector potential in region 2 may be written as

$$\tilde{A}_{z2} = [BI_1(\gamma\rho) + CK_1(\gamma\rho)]\sin\phi.$$

To determine the constants, we apply the tangential field boundary conditions at $\rho = a$ and $\rho = b$. The magnetic field in each region may be found using the curl:

$$\tilde{\mathbf{B}} = \nabla \times \tilde{\mathbf{A}} = \hat{\rho} \frac{1}{\rho} \frac{\partial \tilde{A}_z}{\partial \phi} - \hat{\phi} \frac{\partial \tilde{A}_z}{\partial \rho}.$$

Thus, continuity of tangential $\tilde{\mathbf{H}}$ is equivalent to continuity of $(1/\mu)\partial \tilde{A}_z/\partial \rho$. The electric field may be found from (3.209). However, by the gauge condition (3.210) we see that $\tilde{\phi} = 0$ since $\nabla \cdot \tilde{\mathbf{A}} = \partial \tilde{A}_z/\partial z = 0$. So $\tilde{\mathbf{E}} = -j\omega \tilde{\mathbf{A}}$, and continuity of tangential $\tilde{\mathbf{E}}$ is equivalent to continuity of $\tilde{\mathbf{A}}$. Note that continuity of normal $\tilde{\mathbf{B}}$ is also equivalent to continuity of $\tilde{\mathbf{A}}$.

Continuity of $\tilde{A}_z$ at $\rho = a$ and $\rho = b$ gives

$$Da = BI_1(\gamma a) + CK_1(\gamma a), \qquad \frac{A}{b} + \mu_0 \tilde{H}_0 b = BI_1(\gamma b) + CK_1(\gamma b).$$

Continuity of $(1/\mu)\partial \tilde{A}_z/\partial \rho$ at $\rho = a$ and $\rho = b$ gives

$$Da = \frac{\gamma a}{\mu_r} \left[BI_1'(\gamma a) + CK_1'(\gamma a) \right], \qquad -\frac{A}{b} + \mu_0 \tilde{H}_0 b = \frac{\gamma b}{\mu_r} \left[BI_1'(\gamma b) + CK_1'(\gamma b) \right].$$

Solving these equations simultaneously, we obtain

$$D = \left[\mu_0 \tilde{H}_0 \frac{2\mu_r b}{a} \right] \left\{ [\mu_r K_1(\gamma a) - \gamma a K_1'(\gamma a)] [\mu_r I_1(\gamma b) + \gamma b I_1'(\gamma b)] \right.$$

$$\left. - [\mu_r I_1(\gamma a) - \gamma a I_1'(\gamma a)] [\mu_r K_1(\gamma b) + \gamma b K_1'(\gamma b)] \right\}^{-1}. \tag{3.247}$$

Now, note that the magnetic field in region 3 is just

$$\tilde{\mathbf{H}} = \frac{1}{\mu_0} \nabla \times (\hat{\mathbf{z}} \, D\rho \sin \phi) = \frac{1}{\mu_0} \nabla \times (\hat{\mathbf{z}} \, Dy) = \frac{1}{\mu_0} D \hat{\mathbf{x}},$$

and thus the field interior to the shield is uniform. This allows us to calculate the shielding effectiveness as

$$SE = 20 \log_{10} |\mu_0 \tilde{H}_0 / D| \tag{3.248}$$

where D is given by (3.247).

▶ **Example 3.67:** MQS shielding by a conducting cylindrical shell

Explore the dependence of the cylindrical shell-shielding effectiveness for various shield radii and thicknesses.

Solution: We plot shielding effectiveness as a function of shield thickness for various values of the inner radius a. It is expedient to normalize all distances to the skin depth. Figure 3.43 shows the shielding effectiveness found from (3.248) with $\mu_r = 100$ as a function of $w/\delta = (b - a)/\delta$ for various values of a/δ. Note that even for very small w/δ, a shield can be effective provided μ_r is sufficiently large. This is due to ducting of the magnetic field through the shell, which is most effective for shells of smaller inner radius. For a detailed discussion of MQS shielding by cylindrical shells, see [88] and [209].

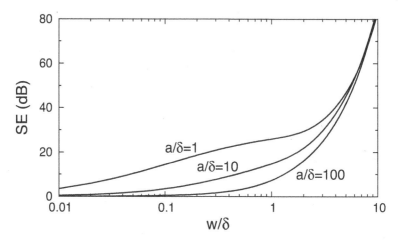

FIGURE 3.43

MQS shielding effectiveness of a conducting cylindrical shell. ◀

3.7.5 Electromagnetic shielding

When the frequency is high enough for displacement current to be important, a quasistatic analysis of shields is no longer accurate, and the wave nature of the electromagnetic field must be taken into account. We shall consider the shielding of electromagnetic waves by planar materials in § 4.11.6, and examine the penetration of electromagnetic waves through a rectangular aperture in a conducting screen in § 7.7.

3.8   Problems

3.1 The z-axis carries a line charge of nonuniform density $\rho_l(z)$. Show that the electric field in the plane $z = 0$ is given by

$$\mathbf{E}(\rho, \phi) = \frac{1}{4\pi\epsilon} \left[\hat{\boldsymbol{\rho}}\rho \int_{-\infty}^{\infty} \frac{\rho_l(z')\,dz'}{(\rho^2 + z'^2)^{3/2}} - \hat{\mathbf{z}} \int_{-\infty}^{\infty} \frac{\rho_l(z')z'\,dz'}{(\rho^2 + z'^2)^{3/2}} \right].$$

Compute $\mathbf{E}$ when $\rho_l = \rho_0 \operatorname{sgn}(z)$, where $\operatorname{sgn}(z)$ is the signum function (A.7).

3.2 The ring $\rho = a$, $z = 0$, carries a line charge of nonuniform density $\rho_l(\phi)$. Show that the electric field at an arbitrary point on the z-axis is given by

$$\mathbf{E}(z) = \frac{-a^2}{4\pi\epsilon(a^2 + z^2)^{3/2}} \left[\hat{\mathbf{x}} \int_0^{2\pi} \rho_l(\phi')\cos\phi'\,d\phi' + \hat{\mathbf{y}} \int_0^{2\pi} \rho_l(\phi')\sin\phi'\,d\phi' \right]$$

$$+ \hat{\mathbf{z}}\frac{az}{4\pi\epsilon(a^2 + z^2)^{3/2}} \int_0^{2\pi} \rho_l(\phi')\,d\phi'.$$

Compute $\mathbf{E}$ when $\rho_l(\phi) = \rho_0 \sin\phi$. Repeat for $\rho_l(\phi) = \rho_0 \cos^2\phi$.

3.3 The plane $z = 0$ carries a surface charge of nonuniform density $\rho_s(\rho, \phi)$. Show that at an arbitrary point on the z-axis, the rectangular components of $\mathbf{E}$ are given by

$$E_x(z) = -\frac{1}{4\pi\epsilon} \int_0^\infty \int_0^{2\pi} \frac{\rho_s(\rho', \phi')\,\rho'^2 \cos\phi'\,d\phi'\,d\rho'}{(\rho'^2 + z^2)^{3/2}},$$

$$E_y(z) = -\frac{1}{4\pi\epsilon} \int_0^\infty \int_0^{2\pi} \frac{\rho_s(\rho', \phi')\,\rho'^2 \sin\phi'\,d\phi'\,d\rho'}{(\rho'^2 + z^2)^{3/2}},$$

$$E_z(z) = \frac{z}{4\pi\epsilon} \int_0^\infty \int_0^{2\pi} \frac{\rho_s(\rho', \phi')\,\rho'\,d\phi'\,d\rho'}{(\rho'^2 + z^2)^{3/2}}.$$

Compute $\mathbf{E}$ when $\rho_s(\rho, \phi) = \rho_0 U(\rho - a)$, where $U(\rho)$ is the unit step function (A.6). Repeat for $\rho_s(\rho, \phi) = \rho_0[1 - U(\rho - a)]$.

3.4 The sphere $r = a$ carries a surface charge of nonuniform density $\rho_s(\theta)$. Show that the electric intensity at an arbitrary point on the z-axis is given by

$$\mathbf{E}(z) = \hat{\mathbf{z}}\frac{a^2}{2\epsilon} \int_0^\pi \frac{\rho_s(\theta')(z - a\cos\theta')\sin\theta'\,d\theta'}{(a^2 + z^2 - 2az\cos\theta')^{3/2}}.$$

Compute $\mathbf{E}(z)$ when $\rho_s(\theta) = \rho_0$, a constant. Repeat for $\rho_s(\theta) = \rho_0\,\mathrm{sgn}(\theta - \frac{\pi}{2})$.

3.5 Beginning with the postulates for the electrostatic field

$$\nabla \times \mathbf{E} = 0, \qquad \nabla \cdot \mathbf{D} = \rho,$$

use the technique of § 2.8.2 to derive the boundary conditions (3.12)–(3.13).

3.6 A material half space of permittivity ϵ_1 occupies the region $z > 0$, while a second material half space of permittivity ϵ_2 occupies $z < 0$. Find the polarization surface charge densities and compute the total induced surface polarization charge for a point charge Q located at $z = h$.

3.7 Consider a point charge between two grounded conducting plates as shown in Figure 3.44. Write the Green's function as the sum of primary and secondary terms and apply the boundary conditions to show that the secondary Green's function is

$$G^s(\mathbf{r}|\mathbf{r}') = \frac{-1}{(2\pi)^2} \int_{-\infty}^\infty \int_{-\infty}^\infty \left[e^{-k_\rho(d-z)} \frac{\sinh k_\rho z'}{\sinh k_\rho d} + e^{-k_\rho z} \frac{\sinh k_\rho(d - z')}{\sinh k_\rho d} \right] \frac{e^{j\mathbf{k}_\rho \cdot (\mathbf{r} - \mathbf{r}')}}{2k_\rho}\,d^2k_\rho.$$

$$(3.249)$$

3.8 Use the expansion

$$\frac{1}{\sinh k_\rho d} = \operatorname{csch} k_\rho d = 2\sum_{n=0}^\infty e^{-(2n+1)k_\rho d}$$

to show that the secondary Green's function for parallel conducting plates (3.249) may be written as an infinite sequence of images of the primary point charge. Identify the geometrical meaning of each image term.

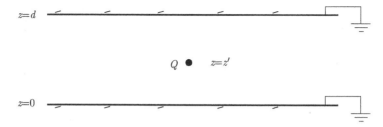

FIGURE 3.44
Geometry for computing Green's function for parallel plates.

3.9 Find the Green's functions for a dielectric slab of thickness d placed over a perfectly conducting ground plane located at $z = 0$.

3.10 Referring to the system of Figure 3.8, find the charge density on the surface of the sphere and integrate to show that the total charge is equal to the image charge.

3.11 Use the method of Green's functions to find the potential inside a conducting sphere for ρ inside the sphere.

3.12 Solve for the total potential and electric field of a grounded conducting sphere centered at the origin within a uniform impressed electric field $\mathbf{E} = E_0\hat{\mathbf{z}}$. Find the total charge induced on the sphere.

3.13 Consider a spherical cavity of radius a centered at the origin within a homogeneous dielectric material of permittivity $\epsilon = \epsilon_0\epsilon_r$. Solve for the total potential and electric field inside the cavity in the presence of an impressed field $\mathbf{E} = E_0\hat{\mathbf{z}}$. Show that the field in the cavity is stronger than the applied field, and explain this using polarization surface charge.

3.14 Find the field of a point charge Q located at $z = d$ above a perfectly conducting ground plane at $z = 0$. Use the boundary condition to find the charge density on the plane and integrate to show that the total charge is $-Q$. Integrate Maxwell's stress tensor over the surface of the ground plane and show that the force on the ground plane is the same as the force on the image charge found from Coulomb's law.

3.15 Consider in free space a point charge $-q$ at $\mathbf{r} = \mathbf{r}_0 + \mathbf{d}$, a point charge $-q$ at $\mathbf{r} = \mathbf{r}_0 - \mathbf{d}$, and a point charge $2q$ at $\mathbf{r}_0$. Find the first three multipole moments and the resulting potential produced by this charge distribution.

3.16 A spherical charge distribution of radius a in free space has the density

$$\rho(\mathbf{r}) = \frac{Q}{\pi a^3} \cos 2\theta.$$

Compute the multipole moments for the charge distribution and find the resulting po-

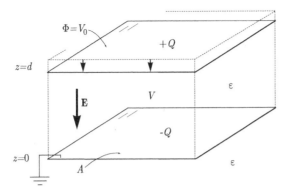

FIGURE 3.45
Parallel plate capacitor.

tential.

3.17 Compute the magnetic flux density **B** for the circular wire loop of Figure 3.17 by (a) using the Biot–Savart law (3.132), and (b) computing the curl of (3.110).

3.18 Two circular current-carrying wires are arranged coaxially along the z-axis. Loop 1 has radius a_1, carries current I_1, and is centered in the $z = 0$ plane. Loop 2 has radius a_2, carries current I_2, and is centered in the $z = d$ plane. Find the force between the loops.

3.19 Obtain (3.136) by integration of Maxwell's stress tensor over the xz-plane.

3.20 Consider two thin conducting parallel plates embedded in a region of permittivity ϵ (Figure 3.45). The bottom plate is connected to ground, and we apply an excess charge $+Q$ to the top plate (and thus $-Q$ is drawn onto the bottom plate). Neglecting fringing, (a) solve Laplace's equation to show that

$$\Phi(z) = \frac{Q}{A\epsilon} z.$$

Use (3.62) to show that

$$W = \frac{Q^2 d}{2A\epsilon}.$$

(b) Verify W using (3.63). (c) Use $\mathbf{F} = -\hat{\mathbf{z}} dW/dz$ to show that the force on the top plate is

$$\mathbf{F} = -\hat{\mathbf{z}} \frac{Q^2}{2A\epsilon}.$$

(d) Verify $\mathbf{F}$ by integrating Maxwell's stress tensor over a closed surface surrounding the top plate.

3.21 Consider two thin conducting parallel plates embedded in a region of permittivity ϵ (Figure 3.45). The bottom plate is connected to ground, and we apply a potential V_0 to the top plate using a battery. Neglecting fringing, (a) solve Laplace's equation to show that

$$\Phi(z) = \frac{V_0}{d} z.$$

Use (3.62) to show that

$$W = \frac{V_0^2 A \epsilon}{2d}.$$

(b) Verify W using (3.63). (c) Use $\mathbf{F} = -\hat{\mathbf{z}} dW/dz$ to show that the force on the top plate is

$$\mathbf{F} = -\hat{\mathbf{z}} \frac{V_0^2 A \epsilon}{2d^2}.$$

(d) Verify $\mathbf{F}$ by integrating Maxwell's stress tensor over a closed surface surrounding the top plate.

3.22 A group of N perfectly conducting bodies is arranged in free space. Body n is held at potential V_n with respect to ground, and charge Q_n is induced upon its surface. By linearity we may write

$$Q_m = \sum_{n=1}^{N} c_{mn} V_n$$

where the c_{mn} are called the *capacitance coefficients*. Using Green's reciprocation theorem, demonstrate that $c_{mn} = c_{nm}$. Hint: Use (3.163). Choose one set of voltages so that $V_k = 0$, $k \neq n$, and place V_n at some potential, say $V_n = V_0$, producing the set of charges $\{Q_k\}$. For the second set choose $V_k' = 0$, $k \neq m$, and $V_m = V_0$, producing $\{Q_k'\}$.

3.23 For the set of conductors of Problem 3.22, show that we may write

$$Q_m = C_{mm} V_m + \sum_{k \neq m} C_{mk}(V_m - V_k)$$

where

$$C_{mn} = -c_{mn}, \quad m \neq n, \qquad C_{mm} = \sum_{k=1}^{N} c_{mk}.$$

Here C_{mm}, called the *self capacitance*, describes the interaction between the mth conductor and ground, while C_{mn}, called the *mutual capacitance*, describes the interaction between the mth and nth conductors.

3.24 For the set of conductors of Problem 3.22, show that the stored electric energy is given by

$$W = \frac{1}{2} \sum_{m=1}^{N} \sum_{n=1}^{N} c_{mn} V_n V_m.$$

3.25 A group of N wires is arranged in free space as shown in Figure 3.46. Wire n carries a steady current I_n, and a flux Ψ_n passes through the surface defined by its contour Γ_n.

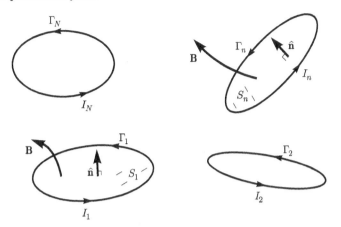

FIGURE 3.46
A system of current-carrying wires.

By linearity we may write

$$\Psi_m = \sum_{n=1}^{N} L_{mn} I_n$$

where the L_{mn} are called the *coefficients of inductance*. Derive *Neumann's formula*

$$L_{mn} = \frac{\mu_0}{4\pi} \oint_{\Gamma_n} \oint_{\Gamma_m} \frac{\mathbf{dl} \cdot \mathbf{dl'}}{|\mathbf{r} - \mathbf{r'}|},$$

and thereby demonstrate the reciprocity relation $L_{mn} = L_{nm}$.

3.26 For the group of wires shown in Figure 3.46, show that the stored magnetic energy is given by

$$W = \frac{1}{2} \sum_{m=1}^{N} \sum_{n=1}^{N} L_{mn} I_n I_m.$$

3.27 Prove the *minimum heat generation theorem*: steady electric currents distribute themselves in a conductor in such a way that the dissipated power is a minimum. Hint: Let $\mathbf{J}$ be the actual distribution of current in a conducting body, and let the power it dissipates be P. Let $\mathbf{J'} = \mathbf{J} + \delta\mathbf{J}$ be any other current distribution, and let the power it dissipates be $P' = P + \delta P$. Show that

$$\delta P = \frac{1}{2} \int_V \frac{1}{\sigma} |\delta\mathbf{J}|^2 \, dV \geq 0.$$

3.28 Establish the following cylindrical harmonic expansion of the line charge potential:

$$-\frac{\lambda}{2\pi\epsilon} \ln R = \begin{cases} \dfrac{\lambda}{2\pi\epsilon} \left[\displaystyle\sum_{n=1}^{\infty} \frac{1}{n} \left(\frac{\rho_0}{\rho} \right)^n (\cos n\phi_0 \cos n\phi + \sin n\phi_0 \sin n\phi) - \ln\rho \right], & \rho > \rho_0, \\[3mm] \dfrac{\lambda}{2\pi\epsilon} \left[\displaystyle\sum_{n=1}^{\infty} \frac{1}{n} \left(\frac{\rho}{\rho_0} \right)^n (\cos n\phi_0 \cos n\phi + \sin n\phi_0 \sin n\phi) - \ln\rho_0 \right], & \rho < \rho_0. \end{cases}$$

The line charge is located at (ρ_0, ϕ_0), and the field point is located at (ρ, ϕ).

3.29 (a) Use spherical harmonics to write down the general solution to Laplace's equation in spherical coordinates. (b) Specialize the result of part (a) for a problem having azimuthal symmetry. (c) Evaluate the result of part (b) for a point on the z-axis. (d) Explain how the result of part (c) can be used to determine the potential at points off the z-axis. (e) Apply this technique, along with the expansion (E.167)–(E.169), to the following problem. Suppose charge Q is distributed uniformly over a ring. The ring is centered on the z-axis, and its radius and height above the $z = 0$ plane are described by (1) the distance c from the origin to the ring (to the ring itself — not to its center!) and (2) the polar angle α measured from the z-axis to the ring. (Note: in (E.168), γ is the angle between $\mathbf{r}$ and $\mathbf{r}'$.)

3.30 The electric field in an anisotropic material is given (in spherical coordinates) by

$$\mathbf{E} = \hat{\mathbf{r}} \frac{2}{r^2} \text{ V/m.}$$

If the permittivity of the material is $\bar{\bar{\epsilon}} = \hat{\boldsymbol{\theta}}\hat{\boldsymbol{\theta}}3\epsilon_0 + \hat{\boldsymbol{\theta}}\hat{\mathbf{r}}4\epsilon_0 + \hat{\mathbf{r}}\hat{\mathbf{r}}\epsilon_0$, calculate (a) $\mathbf{D}$, (b) $\mathbf{P}$, and (c) the equivalent polarization volume charge density.

3.31 Show by direct substitution that the Coulomb field

$$\mathbf{E}(\mathbf{r}) = \frac{1}{4\pi\epsilon_0} \int_V \rho(\mathbf{r}') \frac{\mathbf{r} - \mathbf{r}'}{|\mathbf{r} - \mathbf{r}'|} \, dV'$$

satisfies Gauss's law $\nabla \cdot \mathbf{E} = \rho/\epsilon_0$.

3.32 A parallel-plate capacitor has plates of area A located at $z = 0$ and $z = d$. A voltage V_0 is applied to the plates such that the electric field between the plates is $\mathbf{E} = -\hat{\mathbf{z}}V_0/d$. Find the force on the bottom plate by integrating Maxwell's stress tensor. Neglect fringing.

3.33 A conducting plate of area 10 m^2 is located in the $z = 0$ plane. The field above the plate is given by $\mathbf{E} = \hat{\mathbf{z}}E_0(z^2 + 1)$ V/m, while the field below the plate is zero. Find the force on the plate by integrating Maxwell's stress tensor.

3.34 An infinite static line charge of density $\rho_0\delta(x)\delta(y)$ lies on the interface between two dielectric half spaces: $\epsilon = \epsilon_1$ for $y > 0$, and $\epsilon = \epsilon_2$ for $y < 0$. Using the Fourier transform approach, calculate the potential Φ in each region.

3.35 Calculate the charge distribution that supports the static electric field

$$\mathbf{E} = \frac{e^{-r/a}}{r} \left(\frac{1}{r} + \frac{1}{a} \right) \hat{\mathbf{r}}.$$

Here a is a constant.

3.36 Find the charge distribution that supports the electrostatic potential

$$\Phi = \frac{e^{-ar}}{r}\left(1 + \frac{ar}{2}\right).$$

Here a is a constant.

3.37 A volume charge is distributed throughout an infinite slab of thickness $2a$ symmetrically placed about the xy-plane. Consider the charge density to be uniform in x and y but not necessarily in z. Use the integral form of Coulomb's law to obtain an expression for **E**.

3.38 A surface charge η is distributed over a spherical surface of radius a and centered at the origin. Consider the charge density to be uniform in ϕ but not necessarily in θ, and show that the potential at a point $(0, 0, z)$ is given by

$$\Phi(z) = \frac{a^2}{2\epsilon_0}\int_0^\pi \frac{\eta(\theta')\sin\theta'\,d\theta'}{\sqrt{a^2 + z^2 - 2az\cos\theta'}}.$$

3.39 Given

$$\mathbf{H} = H_0\left(\frac{a}{\rho}\right)e^{-\rho/a}\hat{\boldsymbol{\phi}}$$

in cylindrical coordinates, find **J**.

3.40 A sphere of radius a and permeability μ is centered at the origin in free space, and within a uniform primary magnetic field $\mathbf{H} = H_0\hat{\mathbf{z}}$. Find **H** everywhere.

3.41 A disk of radius a centered in the xy-plane carries a uniform surface charge of density η_0 and rotates with angular velocity ω. Calculate **A** and **B** at the point $(0, 0, z)$.

3.42 In the region $0 \leq r \leq a$ of free space, the electric field is given by

$$\mathbf{E}(\mathbf{r}) = \frac{Q_0}{4\pi\epsilon_0 r}\left(\frac{1}{r} - \frac{1}{a}\right)\hat{\mathbf{r}}$$

where a and Q_0 are constants. Calculate the charge density and total charge within the region $0 \leq r \leq a$.

3.43 A slab of material occupies the region $-a \leq z \leq a$ in otherwise free space. The material is permanently polarized so that

$$\mathbf{P}(\mathbf{r}) = \hat{\mathbf{z}}P_0\left[(z/a)^2 - 1\right]$$

where P_0 is a constant. (a) Calculate the equivalent polarization surface and volume charge densities. (b) Find **E** in the region $-a \leq z \leq a$.

3.44 A hollow, perfectly conducting cylinder of infinite length and radius a is aligned along the z-axis. The cylinder is split into two parts so that the portion $0 \leq \phi \leq \pi$ is held at potential $\Phi = V_0$ while the portion $\pi < \phi < 2\pi$ is held at $\Phi = 0$. Calculate $\Phi(\rho, \phi)$ for $\rho > a$.

3.45 A perfectly conducting sphere of radius a, centered on the origin, is split into halves. The portion $0 \leq \theta \leq \pi/2$ is held at potential $\Phi = V_0$ while the portion $\pi/2 < \phi \leq \pi$ is held at $\Phi = 0$. Calculate $\Phi(r, \theta)$ for $r < a$ and $r > a$.

3.46 An electrostatic dipole having moment $\mathbf{p} = p_0 \hat{\mathbf{z}}$ resides at the center of a spherical dielectric shell. The shell has permittivity ϵ and occupies the region $a \leq r \leq b$. The regions $r > b$ and $r < a$ are free space regions. Calculate the electrostatic potential for $r > 0$.

3.47 A conducting sphere of radius a, centered at the origin, is split into halves described by $0 \leq \theta \leq \pi/2$ and $\pi/2 < \theta \leq \pi$. The sphere is immersed in a uniform electrostatic field $\mathbf{E} = E_0 \hat{\mathbf{z}}$. Find the force tending to separate the two halves of the sphere.

3.48 Consider a PEC screen with a circular aperture, as shown in Figure 3.36. Show that the field on the z-axis as z approaches zero from below is given by $-E_0/2$.

3.49 Consider a PEC screen with a circular aperture, as shown in Figure 3.36. Show that when $|z| \gg a$, the shielding effectiveness may be approximated as

$$SE \approx 13.5 \text{ dB} + 60 \log_{10} \left| \frac{z}{a} \right|.$$

3.50 Consider the cylindrical dielectric shell shown in Figure 3.39. Assume that instead of being x-directed, the applied electric field is z-directed: $\mathbf{E}_0 = \hat{\mathbf{z}} E_0$. Demonstrate that under this condition there is no shielding of the interior region from the applied field.

3.51 Consider the cylindrical shell illuminated by an AC magnetic field shown in Figure 3.42. Solve the system of linear equations resulting from the boundary conditions and show that the amplitude coefficient of the magnetic field internal to the cylindrical shell is given by (3.247). You will need to use the Wronskian relationship for the modified Bessel functions.

4

Temporal and spatial frequency domain representation

4.1 Interpretation of the temporal transform

When a field is represented by a continuous superposition of elemental components, the resulting decomposition can simplify computation and provide physical insight. Such representation is usually accomplished through the use of an integral transform. Although several different transforms are used in electromagnetics, we shall concentrate on the powerful and efficient Fourier transform.

Let us consider the Fourier transform of the electromagnetic field. The field depends on x, y, z, t, and we can transform with respect to any or all of these variables. However, a consideration of units leads us to consider a transform over t separately. Let $\psi(\mathbf{r}, t)$ represent any rectangular component of the electric or magnetic field. Then the temporal transform will be designated by $\tilde{\psi}(\mathbf{r}, \omega)$:

$$\psi(\mathbf{r}, t) \leftrightarrow \tilde{\psi}(\mathbf{r}, \omega).$$

Here ω is the transform variable. The transform field $\tilde{\psi}$ is calculated using (A.1):

$$\tilde{\psi}(\mathbf{r}, \omega) = \int_{-\infty}^{\infty} \psi(\mathbf{r}, t) e^{-j\omega t} \, dt. \tag{4.1}$$

The inverse transform is, by (A.2),

$$\psi(\mathbf{r}, t) = \frac{1}{2\pi} \int_{-\infty}^{\infty} \tilde{\psi}(\mathbf{r}, \omega) e^{j\omega t} \, d\omega. \tag{4.2}$$

Since $\tilde{\psi}$ is complex, it may be written in the amplitude–phase form

$$\tilde{\psi}(\mathbf{r}, \omega) = |\tilde{\psi}(\mathbf{r}, \omega)| e^{j\xi^{\psi}(\mathbf{r}, \omega)} \qquad (-\pi < \xi^{\psi}(\mathbf{r}, \omega) \leq \pi).$$

Since $\psi(\mathbf{r}, t)$ must be real, (4.1) shows that

$$\tilde{\psi}(\mathbf{r}, -\omega) = \tilde{\psi}^{*}(\mathbf{r}, \omega). \tag{4.3}$$

Furthermore, the transform of the derivative of ψ may be found by differentiating (4.2):

$$\frac{\partial}{\partial t} \psi(\mathbf{r}, t) = \frac{1}{2\pi} \int_{-\infty}^{\infty} j\omega \tilde{\psi}(\mathbf{r}, \omega) e^{j\omega t} \, d\omega,$$

hence

$$\frac{\partial}{\partial t} \psi(\mathbf{r}, t) \leftrightarrow j\omega \tilde{\psi}(\mathbf{r}, \omega).$$

By virtue of (4.2), any electromagnetic field component can be decomposed into a continuous, weighted superposition of elemental temporal terms $e^{j\omega t}$. Note that the weighting factor $\tilde{\psi}(\mathbf{r}, \omega)$, often called the *frequency spectrum* of $\psi(\mathbf{r}, t)$, is not arbitrary because $\psi(\mathbf{r}, t)$ must obey a scalar wave equation such as (2.269). For a source-free region of space, we have

$$\left(\nabla^2 - \mu\sigma\frac{\partial}{\partial t} - \mu\epsilon\frac{\partial^2}{\partial t^2}\right)\frac{1}{2\pi}\int_{-\infty}^{\infty}\tilde{\psi}(\mathbf{r}, \omega)\,e^{j\omega t}\,d\omega = 0.$$

Differentiating under the integral sign, we have

$$\frac{1}{2\pi}\int_{-\infty}^{\infty}[(\nabla^2 - j\omega\mu\sigma + \omega^2\mu\epsilon)\tilde{\psi}(\mathbf{r}, \omega)]e^{j\omega t}\,d\omega = 0$$

and hence by the Fourier integral theorem

$$(\nabla^2 + k^2)\tilde{\psi}(\mathbf{r}, \omega) = 0 \tag{4.4}$$

where

$$k = \omega\sqrt{\mu\epsilon}\sqrt{1 - j\frac{\sigma}{\omega\epsilon}}$$

is the *wavenumber*. The *scalar Helmholtz equation* (4.4) represents the wave equation in the temporal frequency domain.

4.2 The frequency-domain Maxwell equations

If the region of interest contains sources, we can return to Maxwell's equations and represent all quantities using the temporal inverse Fourier transform. We have, for example,

$$\mathbf{E}(\mathbf{r}, t) = \frac{1}{2\pi}\int_{-\infty}^{\infty}\tilde{\mathbf{E}}(\mathbf{r}, \omega)\,e^{j\omega t}\,d\omega$$

where

$$\tilde{\mathbf{E}}(\mathbf{r}, \omega) = \sum_{i=1}^{3}\hat{\mathbf{i}}_i\tilde{E}_i(\mathbf{r}, \omega) = \sum_{i=1}^{3}\hat{\mathbf{i}}_i|\tilde{E}_i(\mathbf{r}, \omega)|e^{j\xi_i^E(\mathbf{r}, \omega)}. \tag{4.5}$$

All other field quantities will be written similarly with an appropriate superscript on the phase. Substitution into Ampere's law gives

$$\nabla \times \frac{1}{2\pi}\int_{-\infty}^{\infty}\tilde{\mathbf{H}}(\mathbf{r}, \omega)\,e^{j\omega t}\,d\omega = \frac{\partial}{\partial t}\frac{1}{2\pi}\int_{-\infty}^{\infty}\tilde{\mathbf{D}}(\mathbf{r}, \omega)\,e^{j\omega t}\,d\omega + \frac{1}{2\pi}\int_{-\infty}^{\infty}\tilde{\mathbf{J}}(\mathbf{r}, \omega)\,e^{j\omega t}\,d\omega,$$

hence

$$\frac{1}{2\pi}\int_{-\infty}^{\infty}[\nabla \times \tilde{\mathbf{H}}(\mathbf{r}, \omega) - j\omega\tilde{\mathbf{D}}(\mathbf{r}, \omega) - \tilde{\mathbf{J}}(\mathbf{r}, \omega)]e^{j\omega t}\,d\omega = 0$$

after we differentiate under the integral signs and combine terms. So

$$\nabla \times \tilde{\mathbf{H}} = j\omega\tilde{\mathbf{D}} + \tilde{\mathbf{J}} \tag{4.6}$$

by the Fourier integral theorem. This version of Ampere's law involves only the frequency-domain fields. By similar reasoning, we have

$$\nabla \times \tilde{\mathbf{E}} = -j\omega\tilde{\mathbf{B}}, \tag{4.7}$$

$$\nabla \cdot \tilde{\mathbf{D}} = \tilde{\rho}, \tag{4.8}$$

$$\nabla \cdot \tilde{\mathbf{B}}(\mathbf{r}, \omega) = 0, \tag{4.9}$$

and the continuity equation $\nabla \cdot \tilde{\mathbf{J}} + j\omega\tilde{\rho} = 0$. Relations (4.6)–(4.9) govern the temporal spectra of the electromagnetic fields. We may manipulate them to obtain wave equations, and apply the boundary conditions from the following section. After finding the frequency-domain fields, we may find the temporal fields by Fourier inversion. The frequency-domain equations involve one fewer derivative (the time derivative has been replaced by multiplication by $j\omega$), hence they may be easier to solve. However, the inverse transform may be difficult to compute.

4.3 Boundary conditions on the frequency-domain fields

Several boundary conditions on the source and mediating fields were derived in § 2.8.2. For example, the tangential electric field must obey

$$\hat{\mathbf{n}}_{12} \times \mathbf{E}_1(\mathbf{r}, t) - \hat{\mathbf{n}}_{12} \times \mathbf{E}_2(\mathbf{r}, t) = -\mathbf{J}_{ms}(\mathbf{r}, t).$$

The technique of the previous section yields

$$\hat{\mathbf{n}}_{12} \times [\tilde{\mathbf{E}}_1(\mathbf{r}, \omega) - \tilde{\mathbf{E}}_2(\mathbf{r}, \omega)] = -\tilde{\mathbf{J}}_{ms}(\mathbf{r}, \omega)$$

as the condition satisfied by the frequency-domain electric field. The remaining boundary conditions are treated similarly. Let us summarize the results, including the effects of fictitious magnetic sources:

$$\hat{\mathbf{n}}_{12} \times (\tilde{\mathbf{H}}_1 - \tilde{\mathbf{H}}_2) = \tilde{\mathbf{J}}_s,$$

$$\hat{\mathbf{n}}_{12} \times (\tilde{\mathbf{E}}_1 - \tilde{\mathbf{E}}_2) = -\tilde{\mathbf{J}}_{ms},$$

$$\hat{\mathbf{n}}_{12} \cdot (\tilde{\mathbf{D}}_1 - \tilde{\mathbf{D}}_2) = \tilde{\rho}_s,$$

$$\hat{\mathbf{n}}_{12} \cdot (\tilde{\mathbf{B}}_1 - \tilde{\mathbf{B}}_2) = \tilde{\rho}_{ms},$$

and

$$\hat{\mathbf{n}}_{12} \cdot (\tilde{\mathbf{J}}_1 - \tilde{\mathbf{J}}_2) = -\nabla_s \cdot \tilde{\mathbf{J}}_s - j\omega\tilde{\rho}_s,$$

$$\hat{\mathbf{n}}_{12} \cdot (\tilde{\mathbf{J}}_{m1} - \tilde{\mathbf{J}}_{m2}) = -\nabla_s \cdot \tilde{\mathbf{J}}_{ms} - j\omega\tilde{\rho}_{ms}.$$

Here $\hat{\mathbf{n}}_{12}$ points into region 1 from region 2.

4.4 Constitutive relations in the frequency domain and the Kramers–Kronig relations

All materials are to some extent dispersive. If a field applied to a material undergoes a sufficiently rapid change, there is a time lag in the response of the polarization or magnetization of the atoms. It has been found that such materials have constitutive relations involving products in the frequency domain, and that the frequency-domain constitutive parameters are complex, frequency-dependent quantities. We shall restrict ourselves to the special case of anisotropic materials and refer the reader to Kong [108] and Lindell [120] for the more general case. For anisotropic materials, we write

$$\tilde{\mathbf{P}} = \epsilon_0 \tilde{\bar{\chi}}_e \cdot \tilde{\mathbf{E}}, \tag{4.10}$$

$$\tilde{\mathbf{M}} = \tilde{\bar{\chi}}_m \cdot \tilde{\mathbf{H}}, \tag{4.11}$$

$$\tilde{\mathbf{D}} = \tilde{\bar{\epsilon}} \cdot \tilde{\mathbf{E}} = \epsilon_0 [\bar{\mathbf{I}} + \tilde{\bar{\chi}}_e] \cdot \tilde{\mathbf{E}}, \tag{4.12}$$

$$\tilde{\mathbf{B}} = \tilde{\bar{\mu}} \cdot \tilde{\mathbf{H}} = \mu_0 [\bar{\mathbf{I}} + \tilde{\bar{\chi}}_m] \cdot \tilde{\mathbf{H}}, \tag{4.13}$$

$$\tilde{\mathbf{J}} = \tilde{\bar{\sigma}} \cdot \tilde{\mathbf{E}}. \tag{4.14}$$

By the convolution theorem and the assumption of causality, we immediately obtain the dyadic versions of (2.20)–(2.22):

$$\mathbf{D}(\mathbf{r}, t) = \epsilon_0 \left(\mathbf{E}(\mathbf{r}, t) + \int_{-\infty}^{t} \bar{\chi}_e(\mathbf{r}, t - t') \cdot \mathbf{E}(\mathbf{r}, t') \, dt' \right),$$

$$\mathbf{B}(\mathbf{r}, t) = \mu_0 \left(\mathbf{H}(\mathbf{r}, t) + \int_{-\infty}^{t} \bar{\chi}_m(\mathbf{r}, t - t') \cdot \mathbf{H}(\mathbf{r}, t') \, dt' \right),$$

$$\mathbf{J}(\mathbf{r}, t) = \int_{-\infty}^{t} \bar{\sigma}(\mathbf{r}, t - t') \cdot \mathbf{E}(\mathbf{r}, t') \, dt'.$$

These describe the essential behavior of a dispersive material. The susceptances and conductivity, describing the response of the atomic structure to an applied field, depend not only on the present value of the applied field but on all past values as well.

As $\mathbf{D}(\mathbf{r}, t)$, $\mathbf{B}(\mathbf{r}, t)$, and $\mathbf{J}(\mathbf{r}, t)$ are real, so are the entries in the dyadic matrices $\bar{\epsilon}(\mathbf{r}, t)$, $\bar{\mu}(\mathbf{r}, t)$, and $\bar{\sigma}(\mathbf{r}, t)$. Applying (4.3) to each entry, we must have

$$\tilde{\bar{\chi}}_e(\mathbf{r}, -\omega) = \tilde{\bar{\chi}}_e^*(\mathbf{r}, \omega), \quad \tilde{\bar{\chi}}_m(\mathbf{r}, -\omega) = \tilde{\bar{\chi}}_m^*(\mathbf{r}, \omega), \quad \tilde{\bar{\sigma}}(\mathbf{r}, -\omega) = \tilde{\bar{\sigma}}^*(\mathbf{r}, \omega), \tag{4.15}$$

and hence

$$\tilde{\bar{\epsilon}}(\mathbf{r}, -\omega) = \tilde{\bar{\epsilon}}^*(\mathbf{r}, \omega), \qquad \tilde{\bar{\mu}}(\mathbf{r}, -\omega) = \tilde{\bar{\mu}}^*(\mathbf{r}, \omega). \tag{4.16}$$

In terms of real and imaginary parts, these conditions become

$$\mathrm{Re}\{\tilde{\epsilon}_{ij}(\mathbf{r}, -\omega)\} = \mathrm{Re}\{\tilde{\epsilon}_{ij}(\mathbf{r}, \omega)\}, \qquad \mathrm{Im}\{\tilde{\epsilon}_{ij}(\mathbf{r}, -\omega)\} = -\mathrm{Im}\{\tilde{\epsilon}_{ij}(\mathbf{r}, \omega)\},$$

and so on. The real parts of the constitutive parameters are even functions of frequency, and the imaginary parts are odd functions of frequency.

In most instances, the presence of an imaginary part in the constitutive parameters implies that the material is either *dissipative* (*lossy*), transforming some of the electromagnetic energy in the fields into thermal energy, or *active*, transforming the chemical or mechanical energy of the material into energy in the fields. We investigate this further in § 4.5 and § 4.8.3.

The constitutive equations may also be expressed in amplitude–phase form. Letting

$$\tilde{\epsilon}_{ij} = |\tilde{\epsilon}_{ij}|e^{j\xi_{ij}^\epsilon}, \qquad \tilde{\mu}_{ij} = |\tilde{\mu}_{ij}|e^{j\xi_{ij}^\mu}, \qquad \tilde{\sigma}_{ij} = |\tilde{\sigma}_{ij}|e^{j\xi_{ij}^\sigma},$$

and using the notation (4.5), we can write (4.12)–(4.14) as

$$\tilde{D}_i = |\tilde{D}_i|e^{j\xi_i^D} = \sum_{j=1}^{3} |\tilde{\epsilon}_{ij}||\tilde{E}_j|e^{j[\xi_j^E + \xi_{ij}^\epsilon]}, \tag{4.17}$$

$$\tilde{B}_i = |\tilde{B}_i|e^{j\xi_i^B} = \sum_{j=1}^{3} |\tilde{\mu}_{ij}||\tilde{H}_j|e^{j[\xi_j^H + \xi_{ij}^\mu]}, \tag{4.18}$$

$$\tilde{J}_i = |\tilde{J}_i|e^{j\xi_i^J} = \sum_{j=1}^{3} |\tilde{\sigma}_{ij}||\tilde{E}_j|e^{j[\xi_j^E + \xi_{ij}^\sigma]}. \tag{4.19}$$

Here we remember that the amplitudes and phases may be functions of both $\mathbf{r}$ and ω. For isotropic materials, these reduce to

$$\tilde{D}_i = |\tilde{D}_i|e^{j\xi_i^D} = |\tilde{\epsilon}||\tilde{E}_i|e^{j(\xi_i^E + \xi^\epsilon)}, \tag{4.20}$$

$$\tilde{B}_i = |\tilde{B}_i|e^{j\xi_i^B} = |\tilde{\mu}||\tilde{H}_i|e^{j(\xi_i^H + \xi^\mu)}, \tag{4.21}$$

$$\tilde{J}_i = |\tilde{J}_i|e^{j\xi_i^J} = |\tilde{\sigma}||\tilde{E}_i|e^{j(\xi_i^E + \xi^\sigma)}. \tag{4.22}$$

4.4.1 The complex permittivity

As mentioned above, dissipative effects may be associated with complex entries in the permittivity matrix. Since conduction effects can also lead to dissipation, the permittivity and conductivity matrices are often combined to form a *complex permittivity*. Writing the current as a sum of impressed and secondary conduction terms ($\tilde{\mathbf{J}} = \tilde{\mathbf{J}}^i + \tilde{\mathbf{J}}^c$) and substituting (4.12) and (4.14) into Ampere's law, we find

$$\nabla \times \tilde{\mathbf{H}} = \tilde{\mathbf{J}}^i + \tilde{\bar{\sigma}} \cdot \tilde{\mathbf{E}} + j\omega\tilde{\bar{\epsilon}} \cdot \tilde{\mathbf{E}}.$$

Defining the complex permittivity

$$\tilde{\bar{\epsilon}}^c(\mathbf{r}, \omega) = \frac{\tilde{\bar{\sigma}}(\mathbf{r}, \omega)}{j\omega} + \tilde{\bar{\epsilon}}(\mathbf{r}, \omega), \tag{4.23}$$

we have

$$\nabla \times \tilde{\mathbf{H}} = \tilde{\mathbf{J}}^i + j\omega\tilde{\bar{\epsilon}}^c \cdot \tilde{\mathbf{E}}.$$

Using the complex permittivity, we can include the effects of conduction current by merely replacing the total current with the impressed current. Since Faraday's law is unaffected, any equation (such as the wave equation) derived previously using total current retains its form with the same substitution.

By (4.15) and (4.16) the complex permittivity obeys

$$\tilde{\bar{\epsilon}}^c(\mathbf{r}, -\omega) = \tilde{\bar{\epsilon}}^{c*}(\mathbf{r}, \omega) \tag{4.24}$$

or

$$\text{Re}\{\tilde{\epsilon}_{ij}^c(\mathbf{r}, -\omega)\} = \text{Re}\{\tilde{\epsilon}_{ij}^c(\mathbf{r}, \omega)\}, \qquad \text{Im}\{\tilde{\epsilon}_{ij}^c(\mathbf{r}, -\omega)\} = -\text{Im}\{\tilde{\epsilon}_{ij}^c(\mathbf{r}, \omega)\}. \tag{4.25}$$

For an isotropic material it takes the particularly simple form

$$\tilde{\epsilon}^c = \frac{\tilde{\sigma}}{j\omega} + \tilde{\epsilon} = \frac{\tilde{\sigma}}{j\omega} + \epsilon_0 + \epsilon_0 \tilde{\chi}_e, \tag{4.26}$$

and we have

$$\text{Re}\{\tilde{\epsilon}^c(\mathbf{r}, -\omega)\} = \text{Re}\{\tilde{\epsilon}^c(\mathbf{r}, \omega)\}, \qquad \text{Im}\{\tilde{\epsilon}^c(\mathbf{r}, -\omega)\} = -\text{Im}\{\tilde{\epsilon}^c(\mathbf{r}, \omega)\}. \tag{4.27}$$

4.4.2 High and low frequency behavior of constitutive parameters

At low frequencies the permittivity reduces to the electrostatic permittivity. Since $\text{Re}\{\tilde{\epsilon}\}$ is even in ω and $\text{Im}\{\tilde{\epsilon}\}$ is odd, we have for small ω

$$\text{Re}\,\tilde{\epsilon} \sim \epsilon_0 \epsilon_r, \qquad \text{Im}\,\tilde{\epsilon} \sim \omega.$$

If the material has some dc conductivity σ_0, then for low frequencies the complex permittivity behaves as

$$\text{Re}\,\tilde{\epsilon}^c \sim \epsilon_0 \epsilon_r, \qquad \text{Im}\,\tilde{\epsilon}^c \sim \sigma_0/\omega. \tag{4.28}$$

It is possible for $\mathbf{E}$ or $\mathbf{H}$ to vary so rapidly that polarization or magnetization effects are precluded. Indeed, the atomic structure of a material may be unable to respond to sufficiently rapid oscillations in an applied field. Above some frequency then, we can assume $\tilde{\chi}_e = 0 = \tilde{\chi}_m$ so that $\tilde{\mathbf{P}} = 0 = \tilde{\mathbf{M}}$ and

$$\tilde{\mathbf{D}} = \epsilon_0 \tilde{\mathbf{E}}, \qquad \tilde{\mathbf{B}} = \mu_0 \tilde{\mathbf{H}}.$$

In our simple models of dielectric materials (§ 4.6) we find that as ω becomes large,

$$\text{Re}\,\tilde{\epsilon} - \epsilon_0 \sim 1/\omega^2, \qquad \text{Im}\,\tilde{\epsilon} \sim 1/\omega^3. \tag{4.29}$$

Our assumption of a macroscopic model of matter provides a fairly strict upper frequency limit to the range of validity of the constitutive parameters. We must assume the wavelength of the electromagnetic field is large compared to the size of the atomic structure. This limit suggests that permittivity and permeability might remain meaningful even at optical frequencies, and for dielectrics this is indeed the case since the values of $\tilde{\mathbf{P}}$ remain significant. However, $\tilde{\mathbf{M}}$ becomes insignificant at much lower frequencies, and at optical frequencies we may use $\tilde{\mathbf{B}} = \mu_0 \tilde{\mathbf{H}}$ [113].

4.4.3 The Kramers–Kronig relations

The causality principle is clearly implicit in (2.20)–(2.22). We shall demonstrate that causality leads to explicit relationships between the real and imaginary parts of the frequency-domain constitutive parameters. For simplicity we focus on the isotropic case and merely note that the present analysis holds for all the dyadic components of an anisotropic constitutive parameter. We also focus on the complex permittivity and extend the results to permeability by analogy.

The implications of causality on the behavior of the constitutive parameters in the time domain are easily identified. Writing (2.20) and (2.22) after setting $u = t - t'$ and then $u = t'$, we have

$$\mathbf{D}(\mathbf{r}, t) = \epsilon_0 \mathbf{E}(\mathbf{r}, t) + \epsilon_0 \int_0^\infty \chi_e(\mathbf{r}, t') \mathbf{E}(\mathbf{r}, t - t')\, dt',$$

$$\mathbf{J}(\mathbf{r}, t) = \int_0^\infty \sigma(\mathbf{r}, t') \mathbf{E}(\mathbf{r}, t - t')\, dt'.$$

Clearly the values of $\chi_e(\mathbf{r}, t)$ or $\sigma(\mathbf{r}, t)$ for times $t < 0$ do not contribute. So we can write

$$\mathbf{D}(\mathbf{r}, t) = \epsilon_0 \mathbf{E}(\mathbf{r}, t) + \epsilon_0 \int_{-\infty}^{\infty} \chi_e(\mathbf{r}, t') \mathbf{E}(\mathbf{r}, t - t') \, dt',$$

$$\mathbf{J}(\mathbf{r}, t) = \int_{-\infty}^{\infty} \sigma(\mathbf{r}, t') \mathbf{E}(\mathbf{r}, t - t') \, dt',$$

with the additional assumption

$$\chi_e(\mathbf{r}, t) = 0 \quad (t < 0), \qquad\qquad \sigma(\mathbf{r}, t) = 0 \quad (t < 0). \tag{4.30}$$

By (4.30) the frequency-domain complex permittivity (4.26) is

$$\tilde{\epsilon}^c(\mathbf{r}, \omega) - \epsilon_0 = \frac{1}{j\omega} \int_0^{\infty} \sigma(\mathbf{r}, t') e^{-j\omega t'} \, dt' + \epsilon_0 \int_0^{\infty} \chi_e(\mathbf{r}, t') e^{-j\omega t'} \, dt'. \tag{4.31}$$

To derive the Kramers–Kronig relations, we must understand the behavior of $\tilde{\epsilon}^c(\mathbf{r}, \omega) - \epsilon_0$ in the complex ω-plane. Writing $\omega = \omega_r + j\omega_i$, we require the following two properties.

Property 1: The function $\tilde{\epsilon}^c(\mathbf{r}, \omega) - \epsilon_0$ is analytic in the lower half-plane ($\omega_i < 0$) except for a simple pole at $\omega = 0$.

We can establish the analyticity of $\tilde{\sigma}(\mathbf{r}, \omega)$ by integrating over any closed contour in the lower half-plane. We have

$$\oint_\Gamma \tilde{\sigma}(\mathbf{r}, \omega) \, d\omega = \oint_\Gamma \left[\int_0^{\infty} \sigma(\mathbf{r}, t') e^{-j\omega t'} \, dt' \right] d\omega = \int_0^{\infty} \sigma(\mathbf{r}, t') \left[\oint_\Gamma e^{-j\omega t'} \, d\omega \right] dt'. \tag{4.32}$$

Note that an exchange in the order of integration in the above expression is only valid for ω in the lower half-plane where $\lim_{t' \to \infty} e^{-j\omega t'} = 0$. Since $f(\omega) = e^{-j\omega t'}$ is analytic in the lower half-plane, its closed contour integral vanishes by the Cauchy–Goursat theorem. Hence (4.32) yields

$$\oint_\Gamma \tilde{\sigma}(\mathbf{r}, \omega) \, d\omega = 0.$$

Then, since $\tilde{\sigma}$ may be assumed continuous in the lower half-plane for a physical medium, and since its closed path integral is zero for all possible paths Γ, it is by Morera's theorem [116] analytic in the lower half-plane. Similarly, so is $\chi_e(\mathbf{r}, \omega)$. Because the function $1/\omega$ has a simple pole at $\omega = 0$, the composite function $\tilde{\epsilon}^c(\mathbf{r}, \omega) - \epsilon_0$ given by (4.31) is analytic in the lower half-plane excluding $\omega = 0$ where it has a simple pole.

Property 2: We have
$$\lim_{\omega \to \pm\infty} \tilde{\epsilon}^c(\mathbf{r}, \omega) - \epsilon_0 = 0.$$

Here we require the *Riemann–Lebesgue lemma* [144], which states that if $f(t)$ is absolutely integrable on the interval (a, b) for finite or infinite constants a and b, then

$$\lim_{\omega \to \pm\infty} \int_a^b f(t) e^{-j\omega t} \, dt = 0.$$

This implies

$$\lim_{\omega \to \pm\infty} \frac{\tilde{\sigma}(\mathbf{r}, \omega)}{j\omega} = \lim_{\omega \to \pm\infty} \frac{1}{j\omega} \int_0^{\infty} \sigma(\mathbf{r}, t') e^{-j\omega t'} \, dt' = 0,$$

$$\lim_{\omega \to \pm\infty} \epsilon_0 \chi_e(\mathbf{r}, \omega) = \lim_{\omega \to \pm\infty} \epsilon_0 \int_0^{\infty} \chi_e(\mathbf{r}, t') e^{-j\omega t'} \, dt' = 0,$$

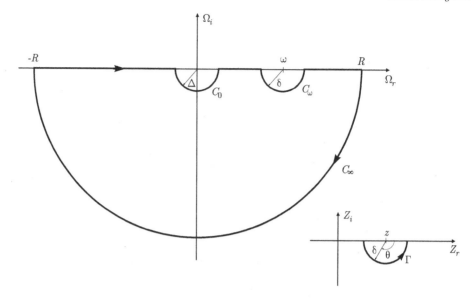

FIGURE 4.1
Complex integration contour used to establish the Kramers–Kronig relations.

and hence

$$\lim_{\omega \to \pm\infty} \tilde{\epsilon}^c(\mathbf{r}, \omega) - \epsilon_0 = 0.$$

To establish the Kramers–Kronig relations, we examine the integral

$$\oint_\Gamma \frac{\tilde{\epsilon}^c(\mathbf{r}, \Omega) - \epsilon_0}{\Omega - \omega} \, d\Omega$$

where Γ is shown in Figure 4.1. Since the points $\Omega = 0, \omega$ are excluded, the integrand is analytic everywhere within and on Γ, and the integral vanishes by the Cauchy–Goursat theorem. By Property 2

$$\lim_{R \to \infty} \int_{C_\infty} \frac{\tilde{\epsilon}^c(\mathbf{r}, \Omega) - \epsilon_0}{\Omega - \omega} \, d\Omega = 0,$$

hence

$$\int_{C_0 + C_\omega} \frac{\tilde{\epsilon}^c(\mathbf{r}, \Omega) - \epsilon_0}{\Omega - \omega} \, d\Omega + \text{P.V.} \int_{-\infty}^{\infty} \frac{\tilde{\epsilon}^c(\mathbf{r}, \Omega) - \epsilon_0}{\Omega - \omega} \, d\Omega = 0. \tag{4.33}$$

Here "P.V." denotes an integral computed in the Cauchy principal value sense (Appendix A). To evaluate the integrals over C_0 and C_ω, consider a function $f(Z)$ analytic in the lower half of the Z-plane ($Z = Z_r + jZ_i$). If the point z lies on the real axis as shown in Figure 4.1, we can calculate the integral

$$F(z) = \lim_{\delta \to 0} \int_\Gamma \frac{f(Z)}{Z - z} \, dZ$$

through the parameterization $Z - z = \delta e^{j\theta}$. Since $dZ = j\delta e^{j\theta} \, d\theta$, we have

$$F(z) = \lim_{\delta \to 0} \int_{-\pi}^{0} \frac{f\left(z + \delta e^{j\theta}\right)}{\delta e^{j\theta}} \left[j\delta e^{j\theta}\right] \, d\theta = jf(z) \int_{-\pi}^{0} d\theta = j\pi f(z).$$

Replacing Z by Ω and z by 0, we can compute

$$
\lim_{\Delta \to 0} \int_{C_0} \frac{\tilde{\epsilon}^c(\mathbf{r}, \Omega) - \epsilon_0}{\Omega - \omega}\, d\Omega
$$

$$
= \lim_{\Delta \to 0} \int_{C_0} \frac{\left[\frac{1}{j} \int_0^\infty \sigma(\mathbf{r}, t') e^{-j\Omega t'}\, dt' + \Omega \epsilon_0 \int_0^\infty \chi_e(\mathbf{r}, t') e^{-j\Omega t'}\, dt' \right] \frac{1}{\Omega - \omega}}{\Omega}\, d\Omega
$$

$$
= -\frac{\pi \int_0^\infty \sigma(\mathbf{r}, t')\, dt'}{\omega}.
$$

We recognize

$$
\int_0^\infty \sigma(\mathbf{r}, t')\, dt' = \sigma_0(\mathbf{r})
$$

as the dc conductivity and write

$$
\lim_{\Delta \to 0} \int_{C_0} \frac{\tilde{\epsilon}^c(\mathbf{r}, \Omega) - \epsilon_0}{\Omega - \omega}\, d\Omega = -\frac{\pi \sigma_0(\mathbf{r})}{\omega}.
$$

If we replace Z by Ω and z by ω we get

$$
\lim_{\delta \to 0} \int_{C_\omega} \frac{\tilde{\epsilon}^c(\mathbf{r}, \Omega) - \epsilon_0}{\Omega - \omega}\, d\Omega = j\pi\tilde{\epsilon}^c(\mathbf{r}, \omega) - j\pi\epsilon_0.
$$

Substituting these into (4.33), we have

$$
\tilde{\epsilon}^c(\mathbf{r}, \omega) - \epsilon_0 = -\frac{1}{j\pi} \,\text{P.V.} \int_{-\infty}^\infty \frac{\tilde{\epsilon}^c(\mathbf{r}, \Omega) - \epsilon_0}{\Omega - \omega}\, d\Omega + \frac{\sigma_0(\mathbf{r})}{j\omega}. \tag{4.34}
$$

By equating real and imaginary parts in (4.34) we find that

$$
\text{Re}\{\tilde{\epsilon}^c(\mathbf{r}, \omega)\} - \epsilon_0 = -\frac{1}{\pi} \,\text{P.V.} \int_{-\infty}^\infty \frac{\text{Im}\{\tilde{\epsilon}^c(\mathbf{r}, \Omega)\}}{\Omega - \omega}\, d\Omega, \tag{4.35}
$$

$$
\text{Im}\{\tilde{\epsilon}^c(\mathbf{r}, \omega)\} = \frac{1}{\pi} \,\text{P.V.} \int_{-\infty}^\infty \frac{\text{Re}\{\tilde{\epsilon}^c(\mathbf{r}, \Omega)\} - \epsilon_0}{\Omega - \omega}\, d\Omega - \frac{\sigma_0(\mathbf{r})}{\omega}. \tag{4.36}
$$

These are the *Kramers–Kronig relations*, named after H.A. Kramers and R. de L. Kronig, who derived them independently. The expressions show that causality requires the real and imaginary parts of the permittivity to depend upon each other through the Hilbert transform pair [144].

It is often more convenient to express the relations purely in terms of positive frequencies. This is accomplished using the even–odd behavior of the real and imaginary parts of $\tilde{\epsilon}^c$. Breaking the integrals in (4.35)–(4.36) into the ranges $(-\infty, 0)$ and $(0, \infty)$, and substituting from (4.27), we find that

$$
\text{Re}\{\tilde{\epsilon}^c(\mathbf{r}, \omega)\} - \epsilon_0 = -\frac{2}{\pi} \,\text{P.V.} \int_0^\infty \frac{\Omega \,\text{Im}\{\tilde{\epsilon}^c(\mathbf{r}, \Omega)\}}{\Omega^2 - \omega^2}\, d\Omega, \tag{4.37}
$$

$$
\text{Im}\{\tilde{\epsilon}^c(\mathbf{r}, \omega)\} = \frac{2\omega}{\pi} \,\text{P.V.} \int_0^\infty \frac{\text{Re}\{\tilde{\epsilon}^c(\mathbf{r}, \Omega)\} - \epsilon_0}{\Omega^2 - \omega^2}\, d\Omega - \frac{\sigma_0(\mathbf{r})}{\omega}. \tag{4.38}
$$

The symbol P.V. in this case indicates that values of the integrand around both $\Omega = 0$ and $\Omega = \omega$ must be excluded from the integration. The details of deriving (4.37)–(4.38) are left as an exercise. We use (4.37) in § 4.6 to demonstrate the Kramers–Kronig relationship for a model of complex permittivity of an actual material.

We cannot specify Re $\tilde{\epsilon}^c$ arbitrarily; for a passive medium Im $\tilde{\epsilon}^c$ must be zero or negative at all values of ω, and (4.36) will not necessarily return these required values. However, if we have a good measurement or physical model for Im $\tilde{\epsilon}^c$, as might come from studies of the absorbing properties of the material, we can approximate the real part of the permittivity using (4.35). We shall demonstrate this using simple models for permittivity in § 4.6.

The Kramers–Kronig properties hold for μ as well. We must for practical reasons consider the fact that magnetization becomes unimportant at a much lower frequency than does polarization, so that the infinite integrals in the Kramers–Kronig relations should be truncated at some upper frequency ω_{max}. If we use a model or measured values of Im $\tilde{\mu}$ to determine Re $\tilde{\mu}$, the form of the relation (4.37) should be [113]

$$\mathrm{Re}\{\tilde{\mu}(\mathbf{r}, \omega)\} - \mu_0 = -\frac{2}{\pi}\,\mathrm{P.V.}\int_0^{\omega_{\mathrm{max}}} \frac{\Omega\,\mathrm{Im}\{\tilde{\mu}(\mathbf{r}, \Omega)\}}{\Omega^2 - \omega^2}\,d\Omega,$$

where ω_{max} is the frequency at which magnetization ceases to be important, and above which $\tilde{\mu} = \mu_0$.

4.5 Dissipated and stored energy in a dispersive medium

Let us write down Poynting's power balance theorem for a dispersive medium. Writing $\mathbf{J} = \mathbf{J}^i + \mathbf{J}^c$, we have (§ 2.9.5)

$$-\mathbf{J}^i \cdot \mathbf{E} = \mathbf{J}^c \cdot \mathbf{E} + \nabla \cdot (\mathbf{E} \times \mathbf{H}) + \left(\mathbf{E} \cdot \frac{\partial \mathbf{D}}{\partial t} + \mathbf{H} \cdot \frac{\partial \mathbf{B}}{\partial t}\right). \tag{4.39}$$

We cannot express this in terms of the time rate of change of a stored energy density because of the difficulty in interpreting the term

$$\mathbf{E} \cdot \frac{\partial \mathbf{D}}{\partial t} + \mathbf{H} \cdot \frac{\partial \mathbf{B}}{\partial t} \tag{4.40}$$

for constitutive parameters of the form (2.20)–(2.22). Physically, this term describes both the energy stored in the electromagnetic field *and* the energy dissipated by the material because of time lags between the application of $\mathbf{E}$ and $\mathbf{H}$ and the polarization or magnetization of the atoms (and thus the response fields $\mathbf{D}$ and $\mathbf{B}$). In principle it can also be used to describe *active* media that transfer mechanical or chemical energy of the material into field energy.

Instead of trying to interpret (4.40), we concentrate on the physical meaning of

$$-\nabla \cdot \mathbf{S}(\mathbf{r}, t) = -\nabla \cdot [\mathbf{E}(\mathbf{r}, t) \times \mathbf{H}(\mathbf{r}, t)].$$

We shall postulate that this term describes the net flow of electromagnetic energy into the point $\mathbf{r}$ at time t. Then (4.39) shows that in the absence of impressed sources the energy flow must act to (1) increase or decrease the stored energy density at $\mathbf{r}$, (2) dissipate energy in ohmic losses through the term involving $\mathbf{J}^c$, or (3) dissipate (or provide) energy through the term (4.40). Assuming linearity, we may write

$$-\nabla \cdot \mathbf{S}(\mathbf{r}, t) = \frac{\partial}{\partial t} w_e(\mathbf{r}, t) + \frac{\partial}{\partial t} w_m(\mathbf{r}, t) + \frac{\partial}{\partial t} w_Q(\mathbf{r}, t), \tag{4.41}$$

where the terms on the right represent the time rates of change of, respectively, stored electric, stored magnetic, and dissipated energies.

4.5.1 Dissipation in a dispersive material

Although we may, in general, be unable to separate the individual terms in (4.41), we can examine these terms under certain conditions. For example, consider a field that builds from zero starting from time $t = -\infty$ and then decays back to zero at $t = \infty$. Then by direct integration[*]

$$-\int_{-\infty}^{\infty} \nabla \cdot \mathbf{S}(t)\, dt = w_{em}(t = \infty) - w_{em}(t = -\infty) + w_Q(t = \infty) - w_Q(t = -\infty)$$

where $w_{em} = w_e + w_m$ is the volume density of stored electromagnetic energy. This stored energy is zero at $t = \pm\infty$ since the fields are zero at those times. So

$$\Delta w_Q = -\int_{-\infty}^{\infty} \nabla \cdot \mathbf{S}(t)\, dt = w_Q(t = \infty) - w_Q(t = -\infty)$$

represents the volume density of the net energy dissipated by a lossy medium (or supplied by an active medium). We may thus classify materials according to the scheme

$$
\begin{aligned}
\Delta w_Q &= 0, \qquad \text{lossless,} \\
\Delta w_Q &> 0, \qquad \text{lossy,} \\
\Delta w_Q &\geq 0, \qquad \text{passive,} \\
\Delta w_Q &< 0, \qquad \text{active.}
\end{aligned}
$$

For an anisotropic material with the constitutive relations

$$\tilde{\mathbf{D}} = \tilde{\bar{\epsilon}} \cdot \tilde{\mathbf{E}}, \qquad \tilde{\mathbf{B}} = \tilde{\bar{\mu}} \cdot \tilde{\mathbf{H}}, \qquad \tilde{\mathbf{J}}^c = \tilde{\bar{\sigma}} \cdot \tilde{\mathbf{E}},$$

we find that dissipation is associated with negative imaginary parts of the constitutive parameters. To see this we write

$$\mathbf{E}(\mathbf{r}, t) = \frac{1}{2\pi} \int_{-\infty}^{\infty} \tilde{\mathbf{E}}(\mathbf{r}, \omega) e^{j\omega t}\, d\omega, \qquad \mathbf{D}(\mathbf{r}, t) = \frac{1}{2\pi} \int_{-\infty}^{\infty} \tilde{\mathbf{D}}(\mathbf{r}, \omega') e^{j\omega' t}\, d\omega',$$

and thus find

$$\mathbf{J}^c \cdot \mathbf{E} + \mathbf{E} \cdot \frac{\partial \mathbf{D}}{\partial t} = \frac{1}{(2\pi)^2} \int_{-\infty}^{\infty} \int_{-\infty}^{\infty} \tilde{\mathbf{E}}(\omega) \cdot \tilde{\bar{\epsilon}}^c(\omega') \cdot \tilde{\mathbf{E}}(\omega') e^{j(\omega + \omega')t} j\omega'\, d\omega\, d\omega'$$

where $\tilde{\bar{\epsilon}}^c$ is the complex dyadic permittivity (4.23). Then

$$
\begin{aligned}
\Delta w_Q = \frac{1}{(2\pi)^2} \int_{-\infty}^{\infty} \int_{-\infty}^{\infty} & \left[\tilde{\mathbf{E}}(\omega) \cdot \tilde{\bar{\epsilon}}^c(\omega') \cdot \tilde{\mathbf{E}}(\omega') + \tilde{\mathbf{H}}(\omega) \cdot \tilde{\bar{\mu}}(\omega') \cdot \tilde{\mathbf{H}}(\omega') \right] \cdot \\
& \cdot \left[\int_{-\infty}^{\infty} e^{j(\omega + \omega')t}\, dt \right] j\omega'\, d\omega\, d\omega'.
\end{aligned}
\tag{4.42}
$$

---

[*]Note that in this section we suppress the $\mathbf{r}$-dependence of most quantities for clarity of presentation.

Using (A.5) and integrating over ω we obtain

$$\Delta w_Q = \frac{1}{2\pi} \int_{-\infty}^{\infty} \left[\tilde{\mathbf{E}}(-\omega') \cdot \bar{\tilde{\epsilon}}^c(\omega') \cdot \tilde{\mathbf{E}}(\omega') + \tilde{\mathbf{H}}(-\omega') \cdot \bar{\tilde{\mu}}(\omega') \cdot \tilde{\mathbf{H}}(\omega') \right] j\omega' \, d\omega'. \quad (4.43)$$

Let us examine (4.43) more closely for the simple case of an isotropic material for which

$$\Delta w_Q = \frac{1}{2\pi} \int_{-\infty}^{\infty} \left\{ [j \operatorname{Re}\{\tilde{\epsilon}^c(\omega')\} - \operatorname{Im}\{\tilde{\epsilon}^c(\omega')\}] \tilde{\mathbf{E}}(-\omega') \cdot \tilde{\mathbf{E}}(\omega') \right.$$
$$\left. + [j \operatorname{Re}\{\tilde{\mu}(\omega')\} - \operatorname{Im}\{\tilde{\mu}(\omega')\}] \tilde{\mathbf{H}}(-\omega') \cdot \tilde{\mathbf{H}}(\omega') \right\} \omega' \, d\omega'.$$

Using the frequency symmetry property for complex permittivity (4.16) (which also holds for permeability), we find that for isotropic materials,

$$\operatorname{Re}\{\tilde{\epsilon}^c(\mathbf{r}, \omega)\} = \operatorname{Re}\{\tilde{\epsilon}^c(\mathbf{r}, -\omega)\}, \qquad \operatorname{Im}\{\tilde{\epsilon}^c(\mathbf{r}, \omega)\} = -\operatorname{Im}\{\tilde{\epsilon}^c(\mathbf{r}, -\omega)\}, \qquad (4.44)$$
$$\operatorname{Re}\{\tilde{\mu}(\mathbf{r}, \omega)\} = \operatorname{Re}\{\tilde{\mu}(\mathbf{r}, -\omega)\}, \qquad \operatorname{Im}\{\tilde{\mu}(\mathbf{r}, \omega)\} = -\operatorname{Im}\{\tilde{\mu}(\mathbf{r}, -\omega)\}. \qquad (4.45)$$

Thus, the products of ω' and the real parts of the constitutive parameters are odd functions, while for the imaginary parts these products are even. Since the dot products of the vector fields are even functions, the integrals of the terms containing the real parts of the constitutive parameters vanish and leave

$$\Delta w_Q = 2\frac{1}{2\pi} \int_0^{\infty} \left[-\operatorname{Im}\{\tilde{\epsilon}^c\} |\tilde{\mathbf{E}}|^2 - \operatorname{Im}\{\tilde{\mu}\} |\tilde{\mathbf{H}}|^2 \right] \omega \, d\omega. \quad (4.46)$$

Here we have used (4.3) in the form

$$\tilde{\mathbf{E}}(\mathbf{r}, -\omega) = \tilde{\mathbf{E}}^*(\mathbf{r}, \omega), \qquad \tilde{\mathbf{H}}(\mathbf{r}, -\omega) = \tilde{\mathbf{H}}^*(\mathbf{r}, \omega). \quad (4.47)$$

Equation (4.46) leads us to associate the imaginary parts of the constitutive parameters with dissipation. Moreover, a lossy isotropic material for which $\Delta w_Q > 0$ must have at least one of $\operatorname{Im} \epsilon^c$ and $\operatorname{Im} \mu$ less than zero over some range of positive frequencies, while an active isotropic medium must have at least one of these greater than zero. In general, we speak of a lossy material as having negative imaginary constitutive parameters:

$$\operatorname{Im} \tilde{\epsilon}^c < 0, \qquad \operatorname{Im} \tilde{\mu} < 0 \qquad (\omega > 0). \quad (4.48)$$

A *lossless* medium must have $\operatorname{Im} \tilde{\epsilon} = \operatorname{Im} \tilde{\mu} = \tilde{\sigma} = 0$ for all ω.

The more general anisotropic case is not as simple. Integration of (4.42) over ω' instead of ω produces

$$\Delta w_Q = -\frac{1}{2\pi} \int_{-\infty}^{\infty} \left[\tilde{\mathbf{E}}(\omega) \cdot \bar{\tilde{\epsilon}}^c(-\omega) \cdot \tilde{\mathbf{E}}(-\omega) + \tilde{\mathbf{H}}(\omega) \cdot \bar{\tilde{\mu}}(-\omega) \cdot \tilde{\mathbf{H}}(-\omega) \right] j\omega \, d\omega.$$

Adding half of this expression to half of (4.43) and using (4.24), (4.16), and (4.47), we obtain

$$\Delta w_Q = \frac{1}{4\pi} \int_{-\infty}^{\infty} \left[\tilde{\mathbf{E}}^* \cdot \bar{\tilde{\epsilon}}^c \cdot \tilde{\mathbf{E}} - \tilde{\mathbf{E}} \cdot \bar{\tilde{\epsilon}}^{c*} \cdot \tilde{\mathbf{E}}^* + \tilde{\mathbf{H}}^* \cdot \bar{\tilde{\mu}} \cdot \tilde{\mathbf{H}} - \tilde{\mathbf{H}} \cdot \bar{\tilde{\mu}}^* \cdot \tilde{\mathbf{H}}^* \right] j\omega \, d\omega.$$

Finally, dyadic identity (A.78) yields

$$\Delta w_Q = \frac{1}{4\pi} \int_{-\infty}^{\infty} \left[\tilde{\mathbf{E}}^* \cdot \left(\bar{\tilde{\epsilon}}^c - \bar{\tilde{\epsilon}}^{c\dagger} \right) \cdot \tilde{\mathbf{E}} + \tilde{\mathbf{H}}^* \cdot \left(\bar{\tilde{\mu}} - \bar{\tilde{\mu}}^{\dagger} \right) \cdot \tilde{\mathbf{H}} \right] j\omega \, d\omega$$

where the dagger (†) denotes the hermitian (conjugate-transpose) operation. The condition for a lossless anisotropic material is

$$\tilde{\bar{\epsilon}}^c = \tilde{\bar{\epsilon}}^{c\dagger}, \qquad \tilde{\bar{\mu}} = \tilde{\bar{\mu}}^\dagger, \tag{4.49}$$

or

$$\tilde{\epsilon}_{ij} = \tilde{\epsilon}_{ji}^*, \qquad \tilde{\mu}_{ij} = \tilde{\mu}_{ji}^*, \qquad \tilde{\sigma}_{ij} = \tilde{\sigma}_{ji}^*. \tag{4.50}$$

These relationships imply that in the lossless case the diagonal entries of the constitutive dyadics are purely real.

Equations (4.50) show that complex entries in a permittivity or permeability matrix do not necessarily imply loss. For example, we will show in § 4.6.2 that an electron plasma exposed to a z-directed dc magnetic field has a permittivity of the form

$$[\tilde{\bar{\epsilon}}] = \begin{bmatrix} \epsilon & -j\delta & 0 \\ j\delta & \epsilon & 0 \\ 0 & 0 & \epsilon_z \end{bmatrix}$$

where $\epsilon, \epsilon_z, \delta$ are real functions of space and frequency. Since $\tilde{\bar{\epsilon}}$ is hermitian it describes a lossless plasma. Similarly, a gyrotropic medium such as a ferrite exposed to a z-directed magnetic field has a permeability dyadic

$$[\tilde{\bar{\mu}}] = \begin{bmatrix} \mu & -j\kappa & 0 \\ j\kappa & \mu & 0 \\ 0 & 0 & \mu_0 \end{bmatrix},$$

which also describes a lossless material.

4.5.2 Energy stored in a dispersive material

In the last section we were able to isolate the dissipative effects for a dispersive material under special circumstances. It is not generally possible, however, to isolate a term describing the stored energy. The Kramers–Kronig relations imply that if the constitutive parameters of a material are frequency-dependent, they must have both real and imaginary parts; such a material, if isotropic, must be lossy. So dispersive materials are generally lossy and must have both dissipative and energy-storage characteristics. However, many materials have frequency ranges called *transparency ranges* over which $\mathrm{Im}\,\tilde{\epsilon}^c$ and $\mathrm{Im}\,\tilde{\mu}$ are small compared to $\mathrm{Re}\,\tilde{\epsilon}^c$ and $\mathrm{Re}\,\tilde{\mu}$. By restricting our interest to these ranges, we may approximate the material as lossless and compute a stored energy. An important case involves a monochromatic field oscillating at a frequency within this range.

To study the energy stored by a monochromatic field in a dispersive material, we must consider the transient period during which energy accumulates in the fields. The assumption of purely sinusoidal field variation would not include the effects described by the temporal constitutive relations (2.20)–(2.21), which show that as the field builds, the energy must be added with a time lag. Instead we shall assume fields with the temporal variation

$$\mathbf{E}(\mathbf{r}, t) = f(t) \sum_{i=1}^{3} \hat{\mathbf{i}}_i |E_i(\mathbf{r})| \cos[\omega_0 t + \xi_i^E(\mathbf{r})] \tag{4.51}$$

where $f(t)$ is an appropriate function describing the build-up of the sinusoidal field. To compute the stored energy of a sinusoidal wave, we must parameterize $f(t)$ so that we

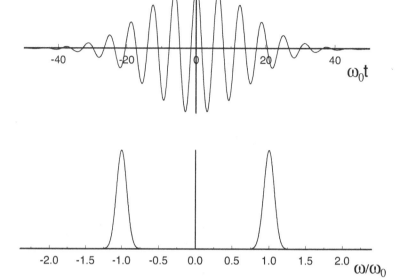

FIGURE 4.2

Temporal (top) and spectral magnitude (bottom) dependences of $\mathbf{E}$ used to compute energy stored in a dispersive material.

may drive it to unity as a limiting case of the parameter. A simple choice is

$$f(t) = e^{-\alpha^2 t^2} \;\leftrightarrow\; \tilde{F}(\omega) = \sqrt{\frac{\pi}{\alpha^2}} e^{-\frac{\omega^2}{4\alpha^2}}. \tag{4.52}$$

Note that since $f(t)$ approaches unity as $\alpha \to 0$, we have the generalized Fourier transform relation

$$\lim_{\alpha \to 0} \tilde{F}(\omega) = 2\pi\delta(\omega). \tag{4.53}$$

Substituting (4.51) into the Fourier transform formula (4.1), we find that

$$\tilde{\mathbf{E}}(\mathbf{r}, \omega) = \frac{1}{2}\sum_{i=1}^{3} \hat{\mathbf{i}}_i |E_i(\mathbf{r})| e^{j\xi_i^E(\mathbf{r})} \tilde{F}(\omega - \omega_0) + \frac{1}{2}\sum_{i=1}^{3} \hat{\mathbf{i}}_i |E_i(\mathbf{r})| e^{-j\xi_i^E(\mathbf{r})} \tilde{F}(\omega + \omega_0).$$

We can simplify this by defining

$$\check{\mathbf{E}}(\mathbf{r}) = \sum_{i=1}^{3} \hat{\mathbf{i}}_i |E_i(\mathbf{r})| e^{j\xi_i^E(\mathbf{r})} \tag{4.54}$$

as the *phasor* vector field, to obtain

$$\tilde{\mathbf{E}}(\mathbf{r}, \omega) = \tfrac{1}{2}[\check{\mathbf{E}}(\mathbf{r})\tilde{F}(\omega - \omega_0) + \check{\mathbf{E}}^*(\mathbf{r})\tilde{F}(\omega + \omega_0)]. \tag{4.55}$$

We shall discuss the phasor concept in detail in § 4.7.

The field $\mathbf{E}(\mathbf{r}, t)$ appears in Figure 4.2 as a function of t, while $\tilde{\mathbf{E}}(\mathbf{r}, \omega)$ appears in Figure 4.2 as a function of ω. As α becomes small, the spectrum of $\mathbf{E}(\mathbf{r}, t)$ concentrates around $\omega = \pm\omega_0$. We assume the material is transparent for all values α of interest so that we may treat ϵ as real. Then, since there is no dissipation, we conclude that the term (4.40) represents the time rate of change of stored energy at time t, including the effects of field build-up. Hence the interpretation[†]

$$\mathbf{E} \cdot \frac{\partial \mathbf{D}}{\partial t} = \frac{\partial w_e}{\partial t}, \qquad \mathbf{H} \cdot \frac{\partial \mathbf{B}}{\partial t} = \frac{\partial w_m}{\partial t}.$$

We focus on the electric field term and infer the magnetic field term by analogy.

Since for periodic signals it is more convenient to deal with the time-averaged stored energy than with the instantaneous stored energy, we compute the time average of $w_e(\mathbf{r}, t)$ over the period of the sinusoid centered at the time origin. That is, we compute

$$\langle w_e \rangle = \frac{1}{T} \int_{-T/2}^{T/2} w_e(t)\, dt \tag{4.56}$$

where $T = 2\pi/\omega_0$. With $\alpha \to 0$, this time-average value is accurate for all periods of the sinusoidal wave.

Because the most expedient approach to the computation of (4.56) is to employ the Fourier spectrum of $\mathbf{E}$, we use

$$\mathbf{E}(\mathbf{r}, t) = \frac{1}{2\pi} \int_{-\infty}^{\infty} \tilde{\mathbf{E}}(\mathbf{r}, \omega)e^{j\omega t}\, d\omega = \frac{1}{2\pi} \int_{-\infty}^{\infty} \tilde{\mathbf{E}}^*(\mathbf{r}, \omega')e^{-j\omega' t}\, d\omega',$$

$$\frac{\partial \mathbf{D}(\mathbf{r}, t)}{\partial t} = \frac{1}{2\pi} \int_{-\infty}^{\infty} (j\omega)\tilde{\mathbf{D}}(\mathbf{r}, \omega)e^{j\omega t}\, d\omega = \frac{1}{2\pi} \int_{-\infty}^{\infty} (-j\omega')\tilde{\mathbf{D}}^*(\mathbf{r}, \omega')e^{-j\omega' t}\, d\omega'.$$

We have obtained the second form of each expression via property (4.3) for the transform of a real function along with the change of variables $\omega' = -\omega$. Multiplying the two forms of the expressions and adding half of each, we have

$$\frac{\partial w_e}{\partial t} = \frac{1}{2} \int_{-\infty}^{\infty} \frac{d\omega}{2\pi} \int_{-\infty}^{\infty} \frac{d\omega'}{2\pi} \left[j\omega \tilde{\mathbf{E}}^*(\omega') \cdot \tilde{\mathbf{D}}(\omega) - j\omega' \tilde{\mathbf{E}}(\omega) \cdot \tilde{\mathbf{D}}^*(\omega') \right] e^{-j(\omega'-\omega)t}. \tag{4.57}$$

Now let us consider a dispersive isotropic medium described by the constitutive relations $\tilde{\mathbf{D}} = \tilde{\epsilon}\tilde{\mathbf{E}}$ and $\tilde{\mathbf{B}} = \tilde{\mu}\tilde{\mathbf{H}}$. Since the imaginary parts of $\tilde{\epsilon}$ and $\tilde{\mu}$ are associated with power dissipation in the medium, we shall approximate $\tilde{\epsilon}$ and $\tilde{\mu}$ as purely real. Then (4.57) becomes

$$\frac{\partial w_e}{\partial t} = \frac{1}{2} \int_{-\infty}^{\infty} \frac{d\omega}{2\pi} \int_{-\infty}^{\infty} \frac{d\omega'}{2\pi} \tilde{\mathbf{E}}^*(\omega') \cdot \tilde{\mathbf{E}}(\omega) \left[j\omega \tilde{\epsilon}(\omega) - j\omega' \tilde{\epsilon}(\omega') \right] e^{-j(\omega'-\omega)t}.$$

Substitution from (4.55) now gives

$$\frac{\partial w_e}{\partial t} = \frac{1}{8} \int_{-\infty}^{\infty} \frac{d\omega}{2\pi} \int_{-\infty}^{\infty} \frac{d\omega'}{2\pi} \left[j\omega \tilde{\epsilon}(\omega) - j\omega' \tilde{\epsilon}(\omega') \right] \cdot$$

$$\cdot \left[\check{\mathbf{E}} \cdot \check{\mathbf{E}}^* \tilde{F}(\omega - \omega_0)\tilde{F}(\omega' - \omega_0) + \check{\mathbf{E}} \cdot \check{\mathbf{E}}^* \tilde{F}(\omega + \omega_0)\tilde{F}(\omega' + \omega_0) \right.$$

$$\left. + \check{\mathbf{E}} \cdot \check{\mathbf{E}} \tilde{F}(\omega - \omega_0)\tilde{F}(\omega' + \omega_0) + \check{\mathbf{E}}^* \cdot \check{\mathbf{E}}^* \tilde{F}(\omega + \omega_0)\tilde{F}(\omega' - \omega_0) \right] e^{-j(\omega'-\omega)t}.$$

[†]Note that in this section we suppress the $\mathbf{r}$-dependence of most quantities for clarity of presentation.

Let $\omega \to -\omega$ wherever the term $\tilde{F}(\omega + \omega_0)$ appears, and $\omega' \to -\omega'$ wherever the term $\tilde{F}(\omega' + \omega_0)$ appears. Since $\tilde{F}(-\omega) = \tilde{F}(\omega)$ and $\tilde{\epsilon}(-\omega) = \tilde{\epsilon}(\omega)$, we find that

$$
\frac{\partial w_e}{\partial t} = \frac{1}{8} \int_{-\infty}^{\infty} \frac{d\omega}{2\pi} \int_{-\infty}^{\infty} \frac{d\omega'}{2\pi} \tilde{F}(\omega - \omega_0)\tilde{F}(\omega' - \omega_0) \cdot
$$
$$
\cdot \left[\check{\mathbf{E}} \cdot \check{\mathbf{E}}^*[j\omega\tilde{\epsilon}(\omega) - j\omega'\tilde{\epsilon}(\omega')]e^{j(\omega-\omega')t} + \check{\mathbf{E}} \cdot \check{\mathbf{E}}^*[j\omega'\tilde{\epsilon}(\omega') - j\omega\tilde{\epsilon}(\omega)]e^{j(\omega'-\omega)t} \right.
$$
$$
\left. + \check{\mathbf{E}} \cdot \check{\mathbf{E}}[j\omega\tilde{\epsilon}(\omega) + j\omega'\tilde{\epsilon}(\omega')]e^{j(\omega+\omega')t} + \check{\mathbf{E}}^* \cdot \check{\mathbf{E}}^*[-j\omega\tilde{\epsilon}(\omega) - j\omega'\tilde{\epsilon}(\omega')]e^{-j(\omega+\omega')t} \right].
$$
$$(4.58)$$

For small α the spectra are concentrated near $\omega = \omega_0$ or $\omega' = \omega_0$. For terms involving the difference in the permittivities, we can expand $g(\omega) = \omega\tilde{\epsilon}(\omega)$ in a Taylor series about ω_0 to obtain the approximation

$$
\omega\tilde{\epsilon}(\omega) \approx \omega_0\tilde{\epsilon}(\omega_0) + (\omega - \omega_0)g'(\omega_0) \quad \text{where} \quad g'(\omega_0) = \left. \frac{\partial[\omega\tilde{\epsilon}(\omega)]}{\partial\omega} \right|_{\omega=\omega_0}.
$$

This is not required for terms involving a sum of permittivities since these will not tend to cancel. For such terms we merely substitute $\omega = \omega_0$ or $\omega' = \omega_0$. With these, (4.58) becomes

$$
\frac{\partial w_e}{\partial t} = \frac{1}{8} \int_{-\infty}^{\infty} \frac{d\omega}{2\pi} \int_{-\infty}^{\infty} \frac{d\omega'}{2\pi} \tilde{F}(\omega - \omega_0)\tilde{F}(\omega' - \omega_0) \cdot
$$
$$
\cdot \left[\check{\mathbf{E}} \cdot \check{\mathbf{E}}^* g'(\omega_0)[j(\omega - \omega')]e^{j(\omega-\omega')t} + \check{\mathbf{E}} \cdot \check{\mathbf{E}}^* g'(\omega_0)[j(\omega' - \omega)]e^{j(\omega'-\omega)t} \right.
$$
$$
\left. + \check{\mathbf{E}} \cdot \check{\mathbf{E}}\tilde{\epsilon}(\omega_0)[j(\omega + \omega')]e^{j(\omega+\omega')t} + \check{\mathbf{E}}^* \cdot \check{\mathbf{E}}^*\tilde{\epsilon}(\omega_0)[-j(\omega + \omega')]e^{-j(\omega+\omega')t} \right].
$$

By integration

$$
w_e(t) = \frac{1}{8} \int_{-\infty}^{\infty} \frac{d\omega}{2\pi} \int_{-\infty}^{\infty} \frac{d\omega'}{2\pi} \tilde{F}(\omega - \omega_0)\tilde{F}(\omega' - \omega_0) \cdot
$$
$$
\cdot \left[\check{\mathbf{E}} \cdot \check{\mathbf{E}}^* g'(\omega_0)e^{j(\omega-\omega')t} + \check{\mathbf{E}} \cdot \check{\mathbf{E}}^* g'(\omega_0)e^{j(\omega'-\omega)t} \right.
$$
$$
\left. + \check{\mathbf{E}} \cdot \check{\mathbf{E}}\tilde{\epsilon}(\omega_0)e^{j(\omega+\omega')t} + \check{\mathbf{E}}^* \cdot \check{\mathbf{E}}^*\tilde{\epsilon}(\omega_0)e^{-j(\omega+\omega')t} \right].
$$

Our last step is to compute the time-average value of w_e and let $\alpha \to 0$. Applying (4.56) we find

$$
\langle w_e \rangle = \frac{1}{8} \int_{-\infty}^{\infty} \frac{d\omega}{2\pi} \int_{-\infty}^{\infty} \frac{d\omega'}{2\pi} \tilde{F}(\omega - \omega_0)\tilde{F}(\omega' - \omega_0) \cdot
$$
$$
\cdot \left[2\check{\mathbf{E}} \cdot \check{\mathbf{E}}^* g'(\omega_0) \operatorname{sinc}\left([\omega - \omega']\frac{\pi}{\omega_0} \right) + \{\check{\mathbf{E}}^* \cdot \check{\mathbf{E}}^* + \check{\mathbf{E}} \cdot \check{\mathbf{E}}\} \tilde{\epsilon}(\omega_0) \operatorname{sinc}\left([\omega + \omega']\frac{\pi}{\omega_0} \right) \right]
$$

where $\operatorname{sinc} x$ is defined in (A.10) and satisfies $\operatorname{sinc}(-x) = \operatorname{sinc} x$. Finally, we let $\alpha \to 0$ and use (4.53) to replace $\tilde{F}(\omega)$ by a δ-function. Upon integration, these δ-functions set $\omega = \omega_0$ and $\omega' = \omega_0$. As $\operatorname{sinc}(0) = 1$ and $\operatorname{sinc}(2\pi) = 0$, the time-average stored electric energy density is simply

$$
\langle w_e \rangle = \frac{1}{4}|\check{\mathbf{E}}|^2 \left. \frac{\partial(\omega\tilde{\epsilon})}{\partial\omega} \right|_{\omega=\omega_0}.
$$
$$(4.59)$$

Similarly

$$\langle w_m \rangle = \tfrac{1}{4}|\check{\mathbf{H}}|^2 \frac{\partial(\omega\tilde{\mu})}{\partial\omega}\bigg|_{\omega=\omega_0}.$$

When applied to anisotropic materials, this approach gives [38]

$$\langle w_e \rangle = \tfrac{1}{4}\check{\mathbf{E}}^* \cdot \frac{\partial(\omega\tilde{\bar{\epsilon}})}{\partial\omega}\bigg|_{\omega=\omega_0} \cdot \check{\mathbf{E}}, \tag{4.60}$$

$$\langle w_m \rangle = \tfrac{1}{4}\check{\mathbf{H}}^* \cdot \frac{\partial(\omega\tilde{\bar{\mu}})}{\partial\omega}\bigg|_{\omega=\omega_0} \cdot \check{\mathbf{H}}. \tag{4.61}$$

For a lossless, nondispersive material where the constitutive parameters are frequency independent, we can use (4.49) and (A.78) to obtain

$$\langle w_e \rangle = \tfrac{1}{4}\check{\mathbf{E}}^* \cdot \bar{\epsilon} \cdot \check{\mathbf{E}} = \tfrac{1}{4}\check{\mathbf{E}} \cdot \check{\mathbf{D}}^*, \tag{4.62}$$

$$\langle w_m \rangle = \tfrac{1}{4}\check{\mathbf{H}}^* \cdot \bar{\mu} \cdot \check{\mathbf{H}} = \tfrac{1}{4}\check{\mathbf{H}} \cdot \check{\mathbf{B}}^*, \tag{4.63}$$

in the anisotropic case, and

$$\langle w_e \rangle = \tfrac{1}{4}\epsilon|\check{\mathbf{E}}|^2 = \tfrac{1}{4}\check{\mathbf{E}} \cdot \check{\mathbf{D}}^*, \tag{4.64}$$

$$\langle w_m \rangle = \tfrac{1}{4}\mu|\check{\mathbf{H}}|^2 = \tfrac{1}{4}\check{\mathbf{H}} \cdot \check{\mathbf{B}}^*, \tag{4.65}$$

in the isotropic case. Here $\check{\mathbf{E}}$, $\check{\mathbf{D}}$, $\check{\mathbf{B}}$, $\check{\mathbf{H}}$ are all phasor fields as defined by (4.54).

4.5.3   The energy theorem

A convenient expression for the time-average stored energies (4.60) and (4.61) is found by manipulating the frequency-domain Maxwell equations. Beginning with the complex conjugates of the two frequency-domain curl equations for anisotropic media,

$$\nabla \times \tilde{\mathbf{E}}^* = j\omega\tilde{\bar{\mu}}^* \cdot \tilde{\mathbf{H}}^*,$$

$$\nabla \times \tilde{\mathbf{H}}^* = \tilde{\mathbf{J}}^* - j\omega\tilde{\bar{\epsilon}}^* \cdot \tilde{\mathbf{E}}^*,$$

we differentiate with respect to frequency:

$$\nabla \times \frac{\partial\tilde{\mathbf{E}}^*}{\partial\omega} = j\frac{\partial(\omega\tilde{\bar{\mu}}^*)}{\partial\omega} \cdot \tilde{\mathbf{H}}^* + j\omega\tilde{\bar{\mu}}^* \cdot \frac{\partial\tilde{\mathbf{H}}^*}{\partial\omega}, \tag{4.66}$$

$$\nabla \times \frac{\partial\tilde{\mathbf{H}}^*}{\partial\omega} = \frac{\partial\tilde{\mathbf{J}}^*}{\partial\omega} - j\frac{\partial(\omega\tilde{\bar{\epsilon}}^*)}{\partial\omega} \cdot \tilde{\mathbf{E}}^* - j\omega\tilde{\bar{\epsilon}}^* \cdot \frac{\partial\tilde{\mathbf{E}}^*}{\partial\omega}. \tag{4.67}$$

These terms also appear as a part of the expansion

$$\nabla \cdot \left(\tilde{\mathbf{E}} \times \frac{\partial\tilde{\mathbf{H}}^*}{\partial\omega} + \frac{\partial\tilde{\mathbf{E}}^*}{\partial\omega} \times \tilde{\mathbf{H}} \right) =$$

$$\frac{\partial\tilde{\mathbf{H}}^*}{\partial\omega} \cdot (\nabla \times \tilde{\mathbf{E}}) - \tilde{\mathbf{E}} \cdot \nabla \times \frac{\partial\tilde{\mathbf{H}}^*}{\partial\omega} + \tilde{\mathbf{H}} \cdot \nabla \times \frac{\partial\tilde{\mathbf{E}}^*}{\partial\omega} - \frac{\partial\tilde{\mathbf{E}}^*}{\partial\omega} \cdot (\nabla \times \tilde{\mathbf{H}})$$

where we have used (B.50). Substituting from (4.66)–(4.67) and eliminating $\nabla \times \tilde{\mathbf{E}}$ and $\nabla \times \tilde{\mathbf{H}}$ by Maxwell's equations, we have

$$\frac{1}{4} \nabla \cdot \left(\tilde{\mathbf{E}} \times \frac{\partial \tilde{\mathbf{H}}^*}{\partial \omega} + \frac{\partial \tilde{\mathbf{E}}^*}{\partial \omega} \times \tilde{\mathbf{H}} \right) =$$

$$j \frac{1}{4} \omega \left(\tilde{\mathbf{E}} \cdot \tilde{\bar{\epsilon}}^* \cdot \frac{\partial \tilde{\mathbf{E}}^*}{\partial \omega} - \frac{\partial \tilde{\mathbf{E}}^*}{\partial \omega} \cdot \tilde{\bar{\epsilon}} \cdot \tilde{\mathbf{E}} \right) + j \frac{1}{4} \omega \left(\tilde{\mathbf{H}} \cdot \tilde{\bar{\mu}}^* \cdot \frac{\partial \tilde{\mathbf{H}}^*}{\partial \omega} - \frac{\partial \tilde{\mathbf{H}}^*}{\partial \omega} \cdot \tilde{\bar{\mu}} \cdot \tilde{\mathbf{H}} \right)$$

$$+ j \frac{1}{4} \left(\tilde{\mathbf{E}} \cdot \frac{\partial(\omega \tilde{\bar{\epsilon}}^*)}{\partial \omega} \cdot \tilde{\mathbf{E}}^* + \tilde{\mathbf{H}} \cdot \frac{\partial(\omega \tilde{\bar{\mu}}^*)}{\partial \omega} \cdot \tilde{\mathbf{H}}^* \right) - \frac{1}{4} \left(\tilde{\mathbf{E}} \cdot \frac{\partial \tilde{\mathbf{J}}^*}{\partial \omega} + \tilde{\mathbf{J}} \cdot \frac{\partial \tilde{\mathbf{E}}^*}{\partial \omega} \right).$$

Let us assume that the sources and fields are narrowband, centered on ω_0, and that ω_0 lies within a transparency range so that within the band the material may be considered lossless. Invoking from (4.49) the facts that $\tilde{\bar{\epsilon}} = \tilde{\bar{\epsilon}}^\dagger$ and $\tilde{\bar{\mu}} = \tilde{\bar{\mu}}^\dagger$, we find that the first two terms on the right are zero. Integrating over a volume and taking the complex conjugate of both sides, we obtain

$$\frac{1}{4} \oint_S \left(\tilde{\mathbf{E}}^* \times \frac{\partial \tilde{\mathbf{H}}}{\partial \omega} + \frac{\partial \tilde{\mathbf{E}}}{\partial \omega} \times \tilde{\mathbf{H}}^* \right) \cdot d\mathbf{S} =$$

$$-j \frac{1}{4} \int_V \left(\tilde{\mathbf{E}}^* \cdot \frac{\partial(\omega \tilde{\bar{\epsilon}})}{\partial \omega} \cdot \tilde{\mathbf{E}} + \tilde{\mathbf{H}}^* \cdot \frac{\partial(\omega \tilde{\bar{\mu}})}{\partial \omega} \cdot \tilde{\mathbf{H}} \right) dV - \frac{1}{4} \int_V \left(\tilde{\mathbf{E}}^* \cdot \frac{\partial \tilde{\mathbf{J}}}{\partial \omega} + \tilde{\mathbf{J}}^* \cdot \frac{\partial \tilde{\mathbf{E}}}{\partial \omega} \right) dV.$$

Evaluating each of the terms at $\omega = \omega_0$ and using (4.60)–(4.61), we have

$$\frac{1}{4} \oint_S \left(\tilde{\mathbf{E}}^* \times \frac{\partial \tilde{\mathbf{H}}}{\partial \omega} + \frac{\partial \tilde{\mathbf{E}}}{\partial \omega} \times \tilde{\mathbf{H}}^* \right) \Bigg|_{\omega = \omega_0} \cdot d\mathbf{S} =$$

$$-j \left[\langle W_e \rangle + \langle W_m \rangle \right] - \frac{1}{4} \int_V \left(\tilde{\mathbf{E}}^* \cdot \frac{\partial \tilde{\mathbf{J}}}{\partial \omega} + \tilde{\mathbf{J}}^* \cdot \frac{\partial \tilde{\mathbf{E}}}{\partial \omega} \right) \Bigg|_{\omega = \omega_0} dV \qquad (4.68)$$

where $\langle W_e \rangle + \langle W_m \rangle$ is the total time-average electromagnetic energy stored in the volume region V. This is known as the *energy theorem*. In § 4.11.3 it will help determine the velocity of energy transport for a plane wave.

4.6 Some simple models for constitutive parameters

Our discussion of electromagnetic fields has been restricted to macroscopic phenomena. Although we recognize that matter is composed of microscopic constituents, we have chosen to describe materials using constitutive relationships whose parameters, such as permittivity, conductivity, and permeability, are viewed in the macroscopic sense. Through experiments on the laboratory scale, we can measure the constitutive parameters to the precision required for engineering applications.

At some point it becomes useful to establish models of the macroscopic behavior of materials based on microscopic considerations, formulating expressions for the constitutive parameters using atomic descriptors such as number density, atomic charge, and molecular dipole moment. These models allow us to predict the behavior of broad classes of materials, such as dielectrics and conductors, over wide ranges of frequency and field strength.

Accurate models for the behavior of materials under the influence of electromagnetic fields must account for many complicated effects, including those best described by quantum mechanics. However, many simple models can be obtained using classical mechanics and field theory. We shall investigate several of the most useful of these, and in the process try to gain a feeling for the relationship between the field applied to a material and the resulting polarization or magnetization of the underlying atomic structure.

For simplicity, we consider only homogeneous materials. The fundamental atomic descriptor of "number density," N, is thus taken to be independent of position and time. The result may be more generally applicable since we may think of an inhomogeneous material in terms of the spatial variation of constitutive parameters originally determined assuming homogeneity. However, we shall not attempt to study the microscopic conditions that give rise to inhomogeneities.

4.6.1   Complex permittivity of a nonmagnetized plasma

A plasma is an ionized gas in which the charged particles are free to move under the influence of an applied field and through particle–particle interactions. A plasma differs from other materials in that there is no atomic lattice restricting particle motion. However, even in a gas the interactions between the particles and the fields give rise to a polarization effect, causing the permittivity of the gas to differ from that of free space. In addition, exposing the gas to an external field will cause a secondary current to flow as a result of the Lorentz force on the particles. As the moving particles collide with one another they relinquish their momentum, an effect describable in terms of a conductivity. In this section we perform a simple analysis to determine the complex permittivity of a nonmagnetized plasma.

To make our analysis tractable, we shall make several assumptions.

1. We assume that the plasma is *neutral*: i.e., that the free electrons and positive ions are of equal number and distributed in like manner. If the particles are sufficiently dense to be considered in the macroscopic sense, then there is no net field produced by the gas and thus no electromagnetic interaction between the particles. We also assume the plasma is homogeneous and that the number density of the electrons N (number of electrons per m^3) is independent of time and position. In contrast to this are *electron beams*, whose properties differ significantly from neutral plasmas because of bunching of electrons by the applied field [152].

2. We ignore the motion of the positive ions in the computation of the secondary current, since the ratio of the mass of an ion to that of an electron is at least as large as the ratio of a proton to an electron ($m_p/m_e = 1836$), and thus the ions accelerate much more slowly.

3. We assume that the applied field is that of an electromagnetic wave. In § 2.10.6 we found that for a wave in free space the ratio of magnetic to electric field is $|\mathbf{H}|/|\mathbf{E}| = \sqrt{\epsilon_0/\mu_0}$, so that $|\mathbf{B}|/|\mathbf{E}| = \mu_0\sqrt{\epsilon_0/\mu_0} = \sqrt{\mu_0\epsilon_0} = 1/c$. So the force on an electron is approximately

$$\mathbf{F} = -e(\mathbf{E} + \mathbf{v} \times \mathbf{B}) \approx -e\mathbf{E}$$

if $v \ll c$. Here e is the *unsigned* charge on an electron, $e = 1.6022 \times 10^{-19}$ C. Note that when an external static magnetic field accompanies the field of the wave, as is the case in the earth's ionosphere, for example, we cannot ignore the magnetic component of the Lorentz force. This case will be considered in § 4.6.2.

4. We assume that the mechanical interactions between particles can be described using a *collision frequency* ν, which describes the rate at which a directed plasma velocity becomes random in the absence of external forces.

With these assumptions, we can write the equation of motion for the plasma medium. Let $\mathbf{v}(\mathbf{r}, t)$ represent the macroscopic velocity of the plasma medium. By Newton's second law, the force acting at each point on the medium is balanced by the time-rate of change in momentum at that point. Because of collisions, the total change in momentum density is described by

$$\mathbf{F}(\mathbf{r}, t) = -Ne\mathbf{E}(\mathbf{r}, t) = \frac{d\wp(\mathbf{r}, t)}{dt} + \nu\wp(\mathbf{r}, t) \tag{4.69}$$

where $\wp(\mathbf{r}, t) = N m_e \mathbf{v}(\mathbf{r}, t)$ is the volume density of momentum. Note that if there is no externally applied electromagnetic force, (4.69) becomes

$$\frac{d\wp(\mathbf{r}, t)}{dt} + \nu\wp(\mathbf{r}, t) = 0.$$

Hence $\wp(\mathbf{r}, t) = \wp_0(\mathbf{r})e^{-\nu t}$, and ν describes the rate at which the electron velocities move toward a random state, producing a macroscopic plasma velocity $\mathbf{v}$ of zero.

The time derivative in (4.69) is the total derivative defined in (A.60):

$$\frac{d\wp(\mathbf{r}, t)}{dt} = \frac{\partial\wp(\mathbf{r}, t)}{\partial t} + (\mathbf{v} \cdot \nabla)\wp(\mathbf{r}, t). \tag{4.70}$$

The second term on the right accounts for the time-rate of change of momentum perceived as the observer moves through regions of spatially changing momentum. Since the electron velocity is induced by the electromagnetic field, we anticipate that for a sinusoidal wave the spatial variation will be on the order of the wavelength of the field: $\lambda = 2\pi c/\omega$. Thus, while the first term in (4.70) is proportional to ω, the second term is proportional to $\omega v/c$ and can be neglected for non-relativistic particle velocities. Then, writing $\mathbf{E}(\mathbf{r}, t)$ and $\mathbf{v}(\mathbf{r}, t)$ as inverse Fourier transforms, we see that (4.69) yields

$$-e\tilde{\mathbf{E}} = j\omega m_e\tilde{\mathbf{v}} + m_e\nu\tilde{\mathbf{v}} \tag{4.71}$$

and thus,

$$\tilde{\mathbf{v}} = -\frac{\frac{e}{m_e}\tilde{\mathbf{E}}}{\nu + j\omega}. \tag{4.72}$$

The secondary current associated with the moving electrons is (since e is unsigned)

$$\tilde{\mathbf{J}}^s = -Ne\tilde{\mathbf{v}} = \frac{\epsilon_0\omega_p^2}{\omega^2 + \nu^2}(\nu - j\omega)\tilde{\mathbf{E}} \tag{4.73}$$

where

$$\omega_p^2 = \frac{Ne^2}{\epsilon_0 m_e} \tag{4.74}$$

is the *plasma frequency*.

The frequency-domain Ampere's law for primary and secondary currents in free space is merely

$$\nabla \times \tilde{\mathbf{H}} = \tilde{\mathbf{J}}^i + \tilde{\mathbf{J}}^s + j\omega\epsilon_0\tilde{\mathbf{E}}.$$

Substitution from (4.73) gives

$$\nabla \times \tilde{\mathbf{H}} = \tilde{\mathbf{J}}^i + \frac{\epsilon_0\omega_p^2\nu}{\omega^2 + \nu^2}\tilde{\mathbf{E}} + j\omega\epsilon_0\left(1 - \frac{\omega_p^2}{\omega^2 + \nu^2}\right)\tilde{\mathbf{E}}.$$

We can determine the material properties of the plasma by realizing that the above expression can be written as

$$\nabla \times \tilde{\mathbf{H}} = \tilde{\mathbf{J}}^i + \tilde{\mathbf{J}}^s + j\omega\tilde{\mathbf{D}}$$

with the constitutive relations $\tilde{\mathbf{J}}^s = \tilde{\sigma}\tilde{\mathbf{E}}$ and $\tilde{\mathbf{D}} = \tilde{\epsilon}\tilde{\mathbf{E}}$. Here we identify the conductivity of the plasma as

$$\tilde{\sigma}(\omega) = \frac{\epsilon_0\omega_p^2\nu}{\omega^2 + \nu^2} \tag{4.75}$$

and the permittivity as

$$\tilde{\epsilon}(\omega) = \epsilon_0\left(1 - \frac{\omega_p^2}{\omega^2 + \nu^2}\right).$$

We can also write Ampere's law as

$$\nabla \times \tilde{\mathbf{H}} = \tilde{\mathbf{J}}^i + j\omega\tilde{\epsilon}^c\tilde{\mathbf{E}}$$

where $\tilde{\epsilon}^c$ is the complex permittivity

$$\tilde{\epsilon}^c(\omega) = \tilde{\epsilon}(\omega) + \frac{\tilde{\sigma}(\omega)}{j\omega} = \epsilon_0\left(1 - \frac{\omega_p^2}{\omega^2 + \nu^2}\right) - j\frac{\epsilon_0\omega_p^2\nu}{\omega(\omega^2 + \nu^2)}. \tag{4.76}$$

To describe the plasma in terms of a polarization vector, we merely use $\tilde{\mathbf{D}} = \epsilon_0\tilde{\mathbf{E}} + \tilde{\mathbf{P}} = \tilde{\epsilon}\tilde{\mathbf{E}}$ to obtain the polarization vector $\tilde{\mathbf{P}} = (\tilde{\epsilon} - \epsilon_0)\tilde{\mathbf{E}} = \epsilon_0\tilde{\chi}_e\tilde{\mathbf{E}}$, where $\tilde{\chi}_e$ is the electric susceptibility

$$\tilde{\chi}_e(\omega) = -\frac{\omega_p^2}{\omega^2 + \nu^2}.$$

Note that $\tilde{\mathbf{P}}$ is directed opposite the applied field $\tilde{\mathbf{E}}$, resulting in $\tilde{\epsilon} < \epsilon_0$.

The plasma is dispersive since both its permittivity and conductivity depend on ω. When a transient plane wave propagates through a dispersive medium, the frequency dependence of the constitutive parameters tends to cause spreading of the waveshape.

▶ **Example 4.1:** Frequency characteristics of a nonmagnetized plasma

Show that the behavior of the permittivity of a nonmagnetized plasma obeys the frequency-symmetry condition (4.27), the low frequency behavior (4.28), and the high-frequency behavior (4.29).

Solution: From (4.76) we have the real and imaginary parts of the plasma permittivity,

$$\text{Re}\,\tilde{\epsilon}^c = \epsilon_0\left(1 - \frac{\omega_p^2}{\omega^2 + \nu^2}\right), \tag{4.77}$$

$$\text{Im}\,\tilde{\epsilon}^c = -\frac{\epsilon_0\omega_p^2\nu}{\omega(\omega^2 + \nu^2)}. \tag{4.78}$$

Since $\text{Re}\{\tilde{\epsilon}^c(-\omega)\} = \text{Re}\{\tilde{\epsilon}^c(\omega)\}$ and $\text{Im}\{\tilde{\epsilon}^c(-\omega)\} = -\text{Im}\{\tilde{\epsilon}^c(\omega)\}$, the frequency symmetry conditions (4.27) are satisfied.

As $\omega \to 0$ we have $\text{Re}\,\tilde{\epsilon}^c \to \epsilon_0\epsilon_r$ where

$$\epsilon_r = 1 - \frac{\omega_p^2}{\nu^2},$$

and $\text{Im}\,\tilde{\epsilon}^c \sim \omega^{-1}$. Hence conditions (4.28) are satisfied. As $\omega \to \infty$ we have $\text{Re}\,\tilde{\epsilon}^c - \epsilon_0 \sim \omega^{-2}$ and $\text{Im}\,\tilde{\epsilon}^c \sim \omega^{-3}$ as required by (4.29). ◄

▶ **Example 4.2:** Permittivity of the ionosphere

The ionosphere may be viewed as a plasma with characteristics highly dependent on altitude, latitude, and time of day. Although the earth's magnetic field plays an important role in the plasma properties, we will postpone its consideration until a later example. For mid latitudes, the D-layer at 80 km altitude has daytime electron density $N_e = 10^{10}$ m^{-3} and collision frequency $\nu = 10^6$ s^{-1} [20]. Plot the real and imaginary parts of the permittivity as functions of frequency.

Solution: The plasma frequency is

$$\omega_p = \sqrt{\frac{N_e e^2}{\epsilon_0 m_e}} = 5.641 \times 10^6 \text{ s}^{-1}.$$

As $\omega \to 0$ we find $\epsilon_r = 1 - \omega_p^2/\nu^2 = -30.82$, so the permittivity is negative at low frequencies.

Figure 4.3 shows a plot of the real and imaginary parts of the complex permittivity vs. frequency. Above about 2 MHz, $\text{Re}\,\tilde{\epsilon}^c \approx \epsilon_0$ and $\text{Im}\,\tilde{\epsilon}^c \approx 0$. Thus, during the day, this region of the ionosphere is transparent to electromagnetic waves above this frequency, and no reflection occurs. Two interesting characteristics of the permittivity can be seen in the plot. The real part of the complex permittivity is zero at $\omega = (\omega_p^2 - \nu^2)^{1/2}$, which occurs at $f = 884$ kHz for the quoted ionospheric parameters. Also, we have $\text{Re}\,\tilde{\epsilon}^c = \text{Im}\,\tilde{\epsilon}^c$ at $\omega = \nu$, which occurs at $f = 159$ kHz.

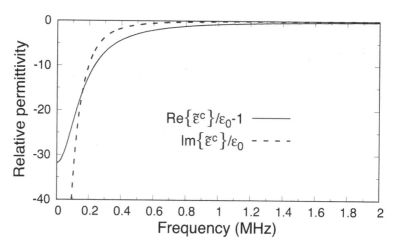

FIGURE 4.3

Permittivity of the ionosphere, neglecting the earth's magnetic field. ◄

We see that the plasma conductivity (4.75) is proportional to the collision frequency ν, and that, since $\operatorname{Im} \tilde{\epsilon}^c < 0$ by the arguments of § 4.5, the plasma must be lossy. Loss arises from the transfer of electromagnetic energy into heat through electron collisions. In the absence of collisions ($\nu = 0$), this mechanism disappears and the conductivity of a lossless (or "collisionless") plasma reduces to zero as expected.

▶ **Example 4.3:** Energy stored in a low-loss nonmagnetized plasma

Determine the time-average electromagnetic energy stored in a low-loss plasma ($\nu \to 0$) for sinusoidal excitation at frequency $\check{\omega}$. Interpret the terms.

Solution: To determine the stored electromagnetic energy, we must be careful to use (4.59), which holds for materials with dispersion. By applying the simpler formula (4.64), we find that for $\nu \to 0$

$$\langle w_e \rangle = \tfrac{1}{4}\epsilon_0 |\check{\mathbf{E}}|^2 - \tfrac{1}{4}\epsilon_0 |\check{\mathbf{E}}|^2 \frac{\omega_p^2}{\check{\omega}^2}.$$

For excitation frequencies $\check{\omega} < \omega_p$, we have $\langle w_e \rangle < 0$, and the material is active. Since there is no mechanism for the plasma to produce energy, this is obviously not valid. But an application of (4.59) gives

$$\langle w_e \rangle = \tfrac{1}{4}|\check{\mathbf{E}}|^2 \frac{\partial}{\partial \omega}\left[\epsilon_0 \omega \left(1 - \frac{\omega_p^2}{\omega^2}\right)\right]\Bigg|_{\omega = \check{\omega}} = \tfrac{1}{4}\epsilon_0 |\check{\mathbf{E}}|^2 + \tfrac{1}{4}\epsilon_0 |\check{\mathbf{E}}|^2 \frac{\omega_p^2}{\check{\omega}^2} \geq 0. \qquad (4.79)$$

The first term represents the time-average energy stored in the vacuum, while the second term represents the energy stored in the kinetic energy of the electrons. For harmonic excitation, the time-average electron kinetic energy density is $\langle w_q \rangle = \tfrac{1}{4}N m_e \check{\mathbf{v}} \cdot \check{\mathbf{v}}^*$. Substituting $\check{\mathbf{v}}$ from (4.72) with $\nu = 0$ we see that

$$\tfrac{1}{4}N m_e \check{\mathbf{v}} \cdot \check{\mathbf{v}}^* = \frac{N e^2}{4 m_e \check{\omega}^2}|\check{\mathbf{E}}|^2$$

$$= \tfrac{1}{4}\epsilon_0 |\check{\mathbf{E}}|^2 \frac{\omega_p^2}{\check{\omega}^2},$$

which matches the second term of (4.79). ◀

▶ **Example 4.4:** Kramers–Kronig relations for a low-loss nonmagnetized plasma

Show that the complex permittivity of a nonmagnetized plasma obeys the Kramers–Kronig relations for a causal material.

Solution: Substituting the imaginary part of the complex plasma permittivity from (4.78) into (4.37), we have

$$\operatorname{Re}\{\tilde{\epsilon}^c(\omega)\} - \epsilon_0 = -\frac{2}{\pi}\,\mathrm{P.V.}\int_0^\infty \left[-\frac{\epsilon_0 \omega_p^2 \nu}{\Omega(\Omega^2 + \nu^2)}\right]\frac{\Omega}{\Omega^2 - \omega^2}\,d\Omega.$$

We can evaluate the principal value integral and thus verify that it produces $\operatorname{Re}\tilde{\epsilon}^c$ by using the contour method of § A.2. Because the integrand is even, we can extend the domain of integration to $(-\infty, \infty)$ and divide the result by two:

$$\operatorname{Re}\{\tilde{\epsilon}^c(\omega)\} - \epsilon_0 = \frac{1}{\pi}\,\mathrm{P.V.}\int_{-\infty}^\infty \frac{\epsilon_0 \omega_p^2 \nu}{(\Omega - j\nu)(\Omega + j\nu)}\frac{d\Omega}{(\Omega - \omega)(\Omega + \omega)}.$$

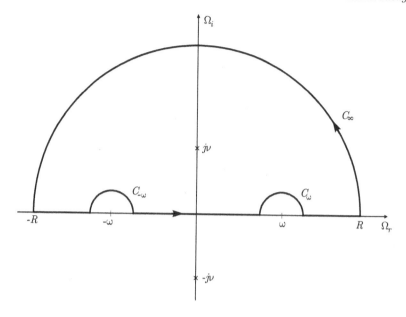

FIGURE 4.4

Integration contour used in Kramers–Kronig relations to find $\mathrm{Re}\,\tilde{\epsilon}^c$ from $\mathrm{Im}\,\tilde{\epsilon}^c$ for a non-magnetized plasma.

We integrate around the closed contour shown in Figure 4.4. Since the integrand decays as Ω^{-4} the contribution from C_∞ is zero. The contributions from the semicircles C_ω and $C_{-\omega}$ are πj times the residues of the integrand at $\Omega = \pm\omega$, which are equal but opposite. These cancel and leave only the contribution from the residue at the upper-half-plane pole $\Omega = j\nu$. Its evaluation yields

$$\mathrm{Re}\{\tilde{\epsilon}^c(\omega)\} - \epsilon_0 = \frac{1}{\pi} 2\pi j \frac{\epsilon_0 \omega_p^2 \nu}{j\nu + j\nu} \frac{1}{(j\nu - \omega)(j\nu + \omega)} = -\frac{\epsilon_0 \omega_p^2}{\nu^2 + \omega^2}$$

and thus

$$\mathrm{Re}\{\tilde{\epsilon}^c(\omega)\} = \epsilon_0 \left(1 - \frac{\omega_p^2}{\nu^2 + \omega^2} \right),$$

which matches (4.77) as expected. ◀

4.6.2   Complex dyadic permittivity of a magnetized plasma

When an electron plasma is exposed to a magnetostatic field, as occurs in the earth's ionosphere, the behavior of the plasma is altered so that the secondary current no longer aligns with the electric field, requiring the constitutive relationships to be written in terms of a complex dyadic permittivity. If the static field is $\mathbf{B}_0$, the velocity field of the plasma is determined by adding the magnetic component of the Lorentz force to (4.71), giving $-e(\tilde{\mathbf{E}} + \tilde{\mathbf{v}} \times \mathbf{B}_0) = \tilde{\mathbf{v}}(j\omega m_e + m_e \nu)$, or equivalently

$$\tilde{\mathbf{v}} - j\frac{e}{m_e(\omega - j\nu)} \tilde{\mathbf{v}} \times \mathbf{B}_0 = j\frac{e}{m_e(\omega - j\nu)} \tilde{\mathbf{E}}. \tag{4.80}$$

Writing this expression generically as

$$\mathbf{v} + \mathbf{v} \times \mathbf{C} = \mathbf{A}, \tag{4.81}$$

we can solve for $\mathbf{v}$ as follows. Dotting both sides of (4.81) with $\mathbf{C}$ we get $\mathbf{C} \cdot \mathbf{v} = \mathbf{C} \cdot \mathbf{A}$. Crossing both sides of (4.81) with $\mathbf{C}$, using (B.7), and substituting $\mathbf{C} \cdot \mathbf{A}$ for $\mathbf{C} \cdot \mathbf{v}$, we have $\mathbf{v} \times \mathbf{C} = \mathbf{A} \times \mathbf{C} + \mathbf{v}(\mathbf{C} \cdot \mathbf{C}) - \mathbf{C}(\mathbf{A} \cdot \mathbf{C})$. Finally, substituting $\mathbf{v} \times \mathbf{C}$ back into (4.81), we obtain

$$\mathbf{v} = \frac{\mathbf{A} - \mathbf{A} \times \mathbf{C} + (\mathbf{A} \cdot \mathbf{C})\mathbf{C}}{1 + \mathbf{C} \cdot \mathbf{C}}. \tag{4.82}$$

Let us first consider a lossless plasma for which $\nu = 0$. We can solve (4.80) for $\tilde{\mathbf{v}}$ by setting

$$\mathbf{C} = -j\frac{\boldsymbol{\omega}_c}{\omega} \quad \text{and} \quad \mathbf{A} = j\frac{\epsilon_0 \omega_p^2}{\omega N e}\tilde{\mathbf{E}} \quad \text{where} \quad \boldsymbol{\omega}_c = \frac{e}{m_e}\mathbf{B}_0.$$

Here $\omega_c = eB_0/m_e = |\boldsymbol{\omega}_c|$ is called the *electron cyclotron frequency* (see § 2.11.4 for its relevance to electron motion in a steady magnetic field). Substituting these into (4.82), we have

$$(\omega^2 - \omega_c^2)\,\tilde{\mathbf{v}} = j\frac{\epsilon_0 \omega \omega_p^2}{Ne}\tilde{\mathbf{E}} + \frac{\epsilon_0 \omega_p^2}{Ne}\boldsymbol{\omega}_c \times \tilde{\mathbf{E}} - j\frac{\omega_c}{\omega}\frac{\epsilon_0 \omega_p^2}{Ne}\boldsymbol{\omega}_c \cdot \tilde{\mathbf{E}}.$$

Since the secondary current produced by the moving electrons is just $\tilde{\mathbf{J}}^s = -Ne\tilde{\mathbf{v}}$, we have

$$\tilde{\mathbf{J}}^s = j\omega\left(-\frac{\epsilon_0 \omega_p^2}{\omega^2 - \omega_c^2}\tilde{\mathbf{E}} + j\frac{\epsilon_0 \omega_p^2}{\omega(\omega^2 - \omega_c^2)}\boldsymbol{\omega}_c \times \tilde{\mathbf{E}} + \frac{\omega_c}{\omega^2}\frac{\epsilon_0 \omega_p^2}{\omega^2 - \omega_c^2}\boldsymbol{\omega}_c \cdot \tilde{\mathbf{E}}\right). \tag{4.83}$$

Now, by the Ampere–Maxwell law, we can write for currents in free space

$$\nabla \times \tilde{\mathbf{H}} = \tilde{\mathbf{J}}^i + \tilde{\mathbf{J}}^s + j\omega\epsilon_0\tilde{\mathbf{E}}. \tag{4.84}$$

Considering the plasma as material implies that we can describe the gas in terms of a complex permittivity dyadic $\tilde{\bar{\boldsymbol{\epsilon}}}^c$ such that the Ampere–Maxwell law is

$$\nabla \times \tilde{\mathbf{H}} = \tilde{\mathbf{J}}^i + j\omega\tilde{\bar{\boldsymbol{\epsilon}}}^c \cdot \tilde{\mathbf{E}}.$$

Substituting (4.83) into (4.84), and defining the dyadic $\bar{\boldsymbol{\omega}}_c$ so that $\bar{\boldsymbol{\omega}}_c \cdot \tilde{\mathbf{E}} = \boldsymbol{\omega}_c \times \tilde{\mathbf{E}}$, we identify the dyadic permittivity

$$\tilde{\bar{\boldsymbol{\epsilon}}}^c(\omega) = \left(\epsilon_0 - \epsilon_0\frac{\omega_p^2}{\omega^2 - \omega_c^2}\right)\bar{\mathbf{I}} + j\frac{\epsilon_0 \omega_p^2}{\omega(\omega^2 - \omega_c^2)}\bar{\boldsymbol{\omega}}_c + \frac{\epsilon_0 \omega_p^2}{\omega^2(\omega^2 - \omega_c^2)}\boldsymbol{\omega}_c\boldsymbol{\omega}_c. \tag{4.85}$$

Note that in rectangular coordinates

$$[\bar{\boldsymbol{\omega}}_c] = \begin{bmatrix} 0 & -\omega_{cz} & \omega_{cy} \\ \omega_{cz} & 0 & -\omega_{cx} \\ -\omega_{cy} & \omega_{cx} & 0 \end{bmatrix}. \tag{4.86}$$

To examine the properties of the dyadic permittivity it is useful to write it in matrix form. To do this we must choose a coordinate system. We shall align $\mathbf{B}_0$ with the z-axis so that $\mathbf{B}_0 = \hat{\mathbf{z}}B_0$ and $\boldsymbol{\omega}_c = \hat{\mathbf{z}}\omega_c$. Then (4.86) becomes

$$[\bar{\boldsymbol{\omega}}_c] = \begin{bmatrix} 0 & -\omega_c & 0 \\ \omega_c & 0 & 0 \\ 0 & 0 & 0 \end{bmatrix} \tag{4.87}$$

and we can write the permittivity dyadic (4.85) as

$$[\bar{\tilde{\epsilon}}(\omega)] = \begin{bmatrix} \epsilon & -j\delta & 0 \\ j\delta & \epsilon & 0 \\ 0 & 0 & \epsilon_z \end{bmatrix} \tag{4.88}$$

where

$$\epsilon = \epsilon_0 \left(1 - \frac{\omega_p^2}{\omega^2 - \omega_c^2} \right), \qquad \epsilon_z = \epsilon_0 \left(1 - \frac{\omega_p^2}{\omega^2} \right), \qquad \delta = \frac{\epsilon_0 \omega_c \omega_p^2}{\omega(\omega^2 - \omega_c^2)}.$$

Note that its form is that for a lossless *gyrotropic* material (2.25).

Since the plasma is lossless, (4.49) shows that the dyadic permittivity must be hermitian. Equation (4.88) confirms this. Moreover, since the sign of $\boldsymbol{\omega}_c$ is determined by the sign of $\mathbf{B}_0$, the dyadic permittivity obeys the symmetry relation

$$\tilde{\epsilon}_{ij}^c(\mathbf{B}_0) = \tilde{\epsilon}_{ji}^c(-\mathbf{B}_0) \tag{4.89}$$

as does the permittivity matrix of any material having anisotropic properties dependent on an externally applied magnetic field [143]. We will find later in this section that the permeability matrix of a magnetized ferrite also obeys such a symmetry condition.

We can let $\omega \to \omega - j\nu$ in (4.83) to obtain the secondary current in a plasma with collisions:

$$\begin{aligned}
\tilde{\mathbf{J}}^s(\mathbf{r}, \omega) = j\omega \Bigg[& -\frac{\epsilon_0 \omega_p^2(\omega - j\nu)}{\omega[(\omega - j\nu)^2 - \omega_c^2]} \tilde{\mathbf{E}}(\mathbf{r}, \omega) \\
& + j\frac{\epsilon_0 \omega_p^2(\omega - j\nu)}{\omega(\omega - j\nu)[(\omega - j\nu)^2 - \omega_c^2)]} \boldsymbol{\omega}_c \times \tilde{\mathbf{E}}(\mathbf{r}, \omega) \\
& + \frac{\boldsymbol{\omega}_c}{(\omega - j\nu)^2} \frac{\epsilon_0 \omega_p^2(\omega - j\nu)}{\omega[(\omega - j\nu)^2 - \omega_c^2]} \boldsymbol{\omega}_c \cdot \tilde{\mathbf{E}}(\mathbf{r}, \omega) \Bigg].
\end{aligned}$$

From this we find

$$\begin{aligned}
\bar{\tilde{\epsilon}}^c(\omega) = & \left[\epsilon_0 - \frac{\epsilon_0 \omega_p^2(\omega - j\nu)}{\omega[(\omega - j\nu)^2 - \omega_c^2]} \right] \bar{\mathbf{I}} + j\frac{\epsilon_0 \omega_p^2}{\omega[(\omega - j\nu)^2 - \omega_c^2)]} \bar{\boldsymbol{\omega}}_c \\
& + \frac{1}{(\omega - j\nu)} \frac{\epsilon_0 \omega_p^2}{\omega[(\omega - j\nu)^2 - \omega_c^2]} \boldsymbol{\omega}_c \boldsymbol{\omega}_c.
\end{aligned}$$

Assuming $\mathbf{B}_0$ is aligned with the z-axis, we can use (4.87) to find the components of the dyadic permittivity matrix:

$$\tilde{\epsilon}_{xx}^c(\omega) = \tilde{\epsilon}_{yy}^c(\omega) = \epsilon_0 \left(1 - \frac{\omega_p^2(\omega - j\nu)}{\omega[(\omega - j\nu)^2 - \omega_c^2]} \right), \tag{4.90}$$

$$\tilde{\epsilon}_{xy}^c(\omega) = -\tilde{\epsilon}_{yx}^c(\omega) = -j\epsilon_0 \frac{\omega_p^2 \omega_c}{\omega[(\omega - j\nu)^2 - \omega_c^2)]}, \tag{4.91}$$

$$\tilde{\epsilon}_{zz}^c(\omega) = \epsilon_0 \left(1 - \frac{\omega_p^2}{\omega(\omega - j\nu)} \right), \tag{4.92}$$

and

$$\tilde{\epsilon}^c_{zx} = \tilde{\epsilon}^c_{xz} = \tilde{\epsilon}^c_{zy} = \tilde{\epsilon}^c_{yz} = 0. \tag{4.93}$$

Each dyadic entry can be separated into real and imaginary parts. The diagonal terms are given by

$$\frac{\mathrm{Re}\{\tilde{\epsilon}^c_{xx}(\omega)\}}{\epsilon_0} - 1 = -\omega_p^2 \frac{\omega^2 + \nu^2 - \omega_c^2}{(\omega^2 + \nu^2 - \omega_c^2)^2 + 4\omega_c^2\nu^2}, \tag{4.94}$$

$$\frac{\mathrm{Im}\{\tilde{\epsilon}^c_{xx}(\omega)\}}{\epsilon_0} = -\omega_p^2 \frac{\nu}{\omega} \frac{\omega^2 + \nu^2 + \omega_c^2}{(\omega^2 + \nu^2 - \omega_c^2)^2 + 4\omega_c^2\nu^2}, \tag{4.95}$$

$$\frac{\mathrm{Re}\{\tilde{\epsilon}^c_{zz}(\omega)\}}{\epsilon_0} - 1 = -\frac{\omega_p^2}{\omega^2 + \nu^2}, \tag{4.96}$$

$$\frac{\mathrm{Im}\{\tilde{\epsilon}^c_{zz}(\omega)\}}{\epsilon_0} = -\frac{\epsilon_0\omega_p^2\nu}{\omega(\omega^2 + \nu^2)}. \tag{4.97}$$

Note that the last two expressions match (4.77) and (4.78). Thus, the zz-component of the permittivity dyadic is identical to the scalar permittivity of the nonmagnetized plasma. Finally, the off-diagonal terms are given by

$$\frac{\mathrm{Re}\{\tilde{\epsilon}^c_{xy}(\omega)\}}{\epsilon_0} = \frac{2\nu\omega_p^2\omega_c}{(\omega^2 + \nu^2 - \omega_c^2)^2 + 4\omega_c^2\nu^2}, \tag{4.98}$$

$$\frac{\mathrm{Im}\{\tilde{\epsilon}^c_{xy}(\omega)\}}{\epsilon_0} = -\omega_p^2 \frac{\omega_c}{\omega} \frac{\omega^2 - \nu^2 - \omega_c^2}{(\omega^2 + \nu^2 - \omega_c^2)^2 + 4\omega_c^2\nu^2}. \tag{4.99}$$

Note that when $B_0 \to 0$ and thus $\omega_c \to 0$, the off-diagonal elements vanish and the diagonal elements reduce to those for a nonmagnetized plasma (4.76).

We see that $[\tilde{\epsilon}^c]$ is not hermitian when $\nu \neq 0$. We expect this since the plasma is lossy when collisions occur. However, we can decompose $[\tilde{\epsilon}^c]$ as the sum of matrices $[\tilde{\epsilon}^c] = [\tilde{\epsilon}] + [\tilde{\sigma}]/j\omega$, where $[\tilde{\epsilon}]$ and $[\tilde{\sigma}]$ are hermitian [143]. The details are left as an exercise. We also note that, as in the case of the lossless plasma, the permittivity dyadic obeys the symmetry condition $\tilde{\epsilon}^c_{ij}(\mathbf{B}_0) = \tilde{\epsilon}^c_{ji}(-\mathbf{B}_0)$.

▶ **Example 4.5:** Frequency characteristics of a magnetized plasma

Show that the behavior of each element of the permittivity dyadic of a magnetized plasma obeys the frequency-symmetry condition (4.25), the low-frequency behavior (4.28) and the high-frequency behavior (4.29).

Solution: By inspection, expressions (4.94), (4.96), and (4.98) are even in ω and thus obey the condition $\mathrm{Re}\{\tilde{\epsilon}^c_{ij}(-\omega)\} = \mathrm{Re}\{\tilde{\epsilon}^c_{ij}(\omega)\}$. Similarly, (4.95), (4.97), and (4.99) are odd in ω and obey $\mathrm{Im}\{\tilde{\epsilon}^c_{ij}(-\omega)\} = -\mathrm{Im}\{\tilde{\epsilon}^c_{ij}(\omega)\}$.

As $\omega \to 0$ we find that $\mathrm{Re}\,\tilde{\epsilon}^c_{ij} \to \epsilon_0\epsilon_{ij,r}$ where

$$\tilde{\epsilon}_{xx,r} = 1 - \frac{\omega_p^2(\nu^2 - \omega_c^2)}{(\nu^2 - \omega_c^2)^2 + 4\omega_c^2\nu^2}, \quad \tilde{\epsilon}_{zz,r} = 1 - \frac{\omega_p^2}{\nu^2}, \quad \tilde{\epsilon}_{xy,r} = \frac{2\nu\omega_p^2\omega_c}{(\nu^2 - \omega_c^2)^2 + 4\omega_c^2\nu^2}.$$

Hence, conditions (4.28) are satisfied for each dyadic element. As $\omega \to \infty$, we find that $\mathrm{Re}\,\tilde{\epsilon}^c_{ij} - \epsilon_0 \sim \omega^{-2}$ and $\mathrm{Im}\,\tilde{\epsilon}^c_{ij} \sim \omega^{-3}$ for each dyadic element, as required by (4.29). ◀

▶ **Example 4.6:** Permittivity of the ionosphere including effects of the earth's magnetic field

Reconsider the earth's ionosphere as described in Example 4.2, including the effects of the earth's magnetic field.

Solution: According to the National Oceanic and Atmospheric Administration, the earth's magnetic field has a value of approximately 52,000 nT at an altitude of 80 km. This field magnetizes the plasma along the direction of the field lines (which is skewed at altitude) and we must regard the plasma permittivity as a dyadic described by (4.94)–(4.99). Here we align the z-direction with the magnetic field lines.

The magnetized plasma has cyclotron frequency

$$\omega_c = \frac{e}{m_e} B_0 = 9.146 \times 10^6 \text{ s}^{-1}$$

or $f_c = 1.46$ MHz. Using values of the plasma and collision frequencies from Example 4.2, we find that as $\omega \to 0$,

$$\tilde{\epsilon}_{xx,r} = 1 - \omega_p^2 \frac{\nu^2 - \omega_c^2}{(\nu^2 - \omega_c^2)^2 + 4\omega_c^2 \nu^2} = 1.367,$$

$$\tilde{\epsilon}_{zz,r} = 1 - \frac{\omega_p^2}{\nu^2} = -30.82,$$

$$\tilde{\epsilon}_{xy,r} = \frac{2\nu \omega_p^2 \omega_c}{(\nu^2 - \omega_c^2)^2 + 4\omega_c^2 \nu^2} = 0.08125.$$

At low frequencies the dyadic permittivity is dominated by the zz-element, and the off-diagonal elements are quite small.

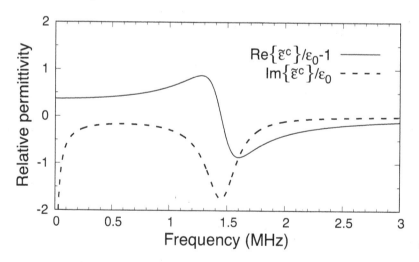

FIGURE 4.5
xx-element of the permittivity dyadic of the ionosphere, including the effects of the earth's magnetic field.

Since $\tilde{\epsilon}_{zz}^c$ is identical to $\tilde{\epsilon}^c$ for the non-magnetized plasma, a plot of this dyadic element appears in Figure 4.3. Figure 4.5 shows a plot of the real and imaginary parts of the diagonal element $\tilde{\epsilon}_{xx}^c$ as functions of frequency. We can clearly see a resonance effect near the cyclotron frequency at 1.46 MHz. The real part of the relative permittivity passes through unity near that frequency and the imaginary part peaks there. In contrast, the zz-component is independent of ω_c. Figure 4.6 shows a plot of the real and imaginary parts of the off-diagonal element $\tilde{\epsilon}_{xy}^c$ as functions of frequency. Again, a resonance can be seen near the cyclotron frequency. In contrast to the diagonal element, the real part peaks near the cyclotron frequency, while the imaginary part passes through zero. Above about 3 MHz, the real parts of both the diagonal and off-diagonal elements approach ϵ_0 while the imaginary parts approach zero. Thus, above this frequency the ionosphere looks transparent.

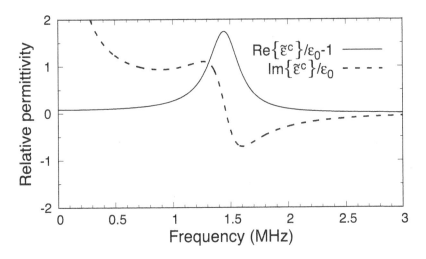

FIGURE 4.6
xy-element of the permittivity dyadic of the ionosphere, including the effects of the earth's magnetic field. ◄

4.6.3 Simple models of dielectrics

We define an isotropic dielectric material (also called an *insulator*) as one that obeys the macroscopic frequency-domain constitutive relationship

$$\tilde{\mathbf{D}}(\mathbf{r}, \omega) = \tilde{\epsilon}(\mathbf{r}, \omega)\tilde{\mathbf{E}}(\mathbf{r}, \omega).$$

Since the polarization vector $\mathbf{P}$ was defined in Chapter 2 as $\mathbf{P}(\mathbf{r}, t) = \mathbf{D}(\mathbf{r}, t) - \epsilon_0 \mathbf{E}(\mathbf{r}, t)$, an isotropic dielectric can also be described through

$$\tilde{\mathbf{P}}(\mathbf{r}, \omega) = [\tilde{\epsilon}(\mathbf{r}, \omega) - \epsilon_0]\tilde{\mathbf{E}}(\mathbf{r}, \omega) = \tilde{\chi}_e(\mathbf{r}, \omega)\epsilon_0\tilde{\mathbf{E}}(\mathbf{r}, \omega)$$

where $\tilde{\chi}_e$ is the susceptibility parameter. In this section we model a homogeneous dielectric consisting of a single, uniform material type.

We found in Chapter 3 that for a dielectric material immersed in a static electric field, $\mathbf{P}$ can be regarded as a volume density of dipole moments. We choose to retain this view as the fundamental link between microscopic dipole moments and the macroscopic polarization vector. Within the framework of our model, we thus describe the polarization through the expression

$$\mathbf{P}(\mathbf{r}, t) = \frac{1}{\Delta V} \sum_{\mathbf{r} - \mathbf{r}_i(t) \in B} \mathbf{p}_i. \tag{4.100}$$

Here $\mathbf{p}_i$ is the dipole moment of the ith elementary microscopic constituent, and we form the macroscopic density function as in § 1.3.1.

We may also write (4.100) as

$$\mathbf{P}(\mathbf{r},t) = \frac{N_B}{\Delta V}\left(\frac{1}{N_B}\sum_{i=1}^{N_B}\mathbf{p}_i\right) = N(\mathbf{r},t)\mathbf{p}(\mathbf{r},t) \tag{4.101}$$

where N_B is the number of constituent particles within ΔV. We identify

$$\mathbf{p}(\mathbf{r},t) = \frac{1}{N_B}\sum_{i=1}^{N_B}\mathbf{p}_i(\mathbf{r},t)$$

as the average dipole moment within ΔV, and $N(\mathbf{r},t) = N_B/\Delta V$ as the dipole moment number density. In this model a dielectric material does not require higher-order multipole moments to describe its behavior. Since we are interested in homogeneous materials in this section, we assume the number density is constant: $N(\mathbf{r},t) = N$.

To understand how dipole moments arise, we choose to adopt the simple idea that matter consists of atomic particles, each of which has a positively charged nucleus surrounded by a negatively charged electron cloud. Isolated, these particles have no net charge or electric dipole moment. However, there are several ways in which individual particles, or aggregates of particles, may take on a dipole moment. When exposed to an external electric field the electron cloud of an individual atom may be displaced, resulting in an *induced* dipole moment that gives rise to *electronic polarization*. When groups of atoms form a molecule, the individual electron clouds may combine to form an asymmetric structure having a *permanent* dipole moment. In some materials these molecules are randomly distributed and no net dipole moment results. However, upon application of an external field the torque acting on the molecules may tend to align them, creating an *induced* dipole moment and *orientation*, or *dipole*, polarization. In other materials, the asymmetric structure of the molecules may be weak until an external field causes the displacement of atoms within each molecule, resulting in an *induced* dipole moment causing *atomic*, or *molecular*, polarization. If a material maintains a permanent polarization without the application of an external field, it is called an *electret* (and is thus similar in behavior to a permanently magnetized magnet).

To describe the constitutive relations, we must establish a link between $\mathbf{P}$ (now describable in microscopic terms) and $\mathbf{E}$. For this we postulate that the average constituent dipole moment is proportional to the *local electric field strength* $\mathbf{E}'$:

$$\mathbf{p} = \alpha\mathbf{E}', \tag{4.102}$$

where α is called the *polarizability* of the elementary constituent. Each polarization effect listed above may have its own polarizability: α_e for electronic polarization, α_a for atomic polarization, and α_d for dipole polarization. The total polarizability is merely the sum $\alpha = \alpha_e + \alpha_a + \alpha_d$.

In a rarefied gas the particles are so far apart that their interaction can be neglected. Here the local field $\mathbf{E}'$ is the same as the applied field $\mathbf{E}$. In liquids and solids where particles are tightly packed, $\mathbf{E}'$ depends on the manner in which the material is polarized and may differ from $\mathbf{E}$. So we need a relationship between $\mathbf{E}'$ and $\mathbf{P}$.

4.6.3.1   The Clausius–Mosotti equation

We seek the local field at an observation point within a polarized material. Let us first assume that the fields are static. We surround the observation point with an artificial spherical surface of radius a and write the field at the observation point as a superposition of the field $\mathbf{E}$ applied, the field $\mathbf{E}_2$ of the polarized molecules external to the sphere, and the field $\mathbf{E}_3$ of the polarized molecules within the sphere. We take a large enough that we may describe the molecules outside the sphere in terms of the macroscopic dipole moment density $\mathbf{P}$, but small enough to assume that $\mathbf{P}$ is uniform over the surface of the sphere. We also assume the major contribution to $\mathbf{E}_2$ comes from the dipoles nearest the observation point. We then approximate $\mathbf{E}_2$ using the electrostatic potential produced by the equivalent polarization surface charge on the sphere $\rho_{Ps} = \hat{\mathbf{n}} \cdot \mathbf{P}$ (where $\hat{\mathbf{n}}$ points toward the center of the sphere). Placing the origin of coordinates at the observation point and orienting the z-axis with the polarization $\mathbf{P}$ so that $\mathbf{P} = P_0 \hat{\mathbf{z}}$, we find that $\hat{\mathbf{n}} \cdot \mathbf{P} = -\cos\theta$ and thus the electrostatic potential at any point $\mathbf{r}$ within the sphere is merely

$$\Phi(\mathbf{r}) = -\frac{1}{4\pi\epsilon_0} \oint_S \frac{P_0 \cos\theta'}{|\mathbf{r} - \mathbf{r}'|} \, dS'.$$

This integral has been computed in § 3.3.8 with the result given by (3.78). Hence

$$\Phi(\mathbf{r}) = -\frac{P_0}{3\epsilon_0} r\cos\theta = -\frac{P_0}{3\epsilon_0} z$$

and

$$\mathbf{E}_2 = \frac{\mathbf{P}}{3\epsilon_0}. \tag{4.103}$$

Note that this is uniform and independent of a.

The assumption that the localized field varies spatially as the electrostatic field, even when $\mathbf{P}$ may depend on frequency, is quite good. In Chapter 5 we will find that for a frequency-dependent source (or, equivalently, a time-varying source), the fields very near the source have a spatial dependence nearly identical to that of the electrostatic case.

We now have the seemingly more difficult task of determining the field $\mathbf{E}_3$ produced by the dipoles within the sphere. This would seem difficult since the field produced by dipoles near the observation point should be highly dependent on the particular dipole arrangement. As mentioned above, there are various mechanisms for polarization, and the distribution of charge near any particular point depends on the molecular arrangement. However, Lorentz showed [122] that for crystalline solids with cubical symmetry, or for a randomly structured gas, the contribution from dipoles within the sphere is zero. Indeed, it is convenient and reasonable to assume that for most dielectrics the effects of the dipoles immediately surrounding the observation point cancel so that $\mathbf{E}_3 = 0$. This was first suggested by O.F. Mosotti in 1850 [46].

With $\mathbf{E}_2$ approximated as (4.103) and $\mathbf{E}_3$ assumed to be zero, we have the value of the resulting local field:

$$\mathbf{E}'(\mathbf{r}) = \mathbf{E}(\mathbf{r}) + \frac{\mathbf{P}(\mathbf{r})}{3\epsilon_0}. \tag{4.104}$$

This is called the *Mosotti field*. Substituting it into (4.102) and using $\mathbf{P} = N\mathbf{p}$, we obtain

$$\mathbf{P}(\mathbf{r}) = N\alpha\mathbf{E}'(\mathbf{r}) = N\alpha\left(\mathbf{E}(\mathbf{r}) + \frac{\mathbf{P}(\mathbf{r})}{3\epsilon_0}\right).$$

This yields

$$\mathbf{P}(\mathbf{r}) = \left(\frac{3\epsilon_0 N\alpha}{3\epsilon_0 - N\alpha}\right)\mathbf{E}(\mathbf{r}) = \chi_e \epsilon_0 \mathbf{E}(\mathbf{r}).$$

So the electric susceptibility of a dielectric may be expressed as

$$\chi_e = \frac{3N\alpha}{3\epsilon_0 - N\alpha}. \tag{4.105}$$

Using $\chi_e = \epsilon_r - 1$ we can rewrite (4.105) as

$$\epsilon = \epsilon_0 \epsilon_r = \epsilon_0 \frac{3 + 2N\alpha/\epsilon_0}{3 - N\alpha/\epsilon_0}, \tag{4.106}$$

which we can rearrange to obtain

$$\alpha = \alpha_e + \alpha_a + \alpha_d = \frac{3\epsilon_0}{N} \frac{\epsilon_r - 1}{\epsilon_r + 2}.$$

This is the *Clausius–Mosotti formula*, named after O.F. Mosotti, who proposed it in 1850, and R. Clausius, who proposed it independently in 1879. When written in terms of the index of refraction n (where $n^2 = \epsilon_r$), it is also known as the *Lorentz–Lorenz formula*, after H. Lorentz and L. Lorenz, who proposed it independently for optical materials in 1880. The formula allows us to determine the dielectric constant from the polarizability and number density of a material. It is reasonably accurate for certain simple gases (with pressures up to 1000 atmospheres) but becomes less reliable for liquids and solids, especially for those with large dielectric constants.

The response of the microscopic structure of matter to an applied field is not instantaneous. When exposed to a rapidly oscillating sinusoidal field, the induced dipole moments may lag in time. The result is a loss mechanism describable macroscopically by a complex permittivity. We can modify the Clausius–Mosotti formula by assuming that both the relative permittivity and polarizability are complex numbers, but this will not model the dependence of these parameters on frequency. Instead we shall (in later sections) model the time response of the dipole moments to the applied field.

4.6.3.2 Maxwell–Garnett and Rayleigh mixing formulas

An interesting application of the Clausius–Mosotti formula is to determine the permittivity of a mixture of dielectrics with different permittivities. Consider the simple case in which many small spheres of permittivity ϵ_2, radius a, and volume V are embedded within a dielectric matrix of permittivity ϵ_1. If we assume a is much smaller than the wavelength of the electromagnetic field and the spheres are sparsely distributed within the matrix, we may ignore any mutual interaction between the spheres. We can use the Clausius–Mosotti formula to define an *effective* permittivity ϵ_e for a material consisting of spheres in a background dielectric by replacing ϵ_0 with ϵ_1 in (4.106) to obtain

$$\epsilon_e = \epsilon_1 \frac{3 + 2N\alpha/\epsilon_1}{3 - N\alpha/\epsilon_1}. \tag{4.107}$$

In this expression, α is the polarizability of a single dielectric sphere embedded in the background dielectric, and N is the number density of dielectric spheres. To find α we use the static field solution for a dielectric sphere immersed in a field (§ 3.3.11). Remembering that $\mathbf{p} = \alpha \mathbf{E}$ and that for a uniform region of volume V we have $\mathbf{p} = V\mathbf{P}$, we can make the replacements $\epsilon_0 \to \epsilon_1$ and $\epsilon \to \epsilon_2$ in (3.90) to get

$$\alpha = 3\epsilon_1 V \frac{\epsilon_2 - \epsilon_1}{\epsilon_2 + 2\epsilon_1}. \tag{4.108}$$

Defining $f = NV$ as the *fractional volume* occupied by the spheres, we can substitute (4.108) into (4.107) to find that

$$\epsilon_e = \epsilon_1 \frac{1 + 2fy}{1 - fy} \quad \text{where} \quad y = \frac{\epsilon_2 - \epsilon_1}{\epsilon_2 + 2\epsilon_1}.$$

This is the *Maxwell–Garnett mixing formula*. Its rearrangement

$$\frac{\epsilon_e - \epsilon_1}{\epsilon_e + 2\epsilon_1} = f \frac{\epsilon_2 - \epsilon_1}{\epsilon_2 + 2\epsilon_1}$$

is the *Rayleigh mixing formula*. As expected, $\epsilon_e \to \epsilon_1$ as $f \to 0$. Even though as $f \to 1$ the formula also reduces to $\epsilon_e = \epsilon_2$, our initial assumption that $f \ll 1$ (sparsely distributed spheres) is violated and the result is inaccurate for non-spherical inhomogeneities [91]. For a discussion of more accurate mixing formulas, see Ishimaru [91] or Sihvola [175].

4.6.3.3 The dispersion formula of classical physics; Lorentz and Sellmeier equations

We may determine the frequency dependence of the permittivity by modeling the time response of induced dipole moments. This was done by H. Lorentz using the simple atomic model we introduced earlier. Consider what happens when a molecule consisting of heavy particles (nuclei) surrounded by clouds of electrons is exposed to a time-harmonic electromagnetic wave. Using the same arguments we made when studying the interactions of fields with a plasma in § 4.6.1, we assume each electron experiences a Lorentz force $\mathbf{F}_e = -e\mathbf{E}'$. We neglect the magnetic component of the force for nonrelativistic charge velocities, and ignore the motion of the much heavier nuclei in favor of studying the motion of the electron cloud. However, several important distinctions exist between the behavior of charges within a plasma and those within a solid or liquid material. Because of the surrounding polarized matter, any molecule responds to the local field $\mathbf{E}'$ instead of the applied field $\mathbf{E}$. Also, as the electron cloud is displaced by the Lorentz force, the attraction from the positive nuclei provides a restoring force $\mathbf{F}_r$. In the absence of loss the restoring force causes the electron cloud (and thus the induced dipole moment) to oscillate in phase with the applied field. In addition, there will be loss due to radiation by the oscillating molecules and collisions between charges that can be modeled using a "frictional force" $\mathbf{F}_s$ in the same manner as for a mechanical harmonic oscillator.

We can express the restoring and frictional forces by the use of a mechanical analogue. The restoring force acting on each electron is taken to be proportional to the displacement from equilibrium $\mathbf{l}$:

$$\mathbf{F}_r(\mathbf{r}, t) = -m_e \omega_r^2 \mathbf{l}(\mathbf{r}, t),$$

where m_e is the mass of an electron and ω_r is a material constant that depends on the molecular structure. The frictional force is similar to the collisional term in § 4.6.1 in that it is assumed to be proportional to the electron momentum $m_e \mathbf{v}$:

$$\mathbf{F}_s(\mathbf{r}, t) = -2\Gamma m_e \mathbf{v}(\mathbf{r}, t)$$

where Γ is a material constant. With these we can apply Newton's second law:

$$\mathbf{F}(\mathbf{r}, t) = -e\mathbf{E}'(\mathbf{r}, t) - m_e \omega_r^2 \mathbf{l}(\mathbf{r}, t) - 2\Gamma m_e \mathbf{v}(\mathbf{r}, t) = m_e \frac{d\mathbf{v}(\mathbf{r}, t)}{dt}.$$

Since $\mathbf{v} = d\mathbf{l}/dt$, the equation of electron motion is

$$\frac{d^2 \mathbf{l}(\mathbf{r}, t)}{dt^2} + 2\Gamma \frac{d\mathbf{l}(\mathbf{r}, t)}{dt} + \omega_r^2 \mathbf{l}(\mathbf{r}, t) = -\frac{e}{m_e} \mathbf{E}'(\mathbf{r}, t). \tag{4.109}$$

We recognize this differential equation as the damped harmonic equation. When $\mathbf{E}' = 0$ we have the homogeneous solution

$$\mathbf{l}(\mathbf{r}, t) = \mathbf{l}_0(\mathbf{r}) e^{-\Gamma t} \cos\left(t\sqrt{\omega_r^2 - \Gamma^2}\right).$$

Thus the electron position is a damped oscillation. The resonant frequency $\sqrt{\omega_r^2 - \Gamma^2}$ is usually only slightly reduced from ω_r since radiation damping is generally quite low.

Since the dipole moment for an electron displaced from equilibrium by $\mathbf{l}$ is $\mathbf{p} = -e\mathbf{l}$, and the polarization density is $\mathbf{P} = N\mathbf{p}$ from (4.101), we can write

$$\mathbf{P}(\mathbf{r}, t) = -Ne\mathbf{l}(\mathbf{r}, t).$$

Multiplying (4.109) by $-Ne$ and substituting the above expression, we have a differential equation for the polarization:

$$\frac{d^2\mathbf{P}}{dt^2} + 2\Gamma\frac{d\mathbf{P}}{dt} + \omega_r^2\mathbf{P} = \frac{Ne^2}{m_e}\mathbf{E}'.$$

To obtain a constitutive equation, we must relate the polarization to the applied field $\mathbf{E}$. We accomplish this by relating the local field $\mathbf{E}'$ to the polarization using the Mosotti field (4.104). Substitution gives

$$\frac{d^2\mathbf{P}}{dt^2} + 2\Gamma\frac{d\mathbf{P}}{dt} + \omega_0^2\mathbf{P} = \frac{Ne^2}{m_e}\mathbf{E} \tag{4.110}$$

where

$$\omega_0 = \sqrt{\omega_r^2 - \frac{Ne^2}{3m_e\epsilon_0}}$$

is the resonance frequency of the dipole moments. It is less than the resonance frequency of the electron oscillation because of the polarization of the surrounding medium.

We can obtain a dispersion equation for the electrical susceptibility by taking the Fourier transform of (4.110). Since

$$-\omega^2\tilde{\mathbf{P}} + j\omega 2\Gamma\tilde{\mathbf{P}} + \omega_0^2\tilde{\mathbf{P}} = \frac{Ne^2}{m_e}\tilde{\mathbf{E}},$$

we have

$$\tilde{\chi}_e(\omega) = \frac{\tilde{\mathbf{P}}}{\epsilon_0\tilde{\mathbf{E}}} = \frac{\omega_p^2}{\omega_0^2 - \omega^2 + j\omega 2\Gamma}$$

where ω_p is the plasma frequency (4.74). Since $\tilde{\epsilon}_r(\omega) = 1 + \tilde{\chi}_e(\omega)$ we also have

$$\tilde{\epsilon}(\omega) = \epsilon_0 + \epsilon_0\frac{\omega_p^2}{\omega_0^2 - \omega^2 + j\omega 2\Gamma}. \tag{4.111}$$

If more than one type of oscillating moment contributes to the permittivity, we may extend (4.111) to

$$\tilde{\epsilon}(\omega) = \epsilon_0 + \sum_i \epsilon_0\frac{\omega_{pi}^2}{\omega_i^2 - \omega^2 + j\omega 2\Gamma_i} \tag{4.112}$$

where $\omega_{pi} = N_i e^2/\epsilon_0 m_i$ is the plasma frequency of the ith resonance component, and ω_i and Γ_i are the oscillation frequency and damping coefficient, respectively, of this component. This expression is the *dispersion formula for classical physics*, so called because it neglects quantum effects. It is also called the *Lorentz permittivity model* of a dielectric in honor of H. Lorentz. Under negligible losses (4.112) reduces to the *Sellmeier equation*:

$$\tilde{\epsilon}(\omega) = \epsilon_0 + \sum_i \epsilon_0 \frac{\omega_{pi}^2}{\omega_i^2 - \omega^2}. \tag{4.113}$$

Let us study the frequency behavior of the dispersion relation (4.112). Splitting the permittivity into real and imaginary parts, we have

$$\mathrm{Re}\{\tilde{\epsilon}(\omega)\} - \epsilon_0 = \epsilon_0 \sum_i \omega_{pi}^2 \frac{\omega_i^2 - \omega^2}{[\omega_i^2 - \omega^2]^2 + 4\omega^2\Gamma_i^2},$$

$$\mathrm{Im}\{\tilde{\epsilon}(\omega)\} = -\epsilon_0 \sum_i \omega_{pi}^2 \frac{2\omega\Gamma_i}{[\omega_i^2 - \omega^2]^2 + 4\omega^2\Gamma_i^2}.$$

As $\omega \to 0$, the permittivity reduces to

$$\epsilon = \epsilon_0 \left(1 + \sum_i \frac{\omega_{pi}^2}{\omega_i^2}\right),$$

which is the static permittivity of the material. As $\omega \to \infty$,

$$\mathrm{Re}\{\tilde{\epsilon}(\omega)\} \to \epsilon_0 \left(1 - \frac{\sum_i \omega_{pi}^2}{\omega^2}\right), \qquad \mathrm{Im}\{\tilde{\epsilon}(\omega)\} \to -\epsilon_0 \frac{2 \sum_i \omega_{pi}^2 \Gamma_i}{\omega^3}.$$

This high-frequency behavior mimics that of a plasma as described by (4.76).

The major characteristic of the dispersion relation (4.112) is the presence of one or more *resonances*. Figure 4.7 shows a plot of a single resonance component, where we have normalized the permittivity as

$$\frac{\mathrm{Re}\{\tilde{\epsilon}(\omega)\} - \epsilon_0}{\epsilon_0 \bar{\omega}_p^2} = \frac{1 - \bar{\omega}^2}{[1 - \bar{\omega}^2]^2 + 4\bar{\omega}^2\bar{\Gamma}^2}, \qquad \frac{-\mathrm{Im}\{\tilde{\epsilon}(\omega)\}}{\epsilon_0 \bar{\omega}_p^2} = \frac{2\bar{\omega}\bar{\Gamma}}{[1 - \bar{\omega}^2]^2 + 4\bar{\omega}^2\bar{\Gamma}^2},$$

with $\bar{\omega} = \omega/\omega_0$, $\bar{\omega}_p = \omega_p/\omega_0$, and $\bar{\Gamma} = \Gamma/\omega_0$. We see a distinct resonance centered at $\omega = \omega_0$. Approaching this resonance through frequencies less than ω_0, we watch $\mathrm{Re}\,\tilde{\epsilon}$ increase slowly and peak at $\omega_{\max} = \omega_0\sqrt{1 - 2\Gamma/\omega_0}$ where it attains a value

$$\mathrm{Re}\{\tilde{\epsilon}\}_{\max} = \epsilon_0 + \tfrac{1}{4}\epsilon_0 \frac{\bar{\omega}_p^2}{\bar{\Gamma}(1 - \bar{\Gamma})}.$$

After peaking, $\mathrm{Re}\,\tilde{\epsilon}$ undergoes a rapid decrease, passing through $\mathrm{Re}\,\tilde{\epsilon} = \epsilon_0$ at $\omega = \omega_0$, and then continuing to decrease until reaching a minimum value

$$\mathrm{Re}\{\tilde{\epsilon}\}_{\min} = \epsilon_0 - \tfrac{1}{4}\epsilon_0 \frac{\bar{\omega}_p^2}{\bar{\Gamma}(1 + \bar{\Gamma})}.$$

at $\omega_{\min} = \omega_0\sqrt{1 + 2\Gamma/\omega_0}$. As ω continues to increase, $\mathrm{Re}\,\tilde{\epsilon}$ again increases slowly toward a final value of $\mathrm{Re}\,\tilde{\epsilon} = \epsilon_0$. The regions of slow variation of $\mathrm{Re}\,\tilde{\epsilon}$ are called regions of *normal dispersion*, while the region where $\mathrm{Re}\,\tilde{\epsilon}$ decreases abruptly is the region of *anomalous dispersion*. Anomalous dispersion is unusual only in the sense that it occurs over a narrower range of frequencies than normal dispersion.

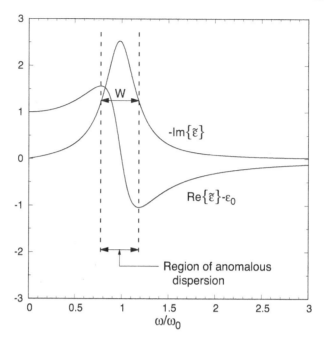

FIGURE 4.7

Real and imaginary parts of permittivity for a single resonance model of a dielectric with $\Gamma/\omega_0 = 0.2$. Permittivity normalized by dividing by $\epsilon_0(\omega_p/\omega_0)^2$.

The imaginary part of the permittivity peaks near the resonant frequency, dropping monotonically in each direction away from the peak. The width of the curve is an important parameter most easily determined by approximating the behavior of $\mathrm{Im}\,\tilde{\epsilon}$ near ω_0. Letting $\Delta\bar{\omega} = (\omega_0 - \omega)/\omega_0$ and using

$$\omega_0^2 - \omega^2 = (\omega_0 - \omega)(\omega_0 + \omega) \approx 2\omega_0^2 \Delta\bar{\omega},$$

we get

$$\mathrm{Im}\{\tilde{\epsilon}(\omega)\} \approx -\tfrac{1}{2}\epsilon_0 \bar{\omega}_p^2 \frac{\bar{\Gamma}}{(\Delta\bar{\omega})^2 + \bar{\Gamma}^2}.$$

This approximation has a maximum value

$$\mathrm{Im}\{\tilde{\epsilon}\}_{\max} = \mathrm{Im}\{\tilde{\epsilon}(\omega_0)\} = -\tfrac{1}{2}\epsilon_0 \bar{\omega}_p^2 \frac{1}{\bar{\Gamma}}$$

located at $\omega = \omega_0$, and has half-amplitude points located at $\Delta\bar{\omega} = \pm\bar{\Gamma}$. So the width of the resonance curve is $W = 2\Gamma$. Note that for a material characterized by a low-loss resonance ($\Gamma \ll \omega_0$), the location of $\mathrm{Re}\{\tilde{\epsilon}\}_{\max}$ can be approximated as $\omega_{\max} = \omega_0\sqrt{1 - 2\Gamma/\omega_0} \approx \omega_0 - \Gamma$ while $\mathrm{Re}\{\tilde{\epsilon}\}_{\min}$ is located at $\omega_{\min} = \omega_0\sqrt{1 + 2\Gamma/\omega_0} \approx \omega_0 + \Gamma$. The region of anomalous dispersion thus lies between the half amplitude points of $\mathrm{Im}\,\tilde{\epsilon}$: $\omega_0 - \Gamma < \omega < \omega_0 + \Gamma$.

As $\Gamma \to 0$, the resonance curve gets more sharply peaked. Thus, a material characterized by a very low-loss resonance may be modeled very simply using $\mathrm{Im}\,\tilde{\epsilon} = A\delta(\omega - \omega_0)$, where A is a constant to be determined. We can find A from the Kramers–Kronig formula (4.37):

$$\text{Re}\{\tilde{\epsilon}(\omega)\} - \epsilon_0 = -\frac{2}{\pi} \, \text{P.V.} \int\limits_0^\infty A\delta(\Omega - \omega_0)\frac{\Omega \, d\Omega}{\Omega^2 - \omega^2} = -\frac{2}{\pi}A\frac{\omega_0}{\omega_0^2 - \omega^2}.$$

Since the material approaches the lossless case, this expression should match the Sellmeier equation (4.113):

$$-\frac{2}{\pi}A\frac{\omega_0}{\omega_0^2 - \omega^2} = \epsilon_0\frac{\omega_p^2}{\omega_0^2 - \omega^2},$$

giving $A = -\pi\epsilon_0\omega_p^2/2\omega_0$. Hence the permittivity of a material characterized by a low-loss resonance may be approximated as

$$\tilde{\epsilon}(\omega) = \epsilon_0\left(1 + \frac{\omega_p^2}{\omega_0^2 - \omega^2}\right) - j\epsilon_0\frac{\pi}{2}\frac{\omega_p^2}{\omega_0}\delta(\omega - \omega_0).$$

The Lorentz resonance formula may be written in a more general form for materials that have conductivity and a static dielectric constant other than unity. For these we write the complex permittivity (4.26) as

$$\tilde{\epsilon}^c(\omega) = \epsilon_\infty + \frac{(\epsilon_s - \epsilon_\infty)\omega_0^2}{\omega_0^2 - \omega^2 + j\omega 2\Gamma} - j\frac{\sigma}{\omega}, \qquad (4.114)$$

where ϵ_s is the low-frequency (static) permittivity, and ϵ_∞ is the high-frequency permittivity. Note that we may also use a multi-resonance model similar to (4.112).

▶ **Example 4.7:** Permittivity of a circuit-board material

The common circuit-board material designated *FR-4* consists of a glass-reinforced epoxy that can be described using the extended Lorentzian model (4.114). A certain sample of the board has parameters $\epsilon_s = 4.307\epsilon_0$, $\epsilon_\infty = 4.181\epsilon_0$, $\omega_0 = 34\pi \times 10^9$ s^{-1}, $\Gamma = 150\pi \times 10^9$ s^{-1}, and $\sigma = 0.00293$ S/m [107]. Plot the real and imaginary parts of the permittivity vs. frequency.

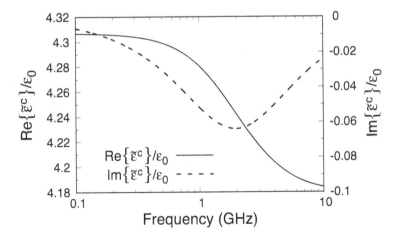

FIGURE 4.8
Relative permittivity of FR-4 circuit-board material. ◀

Solution: The operational band of this circuit-board material is from VHF to low microwave frequencies. Because the resonance frequency is much higher than the operational frequency, the Lorentz model is useful for describing the slow variation of permittivity across the operational band that is observed with FR-4. Figure 4.8 shows the relative permittivity of FR-4 computed using (4.114). The dielectric constant is seen to vary slightly over the operational band from about 4.31 at the low end of the band to 4.18 at the high end. The imaginary part is always fairly small, peaking at -0.06 around 2 GHz.

4.6.3.4 Debye relaxation and the Cole–Cole equation

In solids or liquids consisting of *polar molecules* (those retaining a permanent dipole moment, e.g., water), the resonance effect is replaced by *relaxation*. We can view the molecule as attempting to rotate in response to an applied field within a background medium dominated by the frictional term in (4.109). The rotating molecule experiences many weak collisions that continuously drain off energy, preventing it from accelerating under the force of the applied field. J.W.P. Debye proposed that such materials are described by an exponential damping of their polarization and a complete absence of oscillations. If we neglect the acceleration term in (4.109) we have the equation of motion

$$2\Gamma \frac{d\mathbf{l}(\mathbf{r}, t)}{dt} + \omega_r^2 \mathbf{l}(\mathbf{r}, t) = -\frac{e}{m_e} \mathbf{E}'(\mathbf{r}, t),$$

which has the homogeneous solution

$$\mathbf{l}(\mathbf{r}, t) = \mathbf{l}_0(\mathbf{r}) e^{-\frac{\omega_r^2}{2\Gamma} t} = \mathbf{l}_0(\mathbf{r}) e^{-t/\tau}$$

where τ is Debye's *relaxation time*.

By neglecting the acceleration term in (4.110), we obtain from (4.111) the dispersion equation, or *relaxation spectrum*:

$$\tilde{\epsilon}(\omega) = \epsilon_0 + \epsilon_0 \frac{\omega_p^2}{\omega_0^2 + j\omega 2\Gamma}.$$

Debye proposed a relaxation spectrum a bit more general than this, now called the *Debye equation*:

$$\tilde{\epsilon}(\omega) = \epsilon_\infty + \frac{\epsilon_s - \epsilon_\infty}{1 + j\omega\tau}. \tag{4.115}$$

Here ϵ_s is the real static permittivity obtained when $\omega \to 0$, while ϵ_∞ is the real "optical" permittivity describing the high frequency behavior of $\tilde{\epsilon}$. If we split (4.115) into real and imaginary parts, we find that

$$\text{Re}\{\tilde{\epsilon}(\omega)\} - \epsilon_\infty = \frac{\epsilon_s - \epsilon_\infty}{1 + \omega^2\tau^2},$$

$$\text{Im}\{\tilde{\epsilon}(\omega)\} = -\frac{\omega\tau(\epsilon_s - \epsilon_\infty)}{1 + \omega^2\tau^2}.$$

For a passive material we must have $\text{Im}\,\tilde{\epsilon} < 0$, which requires $\epsilon_s > \epsilon_\infty$. It is straightforward to show that these expressions obey the Kramers–Kronig relationships. The details are left as an exercise.

A plot of $-\text{Im}\,\tilde{\epsilon}$ vs. $\text{Re}\,\tilde{\epsilon}$ traces out a semicircle centered along the real axis at $(\epsilon_s + \epsilon_\infty)/2$ and with radius $(\epsilon_s - \epsilon_\infty)/2$. Such a plot (Figure 4.9) was first described by K.S.

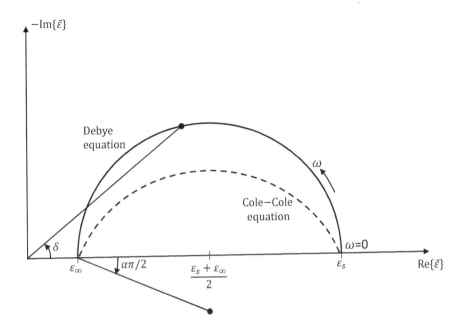

FIGURE 4.9
Arc plots for Debye and Cole–Cole descriptions of a polar material.

Cole and R.H. Cole [37] and is thus called a *Cole–Cole diagram* or "arc plot." We can think of the vector extending from the origin to a point on the semicircle as a phasor whose phase angle δ is described by the *loss tangent* of the material:

$$\tan \delta = -\frac{\mathrm{Im}\,\tilde{\epsilon}}{\mathrm{Re}\,\tilde{\epsilon}} = \frac{\omega \tau (\epsilon_s - \epsilon_\infty)}{\epsilon_s + \epsilon_\infty \omega^2 \tau^2}. \tag{4.116}$$

The Cole–Cole plot shows that the maximum value of $-\mathrm{Im}\,\tilde{\epsilon}$ is $(\epsilon_s - \epsilon_\infty)/2$ and that $\mathrm{Re}\,\tilde{\epsilon} = (\epsilon_s + \epsilon_\infty)/2$ at this point.

A Cole–Cole plot for water, shown in Figure 4.10, displays the typical semicircular nature of the arc plot. However, not all polar materials have a relaxation spectrum that follows the Debye equation as closely as water. Cole and Cole found that for many materials the arc plot traces a circular arc centered *below* the real axis, and that the line through its center makes an angle of $\alpha(\pi/2)$ with the real axis as shown in Figure 4.9. This relaxation spectrum can be described in terms of a modified Debye equation

$$\tilde{\epsilon}(\omega) = \epsilon_\infty + \frac{\epsilon_s - \epsilon_\infty}{1 + (j\omega\tau)^{1-\alpha}},$$

called the *Cole–Cole equation*. A nonzero Cole–Cole parameter α tends to broaden the relaxation spectrum, and results from a spread of relaxation times centered around τ [4]. For water the Cole–Cole parameter is only $\alpha = 0.02$, suggesting that a Debye description is sufficient, but for other materials α may be much higher.

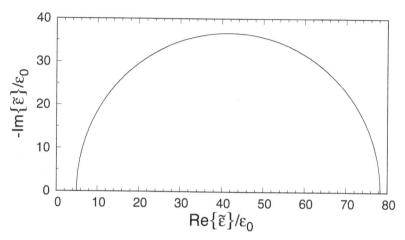

FIGURE 4.10

Cole–Cole diagram for water at 20° C.

▶ **Example 4.8:** Permittivity of water

At $T = 20°$ C the Debye parameters for water are $\epsilon_s = 78.3\epsilon_0$, $\epsilon_\infty = 5\epsilon_0$, and $\tau = 9.6 \times 10^{-12}$ s [45]. Plot the permittivity as a function of frequency and identify any pertinent features of the Debye spectrum.

Solution: Figure 4.11 shows the Debye spectrum over several decades of frequency. We see that $\mathrm{Re}\,\tilde{\epsilon}$ decreases over the entire frequency range.

The frequency dependence of the imaginary part of the permittivity is similar to that found in the resonance model, forming a curve that peaks at the *critical frequency* $\omega_{\max} = 1/\tau = 1.1 \times 10^{11}$ s^{-1} where it attains the value $\mathrm{Im}\{\tilde{\epsilon}(\omega_{\max})\} = (\epsilon_s - \epsilon_\infty)/2 = 36.65$. At this point $\mathrm{Re}\,\tilde{\epsilon}$ takes the average value of ϵ_s and ϵ_∞: $\mathrm{Re}\{\tilde{\epsilon}(\omega_{\max})\} = (\epsilon_s + \epsilon_\infty)/2 = 41.65$.

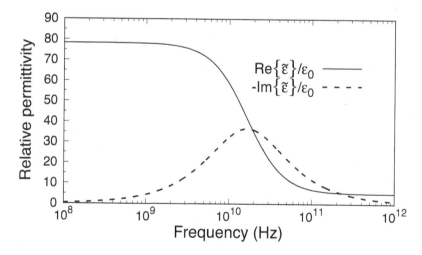

FIGURE 4.11

Relaxation spectrum for water at 20° C via the Debye equation. ◀

▶ **Example 4.9:** Permittivity of transformer oil

A transformer oil has a measured Cole–Cole parameter $\alpha = 0.23$, along with measured relaxation time $\tau = 2.3 \times 10^{-9}$ s, static permittivity $\epsilon_s = 5.9\epsilon_0$, and optical permittivity $\epsilon_\infty = 2.9\epsilon_0$ [4]. Draw the Cole–Cole diagram and plot the permittivity for the oil using both the Debye and Cole–Cole equations.

Solution: Figure 4.12 shows the Cole–Cole plot calculated using both $\alpha = 0$ and $\alpha = 0.23$, demonstrating a significant divergence from the Debye model. Figure 4.13 shows the relaxation spectrum for oil calculated with these same two parameters. The Cole–Cole model produces a more spread-out spectrum, with a reduced imaginary permittivity.

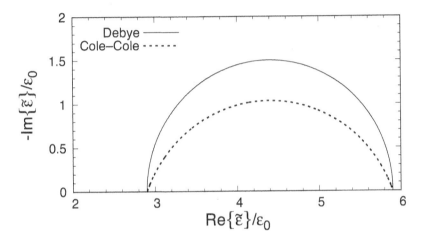

FIGURE 4.12
Cole–Cole diagram for oil via Debye equation and Cole–Cole equation with $\alpha = 0.23$.

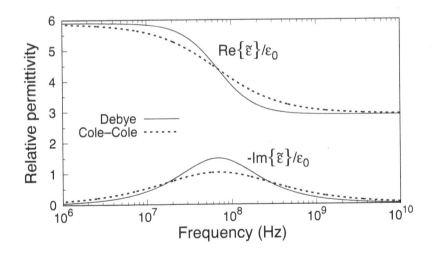

FIGURE 4.13
Relaxation spectrum for oil via Debye equation and Cole–Cole equation with $\alpha = 0.23$. ◀

4.6.4   Permittivity and conductivity of a conductor; the Drude model

The free electrons within a conductor may be considered as an electron gas free to move under the influence of an applied field. Since the electrons are not bound to the atoms of the conductor, no restoring force acts on them. However, there is a damping term associated with electron collisions. The net result is a dispersion equation analogous to (4.111), but with no resonance frequency. Setting $\omega_0 = 0$ gives the permittivity

$$\tilde{\epsilon}(\omega) = \epsilon_0 \left(1 - \frac{\omega_p^2}{\omega^2 - j\omega 2\Gamma} \right).$$

This is the *Drude model* for dispersion in a conductor, named for the German physicist Paul Karl Ludwig Drude, whose work on modeling the electrical properties of materials preceded that of Lorentz.

Alternatively, we may model a conductor as a plasma, but with a very high collision frequency; in a good metallic conductor ν is typically in the range 10^{13}–10^{14} Hz. We therefore have the conductivity of a conductor from (4.75) as

$$\tilde{\sigma}(\omega) = \frac{\epsilon_0 \omega_p^2 \nu}{\omega^2 + \nu^2}$$

and the permittivity as

$$\tilde{\epsilon}(\omega) = \epsilon_0 \left(1 - \frac{\omega_p^2}{\omega^2 + \nu^2} \right).$$

Since ν is so large, the conductivity is approximately

$$\tilde{\sigma}(\omega) \approx \frac{\epsilon_0 \omega_p^2}{\nu} = \frac{Ne^2}{m_e \nu}$$

and the permittivity is $\tilde{\epsilon}(\omega) \approx \epsilon_0$ well past microwave frequencies and into the infrared. Hence the dc conductivity is often employed by engineers throughout the communications bands. When approaching the visible spectrum, the permittivity and conductivity begin to show a strong frequency dependence. In the violet and ultraviolet frequency ranges the free-charge conductivity becomes proportional to $1/\omega$ and is driven toward zero. However, at these frequencies the resonances of the bound electrons of the metal are important and the permittivity behaves more like that of a dielectric. The permittivity is best described using the resonance formula (4.112).

4.6.5   Permeability dyadic of a ferrite

The magnetic properties of materials are complicated and diverse. The formation of accurate models based on atomic behavior requires an understanding of quantum mechanics, but simple models may be constructed using classical mechanics along with very simple quantum-mechanical assumptions, such as the existence of a spin moment. For an excellent review of the magnetic properties of materials, see Elliott [56].

The magnetic properties of matter ultimately result from atomic currents. In our simple microscopic view these currents arise from the spin and orbital motion of negatively charged electrons. These atomic currents potentially give each atom a *magnetic moment* **m**. In *diamagnetic* materials the orbital and spin moments cancel unless the material is exposed to an external magnetic field, in which case the orbital electron velocity changes to produce a net moment opposite the applied field. In *paramagnetic* materials the spin moments are greater than the orbital moments, leaving the atoms with a net permanent magnetic moment. When exposed to an external magnetic field, these moments align in the same direction as an applied field. In either case, the density of magnetic moments **M** is zero in the absence of an applied field.

In most paramagnetic materials the alignment of the permanent moment of neighboring atoms is random. However, in the subsets of paramagnetic materials known as *ferromagnetic, antiferromagnetic,* and *ferrimagnetic* materials, there is a strong coupling between the spin moments of neighboring atoms resulting in either parallel or antiparallel alignment of moments. The most familiar case is the parallel alignment of moments within the domains of ferromagnetic permanent magnets made of iron, nickel, and cobalt. Anti-ferromagnetic materials, such as chromium and manganese, have strongly coupled moments that alternate in direction between small domains, resulting in zero net magnetic moment. Ferrimagnetic materials also have alternating moments, but these are unequal and thus do not cancel completely.

Ferrites form a particularly useful subgroup of ferrimagnetic materials. They were first developed during the 1940s by researchers at the Phillips Laboratories as low-loss magnetic media for supporting electromagnetic waves [56]. Typically, ferrites have conductivities ranging from 10^{-4} to 10^0 S/m (compared to 10^7 for iron), relative permeabilities in the thousands, and dielectric constants in the range 10–15. Their low loss makes them useful for constructing transformer cores and for a variety of microwave applications. Their chemical formula is $XO \cdot Fe_2O_3$, where X is a divalent metal or mixture of metals, such as cadmium, copper, iron, or zinc. When exposed to static magnetic fields, ferrites exhibit gyrotropic magnetic (or *gyromagnetic*) properties and have permeability matrices of the form (2.24). The properties of a wide variety of ferrites are given by von Aulock [207].

To determine the permeability matrix of a ferrite we will model its electrons as simple spinning tops and examine the torque exerted on the magnetic moment by the application of an external field. Each electron has an angular momentum **L** and a magnetic dipole moment **m**, with these two vectors anti-parallel:

$$\mathbf{m}(\mathbf{r}, t) = -\gamma \mathbf{L}(\mathbf{r}, t)$$

where $\gamma = e/m_e = 1.7588 \times 10^{11}$ C/kg is the *gyromagnetic ratio.*

Let us first consider a single spinning electron immersed in an applied static magnetic field $\mathbf{B}_0$. Any torque applied to the electron results in a change of angular momentum:

$$\mathbf{T}(\mathbf{r}, t) = \frac{d\mathbf{L}(\mathbf{r}, t)}{dt}.$$

We found in (3.141) that a very small loop of current in a magnetic field experiences a torque $\mathbf{m} \times \mathbf{B}$. Thus, when first placed into a static magnetic field $\mathbf{B}_0$ an electron's angular momentum obeys the equation

$$\frac{d\mathbf{L}(\mathbf{r}, t)}{dt} = -\gamma \mathbf{L}(\mathbf{r}, t) \times \mathbf{B}_0(\mathbf{r}) = \boldsymbol{\omega}_0(\mathbf{r}) \times \mathbf{L}(\mathbf{r}, t) \tag{4.117}$$

where $\boldsymbol{\omega}_0 = \gamma \mathbf{B}_0$. This equation of motion describes the *precession* of the electron spin axis about the direction of the applied field, which is analogous to the precession of a gyroscope [130]. The spin axis rotates at the *Larmor precessional frequency* $\omega_0 = \gamma B_0 = \gamma \mu_0 H_0$.

We can use this to understand what happens when we place a homogeneous ferrite in a uniform static magnetic field $\mathbf{B}_0 = \mu_0 \mathbf{H}_0$. The internal field $\mathbf{H}_i$ experienced by any magnetic dipole is not the same as the external field $\mathbf{H}_0$, and need not even be in the same direction. In general

$$\mathbf{H}_0(\mathbf{r}, t) - \mathbf{H}_i(\mathbf{r}, t) = \mathbf{H}_d(\mathbf{r}, t)$$

where $\mathbf{H}_d$ is the *demagnetizing field* produced by the magnetic dipole moments of the material. Each electron responds to the internal field by precessing as described above until the precession damps out and the electron moments align with the magnetic field. At this point the ferrite is *saturated*. Because the demagnetizing field depends strongly on the shape of the material, we choose to ignore it as a first approximation, and this allows us to focus on the fundamental atomic properties of the ferrite.

For purposes of understanding its magnetic properties, we view the ferrite as a dense collection of electrons and write $\mathbf{M}(\mathbf{r}, t) = N\mathbf{m}(\mathbf{r}, t)$, where N is the electron number density. Under the assumption that the ferrite is homogeneous, we take N to be independent of time and position. Multiplying (4.117) by $-N\gamma$, we get an equation describing the evolution of $\mathbf{M}$:

$$\frac{d\mathbf{M}(\mathbf{r}, t)}{dt} = -\gamma \mathbf{M}(\mathbf{r}, t) \times \mathbf{B}_i(\mathbf{r}, t). \tag{4.118}$$

To determine the temporal response of the ferrite we must include a time-dependent component of the applied field. We let

$$\mathbf{H}_0(\mathbf{r}, t) = \mathbf{H}_i(\mathbf{r}, t) = \mathbf{H}_T(\mathbf{r}, t) + \mathbf{H}_{dc}$$

where $\mathbf{H}_T$ is the time-dependent component superimposed with the uniform static *biasing field* $\mathbf{H}_{dc}$. Using $\mathbf{B} = \mu_0(\mathbf{H} + \mathbf{M})$ we have from (4.118)

$$\frac{d\mathbf{M}(\mathbf{r}, t)}{dt} = -\gamma \mu_0 \mathbf{M}(\mathbf{r}, t) \times [\mathbf{H}_T(\mathbf{r}, t) + \mathbf{H}_{dc} + \mathbf{M}(\mathbf{r}, t)].$$

With $\mathbf{M} = \mathbf{M}_T(\mathbf{r}, t) + \mathbf{M}_{dc}$ and $\mathbf{M} \times \mathbf{M} = 0$ this becomes

$$\frac{d\mathbf{M}_T(\mathbf{r}, t)}{dt} + \frac{d\mathbf{M}_{dc}}{dt} = -\gamma \mu_0 [\mathbf{M}_T(\mathbf{r}, t) \times \mathbf{H}_T(\mathbf{r}, t) + \mathbf{M}_T(\mathbf{r}, t) \times \mathbf{H}_{dc}$$
$$+ \mathbf{M}_{dc} \times \mathbf{H}_T(\mathbf{r}, t) + \mathbf{M}_{dc} \times \mathbf{H}_{dc}]. \tag{4.119}$$

Suppose the ferrite is saturated. Then $\mathbf{M}_{dc}$ is aligned with $\mathbf{H}_{dc}$ and their cross product vanishes. Suppose further that the spectrum of H_T is small compared to H_{dc} at all frequencies: $|\tilde{H}_T(\mathbf{r}, \omega)| \ll H_{dc}$. This small-signal assumption allows us to neglect $\mathbf{M}_T \times \mathbf{H}_T$. With these and the fact that the time derivative of $\mathbf{M}_{dc}$ is zero, (4.119) reduces to

$$\frac{d\mathbf{M}_T(\mathbf{r}, t)}{dt} = -\gamma \mu_0 [\mathbf{M}_T(\mathbf{r}, t) \times \mathbf{H}_{dc} + \mathbf{M}_{dc} \times \mathbf{H}_T(\mathbf{r}, t)]. \tag{4.120}$$

To determine the frequency response we write (4.120) in terms of inverse Fourier transforms and invoke the Fourier integral theorem to find that

$$j\omega \tilde{\mathbf{M}}_T(\mathbf{r}, \omega) = -\gamma \mu_0 [\tilde{\mathbf{M}}_T(\mathbf{r}, \omega) \times \mathbf{H}_{dc} + \mathbf{M}_{dc} \times \tilde{\mathbf{H}}_T(\mathbf{r}, \omega)].$$

Defining $\gamma \mu_0 \mathbf{M}_{dc} = \boldsymbol{\omega}_M$ where $\omega_M = |\boldsymbol{\omega}_M|$ is the *saturation magnetization frequency*, we find that

$$\tilde{\mathbf{M}}_T + \tilde{\mathbf{M}}_T \times \left(\frac{\boldsymbol{\omega}_0}{j\omega} \right) = -\frac{1}{j\omega} \boldsymbol{\omega}_M \times \tilde{\mathbf{H}}_T, \tag{4.121}$$

where $\boldsymbol{\omega}_0 = \gamma\mu_0\mathbf{H}_{dc}$ with ω_0 now called the *gyromagnetic response frequency*. This has the form $\mathbf{v} + \mathbf{v} \times \mathbf{C} = \mathbf{A}$, which has solution (4.82). Substituting into this expression and remembering that $\boldsymbol{\omega}_0$ is parallel to $\boldsymbol{\omega}_M$, we find that

$$\tilde{\mathbf{M}}_T = \frac{-\frac{1}{j\omega}\boldsymbol{\omega}_M \times \tilde{\mathbf{H}}_T + \frac{1}{\omega^2}[\boldsymbol{\omega}_M(\boldsymbol{\omega}_0 \cdot \tilde{\mathbf{H}}_T) - (\boldsymbol{\omega}_0 \cdot \boldsymbol{\omega}_M)\tilde{\mathbf{H}}_T]}{1 - \omega_0^2/\omega^2}.$$

If we define the dyadic $\bar{\boldsymbol{\omega}}_M$ such that $\bar{\boldsymbol{\omega}}_M \cdot \tilde{\mathbf{H}}_T = \boldsymbol{\omega}_M \times \tilde{\mathbf{H}}_T$, then we identify the dyadic magnetic susceptibility

$$\bar{\tilde{\boldsymbol{\chi}}}_m(\omega) = \frac{j\omega\bar{\boldsymbol{\omega}}_M + \boldsymbol{\omega}_M\boldsymbol{\omega}_0 - \omega_M\omega_0\bar{\mathbf{I}}}{\omega^2 - \omega_0^2} \tag{4.122}$$

with which we can write $\tilde{\mathbf{M}}(\mathbf{r},\omega) = \bar{\tilde{\boldsymbol{\chi}}}_m(\omega) \cdot \tilde{\mathbf{H}}(\mathbf{r},\omega)$. In rectangular coordinates, $\bar{\boldsymbol{\omega}}_M$ is represented by

$$[\bar{\boldsymbol{\omega}}_M] = \begin{bmatrix} 0 & -\omega_{Mz} & \omega_{My} \\ \omega_{Mz} & 0 & -\omega_{Mx} \\ -\omega_{My} & \omega_{Mx} & 0 \end{bmatrix}. \tag{4.123}$$

Finally, using $\tilde{\mathbf{B}} = \mu_0(\tilde{\mathbf{H}} + \tilde{\mathbf{M}}) = \mu_0(\bar{\mathbf{I}} + \bar{\tilde{\boldsymbol{\chi}}}_m) \cdot \tilde{\mathbf{H}} = \bar{\tilde{\boldsymbol{\mu}}} \cdot \tilde{\mathbf{H}}$, we find that

$$\bar{\tilde{\boldsymbol{\mu}}}(\omega) = \mu_0[\bar{\mathbf{I}} + \bar{\tilde{\boldsymbol{\chi}}}_m(\omega)].$$

To examine the properties of the dyadic permeability, it is useful to write it in matrix form. To do this we must choose a coordinate system. We align $\mathbf{H}_{dc}$ with the z-axis so that $\mathbf{H}_{dc} = \hat{\mathbf{z}}H_{dc}$ and thus $\boldsymbol{\omega}_M = \hat{\mathbf{z}}\omega_M$ and $\boldsymbol{\omega}_0 = \hat{\mathbf{z}}\omega_0$. Then (4.123) becomes

$$[\bar{\boldsymbol{\omega}}_M] = \begin{bmatrix} 0 & -\omega_M & 0 \\ \omega_M & 0 & 0 \\ 0 & 0 & 0 \end{bmatrix}$$

and we can write the susceptibility dyadic (4.122) as

$$[\bar{\tilde{\boldsymbol{\chi}}}_m(\omega)] = \frac{\omega_M}{\omega^2 - \omega_0^2}\begin{bmatrix} -\omega_0 & -j\omega & 0 \\ j\omega & -\omega_0 & 0 \\ 0 & 0 & 0 \end{bmatrix}.$$

The permeability dyadic becomes

$$[\bar{\tilde{\boldsymbol{\mu}}}(\omega)] = \begin{bmatrix} \mu & -j\kappa & 0 \\ j\kappa & \mu & 0 \\ 0 & 0 & \mu_0 \end{bmatrix} \tag{4.124}$$

where

$$\mu = \mu_0\left(1 - \frac{\omega_0\omega_M}{\omega^2 - \omega_0^2}\right), \tag{4.125}$$

$$\kappa = \mu_0\frac{\omega\omega_M}{\omega^2 - \omega_0^2}. \tag{4.126}$$

Because its permeability dyadic is that for a lossless *gyrotropic* material (2.25), we call the ferrite *gyromagnetic*.

Since the ferrite is lossless, the dyadic permeability must be hermitian according to (4.49). The form of (4.124) shows this explicitly. We also note that since the sign of ω_M is determined by that of $\mathbf{H}_{dc}$, the dyadic permeability obeys the symmetry relation $\tilde{\mu}_{ij}(\mathbf{H}_{dc}) = \tilde{\mu}_{ji}(-\mathbf{H}_{dc})$, which is the symmetry condition observed for a plasma in (4.89).

► **Example 4.10:** Properties of a representative ferrite

The commercially available ferrite G-1010 manufactured by Trans-Tech [197] has a typical saturation magnetization of $4\pi M_s = 1000$ G. If the applied dc magnetic field is 1000 Oe, find the gyromagnetic response frequency f_0 and the saturation magnetization frequency f_M. Assuming the ferrite is lossless, find μ and κ for $f = 0.8f_0$ and $f = 1.2f_0$.

Solution: Manufacturers often specify the parameters of magnetic materials in units other than SI. Thus, we must make appropriate conversions. The conversion from Oe to A/m is given by $1\,\mathrm{Oe} = 1000/4\pi$ A/m. We have the gyromagnetic response frequency

$$f_0 = \frac{\omega_0}{2\pi} = \frac{\gamma\mu_0 H_{dc}}{2\pi} = 2 \times 10^{-7}\gamma H_{dc} = 3.5184 \times 10^4 H_{dc} \quad \mathrm{Hz}$$

when H_{dc} is in A/m. Converting H_{dc} to Oersteds gives

$$f_0 = 3.5184 \times 10^4 \frac{1000}{4\pi} H_{dc} = 2.800 \times 10^6 H_{dc} \quad \mathrm{Hz}$$

when H_{dc} is in Oe.

Manufacturers further assume that application of the static dc magnetic field saturates the ferrite and thus specify a *saturation magnetization* $4\pi M_s$, which is equivalent to the M_{dc} in the equations above. However, the units are given in terms of magnetic flux density as opposed to magnetic field, which causes some confusion. Here the conversion is $4\pi M_s = 10^4\mu_0 M_{dc} = 10^{-3}(4\pi M_{dc})$ since $1\,\mathrm{T} = 10^4\mathrm{G}$. So $4\pi M_{dc} = 10^3(4\pi M_s)$, where M_{dc} is in A/m and $4\pi M_s$ is in G. We have the saturation magnetization frequency

$$f_M = \frac{\omega_M}{2\pi} = \frac{\gamma\mu_0 M_{dc}}{2\pi} = 3.5184 \times 10^4 M_{dc} = 2.800 \times 10^3(4\pi M_{dc}) \quad \mathrm{Hz}$$

when M_{dc} is in A/m. Conversion to Gauss gives

$$f_M = 2.800 \times 10^3 \times 10^3(4\pi M_s) = 2.800 \times 10^6(4\pi M_s) \quad \mathrm{Hz}$$

when $4\pi M_s$ is in G. Conveniently, with these units we have $f_M/f_0 = 4\pi M_s/H_{dc}$.

So, for $H_{dc} = 1500$ Oe we have $f_0 = 2.800 \times 10^6 \times 1500 = 4.200 \times 10^9$ Hz, or $f_0 = 4.2$ GHz. For $4\pi M_s = 1000$ G, we have $f_M = 2.800 \times 10^6 \times 1000 = 2.8 \times 10^9$ Hz, or $f_M = 2.8$ GHz.

It is useful to normalize (4.125) and (4.126) according to

$$\frac{\mu}{\mu_0} = 1 - \frac{f_M/f_0}{(f/f_0)^2 - 1}, \qquad \frac{\kappa}{\mu_0} = \frac{(f/f_0)(f_M/f_0)}{(f/f_0)^2 - 1}.$$

We can now use $f_M/f_0 = 1000/1500 = 1/1.5$. Then, at $f/f_0 = 0.8$,

$$\frac{\mu}{\mu_0} = 1 - \frac{1/1.5}{0.8^2 - 1} = 2.8518, \qquad \frac{\kappa}{\mu_0} = \frac{0.8/1.5}{0.8^2 - 1} = -1.4815.$$

Similarly, at $f/f_0 = 1.2$,

$$\frac{\mu}{\mu_0} = 1 - \frac{1/1.5}{1.2^2 - 1} = -0.5152, \qquad \frac{\kappa}{\mu_0} = \frac{1.2/1.5}{1.2^2 - 1} = 1.8182. \quad ◄$$

A lossy ferrite material can be modeled by adding a damping term to (4.120):

$$\frac{d\mathbf{M}(\mathbf{r},t)}{dt} = -\gamma\mu_0\left[\mathbf{M}_T(\mathbf{r},t) \times \mathbf{H}_{dc} + \mathbf{M}_{dc} \times \mathbf{H}_T(\mathbf{r},t)\right] + \alpha\frac{\mathbf{M}_{dc}}{M_{dc}} \times \frac{d\mathbf{M}_T(\mathbf{r},t)}{dt},$$

where α is the damping parameter [39, 207]. This term tends to reduce the angle of precession. Fourier transformation gives

$$j\omega\tilde{\mathbf{M}}_T = \boldsymbol{\omega}_0 \times \tilde{\mathbf{M}}_T - \boldsymbol{\omega}_M \times \tilde{\mathbf{H}}_T + \alpha \frac{\boldsymbol{\omega}_M}{\omega_M} \times j\omega\tilde{\mathbf{M}}_T.$$

Remembering that $\boldsymbol{\omega}_0$ and $\boldsymbol{\omega}_M$ are aligned we can write this as

$$\tilde{\mathbf{M}}_T + \tilde{\mathbf{M}}_T \times \left[\frac{\boldsymbol{\omega}_0 \left(1 + j\alpha \frac{\omega}{\omega_0} \right)}{j\omega} \right] = -\frac{1}{j\omega} \boldsymbol{\omega}_M \times \tilde{\mathbf{H}}_T.$$

This is identical to (4.121) with

$$\boldsymbol{\omega}_0 \rightarrow \boldsymbol{\omega}_0 \left(1 + j\alpha \frac{\omega}{\omega_0} \right).$$

Thus, we merely substitute this into (4.122) to find the susceptibility dyadic for a lossy ferrite:

$$\tilde{\bar{\chi}}_m(\omega) = \frac{j\omega\bar{\omega}_M + \boldsymbol{\omega}_M \omega_0 \left(1 + j\alpha\omega/\omega_0 \right) - \omega_M \omega_0 \left(1 + j\alpha\omega/\omega_0 \right) \bar{\mathbf{I}}}{\omega^2(1 + \alpha^2) - \omega_0^2 - 2j\alpha\omega\omega_0}.$$

Making the same substitution into (4.124), we can write the dyadic permeability matrix as

$$[\tilde{\bar{\boldsymbol{\mu}}}(\omega)] = \begin{bmatrix} \tilde{\mu}_{xx} & \tilde{\mu}_{xy} & 0 \\ \tilde{\mu}_{yx} & \tilde{\mu}_{yy} & 0 \\ 0 & 0 & \mu_0 \end{bmatrix} \tag{4.127}$$

where

$$\tilde{\mu}_{xx} = \tilde{\mu}_{yy} = \mu_0 - \mu_0\omega_M \frac{\omega_0 \left[\omega^2(1-\alpha^2) - \omega_0^2 \right] + j\omega\alpha \left[\omega^2(1+\alpha^2) + \omega_0^2 \right]}{\left[\omega^2(1+\alpha^2) - \omega_0^2 \right]^2 + 4\alpha^2\omega^2\omega_0^2} \tag{4.128}$$

and

$$\tilde{\mu}_{xy} = -\tilde{\mu}_{yx} = \frac{2\mu_0\alpha\omega^2\omega_0\omega_M - j\mu_0\omega\omega_M \left[\omega^2(1+\alpha^2) - \omega_0^2 \right]}{\left[\omega^2(1+\alpha^2) - \omega_0^2 \right]^2 + 4\alpha^2\omega^2\omega_0^2}. \tag{4.129}$$

In the case of a lossy ferrite, the hermitian nature of the permeability dyadic is lost.

Ferrite loss may also be described by use of a magnetic field *line width* ΔH. To implement loss, the static biasing field H_{dc} is replaced according to

$$H_{dc} \rightarrow H_{dc} + j\frac{\Delta H}{2}.$$

With this replacement the gyromagnetic response frequency becomes

$$\omega_0 = 2\pi \times 2.8 \times 10^6 H_{dc} \left(1 + j\frac{\Delta H}{2H_{dc}} \right),$$

where H_{dc} is in Oe (see Example 4.10). Now, compare this to the gyromagnetic response frequency in terms of the damping parameter α,

$$\omega_0 = 2\pi \times 2.8 \times 10^6 H_{dc} \left(1 + j\alpha\frac{\omega}{\omega_0} \right).$$

We see that

$$\alpha\omega = \omega_0 \frac{\Delta H}{2H_{dc}}.$$

Substitution into (4.128) and (4.129) gives expressions for the elements of the permeability dyadic in terms of the line width:

$$\tilde{\mu}_{xx} = \tilde{\mu}_{yy} = \mu_0 - \mu_0 \omega_M \omega_0 \frac{\omega^2 - \omega_0^2 \left[1 + \left(\frac{\Delta H}{2H_{dc}}\right)^2\right] + j\frac{\Delta H}{2H_{dc}} \left\{\omega^2 + \omega_0^2 \left[1 + \left(\frac{\Delta H}{2H_{dc}}\right)^2\right]\right\}}{\left\{\omega^2 - \omega_0^2 \left[1 - \left(\frac{\Delta H}{2H_{dc}}\right)^2\right]\right\}^2 + 4\omega_0^4 \left(\frac{\Delta H}{2H_{dc}}\right)^2},$$

(4.130)

and

$$\tilde{\mu}_{xy} = -\tilde{\mu}_{yx} = \mu_0 \omega \omega_M \frac{2\omega_0^2 \frac{\Delta H}{2H_{dc}} - j\left\{\omega^2 - \omega_0^2 \left[1 - \left(\frac{\Delta H}{2H_{dc}}\right)^2\right]\right\}}{\left\{\omega^2 - \omega_0^2 \left[1 - \left(\frac{\Delta H}{2H_{dc}}\right)^2\right]\right\}^2 + 4\omega_0^4 \left(\frac{\Delta H}{2H_{dc}}\right)^2}.$$

(4.131)

▶ **Example 4.11:** Permeability of a lossy ferrite

Consider the ferrite material G-1010 described in Example 4.10. The line width of this ferrite is approximately $\Delta H = 25$ Oe. Plot the entries of the permeability dyadic and compare to the case of a lossless ferrite.

Solution: From Example 4.10 we have $f_0 = 4.2$ GHz and $f_M = 2.8$ GHz. Substituting these into (4.130) and (4.131), and using $\Delta H/(2H_{dc}) = 25/2000 = 0.0125$, we obtain the relative permeabilities shown in Figures 4.14 and 4.15. The resonance is clearly visible at $f_0 = 4.2$ GHz. For $\tilde{\mu}_{xx}$, the real part of the lossless permeability is unbounded at f_0, and the imaginary part is zero. The nonzero line width of the actual lossy ferrite produces a sharp but smooth resonance, with the real part passing from positive to negative, and the imaginary part large and peaking at f_0. Similar behavior is observed for $\tilde{\mu}_{xy}$, with the roles of the real and imaginary parts reversed.

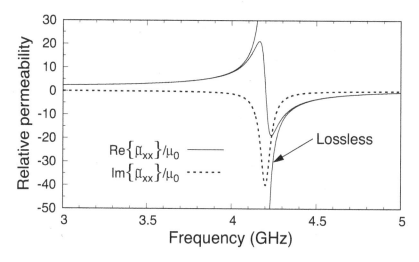

FIGURE 4.14
Relative permeability $\tilde{\mu}_{xx}/\mu_0$ for the ferrite G-1010.

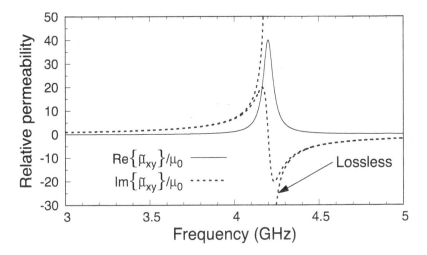

FIGURE 4.15
Relative permeability $\tilde{\mu}_{xy}/\mu_0$ for the ferrite G-1010. ◄

We note that away from resonance the lossless and lossy formulas give nearly the same result, and thus the simpler lossless formulas can be used with good accuracy. For instance, at $f = 0.8f_0 = 3.36$ GHz we find $\mathrm{Re}\{\tilde{\mu}_{xx}\}/\mu_0 = 2.849$, which compares well with the value of 2.852 found in Example 4.10. Similarly, we find $\mathrm{Im}\{\tilde{\mu}_{xy}\}/\mu_0 = 1.4786$, vs. a value of 1.4815 for the lossless case. At this frequency we also find that $\mathrm{Im}\{\tilde{\mu}_{xx}\}/\mu_0 = -0.0702$ and $\mathrm{Re}\{\tilde{\mu}_{xy}\}/\mu_0 = 0.0685$, which are quite small compared to values near resonance.

4.7 Monochromatic fields and the phasor domain

The Fourier transform is very efficient for representing the nearly sinusoidal signals produced by electronic systems such as oscillators. However, we should realize that the elemental term $e^{j\omega t}$ by itself cannot represent any physical quantity; only a continuous superposition of such terms can have physical meaning, because no physical process can be truly monochromatic. All events must have transient periods during which they are established. Even "monochromatic" light appears in bundles called quanta, interpreted as containing finite numbers of oscillations.

Arguments about whether "monochromatic" or "sinusoidal steady-state" fields can actually exist may sound purely academic. After all, a microwave oscillator can create a wave train of 10^{10} oscillations within the first second after being turned on. Such a waveform is surely as close to monochromatic as we would care to measure. But as with all mathematical models of physical systems, we can get into trouble by making non-physical assumptions, in this instance by assuming a physical system has always been in the steady state. Sinusoidal steady-state solutions to Maxwell's equations can lead to troublesome infinities linked to the infinite energy content of each elemental component. For example, an attempt to compute the energy stored within a lossless microwave cavity under steady-state conditions gives an infinite result since the cavity has been building up energy since $t = -\infty$. We handle this by considering time-averaged quantities, but even then must be careful when materials are dispersive (§ 4.5). Nevertheless, the steady-state concept is valuable because of its simplicity and finds widespread application in electromagnetics.

Since the elemental term is complex, we may use its real part, its imaginary part, or some combination of both to represent a monochromatic (or *time-harmonic*) field. We choose the representation

$$\psi(\mathbf{r}, t) = \psi_0(\mathbf{r}) \cos[\breve{\omega}t + \xi(\mathbf{r})], \qquad (4.132)$$

where ξ is the temporal phase angle of the sinusoidal function. The Fourier transform is

$$\tilde{\psi}(\mathbf{r}, \omega) = \int_{-\infty}^{\infty} \psi_0(\mathbf{r}) \cos[\breve{\omega}t + \xi(\mathbf{r})] e^{-j\omega t} \, dt. \qquad (4.133)$$

Here we run into an immediate problem: the transform in (4.133) does not exist in the ordinary sense since $\cos(\breve{\omega}t + \xi)$ is not absolutely integrable on $(-\infty, \infty)$. We should not be surprised by this: the cosine function cannot describe an actual physical process (it extends in time to $\pm\infty$), so it lacks a classical Fourier transform. One way out of this predicament is to extend the meaning of the Fourier transform (§ A.2). Then the monochromatic field (4.132) is viewed as having the generalized transform

$$\tilde{\psi}(\mathbf{r}, \omega) = \psi_0(\mathbf{r}) \pi [e^{j\xi(\mathbf{r})} \delta(\omega - \breve{\omega}) + e^{-j\xi(\mathbf{r})} \delta(\omega + \breve{\omega})]. \qquad (4.134)$$

We can compute the inverse Fourier transform by substituting (4.134) into (4.2):

$$\psi(\mathbf{r}, t) = \frac{1}{2\pi} \int_{-\infty}^{\infty} \psi_0(\mathbf{r}) \pi [e^{j\xi(\mathbf{r})} \delta(\omega - \breve{\omega}) + e^{-j\xi(\mathbf{r})} \delta(\omega + \breve{\omega})] e^{j\omega t} \, d\omega. \qquad (4.135)$$

By our interpretation of the Dirac delta, we see that the decomposition of the cosine function has only two discrete components, located at $\omega = \pm\breve{\omega}$. So we have realized our initial intention of having only a single elemental function present. The sifting property gives

$$\psi(\mathbf{r}, t) = \psi_0(\mathbf{r}) \frac{e^{j\breve{\omega}t} e^{j\xi(\mathbf{r})} + e^{-j\breve{\omega}t} e^{-j\xi(\mathbf{r})}}{2} = \psi_0(\mathbf{r}) \cos[\breve{\omega}t + \xi(\mathbf{r})]$$

as expected.

4.7.1   The time-harmonic EM fields and constitutive relations

The time-harmonic fields are described using the representation (4.132) for each field component. The electric field is

$$\mathbf{E}(\mathbf{r}, t) = \sum_{i=1}^{3} \hat{\mathbf{i}}_i |E_i(\mathbf{r})| \cos[\breve{\omega}t + \xi_i^E(\mathbf{r})],$$

for example. Here $|E_i|$ is the complex magnitude of the ith vector component, and ξ_i^E is the phase angle ($-\pi < \xi_i^E \leq \pi$). Similar terminology is used for the remaining fields.

The frequency-domain constitutive relations (4.10)–(4.14) may be written for the time-harmonic fields by employing (4.135). For instance, for an isotropic material where

$$\tilde{\mathbf{D}}(\mathbf{r}, \omega) = \tilde{\epsilon}(\mathbf{r}, \omega) \tilde{\mathbf{E}}(\mathbf{r}, \omega), \qquad \tilde{\mathbf{B}}(\mathbf{r}, \omega) = \tilde{\mu}(\mathbf{r}, \omega) \tilde{\mathbf{H}}(\mathbf{r}, \omega),$$

with

$$\tilde{\epsilon}(\mathbf{r}, \omega) = |\tilde{\epsilon}(\mathbf{r}, \omega)| e^{\xi^\epsilon(\mathbf{r}, \omega)}, \qquad \tilde{\mu}(\mathbf{r}, \omega) = |\tilde{\mu}(\mathbf{r}, \omega)| e^{\xi^\mu(\mathbf{r}, \omega)},$$

we can write

$$
\mathbf{D}(\mathbf{r},t) = \sum_{i=1}^{3} \hat{\mathbf{i}}_i |D_i(\mathbf{r})| \cos[\breve{\omega}t + \xi_i^D(\mathbf{r})]
$$

$$
= \frac{1}{2\pi} \int_{-\infty}^{\infty} \sum_{i=1}^{3} \hat{\mathbf{i}}_i \tilde{\epsilon}(\mathbf{r},\omega) |E_i(\mathbf{r})| \pi \left[e^{j\xi_i^E(\mathbf{r})} \delta(\omega - \breve{\omega}) + e^{-j\xi_i^E(\mathbf{r})} \delta(\omega + \breve{\omega}) \right] e^{j\omega t} \, d\omega
$$

$$
= \frac{1}{2} \sum_{i=1}^{3} \hat{\mathbf{i}}_i |E_i(\mathbf{r})| \left[\tilde{\epsilon}(\mathbf{r},\breve{\omega}) e^{j(\breve{\omega}t + j\xi_i^E(\mathbf{r}))} + \tilde{\epsilon}(\mathbf{r},-\breve{\omega}) e^{-j(\breve{\omega}t + j\xi_i^E(\mathbf{r}))} \right].
$$

Since (4.24) shows that $\tilde{\epsilon}(\mathbf{r},-\breve{\omega}) = \tilde{\epsilon}^*(\mathbf{r},\breve{\omega})$, we have

$$
\mathbf{D}(\mathbf{r},t) = \frac{1}{2} \sum_{i=1}^{3} \hat{\mathbf{i}}_i |E_i(\mathbf{r})| |\tilde{\epsilon}(\mathbf{r},\breve{\omega})| \left[e^{j(\breve{\omega}t + j\xi_i^E(\mathbf{r}) + j\xi^\epsilon(\mathbf{r},\breve{\omega}))} + e^{-j(\breve{\omega}t + j\xi_i^E(\mathbf{r}) + j\xi^\epsilon(\mathbf{r},\breve{\omega}))} \right]
$$

$$
= \sum_{i=1}^{3} \hat{\mathbf{i}}_i |\tilde{\epsilon}(\mathbf{r},\breve{\omega})| |E_i(\mathbf{r})| \cos[\breve{\omega}t + \xi_i^E(\mathbf{r}) + \xi^\epsilon(\mathbf{r},\breve{\omega})]. \tag{4.136}
$$

Similarly

$$
\mathbf{B}(\mathbf{r},t) = \sum_{i=1}^{3} \hat{\mathbf{i}}_i |B_i(\mathbf{r})| \cos[\breve{\omega}t + \xi_i^B(\mathbf{r})]
$$

$$
= \sum_{i=1}^{3} \hat{\mathbf{i}}_i |\tilde{\mu}(\mathbf{r},\breve{\omega})| |H_i(\mathbf{r})| \cos[\breve{\omega}t + \xi_i^H(\mathbf{r}) + \xi^\mu(\mathbf{r},\breve{\omega})].
$$

4.7.2   The phasor fields and Maxwell's equations

Sinusoidal steady-state computations using the forward and inverse transform formulas are unnecessarily cumbersome. Much more efficient is the *phasor* approach. If we define the complex function

$$
\check{\psi}(\mathbf{r}) = \psi_0(\mathbf{r}) e^{j\xi(\mathbf{r})}
$$

as the *phasor form* of the monochromatic field $\tilde{\psi}(\mathbf{r},\omega)$, then the inverse Fourier transform is easily computed by multiplying $\check{\psi}(\mathbf{r})$ by $e^{j\breve{\omega}t}$ and taking the real part. That is,

$$
\psi(\mathbf{r},t) = \mathrm{Re}\{\check{\psi}(\mathbf{r}) e^{j\breve{\omega}t}\} = \psi_0(\mathbf{r}) \cos[\breve{\omega}t + \xi(\mathbf{r})]. \tag{4.137}
$$

Using the phasor representation of the fields, we can obtain a set of Maxwell equations relating the phasor components. Let

$$
\check{\mathbf{E}}(\mathbf{r}) = \sum_{i=1}^{3} \hat{\mathbf{i}}_i \check{E}_i(\mathbf{r}) = \sum_{i=1}^{3} \hat{\mathbf{i}}_i |E_i(\mathbf{r})| e^{j\xi_i^E(\mathbf{r})}
$$

represent the phasor monochromatic electric field, with similar formulas for the other fields. Then

$$
\mathbf{E}(\mathbf{r},t) = \mathrm{Re}\{\check{\mathbf{E}}(\mathbf{r}) e^{j\breve{\omega}t}\} = \sum_{i=1}^{3} \hat{\mathbf{i}}_i |E_i(\mathbf{r})| \cos[\breve{\omega}t + \xi_i^E(\mathbf{r})].
$$

Substituting these expressions into Ampere's law (2.2), we have

$$\nabla \times \text{Re}\{\check{\mathbf{H}}(\mathbf{r})e^{j\breve{\omega}t}\} = \frac{\partial}{\partial t}\text{Re}\{\check{\mathbf{D}}(\mathbf{r})e^{j\breve{\omega}t}\} + \text{Re}\{\check{\mathbf{J}}(\mathbf{r})e^{j\breve{\omega}t}\}.$$

Since the real part of a sum of complex variables equals the sum of the real parts, we can write

$$\text{Re}\left\{\nabla \times \check{\mathbf{H}}(\mathbf{r})e^{j\breve{\omega}t} - \check{\mathbf{D}}(\mathbf{r})\frac{\partial}{\partial t}e^{j\breve{\omega}t} - \check{\mathbf{J}}(\mathbf{r})e^{j\breve{\omega}t}\right\} = 0. \tag{4.138}$$

If we examine for an arbitrary complex function $F = F_r + jF_i$ the quantity

$$\text{Re}\{(F_r + jF_i)e^{j\breve{\omega}t}\} = \text{Re}\{(F_r\cos\breve{\omega}t - F_i\sin\breve{\omega}t) + j(F_r\sin\breve{\omega}t + F_i\cos\breve{\omega}t)\},$$

we see that both F_r and F_i must be zero for the expression to vanish for all t. Thus (4.138) requires that

$$\nabla \times \check{\mathbf{H}}(\mathbf{r}) = j\breve{\omega}\check{\mathbf{D}}(\mathbf{r}) + \check{\mathbf{J}}(\mathbf{r}), \tag{4.139}$$

which is the phasor Ampere's law. Similarly

$$\nabla \times \check{\mathbf{E}}(\mathbf{r}) = -j\breve{\omega}\check{\mathbf{B}}(\mathbf{r}), \tag{4.140}$$

$$\nabla \cdot \check{\mathbf{D}}(\mathbf{r}) = \check{\rho}(\mathbf{r}), \tag{4.141}$$

$$\nabla \cdot \check{\mathbf{B}}(\mathbf{r}) = 0, \tag{4.142}$$

$$\nabla \cdot \check{\mathbf{J}}(\mathbf{r}) = -j\breve{\omega}\check{\rho}(\mathbf{r}). \tag{4.143}$$

The constitutive relations may be easily handled in the phasor approach. If we use

$$\check{D}_i(\mathbf{r}) = \tilde{\epsilon}(\mathbf{r},\breve{\omega})\check{E}_i(\mathbf{r}) = |\tilde{\epsilon}(\mathbf{r},\breve{\omega})|e^{j\xi^\epsilon(\mathbf{r},\breve{\omega})}|E_i(\mathbf{r})|e^{j\xi_i^E(\mathbf{r})},$$

then forming $D_i(\mathbf{r},t) = \text{Re}\{\check{D}_i(\mathbf{r})e^{j\breve{\omega}t}\}$ we reproduce (4.136). Thus we may write

$$\check{\mathbf{D}}(\mathbf{r}) = \tilde{\epsilon}(\mathbf{r},\breve{\omega})\check{\mathbf{E}}(\mathbf{r}).$$

Note that we never write $\check{\epsilon}$ or refer to a "phasor permittivity" since the permittivity does not vary sinusoidally in the time domain.

An obvious benefit of the phasor method is that we can manipulate field quantities without involving the sinusoidal time dependence. When our manipulations are complete, we return to the time domain via (4.137).

The phasor Maxwell equations (4.139)–(4.142) are identical in form to the temporal frequency-domain Maxwell equations (4.6)–(4.9), except that $\omega = \breve{\omega}$ in the phasor equations. This is sensible, since the phasor fields represent a single component of the complete frequency-domain spectrum of the arbitrary time-varying fields. Thus, if the phasor fields are calculated for some $\breve{\omega}$, we can make the replacements

$$\breve{\omega} \to \omega, \qquad \check{\mathbf{E}}(\mathbf{r}) \to \tilde{\mathbf{E}}(\mathbf{r},\omega), \qquad \check{\mathbf{H}}(\mathbf{r}) \to \tilde{\mathbf{H}}(\mathbf{r},\omega), \qquad \dots,$$

and obtain the general time-domain expressions by performing the inversion (4.2). Similarly, if we evaluate the frequency-domain field $\tilde{\mathbf{E}}(\mathbf{r},\omega)$ at $\omega = \breve{\omega}$, we produce the phasor field $\check{\mathbf{E}}(\mathbf{r}) = \tilde{\mathbf{E}}(\mathbf{r},\breve{\omega})$ for this frequency. That is

$$\text{Re}\{\tilde{\mathbf{E}}(\mathbf{r},\breve{\omega})e^{j\breve{\omega}t}\} = \sum_{i=1}^{3}\hat{\mathbf{i}}_i|\tilde{E}_i(\mathbf{r},\breve{\omega})|\cos\left(\breve{\omega}t + \xi^E(\mathbf{r},\breve{\omega})\right).$$

4.7.3 Boundary conditions on the phasor fields

The boundary conditions developed in § 4.3 for the frequency-domain fields may be adapted for use with the phasor fields by selecting $\omega = \breve{\omega}$. Let us include the effects of fictitious magnetic sources and write

$$\hat{\mathbf{n}}_{12} \times (\breve{\mathbf{H}}_1 - \breve{\mathbf{H}}_2) = \breve{\mathbf{J}}_s,$$

$$\hat{\mathbf{n}}_{12} \times (\breve{\mathbf{E}}_1 - \breve{\mathbf{E}}_2) = -\breve{\mathbf{J}}_{ms},$$

$$\hat{\mathbf{n}}_{12} \cdot (\breve{\mathbf{D}}_1 - \breve{\mathbf{D}}_2) = \breve{\rho}_s,$$

$$\hat{\mathbf{n}}_{12} \cdot (\breve{\mathbf{B}}_1 - \breve{\mathbf{B}}_2) = \breve{\rho}_{ms},$$

and

$$\hat{\mathbf{n}}_{12} \cdot (\breve{\mathbf{J}}_1 - \breve{\mathbf{J}}_2) = -\nabla_s \cdot \breve{\mathbf{J}}_s - j\breve{\omega}\breve{\rho}_s,$$

$$\hat{\mathbf{n}}_{12} \cdot (\breve{\mathbf{J}}_{m1} - \breve{\mathbf{J}}_{m2}) = -\nabla_s \cdot \breve{\mathbf{J}}_{ms} - j\breve{\omega}\breve{\rho}_{ms},$$

where $\hat{\mathbf{n}}_{12}$ points into region 1 from region 2.

4.8 Poynting's theorem for time-harmonic fields

We can specialize Poynting's theorem to time-harmonic form by substituting the time-harmonic field representations. The result depends on whether we use the general form (2.236), which is valid for dispersive materials, or (2.234). For nondispersive materials, (2.234) allows us to interpret the volume integral term as the time rate of change of stored energy. But if the operating frequency lies within the realm of material dispersion and loss, then we can no longer identify an explicit stored energy term.

4.8.1 General form of Poynting's theorem

We begin with (2.236). Substituting the time-harmonic representations, we obtain the term

$$\mathbf{E}(\mathbf{r}, t) \cdot \frac{\partial \mathbf{D}(\mathbf{r}, t)}{\partial t} = \left[\sum_{i=1}^{3} \hat{\mathbf{i}}_i |E_i| \cos[\breve{\omega}t + \xi_i^E] \right] \cdot \frac{\partial}{\partial t} \left[\sum_{i=1}^{3} \hat{\mathbf{i}}_i |D_i| \cos[\breve{\omega}t + \xi_i^D] \right]$$

$$= -\breve{\omega} \sum_{i=1}^{3} |E_i||D_i| \cos[\breve{\omega}t + \xi_i^E] \sin[\breve{\omega}t + \xi_i^D].$$

Since $2 \sin A \cos B \equiv \sin(A + B) + \sin(A - B)$, we have

$$\mathbf{E}(\mathbf{r}, t) \cdot \frac{\partial}{\partial t} \mathbf{D}(\mathbf{r}, t) = -\frac{1}{2} \sum_{i=1}^{3} \breve{\omega} |E_i||D_i| S_{ii}^{DE}(t),$$

where

$$S_{ii}^{DE}(t) = \sin(2\breve{\omega}t + \xi_i^D + \xi_i^E) + \sin(\xi_i^D - \xi_i^E)$$

describes the temporal dependence of the field product. Separating the current into an impressed term $\mathbf{J}^i$ and a secondary term $\mathbf{J}^c$ (assumed to be the conduction current) as $\mathbf{J} = \mathbf{J}^i + \mathbf{J}^c$ and repeating the above steps with the other terms, we obtain

$$
-\frac{1}{2} \int_V \sum_{i=1}^3 |J_i^i||E_i| C_{ii}^{J^iE}(t)\, dV = \frac{1}{2} \oint_S \sum_{i,j=1}^3 |E_i||H_j|(\hat{\mathbf{i}}_i \times \hat{\mathbf{i}}_j) \cdot \hat{\mathbf{n}} C_{ij}^{EH}(t)\, dS
$$

$$
+\frac{1}{2} \int_V \sum_{i=1}^3 \left\{ -\breve{\omega}|D_i||E_i| S_{ii}^{DE}(t) - \breve{\omega}|B_i||H_i| S_{ii}^{BH}(t) + |J_i^c||E_i| C_{ii}^{J^cE}(t) \right\} dV, \tag{4.144}
$$

where

$$
S_{ii}^{BH}(t) = \sin(2\breve{\omega}t + \xi_i^B + \xi_i^H) + \sin(\xi_i^B - \xi_i^H),
$$

$$
C_{ij}^{EH}(t) = \cos(2\breve{\omega}t + \xi_i^E + \xi_j^H) + \cos(\xi_i^E - \xi_j^H),
$$

and so on.

We see that each power term has two temporal components: one oscillating at frequency $2\breve{\omega}$, one constant with time. The oscillating component describes power that cycles through the various mechanisms of energy storage, dissipation, and transfer across the boundary. Dissipation may be produced through conduction processes or through polarization and magnetization phase lag, as described by the volume term on the right side of (4.144). Power may also be delivered to the fields either from the sources, as described by the volume term on the left, or from an active medium, as described by the volume term on the right. The time-average balance of power supplied to the fields and extracted from the fields throughout each cycle, including that transported across the surface S, is given by the constant terms in (4.144):

$$
-\frac{1}{2} \int_V \sum_{i=1}^3 |J_i^i||E_i| \cos(\xi_i^{J^i} - \xi_i^E)\, dV = \frac{1}{2} \int_V \sum_{i=1}^3 \left\{ \breve{\omega}|E_i||D_i| \sin(\xi_i^E - \xi_i^D) \right.
$$

$$
+ \breve{\omega}|B_i||H_i| \sin(\xi_i^H - \xi_i^B) + |J_i^c||E_i| \cos(\xi_i^{J^c} - \xi_i^E) \Big\} dV
$$

$$
+ \frac{1}{2} \oint_S \sum_{i,j=1}^3 |E_i||H_j|(\hat{\mathbf{i}}_i \times \hat{\mathbf{i}}_j) \cdot \hat{\mathbf{n}} \cos(\xi_i^E - \xi_j^H)\, dS. \tag{4.145}
$$

We associate one mechanism for time-average power loss with the phase lag between applied field and resulting polarization or magnetization. We see this more clearly if we use the alternative form of the Poynting theorem (2.237) written in terms of the polarization and magnetization vectors. Writing

$$
\mathbf{P}(\mathbf{r}, t) = \sum_{i=1}^3 |P_i(\mathbf{r})| \cos[\breve{\omega}t + \xi_i^P(\mathbf{r})],
$$

$$
\mathbf{M}(\mathbf{r}, t) = \sum_{i=1}^3 |M_i(\mathbf{r})| \cos[\breve{\omega}t + \xi_i^M(\mathbf{r})],
$$

and substituting the time-harmonic fields, we have

$$-\frac{1}{2}\int_V \sum_{i=1}^3 |J_i||E_i|C_{ii}^{JE}(t)\,dV + \frac{\breve{\omega}}{2}\int_V \sum_{i=1}^3 \left[|P_i||E_i|S_{ii}^{PE}(t) + \mu_0|M_i||H_i|S_{ii}^{MH}(t)\right]dV$$

$$= -\frac{\breve{\omega}}{2}\int_V \sum_{i=1}^3 \left[\epsilon_0|E_i|^2 S_{ii}^{EE}(t) + \mu_0|H_i|^2 S_{ii}^{HH}(t)\right]dV$$

$$+ \frac{1}{2}\oint_S \sum_{i,j=1}^3 |E_i||H_j|(\hat{\mathbf{i}}_i \times \hat{\mathbf{i}}_j)\cdot\hat{\mathbf{n}}C_{ij}^{EH}(t)\,dS.$$

Selection of the constant part gives the balance of time-average power:

$$-\frac{1}{2}\int_V \sum_{i=1}^3 |J_i||E_i|\cos(\xi_i^J - \xi_i^E)\,dV$$

$$= \frac{\breve{\omega}}{2}\int_V \sum_{i=1}^3 \left[|E_i||P_i|\sin(\xi_i^E - \xi_i^P) + \mu_0|H_i||M_i|\sin(\xi_i^H - \xi_i^M)\right]dV$$

$$+ \frac{1}{2}\oint_S \sum_{i,j=1}^3 |E_i||H_j|(\hat{\mathbf{i}}_i \times \hat{\mathbf{i}}_j)\cdot\hat{\mathbf{n}}\cos(\xi_i^E - \xi_j^H)\,dS.$$

Here the power loss associated with the lag in alignment of the electric and magnetic dipoles is easily identified as the volume term on the right side, and is seen to arise through the interaction of the fields with the equivalent sources as described through the phase difference between **E** and **P** and between **H** and **M**. If these pairs are in phase, then the time-average power balance reduces to that for a dispersionless material, Equation (4.148).

4.8.2 Poynting's theorem for nondispersive materials

For nondispersive materials, (2.234) is appropriate. We supply the details here so that we may examine the power-balance implications of nondispersive media. We have, substituting the field expressions,

$$-\frac{1}{2}\int_V \sum_{i=1}^3 |J_i^i||E_i|C_{ii}^{J^iE}(t)\,dV = \frac{1}{2}\int_V \sum_{i=1}^3 |J_i^c||E_i|C_{ii}^{J^cE}(t)\,dV$$

$$+ \frac{\partial}{\partial t}\int_V \sum_{i=1}^3 \left\{\tfrac{1}{4}|D_i||E_i|C_{ii}^{DE}(t) + \tfrac{1}{4}|B_i||H_i|C_{ii}^{BH}(t)\right\}dV$$

$$+ \frac{1}{2}\oint_S \sum_{i,j=1}^3 |E_i||H_j|(\hat{\mathbf{i}}_i \times \hat{\mathbf{i}}_j)\cdot\hat{\mathbf{n}}C_{ij}^{EH}(t)\,dS. \tag{4.146}$$

Here we remember that the conductivity relating **E** to **J**^{c} must also be nondispersive. Note that the electric and magnetic energy densities $w_e(\mathbf{r},t)$ and $w_m(\mathbf{r},t)$ have the time-average values $\langle w_e(\mathbf{r},t)\rangle$ and $\langle w_m(\mathbf{r},t)\rangle$ given by

$$\langle w_e(\mathbf{r},t)\rangle = \frac{1}{T}\int_{-T/2}^{T/2} \tfrac{1}{2}\mathbf{E}(\mathbf{r},t)\cdot\mathbf{D}(\mathbf{r},t)\,dt = \frac{1}{4}\sum_{i=1}^3 |E_i||D_i|\cos(\xi_i^E - \xi_i^D)$$

$$= \tfrac{1}{4}\operatorname{Re}\left\{\breve{\mathbf{E}}(\mathbf{r})\cdot\breve{\mathbf{D}}^*(\mathbf{r})\right\} \tag{4.147}$$

and

$$\langle w_m(\mathbf{r},t)\rangle = \frac{1}{T}\int_{-T/2}^{T/2}\tfrac{1}{2}\mathbf{B}(\mathbf{r},t)\cdot\mathbf{H}(\mathbf{r},t)\,dt = \frac{1}{4}\sum_{i=1}^{3}|B_i||H_i|\cos(\xi_i^H-\xi_i^B)$$

$$= \tfrac{1}{4}\operatorname{Re}\left\{\check{\mathbf{H}}(\mathbf{r})\cdot\check{\mathbf{B}}^*(\mathbf{r})\right\},$$

where $T = 2\pi/\breve{\omega}$. We have already identified the energy stored in a nondispersive material (§ 4.5.2). If (4.147) is to match with (4.62), the phases of $\check{\mathbf{E}}$ and $\check{\mathbf{D}}$ must match: $\xi_i^E = \xi_i^D$. We must also have $\xi_i^H = \xi_i^B$. Since in a dispersionless material σ must be independent of frequency, from $\check{\mathbf{J}}^c = \sigma\check{\mathbf{E}}$ we also see that $\xi_i^{J^c} = \xi_i^E$.

Upon differentiation, the time-average stored energy terms in (4.146) disappear, giving

$$-\frac{1}{2}\int_V\sum_{i=1}^{3}|J_i^i||E_i|C_{ii}^{J^iE}(t)\,dV = \frac{1}{2}\int_V\sum_{i=1}^{3}|J_i^c||E_i|C_{ii}^{EE}(t)\,dV$$

$$-2\breve{\omega}\int_V\sum_{i=1}^{3}\left\{\tfrac{1}{4}|D_i||E_i|S_{ii}^{EE}(t)+\tfrac{1}{4}|B_i||H_i|S_{ii}^{BB}(t)\right\}dV$$

$$+\frac{1}{2}\oint_S\sum_{i,j=1}^{3}|E_i||H_j|(\hat{\mathbf{i}}_i\times\hat{\mathbf{i}}_j)\cdot\hat{\mathbf{n}}C_{ij}^{EH}(t)\,dS.$$

Equating the constant terms, we find the time-average power balance expression

$$-\frac{1}{2}\int_V\sum_{i=1}^{3}|J_i^i||E_i|\cos(\xi_i^{J^i}-\xi_i^E)\,dV = \frac{1}{2}\int_V\sum_{i=1}^{3}|J_i^c||E_i|\,dV$$

$$+\frac{1}{2}\oint_S\sum_{i,j=1}^{3}|E_i||H_j|(\hat{\mathbf{i}}_i\times\hat{\mathbf{i}}_j)\cdot\hat{\mathbf{n}}\cos(\xi_i^E-\xi_j^H)\,dS. \qquad (4.148)$$

This can be written more compactly using phasor notation as

$$\int_V p_J(\mathbf{r})\,dV = \int_V p_\sigma(\mathbf{r})\,dV + \oint_S \mathbf{S}_{av}(\mathbf{r})\cdot\hat{\mathbf{n}}\,dS \qquad (4.149)$$

where

$$p_J(\mathbf{r}) = -\tfrac{1}{2}\operatorname{Re}\{\check{\mathbf{E}}(\mathbf{r})\cdot\check{\mathbf{J}}^{i*}(\mathbf{r})\}$$

is the time-average density of power delivered by the sources to the fields in V,

$$p_\sigma(\mathbf{r}) = \tfrac{1}{2}\check{\mathbf{E}}(\mathbf{r})\cdot\check{\mathbf{J}}^{c*}(\mathbf{r})$$

is the time-average density of power transferred to the conducting material as heat, and

$$\mathbf{S}_{av}(\mathbf{r})\cdot\hat{\mathbf{n}} = \tfrac{1}{2}\operatorname{Re}\{\check{\mathbf{E}}(\mathbf{r})\times\check{\mathbf{H}}^*(\mathbf{r})\}\cdot\hat{\mathbf{n}}$$

is the density of time-average power transferred across the boundary surface S. Here

$$\mathbf{S}^c = \check{\mathbf{E}}(\mathbf{r})\times\check{\mathbf{H}}^*(\mathbf{r})$$

is the *complex Poynting vector* and $\mathbf{S}_{av}$ is the *time-average Poynting vector*.

Comparison of (4.148) with (4.145) shows that nondispersive materials cannot manifest the dissipative (or active) properties determined by the term

$$\frac{1}{2}\int_V\sum_{i=1}^{3}\left\{\breve{\omega}|E_i||D_i|\sin(\xi_i^E-\xi_i^D)+\breve{\omega}|B_i||H_i|\sin(\xi_i^H-\xi_i^B)+|J_i^c||E_i|\cos(\xi_i^{J^c}-\xi_i^E)\right\}dV.$$

This term can be used to classify materials as lossless, lossy, or active, as shown next.

4.8.3   Lossless, lossy, and active media

In § 4.5.1 we classified materials based on whether they dissipate (or provide) energy over the period of a transient event. We can provide the same classification based on their steady-state behavior.

We classify a material as *lossless* if the time-average flow of power entering a homogeneous body is zero when there are sources external to the body, but no sources internal to the body. This implies that the mechanisms within the body either do not dissipate power that enters, or that there is a mechanism that creates energy to exactly balance the dissipation. If the time-average power entering is positive, then the material dissipates power and is termed *lossy*. If the time-average power entering is negative, power must originate from within the body and the material is termed *active*. (Note that the power associated with an active body is not described as arising from sources, but is rather described through the constitutive relations.)

Since materials are generally inhomogeneous we may apply this concept to a vanishingly small volume, thus invoking the point form of Poynting's theorem. From (4.145) we see that the time-average influx of power density is given by

$$-\nabla \cdot \mathbf{S}_{av}(\mathbf{r}) = p_{in}(\mathbf{r}) = \frac{1}{2} \sum_{i=1}^{3} \left\{ \breve{\omega}|E_i||D_i| \sin(\xi_i^E - \xi_i^D) + \breve{\omega}|B_i||H_i| \sin(\xi_i^H - \xi_i^B) \right.$$

$$\left. + |J_i^c||E_i| \cos(\xi_i^{J^c} - \xi_i^E) \right\}.$$

Materials are then classified as follows:

$$p_{in}(\mathbf{r}) = 0, \qquad \text{lossless,}$$
$$p_{in}(\mathbf{r}) > 0, \qquad \text{lossy,}$$
$$p_{in}(\mathbf{r}) \geq 0, \qquad \text{passive,}$$
$$p_{in}(\mathbf{r}) < 0, \qquad \text{active.}$$

We see that if $\xi_i^E = \xi_i^D$, $\xi_i^H = \xi_i^B$, and $\mathbf{J}^c = 0$, then the material is lossless. This implies that $(\mathbf{D},\mathbf{E})$ and $(\mathbf{B},\mathbf{H})$ are exactly in phase and there is no conduction current. If the material is isotropic, we may substitute from the constitutive relations (4.20)–(4.22) to obtain

$$p_{in}(\mathbf{r}) = -\frac{\breve{\omega}}{2} \sum_{i=1}^{3} \left\{ |E_i|^2 \left[|\tilde{\epsilon}| \sin(\xi^\epsilon) - \frac{|\tilde{\sigma}|}{\breve{\omega}} \cos(\xi^\sigma) \right] + |\tilde{\mu}||H_i|^2 \sin(\xi^\mu) \right\}. \qquad (4.150)$$

The first two terms can be regarded as resulting from a single complex permittivity (4.26). Then (4.150) simplifies to

$$p_{in}(\mathbf{r}) = -\frac{\breve{\omega}}{2} \sum_{i=1}^{3} \left\{ |\tilde{\epsilon}^c||E_i|^2 \sin(\xi^{\epsilon^c}) + |\tilde{\mu}||H_i|^2 \sin(\xi^\mu) \right\}. \qquad (4.151)$$

Now we can see that a lossless medium, which requires (4.151) to vanish, has $\xi^{\epsilon^c} = \xi^\mu = 0$ (or perhaps the unlikely condition that dissipative and active effects within the electric and magnetic terms exactly cancel). To have $\xi^\mu = 0$ we need $\mathbf{B}$ and $\mathbf{H}$ to be in phase, hence we need $\tilde{\mu}(\mathbf{r}, \omega)$ to be real. To have $\xi^{\epsilon^c} = 0$ we need $\xi^\epsilon = 0$ ($\tilde{\epsilon}(\mathbf{r}, \omega)$ real) and $\tilde{\sigma}(\mathbf{r}, \omega) = 0$ (or perhaps the unlikely condition that the active and dissipative effects of the permittivity and conductivity exactly cancel).

A lossy medium requires (4.151) to be positive. This occurs when $\xi^\mu < 0$ or $\xi^{\epsilon^c} < 0$, meaning that the imaginary part of the permeability or complex permittivity is negative. The complex permittivity has a negative imaginary part if the imaginary part of $\tilde\epsilon$ is negative or if the real part of $\tilde\sigma$ is positive. Physically, $\xi^\epsilon < 0$ means that $\xi^D < \xi^E$ and thus the phase of the response field **D** lags that of the excitation field **E**. This results from a delay in the polarization alignment of the atoms, and leads to dissipation of power within the material.

An active medium requires (4.151) to be negative. This occurs when $\xi^\mu > 0$ or $\xi^{\epsilon^c} > 0$, meaning that the imaginary part of the permeability or complex permittivity is positive. The complex permittivity has a positive imaginary part if the imaginary part of $\tilde\epsilon$ is positive or if the real part of $\tilde\sigma$ is negative.

In summary, a passive isotropic medium is lossless when the permittivity and permeability are real and when the conductivity is zero. A passive isotropic medium is lossy when one or more of the following holds: the permittivity is complex with negative imaginary part, the permeability is complex with negative imaginary part, or the conductivity has a positive real part. Finally, a complex permittivity or permeability with positive imaginary part or a conductivity with negative real part indicates an *active* medium.

For anisotropic materials the interpretation of p_{in} is not as simple. Here we find that the permittivity or permeability dyadic may be complex, and yet the material may still be lossless. To determine the condition for a lossless medium, let us recompute p_{in} using the constitutive relations (4.17)–(4.19). With these we have

$$\mathbf{E} \cdot \left[\frac{\partial \mathbf{D}}{\partial t} + \mathbf{J}^c \right] + \mathbf{H} \cdot \frac{\partial \mathbf{B}}{\partial t} = \breve\omega \sum_{i,j=1}^{3} |E_i||E_j| \left[-|\tilde\epsilon_{ij}| \sin(\breve\omega t + \xi_j^E + \xi_{ij}^\epsilon) \cos(\breve\omega t + \xi_i^E) \right.$$
$$\left. + \frac{|\tilde\sigma_{ij}|}{\breve\omega} \cos(\breve\omega t + \xi_j^E + \xi_{ij}^\sigma) \cos(\breve\omega t + \xi_i^E) \right]$$
$$+ \breve\omega \sum_{i,j=1}^{3} |H_i||H_j| \left[-|\tilde\mu_{ij}| \sin(\breve\omega t + \xi_j^H + \xi_{ij}^\mu) \cos(\breve\omega t + \xi_i^H) \right].$$

Using the angle-sum formulas and discarding the time-varying quantities, we may obtain the time-average input power density:

$$p_{in}(\mathbf{r}) = -\frac{\breve\omega}{2} \sum_{i,j=1}^{3} |E_i||E_j| \left[|\tilde\epsilon_{ij}| \sin(\xi_j^E - \xi_i^E + \xi_{ij}^\epsilon) - \frac{|\tilde\sigma_{ij}|}{\breve\omega} \cos(\xi_j^E - \xi_i^E + \xi_{ij}^\sigma) \right]$$
$$- \frac{\breve\omega}{2} \sum_{i,j=1}^{3} |H_i||H_j||\tilde\mu_{ij}| \sin(\xi_j^H - \xi_i^H + \xi_{ij}^\mu).$$

It is easily verified that the conditions under which this quantity vanishes, thus describing a lossless material, are

$$|\tilde\epsilon_{ij}| = |\tilde\epsilon_{ji}|, \qquad \xi_{ij}^\epsilon = -\xi_{ji}^\epsilon, \tag{4.152}$$
$$|\tilde\sigma_{ij}| = |\tilde\sigma_{ji}|, \qquad \xi_{ij}^\sigma = -\xi_{ji}^\sigma + \pi, \tag{4.153}$$
$$|\tilde\mu_{ij}| = |\tilde\mu_{ji}|, \qquad \xi_{ij}^\mu = -\xi_{ji}^\mu. \tag{4.154}$$

Note that this requires $\xi_{ii}^\epsilon = \xi_{ii}^\mu = \xi_{ii}^\sigma = 0$.

Condition (4.154) is easily written in the dyadic form

$$\bar{\tilde\mu}(\mathbf{r}, \breve\omega)^\dagger = \bar{\tilde\mu}(\mathbf{r}, \breve\omega)$$

where "†" stands for the conjugate-transpose operation. The dyadic permeability $\tilde{\bar{\mu}}$ is hermitian. The set of conditions (4.152)–(4.153) can also be written quite simply using the complex permittivity dyadic (4.23):

$$\tilde{\bar{\epsilon}}^c(\mathbf{r}, \breve{\omega})^\dagger = \tilde{\bar{\epsilon}}^c(\mathbf{r}, \breve{\omega}).$$

Thus, an anisotropic material is lossless when both the dyadic permeability and the complex dyadic permittivity are hermitian. Since $\breve{\omega}$ is arbitrary, these results are exactly those obtained in § 4.5.1.

4.9   The complex Poynting theorem

An equation having a striking resemblance to Poynting's theorem can be obtained by direct manipulation of the phasor-domain Maxwell equations. The result, although certainly satisfied by the phasor fields, does *not* replace Poynting's theorem as the power-balance equation for time-harmonic fields. We shall be careful to contrast the interpretation of the phasor expression with the actual time-harmonic Poynting theorem.

We begin by dotting both sides of the phasor-domain Faraday's law with $\check{\mathbf{H}}^*$ to obtain

$$\check{\mathbf{H}}^* \cdot (\nabla \times \check{\mathbf{E}}) = -j\breve{\omega}\check{\mathbf{H}}^* \cdot \check{\mathbf{B}}.$$

Taking the complex conjugate of the phasor-domain Ampere's law and dotting with $\check{\mathbf{E}}$, we have

$$\check{\mathbf{E}} \cdot (\nabla \times \check{\mathbf{H}}^*) = \check{\mathbf{E}} \cdot \check{\mathbf{J}}^* - j\breve{\omega}\check{\mathbf{E}} \cdot \check{\mathbf{D}}^*.$$

We subtract these expressions and use (B.50) to write

$$-\check{\mathbf{E}} \cdot \check{\mathbf{J}}^* = \nabla \cdot (\check{\mathbf{E}} \times \check{\mathbf{H}}^*) - j\breve{\omega}[\check{\mathbf{E}} \cdot \check{\mathbf{D}}^* - \check{\mathbf{B}} \cdot \check{\mathbf{H}}^*].$$

Finally, integrating over the volume region V and dividing by two, we have

$$-\frac{1}{2}\int_V \check{\mathbf{E}} \cdot \check{\mathbf{J}}^* \, dV = \frac{1}{2}\oint_S (\check{\mathbf{E}} \times \check{\mathbf{H}}^*) \cdot d\mathbf{S} - 2j\breve{\omega}\int_V \left[\tfrac{1}{4}\check{\mathbf{E}} \cdot \check{\mathbf{D}}^* - \tfrac{1}{4}\check{\mathbf{B}} \cdot \check{\mathbf{H}}^*\right] dV. \qquad (4.155)$$

This is known as the *complex Poynting theorem*, and is an expression that must be obeyed by the phasor fields.

As a power balance theorem, the complex Poynting theorem has meaning only for dispersionless materials. If we let $\mathbf{J} = \mathbf{J}^i + \mathbf{J}^c$ and assume no dispersion, (4.155) becomes

$$-\frac{1}{2}\int_V \check{\mathbf{E}} \cdot \check{\mathbf{J}}^{i*} \, dV = \frac{1}{2}\int_V \check{\mathbf{E}} \cdot \check{\mathbf{J}}^{c*} \, dV + \frac{1}{2}\oint_S (\check{\mathbf{E}} \times \check{\mathbf{H}}^*) \cdot d\mathbf{S}$$
$$- 2j\breve{\omega}\int_V [\langle w_e \rangle - \langle w_m \rangle] \, dV \qquad (4.156)$$

where $\langle w_e \rangle$ and $\langle w_m \rangle$ are the time-average stored electric and magnetic energy densities as described in (4.62)–(4.63). Selection of the real part now gives

$$-\frac{1}{2}\int_V \mathrm{Re}\{\check{\mathbf{E}} \cdot \check{\mathbf{J}}^{i*}\} \, dV = \frac{1}{2}\int_V \check{\mathbf{E}} \cdot \check{\mathbf{J}}^{c*} \, dV + \frac{1}{2}\oint_S \mathrm{Re}\{\check{\mathbf{E}} \times \check{\mathbf{H}}^*\} \cdot d\mathbf{S}, \qquad (4.157)$$

which is identical to (4.149). Thus the real part of the complex Poynting theorem gives the balance of time-average power for a dispersionless material.

Selection of the imaginary part of (4.156) gives the balance of imaginary, or *reactive*, power:

$$-\frac{1}{2}\int_V \text{Im}\{\check{\mathbf{E}}\cdot\check{\mathbf{J}}^{i*}\}\,dV = \frac{1}{2}\oint_S \text{Im}\{\check{\mathbf{E}}\times\check{\mathbf{H}}^*\}\cdot d\mathbf{S} - 2\check{\omega}\int_V [\langle w_e\rangle - \langle w_m\rangle]\,dV.$$

In general, the reactive power balance does not have a simple physical interpretation (it is *not* the balance of the oscillating terms in (4.144)). However, an interesting concept can be gleaned from it. If the source current and electric field are in phase, and there is no reactive power leaving S, then the time-average stored electric energy is equal to the time-average stored magnetic energy:

$$\int_V \langle w_e\rangle\,dV = \int_V \langle w_m\rangle\,dV.$$

This is the condition for "resonance." An example is a series RLC circuit with the source current and voltage in phase. Here the stored energy in the capacitor is equal to the stored energy in the inductor and the input impedance (ratio of voltage to current) is real. Such a resonance occurs at only one value of frequency. In more complicated electromagnetic systems, resonance may occur at many discrete eigenfrequencies.

4.9.1   Boundary condition for the time-average Poynting vector

In § 2.9.5 we developed a boundary condition for the normal component of the time-domain Poynting vector. For time-harmonic fields we can derive a similar condition using the time-average Poynting vector. Consider a surface S across which the electromagnetic sources and constitutive parameters are discontinuous, as shown in Figure 2.6. Let $\hat{\mathbf{n}}_{12}$ be the unit normal to the surface pointing into region 1 from region 2. If we apply the large-scale form of the complex Poynting theorem (4.155) to the two separate surfaces shown in Figure 2.6, we obtain

$$\frac{1}{2}\int_V \left[\check{\mathbf{E}}\cdot\check{\mathbf{J}}^* - 2j\check{\omega}\left(\frac{1}{4}\check{\mathbf{E}}\cdot\check{\mathbf{D}}^* - \frac{1}{4}\check{\mathbf{B}}\cdot\check{\mathbf{H}}^*\right)\right]dV + \frac{1}{2}\oint_S \mathbf{S}^c\cdot\hat{\mathbf{n}}\,dS$$
$$= \frac{1}{2}\int_{S_{10}} \hat{\mathbf{n}}_{12}\cdot(\mathbf{S}_1^c - \mathbf{S}_2^c)\,dS \tag{4.158}$$

where $\mathbf{S}^c = \check{\mathbf{E}}\times\check{\mathbf{H}}^*$ is the complex Poynting vector. If, on the other hand, we apply the large-scale form of Poynting's theorem to the entire volume region including the surface of discontinuity, and include the surface current contribution, we have

$$\frac{1}{2}\int_V \left[\check{\mathbf{E}}\cdot\check{\mathbf{J}}^* - 2j\check{\omega}\left(\frac{1}{4}\check{\mathbf{E}}\cdot\check{\mathbf{D}}^* - \frac{1}{4}\check{\mathbf{B}}\cdot\check{\mathbf{H}}^*\right)\right]dV + \frac{1}{2}\oint_S \mathbf{S}^c\cdot\hat{\mathbf{n}}\,dS$$
$$= -\frac{1}{2}\int_{S_{10}} \check{\mathbf{J}}_s^*\cdot\check{\mathbf{E}}\,dS. \tag{4.159}$$

To have the integrals over V and S in (4.158) and (4.159) produce identical results, we postulate the two conditions

$$\hat{\mathbf{n}}_{12}\times(\check{\mathbf{E}}_1 - \check{\mathbf{E}}_2) = 0$$

and

$$\hat{\mathbf{n}}_{12}\cdot(\mathbf{S}_1^c - \mathbf{S}_2^c) = -\check{\mathbf{J}}_s^*\cdot\check{\mathbf{E}}. \tag{4.160}$$

The first condition is merely the continuity of tangential electric field; it allows us to be nonspecific as to which value of $\mathbf{E}$ we use in the second condition. If we take the real part of the second condition, we have

$$\hat{\mathbf{n}}_{12} \cdot (\mathbf{S}_{av,1} - \mathbf{S}_{av,2}) = p_{J^s},$$

where $\mathbf{S}_{av} = \frac{1}{2} \operatorname{Re}\{\check{\mathbf{E}} \times \check{\mathbf{H}}^*\}$ is the time-average Poynting power flow density and $p_{J^s} = -\frac{1}{2} \operatorname{Re}\{\check{\mathbf{J}}_s^* \cdot \check{\mathbf{E}}\}$ is the time-average density of power delivered by the surface sources. This is the desired boundary condition on the time-average power flow density.

4.10 Fundamental theorems for time-harmonic fields

4.10.1 Uniqueness

If we think of a sinusoidal electromagnetic field as the steady-state culmination of a transient event that has an identifiable starting time, then the conditions for uniqueness established in § 2.2.1 apply. However, a true time-harmonic wave, which has existed since $t = -\infty$ and thus has infinite energy, must be interpreted differently.

Our approach is similar to that of § 2.2.1. Consider a simply connected region of space V bounded by surface S, where both V and S contain only ordinary points. The phasor-domain fields within V are associated with a phasor current distribution $\check{\mathbf{J}}$, which may be internal to V (entirely or in part). We seek conditions under which the phasor electromagnetic fields are uniquely determined. Let the field set $(\check{\mathbf{E}}_1, \check{\mathbf{D}}_1, \check{\mathbf{B}}_1, \check{\mathbf{H}}_1)$ satisfy Maxwell's equations (4.139) and (4.140) associated with the current $\check{\mathbf{J}}$ (along with an appropriate set of constitutive relations), and let $(\check{\mathbf{E}}_2, \check{\mathbf{D}}_2, \check{\mathbf{B}}_2, \check{\mathbf{H}}_2)$ be a second solution. To determine the conditions for uniqueness of the fields, we look for a situation that results in $\check{\mathbf{E}}_1 = \check{\mathbf{E}}_2$, $\check{\mathbf{H}}_1 = \check{\mathbf{H}}_2$, and so on. The electromagnetic fields must obey

$$\nabla \times \check{\mathbf{H}}_1 = j\check{\omega}\check{\mathbf{D}}_1 + \check{\mathbf{J}},$$

$$\nabla \times \check{\mathbf{E}}_1 = -j\check{\omega}\check{\mathbf{B}}_1,$$

$$\nabla \times \check{\mathbf{H}}_2 = j\check{\omega}\check{\mathbf{D}}_2 + \check{\mathbf{J}},$$

$$\nabla \times \check{\mathbf{E}}_2 = -j\check{\omega}\check{\mathbf{B}}_2.$$

Subtracting these and defining the difference fields $\check{\mathbf{E}}_0 = \check{\mathbf{E}}_1 - \check{\mathbf{E}}_2$, $\check{\mathbf{H}}_0 = \check{\mathbf{H}}_1 - \check{\mathbf{H}}_2$, and so on, we find that

$$\nabla \times \check{\mathbf{H}}_0 = j\check{\omega}\check{\mathbf{D}}_0, \tag{4.161}$$

$$\nabla \times \check{\mathbf{E}}_0 = -j\check{\omega}\check{\mathbf{B}}_0. \tag{4.162}$$

Establishing the conditions under which the difference fields vanish throughout V, we shall determine the conditions for uniqueness.

Dotting (4.162) with $\check{\mathbf{H}}_0^*$ and dotting the complex conjugate of (4.161) with $\check{\mathbf{E}}_0$, we have

$$\check{\mathbf{H}}_0^* \cdot (\nabla \times \check{\mathbf{E}}_0) = -j\check{\omega}\check{\mathbf{B}}_0 \cdot \check{\mathbf{H}}_0^*,$$

$$\check{\mathbf{E}}_0 \cdot (\nabla \times \check{\mathbf{H}}_0^*) = -j\check{\omega}\check{\mathbf{D}}_0^* \cdot \check{\mathbf{E}}_0.$$

Subtraction yields

$$\check{\mathbf{H}}_0^* \cdot (\nabla \times \check{\mathbf{E}}_0) - \check{\mathbf{E}}_0 \cdot (\nabla \times \check{\mathbf{H}}_0^*) = -j\check{\omega}\check{\mathbf{B}}_0 \cdot \check{\mathbf{H}}_0^* + j\check{\omega}\check{\mathbf{D}}_0^* \cdot \check{\mathbf{E}}_0$$

which, by (B.50), can be written as

$$\nabla \cdot (\check{\mathbf{E}}_0 \times \check{\mathbf{H}}_0^*) = j\check{\omega}(\check{\mathbf{E}}_0 \cdot \check{\mathbf{D}}_0^* - \check{\mathbf{B}}_0 \cdot \check{\mathbf{H}}_0^*).$$

Adding this expression to its complex conjugate, integrating over V, and using the divergence theorem, we obtain

$$\mathrm{Re}\oint_S (\check{\mathbf{E}}_0 \times \check{\mathbf{H}}_0^*) \cdot d\mathbf{S} = -j\frac{\check{\omega}}{2}\int_V [(\check{\mathbf{E}}_0^* \cdot \check{\mathbf{D}}_0 - \check{\mathbf{E}}_0 \cdot \check{\mathbf{D}}_0^*) + (\check{\mathbf{H}}_0^* \cdot \check{\mathbf{B}}_0 - \check{\mathbf{H}}_0 \cdot \check{\mathbf{B}}_0^*)]\, dV.$$

Breaking S into two arbitrary portions and using (B.6), we obtain

$$\mathrm{Re}\oint_{S_1} \check{\mathbf{H}}_0^* \cdot (\hat{\mathbf{n}} \times \check{\mathbf{E}}_0)\, dS - \mathrm{Re}\oint_{S_2} \check{\mathbf{E}}_0 \cdot (\hat{\mathbf{n}} \times \check{\mathbf{H}}_0^*)\, dS =$$
$$-j\frac{\check{\omega}}{2}\int_V [(\check{\mathbf{E}}_0^* \cdot \check{\mathbf{D}}_0 - \check{\mathbf{E}}_0 \cdot \check{\mathbf{D}}_0^*) + (\check{\mathbf{H}}_0^* \cdot \check{\mathbf{B}}_0 - \check{\mathbf{H}}_0 \cdot \check{\mathbf{B}}_0^*)]\, dV. \qquad (4.163)$$

Now if $\hat{\mathbf{n}} \times \check{\mathbf{E}}_0 = 0$ or $\hat{\mathbf{n}} \times \check{\mathbf{H}}_0 = 0$ over all of S, or some combination of these conditions holds over all of S, then

$$\int_V [(\check{\mathbf{E}}_0^* \cdot \check{\mathbf{D}}_0 - \check{\mathbf{E}}_0 \cdot \check{\mathbf{D}}_0^*) + (\check{\mathbf{H}}_0^* \cdot \check{\mathbf{B}}_0 - \check{\mathbf{H}}_0 \cdot \check{\mathbf{B}}_0^*)]\, dV = 0. \qquad (4.164)$$

This implies a relationship between $\check{\mathbf{E}}_0$, $\check{\mathbf{D}}_0$, $\check{\mathbf{B}}_0$, and $\check{\mathbf{H}}_0$. Since V is arbitrary, one possible relationship is simply to have one of each pair $(\check{\mathbf{E}}_0, \check{\mathbf{D}}_0)$ and $(\check{\mathbf{H}}_0, \check{\mathbf{B}}_0)$ equal to zero. Then, by (4.161) and (4.162), $\check{\mathbf{E}}_0 = 0$ implies $\check{\mathbf{B}}_0 = 0$, and $\check{\mathbf{D}}_0 = 0$ implies $\check{\mathbf{H}}_0 = 0$. Thus $\check{\mathbf{E}}_1 = \check{\mathbf{E}}_2$, etc., and the solution is unique throughout V. However, we cannot in general rule out more complicated relationships. The number of possibilities depends on the additional constraints on the relationship between $\check{\mathbf{E}}_0$, $\check{\mathbf{D}}_0$, $\check{\mathbf{B}}_0$, and $\check{\mathbf{H}}_0$ that we must supply to describe the material supporting the field — i.e., the constitutive relationships. For a simple medium described by $\tilde{\mu}(\omega)$ and $\tilde{\epsilon}^c(\omega)$, Equation (4.164) becomes

$$\int_V \left(|\check{\mathbf{E}}_0|^2 [\tilde{\epsilon}^c(\check{\omega}) - \tilde{\epsilon}^{c*}(\check{\omega})] + |\check{\mathbf{H}}_0|^2 [\tilde{\mu}(\check{\omega}) - \tilde{\mu}^*(\check{\omega})] \right) dV = 0$$

or

$$\int_V \left[|\check{\mathbf{E}}_0|^2 \, \mathrm{Im}\{\tilde{\epsilon}^c(\check{\omega})\} + |\check{\mathbf{H}}_0|^2 \, \mathrm{Im}\{\tilde{\mu}(\check{\omega})\} \right] dV = 0.$$

For a lossy medium, $\mathrm{Im}\,\tilde{\epsilon}^c < 0$ and $\mathrm{Im}\,\tilde{\mu} < 0$ as shown in § 4.5.1. So both terms in the integral must be negative. For the integral to be zero each term must vanish, requiring $\check{\mathbf{E}}_0 = \check{\mathbf{H}}_0 = 0$, and uniqueness is guaranteed.

When establishing more complicated constitutive relations we must be careful to ensure that they lead to a unique solution, and that the condition for uniqueness is understood. In the case above, the assumption $\hat{\mathbf{n}} \times \check{\mathbf{E}}_0|_S = 0$ implies that the tangential components of $\check{\mathbf{E}}_1$ and $\check{\mathbf{E}}_2$ are identical over S — that is, we must give specific values of these quantities on S to ensure uniqueness. A similar statement holds for the condition $\hat{\mathbf{n}} \times \check{\mathbf{H}}_0|_S = 0$.

In summary, the uniqueness conditions for the fields within a region V containing lossy isotropic materials are as follows:

1. the sources within V must be specified;

2. the tangential component of the electric field must be specified over all or part of the bounding surface S;

3. the tangential component of the magnetic field must be specified over the remainder of S.

We may question the requirement of a *lossy* medium to demonstrate uniqueness of the phasor fields. Does this mean that within a vacuum the specification of tangential fields is insufficient? Experience shows that the fields in such a region are indeed properly described by the surface fields, and it is just a case of the mathematical model being slightly out of sync with the physics. As long as we recognize that the sinusoidal steady state requires an initial transient period, we know that specification of the tangential fields is sufficient. We must be careful, however, to understand the restrictions of the mathematical model. Any attempt to describe the fields within a lossless cavity, for instance, is fraught with difficulty if true time-harmonic fields are used to model the actual physical fields. A helpful mathematical strategy is to think of free space as the limit of a lossy medium as the loss recedes to zero. Of course, this does not represent the physical state of "empty" space. Although even interstellar space may have a few particles for every cubic meter to interact with the electromagnetic field, the density of these particles invalidates our initial macroscopic assumptions.

Another important concern is whether we can extend the uniqueness argument to all of space. If we let S recede to infinity, must we continue to specify the fields over S, or is it sufficient to merely specify the sources within S? Since the boundary fields provide information to the internal region about sources that exist outside S, it is sensible to assume that as $S \to \infty$ there are no sources external to S and thus no need for the boundary fields. This is indeed the case. If all sources are localized, the fields they produce behave in just the right manner for the surface integral in (4.163) to vanish, and thus uniqueness is again guaranteed. Later we will find that the electric and magnetic fields produced by a localized source at great distance have the form of a spherical wave:

$$\check{\mathbf{E}} \sim \check{\mathbf{H}} \sim \frac{e^{-jkr}}{r}.$$

If space is taken to be slightly lossy, then k is complex with negative imaginary part, and thus the fields decrease exponentially with distance from the source. As we argued above, it may not be physically meaningful to assume that space is lossy. Sommerfeld postulated that even for lossless space the surface integral in (4.163) vanishes as $S \to \infty$. This has been verified experimentally, and provides the following restrictions on the free-space fields known as the *Sommerfeld radiation condition*:

$$\lim_{r \to \infty} r \left[\eta_0 \hat{\mathbf{r}} \times \check{\mathbf{H}}(\mathbf{r}) + \check{\mathbf{E}}(\mathbf{r}) \right] = 0, \tag{4.165}$$

$$\lim_{r \to \infty} r \left[\hat{\mathbf{r}} \times \check{\mathbf{E}}(\mathbf{r}) - \eta_0 \check{\mathbf{H}}(\mathbf{r}) \right] = 0, \tag{4.166}$$

where $\eta_0 = (\mu_0/\epsilon_0)^{1/2}$. Later we shall see how these expressions arise from the integral solutions to Maxwell's equations.

4.10.2   Reciprocity revisited

In § 2.9.3 we discussed the basic concept of reciprocity, but were unable to examine its real potential since we had not yet developed the theory of time-harmonic fields. In this section we shall apply the reciprocity concept to time-harmonic sources and fields, and investigate the properties a material must display to be reciprocal.

4.10.2.1   The general form of the reciprocity theorem

As in § 2.9.3, we consider a closed surface S enclosing a volume V. Sources of an electromagnetic field are located either inside or outside S. Material media may lie within S, and their properties are described in terms of the constitutive relations. To obtain the time-harmonic (phasor) form of the reciprocity theorem we proceed as in § 2.9.3 but begin with the phasor forms of Maxwell's equations. We find

$$\nabla \cdot (\check{\mathbf{E}}_a \times \check{\mathbf{H}}_b - \check{\mathbf{E}}_b \times \check{\mathbf{H}}_a) = j\check{\omega}[\check{\mathbf{H}}_a \cdot \check{\mathbf{B}}_b - \check{\mathbf{H}}_b \cdot \check{\mathbf{B}}_a] - j\check{\omega}[\check{\mathbf{E}}_a \cdot \check{\mathbf{D}}_b - \check{\mathbf{E}}_b \cdot \check{\mathbf{D}}_a]$$
$$+ [\check{\mathbf{E}}_b \cdot \check{\mathbf{J}}_a - \check{\mathbf{E}}_a \cdot \check{\mathbf{J}}_b - \check{\mathbf{H}}_b \cdot \check{\mathbf{J}}_{ma} + \check{\mathbf{H}}_a \cdot \check{\mathbf{J}}_{mb}], \qquad (4.167)$$

where $(\check{\mathbf{E}}_a, \check{\mathbf{D}}_a, \check{\mathbf{B}}_a, \check{\mathbf{H}}_a)$ are the fields produced by the phasor sources $(\check{\mathbf{J}}_a, \check{\mathbf{J}}_{ma})$, and $(\check{\mathbf{E}}_b, \check{\mathbf{D}}_b, \check{\mathbf{B}}_b, \check{\mathbf{H}}_b)$ are the fields produced by an independent set of sources $(\check{\mathbf{J}}_b, \check{\mathbf{J}}_{mb})$.

As in § 2.9.3, we are interested in the case in which the first two terms on the right side of (4.167) are zero. To see when this might occur, we substitute the constitutive equations for a bianisotropic medium

$$\check{\mathbf{D}} = \tilde{\bar{\xi}} \cdot \check{\mathbf{H}} + \tilde{\bar{\epsilon}} \cdot \check{\mathbf{E}}, \qquad \check{\mathbf{B}} = \tilde{\bar{\mu}} \cdot \check{\mathbf{H}} + \tilde{\bar{\zeta}} \cdot \check{\mathbf{E}},$$

into (4.167), where each of the constitutive parameters is evaluated at $\check{\omega}$. Setting the two terms to zero gives

$$j\check{\omega}\left[\check{\mathbf{H}}_a \cdot \left(\tilde{\bar{\mu}} \cdot \check{\mathbf{H}}_b + \tilde{\bar{\zeta}} \cdot \check{\mathbf{E}}_b\right) - \check{\mathbf{H}}_b \cdot \left(\tilde{\bar{\mu}} \cdot \check{\mathbf{H}}_a + \tilde{\bar{\zeta}} \cdot \check{\mathbf{E}}_a\right)\right]$$
$$- j\check{\omega}\left[\check{\mathbf{E}}_a \cdot \left(\tilde{\bar{\xi}} \cdot \check{\mathbf{H}}_b + \tilde{\bar{\epsilon}} \cdot \check{\mathbf{E}}_b\right) - \check{\mathbf{E}}_b \cdot \left(\tilde{\bar{\xi}} \cdot \check{\mathbf{H}}_a + \tilde{\bar{\epsilon}} \cdot \check{\mathbf{E}}_a\right)\right] = 0,$$

which holds if

$$\check{\mathbf{H}}_a \cdot \tilde{\bar{\mu}} \cdot \check{\mathbf{H}}_b - \check{\mathbf{H}}_b \cdot \tilde{\bar{\mu}} \cdot \check{\mathbf{H}}_a = 0,$$
$$\check{\mathbf{H}}_a \cdot \tilde{\bar{\zeta}} \cdot \check{\mathbf{E}}_b + \check{\mathbf{E}}_b \cdot \tilde{\bar{\xi}} \cdot \check{\mathbf{H}}_a = 0,$$
$$\check{\mathbf{E}}_a \cdot \tilde{\bar{\xi}} \cdot \check{\mathbf{H}}_b + \check{\mathbf{H}}_b \cdot \tilde{\bar{\zeta}} \cdot \check{\mathbf{E}}_a = 0,$$
$$\check{\mathbf{E}}_a \cdot \tilde{\bar{\epsilon}} \cdot \check{\mathbf{E}}_b - \check{\mathbf{E}}_b \cdot \tilde{\bar{\epsilon}} \cdot \check{\mathbf{E}}_a = 0.$$

These in turn hold if

$$\tilde{\bar{\epsilon}} = \tilde{\bar{\epsilon}}^T, \qquad \tilde{\bar{\mu}} = \tilde{\bar{\mu}}^T, \qquad \tilde{\bar{\xi}} = -\tilde{\bar{\zeta}}^T, \qquad \tilde{\bar{\zeta}} = -\tilde{\bar{\xi}}^T. \qquad (4.168)$$

These are the conditions for a *reciprocal medium*. For example, an anisotropic dielectric is a reciprocal medium if its permittivity dyadic is symmetric. An isotropic medium described by scalar quantities μ and ϵ is certainly reciprocal. In contrast, lossless gyrotropic media are nonreciprocal since the constitutive parameters obey $\tilde{\bar{\epsilon}} = \tilde{\bar{\epsilon}}^\dagger$ or $\tilde{\bar{\mu}} = \tilde{\bar{\mu}}^\dagger$ rather than $\tilde{\bar{\epsilon}} = \tilde{\bar{\epsilon}}^T$ or $\tilde{\bar{\mu}} = \tilde{\bar{\mu}}^T$.

For a reciprocal medium, (4.167) reduces to

$$\nabla \cdot (\check{\mathbf{E}}_a \times \check{\mathbf{H}}_b - \check{\mathbf{E}}_b \times \check{\mathbf{H}}_a) = \check{\mathbf{E}}_b \cdot \check{\mathbf{J}}_a - \check{\mathbf{E}}_a \cdot \check{\mathbf{J}}_b - \check{\mathbf{H}}_b \cdot \check{\mathbf{J}}_{ma} + \check{\mathbf{H}}_a \cdot \check{\mathbf{J}}_{mb}. \qquad (4.169)$$

At points where the sources are zero, or are conduction currents described entirely by Ohm's law $\check{\mathbf{J}} = \sigma \check{\mathbf{E}}$, we have

$$\nabla \cdot (\check{\mathbf{E}}_a \times \check{\mathbf{H}}_b - \check{\mathbf{E}}_b \times \check{\mathbf{H}}_a) = 0,$$

known as *Lorentz's lemma*. If we integrate (4.169) over V and use the divergence theorem, we obtain

$$\oint_S (\check{\mathbf{E}}_a \times \check{\mathbf{H}}_b - \check{\mathbf{E}}_b \times \check{\mathbf{H}}_a) \cdot d\mathbf{S} = \int_V (\check{\mathbf{E}}_b \cdot \check{\mathbf{J}}_a - \check{\mathbf{E}}_a \cdot \check{\mathbf{J}}_b - \check{\mathbf{H}}_b \cdot \check{\mathbf{J}}_{ma} + \check{\mathbf{H}}_a \cdot \check{\mathbf{J}}_{mb}) \, dV.$$

(4.170)

This is the general form of the *Lorentz reciprocity theorem*, and is valid when V contains reciprocal media as defined in (4.168).

Note that by an identical set of steps we find that the frequency-domain fields obey an identical Lorentz lemma and reciprocity theorem.

4.10.2.2   The condition for reciprocal systems

The quantity

$$\langle \check{\mathbf{f}}_a, \check{\mathbf{g}}_b \rangle = \int_V (\check{\mathbf{E}}_a \cdot \check{\mathbf{J}}_b - \check{\mathbf{H}}_a \cdot \check{\mathbf{J}}_{mb}) \, dV$$

is called the *reaction* between the source fields $\check{\mathbf{g}}$ of set b and the mediating fields $\check{\mathbf{f}}$ of an independent set a. Note that $\check{\mathbf{E}}_a \cdot \check{\mathbf{J}}_b$ is not quite a power density, since the current lacks a complex conjugate. Using this reaction concept, first introduced by Rumsey [166], we can write (4.170) as

$$\langle \check{\mathbf{f}}_b, \check{\mathbf{g}}_a \rangle - \langle \check{\mathbf{f}}_a, \check{\mathbf{g}}_b \rangle = \oint_S (\check{\mathbf{E}}_a \times \check{\mathbf{H}}_b - \check{\mathbf{E}}_b \times \check{\mathbf{H}}_a) \cdot d\mathbf{S}.$$

(4.171)

If there are no sources within S, then

$$\oint_S (\check{\mathbf{E}}_a \times \check{\mathbf{H}}_b - \check{\mathbf{E}}_b \times \check{\mathbf{H}}_a) \cdot d\mathbf{S} = 0.$$

(4.172)

Whenever (4.172) holds, we call the "system" within S *reciprocal*. For instance, a region of empty space is a reciprocal system.

A system need not be source-free in order for (4.172) to hold. Suppose the relationship between $\check{\mathbf{E}}$ and $\check{\mathbf{H}}$ on S is given by the *impedance boundary condition*

$$\check{\mathbf{E}}_t = -Z(\hat{\mathbf{n}} \times \check{\mathbf{H}}),$$

(4.173)

where $\check{\mathbf{E}}_t$ is the component of $\check{\mathbf{E}}$ tangential to S so that $\hat{\mathbf{n}} \times \mathbf{E} = \hat{\mathbf{n}} \times \mathbf{E}_t$, and the complex *wall impedance* Z may depend on position. By (4.173) we can write

$$(\check{\mathbf{E}}_a \times \check{\mathbf{H}}_b - \check{\mathbf{E}}_b \times \check{\mathbf{H}}_a) \cdot \hat{\mathbf{n}} = \check{\mathbf{H}}_b \cdot (\hat{\mathbf{n}} \times \check{\mathbf{E}}_a) - \check{\mathbf{H}}_a \cdot (\hat{\mathbf{n}} \times \check{\mathbf{E}}_b)$$
$$= -Z \check{\mathbf{H}}_b \cdot [\hat{\mathbf{n}} \times (\hat{\mathbf{n}} \times \check{\mathbf{H}}_a)] + Z \check{\mathbf{H}}_a \cdot [\hat{\mathbf{n}} \times (\hat{\mathbf{n}} \times \check{\mathbf{H}}_b)].$$

Since $\hat{\mathbf{n}} \times (\hat{\mathbf{n}} \times \check{\mathbf{H}}) = \hat{\mathbf{n}}(\hat{\mathbf{n}} \cdot \check{\mathbf{H}}) - \check{\mathbf{H}}$, the right side vanishes. Hence (4.172) still holds even though there are sources within S.

4.10.2.3   The reaction theorem

When sources lie within the surface S and the fields on S obey (4.173), we obtain an important corollary of the Lorentz reciprocity theorem. We have from (4.171) the additional result

$$\langle \check{\mathbf{f}}_a, \check{\mathbf{g}}_b \rangle - \langle \check{\mathbf{f}}_b, \check{\mathbf{g}}_a \rangle = 0.$$

Hence a reciprocal system has

$$\langle \breve{\mathbf{f}}_a, \breve{\mathbf{g}}_b \rangle = \langle \breve{\mathbf{f}}_b, \breve{\mathbf{g}}_a \rangle$$

(which holds even if there are no sources within S, since then the reactions would be identically zero). This condition for reciprocity is sometimes called the *reaction theorem*, and has an important physical meaning, which we explore below in the form of the Rayleigh–Carson reciprocity theorem. Note that in obtaining this relation we must assume that the medium is reciprocal in order to eliminate the terms in (4.167). Thus, in order for a system to be reciprocal, it must involve *both* a reciprocal medium and a boundary over which (4.173) holds.

It is worth noting that the impedance boundary condition (4.173) is widely applicable. If $Z \to 0$, the boundary condition is that for a PEC: $\hat{\mathbf{n}} \times \breve{\mathbf{E}} = 0$. If $Z \to \infty$, a PMC is described: $\hat{\mathbf{n}} \times \breve{\mathbf{H}} = 0$. Suppose S represents a sphere of infinite radius. We know from (4.166) that if the sources and material media within S are spatially finite, the fields far removed from these sources are described by the Sommerfeld radiation condition

$$\hat{\mathbf{r}} \times \breve{\mathbf{E}} = \eta_0 \breve{\mathbf{H}}$$

where $\hat{\mathbf{r}}$ is the radial unit vector of spherical coordinates. This condition is of the type (4.173) since $\hat{\mathbf{r}} = \hat{\mathbf{n}}$ on S, hence the unbounded region that results from S receding to infinity is also reciprocal.

4.10.2.4   Summary of reciprocity for reciprocal systems

We can summarize reciprocity as follows. Unbounded space containing sources and materials of finite size is a reciprocal system if the media are reciprocal; a bounded region of space is a reciprocal system only if the materials within are reciprocal and the boundary fields obey (4.173), or if the region is source-free. In each of these cases

$$\oint_S \left(\breve{\mathbf{E}}_a \times \breve{\mathbf{H}}_b - \breve{\mathbf{E}}_b \times \breve{\mathbf{H}}_a \right) \cdot d\mathbf{S} = 0$$

and

$$\langle \breve{\mathbf{f}}_a, \breve{\mathbf{g}}_b \rangle - \langle \breve{\mathbf{f}}_b, \breve{\mathbf{g}}_a \rangle = 0. \tag{4.174}$$

4.10.2.5   Rayleigh–Carson reciprocity theorem

The physical meaning behind reciprocity can be made clear with a simple example. Consider two electric Hertzian dipoles, each oscillating with frequency $\check{\omega}$ and located within an empty box consisting of PEC walls. These dipoles can be described in terms of volume current density as

$$\breve{\mathbf{J}}_a(\mathbf{r}) = \breve{\mathbf{I}}_a \delta(\mathbf{r} - \mathbf{r}'_a), \qquad \breve{\mathbf{J}}_b(\mathbf{r}) = \breve{\mathbf{I}}_b \delta(\mathbf{r} - \mathbf{r}'_b).$$

Since the fields on the surface obey (4.173) (specifically, $\hat{\mathbf{n}} \times \breve{\mathbf{E}} = 0$), and the medium within the box is empty space (a reciprocal medium), the fields produced by the sources must obey (4.174). We have

$$\int_V \breve{\mathbf{E}}_b(\mathbf{r}) \cdot [\breve{\mathbf{I}}_a \delta(\mathbf{r} - \mathbf{r}'_a)] \, dV = \int_V \breve{\mathbf{E}}_a(\mathbf{r}) \cdot [\breve{\mathbf{I}}_b \delta(\mathbf{r} - \mathbf{r}'_b)] \, dV,$$

hence

$$\breve{\mathbf{I}}_a \cdot \breve{\mathbf{E}}_b(\mathbf{r}'_a) = \breve{\mathbf{I}}_b \cdot \breve{\mathbf{E}}_a(\mathbf{r}'_b).$$

This is the *Rayleigh–Carson reciprocity theorem*. It also holds for two Hertzian dipoles in unbounded free space, as in that case the Sommerfeld radiation condition satisfies (4.173).

As an important application of this principle, consider a closed PEC body located in free space. Reciprocity holds in the region outside the body since we have $\hat{n} \times \tilde{\mathbf{E}} = 0$ at the boundary of the perfect conductor and the Sommerfeld radiation condition on the boundary at infinity. Now let us place dipole a somewhere outside the body, and dipole b adjacent and tangential to the perfectly conducting body. We regard a as the source of an electromagnetic field and b as "sampling" that field. Since the tangential electric field is zero at the conductor surface, the reaction between the dipoles is zero. Now let us switch the roles of the dipoles so that b is regarded as the source and a as the sampler. By reciprocity the reaction is again zero and thus there is no field produced by b at the position of a. Because the position and orientation of a are arbitrary, we conclude that an impressed electric source current placed tangentially to a perfectly conducting body produces no field external to the body. This result is used in Chapter 6 to develop a field equivalence principle useful in the study of antennas and scattering.

4.10.3 Duality

A duality principle analogous to that for time-domain fields (§ 2.9.2) may be established for frequency-domain and time-harmonic fields. Consider a closed surface S enclosing a region of space that includes a frequency-domain electric source current $\tilde{\mathbf{J}}$ and a frequency-domain magnetic source current $\tilde{\mathbf{J}}_m$. The fields $(\tilde{\mathbf{E}}_1, \tilde{\mathbf{D}}_1, \tilde{\mathbf{B}}_1, \tilde{\mathbf{H}}_1)$ in the region (which may also contain arbitrary media) are described by

$$\nabla \times \tilde{\mathbf{E}}_1 = -\tilde{\mathbf{J}}_m - j\omega\tilde{\mathbf{B}}_1, \tag{4.175}$$

$$\nabla \times \tilde{\mathbf{H}}_1 = \tilde{\mathbf{J}} + j\omega\tilde{\mathbf{D}}_1, \tag{4.176}$$

$$\nabla \cdot \tilde{\mathbf{D}}_1 = \tilde{\rho}, \tag{4.177}$$

$$\nabla \cdot \tilde{\mathbf{B}}_1 = \tilde{\rho}_m. \tag{4.178}$$

Suppose we are given a mathematical description of the sources $(\tilde{\mathbf{J}}, \tilde{\mathbf{J}}_m)$ and know the field vectors $(\tilde{\mathbf{E}}_1, \tilde{\mathbf{D}}_1, \tilde{\mathbf{B}}_1, \tilde{\mathbf{H}}_1)$. Of course, we must also be supplied with a set of boundary values and constitutive relations to make the solution unique. Replacing the formula for $\tilde{\mathbf{J}}$ with the formula for $\tilde{\mathbf{J}}_m$ in (4.176) (and $\tilde{\rho}$ with $\tilde{\rho}_m$ in (4.177)) and replacing $\tilde{\mathbf{J}}_m$ with $-\tilde{\mathbf{J}}$ in (4.175) (and $\tilde{\rho}_m$ with $-\tilde{\rho}$ in (4.178)), we get a new problem. But the symmetry of the equations allows us to specify the solution immediately. The new set of curl equations requires

$$\nabla \times \tilde{\mathbf{E}}_2 = \tilde{\mathbf{J}} - j\omega\tilde{\mathbf{B}}_2, \tag{4.179}$$

$$\nabla \times \tilde{\mathbf{H}}_2 = \tilde{\mathbf{J}}_m + j\omega\tilde{\mathbf{D}}_2. \tag{4.180}$$

If we can resolve the question of how the constitutive parameters must be altered to reflect these replacements, then we can conclude, by comparing (4.179) with (4.176) and (4.180) with (4.175), that

$$\tilde{\mathbf{E}}_2 = \tilde{\mathbf{H}}_1, \qquad \tilde{\mathbf{B}}_2 = -\tilde{\mathbf{D}}_1, \qquad \tilde{\mathbf{D}}_2 = \tilde{\mathbf{B}}_1, \qquad \tilde{\mathbf{H}}_2 = -\tilde{\mathbf{E}}_1.$$

The discussion regarding units in § 2.9.2 carries over to the present case. Multiplying Ampere's law by $\eta_0 = (\mu_0/\epsilon_0)^{1/2}$, we have

$$\nabla \times \tilde{\mathbf{E}} = -\tilde{\mathbf{J}}_m - j\omega\tilde{\mathbf{B}}, \qquad \nabla \times (\eta_0\tilde{\mathbf{H}}) = (\eta_0\tilde{\mathbf{J}}) + j\omega(\eta_0\tilde{\mathbf{D}}).$$

Thus if the original problem has solution $(\tilde{\mathbf{E}}_1, \eta_0\tilde{\mathbf{D}}_1, \tilde{\mathbf{B}}_1, \eta_0\tilde{\mathbf{H}}_1)$, then the dual problem with $\tilde{\mathbf{J}}$ replaced by $\tilde{\mathbf{J}}_m/\eta_0$ and $\tilde{\mathbf{J}}_m$ replaced by $-\eta_0\tilde{\mathbf{J}}$ has solution

$$\tilde{\mathbf{E}}_2 = \eta_0\tilde{\mathbf{H}}_1, \tag{4.181}$$

$$\tilde{\mathbf{B}}_2 = -\eta_0\tilde{\mathbf{D}}_1, \tag{4.182}$$

$$\eta_0\tilde{\mathbf{D}}_2 = \tilde{\mathbf{B}}_1, \tag{4.183}$$

$$\eta_0\tilde{\mathbf{H}}_2 = -\tilde{\mathbf{E}}_1. \tag{4.184}$$

As with duality in the time domain, the constitutive parameters for the dual problem must be altered from those of the original problem. For linear anisotropic media, we have by (4.12) and (4.13) the constitutive relationships

$$\tilde{\mathbf{D}}_1 = \bar{\tilde{\boldsymbol{\epsilon}}}_1 \cdot \tilde{\mathbf{E}}_1, \tag{4.185}$$

$$\tilde{\mathbf{B}}_1 = \bar{\tilde{\boldsymbol{\mu}}}_1 \cdot \tilde{\mathbf{H}}_1, \tag{4.186}$$

for the original problem, and

$$\tilde{\mathbf{D}}_2 = \bar{\tilde{\boldsymbol{\epsilon}}}_2 \cdot \tilde{\mathbf{E}}_2, \tag{4.187}$$

$$\tilde{\mathbf{B}}_2 = \bar{\tilde{\boldsymbol{\mu}}}_2 \cdot \tilde{\mathbf{H}}_2, \tag{4.188}$$

for the dual problem. Substitution of (4.181)–(4.184) into (4.185) and (4.186) gives

$$\tilde{\mathbf{D}}_2 = \left(\frac{\bar{\tilde{\boldsymbol{\mu}}}_1}{\eta_0^2}\right) \cdot \tilde{\mathbf{E}}_2, \tag{4.189}$$

$$\tilde{\mathbf{B}}_2 = \left(\eta_0^2\bar{\tilde{\boldsymbol{\epsilon}}}_1\right) \cdot \tilde{\mathbf{H}}_2. \tag{4.190}$$

Comparing (4.189) with (4.187) and (4.190) with (4.188), we conclude that

$$\bar{\tilde{\boldsymbol{\mu}}}_2 = \eta_0^2\bar{\tilde{\boldsymbol{\epsilon}}}_1, \qquad \bar{\tilde{\boldsymbol{\epsilon}}}_2 = \bar{\tilde{\boldsymbol{\mu}}}_1/\eta_0^2. \tag{4.191}$$

For a linear, isotropic medium specified by $\tilde{\epsilon}$ and $\tilde{\mu}$, the dual problem is obtained by replacing $\tilde{\epsilon}_r$ with $\tilde{\mu}_r$ and $\tilde{\mu}_r$ with $\tilde{\epsilon}_r$. The solution to the dual problem is then

$$\tilde{\mathbf{E}}_2 = \eta_0\tilde{\mathbf{H}}_1, \qquad \eta_0\tilde{\mathbf{H}}_2 = -\tilde{\mathbf{E}}_1,$$

as before. The medium in the dual problem must have electric properties numerically equal to the magnetic properties of the medium in the original problem, and magnetic properties numerically equal to the electric properties of the medium in the original problem. Alternatively we may divide Ampere's law by $\eta = (\tilde{\mu}/\tilde{\epsilon})^{1/2}$ instead of η_0. Then the dual problem has $\tilde{\mathbf{J}}$ replaced by $\tilde{\mathbf{J}}_m/\eta$, and $\tilde{\mathbf{J}}_m$ replaced by $-\eta\tilde{\mathbf{J}}$, and the solution is

$$\tilde{\mathbf{E}}_2 = \eta\tilde{\mathbf{H}}_1, \qquad \eta\tilde{\mathbf{H}}_2 = -\tilde{\mathbf{E}}_1. \tag{4.192}$$

There is no need to swap $\tilde{\epsilon}_r$ and $\tilde{\mu}_r$ since information about these parameters is incorporated into the replacement sources.

We may also apply duality to a problem where we have separated the impressed and secondary sources. In a homogeneous, isotropic, conducting medium we may let $\tilde{\mathbf{J}} = \tilde{\mathbf{J}}^i + \tilde{\sigma}\tilde{\mathbf{E}}$. With this the curl equations become

$$\nabla \times \eta\tilde{\mathbf{H}} = \eta\tilde{\mathbf{J}}^i + j\omega\eta\tilde{\epsilon}^c\tilde{\mathbf{E}},$$

$$\nabla \times \tilde{\mathbf{E}} = -\tilde{\mathbf{J}}_m - j\omega\tilde{\mu}\tilde{\mathbf{H}}.$$

The solution to the dual problem is again given by (4.192), except that now $\eta = (\tilde{\mu}/\tilde{\epsilon}^c)^{1/2}$.

As in § 2.9.2, we can consider duality in a source-free region. We let S enclose a source-free region of space and, for simplicity, assume the medium within S is linear, isotropic, and homogeneous. The fields within S are described by

$$\nabla \times \tilde{\mathbf{E}}_1 = -j\omega\tilde{\mu}\tilde{\mathbf{H}}_1, \qquad \nabla \cdot \tilde{\epsilon}\tilde{\mathbf{E}}_1 = 0,$$

$$\nabla \times \eta\tilde{\mathbf{H}}_1 = j\omega\tilde{\epsilon}\eta\tilde{\mathbf{E}}_1, \qquad \nabla \cdot \tilde{\mu}\tilde{\mathbf{H}}_1 = 0.$$

The symmetry of the equations is such that the mathematical form of the solution for $\tilde{\mathbf{E}}$ is the same as that for $\eta\tilde{\mathbf{H}}$. Since the fields

$$\tilde{\mathbf{E}}_2 = \eta\tilde{\mathbf{H}}_1, \qquad \tilde{\mathbf{H}}_2 = -\tilde{\mathbf{E}}_1/\eta,$$

also satisfy Maxwell's equations, the dual problem merely involves replacing $\tilde{\mathbf{E}}$ by $\eta\tilde{\mathbf{H}}$ and $\tilde{\mathbf{H}}$ by $-\tilde{\mathbf{E}}/\eta$.

4.11 The wave nature of the time-harmonic EM field

Time-harmonic electromagnetic waves have been studied in great detail. Narrowband waves are widely used for signal transmission, heating, power transfer, and radar. They share many of the properties of more general transient waves, and the discussions of § 2.10.1 are applicable. Here we shall investigate some of the unique properties of time-harmonic waves and introduce such fundamental quantities as wavelength, the phase and group velocities, and polarization.

4.11.1 The frequency-domain wave equation

We begin by deriving the frequency-domain wave equation for dispersive bianisotropic materials. A solution to this equation may be viewed as the transform of a general time-dependent field. If one specific frequency is considered, the time-harmonic solution is produced.

In § 2.10.2 we derived the time-domain wave equation for bianisotropic materials. There it was necessary to consider only time-independent constitutive parameters. We can overcome this requirement, and thus deal with dispersive materials, through a Fourier transform approach. We solve a frequency-domain wave equation that includes the frequency dependence of the constitutive parameters, then use an inverse transform to return to the time domain.

The derivation of the equation parallels that of § 2.10.2. We substitute the frequency-domain constitutive relationships

$$\tilde{\mathbf{D}} = \tilde{\bar{\epsilon}} \cdot \tilde{\mathbf{E}} + \tilde{\bar{\xi}} \cdot \tilde{\mathbf{H}}, \qquad \tilde{\mathbf{B}} = \tilde{\bar{\zeta}} \cdot \tilde{\mathbf{E}} + \tilde{\bar{\mu}} \cdot \tilde{\mathbf{H}},$$

into Maxwell's curl equations (4.6) and (4.7) to get the coupled differential equations

$$\nabla \times \tilde{\mathbf{E}} = -j\omega[\tilde{\bar{\zeta}} \cdot \tilde{\mathbf{E}} + \tilde{\bar{\mu}} \cdot \tilde{\mathbf{H}}] - \tilde{\mathbf{J}}_m,$$

$$\nabla \times \tilde{\mathbf{H}} = j\omega[\tilde{\bar{\epsilon}} \cdot \tilde{\mathbf{E}} + \tilde{\bar{\xi}} \cdot \tilde{\mathbf{H}}] + \tilde{\mathbf{J}},$$

for $\tilde{\mathbf{E}}$ and $\tilde{\mathbf{H}}$. Here we have included magnetic sources $\tilde{\mathbf{J}}_m$ in Faraday's law. Using the dyadic operator $\bar{\nabla}$ defined in (2.251) we can write these equations as

$$(\bar{\nabla} + j\omega\bar{\tilde{\zeta}}) \cdot \tilde{\mathbf{E}} = -j\omega\bar{\tilde{\mu}} \cdot \tilde{\mathbf{H}} - \tilde{\mathbf{J}}_m, \tag{4.193}$$

$$(\bar{\nabla} - j\omega\bar{\tilde{\xi}}) \cdot \tilde{\mathbf{H}} = j\omega\bar{\tilde{\epsilon}} \cdot \tilde{\mathbf{E}} + \tilde{\mathbf{J}}. \tag{4.194}$$

We can obtain separate equations for $\tilde{\mathbf{E}}$ and $\tilde{\mathbf{H}}$ by defining the inverse dyadics

$$\bar{\tilde{\epsilon}} \cdot \bar{\tilde{\epsilon}}^{-1} = \bar{\mathbf{I}}, \qquad \bar{\tilde{\mu}} \cdot \bar{\tilde{\mu}}^{-1} = \bar{\mathbf{I}}.$$

Using $\bar{\tilde{\mu}}^{-1}$ we can write (4.193) as

$$-j\omega\tilde{\mathbf{H}} = \bar{\tilde{\mu}}^{-1} \cdot (\bar{\nabla} + j\omega\bar{\tilde{\zeta}}) \cdot \tilde{\mathbf{E}} + \bar{\tilde{\mu}}^{-1} \cdot \tilde{\mathbf{J}}_m.$$

Substituting this into (4.194) we get

$$\left[(\bar{\nabla} - j\omega\bar{\tilde{\xi}}) \cdot \bar{\tilde{\mu}}^{-1} \cdot (\bar{\nabla} + j\omega\bar{\tilde{\zeta}}) - \omega^2\bar{\tilde{\epsilon}}\right] \cdot \tilde{\mathbf{E}} = -(\bar{\nabla} - j\omega\bar{\tilde{\xi}}) \cdot \bar{\tilde{\mu}}^{-1} \cdot \tilde{\mathbf{J}}_m - j\omega\tilde{\mathbf{J}}. \tag{4.195}$$

This is the general frequency-domain wave equation for $\tilde{\mathbf{E}}$. Using $\bar{\tilde{\epsilon}}^{-1}$ we can write (4.194) as

$$j\omega\tilde{\mathbf{E}} = \bar{\tilde{\epsilon}}^{-1} \cdot (\bar{\nabla} - j\omega\bar{\tilde{\xi}}) \cdot \tilde{\mathbf{H}} - \bar{\tilde{\epsilon}}^{-1} \cdot \tilde{\mathbf{J}}.$$

Substituting this into (4.193) we get

$$\left[(\bar{\nabla} + j\omega\bar{\tilde{\zeta}}) \cdot \bar{\tilde{\epsilon}}^{-1} \cdot (\bar{\nabla} - j\omega\bar{\tilde{\xi}}) - \omega^2\bar{\tilde{\mu}}\right] \cdot \tilde{\mathbf{H}} = (\bar{\nabla} + j\omega\bar{\tilde{\zeta}}) \cdot \bar{\tilde{\epsilon}}^{-1} \cdot \tilde{\mathbf{J}} - j\omega\tilde{\mathbf{J}}_m. \tag{4.196}$$

This is the general frequency-domain wave equation for $\tilde{\mathbf{H}}$.

4.11.1.1 Wave equation for a homogeneous, lossy, and isotropic medium

We may specialize (4.195) and (4.196) to the case of a homogeneous, lossy, isotropic medium by setting $\bar{\tilde{\zeta}} = \bar{\tilde{\xi}} = 0$, $\bar{\tilde{\mu}} = \tilde{\mu}\bar{\mathbf{I}}$, $\bar{\tilde{\epsilon}} = \tilde{\epsilon}\bar{\mathbf{I}}$, and $\tilde{\mathbf{J}} = \tilde{\mathbf{J}}^i + \tilde{\mathbf{J}}^c$:

$$\nabla \times (\nabla \times \tilde{\mathbf{E}}) - \omega^2\tilde{\mu}\tilde{\epsilon}\tilde{\mathbf{E}} = -\nabla \times \tilde{\mathbf{J}}_m - j\omega\tilde{\mu}(\tilde{\mathbf{J}}^i + \tilde{\mathbf{J}}^c), \tag{4.197}$$

$$\nabla \times (\nabla \times \tilde{\mathbf{H}}) - \omega^2\tilde{\mu}\tilde{\epsilon}\tilde{\mathbf{H}} = \nabla \times (\tilde{\mathbf{J}}^i + \tilde{\mathbf{J}}^c) - j\omega\tilde{\epsilon}\tilde{\mathbf{J}}_m. \tag{4.198}$$

Using (B.53) with Ohm's law $\tilde{\mathbf{J}}^c = \tilde{\sigma}\tilde{\mathbf{E}}$ describing the secondary current, we get from (4.197)

$$\nabla(\nabla \cdot \tilde{\mathbf{E}}) - \nabla^2\tilde{\mathbf{E}} - \omega^2\tilde{\mu}\tilde{\epsilon}\tilde{\mathbf{E}} = -\nabla \times \tilde{\mathbf{J}}_m - j\omega\tilde{\mu}\tilde{\mathbf{J}}^i - j\omega\tilde{\mu}\tilde{\sigma}\tilde{\mathbf{E}},$$

which, through the use of $\nabla \cdot \tilde{\mathbf{E}} = \tilde{\rho}/\tilde{\epsilon}$, can be simplified to

$$(\nabla^2 + k^2)\tilde{\mathbf{E}} = \nabla \times \tilde{\mathbf{J}}_m + j\omega\tilde{\mu}\tilde{\mathbf{J}}^i + \frac{1}{\tilde{\epsilon}}\nabla\tilde{\rho}.$$

This is the *vector Helmholtz equation* for $\tilde{\mathbf{E}}$. Here k is the *complex wavenumber* defined through

$$k^2 = \omega^2\tilde{\mu}\tilde{\epsilon} - j\omega\tilde{\mu}\tilde{\sigma} = \omega^2\tilde{\mu}\left(\tilde{\epsilon} + \frac{\tilde{\sigma}}{j\omega}\right) = \omega^2\tilde{\mu}\tilde{\epsilon}^c \tag{4.199}$$

where $\tilde{\epsilon}^c$ is the complex permittivity (4.26).

By (4.198) we have

$$\nabla(\nabla \cdot \tilde{\mathbf{H}}) - \nabla^2 \tilde{\mathbf{H}} - \omega^2 \tilde{\mu}\tilde{\epsilon}\tilde{\mathbf{H}} = \nabla \times \tilde{\mathbf{J}}^i + \nabla \times \tilde{\mathbf{J}}^c - j\omega\tilde{\epsilon}\tilde{\mathbf{J}}_m.$$

Using

$$\nabla \times \tilde{\mathbf{J}}^c = \nabla \times (\tilde{\sigma}\tilde{\mathbf{E}}) = \tilde{\sigma}\nabla \times \tilde{\mathbf{E}} = \tilde{\sigma}(-j\omega\tilde{\mathbf{B}} - \tilde{\mathbf{J}}_m)$$

and $\nabla \cdot \tilde{\mathbf{H}} = \tilde{\rho}_m/\tilde{\mu}$, we get

$$(\nabla^2 + k^2)\tilde{\mathbf{H}} = -\nabla \times \tilde{\mathbf{J}}^i + j\omega\tilde{\epsilon}^c\tilde{\mathbf{J}}_m + \frac{1}{\tilde{\mu}}\nabla\tilde{\rho}_m,$$

which is the vector Helmholtz equation for $\tilde{\mathbf{H}}$.

4.11.2 Field relationships and the wave equation for two-dimensional fields

Many important canonical problems are two-dimensional in nature, with the sources and fields invariant along one direction. Two-dimensional fields have a simple structure compared to three-dimensional fields, and this often permits decomposition into even simpler field structures.

Consider a homogeneous region of space characterized by permittivity $\tilde{\epsilon}$, permeability $\tilde{\mu}$, and conductivity $\tilde{\sigma}$. We assume that all sources and fields are z-invariant, and wish to find the relationship between the various components of the frequency-domain fields in a source-free region. It is useful to define the transverse vector component of an arbitrary vector $\mathbf{A}$ as the component of $\mathbf{A}$ perpendicular to the axis of invariance:

$$\mathbf{A}_t = \mathbf{A} - \hat{\mathbf{z}}(\hat{\mathbf{z}} \cdot \mathbf{A}).$$

For the position vector $\mathbf{r}$, this component is the transverse position vector $\mathbf{r}_t = \boldsymbol{\rho}$. For instance, we have $\boldsymbol{\rho} = \hat{\mathbf{x}}x + \hat{\mathbf{y}}y$ and $\boldsymbol{\rho} = \hat{\rho}\rho$ in the rectangular and cylindrical coordinate systems, respectively.

Because the region is source-free, $\tilde{\mathbf{E}}$ and $\tilde{\mathbf{H}}$ obey the homogeneous Helmholtz equations

$$(\nabla^2 + k^2)\left\{\begin{matrix}\tilde{\mathbf{E}} \\ \tilde{\mathbf{H}}\end{matrix}\right\} = 0.$$

Writing the fields in terms of rectangular components, we find that each component must obey a homogeneous scalar Helmholtz equation. In particular, the *axial components* $\tilde{E}_z$ and $\tilde{H}_z$ satisfy

$$(\nabla^2 + k^2)\left\{\begin{matrix}\tilde{E}_z \\ \tilde{H}_z\end{matrix}\right\} = 0.$$

But since the fields are independent of z we may also write

$$(\nabla_t^2 + k^2)\left\{\begin{matrix}\tilde{E}_z \\ \tilde{H}_z\end{matrix}\right\} = 0 \tag{4.200}$$

where ∇_t^2 is the transverse Laplacian operator

$$\nabla_t^2 = \nabla^2 - \hat{\mathbf{z}}\frac{\partial^2}{\partial z^2}. \tag{4.201}$$

In rectangular coordinates,

$$\nabla_t^2 = \frac{\partial^2}{\partial x^2} + \frac{\partial^2}{\partial y^2},$$

while in circular cylindrical coordinates,

$$\nabla_t^2 = \frac{\partial^2}{\partial \rho^2} + \frac{1}{\rho}\frac{\partial}{\partial \rho} + \frac{1}{\rho^2}\frac{\partial^2}{\partial \phi^2}. \tag{4.202}$$

With our condition on z-independence we can relate the transverse fields $\tilde{\mathbf{E}}_t$ and $\tilde{\mathbf{H}}_t$ to $\tilde{E}_z$ and $\tilde{H}_z$. By Faraday's law,

$$\nabla \times \tilde{\mathbf{E}}(\boldsymbol{\rho}, \omega) = -j\omega\tilde{\mu}\tilde{\mathbf{H}}(\boldsymbol{\rho}, \omega)$$

and thus

$$\tilde{\mathbf{H}}_t = -\frac{1}{j\omega\tilde{\mu}}\left[\nabla \times \tilde{\mathbf{E}}\right]_t.$$

The transverse portion of the curl is merely

$$\left[\nabla \times \tilde{\mathbf{E}}\right]_t = \hat{\mathbf{x}}\left(\frac{\partial \tilde{E}_z}{\partial y} - \frac{\partial \tilde{E}_y}{\partial z}\right) + \hat{\mathbf{y}}\left(\frac{\partial \tilde{E}_x}{\partial z} - \frac{\partial \tilde{E}_z}{\partial x}\right) = -\hat{\mathbf{z}} \times \left(\hat{\mathbf{x}}\frac{\partial \tilde{E}_z}{\partial x} + \hat{\mathbf{y}}\frac{\partial \tilde{E}_z}{\partial y}\right)$$

since the derivatives with respect to z vanish. The rightmost term in brackets is the transverse gradient of $\tilde{E}_z$, where the transverse gradient operator is

$$\nabla_t = \nabla - \hat{\mathbf{z}}\frac{\partial}{\partial z}.$$

In circular cylindrical coordinates, this operator becomes

$$\nabla_t = \hat{\boldsymbol{\rho}}\frac{\partial}{\partial \rho} + \hat{\boldsymbol{\phi}}\frac{1}{\rho}\frac{\partial}{\partial \phi}. \tag{4.203}$$

Thus

$$\tilde{\mathbf{H}}_t(\boldsymbol{\rho}, \omega) = \frac{1}{j\omega\tilde{\mu}}\hat{\mathbf{z}} \times \nabla_t\tilde{E}_z(\boldsymbol{\rho}, \omega).$$

Similarly, the source-free Ampere's law yields

$$\tilde{\mathbf{E}}_t(\boldsymbol{\rho}, \omega) = -\frac{1}{j\omega\tilde{\epsilon}^c}\hat{\mathbf{z}} \times \nabla_t\tilde{H}_z(\boldsymbol{\rho}, \omega).$$

These results suggest that we can solve a two-dimensional problem by superposition. We first consider the case where $\tilde{E}_z \neq 0$ and $\tilde{H}_z = 0$, called *electric polarization*. This case is also called *TM* or *transverse magnetic* polarization because the magnetic field is transverse to the z-direction (TM_z). We have

$$(\nabla_t^2 + k^2)\tilde{E}_z = 0, \qquad \tilde{\mathbf{H}}_t(\boldsymbol{\rho}, \omega) = \frac{1}{j\omega\tilde{\mu}}\hat{\mathbf{z}} \times \nabla_t\tilde{E}_z(\boldsymbol{\rho}, \omega). \tag{4.204}$$

Once we have solved the Helmholtz equation for $\tilde{E}_z$, the remaining field components follow by differentiation. We next consider the case where $\tilde{H}_z \neq 0$ and $\tilde{E}_z = 0$. This is the case of *magnetic polarization*, also called *TE* or *transverse electric* polarization (TE_z). Here

$$(\nabla_t^2 + k^2)\tilde{H}_z = 0, \qquad \tilde{\mathbf{E}}_t(\boldsymbol{\rho}, \omega) = -\frac{1}{j\omega\tilde{\epsilon}^c}\hat{\mathbf{z}} \times \nabla_t\tilde{H}_z(\boldsymbol{\rho}, \omega). \tag{4.205}$$

A problem involving both $\tilde{E}_z$ and $\tilde{H}_z$ is solved by adding the results for the individual TE$_z$ and TM$_z$ cases.

Note that we can obtain the TE field expressions from the TM field expressions, and vice versa, using duality. For instance, knowing that the TM fields obey (4.204) we may replace $\tilde{\mathbf{H}}_t$ with $\tilde{\mathbf{E}}_t/\eta$ and $\tilde{E}_z$ with $-\eta\tilde{H}_z$ to obtain

$$\frac{\tilde{\mathbf{E}}_t(\boldsymbol{\rho},\omega)}{\eta} = \frac{1}{j\omega\tilde{\mu}}\hat{\mathbf{z}} \times \nabla_t[-\eta\tilde{H}_z(\boldsymbol{\rho},\omega)],$$

which reproduces (4.205).

4.11.3 Plane waves in a homogeneous, isotropic, lossy material

4.11.3.1 The plane-wave field

In later sections we will solve the frequency-domain wave equation with an arbitrary source distribution. At this point we are more interested in the general behavior of EM waves in the frequency domain, so we seek simple solutions to the homogeneous equation

$$(\nabla^2 + k^2)\tilde{\mathbf{E}}(\mathbf{r},\omega) = 0 \tag{4.206}$$

governing the fields in source-free regions of space. Here $[k(\omega)]^2 = \omega^2\tilde{\mu}(\omega)\tilde{\epsilon}^c(\omega)$. Many plane-wave properties are best understood by considering the behavior of a monochromatic field oscillating at a single frequency $\check{\omega}$. In these cases we merely make the replacements

$$\omega \to \check{\omega}, \qquad \tilde{\mathbf{E}}(\mathbf{r},\omega) \to \check{\mathbf{E}}(\mathbf{r}),$$

and apply the rules developed in § 4.7 for manipulating phasor fields.

For our first solutions we choose those that demonstrate rectangular symmetry. *Plane waves* have planar spatial phase loci. That is, the spatial surfaces over which the phase of the complex frequency-domain field is constant are planes. Solutions of this type may be obtained by separating variables in rectangular coordinates. Writing

$$\tilde{\mathbf{E}}(\mathbf{r},\omega) = \hat{\mathbf{x}}\tilde{E}_x(\mathbf{r},\omega) + \hat{\mathbf{y}}\tilde{E}_y(\mathbf{r},\omega) + \hat{\mathbf{z}}\tilde{E}_z(\mathbf{r},\omega)$$

we find that (4.206) reduces to three scalar equations of the form

$$(\nabla^2 + k^2)\tilde{\psi}(\mathbf{r},\omega) = 0$$

where $\tilde{\psi}$ is representative of $\tilde{E}_x$, $\tilde{E}_y$, and $\tilde{E}_z$. This is the homogeneous scalar Helmholtz equation. Product solutions to this equation are considered in § A.5.3. In rectangular coordinates,

$$\tilde{\psi}(\mathbf{r},\omega) = X(x,\omega)Y(y,\omega)Z(z,\omega)$$

where X, Y, and Z are chosen from the list (A.104). Since the exponentials describe propagating wave functions, we choose

$$\tilde{\psi}(\mathbf{r},\omega) = A(\omega)e^{\pm jk_x(\omega)x}e^{\pm jk_y(\omega)y}e^{\pm jk_z(\omega)z}$$

where A is the *amplitude spectrum* of the plane wave and $k_x^2 + k_y^2 + k_z^2 = k^2$. Using this solution to represent each component of $\tilde{\mathbf{E}}$, we have a propagating-wave solution to the homogeneous vector Helmholtz equation:

$$\tilde{\mathbf{E}}(\mathbf{r},\omega) = \tilde{\mathbf{E}}_0(\omega)e^{\pm jk_x(\omega)x}e^{\pm jk_y(\omega)y}e^{\pm jk_z(\omega)z}, \tag{4.207}$$

where $\mathbf{E}_0(\omega)$ is the vector amplitude spectrum. Defining the *wave vector*

$$\mathbf{k}(\omega) = \hat{\mathbf{x}}k_x(\omega) + \hat{\mathbf{y}}k_y(\omega) + \hat{\mathbf{z}}k_z(\omega),$$

we can write (4.207) as

$$\tilde{\mathbf{E}}(\mathbf{r},\omega) = \tilde{\mathbf{E}}_0(\omega)e^{-j\mathbf{k}(\omega)\cdot\mathbf{r}}. \tag{4.208}$$

Note that we adopt a negative sign in the exponential and allow the vector components of $\mathbf{k}$ to be either positive or negative as required by the physical nature of a specific problem. Also note that the magnitude of the wave vector is the wavenumber: $|\mathbf{k}| = k$.

We may always write the wave vector as a sum of real and imaginary vector components

$$\mathbf{k} = \mathbf{k}_r + j\mathbf{k}_i, \tag{4.209}$$

which must obey

$$\mathbf{k}\cdot\mathbf{k} = k^2 = k_r^2 - k_i^2 + 2j\mathbf{k}_r\cdot\mathbf{k}_i. \tag{4.210}$$

When the real and imaginary components are collinear, (4.208) describes a *uniform plane wave* with

$$\mathbf{k} = \hat{\mathbf{k}}(k_r + jk_i).$$

When $\mathbf{k}_r$ and $\mathbf{k}_i$ have different directions, (4.208) describes a *nonuniform plane wave*. We shall find (§ 4.13) that any frequency-domain electromagnetic field in free space may be represented as a continuous superposition of elemental plane-wave components of the type (4.208), but that both uniform and nonuniform terms are required.

4.11.3.2 The TEM nature of a uniform plane wave

Given the plane-wave solution to the wave equation for the electric field, it is straightforward to find the magnetic field. Substitution of (4.208) into Faraday's law gives

$$\nabla \times \left[\tilde{\mathbf{E}}_0(\omega)e^{-j\mathbf{k}(\omega)\cdot\mathbf{r}}\right] = -j\omega\tilde{\mathbf{B}}(\mathbf{r},\omega).$$

Computation of the curl is straightforward and easily done in rectangular coordinates. This and similar derivatives often appear when manipulating plane-wave solutions; see the tabulation in Appendix B. By (B.84) we have

$$\tilde{\mathbf{H}} = \frac{\mathbf{k}\times\tilde{\mathbf{E}}}{\omega\tilde{\mu}}. \tag{4.211}$$

Taking the cross product of this expression with $\mathbf{k}$, we also have

$$\mathbf{k}\times\tilde{\mathbf{H}} = \frac{\mathbf{k}\times(\mathbf{k}\times\tilde{\mathbf{E}})}{\omega\tilde{\mu}} = \frac{\mathbf{k}(\mathbf{k}\cdot\tilde{\mathbf{E}}) - \tilde{\mathbf{E}}(\mathbf{k}\cdot\mathbf{k})}{\omega\tilde{\mu}}. \tag{4.212}$$

We can show that $\mathbf{k}\cdot\tilde{\mathbf{E}} = 0$ by examining Gauss's law and employing (B.83):

$$\nabla\cdot\tilde{\mathbf{E}} = -j\mathbf{k}\cdot\tilde{\mathbf{E}}e^{-j\mathbf{k}\cdot\mathbf{r}} = \tilde{\rho}/\tilde{\epsilon} = 0.$$

Using this and $\mathbf{k}\cdot\mathbf{k} = k^2 = \omega^2\tilde{\mu}\tilde{\epsilon}^c$, we obtain from (4.212)

$$\tilde{\mathbf{E}} = -\frac{\mathbf{k}\times\tilde{\mathbf{H}}}{\omega\tilde{\epsilon}^c}. \tag{4.213}$$

Now, for a uniform plane wave, $\mathbf{k} = \hat{\mathbf{k}}k$, so we can also write (4.211) as

$$\tilde{\mathbf{H}} = \frac{\hat{\mathbf{k}} \times \tilde{\mathbf{E}}}{\eta} = \frac{\hat{\mathbf{k}} \times \tilde{\mathbf{E}}_0}{\eta} e^{-j\mathbf{k}\cdot\mathbf{r}} \qquad (4.214)$$

and (4.213) as

$$\tilde{\mathbf{E}} = -\eta\hat{\mathbf{k}} \times \tilde{\mathbf{H}}.$$

Here

$$\eta = \omega\tilde{\mu}/k = \sqrt{\tilde{\mu}/\tilde{\epsilon}^c}$$

is the complex intrinsic impedance of the medium.

Equations (4.211) and (4.213) show that the electric and magnetic fields and the wave vector are mutually orthogonal; the wave is said to be *transverse electromagnetic* or *TEM* to the direction of propagation.

4.11.3.3 The phase and attenuation constants of a uniform plane wave

For a uniform plane wave we may write

$$\mathbf{k} = k_r\hat{\mathbf{k}} + jk_i\hat{\mathbf{k}} = k\hat{\mathbf{k}} = (\beta - j\alpha)\hat{\mathbf{k}}$$

where $k_r = \beta$ and $k_i = -\alpha$. Here α is the *attenuation constant* and β is the *phase constant*. Since k is defined through (4.199), we have

$$k^2 = (\beta - j\alpha)^2 = \beta^2 - 2j\alpha\beta - \alpha^2 = \omega^2\tilde{\mu}\tilde{\epsilon}^c = \omega^2(\mathrm{Re}\,\tilde{\mu} + j\,\mathrm{Im}\,\tilde{\mu})(\mathrm{Re}\,\tilde{\epsilon}^c + j\,\mathrm{Im}\,\tilde{\epsilon}^c).$$

Equating real and imaginary parts, we have

$$\beta^2 - \alpha^2 = \omega^2[\mathrm{Re}\,\tilde{\mu}\,\mathrm{Re}\,\tilde{\epsilon}^c - \mathrm{Im}\,\tilde{\mu}\,\mathrm{Im}\,\tilde{\epsilon}^c],$$
$$-2\alpha\beta = \omega^2[\mathrm{Im}\,\tilde{\mu}\,\mathrm{Re}\,\tilde{\epsilon}^c + \mathrm{Re}\,\tilde{\mu}\,\mathrm{Im}\,\tilde{\epsilon}^c].$$

We assume the material is passive so that $\mathrm{Im}\,\tilde{\mu} \leq 0$ and $\mathrm{Im}\,\tilde{\epsilon}^c \leq 0$. Letting

$$A = \beta^2 - \alpha^2 = \omega^2[\mathrm{Re}\,\tilde{\mu}\,\mathrm{Re}\,\tilde{\epsilon}^c - \mathrm{Im}\,\tilde{\mu}\,\mathrm{Im}\,\tilde{\epsilon}^c],$$
$$B = 2\alpha\beta = \omega^2[|\,\mathrm{Im}\,\tilde{\mu}|\,\mathrm{Re}\,\tilde{\epsilon}^c + \mathrm{Re}\,\tilde{\mu}|\,\mathrm{Im}\,\tilde{\epsilon}^c|],$$

we may solve simultaneously to get

$$\beta^2 = \tfrac{1}{2}[A + \sqrt{A^2 + B^2}], \qquad \alpha^2 = \tfrac{1}{2}[-A + \sqrt{A^2 + B^2}].$$

Since $A^2 + B^2 = \omega^4(\mathrm{Re}\{\tilde{\epsilon}^c\}^2 + \mathrm{Im}\{\tilde{\epsilon}^c\}^2)(\mathrm{Re}\{\tilde{\mu}\}^2 + \mathrm{Im}\{\tilde{\mu}\}^2)$, we have

$$\beta = \omega\sqrt{\mathrm{Re}\,\tilde{\mu}\,\mathrm{Re}\,\tilde{\epsilon}^c}\sqrt{\frac{1}{2}\left[\sqrt{\left(1 + \tan^2\delta_c\right)\left(1 + \tan^2\delta_\mu\right)} + (1 - \tan\delta_c\tan\delta_\mu)\right]}, \qquad (4.215)$$

$$\alpha = \omega\sqrt{\mathrm{Re}\,\tilde{\mu}\,\mathrm{Re}\,\tilde{\epsilon}^c}\sqrt{\frac{1}{2}\left[\sqrt{\left(1 + \tan^2\delta_c\right)\left(1 + \tan^2\delta_\mu\right)} - (1 - \tan\delta_c\tan\delta_\mu)\right]}, \qquad (4.216)$$

where $\tilde{\epsilon}^c$ and $\tilde{\mu}$ are functions of ω. Here

$$\tan\delta_c = -\frac{\mathrm{Im}\,\tilde{\epsilon}^c}{\mathrm{Re}\,\tilde{\epsilon}^c}, \qquad \tan\delta_\mu = -\frac{\mathrm{Im}\,\tilde{\mu}}{\mathrm{Re}\,\tilde{\mu}}$$

are the electric and magnetic loss tangents, respectively. If $\tilde{\epsilon}(\omega) = \epsilon$, $\tilde{\mu}(\omega) = \mu$, and $\tilde{\sigma}(\omega) = \sigma$ are real and frequency independent, then

$$\alpha = \omega\sqrt{\mu\epsilon}\sqrt{\frac{1}{2}\left[\sqrt{1 + \left(\frac{\sigma}{\omega\epsilon}\right)^2} - 1\right]},$$

$$\beta = \omega\sqrt{\mu\epsilon}\sqrt{\frac{1}{2}\left[\sqrt{1 + \left(\frac{\sigma}{\omega\epsilon}\right)^2} + 1\right]}. \tag{4.217}$$

These values of α and β are valid for $\omega > 0$. For negative frequencies we must be more careful in evaluating the square root in $k = \omega(\tilde{\mu}\tilde{\epsilon}^c)^{1/2}$. Writing

$$\tilde{\mu}(\omega) = |\tilde{\mu}(\omega)|e^{j\xi^\mu(\omega)}, \qquad \tilde{\epsilon}^c(\omega) = |\tilde{\epsilon}^c(\omega)|e^{j\xi^\epsilon(\omega)},$$

we have

$$k(\omega) = \beta(\omega) - j\alpha(\omega) = \omega\sqrt{\tilde{\mu}(\omega)\tilde{\epsilon}^c(\omega)} = \omega\sqrt{|\tilde{\mu}(\omega)||\tilde{\epsilon}^c(\omega)|}e^{j\frac{1}{2}[\xi^\mu(\omega)+\xi^\epsilon(\omega)]}.$$

Now for passive materials we must have, by (4.48), $\operatorname{Im}\tilde{\mu} < 0$ and $\operatorname{Im}\tilde{\epsilon}^c < 0$ for $\omega > 0$. Since $\operatorname{Re}\tilde{\mu} > 0$ and $\operatorname{Re}\tilde{\epsilon}^c > 0$ for $\omega > 0$, we find that $-\pi/2 < \xi^\mu < 0$ and $-\pi/2 < \xi^\epsilon < 0$, and thus $-\pi/2 < (\xi^\mu + \xi^\epsilon)/2 < 0$. Thus we must have $\beta > 0$ and $\alpha > 0$ for $\omega > 0$. For $\omega < 0$ we have by (4.44) and (4.45) that $\operatorname{Im}\tilde{\mu} > 0$, $\operatorname{Im}\tilde{\epsilon}^c > 0$, $\operatorname{Re}\tilde{\mu} > 0$, and $\operatorname{Re}\tilde{\epsilon}^c > 0$. Thus $\pi/2 > (\xi^\mu + \xi^\epsilon)/2 > 0$, and so $\beta < 0$ and $\alpha > 0$ for $\omega < 0$. In summary, $\alpha(\omega)$ is an even function of frequency and $\beta(\omega)$ is an odd function of frequency:

$$\beta(\omega) = -\beta(-\omega), \qquad \alpha(\omega) = \alpha(-\omega), \tag{4.218}$$

where $\beta(\omega) > 0, \alpha(\omega) > 0$ when $\omega > 0$. From this we find a condition on $\tilde{\mathbf{E}}_0$ in (4.208). Since by (4.47) we must have $\tilde{\mathbf{E}}(\omega) = \tilde{\mathbf{E}}^*(-\omega)$, the uniform plane-wave field obeys

$$\tilde{\mathbf{E}}_0(\omega)e^{[-j\beta(\omega)-\alpha(\omega)]\hat{\mathbf{k}}\cdot\mathbf{r}} = \tilde{\mathbf{E}}_0^*(-\omega)e^{[+j\beta(-\omega)-\alpha(-\omega)]\hat{\mathbf{k}}\cdot\mathbf{r}}$$

or

$$\tilde{\mathbf{E}}_0(\omega) = \tilde{\mathbf{E}}_0^*(-\omega),$$

since $\beta(-\omega) = -\beta(\omega)$ and $\alpha(-\omega) = \alpha(\omega)$.

4.11.3.4   Propagation of a uniform plane wave: group and phase velocities

We have derived the plane-wave solution to the wave equation in the frequency domain, but can discover the wave nature of the solution only by examining its behavior in the time domain. Unfortunately, the explicit form of the time-domain field is highly dependent on the frequency behavior of the constitutive parameters. Even the simplest case in which ϵ, μ, and σ are frequency independent is quite complicated (§ 2.10.6). To overcome this difficulty, it is helpful to examine the behavior of a narrowband (but non-monochromatic) signal in a lossy medium with arbitrary constitutive parameters. We will find that the time-domain wave field propagates as a disturbance through the surrounding medium with a velocity determined by the constitutive parameters of the medium. The temporal wave shape does not change as the wave propagates, but the wave amplitude attenuates at a rate dependent on the constitutive parameters.

For simplicity we assume a linearly polarized plane wave (§ 4.11.4.3) with

$$\tilde{\mathbf{E}}(\mathbf{r}, \omega) = \hat{\mathbf{e}}\tilde{E}_0(\omega)e^{-j\mathbf{k}(\omega)\cdot\mathbf{r}}. \tag{4.219}$$

Here $\tilde{E}_0(\omega)$ is the spectrum of the temporal dependence of the wave. For the temporal dependence we choose the narrowband signal

$$E_0(t) = E_0 f(t) \cos \omega_0 t$$

where $f(t)$ has a narrowband spectrum centered on $\omega = 0$ (and is therefore called a *baseband signal*). An appropriate choice for $f(t)$ is the Gaussian function used in (4.52):

$$f(t) = e^{-a^2 t^2} \leftrightarrow \tilde{F}(\omega) = \sqrt{\frac{\pi}{a^2}} e^{-\frac{\omega^2}{4a^2}},$$

producing

$$E_0(t) = E_0 e^{-a^2 t^2} \cos \omega_0 t. \tag{4.220}$$

We regard $f(t)$ as *modulating* the single-frequency cosine *carrier wave*, thus providing the *envelope*. A large value of a yields a narrowband signal having spectrum centered at $\pm \omega_0$. Later we shall let $a \to 0$, driving the width of $f(t)$ to infinity and producing a monochromatic waveform.

By (4.1) we have

$$\tilde{E}_0(\omega) = \tfrac{1}{2} E_0 \left[\tilde{F}(\omega - \omega_0) + \tilde{F}(\omega + \omega_0) \right]$$

where $f(t) \leftrightarrow \tilde{F}(\omega)$. Figure 4.2 shows a plot of this spectrum. We see that the narrowband signal is centered at $\omega = \pm \omega_0$. Substituting into (4.219) and using $\mathbf{k} = (\beta - j\alpha)\hat{\mathbf{k}}$ for a uniform plane wave, we have the frequency-domain field

$$\tilde{\mathbf{E}}(\mathbf{r}, \omega) = \hat{\mathbf{e}} \tfrac{1}{2} E_0 \left[\tilde{F}(\omega - \omega_0) e^{-j[\beta(\omega) - j\alpha(\omega)]\hat{\mathbf{k}} \cdot \mathbf{r}} + \tilde{F}(\omega + \omega_0) e^{-j[\beta(\omega) - j\alpha(\omega)]\hat{\mathbf{k}} \cdot \mathbf{r}} \right].$$

The field at time t and position $\mathbf{r}$ can now be found by inversion:

$$\hat{\mathbf{e}} E(\mathbf{r}, t) = \frac{1}{2\pi} \int_{-\infty}^{\infty} \hat{\mathbf{e}} \tfrac{1}{2} E_0 \left[\tilde{F}(\omega - \omega_0) e^{-j[\beta(\omega) - j\alpha(\omega)]\hat{\mathbf{k}} \cdot \mathbf{r}} \right.$$
$$\left. + \tilde{F}(\omega + \omega_0) e^{-j[\beta(\omega) - j\alpha(\omega)]\hat{\mathbf{k}} \cdot \mathbf{r}} \right] e^{j\omega t} \, d\omega. \tag{4.221}$$

Assuming $\beta(\omega)$ and $\alpha(\omega)$ vary slowly within the band occupied by $\tilde{E}_0(\omega)$, we can expand β and α near $\omega = \omega_0$ as

$$\beta(\omega) = \beta(\omega_0) + \beta'(\omega_0)(\omega - \omega_0) + \tfrac{1}{2}\beta''(\omega_0)(\omega - \omega_0)^2 + \cdots,$$
$$\alpha(\omega) = \alpha(\omega_0) + \alpha'(\omega_0)(\omega - \omega_0) + \tfrac{1}{2}\alpha''(\omega_0)(\omega - \omega_0)^2 + \cdots,$$

where $\beta'(\omega) = d\beta(\omega)/d\omega$, $\beta''(\omega) = d^2\beta(\omega)/d\omega^2$, and so on. We can also expand β and α near $\omega = -\omega_0$:

$$\beta(\omega) = \beta(-\omega_0) + \beta'(-\omega_0)(\omega + \omega_0) + \tfrac{1}{2}\beta''(-\omega_0)(\omega + \omega_0)^2 + \cdots,$$
$$\alpha(\omega) = \alpha(-\omega_0) + \alpha'(-\omega_0)(\omega + \omega_0) + \tfrac{1}{2}\alpha''(-\omega_0)(\omega + \omega_0)^2 + \cdots.$$

Since we are most interested in the propagation velocity, we need not approximate α with great accuracy, hence use $\alpha(\omega) \approx \alpha(\pm \omega_0)$ within the narrow band. We must consider β to greater accuracy to uncover the propagating nature of the wave, and thus use

$$\beta(\omega) \approx \beta(\omega_0) + \beta'(\omega_0)(\omega - \omega_0) \qquad (\omega \approx \omega_0), \tag{4.222}$$
$$\beta(\omega) \approx \beta(-\omega_0) + \beta'(-\omega_0)(\omega + \omega_0) \qquad (\omega \approx -\omega_0).$$

Substitution into (4.221) gives

$$\hat{\mathbf{e}}E(\mathbf{r},t) = \frac{1}{2\pi}\int_{-\infty}^{\infty}\hat{\mathbf{e}}\tfrac{1}{2}E_0\left[\tilde{F}(\omega-\omega_0)e^{-j[\beta(\omega_0)+\beta'(\omega_0)(\omega-\omega_0)]\hat{\mathbf{k}}\cdot\mathbf{r}}e^{-[\alpha(\omega_0)]\hat{\mathbf{k}}\cdot\mathbf{r}}\right.$$
$$\left.+\ \tilde{F}(\omega+\omega_0)e^{-j[\beta(-\omega_0)+\beta'(-\omega_0)(\omega+\omega_0)]\hat{\mathbf{k}}\cdot\mathbf{r}}e^{-[\alpha(-\omega_0)]\hat{\mathbf{k}}\cdot\mathbf{r}}\right]e^{j\omega t}\,d\omega. \qquad (4.223)$$

By (4.218) we know that α is even in ω and β is odd in ω. Since the derivative of an odd function is an even function, we also know that β' is even in ω. We therefore write (4.223) as

$$\hat{\mathbf{e}}E(\mathbf{r},t) = \hat{\mathbf{e}}E_0e^{-\alpha(\omega_0)\hat{\mathbf{k}}\cdot\mathbf{r}}\frac{1}{2\pi}\int_{-\infty}^{\infty}\frac{1}{2}\left[\tilde{F}(\omega-\omega_0)e^{-j\beta(\omega_0)\hat{\mathbf{k}}\cdot\mathbf{r}}e^{-j\beta'(\omega_0)(\omega-\omega_0)\hat{\mathbf{k}}\cdot\mathbf{r}}\right.$$
$$\left.+\ \tilde{F}(\omega+\omega_0)e^{j\beta(\omega_0)\hat{\mathbf{k}}\cdot\mathbf{r}}e^{-j\beta'(\omega_0)(\omega+\omega_0)\hat{\mathbf{k}}\cdot\mathbf{r}}\right]e^{j\omega t}\,d\omega.$$

Multiplying and dividing by $e^{j\omega_0 t}$ and rearranging, we have

$$\hat{\mathbf{e}}E(\mathbf{r},t) = \hat{\mathbf{e}}E_0e^{-\alpha(\omega_0)\hat{\mathbf{k}}\cdot\mathbf{r}}\frac{1}{2\pi}\int_{-\infty}^{\infty}\frac{1}{2}\left[\tilde{F}(\omega-\omega_0)e^{j\phi}e^{j(\omega-\omega_0)[t-\tau]}\right.$$
$$\left.+\ \tilde{F}(\omega+\omega_0)e^{-j\phi}e^{j(\omega+\omega_0)[t-\tau]}\right]d\omega$$

where

$$\phi = \omega_0 t - \beta(\omega_0)\hat{\mathbf{k}}\cdot\mathbf{r}, \qquad \tau = \beta'(\omega_0)\hat{\mathbf{k}}\cdot\mathbf{r}.$$

Setting $u = \omega - \omega_0$ in the first term and $u = \omega + \omega_0$ in the second term, we have

$$\hat{\mathbf{e}}E(\mathbf{r},t) = \hat{\mathbf{e}}E_0e^{-\alpha(\omega_0)\hat{\mathbf{k}}\cdot\mathbf{r}}\cos\phi\frac{1}{2\pi}\int_{-\infty}^{\infty}\tilde{F}(u)e^{ju(t-\tau)}\,du.$$

Finally, the time-shifting theorem (A.3) gives us the time-domain wave field

$$\hat{\mathbf{e}}E(\mathbf{r},t) = \hat{\mathbf{e}}E_0e^{-\alpha(\omega_0)\hat{\mathbf{k}}\cdot\mathbf{r}}\cos\left(\omega_0\left[t-\hat{\mathbf{k}}\cdot\mathbf{r}/v_p(\omega_0)\right]\right)f\left(t-\hat{\mathbf{k}}\cdot\mathbf{r}/v_g(\omega_0)\right) \qquad (4.224)$$

where

$$v_g(\omega) = d\omega/d\beta = [d\beta/d\omega]^{-1} \qquad (4.225)$$

is the *group velocity*, and

$$v_p(\omega) = \omega/\beta$$

is the *phase velocity*.

To interpret (4.224), we note that at any given time t the field is constant over the surface described by

$$\hat{\mathbf{k}}\cdot\mathbf{r} = C \qquad (4.226)$$

where C is some constant. This surface is a plane (Figure 4.16) with normal $\hat{\mathbf{k}}$. It is easily verified that any point $\mathbf{r}$ on this plane satisfies (4.226). Let $\mathbf{r}_0 = r_0\hat{\mathbf{k}}$ describe the point on the plane with position vector in the direction of $\hat{\mathbf{k}}$, and let $\mathbf{d}$ be a displacement from this point to any other point on the plane. Since $\hat{\mathbf{k}}\cdot\mathbf{r} = \hat{\mathbf{k}}\cdot(\mathbf{r}_0+\mathbf{d}) = r_0(\hat{\mathbf{k}}\cdot\hat{\mathbf{k}})+\hat{\mathbf{k}}\cdot\mathbf{d}$ and $\hat{\mathbf{k}}\cdot\mathbf{d} = 0$, we have

$$\hat{\mathbf{k}}\cdot\mathbf{r} = r_0,$$

which is a fixed distance, so (4.226) holds.

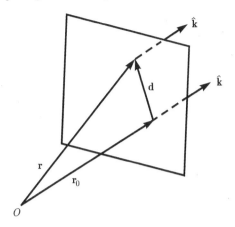

FIGURE 4.16
Surface of constant $\hat{\mathbf{k}} \cdot \mathbf{r}$.

Let us identify the plane over which the envelope f takes a certain value, and follow its motion as time progresses. The value of r_0 associated with this plane must increase with increasing time in such a way that the argument of f remains constant:

$$t - r_0/v_g(\omega_0) = C.$$

Differentiation gives
$$dr_0/dt = v_g = d\omega/d\beta.$$

So the envelope propagates along $\hat{\mathbf{k}}$ at the group velocity v_g. Associated with this propagation is an *attenuation* described by the factor $e^{-\alpha(\omega_0)\hat{\mathbf{k}} \cdot \mathbf{r}}$. This accounts for energy transfer into the lossy medium through Joule heating.

Similarly, we can identify a plane over which the phase of the carrier is constant; this will be parallel to the plane of constant envelope described above. We now set

$$\omega_0[t - \hat{\mathbf{k}} \cdot \mathbf{r}/v_p(\omega_0)] = C$$

and differentiate to get
$$dr_0/dt = v_p = \omega/\beta. \tag{4.227}$$

Surfaces of constant carrier phase propagate along $\hat{\mathbf{k}}$ at velocity v_p.

Caution is required to interpret the velocities v_g and v_p; in particular, we must be careful not to associate propagation speeds of energy or information with v_p. Since envelope propagation represents the actual progression of the disturbance, v_g has the recognizable physical meaning of energy velocity. Kraus and Fleisch [111] suggest we imagine a strolling caterpillar: the speed (v_p) of the undulations along the caterpillar's back (representing the carrier wave) may be much faster than the speed (v_g) of the caterpillar's body (representing the envelope of the disturbance).

In fact, v_g is the speed of energy propagation even for a monochromatic wave (§ 4.11.4). However, for purely monochromatic waves, v_g cannot be identified from the time-domain field, whereas v_p can. This leads to some unfortunate misconceptions, especially when v_p exceeds the speed of light. Since v_p is not the velocity of propagation of a physical quantity, but is rather the rate of change of a phase reference point, Einstein's postulate of c as the limiting velocity is not violated.

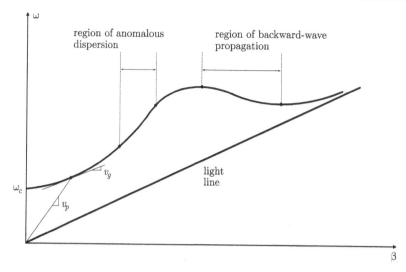

FIGURE 4.17

An ω–β diagram for a fictitious material.

We can obtain interesting relationships between v_p and v_g by manipulating (4.225) and (4.227). For instance, computing

$$\frac{dv_p}{d\omega} = \frac{d}{d\omega}\left(\frac{\omega}{\beta}\right) = \frac{\beta - \omega\frac{d\beta}{d\omega}}{\beta^2}$$

we find that

$$\frac{v_p}{v_g} = 1 - \beta\frac{dv_p}{d\omega}. \tag{4.228}$$

Hence in frequency ranges where v_p decreases with increasing frequency, we have $v_g < v_p$. These are known as regions of *normal dispersion*. In ranges where v_p increases with frequency, we have $v_g > v_p$. These are regions of *anomalous dispersion*. As mentioned in § 4.6.3, the word "anomalous" does not imply that this type of dispersion is unusual.

The propagation of a uniform plane wave through a lossless medium provides a particularly simple example. With

$$\beta(\omega) = \omega\sqrt{\mu\epsilon}, \qquad \alpha(\omega) = 0,$$

relation (4.222) yields $\beta(\omega) = \omega_0\sqrt{\mu\epsilon} + \sqrt{\mu\epsilon}(\omega - \omega_0) = \omega\sqrt{\mu\epsilon}$ and (4.224) becomes

$$\hat{\mathbf{e}}E(\mathbf{r}, t) = \hat{\mathbf{e}}E_0 \cos\left(\omega_0\left[t - \hat{\mathbf{k}}\cdot\mathbf{r}/v_p(\omega_0)\right]\right) f\left(t - \hat{\mathbf{k}}\cdot\mathbf{r}/v_g(\omega_0)\right).$$

Since the linear approximation to the phase constant β is in this case exact, the wave packet truly propagates without distortion, with a group velocity identical to the phase velocity:

$$v_g = \left[\frac{d}{d\omega}\omega\sqrt{\mu\epsilon}\right]^{-1} = \frac{1}{\sqrt{\mu\epsilon}} = \frac{\omega}{\beta} = v_p.$$

4.11.3.5 The ω–β diagram

A plot of ω vs. $\beta(\omega)$ can advantageously display the dispersive properties of a material. Figure 4.17 shows such an ω–β plot, or *dispersion diagram*, for a fictitious material. The slope of the line from the origin (β, ω) is the phase velocity, while the slope of the line tangent to the curve at that point is the group velocity. This plot shows many characteristics of electromagnetic waves (but not necessarily of plane waves). For instance, there may be a minimum frequency ω_c called the *cutoff frequency* at which $\beta = 0$ and below which the wave cannot propagate. This behavior is characteristic of a plane wave propagating in a plasma (as shown below) or of a wave in a hollow pipe waveguide (§ 5.6). Over most values of β we have $v_g < v_p$, so the material demonstrates normal dispersion. However, over a small region we do have anomalous dispersion. In another range the slope of the curve is actually negative and thus $v_g < 0$; here the directions of energy and phase-front propagation are opposite. Such *backward waves* are encountered in certain guided-wave structures used in microwave oscillators. The ω–β plot also includes the *light line* as a reference curve. For all points on this line, $v_g = v_p$; it is generally used to represent propagation within the material under special circumstances, such as when the loss is zero or the material occupies unbounded space. It may also be used to represent propagation within a vacuum.

4.11.3.6 Examples of plane-wave propagation in dispersive media

We consider two examples based on the material properties studied in § 4.6. By examining the properties of plane waves propagating through water and through a nonmagnetized plasma, we gain considerable insight into the behavior of waves in dispersive isotropic media.

▶ **Example 4.12:** Plane-wave propagation in fresh water

Consider a plane wave propagating in fresh water with no dc conductivity. Assuming the Debye parameters for water are $\epsilon_\infty = 5\epsilon_0$, $\epsilon_s = 78.3\epsilon_0$, and $\tau = 9.6 \times 10^{-12}$ s [45], construct the ω–β plot. Plot the attenuation and both the phase and group velocities as functions of frequency.

Solution: By the Debye formula (4.115) we have

$$\tilde{\epsilon}(\omega) = \epsilon_\infty + \frac{\epsilon_s - \epsilon_\infty}{1 + j\omega\tau}.$$

Using the specified Debye parameters, we obtain the relaxation spectrum shown in Figure 4.11. Assuming $\mu = \mu_0$, we may compute β and α as a function of ω from

$$k = \omega\sqrt{\mu_0\tilde{\epsilon}} = \beta - j\alpha.$$

The ω–β plot is shown in Figure 4.18. Since $\mathrm{Re}\,\tilde{\epsilon}$ varies with frequency, we show both the light line for zero frequency found using $\epsilon_s = 78.3\epsilon_0$, and the light line for infinite frequency found using $\epsilon_\infty = 5\epsilon_0$. Over low frequencies the dispersion curve follows the low-frequency light line very closely, and thus $v_p \approx v_g \approx c/\sqrt{78.3}$. As frequency increases, the dispersion curve rises and eventually becomes asymptotic with the high-frequency light line. Plots of v_p and v_g shown in Figure 4.19 verify that the velocities start at $c/\sqrt{78.3} = 0.113c$ for low frequencies, and approach $c/\sqrt{5} = 0.447c$ for high frequencies. Because $v_g > v_p$ at all frequencies, this model of water demonstrates anomalous dispersion.

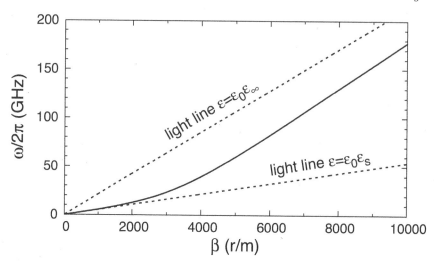

FIGURE 4.18
Dispersion plot for water computed using the Debye relaxation formula.

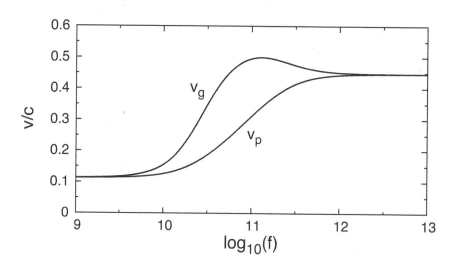

FIGURE 4.19
Phase and group velocities for water computed using the Debye relaxation formula.

The attenuation constant appears in Figure 4.20 as a function of frequency. We see that fresh water has a fairly high attenuation constant for frequencies above several hundred MHz. However, the attenuation constant is far smaller than that for sea water, which has a typical dc conductivity of 4 S/m.

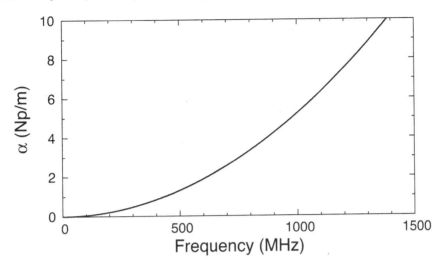

FIGURE 4.20
Attenuation constant for water computed using the Debye relaxation formula. ◄

► **Example 4.13:** Plane-wave propagation in a nonmagnetized plasma

Consider a plane wave propagating in a nonmagnetized plasma. Derive expressions for α and β when there are no collisions and when the collision frequency is small. In particular, consider a plane wave propagating in the earth's ionosphere. Construct the $\omega-\beta$ diagram and plot the phase and group velocities.

Solution: For a collisionless plasma, we may set $\nu = 0$ in (4.76) to find

$$k = \begin{cases} \frac{\omega}{c}\sqrt{1 - \frac{\omega_p^2}{\omega^2}}, & \omega > \omega_p, \\ -j\frac{\omega}{c}\sqrt{\frac{\omega_p^2}{\omega^2} - 1}, & \omega < \omega_p. \end{cases}$$

Thus, when $\omega > \omega_p$ we have $\tilde{\mathbf{E}}(\mathbf{r}, \omega) = \tilde{E}_0(\omega)e^{-j\beta(\omega)\hat{\mathbf{k}}\cdot\mathbf{r}}$ and so

$$\beta = \frac{\omega}{c}\sqrt{1 - \frac{\omega_p^2}{\omega^2}}, \qquad \alpha = 0.$$

In this case a plane wave propagates through the plasma without attenuation. However, when $\omega < \omega_p$ we have $\tilde{\mathbf{E}}(\mathbf{r}, \omega) = \tilde{E}_0(\omega)e^{-\alpha(\omega)\hat{\mathbf{k}}\cdot\mathbf{r}}$ with

$$\alpha = \frac{\omega}{c}\sqrt{\frac{\omega_p^2}{\omega^2} - 1}, \qquad \beta = 0,$$

and a plane wave does not propagate, but only attenuates. Such a wave is said to be *evanescent*. We say that for frequencies below ω_p the wave is *cut off*, and call ω_p the *cutoff frequency*.
   A formula for the phase velocity of a plane wave in a lossless plasma is easily derived:

$$v_p = \frac{\omega}{\beta} = \frac{c}{\sqrt{1 - \frac{\omega_p^2}{\omega^2}}} > c.$$

Similarly,

$$v_g = \left(\frac{d\beta}{d\omega}\right)^{-1}$$

$$= \left(\frac{1}{c}\sqrt{1 - \frac{\omega_p^2}{\omega^2}} + \frac{1}{c}\frac{\omega_p^2/\omega^2}{\sqrt{1 - \frac{\omega_p^2}{\omega^2}}}\right)^{-1}$$

$$= c\sqrt{1 - \frac{\omega_p^2}{\omega^2}} < c.$$

Interestingly, we find that in this case of an unmagnetized collisionless plasma,

$$v_p v_g = c^2.$$

Since $v_p > v_g$, this model of a plasma demonstrates normal dispersion at all frequencies above cutoff.

For the case of a plasma with collisions we retain ν in (4.76) and find that

$$k = \frac{\omega}{c}\sqrt{\left(1 - \frac{\omega_p^2}{\omega^2 + \nu^2}\right) - j\nu\frac{\omega_p^2}{\omega(\omega^2 + \nu^2)}}.$$

When $\nu \neq 0$, a true cutoff effect is absent and the wave may propagate at all frequencies. However, when $\nu \ll \omega_p$ the attenuation for propagating waves of frequency $\omega < \omega_p$ is quite severe, and for practical purposes the wave is cut off. For waves of frequency $\omega > \omega_p$ there is attenuation. Assuming that $\nu \ll \omega_p$ and that $\nu \ll \omega$, we may approximate the square root with the leading terms of a binomial expansion and find that to first order

$$\beta = \frac{\omega}{c}\sqrt{1 - \frac{\omega_p^2}{\omega^2}},$$

$$\alpha = \frac{1}{2}\frac{\nu}{c}\frac{\omega_p^2/\omega^2}{\sqrt{1 - \frac{\omega_p^2}{\omega^2}}}.$$

Hence the phase and group velocities above cutoff are essentially those of a lossless plasma, while the attenuation constant is directly proportional to ν.

As a specific example, consider a plane wave propagating in the earth's ionosphere. Both the electron density and the collision frequency depend on such factors as altitude, time of day, and latitude. However, except at the very lowest altitudes, the collision frequency is low enough that the ionosphere may be considered lossless. For instance, at a height of 200 km (the F_1 layer of the ionosphere), as measured for a mid-latitude region, we find that during the day the electron density is approximately $N_e = 2 \times 10^{11}$ m^{-3}, while the collision frequency is only $\nu = 100$ s^{-1} [20]. The attenuation is so small in this case that the ionosphere may be considered essentially lossless above the cutoff frequency. Figure 4.21 shows the ω–β diagram for the ionosphere assuming $\nu = 0$, along with the light line $v_p = c$. Above the cutoff frequency of

$$f_p = \frac{\omega_p}{2\pi} = 4.0 \text{ MHz},$$

the wave propagates, and $v_g < c$ while $v_p > c$. Below the cutoff frequency the wave does not propagate and the field decays rapidly because α is large.

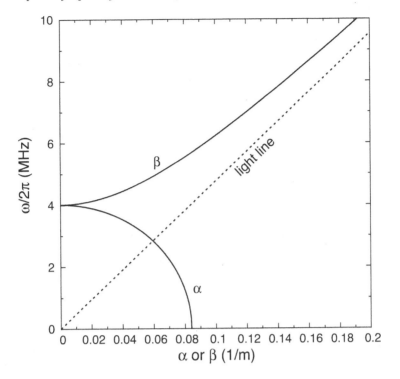

FIGURE 4.21

Dispersion plot for the ionosphere computed using $N_e = 2 \times 10^{11}\,\mathrm{m}^{-3}$, $\nu = 0$. Light line computed using $\epsilon = \epsilon_0$, $\mu = \mu_0$.

Figure 4.22 shows a plot of the phase and group velocities. As the cutoff frequency is approached from above, the group velocity tends to zero while the phase velocity increases without bound. Cutoff behavior also occurs in waveguides (§ 5.6).

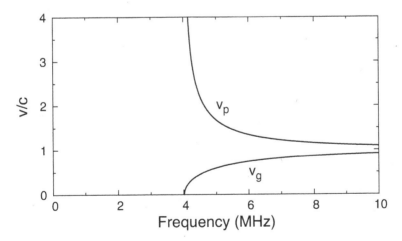

FIGURE 4.22

Phase and group velocities for the ionosphere computed using $N_e = 2 \times 10^{11}\,\mathrm{m}^{-3}$, $\nu = 0$. ◀

4.11.4   Monochromatic plane waves in a lossy medium

Many properties of monochromatic plane waves are particularly simple. In fact, certain properties, such as wavelength, only have meaning for monochromatic fields. And since monochromatic or nearly monochromatic waves are employed extensively in radar, communications, and energy transport, it is worthwhile to examine the results of the preceding section for the special case in which the spectrum of the plane-wave signal consists of a single frequency component. In addition, plane waves of more general time dependence can be viewed as superpositions of individual single-frequency components (through the inverse Fourier transform), and thus we may regard monochromatic waves as building blocks for more complicated plane waves.

We can view the monochromatic field as a specialization of (4.220) for $a \to 0$. This results in $\tilde{F}(\omega) \to \delta(\omega)$, so the linearly polarized plane wave expression (4.221) reduces to

$$\hat{\mathbf{e}} E(\mathbf{r}, t) = \hat{\mathbf{e}} E_0 e^{-\alpha(\omega_0)[\hat{\mathbf{k}} \cdot \mathbf{r}]} \cos(\omega_0 t - j\beta(\omega_0)[\hat{\mathbf{k}} \cdot \mathbf{r}]). \tag{4.229}$$

It is convenient to represent monochromatic fields with frequency $\omega = \check{\omega}$ in phasor form. The phasor form of (4.229) is

$$\check{\mathbf{E}}(\mathbf{r}) = \hat{\mathbf{e}} E_0 e^{-j\beta(\hat{\mathbf{k}} \cdot \mathbf{r})} e^{-\alpha(\hat{\mathbf{k}} \cdot \mathbf{r})} \tag{4.230}$$

where $\beta = \beta(\check{\omega})$ and $\alpha = \alpha(\check{\omega})$. We can identify a surface of constant phase as a locus of points obeying

$$\check{\omega} t - \beta(\hat{\mathbf{k}} \cdot \mathbf{r}) = C_P \tag{4.231}$$

for some constant C_P. This surface is a plane (Figure 4.16) with normal along $\hat{\mathbf{k}}$. It is easily checked that any point $\mathbf{r}$ on this plane satisfies (4.231). Let $\mathbf{r}_0 = r_0 \hat{\mathbf{k}}$ describe the point on the plane with position vector in the $\hat{\mathbf{k}}$ direction, and let $\mathbf{d}$ be a displacement from this point to any other point on the plane. Then $\hat{\mathbf{k}} \cdot \mathbf{r} = \hat{\mathbf{k}} \cdot (\mathbf{r}_0 + \mathbf{d}) = r_0(\hat{\mathbf{k}} \cdot \hat{\mathbf{k}}) + \hat{\mathbf{k}} \cdot \mathbf{d}$ where $\hat{\mathbf{k}} \cdot \mathbf{d} = 0$, so

$$\hat{\mathbf{k}} \cdot \mathbf{r} = r_0, \tag{4.232}$$

which is a spatial constant, hence (4.231) holds for any t. The planar surfaces described by (4.231) are *wavefronts*.

Note that surfaces of constant amplitude are determined by $\alpha(\hat{\mathbf{k}} \cdot \mathbf{r}) = C_A$ for constant C_A. As with the phase term, this requires that $\hat{\mathbf{k}} \cdot \mathbf{r} = $ constant, so surfaces of constant phase and surfaces of constant amplitude are coplanar. This is a property of uniform plane waves. We show later that nonuniform plane waves have planar surfaces that are not parallel.

The cosine term in (4.229) represents a *traveling wave*. As t increases, the argument of the cosine remains unchanged as long as $\hat{\mathbf{k}} \cdot \mathbf{r}$ increases correspondingly. Thus the planar wavefronts propagate along $\hat{\mathbf{k}}$. Simultaneously, the wave is attenuated because of the factor $e^{-\alpha(\hat{\mathbf{k}} \cdot \mathbf{r})}$; this accounts for energy transferred from the propagating wave to the surrounding medium via Joule heating.

4.11.4.1   Phase velocity of a uniform plane wave

The propagation velocity of the progressing wavefront is found by differentiating (4.231):

$$\check{\omega} - \beta\hat{\mathbf{k}} \cdot \frac{d\mathbf{r}}{dt} = 0.$$

By (4.232) we have

$$v_p = dr_0/dt = \check{\omega}/\beta,$$

where the phase velocity v_p represents the propagation speed of the constant-phase surfaces. For a lossy medium with frequency-independent constitutive parameters, (4.217) shows that

$$v_p \leq 1/\sqrt{\mu\epsilon},$$

hence the phase velocity in a conducting medium does not exceed that in a lossless medium with the same parameters μ and ϵ. We cannot draw this conclusion for a medium with frequency-dependent $\tilde{\mu}$ and $\tilde{\epsilon}^c$, since by (4.215) the value of $\breve{\omega}/\beta$ might be greater or less than $1/\sqrt{\operatorname{Re}\tilde{\mu}\operatorname{Re}\tilde{\epsilon}^c}$, depending on the ratios $\operatorname{Im}\tilde{\mu}/\operatorname{Re}\tilde{\mu}$ and $\operatorname{Im}\tilde{\epsilon}^c/\operatorname{Re}\tilde{\epsilon}^c$.

4.11.4.2 Wavelength of a uniform plane wave

Another important property of a uniform plane wave is the distance between adjacent wavefronts that produce the same value of the cosine function in (4.229). Note that the field amplitude may not be the same on these two surfaces because of possible attenuation of the wave. With $\mathbf{r}_1$ and $\mathbf{r}_2$ as two points on adjacent wavefronts, we require $\beta(\hat{\mathbf{k}} \cdot \mathbf{r}_1) = \beta(\hat{\mathbf{k}} \cdot \mathbf{r}_2) - 2\pi$, or

$$\lambda = \hat{\mathbf{k}} \cdot (\mathbf{r}_2 - \mathbf{r}_1) = r_{02} - r_{01} = 2\pi/\beta.$$

We call λ the *wavelength*.

4.11.4.3 Polarization of a uniform plane wave

Plane wave *polarization* describes the temporal evolution of the vector direction of the electric field, which depends on the manner in which the wave is generated. *Completely polarized* waves are produced by antennas or other equipment; these have a deterministic polarization state described completely by three parameters as discussed below. *Randomly polarized* waves are emitted by some natural sources. *Partially polarized* waves, such as those produced by cosmic radio sources, contain both completely polarized and randomly polarized components. We shall concentrate on the description of completely polarized waves.

The polarization ellipse. The polarization state of a completely polarized monochromatic plane wave propagating in a homogeneous, isotropic region may be described by superposing two simpler plane waves that propagate along the same direction, but with different phases and spatially orthogonal electric fields. Without loss of generality we may study propagation along the z-axis and align the orthogonal field directions with $\hat{\mathbf{x}}$ and $\hat{\mathbf{y}}$. So we are interested in the behavior of a wave with electric field

$$\check{\mathbf{E}}(\mathbf{r}) = \hat{\mathbf{x}}E_{x0}e^{j\phi_x}e^{-jkz} + \hat{\mathbf{y}}E_{y0}e^{j\phi_y}e^{-jkz}. \tag{4.233}$$

The time evolution of the direction of $\mathbf{E}$ is examined in the time domain where

$$\mathbf{E}(\mathbf{r},t) = \operatorname{Re}\left\{\check{\mathbf{E}}e^{j\breve{\omega}t}\right\} = \hat{\mathbf{x}}E_{x0}\cos(\breve{\omega}t - kz + \phi_x) + \hat{\mathbf{y}}E_{y0}\cos(\breve{\omega}t - kz + \phi_y)$$

and thus, by the identity $\cos(x+y) \equiv \cos x \cos y - \sin x \sin y$,

$$E_x = E_{x0}\left[\cos(\breve{\omega}t - kz)\cos\phi_x - \sin(\breve{\omega}t - kz)\sin\phi_x\right],$$
$$E_y = E_{y0}\left[\cos(\breve{\omega}t - kz)\cos\phi_y - \sin(\breve{\omega}t - kz)\sin\phi_y\right].$$

The tip of the vector $\mathbf{E}$ moves cyclically in the xy-plane with temporal period $T = \breve{\omega}/2\pi$. Its locus may be found by eliminating the parameter t to obtain a relationship between

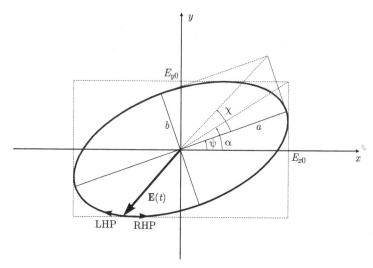

FIGURE 4.23

Polarization ellipse for a monochromatic plane wave.

E_{x0} and E_{y0}. Letting $\delta = \phi_y - \phi_x$ we note that

$$\frac{E_x}{E_{x0}} \sin \phi_y - \frac{E_y}{E_{y0}} \sin \phi_x = \cos(\breve{\omega} t - kz) \sin \delta,$$

$$\frac{E_x}{E_{x0}} \cos \phi_y - \frac{E_y}{E_{y0}} \cos \phi_x = \sin(\breve{\omega} t - kz) \sin \delta;$$

squaring these terms we find that

$$\left(\frac{E_x}{E_{x0}} \right)^2 + \left(\frac{E_y}{E_{y0}} \right)^2 - 2 \frac{E_x}{E_{x0}} \frac{E_y}{E_{y0}} \cos \delta = \sin^2 \delta,$$

which is the equation for the ellipse of Figure 4.23. By (4.214) the magnetic field of the plane wave is

$$\breve{\mathbf{H}} = \frac{\hat{\mathbf{z}} \times \breve{\mathbf{E}}}{\eta},$$

hence its tip also traces an ellipse in the xy-plane.

The tip of the electric vector cycles around the *polarization ellipse* in the xy-plane every T seconds. The sense of rotation is determined by the sign of δ, and is described by the terms *clockwise/counterclockwise* or *right-hand/left-hand*. There is some disagreement about how to do this. We shall adopt the IEEE definitions (IEEE Standard 145-1983 [185]) and associate with $\delta < 0$ rotation in the *right-hand sense*: if the right thumb points in the direction of wave propagation then the fingers curl in the direction of field rotation for increasing time. This is *right-hand polarization (RHP)*. We associate $\delta > 0$ with *left-hand polarization (LHP)*.

The polarization ellipse lies within a rectangle of sides $2E_{x0}$ and $2E_{y0}$, and has its major axis rotated from the x-axis by the *tilt angle* ψ with $0 \leq \psi \leq \pi$. The ratio of E_{y0} to E_{x0} determines an angle α such that $0 \leq \alpha \leq \pi/2$:

$$E_{y0}/E_{x0} = \tan \alpha.$$

The shape of the ellipse is determined by the three parameters E_{x0}, E_{y0}, and δ, while the sense of polarization is described by the sign of δ. These may not, however, be the most convenient parameters for describing wave polarization. We can also inscribe the ellipse within a box measuring $2a$ by $2b$, where a and b are the lengths of the semimajor and semiminor axes. Then b/a determines an angle χ, $-\pi/4 \leq \chi \leq \pi/4$, that is analogous to α:

$$\pm b/a = \tan \chi.$$

Here the algebraic sign of χ is used to indicate the sense of polarization: $\chi > 0$ for LHP, $\chi < 0$ for RHP.

The quantities a, b, ψ can also be used to describe the polarization ellipse. A procedure outlined in Born and Wolf [22] to relate the quantities (a, b, ψ) to (E_{x0}, E_{y0}, δ) yields

$$a^2 + b^2 = E_{x0}^2 + E_{y0}^2,$$

$$\tan 2\psi = (\tan 2\alpha) \cos \delta = \frac{2 E_{x0} E_{y0}}{E_{x0}^2 - E_{y0}^2} \cos \delta,$$

$$\sin 2\chi = (\sin 2\alpha) \sin \delta = \frac{2 E_{x0} E_{y0}}{E_{x0}^2 + E_{y0}^2} \sin \delta.$$

Alternatively, we can describe the polarization ellipse by the angles ψ and χ and one of the amplitudes E_{x0} or E_{y0}.

Stokes parameters. Each of the polarization ellipse parameter sets is somewhat inconvenient, since in each case the units differ among the parameters. In 1852 G. Stokes introduced a system of three independent quantities with identical dimension that can be used to describe plane-wave polarization. Various normalizations of these *Stokes parameters* are employed; when the parameters are chosen to have the dimension of power density we may write them as

$$s_0 = \frac{1}{2\eta} \left[E_{x0}^2 + E_{y0}^2 \right], \tag{4.234}$$

$$s_1 = \frac{1}{2\eta} \left[E_{x0}^2 - E_{y0}^2 \right] = s_0 \cos(2\chi) \cos(2\psi), \tag{4.235}$$

$$s_2 = \frac{1}{\eta} E_{x0} E_{y0} \cos \delta = s_0 \cos(2\chi) \sin(2\psi), \tag{4.236}$$

$$s_3 = \frac{1}{\eta} E_{x0} E_{y0} \sin \delta = s_0 \sin(2\chi). \tag{4.237}$$

Only three of these four parameters are independent, since $s_0^2 = s_1^2 + s_2^2 + s_3^2$. Often the Stokes parameters are designated (I, Q, U, V) rather than (s_0, s_1, s_2, s_3).

Figure 4.24 summarizes various polarization states as a function of the angles ψ and χ. Interesting cases occur when $\chi = 0$ and $\chi = \pm\pi/4$. The value $\chi = 0$ corresponds to $b = 0$ and thus $\delta = 0$. In this case the electric vector traces out a straight line and we call the polarization *linear*. Here

$$\mathbf{E} = (\hat{\mathbf{x}} E_{x0} + \hat{\mathbf{y}} E_{y0}) \cos(\breve{\omega} t - kz + \phi_x).$$

When $\psi = 0$ we have $E_{y0} = 0$ and call this *horizontal linear polarization* (HLP); when $\psi = \pi/2$ we have $E_{x0} = 0$ and *vertical linear polarization* (VLP).

The case $\chi = \pm\pi/4$ corresponds to $b = a$ and $\delta = \pm\pi/2$. Thus $E_{x0} = E_{y0}$, and $\mathbf{E}$ traces out a circle regardless of the value of ψ. If $\chi = -\pi/4$ we have right-hand rotation

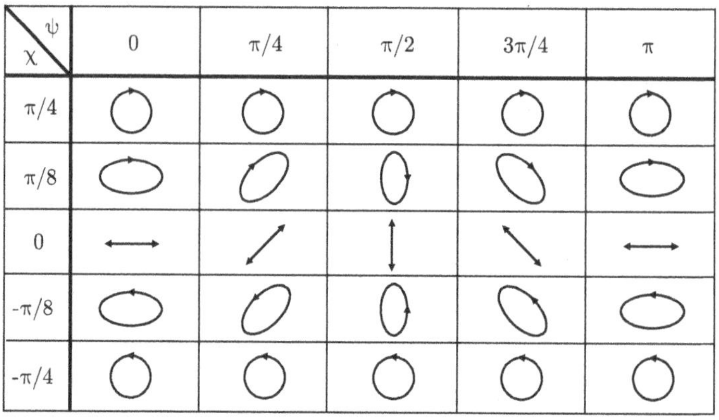

χ \ ψ	0	$\pi/4$	$\pi/2$	$3\pi/4$	π
$\pi/4$					
$\pi/8$					
0					
$-\pi/8$					
$-\pi/4$					

FIGURE 4.24

Polarization states as a function of tilt angle ψ and ellipse aspect ratio angle χ. Left-hand polarization for $\chi > 0$, right-hand for $\chi < 0$.

of **E** and thus refer to *right-hand circular polarization* (RHCP). If $\chi = \pi/4$ we have *left-hand circular polarization* (LHCP). For these cases,

$$\mathbf{E} = E_{x0} \left[\hat{\mathbf{x}} \cos(\breve{\omega}t - kz) \mp \hat{\mathbf{y}} \sin(\breve{\omega}t - kz) \right],$$

where the upper and lower signs correspond to LHCP and RHCP, respectively. All other values of χ result in the general cases of left-hand or right-hand *elliptical polarization*.

The Poincaré sphere. The French mathematician H. Poincaré realized that the Stokes parameters (s_1, s_2, s_3) describe a point on a sphere of radius s_0, and that this *Poincaré sphere* is useful for visualizing the various polarization states. Each state corresponds uniquely to one point on the sphere, and by (4.235)–(4.237) the angles 2χ and 2ψ are the spherical angular coordinates of the point as shown in Figure 4.25. We may therefore map the polarization states shown in Figure 4.24 directly onto the sphere: left- and right-hand polarizations appear in the upper and lower hemispheres, respectively; circular polarization appears at the poles $(2\chi = \pm\pi/2)$; linear polarization appears on the equator $(2\chi = 0)$, with HLP at $2\psi = 0$ and VLP at $2\psi = \pi$. The angles α and δ also have geometrical interpretations on the Poincaré sphere. The spherical angle of the great-circle route between the point of HLP and a point on the sphere is 2α, while the angle between the great-circle path and the equator is δ.

4.11.4.4 Uniform plane waves in a good dielectric

We may base some useful plane-wave approximations on whether the real or imaginary part of $\tilde{\epsilon}^c$ dominates at the frequency of operation. We assume that $\tilde{\mu}(\omega) = \mu$ is independent of frequency and use the notation $\epsilon^c = \tilde{\epsilon}^c(\breve{\omega})$, $\sigma = \tilde{\sigma}(\breve{\omega})$, etc. Remember that

$$\epsilon^c = (\operatorname{Re}\epsilon + j\operatorname{Im}\epsilon) + \frac{\sigma}{j\breve{\omega}} = \operatorname{Re}\epsilon + j\left(\operatorname{Im}\epsilon - \frac{\sigma}{\breve{\omega}}\right) = \operatorname{Re}\epsilon^c + j\operatorname{Im}\epsilon^c.$$

By definition, a "good dielectric" obeys

$$\tan\delta_c = -\frac{\operatorname{Im}\epsilon^c}{\operatorname{Re}\epsilon^c} = \frac{\sigma}{\breve{\omega}\operatorname{Re}\epsilon} - \frac{\operatorname{Im}\epsilon}{\operatorname{Re}\epsilon} \ll 1. \tag{4.238}$$

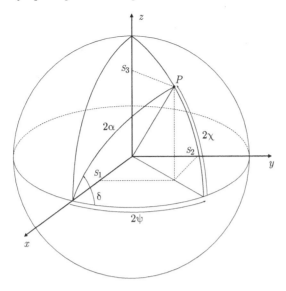

FIGURE 4.25
Graphical representation of the polarization of a monochromatic plane wave using the Poincaré sphere.

Here $\tan \delta_c$ is the *loss tangent* of the material, as described in (4.116) for a material without conductivity. For a good dielectric, we have

$$k = \beta - j\alpha = \breve{\omega}\sqrt{\mu \epsilon^c} = \breve{\omega}\sqrt{\mu(\operatorname{Re}\epsilon + j\operatorname{Im}\epsilon^c)} = \breve{\omega}\sqrt{\mu \operatorname{Re}\epsilon}\sqrt{1 - j\tan\delta_c}, \qquad (4.239)$$

hence,

$$k \approx \breve{\omega}\sqrt{\mu \operatorname{Re}\epsilon}\left(1 - j\tfrac{1}{2}\tan\delta_c\right)$$

by the binomial approximation for the square root. Therefore

$$\beta \approx \breve{\omega}\sqrt{\mu \operatorname{Re}\epsilon}$$

and

$$\alpha \approx \frac{\beta}{2}\tan\delta_c = \frac{\sigma}{2}\sqrt{\frac{\mu}{\operatorname{Re}\epsilon}}\left(1 - \frac{\breve{\omega}\operatorname{Im}\epsilon}{\sigma}\right).$$

We conclude that $\alpha \ll \beta$. Using this and the binomial approximation, we establish

$$\eta = \frac{\breve{\omega}\mu}{k} = \frac{\breve{\omega}\mu}{\beta}\frac{1}{1 - j\alpha/\beta} \approx \frac{\breve{\omega}\mu}{\beta}\left(1 + j\frac{\alpha}{\beta}\right).$$

Finally,

$$v_p = \frac{\breve{\omega}}{\beta} \approx \frac{1}{\sqrt{\mu \operatorname{Re}\epsilon}} \quad \text{and} \quad v_g = \left[\frac{d\beta}{d\omega}\right]^{-1} \approx \frac{1}{\sqrt{\mu \operatorname{Re}\epsilon}}.$$

To first order, the phase constant, phase velocity, and group velocity are those of a lossless medium.

4.11.4.5 Uniform plane waves in a good conductor

We classify a material as a "good conductor" if

$$\tan \delta_c \approx \frac{\sigma}{\check{\omega}\epsilon} \gg 1.$$

In a good conductor the conduction current $\sigma\check{\mathbf{E}}$ is much greater than the displacement current $j\check{\omega}\,\mathrm{Re}\,\epsilon\,\check{\mathbf{E}}$, and $\mathrm{Im}\,\epsilon$ is usually ignored. Now we may approximate

$$k = \beta - j\alpha = \check{\omega}\sqrt{\mu\,\mathrm{Re}\,\epsilon}\sqrt{1 - j\tan\delta_c} \approx \check{\omega}\sqrt{\mu\,\mathrm{Re}\,\epsilon}\sqrt{-j\tan\delta_c}.$$

Since $\sqrt{-j} = (1 - j)/\sqrt{2}$, we find that

$$\beta = \alpha \approx \sqrt{\pi f \mu \sigma}.$$

Hence,

$$v_p = \frac{\check{\omega}}{\beta} \approx \sqrt{\frac{2\check{\omega}}{\mu\sigma}} = \frac{1}{\sqrt{\mu\,\mathrm{Re}\,\epsilon}}\sqrt{\frac{2}{\tan\delta_c}}.$$

To find v_g we must replace $\check{\omega}$ by ω and differentiate, obtaining

$$v_g = \left[\frac{d\beta}{d\omega}\right]^{-1}\bigg|_{\omega=\check{\omega}} \approx \left[\frac{1}{2}\sqrt{\frac{\mu\sigma}{2\check{\omega}}}\right]^{-1} = 2\sqrt{\frac{2\check{\omega}}{\mu\sigma}} = 2v_p.$$

In a good conductor the group velocity is approximately twice the phase velocity. We could have found this relation from the phase velocity using (4.228). Indeed, noting that

$$\frac{dv_p}{d\omega} = \frac{d}{d\omega}\sqrt{\frac{2\omega}{\mu\sigma}} = \frac{1}{2}\sqrt{\frac{2}{\omega\mu\sigma}} \quad\text{and}\quad \beta\frac{dv_p}{d\omega} = \sqrt{\frac{\omega\mu\sigma}{2}}\frac{1}{2}\sqrt{\frac{2}{\omega\mu\sigma}} = \frac{1}{2},$$

we see that

$$\frac{v_p}{v_g} = 1 - \frac{1}{2} = \frac{1}{2}.$$

Note that the phase and group velocities may be only small fractions of the free-space light velocity. For example, in copper ($\sigma = 5.8 \times 10^7$ S/m, $\mu = \mu_0$, $\epsilon = \epsilon_0$) at 1 MHz, we have $v_p = 415$ m/s.

A factor often used to judge the quality of a conductor is the distance required for a propagating uniform plane wave to decrease in amplitude by the factor $1/e$. By (4.230) this distance is given by

$$\delta = \frac{1}{\alpha} = \frac{1}{\sqrt{\pi f \mu \sigma}}.$$

We call δ the *skin depth*. A good conductor is characterized by a small skin depth. For example, copper at 1 MHz has $\delta = 0.066$ mm. See § 3.6.2.3 for a detailed discussion of the importance of skin depth in quasistatics.

4.11.4.6 Power carried by a uniform plane wave

Since a plane wavefront is infinite in extent, we usually speak of the *power density* carried by the wave. This is identical to the time-average Poynting flux. Substitution from (4.214) and (4.230) gives

$$\mathbf{S}_{av} = \tfrac{1}{2}\,\mathrm{Re}[\check{\mathbf{E}} \times \check{\mathbf{H}}^*] = \tfrac{1}{2}\,\mathrm{Re}\left[\check{\mathbf{E}} \times \left(\frac{\hat{\mathbf{k}} \times \check{\mathbf{E}}}{\eta}\right)^*\right].$$

Expanding and remembering that $\mathbf{k} \cdot \check{\mathbf{E}} = 0$, we get

$$\mathbf{S}_{av} = \tfrac{1}{2}\hat{\mathbf{k}} \operatorname{Re}\left(\frac{|\check{\mathbf{E}}|^2}{\eta^*}\right) = \hat{\mathbf{k}} \operatorname{Re}\left(\frac{E_0^2}{2\eta^*}\right) e^{-2\alpha\hat{\mathbf{k}}\cdot\mathbf{r}}.$$

Hence a uniform plane wave propagating in an isotropic medium carries power in the direction of wavefront propagation.

4.11.4.7 Velocity of energy transport

The group velocity (4.225) has an additional interpretation as the velocity of energy transport. If the time-average volume density of energy is given by

$$\langle w_{em} \rangle = \langle w_e \rangle + \langle w_m \rangle$$

and the time-average volume density of energy flow is given by the Poynting flux density

$$\mathbf{S}_{av} = \tfrac{1}{2}\operatorname{Re}[\check{\mathbf{E}}(\mathbf{r}) \times \check{\mathbf{H}}^*(\mathbf{r})] = \tfrac{1}{4}[\check{\mathbf{E}}(\mathbf{r}) \times \check{\mathbf{H}}^*(\mathbf{r}) + \check{\mathbf{E}}^*(\mathbf{r}) \times \check{\mathbf{H}}(\mathbf{r})], \qquad (4.240)$$

then the velocity of energy flow, $\mathbf{v}_e$, is defined by

$$\mathbf{S}_{av} = \langle w_{em} \rangle \mathbf{v}_e. \qquad (4.241)$$

Let us calculate $\mathbf{v}_e$ for a plane wave propagating in a lossless, source-free medium where $\mathbf{k} = \hat{\mathbf{k}}\omega\sqrt{\mu\epsilon}$. By (4.208) and (4.214), we have

$$\tilde{\mathbf{E}}(\mathbf{r}, \omega) = \tilde{\mathbf{E}}_0(\omega)e^{-j\beta\hat{\mathbf{k}}\cdot\mathbf{r}}, \qquad (4.242)$$

$$\tilde{\mathbf{H}}(\mathbf{r}, \omega) = \left(\frac{\hat{\mathbf{k}} \times \tilde{\mathbf{E}}_0(\omega)}{\eta}\right)e^{-j\beta\hat{\mathbf{k}}\cdot\mathbf{r}} = \tilde{\mathbf{H}}_0(\omega)e^{-j\beta\hat{\mathbf{k}}\cdot\mathbf{r}}. \qquad (4.243)$$

We compute the time-average stored energy density using the energy theorem (4.68). In point form,

$$-\nabla \cdot \left(\tilde{\mathbf{E}}^* \times \frac{\partial\tilde{\mathbf{H}}}{\partial\omega} + \frac{\partial\tilde{\mathbf{E}}}{\partial\omega} \times \tilde{\mathbf{H}}^*\right)\bigg|_{\omega=\check{\omega}} = 4j\langle w_{em} \rangle. \qquad (4.244)$$

Upon substitution of (4.242) and (4.243) we find that we must compute the frequency derivatives of $\tilde{\mathbf{E}}$ and $\tilde{\mathbf{H}}$. Using

$$\frac{\partial}{\partial\omega}e^{-j\beta\hat{\mathbf{k}}\cdot\mathbf{r}} = \left(\frac{\partial}{\partial\beta}e^{-j\beta\hat{\mathbf{k}}\cdot\mathbf{r}}\right)\frac{d\beta}{d\omega} = -j\hat{\mathbf{k}} \cdot \mathbf{r}\frac{d\beta}{d\omega}e^{-j\beta\hat{\mathbf{k}}\cdot\mathbf{r}}$$

and remembering that $\mathbf{k} = \hat{\mathbf{k}}\beta$, we have

$$\frac{\partial\tilde{\mathbf{E}}(\mathbf{r}, \omega)}{\partial\omega} = \frac{d\tilde{\mathbf{E}}_0(\omega)}{d\omega}e^{-j\mathbf{k}\cdot\mathbf{r}} + \tilde{\mathbf{E}}_0(\omega)\left(-j\mathbf{r} \cdot \frac{d\mathbf{k}}{d\omega}\right)e^{-j\mathbf{k}\cdot\mathbf{r}},$$

$$\frac{\partial\tilde{\mathbf{H}}(\mathbf{r}, \omega)}{\partial\omega} = \frac{d\tilde{\mathbf{H}}_0(\omega)}{d\omega}e^{-j\mathbf{k}\cdot\mathbf{r}} + \tilde{\mathbf{H}}_0(\omega)\left(-j\mathbf{r} \cdot \frac{d\mathbf{k}}{d\omega}\right)e^{-j\mathbf{k}\cdot\mathbf{r}}.$$

Equation (4.244) becomes

$$-\nabla \cdot \left\{\tilde{\mathbf{E}}_0^*(\omega) \times \frac{d\tilde{\mathbf{H}}_0(\omega)}{d\omega} + \frac{d\tilde{\mathbf{E}}_0(\omega)}{d\omega} \times \tilde{\mathbf{H}}_0^*(\omega)\right.$$

$$\left. - j\mathbf{r} \cdot \frac{d\mathbf{k}}{d\omega}[\tilde{\mathbf{E}}_0^*(\omega) \times \tilde{\mathbf{H}}_0(\omega) + \tilde{\mathbf{E}}_0(\omega) \times \tilde{\mathbf{H}}_0^*(\omega)]\right\}\bigg|_{\omega=\check{\omega}} = 4j\langle w_{em} \rangle.$$

The first two terms on the left side have zero divergence, since these terms do not depend on $\mathbf{r}$. By the product rule (B.48) we have

$$[\tilde{\mathbf{E}}_0^*(\breve{\omega}) \times \tilde{\mathbf{H}}_0(\breve{\omega}) + \tilde{\mathbf{E}}_0(\breve{\omega}) \times \tilde{\mathbf{H}}_0^*(\breve{\omega})] \cdot \nabla \left(\mathbf{r} \cdot \frac{d\mathbf{k}}{d\omega} \right) \bigg|_{\omega=\breve{\omega}} = 4 \langle w_{em} \rangle.$$

The gradient term is merely

$$\nabla \left(\mathbf{r} \cdot \frac{d\mathbf{k}}{d\omega} \right) \bigg|_{\omega=\breve{\omega}} = \nabla \left(x \frac{dk_x}{d\omega} + y \frac{dk_y}{d\omega} + z \frac{dk_z}{d\omega} \right) \bigg|_{\omega=\breve{\omega}} = \frac{d\mathbf{k}}{d\omega} \bigg|_{\omega=\breve{\omega}},$$

hence

$$[\tilde{\mathbf{E}}_0^*(\breve{\omega}) \times \tilde{\mathbf{H}}_0(\breve{\omega}) + \tilde{\mathbf{E}}_0(\breve{\omega}) \times \tilde{\mathbf{H}}_0^*(\breve{\omega})] \cdot \frac{d\mathbf{k}}{d\omega} \bigg|_{\omega=\breve{\omega}} = 4 \langle w_{em} \rangle. \qquad (4.245)$$

Finally, the left side of this expression can be written in terms of the time-average Poynting vector. By (4.240) we have

$$\mathbf{S}_{av} = \tfrac{1}{2} \operatorname{Re}[\breve{\mathbf{E}} \times \breve{\mathbf{H}}^*] = \tfrac{1}{4} [\tilde{\mathbf{E}}_0(\breve{\omega}) \times \tilde{\mathbf{H}}_0^*(\breve{\omega}) + \tilde{\mathbf{E}}_0^*(\breve{\omega}) \times \tilde{\mathbf{H}}_0(\breve{\omega})]$$

and thus we can write (4.245) as

$$\mathbf{S}_{av} \cdot \frac{d\mathbf{k}}{d\omega} \bigg|_{\omega=\breve{\omega}} = \langle w_{em} \rangle.$$

Since for a uniform plane wave in an isotropic medium $\mathbf{k}$ and $\mathbf{S}_{av}$ are in the same direction, we have

$$\mathbf{S}_{av} = \hat{\mathbf{k}} \frac{d\omega}{d\beta} \bigg|_{\omega=\breve{\omega}} \langle w_{em} \rangle$$

and the velocity of energy transport for a plane wave of frequency $\breve{\omega}$ is then

$$\mathbf{v}_e = \hat{\mathbf{k}} \frac{d\omega}{d\beta} \bigg|_{\omega=\breve{\omega}}.$$

Thus, for a uniform plane wave in a lossless medium, the velocity of energy transport is identical to the group velocity.

4.11.4.8   Nonuniform plane waves

A nonuniform plane wave has the same form (4.208) as a uniform plane wave, but the vectors $\mathbf{k}_r$ and $\mathbf{k}_i$ described in (4.209) are not aligned. Thus, under linear polarization

$$\breve{\mathbf{E}}(\mathbf{r}) = \hat{\mathbf{e}} E_0 e^{-j\mathbf{k}_r \cdot \mathbf{r}} e^{\mathbf{k}_i \cdot \mathbf{r}}.$$

In the time domain this becomes

$$\mathbf{E}(\mathbf{r}, t) = \hat{\mathbf{e}} E_0 e^{\mathbf{k}_i \cdot \mathbf{r}} \cos[\breve{\omega} t - k_r (\hat{\mathbf{k}}_r \cdot \mathbf{r})]$$

where $\mathbf{k}_r = \hat{\mathbf{k}}_r k_r$. The surfaces of constant phase are planes perpendicular to $\mathbf{k}_r$ and propagating in the direction of $\hat{\mathbf{k}}_r$. The phase velocity is now

$$v_p = \breve{\omega} / k_r$$

and the wavelength is

$$\lambda = 2\pi / k_r.$$

In contrast, surfaces of constant amplitude must obey $\mathbf{k}_i \cdot \mathbf{r} = C$ and hence are planes normal to $\mathbf{k}_i$.

In a nonuniform plane wave the TEM nature of the fields is lost. This is easily seen by calculating $\check{\mathbf{H}}$ from (4.211):

$$\check{\mathbf{H}}(\mathbf{r}) = \frac{\mathbf{k} \times \check{\mathbf{E}}(\mathbf{r})}{\tilde{\omega}\mu} = \frac{\mathbf{k}_r \times \check{\mathbf{E}}(\mathbf{r})}{\tilde{\omega}\mu} + j\frac{\mathbf{k}_i \times \check{\mathbf{E}}(\mathbf{r})}{\tilde{\omega}\mu}.$$

Thus, $\check{\mathbf{H}}$ is no longer perpendicular to the direction of propagation of the phase front. The power carried by the wave also differs from that of the uniform case. The time-average Poynting vector

$$\mathbf{S}_{av} = \tfrac{1}{2} \operatorname{Re}\left[\check{\mathbf{E}} \times \left(\frac{\mathbf{k} \times \check{\mathbf{E}}}{\tilde{\omega}\mu}\right)^*\right]$$

can be expanded using (B.7):

$$\mathbf{S}_{av} = \tfrac{1}{2} \operatorname{Re}\left\{\frac{1}{\tilde{\omega}\mu^*}[\mathbf{k}^*(\check{\mathbf{E}} \cdot \check{\mathbf{E}}^*) - \check{\mathbf{E}}^*(\mathbf{k}^* \cdot \check{\mathbf{E}})]\right\}.$$

Since we still have $\mathbf{k} \cdot \check{\mathbf{E}} = 0$, we may write

$$\begin{aligned}
\mathbf{S}_{av} &= \frac{1}{2} \operatorname{Re}\left\{\frac{E_0^2}{\tilde{\omega}\mu^*}\mathbf{k}^*\right\} \\
&= \frac{E_0^2}{2\tilde{\omega}} \operatorname{Re}\left\{\frac{(\mathbf{k}_r \operatorname{Re}\mu - \mathbf{k}_i \operatorname{Im}\mu) - j(\mathbf{k}_r \operatorname{Im}\mu + \mathbf{k}_i \operatorname{Re}\mu)}{|\mu|^2}\right\} \\
&= \frac{E_0^2}{2\tilde{\omega}}\left\{\frac{\mathbf{k}_r \operatorname{Re}\mu - \mathbf{k}_i \operatorname{Im}\mu}{|\mu|^2}\right\}.
\end{aligned} \tag{4.246}$$

Thus the vector direction of $\mathbf{S}_{av}$ is not generally in the direction of propagation of the plane wavefronts.

Let us examine the special case of nonuniform plane waves propagating in a lossless material. It is intriguing that $\mathbf{k}$ may be complex when k is real, and the implication is important for the plane-wave expansion of complicated fields in free space. By (4.210), real k requires that if $\mathbf{k}_i \neq 0$ then $\mathbf{k}_r \cdot \mathbf{k}_i = 0$. Thus, for a nonuniform plane wave in a lossless material the surfaces of constant phase and the surfaces of constant amplitude are orthogonal. To specialize the time-average power to the lossless case we note that μ is purely real and (4.246) becomes

$$\mathbf{S}_{av} = \frac{E_0^2}{2\tilde{\omega}\mu}\mathbf{k}_r.$$

We see that in a lossless medium, the direction of energy propagation for a linearly polarized plane wave is perpendicular to surfaces of constant phase and parallel to surfaces of constant amplitude.

We shall encounter nonuniform plane waves when studying the reflection and refraction of a plane wave from a planar interface in the next section. We shall also find (§ 4.13) that nonuniform plane waves are a necessary constituent of the angular spectrum representation of an arbitrary wave field.

4.11.5 Plane waves in layered media

A useful canonical problem in wave propagation involves the reflection of plane waves by planar interfaces between differing material regions. This has many direct applications, from the design of optical coatings and microwave absorbers to the probing of underground oil-bearing rock layers. We begin by studying the reflection of a plane wave at a single interface and then extend the results to arbitrarily many material layers.

4.11.5.1 Reflection of a uniform plane wave at a planar material interface

Consider two lossy media separated by the $z = 0$ plane as shown in Figure 4.26. The media are assumed to be isotropic and homogeneous with permeability $\tilde{\mu}(\omega)$ and complex permittivity $\tilde{\epsilon}^c(\omega)$. Both $\tilde{\mu}$ and $\tilde{\epsilon}^c$ may be complex numbers describing magnetic and dielectric loss, respectively. We assume that a linearly polarized plane-wave field of the form (4.208) is created within region 1 by a process not considered here. We take this field as the known "incident wave" produced by an impressed source, and seek the total fields in regions 1 and 2. Here we assume that the incident field is that of a uniform plane wave, and subsequently extend the analysis to certain types of nonuniform plane waves.

Since the incident field is uniform, we may write the wave vector associated with this field as

$$\mathbf{k}^i = \hat{\mathbf{k}}^i k^i = \hat{\mathbf{k}}^i (k_r^i + j k_i^i) \quad \text{where} \quad [k^i(\omega)]^2 = \omega^2 \tilde{\mu}_1(\omega) \tilde{\epsilon}_1^c(\omega).$$

We can assume without loss of generality that $\hat{\mathbf{k}}^i$ lies in the xz-plane and makes an angle θ_i with the interface normal as in Figure 4.26. We refer to θ_i as the *incidence angle* of the incident field, and note that it is the angle between the direction of propagation of the planar phase fronts and the interface normal. With this we have

$$\mathbf{k}^i = \hat{\mathbf{x}} k_1 \sin \theta_i + \hat{\mathbf{z}} k_1 \cos \theta_i = \hat{\mathbf{x}} k_x^i + \hat{\mathbf{z}} k_z^i.$$

Using $k_1 = \beta_1 - j\alpha_1$ we also have

$$k_x^i = (\beta_1 - j\alpha_1) \sin \theta_i.$$

The term k_z^i is written in a somewhat different form in order to make the result easily applicable to reflections from multiple interfaces. We write

$$k_z^i = (\beta_1 - j\alpha_1) \cos \theta_i = \tau^i e^{-j\gamma^i} = \tau^i \cos \gamma^i - j \tau^i \sin \gamma^i.$$

Thus,

$$\tau^i = \sqrt{\beta_1^2 + \alpha_1^2} \cos \theta_i, \qquad \gamma^i = \tan^{-1}(\alpha_1/\beta_1).$$

We solve for the fields in each region of space directly in the frequency domain. The incident electric field has the form of (4.208),

$$\tilde{\mathbf{E}}^i(\mathbf{r}, \omega) = \tilde{\mathbf{E}}_0^i(\omega) e^{-j \mathbf{k}^i(\omega) \cdot \mathbf{r}}, \tag{4.247}$$

while the magnetic field is by (4.211)

$$\tilde{\mathbf{H}}^i = \frac{\mathbf{k}^i \times \tilde{\mathbf{E}}^i}{\omega \tilde{\mu}_1}. \tag{4.248}$$

The incident field may be decomposed into two orthogonal components, one parallel to the plane of incidence (the plane containing $\hat{\mathbf{k}}$ and the interface normal $\hat{\mathbf{z}}$) and one

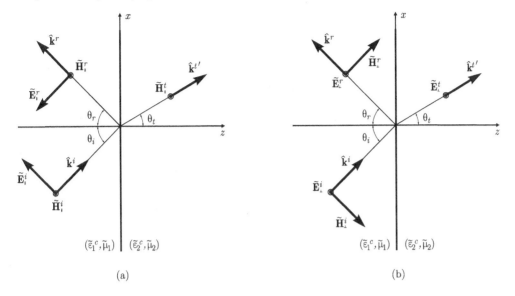

FIGURE 4.26
Uniform plane wave incident on planar interface between two lossy regions of space. (a) TM polarization, (b) TE polarization.

perpendicular to this plane. We seek unique solutions for the fields in both regions, first for the case in which the incident electric field has only a parallel component, and then for the case in which it has only a perpendicular component. The total field is determined by superposing the individual solutions. For perpendicular polarization we have from (4.247) and (4.248)

$$\tilde{\mathbf{E}}^i_\perp = \hat{\mathbf{y}} \tilde{E}^i_\perp e^{-j(k^i_x x + k^i_z z)},$$

$$\tilde{\mathbf{H}}^i_\perp = \frac{-\hat{\mathbf{x}} k^i_z + \hat{\mathbf{z}} k^i_x}{k_1} \frac{\tilde{E}^i_\perp}{\eta_1} e^{-j(k^i_x x + k^i_z z)},$$

as shown graphically in Figure 4.26. Here $\eta_1 = (\tilde{\mu}_1 / \tilde{\epsilon}^c_1)^{1/2}$ is the intrinsic impedance of medium 1. For parallel polarization, the direction of $\tilde{\mathbf{E}}$ is found by remembering that the wave must be TEM. Thus $\tilde{\mathbf{E}}_\parallel$ is perpendicular to $\mathbf{k}^i$. Since $\tilde{\mathbf{E}}_\parallel$ must also be perpendicular to $\tilde{\mathbf{E}}_\perp$, we have two possible directions for $\tilde{\mathbf{E}}_\parallel$. By convention we choose the one for which $\tilde{\mathbf{H}}$ lies in the same direction as did $\tilde{\mathbf{E}}$ for perpendicular polarization. Thus for parallel polarization,

$$\tilde{\mathbf{H}}^i_\parallel = \hat{\mathbf{y}} \frac{\tilde{E}^i_\parallel}{\eta_1} e^{-j(k^i_x x + k^i_z z)},$$

$$\tilde{\mathbf{E}}^i_\parallel = \frac{\hat{\mathbf{x}} k^i_z - \hat{\mathbf{z}} k^i_x}{k_1} \tilde{E}^i_\parallel e^{-j(k^i_x x + k^i_z z)},$$

as shown in Figure 4.26. Because $\tilde{\mathbf{E}}$ lies transverse (normal) to the plane of incidence under perpendicular polarization, the field set is often described as *transverse electric* or *TE*. Because $\tilde{\mathbf{H}}$ lies transverse to the plane of incidence under parallel polarization, the fields in that case are *transverse magnetic* or *TM*.

Uniqueness requires that the total field obey the boundary conditions at the planar interface. We hypothesize that the total field within region 1 consists of the incident field superposed with a "reflected" plane-wave field having wave vector $\mathbf{k}^r$, while the field in region 2 consists of a single "transmitted" plane-wave field having wave vector $\mathbf{k}^t$. We cannot at the outset make any assumption regarding whether either of these fields are uniform plane waves. However, we do note that the reflected and transmitted fields cannot have vector components that are not present in the incident field; extra components would prevent satisfaction of the boundary conditions. Letting $\tilde{E}^r$ be the amplitude of the reflected plane-wave field, we may write

$$\tilde{\mathbf{E}}_\perp^r = \hat{\mathbf{y}}\tilde{E}_\perp^r e^{-j(k_x^r x + k_z^r z)}, \qquad \tilde{\mathbf{H}}_\perp^r = \frac{-\hat{\mathbf{x}}k_z^r + \hat{\mathbf{z}}k_x^r}{k_1}\frac{\tilde{E}_\perp^r}{\eta_1}e^{-j(k_x^r x + k_z^r z)},$$

$$\tilde{\mathbf{H}}_\parallel^r = \hat{\mathbf{y}}\frac{\tilde{E}_\parallel^r}{\eta_1}e^{-j(k_x^r x + k_z^r z)}, \qquad \tilde{\mathbf{E}}_\parallel^r = \frac{\hat{\mathbf{x}}k_z^r - \hat{\mathbf{z}}k_x^r}{k_1}\tilde{E}_\parallel^r e^{-j(k_x^r x + k_z^r z)},$$

where $(k_x^r)^2 + (k_z^r)^2 = k_1^2$. Similarly, letting $\tilde{E}^t$ be the amplitude of the transmitted field, we have

$$\tilde{\mathbf{E}}_\perp^t = \hat{\mathbf{y}}\tilde{E}_\perp^t e^{-j(k_x^t x + k_z^t z)}, \qquad \tilde{\mathbf{H}}_\perp^t = \frac{-\hat{\mathbf{x}}k_z^t + \hat{\mathbf{z}}k_x^t}{k_2}\frac{\tilde{E}_\perp^t}{\eta_2}e^{-j(k_x^t x + k_z^t z)},$$

$$\tilde{\mathbf{H}}_\parallel^t = \hat{\mathbf{y}}\frac{\tilde{E}_\parallel^t}{\eta_2}e^{-j(k_x^t x + k_z^t z)}, \qquad \tilde{\mathbf{E}}_\parallel^t = \frac{\hat{\mathbf{x}}k_z^t - \hat{\mathbf{z}}k_x^t}{k_2}\tilde{E}_\parallel^t e^{-j(k_x^t x + k_z^t z)},$$

where $(k_x^t)^2 + (k_z^t)^2 = k_2^2$.

The relationships between the field amplitudes $\tilde{E}^i$, $\tilde{E}^r$, $\tilde{E}^t$, and between the components of the reflected and transmitted wave vectors $\mathbf{k}^r$ and $\mathbf{k}^t$, can be found by applying the boundary conditions. The tangential electric and magnetic fields are continuous across the interface at $z = 0$:

$$\hat{\mathbf{z}} \times (\tilde{\mathbf{E}}^i + \tilde{\mathbf{E}}^r)|_{z=0} = \hat{\mathbf{z}} \times \tilde{\mathbf{E}}^t|_{z=0},$$

$$\hat{\mathbf{z}} \times (\tilde{\mathbf{H}}^i + \tilde{\mathbf{H}}^r)|_{z=0} = \hat{\mathbf{z}} \times \tilde{\mathbf{H}}^t|_{z=0}.$$

Substituting the field expressions we find that for perpendicular polarization, the two boundary conditions require

$$\tilde{E}_\perp^i e^{-jk_x^i x} + \tilde{E}_\perp^r e^{-jk_x^r x} = \tilde{E}_\perp^t e^{-jk_x^t x}, \tag{4.249}$$

$$\frac{k_z^i}{k_1}\frac{\tilde{E}_\perp^i}{\eta_1}e^{-jk_x^i x} + \frac{k_z^r}{k_1}\frac{\tilde{E}_\perp^r}{\eta_1}e^{-jk_x^r x} = \frac{k_z^t}{k_2}\frac{\tilde{E}_\perp^t}{\eta_2}e^{-jk_x^t x}, \tag{4.250}$$

while for parallel polarization, they require

$$\frac{k_z^i}{k_1}\tilde{E}_\parallel^i e^{-jk_x^i x} + \frac{k_z^r}{k_1}\tilde{E}_\parallel^r e^{-jk_x^r x} = \frac{k_z^t}{k_2}\tilde{E}_\parallel^t e^{-jk_x^t x}, \tag{4.251}$$

$$\frac{\tilde{E}_\parallel^i}{\eta_1}e^{-jk_x^i x} + \frac{\tilde{E}_\parallel^r}{\eta_1}e^{-jk_x^r x} = \frac{\tilde{E}_\parallel^t}{\eta_2}e^{-jk_x^t x}. \tag{4.252}$$

For the above to hold for all x the exponential terms must match. This requires

$$k_x^i = k_x^r = k_x^t,$$

and also establishes a relation between k_z^i, k_z^r, and k_z^t. Since $(k_x^i)^2 + (k_z^i)^2 = (k_x^r)^2 + (k_z^r)^2 = k_1^2$, we require $k_z^r = \pm k_z^i$. In order to make the reflected wavefronts propagate away from the interface, we select $k_z^r = -k_z^i$. Letting $k_x^i = k_x^r = k_x^t = k_{1x}$ and $k_z^i = -k_z^r = k_{1z}$, we may write the wave vectors in region 1 as

$$\mathbf{k}^i = \hat{\mathbf{x}} k_{1x} + \hat{\mathbf{z}} k_{1z}, \qquad \mathbf{k}^r = \hat{\mathbf{x}} k_{1x} - \hat{\mathbf{z}} k_{1z}.$$

Since $(k_x^t)^2 + (k_z^t)^2 = k_2^2$, letting $k_2 = \beta_2 - j\alpha_2$ we have

$$k_z^t = \sqrt{k_2^2 - k_{1x}^2} = \sqrt{(\beta_2 - j\alpha_2)^2 - (\beta_1 - j\alpha_1)^2 \sin^2\theta_i} = \tau^t e^{-j\gamma^t}.$$

Squaring the above relation, we have

$$A - jB = (\tau^t)^2 \cos 2\gamma^t - j(\tau^t)^2 \sin 2\gamma^t$$

where

$$A = \beta_2^2 - \alpha_2^2 - (\beta_1^2 - \alpha_1^2)\sin^2\theta_i, \qquad B = 2(\beta_2\alpha_2 - \beta_1\alpha_1 \sin^2\theta_i). \tag{4.253}$$

Thus

$$\tau^t = (A^2 + B^2)^{1/4}, \qquad \gamma^t = \tfrac{1}{2}\tan^{-1} B/A. \tag{4.254}$$

Renaming k_z^t as k_{2z}, we may write the transmitted wave vector as

$$\mathbf{k}^t = \hat{\mathbf{x}} k_{1x} + \hat{\mathbf{z}} k_{2z} = \mathbf{k}_{2r} + j\mathbf{k}_{2i}$$

where

$$\mathbf{k}_{2r} = \hat{\mathbf{x}}\beta_1 \sin\theta_i + \hat{\mathbf{z}}\tau^t \cos\gamma^t, \qquad \mathbf{k}_{2i} = -\hat{\mathbf{x}}\alpha_1 \sin\theta_i - \hat{\mathbf{z}}\tau^t \sin\gamma^t.$$

Since the direction of propagation of the transmitted field phase fronts is perpendicular to $\mathbf{k}_{2r}$, a unit vector in the direction of propagation is

$$\hat{\mathbf{k}}_{2r} = \frac{\hat{\mathbf{x}}\beta_1 \sin\theta_i + \hat{\mathbf{z}}\tau^t \cos\gamma^t}{\sqrt{\beta_1^2 \sin^2\theta_i + (\tau^t)^2 \cos^2\theta_i}}. \tag{4.255}$$

Similarly, a unit vector perpendicular to planar surfaces of constant amplitude is

$$\hat{\mathbf{k}}_{2i} = \frac{\hat{\mathbf{x}}\alpha_1 \sin\theta_i + \hat{\mathbf{z}}\tau^t \sin\gamma^t}{\sqrt{\alpha_1^2 \sin^2\theta_i + (\tau^t)^2 \sin^2\gamma^t}}. \tag{4.256}$$

In general, $\hat{\mathbf{k}}_r$ is not aligned with $\hat{\mathbf{k}}_i$ and the transmitted field is a nonuniform plane wave.

With these definitions of k_{1x}, k_{1z}, k_{2z}, Equations (4.249) and (4.250) can be solved simultaneously, and we have

$$\tilde{E}_\perp^r = \tilde{\Gamma}_\perp \tilde{E}_\perp^i, \qquad \tilde{E}_\perp^t = \tilde{T}_\perp \tilde{E}_\perp^i,$$

where

$$\tilde{\Gamma}_\perp = \frac{Z_{2\perp} - Z_{1\perp}}{Z_{2\perp} + Z_{1\perp}}, \qquad \tilde{T}_\perp = 1 + \tilde{\Gamma}_\perp = \frac{2Z_{2\perp}}{Z_{2\perp} + Z_{1\perp}}, \tag{4.257}$$

with

$$Z_{1\perp} = \frac{k_1 \eta_1}{k_{1z}}, \qquad Z_{2\perp} = \frac{k_2 \eta_2}{k_{2z}}. \tag{4.258}$$

Here $\tilde{\Gamma}$ is a frequency-dependent *reflection coefficient* relating the tangential components of the incident and reflected electric fields, and $\tilde{T}$ is a frequency-dependent *transmission coefficient* relating the tangential components of the incident and transmitted electric fields. These coefficients are also called the *Fresnel coefficients*.

For the case of parallel polarization, we solve (4.251) and (4.252) to find

$$\frac{\tilde{E}^r_{\parallel,x}}{\tilde{E}^i_{\parallel,x}} = \frac{k^r_z}{k^i_z}\frac{\tilde{E}^r_\parallel}{\tilde{E}^i_\parallel} = -\frac{\tilde{E}^r_\parallel}{\tilde{E}^i_\parallel} = \tilde{\Gamma}_\parallel, \qquad \frac{\tilde{E}^t_{\parallel,x}}{\tilde{E}^i_{\parallel,x}} = \frac{(k^t_z/k_2)\tilde{E}^t_\parallel}{(k^i_z/k_1)\tilde{E}^i_\parallel} = \tilde{T}_\parallel.$$

Here

$$\tilde{\Gamma}_\parallel = \frac{Z_{2\parallel} - Z_{1\parallel}}{Z_{2\parallel} + Z_{1\parallel}}, \qquad \tilde{T}_\parallel = 1 + \tilde{\Gamma}_\parallel = \frac{2Z_{2\parallel}}{Z_{2\parallel} + Z_{1\parallel}}, \tag{4.259}$$

with

$$Z_{1\parallel} = \frac{k_{1z}\eta_1}{k_1}, \qquad Z_{2\parallel} = \frac{k_{2z}\eta_2}{k_2}. \tag{4.260}$$

Note that we may also write

$$\tilde{E}^r_\parallel = -\tilde{\Gamma}_\parallel \tilde{E}^i_\parallel, \qquad \tilde{E}^t_\parallel = \tilde{T}_\parallel \tilde{E}^i_\parallel \left(\frac{k^i_z}{k_1} \frac{k_2}{k^t_z} \right).$$

Let us summarize the fields in each region. For perpendicular polarization, we have

$$\begin{aligned}
\tilde{\mathbf{E}}^i_\perp &= \hat{\mathbf{y}}\tilde{E}^i_\perp e^{-j\mathbf{k}^i\cdot\mathbf{r}}, \\
\tilde{\mathbf{E}}^r_\perp &= \hat{\mathbf{y}}\tilde{\Gamma}_\perp \tilde{E}^i_\perp e^{-j\mathbf{k}^r\cdot\mathbf{r}}, \\
\tilde{\mathbf{E}}^t_\perp &= \hat{\mathbf{y}}\tilde{T}_\perp \tilde{E}^i_\perp e^{-j\mathbf{k}^t\cdot\mathbf{r}},
\end{aligned} \tag{4.261}$$

and

$$\tilde{\mathbf{H}}^i_\perp = \frac{\mathbf{k}^i \times \tilde{\mathbf{E}}^i_\perp}{k_1\eta_1}, \quad \tilde{\mathbf{H}}^r_\perp = \frac{\mathbf{k}^r \times \tilde{\mathbf{E}}^r_\perp}{k_1\eta_1}, \quad \tilde{\mathbf{H}}^t_\perp = \frac{\mathbf{k}^t \times \tilde{\mathbf{E}}^t_\perp}{k_2\eta_2}.$$

For parallel polarization, we have

$$\tilde{\mathbf{E}}^i_\parallel = -\eta_1 \frac{\mathbf{k}^i \times \tilde{\mathbf{H}}^i_\parallel}{k_1} e^{-j\mathbf{k}^i\cdot\mathbf{r}},$$

$$\tilde{\mathbf{E}}^r_\parallel = -\eta_1 \frac{\mathbf{k}^r \times \tilde{\mathbf{H}}^r_\parallel}{k_1} e^{-j\mathbf{k}^r\cdot\mathbf{r}},$$

$$\tilde{\mathbf{E}}^t_\parallel = -\eta_2 \frac{\mathbf{k}^t \times \tilde{\mathbf{H}}^t_\parallel}{k_2} e^{-j\mathbf{k}^t\cdot\mathbf{r}},$$

and

$$\tilde{\mathbf{H}}^i_\parallel = \hat{\mathbf{y}}\frac{\tilde{E}^i_\parallel}{\eta_1} e^{-j\mathbf{k}^i\cdot\mathbf{r}},$$

$$\tilde{\mathbf{H}}^r_\parallel = -\hat{\mathbf{y}}\frac{\tilde{\Gamma}_\parallel \tilde{E}^i_\parallel}{\eta_1} e^{-j\mathbf{k}^r\cdot\mathbf{r}},$$

$$\tilde{\mathbf{H}}^t_\parallel = \hat{\mathbf{y}}\frac{\tilde{T}_\parallel \tilde{E}^i_\parallel}{\eta_2} \left(\frac{k^i_z}{k_1} \frac{k_2}{k^t_z} \right) e^{-j\mathbf{k}^t\cdot\mathbf{r}}.$$

The wave vectors are given by

$$\mathbf{k}^i = (\hat{\mathbf{x}}\beta_1 \sin\theta_i + \hat{\mathbf{z}}\tau^i \cos\gamma^i) - j(\hat{\mathbf{x}}\alpha_1 \sin\theta_i + \hat{\mathbf{z}}\tau^i \sin\gamma^i), \tag{4.262}$$

$$\mathbf{k}^r = (\hat{\mathbf{x}}\beta_1 \sin\theta_i - \hat{\mathbf{z}}\tau^i \cos\gamma^i) - j(\hat{\mathbf{x}}\alpha_1 \sin\theta_i - \hat{\mathbf{z}}\tau^i \sin\gamma^i), \tag{4.263}$$

$$\mathbf{k}^t = (\hat{\mathbf{x}}\beta_1 \sin\theta_i + \hat{\mathbf{z}}\tau^t \cos\gamma^t) - j(\hat{\mathbf{x}}\alpha_1 \sin\theta_i + \hat{\mathbf{z}}\tau^t \sin\gamma^t). \tag{4.264}$$

We see that the reflected wave must, like the incident wave, be a uniform plane wave. We define the unsigned *reflection angle* θ_r as the angle between the surface normal and the direction of propagation of the reflected wavefronts (Figure 4.26). Since

$$\mathbf{k}^i \cdot \hat{\mathbf{z}} = k_1 \cos\theta_i = -\mathbf{k}^r \cdot \hat{\mathbf{z}} = k_1 \cos\theta_r,$$

$$\mathbf{k}^i \cdot \hat{\mathbf{x}} = k_1 \sin\theta_i = \mathbf{k}^r \cdot \hat{\mathbf{x}} = k_1 \sin\theta_r,$$

we must have $\theta_i = \theta_r$. This is *Snell's law of reflection*. We can similarly define the *transmission angle* as the angle between the direction of propagation of the transmitted wavefronts and the interface normal. Noting that $\hat{\mathbf{k}}_{2r} \cdot \hat{\mathbf{z}} = \cos\theta_t$ and $\hat{\mathbf{k}}_{2r} \cdot \hat{\mathbf{x}} = \sin\theta_t$, we have from (4.255) and (4.256)

$$\cos\theta_t = \frac{\tau^t \cos\gamma^t}{\sqrt{\beta_1^2 \sin^2\theta_i + (\tau^t)^2 \cos^2\gamma^t}},$$

$$\sin\theta_t = \frac{\beta_1 \sin\theta_i}{\sqrt{\beta_1^2 \sin^2\theta_i + (\tau^t)^2 \cos^2\gamma^t}}, \tag{4.265}$$

and thus

$$\theta_t = \tan^{-1}\left(\frac{\beta_1}{\tau^t} \frac{\sin\theta_i}{\cos\gamma^t}\right).$$

Depending on the properties of the media, at a certain incidence angle θ_c, called the *critical angle*, the angle of transmission becomes $\pi/2$. Under this condition $\hat{\mathbf{k}}_{2r}$ has only an x-component. Thus, surfaces of constant phase propagate parallel to the interface. Later we shall see that for low-loss (or lossless) media, this implies that no time-average power is carried by a monochromatic transmitted wave into the second medium.

We also see that although the transmitted field may be a nonuniform plane wave, its mathematical form is that of the incident plane wave. This allows us to easily generalize the single-interface reflection problem to one involving many layers.

4.11.5.2 Uniform plane-wave reflection for lossless media

We can specialize the preceding results to the case in which both regions are lossless, with $\tilde{\mu} = \mu$ and $\tilde{\epsilon}^c = \epsilon$ real and frequency-independent. By (4.215) we have

$$\beta = \omega\sqrt{\mu\epsilon},$$

while (4.216) gives $\alpha = 0$. We can easily show that the transmitted wave must be uniform unless the incidence angle exceeds the critical angle. By (4.253) we have

$$A = \beta_2^2 - \beta_1^2 \sin^2\theta_i, \qquad B = 0, \tag{4.266}$$

while (4.254) gives

$$\tau = (A^2)^{1/4} = \sqrt{|\beta_2^2 - \beta_1^2 \sin^2\theta_i|} \quad \text{and} \quad \gamma^t = \tfrac{1}{2}\tan^{-1}0.$$

There are several possible choices for γ^t. To choose properly we note that γ^t represents the negative of the phase of the quantity $k_z^t = \sqrt{A}$. If $A > 0$ the phase of the square root is 0. If $A < 0$ the phase of the square root is $-\pi/2$ and thus $\gamma^t = +\pi/2$. Here we choose the plus sign on γ^t to ensure that the transmitted field decays as z increases. We note that if $A = 0$ then $\tau^t = 0$ and (4.265) gives $\theta_t = \pi/2$. This defines the critical angle, which from (4.266) is

$$\theta_c = \sin^{-1} \sqrt{\beta_2^2/\beta_1^2} = \sin^{-1} \sqrt{\frac{\mu_2 \epsilon_2}{\mu_1 \epsilon_1}}.$$

Therefore

$$\gamma^t = \begin{cases} 0, & \theta_i < \theta_c, \\ \pi/2, & \theta_i > \theta_c. \end{cases}$$

Using these we can write down the transmitted wave vector from (4.264):

$$\mathbf{k}^t = \mathbf{k}_r^t + j\mathbf{k}_i^t = \begin{cases} \hat{\mathbf{x}}\beta_1 \sin\theta_i + \hat{\mathbf{z}}\sqrt{|A|}, & \theta_i < \theta_c, \\ \hat{\mathbf{x}}\beta_1 \sin\theta_i - j\hat{\mathbf{z}}\sqrt{|A|}, & \theta_i > \theta_c. \end{cases} \tag{4.267}$$

By (4.265) we have

$$\sin\theta_t = \frac{\beta_1 \sin\theta_i}{\sqrt{\beta_1^2 \sin^2\theta_i + \beta_2^2 - \beta_1^2 \sin^2\theta_i}} = \frac{\beta_1 \sin\theta_i}{\beta_2}$$

or

$$\beta_2 \sin\theta_t = \beta_1 \sin\theta_i. \tag{4.268}$$

This is *Snell's law of refraction*. With this we can write for $\theta_i < \theta_c$

$$A = \beta_2^2 - \beta_1^2 \sin^2\theta_i = \beta_2^2 \cos^2\theta_t.$$

Using this and substituting $\beta_2 \sin\theta_t$ for $\beta_1 \sin\theta_i$, we may rewrite (4.267) for $\theta_i < \theta_c$ as

$$\mathbf{k}^t = \mathbf{k}_r^t + j\mathbf{k}_i^t = \hat{\mathbf{x}}\beta_2 \sin\theta_t + \hat{\mathbf{z}}\beta_2 \cos\theta_t. \tag{4.269}$$

So the transmitted plane wave is uniform with $\mathbf{k}_i^t = 0$. When $\theta_i > \theta_c$ we have from (4.267)

$$\mathbf{k}_r^t = \hat{\mathbf{x}}\beta_1 \sin\theta_i, \qquad \mathbf{k}_i^t = -\hat{\mathbf{z}}\sqrt{\beta_1^2 \sin^2\theta_i - \beta_2^2}.$$

Since $\mathbf{k}_r^t$ and $\mathbf{k}_i^t$ are not collinear, the plane wave is nonuniform. Let us examine the cases $\theta_i < \theta_c$ and $\theta_i > \theta_c$ in greater detail.

Case 1: $\theta_i < \theta_c$. By (4.262)–(4.263) and (4.269) the wave vectors are

$$\mathbf{k}^i = \hat{\mathbf{x}}\beta_1 \sin\theta_i + \hat{\mathbf{z}}\beta_1 \cos\theta_i, \qquad \mathbf{k}^r = \hat{\mathbf{x}}\beta_1 \sin\theta_i - \hat{\mathbf{z}}\beta_1 \cos\theta_i, \qquad \mathbf{k}^t = \hat{\mathbf{x}}\beta_2 \sin\theta_t + \hat{\mathbf{z}}\beta_2 \cos\theta_t,$$

and the wave impedances are

$$Z_{1\perp} = \frac{\eta_1}{\cos\theta_i}, \qquad Z_{2\perp} = \frac{\eta_2}{\cos\theta_t}, \qquad Z_{1\parallel} = \eta_1 \cos\theta_i, \qquad Z_{2\parallel} = \eta_2 \cos\theta_t.$$

The reflection coefficients are

$$\tilde{\Gamma}_\perp = \frac{\eta_2 \cos\theta_i - \eta_1 \cos\theta_t}{\eta_2 \cos\theta_i + \eta_1 \cos\theta_t}, \qquad \tilde{\Gamma}_\parallel = \frac{\eta_2 \cos\theta_t - \eta_1 \cos\theta_i}{\eta_2 \cos\theta_t + \eta_1 \cos\theta_i}.$$

So the reflection coefficients are purely real, with signs dependent on the constitutive parameters of the media. We can write

$$\tilde{\Gamma}_{\perp} = \rho_{\perp} e^{j\phi_{\perp}}, \qquad \tilde{\Gamma}_{\|} = \rho_{\|} e^{j\phi_{\|}},$$

where ρ and ϕ are real, and where $\phi = 0$ or π.

Under certain conditions the reflection coefficients vanish. For a given set of constitutive parameters we may achieve $\tilde{\Gamma} = 0$ at an incidence angle θ_B, known as the *Brewster* or *polarizing angle*. A wave with an arbitrary combination of perpendicular and parallel polarized components incident at this angle produces a reflected field with a single component. A wave incident with only the appropriate single component produces no reflected field, regardless of its amplitude.

For perpendicular polarization we set $\tilde{\Gamma}_{\perp} = 0$, requiring

$$\eta_2 \cos\theta_i - \eta_1 \cos\theta_t = 0$$

or equivalently

$$\frac{\mu_2}{\epsilon_2}(1 - \sin^2\theta_i) = \frac{\mu_1}{\epsilon_1}(1 - \sin^2\theta_t).$$

By (4.268) we may put

$$\sin^2\theta_t = \frac{\mu_1\epsilon_1}{\mu_2\epsilon_2}\sin^2\theta_i,$$

resulting in

$$\sin^2\theta_i = \frac{\mu_2}{\epsilon_1}\frac{\epsilon_2\mu_1 - \epsilon_1\mu_2}{\mu_1^2 - \mu_2^2}.$$

The value of θ_i that satisfies this equation must be the Brewster angle, and thus

$$\theta_{B\perp} = \sin^{-1}\sqrt{\frac{\mu_2}{\epsilon_1}\frac{\epsilon_2\mu_1 - \epsilon_1\mu_2}{\mu_1^2 - \mu_2^2}}.$$

When $\mu_1 = \mu_2$ there is no solution to this equation, hence the reflection coefficient cannot vanish. When $\epsilon_1 = \epsilon_2$ we have

$$\theta_{B\perp} = \sin^{-1}\sqrt{\frac{\mu_2}{\mu_1 + \mu_2}} = \tan^{-1}\sqrt{\frac{\mu_2}{\mu_1}}.$$

For parallel polarization, we set $\tilde{\Gamma}_{\|} = 0$ and have

$$\eta_2 \cos\theta_t = \eta_1 \cos\theta_i.$$

Proceeding as above, we find that

$$\theta_{B\|} = \sin^{-1}\sqrt{\frac{\epsilon_2}{\mu_1}\frac{\epsilon_1\mu_2 - \epsilon_2\mu_1}{\epsilon_1^2 - \epsilon_2^2}}.$$

This expression has no solution when $\epsilon_1 = \epsilon_2$, and thus the reflection coefficient cannot vanish under this condition. When $\mu_1 = \mu_2$ we have

$$\theta_{B\|} = \sin^{-1}\sqrt{\frac{\epsilon_2}{\epsilon_1 + \epsilon_2}} = \tan^{-1}\sqrt{\frac{\epsilon_2}{\epsilon_1}}.$$

We find that when $\theta_i < \theta_c$ the total field in region 1 behaves as a traveling wave along x, but has characteristics of both a standing wave and a traveling wave along z (Problem 4.7). The traveling-wave component is associated with a Poynting power flux, while the standing-wave component is not. This flux is carried across the boundary into region 2 where the transmitted field consists only of a traveling wave. By (4.160) the normal component of time-average Poynting flux is continuous across the boundary, demonstrating that the time-average power carried by the wave into the interface from region 1 passes out through the interface into region 2 (Problem 4.8).

Case 2: $\theta_i > \theta_c$. The wave vectors are, from (4.262)–(4.263) and (4.267),

$$\mathbf{k}^i = \hat{\mathbf{x}}\beta_1 \sin\theta_i + \hat{\mathbf{z}}\beta_1 \cos\theta_i, \qquad \mathbf{k}^r = \hat{\mathbf{x}}\beta_1 \sin\theta_i - \hat{\mathbf{z}}\beta_1 \cos\theta_i, \qquad \mathbf{k}^t = \hat{\mathbf{x}}\beta_1 \sin\theta_i - j\hat{\mathbf{z}}\alpha_c,$$

where

$$\alpha_c = \sqrt{\beta_1^2 \sin^2\theta_i - \beta_2^2}$$

is the *critical angle attenuation constant*. The wave impedances are

$$Z_{1\perp} = \frac{\eta_1}{\cos\theta_i}, \qquad Z_{2\perp} = j\frac{\beta_2\eta_2}{\alpha_c}, \qquad Z_{1\parallel} = \eta_1 \cos\theta_i, \qquad Z_{2\parallel} = -j\frac{\alpha_c\eta_2}{\beta_2}.$$

Substituting these into (4.257) and (4.259), we find that the reflection coefficients are the complex quantities

$$\tilde{\Gamma}_\perp = \frac{\beta_2\eta_2 \cos\theta_i + j\eta_1\alpha_c}{\beta_2\eta_2 \cos\theta_i - j\eta_1\alpha_c} = e^{j\phi_\perp}, \qquad \tilde{\Gamma}_\parallel = -\frac{\beta_2\eta_1 \cos\theta_i + j\eta_2\alpha_c}{\beta_2\eta_1 \cos\theta_i - j\eta_2\alpha_c} = e^{j\phi_\parallel},$$

where

$$\phi_\perp = 2\tan^{-1}\left(\frac{\eta_1\alpha_c}{\beta_2\eta_2 \cos\theta_i}\right), \qquad \phi_\parallel = \pi + 2\tan^{-1}\left(\frac{\eta_2\alpha_c}{\beta_2\eta_1 \cos\theta_i}\right).$$

We note with interest that $\rho_\perp = \rho_\parallel = 1$. So the amplitudes of the reflected waves are identical to those of the incident waves, and we call this the case of *total internal reflection*. The phase of the reflected wave at the interface is changed from that of the incident wave by an amount $\phi_\perp$ or $\phi_\parallel$. The phase shift incurred by the reflected wave upon total internal reflection is called the *Goos–Hänchen shift*.

In the case of total internal reflection, the field in region 1 is a pure standing wave while the field in region 2 decays exponentially in the z-direction and is evanescent (Problem 4.9). Since a standing wave transports no power, there is no Poynting flux into region 2. We find that the evanescent wave also carries no power and thus the boundary condition on power flux at the interface is satisfied (Problem 4.10). We note that for any incident angle except $\theta_i = 0$ (normal incidence) the wave in region 1 does transport power in the x-direction.

4.11.5.3   Reflection of time-domain uniform plane waves

Solution for the fields reflected and transmitted at an interface shows us the properties of the fields for a certain single excitation frequency and allows us to obtain time-domain fields by Fourier inversion. Under certain conditions it is possible to do the inversion analytically, providing physical insight into the temporal behavior of the fields.

As a simple example, consider a perpendicularly polarized, uniform plane wave incident from free space at an angle θ_i on the planar surface of a conducting material (Figure

4.26). The material is assumed to have frequency-independent constitutive parameters $\tilde{\mu} = \mu_0$, $\tilde{\epsilon} = \epsilon$, and $\tilde{\sigma} = \sigma$. By (4.261) we have the reflected field

$$\tilde{\mathbf{E}}_\perp^r(\mathbf{r}, \omega) = \hat{\mathbf{y}}\tilde{\Gamma}_\perp(\omega)\tilde{E}_\perp^i(\omega)e^{-j\mathbf{k}^r(\omega)\cdot\mathbf{r}} = \hat{\mathbf{y}}\tilde{E}^r(\omega)e^{-j\omega\frac{\hat{\mathbf{k}}^r\cdot\mathbf{r}}{c}}$$

where $\tilde{E}^r = \tilde{\Gamma}_\perp\tilde{E}_\perp^i$. We can use the time-shifting theorem (A.3) to invert the transform and obtain

$$\mathbf{E}_\perp^r(\mathbf{r}, t) = \mathcal{F}^{-1}\{\tilde{\mathbf{E}}_\perp^r(\mathbf{r}, \omega)\} = \hat{\mathbf{y}}E^r\left(t - \frac{\hat{\mathbf{k}}^r\cdot\mathbf{r}}{c}\right) \tag{4.270}$$

where by the convolution theorem (A.4)

$$E^r(t) = \mathcal{F}^{-1}\{\tilde{E}^r(\omega)\} = \Gamma_\perp(t) * E_\perp(t).$$

Here $E_\perp(t) = \mathcal{F}^{-1}\{\tilde{E}_\perp^i(\omega)\}$ is the time waveform of the incident plane wave, while $\Gamma_\perp(t) = \mathcal{F}^{-1}\{\tilde{\Gamma}_\perp(\omega)\}$ is the time-domain reflection coefficient.

By (4.270) the reflected time-domain field propagates along the direction $\hat{\mathbf{k}}^r$ at the speed of light. The time waveform of the field is the convolution of the waveform of the incident field with the time-domain reflection coefficient $\Gamma_\perp(t)$. In the lossless case ($\sigma = 0$), $\Gamma_\perp(t)$ is a δ-function, so the waveforms of the reflected and incident fields are identical. With losses present, $\Gamma_\perp(t)$ broadens and the reflected field waveform becomes a convolution-broadened version of the incident field waveform. To understand the waveform of the reflected field we must compute $\Gamma_\perp(t)$. Note that by choosing the permittivity of region 2 larger than that of region 1 we prevent total internal reflection.

We specialize the frequency-domain reflection coefficient (4.257) for our problem by noting that

$$k_{1z} = \beta_1\cos\theta_i, \qquad k_{2z} = \sqrt{k_2^2 - k_{1x}^2} = \omega\sqrt{\mu_0}\sqrt{\epsilon + \frac{\sigma}{j\omega} - \epsilon_0\sin^2\theta_i},$$

and thus

$$Z_{1\perp} = \frac{\eta_0}{\cos\theta_i}, \qquad Z_{2\perp} = \frac{\eta_0}{\sqrt{\epsilon_r + \frac{\sigma}{j\omega\epsilon_0} - \sin^2\theta_i}},$$

where $\epsilon_r = \epsilon/\epsilon_0$ and $\eta_0 = \sqrt{\mu_0/\epsilon_0}$. So

$$\tilde{\Gamma}_\perp = \frac{\sqrt{s} - \sqrt{Ds + B}}{\sqrt{s} + \sqrt{Ds + B}} \tag{4.271}$$

where $s = j\omega$ and

$$D = \frac{\epsilon_r - \sin^2\theta_i}{\cos^2\theta_i}, \qquad B = \frac{\sigma}{\epsilon_0\cos^2\theta_i}.$$

We can put (4.271) into a better form for inversion. We begin by subtracting $\Gamma_{\perp\infty}$, the high-frequency limit of $\tilde{\Gamma}_\perp$. Noting that

$$\lim_{\omega\to\infty}\tilde{\Gamma}_\perp(\omega) = \Gamma_{\perp\infty} = \frac{1 - \sqrt{D}}{1 + \sqrt{D}},$$

we can form

$$\tilde{\Gamma}_\perp^0(\omega) = \tilde{\Gamma}_\perp(\omega) - \Gamma_{\perp\infty} = \frac{\sqrt{s} - \sqrt{Ds + B}}{\sqrt{s} + \sqrt{Ds + B}} - \frac{1 - \sqrt{D}}{1 + \sqrt{D}}$$

$$= 2\frac{\sqrt{D}}{1 + \sqrt{D}}\left[\frac{\sqrt{s} - \sqrt{s + B/D}}{\sqrt{s} + \sqrt{D}\sqrt{s + D/B}}\right].$$

With a bit of algebra this becomes

$$\tilde{\Gamma}^0_\perp(\omega) = -\frac{2\sqrt{D}}{D-1}\left(\frac{s}{s+\frac{B}{D-1}}\right)\left(1 - \sqrt{\frac{s+\frac{B}{D}}{s}}\right) - \frac{2B}{\left(1+\sqrt{D}\right)(D-1)}\left(\frac{1}{s+\frac{B}{D-1}}\right).$$

Now we can apply (C.12), (C.18), and (C.19) to obtain

$$\Gamma^0_\perp(t) = \mathcal{F}^{-1}\{\tilde{\Gamma}^0_\perp(\omega)\} = f_1(t) + f_2(t) + f_3(t)$$

where

$$f_1(t) = -\frac{2B}{(1+\sqrt{D})(D-1)}e^{-\frac{Bt}{D-1}}U(t),$$

$$f_2(t) = -\frac{B^2}{\sqrt{D}(D-1)^2}U(t)\int_0^t e^{-\frac{B(t-x)}{D-1}}I\left(\frac{Bx}{2D}\right)dx,$$

$$f_3(t) = \frac{B}{\sqrt{D}(D-1)}I\left(\frac{Bt}{2D}\right)U(t).$$

Here

$$I(x) = e^{-x}\left[I_0(x) + I_1(x)\right]$$

where $I_0(x)$ and $I_1(x)$ are modified Bessel functions of the first kind. Setting $u = Bx/2D$ we can also write

$$f_2(t) = -\frac{2B\sqrt{D}}{(D-1)^2}U(t)\int_0^{\frac{Bt}{2D}} e^{-\frac{Bt-2Du}{D-1}}I(u)\,du.$$

Polynomial approximations for $I(x)$ found in [1] and exponential approximations found in [165], make the computation of $\Gamma^0_\perp(t)$ straightforward.

The complete time-domain reflection coefficient is

$$\Gamma_\perp(t) = \frac{1-\sqrt{D}}{1+\sqrt{D}}\delta(t) + \Gamma^0_\perp(t).$$

If $\sigma = 0$ then $\Gamma^0_\perp(t) = 0$ and the reflection coefficient reduces to a single δ-function. Since convolution with this term does not alter wave shape, the reflected field has the same waveform as the incident field.

▶ **Example 4.14:** Transient plane wave reflected from water

Consider a perpendicularly polarized, uniform plane wave normally incident from free space on the planar surface of dry water ice ($\epsilon_r = 3$, $\sigma = 0.01$ S/m). Plot the time-domain reflection coefficient for $0 \le t \le 15$ ns. Repeat for a wave incident on sea water ($\epsilon_r = 80$, $\sigma = 4$ S/m).

Solution: Figure 4.27 shows a plot of $\Gamma^0_\perp(t)$ for normal incidence ($\theta_i = 0^0$). Note that a pulse waveform experiences more temporal spreading upon reflection from ice than from sea water, but that the amplitude of the dispersive component is less than that for sea water.

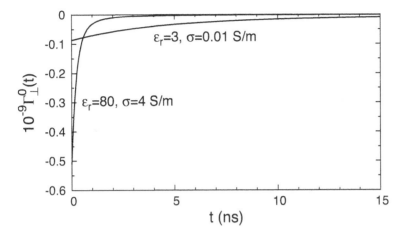

FIGURE 4.27
Time-domain reflection coefficients. ◀

4.11.5.4 Reflection of a nonuniform plane wave from a planar interface

Describing the interaction of a general nonuniform plane wave with a planar interface is problematic because of the non-TEM behavior of the incident wave. We cannot decompose the fields into two mutually orthogonal cases as we did with uniform waves. However, we found in the last section that when a uniform wave is incident on a planar interface, the transmitted wave, even if nonuniform in nature, takes the same mathematical form and may be decomposed in the same manner as the incident wave. Thus, we may study the case in which this refracted wave is incident on a successive interface using exactly the same analysis as with a uniform incident wave. This is helpful in the case of multi-layered media, examined next.

4.11.5.5 Interaction of a plane wave with multi-layered, planar materials

Consider $N+1$ regions of space separated by N planar interfaces as shown in Figure 4.28, and suppose a uniform plane wave is incident on the first interface at angle θ_i. Assume each region is isotropic and homogeneous with a frequency-dependent complex permittivity and permeability. We can easily generalize the previous analysis regarding reflection from a single interface by realizing that in order to satisfy the boundary conditions, each region, except region N, contains incident- and reflected-type waves of the forms

$$\tilde{\mathbf{E}}^i(\mathbf{r}, \omega) = \tilde{\mathbf{E}}_0^i e^{-j\mathbf{k}^i \cdot \mathbf{r}}, \qquad \tilde{\mathbf{E}}^r(\mathbf{r}, \omega) = \tilde{\mathbf{E}}_0^r e^{-j\mathbf{k}^r \cdot \mathbf{r}}.$$

In region n we may write the wave vectors describing these waves as

$$\mathbf{k}_n^i = \hat{\mathbf{x}} k_{x,n} + \hat{\mathbf{z}} k_{z,n}, \qquad \mathbf{k}_n^r = \hat{\mathbf{x}} k_{x,n} - \hat{\mathbf{z}} k_{z,n},$$

where

$$k_{x,n}^2 + k_{z,n}^2 = k_n^2, \qquad k_n^2 = \omega^2 \tilde{\mu}_n \tilde{\epsilon}_n^c = (\beta_n - j\alpha_n)^2.$$

We note at the outset that, as with the single interface case, the boundary conditions are only satisfied when Snell's law of reflection holds, and thus

$$k_{x,n} = k_{x,0} = k_0 \sin \theta_i$$

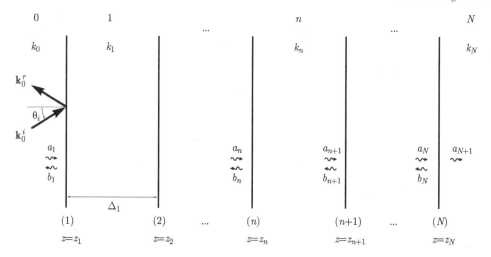

FIGURE 4.28

Interaction of a uniform plane wave with a multi-layered material.

where $k_0 = \omega(\tilde{\mu}_0 \tilde{\epsilon}_0^c)^{1/2}$ is the wavenumber of the 0th region (not necessarily free space). From this condition we have

$$k_{z,n} = \sqrt{k_n^2 - k_{x,0}^2} = \tau_n e^{-j\gamma_n}$$

where

$$\tau_n = (A_n^2 + B_n^2)^{1/4}, \qquad \gamma_n = \tfrac{1}{2} \tan^{-1}(B_n/A_n),$$

and

$$A_n = \beta_n^2 - \alpha_n^2 - (\beta_0^2 - \alpha_0^2)\sin^2\theta_i, \qquad B_n = 2(\beta_n\alpha_n - \beta_0\alpha_0\sin^2\theta_i).$$

Provided the incident wave is uniform, we can decompose the fields in every region into cases of perpendicular and parallel polarization. This is true even when the waves in certain layers are nonuniform. For the case of perpendicular polarization we can write the electric field in region n, $0 \le n \le N-1$, as $\tilde{\mathbf{E}}_{\perp n} = \tilde{\mathbf{E}}_{\perp n}^i + \tilde{\mathbf{E}}_{\perp n}^r$ where

$$\tilde{\mathbf{E}}_{\perp n}^i = \hat{\mathbf{y}} a_{n+1} e^{-jk_{x,n}x} e^{-jk_{z,n}(z-z_{n+1})},$$
$$\tilde{\mathbf{E}}_{\perp n}^r = \hat{\mathbf{y}} b_{n+1} e^{-jk_{x,n}x} e^{+jk_{z,n}(z-z_{n+1})},$$

and the magnetic field as $\tilde{\mathbf{H}}_{\perp n} = \tilde{\mathbf{H}}_{\perp n}^i + \mathbf{H}_{\perp n}^r$ where

$$\tilde{\mathbf{H}}_{\perp n}^i = \frac{-\hat{\mathbf{x}}k_{z,n} + \hat{\mathbf{z}}k_{x,n}}{k_n\eta_n} a_{n+1} e^{-jk_{x,n}x} e^{-jk_{z,n}(z-z_{n+1})},$$
$$\tilde{\mathbf{H}}_{\perp n}^r = \frac{+\hat{\mathbf{x}}k_{z,n} + \hat{\mathbf{z}}k_{x,n}}{k_n\eta_n} b_{n+1} e^{-jk_{x,n}x} e^{+jk_{z,n}(z-z_{n+1})}.$$

When $n = N$ there is no reflected wave; in this region we write

$$\tilde{\mathbf{E}}_{\perp N} = \hat{\mathbf{y}} a_{N+1} e^{-jk_{x,N}x} e^{-jk_{z,N}(z-z_N)},$$
$$\tilde{\mathbf{H}}_{\perp N} = \frac{-\hat{\mathbf{x}}k_{z,N} + \hat{\mathbf{z}}k_{x,N}}{k_N\eta_N} a_{N+1} e^{-jk_{x,N}x} e^{-jk_{z,N}(z-z_N)}.$$

Since a_1 is the known amplitude of the incident wave, there are $2N$ unknown wave amplitudes. We obtain the needed $2N$ simultaneous equations by applying the boundary conditions at each of the interfaces. At interface n located at $z = z_n$ $(1 \le n \le N - 1)$, we have by continuity of the tangential electric field,

$$a_n + b_n = a_{n+1}e^{-jk_{z,n}(z_n - z_{n+1})} + b_{n+1}e^{+jk_{z,n}(z_n - z_{n+1})}$$

while by continuity of the magnetic field,

$$-a_n \frac{k_{z,n-1}}{k_{n-1}\eta_{n-1}} + b_n \frac{k_{z,n-1}}{k_{n-1}\eta_{n-1}} = -a_{n+1}\frac{k_{z,n}}{k_n\eta_n}e^{-jk_{z,n}(z_n - z_{n+1})} + b_{n+1}\frac{k_{z,n}}{k_n\eta_n}e^{+jk_{z,n}(z_n - z_{n+1})}.$$

Noting that the wave impedance of region n is $Z_{\perp n} = k_n \eta_n / k_{z,n}$ and defining the region n propagation factor as

$$\tilde{P}_n = e^{-jk_{z,n}\Delta_n} \tag{4.272}$$

where $\Delta_n = z_{n+1} - z_n$, we can write

$$a_n \tilde{P}_n + b_n \tilde{P}_n = a_{n+1} + b_{n+1}\tilde{P}_n^2, \tag{4.273}$$

$$-a_n \tilde{P}_n + b_n \tilde{P}_n = -a_{n+1}\frac{Z_{\perp n-1}}{Z_{\perp n}} + b_{n+1}\frac{Z_{\perp n-1}}{Z_{\perp n}}\tilde{P}_n^2. \tag{4.274}$$

We must still apply the boundary conditions at $z = z_N$. Proceeding as above, we find that (4.273) and (4.274) hold for $n = N$ if we set $b_{N+1} = 0$ and $\tilde{P}_N = 1$.

Equations (4.273) and (4.274) may be put into a somewhat more convenient form through addition and subtraction, giving

$$2a_n = a_{n+1}\tilde{P}_n^{-1}\left(1 + \frac{Z_{\perp n-1}}{Z_{\perp n}}\right) + b_{n+1}\tilde{P}_n\left(1 - \frac{Z_{\perp n-1}}{Z_{\perp n}}\right), \tag{4.275}$$

$$2b_n = a_{n+1}\tilde{P}_n^{-1}\left(1 - \frac{Z_{\perp n-1}}{Z_{\perp n}}\right) + b_{n+1}\tilde{P}_n\left(1 + \frac{Z_{\perp n-1}}{Z_{\perp n}}\right). \tag{4.276}$$

These form a system of $2N$ simultaneous equations solvable by standard matrix methods. However, two alternative approaches may prove advantageous, depending on the application. We consider these next.

4.11.5.6 Analysis of multi-layered planar materials using a recursive approach

A recursive approach to (4.275)–(4.276) offers a nice physical picture of the multiple reflections in the layered medium. We begin by defining

$$\tilde{\Gamma}_n = \frac{Z_{\perp n} - Z_{\perp n-1}}{Z_{\perp n} + Z_{\perp n-1}} \tag{4.277}$$

as the *interfacial reflection coefficient* for interface n (i.e., the reflection coefficient assuming a single interface as in (4.257)), and

$$\tilde{T}_n = \frac{2Z_{\perp n}}{Z_{\perp n} + Z_{\perp n-1}} = 1 + \tilde{\Gamma}_n$$

as the *interfacial transmission coefficient* for interface n. Then (4.275) becomes

$$a_{n+1} = a_n \tilde{T}_n \tilde{P}_n + b_{n+1}\tilde{P}_n(-\tilde{\Gamma}_n)\tilde{P}_n.$$

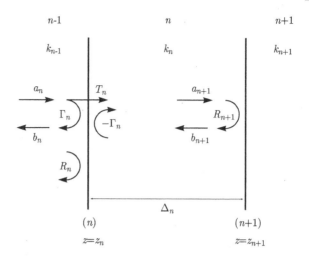

FIGURE 4.29
Wave flow diagram showing interaction of incident and reflected waves for region n.

Next, defining the *global* reflection coefficient R_n for region n as the ratio of the amplitudes of the reflected and incident waves,

$$\tilde{R}_n = b_n/a_n,$$

we can write

$$a_{n+1} = a_n \tilde{T}_n \tilde{P}_n + a_{n+1} \tilde{R}_{n+1} \tilde{P}_n (-\tilde{\Gamma}_n) \tilde{P}_n. \qquad (4.278)$$

For $n = N$ we note that $\tilde{R}_{N+1} = 0$, and $\tilde{P}_N = 1$ since a_N and a_{N+1} are assigned to opposite sides of the same interface. This gives

$$a_{N+1} = a_N \tilde{T}_N. \qquad (4.279)$$

If we choose to eliminate a_{n+1} from (4.273) and (4.274) we find that

$$b_n = a_n \tilde{\Gamma}_n + \tilde{R}_{n+1} \tilde{P}_n (1 - \tilde{\Gamma}_n) a_{n+1}. \qquad (4.280)$$

For $n = N$ this reduces to

$$b_N = a_N \tilde{\Gamma}_N. \qquad (4.281)$$

Equations (4.278) and (4.280) have nice physical interpretations. Consider Figure 4.29, which shows the wave amplitudes for region n. We may think of the wave incident on interface $n + 1$ with amplitude a_{n+1} as consisting of two terms. The first term is the wave transmitted through interface n (at $z = z_n$). This wave must propagate through a distance Δ_n to reach interface $n+1$ and thus has an amplitude $a_n \tilde{T}_n \tilde{P}_n$. The second term is the reflection at interface n of the wave traveling in the $-z$ direction within region n. The amplitude of the wave before reflection is merely $b_{n+1} \tilde{P}_n$, where the term $\tilde{P}_n$ results from the propagation of the negatively traveling wave from interface $n + 1$ to interface n. Now, since the interfacial reflection coefficient at interface n for a wave incident from region n is the negative of that for a wave incident from region $n - 1$ (since the wave is traveling in the reverse direction), and since the reflected wave must travel through a distance Δ_n from interface n back to interface $n + 1$, the amplitude of the second term is $b_{n+1}\tilde{P}_n(-\Gamma_n)\tilde{P}_n$. Finally, remembering that $b_{n+1} = \tilde{R}_{n+1}a_{n+1}$, we can write

$$a_{n+1} = a_n\tilde{T}_n\tilde{P}_n + a_{n+1}\tilde{R}_{n+1}\tilde{P}_n(-\tilde{\Gamma}_n)\tilde{P}_n.$$

This equation replicates (4.278), which was found using the boundary conditions. By similar reasoning, we may say that the wave traveling in the $-z$ direction in region $n - 1$ consists of a term reflected from the interface and a term transmitted through the interface. The amplitude of the reflected term is merely $a_n\tilde{\Gamma}_n$. The amplitude of the transmitted term is found by considering $b_{n+1} = \tilde{R}_{n+1}a_{n+1}$ propagated through a distance Δ_n and then transmitted backwards through interface n. Since the transmission coefficient for a wave going from region n to region $n - 1$ is $1 + (-\tilde{\Gamma}_n)$, the amplitude of the transmitted term is $\tilde{R}_{n+1}\tilde{P}_n(1 - \tilde{\Gamma}_n)a_{n+1}$. Thus

$$b_n = \tilde{\Gamma}_na_n + \tilde{R}_{n+1}\tilde{P}_n(1 - \tilde{\Gamma}_n)a_{n+1},$$

which replicates (4.280).

We are left with determining the various field amplitudes. This can be done using a simple recursive technique. Using $\tilde{T}_n = 1 + \tilde{\Gamma}_n$ we find from (4.278) that

$$a_{n+1} = \tilde{\tau}_na_n, \tag{4.282}$$

where

$$\tilde{\tau}_n = \frac{(1 + \tilde{\Gamma}_n)\tilde{P}_n}{1 + \tilde{\Gamma}_n\tilde{R}_{n+1}\tilde{P}_n^2}. \tag{4.283}$$

Substituting this into (4.280), we find

$$b_n = [\tilde{\Gamma}_n + \tilde{P}_n(1 - \tilde{\Gamma}_n)\tilde{R}_{n+1}\tilde{\tau}_n]a_n.$$

Using this expression we find a recursive relationship for the global reflection coefficient:

$$\tilde{R}_n = b_n/a_n = \tilde{\Gamma}_n + \tilde{P}_n(1 - \tilde{\Gamma}_n)\tilde{R}_{n+1}\tilde{\tau}_n. \tag{4.284}$$

The procedure is now as follows. Start with (4.279) to find

$$\tilde{\tau}_N = \tilde{T}_N. \tag{4.285}$$

Then find the global reflection coefficient for interface N, which by (4.281) is

$$\tilde{R}_N = b_N/a_N = \tilde{\Gamma}_N. \tag{4.286}$$

This is also obtained from (4.284) with $\tilde{R}_{N+1} = 0$. Next, compute $\tilde{\tau}_{N-1}$ from (4.283):

$$\tilde{\tau}_{N-1} = \frac{(1 + \tilde{\Gamma}_{N-1})\tilde{P}_{N-1}}{1 + \tilde{\Gamma}_{N-1}\tilde{R}_N\tilde{P}_{N-1}^2}.$$

Lastly use (4.284) to find $\tilde{R}_{N-1}$:

$$\tilde{R}_{N-1} = \tilde{\Gamma}_{N-1} + \tilde{P}_{N-1}(1 - \tilde{\Gamma}_{N-1})\tilde{R}_N\tilde{\tau}_{N-1}.$$

This process is repeated until reaching $\tilde{R}_1$, whereupon all the values of $\tilde{R}_n$ and $\tilde{\tau}_n$ are known. We then find the amplitudes beginning with a_1, which is the known incident field amplitude. From (4.284) we find

$$b_1 = a_1\tilde{R}_1,$$

and from (4.282) we find

$$a_2 = \tilde{\tau}_1 a_1.$$

This process is repeated until all field amplitudes are known.

Often we wish to know the transmission and reflection coefficients for the entire structure. In this case we write the structure reflection coefficient as

$$\tilde{\mathbb{R}} = b_1/a_1 = \tilde{R}_1,$$

and the structure transmission coefficient as

$$\tilde{\mathbb{T}} = a_{N+1}/a_1 = \tilde{\tau}_1 \tilde{\tau}_2 \cdots \tilde{\tau}_N.$$

To find the reflection coefficient for the opposite side of the structure, or for transmission in the opposite direction, we merely reverse the structure.

We note that the process outlined above holds for parallel polarization, provided we use the parallel wave impedances

$$Z_{\|n} = k_{z,n} \eta_n / k_n$$

when computing the interfacial reflection coefficients.

▶ **Example 4.15:** Single-layer reflection and transmission using recursion

Consider a slab of material of thickness Δ, sandwiched between two lossless dielectrics. A uniform plane wave of frequency ω impinges on interface 1. Compute the structure reflection and transmission coefficients.

Solution: Here $N = 2$ (two interfaces and three regions). By (4.285) we have $\tilde{\tau}_2 = \tilde{T}_2$ and by (4.286) we have $\tilde{R}_2 = \tilde{\Gamma}_2$. Then (4.283) gives

$$\tilde{\tau}_1 = \frac{(1 + \tilde{\Gamma}_1)\tilde{P}_1}{1 + \tilde{\Gamma}_1 \tilde{R}_2 \tilde{P}_1^2} = \frac{(1 + \tilde{\Gamma}_1)\tilde{P}_1}{1 + \tilde{\Gamma}_1 \tilde{\Gamma}_2 \tilde{P}_1^2},$$

while (4.284) gives

$$\tilde{R}_1 = \tilde{\Gamma}_1 + \tilde{P}_1(1 - \tilde{\Gamma}_1)\tilde{R}_2 \tilde{\tau}_1 = \tilde{\Gamma}_1 + \frac{(1 - \tilde{\Gamma}_1^2)\tilde{P}_1^2 \tilde{\Gamma}_2}{1 + \tilde{\Gamma}_1 \tilde{\Gamma}_2 \tilde{P}_1^2} = \frac{\tilde{\Gamma}_1 + \tilde{\Gamma}_2 \tilde{P}_1^2}{1 + \tilde{\Gamma}_1 \tilde{\Gamma}_2 \tilde{P}_1^2}.$$

The structure reflection coefficient is

$$\tilde{\mathbb{R}} = \tilde{R}_1 = \frac{\tilde{\Gamma}_1 + \tilde{\Gamma}_2 \tilde{P}_1^2}{1 + \tilde{\Gamma}_1 \tilde{\Gamma}_2 \tilde{P}_1^2} \tag{4.287}$$

and the structure transmission coefficient is

$$\tilde{\mathbb{T}} = \tilde{\tau}_1 \tilde{\tau}_2 = \frac{(1 + \tilde{\Gamma}_1)(1 + \tilde{\Gamma}_2)\tilde{P}_1}{1 + \tilde{\Gamma}_1 \tilde{\Gamma}_2 \tilde{P}_1^2}. \quad \blacktriangleleft \tag{4.288}$$

▶ **Example 4.16:** Quarter-wave transformer

Consider a lossless slab of material of thickness Δ sandwiched between two lossless dielectrics. A time-harmonic uniform plane wave of frequency $\omega = \breve{\omega}$ is normally incident onto interface 1. Determine the conditions under which the reflected wave vanishes.

Solution: Let $\mathbb{R} = \tilde{\mathbb{R}}(\breve{\omega})$, etc. From (4.287) we see that $\mathbb{R} = 0$ and thus the wave reflected by a material slab vanishes when

$$\Gamma_1 + \Gamma_2 P_1^2 = 0.$$

Since the field in region 0 is normally incident, we have

$$k_{z,n} = k_n = \beta_n = \breve{\omega}\sqrt{\mu_n \epsilon_n}.$$

If we choose $P_1^2 = -1$, then $\Gamma_1 = \Gamma_2$ results in no reflected wave. This requires

$$\frac{Z_1 - Z_0}{Z_1 + Z_0} = \frac{Z_2 - Z_1}{Z_2 + Z_1}.$$

Clearing the denominator, we find that $2Z_1^2 = 2Z_0 Z_2$, or

$$Z_1 = \sqrt{Z_0 Z_2}.$$

This condition makes the reflected field vanish if we can ensure that $P_1^2 = -1$. To do this we need $e^{-j\beta_1 2\Delta} = -1$. The minimum thickness that satisfies this condition is $\beta_1 2\Delta = \pi$. Since $\beta_1 = 2\pi/\lambda_1$, this is equivalent to $\Delta = \lambda_1/4$.

A layer of this type is called a *quarter-wave transformer*. Since no wave is reflected from the initial interface, and since all the regions are assumed lossless, all of the power carried by the incident wave in the first region is transferred into the third region. Thus, two regions of differing materials may be "matched" by inserting an appropriate slab between them. This technique finds use in optical coatings for lenses and for reducing the radar reflectivity of objects. ◄

▶ **Example 4.17:** Reflection from a conductor-backed dielectric layer

A lossless dielectric slab with $\tilde{\epsilon} = \epsilon_1 = \epsilon_{1r}\epsilon_0$ and $\tilde{\mu} = \mu_0$ is backed by a perfect conductor and immersed in free space, as shown in Figure 4.30. A perpendicularly polarized uniform plane wave is incident on the slab from free space. If the incident electric field is given by $\tilde{\mathbf{E}}^i(\omega) = \hat{\mathbf{y}}\tilde{E}_\perp^i(\omega)$, determine the frequency-domain reflected field.

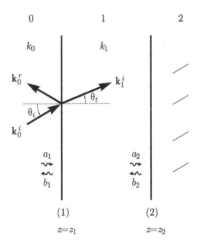

FIGURE 4.30
Interaction of a uniform plane wave with a conductor-backed dielectric slab.

Solution: Since $\epsilon_0 < \epsilon_1$, total internal reflection cannot occur. Thus the wave vectors in region 1 have real components and can be written as

$$\mathbf{k}_1^i = k_{x,1}\hat{\mathbf{x}} + k_{z,1}\hat{\mathbf{z}}, \qquad \mathbf{k}_1^r = k_{x,1}\hat{\mathbf{x}} - k_{z,1}\hat{\mathbf{z}}.$$

By Snell's law, $k_{x,1} = k_0 \sin\theta_i = k_1 \sin\theta_t$, and so

$$k_{z,1} = \sqrt{k_1^2 - k_{x,1}^2} = \frac{\omega}{c}\sqrt{\epsilon_{1r} - \sin^2\theta_i} = k_1 \cos\theta_t$$

where θ_t is the transmission angle in region 1. Since region 2 is a perfect conductor, we have $\tilde{R}_2 = -1$. By (4.283) we have

$$\tilde{\tau}_1(\omega) = \frac{(1+\tilde{\Gamma}_1)\tilde{P}_1(\omega)}{1 - \tilde{\Gamma}_1 \tilde{P}_1^2(\omega)},$$

where from (4.277),

$$\tilde{\Gamma}_1 = \frac{Z_1 - Z_0}{Z_1 + Z_0}$$

is not a function of frequency. Using this in (4.284) gives the structure reflection coefficient

$$\tilde{\mathbb{R}}(\omega) = \tilde{R}_1(\omega) = \tilde{\Gamma}_1 - \frac{(1-\tilde{\Gamma}_1^2)\tilde{P}_1^2(\omega)}{1 - \tilde{\Gamma}_1 \tilde{P}_1^2(\omega)} = \frac{\tilde{\Gamma}_1 - \tilde{P}_1^2(\omega)}{1 - \tilde{\Gamma}_1 \tilde{P}_1^2(\omega)}. \qquad (4.289)$$

From this we have the frequency-domain reflected field

$$\tilde{\mathbf{E}}_\perp^r(\mathbf{r},\omega) = \hat{\mathbf{y}}\tilde{\mathbb{R}}(\omega)\tilde{E}_\perp^i(\omega)e^{-j\mathbf{k}_1^r(\omega)\cdot\mathbf{r}}. \quad \blacktriangleleft \qquad (4.290)$$

In many layered-media problems it is possible to obtain the time-domain fields from the frequency-domain fields through the inverse Fourier transform. Calculation of the transform is facilitated by a clever expansion of the structure reflection coefficient or structure transmission coefficient, which allows an interpretation in terms of multiple reflections within the layers. The following example applies this approach to the conductor-backed dielectric layer considered above.

▶ **Example 4.18:** Time-domain reflection from a conductor-backed dielectric layer

Compute the time-domain electric field reflected from a conductor-backed dielectric layer by inverse transforming the frequency-domain field (4.290). Interpret the result in terms of multiple reflections within the dielectric layer.

Solution: By the approach used to obtain (4.270), we write

$$\mathbf{E}_\perp^r(\mathbf{r},t) = \hat{\mathbf{y}}E^r\left(t - \frac{\hat{\mathbf{k}}_1^r \cdot \mathbf{r}}{c}\right),$$

where by the convolution theorem,

$$E^r(t) = \mathbb{R}(t) * E_\perp^i(t). \qquad (4.291)$$

Here $E_\perp^i(t) = \mathcal{F}^{-1}\{\tilde{E}_\perp^i(\omega)\}$ is the time waveform of the incident plane wave and $\mathbb{R}(t) = \mathcal{F}^{-1}\{\tilde{\mathbb{R}}(\omega)\}$ is the time-domain reflection coefficient of the structure. To invert $\tilde{\mathbb{R}}(\omega)$ we apply the binomial expansion

$$(1-x)^{-1} = 1 + x + x^2 + x^3 + \cdots$$

to the denominator of (4.289):

$$\tilde{\mathbb{R}}(\omega) = [\Gamma_1 - \tilde{P}_1^2(\omega)] \left\{ 1 + [\Gamma_1 \tilde{P}_1^2(\omega)] + [\Gamma_1 \tilde{P}_1^2(\omega)]^2 + [\Gamma_1 \tilde{P}_1^2(\omega)]^3 + \ldots \right\}$$
$$= \Gamma_1 - [1 - \Gamma_1^2]\tilde{P}_1^2(\omega) - [1 - \Gamma_1^2]\Gamma_1 \tilde{P}_1^4(\omega)$$
$$- [1 - \Gamma_1^2]\Gamma_1^2 \tilde{P}_1^6(\omega) - \cdots . \tag{4.292}$$

Thus we need the inverse transform of

$$\tilde{P}_1^{2n}(\omega) = e^{-j2nk_{z,1}\Delta_1}$$
$$= e^{-j2nk_1\Delta_1 \cos\theta_t}.$$

Writing $k_1 = \omega/v_1$, where $v_1 = 1/(\mu_0 \epsilon_1)^{1/2}$ is the phase velocity of the wave in region 1, and using $1 \leftrightarrow \delta(t)$ along with the time-shifting theorem (A.3), we have

$$\tilde{P}_1^{2n}(\omega) = e^{-j\omega 2n\mathcal{T}} \leftrightarrow \delta(t - 2n\mathcal{T})$$

where $\mathcal{T} = \Delta_1 \cos\theta_t/v_1$. With this the inverse transform of $\tilde{\mathbb{R}}$ in (4.292) is

$$\mathbb{R}(t) = \Gamma_1 \delta(t) - (1 + \Gamma_1)(1 - \Gamma_1)\delta(t - 2\mathcal{T}) - (1 + \Gamma_1)(1 - \Gamma_1)\Gamma_1 \delta(t - 4\mathcal{T}) - \cdots$$

and thus from (4.291)

$$E^r(t) = \Gamma_1 E_\perp^i(t) - (1 + \Gamma_1)(1 - \Gamma_1)E_\perp^i(t - 2\mathcal{T})$$
$$- (1 + \Gamma_1)(1 - \Gamma_1)\Gamma_1 E_\perp^i(t - 4\mathcal{T}) - \cdots .$$

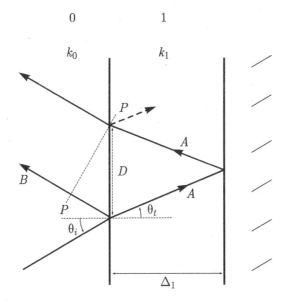

FIGURE 4.31
Timing diagram for multiple reflections from a conductor-backed dielectric slab.

The reflected field consists of time-shifted and amplitude-scaled versions of the incident field waveform. These terms can be interpreted as multiple reflections of the incident wave.

Consider Figure 4.31. The first term is the direct reflection from interface 1 and thus has its amplitude multiplied by Γ_1. The next term represents a wave that penetrates the interface (and thus has its amplitude multiplied by the transmission coefficient $1 + \Gamma_1$), propagates to and reflects from the conductor (and thus has its amplitude multiplied by -1), and then propagates back to the interface and passes through in the opposite direction (and thus has its amplitude multiplied by the transmission coefficient for passage from region 1 to region 0, $1 - \Gamma_1$). The time delay between this wave and the initially reflected wave is given by $2\mathcal{T}$, as discussed in detail below. The third term represents a wave that penetrates the interface, reflects from the conductor, returns to and reflects from the interface a second time, again reflects from the conductor, and then passes through the interface in the opposite direction. Its amplitude has an additional multiplicative factor of $-\Gamma_1$ to account for reflection from the interface and an additional factor of -1 to account for the second reflection from the conductor, and is time-delayed by an additional $2\mathcal{T}$. Subsequent terms account for additional reflections; the nth reflected wave amplitude is multiplied by an additional $(-1)^n$ and $(-\Gamma_1)^n$ and is time-delayed by an additional $2n\mathcal{T}$.

It is important to understand that the time delay $2\mathcal{T}$ is *not* just the propagation time for the wave to travel through the slab. To properly describe the timing between the initially reflected wave and the waves that reflect from the conductor, we must consider the field over identical observation planes as shown in Figure 4.31. For example, consider the observation plane designated P-P, intersecting the first "exit point" on interface 1. To arrive at this plane the initially reflected wave takes the path labeled B, arriving at a time

$$\frac{D \sin \theta_i}{v_0}$$

after the time of initial reflection, where $v_0 = c$ is the velocity in region 0. To arrive at this same plane the wave that penetrates the surface takes the path labeled A, arriving at a time

$$\frac{2\Delta_1}{v_1 \cos \theta_t}$$

where v_1 is the wave velocity in region 1 and θ_t is the transmission angle. Noting that $D = 2\Delta_1 \tan \theta_t$, the time delay between the arrival of the two waves at the plane P-P is

$$
\begin{aligned}
T &= \frac{2\Delta_1}{v_1 \cos \theta_t} - \frac{D \sin \theta_i}{v_0} \\
&= \frac{2\Delta_1}{v_1 \cos \theta_t} \left(1 - \frac{\sin \theta_t \sin \theta_i}{v_0/v_1} \right).
\end{aligned}
$$

By Snell's law of refraction (4.268) we can write

$$\frac{v_0}{v_1} = \frac{\sin \theta_i}{\sin \theta_t},$$

which, upon substitution, gives

$$T = 2\frac{\Delta_1 \cos \theta_t}{v_1}.$$

This is exactly the time delay $2\mathcal{T}$. ◄

4.11.5.7   Analysis of multi-layered planar materials using cascaded matrices

Equations (4.275) and (4.276) can be recast in the matrix form

$$
\begin{bmatrix} T_{11}^{(n)} & T_{12}^{(n)} \\ T_{21}^{(n)} & T_{22}^{(n)} \end{bmatrix}
\begin{bmatrix} a_{n+1} \\ b_{n+1} \end{bmatrix}
=
\begin{bmatrix} a_n \\ b_n \end{bmatrix}
$$

where

$$T_{11}^{(n)} = \frac{1}{2}\frac{Z_n + Z_{n-1}}{Z_n}\tilde{P}_n^{-1} = \frac{1}{\tilde{T}_n\tilde{P}_n},$$ (4.293)

$$T_{12}^{(n)} = \frac{1}{2}\frac{Z_n - Z_{n-1}}{Z_n}\tilde{P}_n = \frac{\tilde{\Gamma}_n\tilde{P}_n}{\tilde{T}_n},$$ (4.294)

$$T_{21}^{(n)} = \frac{1}{2}\frac{Z_n - Z_{n-1}}{Z_n}\tilde{P}_n^{-1} = \frac{\tilde{\Gamma}_n}{\tilde{T}_n\tilde{P}_n},$$ (4.295)

$$T_{22}^{(n)} = \frac{1}{2}\frac{Z_n + Z_{n-1}}{Z_n}\tilde{P}_n = \frac{\tilde{P}_n}{\tilde{T}_n}.$$ (4.296)

Here Z_n represents $Z_{n\perp}$ for perpendicular polarization and $Z_{n\parallel}$ for parallel polarization. The matrix entries, often called *transmission parameters*, are similar to the parameters used to describe microwave networks except that in network theory the wave amplitudes are often normalized using the wave impedances.

We may use the transmission parameters to describe the cascaded system of any number of layers by simply multiplying transmission matrices. For example, a single layer of material may be described using the two transmission matrices

$$\begin{bmatrix} T_{11}^{(1)} & T_{12}^{(1)} \\ T_{21}^{(1)} & T_{22}^{(1)} \end{bmatrix}\begin{bmatrix} a_2 \\ b_2 \end{bmatrix} = \begin{bmatrix} a_1 \\ b_1 \end{bmatrix}$$ (4.297)

and

$$\begin{bmatrix} T_{11}^{(2)} & T_{12}^{(2)} \\ T_{21}^{(2)} & T_{22}^{(2)} \end{bmatrix}\begin{bmatrix} a_3 \\ b_3 \end{bmatrix} = \begin{bmatrix} a_2 \\ b_2 \end{bmatrix}.$$ (4.298)

Substituting (4.298) into (4.297), we get

$$\begin{bmatrix} T_{11}^{(1)} & T_{12}^{(1)} \\ T_{21}^{(1)} & T_{22}^{(1)} \end{bmatrix}\begin{bmatrix} T_{11}^{(2)} & T_{12}^{(2)} \\ T_{21}^{(2)} & T_{22}^{(2)} \end{bmatrix}\begin{bmatrix} a_3 \\ b_3 \end{bmatrix} = \begin{bmatrix} a_1 \\ b_1 \end{bmatrix}.$$

Here we also use $\tilde{P}_2 = 1$. Extending this to the system of $N+1$ regions shown in Figure 4.28, we have

$$\begin{bmatrix} T_{11} & T_{12} \\ T_{21} & T_{22} \end{bmatrix}\begin{bmatrix} a_{N+1} \\ b_{N+1} \end{bmatrix} = \begin{bmatrix} a_1 \\ b_1 \end{bmatrix}$$

where

$$\begin{bmatrix} T_{11} & T_{12} \\ T_{21} & T_{22} \end{bmatrix} = \begin{bmatrix} T_{11}^{(1)} & T_{12}^{(1)} \\ T_{21}^{(1)} & T_{22}^{(1)} \end{bmatrix}\begin{bmatrix} T_{11}^{(2)} & T_{12}^{(2)} \\ T_{21}^{(2)} & T_{22}^{(2)} \end{bmatrix}\cdots\begin{bmatrix} T_{11}^{(N)} & T_{12}^{(N)} \\ T_{21}^{(2)} & T_{22}^{(2)} \end{bmatrix},$$

with $\tilde{P}_N = 1$. To find the reflection and transmission coefficients for the entire structure, set $b_{N+1} = 0$:

$$\begin{bmatrix} T_{11} & T_{12} \\ T_{21} & T_{22} \end{bmatrix}\begin{bmatrix} a_{N+1} \\ 0 \end{bmatrix} = \begin{bmatrix} a_1 \\ b_1 \end{bmatrix}.$$

Then $a_1 = T_{11}a_{N+1}$ and

$$\tilde{\mathbb{T}} = a_{N+1}/a_1 = 1/T_{11}.$$ (4.299)

Also, $b_1 = T_{21}a_{N+1}$, which along with (4.299) gives $\tilde{\mathbb{R}} = b_1/a_1 = T_{21}/T_{11}$.

▶ **Example 4.19:** Single layer reflection and transmission using transmission matrices

Repeat Example 4.15 using transmission matrices.

Solution: Since $N = 2$, the transmission matrix for the structure is

$$
\begin{bmatrix} T_{11} & T_{12} \\ T_{21} & T_{22} \end{bmatrix} = \begin{bmatrix} \frac{1}{\tilde{T}_1 \tilde{P}_1} & \frac{\tilde{\Gamma}_1 \tilde{P}_1}{\tilde{T}_1} \\ \frac{\tilde{\Gamma}_1}{\tilde{T}_1 \tilde{P}_1} & \frac{\tilde{P}_1}{\tilde{T}_1} \end{bmatrix} \begin{bmatrix} \frac{1}{\tilde{T}_2} & \frac{\tilde{\Gamma}_2}{\tilde{T}_2} \\ \frac{\tilde{\Gamma}_2}{\tilde{T}_2} & \frac{1}{\tilde{T}_2} \end{bmatrix}
$$

$$
= \begin{bmatrix} \frac{1 + \tilde{\Gamma}_1 \tilde{\Gamma}_2 \tilde{P}_1^2}{\tilde{T}_1 \tilde{T}_2 \tilde{P}_1} & \frac{\tilde{\Gamma}_2 + \tilde{\Gamma}_1 \tilde{P}_1^2}{\tilde{T}_1 \tilde{T}_2 \tilde{P}_1} \\ \frac{\tilde{\Gamma}_1 + \tilde{\Gamma}_2 \tilde{P}_1^2}{\tilde{T}_1 \tilde{T}_2 \tilde{P}_1} & \frac{\tilde{\Gamma}_1 \tilde{\Gamma}_2 + \tilde{P}_1^2}{\tilde{T}_1 \tilde{T}_2 \tilde{P}_1} \end{bmatrix}. \tag{4.300}
$$

Here we have used $\tilde{P}_2 = 1$. The structure reflection coefficient is

$$
\begin{aligned}
\tilde{\mathbb{R}} &= \frac{T_{21}}{T_{11}} \\
&= \frac{\tilde{\Gamma}_1 + \tilde{\Gamma}_2 \tilde{P}_1^2}{1 + \tilde{\Gamma}_1 \tilde{\Gamma}_2 \tilde{P}_1^2}
\end{aligned} \tag{4.301}
$$

as in (4.287). The structure transmission coefficient expression,

$$
\begin{aligned}
\tilde{\mathbb{T}} &= \frac{1}{T_{11}} = \frac{\tilde{T}_1 \tilde{T}_2 \tilde{P}_1}{1 + \tilde{\Gamma}_1 \tilde{\Gamma}_2 \tilde{P}_1^2} \\
&= \frac{(1 + \tilde{\Gamma}_1)(1 + \tilde{\Gamma}_2) \tilde{P}_1}{1 + \tilde{\Gamma}_1 \tilde{\Gamma}_2 \tilde{P}_1^2},
\end{aligned}
$$

is the same as (4.288). ◀

▶ **Example 4.20:** Reflection from a conductor-backed material layer using transmission matrices

A material slab is backed by a perfect conductor. Determine the structure reflection coefficient using transmission matrices.

Solution: Since $N = 2$, the transmission matrix is (4.300). The presence of the conductor backing requires $\tilde{\Gamma}_2 = -1$. Substitution into (4.301) gives

$$
\tilde{\mathbb{R}} = \frac{\tilde{\Gamma}_1 - \tilde{P}_1^2}{1 - \tilde{\Gamma}_1 \tilde{P}_1^2}.
$$

Compare with (4.289) of Example 4.17. ◀

▶ **Example 4.21:** Reflection from a double material layer using transmission matrices

Consider the two material layers of Figure 4.32. This configuration is used for the characterization of material properties (cf., § 4.15). Determine the structure reflection coefficient using transmission matrices. Consider the cases when the second layer is backed by (1) air and (2) a conductor.

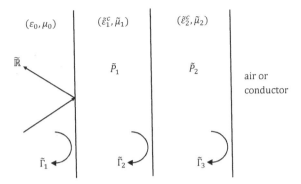

FIGURE 4.32
Reflection from two material layers.

Solution: With $N = 3$, the cascaded system is described by a product of three transmission matrices:

$$
\begin{bmatrix} T_{11} & T_{12} \\ T_{21} & T_{22} \end{bmatrix} = \begin{bmatrix} \frac{1}{\tilde{T}_1 \tilde{P}_1} & \frac{\tilde{\Gamma}_1 \tilde{P}_1}{\tilde{T}_1} \\ \frac{\tilde{\Gamma}_1}{\tilde{T}_1 \tilde{P}_1} & \frac{\tilde{P}_1}{\tilde{T}_1} \end{bmatrix} \begin{bmatrix} \frac{1}{\tilde{T}_2 \tilde{P}_2} & \frac{\tilde{\Gamma}_2 \tilde{P}_2}{\tilde{T}_2} \\ \frac{\tilde{\Gamma}_2}{\tilde{T}_2 \tilde{P}_2} & \frac{\tilde{P}_2}{\tilde{T}_2} \end{bmatrix} \begin{bmatrix} \frac{1}{\tilde{T}_3} & \frac{\tilde{\Gamma}_3}{\tilde{T}_3} \\ \frac{\tilde{\Gamma}_3}{\tilde{T}_3} & \frac{1}{\tilde{T}_3} \end{bmatrix}
$$

$$
= \begin{bmatrix} \frac{1 + \tilde{\Gamma}_2 \tilde{\Gamma}_3 \tilde{P}_2^2 + \tilde{\Gamma}_1 \tilde{\Gamma}_2 \tilde{P}_1^2 + \tilde{\Gamma}_1 \tilde{\Gamma}_3 \tilde{P}_1^2 \tilde{P}_2^2}{\tilde{T}_1 \tilde{T}_2 \tilde{T}_3 \tilde{P}_1 \tilde{P}_2} & \frac{\tilde{\Gamma}_3 + \tilde{\Gamma}_2 \tilde{P}_2^2 + \tilde{\Gamma}_1 \tilde{\Gamma}_2 \tilde{\Gamma}_3 \tilde{P}_1^2 + \tilde{\Gamma}_1 \tilde{P}_1^2 \tilde{P}_2^2}{\tilde{T}_1 \tilde{T}_2 \tilde{T}_3 \tilde{P}_1 \tilde{P}_2} \\ \frac{\tilde{\Gamma}_1 + \tilde{\Gamma}_1 \tilde{\Gamma}_2 \tilde{\Gamma}_3 \tilde{P}_2^2 + \tilde{\Gamma}_2 \tilde{P}_1^2 + \tilde{\Gamma}_3 \tilde{P}_1^2 \tilde{P}_2^2}{\tilde{T}_1 \tilde{T}_2 \tilde{T}_3 \tilde{P}_1 \tilde{P}_2} & \frac{\tilde{\Gamma}_1 \tilde{\Gamma}_3 + \tilde{\Gamma}_1 \tilde{\Gamma}_2 \tilde{P}_2^2 + \tilde{\Gamma}_2 \tilde{\Gamma}_3 \tilde{P}_1^2 + \tilde{P}_1^2 \tilde{P}_2^2}{\tilde{T}_1 \tilde{T}_2 \tilde{T}_3 \tilde{P}_1 \tilde{P}_2} \end{bmatrix}.
$$

Here we have used $\tilde{P}_3 = 1$. The structure reflection coefficient is given by $\tilde{\mathbb{R}} = T_{21}/T_{11}$. When the second layer is backed by air (free space) we have

$$
\tilde{\mathbb{R}} = \frac{\tilde{\Gamma}_1 + \tilde{\Gamma}_1 \tilde{\Gamma}_2 \tilde{\Gamma}_3 \tilde{P}_2^2 + \tilde{\Gamma}_2 \tilde{P}_1^2 + \tilde{\Gamma}_3 \tilde{P}_1^2 \tilde{P}_2^2}{1 + \tilde{\Gamma}_2 \tilde{\Gamma}_3 \tilde{P}_2^2 + \tilde{\Gamma}_1 \tilde{\Gamma}_2 \tilde{P}_1^2 + \tilde{\Gamma}_1 \tilde{\Gamma}_3 \tilde{P}_1^2 \tilde{P}_2^2} \quad \text{where} \quad \tilde{\Gamma}_3 = \frac{Z_0 - Z_2}{Z_0 + Z_2}. \tag{4.302}
$$

When the final layer is a conductor, $\tilde{\Gamma}_3 = -1$ and

$$
\tilde{\mathbb{R}} = \frac{\tilde{\Gamma}_1 - \tilde{\Gamma}_1 \tilde{\Gamma}_2 \tilde{P}_2^2 + \tilde{\Gamma}_2 \tilde{P}_1^2 - \tilde{P}_1^2 \tilde{P}_2^2}{1 - \tilde{\Gamma}_2 \tilde{P}_2^2 + \tilde{\Gamma}_1 \tilde{\Gamma}_2 \tilde{P}_1^2 - \tilde{\Gamma}_1 \tilde{P}_1^2 \tilde{P}_2^2}. \tag{4.303}
$$

We can check this by specializing it to the single layer case considered in Example 4.20. Setting $\tilde{P}_2 = 1$ and $\tilde{\Gamma}_2 = -1$ in (4.303), we get

$$
\tilde{\mathbb{R}} = \frac{\tilde{\Gamma}_1 + \tilde{\Gamma}_1 - \tilde{P}_1^2 - \tilde{P}_1^2}{1 + 1 - \tilde{\Gamma}_1 \tilde{P}_1^2 - \tilde{\Gamma}_1 \tilde{P}_1^2} = \frac{\tilde{\Gamma}_1 - \tilde{P}_1^2}{1 - \tilde{\Gamma}_1 \tilde{P}_1^2},
$$

which agrees with (4.289). ◄

4.11.6 Electromagnetic shielding

The concept of electromagnetic shielding was introduced in § 3.7, where we defined shielding effectiveness and examined several canonical problems for static and quasistatic fields. Canonical problems enable us to estimate shielding effectiveness for practical situations and also promote insight into the dependence of shielding effectiveness on a wide variety of parameters. As this approach is substantially more difficult for general electromagnetic fields, however, engineers often resort to numerical tools for complicated practical problems. For instance, we will explore the penetration of fields through a narrow rectangular slot in a conducting ground plane in § 7.7; this will help us understand how apertures can reduce the effectiveness of a shield. But the integral equation approach is not easily extended to more complicated situations.

There is, however, one canonical problem that finds widespread use: the planar conducting layer. Estimation of fields penetrating a planar shield relies on simple formulas; even simpler formulas are available through suitable approximations.

Consider a planar shield of thickness Δ, immersed in free space and illuminated by a plane wave at incidence angle θ_i (Figure 4.33). Although the shielding effectiveness can be due to both electric and magnetic losses in the shield, we will restrict ourselves to metallic shields, where the loss is due to conductivity. Thus we assume that the shield has permeability $\mu = \mu_0 \mu_r$ and complex permittivity $\tilde{\epsilon}^c$, where by (4.26)

$$\tilde{\epsilon}^c = \epsilon_0 \epsilon_r + \frac{\sigma}{j\omega},$$

with σ the conductivity of the shield. For most metals σ may be assumed constant over frequency bands where shielding is desired. We can find the transmission coefficient of the shield by specializing (4.288) to $\tilde{\Gamma}_1 = -\tilde{\Gamma}_2 = \tilde{\Gamma}$, giving

$$\tilde{\mathbb{T}} = \frac{(1 - \tilde{\Gamma}^2)\tilde{P}}{1 - \tilde{\Gamma}^2 \tilde{P}^2}, \tag{4.304}$$

where the interfacial reflection coefficient is

$$\tilde{\Gamma} = \frac{Z - Z_0}{Z + Z_0}$$

and the propagation term is $\tilde{P} = e^{-jk_z \Delta}$. Here

$$Z = Z_\parallel = \frac{k_z \eta}{k}, \qquad Z_0 = Z_{0\parallel} = \eta_0 \cos \theta_i$$

for parallel polarization, and

$$Z = Z_\perp = \frac{k\eta}{k_z}, \qquad Z_0 = Z_{0\perp} = \frac{\eta_0}{\cos \theta_i}$$

for perpendicular polarization. In both cases $k_z = \sqrt{k^2 - k_0^2 \sin^2 \theta_i}$, $\eta = \sqrt{\mu/\tilde{\epsilon}^c}$, and $k = \omega\sqrt{\mu\tilde{\epsilon}^c}$. For the case of normal incidence, $k_z = k$ and the interfacial reflection coefficient reduces to

$$\tilde{\Gamma} = \frac{\eta - \eta_0}{\eta + \eta_0}.$$

For either polarization, the shielding effectiveness is given by

$$SE = 20 \log_{10} |1/\tilde{\mathbb{T}}|, \tag{4.305}$$

which is dependent on the incidence angle θ_i. Substituting the transmission coefficient (4.304) into (4.305), we find that the shielding effectiveness can be written as the sum

$$SE = R + A + M. \tag{4.306}$$

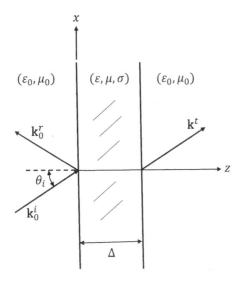

FIGURE 4.33
Planar shield obliquely illuminated by a plane wave.

Here

$$R = -20 \log_{10} |1 - \tilde{\Gamma}^2|, \tag{4.307}$$

$$A = -20 \log_{10} |\tilde{P}|, \tag{4.308}$$

$$M = 20 \log_{10} |1 - \tilde{\Gamma}^2 \tilde{P}^2|. \tag{4.309}$$

Each term has an important physical interpretation. The term A is independent of $\tilde{\Gamma}$ and represents the shielding effectiveness provided by attenuation of the wave as it passes through the shield. The term R is independent of $\tilde{P}$ and represents the shielding provided by the reflection of the incident wave from the initial air-shield interface. Note that since $1 - \tilde{\Gamma}^2 = (1 + \tilde{\Gamma})(1 - \tilde{\Gamma})$ the term R is the product of the interfacial transmission coefficient for passage into the shield with the interfacial transmission coefficient for passage out of the shield. When the shield is thick, $|\tilde{P}| \ll 1$, and $M \approx 0$. In this case the shielding effectiveness is completely determined by initial reflection and attenuation, i.e., by R and A. When the shield is thin, a significant contribution to shielding is provided by the term M, which represents the contribution due to reflections from the second (shield-air) interface, and any subsequent multiple reflections that lead to additional attenuation and transmission back into the region of the incident wave. While R is independent of the shield thickness, A and M are highly dependent on the thickness. Often the thick shield assumption is invoked and the term M is ignored.

Because the shield is assumed to be a good conductor, it is possible to provide useful approximations to the terms in (4.306). Note that

$$k = \omega \sqrt{\mu \tilde{\epsilon}^c} = \omega \sqrt{\mu \epsilon_0 \epsilon_r \left(1 - j \frac{\sigma}{\omega \epsilon_0 \epsilon_r}\right)}.$$

We define a good conductor as one for which

$$\frac{\sigma}{\omega \epsilon_0 \epsilon_r} \gg 1$$

and thus

$$k \approx \omega \sqrt{-j\mu\sigma\omega} = \frac{1-j}{\delta},$$

where

$$\delta = \frac{1}{\sqrt{\pi f \mu \sigma}} \tag{4.310}$$

is the skin depth. Note that

$$\frac{k}{k_0} = \sqrt{\mu_r \epsilon_r} \sqrt{1 - j\frac{\sigma}{\omega \epsilon_0 \epsilon_r}},$$

and thus $|k| \gg k_0$. Then

$$k_z = \sqrt{k^2 - k_0^2 \sin^2 \theta_i} \approx k.$$

Hence, in the shield, $\mathbf{k} = \hat{\mathbf{x}}k_0 \sin \theta_i + \hat{\mathbf{z}}k_z \approx \hat{\mathbf{z}}k_z$, meaning that the wave travels in the shield primarily normal to the interface regardless of the incidence angle. Therefore

$$\tilde{P} = e^{-(1-j)\frac{\Delta}{\delta}} \tag{4.311}$$

and

$$A = -20 \log_{10} \left| e^{-\frac{\Delta}{\delta}} \right| = 8.686 \frac{\Delta}{\delta} \ \text{dB}. \tag{4.312}$$

Because $k_z \approx k$ when the shield is a good conductor, we also have

$$Z_\perp \approx Z_\parallel \approx \eta \quad \text{and} \quad \tilde{\Gamma} \approx \frac{\eta - Z_0}{\eta + Z_0}.$$

Note that

$$\frac{\eta}{\eta_0} = \sqrt{\frac{\mu_r}{\epsilon_r + j\frac{\sigma}{\omega \epsilon_0}}} \tag{4.313}$$

so that $|\eta| \ll \eta_0$. Thus, at all incidence angles except $\theta_i \approx 90°$,

$$\tilde{\Gamma} \approx -\frac{Z_0 - \eta}{Z_0 + \eta} = -\frac{1 - \frac{\eta}{Z_0}}{1 + \frac{\eta}{Z_0}} \approx -\left(1 - \frac{\eta}{Z_0}\right)\left(1 - \frac{\eta}{Z_0}\right) \approx -1 + 2\frac{\eta}{Z_0}, \tag{4.314}$$

and so

$$1 - \tilde{\Gamma}^2 \approx 1 - \left(1 - 4\frac{\eta}{Z_0}\right) = 4\frac{\eta}{Z_0}.$$

This gives

$$R \approx 20 \log_{10} \left| \frac{Z_0}{4\eta} \right| \ \text{dB}.$$

Finally, use of

$$\eta = \sqrt{\frac{\mu}{\epsilon_0 \epsilon_r + j\frac{\sigma}{\omega}}} \approx \sqrt{\frac{\mu}{j\frac{\sigma}{\omega}}} = \frac{1+j}{\sigma\delta}$$

gives

$$R \approx 20 \log_{10} \left[\frac{\sigma \delta \eta_0}{4\sqrt{2}} (\cos \theta_i)^{\pm 1} \right] \ \text{dB}, \tag{4.315}$$

where the upper sign corresponds to parallel polarization and the lower sign to perpendicular polarization.

The term M can be approximated as well. Use of (4.314) gives

$$1 - \tilde{\Gamma}^2 \tilde{P}^2 \approx 1 - \left(1 - 4\frac{\eta}{Z_0}\right)\tilde{P}^2 \approx 1 - \tilde{P}^2$$

since $|\eta/Z_0| \ll 1$. Use of (4.311) then gives

$$M \approx 10 \log_{10}\left[1 - 2e^{-2\frac{\Delta}{\delta}}\cos 2\frac{\Delta}{\delta} + e^{-4\frac{\Delta}{\delta}}\right] \text{ dB}. \qquad (4.316)$$

Assembling the approximations for R, A, and M, we have a final approximation for the shielding effectiveness of a planar conducting screen:

$$SE \approx 20 \log_{10}\left[\frac{\sigma\delta\eta_0}{4\sqrt{2}}(\cos\theta_i)^{\pm 1}\right] + 8.686\frac{\Delta}{\delta}$$

$$+ 10 \log_{10}\left[1 - 2e^{-2\frac{\Delta}{\delta}}\cos 2\frac{\Delta}{\delta} + e^{-4\frac{\Delta}{\delta}}\right] \text{ dB}. \qquad (4.317)$$

Note that to the order of this approximation, only the R term depends on the incidence angle.

▶ **Example 4.22:** Low-frequency shielding effectiveness

It turns out that at low frequency, the shielding effectiveness is dominated by a constant-frequency term. Derive a simple approximation for this term by assuming that the thickness is much smaller than a skin depth ($\Delta/\delta \ll 1$).

Solution: We first approximate M. Use of

$$\cos x \approx 1 - \tfrac{1}{2}x^2, \qquad e^x \approx 1 + x + \tfrac{1}{2}x^2,$$

in (4.316) gives

$$1 - 2e^{-2\frac{\Delta}{\delta}}\cos 2\frac{\Delta}{\delta} + e^{-4\frac{\Delta}{\delta}} \approx 1 - 2\left(1 - 2\frac{\Delta}{\delta} + 2\frac{\Delta^2}{\delta^2}\right)\left(1 - 2\frac{\Delta^2}{\delta^2}\right) + 1 - 4\frac{\Delta}{\delta} + 8\frac{\Delta^2}{\delta^2}$$

$$\approx 8\frac{\Delta^2}{\delta^2}.$$

Thus, for $\frac{\Delta}{\delta} \ll 1$,

$$M \approx 20 \log_{10}\left[2\sqrt{2}\frac{\Delta}{\delta}\right].$$

Adding this to R from (4.315), we find

$$M + R \approx 20 \log_{10}\left[\frac{\sigma\delta\eta_0}{4\sqrt{2}}(\cos\theta_i)^{\pm 1}\right] + 20 \log_{10}\left[2\sqrt{2}\frac{\Delta}{\delta}\right]$$

$$= 20 \log_{10}\left[\frac{\sigma\Delta\eta_0}{2}(\cos\theta_i)^{\pm 1}\right].$$

The propagation term is $|\tilde{P}| = |e^{-jk_z\Delta}| \approx e^{-\Delta/\delta} \approx 1 - \Delta/\delta$. Thus, to lowest order we may set $|P| \approx 1$ and conclude that $A \approx 0$ dB. This gives a simple approximation for the shielding effectiveness at low frequencies:

$$SE \approx 20\log_{10}\left[\left(\frac{\eta_0}{R_s}\right)\frac{(\cos\theta_i)^{\pm 1}}{2}\right],\qquad(4.318)$$

where

$$R_s = \frac{1}{\sigma\Delta}$$

is the *surface resistance*.

At low frequency, the expression determining M is proportional to skin depth, whereas the expression determining R is inversely proportional to skin depth. This dependence cancels, leaving a constant value as the dominant contribution to shielding effectiveness at low frequency. ◀

▶ **Example 4.23:** Skin depth of silver paint

Silver paint is often applied to the plastic cabinets of electronic devices to provide RF shielding. A typical silver paint has $\mu_r = 1$, $\epsilon_r = 1$, and $\sigma \approx 6000$ S/cm. Plot the skin depth of the paint and the ratio $|\eta/\eta_0|$ for the frequency range 1 MHz to 10 GHz.

Solution: From (4.313) we have

$$\frac{\eta}{\eta_0} = \sqrt{\frac{1}{1 + j\frac{\sigma}{\omega\epsilon_0}}}.$$

Substituting $\sigma = 6 \times 10^5$ S/m and plotting vs. frequency gives the result shown in Figure 4.34. Also shown is the skin depth, computed using (4.310), which is quite small as expected for a good conductor at these frequencies. The intrinsic impedance of the shield is also small next to that of free space, suggesting that the approximate formulas used to find R and M should be quite accurate.

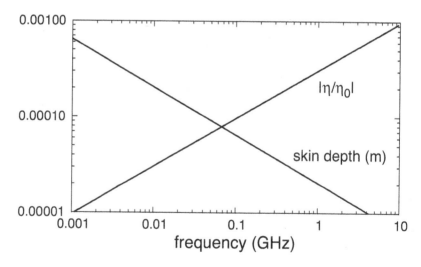

FIGURE 4.34

Skin depth and normalized intrinsic impedance for a shield made from silver paint. ◀

► **Example 4.24:** Shielding with silver paint

Consider the silver paint shield described in Example 4.23. A typical application has a thickness of 0.025 mm. For the frequency range 1 MHz to 10 GHz, compute the terms in the shielding effectiveness (A, R, and M) using the exact formulas and add them to find the shielding effectiveness (SE). Compare with the result from (4.317). Assume incidence angle $\theta_i = 60°$ and perpendicular polarization.

Solution: Based on Example 4.22 we expect that the shielding effectiveness will not be significantly dependent on frequency at lower frequencies, and should be given approximately by (4.318). We first compute the surface resistance

$$R_s = \frac{1}{\sigma\Delta} = \frac{1}{15} \ \Omega/\text{square}.$$

With this, we have from (4.318) a low-frequency shielding effectiveness

$$SE = 20\log_{10}(15\eta_0) = 75.04 \ \text{dB}.$$

Figure 4.35 shows R, M, A and the total shielding effectiveness SE as functions of frequency, computed from the exact formulas (4.307)–(4.309). As expected, A is very small for low frequency. Below about 100 MHz, M increases and R decreases in such a way that the total shielding effectiveness is nearly frequency independent. In fact, SE matches the low-frequency value of 75.04 dB to four digits until about 100 MHz. Above that SE begins to increase along with A. When the approximate formulas are used — (4.315) for R, (4.316) for M, and (4.312) for A — agreement to four significant figures is observed across the band of interest.

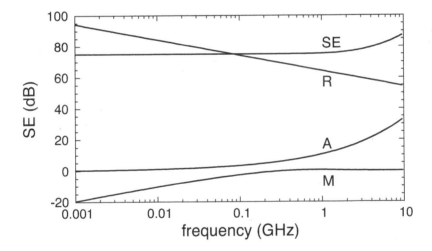

FIGURE 4.35
Components of shielding effectiveness for a 0.025 mm thick shield made from silver paint. $\theta_i = 60°$. ◄

4.11.7 Plane-wave propagation in an anisotropic ferrite medium

Several interesting properties of plane waves, such as Faraday rotation and the existence of stopbands, appear only when the waves propagate through anisotropic media. We shall study waves propagating in a magnetized ferrite medium, noting that their behavior is shared by waves propagating in a magnetized plasma because of the similarity in the dyadic constitutive parameters of the two media.

Consider a uniform ferrite material having scalar permittivity $\tilde{\epsilon} = \epsilon$ and dyadic permeability $\bar{\tilde{\mu}}$. We assume the ferrite is lossless and magnetized along the z-direction. By (4.124)–(4.126) the permeability of the medium is

$$[\bar{\tilde{\mu}}(\omega)] = \begin{bmatrix} \mu_1 & j\mu_2 & 0 \\ -j\mu_2 & \mu_1 & 0 \\ 0 & 0 & \mu_0 \end{bmatrix}$$

where

$$\mu_1 = \mu_0 \left(1 + \frac{\omega_M \omega_0}{\omega_0^2 - \omega^2} \right), \qquad \mu_2 = \mu_0 \frac{\omega \omega_M}{\omega_0^2 - \omega^2}.$$

The source-free frequency-domain wave equation can be found using (4.195) with $\bar{\tilde{\zeta}} = \bar{\tilde{\xi}} = 0$ and $\tilde{\epsilon} = \epsilon \bar{\mathbf{I}}$:

$$\left[\bar{\nabla} \cdot \left(\bar{\mathbf{I}} \frac{1}{\epsilon} \right) \cdot \bar{\nabla} - \omega^2 \bar{\tilde{\mu}} \right] \cdot \tilde{\mathbf{H}} = 0$$

or, since $\bar{\nabla} \cdot \mathbf{A} = \nabla \times \mathbf{A}$,

$$\frac{1}{\epsilon} \nabla \times (\nabla \times \tilde{\mathbf{H}}) - \omega^2 \bar{\tilde{\mu}} \cdot \tilde{\mathbf{H}} = 0. \tag{4.319}$$

The simplest solutions to the wave equation for this anisotropic medium are TEM plane waves propagating along the applied dc magnetic field. So we seek solutions of the form

$$\tilde{\mathbf{H}}(\mathbf{r}, \omega) = \tilde{\mathbf{H}}_0(\omega) e^{-j\mathbf{k} \cdot \mathbf{r}} \tag{4.320}$$

where $\mathbf{k} = \hat{\mathbf{z}}\beta$ and $\hat{\mathbf{z}} \cdot \tilde{\mathbf{H}}_0 = 0$. We can find β by enforcing (4.319). From (B.84) we find that

$$\nabla \times \tilde{\mathbf{H}} = -j\beta \hat{\mathbf{z}} \times \tilde{\mathbf{H}}_0 e^{-j\beta z}.$$

By Ampere's law

$$\tilde{\mathbf{E}} = \frac{\nabla \times \tilde{\mathbf{H}}}{j\omega\epsilon} = -Z_{TEM} \hat{\mathbf{z}} \times \tilde{\mathbf{H}}, \tag{4.321}$$

where $Z_{TEM} = \beta / \omega\epsilon$ is the wave impedance. Note that the wave is indeed TEM. The second curl is found to be

$$\nabla \times (\nabla \times \tilde{\mathbf{H}}) = -j\beta \nabla \times [\hat{\mathbf{z}} \times \tilde{\mathbf{H}}_0 e^{-j\beta z}].$$

After an application of (B.49) this becomes

$$\nabla \times (\nabla \times \tilde{\mathbf{H}}) = -j\beta [e^{-j\beta z} \nabla \times (\hat{\mathbf{z}} \times \tilde{\mathbf{H}}_0) - (\hat{\mathbf{z}} \times \tilde{\mathbf{H}}_0) \times \nabla e^{-j\beta z}].$$

The first term on the right side is zero, so by (B.82) we have

$$\nabla \times (\nabla \times \tilde{\mathbf{H}}) = [-j\beta e^{-j\beta z} \hat{\mathbf{z}} \times (\hat{\mathbf{z}} \times \tilde{\mathbf{H}}_0)](-j\beta)$$

or, using (B.7),

$$\nabla \times (\nabla \times \tilde{\mathbf{H}}) = \beta^2 e^{-j\beta z} \tilde{\mathbf{H}}_0$$

since $\hat{\mathbf{z}} \cdot \tilde{\mathbf{H}}_0 = 0$. With this (4.319) becomes

$$\beta^2 \tilde{\mathbf{H}}_0 = \omega^2 \epsilon \bar{\tilde{\mu}} \cdot \tilde{\mathbf{H}}_0. \tag{4.322}$$

We can solve (4.322) for β by writing the vector equation in component form:

$$\beta^2 H_{0x} = \omega^2 \epsilon(\mu_1 H_{0x} + j\mu_2 H_{0y}),$$
$$\beta^2 H_{0y} = \omega^2 \epsilon(-j\mu_2 H_{0x} + \mu_1 H_{0y}).$$

In matrix form these are

$$\begin{bmatrix} \beta^2 - \omega^2\epsilon\mu_1 & -j\omega^2\epsilon\mu_2 \\ j\omega^2\epsilon\mu_2 & \beta^2 - \omega^2\epsilon\mu_1 \end{bmatrix} \begin{bmatrix} H_{0x} \\ H_{0y} \end{bmatrix} = \begin{bmatrix} 0 \\ 0 \end{bmatrix}, \tag{4.323}$$

and nontrivial solutions occur only if

$$\begin{vmatrix} \beta^2 - \omega^2\epsilon\mu_1 & -j\omega^2\epsilon\mu_2 \\ j\omega^2\epsilon\mu_2 & \beta^2 - \omega^2\epsilon\mu_1 \end{vmatrix} = 0.$$

Expansion yields the two solutions

$$\beta_\pm = \omega\sqrt{\epsilon\mu_\pm} \tag{4.324}$$

where

$$\mu_\pm = \mu_1 \pm \mu_2 = \mu_0\left(1 + \frac{\omega_M}{\omega_0 \mp \omega}\right). \tag{4.325}$$

So the propagation properties of the plane wave are the same as those in a medium with an equivalent scalar permeability given by $\mu_\pm$.

Associated with each of these solutions is a relationship between H_{0x} and H_{0y} that can be found from (4.323). Substituting β_+ into the first equation, we have

$$\omega^2\epsilon\mu_2 H_{0x} - j\omega^2\epsilon\mu_2 H_{0y} = 0$$

or $H_{0x} = jH_{0y}$. Similarly, substitution of β_- produces $H_{0x} = -jH_{0y}$. Thus, by (4.320) the magnetic field may be expressed as

$$\tilde{\mathbf{H}}(\mathbf{r}, \omega) = H_{0y}(\pm j\hat{\mathbf{x}} + \hat{\mathbf{y}})e^{-j\beta_\pm z}.$$

By (4.321) we also have the electric field

$$\tilde{\mathbf{E}}(\mathbf{r}, \omega) = Z_{TEM} H_{0y}(\hat{\mathbf{x}} + e^{\mp j\frac{\pi}{2}}\hat{\mathbf{y}})e^{-j\beta_\pm z}.$$

This has the form of (4.233). For β_+ we have $\phi_y - \phi_x = -\pi/2$ and thus the wave exhibits RHCP. For β_- we have $\phi_y - \phi_x = \pi/2$ and LHCP.

The dispersion diagram for each polarization is shown in Figure 4.36, where we have arbitrarily chosen $\omega_M = 2\omega_0$. Here we have combined (4.324) and (4.325) to produce the normalized expression

$$\frac{\beta_\pm}{\omega_0/v_c} = \frac{\omega}{\omega_0}\sqrt{1 + \frac{\omega_M/\omega_0}{1 \mp \omega/\omega_0}}$$

where $v_c = 1/(\mu_0\epsilon)^{1/2}$. Except at low frequencies, an LHCP plane wave passes through the ferrite as if the permeability is close to that of free space. Over all frequencies we have $v_p < v_c$ and $v_g < v_c$. In contrast, an RHCP wave excites the electrons in the ferrite and a resonance occurs at $\omega = \omega_0$. For all frequencies below ω_0, we have $v_p < v_c$ and $v_g < v_c$, and both v_p and v_g reduce to zero as $\omega \to \omega_0$. Because the ferrite is lossless, frequencies between $\omega = \omega_0$ and $\omega = \omega_0 + \omega_M$ result in β being purely imaginary and thus the wave being evanescent. We thus call the frequency range $\omega_0 < \omega < \omega_0 + \omega_M$ a *stopband*; within this band the plane wave cannot transport energy. For frequencies above $\omega_0 + \omega_M$ the RHCP wave propagates as if it is in a medium with permeability less than that of free space. Here we have $v_p > v_c$ and $v_g < v_c$, with $v_p \to v_c$ and $v_g \to v_c$ as $\omega \to \infty$.

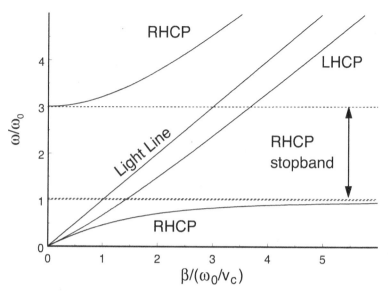

FIGURE 4.36

Dispersion plot for unmagnetized ferrite with $\omega_M = 2\omega_0$. Light line shows $\omega/\beta = v_c = 1/(\mu_0\epsilon)^{1/2}$.

4.11.7.1 Faraday rotation

The solutions to the wave equation found above do not allow the existence of linearly polarized plane waves. However, by superposing LHCP and RHCP waves we can obtain a wave with the appearance of linear polarization. That is, over any z-plane the electric field vector may be written as $\hat{\mathbf{E}} = K(E_{x0}\hat{\mathbf{x}} + E_{y0}\hat{\mathbf{y}})$ where E_{x0} and E_{y0} are real (although K may be complex). To see this let us examine

$$\tilde{\mathbf{E}} = \tilde{\mathbf{E}}^+ + \tilde{\mathbf{E}}^- = \tfrac{1}{2}E_0(\hat{\mathbf{x}} - j\hat{\mathbf{y}})e^{-j\beta_+ z} + \tfrac{1}{2}E_0(\hat{\mathbf{x}} + j\hat{\mathbf{y}})e^{-j\beta_- z}$$

$$= \tfrac{1}{2}E_0\left[\hat{\mathbf{x}}\left(e^{-j\beta_+ z} + e^{-j\beta_- z}\right) + j\hat{\mathbf{y}}\left(-e^{-j\beta_+ z} + e^{-j\beta_- z}\right)\right]$$

$$= E_0 e^{-j\frac{1}{2}(\beta_+ + \beta_-)z}\left[\hat{\mathbf{x}}\cos\tfrac{1}{2}(\beta_+ - \beta_-)z + \hat{\mathbf{y}}\sin\tfrac{1}{2}(\beta_+ - \beta_-)z\right]$$

or

$$\tilde{\mathbf{E}} = E_0 e^{-j\frac{1}{2}(\beta_+ + \beta_-)z}\left[\hat{\mathbf{x}}\cos\theta(z) + \hat{\mathbf{y}}\sin\theta(z)\right]$$

where $\theta(z) = (\beta_+ - \beta_-)z/2$. Because $\beta_+ \neq \beta_-$, the velocities of the two circularly polarized waves differ and the waves superpose to form a linearly polarized wave with a polarization that depends on the observation plane z-value. We may think of the wave as undergoing a phase shift of $(\beta_+ + \beta_-)z/2$ radians as it propagates, while the direction of $\tilde{\mathbf{E}}$ rotates to an angle $\theta(z) = (\beta_+ - \beta_-)z/2$ as the wave propagates. *Faraday rotation* can only occur at frequencies where both the LHCP and RHCP waves propagate, and therefore not within the stopband $\omega_0 < \omega < \omega_0 + \omega_M$.

Faraday rotation is nonreciprocal. That is, if a wave that has undergone a rotation of θ_0 radians by propagating through a distance z_0 is made to propagate an equal distance back in the direction from whence it came, the polarization does not return to its initial state but rather incurs an additional rotation of θ_0. Thus, the polarization angle of the wave when it returns to the starting point is not zero, but $2\theta_0$. This effect is employed in a number of microwave devices including gyrators, isolators, and circulators. The interested reader should see Collin [39], Elliott [58], or Liao [118] for details. We note that for $\omega \gg \omega_M$ we can approximate the rotation angle as

$$\theta(z) = (\beta_+ - \beta_-)z/2 = \tfrac{1}{2}\omega z\sqrt{\epsilon\mu_0}\left[\sqrt{1 + \frac{\omega_M}{\omega_0 - \omega}} - \sqrt{1 + \frac{\omega_M}{\omega_0 + \omega}}\right] \approx -\tfrac{1}{2}z\omega_M\sqrt{\epsilon\mu_0},$$

which is independent of frequency. So it is possible to construct Faraday rotation-based ferrite devices that maintain their properties over wide bandwidths.

It is straightforward to extend the above analysis to the case of a lossy ferrite. We find that for typical ferrites the attenuation constant associated with μ_- is small for all frequencies, but the attenuation constant associated with μ_+ is large near the resonant frequency ($\omega \approx \omega_0$) [39]. See Problem 4.14.

4.11.8 Propagation of cylindrical waves

By studying plane waves we have gained insight into the basic behavior of frequency-domain and time-harmonic waves. However, these solutions do not display the fundamental property that waves in space must diverge from their sources. To understand this behavior we shall treat waves having cylindrical and spherical symmetries.

4.11.8.1 Uniform cylindrical waves

In § 2.10.7 we studied the temporal behavior of cylindrical waves in a homogeneous, lossless medium, and found that they diverge from a line source located along the z-axis. Here we shall extend the analysis to lossy media and investigate the waves in the frequency domain.

Consider a homogeneous region of space described by permittivity $\tilde{\epsilon}(\omega)$, permeability $\tilde{\mu}(\omega)$, and conductivity $\tilde{\sigma}(\omega)$. We seek solutions that are invariant over a cylindrical surface: $\tilde{\mathbf{E}}(\mathbf{r}, \omega) = \tilde{\mathbf{E}}(\rho, \omega)$, $\tilde{\mathbf{H}}(\mathbf{r}, \omega) = \tilde{\mathbf{H}}(\rho, \omega)$. Such waves are called *uniform cylindrical waves*. Since the fields are z-independent we may decompose them into TE and TM sets (§ 4.11.2). For TM polarization we may insert (4.203) into (4.204) to find

$$\tilde{H}_\phi(\rho, \omega) = \frac{1}{j\omega\tilde{\mu}(\omega)}\frac{\partial \tilde{E}_z(\rho, \omega)}{\partial \rho}. \tag{4.326}$$

For TE polarization we have from (4.205)

$$\tilde{E}_\phi(\rho, \omega) = -\frac{1}{j\omega\tilde{\epsilon}^c(\omega)}\frac{\partial \tilde{H}_z(\rho, \omega)}{\partial \rho}, \tag{4.327}$$

where $\tilde{\epsilon}^c = \tilde{\epsilon} + \tilde{\sigma}/j\omega$ is the complex permittivity of § 4.4.1. Since $\tilde{\mathbf{E}} = \hat{\phi}\tilde{E}_\phi + \hat{z}\tilde{E}_z$ and $\tilde{\mathbf{H}} = \hat{\phi}\tilde{H}_\phi + \hat{z}\tilde{H}_z$, we can always decompose a cylindrical electromagnetic wave into cases of electric and magnetic polarization. In each case the resulting field is TEM_ρ since $\tilde{\mathbf{E}}$, $\tilde{\mathbf{H}}$, and $\hat{\rho}$ are mutually orthogonal.

Wave equations for $\tilde{E}_z$ in the electric polarization case and for $\tilde{H}_z$ in the magnetic polarization case can be derived by substituting (4.202) into (4.200):

$$\left(\frac{\partial^2}{\partial \rho^2} + \frac{1}{\rho}\frac{\partial}{\partial \rho} + k^2\right)\left\{\begin{matrix} \tilde{E}_z \\ \tilde{H}_z \end{matrix}\right\} = 0.$$

Thus the electric field must satisfy Bessel's equation (A.126)

$$\frac{d^2\tilde{E}_z}{d\rho^2} + \frac{1}{\rho}\frac{d\tilde{E}_z}{d\rho} + k^2\tilde{E}_z = 0. \tag{4.328}$$

This is a second-order equation with two independent solutions chosen from the list

$$J_0(k\rho), \quad Y_0(k\rho), \quad H_0^{(1)}(k\rho), \quad H_0^{(2)}(k\rho).$$

We find that $J_0(k\rho)$ and $Y_0(k\rho)$ are useful for describing standing waves between boundaries, while $H_0^{(1)}(k\rho)$ and $H_0^{(2)}(k\rho)$ are useful for describing waves propagating in the ρ-direction. Of these, $H_0^{(1)}(k\rho)$ represents waves traveling inward while $H_0^{(2)}(k\rho)$ represents waves traveling outward. At this point we are interested in studying the behavior of outward propagating waves, and so we choose

$$\tilde{E}_z(\rho,\omega) = -\frac{j}{4}\tilde{E}_{z0}(\omega)H_0^{(2)}(k\rho). \tag{4.329}$$

As explained in § 2.10.7, $\tilde{E}_{z0}(\omega)$ is the amplitude spectrum of the wave, while the factor $-j/4$ is included to make the conversion to the time domain more convenient. By (4.326) we have

$$\tilde{H}_\phi = \frac{1}{j\omega\tilde{\mu}}\frac{\partial\tilde{E}_z}{\partial\rho} = \frac{1}{j\omega\tilde{\mu}}\frac{\partial}{\partial\rho}\left[-\frac{j}{4}\tilde{E}_{z0}H_0^{(2)}(k\rho)\right]. \tag{4.330}$$

Using $dH_0^{(2)}(x)/dx = -H_1^{(2)}(x)$ we find that

$$\tilde{H}_\phi = \frac{1}{Z_{TM}}\frac{\tilde{E}_{z0}}{4}H_1^{(2)}(k\rho) \tag{4.331}$$

where $Z_{TM} = \omega\tilde{\mu}/k$ is the *TM wave impedance*.

For the case of magnetic polarization, the field $\tilde{H}_z$ must satisfy Bessel's equation (4.328). Thus we choose

$$\tilde{H}_z(\rho,\omega) = -\frac{j}{4}\tilde{H}_{z0}(\omega)H_0^{(2)}(k\rho). \tag{4.332}$$

From (4.327) we find the electric field associated with the wave:

$$\tilde{E}_\phi = -Z_{TE}\frac{\tilde{H}_{z0}}{4}H_1^{(2)}(k\rho), \tag{4.333}$$

where $Z_{TE} = k/(\omega\tilde{\epsilon}^c)$ is the *TE wave impedance*.

It is not readily apparent that the terms $H_0^{(2)}(k\rho)$ or $H_1^{(2)}(k\rho)$ describe outward propagating waves. We shall see later that the cylindrical wave may be written as a superposition of plane waves, both uniform and evanescent, propagating in all possible directions. Each of these components does have the expected wave behavior, but it is still not obvious that the sum of such waves is outward propagating.

We saw in § 2.10.7 that when examined in the time domain, a cylindrical wave of the form $H_0^{(2)}(k\rho)$ does indeed propagate outward, and that for lossless media the velocity of propagation of its wavefronts is $v = 1/(\mu\epsilon)^{1/2}$. For time-harmonic fields, the cylindrical wave takes on a familiar behavior when the observation point is sufficiently removed from the source. We may specialize (4.329) to the time-harmonic case by setting $\omega = \check{\omega}$ and using phasors, giving

$$\check{E}_z(\rho) = -\frac{j}{4}\check{E}_{z0}H_0^{(2)}(k\rho).$$

If $|k\rho| \gg 1$ we can use the asymptotic representation (E.64)

$$H_\nu^{(2)}(z) \sim \sqrt{\frac{2}{\pi z}} e^{-j(z-\pi/4-\nu\pi/2)} \qquad (|z| \gg 1,\ -2\pi < \arg z < \pi)$$

to obtain

$$\check{E}_z(\rho) \sim \check{E}_{z0} \frac{e^{-jk\rho}}{\sqrt{8j\pi k\rho}} \tag{4.334}$$

and

$$\check{H}_\phi(\rho) \sim -\check{E}_{z0} \frac{1}{Z_{TM}} \frac{e^{-jk\rho}}{\sqrt{8j\pi k\rho}} \tag{4.335}$$

for $|k\rho| \gg 1$. Except for the $\sqrt{\rho}$ term in the denominator, the wave has very much the same form as the plane waves encountered earlier. For the case of magnetic polarization, we can approximate (4.332) and (4.333) to obtain

$$\check{H}_z(\rho) \sim \check{H}_{z0} \frac{e^{-jk\rho}}{\sqrt{8j\pi k\rho}} \tag{4.336}$$

and

$$\check{E}_\phi(\rho) \sim Z_{TE} \check{H}_{z0} \frac{e^{-jk\rho}}{\sqrt{8j\pi k\rho}} \tag{4.337}$$

for $|k\rho| \gg 1$.

To interpret the wave nature of the field (4.334), let us substitute $k = \beta - j\alpha$ into the exponential function, where β is the phase constant (4.215) and α is the attenuation constant (4.216). Then

$$\check{E}_z(\rho) \sim \check{E}_{z0} \frac{1}{\sqrt{8j\pi k\rho}} e^{-\alpha\rho} e^{-j\beta\rho}.$$

Assuming $\check{E}_{z0} = |E_{z0}|e^{j\xi^E}$, the time-domain representation is found from (4.137):

$$E_z(\rho, t) = \frac{|E_{z0}|}{\sqrt{8\pi k\rho}} e^{-\alpha\rho} \cos[\check{\omega}t - \beta\rho - \pi/4 + \xi^E]. \tag{4.338}$$

We can identify a surface of constant phase as a locus of points obeying

$$\check{\omega}t - \beta\rho - \pi/4 + \xi^E = C_P \tag{4.339}$$

where C_P is some constant. These surfaces are cylinders coaxial with the z-axis, and are called *cylindrical wavefronts*. Note that surfaces of constant amplitude, as determined by $e^{-\alpha\rho}/\sqrt{\rho} = C_A$ for constant C_A, are also cylinders.

The cosine term in (4.338) represents a traveling wave. As t is increased the argument of the cosine function remains fixed as long as ρ is increased correspondingly. Hence the cylindrical wavefronts propagate outward as time progresses. As the wavefront travels outward, the field is attenuated because of the factor $e^{-\alpha\rho}$. The velocity of propagation of the phase fronts may be computed by a now-familiar technique. Differentiating (4.339) with respect to t we find that

$$\check{\omega} - \beta \frac{d\rho}{dt} = 0,$$

and thus have the phase velocity v_p of the outward expanding phase fronts:

$$v_p = \frac{d\rho}{dt} = \frac{\check{\omega}}{\beta}.$$

Calculation of wavelength also proceeds as before. Examining the two adjacent wavefronts that produce the same value of the cosine function in (4.338), we find $\beta\rho_1 = \beta\rho_2 - 2\pi$, or

$$\lambda = \rho_2 - \rho_1 = 2\pi/\beta.$$

Computation of the power carried by a cylindrical wave is straightforward. Since a cylindrical wavefront is infinite in extent, we usually speak of the *power per unit length* carried by the wave. This is found by integrating the time-average Poynting flux (4.157). For electric polarization we find the time-average power flux density using (4.329) and (4.330):

$$\mathbf{S}_{av} = \tfrac{1}{2}\,\mathrm{Re}\{\check{E}_z\hat{\mathbf{z}} \times \check{H}_\phi^*\hat{\boldsymbol{\phi}}\} = \tfrac{1}{2}\,\mathrm{Re}\left\{\hat{\boldsymbol{\rho}}\frac{j}{16Z_{TM}^*}|\check{E}_{z0}|^2 H_0^{(2)}(k\rho)H_1^{(2)*}(k\rho)\right\}. \qquad (4.340)$$

For magnetic polarization we use (4.332) and (4.333):

$$\mathbf{S}_{av} = \tfrac{1}{2}\,\mathrm{Re}\{\check{E}_\phi\hat{\boldsymbol{\phi}} \times \check{H}_z^*\hat{\mathbf{z}}\} \doteq \tfrac{1}{2}\,\mathrm{Re}\left\{-\hat{\boldsymbol{\rho}}\frac{jZ_{TE}}{16}|\check{H}_{z0}|^2 H_0^{(2)*}(k\rho)H_1^{(2)}(k\rho)\right\}.$$

For a lossless medium these expressions can be greatly simplified. By (E.5)

$$jH_0^{(2)}(k\rho)H_1^{(2)*}(k\rho) = j[J_0(k\rho) - jN_0(k\rho)][J_1(k\rho) + jN_1(k\rho)],$$

hence

$$jH_0^{(2)}(k\rho)H_1^{(2)*}(k\rho) = [N_0(k\rho)J_1(k\rho) - J_0(k\rho)N_1(k\rho)] + j[J_0(k\rho)J_1(k\rho) + N_0(k\rho)N_1(k\rho)].$$

Substituting this into (4.340) and remembering that $Z_{TM} = \eta = (\mu/\epsilon)^{1/2}$ is real for lossless media, we have

$$\mathbf{S}_{av} = \hat{\boldsymbol{\rho}}\,\frac{1}{32\eta}|\check{E}_{z0}|^2[N_0(k\rho)J_1(k\rho) - J_0(k\rho)N_1(k\rho)] = \hat{\boldsymbol{\rho}}\frac{|\check{E}_{z0}|^2}{16\pi k\rho\eta}$$

by the Wronskian relation (E.90).

The power density is inversely proportional to ρ. When we compute the total time-average power per unit length passing through a cylinder of radius ρ, this factor cancels with the ρ-dependence of the surface area to give a result independent of radius:

$$P_{av}/l = \int_0^{2\pi} \mathbf{S}_{av} \cdot \hat{\boldsymbol{\rho}}\rho\,d\phi = \frac{|\check{E}_{z0}|^2}{8k\eta}. \qquad (4.341)$$

For a lossless medium there is no mechanism to dissipate the power and so the wave propagates unabated. A similar calculation for the case of magnetic polarization (Problem 4.15) gives

$$\mathbf{S}_{av} = \hat{\boldsymbol{\rho}}\frac{\eta|\check{H}_{z0}|^2}{16\pi k\rho} \quad \text{and} \quad P_{av}/l = \frac{\eta|\check{H}_{z0}|^2}{8k}.$$

For a lossy medium the expressions are more difficult to evaluate. In this case we expect the total power passing through a cylinder to depend on the radius of the cylinder, since the fields decay exponentially with distance and thus give up power as they propagate. Assuming the observation point is far from the z-axis with $|k\rho| \gg 1$, we can use (4.334) and (4.335) for the electric polarization case to obtain

$$\mathbf{S}_{av} = \tfrac{1}{2}\,\mathrm{Re}\{\check{E}_z\hat{\mathbf{z}} \times \check{H}_\phi^*\hat{\boldsymbol{\phi}}\} = \tfrac{1}{2}\,\mathrm{Re}\left\{\hat{\boldsymbol{\rho}}\frac{e^{-2\alpha\rho}}{8\pi\rho|k|Z_{TM}^*}|\check{E}_{z0}|^2\right\}.$$

So

$$P_{av}/l = \int_0^{2\pi} \mathbf{S}_{av} \cdot \hat{\boldsymbol{\rho}} \rho \, d\phi = \mathrm{Re}\left\{\frac{1}{Z_{TM}^*}\right\} |\check{E}_{z0}|^2 \frac{e^{-2\alpha\rho}}{8|k|}.$$

For a lossless material, $Z_{TM} = \eta$ and $\alpha = 0$, and the expression reduces to (4.341) as expected. Thus for lossy materials the power depends on the radius of the cylinder. In the case of magnetic polarization, we use (4.336) and (4.337) to get

$$\mathbf{S}_{av} = \tfrac{1}{2} \mathrm{Re}\{\check{E}_\phi \hat{\boldsymbol{\phi}} \times \check{H}_z^* \hat{\mathbf{z}}\} = \tfrac{1}{2} \mathrm{Re}\left\{\hat{\boldsymbol{\rho}} Z_{TE}^* \frac{e^{-2\alpha\rho}}{8\pi\rho|k|} |\check{H}_{z0}|^2\right\}$$

and

$$P_{av}/l = \mathrm{Re}\left\{Z_{TE}^*\right\} |\check{H}_{z0}|^2 \frac{e^{-2\alpha\rho}}{8|k|}.$$

4.11.8.2   Fields of a line source

The simplest example of a uniform cylindrical wave is that produced by an electric or magnetic line source. Consider first an infinite electric line current of amplitude $\tilde{I}(\omega)$ on the z-axis, immersed within a medium of permittivity $\tilde{\epsilon}(\omega)$, permeability $\tilde{\mu}(\omega)$, and conductivity $\tilde{\sigma}(\omega)$. We assume the current does not vary in the z-direction, so the problem is two-dimensional. We can decompose the field produced by the line source into TE and TM cases according to § 4.11.2. It turns out that an electric line source only excites TM fields, as we shall show in § 5.4, and thus we need only $\tilde{E}_z$ to completely describe the fields.

By symmetry the fields are ϕ-independent and thus the wave produced by the line source is a uniform cylindrical wave. Since the wave propagates outward from the line source we have the electric field from (4.329),

$$\tilde{E}_z(\rho, \omega) = -\frac{j}{4} \tilde{E}_{z0}(\omega) H_0^{(2)}(k\rho),$$

and the magnetic field from (4.331),

$$\tilde{H}_\phi(\rho, \omega) = \frac{k}{\omega\tilde{\mu}} \frac{\tilde{E}_{z0}(\omega)}{4} H_1^{(2)}(k\rho).$$

We can find $\tilde{E}_{z0}$ by using Ampere's law:

$$\oint_\Gamma \tilde{\mathbf{H}} \cdot \mathbf{dl} = \int_S \tilde{\mathbf{J}} \cdot \mathbf{dS} + j\omega \int_S \tilde{\mathbf{D}} \cdot \mathbf{dS}.$$

Since $\tilde{\mathbf{J}}$ is the sum of the impressed current $\tilde{I}$ and the secondary conduction current $\tilde{\sigma}\tilde{\mathbf{E}}$, we can also write

$$\oint_\Gamma \tilde{\mathbf{H}} \cdot \mathbf{dl} = \tilde{I} + \int_S (\tilde{\sigma} + j\omega\tilde{\epsilon})\tilde{\mathbf{E}} \cdot \mathbf{dS} = \tilde{I} + j\omega\tilde{\epsilon}^c \int_S \tilde{\mathbf{E}} \cdot \mathbf{dS}.$$

Choosing our path of integration as a circle of radius a in the $z = 0$ plane and substituting for $\tilde{E}_z$ and $\tilde{H}_\phi$, we find that

$$\frac{k}{\omega\tilde{\mu}} \frac{\tilde{E}_{z0}}{4} H_1^{(2)}(ka) 2\pi a = \tilde{I} + j\omega\tilde{\epsilon}^c 2\pi \frac{-j\tilde{E}_{z0}}{4} \lim_{\delta \to 0} \int_\delta^a H_0^{(2)}(k\rho) \rho \, d\rho. \tag{4.342}$$

The limit operation is required because $H_0^{(2)}(k\rho)$ diverges as $\rho \to 0$. By (E.106) the integral is

$$\lim_{\delta \to 0} \int_\delta^a H_0^{(2)}(k\rho)\rho \, d\rho = \frac{a}{k}H_1^{(2)}(ka) - \frac{1}{k}\lim_{\delta \to 0}\delta H_1^{(2)}(k\delta).$$

The limit may be found by using $H_1^{(2)}(x) = J_1(x) - jN_1(x)$ and the small argument approximations (E.50) and (E.53):

$$\lim_{\delta \to 0}\delta H_1^{(2)}(\delta) = \lim_{\delta \to 0}\delta\left[\frac{k\delta}{2} - j\left(-\frac{1}{\pi}\frac{2}{k\delta}\right)\right] = j\frac{2}{\pi k}.$$

Substituting these expressions into (4.342), we obtain

$$\frac{k}{\omega\tilde{\mu}}\frac{\tilde{E}_{z0}}{4}H_1^{(2)}(ka)2\pi a = \tilde{I} + j\omega\tilde{\epsilon}^c 2\pi\frac{-j\tilde{E}_{z0}}{4}\left[\frac{a}{k}H_1^{(2)}(ka) - j\frac{2}{\pi k^2}\right].$$

Using $k^2 = \omega^2\tilde{\mu}\tilde{\epsilon}^c$ we find that the two Hankel function terms cancel. Solving for $\tilde{E}_{z0}$ we have $\tilde{E}_{z0} = -j\omega\tilde{\mu}\tilde{I}$ and therefore

$$\tilde{E}_z(\rho, \omega) = -\frac{\omega\tilde{\mu}}{4}\tilde{I}(\omega)H_0^{(2)}(k\rho) = -j\omega\tilde{\mu}\tilde{I}(\omega)\tilde{G}(x, y|0, 0; \omega). \tag{4.343}$$

Here $\tilde{G}$ is the *two-dimensional Green's function*

$$\tilde{G}(x, y|x', y'; \omega) = \frac{1}{4j}H_0^{(2)}\left(k\sqrt{(x - x')^2 + (y - y')^2}\right). \tag{4.344}$$

Green's functions are examined in greater detail in Chapter 5.

It is also possible to determine the field amplitude by evaluating

$$\lim_{a \to 0}\oint_C \tilde{\mathbf{H}} \cdot \mathbf{dl}.$$

This produces an identical result and is a bit simpler since it can be argued that the surface integral of $\tilde{E}_z$ vanishes as $a \to 0$ without having to perform the calculation directly [82, 12].

For a magnetic line source $\tilde{I}_m(\omega)$ aligned along the z-axis, we proceed as above, but note that the source only produces TE fields. By (4.332) and (4.333) we have

$$\tilde{H}_z(\rho, \omega) = -\frac{j}{4}\tilde{H}_{z0}(\omega)H_0^{(2)}(k\rho), \qquad \tilde{E}_\phi = -\frac{k}{\omega\tilde{\epsilon}^c}\frac{\tilde{H}_{0z}}{4}H_1^{(2)}(k\rho).$$

We can find $\tilde{H}_{z0}$ by applying Faraday's law

$$\oint_C \tilde{\mathbf{E}} \cdot \mathbf{dl} = -\int_S \tilde{\mathbf{J}}_m \cdot \mathbf{dS} - j\omega\int_S \tilde{\mathbf{B}} \cdot \mathbf{dS}$$

about a circle of radius a in the $z = 0$ plane. We have

$$-\frac{k}{\omega\tilde{\epsilon}^c}\frac{\tilde{H}_{z0}}{4}H_1^{(2)}(ka)2\pi a = -\tilde{I}_m - j\omega\tilde{\mu}\left[-\frac{j}{4}\right]\tilde{H}_{z0}2\pi\lim_{\delta \to 0}\int_\delta^a H_0^{(2)}(k\rho)\rho \, d\rho.$$

Proceeding as above we find that $\tilde{H}_{z0} = j\omega\tilde{\epsilon}^c\tilde{I}_m$ and

$$\tilde{H}_z(\rho, \omega) = -\frac{\omega\tilde{\epsilon}^c}{4}\tilde{I}_m(\omega)H_0^{(2)}(k\rho) = -j\omega\tilde{\epsilon}^c\tilde{I}_m(\omega)\tilde{G}(x, y|0, 0; \omega). \tag{4.345}$$

We can find the magnetic field of a magnetic line current via the field of an electric line current and the duality principle. Letting the magnetic current equal $-\eta$ times the electric current and using (4.192), we have

$$\tilde{H}_{z0} = \left(-\frac{1}{\eta}\frac{\tilde{I}_m(\omega)}{\tilde{I}(\omega)}\right)\left(-\frac{1}{\eta}\left[-\frac{\omega\tilde{\mu}}{4}\tilde{I}(\omega)H_0^{(2)}(k\rho)\right]\right) = -\tilde{I}_m(\omega)\frac{\omega\tilde{\epsilon}^c}{4}H_0^{(2)}(k\rho)$$

as in (4.345).

4.11.8.3 Nonuniform cylindrical waves

When we solve two-dimensional boundary value problems we encounter cylindrical waves that are z-independent but ϕ-dependent. Although such waves propagate outward, they have a more complicated structure than those considered above.

For the case of TM polarization, we have, by (4.204),

$$\tilde{H}_\rho = \frac{j}{Z_{TM}k}\frac{1}{\rho}\frac{\partial\tilde{E}_z}{\partial\phi}, \tag{4.346}$$

$$\tilde{H}_\phi = -\frac{j}{Z_{TM}k}\frac{\partial\tilde{E}_z}{\partial\rho}, \tag{4.347}$$

where $Z_{TM} = \omega\tilde{\mu}/k$. For the TE case we have, by (4.205),

$$\tilde{E}_\rho = -\frac{jZ_{TE}}{k}\frac{1}{\rho}\frac{\partial\tilde{H}_z}{\partial\phi}, \tag{4.348}$$

$$\tilde{E}_\phi = \frac{jZ_{TE}}{k}\frac{\partial\tilde{H}_z}{\partial\rho},$$

where $Z_{TE} = k/\omega\tilde{\epsilon}^c$. By (4.200) the wave equations are

$$\left(\frac{\partial^2}{\partial\rho^2} + \frac{1}{\rho}\frac{\partial}{\partial\rho} + \frac{1}{\rho^2}\frac{\partial^2}{\partial\phi^2} + k^2\right)\left\{\begin{array}{c}\tilde{E}_z \\ \tilde{H}_z\end{array}\right\} = 0.$$

Because this has the form of (A.119) with $\partial/\partial z \to 0$, we have

$$\left\{\begin{array}{c}\tilde{E}_z(\rho,\phi,\omega) \\ \tilde{H}_z(\rho,\phi,\omega)\end{array}\right\} = P(\rho,\omega)\Phi(\phi,\omega) \tag{4.349}$$

where

$$\Phi(\phi,\omega) = A_\phi(\omega)\sin k_\phi\phi + B_\phi(\omega)\cos k_\phi\phi,$$

$$P(\rho) = A_\rho(\omega)B_{k_\phi}^{(1)}(k\rho) + B_\rho(\omega)B_{k_\phi}^{(2)}(k\rho), \tag{4.350}$$

and where $B_\nu^{(1)}(z)$ and $B_\nu^{(2)}(z)$ are any two independent Bessel functions chosen from the set

$$J_\nu(z), \quad N_\nu(z), \quad H_\nu^{(1)}(z), \quad H_\nu^{(2)}(z).$$

In bounded regions we generally use the oscillatory functions $J_\nu(z)$ and $N_\nu(z)$ to represent standing waves. In unbounded regions we generally use $H_\nu^{(2)}(z)$ and $H_\nu^{(1)}(z)$ to represent outward and inward propagating waves, respectively.

▶ **Example 4.25:** Expansion of a plane wave in term of cylindrical waves

Consider a TM plane wave in free space propagating along the x-axis with the electric field

$$\tilde{\mathbf{E}}(\mathbf{r}) = \hat{\mathbf{z}}\tilde{E}_0 e^{-jk_0 x}.$$

Find a representation for $\tilde{\mathbf{E}}$ in terms of a superposition of cylindrical waves.

Solution: Since the plane-wave field is z-independent and obeys the homogeneous Helmholtz equation, we may represent it in terms of nonuniform cylindrical waves:

$$\tilde{E}_z = \tilde{E}_0 e^{-jk_0\rho\cos\phi} = \sum_{n=0}^{\infty} \left[E_n \sin n\phi + F_n \cos n\phi \right] J_n(k_0\rho).$$

Here we have chosen the Bessel function $J_n(k_0\rho)$ since the origin is included and the functions $N_\nu(z)$, $H_\nu^{(1)}(z)$, and $H_\nu^{(2)}(z)$ are all singular at $z = 0$. We also note that $\nu = n$ must be an integer since the plane-wave field is periodic in ϕ. Applying orthogonality, we see immediately that $E_n = 0$ and that

$$\frac{2\pi}{\epsilon_m} F_m J_m(k_0\rho) = \tilde{E}_0 \int_{-\pi}^{\pi} \cos m\phi e^{-jk_0\rho\cos\phi}\, d\phi = \tilde{E}_0 2\pi j^{-m} J_m(k_0\rho),$$

where ϵ_n is Neumann's number (A.133) and where we have used (E.85) and (E.39) to evaluate the integral. Thus, $F_n = \tilde{E}_0 \epsilon_n j^{-n}$ and

$$\tilde{E}_z = \sum_{n=0}^{\infty} \tilde{E}_0 \epsilon_n j^{-n} J_n(k_0\rho) \cos n\phi. \quad \blacktriangleleft \qquad (4.351)$$

4.11.8.4 Scattering by a material cylinder

A variety of boundary value problems can be solved using nonuniform cylindrical waves. We shall examine two interesting cases in which an external field is impressed on a two-dimensional object. The impressed field creates secondary sources within or on the object, and these in turn create a secondary field. Our goal is to determine the latter by applying appropriate boundary conditions.

As a first example, consider a material cylinder of radius a, complex permittivity $\tilde{\epsilon}^c$, and permeability $\tilde{\mu}$, aligned along the z-axis in free space (Figure 4.37). An incident plane wave propagating in the x-direction is impressed on the cylinder, inducing secondary polarization and conduction currents within the cylinder. These in turn produce secondary or scattered fields, which are standing waves within the cylinder and outward traveling waves outside the cylinder. Although we have not yet learned how to write the secondary fields in terms of the impressed sources, we can solve for the fields as a boundary value problem. The total field must obey the boundary conditions on tangential components at the interface between the cylinder and surrounding free space. We need not worry about the effect of the secondary sources on the source of the primary field, since by definition impressed sources cannot be influenced by secondary fields.

The scattered field can be found using superposition. When excited by a TM impressed field, the secondary field is also TM. The situation for TE excitation is similar. By decomposing the impressed field into TE and TM components, we may solve for the scattered field in each case and then superpose the results to determine the complete solution.

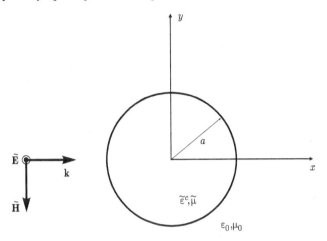

FIGURE 4.37
TM plane-wave field incident on a material cylinder.

We first consider the TM case. The impressed electric field may be written as

$$\tilde{\mathbf{E}}^i(\mathbf{r},\omega) = \hat{\mathbf{z}}\tilde{E}_0(\omega)e^{-jk_0x} = \hat{\mathbf{z}}\tilde{E}_0(\omega)e^{-jk_0\rho\cos\phi} \qquad (4.352)$$

while the magnetic field is, by (4.214),

$$\tilde{\mathbf{H}}^i(\mathbf{r},\omega) = -\hat{\mathbf{y}}\frac{\tilde{E}_0(\omega)}{\eta_0}e^{-jk_0x} = -(\hat{\boldsymbol{\rho}}\sin\phi + \hat{\boldsymbol{\phi}}\cos\phi)\frac{\tilde{E}_0(\omega)}{\eta_0}e^{-jk_0\rho\cos\phi}.$$

Here $k_0 = \omega(\mu_0\epsilon_0)^{1/2}$ and $\eta_0 = (\mu_0/\epsilon_0)^{1/2}$. The scattered electric field takes the form of a nonuniform cylindrical wave (4.349). Periodicity in ϕ implies that k_ϕ is an integer, say $k_\phi = n$. Within the cylinder we cannot use any of the functions $N_n(k\rho)$, $H_n^{(2)}(k\rho)$, or $H_n^{(1)}(k\rho)$ to represent the radial dependence of the field, since each is singular at the origin. So we choose $B_n^{(1)}(k\rho) = J_n(k\rho)$ and $B_\rho(\omega) = 0$ in (4.350). Physically, $J_n(k\rho)$ represents the standing wave created by the interaction of outward and inward propagating waves. External to the cylinder we use $H_n^{(2)}(k\rho)$ to represent the radial dependence of the secondary field components: we avoid $N_n(k\rho)$ and $J_n(k\rho)$ since these represent standing waves, and avoid $H_n^{(1)}(k\rho)$ since there are no external secondary sources to create an inward traveling wave.

Any attempt to satisfy the boundary conditions by using a single nonuniform wave fails. This is because the sinusoidal dependence on ϕ of each individual nonuniform wave cannot match the more complicated dependence of the impressed field (4.352). Since the sinusoids are complete, an infinite series of the functions (4.349) can be used to represent the scattered field. So we have inside the cylinder the total field

$$\tilde{E}_z(\mathbf{r},\omega) = \sum_{n=0}^{\infty} \left[A_n(\omega)\sin n\phi + B_n(\omega)\cos n\phi\right] J_n(k\rho)$$

where $k = \omega(\tilde{\mu}\tilde{\epsilon}^c)^{1/2}$. Outside the cylinder we have free space and thus have the scattered field

$$\tilde{E}_z^s(\mathbf{r},\omega) = \sum_{n=0}^{\infty} \left[C_n(\omega)\sin n\phi + D_n(\omega)\cos n\phi\right] H_n^{(2)}(k_0\rho). \qquad (4.353)$$

Equations (4.346) and (4.347) yield the total magnetic field inside the cylinder:

$$\tilde{H}_\rho = \sum_{n=0}^\infty \frac{jn}{Z_{TM}k}\frac{1}{\rho}\left[A_n(\omega)\cos n\phi - B_n(\omega)\sin n\phi\right] J_n(k\rho),$$

$$\tilde{H}_\phi = -\sum_{n=0}^\infty \frac{j}{Z_{TM}}\left[A_n(\omega)\sin n\phi + B_n(\omega)\cos n\phi\right] J_n'(k\rho),$$

where $Z_{TM} = \omega\tilde{\mu}/k$. Outside the cylinder

$$\tilde{H}_\rho^s = \sum_{n=0}^\infty \frac{jn}{\eta_0 k_0}\frac{1}{\rho}\left[C_n(\omega)\cos n\phi - D_n(\omega)\sin n\phi\right] H_n^{(2)}(k_0\rho),$$

$$\tilde{H}_\phi^s = -\sum_{n=0}^\infty \frac{j}{\eta_0}\left[C_n(\omega)\sin n\phi + D_n(\omega)\cos n\phi\right] H_n^{(2)\prime}(k_0\rho).$$

Here $J_n'(z) = dJ_n(z)/dz$ and $H_n^{(2)\prime}(z) = dH_n^{(2)}(z)/dz$.

We have two sets of unknown spectral amplitudes, (A_n, B_n) and (C_n, D_n). These can be determined by applying the boundary conditions at the interface. Since the total field outside the cylinder is the sum of the impressed and scattered terms, an application of continuity of the tangential electric field at $\rho = a$ gives us

$$\sum_{n=0}^\infty \left[A_n\sin n\phi + B_n\cos n\phi\right] J_n(ka)$$

$$= \sum_{n=0}^\infty \left[C_n\sin n\phi + D_n\cos n\phi\right] H_n^{(2)}(k_0a) + \tilde{E}_0 e^{-jk_0a\cos\phi},$$

which must hold for all $-\pi \le \phi \le \pi$. To remove the coefficients from the sum, we apply orthogonality. Multiplying both sides by $\sin m\phi$, integrating over $[-\pi, \pi]$, and using the orthogonality conditions (A.130)–(A.132), we obtain

$$\pi A_m J_m(ka) - \pi C_m H_m^{(2)}(k_0a) = \tilde{E}_0 \int_{-\pi}^\pi \sin m\phi e^{-jk_0a\cos\phi}\, d\phi = 0. \qquad (4.354)$$

Multiplying by $\cos m\phi$ and integrating, we find that

$$2\pi B_m J_m(ka) - 2\pi D_m H_m^{(2)}(k_0a) = \tilde{E}_0\epsilon_m \int_{-\pi}^\pi \cos m\phi e^{-jk_0a\cos\phi}\, d\phi$$

$$= 2\pi\tilde{E}_0\epsilon_m j^{-m} J_m(k_0a) \qquad (4.355)$$

where ϵ_n is Neumann's number (A.133) and where we have used (E.85) and (E.39) to evaluate the integral.

We must also have continuity of the tangential magnetic field $\tilde{H}_\phi$ at $\rho = a$. Thus

$$-\sum_{n=0}^\infty \frac{j}{Z_{TM}}\left[A_n\sin n\phi + B_n\cos n\phi\right] J_n'(ka) =$$

$$-\sum_{n=0}^\infty \frac{j}{\eta_0}\left[C_n\sin n\phi + D_n\cos n\phi\right] H_n^{(2)\prime}(k_0a) - \cos\phi\frac{\tilde{E}_0}{\eta_0}e^{-jk_0a\cos\phi}$$

must hold for all $-\pi \leq \phi \leq \pi$. By orthogonality,

$$\pi \frac{j}{Z_{TM}} A_m J'_m(ka) - \pi \frac{j}{\eta_0} C_m H_m^{(2)\prime}(k_0 a) = \frac{\tilde{E}_0}{\eta_0} \int_{-\pi}^{\pi} \sin m\phi \cos \phi \, e^{-jk_0 a \cos \phi} \, d\phi = 0 \quad (4.356)$$

and

$$2\pi \frac{j}{Z_{TM}} B_m J'_m(ka) - 2\pi \frac{j}{\eta_0} D_m H_m^{(2)\prime}(k_0 a) = \epsilon_m \frac{\tilde{E}_0}{\eta_0} \int_{-\pi}^{\pi} \cos m\phi \cos \phi \, e^{-jk_0 a \cos \phi} \, d\phi.$$

The integral may be computed as

$$\int_{-\pi}^{\pi} \cos m\phi \cos \phi \, e^{-jk_0 a \cos \phi} \, d\phi = j \frac{d}{d(k_0 a)} \int_{-\pi}^{\pi} \cos m\phi \, e^{-jk_0 a \cos \phi} \, d\phi = j 2\pi j^{-m} J'_m(k_0 a)$$

and thus

$$\frac{1}{Z_{TM}} B_m J'_m(ka) - \frac{1}{\eta_0} D_m H_m^{(2)\prime}(k_0 a) = \frac{\tilde{E}_0}{\eta_0} \epsilon_m j^{-m} J'_m(k_0 a). \quad (4.357)$$

We now have four equations for the coefficients A_n, B_n, C_n, D_n. We may write (4.354) and (4.356) as

$$\begin{bmatrix} J_m(ka) & -H_m^{(2)}(k_0 a) \\ \frac{\eta_0}{Z_{TM}} J'_m(ka) & -H_m^{(2)\prime}(k_0 a) \end{bmatrix} \begin{bmatrix} A_m \\ C_m \end{bmatrix} = 0, \quad (4.358)$$

and (4.355) and (4.357) as

$$\begin{bmatrix} J_m(ka) & -H_m^{(2)}(k_0 a) \\ \frac{\eta_0}{Z_{TM}} J'_m(ka) & -H_m^{(2)\prime}(k_0 a) \end{bmatrix} \begin{bmatrix} B_m \\ D_m \end{bmatrix} = \begin{bmatrix} \tilde{E}_0 \epsilon_m j^{-m} J_m(k_0 a) \\ \tilde{E}_0 \epsilon_m j^{-m} J'_m(k_0 a) \end{bmatrix}. \quad (4.359)$$

Matrix equations (4.358) and (4.359) cannot hold simultaneously unless $A_m = C_m = 0$. Then the solution to (4.359) is

$$B_m = \tilde{E}_0 \epsilon_m j^{-m} \left[\frac{H_m^{(2)}(k_0 a) J'_m(k_0 a) - J_m(k_0 a) H_m^{(2)\prime}(k_0 a)}{\frac{\eta_0}{Z_{TM}} J'_m(ka) H_m^{(2)}(k_0 a) - H_m^{(2)\prime}(k_0 a) J_m(ka)} \right],$$

$$D_m = -\tilde{E}_0 \epsilon_m j^{-m} \left[\frac{\frac{\eta_0}{Z_{TM}} J'_m(ka) J_m(k_0 a) - J'_m(k_0 a) J_m(ka)}{\frac{\eta_0}{Z_{TM}} J'_m(ka) H_m^{(2)}(k_0 a) - H_m^{(2)\prime}(k_0 a) J_m(ka)} \right].$$

With these coefficients we can calculate the field inside the cylinder ($\rho \leq a$) from

$$\tilde{E}_z(\mathbf{r}, \omega) = \sum_{n=0}^{\infty} B_n(\omega) J_n(k\rho) \cos n\phi, \quad (4.360)$$

$$\tilde{H}_\rho(\mathbf{r}, \omega) = -\sum_{n=0}^{\infty} \frac{jn}{Z_{TM}k} \frac{1}{\rho} B_n(\omega) J_n(k\rho) \sin n\phi,$$

$$\tilde{H}_\phi(\mathbf{r}, \omega) = -\sum_{n=0}^{\infty} \frac{j}{Z_{TM}} B_n(\omega) J'_n(k\rho) \cos n\phi,$$

and the field outside the cylinder $(\rho > a)$ from

$$\tilde{E}_z(\mathbf{r},\omega) = \tilde{E}_0(\omega)e^{-jk_0\rho\cos\phi} + \sum_{n=0}^{\infty} D_n(\omega)H_n^{(2)}(k_0\rho)\cos n\phi, \qquad (4.361)$$

$$\tilde{H}_\rho(\mathbf{r},\omega) = -\sin\phi\frac{\tilde{E}_0(\omega)}{\eta_0}e^{-jk_0\rho\cos\phi} - \sum_{n=0}^{\infty}\frac{jn}{\eta_0 k_0}\frac{1}{\rho}D_n(\omega)H_n^{(2)}(k_0\rho)\sin n\phi,$$

$$\tilde{H}_\phi(\mathbf{r},\omega) = -\cos\phi\frac{\tilde{E}_0(\omega)}{\eta_0}e^{-jk_0\rho\cos\phi} - \sum_{n=0}^{\infty}\frac{j}{\eta_0}D_n(\omega)H_n^{(2)\prime}(k_0\rho)\cos n\phi.$$

▶ **Example 4.26:** Total field for plane-wave scattering by a dielectric cylinder

Consider a TM plane wave incident along the x-axis onto a dielectric cylinder immersed in free space. The cylinder has dielectric constant $\epsilon_r = 3$ and radius $a = 2\lambda_0$ where λ_0 is the free-space wavelength. Use the series solution to compute and plot the magnitude of the total field $|\tilde{E}_z/\tilde{E}_0|$ within a square of side $4a$ centered on the cylinder.

Solution: The total axial electric field may be computed by evaluating the series (4.360) inside the cylinder and the series (4.361) outside the cylinder. We normalize distance to the cylinder radius such that

$$k_0\rho = 2\pi\frac{\rho}{a}\frac{a}{\lambda_0}, \qquad k\rho = 2\pi\sqrt{\epsilon_r}\frac{\rho}{a}\frac{a}{\lambda_0},$$

with $a/\lambda_0 = 2$ and $k_0 a = 2\pi(a/\lambda_0)$, and plot the total field for values of $(\rho/a, \phi)$. The field magnitude is shown in Figure 4.38, where the series has been terminated at $n = 150$.

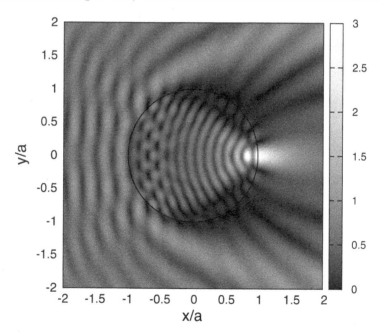

FIGURE 4.38
Total axial electric field $|\tilde{E}_z/\tilde{E}_0|$ for a dielectric cylinder of radius $2\lambda_0$ and permittivity $3\epsilon_0$ found using the series solution.

A standing wave pattern appears outside the cylinder along the incidence direction, the incident and scattered fields combining to create regions of constructive and destructive interference with a period of a half wavelength ($a/4$). A standing wave pattern also appears inside the cylinder, where internal reflections cause constructive and destructive interference. The internal interference pattern has a smaller period of $a/4\sqrt{3}$ corresponding to the shorter wavelength in the dielectric. Note that the cylinder produces a lens effect, with the wave inside converging to a point at the far side of the cylinder where the field is about 3.1 times the strength of the incident field. ◀

We can specialize these results for a perfectly conducting cylinder by allowing $\tilde{\sigma} \to \infty$:

$$\frac{\eta_0}{Z_{TM}} = \sqrt{\frac{\mu_0 \tilde{\epsilon}^c}{\tilde{\mu} \epsilon_0}} \to \infty, \qquad B_n \to 0, \qquad D_n \to -\tilde{E}_0 \epsilon_n j^{-n} \frac{J_n(k_0 a)}{H_n^{(2)}(k_0 a)}.$$

In this case it is convenient to combine the formulas for the impressed and scattered fields when forming the total fields. Expanding the impressed field as in (4.351) gives

$$\tilde{E}_z^i = \sum_{n=0}^{\infty} \tilde{E}_0 \epsilon_n j^{-n} J_n(k_0 \rho) \cos n\phi.$$

Adding this field to the scattered field, we have the total field outside the cylinder,

$$\tilde{E}_z = \tilde{E}_0 \sum_{n=0}^{\infty} \frac{\epsilon_n j^{-n}}{H_n^{(2)}(k_0 a)} \left[J_n(k_0 \rho) H_n^{(2)}(k_0 a) - J_n(k_0 a) H_n^{(2)}(k_0 \rho) \right] \cos n\phi, \qquad (4.362)$$

while the field within the cylinder vanishes. Then, by (4.347),

$$\tilde{H}_\phi = -\frac{j}{\eta_0} \tilde{E}_0 \sum_{n=0}^{\infty} \frac{\epsilon_n j^{-n}}{H_n^{(2)}(k_0 a)} \left[J_n'(k_0 \rho) H_n^{(2)}(k_0 a) - J_n(k_0 a) H_n^{(2)\prime}(k_0 \rho) \right] \cos n\phi.$$

This in turn gives us the surface current induced on the cylinder. From the boundary condition $\tilde{\mathbf{J}}_s = \hat{\mathbf{n}} \times \tilde{\mathbf{H}}|_{\rho=a} = \hat{\boldsymbol{\rho}} \times [\hat{\boldsymbol{\rho}} \tilde{H}_\rho + \hat{\boldsymbol{\phi}} \tilde{H}_\phi]|_{\rho=a} = \hat{\mathbf{z}} \tilde{H}_\phi|_{\rho=a}$ and an application of (E.95) we have

$$\mathbf{J}_s(\phi, \omega) = \hat{\mathbf{z}} \frac{2\tilde{E}_0}{\eta_0 k_0 \pi a} \sum_{n=0}^{\infty} \frac{\epsilon_n j^{-n}}{H_n^{(2)}(k_0 a)} \cos n\phi. \qquad (4.363)$$

Plots of the surface current may be found in Chapter 7, where results from this section are compared to results obtained by solving an integral equation.

▶ **Example 4.27:** Total field for plane-wave scattering by a conducting cylinder

Consider a TM plane wave incident along the x-axis onto a perfectly conducting cylinder of radius $2\lambda_0$ in free space. Compute and plot the magnitude of the total field $|\tilde{E}_z/\tilde{E}_0|$ within a square of side $4a$ centered on the cylinder using the series solution.

Solution: The total axial electric field may be computed by evaluating the series (4.362). We normalize distance to the cylinder radius and compute

$$\frac{\tilde{E}_z}{\tilde{E}_0} = \sum_{n=0}^{\infty} \frac{\epsilon_n j^{-n} \cos n\phi}{H_n^{(2)}\left(2\pi \frac{a}{\lambda_0}\right)} \left[J_n\left(2\pi \frac{a}{\lambda_0} \frac{\rho}{a}\right) H_n^{(2)}\left(2\pi \frac{a}{\lambda_0}\right) - J_n\left(2\pi \frac{a}{\lambda_0}\right) H_n^{(2)}\left(2\pi \frac{a}{\lambda_0} \frac{\rho}{a}\right) \right]$$

with $a/\lambda_0 = 2$, for values of $(\rho/a, \phi)$ surrounding the cylinder. The field magnitude is shown in Figure 4.39 for $n = 150$ terms. First note that the field vanishes at the surface of the cylinder (and has been set to zero within), as expected from the boundary condition.

Second, a standing wave pattern is seen along the incidence direction, where the incident and scattered fields combine to create regions of constructive and destructive interference with a period of a half wavelength ($a/4$). The maximum amplitude at the points of constructive interference is twice the incident field amplitude, as expected. Finally, a shadow region appears on the side opposite the incidence direction (the direction of forward scattering). The field is not identically zero within this region, because of diffraction produced by creeping waves of current that travel around the cylinder and radiate as they travel.

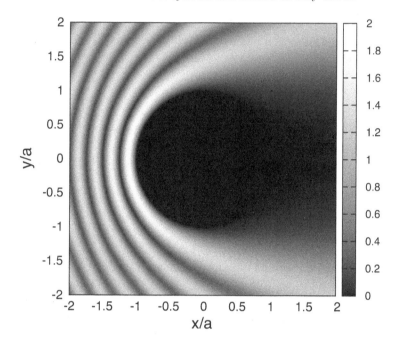

FIGURE 4.39
Total axial electric field $|\tilde{E}_z/\tilde{E}_0|$ around a PEC cylinder of radius $2\lambda_0$, found using the series solution. ◄

▶ **Example 4.28:** Total field for plane-wave scattering by a conducting cylinder using physical optics

Consider a TM plane wave incident along the x-axis onto a perfectly conducting cylinder of radius $2\lambda_0$ in free space. Compute and plot the magnitude of the total field $|\tilde{E}_z/\tilde{E}_0|$ within a square of side $4a$ centered on the cylinder using the physical optics approximation.

Solution: The physical optics (PO) current is given by $\tilde{\mathbf{J}}_s \approx 2\hat{n} \times \tilde{\mathbf{H}}^i$ in regions of illumination and by $\tilde{\mathbf{J}}_s = 0$ in regions of shadow. Use of

$$\tilde{\mathbf{H}}^i = -\hat{\mathbf{y}}\frac{\tilde{E}_0}{\eta_0}e^{jk_0 x}$$

gives

$$\tilde{J}_{sz}(\phi) = -2\frac{\tilde{E}_0}{\eta_0}\cos\phi\, e^{-jk_0 a\cos\phi} \qquad (4.364)$$

on the surface at $\rho = a$. Figure 7.51 compares the PO current with that found using the series solution. The currents match closely over much of the illuminated region, although the series solution reveals a small but significant current in the shadow region. Hence we might expect the scattered field produced using the PO approximation to work well in the illuminated region, but not necessarily in the shadow region.

The scattered field for the PO approximation may be computed using the two-dimensional Green's function:

$$\tilde{E}_z^s(x,y,\omega) = -j\omega\mu_0 \int J_{sz}(x',y',\omega)\tilde{G}(x,y|x',y';\omega)\,dl'.$$

The Green's function is given by (4.344), where $x = \rho\cos\phi$, $y = \rho\sin\phi$, $x' = a\cos\phi$, and $y' = a\sin\phi'$. Noting that

$$(x - x')^2 + (y - y')^2 = \rho^2 + a^2 - 2a\rho\cos(\phi - \phi'),$$

substituting the PO current from (4.364), and writing in terms of normalized distances, we obtain

$$\tilde{E}_z^s(\rho,\phi) = \pi\tilde{E}_0\frac{a}{\lambda_0}\int_{\pi/2}^{3\pi/2}\cos\phi' e^{-j2\pi\frac{a}{\lambda_0}\cos\phi'}$$
$$\cdot H_0^{(2)}\left(2\pi\frac{a}{\lambda_0}\sqrt{1 + \left(\frac{\rho}{a}\right)^2 - 2\frac{\rho}{a}\cos(\phi - \phi')}\right)d\phi'.$$

Figure 4.40 shows the magnitude of the normalized total electric field around the cylinder found with $a/\lambda_0 = 2$ using PO. The field on the illuminated side matches well with the series solution shown in Figure 4.39, revealing a clear standing wave pattern. The field on the opposite side does not match as well, with a much less distinct shadow region than is seen using the series solution. Thus, ignoring the surface current in the shadow region leads to an error most pronounced in the shadow region.

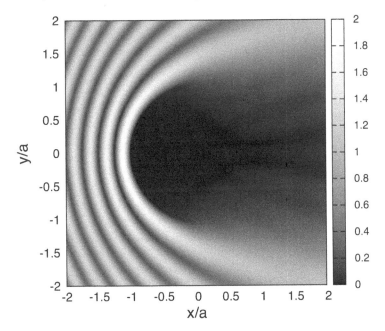

FIGURE 4.40
Total axial electric field $|\tilde{E}_z/\tilde{E}_0|$ around a PEC cylinder of radius $2\lambda_0$ found using PO. ◄

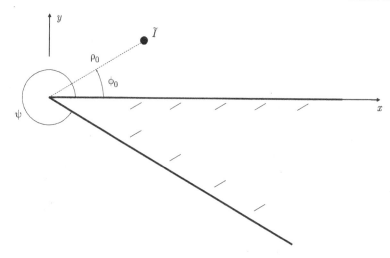

FIGURE 4.41
Geometry of a perfectly conducting wedge illuminated by a line source.

Computation of the scattered field for a magnetically polarized impressed field proceeds similarly. The impressed electric and magnetic fields are assumed to be

$$\tilde{\mathbf{E}}^i(\mathbf{r},\omega) = \hat{\mathbf{y}}\tilde{E}_0(\omega)e^{-jk_0 x} = (\hat{\boldsymbol{\rho}}\sin\phi + \hat{\boldsymbol{\phi}}\cos\phi)\tilde{E}_0(\omega)e^{-jk_0\rho\cos\phi},$$

$$\tilde{\mathbf{H}}^i(\mathbf{r},\omega) = \hat{\mathbf{z}}\frac{\tilde{E}_0(\omega)}{\eta_0}e^{-jk_0 x} = \hat{\mathbf{z}}\frac{\tilde{E}_0(\omega)}{\eta_0}e^{-jk_0\rho\cos\phi}.$$

For a perfectly conducting cylinder, the total magnetic field is

$$\tilde{H}_z = \frac{\tilde{E}_0}{\eta_0}\sum_{n=0}^{\infty}\frac{\epsilon_n j^{-n}}{H_n^{(2)\prime}(k_0 a)}\left[J_n(k_0\rho)H_n^{(2)\prime}(k_0 a) - J_n'(k_0 a)H_n^{(2)}(k_0\rho)\right]\cos n\phi. \qquad (4.365)$$

The details are left as an exercise.

4.11.8.5 Scattering by a perfectly conducting wedge

As a second example, consider a perfectly conducting wedge immersed in free space and illuminated by a line source (Figure 4.41) carrying current $\tilde{I}(\omega)$ and located at (ρ_0, ϕ_0). The current, assumed z-invariant, induces a secondary current on the surface of the wedge, which in turn produces a secondary (scattered) field. This scattered field, also z-invariant, can be found by solving a boundary value problem. We do this by separating space into the two regions $\rho < \rho_0$ and $\rho > \rho_0$, $0 < \phi < \psi$. Each of these is source-free, so we can represent the total field using nonuniform cylindrical waves of the type (4.349). The line source is brought into the problem by applying the boundary condition on the tangential magnetic field across the cylindrical surface $\rho = \rho_0$.

As the impressed electric field has only a z-component, so do the scattered and total electric fields. We wish to represent the total field $\tilde{E}_z$ in terms of nonuniform cylindrical waves of the type (4.349). Since the field is not periodic in ϕ, the separation constant k_ϕ need not be an integer; instead, its value is determined by the positions of the wedge boundaries. For the region $\rho < \rho_0$ we represent the radial dependence of the field using the functions J_ν since the field must be finite at the origin. For $\rho > \rho_0$ we use the outward-propagating wave functions $H_\delta^{(2)}$. Thus

$$\tilde{E}_z(\rho, \phi, \omega) = \begin{cases} \sum_\nu \left[A_\nu \sin \nu\phi + B_\nu \cos \nu\phi\right] J_\nu(k_0\rho), & \rho < \rho_0, \\ \sum_\delta \left[C_\delta \sin \delta\phi + D_\delta \cos \delta\phi\right] H_\delta^{(2)}(k_0\rho), & \rho > \rho_0. \end{cases} \quad (4.366)$$

The coefficients $A_\nu, B_\nu, C_\delta, D_\delta$ and separation constants ν, δ may be found by applying the boundary conditions on the fields at the surface of the wedge and across the surface $\rho = \rho_0$. On the wedge face at $\phi = 0$ we must have $\tilde{E}_z = 0$, hence $B_\nu = D_\delta = 0$. On the wedge face at $\phi = \psi$ we must also have $\tilde{E}_z = 0$, requiring $\sin \nu\psi = \sin \delta\psi = 0$ and hence $\nu = \delta = \nu_n = n\pi/\psi$ for $n = 1, 2, \ldots$. So

$$\tilde{E}_z = \begin{cases} \sum_{n=0}^\infty A_n \sin \nu_n\phi J_{\nu_n}(k_0\rho), & \rho < \rho_0, \\ \sum_{n=0}^\infty C_n \sin \nu_n\phi H_{\nu_n}^{(2)}(k_0\rho), & \rho > \rho_0. \end{cases} \quad (4.367)$$

The magnetic field can be found from (4.346)–(4.347):

$$\tilde{H}_\rho = \begin{cases} \sum_{n=0}^\infty A_n \frac{j}{\eta_0 k_0} \frac{\nu_n}{\rho} \cos \nu_n\phi J_{\nu_n}(k_0\rho), & \rho < \rho_0, \\ \sum_{n=0}^\infty C_n \frac{j}{\eta_0 k_0} \frac{\nu_n}{\rho} \cos \nu_n\phi H_{\nu_n}^{(2)}(k_0\rho), & \rho > \rho_0, \end{cases} \quad (4.368)$$

$$\tilde{H}_\phi = \begin{cases} -\sum_{n=0}^\infty A_n \frac{j}{\eta_0} \sin \nu_n\phi J'_{\nu_n}(k_0\rho), & \rho < \rho_0, \\ -\sum_{n=0}^\infty C_n \frac{j}{\eta_0} \sin \nu_n\phi H_{\nu_n}^{(2)\prime}(k_0\rho), & \rho > \rho_0. \end{cases} \quad (4.369)$$

The coefficients A_n and C_n are found by applying the boundary conditions at $\rho = \rho_0$. By continuity of the tangential electric field,

$$\sum_{n=0}^\infty A_n \sin \nu_n\phi J_{\nu_n}(k_0\rho_0) = \sum_{n=0}^\infty C_n \sin \nu_n\phi H_{\nu_n}^{(2)}(k_0\rho_0).$$

We now apply orthogonality over the interval $[0, \psi]$. Multiplying by $\sin \nu_m\phi$ and integrating, we have

$$\sum_{n=0}^\infty A_n J_{\nu_n}(k_0\rho_0) \int_0^\psi \sin \nu_n\phi \sin \nu_m\phi \, d\phi = \sum_{n=0}^\infty C_n H_{\nu_n}^{(2)}(k_0\rho_0) \int_0^\psi \sin \nu_n\phi \sin \nu_m\phi \, d\phi.$$

Setting $u = \phi\pi/\psi$, we have

$$\int_0^\psi \sin \nu_n\phi \sin \nu_m\phi \, d\phi = \frac{\psi}{\pi} \int_0^\pi \sin nu \sin mu \, du = \frac{\psi}{2} \delta_{mn},$$

thus

$$A_m J_{\nu_m}(k_0\rho_0) = C_m H_{\nu_m}^{(2)}(k_0\rho_0).$$

The boundary condition $\hat{n}_{12} \times (\tilde{\mathbf{H}}_1 - \tilde{\mathbf{H}}_2) = \tilde{\mathbf{J}}_s$ requires the surface current at $\rho = \rho_0$. We can write the line current in terms of a surface current density using the δ-function:

$$\tilde{\mathbf{J}}_s = \hat{\mathbf{z}} \tilde{I} \frac{\delta(\phi - \phi_0)}{\rho_0}.$$

This is easily verified as the correct expression since the integral of this density along the circular arc at $\rho = \rho_0$ returns the correct value $\tilde{I}$ for the total current. Thus the boundary condition requires

$$\tilde{H}_\phi(\rho_0^+, \phi, \omega) - \tilde{H}_\phi(\rho_0^-, \phi, \omega) = \tilde{I} \frac{\delta(\phi - \phi_0)}{\rho_0}.$$

By (4.369) we have

$$-\sum_{n=0}^{\infty} C_n \frac{j}{\eta_0} \sin \nu_n \phi H_{\nu_n}^{(2)\prime}(k_0\rho_0) + \sum_{n=0}^{\infty} A_n \frac{j}{\eta_0} \sin \nu_n \phi J_{\nu_n}'(k_0\rho_0) = \tilde{I} \frac{\delta(\phi - \phi_0)}{\rho_0}$$

and orthogonality yields

$$-C_m \frac{\psi}{2} \frac{j}{\eta_0} H_{\nu_m}^{(2)\prime}(k_0\rho_0) + A_m \frac{\psi}{2} \frac{j}{\eta_0} J_{\nu_m}'(k_0\rho_0) = \tilde{I} \frac{\sin \nu_m \phi_0}{\rho_0}.$$

The coefficients A_m and C_m thus obey the matrix equation

$$\begin{bmatrix} J_{\nu_m}(k_0\rho_0) & -H_{\nu_m}^{(2)}(k_0\rho_0) \\ J_{\nu_m}'(k_0\rho_0) & -H_{\nu_m}^{(2)\prime}(k_0\rho_0) \end{bmatrix} \begin{bmatrix} A_m \\ C_m \end{bmatrix} = \begin{bmatrix} 0 \\ -j2\tilde{I} \frac{\eta_0}{\psi} \frac{\sin \nu_m \phi_0}{\rho_0} \end{bmatrix}$$

and are

$$A_m = \frac{j2\tilde{I} \frac{\eta_0}{\psi} \frac{\sin \nu_m \phi_0}{\rho_0} H_{\nu_m}^{(2)}(k_0\rho_0)}{H_{\nu_m}^{(2)\prime}(k_0\rho_0) J_{\nu_m}(k_0\rho_0) - J_{\nu_m}'(k_0\rho_0) H_{\nu_m}^{(2)}(k_0\rho_0)},$$

$$C_m = \frac{j2\tilde{I} \frac{\eta_0}{\psi} \frac{\sin \nu_m \phi_0}{\rho_0} J_{\nu_m}(k_0\rho_0)}{H_{\nu_m}^{(2)\prime}(k_0\rho_0) J_{\nu_m}(k_0\rho_0) - J_{\nu_m}'(k_0\rho_0) H_{\nu_m}^{(2)}(k_0\rho_0)}.$$

Using the Wronskian relation (E.95), we replace the denominators in these expressions by $2/(j\pi k_0\rho_0)$:

$$A_m = -\tilde{I} \frac{\eta_0}{\psi} \pi k_0 \sin \nu_m \phi_0 H_{\nu_m}^{(2)}(k_0\rho_0),$$

$$C_m = -\tilde{I} \frac{\eta_0}{\psi} \pi k_0 \sin \nu_m \phi_0 J_{\nu_m}(k_0\rho_0).$$

Hence (4.367) gives

$$\tilde{E}_z(\rho, \phi, \omega) = \begin{cases} -\sum_{n=0}^{\infty} \tilde{I} \frac{\eta_0}{2\psi} \pi k_0 \epsilon_n J_{\nu_n}(k_0\rho) H_{\nu_n}^{(2)}(k_0\rho_0) \sin \nu_n \phi \sin \nu_n \phi_0, & \rho < \rho_0, \\ -\sum_{n=0}^{\infty} \tilde{I} \frac{\eta_0}{2\psi} \pi k_0 \epsilon_n H_{\nu_n}^{(2)}(k_0\rho) J_{\nu_n}(k_0\rho_0) \sin \nu_n \phi \sin \nu_n \phi_0, & \rho > \rho_0, \end{cases}$$

$$(4.370)$$

where ϵ_n is Neumann's number (A.133). The magnetic fields can also be found by substituting the coefficients into (4.368) and (4.369).

The fields produced by an impressed plane wave may now be obtained by letting the line source recede to infinity. For large ρ_0 we use the asymptotic form (E.64) and find that

$$\tilde{E}_z(\rho, \phi, \omega) = -\sum_{n=0}^{\infty} \tilde{I} \frac{\eta_0}{2\psi} \pi k_0 \epsilon_n J_{\nu_n}(k_0\rho) \left[\sqrt{\frac{2j}{\pi k_0\rho_0}} j^{\nu_n} e^{-jk_0\rho_0} \right] \sin \nu_n \phi \sin \nu_n \phi_0 \quad (\rho < \rho_0).$$

$$(4.371)$$

Since the field of a line source decays as $\rho_0^{-1/2}$, the amplitude of the impressed field approaches zero as $\rho_0 \to \infty$. We must compensate for the reduction in the impressed field by scaling the amplitude of the current source. To obtain the proper scale factor, we note that the electric field produced at a point $\boldsymbol{\rho}$ by a line source located at $\boldsymbol{\rho}_0$ may be found from (4.343):

$$\tilde{E}_z = -\tilde{I} \frac{k_0\eta_0}{4} H_0^{(2)}(k_0|\boldsymbol{\rho} - \boldsymbol{\rho}_0|) \approx -\tilde{I} \frac{k_0\eta_0}{4} \sqrt{\frac{2j}{\pi k_0\rho_0}} e^{-jk_0\rho_0} e^{jk\rho\cos(\phi-\phi_0)} \quad (k_0\rho_0 \gg 1).$$

But if we write this as $\tilde{E}_z \approx \tilde{E}_0 e^{j\mathbf{k}\cdot\boldsymbol{\rho}}$, then the field resembles that of a plane wave with amplitude $\tilde{E}_0$ traveling along the wave vector $\mathbf{k} = -k_0\hat{\mathbf{x}}\cos\phi_0 - k_0\hat{\mathbf{y}}\sin\phi_0$. Solving for $\tilde{I}$ in terms of $\tilde{E}_0$ and substituting into (4.371), we get the total electric field scattered from a wedge with an impressed TM plane-wave field:

$$\tilde{E}_z(\rho, \phi, \omega) = \frac{2\pi}{\psi}\tilde{E}_0 \sum_{n=0}^{\infty} \epsilon_n j^{\nu_n} J_{\nu_n}(k_0\rho) \sin\nu_n\phi \sin\nu_n\phi_0. \qquad (4.372)$$

Here we interpret the angle ϕ_0 as the incidence angle of the plane wave.

▶ **Example 4.29:** Total field for plane-wave scattering by a conducting wedge

A TM plane wave is incident at angle ϕ_0 on the perfectly conducting wedge shown in Figure 4.41. The wedge is defined by a right angle ($\psi = 3\pi/2$). Compute and plot the magnitude of the total field $|\tilde{E}_z|$ within a square of side $8\lambda_0$ centered on the edge, treating the cases $\phi_0 = 45°$ and $\phi_0 = 135°$.

Solution: The total axial electric field may be computed by evaluating (4.372). We normalize the distance from the edge so that $k\rho = 2\pi\rho/\lambda_0$. The magnitude of $\tilde{E}_z/\tilde{E}_0$ is shown in Figure 4.42 for $\phi_0 = 45°$. Here we have truncated the series at 200 terms for computational expediency.

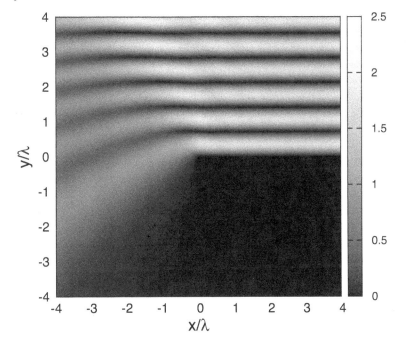

FIGURE 4.42
Total axial electric field $|\tilde{E}_z/\tilde{E}_0|$ for a TM plane wave incident at angle $\phi_0 = 45°$ on a perfectly conducting right-angle wedge.

Several interesting characteristics can be seen. First, the field vanishes at both surfaces of the wedge, as expected from the boundary condition. Second, a standing wave pattern exists above the wedge, where the incident field and the field reflected from the top surface combine to create regions of constructive and destructive interference with a period of a half wavelength. The *reflection boundary* at $\phi = 180° - \phi_0 = 135°$ separates the region illuminated by both the incident and reflected waves from the region illuminated only by the incident wave, according to the laws of geometrical optics. Finally, a shadow is seen to the left of the wedge. The *shadow boundary* at $\phi = 180° + \phi_0 = 225°$ separates the region illuminated by the incident wave from the *shadow zone*, which the incident wave cannot reach according to the laws of geometrical optics. Geometrical optics predicts that the fields should be discontinuous across the reflection and shadow boundaries, but the presence of a diffracted field provides for the expected continuity. Because of diffraction, a remnant of the standing wave pattern can be seen outside the reflection zone, and a nonzero (but small) field in the shadow region.

The magnitude of $\tilde{E}_z / \tilde{E}_0$ is shown in Figure 4.43 for $\phi_0 = 135°$. In this case the incident field illuminates both surfaces of the wedge, and a standing wave can be seen parallel to each surface. The reflection boundary for the top of the wedge is along $\phi = 45°$, while the reflection boundary for the side of the wedge is along $\phi = 225°$. Because both surfaces are illuminated, there are no shadow regions. Note that diffraction produces a very strong standing wave pattern outside the two reflection zones, and that the standing waves due to reflection from the surfaces merge into a symmetric and continuous pattern, as is expected from the symmetry of illumination.

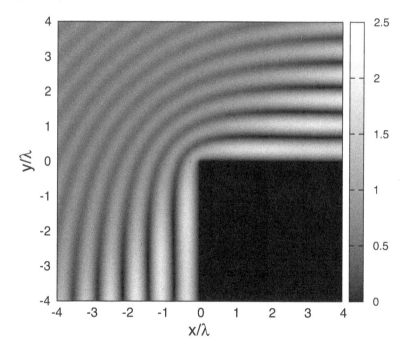

FIGURE 4.43
Total axial electric field $|\tilde{E}_z / \tilde{E}_0|$ for a TM plane wave incident at angle $\phi_0 = 135°$ on a perfectly conducting right-angle wedge. ◄

To determine the field produced by an impressed TE plane-wave field, we use a magnetic line source $\tilde{I}_m$ located at ρ_0, ϕ_0 and proceed as above. By analogy with (4.366) we write

$$\tilde{H}_z(\rho, \phi, \omega) = \begin{cases} \sum_\nu [A_\nu \sin \nu\phi + B_\nu \cos \nu\phi] \, J_\nu(k_0\rho), & \rho < \rho_0, \\ \sum_\delta [C_\delta \sin \delta\phi + D_\delta \cos \delta\phi] \, H_\delta^{(2)}(k_0\rho), & \rho > \rho_0. \end{cases}$$

By (4.348) the tangential electric field is

$$\tilde{E}_\rho(\rho, \phi, \omega) = \begin{cases} -\sum_\nu [A_\nu \cos \nu\phi - B_\nu \sin \nu\phi] \, j\frac{Z_{TE}}{k} \frac{1}{\rho} \nu J_\nu(k_0\rho), & \rho < \rho_0, \\ -\sum_\delta [C_\delta \cos \delta\phi - D_\delta \sin \delta\phi] \, j\frac{Z_{TE}}{k} \frac{1}{\rho} \delta H_\delta^{(2)}(k_0\rho), & \rho > \rho_0. \end{cases}$$

Application of the boundary conditions on the tangential electric field at $\phi = 0, \psi$ results in $A_\nu = C_\delta = 0$ and $\nu = \delta = \nu_n = n\pi/\psi$, and thus $\tilde{H}_z$ becomes

$$\tilde{H}_z(\rho, \phi, \omega) = \begin{cases} \sum_{n=0}^\infty B_n \cos \nu_n\phi J_{\nu_n}(k_0\rho), & \rho < \rho_0, \\ \sum_{n=0}^\infty D_n \cos \nu_n\phi H_{\nu_n}^{(2)}(k_0\rho), & \rho > \rho_0. \end{cases} \tag{4.373}$$

Application of the boundary conditions on tangential electric and magnetic fields across the magnetic line source then leads directly to

$$\tilde{H}_z(\rho, \phi, \omega) = \begin{cases} -\sum_{n=0}^\infty \tilde{I}_m \frac{\eta_0}{2\psi} \pi k_0 \epsilon_n J_{\nu_n}(k_0\rho) H_{\nu_n}^{(2)}(k_0\rho_0) \cos \nu_n\phi \cos \nu_n\phi_0, & \rho < \rho_0, \\ -\sum_{n=0}^\infty \tilde{I}_m \frac{\eta_0}{2\psi} \pi k_0 \epsilon_n H_{\nu_n}^{(2)}(k_0\rho) J_{\nu_n}(k_0\rho_0) \cos \nu_n\phi \cos \nu_n\phi_0, & \rho > \rho_0. \end{cases}$$

For a plane-wave impressed field this reduces to

$$\tilde{H}_z(\rho, \phi, \omega) = \frac{2\pi}{\psi} \frac{\tilde{E}_0}{\eta_0} \sum_{n=0}^\infty \epsilon_n j^{\nu_n} J_{\nu_n}(k_0\rho) \cos \nu_n\phi \cos \nu_n\phi_0.$$

4.11.8.6 Behavior of current near a sharp edge

In § 3.3.10 we studied the behavior of static charge near a sharp conducting edge by modeling the latter as a wedge. We can follow the same procedure for frequency-domain fields. Assume the perfectly conducting wedge shown in Figure 4.41 is immersed in a finite, z-independent impressed field of a sort that will not concern us. A current is induced on the surface of the wedge and we wish to study its behavior as we approach the edge.

Because the field is z-independent, we may consider the superposition of TM and TE fields as was done above to solve for the field scattered by a wedge. For TM polarization, if the source is not located near the edge we may write the total field (impressed plus scattered) in terms of nonuniform cylindrical waves. The form of the field that obeys the boundary conditions at $\phi = 0$ and $\phi = \psi$ is given by (4.367):

$$\tilde{E}_z = \sum_{n=0}^\infty A_n \sin \nu_n\phi J_{\nu_n}(k_0\rho),$$

where $\nu_n = n\pi/\psi$. Although the A_n depend on the impressed source, the general behavior of the current near the edge is determined by the properties of the Bessel functions. The current on the wedge face at $\phi = 0$ has density given by

$$\tilde{\mathbf{J}}_s(\rho, \omega) = \hat{\phi} \times [\hat{\phi}\tilde{H}_\phi + \hat{\rho}\tilde{H}_\rho]|_{\phi=0} = -\hat{z}\tilde{H}_\rho(\rho, 0, \omega).$$

By (4.346) we have the surface current density

$$\tilde{\mathbf{J}}_s(\rho, \omega) = -\hat{z} \frac{1}{Z_{TM}k_0} \sum_{n=0}^\infty A_n \frac{\nu_n}{\rho} J_{\nu_n}(k_0\rho).$$

For $\rho \to 0$ the small-argument approximation (E.51) yields

$$\tilde{\mathbf{J}}_s(\rho, \omega) \approx -\hat{\mathbf{z}} \frac{1}{Z_{TM}k_0} \sum_{n=0}^{\infty} A_n \nu_n \frac{1}{\Gamma(\nu_n + 1)} \left(\frac{k_0}{2}\right)^{\nu_n} \rho^{\nu_n - 1}.$$

The sum is dominated by the smallest power of ρ. Since the $n = 0$ term vanishes we have

$$\tilde{J}_s(\rho, \omega) \sim \rho^{\frac{\pi}{\psi} - 1} \qquad (\rho \to 0).$$

For $\psi < \pi$ the current density, which runs parallel to the edge, is unbounded as $\rho \to 0$. A right-angle wedge ($\psi = 3\pi/2$) carries $\tilde{J}_s(\rho, \omega) \sim \rho^{-1/3}$. Another important case is that of a half-plane ($\psi = 2\pi$) where

$$\tilde{J}_s(\rho, \omega) \sim 1/\sqrt{\rho}. \tag{4.374}$$

This square-root edge singularity dominates the behavior of the current flowing parallel to any flat edge, either straight or with curvature large compared to a wavelength, and is useful for modeling currents on complicated structures.

▶ **Example 4.30:** Current induced on the surface of a conducting wedge by a plane wave

A TM plane wave is incident at angle ϕ_0 on a perfectly conducting wedge (Figure 4.41). Derive a formula for the density of the surface current induced on each face, and approximate the current density near the edge. Assuming the wedge is defined by a right angle ($\psi = 3\pi/2$) and illumination occurs along $\phi_0 = 45°$, compute and plot the magnitude of the current density on each face. Use the exact and approximate formulas.

Solution: The surface current density on the face located at $\phi = \phi_w$ is

$$\tilde{\mathbf{J}}_s(\rho, \omega) = \hat{\boldsymbol{\phi}} \times [\hat{\boldsymbol{\phi}} \tilde{H}_\phi + \hat{\boldsymbol{\rho}} \tilde{H}_\rho]|_{\phi=\phi_w} = -\hat{\mathbf{z}} \tilde{H}_\rho(\rho, \phi_w, \omega),$$

where $\phi_w = 0$ for the top face and $\phi_w = \psi$ for the bottom face. Substituting (4.372) into (4.346) we get the surface current density

$$\tilde{\mathbf{J}}_s(\rho, \omega) = -\hat{\mathbf{z}} \frac{2\pi}{\psi} \tilde{E}_0 \frac{j}{Z_{TM}k_0\rho} \sum_{n=0}^{\infty} \epsilon_n j^{\nu_n} J_{\nu_n}(k_0\rho) \nu_n \cos \nu_n \phi_w \sin \nu_n \phi_0,$$

or

$$\frac{\tilde{J}_{sz}}{\tilde{E}_0/\eta_0} = -\frac{2\pi}{\psi} \frac{j}{k_0\rho} \sum_{n=0}^{\infty} \epsilon_n j^{\nu_n} \nu_n J_{\nu_n}(k_0\rho) \cos \nu_n \phi_w \sin \nu_n \phi_0.$$

When $k_0\rho \ll 1$ we have

$$J_{\nu_n}(k_0\rho) \approx \frac{1}{\Gamma(\nu_n + 1)} \left(\frac{k_0\rho}{2}\right)^{\nu_n}.$$

Since the series is dominated by the first nonzero term, retention of the $n = 1$ term gives a simple approximation for the surface current density near the edge:

$$\frac{\tilde{J}_{sz}}{\tilde{E}_0/\eta_0} = -\left(\frac{2\pi}{\psi}\right)^2 2^{-\pi/\psi} (k_0\rho)^{\pi/\psi - 1} \frac{j^{\pi/\psi+1}}{\Gamma(\pi/\psi + 1)} \cos \pi\frac{\phi_w}{\psi} \sin \pi\frac{\phi_0}{\psi}. \tag{4.375}$$

Under this approximation the currents on the faces $\phi = 0$ and $\phi = \psi$ differ only in sign.

According to the physical optics (PO) approximation, we have $\tilde{\mathbf{J}}_s \approx 2\hat{\mathbf{n}} \times \tilde{\mathbf{H}}^i$ in regions of illumination and $\tilde{\mathbf{J}}_s = 0$ in regions of shadow. Use of

$$\tilde{\mathbf{H}}^i = (-\hat{\mathbf{x}} \sin \phi_0 + \hat{\mathbf{y}} \cos \phi_0) \frac{\tilde{E}_0}{\eta_0} e^{jk_0(x \cos \phi_0 + y \sin \phi_0)}$$

gives

$$\frac{\tilde{J}_{sz}}{\tilde{E}_0/\eta_0} \approx 2\sin\phi_0 e^{jk_0 x\cos\phi_0}.$$

As a specific example, take $\psi = 3\pi/2$ and $\phi_0 = 45°$. We normalize the distance from the edge so that $k\rho = 2\pi\rho/\lambda$. Figure 4.44 shows the normalized current densities on both faces of the wedge, plotted using a logarithmic scale. Approaching the edge, both currents vary as $\tilde{J}_{sz} \sim \rho^{\pi/\psi-1} \sim \rho^{-1/3}$, and (4.375) yields good results. On the top face, as the distance from the edge tends to a wavelength, the PO approximation

$$\left|\frac{\tilde{J}_{sz}}{\tilde{E}_0/\eta_0}\right| \approx 2\sin\phi_0 = \sqrt{2}$$

matches the exact result very well. On the side face, in the shadow of the incident wave, the current decays rapidly away from the edge. Ten wavelengths away, the current on the side face is less than 0.2% of the PO value on the top face.

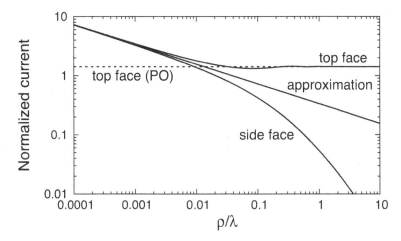

FIGURE 4.44
Normalized surface current density $|\tilde{J}_{sz}/(\tilde{E}_0/\eta_0)|$ for a plane wave incident at angle $\phi_0 = 45°$ on a perfectly conducting right-angle wedge. ◄

In the case of TE polarization, the magnetic field near the edge is, by (4.373),

$$\tilde{H}_z(\rho,\phi,\omega) = \sum_{n=0}^{\infty} B_n \cos\nu_n\phi J_{\nu_n}(k_0\rho) \qquad (\rho < \rho_0).$$

The current density at $\phi = 0$ is

$$\tilde{\mathbf{J}}_s(\rho,\omega) = \hat{\boldsymbol{\phi}} \times \hat{\mathbf{z}}\tilde{H}_z|_{\phi=0} = \hat{\boldsymbol{\rho}}\tilde{H}_z(\rho,0,\omega)$$

$$= \hat{\boldsymbol{\rho}} \sum_{n=0}^{\infty} B_n J_{\nu_n}(k_0\rho).$$

For $\rho \to 0$ we use (E.51) to write

$$\tilde{\mathbf{J}}_s(\rho,\omega) = \hat{\boldsymbol{\rho}} \sum_{n=0}^{\infty} B_n \frac{1}{\Gamma(\nu_n+1)} \left(\frac{k_0}{2}\right)^{\nu_n} \rho^{\nu_n}.$$

The $n = 0$ term gives a constant contribution, so we keep the first two terms to see how the current behaves near $\rho = 0$:

$$\tilde{J}_s \sim b_0 + b_1 \rho^{\frac{\pi}{\psi}}.$$

Here b_0 and b_1 depend on the form of the impressed field. For a thin plate where $\psi = 2\pi$ this becomes $\tilde{J}_s \sim b_0 + b_1 \sqrt{\rho}$. This is the companion square-root behavior to (4.374). When perpendicular to a sharp edge, the current grows away from the edge as $\rho^{1/2}$. In most cases $b_0 = 0$ since there is no mechanism to store charge along a sharp edge.

▶ **Example 4.31:** Total field for plane-wave scattering by a conducting wedge using physical optics

A TM plane wave is incident at angle ϕ_0 on the perfectly conducting wedge shown in Figure 4.41. The wedge is defined by a right angle ($\psi = 3\pi/2$). Compute and plot the magnitude of the total field $|\tilde{E}_z|$ within a square of side $8\lambda_0$ centered on the edge. Treat the cases $\phi_0 = 45°$ and $\phi_0 = 135°$. Assume the current on the wedge surface is well described by physical optics.

Solution: Figure 4.44 shows that the current on the top surface of a wedge illuminated by a plane wave from above is very close to the PO current, except for points near the edge (where there is a singularity for TM-polarization). So the PO current should lead to a reasonable approximation for the total field around the wedge if used to compute the scattered field in place of the series solution.

The physical optics current density is given by

$$\tilde{\mathbf{J}}_s \approx 2\hat{\mathbf{n}} \times \tilde{\mathbf{H}}^i$$

in regions of illumination, and $\tilde{\mathbf{J}}_s = 0$ in regions of shadow. Use of

$$\tilde{\mathbf{H}}^i = (-\hat{\mathbf{x}} \sin \phi_0 + \hat{\mathbf{y}} \cos \phi_0) \frac{\tilde{E}_0}{\eta_0} e^{jk_0(x \cos \phi_0 + y \sin \phi_0)}$$

gives

$$\tilde{J}_{sz}(x) = \frac{\tilde{E}_0}{\eta_0} 2 \sin \phi_0 e^{jk_0 x \cos \phi_0}$$

on the surface at $y = 0$. The surface at $x = 0$ will carry no current if it is in shadow ($\phi_0 < 90°$) and

$$\tilde{J}_{sz}(y) = -\frac{\tilde{E}_0}{\eta_0} 2 \cos \phi_0 e^{jky \sin \phi_0}$$

if it is illuminated.

The scattered field may be computed from

$$\tilde{E}_z^s(x, y, \omega) = -j\omega\mu_0 \int J_{sz}(x', y', \omega) \tilde{G}(x, y | x', y'; \omega) \, dl'.$$

For this purpose we truncate the size of the wedge at a large value W. Substitution from (4.344) gives

$$\tilde{E}_z^s = -\frac{\tilde{E}_0}{2\eta_0} \omega\mu_0 \sin \phi_0 \int_0^W e^{jk_0 x' \cos\phi_0} H_0^{(2)} \left(k_0 \sqrt{(x - x')^2 + y^2} \right) dx'$$

$$+ \frac{\tilde{E}_0}{2\eta_0} \omega\mu_0 \cos \phi_0 \int_{-W}^0 e^{jk_0 y' \sin \phi_0} H_0^{(2)} \left(k_0 \sqrt{x^2 + (y - y')^2} \right) dy',$$

where the second integral is included only when the surface at $x = 0$ is illuminated. Using the change of variables $u = k_0 x'$ in the first integral and $u = -k_0 y'$ in the second integral, we obtain the field in terms of normalized distance:

$$\tilde{E}_z^s = -\frac{\tilde{E}_0}{2} \sin \phi_0 \int_0^{2\pi \frac{W}{\lambda_0}} e^{ju \cos \phi_0} H_0^{(2)} \left(\sqrt{\left(2\pi \frac{x}{\lambda_0} - u\right)^2 + \left(2\pi \frac{y}{\lambda_0}\right)^2} \right) du$$

$$+ \frac{\tilde{E}_0}{2} \cos \phi_0 \int_0^{2\pi \frac{W}{\lambda_0}} e^{-ju \sin \phi_0} H_0^{(2)} \left(\sqrt{\left(2\pi \frac{x}{\lambda_0}\right)^2 + \left(2\pi \frac{y}{\lambda_0} + u\right)^2} \right) du. \quad (4.376)$$

Figure 4.45 shows the total field $\tilde{E}_z / \tilde{E}_0$ (incident plus scattered) for incidence angle $\phi_0 = 45°$. Here the wedge has been truncated at $W = 100\lambda_0$, and only the first integral in (4.376) has been used since only the top surface is illuminated. The PO current gives a result very similar to the exact result of Figure 4.42, including the standing wave pattern above the wedge. Because the PO current is approximate, the boundary condition at the wedge surface is not satisfied exactly, especially near the edge since the PO current lacks the expected singularity. Moreover, the shadow region is not as pronounced, and there is a small field within the wedge region due to truncation and inaccuracy of the current.

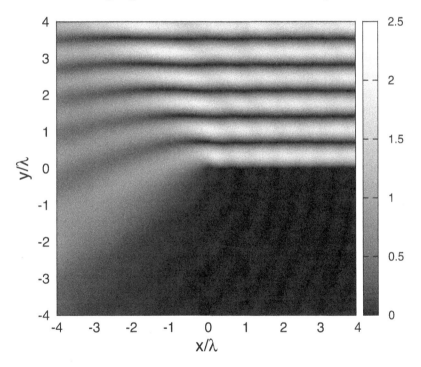

FIGURE 4.45
Total axial electric field $|\tilde{E}_z / \tilde{E}_0|$ for a plane wave incident at angle $\phi_0 = 45°$ on a perfectly conducting right-angle wedge, computed using the physical optics current.

Figure 4.46 shows the total field $\tilde{E}_z / \tilde{E}_0$ (incident plus scattered) for an incidence angle of $\phi_0 = 135°$. Here the wedge has been truncated at $W = 100\lambda_0$, and both of the integrals in (4.376) have been used since both wedge surfaces are illuminated. Once again the PO current gives a result very similar to the exact result; compare to Figure 4.43. Standing wave patterns are clearly seen both above and to the side of the wedge, just as in the exact solution. Again, because the PO current is approximate, the boundary condition at the surface of the wedge is not satisfied exactly and there is a small field within the wedge region due to truncation and inaccuracy of the current.

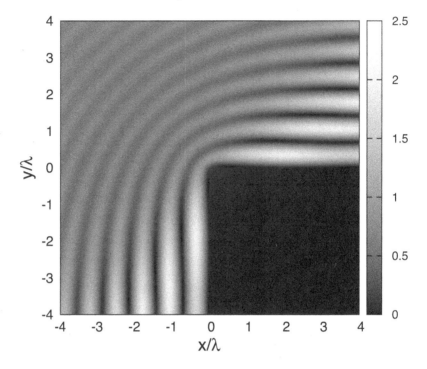

FIGURE 4.46
Total axial electric field $|\tilde{E}_z/\tilde{E}_0|$ for a plane wave incident at angle $\phi_0 = 135°$ on a perfectly conducting right-angle wedge, computed using the physical optics current. ◄

4.11.9   Propagation of spherical waves in a conducting medium

We cannot obtain uniform spherical wave solutions to Maxwell's equations. Any field dependent only on r produces the null field external to the source region (§ 4.11.10). Nonuniform spherical waves are in general complicated and most easily handled using potentials. We consider here only the simple problem of fields dependent on r and θ. These waves display the fundamental properties of all spherical waves: they diverge from a localized source and expand with finite velocity.

Consider a homogeneous, source-free region characterized by $\tilde{\epsilon}(\omega)$, $\tilde{\mu}(\omega)$, and $\tilde{\sigma}(\omega)$. We seek wave solutions that are TEM_r in spherical coordinates ($\tilde{H}_r = \tilde{E}_r = 0$) and ϕ-independent. Thus we write

$$\tilde{\mathbf{E}}(\mathbf{r}, \omega) = \hat{\theta}\tilde{E}_\theta(r, \theta, \omega) + \hat{\phi}\tilde{E}_\phi(r, \theta, \omega),$$
$$\tilde{\mathbf{H}}(\mathbf{r}, \omega) = \hat{\theta}\tilde{H}_\theta(r, \theta, \omega) + \hat{\phi}\tilde{H}_\phi(r, \theta, \omega).$$

To determine the behavior of these fields, we first examine Faraday's law. Expanding the curl in spherical coordinates, we have

$$\nabla \times \tilde{\mathbf{E}}(r,\theta,\omega) = \hat{\mathbf{r}}\frac{1}{r\sin\theta}\frac{\partial}{\partial\theta}[\sin\theta \tilde{E}_\phi(r,\theta,\omega)] - \hat{\boldsymbol{\theta}}\frac{1}{r}\frac{\partial}{\partial r}[r\tilde{E}_\phi(r,\theta,\omega)] + \hat{\boldsymbol{\phi}}\frac{1}{r}\frac{\partial}{\partial r}[r\tilde{E}_\theta(r,\theta,\omega)]$$

$$= -j\omega\tilde{\mu}\tilde{\mathbf{H}}(r,\theta,\omega). \tag{4.377}$$

Since we require $\tilde{H}_r = 0$ we must have

$$\frac{\partial}{\partial\theta}[\sin\theta\tilde{E}_\phi(r,\theta,\omega)] = 0.$$

This implies that either $\tilde{E}_\phi \sim 1/\sin\theta$ or $\tilde{E}_\phi = 0$. We choose $\tilde{E}_\phi = 0$ and investigate whether the resulting fields satisfy the remaining Maxwell equations.

In a source-free, homogeneous region of space we have $\nabla \cdot \tilde{\mathbf{D}} = 0$ and hence $\nabla \cdot \tilde{\mathbf{E}} = 0$. Since we have only a θ-component of the electric field, this requires

$$\frac{1}{r}\frac{\partial}{\partial\theta}\tilde{E}_\theta(r,\theta,\omega) + \frac{\cot\theta}{r}\tilde{E}_\theta(r,\theta,\omega) = 0.$$

So when $\tilde{E}_\phi = 0$, the field $\tilde{E}_\theta$ must obey

$$\tilde{E}_\theta(r,\theta,\omega) = \frac{\tilde{f}_E(r,\omega)}{\sin\theta}.$$

By (4.377) there is only a ϕ-component of magnetic field, which obeys

$$\tilde{H}_\phi(r,\theta,\omega) = \frac{\tilde{f}_H(r,\omega)}{\sin\theta}$$

where

$$-j\omega\tilde{\mu}\tilde{f}_H(r,\omega) = \frac{1}{r}\frac{\partial}{\partial r}[r\tilde{f}_E(r,\omega)]. \tag{4.378}$$

So the spherical wave is TEM to the r-direction.

We can obtain a wave equation for $\tilde{f}_E$ by taking the curl of (4.377) and substituting from Ampere's law:

$$\nabla \times (\nabla \times \tilde{\mathbf{E}}) = -\hat{\boldsymbol{\theta}}\frac{1}{r}\frac{\partial^2}{\partial r^2}(r\tilde{E}_\theta) = \nabla \times (-j\omega\tilde{\mu}\tilde{\mathbf{H}}) = -j\omega\tilde{\mu}(\tilde{\sigma}\tilde{\mathbf{E}} + j\omega\tilde{\epsilon}\tilde{\mathbf{E}}),$$

hence

$$\frac{d^2}{dr^2}[r\tilde{f}_E(r,\omega)] + k^2[r\tilde{f}_E(r,\omega)] = 0. \tag{4.379}$$

Here $k = \omega(\tilde{\mu}\tilde{\epsilon}^c)^{1/2}$ is the complex wavenumber and $\tilde{\epsilon}^c = \tilde{\epsilon} + \tilde{\sigma}/j\omega$ is the complex permittivity. The equation for $\tilde{f}_H$ is identical.

The wave equation (4.379) is merely the second-order harmonic differential equation, with two independent solutions chosen from the list

$$\sin kr, \quad \cos kr, \quad e^{-jkr}, \quad e^{jkr}.$$

We find $\sin kr$ and $\cos kr$ useful for describing standing waves between boundaries, and e^{jkr} and e^{-jkr} useful for describing waves propagating in the r-direction. Of these, e^{jkr} represents waves traveling inward while e^{-jkr} represents waves traveling outward. At this point we choose $r\tilde{f}_E = e^{-jkr}$ and thus

$$\tilde{\mathbf{E}}(r,\theta,\omega) = \hat{\boldsymbol{\theta}}\tilde{E}_0(\omega)\frac{e^{-jkr}}{r\sin\theta}. \tag{4.380}$$

By (4.378) we have

$$\tilde{\mathbf{H}}(r,\theta,\omega) = \hat{\boldsymbol{\phi}} \frac{\tilde{E}_0(\omega)}{Z_{TEM}} \frac{e^{-jkr}}{r\sin\theta} \tag{4.381}$$

where $Z_{TEM} = (\tilde{\mu}/\tilde{\epsilon}^c)^{1/2}$ is the complex wave impedance. Since we can also write

$$\tilde{\mathbf{H}}(r,\theta,\omega) = \frac{\hat{\mathbf{r}} \times \tilde{\mathbf{E}}(r,\theta,\omega)}{Z_{TEM}},$$

the field is TEM to the r-direction, which is the direction of wave propagation as shown below.

The wave nature of the field is easily identified by considering the fields in the phasor domain. Letting $\omega \to \breve{\omega}$ and setting $k = \beta - j\alpha$ in the exponential function, we find that

$$\breve{\mathbf{E}}(r,\theta) = \hat{\boldsymbol{\theta}} \breve{E}_0 e^{-\alpha r} \frac{e^{-j\beta r}}{r\sin\theta}$$

where $\breve{E}_0 = E_0 e^{j\xi^E}$. The time-domain representation may be found using (4.137):

$$\mathbf{E}(r,\theta,t) = \hat{\boldsymbol{\theta}} E_0 \frac{e^{-\alpha r}}{r\sin\theta} \cos(\breve{\omega}t - \beta r + \xi^E). \tag{4.382}$$

We can identify a surface of constant phase as a locus of points satisfying

$$\breve{\omega}t - \beta r + \xi^E = C_P$$

where C_P is some constant. These surfaces, which are spheres centered on the origin, are called *spherical wavefronts*. Note that surfaces of constant amplitude, determined by $e^{-\alpha r}/r = C_A$ for constant C_A, are also spheres.

The cosine term in (4.382) represents a traveling wave with spherical wavefronts that propagate outward as time progresses. Attenuation is caused by the factor $e^{-\alpha r}$. By differentiation we find that the phase velocity is

$$v_p = \breve{\omega}/\beta.$$

The wavelength is given by $\lambda = 2\pi/\beta$.

Our solution is not appropriate for unbounded space since the fields have a singularity at $\theta = 0$. To exclude the z-axis we add conducting cones as mentioned on page 119. This results in a biconical structure that can be used as a transmission line or antenna.

To compute the power carried by a spherical wave, we use (4.380) and (4.381) to obtain the time-average Poynting flux

$$\mathbf{S}_{av} = \tfrac{1}{2} \operatorname{Re}\{\breve{E}_\theta \hat{\boldsymbol{\theta}} \times \breve{H}_\phi^* \hat{\boldsymbol{\phi}}\} = \tfrac{1}{2}\hat{\mathbf{r}} \operatorname{Re}\left\{\frac{1}{Z_{TEM}^*}\right\} \frac{E_0^2}{r^2\sin^2\theta} e^{-2\alpha r}.$$

The power flux is radial and has density inversely proportional to r^2. The time-average power carried by the wave through a spherical surface at r sandwiched between the cones at θ_1 and θ_2 is

$$P_{av}(r) = \tfrac{1}{2}\operatorname{Re}\left\{\frac{1}{Z_{TEM}^*}\right\} E_0^2 e^{-2\alpha r} \int_0^{2\pi} d\phi \int_{\theta_1}^{\theta_2} \frac{d\theta}{\sin\theta} = \pi F \operatorname{Re}\left\{\frac{1}{Z_{TEM}^*}\right\} E_0^2 e^{-2\alpha r}$$

where

$$F = \ln\left[\frac{\tan(\theta_2/2)}{\tan(\theta_1/2)}\right]. \tag{4.383}$$

This is independent of r when $\alpha = 0$. For lossy media the power decays exponentially because of Joule heating.

We can write the phasor electric field in terms of the transverse gradient of a scalar potential function $\check{\Phi}$:

$$\check{\mathbf{E}}(r,\theta) = \hat{\boldsymbol{\theta}}\check{E}_0 \frac{e^{-jkr}}{r\sin\theta} = -\nabla_t\check{\Phi}(\theta) \quad \text{where} \quad \check{\Phi}(\theta) = -\check{E}_0 e^{-jkr}\ln\left(\tan\frac{\theta}{2}\right).$$

By ∇_t we mean the gradient with the r-component excluded. It is easily verified that

$$\check{\mathbf{E}}(r,\theta) = -\nabla_t\check{\Phi}(\theta) = -\hat{\boldsymbol{\theta}}\check{E}_0\frac{1}{r}\frac{\partial\check{\Phi}(\theta)}{\partial\theta} = \hat{\boldsymbol{\theta}}\check{E}_0\frac{e^{-jkr}}{r\sin\theta}.$$

Because $\check{\mathbf{E}}$ and $\check{\Phi}$ are related by the gradient, we can define a unique potential difference between the two cones at any radial position r:

$$\check{V}(r) = -\int_{\theta_1}^{\theta_2}\check{\mathbf{E}}\cdot\mathbf{dl} = \check{\Phi}(\theta_2) - \check{\Phi}(\theta_1) = \check{E}_0 F e^{-jkr},$$

where F is given in (4.383). The existence of a unique voltage difference is a property of all transmission line structures operated in the TEM mode. We can similarly compute the current flowing outward on the cone surfaces. The surface current on the cone at $\theta = \theta_1$ is $\check{\mathbf{J}}_s = \hat{\mathbf{n}}\times\check{\mathbf{H}} = \hat{\boldsymbol{\theta}}\times\hat{\boldsymbol{\phi}}\check{H}_\phi = \hat{\mathbf{r}}\check{H}_\phi$, hence

$$\check{I}(r) = \int_0^{2\pi}\check{\mathbf{J}}_s\cdot\hat{\mathbf{r}}r\sin\theta d\phi = 2\pi\frac{\check{E}_0}{Z_{TEM}}e^{-jkr}.$$

The ratio of voltage to current at any radius r is the *characteristic impedance* of the biconical transmission line (or, equivalently, the *input impedance* of the bicone antenna):

$$Z = \frac{\check{V}(r)}{\check{I}(r)} = \frac{Z_{TEM}}{2\pi}F.$$

If the material between the cones is lossless (so $\tilde{\mu} = \mu$ and $\tilde{\epsilon}^c = \epsilon$ are real), this becomes

$$Z = \frac{\eta}{2\pi}F$$

where $\eta = (\mu/\epsilon)^{1/2}$. The frequency independence of this quantity makes biconical antennas (or their approximate representations) useful in broadband applications.

Finally, the time-average power carried by the wave may be found from

$$P_{av}(r) = \tfrac{1}{2}\operatorname{Re}\left\{\check{V}(r)\check{I}^*(r)\right\} = \pi F\operatorname{Re}\left\{\frac{1}{Z_{TEM}^*}\right\}E_0^2 e^{-2\alpha r}.$$

The complex power relationship $P = VI^*$ is also a property of TEM guided-wave structures. We consider the biconical transmission line again in § 5.6.5.2 in the context of transmission line theory, using a TE-TM decomposition of the fields in spherical coordinates.

4.11.10 Nonradiating sources

We showed in § 2.10.10 that not all time-varying sources produce electromagnetic waves. In fact, a subset of localized sources known as *nonradiating sources* produce no field external to the source region. Devaney and Wolf [48] have shown that all nonradiating time-harmonic sources in an unbounded homogeneous medium can be represented in the form

$$\check{\mathbf{J}}^{nr}(\mathbf{r}) = -\nabla \times [\nabla \times \check{\mathbf{f}}(\mathbf{r})] + k^2 \check{\mathbf{f}}(\mathbf{r}) \tag{4.384}$$

where $\check{\mathbf{f}}$ is any vector field that is continuous, has partial derivatives up to third order, and vanishes outside some localized region V_s. In fact, $\check{\mathbf{E}}(\mathbf{r}) = j\breve{\omega}\mu\check{\mathbf{f}}(\mathbf{r})$ is precisely the phasor electric field produced by $\check{\mathbf{J}}^{nr}(\mathbf{r})$. The reasoning is straightforward. Consider the Helmholtz equation (4.197):

$$\nabla \times (\nabla \times \check{\mathbf{E}}) - k^2 \check{\mathbf{E}} = -j\breve{\omega}\mu\check{\mathbf{J}}.$$

By (4.384) we have

$$(\nabla \times \nabla \times -k^2)[\check{\mathbf{E}} - j\breve{\omega}\mu\check{\mathbf{f}}] = 0.$$

Since $\check{\mathbf{f}}$ is zero outside the source region, it must vanish at infinity. $\check{\mathbf{E}}$ also vanishes at infinity by the radiation condition, and thus the quantity $\check{\mathbf{E}} - j\breve{\omega}\mu\check{\mathbf{f}}$ obeys the radiation condition and is a unique solution to the Helmholtz equation throughout all space. Since the Helmholtz equation is homogeneous, we have

$$\check{\mathbf{E}} - j\breve{\omega}\mu\check{\mathbf{f}} = 0$$

everywhere; since $\check{\mathbf{f}}$ is zero outside the source region, so is $\check{\mathbf{E}}$ (and so is $\check{\mathbf{H}}$).

An interesting special case of nonradiating sources is

$$\check{\mathbf{f}} = \frac{\nabla\check{\Phi}}{k^2}$$

so that

$$\check{\mathbf{J}}^{nr} = -(\nabla \times \nabla \times -k^2)\frac{\nabla\check{\Phi}}{k^2} = \nabla\check{\Phi}.$$

Using $\check{\Phi}(\mathbf{r}) = \check{\Phi}(r)$, we see that this source describes the current produced by an oscillating spherical balloon of charge (cf., § 2.10.10). Radially directed, spherically symmetric sources cannot produce uniform spherical waves, since these sources are of the nonradiating type.

4.12   Interpretation of the spatial transform

Now that we understand the meaning of a Fourier transform on the time variable, let us consider a single transform involving one of the spatial variables. For a transform over z we shall use the notation

$$\psi^z(x, y, k_z, t) \leftrightarrow \psi(x, y, z, t).$$

Here the spatial frequency transform variable k_z has units of m^{-1}. The forward and inverse transform expressions are

$$\psi^z(x, y, k_z, t) = \int_{-\infty}^{\infty} \psi(x, y, z, t)e^{-jk_z z}\, dz,$$

$$\psi(x, y, z, t) = \frac{1}{2\pi}\int_{-\infty}^{\infty} \psi^z(x, y, k_z, t)e^{jk_z z}\, dk_z, \tag{4.385}$$

by (A.1) and (A.2).

We interpret (4.385) much as we interpreted the temporal inverse transform (4.2). Any vector component of the electromagnetic field can be decomposed into a continuous superposition of elemental spatial terms $e^{jk_z z}$ with weighting factors $\psi^z(x, y, k_z, t)$. In this case ψ^z is the *spatial frequency spectrum* of ψ. The elemental terms are spatial sinusoids along z with rapidity of variation described by k_z.

As with the temporal transform, ψ^z cannot be arbitrary since ψ must obey a scalar wave equation such as (2.269). For instance, in a source-free region of free space we must have

$$\left(\nabla^2 - \frac{1}{c^2}\frac{\partial}{\partial t^2}\right)\frac{1}{2\pi}\int_{-\infty}^{\infty}\psi^z(x, y, k_z, t)e^{jk_z z}\,dk_z = 0.$$

Decomposing the Laplacian operator as $\nabla^2 = \nabla_t^2 + \partial^2/\partial z^2$ and taking the derivatives into the integrand, we have

$$\frac{1}{2\pi}\int_{-\infty}^{\infty}\left[\left(\nabla_t^2 - k_z^2 - \frac{1}{c^2}\frac{\partial^2}{\partial t^2}\right)\psi^z(x, y, k_z, t)\right]e^{jk_z z}\,dk_z = 0.$$

Hence

$$\left(\nabla_t^2 - k_z^2 - \frac{1}{c^2}\frac{\partial^2}{\partial t^2}\right)\psi^z(x, y, k_z, t) = 0$$

by the Fourier integral theorem.

The elemental component $e^{jk_z z}$ is spatially sinusoidal and occupies all of space. Because such an element could only be created by a source that spans all of space, it is nonphysical when taken by itself. Nonetheless it is often used to represent more complicated fields. If the elemental spatial term is to be used alone, it is best interpreted physically when combined with a temporal decomposition. That is, we consider a two-dimensional transform, with transforms over both time and space. Then the time-domain representation of the elemental component is

$$\phi(z, t) = \frac{1}{2\pi}\int_{-\infty}^{\infty}e^{jk_z z}e^{j\omega t}\,d\omega. \tag{4.386}$$

Before attempting to compute this transform, we should note that if the elemental term is to describe an EM field ψ in a source-free region, it must obey the homogeneous scalar wave equation. Substituting (4.386) into the latter, we have

$$\left(\nabla^2 - \frac{1}{c^2}\frac{\partial^2}{\partial t^2}\right)\frac{1}{2\pi}\int_{-\infty}^{\infty}e^{jk_z z}e^{j\omega t}\,d\omega = 0.$$

Differentiation under the integral sign gives

$$\frac{1}{2\pi}\int_{-\infty}^{\infty}\left[\left(-k_z^2 + \frac{\omega^2}{c^2}\right)e^{jk_z z}\right]e^{j\omega t}\,d\omega = 0$$

and thus

$$k_z^2 = \frac{\omega^2}{c^2} = k^2.$$

Substitution of $k_z = k$ into (4.386) gives the time-domain representation of the elemental component

$$\phi(z, t) = \frac{1}{2\pi}\int_{-\infty}^{\infty}e^{j\omega(t+z/c)}\,d\omega.$$

Finally, by the shifting theorem (A.3) and (A.5) we have

$$\phi(z,t) = \delta\left(t + \frac{z}{c}\right), \tag{4.387}$$

which is a uniform plane wave propagating in the $-z$-direction with velocity c. There is no variation transverse to the direction of propagation and the surface describing a constant argument of the δ-function at any time t is a plane perpendicular to that direction.

We can also consider the elemental spatial component in tandem with a single sinusoidal steady-state elemental component. The phasor representation of the elemental spatial component is

$$\check{\phi}(z) = e^{jk_z z} = e^{jkz}.$$

This elemental term is a time-harmonic plane wave propagating in the $-z$-direction. Indeed, multiplying by $e^{j\check{\omega}t}$ and taking the real part, we get

$$\phi(z,t) = \cos(\check{\omega}t + kz),$$

which is the sinusoidal steady-state analogue of (4.387).

Many authors define the temporal and spatial transforms using differing sign conventions. The temporal transform is defined as in (4.1) and (4.2), but the spatial transform is defined through

$$\psi^z(x,y,k_z,t) = \int_{-\infty}^{\infty} \psi(x,y,z,t)e^{jk_z z}\,dz, \tag{4.388}$$

$$\psi(x,y,z,t) = \frac{1}{2\pi}\int_{-\infty}^{\infty} \psi^z(x,y,k_z,t)e^{-jk_z z}\,dk_z.$$

This employs a wave traveling in the positive z-direction as the elemental spatial component, which is quite useful for physical interpretation. We shall adopt this notation in § 4.13. The drawback is that we must alter the formulas from standard Fourier transform tables (replacing k by $-k$) accordingly.

In the following sections we show how a spatial Fourier decomposition can be used to find the electromagnetic fields in a source-free region of space. By employing the spatial transform we eliminate one or more spatial variables from Maxwell's equations, making the wave equation easier to solve. In the end we must perform an inversion to return to the space domain. This may be difficult or impossible to do analytically, requiring a numerical Fourier inversion.

4.13   Spatial Fourier decomposition of two-dimensional fields

Consider a homogeneous, source-free region characterized by $\tilde{\epsilon}(\omega)$, $\tilde{\mu}(\omega)$, and $\tilde{\sigma}(\omega)$. We seek z-independent solutions to the frequency-domain Maxwell's equations, using the Fourier transform to represent the spatial dependence. By § 4.11.2 a general two-dimensional field may be decomposed into fields TE and TM to the z-direction. In the TM case, $\tilde{H}_z = 0$, and $\tilde{E}_z$ obeys the homogeneous scalar Helmholtz equation (4.200). In the TE case, $\tilde{E}_z = 0$, and $\tilde{H}_z$ obeys the homogeneous scalar Helmholtz equation. Since each field component obeys the same equation, we let $\tilde{\psi}(x,y,\omega)$ represent either $\tilde{E}_z(x,y,\omega)$ or $\tilde{H}_z(x,y,\omega)$. Then $\tilde{\psi}$ obeys

$$(\nabla_t^2 + k^2)\tilde{\psi}(x,y,\omega) = 0 \tag{4.389}$$

where ∇_t^2 is the transverse Laplacian (4.201) and $k = \omega(\tilde{\mu}\tilde{\epsilon}^c)^{1/2}$ with $\tilde{\epsilon}^c$ the complex permittivity.

We may represent $\tilde{\psi}(x, y, \omega)$ using Fourier transforms over one or both spatial variables. For problems in which boundary values or boundary conditions are specified at a constant value of a single variable (e.g., over a plane), one transform suffices. For instance, we may know the values of the field in the $y = 0$ plane (as we will, for example, with the boundary value problems of § 4.13.1). Then we may transform over x and leave the y variable intact so that we may substitute the boundary values.

We adopt (4.388) since the result is more readily interpreted in terms of propagating plane waves. Choosing to transform over x, we have

$$\tilde{\psi}^x(k_x, y, \omega) = \int_{-\infty}^{\infty} \tilde{\psi}(x, y, \omega)e^{jk_x x}\, dx,$$

$$\tilde{\psi}(x, y, \omega) = \frac{1}{2\pi}\int_{-\infty}^{\infty} \psi^x(k_x, y, \omega)e^{-jk_x x}\, dk_x.$$

For convenience in computation or interpretation of the inverse transform, we often regard k_x as a complex variable and perturb the inversion contour into the complex $k_x = k_{xr} + jk_{xi}$ plane. The integral is not altered if the contour is not moved past singularities such as poles or branch points. If the function being transformed has exponential (wave) behavior, then a pole exists in the complex plane; if we move the inversion contour across this pole, the inverse transform does not return the original function. We generally indicate the desire to interpret k_x as complex by indicating that the inversion contour is parallel to the real axis but located in the complex plane at $k_{xi} = \Delta$:

$$\tilde{\psi}(x, y, \omega) = \frac{1}{2\pi}\int_{-\infty+j\Delta}^{\infty+j\Delta} \tilde{\psi}^x(k_x, y, \omega)e^{-jk_x x}\, dk_x. \tag{4.390}$$

Additional perturbations of the contour are allowed, provided that the contour is not moved through singularities.

As an example, consider the function

$$u(x) = \begin{cases} 0, & x < 0, \\ e^{-jkx}, & x > 0, \end{cases} \tag{4.391}$$

where $k = k_r + jk_i$ represents a wavenumber. This function has the form of a plane wave propagating in the x-direction and is thus relevant to our studies. If the material through which the wave is propagating is lossy, then $k_i < 0$. The Fourier transform of the function is

$$u^x(k_x) = \int_0^{\infty} e^{-jkx}e^{jk_x x}\, dx = \frac{1}{j(k_x - k)}\left[e^{j(k_{xr}-k_r)x}e^{-(k_{xi}-k_i)x}\right]\Big|_0^{\infty}.$$

The integral converges if $k_{xi} > k_i$, and the transform is

$$u^x(k_x) = -\frac{1}{j(k_x - k)}.$$

Since $u(x)$ is an exponential function, $u^x(k_x)$ has a pole at $k_x = k$ as anticipated.

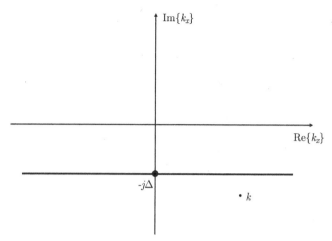

FIGURE 4.47

Inversion contour for evaluating the spectral integral for a plane wave.

To compute the inverse transform, we use (4.390):

$$u(x) = \frac{1}{2\pi} \int\limits_{-\infty+j\Delta}^{\infty+j\Delta} \left[-\frac{1}{j(k_x - k)} \right] e^{-jk_x x} \, dk_x. \qquad (4.392)$$

We must be careful to choose Δ in such a way that all values of k_x along the inversion contour lead to a convergent forward Fourier transform. Since we must have $k_{xi} > k_i$, choosing $\Delta > k_i$ ensures proper convergence. This gives the inversion contour shown in Figure 4.47, a special case of which is the real axis. We compute the inversion integral using contour integration (§ A.2). We close the contour in the complex plane and use Cauchy's residue theorem (A.15). For $x > 0$ we take $0 > \Delta > k_i$ and close the contour in the lower half-plane using a semicircular contour C_R of radius R. Then the closed contour integral is equal to $-2\pi j$ times the residue at the pole $k_x = k$. As $R \to \infty$ we find that $k_{xi} \to -\infty$ at all points on the contour C_R. Thus the integrand, which varies as $e^{k_{xi} x}$, vanishes on C_R and there is no contribution to the integral. The inversion integral (4.392) is found from the residue at the pole:

$$u(x) = (-2\pi j) \frac{1}{2\pi} \mathrm{Res}_{k_x = k} \left[-\frac{1}{j(k_x - k)} e^{-jk_x x} \right].$$

Since the residue is merely je^{-jkx} we have $u(x) = e^{-jkx}$. When $x < 0$ we choose $\Delta > 0$ and close the contour along a semicircle C_R of radius R in the upper half-plane. Again we find that on C_R the integrand vanishes as $R \to \infty$, and thus the inversion integral (4.392) is given by $2\pi j$ times the residues of the integrand at any poles within the closed contour. This time, however, there are no poles enclosed and thus $u(x) = 0$. We have recovered the original function (4.391) for both $x > 0$ and $x < 0$. Note that if we had erroneously chosen $\Delta < k_i$ we would not have properly enclosed the pole and would have obtained an incorrect inverse transform.

   Now that we know how to represent the Fourier transform pair, let us apply the transform to solve (4.389). Our hope is that by representing $\tilde{\psi}$ in terms of a spatial

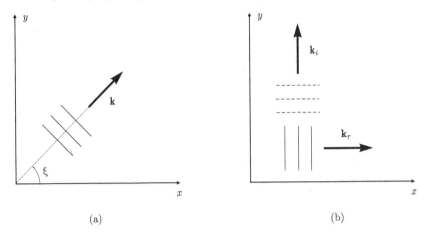

FIGURE 4.48
Propagation behavior of the angular spectrum for (a) $k_x^2 \le k^2$, (b) $k_x^2 > k^2$.

Fourier integral we will make the equation easier to solve. We have

$$(\nabla_t^2 + k^2) \frac{1}{2\pi} \int\limits_{-\infty+j\Delta}^{\infty+j\Delta} \tilde{\psi}^x(k_x, y, \omega) e^{-jk_x x} \, dk_x = 0.$$

Differentiation under the integral sign with subsequent application of the Fourier integral theorem implies that $\tilde{\psi}$ must obey the second-order harmonic differential equation

$$\left[\frac{d^2}{dy^2} + k_y^2 \right] \tilde{\psi}^x(k_x, y, \omega) = 0$$

where we have defined the dependent parameter $k_y = k_{yr} + jk_{yi}$ through $k_x^2 + k_y^2 = k^2$. Two independent solutions to the differential equation are $e^{\mp jk_y y}$ and thus

$$\tilde{\psi}(k_x, y, \omega) = A(k_x, \omega) e^{\mp jk_y y}.$$

Substituting this into the inversion integral, we have the solution to the Helmholtz equation:

$$\tilde{\psi}(x, y, \omega) = \frac{1}{2\pi} \int\limits_{-\infty+j\Delta}^{\infty+j\Delta} A(k_x, \omega) e^{-jk_x x} e^{\mp jk_y y} \, dk_x. \tag{4.393}$$

Defining the wave vector $\mathbf{k} = \hat{\mathbf{x}} k_x \pm \hat{\mathbf{y}} k_y$, we can also write the solution in the form

$$\tilde{\psi}(x, y, \omega) = \frac{1}{2\pi} \int\limits_{-\infty+j\Delta}^{\infty+j\Delta} A(k_x, \omega) e^{-j\mathbf{k}\cdot\boldsymbol{\rho}} \, dk_x \tag{4.394}$$

where $\boldsymbol{\rho} = \hat{\mathbf{x}} x + \hat{\mathbf{y}} y$ is the two-dimensional position vector.

The solution (4.394) has an important physical interpretation. The exponential term looks exactly like a plane wave with its wave vector lying in the xy-plane. For lossy media the plane wave is nonuniform, and the surfaces of constant phase may not be aligned with the surfaces of constant amplitude (see § 4.11.4). For the special case of a lossless medium we have $k_i \to 0$ and can let $\Delta \to 0$ as long as $\Delta > k_i$. As we perform the inverse transform integral over k_x from $-\infty$ to ∞ we will encounter both the condition $k_x^2 > k^2$ and $k_x^2 \leq k^2$. For $k_x^2 \leq k^2$ we have

$$e^{-jk_x x} e^{\mp jk_y y} = e^{-jk_x x} e^{\mp j\sqrt{k^2 - k_x^2} y}$$

where the choice of upper sign for $y > 0$ and lower sign for $y < 0$ ensures waves propagating in the $\pm y$-directions, respectively. Thus, in this regime the exponential represents a propagating wave that travels into the half-plane $y > 0$ along a direction that depends on k_x, making an angle ξ with the x-axis as shown in Figure 4.48. For k_x in $[-k, k]$, all possible wave directions are covered, and we may think of the inversion integral as constructing the solution to the two-dimensional Helmholtz equation from a continuous superposition of plane waves. The amplitude of each plane wave component is given by $A(k_x, \omega)$, which is often called the *angular spectrum* of the plane waves and is determined by the values of the field over the boundaries of the solution region. But this is not the whole picture. The inverse transform integral also requires values of k_x in the intervals $[-\infty, k]$ and $[k, \infty]$. Here we have $k_x^2 > k^2$ and thus

$$e^{-jk_x x} e^{-jk_y y} = e^{-jk_x x} e^{\mp \sqrt{k_x^2 - k^2} y},$$

where we choose the upper sign for $y > 0$ and the lower sign for $y < 0$ to obtain a field decaying along the y-direction. In these regimes we have an evanescent wave, propagating along x but decaying along y, with surfaces of constant phase and amplitude mutually perpendicular (Figure 4.48). As k_x ranges out to ∞, evanescent waves of all possible decay constants also contribute to the plane-wave superposition.

We may summarize the plane-wave contributions by letting $\mathbf{k} = \hat{\mathbf{x}} k_x + \hat{\mathbf{y}} k_y = \mathbf{k}_r + j\mathbf{k}_i$ where

$$\mathbf{k}_r = \begin{cases} \hat{\mathbf{x}} k_x \pm \hat{\mathbf{y}}\sqrt{k^2 - k_x^2}, & k_x^2 < k^2, \\ \hat{\mathbf{x}} k_x, & k_x^2 > k^2, \end{cases} \qquad \mathbf{k}_i = \begin{cases} 0, & k_x^2 < k^2, \\ \mp\hat{\mathbf{y}}\sqrt{k_x^2 - k^2}, & k_x^2 > k^2. \end{cases}$$

The upper sign applies for $y > 0$ and the lower sign for $y < 0$.

In many applications, including the half-plane example considered later, it is useful to write the inversion integral in polar coordinates. Letting $k_x = k\cos\xi$ and $k_y = \pm k\sin\xi$ where $\xi = \xi_r + j\xi_i$ is a new complex variable, we have $\mathbf{k} \cdot \boldsymbol{\rho} = kx\cos\xi \pm ky\sin\xi$ and $dk_x = -k\sin\xi\, d\xi$. With this change of variables (4.394) becomes

$$\tilde{\psi}(x, y, \omega) = \frac{k}{2\pi} \int_C A(k\cos\xi, \omega) e^{-jkx\cos\xi} e^{\pm jky\sin\xi} \sin\xi\, d\xi. \qquad (4.395)$$

Since $A(k_x, \omega)$ is a function to be determined, we may introduce a new function

$$f(\xi, \omega) = \frac{k}{2\pi} A(k_x, \omega) \sin\xi$$

so that (4.395) becomes

$$\tilde{\psi}(x, y, \omega) = \int_C f(\xi, \omega) e^{-jk\rho\cos(\phi \pm \xi)}\, d\xi \qquad (4.396)$$

where $x = \rho\cos\phi$, $y = \rho\sin\phi$, and where the upper sign corresponds to $0 < \phi < \pi$ ($y > 0$) and the lower sign to $\pi < \phi < 2\pi$ ($y < 0$). In these expressions C is a contour in

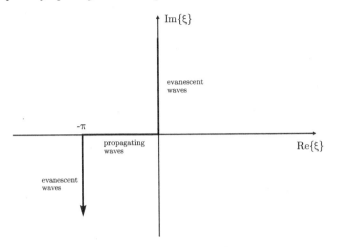

FIGURE 4.49

Inversion contour for the polar coordinate representation of the inverse Fourier transform.

the complex ξ-plane to be determined. Values along this contour must produce identical values of the integrand as did the values of k_x over $[-\infty, \infty]$ in the original inversion integral. By the identities

$$\cos z = \cos(u + jv) = \cos u \cosh v - j \sin u \sinh v,$$
$$\sin z = \sin(u + jv) = \sin u \cosh v + j \cos u \sinh v,$$

we find that the contour shown in Figure 4.49 provides identical values of the integrand (Problem 4.20). The portions of the contour $[0 + j\infty, 0]$ and $[-\pi, -\pi - j\infty]$ together correspond to the regime of evanescent waves ($k < k_x < \infty$ and $-\infty < k_x < k$), while the segment $[0, -\pi]$ along the real axis corresponds to $-k < k_x < k$ and thus describes contributions from propagating plane waves. In this case ξ represents the propagation angle of the waves.

4.13.1 Boundary value problems using the spatial Fourier transform representation

4.13.1.1 The field of a line source

As a first example we calculate the Fourier representation of the field of an electric line source. Assume a uniform line current $\tilde{I}(\omega)$ along the z-axis in a medium characterized by complex permittivity $\tilde{\epsilon}^c(\omega)$ and permeability $\tilde{\mu}(\omega)$. We separate space into two source-free portions, $y > 0$ and $y < 0$, and write the field in each region in terms of an inverse spatial Fourier transform. Then, by applying the boundary conditions in the $y = 0$ plane, we determine the angular spectrum of the line source.

Since this is a two-dimensional problem, we may decompose the fields into TE and TM sets. For an electric line source we need only the TM set, and write $\tilde{E}_z$ as a superposition of plane waves using (4.393). For $y \gtrless 0$ we represent the field in terms of plane waves traveling in the $\pm y$-direction. Thus

$$\tilde{E}_z(x,y,\omega) = \frac{1}{2\pi} \int_{-\infty+j\Delta}^{\infty+j\Delta} A^+(k_x,\omega) e^{-jk_x x} e^{-jk_y y} \, dk_x \qquad (y > 0),$$

$$\tilde{E}_z(x,y,\omega) = \frac{1}{2\pi} \int_{-\infty+j\Delta}^{\infty+j\Delta} A^-(k_x,\omega) e^{-jk_x x} e^{+jk_y y} \, dk_x \qquad (y < 0).$$

The transverse magnetic field may be found from the axial electric field using (4.204):

$$\tilde{H}_x = -\frac{1}{j\omega\tilde{\mu}} \frac{\partial \tilde{E}_z}{\partial y}, \qquad (4.397)$$

thus

$$\tilde{H}_x(x,y,\omega) = \frac{1}{2\pi} \int_{-\infty+j\Delta}^{\infty+j\Delta} A^+(k_x,\omega) \left[\frac{k_y}{\omega\tilde{\mu}}\right] e^{-jk_x x} e^{-jk_y y} \, dk_x \qquad (y > 0),$$

$$\tilde{H}_x(x,y,\omega) = \frac{1}{2\pi} \int_{-\infty+j\Delta}^{\infty+j\Delta} A^-(k_x,\omega) \left[-\frac{k_y}{\omega\tilde{\mu}}\right] e^{-jk_x x} e^{+jk_y y} \, dk_x \qquad (y < 0).$$

To find the spectra $A^\pm(k_x,\omega)$ we apply the boundary conditions at $y = 0$. Since tangential $\tilde{\mathbf{E}}$ is continuous, we have, after combining the integrals,

$$\frac{1}{2\pi} \int_{-\infty+j\Delta}^{\infty+j\Delta} [A^+(k_x,\omega) - A^-(k_x,\omega)] e^{-jk_x x} \, dk_x = 0,$$

and, hence, by the Fourier integral theorem

$$A^+(k_x,\omega) - A^-(k_x,\omega) = 0. \qquad (4.398)$$

We must also apply $\hat{\mathbf{n}}_{12} \times (\tilde{\mathbf{H}}_1 - \tilde{\mathbf{H}}_2) = \tilde{\mathbf{J}}_s$. The line current may be written as a surface current density using the δ-function, giving

$$-[\tilde{H}_x(x,0^+,\omega) - \tilde{H}_x(x,0^-,\omega)] = \tilde{I}(\omega)\delta(x).$$

By (A.5)

$$\delta(x) = \frac{1}{2\pi} \int_{-\infty}^{\infty} e^{-jk_x x} \, dk_x.$$

Then, substituting for the fields and combining the integrands, we have

$$\frac{1}{2\pi} \int_{-\infty+j\Delta}^{\infty+j\Delta} \left[A^+(k_x,\omega) + A^-(k_x,\omega) + \frac{\omega\tilde{\mu}}{k_y}\tilde{I}(\omega)\right] e^{-jk_x x} \, dk_x = 0,$$

hence,

$$A^+(k_x,\omega) + A^-(k_x,\omega) = -\frac{\omega\tilde{\mu}}{k_y}\tilde{I}(\omega). \qquad (4.399)$$

Solution of (4.398) and (4.399) gives the angular spectra

$$A^+(k_x, \omega) = A^-(k_x, \omega) = -\frac{\omega\tilde{\mu}}{2k_y}\tilde{I}(\omega).$$

Substituting this into the field expressions and combining the cases for $y > 0$ and $y < 0$, we find

$$\tilde{E}_z(x, y, \omega) = -\frac{\omega\tilde{\mu}\tilde{I}(\omega)}{2\pi}\int\limits_{-\infty+j\Delta}^{\infty+j\Delta}\frac{e^{-jk_y|y|}}{2k_y}e^{-jk_x x}\,dk_x = -j\omega\tilde{\mu}\tilde{I}(\omega)\tilde{G}(x, y|0, 0; \omega). \quad (4.400)$$

Here $\tilde{G}$ is the spectral representation of the two-dimensional Green's function first found in § 4.11.8, and is given by

$$\tilde{G}(x, y|x', y'; \omega) = \frac{1}{2\pi j}\int\limits_{-\infty+j\Delta}^{\infty+j\Delta}\frac{e^{-jk_y|y-y'|}}{2k_y}e^{-jk_x(x-x')}\,dk_x. \quad (4.401)$$

By duality,

$$\tilde{H}_z(x, y, \omega) = -\frac{\omega\tilde{\epsilon}^c\tilde{I}_m(\omega)}{2\pi}\int\limits_{-\infty+j\Delta}^{\infty+j\Delta}\frac{e^{-jk_y|y|}}{2k_y}e^{-jk_x x}\,dk_x = -j\omega\tilde{\epsilon}^c\tilde{I}_m(\omega)G(x, y|0, 0; \omega)$$

$$(4.402)$$

for a magnetic line current $\tilde{I}_m(\omega)$ on the z-axis.

Since (4.344) and (4.401) must be equivalent, we have the well-known identity [35]

$$\frac{1}{\pi}\int\limits_{-\infty+j\Delta}^{\infty+j\Delta}\frac{e^{-jk_y|y|}}{k_y}e^{-jk_x x}\,dk_x = H_0^{(2)}(k\rho).$$

We have not yet specified the contour appropriate for calculating the inverse transform (4.400). Caution is required, as the denominator of (4.400) has branch points at $k_y = \sqrt{k^2 - k_x^2} = 0$, or equivalently, $k_x = \pm k = \pm(k_r + jk_i)$. For lossy materials, $k_i < 0$ and $k_r > 0$, so the branch points appear as in Figure 4.50. We may take the branch cuts outward from these points, with an inversion contour lying between the branch points so that the latter are not traversed. This requires $k_i < \Delta < -k_i$. It is natural to choose $\Delta = 0$ and use the real axis as the inversion contour. But care is needed to extend these arguments to the lossless case. Taking the lossless case as the limit of the lossy case as $k_i \to 0$, we find that the branch points migrate to the real axis and thus lie on the inversion contour. We can eliminate this problem by realizing that the inversion contour may be perturbed without affecting the value of the integral, provided the contour does not cross the branch cuts. If we perturb the contour as shown in Figure 4.50, then as $k_i \to 0$ the branch points do not fall on the contour.

Many interesting techniques may be used to compute the inversion integral appearing in (4.400) and in the other expressions obtained in this section. These include direct real-axis integration and closed contour methods using Cauchy's residue theorem to capture poles of the integrand (which often describe the properties of waves guided by surfaces). Often it is necessary to integrate around branch cuts to satisfy the hypotheses of the residue theorem. When the observation point is far from the source, we may use the method of steepest descents to obtain asymptotic forms for the fields. The interested reader should consult Chew [35], Kong [108], or Sommerfeld [179].

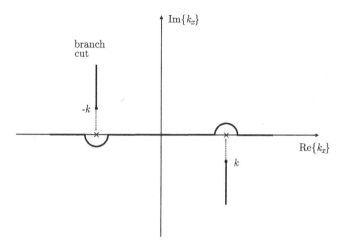

FIGURE 4.50

Inversion contour in complex k_x-plane for a line source. Dotted arrow shows migration of branch points to real axis as loss goes to zero.

4.13.1.2   Field of a line source above an interface

Consider a z-directed electric line current located at $y = h$ within a medium having parameters $\tilde{\mu}_1(\omega)$ and $\tilde{\epsilon}_1^c(\omega)$. The $y = 0$ plane separates this region from a region having parameters $\tilde{\mu}_2(\omega)$ and $\tilde{\epsilon}_2^c(\omega)$. See Figure 4.51. The impressed line current source creates an electromagnetic field that induces secondary polarization and conduction currents in both regions. This current in turn produces a secondary field that adds to the primary field of the line source to satisfy the boundary conditions at the interface. We would like to obtain the secondary field and give its sources an image interpretation.

Since the fields are z-independent we may decompose the fields into sets TE and TM to z. For a z-directed impressed source there is a z-component of $\tilde{\mathbf{E}}$ but no z-component of $\tilde{\mathbf{H}}$, so the fields are entirely specified by the TM set. The impressed source is unaffected by the secondary field, and we may represent the impressed electric field using (4.400):

$$\tilde{E}_z^i(x,y,\omega) = -\frac{\omega\tilde{\mu}_1\tilde{I}(\omega)}{2\pi} \int\limits_{-\infty+j\Delta}^{\infty+j\Delta} \frac{e^{-jk_{y1}|y-h|}}{2k_{y1}} e^{-jk_x x}\, dk_x \qquad (y \geq 0) \qquad (4.403)$$

where $k_{y1} = \sqrt{k_1^2 - k_x^2}$ and $k_1 = \omega(\tilde{\mu}_1\tilde{\epsilon}_1^c)^{1/2}$. From (4.397) we find that

$$\tilde{H}_x^i = -\frac{1}{j\omega\tilde{\mu}_1}\frac{\partial \tilde{E}_z^i}{\partial y} = \frac{\tilde{I}(\omega)}{2\pi} \int\limits_{-\infty+j\Delta}^{\infty+j\Delta} \frac{e^{jk_{y1}(y-h)}}{2} e^{-jk_x x}\, dk_x \qquad (0 \leq y < h).$$

The scattered field obeys the homogeneous Helmholtz equation for all $y > 0$, and thus

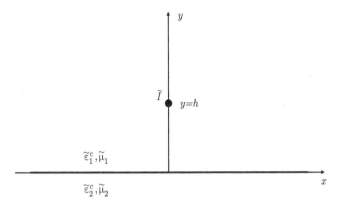

FIGURE 4.51
Geometry of a z-directed line source above an interface between two material regions.

may be written using (4.393) as a superposition of upward-traveling waves:

$$\tilde{E}_{z1}^{s}(x,y,\omega) = \frac{1}{2\pi} \int_{-\infty+j\Delta}^{\infty+j\Delta} A_1(k_x,\omega) e^{-jk_{y1}y} e^{-jk_x x} \, dk_x,$$

$$\tilde{H}_{x1}^{s}(x,y,\omega) = \frac{1}{2\pi} \int_{-\infty+j\Delta}^{\infty+j\Delta} \frac{k_{y1}}{\omega\tilde{\mu}_1} A_1(k_x,\omega) e^{-jk_{y1}y} e^{-jk_x x} \, dk_x.$$

Similarly, in region 2 the scattered field may be written as a superposition of downward-traveling waves:

$$\tilde{E}_{z2}^{s}(x,y,\omega) = \frac{1}{2\pi} \int_{-\infty+j\Delta}^{\infty+j\Delta} A_2(k_x,\omega) e^{jk_{y2}y} e^{-jk_x x} \, dk_x,$$

$$\tilde{H}_{x2}^{s}(x,y,\omega) = -\frac{1}{2\pi} \int_{-\infty+j\Delta}^{\infty+j\Delta} \frac{k_{y2}}{\omega\tilde{\mu}_2} A_2(k_x,\omega) e^{jk_{y2}y} e^{-jk_x x} \, dk_x,$$

where $k_{y2} = \sqrt{k_2^2 - k_x^2}$ and $k_2 = \omega(\tilde{\mu}_2 \tilde{\epsilon}_2^c)^{1/2}$.

We can determine the angular spectra A_1 and A_2 by applying the boundary conditions at the interface between the two media. Continuity of the total tangential electric field implies

$$\frac{1}{2\pi} \int_{-\infty+j\Delta}^{\infty+j\Delta} \left[-\frac{\omega\tilde{\mu}_1 \tilde{I}(\omega)}{2k_{y1}} e^{-jk_{y1}h} + A_1(k_x,\omega) - A_2(k_x,\omega) \right] e^{-jk_x x} \, dk_x = 0,$$

hence, by the Fourier integral theorem,

$$A_1(k_x,\omega) - A_2(k_x,\omega) = \frac{\omega\tilde{\mu}_1 \tilde{I}(\omega)}{2k_{y1}} e^{-jk_{y1}h}.$$

Continuity of $\tilde{H}_x$ yields

$$-\frac{\tilde{I}(\omega)}{2}e^{-jk_{y1}h} = \frac{k_{y1}}{\omega\tilde{\mu}_1}A_1(k_x,\omega) + \frac{k_{y2}}{\omega\tilde{\mu}_2}A_2(k_x,\omega)$$

and we obtain

$$A_1(k_x,\omega) = \frac{\omega\tilde{\mu}_1\tilde{I}(\omega)}{2k_{y1}}R_{TM}(k_x,\omega)e^{-jk_{y1}h},$$

$$A_2(k_x,\omega) = -\frac{\omega\tilde{\mu}_2\tilde{I}(\omega)}{2k_{y2}}T_{TM}(k_x,\omega)e^{-jk_{y1}h},$$

where R_{TM} and $T_{TM} = 1 + R_{TM}$ are the reflection and transmission coefficients

$$R_{TM}(k_x,\omega) = \frac{\tilde{\mu}_1 k_{y2} - \tilde{\mu}_2 k_{y1}}{\tilde{\mu}_1 k_{y2} + \tilde{\mu}_2 k_{y1}}, \qquad T_{TM}(k_x,\omega) = \frac{2\tilde{\mu}_1 k_{y2}}{\tilde{\mu}_1 k_{y2} + \tilde{\mu}_2 k_{y1}}.$$

These describe the reflection and transmission of each component of the plane-wave spectrum of the impressed field, and thus depend on the parameter k_x. The scattered fields are

$$\tilde{E}_{z1}^s(x,y,\omega) = \frac{\omega\tilde{\mu}_1\tilde{I}(\omega)}{2\pi} \int_{-\infty+j\Delta}^{\infty+j\Delta} \frac{e^{-jk_{y1}(y+h)}}{2k_{y1}}R_{TM}(k_x,\omega)e^{-jk_x x}\,dk_x, \qquad (4.404)$$

$$\tilde{E}_{z2}^s(x,y,\omega) = -\frac{\omega\tilde{\mu}_2\tilde{I}(\omega)}{2\pi} \int_{-\infty+j\Delta}^{\infty+j\Delta} \frac{e^{jk_{y2}(y-hk_{y1}/k_{y2})}}{2k_{y2}}T_{TM}(k_x,\omega)e^{-jk_x x}\,dk_x. \qquad (4.405)$$

We may now obtain the field produced by an electric line source above a perfect conductor. As $\tilde{\sigma}_2 \to \infty$ we get $k_{y2} = \sqrt{k_2^2 - k_x^2} \to \infty$ so that $R_{TM} \to 1$ and $T_{TM} \to 2$. The scattered fields (4.404) and (4.405) become

$$\tilde{E}_{z1}^s(x,y,\omega) = \frac{\omega\tilde{\mu}_1\tilde{I}(\omega)}{2\pi} \int_{-\infty+j\Delta}^{\infty+j\Delta} \frac{e^{-jk_{y1}(y+h)}}{2k_{y1}}e^{-jk_x x}\,dk_x, \qquad (4.406)$$

$$\tilde{E}_{z2}^s(x,y,\omega) = 0.$$

Comparing (4.406) to (4.403) we see that the scattered field is the same as that produced by a line source of amplitude $-\tilde{I}(\omega)$ located at $y = -h$. We call this line source the image of the impressed source, and say that the problem of two line sources located symmetrically on the y-axis is equivalent for $y > 0$ to the problem of the line source above a ground plane. The total field is the sum of the impressed and scattered fields:

$$\tilde{E}_z(x,y,\omega) = -\frac{\omega\tilde{\mu}_1\tilde{I}(\omega)}{2\pi} \int_{-\infty+j\Delta}^{\infty+j\Delta} \frac{e^{-jk_{y1}|y-h|} - e^{-jk_{y1}(y+h)}}{2k_{y1}}e^{-jk_x x}\,dk_x \qquad (y \geq 0).$$

We can write this in another form using the Hankel-function representation of the line source (4.343):

$$\tilde{E}_z(x,y,\omega) = -\frac{\omega\tilde{\mu}}{4}\tilde{I}(\omega)H_0^{(2)}(k|\boldsymbol{\rho} - \hat{\mathbf{y}}h|) + \frac{\omega\tilde{\mu}}{4}\tilde{I}(\omega)H_0^{(2)}(k|\boldsymbol{\rho} + \hat{\mathbf{y}}h|)$$

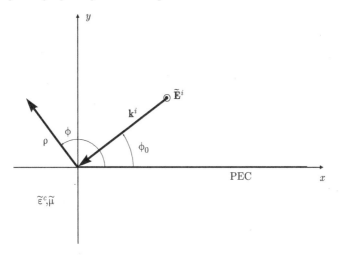

FIGURE 4.52
Geometry for scattering of a TM plane wave by a conducting half-plane.

where $|\boldsymbol{\rho} \pm \hat{\mathbf{y}}h| = |\rho\hat{\boldsymbol{\rho}} \pm \hat{\mathbf{y}}h| = \sqrt{x^2 + (y \pm h)^2}$.

Interpreting the general case in terms of images is more difficult. Comparing (4.404) and (4.405) with (4.403), we see that each spectral component of the field in region 1 has the form of an image line source located at $y = -h$ in region 2, but that the amplitude of the line source, $R_{TM}\tilde{I}$, depends on k_x. Similarly, the field in region 2 is composed of spectral components that seem to originate from line sources with amplitudes $-T_{TM}\tilde{I}$ located at $y = hk_{y1}/k_{y2}$ in region 1. In this case the amplitude and position of the image line source producing a spectral component are both dependent on k_x.

4.13.1.3   The field scattered by a half-plane

Consider a thin planar conductor that occupies the half-plane $y = 0$, $x > 0$. We assume the half-plane lies within a slightly lossy medium having parameters $\tilde{\mu}(\omega)$ and $\tilde{\epsilon}^c(\omega)$, and may consider the case of free space as a lossless limit. The half-plane is illuminated by an impressed uniform plane wave with a z-directed electric field (Figure 4.52). The primary field induces a secondary current on the conductor and this in turn produces a secondary field. The total field must obey the boundary conditions at $y = 0$.

Because the z-directed incident field induces a z-directed secondary current, the fields may be described entirely in terms of a TM set. The impressed plane wave may be written as
$$\tilde{\mathbf{E}}^i(\mathbf{r}, \omega) = \hat{\mathbf{z}}\tilde{E}_0(\omega)e^{jk(x\cos\phi_0 + y\sin\phi_0)}$$
where ϕ_0 is the angle between the incident wave vector and the x-axis. By (4.214) we also have
$$\tilde{\mathbf{H}}^i(\mathbf{r}, \omega) = \frac{\tilde{E}_0(\omega)}{\eta}(\hat{\mathbf{y}}\cos\phi_0 - \hat{\mathbf{x}}\sin\phi_0)e^{jk(x\cos\phi_0 + y\sin\phi_0)}.$$

The scattered fields may be written in terms of the Fourier transform solution to the Helmholtz equation. It is convenient to use the polar coordinate representation (4.396) to develop the necessary equations. Thus, for the scattered electric field we can write
$$\tilde{E}_z^s(x, y, \omega) = \int_C f(\xi, \omega)e^{-jk\rho\cos(\phi\pm\xi)}\, d\xi. \qquad (4.407)$$

By (4.397) the x-component of the magnetic field is

$$\tilde{H}_x^s(x, y, \omega) = -\frac{1}{j\omega\tilde{\mu}}\frac{\partial \tilde{E}_z^s}{\partial y} = -\frac{1}{j\omega\tilde{\mu}}\int_C f(\xi, \omega)\frac{\partial}{\partial y}\left(e^{-jkx\cos\xi}e^{\pm jky\sin\xi}\right)d\xi$$

$$= -\frac{1}{j\omega\tilde{\mu}}(\pm jk)\int_C f(\xi, \omega)\sin\xi\, e^{-jk\rho\cos(\phi\pm\xi)}\, d\xi.$$

To find the angular spectrum $f(\xi, \omega)$ and ensure uniqueness of solution, we must apply the boundary conditions over the entire $y = 0$ plane. For $x > 0$ where the conductor resides, the total tangential electric field must vanish. Setting the sum of the incident and scattered fields to zero at $\phi = 0$, we have

$$\int_C f(\xi, \omega)e^{-jkx\cos\xi}\, d\xi = -\tilde{E}_0 e^{jkx\cos\phi_0} \qquad (x > 0). \tag{4.408}$$

To find the boundary condition for $x < 0$, we note that by symmetry, $\tilde{E}_z^s$ is even about $y = 0$, while $\tilde{H}_x^s$, as the y-derivative of $\tilde{E}_z^s$, is odd. Since no current can be induced in the $y = 0$ plane for $x < 0$, the x-directed scattered magnetic field must be continuous and thus equal to zero there. Hence our second condition is

$$\int_C f(\xi, \omega)\sin\xi\, e^{-jkx\cos\xi}\, d\xi = 0 \qquad (x < 0). \tag{4.409}$$

Now that we have developed the two equations that describe $f(\xi, \omega)$, it is convenient to return to a rectangular-coordinate-based spectral integral to analyze them. Writing $\xi = \cos^{-1}(k_x/k)$, we have

$$\frac{d}{d\xi}(k\cos\xi) = -k\sin\xi = \frac{dk_x}{d\xi}$$

and

$$d\xi = -\frac{dk_x}{k\sin\xi} = -\frac{dk_x}{k\sqrt{1 - \cos^2\xi}} = -\frac{dk_x}{\sqrt{k^2 - k_x^2}}.$$

Upon substitution of these relations, the inversion contour returns to the real k_x axis (which may then be perturbed by $j\Delta$). Thus, (4.408) and (4.409) may be written as

$$\int_{-\infty+j\Delta}^{\infty+j\Delta} \frac{f\left(\cos^{-1}\frac{k_x}{k}\right)}{\sqrt{k^2 - k_x^2}}e^{-jk_x x}\, dk_x = -\tilde{E}_0 e^{jk_{x0} x} \qquad (x > 0), \tag{4.410}$$

$$\int_{-\infty+j\Delta}^{\infty+j\Delta} f\left(\cos^{-1}\frac{k_x}{k}\right)e^{-jk_x x}\, dk_x = 0 \qquad (x < 0), \tag{4.411}$$

where $k_{x0} = k\cos\phi_0$. Equations (4.410) and (4.411) form a set of *dual integral equations* for f. They may be treated by the *Wiener–Hopf technique*.

We begin by considering (4.411). If we close the integration contour in the upper half-plane using a semicircle C_R of radius R where $R \to \infty$, we find that the contribution from the semicircle is

$$\lim_{R\to\infty}\int_{C_R} f\left(\cos^{-1}\frac{k_x}{k}\right)e^{-|x|k_{xi}}e^{j|x|k_{xr}}\, dk_x = 0$$

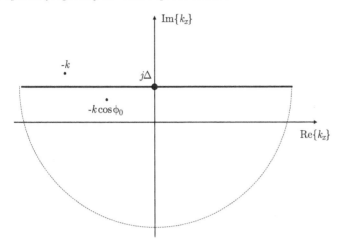

FIGURE 4.53
Integration contour used to evaluate the function $F(x)$.

since $x < 0$. This assumes f does not grow exponentially with R. Thus

$$\oint_C f\left(\cos^{-1}\frac{k_x}{k}\right) e^{-jk_x x}\, dk_x = 0$$

where C now encloses the portion of the upper half-plane $k_{xi} > \Delta$. By Morera's theorem [116], the above relation holds if f is regular (contains no singularities or branch points) in this portion of the upper half-plane. We shall assume this and investigate the other properties of f implied by (4.410).

In (4.410) we have an integral equated to an exponential function. To understand the implications of this it is helpful to write the exponential function as an integral as well. Consider the integral

$$F(x) = \frac{1}{2j\pi} \int\limits_{-\infty+j\Delta}^{\infty+j\Delta} \frac{h(k_x)}{h(-k_{x0})} \frac{1}{k_x + k_{x0}} e^{-jk_x x}\, dk_x.$$

Here $h(k_x)$ is some function regular in the region $k_{xi} < \Delta$, with $h(k_x) \to 0$ as $k_x \to \infty$. If we choose Δ so that $-k_{xi} > \Delta > -k_{xi}\cos\theta_0$ and close the contour with a semicircle in the lower half-plane (Figure 4.53), then the contribution from the semicircle vanishes for large radius and thus, by Cauchy's residue theorem, $F(x) = -e^{jk_{x0}x}$. Hence we can rewrite (4.410) as

$$\int\limits_{-\infty+j\Delta}^{\infty+j\Delta} \left[\frac{f\left(\cos^{-1}\frac{k_x}{k}\right)}{\sqrt{k^2 - k_x^2}} - \frac{\tilde{E}_0}{2j\pi} \frac{h(k_x)}{h(-k_{x0})} \frac{1}{k_x + k_{x0}} \right] e^{-jk_x x}\, dk_x = 0.$$

Setting the integrand to zero and using $\sqrt{k^2 - k_x^2} = \sqrt{k - k_x}\sqrt{k + k_x}$, we have

$$\frac{f\left(\cos^{-1}\frac{k_x}{k}\right)}{\sqrt{k - k_x}}(k_x + k_{x0}) = \frac{\tilde{E}_0}{2j\pi}\sqrt{k + k_x}\frac{h(k_x)}{h(-k_{x0})}. \tag{4.412}$$

The left member has a branch point at $k_x = k$ while the right member has a branch point at $k_x = -k$. If we choose the branch cuts as in Figure 4.50, then since f is regular in the region $k_{xi} > \Delta$, the left side of (4.412) is regular there. Also, since $h(k_x)$ is regular in the region $k_{xi} < \Delta$, the right side is regular there. We assert that since the two sides are equal, both sides must be regular in the entire complex plane. By Liouville's theorem [36], if a function is entire (regular in the entire plane) and bounded, it is constant. So

$$\frac{f\left(\cos^{-1}\frac{k_x}{k}\right)}{\sqrt{k - k_x}}(k_x + k_{x0}) = \frac{\tilde{E}_0}{2j\pi}\sqrt{k + k_x}\frac{h(k_x)}{h(-k_{x0})} = \text{constant.}$$

We may evaluate the constant by inserting any value of k_x. Using $k_x = -k_{x0}$ on the right, we find that

$$\frac{f\left(\cos^{-1}\frac{k_x}{k}\right)}{\sqrt{k - k_x}}(k_x + k_{x0}) = \frac{\tilde{E}_0}{2j\pi}\sqrt{k - k_{x0}}.$$

Substituting $k_x = k\cos\xi$ and $k_{x0} = k\cos\phi_0$, we have

$$f(\xi) = \frac{\tilde{E}_0}{2j\pi}\frac{\sqrt{1 - \cos\phi_0}\sqrt{1 - \cos\xi}}{\cos\xi + \cos\phi_0}.$$

Since $\sin(x/2) = \sqrt{(1 - \cos x)/2}$, we may also write

$$f(\xi) = \frac{\tilde{E}_0}{j\pi}\frac{\sin\frac{\phi_0}{2}\sin\frac{\xi}{2}}{\cos\xi + \cos\phi_0}.$$

Finally, substituting this into (4.407), we have the spectral representation for the field scattered by a half-plane:

$$\tilde{E}_z^s(\rho, \phi, \omega) = \frac{\tilde{E}_0(\omega)}{j\pi}\int_C \frac{\sin\frac{\phi_0}{2}\sin\frac{\xi}{2}}{\cos\xi + \cos\phi_0}e^{-jk\rho\cos(\phi\pm\xi)}\,d\xi. \tag{4.413}$$

The scattered field inversion integral in (4.413) may be rewritten in such a way as to separate geometrical optics (plane-wave) terms from diffraction terms. First, a change of variables $\xi' = \xi + \phi$ is used, resulting in a shift of the inversion contour and a pole of the integrand at $\xi' = \phi + \phi_0 - \pi$. The inversion contour is then shifted to the contour defined by $\cos\xi_r'\cosh\xi_i' = 1$ where ξ_r and ξ_i are the real and imaginary parts of ξ, respectively. The contour passes through the origin and remains in the region defined by $\sin\xi_r'\sinh\xi_i' \geq 0$. Note that moving the contour may require adding a contribution to the integral because the pole on the real axis may lie within the closed contour created by appending the shifted and original contours. Whether the pole is implicated depends on the observation angle ϕ. Finally, a change of variables

$$\nu = \sqrt{2}e^{-j\frac{\pi}{4}}\sin\frac{\xi'}{2}$$

allows the contour integral to be written in terms of standard functions. Addition of the the incident field gives the total field:

$$\tilde{E}_z(\rho, \phi, \omega) = U(\epsilon^i)\tilde{E}_0 e^{jk\rho\cos(\phi - \phi_0)} - U(\epsilon^r)\tilde{E}_0 e^{jk\rho\cos(\phi + \phi_0)}$$

$$- \tilde{E}_0\epsilon^i K_-\left(|a_i|\sqrt{k\rho}\right)e^{-jk\rho} + \tilde{E}_0\epsilon^r K_-\left(|a_r|\sqrt{k\rho}\right)e^{-jk\rho} \qquad (0 \leq \phi < 2\pi). \tag{4.414}$$

Here $U(x)$ is the unit step function (A.6), and $\epsilon^{i,r}$ are the functions (not to be confused with permittivity)

$$\epsilon^i = \text{sgn}(a^i), \qquad \epsilon^r = \text{sgn}(a^r),$$

where

$$a^i = \sqrt{2} \cos \tfrac{1}{2}(\phi - \phi_0), \qquad a^r = \sqrt{2} \cos \tfrac{1}{2}(\phi + \phi_0),$$

and $\text{sgn}(x)$ is the signum function (A.7). In addition, $K_-(x)$ is the modified Fresnel integral defined by

$$K_-(x) = \frac{1}{\sqrt{\pi}} e^{j\left(x^2 + \frac{\pi}{4}\right)} \int_x^\infty e^{-jt^2} \, dt.$$

The Fresnel integral is a standard handbook function available in many numerical libraries. See James [94] for details of the derivation.

The result (4.414) has an important interpretation in terms of geometrical optics and diffraction. The first term represents the incident plane-wave field. It is nonzero only when $0 \leq \phi \leq \pi + \phi_0$. This is the region directly illuminated by the incident plane wave according to geometrical optics. It is not present in the *shadow zone* where the geometrical optics rays are blocked by the conductor. The second term is a plane wave representing reflection from the half-plane, with angle of reflection equal to angle of incidence. It is nonzero only when $0 \leq \phi \leq \pi - \phi_0$, which is the region illuminated by rays reflected from the half-plane. The third and fourth terms represent the diffracted field, with one term associated with the incident field and the other associated with the reflected field. The factor $\exp(-jk\rho)$ in each term suggests that the diffracted field may behave similar to a cylindrical wave emanating from the edge of the half-plane; this is true for angles sufficiently removed from the geometrical optics boundaries. The multiplicative factor of the Fresnel integral determines the dependence on angle and represents a *pattern factor* of diffraction, much as with the pattern of antennas; it is called a *diffraction coefficient*. Diffraction coefficients form an important part of the *geometrical theory of diffraction* in which the principles of geometrical optics are used to interpret the nature of diffracted fields and model their interactions with nearby objects.

▶ **Example 4.32:** Total field for plane-wave scattering by a half-plane

A TM plane wave is incident at angle $\phi_0 = 45°$ on a perfectly conducting half-plane located in free space, as shown in Figure 4.52. Investigate the magnitude of the total field $|\tilde{E}_z|$ within a square of side $8\lambda_0$ centered on the edge of the half-plane.

Solution: The total axial electric field may be computed using (4.414). We normalize the distance from the edge of the half plane so that $k_0\rho = 2\pi\rho/\lambda_0$. The magnitude of $\tilde{E}_z/\tilde{E}_0$ is shown in Figure 4.54. Several interesting characteristics can be seen, similar to those observed for the case of the conducting wedge considered in Example 4.29. First, the field vanishes at the half-plane surface, as expected from the boundary condition. Second, a standing wave pattern exists above the half-plane, where the incident and scattered fields combine to create regions of constructive and destructive interference with a period of a half wavelength. The *reflection boundary* at $\phi = 180° - \phi_0 = 135°$ separates the region illuminated by both the incident and reflected waves from the region illuminated only by the incident wave, according to the laws of geometrical optics. A clear shadow is seen below the half-plane. The *shadow boundary* at $\phi = 180° + \phi_0 = 225°$ separates the region illuminated by the incident wave from the *shadow zone*, where the incident wave cannot reach according to the laws of geometrical optics. Geometrical optics predicts fields discontinuous across the reflection and shadow boundaries, but the diffraction terms in (4.414) provide for the smooth transition that is observed in practice. Thus, a remnant of the standing wave pattern can be seen outside the reflection zone, and a nonzero (but small) field is present in the shadow region.

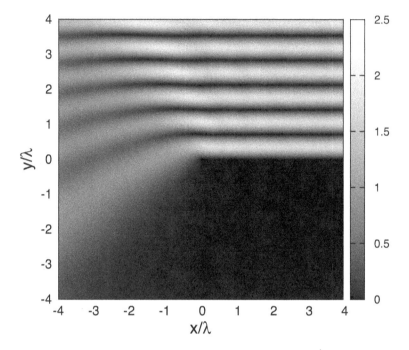

FIGURE 4.54
Total axial electric field $|\tilde{E}_z/\tilde{E}_0|$ for a plane wave incident at angle $\phi_0 = 45°$ on a perfectly conducting half-plane. ◄

4.14 Periodic fields and Floquet's theorem

In several practical situations, EM waves interact with, or are radiated by, structures spatially periodic along one or more directions. Periodic symmetry simplifies field computation, since boundary conditions need only be applied within one period, or *cell*, of the structure. Examples of situations that lead to periodic fields include the guiding of waves in slow-wave structures such as helices and meander lines, the scattering of plane waves from gratings, and the radiation of waves by antenna arrays. In this section we will study the representation of fields with infinite periodicity as spatial Fourier series.

4.14.1 Floquet's theorem

Consider an environment having spatial periodicity along the z-direction. In this environment the frequency-domain field may be represented in terms of a periodic function $\tilde{\psi}_p$ that obeys

$$\tilde{\psi}_p(x, y, z \pm mL, \omega) = \tilde{\psi}_p(x, y, z, \omega)$$

where m is an integer and L is the spatial period. According to *Floquet's theorem*, if $\tilde{\psi}$ represents some vector component of the field, then the field obeys

$$\tilde{\psi}(x,y,z,\omega) = e^{-j\kappa z}\tilde{\psi}_p(x,y,z,\omega). \qquad (4.415)$$

Here $\kappa = \beta - j\alpha$ is a complex wavenumber describing the phase shift and attenuation of the field between the various cells of the environment. The phase shift and attenuation may arise from a wave propagating through a lossy periodic medium (see example below) or may be impressed by a plane wave as it scatters from a periodic surface, or may be produced by the excitation of an antenna array by a distributed terminal voltage. It is also possible to have $\kappa = 0$ as when, for example, a periodic antenna array is driven with all elements in phase.

Because $\tilde{\psi}_p$ is periodic, we may expand it in a Fourier series

$$\tilde{\psi}_p(x,y,z,\omega) = \sum_{n=-\infty}^{\infty} \tilde{\psi}_n(x,y,\omega)e^{-j2\pi nz/L}$$

where the $\tilde{\psi}_n$ are found by orthogonality:

$$\tilde{\psi}_n(x,y,\omega) = \frac{1}{L}\int_{-L/2}^{L/2} \tilde{\psi}_p(x,y,z,\omega)e^{j2\pi nz/L}\,dz.$$

Substituting this into (4.415), we have a representation for the field as a Fourier series:

$$\tilde{\psi}(x,y,z,\omega) = \sum_{n=-\infty}^{\infty} \tilde{\psi}_n(x,y,\omega)e^{-j\kappa_n z}$$

where

$$\kappa_n = \beta + 2\pi n/L + j\alpha = \beta_n - j\alpha.$$

We see that within each cell the field consists of a number of constituents called *space harmonics* or *Hartree harmonics*, each with the property of a propagating or evanescent wave. Each has phase velocity

$$v_{pn} = \frac{\omega}{\beta_n} = \frac{\omega}{\beta + 2\pi n/L}.$$

A number of the space harmonics have phase velocities in the $+z$-direction while the remainder have phase velocities in the $-z$-direction, depending on the value of β. However, all of the space harmonics have the same group velocity

$$v_{gn} = \frac{d\omega}{d\beta} = \left(\frac{d\beta_n}{d\omega}\right)^{-1} = \left(\frac{d\beta}{d\omega}\right)^{-1} = v_g.$$

Those space harmonics having group and phase velocities oppositely directed are called *backward waves*, and form the basis of operation of microwave tubes known as "backward wave oscillators."

4.14.2   Examples of periodic systems

4.14.2.1   Plane-wave propagation within a periodically stratified medium

As an example of wave propagation in a periodic structure, let us consider a plane wave propagating within a layered medium consisting of two material layers repeated

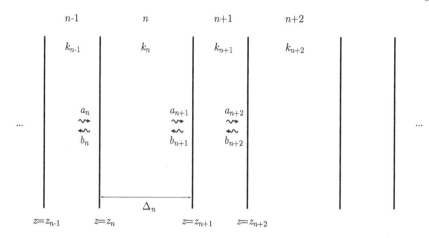

FIGURE 4.55

Geometry of a periodic stratified medium with each cell consisting of two material layers.

periodically, as shown in Figure 4.55. Each section of two layers is a cell within the periodic medium, and we seek an expression for the propagation constant within the cells, κ.

Recall (§ 4.11.5.7) that the wave amplitudes in any region in terms of the amplitudes in the region immediately preceding it may be written in terms of a transmission matrix as

$$\begin{bmatrix} T_{11}^{(n)} & T_{12}^{(n)} \\ T_{21}^{(n)} & T_{22}^{(n)} \end{bmatrix} \begin{bmatrix} a_{n+1} \\ b_{n+1} \end{bmatrix} = \begin{bmatrix} a_n \\ b_n \end{bmatrix}$$

where $T_{11}^{(n)}$, etc., are defined in (4.293)–(4.296). We may use these parameters to describe the cascaded system of two layers:

$$\begin{bmatrix} T_{11}^{(n)} & T_{12}^{(n)} \\ T_{21}^{(n)} & T_{22}^{(n)} \end{bmatrix} \begin{bmatrix} T_{11}^{(n+1)} & T_{12}^{(n+1)} \\ T_{21}^{(n+1)} & T_{22}^{(n+1)} \end{bmatrix} \begin{bmatrix} a_{n+2} \\ b_{n+2} \end{bmatrix} = \begin{bmatrix} a_n \\ b_n \end{bmatrix}.$$

Since for a periodic layered medium the wave amplitudes should obey (4.415), we have

$$\begin{bmatrix} T_{11} & T_{12} \\ T_{21} & T_{22} \end{bmatrix} \begin{bmatrix} a_{n+2} \\ b_{n+2} \end{bmatrix} = \begin{bmatrix} a_n \\ b_n \end{bmatrix} = e^{j\kappa L} \begin{bmatrix} a_{n+2} \\ b_{n+2} \end{bmatrix} \qquad (4.416)$$

where $L = \Delta_n + \Delta_{n+1}$ is the period of the structure, and

$$\begin{bmatrix} T_{11} & T_{12} \\ T_{21} & T_{22} \end{bmatrix} = \begin{bmatrix} T_{11}^{(n)} & T_{12}^{(n)} \\ T_{21}^{(n)} & T_{22}^{(n)} \end{bmatrix} \begin{bmatrix} T_{11}^{(n+1)} & T_{12}^{(n+1)} \\ T_{21}^{(n+1)} & T_{22}^{(n+1)} \end{bmatrix}.$$

Equation (4.416) is an eigenvalue equation for κ and can be rewritten as

$$\begin{bmatrix} T_{11} - e^{j\kappa L} & T_{12} \\ T_{21} & T_{22} - e^{j\kappa L} \end{bmatrix} \begin{bmatrix} a_{n+2} \\ b_{n+2} \end{bmatrix} = \begin{bmatrix} 0 \\ 0 \end{bmatrix}.$$

This equation has nontrivial solutions only when the determinant of the matrix vanishes. Expansion of the determinant gives

$$T_{11}T_{22} - T_{12}T_{21} - e^{j\kappa L}(T_{11} + T_{22}) + e^{j2\kappa L} = 0. \qquad (4.417)$$

The first two terms are merely

$$T_{11}T_{22} - T_{12}T_{21} = \begin{vmatrix} T_{11} & T_{12} \\ T_{21} & T_{22} \end{vmatrix} = \begin{vmatrix} T_{11}^{(n)} & T_{12}^{(n)} \\ T_{21}^{(n)} & T_{22}^{(n)} \end{vmatrix} \begin{vmatrix} T_{11}^{(n+1)} & T_{12}^{(n+1)} \\ T_{21}^{(n+1)} & T_{22}^{(n+1)} \end{vmatrix}.$$

Since we can show that

$$\begin{vmatrix} T_{11}^{(n)} & T_{12}^{(n)} \\ T_{21}^{(n)} & T_{22}^{(n)} \end{vmatrix} = \frac{Z_{n-1}}{Z_n},$$

we have

$$T_{11}T_{22} - T_{12}T_{21} = \frac{Z_{n-1}}{Z_n}\frac{Z_n}{Z_{n+1}} = 1$$

where we have used $Z_{n-1} = Z_{n+1}$ because of the periodicity of the medium. With this, (4.417) becomes

$$\cos \kappa L = \frac{T_{11} + T_{22}}{2}.$$

Finally, computing the matrix product and simplifying to find $T_{11} + T_{22}$, we have

$$\cos \kappa L = \cos(k_{z,n}\Delta_n)\cos(k_{k,n-1}\Delta_{n-1})$$
$$- \frac{1}{2}\left(\frac{Z_{n-1}}{Z_n} + \frac{Z_n}{Z_{n-1}}\right)\sin(k_{z,n}\Delta_n)\sin(k_{z,n-1}\Delta_{n-1}) \qquad (4.418)$$

or equivalently

$$\cos \kappa L = \frac{1}{4}\frac{(Z_{n-1} + Z_n)^2}{Z_n Z_{n-1}}\cos(k_{z,n}\Delta_n + k_{z,n-1}\Delta_{n-1})$$
$$- \frac{1}{4}\frac{(Z_{n-1} - Z_n)^2}{Z_n Z_{n-1}}\cos(k_{z,n}\Delta_n - k_{z,n-1}\Delta_{n-1}). \qquad (4.419)$$

Note that both $\pm\kappa$ satisfy this equation, allowing waves with phase-front propagation in both the $\pm z$-directions.

We see in (4.418) that even for lossless materials, certain values of ω result in $\cos \kappa L > 1$, causing κL to be imaginary and producing evanescent waves. We refer to the frequency ranges over which $\cos \kappa L > 1$ as *stopbands*, and those over which $\cos \kappa L < 1$ as *passbands*. This terminology is used in filter analysis and, indeed, waves propagating in periodic media experience effects similar to those experienced by signals passing through filters.

4.14.2.2 Field produced by an infinite array of line sources

As a second example, consider an infinite number of z-directed line sources within a homogeneous medium of complex permittivity $\tilde{\epsilon}^c(\omega)$ and permeability $\tilde{\mu}(\omega)$, aligned along the x-axis with separation L such that

$$\tilde{\mathbf{J}}(\mathbf{r}, \omega) = \sum_{n=-\infty}^{\infty} \hat{z}\tilde{I}_n \delta(y)\delta(x - nL).$$

The current on each element is allowed to show a progressive phase shift and attenuation. (Such progression may result from a particular method of driving primary currents on successive elements, or, if the currents are secondary, from their excitation by an impressed field such as a plane wave.) Thus we write

$$\tilde{I}_n = \tilde{I}_0 e^{-j\kappa n L} \qquad (4.420)$$

where κ is a complex constant.

We may represent the field produced by the source array as a superposition of the fields of individual line sources found earlier. In particular we may use the Hankel function representation (4.343) or the Fourier transform representation (4.400). Using the latter, we have

$$\tilde{E}_z(x,y,\omega) = \sum_{n=-\infty}^{\infty} e^{-j\kappa nL} \left[-\frac{\omega\tilde{\mu}\tilde{I}_0(\omega)}{2\pi} \int_{-\infty+j\Delta}^{\infty+j\Delta} \frac{e^{-jk_y|y|}}{2k_y} e^{-jk_x(x-nL)}\, dk_x \right].$$

Interchanging the order of summation and integration, we have

$$\tilde{E}_z(x,y,\omega) = -\frac{\omega\tilde{\mu}\tilde{I}_0(\omega)}{2\pi} \int_{-\infty+j\Delta}^{\infty+j\Delta} \frac{e^{-jk_y|y|}}{2k_y} \left[\sum_{n=-\infty}^{\infty} e^{jn(k_x-\kappa)L} \right] e^{-jk_x x}\, dk_x. \qquad (4.421)$$

We can rewrite the sum in this expression using Poisson's sum formula [144],

$$\sum_{n=-\infty}^{\infty} f(x-nD) = \frac{1}{D} \sum_{n=-\infty}^{\infty} F(nk_0)e^{jnk_0 x},$$

where $k_0 = 2\pi/D$. Letting $f(x) = \delta(x-x_0)$ in that expression, we have

$$\sum_{n=-\infty}^{\infty} \delta\left(x-x_0-n\frac{2\pi}{L}\right) = \frac{L}{2\pi} \sum_{n=-\infty}^{\infty} e^{jnL(x-x_0)}.$$

Substituting this into (4.421), we have

$$\tilde{E}_z(x,y,\omega) = -\frac{\omega\tilde{\mu}\tilde{I}_0(\omega)}{2\pi} \int_{-\infty+j\Delta}^{\infty+j\Delta} \frac{e^{-jk_y|y|}}{2k_y} \left[\sum_{n=-\infty}^{\infty} \frac{2\pi}{L}\delta\left(k_x-\kappa-n\frac{2\pi}{L}\right) \right] e^{-jk_x x}\, dk_x.$$

Carrying out the integral, we replace k_x with $\kappa_n = \kappa + 2n\pi/L$, giving

$$\tilde{E}_z(x,y,\omega) = -\omega\tilde{\mu}\tilde{I}_0(\omega) \sum_{n=-\infty}^{\infty} \frac{e^{-jk_{y,n}|y|}e^{-j\kappa_n x}}{2Lk_{y,n}}$$

$$= -j\omega\tilde{\mu}\tilde{I}_0(\omega)\tilde{G}_\infty(x,y\,|\,0,0,\omega)$$

where $k_{y,n} = \sqrt{k^2-\kappa_n^2}$, and where

$$\tilde{G}_\infty(x,y\,|\,x',y',\omega) = \sum_{n=-\infty}^{\infty} \frac{e^{-jk_{y,n}|y-y'|}e^{-j\kappa_n(x-x')}}{2jLk_{y,n}} \qquad (4.422)$$

is called the *periodic Green's function.*

We may also find the field produced by an infinite array of line sources in terms of the Hankel function representation of a single line source (4.343). Using the current representation (4.420) and summing over the sources, we obtain

$$\tilde{E}_z(\rho,\omega) = -\frac{\omega\tilde{\mu}}{4} \sum_{n=-\infty}^{\infty} \tilde{I}_0(\omega)e^{-j\kappa nL}H_0^{(2)}(k|\boldsymbol{\rho}-\boldsymbol{\rho}_n|) = -j\omega\tilde{\mu}\tilde{I}_0(\omega)\tilde{G}_\infty(x,y\,|\,0,0,\omega)$$

where

$$|\boldsymbol{\rho} - \boldsymbol{\rho}_n| = |\hat{\mathbf{y}}y + \hat{\mathbf{x}}(x - nL)| = \sqrt{y^2 + (x - nL)^2}$$

and where $\tilde{G}_\infty$ is an alternative form of the periodic Green's function

$$\tilde{G}_\infty(x, y \mid x', y', \omega) = \frac{1}{4j} \sum_{n=-\infty}^{\infty} e^{-j\kappa nL} H_0^{(2)} \left(k\sqrt{(y - y')^2 + (x - nL - x')^2} \right). \quad (4.423)$$

The periodic Green's functions (4.422) and (4.423) produce identical results, but are each appropriate for certain applications. For example, (4.422) is useful for situations in which boundary conditions at constant values of y are to be applied. Both forms are difficult to compute under certain circumstances, and variants of these forms have been introduced in the literature [205].

▶ **Example 4.33**: Computing the periodic Green's function

Assume that $\kappa = 0$, $x' = y' = 0$, $x = 0$, $y = L$, and $kL = \pi$. Specialize each of the two forms of the periodic Green's function, (4.422) and (4.423). Investigate the convergence rates of these series.

Solution: First examine (4.422). Note that

$$\kappa_n = 2n\frac{\pi}{L}.$$

For $n = 0$, $k_{y,n} = k$. For $n \neq 0$,

$$k_{y,n} = -j\sqrt{\left(2n\frac{\pi}{L}\right)^2 - k^2}.$$

Substitution gives

$$\tilde{G}_{\infty,1}(0, L \mid 0, 0, \omega) = \sum_{n=-\infty}^{\infty} \frac{e^{-jk_{y,n}L}}{2jk_{y,n}L}$$

$$= -j\frac{1}{2\pi}e^{-j\pi} + 2\sum_{n=1}^{\infty} \frac{e^{-L\sqrt{\left(2n\frac{\pi}{L}\right)^2 - \left(\frac{\pi}{L}\right)^2}}}{2L\sqrt{\left(2n\frac{\pi}{L}\right)^2 - \left(\frac{\pi}{L}\right)^2}}$$

$$= j\frac{1}{2\pi} + \frac{1}{\pi}\sum_{n=1}^{\infty} \frac{e^{-\pi\sqrt{4n^2 - 1}}}{\sqrt{4n^2 - 1}}. \quad (4.424)$$

Next, examine (4.423). Substituting the values of the parameters gives

$$\tilde{G}_{\infty,2}(0, L \mid 0, 0, \omega) = \frac{1}{4j} \sum_{n=-\infty}^{\infty} H_0^{(2)} \left(\frac{\pi}{L}\sqrt{L^2 + n^2 L^2} \right)$$

$$= \frac{1}{4j} H_0^{(2)}(\pi) + \frac{1}{2j} \sum_{n=1}^{\infty} H_0^{(2)} \left(\pi\sqrt{n^2 + 1} \right). \quad (4.425)$$

The table below shows values of the periodic Green's function computed from (4.424) and (4.425), found by summing the series to an upper limit of N using 15-digit precision. It is clear that for this combination of parameters, computing (4.424) is far more efficient than computing (4.425). With (4.424), the imaginary part is computed exactly in the first term and the real part has been computed to 10 significant digits by the fifth term. With (4.425),

even after a billion terms the real part is only accurate to two digits and the imaginary part
to 5 digits!

N	$\mathrm{Re}\{\tilde{G}_{\infty,1}\}$	$\mathrm{Re}\{\tilde{G}_{\infty,2}\}$	$\mathrm{Im}\{\tilde{G}_{\infty,1}\}$	$\mathrm{Im}\{\tilde{G}_{\infty,2}\}$
1	0.0007963799811	0.006455135089	0.1591549431	0.2427066943
2	0.0007968071480	0.01567773939	0.1591549431	0.09265520444
3	0.0007968076041	-0.02026337805	0.1591549431	0.2139340659
4	0.0007968076047	0.02372911342	0.1591549431	0.1122355519
5	0.0007968076047	-0.02474289964	0.1591549431	0.2005293266
10		0.02148396207		0.1313924649
10^2		0.008612629308		0.1510973026
10^3		0.003308785678		0.1566352654
10^4		0.001592440590		0.1583590665
10^5		0.001048449180		0.1589032938
10^6		0.0008763849212		0.1590753655
10^7		0.0008219721912		0.1591297785
10^8		0.0008047653455		0.1591469853
10^9		0.0007993240799		0.1591524266

4.15 Application: electromagnetic characterization of materials

Material characterization uses measurements of the interaction of electromagnetic fields
with material samples to determine the intrinsic electromagnetic properties of the materials. For isotropic materials, this entails finding complex permittivity and permeability.
While the properties of some materials with simple chemical composition may be pre-
dictable theoretically, many complicated materials, especially engineered materials such
as chiral materials [26], nanotube composites [60], and metamaterials [178] can be characterized only through measurements. Material characterization can be done accurately,
but care is required since material characterization is in essence an inversion process and
can be fraught with measurement sensitivity issues.

An astonishing number of characterization methods have been devised. Simple techniques using measurements of capacitance or inductance are useful at lower frequencies
where field fringing may be easily compensated. More sophisticated methods are required in the microwave, millimeter-wave, and THz frequency bands, and these are often
categorized by the width of the frequency band over which they are applied. Wideband
techniques include: (1) free space methods, in which a planar layered sample is illuminated by a focused beam with planar phase fronts [70]; (2) probe methods, in which
the open-ended aperture of a guided wave structure is placed against a material sample
[10]; (3) guided wave techniques where an open applicator, such as a microstrip line, is
placed against a sample [84]; and (4) closed-boundary guided wave systems in which a
sample is placed into the waveguiding system [212]. Narrowband techniques often involve placing a sample in a cavity [15] or constructing a resonator out of the sample
itself [76]. Although resonator methods are narrowband, they are often the most reliable for characterizing low loss materials that produce little attenuation of propagating
waves. Characterization methods may also be categorized as destructive, where a sample must be excised or fabricated to a certain shape to fit into a field applicator, and
non-destructive, in which fields are applied to a sample without altering its physical con-

struction. Probes and open guided-wave applicators provide a means for non-destructive characterization, while cavity methods and closed guided-wave systems require specially prepared samples. Finally, guided-wave and free-space techniques may be categorized as reflection/transmission methods, which use both the field reflected by the sample and the field transmitted through the sample, and reflection-only methods that do not utilize the transmitted field [62]. A useful overview is provided in [30].

We will describe a subset of techniques that use measured reflection and transmission data for waves incident on samples with planar surfaces. This includes free space systems, where we model both the incident field and the sample as being infinite in extent. (This model works well for a finite-sized sample illuminated by a focused beam if the sample is larger than the spot size of the beam, and both the phase and amplitude of the incident field are predominantly uniform across the spot.) Also included are guided wave techniques where a sample is placed into the waveguiding structure. Both TEM systems (such as coaxial lines and striplines) and dominant-mode waveguide systems (such as rectangular guides operating in the TE_{10} mode) are appropriate provided the sample fills the guide cross-section and the surfaces of the sample are planar and aligned perpendicular to the direction of propagation.

Some techniques have been specifically developed to characterize dielectric materials, where it is known *a priori* that $\tilde{\mu} = \mu_0$. In this case measurement of either the reflected or transmitted signal suffices. We assume both $\tilde{\mu}$ and $\tilde{\epsilon}^c$ are desired. This requires two independent measurements, which could be the reflected and transmitted signals, or the reflected signal under two conditions that make the measurements independent. Remarkably, many techniques share the common approach of determining the interfacial reflection coefficient and the propagation term for the material sample. From these quantities both $\tilde{\epsilon}^c$ and $\tilde{\mu}$ may be found, as described next.

The propagation of plane waves in layered media is considered in § 4.11.5. Many concepts from that chapter can be extended to layered media in waveguiding systems, including rectangular waveguides, coaxial guides, and striplines. We use the terminology for plane waves in the following sections, providing relationships to waveguide terminology when needed.

4.15.1 $\tilde{\Gamma}$-$\tilde{P}$ methods

Two independent measurements are required to determine both $\tilde{\epsilon}^c$ and $\tilde{\mu}$. Knowing both the product $\tilde{\mu}\tilde{\epsilon}^c$ and the quotient $\tilde{\mu}/\tilde{\epsilon}^c$, we can find $\tilde{\epsilon}^c$ and $\tilde{\mu}$ via multiplication and division. The product appears in the wavenumber k, which appears in k_z, which appears in $\tilde{P}$. Thus, finding $\tilde{P}$ allows to find $\tilde{\mu}\tilde{\epsilon}^c$. The quotient appears in the intrinsic impedance η, which appears in the wave impedance $\tilde{Z}$ (along with k_z), which appears in the interfacial reflection coefficient $\tilde{\Gamma}$. Thus, if k_z has been found, the quotient may be found from $\tilde{\Gamma}$. We refer to techniques that provide means to find $\tilde{\Gamma}$ and $\tilde{P}$ as $\tilde{\Gamma}$-$\tilde{P}$ *methods*.

A variety of approaches give rise to measurements of $\tilde{\Gamma}$ and $\tilde{P}$, and several of these are described below. In each case, the same procedure may be used to find $\tilde{\epsilon}^c$ and $\tilde{\mu}$ after finding $\tilde{\Gamma}$ and $\tilde{P}$. Although methods exist where the sample is sandwiched among known materials, we assume here that the excitation field originates from a free-space region and is incident on a sample of thickness Δ. Thus, the first interface is between free space and the sample material. Recall from (4.272) that $\tilde{P} = e^{-jk_z\Delta}$. If $\tilde{P}$ is determined from measurements, k_z may be found using the logarithm:

$$k_z = \frac{\ln \tilde{P} + j2n\pi}{-j\Delta} \qquad (n = 0, \pm1, \pm2, \ldots). \tag{4.426}$$

But the multivalued nature of the logarithm leaves us with having to choose n. If we can choose the sample thickness to be less than a half-wavelength, the phase cannot progress more than $-180°$ as the wave passes through the sample, and thus $n = 0$. However, without a good estimate of the material properties we may not be able to estimate the proper sample thickness to satisfy this requirement. In other scenarios we may be given a sample whose thickness is predetermined, or we may need to measure across a wide band such that the electrical thickness of the sample is wide ranging. Several authors have suggested techniques for choosing the value of n; see [7].

The interfacial reflection coefficient may be written as

$$\tilde{\Gamma} = \frac{Z - Z_0}{Z + Z_0},$$

where Z is the wave impedance of the sample and Z_0 is the impedance of the free-space region immediately before the sample. From this we find

$$Z = Z_0 \frac{1 + \tilde{\Gamma}}{1 - \tilde{\Gamma}}.$$

The formulas for k_z, Z and Z_0 depend on the form of the excitation, and hence so do the final expressions for $\tilde{\mu}$ and $\tilde{\epsilon}^c$. Three situations are considered here.

Incident plane wave. Suppose a plane wave is incident at an angle θ_0 on an infinite planar interface. First we have $k_z^2 = k^2 - k_0^2 \sin^2 \theta_0$ where $k^2 = \omega^2 \tilde{\mu} \tilde{\epsilon}^c$. Thus,

$$k = \sqrt{k_z^2 + k_0^2 \sin^2 \theta_0},$$

which is known, since k_z is known from (4.426). For perpendicular polarization we also have from (4.258),

$$Z = \frac{k\eta}{k_z}, \qquad Z_0 = \frac{\eta_0}{\cos \theta_0},$$

and thus

$$\eta = Z \frac{k_z}{k} = \frac{\eta_0}{\cos \theta_0} \frac{1 + \tilde{\Gamma}}{1 - \tilde{\Gamma}} \frac{k_z}{k}.$$

With η and k known, we find

$$\tilde{\mu} = \frac{\eta k}{\omega}, \qquad \tilde{\epsilon}^c = \frac{k}{\omega \eta}.$$

For parallel polarization, we have from (4.260),

$$Z = \frac{k_z \eta}{k}, \qquad Z_0 = \eta_0 \cos \theta_0,$$

and thus

$$\eta = Z \frac{k}{k_z} = \eta_0 \cos \theta_0 \frac{1 + \tilde{\Gamma}}{1 - \tilde{\Gamma}} \frac{k}{k_z}.$$

With η and k known, we again find

$$\tilde{\mu} = \frac{\eta k}{\omega}, \qquad \tilde{\epsilon}^c = \frac{k}{\omega \eta}.$$

TEM waveguiding system. In this case the sample is placed into a TEM waveguiding system such as a coaxial cable or a stripline. We then have simply $k = k_z$, $Z = \eta$, and $Z_0 = \eta_0$. Thus

$$\eta = \eta_0 \frac{1 + \tilde{\Gamma}}{1 - \tilde{\Gamma}}.$$

With η and k known, we find

$$\tilde{\mu} = \frac{\eta k_z}{\omega}, \qquad \tilde{\epsilon}^c = \frac{k_z}{\omega \eta}.$$

Note that this result is identical to the case of a plane-wave normally incident on a planar sample ($\theta_0 = 0$).

Rectangular waveguide with TE$_{10}$ mode incident. Here the sample is placed in the cross-section of a rectangular waveguide. We have (§ 5.6)

$$k_z^2 = k^2 - \left(\frac{\pi}{a}\right)^2$$

where a is the width of the guide. Thus

$$k = \sqrt{k_z^2 + \left(\frac{\pi}{a}\right)^2}.$$

We also have from (5.201)

$$Z = \frac{\omega \tilde{\mu}}{k_z}, \qquad Z_0 = \frac{\omega \mu_0}{k_{z0}}$$

where

$$k_{z0} = \sqrt{k_0^2 - \left(\frac{\pi}{a}\right)^2}.$$

Thus

$$\tilde{\mu} = Z\frac{k_z}{\omega} = \mu_0 \frac{1 + \tilde{\Gamma}}{1 - \tilde{\Gamma}} \frac{k_z}{k_{z0}} \quad \text{and} \quad \tilde{\epsilon}^c = \frac{k^2}{\omega^2 \tilde{\mu}}.$$

4.15.1.1 Reflection-transmission (Nicolson–Ross–Weir) method

Probably the most widely applied material characterization technique is the reflection-transmission technique proposed by Nicolson and Ross [140] and by Weir [212] (commonly called the *Nicolson–Ross–Weir method*, or simply NRW). Here we make measurements of the reflection from, and transmission through, a sample with a matched termination (or equivalently, with a region behind the sample that is infinite in extent). The measured reflection coefficient is described by (4.287) from Example 4.15. However, since free space exists on both sides of the sample, we have $\tilde{\Gamma}_1 = \tilde{\Gamma} = -\tilde{\Gamma}_2$. With this, the reflection coefficient becomes

$$\tilde{\mathbb{R}} = \frac{(1 - \tilde{P}^2)\tilde{\Gamma}}{1 - \tilde{\Gamma}^2 \tilde{P}^2},$$

where $\tilde{P}$ is the propagation term for the sample. Similarly, the measured transmission coefficient is described by (4.288) from Example 4.15, again with $\tilde{\Gamma}_1 = \tilde{\Gamma} = -\tilde{\Gamma}_2$:

$$\tilde{\mathbb{T}} = \frac{(1 - \tilde{\Gamma}^2)\tilde{P}}{1 - \tilde{\Gamma}^2 \tilde{P}^2}.$$

We must solve these two equations simultaneously to determine $\tilde{\Gamma}$ and $\tilde{P}$.

The procedure outlined by Nicolson and Ross and by Weir is as follows. Define the intermediate quantities

$$V_1 = \tilde{\mathbb{T}} + \tilde{\mathbb{R}} = \frac{\tilde{P} - \tilde{\Gamma}^2\tilde{P} + \tilde{\Gamma} - \tilde{P}^2\tilde{\Gamma}}{1 - \tilde{\Gamma}^2\tilde{P}^2} = \frac{\tilde{P} + \tilde{\Gamma}}{1 + \tilde{\Gamma}\tilde{P}}, \tag{4.427}$$

$$V_2 = \tilde{\mathbb{T}} - \tilde{\mathbb{R}} = \frac{\tilde{P} - \tilde{\Gamma}^2\tilde{P} - \tilde{\Gamma} + \tilde{P}^2\tilde{\Gamma}}{1 - \tilde{\Gamma}^2\tilde{P}^2} = \frac{\tilde{P} - \tilde{\Gamma}}{1 - \tilde{\Gamma}\tilde{P}}. \tag{4.428}$$

We solve for $\tilde{P}$ using (4.427),

$$\tilde{P} = \frac{V_1 - \tilde{\Gamma}}{1 - V_1\tilde{\Gamma}}, \tag{4.429}$$

and using (4.428),

$$\tilde{P} = \frac{V_2 + \tilde{\Gamma}}{1 + V_2\tilde{\Gamma}}. \tag{4.430}$$

Equating, we find $(V_1 - \tilde{\Gamma})(1 + V_2\tilde{\Gamma}) = (V_2 + \tilde{\Gamma})(1 - V_1\tilde{\Gamma})$, which can be rearranged as

$$(V_1 - V_2)\tilde{\Gamma}^2 - 2(1 - V_1V_2)\tilde{\Gamma} + (V_1 - V_2) = 0.$$

This is a quadratic equation of the form $\tilde{\Gamma}^2 - 2X\tilde{\Gamma} + 1 = 0$, with solution

$$\tilde{\Gamma} = X \pm \sqrt{X^2 - 1} \quad \text{where} \quad X = \frac{1 - V_1V_2}{V_1 - V_2} = \frac{1 - \tilde{\mathbb{T}}^2 + \tilde{\mathbb{R}}^2}{2\tilde{\mathbb{R}}}.$$

The sign ambiguity is resolved by recognizing that only one choice results in $|\tilde{\Gamma}| \leq 1$, which must hold for passive materials. Finally, with $\tilde{\Gamma}$ determined, $\tilde{P}$ is found from either (4.429) or (4.430).

There is an obvious alternative approach to finding $\tilde{\Gamma}$ and $\tilde{P}$. We solve for $\tilde{\Gamma}$ from (4.427) as

$$\tilde{\Gamma} = \frac{V_1 - \tilde{P}}{1 - V_1\tilde{P}}. \tag{4.431}$$

Similarly, we solve for $\tilde{\Gamma}$ from (4.428) as

$$\tilde{\Gamma} = \frac{\tilde{P} - V_2}{1 - V_2\tilde{P}}. \tag{4.432}$$

Equating, we obtain $(V_1 - \tilde{P})(1 - V_2\tilde{P}) = (\tilde{P} - V_2)(1 - V_1\tilde{P})$ or

$$(V_1 + V_2)\tilde{P}^2 - 2(1 + V_1V_2)\tilde{P} + (V_1 + V_2) = 0.$$

This is a quadratic equation of the form $\tilde{P}^2 - 2Y\tilde{P} + 1 = 0$ with solution

$$\tilde{P} = Y \pm \sqrt{Y^2 - 1} \quad \text{where} \quad Y = \frac{1 + V_1V_2}{V_1 + V_2} = \frac{1 + \tilde{\mathbb{T}}^2 - \tilde{\mathbb{R}}^2}{2\tilde{\mathbb{T}}}.$$

With $\tilde{P}$ determined, $\tilde{\Gamma}$ is found from either (4.431) or (4.432). The sign ambiguity is again resolved by recognizing that only one choice results in $|\tilde{\Gamma}| \leq 1$. It is interesting that X and Y satisfy $\tilde{\mathbb{R}}X + \tilde{\mathbb{T}}Y = 1$.

4.15.1.2 Two-thickness method

The two-thickness method [9] is one of several that rely on reflection-only measurements. It is often more convenient to measure reflection than transmission, since no receiver is required on the opposite side of the sample. In some cases, a transmission measurement may not be possible, as when the sample is glued to a conductor backing that cannot be removed [62].

The two-thickness method relies on either having two samples of the material with two different thicknesses, or having one sample that can be cut into two pieces. In the former situation, a reflection measurement is made with each of the two samples. In the latter, the reflection from the sample is measured, then the sample is split, and a second reflection measurement is made using one of the pieces.

With conductor backing. If the sample is backed by a conductor, the measured reflection coefficients are given by (4.287) with $\tilde{\Gamma}_2 = -1$. Let the thickness of the sample for the first measurement be Δ, and for the second measurement be $\kappa\Delta$, where $\kappa > 1$. See Figure 4.56. The reflection coefficient for the first measurement is

$$\tilde{\mathbb{R}}_A = \frac{\tilde{\Gamma} - \tilde{P}_A^2}{1 - \tilde{\Gamma}\tilde{P}_A^2} = \frac{\tilde{\Gamma} - \tilde{Q}}{1 - \tilde{\Gamma}\tilde{Q}} \tag{4.433}$$

where $\tilde{P}_A^2 = e^{-2jk_z\Delta} = \tilde{Q}$. Similarly, the reflection coefficient for the second measurement is

$$\tilde{\mathbb{R}}_B = \frac{\tilde{\Gamma} - P_B^2}{1 - \tilde{\Gamma}P_B^2} = \frac{\tilde{\Gamma} - Q^\kappa}{1 - \tilde{\Gamma}Q^\kappa} \tag{4.434}$$

where $P_B^2 = e^{-2jk_z(\kappa\Delta)} = \tilde{Q}^\kappa$. Simultaneous solution of (4.433) and (4.434) will yield $\tilde{\Gamma}$ and $\tilde{Q}$. We rearrange (4.433) to get

$$\tilde{\Gamma} = \frac{\tilde{\mathbb{R}}_A + \tilde{Q}}{1 + \tilde{\mathbb{R}}_A\tilde{Q}} \tag{4.435}$$

and (4.434) to get

$$\tilde{\Gamma} = \frac{\tilde{\mathbb{R}}_B + \tilde{Q}^\kappa}{1 + \tilde{\mathbb{R}}_B\tilde{Q}^\kappa}. \tag{4.436}$$

Equating, we find $(\tilde{\mathbb{R}}_A + \tilde{Q})(1 + \tilde{\mathbb{R}}_B\tilde{Q}^\kappa) - (\tilde{\mathbb{R}}_B + \tilde{Q}^\kappa)(1 + \tilde{\mathbb{R}}_A\tilde{Q}) = 0$ or

$$\tilde{Q}^{\kappa+1} - A\tilde{Q}^\kappa + A\tilde{Q} - 1 = 0 \quad \text{where} \quad A = \frac{\tilde{\mathbb{R}}_A\tilde{\mathbb{R}}_B - 1}{\tilde{\mathbb{R}}_A - \tilde{\mathbb{R}}_B}. \tag{4.437}$$

Equation (4.437) may be solved for $\tilde{Q}$, which in turn determines $\tilde{P}$. Once $\tilde{P}$ is found, $\tilde{\Gamma}$ may be found from either (4.435) or (4.436).

If κ is irrational, (4.437) has infinitely many solutions. If κ is rational, there may be many solutions. Neither possibility is desirable. For $\kappa = 2$, however (which corresponds to the thickness of the second sample being exactly twice that of the first sample), (4.437) becomes $\tilde{Q}^3 - A\tilde{Q}^2 + A\tilde{Q} - 1 = 0$, which may be factored as

$$(\tilde{Q} - 1)[\tilde{Q}^2 - (A - 1)\tilde{Q} + 1] = 0.$$

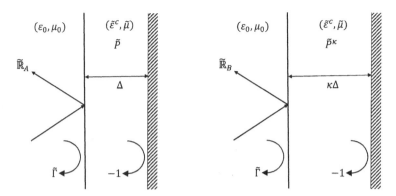

FIGURE 4.56

Measurements used in the two-thickness method (with conductor backing).

This has solutions $\tilde{Q} = 1$ and

$$\tilde{Q} = \frac{A-1}{2} \pm \sqrt{\left(\frac{A-1}{2}\right)^2 - 1}.$$

The solution $\tilde{Q} = 1$ would only be permissible for a lossless sample exactly a half wavelength thick. The remaining sign choice is made by requiring $|\tilde{\Gamma}| \leq 1$, where $\tilde{\Gamma}$ is found using either (4.435) or (4.436).

As with the reflection-transmission method, an alternative solution approach is possible in which $\tilde{\Gamma}$ is first found. Consider the case with $\kappa = 2$. Rearrangement of (4.433) gives

$$\tilde{Q} = \frac{\tilde{\mathbb{R}}_A - \tilde{\Gamma}}{\tilde{\mathbb{R}}_A \tilde{\Gamma} - 1}, \tag{4.438}$$

while (4.434) gives

$$\tilde{Q}^2 = \frac{\tilde{\mathbb{R}}_B - \tilde{\Gamma}}{\tilde{\mathbb{R}}_B \tilde{\Gamma} - 1}. \tag{4.439}$$

Squaring (4.438) and equating with (4.439), we have

$$(\tilde{\mathbb{R}}_A - \tilde{\Gamma})^2 (\tilde{\mathbb{R}}_B \tilde{\Gamma} - 1) - (\tilde{\mathbb{R}}_A \tilde{\Gamma} - 1)^2 (\tilde{\mathbb{R}}_B - \tilde{\Gamma}) = 0$$

or

$$\tilde{\Gamma}^3 - B\tilde{\Gamma}^2 + B\tilde{\Gamma} - 1 = 0 \quad \text{where} \quad B = \frac{1 + \tilde{\mathbb{R}}_A^2 \tilde{\mathbb{R}}_B + 2\tilde{\mathbb{R}}_A \tilde{\mathbb{R}}_B + 2\tilde{\mathbb{R}}_A}{\tilde{\mathbb{R}}_A^2 + \tilde{\mathbb{R}}_B}. \tag{4.440}$$

We may factor (4.440) as

$$(\tilde{\Gamma} - 1)[\tilde{\Gamma}^2 - (B-1)\tilde{\Gamma} + 1] = 0.$$

This has solutions $\tilde{\Gamma} = 1$ and

$$\tilde{\Gamma} = \frac{B-1}{2} \pm \sqrt{\left(\frac{B-1}{2}\right)^2 - 1}.$$

The solution $\tilde{\Gamma} = 1$ would correspond to the unlikely case of total interfacial reflection. The remaining sign choice is made by requiring $|\tilde{\Gamma}| \leq 1$. Once $\tilde{\Gamma}$ is determined, $\tilde{Q}$, and thus $\tilde{P}$, may be found from either (4.438) or (4.439).

With air backing. If the sample is backed by air, then the measured reflection coefficients are given by (4.287) with $\tilde{\Gamma}_1 = \tilde{\Gamma} = -\tilde{\Gamma}_2$. This is a more complicated expression than for the case of a conductor-backed sample considered above. Thus, we will restrict ourselves to the case in which one sample is exactly twice the thickness of the other sample.

Let the thickness of the sample for the first measurement be Δ, and for the second measurement be 2Δ. Then the reflection coefficient for the first measurement is

$$\tilde{\mathbb{R}}_A = \frac{\tilde{\Gamma}(1 - \tilde{P}^2)}{1 - \tilde{\Gamma}^2 \tilde{P}^2} = \frac{\tilde{\Gamma}(1 - \tilde{Q})}{1 - \tilde{\Gamma}^2 \tilde{Q}} \tag{4.441}$$

where $\tilde{P}^2 = e^{-2jk_z\Delta} = \tilde{Q}$. Similarly, the reflection coefficient for the second measurement is

$$\tilde{\mathbb{R}}_B = \frac{\tilde{\Gamma}(1 - \tilde{P}^4)}{1 - \tilde{\Gamma}^2 \tilde{P}^4} = \frac{\tilde{\Gamma}(1 - \tilde{Q}^2)}{1 - \tilde{\Gamma}^2 \tilde{Q}^2} \tag{4.442}$$

where $P^4 = e^{-2jk_z(2\Delta)} = \tilde{Q}^2$. Simultaneous solution of (4.441) and (4.442) yields $\tilde{\Gamma}$ and $\tilde{Q}$. Rearrangement of (4.441) gives

$$\tilde{Q} = \frac{\tilde{\Gamma} - \tilde{\mathbb{R}}_A}{\tilde{\Gamma}(1 - \tilde{\mathbb{R}}_A \tilde{\Gamma})} \tag{4.443}$$

while (4.442) gives

$$\tilde{Q}^2 = \frac{\tilde{\Gamma} - \tilde{\mathbb{R}}_B}{\tilde{\Gamma}(1 - \tilde{\mathbb{R}}_B \tilde{\Gamma})}. \tag{4.444}$$

Squaring (4.443) and equating with (4.444), we get

$$(\tilde{\Gamma} - \tilde{\mathbb{R}}_A)^2 \tilde{\Gamma}(1 - \tilde{\mathbb{R}}_B \tilde{\Gamma}) - \tilde{\Gamma}^2 (1 - \tilde{\mathbb{R}}_A \tilde{\Gamma})^2 (\tilde{\Gamma} - \tilde{\mathbb{R}}_B) = 0,$$

or

$$\tilde{\Gamma}^4 - C\tilde{\Gamma}^3 + C\tilde{\Gamma} - 1 = 0 \quad \text{where} \quad C = \frac{\tilde{\mathbb{R}}_A^2 \tilde{\mathbb{R}}_B + 2\tilde{\mathbb{R}}_A - \tilde{\mathbb{R}}_B}{\tilde{\mathbb{R}}_A^2}. \tag{4.445}$$

We may factor (4.445) as

$$(\tilde{\Gamma} - 1)(\tilde{\Gamma} + 1)(\tilde{\Gamma}^2 - C\tilde{\Gamma} + 1) = 0.$$

This has solutions $\tilde{\Gamma} = 1$, $\tilde{\Gamma} = -1$, and

$$\tilde{\Gamma} = \frac{C}{2} \pm \sqrt{\left(\frac{C}{2}\right)^2 - 1}.$$

The first two solutions, $\tilde{\Gamma} = \pm 1$, represent the unlikely cases of total interfacial reflection. The remaining sign choice is made by requiring $|\tilde{\Gamma}| \leq 1$. Once $\tilde{\Gamma}$ is determined, $\tilde{P}$ may be found from either (4.443) or (4.444).

The two-thickness method depends on changing the structure of the sample. Several other methods depend on changing the structure of the region immediately behind the sample. These include the conductor-backed/air-backed method, the layer-shift method, and the two-backing method. These are considered next. It is important to note that changing the structure of the region in front of the sample does not provide any new information about the sample, and thus cannot be used to provide an independent second measurement. For instance, measuring the reflection from a layer of known material placed immediately in front of the sample provides no more information than reflection from the sample itself. This is discussed in some detail in [62].

4.15.1.3   Conductor-backed/air-backed method

This method uses measurements of the reflection from the sample backed first by a conductor, then backed by air [30]. See Figure 4.57. From (4.433) we have the conductor-backed reflection coefficient

$$\mathbb{R}_A = \frac{\tilde{\Gamma} - \tilde{P}^2}{1 - \tilde{\Gamma}\tilde{P}^2},$$

(4.446)

while from (4.441) we have the air-backed reflection coefficient

$$\mathbb{R}_B = \frac{\tilde{\Gamma}(1 - \tilde{P}^2)}{1 - \tilde{\Gamma}^2\tilde{P}^2}.$$

(4.447)

We must solve (4.446) and (4.447) simultaneously to find $\tilde{\Gamma}$ and $\tilde{P}$. Rearrangement of (4.446) gives

$$\tilde{P}^2 = \frac{\tilde{\Gamma} - \mathbb{R}_A}{1 - \mathbb{R}_A\tilde{\Gamma}},$$

(4.448)

while (4.447) gives

$$\tilde{P}^2 = \frac{\tilde{\Gamma} - \mathbb{R}_B}{\tilde{\Gamma}(1 - \mathbb{R}_B\tilde{\Gamma})}.$$

(4.449)

Equating, we have $(\tilde{\Gamma} - \mathbb{R}_A)\tilde{\Gamma}(1 - \mathbb{R}_B\tilde{\Gamma}) - (1 - \mathbb{R}_A\tilde{\Gamma})(\tilde{\Gamma} - \mathbb{R}_B) = 0$, or

$$\tilde{\Gamma}^3 - A\tilde{\Gamma}^2 + A\tilde{\Gamma} - 1 = 0 \quad \text{where} \quad A = \frac{1 + \mathbb{R}_A\mathbb{R}_B + \mathbb{R}_A}{\mathbb{R}_B}.$$

(4.450)

We may factor (4.450) as

$$(\tilde{\Gamma} - 1)[\tilde{\Gamma}^2 - (A - 1)\tilde{\Gamma} + 1] = 0.$$

This has solutions $\tilde{\Gamma} = 1$ and

$$\tilde{\Gamma} = \frac{A - 1}{2} \pm \sqrt{\left(\frac{A - 1}{2}\right)^2 - 1}.$$

The first solution, $\tilde{\Gamma} = 1$, corresponds to the unlikely case of total interfacial reflection. The remaining sign choice is made by requiring $|\tilde{\Gamma}| \leq 1$. Once $\tilde{\Gamma}$ is determined, $\tilde{Q}$, and thus $\tilde{P}$, may be found from either (4.448) or (4.449).

Alternatively, we may solve for $\tilde{P}$ and then $\tilde{\Gamma}$. Rearranging (4.446) gives

$$\tilde{\Gamma} = \frac{\mathbb{R}_A + \tilde{Q}}{1 + \mathbb{R}_A\tilde{Q}},$$

(4.451)

where $\tilde{Q} = \tilde{P}^2$. Substituting this into (4.447) then gives

$$(1 + \mathbb{R}_A\tilde{Q})^2\mathbb{R}_B - \mathbb{R}_B\tilde{Q}(\mathbb{R}_A + \tilde{Q})^2 - (\mathbb{R}_A + \tilde{Q})(1 - \tilde{Q})(1 + \mathbb{R}_A\tilde{Q}) = 0,$$

or

$$\tilde{Q}^3 - B\tilde{Q}^2 + B\tilde{Q} - 1 = 0,$$

(4.452)

where

$$B = \frac{(\mathbb{R}_A\mathbb{R}_B - 1)(1 - \mathbb{R}_A) + \mathbb{R}_A(\mathbb{R}_B - \mathbb{R}_A)}{\mathbb{R}_A - \mathbb{R}_B}.$$

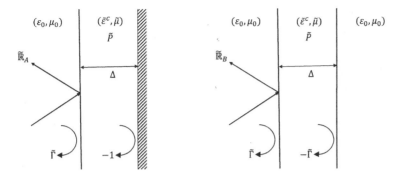

FIGURE 4.57
Measurements used in the conductor-backed/air-backed method.

We may factor (4.452) as

$$(\tilde{Q} - 1)[\tilde{Q}^2 - (B - 1)\tilde{Q} + 1] = 0.$$

This has solutions $\tilde{Q} = 1$ and

$$\tilde{Q} = \frac{B - 1}{2} \pm \sqrt{\left(\frac{B - 1}{2}\right)^2 - 1}.$$

The first solution, $\tilde{Q} = 1$, is only possible if the sample is lossless and exactly a half wavelength thick, which is unlikely. The remaining sign choice is made by requiring $|\tilde{\Gamma}| \leq 1$ where $\tilde{\Gamma}$ is found using (4.451).

4.15.1.4 Layer-shift method

In the layer-shift method a measurement is first taken of the reflection from the sample backed by a conductor, then with the conductor removed a distance d from the sample. In the second measurement, air is assumed to occupy the region between the sample and the conductor. See Figure 4.58. A variant of this method uses measurements with the conductor placed in two different positions behind the sample [100, 62]; see Problem 4.37.

From (4.433) we have the conductor-backed reflection coefficient

$$\mathbb{\tilde{R}}_A = \frac{\tilde{\Gamma} - \tilde{P}^2}{1 - \tilde{\Gamma}\tilde{P}^2}. \tag{4.453}$$

The case of an offset conductor was considered in Example 4.21. Using (4.303) with $\tilde{\Gamma}_2 = -\tilde{\Gamma}_1$, we get

$$\mathbb{\tilde{R}}_B = \frac{\tilde{\Gamma} + \tilde{\Gamma}^2\tilde{P}_B^2 - \tilde{\Gamma}\tilde{P}^2 - \tilde{P}^2\tilde{P}_B^2}{1 + \tilde{\Gamma}\tilde{P}_B^2 - \tilde{\Gamma}^2\tilde{P}^2 - \tilde{\Gamma}\tilde{P}^2\tilde{P}_B^2}. \tag{4.454}$$

Here $\tilde{P}_B = e^{-jk_{z0}d}$ is known. Simultaneous solution of (4.446) and (4.447) yields $\tilde{\Gamma}$ and $\tilde{P}$. Rearranging (4.453), we have

$$\tilde{P}^2 = \frac{\tilde{\Gamma} - \mathbb{\tilde{R}}_A}{1 - \tilde{\Gamma}\mathbb{\tilde{R}}_A}, \tag{4.455}$$

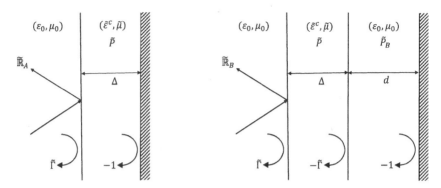

FIGURE 4.58
Measurements used in the layer-shift method.

and substitution into (4.454) gives

$$\tilde{\mathbb{R}}_B - \tilde{\Gamma}^2 \tilde{\mathbb{R}}_B \frac{\tilde{\Gamma} - \tilde{\mathbb{R}}_A}{1 - \tilde{\Gamma}\tilde{\mathbb{R}}_A} - \tilde{\mathbb{R}}_B \tilde{\Gamma} \tilde{P}_B^2 \frac{\tilde{\Gamma} - \tilde{\mathbb{R}}_A}{1 - \tilde{\Gamma}\tilde{\mathbb{R}}_A} + \tilde{\Gamma}\tilde{\mathbb{R}}_B \tilde{P}_B^2$$

$$= \tilde{\Gamma} - \tilde{\Gamma} \frac{\tilde{\Gamma} - \tilde{\mathbb{R}}_A}{1 - \tilde{\Gamma}\tilde{\mathbb{R}}_A} - \tilde{P}_B^2 \frac{\tilde{\Gamma} - \tilde{\mathbb{R}}_A}{1 - \tilde{\Gamma}\tilde{\mathbb{R}}_A} + \tilde{\Gamma}^2 \tilde{P}_B^2$$

or

$$\tilde{\Gamma}^3 - A\tilde{\Gamma}^2 + A\tilde{\Gamma} - 1 = 0 \qquad (4.456)$$

where

$$A = \frac{(\tilde{\mathbb{R}}_A \tilde{\mathbb{R}}_B + 1)(1 - \tilde{P}_B^2) + \tilde{\mathbb{R}}_A - \tilde{\mathbb{R}}_B \tilde{P}_B^2}{\tilde{\mathbb{R}}_B - \tilde{\mathbb{R}}_A \tilde{P}_B^2}.$$

We may factor (4.456) as

$$(\tilde{\Gamma} - 1)[\tilde{\Gamma}^2 - (A - 1)\tilde{\Gamma} + 1] = 0.$$

This has solutions $\tilde{\Gamma} = 1$ and

$$\tilde{\Gamma} = \frac{A - 1}{2} \pm \sqrt{\left(\frac{A - 1}{2}\right)^2 - 1}.$$

The solution $\tilde{\Gamma} = 1$ corresponds to the unlikely case of total interfacial reflection. The remaining sign choice is made by requiring $|\tilde{\Gamma}| \leq 1$. Once $\tilde{\Gamma}$ is determined, $\tilde{P}$ may be found from (4.455).

4.15.1.5   Two-backing method

In the two-backing method a measurement is first taken of the reflection from the sample backed by air (free space), then with a known layer of material placed behind the sample and backed by air [62]. See Figure 4.59. Let the propagation term for the known backing region be $\tilde{P}_B$, and the impedance be Z_B.

From (4.287) we have the reflection coefficient for an air-backed sample,

$$\tilde{\mathbb{R}}_A = \frac{\tilde{\Gamma}(1 - \tilde{P}^2)}{1 - \tilde{\Gamma}^2 \tilde{P}^2}, \qquad (4.457)$$

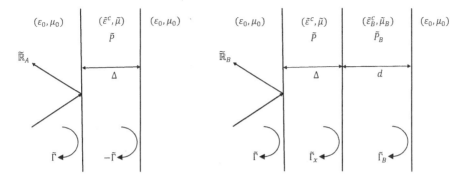

FIGURE 4.59
Measurements used in the two-backing method.

where we have used $\tilde{\Gamma}_1 = -\tilde{\Gamma}_2 = \tilde{\Gamma}$ and $\tilde{P}_1 = \tilde{P}$.

The structure reflection coefficient for the sample backed by a known layer is given by (4.302). Letting $\tilde{\Gamma}_1 = \tilde{\Gamma}$, $\tilde{\Gamma}_2 = \tilde{\Gamma}_x$, $\tilde{\Gamma}_3 = \tilde{\Gamma}_B$, $\tilde{P}_1 = \tilde{P}$, and $\tilde{P}_2 = \tilde{P}_B$, we have

$$\tilde{\mathbb{R}}_B = \frac{\tilde{\Gamma} + \tilde{\Gamma}\tilde{\Gamma}_x\tilde{\Gamma}_B\tilde{P}_B^2 + \tilde{\Gamma}_x\tilde{P}^2 + \tilde{\Gamma}_B\tilde{P}^2\tilde{P}_B^2}{1 + \tilde{\Gamma}_x\tilde{\Gamma}_B\tilde{P}_B^2 + \tilde{\Gamma}\tilde{\Gamma}_x\tilde{P}^2 + \tilde{\Gamma}\tilde{\Gamma}_B\tilde{P}^2\tilde{P}_B^2}. \tag{4.458}$$

Simultaneous solution of (4.457) and (4.458) yields $\tilde{\Gamma}$ and $\tilde{P}$.

Note that the interfacial reflection coefficient $\tilde{\Gamma}_x$ can be written in terms of $\tilde{\Gamma}$ and $\tilde{\Gamma}_B$. Since

$$\tilde{\Gamma} = \frac{Z - Z_0}{Z + Z_0},$$

we have

$$Z = Z_0\frac{1 + \tilde{\Gamma}}{1 - \tilde{\Gamma}}.$$

Substituting this into

$$\tilde{\Gamma}_x = \frac{Z_B - Z}{Z_B + Z}$$

and rearranging, we find that

$$\tilde{\Gamma}_x = \frac{\tilde{\Gamma} + \frac{Z_0 - Z_B}{Z_0 + Z_B}}{-\tilde{\Gamma}\frac{Z_0 - Z_B}{Z_0 + Z_B} - 1} = -\frac{\tilde{\Gamma} + \tilde{\Gamma}_B}{\tilde{\Gamma}\tilde{\Gamma}_B + 1}. \tag{4.459}$$

We can rearrange (4.457) to obtain $\tilde{P}^2$ as

$$\tilde{P}^2 = \frac{\tilde{\mathbb{R}}_A - \tilde{\Gamma}}{\tilde{\Gamma}^2\tilde{\mathbb{R}}_A - \tilde{\Gamma}}. \tag{4.460}$$

Substituting this and $\tilde{\Gamma}_x$ from (4.459) into (4.458) then gives, after some tedious algebra,

$$\tilde{\Gamma}^4 - C\tilde{\Gamma}^3 + C\tilde{\Gamma} - 1 = 0, \tag{4.461}$$

where

$$C = \frac{\tilde{\Gamma}_B(\tilde{P}_B^2 - 1)(\tilde{\mathbb{R}}_A\tilde{\mathbb{R}}_B - \tilde{\mathbb{R}}_A + 1) + (\tilde{\Gamma}_B^2\tilde{P}_B^2 - 1)(\tilde{\mathbb{R}}_B - \tilde{\mathbb{R}}_A)}{\tilde{\mathbb{R}}_A\tilde{\Gamma}_B(\tilde{P}_B^2 - 1)}.$$

We may factor (4.461) as

$$(\tilde{\Gamma} - 1)(\tilde{\Gamma} + 1)(\tilde{\Gamma}^2 - C\tilde{\Gamma} + 1) = 0.$$

This has solutions $\tilde{\Gamma} = 1$, $\tilde{\Gamma} = -1$, and

$$\tilde{\Gamma} = \frac{C}{2} \pm \sqrt{\left(\frac{C}{2}\right)^2 - 1}.$$

The solutions $\tilde{\Gamma} = \pm 1$ correspond to the unlikely case of total interfacial reflection. The remaining sign choice is made by requiring $|\tilde{\Gamma}| \leq 1$. Once $\tilde{\Gamma}$ is determined, $\tilde{P}$ may be found most easily from (4.460).

4.15.2   Other $\tilde{\Gamma}$-$\tilde{P}$ methods

There are other ways to obtain two independent measurements and thereby determine $\tilde{\Gamma}$ and $\tilde{P}$. For instance, one may make measurements of the transmission coefficients for two different sample thicknesses, rather than two reflection measurements (see Problem 4.38). Interestingly, all reflection-only methods that generate two independent measurements by altering the structure behind the sample (such as the conductor/air-backed method, the layer-shift method, and the two-backing method) may be handled using one unified technique. This approach, described in [62], uses an impedance transformation to give closed-form answers for $\tilde{\epsilon}^c$ and $\tilde{\mu}$ without the need to find the roots of a polynomial equation (although the need to find the proper value of n, and a square root ambiguity, remain). It is a bit less transparent than the $\tilde{\Gamma}$-$\tilde{P}$ approach, but those interested in applying reflection-only techniques might find it worth pursuing. But the reader is warned again that altering the region in *front* of the sample may not provide independent reflection measurements [62].

4.15.3   Non $\tilde{\Gamma}$-$\tilde{P}$ methods

The $\tilde{\Gamma}$-$\tilde{P}$ methods described above meet the requirement for two independent measurements either by using both transmission and reflection, or by altering the sample environment. Alternative methods use two measurements made with different excitations. For example, the free space method may use two reflection measurements made at different incidence angles. A method that is appealing on the surface is the *two-polarization method*, which uses a free space system with reflection-only measurements made using orthogonal (perpendicular and parallel) polarizations of the incident plane wave. Closed form expressions for $\tilde{\epsilon}^c$ and $\tilde{\mu}$ may be found without any ambiguities regarding branches of logarithms. Unfortunately, for materials with high values of $\tilde{\epsilon}^c$ or $\tilde{\mu}$, the technique is extremely sensitive to uncertainty in the value of the incidence angle. This makes the technique problematic for high-loss or high-permittivity materials. However, it may find application for materials in which the product of relative permittivity and relative permeability is not too large.

4.15.3.1   Two-polarization method

The two-polarization method is a reflection-only free-space method that uses the measured structure reflection coefficient under both parallel and perpendicular polarizations. It is convenient to use only if the sample is backed by a conductor.

Consider a conductor-backed sample of thickness Δ illuminated by a plane wave at an angle θ_0 under either parallel or perpendicular polarization. Recall that for either polarization, the structure reflection coefficient is

$$\tilde{\mathbb{R}} = \frac{\tilde{\Gamma} - \tilde{P}^2}{1 - \tilde{\Gamma}\tilde{P}^2} \quad \text{where} \quad \tilde{\Gamma} = \frac{Z - Z_0}{Z + Z_0} = \frac{z - 1}{z + 1}$$

with $z = Z/Z_0$. Substitution gives

$$\tilde{\mathbb{R}} = \frac{z - 1 - \tilde{P}^2 z - \tilde{P}^2}{z + 1 - \tilde{P}^2 z + \tilde{P}^2}.$$

Rearranging and solving for z, we have

$$z = \left(\frac{1 + \tilde{\mathbb{R}}}{1 - \tilde{\mathbb{R}}}\right)\left(\frac{1 + \tilde{P}^2}{1 - \tilde{P}^2}\right) = -j\left(\frac{1 + \tilde{\mathbb{R}}}{1 - \tilde{\mathbb{R}}}\right)\cot(k_z\Delta).$$

Now assume two measurements are made: one with perpendicular polarization and one with parallel polarization, both at the same incidence angle θ_0. Then

$$z_\perp = -j\left(\frac{1 + \tilde{\mathbb{R}}_\perp}{1 - \tilde{\mathbb{R}}_\perp}\right)\cot(k_z\Delta), \qquad z_\| = -j\left(\frac{1 + \tilde{\mathbb{R}}_\|}{1 - \tilde{\mathbb{R}}_\|}\right)\cot(k_z\Delta).$$

By (4.258) and (4.260),

$$z_\perp = \frac{k\eta/k_z}{\eta_0/\cos\theta_0}, \qquad z_\| = \frac{k_z\eta/k}{\eta_0\cos\theta_0}.$$

Taking the ratio $z_\|/z_\perp$ we find that

$$\frac{k_z^2}{k^2\cos^2\theta_0} = Y,$$

which only depends on the product $\tilde{\mu}\tilde{\epsilon}^c$, and not the ratio $\tilde{\mu}/\tilde{\epsilon}^c$. Here

$$Y = \left(\frac{1 - \tilde{\mathbb{R}}_\perp}{1 + \tilde{\mathbb{R}}_\perp}\right)\left(\frac{1 + \tilde{\mathbb{R}}_\|}{1 - \tilde{\mathbb{R}}_\|}\right).$$

Using $k_z^2 = k^2 - k_0^2\sin^2\theta_0$ we then have $k^2 - k_0^2\sin^2\theta_0 = Yk^2\cos^2\theta_0$ so that

$$\frac{k^2}{k_0^2} = \frac{\tilde{\mu}}{\mu_0}\frac{\tilde{\epsilon}^c}{\epsilon_0} = k_r^2 = \frac{\sin^2\theta_0}{1 - Y\cos^2\theta_0}, \tag{4.462}$$

where k_r is the relative wavenumber (complex refractive index) of the sample. Formation of the product $z_\| z_\perp$ gives

$$\frac{\eta^2}{\eta_0^2} = \eta_r^2 = \frac{\tilde{\mu}}{\mu_0}\frac{\epsilon_0}{\tilde{\epsilon}^c} = X\cot^2(k_z\Delta),$$

where η_r is the complex relative intrinsic impedance of the sample, and

$$X = -\left(\frac{1 + \tilde{\mathbb{R}}_\perp}{1 - \tilde{\mathbb{R}}_\perp}\right)\left(\frac{1 + \tilde{\mathbb{R}}_\|}{1 - \tilde{\mathbb{R}}_\|}\right).$$

Since $\tilde{\mu}\tilde{\epsilon}^c$ is known from (4.462), we can compute k_z and thus η_r. From k_r and η_r we find $\tilde{\mu}$ and $\tilde{\epsilon}^c$ by multiplication and division:

$$\frac{\tilde{\mu}}{\mu_0} = k_r \eta_r = \pm\sqrt{\frac{\sin^2 \theta_0}{1 - Y \cos^2 \theta_0} X \cot^2(k_z \Delta)}, \tag{4.463}$$

$$\frac{\tilde{\epsilon}^c}{\epsilon_0} = \frac{k_r}{\eta_r} = \pm\sqrt{\frac{\sin^2 \theta_0}{1 - Y \cos^2 \theta_0} \frac{\tan^2(k_z \Delta)}{X}}. \tag{4.464}$$

These can be written more compactly by noting that

$$k_z = k_0\sqrt{k_r^2 - \sin^2 \theta_0} = k_0 \sin \theta_0 \sqrt{\frac{Y \cos^2 \theta_0}{1 - Y \cos^2 \theta_0}}$$

so

$$\frac{\sin^2 \theta_0}{1 - Y \cos^2 \theta_0} = \frac{k_z^2}{Y k_0^2 \cos^2 \theta_0}.$$

With this, (4.463)–(4.464) become

$$\frac{\tilde{\mu}}{\mu_0} = \pm j \frac{1 + \tilde{\mathbb{R}}_\perp}{1 - \tilde{\mathbb{R}}_\perp} \frac{k_z}{k_0} \frac{\cot(k_z \Delta)}{\cos \theta_0}, \qquad \frac{\tilde{\epsilon}^c}{\epsilon_0} = \mp j \frac{1 - \tilde{\mathbb{R}}_\parallel}{1 + \tilde{\mathbb{R}}_\parallel} \frac{k_z}{k_0} \frac{\tan(k_z \Delta)}{\cos \theta_0}.$$

The sign ambiguity is resolved by requiring $\mathrm{Im}\{\tilde{\epsilon}^c\} < 0$ and $\mathrm{Im}\{\tilde{\mu}\} < 0$ for passive materials.

The benefit of the two-polarization method over $\tilde{\Gamma}$-$\tilde{P}$ methods is that k_z is found without the logarithm and its branch ambiguity. However, the expression (4.462) for the product $\tilde{\mu}\tilde{\epsilon}^c$ suggests that this method can be sensitive to uncertainty in angle. If k_r^2 is large, the denominator of the expression on the right-hand side must be small so that $Y \cos^2 \theta_0$ is near unity. Thus, an inaccurate specification of θ_0 will not allow proper cancellation and produce error in the expression. This will then propagate to the ratio $\tilde{\mu}/\tilde{\epsilon}^c$ through k_z. This is examined in more detail next.

4.15.4 Uncertainty analysis in material characterization

Uncertainty analysis is essential to any experimental undertaking. The uncertainties inherent both in the measurement equipment and in the process are quantified, and the resulting error in the material parameters $\tilde{\epsilon}^c$ and $\tilde{\mu}$ are predicted. Sources of error arise from uncertainty in the parameters specified in the extraction equations, such as the sample thickness, the incidence angle, and the measured structure reflection coefficients. Because these parameters appear directly in the equations, we can explore how uncertainty in their values propagates through to become errors in the extracted values of $\tilde{\epsilon}^c$ and $\tilde{\mu}$. Many other factors, however, are omitted as parameters from our simple model used to produce the reflection and transmission coefficients. The sample may have surface roughness, for instance, or nonuniform thickness. It may be bowed or tilted, or if placed in a waveguide there may be gaps between the sample and the waveguide walls. In the free-space method the incident field may not have a planar wavefront. Since we did not account for these situations in our model, we can't examine how these effects create errors in the extracted values of $\tilde{\epsilon}^c$ and $\tilde{\mu}$. So we are confined to examinations of uncertainty in thickness and incidence angle.

Assume that the sample thickness is measured many times using some appropriate process, and that the mean value is $\bar{\Delta}$ and the standard deviation is σ_Δ. Then $\bar{\Delta}$ is the estimate of the thickness that we use in the extraction equations, and σ_Δ characterizes the uncertainty of the thickness. Similarly, assume that the incidence angle is measured many times, and its mean and standard deviation are $\bar{\theta}_0$ and σ_{θ_0}, respectively. Then the propagation of the uncertainties in the values of Δ and θ_0 into errors in the extracted values of $\tilde{\epsilon}^c$ and $\tilde{\mu}$ is described using the propagation of errors formulas [192],

$$\sigma^2_{\mathrm{Re}\,\tilde{\epsilon}^c} = \left(\frac{\partial\,\mathrm{Re}\,\tilde{\epsilon}^c}{\partial\Delta}\right)^2 \sigma^2_\Delta + \left(\frac{\partial\,\mathrm{Re}\,\tilde{\epsilon}^c}{\partial\theta_0}\right)^2 \sigma^2_{\theta_0},$$

$$\sigma^2_{\mathrm{Im}\,\tilde{\epsilon}^c} = \left(\frac{\partial\,\mathrm{Im}\,\tilde{\epsilon}^c}{\partial\Delta}\right)^2 \sigma^2_\Delta + \left(\frac{\partial\,\mathrm{Im}\,\tilde{\epsilon}^c}{\partial\theta_0}\right)^2 \sigma^2_{\theta_0},$$

$$\sigma^2_{\mathrm{Re}\,\tilde{\mu}} = \left(\frac{\partial\,\mathrm{Re}\,\tilde{\mu}}{\partial\Delta}\right)^2 \sigma^2_\Delta + \left(\frac{\partial\,\mathrm{Re}\,\tilde{\mu}}{\partial\theta_0}\right)^2 \sigma^2_{\theta_0},$$

$$\sigma^2_{\mathrm{Im}\,\tilde{\mu}} = \left(\frac{\partial\,\mathrm{Im}\,\tilde{\mu}}{\partial\Delta}\right)^2 \sigma^2_\Delta + \left(\frac{\partial\,\mathrm{Im}\,\tilde{\mu}}{\partial\theta_0}\right)^2 \sigma^2_{\theta_0},$$

where it is assumed that there is no correlation between uncertainties (i.e., a change in the knowledge of Δ does not affect the known value of θ_0, and vice versa). Here $\sigma^2_{\mathrm{Re}\,\tilde{\epsilon}^c}$ is the predicted variance in the extracted value of $\mathrm{Re}\,\tilde{\epsilon}^c$, etc. The partial derivative quantities, $\partial\,\mathrm{Re}\,\tilde{\epsilon}^c/\partial\Delta$, etc., are sometimes called the *sensitivity coefficients* or *amplification factors* because they describe how extensively the uncertainties in the measured quantities are transferred to errors in the extracted parameters. These coefficients provide the crucial link between uncertainty and error. When they are small, uncertainties in thickness and angle may introduce little error into the calculation of $\tilde{\mu}$ and $\tilde{\epsilon}^c$. When they are large, small uncertainties may be amplified into large errors. In any case, note that these formulas are only accurate when the uncertainties are small compared to the sensitivity coefficients.

We illustrate with some examples.

▶ **Example 4.34:** Thickness sensitivity coefficients for free space Γ-P methods

Assume that some free-space method has been used to obtain the interfacial reflection coefficient and propagation factor of a material sample of thickness Δ. This method could be, for instance, the reflection/transmission method of Nicolson, Ross, and Weir, or one of the reflection-only methods. Determine the sensitivity coefficients for uncertainty in sample thickness.

Solution: From § 4.15.1 we have, for perpendicular polarization,

$$\tilde{\mu} = \frac{\eta_0}{\cos\theta_0}\frac{1+\tilde{\Gamma}}{1-\tilde{\Gamma}}\frac{k_z}{\omega},$$

$$\tilde{\epsilon}^c = \frac{\cos\theta_0}{\eta_0}\frac{1-\tilde{\Gamma}}{1+\tilde{\Gamma}}\frac{k^2}{\omega k_z},$$

where $k^2 = k_z^2 + k_0^2 \sin^2\theta_0$ and

$$k_z = \frac{\ln P + j2n\pi}{-j\Delta}.$$

Computing the derivative of $\tilde{\mu}$, we find

$$\frac{\partial \tilde{\mu}}{\partial \Delta} = \frac{\eta_0}{\cos \theta_0} \frac{1 + \tilde{\Gamma}}{1 - \tilde{\Gamma}} \frac{1}{\omega} \frac{\partial k_z}{\partial \Delta}.$$

But

$$\frac{\partial k_z}{\partial \Delta} = -\frac{k_z}{\Delta},$$

so we have the simple expression

$$\frac{\partial \tilde{\mu}}{\partial \Delta} = -\frac{\tilde{\mu}}{\Delta}.$$

This yields the sensitivity coefficients

$$\frac{\partial \operatorname{Re} \tilde{\mu}}{\partial \Delta} = -\frac{\operatorname{Re} \tilde{\mu}}{\Delta}, \qquad \frac{\partial \operatorname{Im} \tilde{\mu}}{\partial \Delta} = -\frac{\operatorname{Im} \tilde{\mu}}{\Delta}.$$

The derivative of $\tilde{\epsilon}^c$ is a bit more complicated. Write

$$\tilde{\epsilon}^c = \frac{\cos \theta_0}{\eta_0} \frac{1 - \tilde{\Gamma}}{1 + \tilde{\Gamma}} \frac{k_z^2 + k_0^2 \sin^2 \theta_0}{\omega k_z}.$$

Then

$$\frac{\partial \tilde{\epsilon}^c}{\partial \Delta} = \frac{\cos \theta_0}{\eta_0} \frac{1 - \tilde{\Gamma}}{1 + \tilde{\Gamma}} \frac{k_z^2 - k_0^2 \sin^2 \theta_0}{\omega k_z^2} \frac{\partial k_z}{\partial \Delta} = -\frac{\tilde{\epsilon}^c}{\Delta} \left(\frac{k_z^2 - k_0^2 \sin^2 \theta_0}{k_z^2 + k_0^2 \sin^2 \theta_0} \right),$$

which at normal incidence ($\theta_0 = 0$) reduces to

$$\frac{\partial \tilde{\epsilon}^c}{\partial \Delta} = -\frac{\tilde{\epsilon}^c}{\Delta}.$$

Thus, we have the sensitivity coefficients

$$\frac{\partial \operatorname{Re} \tilde{\epsilon}^c}{\partial \Delta} = -\operatorname{Re}\left\{ \frac{\tilde{\epsilon}^c}{\Delta} \left(\frac{k_z^2 - k_0^2 \sin^2 \theta_0}{k_z^2 + k_0^2 \sin^2 \theta_0} \right) \right\},$$

$$\frac{\partial \operatorname{Im} \tilde{\epsilon}^c}{\partial \Delta} = -\operatorname{Im}\left\{ \frac{\tilde{\epsilon}^c}{\Delta} \left(\frac{k_z^2 - k_0^2 \sin^2 \theta_0}{k_z^2 + k_0^2 \sin^2 \theta_0} \right) \right\}.$$

Note that the sensitivity coefficient for $\tilde{\mu}$ is independent of incidence angle, while the sensitivity coefficient for $\tilde{\epsilon}^c$ is dependent on incidence angle, although this dependence may be modest.

For parallel polarization we have

$$\tilde{\mu} = \eta_0 \cos \theta_0 \frac{1 + \tilde{\Gamma}}{1 - \tilde{\Gamma}} \frac{k^2}{\omega k_z},$$

$$\tilde{\epsilon}^c = \frac{1}{\eta_0 \cos \theta_0} \frac{1 - \tilde{\Gamma}}{1 + \tilde{\Gamma}} \frac{k_z}{\omega}.$$

Comparing to the case of perpendicular polarization, we quickly see that

$$\frac{\partial \operatorname{Re} \tilde{\epsilon}^c}{\partial \Delta} = -\frac{\operatorname{Re} \tilde{\epsilon}^c}{\Delta}, \qquad \frac{\partial \operatorname{Im} \tilde{\epsilon}^c}{\partial \Delta} = -\frac{\operatorname{Im} \tilde{\epsilon}^c}{\Delta},$$

and

$$\frac{\partial \operatorname{Re} \tilde{\mu}_r}{\partial \Delta} = -\operatorname{Re}\left\{ \frac{\tilde{\mu}_r}{\Delta} \left(\frac{k_z^2 - k_0^2 \sin^2 \theta_0}{k_z^2 + k_0^2 \sin^2 \theta_0} \right) \right\},$$

$$\frac{\partial \operatorname{Im} \tilde{\mu}_r}{\partial \Delta} = -\operatorname{Im}\left\{ \frac{\tilde{\mu}_r}{\Delta} \left(\frac{k_z^2 - k_0^2 \sin^2 \theta_0}{k_z^2 + k_0^2 \sin^2 \theta_0} \right) \right\}.$$

These results show that as the thickness of the sample decreases, the sensitivity to uncertainty in thickness increases in inverse proportion. Thus, if the experimenter's ability to measure thickness is independent of the thickness, then thin samples should be avoided to minimize errors in $\tilde{\epsilon}^c$ and $\tilde{\mu}$. That is, the ratio of the thickness uncertainty to the thickness should be kept as small as possible. ◀

▶ **Example 4.35:** Angle sensitivity coefficients for free space Γ-P methods

Assume that some free-space method has been used to obtain the interfacial reflection coefficient and propagation factor of a material sample of thickness Δ. This method could be, for instance, the reflection/transmission method of Nicolson, Ross, and Weir, or one of the reflection-only methods. Determine the sensitivity coefficients for uncertainty in measurement angle.

Solution: From § 4.15.1 we have, for perpendicular polarization,

$$\tilde{\mu} = \frac{\eta_0}{\cos\theta_0} \frac{1 + \tilde{\Gamma}}{1 - \tilde{\Gamma}} \frac{k_z}{\omega}, \qquad \tilde{\epsilon} = \frac{\cos\theta_0}{\eta_0} \frac{1 - \tilde{\Gamma}}{1 + \tilde{\Gamma}} \frac{k^2}{\omega k_z},$$

where $k^2 = k_z^2 + k_0^2 \sin^2 \theta_0$ and

$$k_z = \frac{\ln P + j2n\pi}{-j\Delta}.$$

Computing the derivative of $\tilde{\mu}$, we find

$$\frac{\partial \tilde{\mu}}{\partial \theta_0} = \eta_0 \frac{1 + \tilde{\Gamma}}{1 - \tilde{\Gamma}} \frac{k_z}{\omega} \frac{\partial}{\partial \theta_0} \left(\frac{1}{\cos\theta_0} \right) = \tilde{\mu} \tan\theta_0.$$

This yields the sensitivity coefficients

$$\frac{\partial \operatorname{Re}\tilde{\mu}}{\partial \theta_0} = \operatorname{Re}\{\tilde{\mu}\} \tan\theta_0, \qquad \frac{\partial \operatorname{Im}\tilde{\mu}}{\partial \theta_0} = \operatorname{Im}\{\tilde{\mu}\} \tan\theta_0.$$

Differentiation yields

$$\frac{\partial \tilde{\epsilon}^c}{\partial \theta_0} = \frac{\partial}{\partial \theta_0} \left(\frac{\cos\theta_0}{\eta_0} \frac{1 - \tilde{\Gamma}}{1 + \tilde{\Gamma}} \frac{k_z^2 + k_0^2 \sin^2 \theta_0}{\omega k_z} \right)$$

$$= \frac{1 - \tilde{\Gamma}}{1 + \tilde{\Gamma}} \frac{1}{\eta_0 \omega k_z} \left(\cos\theta_0 \left[2k_0^2 \sin\theta_0 \cos\theta_0 \right] - \sin\theta_0 \left[k_z^2 + k_0^2 \sin^2 \theta_0 \right] \right)$$

$$= \tilde{\epsilon}^c \left(\frac{k_0^2 \sin 2\theta_0}{k_z^2 + k_0^2 \sin^2 \theta_0} - \tan\theta_0 \right).$$

Therefore

$$\frac{\partial \operatorname{Re}\tilde{\epsilon}^c}{\partial \theta_0} = \operatorname{Re}\left\{ \tilde{\epsilon}^c \left(\frac{k_0^2 \sin 2\theta_0}{k_z^2 + k_0^2 \sin^2 \theta_0} - \tan\theta_0 \right) \right\},$$

$$\frac{\partial \operatorname{Im}\tilde{\epsilon}^c}{\partial \theta_0} = \operatorname{Im}\left\{ \tilde{\epsilon}^c \left(\frac{k_0^2 \sin 2\theta_0}{k_z^2 + k_0^2 \sin^2 \theta_0} - \tan\theta_0 \right) \right\}.$$

For parallel polarization, we have

$$\tilde{\mu} = \eta_0 \cos\theta_0 \frac{1 + \tilde{\Gamma}}{1 - \tilde{\Gamma}} \frac{k^2}{\omega k_z},$$

$$\tilde{\epsilon}^c = \frac{1}{\eta_0 \cos\theta_0} \frac{1 - \tilde{\Gamma}}{1 + \tilde{\Gamma}} \frac{k_z}{\omega}.$$

Comparing to the case of perpendicular polarization, we quickly see that

$$\frac{\partial \operatorname{Re} \tilde{\epsilon}^c}{\partial \theta_0} = \operatorname{Re}\{\tilde{\epsilon}^c\} \tan \theta_0, \qquad \frac{\partial \operatorname{Im} \tilde{\epsilon}^c}{\partial \theta_0} = \operatorname{Im}\{\tilde{\epsilon}^c\} \tan \theta_0.$$

and

$$\frac{\partial \operatorname{Re} \tilde{\mu}}{\partial \theta_0} = \operatorname{Re}\left\{\tilde{\mu}\left(\frac{k_0^2 \sin 2\theta_0}{k_z^2 + k_0^2 \sin^2 \theta_0} - \tan \theta_0\right)\right\},$$

$$\frac{\partial \operatorname{Im} \tilde{\mu}}{\partial \theta_0} = -\operatorname{Im}\left\{\tilde{\mu}\left(\frac{k_0^2 \sin 2\theta_0}{k_z^2 + k_0^2 \sin^2 \theta_0} - \tan \theta_0\right)\right\}.$$

Note that the sensitivity coefficients are smallest at $\theta_0 = 0$. Thus, illuminating the sample at normal incidence results in the least propagated error from angle uncertainty. In contrast, as the incidence angle approaches grazing ($\theta_0 = \pi/2$), the propagated error increases without bound, with sensitivity coefficients

$$\left|\frac{\partial \operatorname{Re} \tilde{\epsilon}^c}{\partial \theta_0}\right|, \left|\frac{\partial \operatorname{Im} \tilde{\epsilon}^c}{\partial \theta_0}\right|, \left|\frac{\partial \operatorname{Re} \tilde{\mu}}{\partial \theta_0}\right|, \left|\frac{\partial \operatorname{Im} \tilde{\mu}}{\partial \theta_0}\right| \sim \frac{1}{\frac{\pi}{2} - \theta_0}. \quad \blacktriangleleft$$

▶ **Example 4.36:** Angle sensitivity coefficients for two-polarization method

Determine the sensitivity coefficients for uncertainty in measurement angle for the two-polarization method. Compare the coefficients as functions of angle at 10 GHz for a sample of Eccosorb® FGM-125 and a 1/4-wavelength-thick sample of Teflon®.

Solution: From § 4.15.3.1 we have

$$\frac{\tilde{\mu}}{\mu_0} = \pm j \frac{1 + \tilde{\mathbb{R}}_\perp}{1 - \tilde{\mathbb{R}}_\perp} \frac{k_z}{k_0} \frac{\cot(k_z \Delta)}{\cos \theta_0}, \qquad \frac{\tilde{\epsilon}^c}{\epsilon_0} = \mp j \frac{1 - \tilde{\mathbb{R}}_\parallel}{1 + \tilde{\mathbb{R}}_\parallel} \frac{k_z}{k_0} \frac{\tan(k_z \Delta)}{\cos \theta_0}.$$

Differentiation of $\tilde{\epsilon}^c$ with respect to θ_0 gives

$$\frac{\partial \tilde{\epsilon}^c}{\partial \theta_0} = \mp j \epsilon_0 \frac{1 - \tilde{\mathbb{R}}_\parallel}{1 + \tilde{\mathbb{R}}_\parallel} \frac{1}{k_0} \left[k_z \tan(k_z \Delta) \frac{\sin \theta_0}{\cos^2 \theta_0}\right]$$

$$\mp j \epsilon_0 \frac{1 - \tilde{\mathbb{R}}_\parallel}{1 + \tilde{\mathbb{R}}_\parallel} \frac{1}{k_0 \cos \theta_0} \left[\frac{k_z \Delta}{\cos^2(k_z \Delta)} + \tan(k_z \Delta)\right] \frac{dk_z}{d\theta_0}$$

$$= \tilde{\epsilon}^c \left\{\tan \theta_0 + \left[\frac{2 k_z \Delta}{\sin(2 k_z \Delta)} + 1\right] \frac{1}{k_z} \frac{dk_z}{d\theta_0}\right\}. \qquad (4.465)$$

Now,

$$k_z^2 = \frac{Y k_0^2 \sin^2 \theta_0 \cos^2 \theta_0}{1 - Y \cos^2 \theta_0} = \frac{Y k_0^2}{4} \frac{\sin^2(2\theta_0)}{1 - Y \cos^2 \theta_0}.$$

Thus,

$$2 k_z \frac{\partial k_z}{\partial \theta_0} = \frac{Y k_0^2}{4} \frac{(1 - Y \cos^2 \theta_0) 4 \sin(2\theta_0) \cos(2\theta_0) - 2Y \sin^2(2\theta_0) \sin \theta_0 \cos \theta_0}{(1 - Y \cos^2 \theta_0)^2},$$

which, after much simplification, leads to

$$\frac{\partial k_z}{\partial \theta_0} = k_z \cot \theta_0 \left(1 - \frac{k_z^2}{k_0^2} \frac{1}{Y \cos^4 \theta_0}\right).$$

Use of this in (4.465) gives

$$\frac{\partial \tilde{\epsilon}^c}{\partial \theta_0} = \tilde{\epsilon}^c \left\{ \tan \theta_0 + \cot \theta_0 \left(1 + \frac{2k_z \Delta}{\sin(2k_z \Delta)} \right) \left(1 - \frac{k_z^2}{k_0^2} \frac{1}{Y \cos^4 \theta_0} \right) \right\}.$$

From these the expressions for $\partial \operatorname{Re} \tilde{\epsilon}^c / \partial \theta_0$ and $\partial \operatorname{Im} \tilde{\epsilon}^c / \partial \theta_0$ can be obtained by taking real and imaginary parts. An analogous set of steps gives

$$\frac{\partial \tilde{\mu}}{\partial \theta_0} = \tilde{\mu} \left\{ \tan \theta_0 + \cot \theta_0 \left(1 - \frac{2k_z \Delta}{\sin(2k_z \Delta)} \right) \left(1 - \frac{k_z^2}{k_0^2} \frac{1}{Y \cos^4 \theta_0} \right) \right\}.$$

Note the term $\cot \theta_0$ in each of these expressions. This term shows that the sensitivity coefficients approach infinity as normal incidence is approached. This is reasonable, since at normal incidence the structure reflection coefficients for parallel and perpendicular polarization are identical, and thus the inverse problem is underdetermined.

To get a grasp on the magnitude of the sensitivity coefficients, consider two samples of material. The first is a commercially available magnetic radar absorbing material (magRAM), Eccosorb FGM-125. This material is 125 one-thousandths of an inch (3.175 mm) thick, and at 10 GHz has representative material parameters $\tilde{\epsilon}^c = (7.32 - j0.0464)\epsilon_0$ and $\tilde{\mu} = (0.576 - j0.484)\mu_0$ [47]. Figure 4.60 plots the absolute values of the sensitivity coefficients vs. incidence angle. As expected, the coefficients approach infinity at normal incidence. They also become large near grazing incidence. The smallest sensitivity coefficients appear for an angle in the vicinity of 60° (although each coefficient has a different angle at which it is minimum). At this angle,

$$\left| \frac{\partial \operatorname{Re} \tilde{\epsilon}^c}{\partial \theta_0} \right| \approx 55\epsilon_0, \quad \left| \frac{\partial \operatorname{Im} \tilde{\epsilon}^c}{\partial \theta_0} \right| \approx 203\epsilon_0, \quad \left| \frac{\partial \operatorname{Re} \tilde{\mu}}{\partial \theta_0} \right| \approx 18\mu_0, \quad \left| \frac{\partial \operatorname{Im} \tilde{\mu}}{\partial \theta_0} \right| \approx 4.2\mu_0.$$

These numbers are surprisingly large. Assume the uncertainty in the angle is $\sigma_{\theta_0} = 0.02$ rad, which is just over one degree. Then, the expected error in $\operatorname{Re} \tilde{\epsilon}^c$ is approximately $\sigma_{\operatorname{Re}\{\tilde{\epsilon}^c\}} = 55\epsilon_0 \times 0.02 = 1.1\epsilon_0$. Recall that $\operatorname{Re} \tilde{\epsilon}^c = 7.32\epsilon_0$. This means that an uncertainty in angle of just one degree leads to roughly a 15% error in $\operatorname{Re} \tilde{\epsilon}^c$. Even worse is the error in $\operatorname{Im} \tilde{\epsilon}^c$, which at $\sigma_{\operatorname{Im} \tilde{\epsilon}^c} = 203\epsilon_0 \times 0.02 = 4.06\epsilon_0$ is 10,000%! The errors in $\operatorname{Re} \tilde{\mu}$ and $\operatorname{Im} \tilde{\mu}$ are around 30% and 9%, respectively. These suggest that it might be difficult to accurately employ the two-polarization method because of the need to precisely control the incidence angle.

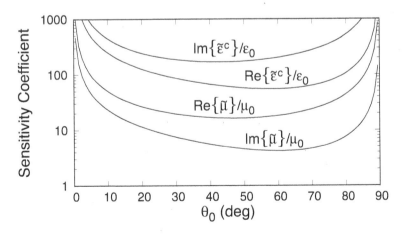

FIGURE 4.60
Sensitivity coefficients for FGM-125 at 10 GHz.

The second material sample consists of a quarter-wavelength-thick piece of Teflon, which is a nonmagnetic dielectric with representative material parameters $\tilde{\epsilon}^c = (2.08 - j0.0008)\epsilon_0$ and $\tilde{\mu} = (1 - j0)\mu_0$ at 10 GHz [206]. Figure 4.61 plots the absolute values of the sensitivity coefficients vs. incidence angle. As with the magRAM, the smallest sensitivity coefficients appear for $\theta_0 \approx 60°$. At this angle,

$$\left|\frac{\partial \operatorname{Re}\tilde{\epsilon}^c}{\partial \theta_0}\right| \approx 0.090\epsilon_0, \quad \left|\frac{\partial \operatorname{Im}\tilde{\epsilon}^c}{\partial \theta_0}\right| \approx 43\epsilon_0, \quad \left|\frac{\partial \operatorname{Re}\tilde{\mu}}{\partial \theta_0}\right| \approx 15.5\mu_0, \quad \left|\frac{\partial \operatorname{Im}\tilde{\mu}}{\partial \theta_0}\right| \approx 0.031\mu_0.$$

Clearly, the errors due to angle uncertainty are much less for Teflon than for magRAM. Indeed, the error is dependent on the contrast of the material parameters of the sample with the parameters of the air region outside the sample. In general, the larger this contrast, the larger the error.

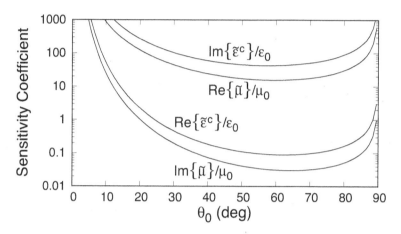

FIGURE 4.61
Sensitivity coefficients for Teflon at 10 GHz. ◄

4.16 Problems

4.1 Beginning with the Kramers–Kronig formulas (4.35)–(4.36), use the even–odd behavior of the real and imaginary parts of $\tilde{\epsilon}^c$ to derive the alternative relations (4.37)–(4.38).

4.2 Consider the complex permittivity dyadic of a magnetized plasma given by (4.90)–(4.93). Show that we may decompose $[\tilde{\bar{\epsilon}}^c]$ as the sum of two matrices

$$[\tilde{\bar{\epsilon}}^c] = [\tilde{\bar{\epsilon}}] + \frac{[\tilde{\bar{\sigma}}]}{j\omega}$$

where $[\tilde{\bar{\epsilon}}]$ and $[\tilde{\bar{\sigma}}]$ are hermitian.

4.3 Show that the Debye permittivity formulas

$$\operatorname{Re}\{\tilde{\epsilon}(\omega)\} - \epsilon_\infty = \frac{\epsilon_s - \epsilon_\infty}{1 + \omega^2\tau^2}, \qquad \operatorname{Im}\{\tilde{\epsilon}(\omega)\} = -\frac{\omega\tau(\epsilon_s - \epsilon_\infty)}{1 + \omega^2\tau^2},$$

obey the Kramers–Kronig relations.

4.4 The frequency-domain duality transformations for the constitutive parameters of an anisotropic medium are given in (4.191). Determine the analogous transformations for the constitutive parameters of a bianisotropic medium.

4.5 Establish the plane-wave identities (B.82)–(B.85) by direct differentiation in rectangular coordinates.

4.6 Assume that sea water has the parameters $\epsilon = 80\epsilon_0$, $\mu = \mu_0$, $\sigma = 4$ S/m, and that these parameters are frequency-independent. Plot the ω–β diagram for a plane wave propagating in this medium and compare to Figure 4.18. Describe the dispersion: is it normal or anomalous? Also plot the phase and group velocities and compare to Figure 4.19. How does the relaxation phenomenon affect the velocity of a wave in this medium?

4.7 Consider a uniform plane wave incident at angle θ_i onto an interface separating two lossless media (Figure 4.26). Assuming perpendicular polarization, write the explicit forms of the total fields in each region under the condition $\theta_i < \theta_c$, where θ_c is the critical angle. Show that the total field in region 1 can be decomposed into a portion that is a pure standing wave in the z-direction and a portion that is a pure traveling wave in the z-direction. Also show that the field in region 2 is a pure traveling wave. Repeat for parallel polarization.

4.8 Consider a uniform plane wave incident at angle θ_i onto an interface separating two lossless media (Figure 4.26). Assuming perpendicular polarization, use the total fields from Problem 4.7 to show that under the condition $\theta_i < \theta_c$ the normal component of the time-average Poynting vector is continuous across the interface. Here θ_c is the critical angle. Repeat for parallel polarization.

4.9 Consider a uniform plane wave incident at angle θ_i onto an interface separating two lossless media (Figure 4.26). Assuming perpendicular polarization, write the explicit forms of the total fields in each region under the condition $\theta_i > \theta_c$, where θ_c is the critical angle. Show that the field in region 1 is a pure standing wave in the z-direction and that the field in region 2 is an evanescent wave. Repeat for parallel polarization.

4.10 Consider a uniform plane wave incident at angle θ_i onto an interface separating two lossless media (Figure 4.26). Assuming perpendicular polarization, use the fields from Problem 4.9 to show that under the condition $\theta_i > \theta_c$ neither the fields in region 1 nor the fields in region 2 carry time-average power in the z-direction. Here θ_c is the critical angle. Repeat for parallel polarization.

4.11 Consider a uniform plane wave incident at angle θ_i from a lossless material onto a good conductor (Figure 4.26). The conductor has permittivity ϵ_0, permeability μ_0, and conductivity σ. Show that the transmission angle is $\theta_t \approx 0$ and thus the wave in the conductor propagates normal to the interface. Also show that for perpendicular polarization the current per unit width induced by the wave in region 2 is

$$\tilde{\mathbf{K}}(\omega) = \hat{\mathbf{y}}\sigma\tilde{T}_\perp(\omega)\tilde{E}_\perp(\omega)\frac{1-j}{2\beta_2}$$

and that this is identical to the tangential magnetic field at the surface:

$$\tilde{\mathbf{K}}(\omega) = -\hat{\mathbf{z}} \times \tilde{\mathbf{H}}^t|_{z=0}.$$

If we define the *surface impedance* $Z_s(\omega)$ of the conductor as the ratio of tangential electric and magnetic fields at the interface, show that

$$Z_s(\omega) = \frac{1+j}{\sigma\delta} = R_s(\omega) + jX_s(\omega).$$

Then show that the time-average power flux entering region 2 for a monochromatic wave of frequency $\breve{\omega}$ is simply

$$\mathbf{S}_{av,2} = \hat{\mathbf{z}}\frac{1}{2}(\breve{\mathbf{K}} \cdot \breve{\mathbf{K}}^*)R_s.$$

Note that the since the surface impedance is also the ratio of tangential electric field to induced current per unit width in region 2, it is also called the *internal impedance*.

4.12 Consider a slab of lossless material with permittivity $\epsilon = \epsilon_r\epsilon_0$ and permeability μ_0 located in free space between the planes $z = z_1$ and $z = z_2$. A transient, perpendicularly polarized plane wave is obliquely incident on the slab as shown in Figure 4.30. If the temporal waveform of the incident wave is $E_\perp^i(t)$, find the transient reflected field in region 0 and the transient transmitted field in region 2 in terms of an infinite superposition of amplitude-scaled, time-shifted versions of the incident wave. Interpret each of the first four terms in the reflected and transmitted fields in terms of multiple reflection within the slab.

4.13 Consider a free-space gap embedded between the planes $z = z_1$ and $z = z_2$ in an infinite, lossless dielectric medium of permittivity $\epsilon_r\epsilon_0$ and permeability μ_0. A perpendicularly polarized plane wave is incident on the gap at angle $\theta_i > \theta_c$ as shown in Figure 4.30. Here θ_c is the critical angle for a plane wave incident on the *single* interface between a lossless dielectric of permittivity $\epsilon_r\epsilon_0$ and free space. Apply the boundary conditions and find the fields in each of the three regions. Find the time-average Poynting vector in region 0 at $z = z_1$, in region 1 at $z = z_2$, and in region 2 at $z = z_2$. Is conservation of energy obeyed?

4.14 A uniform ferrite material has scalar permittivity $\tilde{\epsilon} = \epsilon$ and dyadic permeability $\bar{\tilde{\mu}}$. Assume the ferrite is magnetized along the z-direction and has losses so that its permeability dyadic is given by (4.127). Show that the wave equation for a TEM plane wave of the form

$$\tilde{\mathbf{H}}(\mathbf{r},\omega) = \tilde{\mathbf{H}}_0(\omega)e^{-jk_z z}$$

is

$$k_z^2\tilde{\mathbf{H}}_0 = \omega^2\epsilon\bar{\tilde{\mu}} \cdot \tilde{\mathbf{H}}_0$$

where $k_z = \beta - j\alpha$. Find explicit formulas for the two solutions $k_{z\pm} = \beta_\pm - j\alpha_\pm$. Show that when the damping parameter $\alpha \ll 1$, near resonance $\alpha_+ \gg \alpha_-$.

4.15 A time-harmonic, TE-polarized, uniform cylindrical wave propagates in a lossy medium. Assuming $|k\rho| \gg 1$, show that the power per unit length passing through a cylinder of radius ρ is given by

$$P_{av}/l = \mathrm{Re}\,\{Z_{TE}^*\}\,|\check{H}_{z0}|^2 \frac{e^{-2\alpha\rho}}{8|k|}.$$

If the material is lossless, show that the power per unit length passing through a cylinder is independent of the radius and is given by

$$P_{av}/l = \frac{\eta|\check{H}_{z0}|^2}{8k}.$$

4.16 The *radar cross-section* of a two-dimensional object illuminated by a TM-polarized plane wave is defined by

$$\sigma_{2D}(\omega, \phi) = \lim_{\rho \to \infty} 2\pi\rho \frac{|\tilde{E}_z^s|^2}{|\tilde{E}_z^i|^2}.$$

This quantity has units of meters and is sometimes called the "scattering width" of the object. Using the asymptotic form of the Hankel function, determine the formula for the radar cross-section of a TM-illuminated cylinder made of perfect electric conductor. Show that when the cylinder radius is small compared to a wavelength, the radar cross-section may be approximated as

$$\sigma_{2D}(\omega, \phi) = a\frac{\pi^2}{k_0 a}\frac{1}{\ln^2(0.89k_0 a)}$$

and is thus independent of the observation angle ϕ.

4.17 A TE-polarized plane wave is incident on a material cylinder with complex permittivity $\tilde{\epsilon}^c(\omega)$ and permeability $\tilde{\mu}(\omega)$, aligned along the z-axis in free space. Apply the boundary conditions on the surface of the cylinder and determine the total field both internal and external to the cylinder. Show that as $\tilde{\sigma} \to \infty$ the magnetic field external to the cylinder reduces to (4.365).

4.18 A TM-polarized plane wave is incident on a PEC cylinder of radius a aligned along the z-axis in free space. The cylinder is coated with a material layer of radius b with complex permittivity $\tilde{\epsilon}^c(\omega)$ and permeability $\tilde{\mu}(\omega)$. Apply the boundary conditions on the surface of the cylinder and across the interface between the material and free space and determine the total field both internal and external to the material layer.

4.19 A PEC cylinder of radius a, aligned along the z-axis in free space, is illuminated by a z-directed electric line source $\tilde{I}(\omega)$ located at (ρ_0, ϕ_0). Expand the fields in the regions $a < \rho < \rho_0$ and $\rho > \rho_0$ in terms of nonuniform cylindrical waves, and apply the boundary conditions at $\rho = a$ and $\rho = \rho_0$ to determine the fields everywhere.

4.20 Assuming

$$f(\xi, \omega) = \frac{k}{2\pi}A(k_x, \omega)\sin\xi,$$

use the relations

$$\cos z = \cos(u + jv) = \cos u \cosh v - j \sin u \sinh v,$$
$$\sin z = \sin(u + jv) = \sin u \cosh v + j \cos u \sinh v,$$

to show that the contour in Figure 4.49 provides identical values of the integrand in

$$\tilde{\psi}(x, y, \omega) = \int_C f(\xi, \omega) e^{-jk\rho \cos(\phi \pm \xi)} \, d\xi$$

as does the contour $[-\infty + j\Delta, \infty + j\Delta]$ in

$$\tilde{\psi}(x, y, \omega) = \frac{1}{2\pi} \int\limits_{-\infty+j\Delta}^{\infty+j\Delta} A(k_x, \omega) e^{-jk_x x} e^{\mp jk_y y} \, dk_x. \tag{4.466}$$

4.21 Verify (4.402) by writing the TE fields in terms of Fourier transforms and applying boundary conditions.

4.22 Consider a z-directed electric line source $\tilde{I}(\omega)$ located on the y-axis at $y = h$. The region $y < 0$ contains a perfect electric conductor. Write the fields in the regions $0 < y < h$ and $y > h$ in terms of the Fourier transform solution to the homogeneous Helmholtz equation. Note that in the region $0 < y < h$ terms representing waves traveling in *both* the $\pm y$-directions are needed, while in the region $y > h$ only terms traveling in the y-direction are needed. Apply the boundary conditions at $y = 0, h$ to determine the spectral amplitudes. Show that the total field may be decomposed into an impressed term identical to (4.403) and a scattered term identical to (4.406).

4.23 Consider a z-directed magnetic line source $\tilde{I}_m(\omega)$ located on the y-axis at $y = h$. The region $y > 0$ contains a material with parameters $\tilde{\epsilon}_1^c(\omega)$ and $\tilde{\mu}_1(\omega)$, while the region $y < 0$ contains a material with parameters $\tilde{\epsilon}_2^c(\omega)$ and $\tilde{\mu}_2(\omega)$. Using the Fourier transform solution to the Helmholtz equation, write the total field for $y > 0$ as the sum of an impressed field of the magnetic line source and a scattered field, and write the field for $y < 0$ as a scattered field. Apply the boundary conditions at $y = 0$ to determine the spectral amplitudes.

4.24 Consider a TE-polarized plane wave incident on a PEC plane located at $x = 0$. If the incident magnetic field is given by

$$\tilde{\mathbf{H}}^i(\mathbf{r}, \omega) = \hat{\mathbf{z}} \tilde{H}_0(\omega) e^{jk(x \cos \phi_0 + y \sin \phi_0)},$$

solve for the scattered magnetic field using the Fourier transform approach.

4.25 Consider the layered medium of Figure 4.55 with alternating layers of free space and perfect dielectric. The dielectric layer has permittivity $4\epsilon_0$ and thickness Δ while the free space layer has thickness 2Δ. Assuming a normally incident plane wave, solve for $k_0\Delta$ in terms of $\kappa\Delta$, and plot k_0 vs. κ, identifying the stop and pass bands. This type of ω–β plot for a periodic medium is named a *Brillouin diagram*, after L. Brillouin who investigated energy bands in periodic crystal lattices [25].

4.26 Consider a periodic layered medium as in Figure 4.55, but with each cell consisting of three different layers. Derive an eigenvalue equation similar to (4.419) for the propagation constant.

4.27 The ionosphere may be considered as an ionized gas. A typical electron concentration is $N = 2 \times 10^{11}$ electrons per cubic meter, and a typical collision frequency is $\nu = 100$ Hz. (a) Neglecting the earth's magnetic field, compute the permittivity and conductivity of the ionosphere at 5 MHz and at 100 MHz. (b) Neglecting collisions, compute the entries in the permittivity dyadic assuming the earth's magnetic field is 0.5 Gauss.

4.28 Consider a dielectric material with a single resonance. (a) Show that the maximum value of $\mathrm{Re}\{\epsilon\}$ occurs at $\omega_{\max} = \omega_0\sqrt{1 - 2\Gamma/\omega_0}$ and has the value

$$\mathrm{Re}\{\epsilon\}_{\max} = \epsilon_0 + \frac{1}{4}\epsilon_0\frac{\omega_p^2}{\Gamma(\omega_0 - \Gamma)}.$$

(b) If the static permittivity of the material is $\epsilon = 5\epsilon_0$ and the maximum value of $\mathrm{Im}\{\epsilon\}$ is $\mathrm{Im}\{\epsilon\}_{\max} = -16\epsilon_0$, find the width of the resonance curve as a percentage of the resonant frequency.

4.29 A spherical time-harmonic wave in free space has the form

$$\mathbf{E}(\mathbf{r}, t) = \hat{\boldsymbol{\theta}}\frac{E_0}{r\sin\theta}\cos(\breve{\omega}t - \beta r + \xi^E).$$

(a) Write the field in phasor form. (b) Find the magnetic field using

$$\breve{\mathbf{H}} = \frac{\hat{\mathbf{r}} \times \breve{\mathbf{E}}}{\eta_0}.$$

(c) Show that the fields obey the Sommerfeld radiation condition.

4.30 Consider sinusoidally varying fields $\mathbf{E}(\mathbf{r}, t)$ and $\mathbf{D}(\mathbf{r}, t)$. Compute the time-average integral to show that

$$\langle w_e(\mathbf{r}, t)\rangle = \tfrac{1}{4}\mathrm{Re}\left\{\breve{\mathbf{E}}(\mathbf{r}) \cdot \breve{\mathbf{D}}^*(\mathbf{r})\right\}.$$

4.31 A dielectric material has the constitutive parameters $\tilde{\mu} = \mu_0$ and $\tilde{\epsilon}^c = \mathrm{Re}\,\tilde{\epsilon}^c + j\,\mathrm{Im}\,\tilde{\epsilon}^c$. Derive the formulas for α and β.

4.32 The instantaneous fields in a source-free region of free space are given by

$$\mathbf{E}(\mathbf{r}, t) = \hat{\mathbf{x}}E_0\cos(\omega t - \beta z), \qquad \mathbf{H}(\mathbf{r}, t) = \hat{\mathbf{y}}\frac{E_0}{\eta_0}\cos(\omega t - \beta z).$$

Verify the complex Poynting theorem over the cube $0 \le x, y, z \le a$.

4.33 A plane wave propagates through a homogeneous material having permeability $\mu = \mu_0$ and permittivity

$$\tilde{\epsilon} = \frac{1}{\mu_0 c^2}\left(2 + 10\frac{c}{\omega}\right)^2$$

where c is the speed of light in free space. Calculate the group velocity and phase velocity.

4.34 The phasor electric field of a propagating plane wave is given by

$$\check{\mathbf{E}}(\mathbf{r}) = \hat{\mathbf{x}} E_0 e^{-j\beta z} + j\hat{\mathbf{y}} E_0 e^{-j\beta z}.$$

By converting to the time domain, derive the equation for the locus traced out by the tip of the electric field vector. What type of wave polarization is represented here?

4.35 The source of an EM field in free space is contained within a sphere of radius r. The fields on the surface of the sphere are given by

$$\check{\mathbf{E}}(\mathbf{r}) = \check{E}_0 e^{-j\beta r} \left[\hat{\boldsymbol{\theta}} \frac{1}{\beta r} + \hat{\mathbf{r}} \frac{j}{(\beta r)^2} \right] \sin\theta,$$

$$\check{\mathbf{H}}(\mathbf{r}) = \frac{\check{E}_0}{\eta_0} e^{-j\beta r} \left[\hat{\boldsymbol{\phi}} \frac{2}{\beta r} - \hat{\mathbf{r}} \frac{3j}{(\beta r)^2} \right] \sin\theta.$$

Compute the time-average power supplied by the source. Assume that β and η_0 are real and that $\check{E}_0$ is complex.

4.36 Consider a source-free region of space V, described by $a \le \rho \le b$, $0 \le \phi < 2\pi$, $0 \le z \le L$. The parameters of the medium are (σ, ϵ, μ). The fields within the region are given by

$$\check{\mathbf{E}}(\mathbf{r}) = \hat{\boldsymbol{\rho}} \frac{\check{E}_0 a}{\rho} e^{-j\beta z} e^{-\alpha z}, \qquad \check{\mathbf{H}}(\mathbf{r}) = \hat{\boldsymbol{\phi}} \frac{\check{E}_0 a}{\eta \rho} e^{-j\beta z} e^{-\alpha z}.$$

Here $\check{E}_0$ and η are complex numbers, while α and β are real. (a) Compute the total electric current passing through the disk of radius $(a + b)/2$ located at $z = 0$. (b) Compute the power dissipated within V. (c) Compute the power flowing through the cross-section of V located at z. (d) Using Poynting's theorem, find a formula for α in terms of σ and η.

4.37 In the layer-shift method described in Section 4.15.1.4, a measurement is first taken of the reflection from a material sample backed by a conductor, then with the conductor removed a distance d from the sample. A variant of this method is to make the first measurement with the conductor a distance d_A behind the sample, and then to make a second measurement with the conductor a distance d_B behind the sample. Determine expressions for $\tilde{\Gamma}$ and $\tilde{P}$ for this variant. Assume that in both measurements the region between the sample and the conductor is free space.

4.38 In the two-thickness method described in Section 4.15.1.2, measurements are taken of the reflection coefficient of a material sample with two different thicknesses. A variant of this method is to make measurements of the transmission coefficient of a sample with two thicknesses. If the samples have thicknesses Δ and 2Δ, determine expressions for $\tilde{\Gamma}$ and $\tilde{P}$ for this variant.

5

Field decompositions and the EM potentials

5.1 Spatial symmetry decompositions

Spatial symmetry can often be exploited to solve electromagnetics problems. For analytic solutions, symmetry can be used to reduce the number of boundary conditions that must be applied. For computer solutions, the storage requirements can be reduced. Typical symmetries include rotation about a point or axis, and reflection through a plane, along an axis, or through a point. We shall consider the common case of reflection through a plane.

Spatial symmetry decompositions may be applied even if the sources and fields possess no spatial symmetry. As long as the boundaries and material media are symmetric, the sources and fields may be decomposed into constituents that individually mimic the symmetry of the environment.

Consider a region of space consisting of linear, isotropic, time-invariant media having material parameters $\epsilon(\mathbf{r})$, $\mu(\mathbf{r})$, and $\sigma(\mathbf{r})$. The electromagnetic fields $(\mathbf{E}, \mathbf{H})$ within this region are related to their impressed sources $(\mathbf{J}^i, \mathbf{J}^i_m)$ and their secondary sources $\mathbf{J}^s = \sigma \mathbf{E}$ through Maxwell's curl equations:

$$\frac{\partial E_z}{\partial y} - \frac{\partial E_y}{\partial z} = -\mu \frac{\partial H_x}{\partial t} - J^i_{mx}, \tag{5.1}$$

$$\frac{\partial E_x}{\partial z} - \frac{\partial E_z}{\partial x} = -\mu \frac{\partial H_y}{\partial t} - J^i_{my}, \tag{5.2}$$

$$\frac{\partial E_y}{\partial x} - \frac{\partial E_x}{\partial y} = -\mu \frac{\partial H_z}{\partial t} - J^i_{mz}, \tag{5.3}$$

$$\frac{\partial H_z}{\partial y} - \frac{\partial H_y}{\partial z} = \epsilon \frac{\partial E_x}{\partial t} + \sigma E_x + J^i_x, \tag{5.4}$$

$$\frac{\partial H_x}{\partial z} - \frac{\partial H_z}{\partial x} = \epsilon \frac{\partial E_y}{\partial t} + \sigma E_y + J^i_y, \tag{5.5}$$

$$\frac{\partial H_y}{\partial x} - \frac{\partial H_x}{\partial y} = \epsilon \frac{\partial E_z}{\partial t} + \sigma E_z + J^i_z. \tag{5.6}$$

We assume the material constants are symmetric about some plane, say $z = 0$. Then

$$\epsilon(x, y, -z) = \epsilon(x, y, z), \qquad \mu(x, y, -z) = \mu(x, y, z), \qquad \sigma(x, y, -z) = \sigma(x, y, z).$$

That is, with respect to z the material constants are even functions. We further assume that the boundaries and boundary conditions, which guarantee uniqueness of solution, are also symmetric about the $z = 0$ plane. Then we define two cases of reflection symmetry.

5.1.1   Conditions for even symmetry

We claim that if the sources obey

$$J_x^i(x,y,z) = J_x^i(x,y,-z), \qquad J_{mx}^i(x,y,z) = -J_{mx}^i(x,y,-z),$$
$$J_y^i(x,y,z) = J_y^i(x,y,-z), \qquad J_{my}^i(x,y,z) = -J_{my}^i(x,y,-z),$$
$$J_z^i(x,y,z) = -J_z^i(x,y,-z), \qquad J_{mz}^i(x,y,z) = J_{mz}^i(x,y,-z),$$

then the fields obey

$$E_x(x,y,z) = E_x(x,y,-z), \qquad H_x(x,y,z) = -H_x(x,y,-z),$$
$$E_y(x,y,z) = E_y(x,y,-z), \qquad H_y(x,y,z) = -H_y(x,y,-z),$$
$$E_z(x,y,z) = -E_z(x,y,-z), \qquad H_z(x,y,z) = H_z(x,y,-z).$$

The electric field shares the symmetry of the electric source: components parallel to the $z = 0$ plane are even in z, and the component perpendicular is odd. The magnetic field shares the symmetry of the magnetic source: components parallel to the $z = 0$ plane are odd in z, and the component perpendicular is even.

We can verify our claim by showing that the symmetric fields and sources obey Maxwell's equations. At an arbitrary point $z = a > 0$, Equation (5.1) requires

$$\left.\frac{\partial E_z}{\partial y}\right|_{z=a} - \left.\frac{\partial E_y}{\partial z}\right|_{z=a} = -\mu|_{z=a}\left.\frac{\partial H_x}{\partial t}\right|_{z=a} - J_{mx}^i|_{z=a}.$$

By the assumed symmetry condition on source and material constant, we get

$$\left.\frac{\partial E_z}{\partial y}\right|_{z=a} - \left.\frac{\partial E_y}{\partial z}\right|_{z=a} = -\mu|_{z=-a}\left.\frac{\partial H_x}{\partial t}\right|_{z=a} + J_{mx}^i|_{z=-a}.$$

If our claim holds regarding the field behavior, then

$$\left.\frac{\partial E_z}{\partial y}\right|_{z=-a} = -\left.\frac{\partial E_z}{\partial y}\right|_{z=a}, \qquad \left.\frac{\partial E_y}{\partial z}\right|_{z=-a} = -\left.\frac{\partial E_y}{\partial z}\right|_{z=a}, \qquad \left.\frac{\partial H_x}{\partial t}\right|_{z=-a} = -\left.\frac{\partial H_x}{\partial t}\right|_{z=a},$$

and we have

$$-\left.\frac{\partial E_z}{\partial y}\right|_{z=-a} + \left.\frac{\partial E_y}{\partial z}\right|_{z=-a} = \mu|_{z=-a}\left.\frac{\partial H_x}{\partial t}\right|_{z=-a} + J_{mx}^i|_{z=-a}.$$

So this component of Faraday's law is satisfied. With similar reasoning we can show that the symmetric sources and fields satisfy (5.2)–(5.6) as well.

5.1.2   Conditions for odd symmetry

We can also show that if the sources obey

$$J_x^i(x,y,z) = -J_x^i(x,y,-z), \qquad J_{mx}^i(x,y,z) = J_{mx}^i(x,y,-z),$$
$$J_y^i(x,y,z) = -J_y^i(x,y,-z), \qquad J_{my}^i(x,y,z) = J_{my}^i(x,y,-z),$$
$$J_z^i(x,y,z) = J_z^i(x,y,-z), \qquad J_{mz}^i(x,y,z) = -J_{mz}^i(x,y,-z),$$

then the fields obey

$$E_x(x,y,z) = -E_x^i(x,y,-z), \qquad H_x(x,y,z) = H_x(x,y,-z),$$
$$E_y(x,y,z) = -E_y(x,y,-z), \qquad H_y(x,y,z) = H_y(x,y,-z),$$
$$E_z(x,y,z) = E_z(x,y,-z), \qquad H_z(x,y,z) = -H_z(x,y,-z).$$

Again the electric field has the same symmetry as the electric source. However, in this case components parallel to the $z = 0$ plane are odd in z and the component perpendicular is even. Similarly, the magnetic field has the same symmetry as the magnetic source. Here components parallel to the $z = 0$ plane are even in z and the component perpendicular is odd.

5.1.3 Field symmetries and the concept of source images

In the case of odd symmetry, the electric field parallel to the $z = 0$ plane is an odd function of z. If we assume that the field is also continuous across this plane, then the electric field tangential to $z = 0$ must vanish: the condition required at the surface of a perfect electric conductor (PEC). We may regard the problem of sources above a perfect conductor in the $z = 0$ plane as *equivalent* to the problem of sources odd about this plane, provided the sources in both cases are identical for $z > 0$. We call the source in the region $z < 0$ the *image* of the source in the region $z > 0$. So the image source $(\mathbf{J}^I, \mathbf{J}_m^I)$ obeys

$$J_x^I(x,y,-z) = -J_x^i(x,y,z), \qquad J_{mx}^I(x,y,-z) = J_{mx}^i(x,y,z),$$
$$J_y^I(x,y,-z) = -J_y^i(x,y,z), \qquad J_{my}^I(x,y,-z) = J_{my}^i(x,y,z),$$
$$J_z^I(x,y,-z) = J_z^i(x,y,z), \qquad J_{mz}^I(x,y,-z) = -J_{mz}^i(x,y,z).$$

That is, parallel components of electric current image in the opposite direction, and the perpendicular component images in the same direction; parallel components of the magnetic current image in the same direction, while the perpendicular component images in the opposite direction.

In the case of even symmetry, the magnetic field parallel to the $z = 0$ plane is odd, and thus the magnetic field tangential to the $z = 0$ plane must be zero. We therefore have an equivalence between the problem of a source above a plane of perfect magnetic conductor (PMC) and the problem of sources even about that plane. In this case we identify image sources that obey

$$J_x^I(x,y,-z) = J_x^i(x,y,z), \qquad J_{mx}^I(x,y,-z) = -J_{mx}^i(x,y,z),$$
$$J_y^I(x,y,-z) = J_y^i(x,y,z), \qquad J_{my}^I(x,y,-z) = -J_{my}^i(x,y,z),$$
$$J_z^I(x,y,-z) = -J_z^i(x,y,z), \qquad J_{mz}^I(x,y,-z) = J_{mz}^i(x,y,z).$$

Parallel components of electric current image in the same direction, and the perpendicular component images in the opposite direction; parallel components of magnetic current image in the opposite direction, and the perpendicular component images in the same direction.

In the case of odd symmetry, we sometimes say that an "electric wall" exists at $z = 0$. The term "magnetic wall" can be used in the case of even symmetry. These terms are particularly common in the description of waveguide fields.

5.1.4   Symmetric field decomposition

Field symmetries may be applied to arbitrary source distributions through a symmetry decomposition of the sources and fields. Consider the general impressed source distributions $(\mathbf{J}^i, \mathbf{J}^i_m)$. The source set

$$J_x^{ie}(x,y,z) = \tfrac{1}{2}\left[J_x^i(x,y,z) + J_x^i(x,y,-z)\right],$$

$$J_y^{ie}(x,y,z) = \tfrac{1}{2}\left[J_y^i(x,y,z) + J_y^i(x,y,-z)\right],$$

$$J_z^{ie}(x,y,z) = \tfrac{1}{2}\left[J_z^i(x,y,z) - J_z^i(x,y,-z)\right],$$

$$J_{mx}^{ie}(x,y,z) = \tfrac{1}{2}\left[J_{mx}^i(x,y,z) - J_{mx}^i(x,y,-z)\right],$$

$$J_{my}^{ie}(x,y,z) = \tfrac{1}{2}\left[J_{my}^i(x,y,z) - J_{my}^i(x,y,-z)\right],$$

$$J_{mz}^{ie}(x,y,z) = \tfrac{1}{2}\left[J_{mz}^i(x,y,z) + J_{mz}^i(x,y,-z)\right]$$

is clearly of even symmetric type, while the source set

$$J_x^{io}(x,y,z) = \tfrac{1}{2}\left[J_x^i(x,y,z) - J_x^i(x,y,-z)\right],$$

$$J_y^{io}(x,y,z) = \tfrac{1}{2}\left[J_y^i(x,y,z) - J_y^i(x,y,-z)\right],$$

$$J_z^{io}(x,y,z) = \tfrac{1}{2}\left[J_z^i(x,y,z) + J_z^i(x,y,-z)\right],$$

$$J_{mx}^{io}(x,y,z) = \tfrac{1}{2}\left[J_{mx}^i(x,y,z) + J_{mx}^i(x,y,-z)\right],$$

$$J_{my}^{io}(x,y,z) = \tfrac{1}{2}\left[J_{my}^i(x,y,z) + J_{my}^i(x,y,-z)\right],$$

$$J_{mz}^{io}(x,y,z) = \tfrac{1}{2}\left[J_{mz}^i(x,y,z) - J_{mz}^i(x,y,-z)\right]$$

is of the odd symmetric type. Since $\mathbf{J}^i = \mathbf{J}^{ie} + \mathbf{J}^{io}$ and $\mathbf{J}^i_m = \mathbf{J}^{ie}_m + \mathbf{J}^{io}_m$, we can decompose any source into constituents having, respectively, even and odd symmetry with respect to a plane. The source with even symmetry produces an even field set, while the source with odd symmetry produces an odd field set. The total field is the sum of the fields from each field set.

5.1.5   Planar symmetry for frequency-domain fields

The symmetry conditions introduced above for the time-domain fields also hold for the frequency-domain fields. Because both the conductivity and permittivity must be even functions, we combine their effects and require the complex permittivity to be even. Otherwise the field symmetries and source decompositions are identical.

▶ **Example 5.1:** Line source between conducting planes

Consider a z-directed electric line source $\tilde{I}_0$ located at $y = h, x = 0$ between conducting planes at $y = \pm d$, $d > h$. The material between the plates has permeability $\tilde{\mu}(\omega)$ and complex permittivity $\tilde{\epsilon}^c(\omega)$. Find the electric field everywhere between the plates by decomposing the source into one of even symmetric type with line sources $\tilde{I}_0/2$ located at $y = \pm h$, and one of odd symmetric type with a line source $\tilde{I}_0/2$ located at $y = h$ and a line source $-\tilde{I}_0/2$ located at $y = -h$. Find the electric field for each of these problems by exploiting the appropriate symmetry, and superpose the results to find the field for the original problem.

Solution: For the even-symmetric case, we begin by using (4.400) to represent the impressed field:

$$\tilde{E}_z^i(x,y,\omega) = -\frac{\omega\tilde{\mu}\frac{\tilde{I}_0(\omega)}{2}}{2\pi} \int\limits_{-\infty+j\Delta}^{\infty+j\Delta} \frac{e^{-jk_y|y-h|} + e^{-jk_y|y+h|}}{2k_y} e^{-jk_x x}\, dk_x.$$

For $y > h$ this becomes

$$\tilde{E}_z^i(x,y,\omega) = -\frac{\omega\tilde{\mu}\frac{\tilde{I}_0(\omega)}{2}}{2\pi} \int\limits_{-\infty+j\Delta}^{\infty+j\Delta} \frac{2\cos k_y h}{2k_y} e^{-jk_y y} e^{-jk_x x}\, dk_x.$$

The secondary (scattered) field consists of waves propagating in both the $\pm y$-directions:

$$\tilde{E}_z^s(x,y,\omega) = \frac{1}{2\pi} \int\limits_{-\infty+j\Delta}^{\infty+j\Delta} \left[A^+(k_x,\omega)e^{-jk_y y} + A^-(k_x,\omega)e^{jk_y y} \right] e^{-jk_x x}\, dk_x. \qquad (5.7)$$

The impressed field is even about $y = 0$. Since the total field $E_z = E_z^i + E_z^s$ must be even in y (E_z is parallel to the plane $y = 0$), the scattered field must also be even. Thus, $A^+ = A^-$ and the total field is for $y > h$

$$\tilde{E}_z(x,y,\omega) = \frac{1}{2\pi} \int\limits_{-\infty+j\Delta}^{\infty+j\Delta} \left[2A^+(k_x,\omega)\cos k_y y \right.$$

$$\left. - \omega\tilde{\mu}\frac{\tilde{I}_0(\omega)}{2}\frac{2\cos k_y h}{2k_y} e^{-jk_y y} \right] e^{-jk_x x}\, dk_x.$$

Now the electric field must obey the boundary condition $\tilde{E}_z = 0$ at $y = \pm d$. However, since $\tilde{E}_z$ is even the satisfaction of this condition at $y = d$ automatically implies its satisfaction at $y = -d$. So we set

$$\frac{1}{2\pi} \int\limits_{-\infty+j\Delta}^{\infty+j\Delta} \left[2A^+(k_x,\omega)\cos k_y d - \omega\tilde{\mu}\frac{\tilde{I}_0(\omega)}{2}\frac{2\cos k_y h}{2k_y} e^{-jk_y d} \right] e^{-jk_x x}\, dk_x = 0$$

and invoke the Fourier integral theorem to get

$$A^+(k_x,\omega) = \omega\tilde{\mu}\frac{\tilde{I}_0(\omega)}{2}\frac{\cos k_y h}{2k_y}\frac{e^{-jk_y d}}{\cos k_y d}.$$

The total field for this case is

$$\tilde{E}_z(x,y,\omega) = -\frac{\omega\tilde{\mu}\frac{\tilde{I}_0(\omega)}{2}}{2\pi} \int\limits_{-\infty+j\Delta}^{\infty+j\Delta} \left[\frac{e^{-jk_y|y-h|} + e^{-jk_y|y+h|}}{2k_y} \right.$$

$$\left. - \frac{2\cos k_y h}{2k_y}\frac{e^{-jk_y d}}{\cos k_y d}\cos k_y y \right] e^{-jk_x x}\, dk_x.$$

For the odd-symmetric case the impressed field is

$$\tilde{E}_z^i(x,y,\omega) = -\frac{\omega\tilde{\mu}\frac{\tilde{I}_0(\omega)}{2}}{2\pi} \int\limits_{-\infty+j\Delta}^{\infty+j\Delta} \frac{e^{-jk_y|y-h|} - e^{-jk_y|y+h|}}{2k_y} e^{-jk_x x}\, dk_x,$$

which for $y > h$ is

$$\tilde{E}_z^i(x, y, \omega) = -\frac{\omega\tilde{\mu}\frac{\tilde{I}_0(\omega)}{2}}{2\pi} \int\limits_{-\infty+j\Delta}^{\infty+j\Delta} \frac{2j\sin k_y h}{2k_y} e^{-jk_y y} e^{-jk_x x}\, dk_x.$$

The scattered field has the form of (5.7) but must be odd. Thus $A^+ = -A^-$ and the total field for $y > h$ is

$$\tilde{E}_z(x, y, \omega) = \frac{1}{2\pi} \int\limits_{-\infty+j\Delta}^{\infty+j\Delta} \Big[2jA^+(k_x, \omega)\sin k_y y$$

$$- \omega\tilde{\mu}\frac{\tilde{I}_0(\omega)}{2}\frac{2j\sin k_y h}{2k_y} e^{-jk_y y} \Big] e^{-jk_x x}\, dk_x.$$

Setting $\tilde{E}_z = 0$ at $z = d$ and solving for A^+, we find that the total field for this case is

$$\tilde{E}_z(x, y, \omega) = -\frac{\omega\tilde{\mu}\frac{\tilde{I}_0(\omega)}{2}}{2\pi} \int\limits_{-\infty+j\Delta}^{\infty+j\Delta} \Big[\frac{e^{-jk_y|y-h|} - e^{-jk_y|y+h|}}{2k_y}$$

$$- \frac{2j\sin k_y h}{2k_y}\frac{e^{-jk_y d}}{\sin k_y d}\sin k_y y \Big] e^{-jk_x x}\, dk_x.$$

Adding the fields for the two cases, we find that

$$\tilde{E}_z(x, y, \omega) = -\frac{\omega\tilde{\mu}\tilde{I}_0(\omega)}{2\pi} \int\limits_{-\infty+j\Delta}^{\infty+j\Delta} \frac{e^{-jk_y|y-h|}}{2k_y} e^{-jk_x x}\, dk_x$$

$$+ \frac{\omega\tilde{\mu}\tilde{I}_0(\omega)}{2\pi} \int\limits_{-\infty+j\Delta}^{\infty+j\Delta} \Big[\frac{\cos k_y h \cos k_y y}{\cos k_y d} + j\frac{\sin k_y h \sin k_y y}{\sin k_y d} \Big] \frac{e^{-jk_y d}}{2k_y} e^{-jk_x x}\, dk_x,$$

$$(5.8)$$

which is a superposition of impressed and scattered fields. ◄

▶ **Example 5.2:** Line source centered between conducting planes

Consider a z-directed electric line source $\tilde{I}_0$ centered between conducting planes at $y = \pm d$. The material between the plates is Teflon® with permeability μ_0 and complex permittivity $\tilde{\epsilon}^c = (2.08 - j0.0008)\epsilon_0$. Plot the magnitude of $\tilde{E}_z$ in the region $-2d \leq x \leq 2d$, $-d \leq y \leq d$ for $d = 2\lambda_0$.

Solution: We can adapt the solution of Example 5.1 to the present problem by setting $h = 0$. Since the line source is centered, both the impressed and scattered fields will be even about $y = 0$. For computation expediency, we use the Hankel function representation of impressed field, and normalize all distances to d. Thus we can write,

$$\frac{\tilde{E}_z^i(x, y, \omega)d}{\eta_0\tilde{I}_0} = -\frac{k_0 d}{4} H_0^{(2)}\left(k_0 d\tilde{\epsilon}_r^c \sqrt{\bar{x}^2 + \bar{y}^2} \right)$$

which is a dimensionless quantity. Here $\bar{x} = x/d$, $\bar{y} = y/d$, and $\tilde{\epsilon}_r^c = \tilde{\epsilon}^c/\epsilon_0$. Since the scattered field has only an even contribution, we have

$$\frac{\tilde{E}_z^s(x,y,\omega)d}{\eta_0 \tilde{I}_0} = \frac{k_0 d}{2\pi} \int_0^\infty \frac{\cos{(k_0 d \kappa \bar{y})}}{\cos{(k_0 d \kappa)}} e^{-jk_0 d\kappa} \frac{\cos{(k_0 d \xi \bar{x})}}{\kappa} d\xi$$

where $\xi = k_x/k_0$ and $\kappa = \sqrt{\tilde{\epsilon}_r^c - \xi^2}$, with $\text{Im}\{\kappa\} < 0$. Note that we have used the even and odd behavior of the integrand about $x = 0$ to reduce the domain of integration to $[0, \infty)$. Because the trigonometric functions diverge as $\xi \to \infty$, it is useful to rewrite the integrand in terms of exponential functions:

$$\frac{\tilde{E}_z^s(x,y,\omega)d}{\eta_0 \tilde{I}_0} = \frac{k_0 d}{2\pi} \int_0^\infty \frac{e^{-jk_0 d\kappa(2-\bar{y})} + e^{-jk_0 d\kappa(2+\bar{y})}}{1 + e^{-2jk_0 d\kappa}} \frac{\cos{(k_0 d \xi \bar{x})}}{\kappa} d\xi.$$

Letting $d = 2\lambda_0$ so that $k_0 d = 4\pi$, the magnitude of the total electric field is computed and plotted in Figure 5.1. Since the impressed field diverges on the z-axis, the normalized total field is cropped at a value of 2. The impressed field sets up a traveling wave between the plates that interferes with the scattered field to produce a standing wave pattern, with nulls separated by a half-wavelength in the material. Note that the total field is zero at $y = \pm d$, as required by the boundary conditions.

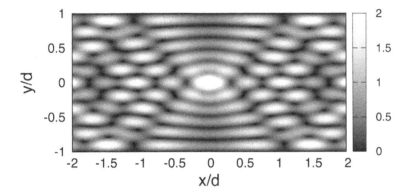

FIGURE 5.1
Magnitude of the total field $|\tilde{E}_z d/(\eta_0 \tilde{I}_0)|$ produced by a line source centered between parallel plates. ◀

5.2 Solenoidal–lamellar decomposition and the electromagnetic potentials

We now discuss the decomposition of a general vector field into a *lamellar* component having zero curl and a *solenoidal* component having zero divergence. This is known as a *Helmholtz decomposition*. If **V** is any vector field, then we wish to write

$$\mathbf{V} = \mathbf{V}_s + \mathbf{V}_l, \tag{5.9}$$

where $\mathbf{V}_s$ and $\mathbf{V}_l$ are the solenoidal and lamellar components of **V**. Formulas expressing these components in terms of **V** are obtained as follows. We first write $\mathbf{V}_s$ in terms of a "vector potential" $\mathbf{A}$ as

$$\mathbf{V}_s = \nabla \times \mathbf{A}. \tag{5.10}$$

This is possible by virtue of (B.55). Similarly, we write $\mathbf{V}_l$ in terms of a "scalar potential" ϕ as

$$\mathbf{V}_l = \nabla \phi. \tag{5.11}$$

To obtain a formula for $\mathbf{V}_l$ we take the divergence of (5.9) and use (5.11) to get

$$\nabla \cdot \mathbf{V} = \nabla \cdot \mathbf{V}_l = \nabla \cdot \nabla \phi = \nabla^2 \phi.$$

The result,

$$\nabla^2 \phi = \nabla \cdot \mathbf{V},$$

may be regarded as Poisson's equation for the unknown ϕ. This equation is solved in Chapter 3. By (3.41) we have

$$\phi(\mathbf{r}) = -\int_V \frac{\nabla' \cdot \mathbf{V}(\mathbf{r}')}{4\pi R} \, dV',$$

where $R = |\mathbf{r} - \mathbf{r}'|$, and we have

$$\mathbf{V}_l(\mathbf{r}) = -\nabla \int_V \frac{\nabla' \cdot \mathbf{V}(\mathbf{r}')}{4\pi R} \, dV'. \tag{5.12}$$

Similarly, a formula for $\mathbf{V}_s$ can be obtained by taking the curl of (5.9) to get

$$\nabla \times \mathbf{V} = \nabla \times \mathbf{V}_s.$$

Substituting (5.10) we have

$$\nabla \times \mathbf{V} = \nabla \times (\nabla \times \mathbf{A}) = \nabla(\nabla \cdot \mathbf{A}) - \nabla^2 \mathbf{A}.$$

We may choose any value we wish for $\nabla \cdot \mathbf{A}$, since this does not alter $\mathbf{V}_s = \nabla \times \mathbf{A}$. (We discuss such "gauge transformations" in greater detail later in this chapter.) With $\nabla \cdot \mathbf{A} = 0$ we obtain

$$-\nabla \times \mathbf{V} = \nabla^2 \mathbf{A}.$$

This is Poisson's equation for each rectangular component of $\mathbf{A}$; therefore

$$\mathbf{A}(\mathbf{r}) = \int_V \frac{\nabla' \times \mathbf{V}(\mathbf{r}')}{4\pi R} \, dV',$$

and we have

$$\mathbf{V}_s(\mathbf{r}) = \nabla \times \int_V \frac{\nabla' \times \mathbf{V}(\mathbf{r}')}{4\pi R} \, dV'.$$

Summing the results, we obtain the Helmholtz decomposition

$$\mathbf{V} = \mathbf{V}_l + \mathbf{V}_s = -\nabla \int_V \frac{\nabla' \cdot \mathbf{V}(\mathbf{r}')}{4\pi R} \, dV' + \nabla \times \int_V \frac{\nabla' \times \mathbf{V}(\mathbf{r}')}{4\pi R} \, dV'. \tag{5.13}$$

5.2.1 Identification of the electromagnetic potentials

Let us write the electromagnetic fields as a general superposition of solenoidal and lamellar components:

$$\mathbf{E} = \nabla \times \mathbf{A}_E + \nabla \phi_E, \tag{5.14}$$

$$\mathbf{B} = \nabla \times \mathbf{A}_B + \nabla \phi_B. \tag{5.15}$$

One possible form of the potentials $\mathbf{A}_E$, $\mathbf{A}_B$, ϕ_E, and ϕ_B appears in (5.13). However, because $\mathbf{E}$ and $\mathbf{B}$ are related by Maxwell's equations, the potentials should be related to the sources. We can determine the explicit relationship by substituting (5.14) and (5.15) into Ampere's and Faraday's laws. It is most convenient to analyze the relationships using superposition of the cases for which $\mathbf{J}_m = 0$ and $\mathbf{J} = 0$.

With $\mathbf{J}_m = 0$, Faraday's law is

$$\nabla \times \mathbf{E} = -\frac{\partial \mathbf{B}}{\partial t}. \tag{5.16}$$

Since $\nabla \times \mathbf{E}$ is solenoidal, $\mathbf{B}$ must be solenoidal and thus $\nabla \phi_B = 0$. This implies that $\phi_B = 0$, which is equivalent to the auxiliary Maxwell equation $\nabla \cdot \mathbf{B} = 0$. Now, substitution of (5.14) and (5.15) into (5.16) gives

$$\nabla \times (\nabla \times \mathbf{A}_E + \nabla \phi_E) = -\frac{\partial}{\partial t} (\nabla \times \mathbf{A}_B).$$

Using $\nabla \times (\nabla \phi_E) = 0$ and combining the terms, we get

$$\nabla \times \left(\nabla \times \mathbf{A}_E + \frac{\partial \mathbf{A}_B}{\partial t} \right) = 0,$$

hence

$$\nabla \times \mathbf{A}_E = -\frac{\partial \mathbf{A}_B}{\partial t} + \nabla \xi.$$

Substitution into (5.14) gives

$$\mathbf{E} = -\frac{\partial \mathbf{A}_B}{\partial t} + \nabla \phi_E + \nabla \xi.$$

Combining the two gradient functions together, we see that we can write both $\mathbf{E}$ and $\mathbf{B}$ in terms of two potentials:

$$\mathbf{E} = -\frac{\partial \mathbf{A}_e}{\partial t} - \nabla \phi_e, \tag{5.17}$$

$$\mathbf{B} = \nabla \times \mathbf{A}_e, \tag{5.18}$$

where the negative sign on the gradient term is introduced by convention.

5.2.2 Gauge transformations

5.2.2.1 The Coulomb gauge

We pay a price for the simplicity of using only two potentials to represent $\mathbf{E}$ and $\mathbf{B}$. While $\nabla \times \mathbf{A}_e$ is definitely solenoidal, $\mathbf{A}_e$ itself may not be; because of this, (5.17) may not be a decomposition into solenoidal and lamellar components. However, a corollary of the Helmholtz theorem states that a vector field is uniquely specified only when *both* its curl and divergence are specified. Here there is an ambiguity in the representation of $\mathbf{E}$ and $\mathbf{B}$; we may remove this ambiguity and define $\mathbf{A}_e$ uniquely by requiring that

$$\nabla \cdot \mathbf{A}_e = 0. \tag{5.19}$$

Then $\mathbf{A}_e$ is solenoidal and the decomposition (5.17) is solenoidal–lamellar. This requirement on $\mathbf{A}_e$ is called the *Coulomb gauge*.

The ambiguity implied by the nonuniqueness of $\nabla \cdot \mathbf{A}_e$ can also be expressed by the observation that a transformation of the type

$$\mathbf{A}_e \rightarrow \mathbf{A}_e + \nabla\Gamma, \tag{5.20}$$

$$\phi_e \rightarrow \phi_e - \frac{\partial\Gamma}{\partial t},$$

leaves expressions (5.17) and (5.18) unchanged. This is called a *gauge transformation*, and the choice of a certain Γ alters the specification of $\nabla \cdot \mathbf{A}_e$. Thus we may begin with the Coulomb gauge as our baseline, and allow any alteration of $\mathbf{A}_e$ according to (5.20) as long as we augment $\nabla \cdot \mathbf{A}_e$ by $\nabla \cdot \nabla\Gamma = \nabla^2\Gamma$.

Once $\nabla \cdot \mathbf{A}_e$ is specified, the relationship between the potentials and the current $\mathbf{J}$ can be found by substitution of (5.17) and (5.18) into Ampere's law. At this point we assume media that are linear, homogeneous, isotropic, and described by the time-invariant parameters μ, ϵ, and σ. Writing $\mathbf{J} = \mathbf{J}^i + \sigma\mathbf{E}$, we have

$$\frac{1}{\mu}\nabla \times (\nabla \times \mathbf{A}_e) = \mathbf{J}^i - \sigma\frac{\partial \mathbf{A}_e}{\partial t} - \sigma\nabla\phi_e - \epsilon\frac{\partial^2 \mathbf{A}_e}{\partial t^2} - \epsilon\frac{\partial}{\partial t}\nabla\phi_e. \tag{5.21}$$

Taking the divergence of both sides of (5.21), we get

$$0 = \nabla \cdot \mathbf{J}^i - \sigma\frac{\partial}{\partial t}\nabla \cdot \mathbf{A} - \sigma\nabla \cdot \nabla\phi_e - \epsilon\frac{\partial^2}{\partial t^2}\nabla \cdot \mathbf{A}_e - \epsilon\frac{\partial}{\partial t}\nabla \cdot \nabla\phi_e. \tag{5.22}$$

Then, by substitution from the continuity equation and use of (5.19) along with $\nabla \cdot \nabla\phi_e = \nabla^2\phi_e$, we obtain

$$\frac{\partial}{\partial t}\left(\rho^i + \epsilon\nabla^2\phi_e\right) = -\sigma\nabla^2\phi_e.$$

For a lossless medium, this reduces to

$$\nabla^2\phi_e = -\rho^i/\epsilon \tag{5.23}$$

and we have

$$\phi_e(\mathbf{r}, t) = \int_V \frac{\rho^i(\mathbf{r}', t)}{4\pi\epsilon R}\, dV'. \tag{5.24}$$

We can obtain an equation for $\mathbf{A}_e$ by expanding the left side of (5.21) to get

$$\nabla(\nabla \cdot \mathbf{A}_e) - \nabla^2\mathbf{A}_e = \mu\mathbf{J}^i - \sigma\mu\frac{\partial \mathbf{A}_e}{\partial t} - \sigma\mu\nabla\phi_e - \mu\epsilon\frac{\partial^2 \mathbf{A}_e}{\partial t^2} - \mu\epsilon\frac{\partial}{\partial t}\nabla\phi_e, \tag{5.25}$$

hence

$$\nabla^2\mathbf{A}_e - \mu\epsilon\frac{\partial^2 \mathbf{A}_e}{\partial t^2} = -\mu\mathbf{J}^i + \sigma\mu\frac{\partial \mathbf{A}_e}{\partial t} + \sigma\mu\nabla\phi_e + \mu\epsilon\frac{\partial}{\partial t}\nabla\phi_e$$

under the Coulomb gauge. For lossless media this becomes

$$\nabla^2\mathbf{A}_e - \mu\epsilon\frac{\partial^2 \mathbf{A}_e}{\partial t^2} = -\mu\mathbf{J}^i + \mu\epsilon\frac{\partial}{\partial t}\nabla\phi_e. \tag{5.26}$$

Observe that the left side of (5.26) is solenoidal (since the Laplacian term came from the curl-curl, and $\nabla \cdot \mathbf{A}_e = 0$), while the right side contains a general vector field $\mathbf{J}^i$ and a lamellar term. We might expect the $\nabla \phi_e$ term to cancel the lamellar portion of $\mathbf{J}^i$, and this does happen [92]. By (5.12) and the continuity equation we can write the lamellar component of the current as

$$\mathbf{J}_l^i(\mathbf{r},t) = -\nabla \int_V \frac{\nabla' \cdot \mathbf{J}^i(\mathbf{r}',t)}{4\pi R} \, dV' = \frac{\partial}{\partial t}\nabla \int_V \frac{\rho^i(\mathbf{r}',t)}{4\pi R} \, dV' = \epsilon\frac{\partial}{\partial t}\nabla \phi_e.$$

Thus (5.26) becomes

$$\nabla^2 \mathbf{A}_e - \mu\epsilon\frac{\partial^2 \mathbf{A}_e}{\partial t^2} = -\mu\mathbf{J}_s^i. \tag{5.27}$$

Therefore, the vector potential $\mathbf{A}_e$, which describes the solenoidal portion of both $\mathbf{E}$ and $\mathbf{B}$, is found from just the solenoidal portion of the current. On the other hand, the scalar potential, which describes the lamellar portion of $\mathbf{E}$, is found from ρ^i, which arises from $\nabla \cdot \mathbf{J}^i$, the lamellar portion of the current.

From the perspective of field computation, we see that the introduction of potential functions has reoriented the solution process from dealing with two coupled first-order partial differential equations (Maxwell's equations), to two uncoupled second-order equations (the potential equations (5.23) and (5.27)). The decoupling of the equations is often worth the added complexity of dealing with potentials, and, in fact, is the solution technique of choice in such areas as radiation and guided waves. It is worth pausing to examine the form of these equations. We see that the scalar potential obeys Poisson's equation with the solution (5.24), while the vector potential obeys the wave equation. As a wave, the vector potential must propagate away from the source with finite velocity. However, the solution for the scalar potential (5.24) shows no such behavior. In fact, any change to the charge distribution instantaneously permeates all of space. This apparent violation of Einstein's postulate shows that we must be careful when interpreting the potentials physically. Once the computations (5.17) and (5.18) are undertaken, we find that both $\mathbf{E}$ and $\mathbf{B}$ behave as waves, and thus propagate at finite velocity. Mathematically, the conundrum can be resolved by realizing that individually the solenoidal and lamellar components of current must occupy all of space, even if their sum, the actual current $\mathbf{J}^i$, is localized [92].

5.2.2.2 The Lorenz gauge

A different choice of gauge condition can allow both the vector and scalar potentials to act as waves. In this case $\mathbf{E}$ may be written as a sum of two terms: one purely solenoidal, and the other a superposition of lamellar and solenoidal parts.

Let us examine the effect of choosing the *Lorenz gauge*

$$\nabla \cdot \mathbf{A}_e = -\mu\epsilon\frac{\partial \phi_e}{\partial t} - \mu\sigma\phi_e. \tag{5.28}$$

Substituting this expression into (5.25) we find that the gradient terms cancel, giving

$$\nabla^2 \mathbf{A}_e - \mu\sigma\frac{\partial \mathbf{A}_e}{\partial t} - \mu\epsilon\frac{\partial^2 \mathbf{A}_e}{\partial t^2} = -\mu\mathbf{J}^i. \tag{5.29}$$

For lossless media,

$$\nabla^2 \mathbf{A}_e - \mu\epsilon\frac{\partial^2 \mathbf{A}_e}{\partial t^2} = -\mu\mathbf{J}^i, \tag{5.30}$$

and (5.22) becomes

$$\nabla^2 \phi_e - \mu\epsilon \frac{\partial^2 \phi_e}{\partial t^2} = -\frac{\rho^i}{\epsilon}. \tag{5.31}$$

For lossy media we have obtained a second-order differential equation for $\mathbf{A}_e$, but ϕ_e must be found through the somewhat cumbersome relation (5.28). For lossless media the coupled Maxwell equations have been decoupled into two second-order equations, one involving $\mathbf{A}_e$ and one involving ϕ_e. Both (5.30) and (5.31) are wave equations, with $\mathbf{J}^i$ as the source for $\mathbf{A}_e$ and ρ^i as the source for ϕ_e. Thus the expected finite-velocity wave nature of the electromagnetic fields is also manifested in each of the potential functions. The drawback is that, even though we can still use (5.17) and (5.18), the expression for $\mathbf{E}$ is no longer a decomposition into solenoidal and lamellar components. Nevertheless, the choice of the Lorenz gauge is popular in the study of radiated and guided waves.

5.2.3   The Hertzian potentials

With a little manipulation and the introduction of a new notation, we can maintain the wave nature of the potential functions and still provide a decomposition into purely lamellar and solenoidal components. In this analysis we shall assume lossless media only.

When we chose the Lorenz gauge to remove the arbitrariness of the divergence of the vector potential, we established a relationship between $\mathbf{A}_e$ and ϕ_e. Thus we should be able to write both the electric and magnetic fields in terms of a single potential function. From the Lorenz gauge we can write ϕ_e as

$$\phi_e(\mathbf{r}, t) = -\frac{1}{\mu\epsilon} \int_{-\infty}^{t} \nabla \cdot \mathbf{A}_e(\mathbf{r}, t) \, dt.$$

By (5.17) and (5.18) we can thus write the EM fields as

$$\mathbf{E} = \frac{1}{\mu\epsilon} \nabla \int_{-\infty}^{t} \nabla \cdot \mathbf{A}_e dt - \frac{\partial \mathbf{A}_e}{\partial t}, \tag{5.32}$$

$$\mathbf{B} = \nabla \times \mathbf{A}_e. \tag{5.33}$$

The integro-differential representation of $\mathbf{E}$ in (5.32) is somewhat clumsy in appearance. We can make it easier to manipulate by defining the *Hertzian potential*

$$\mathbf{\Pi}_e = \frac{1}{\mu\epsilon} \int_{-\infty}^{t} \mathbf{A}_e \, dt.$$

In differential form,

$$\mathbf{A}_e = \mu\epsilon \frac{\partial \mathbf{\Pi}_e}{\partial t}. \tag{5.34}$$

With this, (5.32) and (5.33) become

$$\mathbf{E} = \nabla(\nabla \cdot \mathbf{\Pi}_e) - \mu\epsilon \frac{\partial^2}{\partial t^2} \mathbf{\Pi}_e, \tag{5.35}$$

$$\mathbf{B} = \mu\epsilon \nabla \times \frac{\partial \mathbf{\Pi}_e}{\partial t}.$$

An equation for $\mathbf{\Pi}_e$ in terms of the source current can be found by substituting (5.34) into (5.30):

$$\mu\epsilon \frac{\partial}{\partial t} \left(\nabla^2 \mathbf{\Pi}_e - \mu\epsilon \frac{\partial^2}{\partial t^2} \mathbf{\Pi}_e \right) = -\mu\mathbf{J}^i.$$

Let us define

$$\mathbf{J}^i = \frac{\partial \mathbf{P}^i}{\partial t}. \tag{5.36}$$

For general impressed current sources, (5.36) is just a convenient notation. However, we can conceive of an *impressed polarization current* that is independent of $\mathbf{E}$ and defined through the relation $\mathbf{D} = \epsilon_0 \mathbf{E} + \mathbf{P} + \mathbf{P}^i$. Then (5.36) has a physical interpretation as described in (2.101). We now have

$$\nabla^2 \mathbf{\Pi}_e - \mu\epsilon \frac{\partial^2}{\partial t^2} \mathbf{\Pi}_e = -\frac{1}{\epsilon} \mathbf{P}^i, \tag{5.37}$$

which is a wave equation for $\mathbf{\Pi}_e$. Thus the Hertzian potential has the same wave behavior as the vector potential under the Lorenz gauge.

We can use (5.37) to perform one final simplification of the EM field representation. By the vector identity $\nabla(\nabla \cdot \mathbf{\Pi}) = \nabla \times (\nabla \times \mathbf{\Pi}) + \nabla^2 \mathbf{\Pi}$ we get

$$\nabla(\nabla \cdot \mathbf{\Pi}_e) = \nabla \times (\nabla \times \mathbf{\Pi}_e) - \frac{1}{\epsilon} \mathbf{P}^i + \mu\epsilon \frac{\partial^2}{\partial t^2} \mathbf{\Pi}_e.$$

Substituting this into (5.35), we obtain

$$\mathbf{E} = \nabla \times (\nabla \times \mathbf{\Pi}_e) - \frac{\mathbf{P}^i}{\epsilon}, \tag{5.38}$$

$$\mathbf{B} = \mu\epsilon \nabla \times \frac{\partial \mathbf{\Pi}_e}{\partial t}.$$

Let us examine these closely. We know that $\mathbf{B}$ is solenoidal since it is written as the curl of another vector (this is also clear from the auxiliary Maxwell equation $\nabla \cdot \mathbf{B} = 0$). The first term in the expression for $\mathbf{E}$ is also solenoidal. So the lamellar part of $\mathbf{E}$ must be contained within the source term $\mathbf{P}^i$. If we write $\mathbf{P}^i$ in terms of its lamellar and solenoidal components by using

$$\mathbf{J}^i_s = \frac{\partial \mathbf{P}^i_s}{\partial t}, \qquad \mathbf{J}^i_l = \frac{\partial \mathbf{P}^i_l}{\partial t},$$

then (5.38) becomes

$$\mathbf{E} = \left[\nabla \times (\nabla \times \mathbf{\Pi}_e) - \frac{\mathbf{P}^i_s}{\epsilon} \right] - \frac{\mathbf{P}^i_l}{\epsilon}.$$

So we have again succeeded in dividing $\mathbf{E}$ into lamellar and solenoidal components.

5.2.4 Potential functions for magnetic current

We can proceed as above to derive the field–potential relationships when $\mathbf{J}^i = 0$ but $\mathbf{J}^i_m \neq 0$. We assume a homogeneous, lossless, isotropic medium with permeability μ and permittivity ϵ, and begin with Faraday's and Ampere's laws

$$\nabla \times \mathbf{E} = -\mathbf{J}^i_m - \frac{\partial \mathbf{B}}{\partial t}, \tag{5.39}$$

$$\nabla \times \mathbf{H} = \frac{\partial \mathbf{D}}{\partial t}.$$

We write $\mathbf{H}$ and $\mathbf{D}$ in terms of two potential functions $\mathbf{A}_h$ and ϕ_h as

$$\mathbf{H} = -\frac{\partial \mathbf{A}_h}{\partial t} - \nabla\phi_h, \qquad \mathbf{D} = -\nabla \times \mathbf{A}_h,$$

and the differential equation for the potentials is found by substitution into (5.39):

$$\nabla \times (\nabla \times \mathbf{A}_h) = \epsilon \mathbf{J}_m^i - \mu\epsilon \frac{\partial^2 \mathbf{A}_h}{\partial t^2} - \mu\epsilon \frac{\partial}{\partial t} \nabla \phi_h. \tag{5.40}$$

Taking the divergence of this equation and substituting from the magnetic continuity equation, we obtain

$$\mu\epsilon \frac{\partial^2}{\partial t^2} \nabla \cdot \mathbf{A}_h + \mu\epsilon \frac{\partial}{\partial t} \nabla^2 \phi_h = -\epsilon \frac{\partial \rho_m^i}{\partial t}.$$

Under the Lorenz gauge condition

$$\nabla \cdot \mathbf{A}_h = -\mu\epsilon \frac{\partial \phi_h}{\partial t},$$

this reduces to

$$\nabla^2 \phi_h - \mu\epsilon \frac{\partial^2 \phi_h}{\partial t^2} = -\frac{\rho_m^i}{\mu}.$$

Expanding the curl-curl operation in (5.40) we have

$$\nabla(\nabla \cdot \mathbf{A}_h) - \nabla^2 \mathbf{A}_h = \epsilon \mathbf{J}_m^i - \mu\epsilon \frac{\partial^2 \mathbf{A}_h}{\partial t^2} - \mu\epsilon \frac{\partial}{\partial t} \nabla \phi_h,$$

which, upon substitution of the Lorenz gauge condition, gives

$$\nabla^2 \mathbf{A}_h - \mu\epsilon \frac{\partial^2 \mathbf{A}_h}{\partial t^2} = -\epsilon \mathbf{J}_m^i. \tag{5.41}$$

We can also derive a Hertzian potential for the case of magnetic current. Letting

$$\mathbf{A}_h = \mu\epsilon \frac{\partial \mathbf{\Pi}_h}{\partial t} \tag{5.42}$$

and employing the Lorenz condition, we have

$$\mathbf{D} = -\mu\epsilon \nabla \times \frac{\partial \mathbf{\Pi}_h}{\partial t}, \qquad \mathbf{H} = \nabla(\nabla \cdot \mathbf{\Pi}_h) - \mu\epsilon \frac{\partial^2 \mathbf{\Pi}_h}{\partial t^2}.$$

The wave equation for $\mathbf{\Pi}_h$ is found by substituting (5.42) into (5.41) to give

$$\frac{\partial}{\partial t} \left[\nabla^2 \mathbf{\Pi}_h - \mu\epsilon \frac{\partial^2 \mathbf{\Pi}_h}{\partial t^2} \right] = -\frac{1}{\mu} \mathbf{J}_m^i. \tag{5.43}$$

Defining $\mathbf{M}^i$ through

$$\mathbf{J}_m^i = \mu \frac{\partial \mathbf{M}^i}{\partial t},$$

we write the wave equation as

$$\nabla^2 \mathbf{\Pi}_h - \mu\epsilon \frac{\partial^2 \mathbf{\Pi}_h}{\partial t^2} = -\mathbf{M}^i.$$

We can think of $\mathbf{M}^i$ as a convenient way of representing $\mathbf{J}_m^i$, or we can conceive of an *impressed magnetization current* that is independent of $\mathbf{H}$ and defined through $\mathbf{B} = \mu_0(\mathbf{H} + \mathbf{M} + \mathbf{M}^i)$. With the help of (5.43) we can also write the fields as

$$\mathbf{H} = \nabla \times (\nabla \times \mathbf{\Pi}_h) - \mathbf{M}_i, \qquad \mathbf{D} = -\mu\epsilon \nabla \times \frac{\partial \mathbf{\Pi}_h}{\partial t}.$$

5.2.5 Summary of potential relations for lossless media

When both electric and magnetic sources are present, we may superpose the potential representations derived above. We assume a homogeneous, lossless medium with time-invariant parameters μ and ϵ. For the scalar/vector potential representation, we have

$$\mathbf{E} = -\frac{\partial \mathbf{A}_e}{\partial t} - \nabla \phi_e - \frac{1}{\epsilon} \nabla \times \mathbf{A}_h, \tag{5.44}$$

$$\mathbf{H} = \frac{1}{\mu} \nabla \times \mathbf{A}_e - \frac{\partial \mathbf{A}_h}{\partial t} - \nabla \phi_h. \tag{5.45}$$

Here the potentials satisfy the wave equations

$$\left(\nabla^2 - \mu\epsilon \frac{\partial^2}{\partial t^2} \right) \left\{ \begin{matrix} \mathbf{A}_e \\ \phi_e \end{matrix} \right\} = \left\{ \begin{matrix} -\mu \mathbf{J}^i \\ -\frac{\rho^i}{\epsilon} \end{matrix} \right\}, \tag{5.46}$$

$$\left(\nabla^2 - \mu\epsilon \frac{\partial^2}{\partial t^2} \right) \left\{ \begin{matrix} \mathbf{A}_h \\ \phi_h \end{matrix} \right\} = \left\{ \begin{matrix} -\epsilon \mathbf{J}_m^i \\ -\frac{\rho_m^i}{\mu} \end{matrix} \right\},$$

and are linked by the Lorenz conditions

$$\nabla \cdot \mathbf{A}_e = -\mu\epsilon \frac{\partial \phi_e}{\partial t}, \tag{5.47}$$

$$\nabla \cdot \mathbf{A}_h = -\mu\epsilon \frac{\partial \phi_h}{\partial t}.$$

We also have the Hertz potential representation

$$\mathbf{E} = \nabla(\nabla \cdot \mathbf{\Pi}_e) - \mu\epsilon \frac{\partial^2 \mathbf{\Pi}_e}{\partial t^2} - \mu \nabla \times \frac{\partial \mathbf{\Pi}_h}{\partial t}$$

$$= \nabla \times (\nabla \times \mathbf{\Pi}_e) - \frac{\mathbf{P}^i}{\epsilon} - \mu \nabla \times \frac{\partial \mathbf{\Pi}_h}{\partial t},$$

$$\mathbf{H} = \epsilon \nabla \times \frac{\partial \mathbf{\Pi}_e}{\partial t} + \nabla(\nabla \cdot \mathbf{\Pi}_h) - \mu\epsilon \frac{\partial^2 \mathbf{\Pi}_h}{\partial t^2}$$

$$= \epsilon \nabla \times \frac{\partial \mathbf{\Pi}_e}{\partial t} + \nabla \times (\nabla \times \mathbf{\Pi}_h) - \mathbf{M}_i.$$

The Hertz potentials satisfy the wave equations

$$\left(\nabla^2 - \mu\epsilon \frac{\partial^2}{\partial t^2} \right) \left\{ \begin{matrix} \mathbf{\Pi}_e \\ \mathbf{\Pi}_h \end{matrix} \right\} = \left\{ \begin{matrix} -\frac{1}{\epsilon} \mathbf{P}^i \\ -\mathbf{M}^i \end{matrix} \right\}.$$

5.2.6 Potential functions for the frequency-domain fields

In the frequency domain it is much easier to handle lossy media. Consider a lossy, isotropic, homogeneous medium described by the frequency-dependent parameters $\tilde{\mu}$, $\tilde{\epsilon}$, and $\tilde{\sigma}$. Maxwell's curl equations are

$$\nabla \times \tilde{\mathbf{E}} = -\tilde{\mathbf{J}}_m^i - j\omega\tilde{\mu}\tilde{\mathbf{H}}, \tag{5.48}$$

$$\nabla \times \tilde{\mathbf{H}} = \tilde{\mathbf{J}}^i + j\omega\tilde{\epsilon}^c\tilde{\mathbf{E}}. \tag{5.49}$$

Here we have separated the primary and secondary currents through $\tilde{\mathbf{J}} = \tilde{\mathbf{J}}^i + \tilde{\sigma}\tilde{\mathbf{E}}$, and used the complex permittivity $\tilde{\epsilon}^c = \tilde{\epsilon} + \tilde{\sigma}/j\omega$. As with the time-domain equations, we introduce the potential functions using superposition. If $\tilde{\mathbf{J}}_m^i = 0$ and $\tilde{\mathbf{J}}^i \neq 0$, then we may introduce the electric potentials through the relationships

$$\tilde{\mathbf{E}} = -\nabla\tilde{\phi}_e - j\omega\tilde{\mathbf{A}}_e, \tag{5.50}$$

$$\tilde{\mathbf{H}} = \frac{1}{\tilde{\mu}}\nabla \times \tilde{\mathbf{A}}_e. \tag{5.51}$$

Assuming the Lorenz condition

$$\nabla \cdot \tilde{\mathbf{A}}_e = -j\omega\tilde{\mu}\tilde{\epsilon}^c\tilde{\phi}_e,$$

we find that upon substitution of (5.50)–(5.51) into (5.48)–(5.49) the potentials must obey the Helmholtz equation

$$(\nabla^2 + k^2)\left\{\begin{matrix} \tilde{\phi}_e \\ \tilde{\mathbf{A}}_e \end{matrix}\right\} = \left\{\begin{matrix} -\tilde{\rho}^i/\tilde{\epsilon}^c \\ -\tilde{\mu}\tilde{\mathbf{J}}^i \end{matrix}\right\}.$$

If $\tilde{\mathbf{J}}_m^i \neq 0$ and $\tilde{\mathbf{J}}^i = 0$, then we may introduce the magnetic potentials through

$$\tilde{\mathbf{E}} = -\frac{1}{\tilde{\epsilon}^c}\nabla \times \tilde{\mathbf{A}}_h, \tag{5.52}$$

$$\tilde{\mathbf{H}} = -\nabla\tilde{\phi}_h - j\omega\tilde{\mathbf{A}}_h. \tag{5.53}$$

Assuming $\nabla \cdot \tilde{\mathbf{A}}_h = -j\omega\tilde{\mu}\tilde{\epsilon}^c\tilde{\phi}_h$, we find that upon substitution of (5.52)–(5.53) into (5.48)–(5.49), the potentials must obey

$$(\nabla^2 + k^2)\left\{\begin{matrix} \tilde{\phi}_h \\ \tilde{\mathbf{A}}_h \end{matrix}\right\} = \left\{\begin{matrix} -\tilde{\rho}_m^i/\tilde{\mu} \\ -\tilde{\epsilon}^c\tilde{\mathbf{J}}_m^i \end{matrix}\right\}.$$

When both electric and magnetic sources are present, we use superposition:

$$\tilde{\mathbf{E}} = -\nabla\tilde{\phi}_e - j\omega\tilde{\mathbf{A}}_e - \frac{1}{\tilde{\epsilon}^c}\nabla \times \tilde{\mathbf{A}}_h,$$

$$\tilde{\mathbf{H}} = \frac{1}{\tilde{\mu}}\nabla \times \tilde{\mathbf{A}}_e - \nabla\tilde{\phi}_h - j\omega\tilde{\mathbf{A}}_h.$$

Using the Lorenz conditions, we can also write the fields in terms of the vector potentials alone:

$$\tilde{\mathbf{E}} = -\frac{j\omega}{k^2}\nabla(\nabla \cdot \tilde{\mathbf{A}}_e) - j\omega\tilde{\mathbf{A}}_e - \frac{1}{\tilde{\epsilon}^c}\nabla \times \tilde{\mathbf{A}}_h, \tag{5.54}$$

$$\tilde{\mathbf{H}} = \frac{1}{\tilde{\mu}}\nabla \times \tilde{\mathbf{A}}_e - \frac{j\omega}{k^2}\nabla(\nabla \cdot \tilde{\mathbf{A}}_h) - j\omega\tilde{\mathbf{A}}_h. \tag{5.55}$$

We can also define Hertzian potentials for the frequency-domain fields. When $\tilde{\mathbf{J}}_m^i = 0$ and $\tilde{\mathbf{J}}^i \neq 0$, we let

$$\tilde{\mathbf{A}}_e = j\omega\tilde{\mu}\tilde{\epsilon}^c\tilde{\mathbf{\Pi}}_e$$

and find

$$\tilde{\mathbf{E}} = \nabla(\nabla \cdot \tilde{\mathbf{\Pi}}_e) + k^2\tilde{\mathbf{\Pi}}_e = \nabla \times (\nabla \times \tilde{\mathbf{\Pi}}_e) - \frac{\tilde{\mathbf{J}}^i}{j\omega\tilde{\epsilon}^c} \tag{5.56}$$

and

$$\tilde{\mathbf{H}} = j\omega\tilde{\epsilon}^c\nabla \times \tilde{\mathbf{\Pi}}_e. \tag{5.57}$$

Here $\tilde{\mathbf{J}}^i$ can represent either an impressed electric current source or an impressed polarization current source $\tilde{\mathbf{J}}^i = j\omega\tilde{\mathbf{P}}^i$. The electric Hertzian potential obeys

$$(\nabla^2 + k^2)\tilde{\boldsymbol{\Pi}}_e = -\frac{\tilde{\mathbf{J}}^i}{j\omega\tilde{\epsilon}^c}. \tag{5.58}$$

When $\tilde{\mathbf{J}}^i_m \neq 0$ and $\tilde{\mathbf{J}}^i = 0$, we let

$$\tilde{\mathbf{A}}_h = j\omega\tilde{\mu}\tilde{\epsilon}^c\tilde{\boldsymbol{\Pi}}_h$$

and find

$$\tilde{\mathbf{E}} = -j\omega\tilde{\mu}\nabla \times \tilde{\boldsymbol{\Pi}}_h \tag{5.59}$$

and

$$\tilde{\mathbf{H}} = \nabla(\nabla \cdot \tilde{\boldsymbol{\Pi}}_h) + k^2\tilde{\boldsymbol{\Pi}}_h = \nabla \times (\nabla \times \tilde{\boldsymbol{\Pi}}_h) - \frac{\tilde{\mathbf{J}}^i_m}{j\omega\tilde{\mu}}. \tag{5.60}$$

Here $\tilde{\mathbf{J}}^i_m$ can represent either an impressed magnetic current source or an impressed magnetization current source $\tilde{\mathbf{J}}^i_m = j\omega\tilde{\mu}\tilde{\mathbf{M}}^i$. The magnetic Hertzian potential obeys

$$(\nabla^2 + k^2)\tilde{\boldsymbol{\Pi}}_h = -\frac{\tilde{\mathbf{J}}^i_m}{j\omega\tilde{\mu}}.$$

If both electric and magnetic sources are present, then by superposition,

$$\tilde{\mathbf{E}} = \nabla(\nabla \cdot \tilde{\boldsymbol{\Pi}}_e) + k^2\tilde{\boldsymbol{\Pi}}_e - j\omega\tilde{\mu}\nabla \times \tilde{\boldsymbol{\Pi}}_h$$

$$= \nabla \times (\nabla \times \tilde{\boldsymbol{\Pi}}_e) - \frac{\tilde{\mathbf{J}}^i}{j\omega\tilde{\epsilon}^c} - j\omega\tilde{\mu}\nabla \times \tilde{\boldsymbol{\Pi}}_h$$

and

$$\tilde{\mathbf{H}} = j\omega\tilde{\epsilon}^c\nabla \times \tilde{\boldsymbol{\Pi}}_e + \nabla(\nabla \cdot \tilde{\boldsymbol{\Pi}}_h) + k^2\tilde{\boldsymbol{\Pi}}_h$$

$$= j\omega\tilde{\epsilon}^c\nabla \times \tilde{\boldsymbol{\Pi}}_e + \nabla \times (\nabla \times \tilde{\boldsymbol{\Pi}}_h) - \frac{\tilde{\mathbf{J}}^i_m}{j\omega\tilde{\mu}}.$$

5.2.7 Solution for potentials in an unbounded medium: the retarded potentials

Under the Lorenz condition, each of the potential functions obeys the wave equation. This equation can be solved using the method of Green's functions to determine the potentials, and the electromagnetic fields can therefore be determined. We now examine the solution for an unbounded medium. Solutions for bounded regions are considered in § 5.2.9.

Consider a linear operator $\mathcal{L}$ that operates on a function of $\mathbf{r}$ and t. If we wish to solve the equation

$$\mathcal{L}\{\psi(\mathbf{r}, t)\} = S(\mathbf{r}, t), \tag{5.61}$$

we first solve

$$\mathcal{L}\{G(\mathbf{r}, t|\mathbf{r}', t')\} = \delta(\mathbf{r} - \mathbf{r}')\delta(t - t')$$

and determine the Green's function G for the operator $\mathcal{L}$. Provided that S resides within V, we have

$$\mathcal{L}\left\{\int_V \int_{-\infty}^{\infty} S(\mathbf{r}',t')G(\mathbf{r},t|\mathbf{r}',t')\,dt'\,dV'\right\} = \int_V \int_{-\infty}^{\infty} S(\mathbf{r}',t')\,\mathcal{L}\{G(\mathbf{r},t|\mathbf{r}',t')\}\,dt'\,dV'$$

$$= \int_V \int_{-\infty}^{\infty} S(\mathbf{r}',t')\delta(\mathbf{r}-\mathbf{r}')\delta(t-t')\,dt'\,dV'$$

$$= S(\mathbf{r},t),$$

hence

$$\psi(\mathbf{r},t) = \int_V \int_{-\infty}^{\infty} S(\mathbf{r}',t')G(\mathbf{r},t|\mathbf{r}',t')\,dt'\,dV' \qquad (5.62)$$

by comparison with (5.61).

We can also apply this idea in the frequency domain. The solution to

$$\mathcal{L}\{\tilde{\psi}(\mathbf{r},\omega)\} = \tilde{S}(\mathbf{r},\omega) \qquad (5.63)$$

is

$$\tilde{\psi}(\mathbf{r},\omega) = \int_V \tilde{S}(\mathbf{r}',\omega)\tilde{G}(\mathbf{r}|\mathbf{r}';\omega)\,dV'$$

where the Green's function $\tilde{G}$ satisfies

$$\mathcal{L}\{\tilde{G}(\mathbf{r}|\mathbf{r}';\omega)\} = \delta(\mathbf{r}-\mathbf{r}').$$

Equation (5.62) is the basic superposition integral that allows us to find the potentials in an infinite, unbounded medium. If the medium is bounded, we must use Green's theorem to include the effects of sources that reside external to the boundaries. These are manifested in terms of the values of the potentials on the boundaries in the same manner as with the static potentials in Chapter 3. In order to determine whether (5.62) is the unique solution to the wave equation, we must also examine the behavior of the fields on the boundary as the boundary recedes to infinity. In the frequency domain we find that an additional "radiation condition" is required to ensure uniqueness.

5.2.7.1   The retarded potentials in the time domain

Consider an unbounded, homogeneous, lossy, isotropic medium described by parameters μ,ϵ,σ. In the time domain the vector potential $\mathbf{A}_e$ satisfies (5.29). The scalar components of $\mathbf{A}_e$ must obey

$$\nabla^2 A_{e,n}(\mathbf{r},t) - \mu\sigma\frac{\partial A_{e,n}(\mathbf{r},t)}{\partial t} - \mu\epsilon\frac{\partial^2 A_{e,n}(\mathbf{r},t)}{\partial t^2} = -\mu J_n^i(\mathbf{r},t) \qquad (n=x,y,z).$$

We may write this in the form

$$\left(\nabla^2 - \frac{2\Omega}{v^2}\frac{\partial}{\partial t} - \frac{1}{v^2}\frac{\partial^2}{\partial t^2}\right)\psi(\mathbf{r},t) = -S(\mathbf{r},t) \qquad (5.64)$$

where $\psi = A_{e,n}$, $v^2 = 1/\mu\epsilon$, $\Omega = \sigma/2\epsilon$, and $S = \mu J_n^i$. The solution is

$$\psi(\mathbf{r},t) = \int_V \int_{-\infty}^{\infty} S(\mathbf{r}',t')G(\mathbf{r},t|\mathbf{r}',t')\,dt'\,dV'$$

where G satisfies

$$\left(\nabla^2 - \frac{2\Omega}{v^2}\frac{\partial}{\partial t} - \frac{1}{v^2}\frac{\partial^2}{\partial t^2}\right)G(\mathbf{r}, t|\mathbf{r}', t') = -\delta(\mathbf{r} - \mathbf{r}')\delta(t - t').$$

In § A.2.6, we find that

$$G(\mathbf{r}, t|\mathbf{r}', t') = e^{-\Omega(t-t')}\frac{\delta(t - t' - R/v)}{4\pi R}$$

$$+ \frac{\Omega^2}{4\pi v}e^{-\Omega(t-t')}\frac{I_1\left(\Omega\sqrt{(t - t')^2 - (R/v)^2}\right)}{\Omega\sqrt{(t - t')^2 - (R/v)^2}} \qquad (t - t' > R/v),$$

where $R = |\mathbf{r} - \mathbf{r}'|$. For lossless media where $\sigma = 0$, this becomes

$$G(\mathbf{r}, t|\mathbf{r}', t') = \frac{\delta(t - t' - R/v)}{4\pi R}$$

and thus

$$\psi(\mathbf{r}, t) = \int_V \int_{-\infty}^{\infty} S(\mathbf{r}', t')\frac{\delta(t - t' - R/v)}{4\pi R}\,dt'\,dV'$$

$$= \int_V \frac{S(\mathbf{r}', t - R/v)}{4\pi R}\,dV'. \tag{5.65}$$

For lossless media, the scalar potentials and all rectangular components of the vector potentials obey the same wave equation. Thus, for instance, the solutions to (5.46) are

$$\mathbf{A}_e(\mathbf{r}, t) = \frac{\mu}{4\pi}\int_V \frac{\mathbf{J}^i(\mathbf{r}', t - R/v)}{R}\,dV', \tag{5.66}$$

$$\phi_e(\mathbf{r}, t) = \frac{1}{4\pi\epsilon}\int_V \frac{\rho^i(\mathbf{r}', t - R/v)}{R}\,dV'. \tag{5.67}$$

These are called the *retarded potentials*, since their values at time t are determined by the values of the sources at an earlier (or retardation) time $t - R/v$. The retardation time is determined by the propagation velocity v of the potential waves.

The fields are determined by the potentials:

$$\mathbf{E}(\mathbf{r}, t) = -\nabla\frac{1}{4\pi\epsilon}\int_V \frac{\rho^i(\mathbf{r}', t - R/v)}{R}\,dV' - \frac{\partial}{\partial t}\frac{\mu}{4\pi}\int_V \frac{\mathbf{J}^i(\mathbf{r}', t - R/v)}{R}\,dV',$$

$$\mathbf{H}(\mathbf{r}, t) = \nabla \times \frac{1}{4\pi}\int_V \frac{\mathbf{J}^i(\mathbf{r}', t - R/v)}{R}\,dV'.$$

The derivatives may be brought inside the integrals, but some care must be taken when the observation point $\mathbf{r}$ lies within the source region. In this case the integrals must be performed in a principal value sense by excluding a small volume around the observation point. We discuss this in more detail below for the frequency-domain fields. For details regarding this procedure in the time domain, the reader may see Hansen [79].

5.2.7.2 The retarded potentials in the frequency domain

Consider an unbounded, homogeneous, isotropic medium described by $\tilde{\mu}(\omega)$ and $\tilde{\epsilon}^c(\omega)$. If $\tilde{\psi}(\mathbf{r}, \omega)$ represents a scalar potential or any rectangular component of a vector or Hertzian potential, then it must satisfy

$$(\nabla^2 + k^2)\tilde{\psi}(\mathbf{r}, \omega) = -\tilde{S}(\mathbf{r}, \omega) \tag{5.68}$$

where $k = \omega(\tilde{\mu}\tilde{\epsilon}^c)^{1/2}$. This Helmholtz equation has the form of (5.63), and thus

$$\tilde{\psi}(\mathbf{r}, \omega) = \int_V \tilde{S}(\mathbf{r}', \omega)\tilde{G}(\mathbf{r}|\mathbf{r}'; \omega)\, dV'$$

where

$$(\nabla^2 + k^2)\tilde{G}(\mathbf{r}|\mathbf{r}'; \omega) = -\delta(\mathbf{r} - \mathbf{r}'). \tag{5.69}$$

This is Equation (A.47) and its solution, as given by (A.50), is

$$\tilde{G}(\mathbf{r}|\mathbf{r}'; \omega) = \frac{e^{-jkR}}{4\pi R}. \tag{5.70}$$

Here we use $v^2 = 1/\tilde{\mu}\tilde{\epsilon}$ and $\Omega = \tilde{\sigma}/2\epsilon$ in (A.48):

$$k = \frac{1}{v}\sqrt{\omega^2 - j2\omega\Omega} = \omega\sqrt{\tilde{\mu}\left(\tilde{\epsilon} - j\frac{\tilde{\sigma}}{\omega}\right)} = \omega\sqrt{\tilde{\mu}\tilde{\epsilon}^c}.$$

The solution to (5.68) is therefore

$$\tilde{\psi}(\mathbf{r}, \omega) = \int_V \tilde{S}(\mathbf{r}', \omega)\frac{e^{-jkR}}{4\pi R}\, dV'. \tag{5.71}$$

When the medium is lossless, the potential must also satisfy the *radiation condition*

$$\lim_{r\to\infty} r\left(\frac{\partial}{\partial r} + jk\right)\tilde{\psi}(\mathbf{r}) = 0 \tag{5.72}$$

to guarantee uniqueness of solution. In § 5.2.9 we shall show how this requirement arises from the solution within a bounded region. For a uniqueness proof for the Helmholtz equation, the reader may consult Chew [35].

We may use (5.71) to find that

$$\tilde{\mathbf{A}}_e(\mathbf{r}, \omega) = \frac{\tilde{\mu}}{4\pi} \int_V \tilde{\mathbf{J}}^i(\mathbf{r}', \omega)\frac{e^{-jkR}}{R}\, dV'.$$

Comparison with (5.65) shows that in the frequency domain, time retardation takes the form of a phase shift. Similarly,

$$\tilde{\phi}(\mathbf{r}, \omega) = \frac{1}{4\pi\tilde{\epsilon}^c} \int_V \tilde{\rho}^i(\mathbf{r}', \omega)\frac{e^{-jkR}}{R}\, dV'.$$

▶ **Example 5.3:** Fields of a Hertzian dipole

Figure 5.2 depicts a short electric line current of length $l \ll \lambda$ at position $\mathbf{r}_p$, oriented along a direction $\hat{\mathbf{p}}$ in a medium with constitutive parameters $\tilde{\mu}(\omega)$, $\tilde{\epsilon}^c(\omega)$. Use the vector potential to find the electric and magnetic fields produced by the source. Repeat for a short magnetic current.

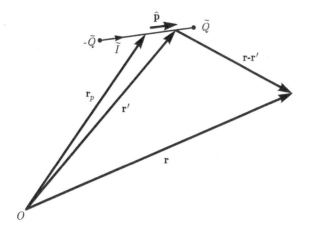

FIGURE 5.2
Geometry of an electric Hertzian dipole.

Solution: We assume that the frequency-domain current $\tilde{I}(\omega)$ is independent of position, and therefore this *Hertzian dipole* must be terminated by point charges

$$\tilde{Q}(\omega) = \pm\frac{\tilde{I}(\omega)}{j\omega}$$

as required by the continuity equation. The electric vector potential produced by this short current element is

$$\tilde{\mathbf{A}}_e = \frac{\tilde{\mu}}{4\pi}\int_\Gamma \tilde{I}\hat{\mathbf{p}}\frac{e^{-jkR}}{R}\,dl'.$$

At observation points far from the dipole (compared to its length) such that $|\mathbf{r} - \mathbf{r}_p| \gg l$, we may approximate

$$\frac{e^{-jkR}}{R} \approx \frac{e^{-jk|\mathbf{r}-\mathbf{r}_p|}}{|\mathbf{r} - \mathbf{r}_p|}.$$

Then

$$\tilde{\mathbf{A}}_e = \hat{\mathbf{p}}\tilde{\mu}\tilde{I}\tilde{G}(\mathbf{r}|\mathbf{r}_p;\omega)\int_\Gamma dl' = \hat{\mathbf{p}}\tilde{\mu}\tilde{I}l\tilde{G}(\mathbf{r}|\mathbf{r}_p;\omega). \qquad (5.73)$$

Note that we obtain the same answer if we let the current density of the dipole be

$$\tilde{\mathbf{J}} = j\omega\tilde{\mathbf{p}}\delta(\mathbf{r} - \mathbf{r}_p)$$

where $\tilde{\mathbf{p}}$ is the *dipole moment* defined by

$$\tilde{\mathbf{p}} = \tilde{Q}l\hat{\mathbf{p}} = \frac{\tilde{I}l}{j\omega}\hat{\mathbf{p}}.$$

That is, we consider a Hertzian dipole to be a "point source" of electromagnetic radiation. With this notation we have

$$\tilde{\mathbf{A}}_e = \tilde{\mu}\int_V \left[j\omega\tilde{\mathbf{p}}\delta(\mathbf{r}' - \mathbf{r}_p)\right]\tilde{G}(\mathbf{r}|\mathbf{r}';\omega)\,dV' = j\omega\tilde{\mu}\tilde{\mathbf{p}}\tilde{G}(\mathbf{r}|\mathbf{r}_p;\omega),$$

which is identical to (5.73). The electromagnetic fields are then

$$\tilde{\mathbf{H}}(\mathbf{r},\omega) = j\omega\nabla \times [\tilde{\mathbf{p}}\tilde{G}(\mathbf{r}|\mathbf{r}_p;\omega)], \qquad (5.74)$$

and

$$\tilde{\mathbf{E}}(\mathbf{r}, \omega) = \frac{1}{\tilde{\epsilon}^c} \nabla \times \nabla \times [\tilde{\mathbf{p}} \tilde{G}(\mathbf{r}|\mathbf{r}_p; \omega)]. \tag{5.75}$$

Here we have obtained $\tilde{\mathbf{E}}$ from $\tilde{\mathbf{H}}$ outside the source region by applying Ampere's law. By duality we may obtain the fields produced by a magnetic Hertzian dipole of moment

$$\tilde{\mathbf{p}}_m = \frac{\tilde{I}_m l}{j\omega} \hat{\mathbf{p}}$$

located at $\mathbf{r} = \mathbf{r}_p$ as

$$\tilde{\mathbf{E}}(\mathbf{r}, \omega) = -j\omega \nabla \times [\tilde{\mathbf{p}}_m \tilde{G}(\mathbf{r}|\mathbf{r}_p; \omega)],$$

$$\tilde{\mathbf{H}}(\mathbf{r}, \omega) = \frac{1}{\tilde{\mu}} \nabla \times \nabla \times [\tilde{\mathbf{p}}_m \tilde{G}(\mathbf{r}|\mathbf{r}_p; \omega)]. \ \blacktriangleleft$$

▶ **Example 5.4:** Near-zone and far-zone fields of a Hertzian dipole

Consider the special case of a Hertzian dipole located on the z-axis and centered on the origin. Identify the terms in the field expressions that are dominant very near to the dipole, and those that are dominant very far from the dipole.

Solution: Using $\hat{\mathbf{p}} = \hat{\mathbf{z}}$ and $\mathbf{r}_p = 0$ in (5.74), we find that

$$\tilde{\mathbf{H}}(\mathbf{r}, \omega) = j\omega \nabla \times \left[\hat{\mathbf{z}} \frac{\tilde{I}}{j\omega} l \frac{e^{-jkr}}{4\pi r} \right] = \hat{\boldsymbol{\phi}} \frac{1}{4\pi} \tilde{I} l k^2 \left[\frac{1}{(kr)^2} + j \frac{1}{kr} \right] \sin\theta e^{-jkr}. \tag{5.76}$$

By Ampere's law,

$$\tilde{\mathbf{E}}(\mathbf{r}, \omega) = \frac{1}{j\omega\tilde{\epsilon}^c} \nabla \times \tilde{\mathbf{H}}(\mathbf{r}, \omega) = \hat{\mathbf{r}} \frac{\eta}{4\pi} \tilde{I} l k^2 \left[\frac{2}{(kr)^2} - j \frac{2}{(kr)^3} \right] \cos\theta e^{-jkr}$$

$$+ \hat{\boldsymbol{\theta}} \frac{\eta}{4\pi} \tilde{I} l k^2 \left[j \frac{1}{kr} + \frac{1}{(kr)^2} - j \frac{1}{(kr)^3} \right] \sin\theta e^{-jkr}. \tag{5.77}$$

The fields involve various inverse powers of r, with the $1/r$ and $1/r^3$ terms $90°$ out of phase from the $1/r^2$ term. Some terms dominate the field close to the source, while others dominate far away. Assume the dipole is so small that $r \gg l$ at all points of interest. We then say that an observation point is in the *near-zone* region of the dipole when $kr \ll 1$. The terms that are dominant near the dipole are the *near-zone fields* or *induction-zone fields*:

$$\tilde{\mathbf{H}}^{NZ}(\mathbf{r}, \omega) = \hat{\boldsymbol{\phi}} \frac{\tilde{I} l k^2}{4\pi} \frac{e^{-jkr}}{(kr)^2} \sin\theta,$$

$$\tilde{\mathbf{E}}^{NZ}(\mathbf{r}, \omega) = -j\eta \frac{\tilde{I} l k^2}{4\pi} \frac{e^{-jkr}}{(kr)^3} \left[2\hat{\mathbf{r}} \cos\theta + \hat{\boldsymbol{\theta}} \sin\theta \right].$$

Note that $\tilde{\mathbf{H}}^{NZ}$ and $\tilde{\mathbf{E}}^{NZ}$ are $90°$ out of phase. Also, the electric field has the same spatial dependence as the field of a static electric dipole. The terms that dominate far from the source ($kr \gg 1$) are called the *far-zone* or *radiation fields*:

$$\tilde{\mathbf{H}}^{FZ}(\mathbf{r}, \omega) = \hat{\boldsymbol{\phi}} \frac{jk^2 \tilde{I} l}{4\pi} \frac{e^{-jkr}}{kr} \sin\theta,$$

$$\tilde{\mathbf{E}}^{FZ}(\mathbf{r}, \omega) = \hat{\boldsymbol{\theta}} \eta \frac{jk^2 \tilde{I} l}{4\pi} \frac{e^{-jkr}}{kr} \sin\theta.$$

The far-zone fields are in-phase and in fact form a TEM spherical wave with

$$\tilde{\mathbf{H}}^{FZ} = \frac{\hat{\mathbf{r}} \times \tilde{\mathbf{E}}^{FZ}}{\eta}. \qquad (5.78)$$

The dipole field is the first term in a general expansion of the electromagnetic fields in terms of the multipole moments of the sources. Either a Taylor expansion or a spherical-harmonic expansion may be used. See Papas [143] for details. ◄

▶ **Example 5.5:** Power radiated by a Hertzian dipole

Compute the power radiated by a Hertzian dipole in free space.

Solution: We speak of the time-average power *radiated* by a time-harmonic source as the integral of the time-average power density over a large sphere. Thus *radiated power* is the power delivered by the sources to infinity. If the dipole is situated within a lossy medium, all of the time-average power delivered by the sources is dissipated by the medium. If the medium is lossless then all the time-average power is delivered to infinity.

For a time-harmonic Hertzian dipole immersed in a lossless medium, we write (5.76) and (5.77) in terms of phasors and compute the complex Poynting vector

$$\mathbf{S}^c(\mathbf{r}) = \check{\mathbf{E}}(\mathbf{r}) \times \check{\mathbf{H}}^*(\mathbf{r})$$

$$= \hat{\boldsymbol{\theta}}\eta \left(\frac{|\check{I}|l}{4\pi} \right)^2 j \frac{2}{kr^5} \left[k^2 r^2 + 1 \right] \cos\theta \sin\theta$$

$$+ \hat{\mathbf{r}}\eta \left(\frac{|\check{I}|l}{4\pi} \right)^2 \frac{k^2}{r^2} \left[1 - j \frac{1}{k^3 r^5} \right] \sin^2\theta.$$

The θ-component of $\mathbf{S}^c$ is purely imaginary and gives rise to no time-average power flux; it decays as $1/r^3$ for large r and produces no net flux through a sphere of radius $r \to \infty$. Additionally, the angular variation $\cos\theta \sin\theta$ integrates to zero over a sphere. In contrast, the r-component has a real part that varies as $1/r^2$ and as $\sin^2\theta$. Hence the total time-average power passing through a sphere expanding to infinity is nonzero:

$$P_{av} = \lim_{r \to \infty} \int_0^{2\pi} \int_0^{\pi} \tfrac{1}{2}\mathrm{Re}\left\{ \hat{\mathbf{r}}\eta \left(\frac{|\check{I}|l}{4\pi} \right)^2 \frac{k^2}{r^2} \sin^2\theta \right\} \cdot \hat{\mathbf{r}} r^2 \sin\theta \, d\theta \, d\phi = \eta \frac{\pi}{3} |\check{I}|^2 \left(\frac{l}{\lambda} \right)^2$$

where $\lambda = 2\pi/k$ is the wavelength in the lossless medium. This is the power radiated by the Hertzian dipole.

Note that the radiated power is proportional to $|\check{I}|^2$ as it is in a circuit, and thus we may define a *radiation resistance*

$$R_r = \frac{2P_{av}}{|\check{I}|^2} = \eta \frac{2\pi}{3} \left(\frac{l}{\lambda} \right)^2$$

representing the resistance of a lumped element that would absorb the same power as radiated by the Hertzian dipole when carrying the same current. We also note that the power radiated by a Hertzian dipole (and, in fact, by any source of finite extent) may be calculated directly from its far-zone fields. In fact, from (5.78) we have the simple formula for the time-average power density in lossless media

$$\mathbf{S}_{av} = \tfrac{1}{2}\mathrm{Re}\left\{ \check{\mathbf{E}}^{FZ} \times \check{\mathbf{H}}^{FZ*} \right\} = \hat{\mathbf{r}} \frac{|\check{\mathbf{E}}^{FZ}|^2}{2\eta}.$$

The concepts of radiated power and radiation resistance are treated in detail in § 6.4. ◄

5.2.8   The electric and magnetic dyadic Green's functions

The frequency-domain electromagnetic fields may be found for electric sources from the electric vector potential using (5.54) and (5.55):

$$\tilde{\mathbf{E}}(\mathbf{r},\omega) = -j\omega\tilde{\mu}(\omega)\int_V \tilde{\mathbf{J}}^i(\mathbf{r}',\omega)\tilde{G}(\mathbf{r}|\mathbf{r}';\omega)\,dV' - \frac{j\omega\tilde{\mu}(\omega)}{k^2}\nabla\nabla\cdot\int_V \tilde{\mathbf{J}}^i(\mathbf{r}',\omega)\tilde{G}(\mathbf{r}|\mathbf{r}';\omega)\,dV',$$

$$\tilde{\mathbf{H}} = \nabla\times\int_V \tilde{\mathbf{J}}^i(\mathbf{r}',\omega)\tilde{G}(\mathbf{r}|\mathbf{r}';\omega)\,dV'. \tag{5.79}$$

Provided the observation point $\mathbf{r}$ does not lie within the source region, we may take the derivatives inside the integrals. Using

$$\nabla\cdot\left[\tilde{\mathbf{J}}^i(\mathbf{r}',\omega)\tilde{G}(\mathbf{r}|\mathbf{r}';\omega)\right] = \tilde{\mathbf{J}}^i(\mathbf{r}',\omega)\cdot\nabla\tilde{G}(\mathbf{r}|\mathbf{r}';\omega) + \tilde{G}(\mathbf{r}|\mathbf{r}';\omega)\nabla\cdot\tilde{\mathbf{J}}(\mathbf{r}',\omega)$$

$$= \nabla\tilde{G}(\mathbf{r}|\mathbf{r}';\omega)\cdot\tilde{\mathbf{J}}^i(\mathbf{r}',\omega)$$

we have

$$\tilde{\mathbf{E}}(\mathbf{r},\omega) = -j\omega\tilde{\mu}(\omega)\int_V\left\{\tilde{\mathbf{J}}^i(\mathbf{r}',\omega)\tilde{G}(\mathbf{r}|\mathbf{r}';\omega) + \frac{1}{k^2}\nabla\left[\nabla\tilde{G}(\mathbf{r}|\mathbf{r}';\omega)\cdot\mathbf{J}^i(\mathbf{r}',\omega)\right]\right\}dV'.$$

This can be written more compactly as

$$\tilde{\mathbf{E}}(\mathbf{r},\omega) = -j\omega\tilde{\mu}(\omega)\int_V \bar{\mathbf{G}}_e(\mathbf{r}|\mathbf{r}';\omega)\cdot\tilde{\mathbf{J}}^i(\mathbf{r}',\omega)\,dV'$$

where

$$\bar{\mathbf{G}}_e(\mathbf{r}|\mathbf{r}';\omega) = \left[\bar{\mathbf{I}} + \frac{\nabla\nabla}{k^2}\right]\tilde{G}(\mathbf{r}|\mathbf{r}';\omega) \tag{5.80}$$

is called the *electric dyadic Green's function*. Using

$$\nabla\times[\tilde{\mathbf{J}}^i\tilde{G}] = \nabla\tilde{G}\times\tilde{\mathbf{J}}^i + \tilde{G}\nabla\times\tilde{\mathbf{J}}^i = \nabla\tilde{G}\times\tilde{\mathbf{J}}^i,$$

we have for the magnetic field

$$\tilde{\mathbf{H}}(\mathbf{r},\omega) = \int_V \nabla\tilde{G}(\mathbf{r}|\mathbf{r}';\omega)\times\tilde{\mathbf{J}}^i(\mathbf{r}',\omega)\,dV'.$$

Now, using the dyadic identity (B.16), we may show that

$$\tilde{\mathbf{J}}^i\times\nabla\tilde{G} = (\tilde{\mathbf{J}}^i\times\nabla\tilde{G})\cdot\bar{\mathbf{I}} = (\nabla\tilde{G}\times\bar{\mathbf{I}})\cdot\mathbf{J}^i.$$

So

$$\tilde{\mathbf{H}}(\mathbf{r},\omega) = -\int_V \bar{\mathbf{G}}_m(\mathbf{r}|\mathbf{r}';\omega)\cdot\tilde{\mathbf{J}}^i(\mathbf{r}',\omega)\,dV'$$

where

$$\bar{\mathbf{G}}_m(\mathbf{r}|\mathbf{r}';\omega) = \nabla\tilde{G}(\mathbf{r}|\mathbf{r}';\omega)\times\bar{\mathbf{I}} \tag{5.81}$$

is called the *magnetic dyadic Green's function*.

Proceeding similarly for magnetic sources (or using duality), we have

$$\tilde{\mathbf{H}}(\mathbf{r}) = -j\omega\tilde{\epsilon}^c\int_V \bar{\mathbf{G}}_e(\mathbf{r}|\mathbf{r}';\omega)\cdot\tilde{\mathbf{J}}_m^i(\mathbf{r}',\omega)\,dV',$$

$$\tilde{\mathbf{E}}(\mathbf{r}) = \int_V \bar{\mathbf{G}}_m(\mathbf{r}|\mathbf{r}';\omega)\cdot\tilde{\mathbf{J}}_m^i(\mathbf{r}',\omega)\,dV'.$$

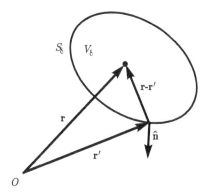

FIGURE 5.3

Geometry of excluded region used to compute the electric field within a source region.

When both electric and magnetic sources are present, we simply use superposition and add the fields.

When the observation point lies within the source region, we must be more careful when formulating the dyadic Green's functions. In (5.79) we encounter the integral

$$\int_V \tilde{\mathbf{J}}^i(\mathbf{r}',\omega)\tilde{G}(\mathbf{r}|\mathbf{r}';\omega)\,dV'.$$

If $\mathbf{r}$ lies within the source region, then $\tilde{G}$ is singular, since $R \to 0$ when $\mathbf{r} \to \mathbf{r}'$. However, the integral converges and the potentials exist within the source region. While trouble arises from passing both derivatives in the operator $\nabla\nabla\cdot$ through the integral and allowing them to operate on $\tilde{G}$, as differentiation of $\tilde{G}$ increases the order of the singularity, we may safely take one derivative of $\tilde{G}$.

Even when applying just one derivative to $\tilde{G}$, we must compute the integral carefully. We exclude the point $\mathbf{r}$ by surrounding it with a small volume element V_δ as shown in Figure 5.3, and write

$$\nabla\nabla \cdot \int_V \tilde{\mathbf{J}}^i(\mathbf{r}',\omega)\tilde{G}(\mathbf{r}|\mathbf{r}';\omega)\,dV' =$$
$$\lim_{V_\delta \to 0} \int_{V-V_\delta} \nabla\left[\nabla\tilde{G}(\mathbf{r}|\mathbf{r}';\omega)\cdot\tilde{\mathbf{J}}^i(\mathbf{r}',\omega)\right]\,dV' + \lim_{V_\delta \to 0} \nabla\int_{V_\delta}\nabla\tilde{G}(\mathbf{r}|\mathbf{r}';\omega)\cdot\tilde{\mathbf{J}}^i(\mathbf{r}',\omega)\,dV'.$$

The first integral on the right is called the *principal value integral* and is usually abbreviated

$$\text{P.V.} \int_V \nabla\left[\nabla\tilde{G}(\mathbf{r}|\mathbf{r}';\omega)\cdot\tilde{\mathbf{J}}^i(\mathbf{r}',\omega)\right]\,dV'.$$

It converges to a value dependent on the shape of the excluded region V_δ, as does the second integral. However, the sum of these two integrals produces a unique result. Using $\nabla\tilde{G} = -\nabla'\tilde{G}$, the identity $\nabla' \cdot (\tilde{\mathbf{J}}\tilde{G}) = \tilde{\mathbf{J}}\cdot\nabla'\tilde{G} + \tilde{G}\nabla'\cdot\tilde{\mathbf{J}}$, and the divergence theorem, we can write

$$-\int_{V_\delta}\nabla'\tilde{G}(\mathbf{r}|\mathbf{r}';\omega)\cdot\tilde{\mathbf{J}}^i(\mathbf{r}',\omega)\,dV' =$$
$$-\oint_{S_\delta}\tilde{G}(\mathbf{r}|\mathbf{r}';\omega)\tilde{\mathbf{J}}^i(\mathbf{r}',\omega)\cdot\hat{\mathbf{n}}'\,dS' + \int_{V_\delta}\tilde{G}(\mathbf{r}|\mathbf{r}';\omega)\nabla'\cdot\tilde{\mathbf{J}}^i(\mathbf{r}',\omega)\,dV'$$

where S_δ is the surface surrounding V_δ. By the continuity equation, the second integral on the right is proportional to the scalar potential produced by the charge within V_δ, and thus vanishes as $V_\delta \to 0$. The first term is proportional to the field at $\mathbf{r}$ produced by surface charge on S_δ, which results in a value proportional to $\mathbf{J}^i$. Thus

$$\lim_{V_\delta \to 0} \nabla \int_{V_\delta} \nabla \tilde{G}(\mathbf{r}|\mathbf{r}';\omega) \cdot \tilde{\mathbf{J}}^i(\mathbf{r}',\omega)\, dV' = -\lim_{V_\delta \to 0} \nabla \oint_{S_\delta} \tilde{G}(\mathbf{r}|\mathbf{r}';\omega) \tilde{\mathbf{J}}^i(\mathbf{r}',\omega) \cdot \hat{\mathbf{n}}'\, dS'$$

$$= -\bar{\mathbf{L}} \cdot \tilde{\mathbf{J}}^i(\mathbf{r},\omega), \qquad (5.82)$$

so

$$\nabla\nabla \cdot \int_V \tilde{\mathbf{J}}^i(\mathbf{r}',\omega) \tilde{G}(\mathbf{r}|\mathbf{r}';\omega)\, dV' = \mathrm{P.V.} \int_V \nabla \left[\nabla \tilde{G}(\mathbf{r}|\mathbf{r}';\omega) \cdot \tilde{\mathbf{J}}^i(\mathbf{r}',\omega) \right] dV' - \bar{\mathbf{L}} \cdot \tilde{\mathbf{J}}^i(\mathbf{r},\omega).$$

Here $\bar{\mathbf{L}}$ is usually called the *depolarizing dyadic* [120]. Its value depends on the shape of V_δ, as considered below.

We may now write

$$\tilde{\mathbf{E}}(\mathbf{r},\omega) = -j\omega\tilde{\mu}(\omega)\,\mathrm{P.V.} \int_V \bar{\mathbf{G}}_e(\mathbf{r}|\mathbf{r}';\omega) \cdot \tilde{\mathbf{J}}(\mathbf{r}',\omega)\, dV' - \frac{1}{j\omega\tilde{\epsilon}^c(\omega)} \bar{\mathbf{L}} \cdot \tilde{\mathbf{J}}^i(\mathbf{r},\omega). \qquad (5.83)$$

We may also incorporate both terms into a single dyadic Green's function using the notation

$$\bar{\mathbf{G}}(\mathbf{r}|\mathbf{r}';\omega) = \mathrm{P.V.}\,\bar{\mathbf{G}}_e(\mathbf{r}|\mathbf{r}';\omega) - \frac{1}{k^2}\bar{\mathbf{L}}\delta(\mathbf{r}-\mathbf{r}').$$

Hence, when we compute

$$\tilde{\mathbf{E}}(\mathbf{r},\omega) = -j\omega\tilde{\mu}(\omega) \int_V \bar{\mathbf{G}}(\mathbf{r}|\mathbf{r}';\omega) \cdot \tilde{\mathbf{J}}^i(\mathbf{r}',\omega)\, dV'$$

$$= -j\omega\tilde{\mu}(\omega) \int_V \left[\mathrm{P.V.}\,\bar{\mathbf{G}}_e(\mathbf{r}|\mathbf{r}';\omega) - \frac{1}{k^2}\bar{\mathbf{L}}\delta(\mathbf{r}-\mathbf{r}') \right] \cdot \tilde{\mathbf{J}}^i(\mathbf{r}',\omega)\, dV',$$

we reproduce (5.83). That is, the symbol P.V. on $\bar{\mathbf{G}}_e$ indicates that a principal value integral must be performed.

Our final task is to compute $\bar{\mathbf{L}}$ from (5.82). Removal of the excluded region from the principal value computation leaves behind a hole in the source region. The contribution to the field at $\mathbf{r}$ by the sources in the excluded region is found from the scalar potential produced by the surface distribution $\hat{\mathbf{n}} \cdot \mathbf{J}^i$. The value of this *correction term* depends on the shape of the excluded volume. However, the correction term always adds to the principal value integral to give the true field at $\mathbf{r}$, regardless of the shape of the volume. So we must always match the shape of the excluded region used to compute the principal value integral with that used to compute the correction term so that the true field is obtained. Note that as $V_\delta \to 0$, the phase factor in the Green's function becomes insignificant, and the values of the current on the surface approach the value at $\mathbf{r}$ (assuming $\mathbf{J}^i$ is continuous at $\mathbf{r}$). Thus we may write

$$\lim_{V_\delta \to 0} \nabla \oint_{S_\delta} \frac{\tilde{\mathbf{J}}^i(\mathbf{r},\omega) \cdot \hat{\mathbf{n}}'}{4\pi|\mathbf{r}-\mathbf{r}'|}\, dS' = \bar{\mathbf{L}} \cdot \tilde{\mathbf{J}}^i(\mathbf{r},\omega).$$

This has the form of a static field integral. For a spherical excluded region, we may compute the above quantity quite simply by assuming the current to be uniform throughout

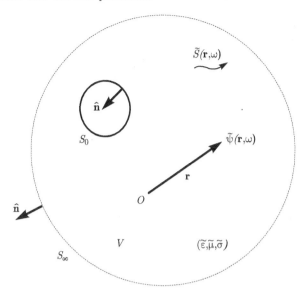

FIGURE 5.4
Geometry for solution to the frequency-domain Helmholtz equation.

V_δ and by aligning the current with the z-axis and placing the center of the sphere at the origin. We then compute the integral at a point $\mathbf{r}$ within the sphere, take the gradient, and allow $\mathbf{r} \to 0$. We thus have for a sphere

$$\lim_{V_\delta \to 0} \nabla \oint_S \frac{\tilde{J}^i \cos\theta'}{4\pi|\mathbf{r} - \mathbf{r}'|} \, dS' = \bar{\mathbf{L}} \cdot [\hat{\mathbf{z}} \tilde{J}^i(\mathbf{r}, \omega)].$$

This integral has been computed in § 3.3.8 with the result given by (3.78). Using this we find

$$\lim_{V_\delta \to 0} \left[\nabla \left(\tfrac{1}{3} \tilde{J}^i z \right) \right]\Big|_{\mathbf{r}=0} = \hat{\mathbf{z}} \tfrac{1}{3} \tilde{J}^i = \bar{\mathbf{L}} \cdot [\hat{\mathbf{z}} \tilde{J}^i(\mathbf{r}, \omega)]$$

and thus

$$\bar{\mathbf{L}} = \tfrac{1}{3}\bar{\mathbf{I}}.$$

We leave it as an exercise to show that for a cubical excluded volume, the depolarizing dyadic has this same value. Values for other shapes may be found in Yaghjian [215].

The theory of dyadic Green's functions is well developed and there exist techniques for their construction under a variety of conditions. For an excellent overview, the reader may see Tai [189].

5.2.9 Solution for potential functions in a bounded medium

In the previous section, we solved for the frequency-domain potential functions in an unbounded region of space. Here we extend the solution to a bounded region and identify the physical meaning of the radiation condition (5.72).

Consider a bounded region of space V containing a linear, homogeneous, isotropic medium characterized by $\tilde{\mu}(\omega)$ and $\tilde{\epsilon}^c(\omega)$. As shown in Figure 5.4, we decompose the multiply connected boundary into a closed "excluding surface" S_0 and a closed "encompassing surface" S_∞ permitted to expand outward to infinity. S_0 may consist of more than one closed surface and is often used to exclude unknown sources from V. We wish to solve the Helmholtz equation (5.68) for $\tilde{\psi}$ within V in terms of the sources within V and the values of $\tilde{\psi}$ on S_0. The actual sources of $\tilde{\psi}$ lie entirely with S_∞ but may lie partly, or entirely, within S_0.

The approach parallels that of § 3.3.5.2. We begin with Green's second identity, written in terms of the source point (primed) variables and applied to the region V:

$$\int_V [\psi(\mathbf{r}',\omega)\nabla'^2\tilde{G}(\mathbf{r}|\mathbf{r}';\omega) - \tilde{G}(\mathbf{r}|\mathbf{r}';\omega)\nabla'^2\psi(\mathbf{r}',\omega)]\, dV' =$$

$$\oint_{S_0+S_\infty} \left[\psi(\mathbf{r}',\omega)\frac{\partial\tilde{G}(\mathbf{r}|\mathbf{r}';\omega)}{\partial n'} - \tilde{G}(\mathbf{r}|\mathbf{r}';\omega)\frac{\partial\psi(\mathbf{r}',\omega)}{\partial n'} \right] dS'.$$

Here $\hat{\mathbf{n}}$ points outward from V, and $\tilde{G}$ is the Green's function (5.70). By inspection, the latter obeys the reciprocity condition

$$\tilde{G}(\mathbf{r}|\mathbf{r}';\omega) = \tilde{G}(\mathbf{r}'|\mathbf{r};\omega)$$

and satisfies

$$\nabla^2\tilde{G}(\mathbf{r}|\mathbf{r}';\omega) = \nabla'^2\tilde{G}(\mathbf{r}|\mathbf{r}';\omega).$$

Substituting $\nabla'^2\tilde{\psi} = -k^2\tilde{\psi} - \tilde{S}$ from (5.68) and $\nabla'^2\tilde{G} = -k^2\tilde{G} - \delta(\mathbf{r}-\mathbf{r}')$ from (5.69) we get

$$\tilde{\psi}(\mathbf{r},\omega) = \int_V \tilde{S}(\mathbf{r}',\omega)\tilde{G}(\mathbf{r}|\mathbf{r}';\omega)\, dV'$$

$$- \oint_{S_0+S_\infty} \left[\tilde{\psi}(\mathbf{r}',\omega)\frac{\partial\tilde{G}(\mathbf{r}|\mathbf{r}';\omega)}{\partial n'} - \tilde{G}(\mathbf{r}|\mathbf{r}';\omega)\frac{\partial\tilde{\psi}(\mathbf{r}',\omega)}{\partial n'} \right] dS'.$$

Hence $\tilde{\psi}$ within V may be written in terms of the sources within V and the values of $\tilde{\psi}$ and its normal derivative over $S_0 + S_\infty$. The surface contributions account for sources excluded by S_0.

Let us examine the integral over S_∞ more closely. If we let S_∞ recede to infinity, we expect no contribution to the potential at $\mathbf{r}$ from the fields on S_∞. Choosing a sphere centered at the origin, we note that $\hat{\mathbf{n}}' = \hat{\mathbf{r}}'$ and that as $r' \to \infty$,

$$\tilde{G}(\mathbf{r}|\mathbf{r}';\omega) = \frac{e^{-jk|\mathbf{r}-\mathbf{r}'|}}{4\pi|\mathbf{r}-\mathbf{r}'|} \approx \frac{e^{-jkr'}}{4\pi r'},$$

$$\frac{\partial\tilde{G}(\mathbf{r}|\mathbf{r}';\omega)}{\partial n'} = \hat{\mathbf{n}}' \cdot \nabla'\tilde{G}(\mathbf{r}|\mathbf{r}';\omega) \approx \frac{\partial}{\partial r'}\frac{e^{-jkr'}}{4\pi r'} = -(1+jkr')\frac{e^{-jkr'}}{4\pi r'}.$$

Substituting these, we find that as $r' \to \infty$,

$$\oint_{S_\infty} \left[\tilde{\psi}\frac{\partial\tilde{G}}{\partial n'} - \tilde{G}\frac{\partial\tilde{\psi}}{\partial n'} \right] dS' \approx \int_0^{2\pi}\int_0^\pi \left[-\frac{1+jkr'}{r'^2}\tilde{\psi} - \frac{1}{r'}\frac{\partial\tilde{\psi}}{\partial r'} \right] \frac{e^{-jkr'}}{4\pi} r'^2 \sin\theta'\, d\theta'\, d\phi'$$

$$\approx -\int_0^{2\pi}\int_0^\pi \left[\tilde{\psi} + r'\left(jk\tilde{\psi} + \frac{\partial\tilde{\psi}}{\partial r'} \right) \right] \frac{e^{-jkr}}{4\pi} \sin\theta'\, d\theta'\, d\phi'.$$

Since this gives the contribution to the field in V from the fields on the surface receding to infinity, we expect that this term should be zero. For a lossy medium the exponential term decays and drives the contribution to zero. For a lossless medium the contribution is zero if

$$\lim_{r\to\infty} \tilde{\psi}(\mathbf{r},\omega) = 0, \tag{5.84}$$

$$\lim_{r\to\infty} r\left[jk\tilde{\psi}(\mathbf{r},\omega) + \frac{\partial\tilde{\psi}(\mathbf{r},\omega)}{\partial r} \right] = 0. \tag{5.85}$$

This is the *radiation condition* for the Helmholtz equation. It is also called the *Sommerfeld radiation condition* after the German physicist A. Sommerfeld. Note that we have not derived this condition: we have merely postulated it. As with all postulates it is subject to experimental verification.

The radiation condition implies that for points far from the source, the potentials behave as spherical waves:

$$\tilde{\psi}(\mathbf{r},\omega) \sim \frac{e^{-jkr}}{r} \qquad (r\to\infty).$$

Substituting this into (5.84) and (5.85), we find that the radiation condition is satisfied. With $S_\infty \to \infty$ we have

$$\tilde{\psi}(\mathbf{r},\omega) = \int_V \tilde{S}(\mathbf{r}',\omega)\tilde{G}(\mathbf{r}|\mathbf{r}';\omega)\,dV'$$

$$- \oint_{S_0} \left[\tilde{\psi}(\mathbf{r}',\omega)\frac{\partial\tilde{G}(\mathbf{r}|\mathbf{r}';\omega)}{\partial n'} - \tilde{G}(\mathbf{r}|\mathbf{r}';\omega)\frac{\partial\tilde{\psi}(\mathbf{r}',\omega)}{\partial n'} \right] dS',$$

which is the expression for the potential within an infinite medium having source-excluding regions. As $S_0 \to 0$, we obtain the expression for the potential in an unbounded medium:

$$\tilde{\psi}(\mathbf{r},\omega) = \int_V \tilde{S}(\mathbf{r}',\omega)\tilde{G}(\mathbf{r}|\mathbf{r}';\omega)\,dV',$$

as expected.

The time-domain equation (5.64) may also be solved (at least for the lossless case) in a bounded region of space. The interested reader should see Pauli [147] for details.

5.3 Transverse–longitudinal decomposition

We have seen that when only electric sources are present, the electromagnetic fields in a homogeneous, isotropic region can be represented by a single Hertzian potential $\boldsymbol{\Pi}_e$. Similarly, when only magnetic sources are present, the fields can be represented by a single Hertzian potential $\boldsymbol{\Pi}_h$. Hence two vector potentials may be used to represent the field if both electric and magnetic sources are present.

We may also represent the electromagnetic field in a homogeneous, isotropic region using two scalar functions and the sources. This follows naturally from another important field decomposition: a splitting of each field vector into (1) a component along a certain pre-chosen constant direction, and (2) a component transverse to this direction. Depending on the geometry of the sources, it is possible that only one of these components will be present. A special case of this decomposition, the *TE–TM field decomposition*, holds for a source-free region and will be discussed in the next section.

5.3.1   Transverse–longitudinal decomposition for isotropic media

Consider a direction defined by a constant unit vector $\hat{\mathbf{u}}$. We define the *longitudinal component* of $\mathbf{A}$ as $\hat{\mathbf{u}}A_u$ where

$$A_u = \hat{\mathbf{u}} \cdot \mathbf{A},$$

and the *transverse component* of $\mathbf{A}$ as

$$\mathbf{A}_t = \mathbf{A} - \hat{\mathbf{u}}A_u.$$

We may thus decompose any vector into a sum of longitudinal and transverse parts. An important consequence of Maxwell's equations is that the transverse fields may be written entirely in terms of the longitudinal fields and the sources. This holds in both the time and frequency domains; we derive the decomposition in the frequency domain and leave the derivation of the time-domain expressions as exercises. We begin by decomposing the operators in Maxwell's equations into longitudinal and transverse components. We note that

$$\frac{\partial}{\partial u} \equiv \hat{\mathbf{u}} \cdot \nabla$$

and define a *transverse del operator* as

$$\nabla_t \equiv \nabla - \hat{\mathbf{u}}\frac{\partial}{\partial u}.$$

Using these basic definitions, the identities listed in Appendix B may be derived. We shall find it helpful to express the vector curl and Laplacian operations in terms of their longitudinal and transverse components. Using (B.99) and (B.102), we find that the transverse component of the curl is given by

$$(\nabla \times \mathbf{A})_t = -\hat{\mathbf{u}} \times \hat{\mathbf{u}} \times (\nabla \times \mathbf{A})$$

$$= -\hat{\mathbf{u}} \times \hat{\mathbf{u}} \times (\nabla_t \times \mathbf{A}_t) - \hat{\mathbf{u}} \times \hat{\mathbf{u}} \times \left(\hat{\mathbf{u}} \times \left[\frac{\partial \mathbf{A}_t}{\partial u} - \nabla_t A_u \right] \right).$$

The first term in the right member is zero by property (B.97). Using (B.7) we can replace the second term by

$$-\hat{\mathbf{u}}\left\{ \hat{\mathbf{u}} \cdot \left(\hat{\mathbf{u}} \times \left[\frac{\partial \mathbf{A}_t}{\partial u} - \nabla_t A_u \right] \right) \right\} + (\hat{\mathbf{u}} \cdot \hat{\mathbf{u}})\left(\hat{\mathbf{u}} \times \left[\frac{\partial \mathbf{A}_t}{\partial u} - \nabla_t A_u \right] \right).$$

The first of these terms is zero since

$$\hat{\mathbf{u}} \cdot \left(\hat{\mathbf{u}} \times \left[\frac{\partial \mathbf{A}_t}{\partial u} - \nabla_t A_u \right] \right) = \left[\frac{\partial \mathbf{A}_t}{\partial u} - \nabla_t A_u \right] \cdot (\hat{\mathbf{u}} \times \hat{\mathbf{u}}) = 0,$$

hence

$$(\nabla \times \mathbf{A})_t = \hat{\mathbf{u}} \times \left[\frac{\partial \mathbf{A}_t}{\partial u} - \nabla_t A_u \right]. \tag{5.86}$$

The longitudinal part is then, by property (B.86), merely the difference between the curl and its transverse part, or

$$\hat{\mathbf{u}}\,(\hat{\mathbf{u}} \cdot \nabla \times \mathbf{A}) = \nabla_t \times \mathbf{A}_t. \tag{5.87}$$

A similar set of steps gives the transverse component of the Laplacian as

$$(\nabla^2 \mathbf{A})_t = \left[\nabla_t(\nabla_t \cdot \mathbf{A}_t) + \frac{\partial^2 \mathbf{A}_t}{\partial u^2} - \nabla_t \times \nabla_t \times \mathbf{A}_t \right],$$

and the longitudinal part as

$$\hat{\mathbf{u}}\left(\hat{\mathbf{u}} \cdot \nabla^2 \mathbf{A}\right) = \hat{\mathbf{u}} \nabla^2 A_u. \tag{5.88}$$

Verification is left as an exercise.

Now we are ready to give a transverse–longitudinal decomposition of the fields in a lossy, homogeneous, isotropic region in terms of the direction $\hat{\mathbf{u}}$. We write Maxwell's equations as

$$\nabla \times \tilde{\mathbf{E}} = -j\omega\tilde{\mu}\tilde{\mathbf{H}}_t - j\omega\tilde{\mu}\hat{\mathbf{u}}\tilde{H}_u - \tilde{\mathbf{J}}^i_{mt} - \hat{\mathbf{u}}\tilde{J}^i_{mu},$$

$$\nabla \times \tilde{\mathbf{H}} = j\omega\tilde{\epsilon}^c\tilde{\mathbf{E}}_t + j\omega\tilde{\epsilon}^c\hat{\mathbf{u}}\tilde{E}_u + \tilde{\mathbf{J}}^i_t + \hat{\mathbf{u}}\tilde{J}^i_u,$$

where we have split the right-hand sides into longitudinal and transverse parts. Then, using (5.86) and (5.87), we can equate the transverse and longitudinal parts of each equation to obtain

$$\nabla_t \times \tilde{\mathbf{E}}_t = -j\omega\tilde{\mu}\hat{\mathbf{u}}\tilde{H}_u - \hat{\mathbf{u}}\tilde{J}^i_{mu}, \tag{5.89}$$

$$-\hat{\mathbf{u}} \times \nabla_t \tilde{E}_u + \hat{\mathbf{u}} \times \frac{\partial \tilde{\mathbf{E}}_t}{\partial u} = -j\omega\tilde{\mu}\tilde{\mathbf{H}}_t - \tilde{\mathbf{J}}^i_{mt}, \tag{5.90}$$

$$\nabla_t \times \tilde{\mathbf{H}}_t = j\omega\tilde{\epsilon}^c\hat{\mathbf{u}}\tilde{E}_u + \hat{\mathbf{u}}\tilde{J}^i_u, \tag{5.91}$$

$$-\hat{\mathbf{u}} \times \nabla_t \tilde{H}_u + \hat{\mathbf{u}} \times \frac{\partial \tilde{\mathbf{H}}_t}{\partial u} = j\omega\tilde{\epsilon}^c\tilde{\mathbf{E}}_t + \tilde{\mathbf{J}}^i_t. \tag{5.92}$$

Let us isolate the transverse fields in terms of the longitudinal fields. Forming the cross product of $\hat{\mathbf{u}}$ and the partial derivative of (5.92) with respect to u, we have

$$-\hat{\mathbf{u}} \times \hat{\mathbf{u}} \times \nabla_t \frac{\partial \tilde{H}_u}{\partial u} + \hat{\mathbf{u}} \times \hat{\mathbf{u}} \times \frac{\partial^2 \tilde{\mathbf{H}}_t}{\partial u^2} = j\omega\tilde{\epsilon}^c\hat{\mathbf{u}} \times \frac{\partial \tilde{\mathbf{E}}_t}{\partial u} + \hat{\mathbf{u}} \times \frac{\partial \tilde{\mathbf{J}}^i_t}{\partial u}.$$

Using (B.7) and (B.86) we find that

$$\nabla_t \frac{\partial \tilde{H}_u}{\partial u} - \frac{\partial^2 \tilde{\mathbf{H}}_t}{\partial u^2} = j\omega\tilde{\epsilon}^c\hat{\mathbf{u}} \times \frac{\partial \tilde{\mathbf{E}}_t}{\partial u} + \hat{\mathbf{u}} \times \frac{\partial \tilde{\mathbf{J}}^i_t}{\partial u}. \tag{5.93}$$

Multiplying (5.90) by $j\omega\tilde{\epsilon}^c$, we have

$$-j\omega\tilde{\epsilon}^c\hat{\mathbf{u}} \times \nabla_t \tilde{E}_u + j\omega\tilde{\epsilon}^c\hat{\mathbf{u}} \times \frac{\partial \tilde{\mathbf{E}}_t}{\partial u} = \omega^2\tilde{\mu}\tilde{\epsilon}^c\tilde{\mathbf{H}}_t - j\omega\tilde{\epsilon}^c\tilde{\mathbf{J}}^i_{mt}. \tag{5.94}$$

We now add (5.93) to (5.94) and eliminate $\tilde{\mathbf{E}}_t$ to get

$$\left(\frac{\partial^2}{\partial u^2} + k^2\right)\tilde{\mathbf{H}}_t = \nabla_t \frac{\partial \tilde{H}_u}{\partial u} - j\omega\tilde{\epsilon}^c\hat{\mathbf{u}} \times \nabla_t \tilde{E}_u + j\omega\tilde{\epsilon}^c\tilde{\mathbf{J}}^i_{mt} - \hat{\mathbf{u}} \times \frac{\partial \tilde{\mathbf{J}}^i_t}{\partial u}. \tag{5.95}$$

This one-dimensional Helmholtz equation can be used to express the transverse magnetic field in terms of the longitudinal components of $\tilde{\mathbf{E}}$ and $\tilde{\mathbf{H}}$. Similar steps lead to a formula for the transverse component of $\tilde{\mathbf{E}}$:

$$\left(\frac{\partial^2}{\partial u^2} + k^2\right)\tilde{\mathbf{E}}_t = \nabla_t \frac{\partial \tilde{E}_u}{\partial u} + j\omega\tilde{\mu}\hat{\mathbf{u}} \times \nabla_t \tilde{H}_u + \hat{\mathbf{u}} \times \frac{\partial \tilde{\mathbf{J}}^i_{mt}}{\partial u} + j\omega\tilde{\mu}\tilde{\mathbf{J}}^i_t. \tag{5.96}$$

We find the longitudinal components from the wave equation for $\tilde{\mathbf{E}}$ and $\tilde{\mathbf{H}}$. Recall that the fields satisfy

$$(\nabla^2 + k^2)\tilde{\mathbf{E}} = \frac{1}{\tilde{\epsilon}^c}\nabla\tilde{\rho}^i + j\omega\tilde{\mu}\tilde{\mathbf{J}}^i + \nabla \times \tilde{\mathbf{J}}_m^i,$$

$$(\nabla^2 + k^2)\tilde{\mathbf{H}} = \frac{1}{\tilde{\mu}}\nabla\tilde{\rho}_m^i + j\omega\tilde{\epsilon}^c\tilde{\mathbf{J}}_m^i - \nabla \times \tilde{\mathbf{J}}^i.$$

Splitting the vectors into longitudinal and transverse parts, and using (5.87) and (5.88), we equate the longitudinal components of the wave equations to obtain

$$\left(\nabla^2 + k^2\right)\tilde{E}_u = \frac{1}{\tilde{\epsilon}^c}\frac{\partial\tilde{\rho}^i}{\partial u} + j\omega\tilde{\mu}\tilde{J}_u^i + \hat{\mathbf{u}} \cdot (\nabla_t \times \tilde{\mathbf{J}}_{mt}^i), \qquad (5.97)$$

$$\left(\nabla^2 + k^2\right)\tilde{H}_u = \frac{1}{\tilde{\mu}}\frac{\partial\tilde{\rho}_m^i}{\partial u} + j\omega\tilde{\epsilon}^c\tilde{J}_{mu}^i - \hat{\mathbf{u}} \cdot (\nabla_t \times \tilde{\mathbf{J}}_t^i). \qquad (5.98)$$

Note that if $\tilde{\mathbf{J}}_m^i = \tilde{\mathbf{J}}_t^i = 0$, then $\tilde{H}_u = 0$ and the fields are TM to the u-direction; these fields may be determined completely from $\tilde{E}_u$. Similarly, if $\tilde{\mathbf{J}}^i = \tilde{\mathbf{J}}_{mt}^i = 0$, then $\tilde{E}_u = 0$ and the fields are TE to the u-direction; these fields may be determined completely from $\tilde{H}_u$. These properties are used in § 4.11.8.2, where the fields of z-directed electric and magnetic line sources are assumed to be purely TM$_z$ or TE$_z$, respectively.

5.3.2   Transverse–longitudinal decomposition for anisotropic media

The transverse–longitudinal decomposition presented in Section 5.3.1 for a homogeneous isotropic medium may also be formulated for a homogeneous anisotropic medium, with an expected increase in complexity. In fact, the decomposition may even be formulated for a homogeneous bianisotropic medium, but we restrict ourselves to anisotropic media for expediency. The reader interested in the bianisotropic case should see [42].

Consider a homogeneous anisotropic medium with constitutive relations $\tilde{\mathbf{D}} = \bar{\tilde{\epsilon}} \cdot \tilde{\mathbf{E}}$ and $\tilde{\mathbf{B}} = \bar{\tilde{\mu}} \cdot \tilde{\mathbf{H}}$. Maxwell's equations can be written as

$$\nabla \times \tilde{\mathbf{E}} = -j\omega\tilde{\mathbf{B}}_t - j\omega\hat{\mathbf{u}}\tilde{B}_u - \tilde{\mathbf{J}}_{mt}^i - \hat{\mathbf{u}}\tilde{J}_{mu}^i,$$

$$\nabla \times \tilde{\mathbf{H}} = j\omega\tilde{\mathbf{D}}_t + j\omega\hat{\mathbf{u}}\tilde{D}_u + \tilde{\mathbf{J}}_t^i + \hat{\mathbf{u}}\tilde{J}_u^i,$$

where we have split the right-hand sides into longitudinal and transverse parts. Using (5.86) and (5.87), we can equate the transverse and longitudinal parts of each equation to obtain

$$\nabla_t \times \tilde{\mathbf{E}}_t = -j\omega\hat{\mathbf{u}}\tilde{B}_u - \hat{\mathbf{u}}\tilde{J}_{mu}^i, \qquad (5.99)$$

$$-\hat{\mathbf{u}} \times \nabla_t\tilde{E}_u + \hat{\mathbf{u}} \times \frac{\partial\tilde{\mathbf{E}}_t}{\partial u} = -j\omega\tilde{\mathbf{B}}_t - \tilde{\mathbf{J}}_{mt}^i, \qquad (5.100)$$

$$\nabla_t \times \tilde{\mathbf{H}}_t = j\omega\hat{\mathbf{u}}\tilde{D}_u + \hat{\mathbf{u}}\tilde{J}_u^i, \qquad (5.101)$$

$$-\hat{\mathbf{u}} \times \nabla_t\tilde{H}_u + \hat{\mathbf{u}} \times \frac{\partial\tilde{\mathbf{H}}_t}{\partial u} = j\omega\tilde{\mathbf{D}}_t + \tilde{\mathbf{J}}_t^i. \qquad (5.102)$$

To proceed, we must write $\tilde{B}_u$ and $\tilde{\mathbf{B}}_t$ in terms of $\tilde{H}_u$ and $\tilde{\mathbf{H}}_t$, and $\tilde{D}_u$ and $\tilde{\mathbf{D}}_t$ in terms of $\tilde{E}_u$ and $\tilde{\mathbf{E}}_t$.

First note that

$$\tilde{B}_u = \hat{\mathbf{u}} \cdot \tilde{\mathbf{B}} = \hat{\mathbf{u}} \cdot (\tilde{\boldsymbol{\mu}} \cdot \tilde{\mathbf{H}}) = (\hat{\mathbf{u}} \cdot \tilde{\boldsymbol{\mu}}) \cdot \tilde{\mathbf{H}} = \tilde{\boldsymbol{\mu}} \cdot \tilde{\mathbf{H}},$$

where we have used (B.11), and where we define $\tilde{\boldsymbol{\mu}} = \hat{\mathbf{u}} \cdot \tilde{\boldsymbol{\mu}}$. Then

$$\tilde{B}_u = \tilde{\boldsymbol{\mu}} \cdot (\tilde{\mathbf{H}}_t + \hat{\mathbf{u}}\tilde{H}_u) = \tilde{\boldsymbol{\mu}} \cdot \tilde{\mathbf{H}}_t + \tilde{\mu}_{uu}\tilde{H}_u, \tag{5.103}$$

where $\tilde{\mu}_{uu} = \hat{\mathbf{u}} \cdot \tilde{\boldsymbol{\mu}} \cdot \hat{\mathbf{u}} = \tilde{\boldsymbol{\mu}} \cdot \hat{\mathbf{u}}$. Next note that

$$\tilde{\mathbf{B}}_t = \tilde{\mathbf{B}} - \hat{\mathbf{u}}\tilde{B}_u = \tilde{\boldsymbol{\mu}} \cdot (\tilde{\mathbf{H}}_t + \hat{\mathbf{u}}\tilde{H}_u) - \hat{\mathbf{u}}(\tilde{\boldsymbol{\mu}} \cdot \tilde{\mathbf{H}}_t + \tilde{\mu}_{uu}\tilde{H}_u) = \tilde{\boldsymbol{\mu}}_t \cdot \tilde{\mathbf{H}}_t + \tilde{\boldsymbol{\mu}}_t\tilde{H}_u, \tag{5.104}$$

where we have defined $\tilde{\boldsymbol{\mu}}_t = \tilde{\boldsymbol{\mu}} - \hat{\mathbf{u}}\tilde{\boldsymbol{\mu}}$ and $\tilde{\boldsymbol{\mu}}_t = (\bar{\mathbf{I}} - \hat{\mathbf{u}}\hat{\mathbf{u}}) \cdot \tilde{\boldsymbol{\mu}} \cdot \hat{\mathbf{u}}$. Similarly,

$$\tilde{D}_u = \tilde{\boldsymbol{\epsilon}} \cdot \tilde{\mathbf{E}}_t + \tilde{\epsilon}_{uu}\tilde{E}_u, \tag{5.105}$$

$$\tilde{\mathbf{D}}_t = \tilde{\boldsymbol{\epsilon}}_t \cdot \tilde{\mathbf{E}}_t + \tilde{\boldsymbol{\epsilon}}_t\tilde{E}_u, \tag{5.106}$$

where $\tilde{\boldsymbol{\epsilon}} = \hat{\mathbf{u}} \cdot \tilde{\boldsymbol{\epsilon}}$, $\tilde{\epsilon}_{uu} = \hat{\mathbf{u}} \cdot \tilde{\boldsymbol{\epsilon}} \cdot \hat{\mathbf{u}} = \tilde{\boldsymbol{\epsilon}} \cdot \hat{\mathbf{u}}$, $\tilde{\boldsymbol{\epsilon}}_t = \tilde{\boldsymbol{\epsilon}} - \hat{\mathbf{u}}\tilde{\boldsymbol{\epsilon}}$, and $\tilde{\boldsymbol{\epsilon}}_t = (\bar{\mathbf{I}} - \hat{\mathbf{u}}\hat{\mathbf{u}}) \cdot \tilde{\boldsymbol{\epsilon}} \cdot \hat{\mathbf{u}}$.

Substituting (5.103), (5.104), (5.105), and (5.106) into (5.100) and (5.102), we obtain

$$-\hat{\mathbf{u}} \times \nabla_t\tilde{E}_u + \hat{\mathbf{u}} \times \frac{\partial\tilde{\mathbf{E}}_t}{\partial u} = -j\omega(\tilde{\boldsymbol{\mu}}_t \cdot \tilde{\mathbf{H}}_t + \tilde{\boldsymbol{\mu}}_t\tilde{H}_u) - \tilde{\mathbf{J}}_{mt}^i, \tag{5.107}$$

$$-\hat{\mathbf{u}} \times \nabla_t\tilde{H}_u + \hat{\mathbf{u}} \times \frac{\partial\tilde{\mathbf{H}}_t}{\partial u} = j\omega(\tilde{\boldsymbol{\epsilon}}_t \cdot \tilde{\mathbf{E}}_t + \tilde{\boldsymbol{\epsilon}}_t\tilde{E}_u) + \tilde{\mathbf{J}}_t^i. \tag{5.108}$$

Next, crossing $\hat{\mathbf{u}}$ with (5.107) and using (B.99), we get

$$\frac{\partial\tilde{\mathbf{E}}_t}{\partial u} = \nabla_t\tilde{E}_u + j\omega\hat{\mathbf{u}} \times (\tilde{\boldsymbol{\mu}}_t \cdot \tilde{\mathbf{H}}_t) + j\omega\hat{\mathbf{u}} \times (\tilde{\boldsymbol{\mu}}_t\tilde{H}_u) + \hat{\mathbf{u}} \times \tilde{\mathbf{J}}_{mt}^i. \tag{5.109}$$

Taking $\partial/\partial u$ of (5.108) and using (5.109) to eliminate $\tilde{\mathbf{E}}_t$, we have

$$-\hat{\mathbf{u}} \times \nabla_t\frac{\partial\tilde{H}_u}{\partial u} + \hat{\mathbf{u}} \times \frac{\partial^2\tilde{\mathbf{H}}_t}{\partial u^2} =$$
$$j\omega\tilde{\boldsymbol{\epsilon}}_t \cdot \left[\nabla_t\tilde{E}_u + j\omega\hat{\mathbf{u}} \times (\tilde{\boldsymbol{\mu}}_t \cdot \tilde{\mathbf{H}}_t) + j\omega\hat{\mathbf{u}} \times (\tilde{\boldsymbol{\mu}}_t\tilde{H}_u) + \hat{\mathbf{u}} \times \tilde{\mathbf{J}}_{mt}^i\right] + j\omega\tilde{\boldsymbol{\epsilon}}_t\frac{\partial\tilde{E}_u}{\partial u} + \frac{\partial\tilde{\mathbf{J}}_t^i}{\partial u}.$$

This can be written more conveniently by forming $\hat{\mathbf{u}}$ crossed with this expression:

$$\nabla_t\frac{\partial\tilde{H}_u}{\partial u} - \frac{\partial^2\tilde{\mathbf{H}}_t}{\partial u^2} = j\omega\hat{\mathbf{u}} \times (\tilde{\boldsymbol{\epsilon}}_t \cdot \nabla_t\tilde{E}_u) - \omega^2\hat{\mathbf{u}} \times \left\{\tilde{\boldsymbol{\epsilon}}_t \cdot [\hat{\mathbf{u}} \times (\tilde{\boldsymbol{\mu}}_t \cdot \tilde{\mathbf{H}}_t)]\right\}$$
$$-\omega^2\hat{\mathbf{u}} \times \left\{\tilde{\boldsymbol{\epsilon}}_t \cdot [\hat{\mathbf{u}} \times (\tilde{\boldsymbol{\mu}}_t\tilde{H}_u)]\right\} + j\omega(\hat{\mathbf{u}} \times \tilde{\boldsymbol{\epsilon}}_t)\frac{\partial\tilde{E}_u}{\partial u} + j\omega\hat{\mathbf{u}} \times [\tilde{\boldsymbol{\epsilon}}_t \cdot (\hat{\mathbf{u}} \times \tilde{\mathbf{J}}_{mt}^i)] + \hat{\mathbf{u}} \times \frac{\partial\tilde{\mathbf{J}}_t^i}{\partial u}. \tag{5.110}$$

Several terms in (5.110) may be simplified. Use of (B.12) gives

$$\hat{\mathbf{u}} \times \tilde{\boldsymbol{\epsilon}}_t = \hat{\mathbf{u}} \times (\bar{\mathbf{I}} - \hat{\mathbf{u}}\hat{\mathbf{u}}) \cdot \tilde{\boldsymbol{\epsilon}} \cdot \hat{\mathbf{u}} = \hat{\mathbf{u}} \times (\tilde{\boldsymbol{\epsilon}} \cdot \hat{\mathbf{u}}) = (\hat{\mathbf{u}} \times \tilde{\boldsymbol{\epsilon}}) \cdot \hat{\mathbf{u}}.$$

With use of the additional relations

$$\hat{\mathbf{u}} \times (\tilde{\boldsymbol{\epsilon}}_t \cdot \nabla_t\tilde{E}_u) = \hat{\mathbf{u}} \times (\tilde{\boldsymbol{\epsilon}} \cdot \nabla_t\tilde{E}_u - \hat{\mathbf{u}}\tilde{\boldsymbol{\epsilon}} \cdot \nabla_t\tilde{E}_u) = (\hat{\mathbf{u}} \times \tilde{\boldsymbol{\epsilon}}) \cdot \nabla_t\tilde{E}_u,$$

$$\hat{\mathbf{u}} \times \left\{\tilde{\boldsymbol{\epsilon}}_t \cdot [\hat{\mathbf{u}} \times (\tilde{\boldsymbol{\mu}}_t \cdot \tilde{\mathbf{H}}_t)]\right\} = (\hat{\mathbf{u}} \times \tilde{\boldsymbol{\epsilon}}_t) \cdot [\hat{\mathbf{u}} \times (\tilde{\boldsymbol{\mu}}_t \cdot \tilde{\mathbf{H}}_t)] = [(\hat{\mathbf{u}} \times \tilde{\boldsymbol{\epsilon}}) \cdot (\hat{\mathbf{u}} \times \tilde{\boldsymbol{\mu}})] \cdot \tilde{\mathbf{H}}_t,$$

$$\hat{\mathbf{u}} \times \left\{\tilde{\boldsymbol{\epsilon}}_t \cdot [\hat{\mathbf{u}} \times (\tilde{\boldsymbol{\mu}}_t\tilde{H}_u)]\right\} = (\hat{\mathbf{u}} \times \tilde{\boldsymbol{\epsilon}}_t) \cdot [\hat{\mathbf{u}} \times (\tilde{\boldsymbol{\mu}}_t\tilde{H}_u)] = [(\hat{\mathbf{u}} \times \tilde{\boldsymbol{\epsilon}}) \cdot (\hat{\mathbf{u}} \times \tilde{\boldsymbol{\mu}})] \cdot (\hat{\mathbf{u}}\tilde{H}_u),$$

$$\hat{\mathbf{u}} \times [\tilde{\boldsymbol{\epsilon}}_t \cdot (\hat{\mathbf{u}} \times \tilde{\mathbf{J}}_{mt}^i)] = (\hat{\mathbf{u}} \times \tilde{\boldsymbol{\epsilon}}) \cdot (\hat{\mathbf{u}} \times \tilde{\mathbf{J}}_{mt}^i),$$

(5.110) becomes

$$\frac{\partial^2 \tilde{\mathbf{H}}_t}{\partial u^2} - \omega^2 \left[(\hat{\mathbf{u}} \times \tilde{\bar{\boldsymbol{\epsilon}}}) \cdot (\hat{\mathbf{u}} \times \tilde{\bar{\boldsymbol{\mu}}}) \right] \cdot \tilde{\mathbf{H}}_t = \nabla_t \frac{\partial \tilde{H}_u}{\partial u} - j\omega (\hat{\mathbf{u}} \times \tilde{\bar{\boldsymbol{\epsilon}}}) \cdot \nabla_t \tilde{E}_u$$

$$+ \omega^2 \left[(\hat{\mathbf{u}} \times \tilde{\bar{\boldsymbol{\epsilon}}}) \cdot (\hat{\mathbf{u}} \times \tilde{\bar{\boldsymbol{\mu}}}) \right] \cdot (\hat{\mathbf{u}} \tilde{H}_u) - j\omega (\hat{\mathbf{u}} \times \tilde{\bar{\boldsymbol{\epsilon}}}) \cdot (\hat{\mathbf{u}} \times \tilde{\mathbf{J}}^i_{mt}) - \hat{\mathbf{u}} \times \frac{\partial \tilde{\mathbf{J}}^i_t}{\partial u}. \qquad (5.111)$$

This expresses the transverse component of $\tilde{\mathbf{H}}$ in terms of the longitudinal components of $\tilde{\mathbf{H}}$ and $\tilde{\mathbf{E}}$ and the sources. Similar steps yield the transverse component of $\tilde{\mathbf{E}}$ as

$$\frac{\partial^2 \tilde{\mathbf{E}}_t}{\partial u^2} - \omega^2 \left[(\hat{\mathbf{u}} \times \tilde{\bar{\boldsymbol{\mu}}}) \cdot (\hat{\mathbf{u}} \times \tilde{\bar{\boldsymbol{\epsilon}}}) \right] \cdot \tilde{\mathbf{E}}_t = \nabla_t \frac{\partial \tilde{E}_u}{\partial u} + j\omega (\hat{\mathbf{u}} \times \tilde{\bar{\boldsymbol{\mu}}}) \cdot \nabla_t \tilde{H}_u$$

$$+ \omega^2 \left[(\hat{\mathbf{u}} \times \tilde{\bar{\boldsymbol{\mu}}}) \cdot (\hat{\mathbf{u}} \times \tilde{\bar{\boldsymbol{\epsilon}}}) \right] \cdot (\hat{\mathbf{u}} \tilde{E}_u) - j\omega (\hat{\mathbf{u}} \times \tilde{\bar{\boldsymbol{\mu}}}) \cdot (\hat{\mathbf{u}} \times \tilde{\mathbf{J}}^i_t) + \hat{\mathbf{u}} \times \frac{\partial \tilde{\mathbf{J}}^i_{mt}}{\partial u}. \qquad (5.112)$$

This is left as an exercise.

▶ **Example 5.6:** Transverse–longitudinal decomposition for isotropic media

Show that (5.111)–(5.112) reduce to (5.95)–(5.96) when the medium is isotropic.

Solution: For an isotropic medium, we have

$$\tilde{\bar{\boldsymbol{\epsilon}}} = \bar{\mathbf{I}} \tilde{\epsilon}^c, \qquad (5.113)$$

$$\tilde{\bar{\boldsymbol{\mu}}} = \bar{\mathbf{I}} \tilde{\mu}. \qquad (5.114)$$

Using (B.12) we have

$$[(\hat{\mathbf{u}} \times \tilde{\bar{\boldsymbol{\epsilon}}}) \cdot (\hat{\mathbf{u}} \times \tilde{\bar{\boldsymbol{\mu}}})] \cdot \tilde{\mathbf{H}}_t = (\hat{\mathbf{u}} \times \tilde{\bar{\boldsymbol{\epsilon}}}) \cdot [\hat{\mathbf{u}} \times (\tilde{\bar{\boldsymbol{\mu}}} \cdot \tilde{\mathbf{H}}_t)].$$

By (5.114),

$$[(\hat{\mathbf{u}} \times \tilde{\bar{\boldsymbol{\epsilon}}}) \cdot (\hat{\mathbf{u}} \times \tilde{\bar{\boldsymbol{\mu}}})] \cdot \tilde{\mathbf{H}}_t = (\hat{\mathbf{u}} \times \tilde{\bar{\boldsymbol{\epsilon}}}) \cdot [\hat{\mathbf{u}} \times (\tilde{\mu} \bar{\mathbf{I}} \cdot \tilde{\mathbf{H}}_t)] = \tilde{\mu} (\hat{\mathbf{u}} \times \tilde{\bar{\boldsymbol{\epsilon}}}) \cdot (\hat{\mathbf{u}} \times \tilde{\mathbf{H}}_t).$$

By (5.113),

$$[(\hat{\mathbf{u}} \times \tilde{\bar{\boldsymbol{\epsilon}}}) \cdot (\hat{\mathbf{u}} \times \tilde{\bar{\boldsymbol{\mu}}})] \cdot \tilde{\mathbf{H}}_t = \tilde{\mu} \tilde{\epsilon}^c (\hat{\mathbf{u}} \times \bar{\mathbf{I}}) \cdot (\hat{\mathbf{u}} \times \tilde{\mathbf{H}}_t) = \tilde{\mu} \tilde{\epsilon}^c \hat{\mathbf{u}} \times [\bar{\mathbf{I}} \cdot (\hat{\mathbf{u}} \times \tilde{\mathbf{H}}_t)]$$

$$= \tilde{\mu} \tilde{\epsilon}^c \hat{\mathbf{u}} \times (\hat{\mathbf{u}} \times \tilde{\mathbf{H}}_t) = -\tilde{\mu} \tilde{\epsilon}^c \tilde{\mathbf{H}}_t.$$

Similarly

$$[(\hat{\mathbf{u}} \times \tilde{\bar{\boldsymbol{\epsilon}}}) \cdot (\hat{\mathbf{u}} \times \tilde{\bar{\boldsymbol{\mu}}})] \cdot (\hat{\mathbf{u}} \tilde{H}_u) = (\hat{\mathbf{u}} \times \tilde{\bar{\boldsymbol{\epsilon}}}) \cdot [\hat{\mathbf{u}} \times (\tilde{\bar{\boldsymbol{\mu}}} \cdot \hat{\mathbf{u}} \tilde{H}_u)] = (\hat{\mathbf{u}} \times \tilde{\bar{\boldsymbol{\epsilon}}}) \cdot [\tilde{\mu} (\hat{\mathbf{u}} \times \hat{\mathbf{u}} \tilde{H}_u)] = 0.$$

We also have

$$(\hat{\mathbf{u}} \times \tilde{\bar{\boldsymbol{\epsilon}}}) \cdot \nabla_t \tilde{E}_u = \hat{\mathbf{u}} \times (\tilde{\epsilon}^c \bar{\mathbf{I}} \cdot \nabla_t \tilde{E}_u) = \tilde{\epsilon}^c \hat{\mathbf{u}} \times \nabla_t \tilde{E}_u,$$

and

$$(\hat{\mathbf{u}} \times \tilde{\bar{\boldsymbol{\epsilon}}}) \cdot (\hat{\mathbf{u}} \times \tilde{\mathbf{J}}^i_{mt}) = \hat{\mathbf{u}} \times [\tilde{\bar{\boldsymbol{\epsilon}}} \cdot (\hat{\mathbf{u}} \times \tilde{\mathbf{J}}^i_{mt})] = \tilde{\epsilon}^c \hat{\mathbf{u}} \times (\hat{\mathbf{u}} \times \tilde{\mathbf{J}}^i_{mt}) = -\tilde{\epsilon}^c \tilde{\mathbf{J}}^i_{mt}.$$

By these and (5.111), we obtain

$$\frac{\partial^2 \tilde{\mathbf{H}}_t}{\partial u^2} + \omega^2 \tilde{\epsilon}^c \tilde{\mu} \tilde{\mathbf{H}}_t = \nabla_t \frac{\partial \tilde{H}_u}{\partial u} - j\omega \tilde{\epsilon}^c \hat{\mathbf{u}} \times \nabla_t \tilde{E}_u + j\omega \tilde{\epsilon}^c \tilde{\mathbf{J}}^i_{mt} - \hat{\mathbf{u}} \times \frac{\partial \tilde{\mathbf{J}}^i_t}{\partial u}.$$

Since $k^2 = \omega^2 \tilde{\mu} \tilde{\epsilon}^c$, this is identical to (5.95). Similar steps show that (5.112) reduces to (5.96). ◀

It would be convenient to solve (5.111)–(5.112) explicitly for the transverse fields in terms of the longitudinal fields and the sources. This is done most expediently by applying a spatial Fourier transform to eliminate the u-derivatives. A Fourier transform approach is particularly useful when the geometry of a problem is invariant in the longitudinal direction (such as in the case of waveguides or transmission lines). Define the Fourier transform pair $f(u) \leftrightarrow f^u(k_u)$ according to

$$f^u(k_u) = \int_{-\infty}^{\infty} f(u)e^{-jk_u u}\, du, \qquad f(u) = \frac{1}{2\pi}\int_{-\infty}^{\infty} f^u(k_u)e^{k_u u}\, dk_u.$$

Writing each term in (5.111) as an inverse Fourier transform and invoking the Fourier integral theorem, we obtain

$$k_u^2 \tilde{\mathbf{H}}_t^u + \omega^2\left[(\hat{\mathbf{u}} \times \tilde{\bar{\boldsymbol{\epsilon}}}) \cdot (\hat{\mathbf{u}} \times \tilde{\bar{\boldsymbol{\mu}}})\right] \cdot \tilde{\mathbf{H}}_t^u = -jk_u\nabla_t\tilde{H}_u^u + j\omega(\hat{\mathbf{u}} \times \tilde{\bar{\boldsymbol{\epsilon}}}) \cdot \left[\nabla_t\tilde{E}_u^u + \hat{\mathbf{u}}jk_u\tilde{E}_u^u\right]$$

$$-\omega^2\left[(\hat{\mathbf{u}} \times \tilde{\bar{\boldsymbol{\epsilon}}}) \cdot (\hat{\mathbf{u}} \times \tilde{\bar{\boldsymbol{\mu}}})\right] \cdot \hat{\mathbf{u}}\tilde{H}_u^u + j\omega(\hat{\mathbf{u}} \times \tilde{\bar{\boldsymbol{\epsilon}}}) \cdot \left[\hat{\mathbf{u}} \times \tilde{\mathbf{J}}_{mt}^{iu}\right] + jk_u\hat{\mathbf{u}} \times \tilde{\mathbf{J}}_t^{iu}. \tag{5.115}$$

Defining the dyadic

$$\bar{\boldsymbol{\Omega}} = k_u^2\bar{\mathbf{I}} + \omega^2(\hat{\mathbf{u}} \times \tilde{\bar{\boldsymbol{\epsilon}}}) \cdot (\hat{\mathbf{u}} \times \tilde{\bar{\boldsymbol{\mu}}})$$

we can solve (5.115) for $\tilde{\mathbf{H}}_t^u$:

$$\tilde{\mathbf{H}}_t^u = \bar{\boldsymbol{\Omega}}^{-1} \cdot \left\{ j\omega(\hat{\mathbf{u}} \times \tilde{\bar{\boldsymbol{\epsilon}}}) \cdot \left[\nabla_t\tilde{E}_u^u + \hat{\mathbf{u}}jk_u\tilde{E}_u^u\right] - jk_u\nabla_t\tilde{H}_u^u \right.$$

$$\left. -\omega^2\left[(\hat{\mathbf{u}} \times \tilde{\bar{\boldsymbol{\epsilon}}}) \cdot (\hat{\mathbf{u}} \times \tilde{\bar{\boldsymbol{\mu}}})\right] \cdot \hat{\mathbf{u}}\tilde{H}_u^u + j\omega(\hat{\mathbf{u}} \times \tilde{\bar{\boldsymbol{\epsilon}}}) \cdot \left[\hat{\mathbf{u}} \times \tilde{\mathbf{J}}_{mt}^{iu}\right] + jk_u\hat{\mathbf{u}} \times \tilde{\mathbf{J}}_t^{iu} \right\}. \tag{5.116}$$

Similarly, we can transform the entities in (5.112) and solve for $\tilde{\mathbf{E}}_t^u$:

$$\tilde{\mathbf{E}}_t^u = \bar{\boldsymbol{\Phi}}^{-1} \cdot \left\{ -j\omega(\hat{\mathbf{u}} \times \tilde{\bar{\boldsymbol{\mu}}}) \cdot \left[\nabla_t\tilde{H}_u^u + \hat{\mathbf{u}}jk_u\tilde{H}_u^u\right] - jk_u\nabla_t\tilde{E}_u^u \right.$$

$$\left. -\omega^2\left[(\hat{\mathbf{u}} \times \tilde{\bar{\boldsymbol{\mu}}}) \cdot (\hat{\mathbf{u}} \times \tilde{\bar{\boldsymbol{\epsilon}}})\right] \cdot \hat{\mathbf{u}}\tilde{E}_u^u + j\omega(\hat{\mathbf{u}} \times \tilde{\bar{\boldsymbol{\mu}}}) \cdot \left[\hat{\mathbf{u}} \times \tilde{\mathbf{J}}_t^{iu}\right] - jk_u\hat{\mathbf{u}} \times \tilde{\mathbf{J}}_{mt}^{iu} \right\}, \tag{5.117}$$

where

$$\bar{\boldsymbol{\Phi}} = k_u^2\bar{\mathbf{I}} + \omega^2(\hat{\mathbf{u}} \times \tilde{\bar{\boldsymbol{\mu}}}) \cdot (\hat{\mathbf{u}} \times \tilde{\bar{\boldsymbol{\epsilon}}}).$$

▶ **Example 5.7:** Transverse field relations for isotropic media

Specialize (5.116)–(5.117) for the case of an isotropic medium.

Solution: The dyadic permeability and permittivity for an isotropic medium are given by (5.114) and (5.113). Proceeding as in Example 5.6, we have

$$(\hat{\mathbf{u}} \times \tilde{\bar{\boldsymbol{\epsilon}}}) \cdot \left[\hat{\mathbf{u}} \times \tilde{\mathbf{H}}_t^u\right] = \tilde{\epsilon}^c\hat{\mathbf{u}} \times \left[\bar{\mathbf{I}} \cdot (\hat{\mathbf{u}} \times \tilde{\mathbf{H}}_t^u)\right] = \tilde{\epsilon}^c\hat{\mathbf{u}} \times (\hat{\mathbf{u}} \times \tilde{\mathbf{H}}_t^u) = -\tilde{\epsilon}^c\tilde{\mathbf{H}}_t^u,$$

$$(\hat{\mathbf{u}} \times \tilde{\bar{\boldsymbol{\epsilon}}}) \cdot \left[\hat{\mathbf{u}} \times \tilde{\mathbf{J}}_{mt}^{iu}\right] = \tilde{\epsilon}^c\hat{\mathbf{u}} \times \left[\bar{\mathbf{I}} \cdot (\hat{\mathbf{u}} \times \tilde{\mathbf{J}}_{mt}^{iu})\right] = \tilde{\epsilon}^c\hat{\mathbf{u}} \times (\hat{\mathbf{u}} \times \tilde{\mathbf{J}}_{mt}^{iu}) = -\tilde{\epsilon}^c\tilde{\mathbf{J}}_{mt}^{iu},$$

$$(\hat{\mathbf{u}} \times \tilde{\bar{\boldsymbol{\epsilon}}}) \cdot \nabla_t\tilde{E}_u^u = \hat{\mathbf{u}} \times (\tilde{\epsilon}^c\bar{\mathbf{I}} \cdot \nabla_t\tilde{E}_u^u) = \tilde{\epsilon}^c\hat{\mathbf{u}} \times \nabla_t\tilde{E}_u^u,$$

$$(\hat{\mathbf{u}} \times \tilde{\bar{\boldsymbol{\epsilon}}}) \cdot \hat{\mathbf{u}}\tilde{E}_u^u = \hat{\mathbf{u}} \times (\tilde{\epsilon}^c\bar{\mathbf{I}} \cdot \hat{\mathbf{u}}\tilde{E}_u^u) = \tilde{\epsilon}^c(\hat{\mathbf{u}} \times \hat{\mathbf{u}})\tilde{E}_u^u = 0,$$

and

$$[(\hat{\mathbf{u}} \times \tilde{\boldsymbol{\epsilon}}) \cdot (\hat{\mathbf{u}} \times \tilde{\boldsymbol{\mu}})] \cdot (\hat{\mathbf{u}}\tilde{H}_u^u) = (\hat{\mathbf{u}} \times \tilde{\boldsymbol{\epsilon}}) \cdot [\hat{\mathbf{u}} \times (\tilde{\boldsymbol{\mu}} \cdot \hat{\mathbf{u}}\tilde{H}_u^u)] = (\hat{\mathbf{u}} \times \tilde{\boldsymbol{\epsilon}}) \cdot [\tilde{\mu}(\hat{\mathbf{u}} \times \hat{\mathbf{u}}\tilde{H}_u^u)] = 0.$$

Substituting these into (5.116), we get the transverse magnetic field

$$(k_u^2 - \omega^2\tilde{\mu}\tilde{\epsilon}^c)\tilde{\mathbf{H}}_t = -jk_u\nabla_t\tilde{H}_u + j\omega\tilde{\epsilon}^c\hat{\mathbf{u}} \times \nabla_t\tilde{E}_u - j\omega\tilde{\epsilon}^c\tilde{\mathbf{J}}_{mt}^{iu} + jk_u\hat{\mathbf{u}} \times \tilde{\mathbf{J}}_t^{iu}.$$

Similar steps allow (5.117) to be written as

$$(k_u^2 - \omega^2\tilde{\mu}\tilde{\epsilon}^c)\tilde{\mathbf{E}}_t = -jk_u\nabla_t\tilde{E}_u - j\omega\tilde{\mu}\hat{\mathbf{u}} \times \nabla_t\tilde{H}_u - j\omega\tilde{\mu}\tilde{\mathbf{J}}_t^{iu} - jk_u\hat{\mathbf{u}} \times \tilde{\mathbf{J}}_{mt}^{iu}. \quad \blacktriangleleft$$

5.4 TE–TM decomposition

5.4.1 TE–TM decomposition in terms of fields

A particularly useful field decomposition results for a source-free region. With $\tilde{\mathbf{J}}^i = \tilde{\mathbf{J}}_m^i = 0$ in (5.95)–(5.96) we obtain

$$\left(\frac{\partial^2}{\partial u^2} + k^2\right)\tilde{\mathbf{H}}_t = \nabla_t\frac{\partial\tilde{H}_u}{\partial u} - j\omega\tilde{\epsilon}^c\hat{\mathbf{u}} \times \nabla_t\tilde{E}_u, \tag{5.118}$$

$$\left(\frac{\partial^2}{\partial u^2} + k^2\right)\tilde{\mathbf{E}}_t = \nabla_t\frac{\partial\tilde{E}_u}{\partial u} + j\omega\tilde{\mu}\hat{\mathbf{u}} \times \nabla_t\tilde{H}_u. \tag{5.119}$$

Setting the sources to zero in (5.97)–(5.98) we get

$$\left(\nabla^2 + k^2\right)\tilde{E}_u = 0,$$

$$\left(\nabla^2 + k^2\right)\tilde{H}_u = 0.$$

Hence the longitudinal field components satisfy the homogeneous Helmholtz equation, and the transverse components are specified solely in terms of the longitudinal components. The electromagnetic field is completely specified by the two scalar fields $\tilde{E}_u$ and $\tilde{H}_u$ (and, of course, appropriate boundary values).

Superposition aids in the task of solving (5.118)–(5.119). Since each equation has two forcing terms on the right, we can solve the equations using one forcing term at a time, and add the results. That is, let $\tilde{\mathbf{E}}_1$ and $\tilde{\mathbf{H}}_1$ be the solutions to (5.118)–(5.119) with $\tilde{E}_u = 0$, and $\tilde{\mathbf{E}}_2$ and let $\tilde{\mathbf{H}}_2$ be the solutions with $\tilde{H}_u = 0$. This results in a decomposition

$$\tilde{\mathbf{E}} = \tilde{\mathbf{E}}_1 + \tilde{\mathbf{E}}_2,$$

$$\tilde{\mathbf{H}} = \tilde{\mathbf{H}}_1 + \tilde{\mathbf{H}}_2,$$

with

$$\tilde{\mathbf{E}}_1 = \tilde{\mathbf{E}}_{1t}, \qquad \tilde{\mathbf{H}}_1 = \tilde{\mathbf{H}}_{1t} + \tilde{H}_{1u}\hat{\mathbf{u}},$$

$$\tilde{\mathbf{H}}_2 = \tilde{\mathbf{H}}_{2t}, \qquad \tilde{\mathbf{E}}_2 = \tilde{\mathbf{E}}_{2t} + \tilde{E}_{2u}\hat{\mathbf{u}}.$$

Because $\tilde{\mathbf{E}}_1$ has no u-component, $\tilde{\mathbf{E}}_1$ and $\tilde{\mathbf{H}}_1$ are termed *transverse electric* (or *TE*) to the u-direction; $\tilde{\mathbf{H}}_2$ has no u-component, and $\tilde{\mathbf{E}}_2$ and $\tilde{\mathbf{H}}_2$ are termed *transverse magnetic* (or *TM*) to the *u*-direction.* We see that in a source-free region any electromagnetic field can be decomposed into a set of two fields that are TE and TM, respectively, to some fixed *u*-direction. This is useful for boundary value (e.g., waveguide and scattering) problems where external source information is easily specified using field values on the boundary of the source-free region. In that case $\tilde{E}_u$ and $\tilde{H}_u$ are determined by solving the homogeneous wave equation in an appropriate coordinate system, and the other field components are found from (5.118)–(5.119). Often the boundary conditions can be satisfied by the TM fields or the TE fields alone. This simplifies the analysis of many types of EM systems.

5.4.2 TE–TM decomposition in terms of Hertzian potentials

We are free to represent $\tilde{\mathbf{E}}$ and $\tilde{\mathbf{H}}$ in terms of scalar fields other than $\tilde{E}_u$ and $\tilde{H}_u$. In doing so, it is helpful to retain the wave nature of the solution so that a meaningful physical interpretation is still possible; we thus use Hertzian potentials since they obey the wave equation.

For the TM case let $\tilde{\boldsymbol{\Pi}}_h = 0$ and $\tilde{\boldsymbol{\Pi}}_e = \hat{\mathbf{u}}\tilde{\Pi}_e$. Setting $\tilde{\mathbf{J}}^i = 0$ in (5.58) we have

$$(\nabla^2 + k^2)\tilde{\boldsymbol{\Pi}}_e = 0.$$

Since $\tilde{\boldsymbol{\Pi}}_e$ is purely longitudinal, we can use (B.105) to obtain the scalar Helmholtz equation for $\tilde{\Pi}_e$:

$$(\nabla^2 + k^2)\tilde{\Pi}_e = 0. \tag{5.120}$$

Once a solution $\tilde{\Pi}_e$ has been found, the fields are determined by (5.56)–(5.57) with $\tilde{\mathbf{J}}^i = 0$:

$$\tilde{\mathbf{E}} = \nabla \times (\nabla \times \tilde{\boldsymbol{\Pi}}_e),$$
$$\tilde{\mathbf{H}} = j\omega\tilde{\epsilon}^c \nabla \times \tilde{\boldsymbol{\Pi}}_e. \tag{5.121}$$

We can evaluate $\tilde{\mathbf{E}}$ by noting that $\tilde{\boldsymbol{\Pi}}_e$ is purely longitudinal. Use of (B.104) gives

$$\nabla \times \nabla \times \tilde{\boldsymbol{\Pi}}_e = \nabla_t \frac{\partial \tilde{\Pi}_e}{\partial u} - \hat{\mathbf{u}}\nabla_t^2 \tilde{\Pi}_e.$$

Then, by (B.103),

$$\nabla \times \nabla \times \tilde{\boldsymbol{\Pi}}_e = \nabla_t \frac{\partial \tilde{\Pi}_e}{\partial u} - \hat{\mathbf{u}}\left[\nabla^2 \tilde{\Pi}_e - \frac{\partial^2 \tilde{\Pi}_e}{\partial u^2}\right].$$

By (5.120) then,

$$\tilde{\mathbf{E}} = \nabla_t \frac{\partial \tilde{\Pi}_e}{\partial u} + \hat{\mathbf{u}}\left(\frac{\partial^2}{\partial u^2} + k^2\right)\tilde{\Pi}_e. \tag{5.122}$$

The field $\tilde{\mathbf{H}}$ can be found by noting that $\tilde{\boldsymbol{\Pi}}_e$ is purely longitudinal. Use of (B.102) in (5.121) gives

$$\tilde{\mathbf{H}} = -j\omega\tilde{\epsilon}^c \hat{\mathbf{u}} \times \nabla_t \tilde{\Pi}_e. \tag{5.123}$$

---

*Some authors prefer the terminology *E mode* instead of TM, and *H mode* instead of TE, indicating the presence of a *u*-directed electric or magnetic field component.

Similar steps yield the TE representation. Substitution of $\tilde{\mathbf{\Pi}}_e = 0$ and $\tilde{\mathbf{\Pi}}_h = \hat{\mathbf{u}}\tilde{\Pi}_h$ into (5.59)–(5.60) gives

$$\tilde{\mathbf{E}} = j\omega\tilde{\mu}\hat{\mathbf{u}} \times \nabla_t\tilde{\Pi}_h, \tag{5.124}$$

$$\tilde{\mathbf{H}} = \nabla_t\frac{\partial\tilde{\Pi}_h}{\partial u} + \hat{\mathbf{u}}\left(\frac{\partial^2}{\partial u^2} + k^2\right)\tilde{\Pi}_h, \tag{5.125}$$

while $\tilde{\Pi}_h$ must satisfy

$$(\nabla^2 + k^2)\tilde{\Pi}_h = 0.$$

5.4.2.1   Hertzian potential representation of TEM fields

An interesting situation occurs when a field is both TE and TM to a particular direction. Such a field is said to be *transverse electromagnetic* (or *TEM*) to that direction. Unfortunately, with $\tilde{E}_u = \tilde{H}_u = 0$ we cannot obtain the transverse field components from (5.118) or (5.119). It turns out that a single scalar potential function suffices to represent the field, and we may use either $\tilde{\Pi}_e$ or $\tilde{\Pi}_h$.

For the TM case, Equations (5.122)–(5.123) show that we can represent the electromagnetic fields completely with $\tilde{\Pi}_e$. Unfortunately, (5.122) has a longitudinal component and cannot describe a TEM field. But if we require that $\tilde{\Pi}_e$ obey the additional equation

$$\left(\frac{\partial^2}{\partial u^2} + k^2\right)\tilde{\Pi}_e = 0, \tag{5.126}$$

then both $\tilde{\mathbf{E}}$ and $\tilde{\mathbf{H}}$ are transverse to u and thus describe a TEM field. Since $\tilde{\Pi}_e$ must also obey

$$\left(\nabla^2 + k^2\right)\tilde{\Pi}_e = 0,$$

using (B.103) we can write (5.126) as

$$\nabla_t^2\tilde{\Pi}_e = 0.$$

Similarly, for the TE case the EM fields were completely described in (5.124)–(5.125) by $\tilde{\Pi}_h$. Here $\tilde{\mathbf{H}}$ has a longitudinal component. Thus, if we require

$$\left(\frac{\partial^2}{\partial u^2} + k^2\right)\tilde{\Pi}_h = 0, \tag{5.127}$$

then both $\tilde{\mathbf{E}}$ and $\tilde{\mathbf{H}}$ are transverse to u. Equation (5.127) is equivalent to

$$\nabla_t^2\tilde{\Pi}_h = 0.$$

We can therefore describe a TEM field using either $\tilde{\Pi}_e$ or $\tilde{\Pi}_h$, since a TEM field is both TE and TM to the longitudinal direction. If we choose $\tilde{\Pi}_e$ we can use (5.122)–(5.123) to obtain the expressions

$$\tilde{\mathbf{E}} = \nabla_t\frac{\partial\tilde{\Pi}_e}{\partial u},$$

$$\tilde{\mathbf{H}} = -j\omega\tilde{\epsilon}^c\hat{\mathbf{u}} \times \nabla_t\tilde{\Pi}_e, \tag{5.128}$$

where $\tilde{\Pi}_e$ must obey

$$\nabla_t^2\tilde{\Pi}_e = 0, \qquad \left(\frac{\partial^2}{\partial u^2} + k^2\right)\tilde{\Pi}_e = 0.$$

If we choose $\tilde{\Pi}_h$ we can use (5.124)–(5.125) to obtain

$$\tilde{\mathbf{E}} = j\omega\tilde{\mu}\hat{\mathbf{u}} \times \nabla_t\tilde{\Pi}_h,$$

$$\tilde{\mathbf{H}} = \nabla_t\frac{\partial\tilde{\Pi}_h}{\partial u},$$

where $\tilde{\Pi}_h$ must obey

$$\nabla_t^2\tilde{\Pi}_h = 0, \qquad \left(\frac{\partial^2}{\partial u^2} + k^2\right)\tilde{\Pi}_h = 0.$$

5.4.3 TE–TM decomposition in spherical coordinates

It is not necessary for the longitudinal direction to be constant to achieve a TE–TM decomposition. It is possible, for instance, to represent the electromagnetic field in terms of components either TE or TM to the radial direction of spherical coordinates. This may be shown using a procedure identical to that used for the longitudinal–transverse decomposition in rectangular coordinates. We carry out the decomposition in the frequency domain and leave the time-domain decomposition as an exercise.

5.4.3.1 TE–TM decomposition in terms of the radial fields

Consider a source-free region of space filled with a homogeneous, isotropic material described by parameters $\tilde{\mu}(\omega)$ and $\tilde{\epsilon}^c(\omega)$. We substitute the spherical coordinate representation of the curl into Faraday's and Ampere's laws with source terms $\tilde{\mathbf{J}}$ and $\tilde{\mathbf{J}}_m$ set to zero. Equating vector components, we have, in particular,

$$\frac{1}{r}\left[\frac{1}{\sin\theta}\frac{\partial\tilde{E}_r}{\partial\phi} - \frac{\partial}{\partial r}(r\tilde{E}_\phi)\right] = -j\omega\tilde{\mu}\tilde{H}_\theta \tag{5.129}$$

and

$$\frac{1}{r}\left[\frac{\partial}{\partial r}(r\tilde{H}_\theta) - \frac{\partial\tilde{H}_r}{\partial\theta}\right] = j\omega\tilde{\epsilon}^c\tilde{E}_\phi. \tag{5.130}$$

We seek to isolate the transverse components of the fields in terms of the radial components. Multiplying (5.129) by $j\omega\tilde{\epsilon}^c r$ we get

$$j\omega\tilde{\epsilon}^c\frac{1}{\sin\theta}\frac{\partial\tilde{E}_r}{\partial\phi} - j\omega\tilde{\epsilon}^c\frac{\partial(r\tilde{E}_\phi)}{\partial r} = k^2 r\tilde{H}_\theta;$$

next, multiplying (5.130) by r and then differentiating with respect to r we get

$$\frac{\partial^2}{\partial r^2}(r\tilde{H}_\theta) - \frac{\partial^2\tilde{H}_r}{\partial\theta\partial r} = j\omega\tilde{\epsilon}^c\frac{\partial(r\tilde{E}_\phi)}{\partial r}.$$

Subtracting these two equations and rearranging, we obtain

$$\left(\frac{\partial^2}{\partial r^2} + k^2\right)(r\tilde{H}_\theta) = j\omega\tilde{\epsilon}^c\frac{1}{\sin\theta}\frac{\partial\tilde{E}_r}{\partial\phi} + \frac{\partial^2\tilde{H}_r}{\partial r\partial\theta}.$$

This is a one-dimensional wave equation for the product of r with the transverse field component $\tilde{H}_\theta$. Similarly

$$\left(\frac{\partial^2}{\partial r^2} + k^2\right)(r\tilde{H}_\phi) = -j\omega\tilde{\epsilon}^c\frac{\partial\tilde{E}_r}{\partial\theta} + \frac{1}{\sin\theta}\frac{\partial^2\tilde{H}_r}{\partial r\partial\phi},$$

and

$$\left(\frac{\partial^2}{\partial r^2} + k^2\right)(r\tilde{E}_\phi) = \frac{1}{\sin\theta}\frac{\partial^2 \tilde{E}_r}{\partial\phi\partial r} + j\omega\tilde{\mu}\frac{\partial \tilde{H}_r}{\partial\theta},$$

$$\left(\frac{\partial^2}{\partial r^2} + k^2\right)(r\tilde{E}_\theta) = \frac{\partial^2 \tilde{E}_r}{\partial\theta\partial r} + j\omega\tilde{\mu}\frac{1}{\sin\theta}\frac{\partial \tilde{H}_r}{\partial\phi}.$$

Hence we can represent the electromagnetic field in a source-free region in terms of the scalar quantities $\tilde{E}_r$ and $\tilde{H}_r$. Superposition allows us to solve the TE case with $\tilde{E}_r = 0$, solve the TM case with $\tilde{H}_r = 0$, and combine the results for the general expansion of the field.

5.4.3.2 TE–TM decomposition in terms of potential functions

If we allow the vector potential (or Hertzian potential) to have only an r-component, the resulting fields are TE or TM to the r-direction. Unfortunately, this scalar component does not satisfy the Helmholtz equation. To employ a potential component that satisfies the Helmholtz equation, we must discard the Lorenz condition in favor of a different relation between the vector and scalar potentials.

1. TM fields. To generate fields TM to r we recall that the electromagnetic fields may be written in terms of electric vector and scalar potentials as

$$\tilde{\mathbf{E}} = -j\omega\tilde{\mathbf{A}}_e - \nabla\tilde{\phi}_e, \qquad\qquad (5.131)$$

$$\tilde{\mathbf{B}} = \nabla \times \tilde{\mathbf{A}}_e. \qquad\qquad (5.132)$$

In a source-free region, we have by Ampere's law,

$$\tilde{\mathbf{E}} = \frac{1}{j\omega\tilde{\mu}\tilde{\epsilon}^c}\nabla \times \tilde{\mathbf{B}} = \frac{1}{j\omega\tilde{\mu}\tilde{\epsilon}^c}\nabla \times (\nabla \times \tilde{\mathbf{A}}_e).$$

Here $\tilde{\phi}_e$ and $\tilde{\mathbf{A}}_e$ must satisfy a differential equation that may be derived by examining

$$\nabla \times (\nabla \times \tilde{\mathbf{E}}) = -j\omega\nabla \times \tilde{\mathbf{B}} = -j\omega(j\omega\tilde{\mu}\tilde{\epsilon}^c\tilde{\mathbf{E}}) = k^2\tilde{\mathbf{E}},$$

where $k^2 = \omega^2\tilde{\mu}\tilde{\epsilon}^c$. Substitution from (5.131) gives

$$\nabla \times \left(\nabla \times [-j\omega\tilde{\mathbf{A}}_e - \nabla\tilde{\phi}_e]\right) = k^2[-j\omega\tilde{\mathbf{A}}_e - \nabla\tilde{\phi}_e]$$

or

$$\nabla \times (\nabla \times \tilde{\mathbf{A}}_e) - k^2\tilde{\mathbf{A}}_e = \frac{k^2}{j\omega}\nabla\tilde{\phi}_e. \qquad\qquad (5.133)$$

We are still free to specify $\nabla \cdot \tilde{\mathbf{A}}_e$.

At this point let us examine the effect of choosing a vector potential with only an r-component: $\tilde{\mathbf{A}}_e = \hat{\mathbf{r}}\tilde{A}_e$. Since

$$\nabla \times (\hat{\mathbf{r}}\tilde{A}_e) = \frac{\hat{\boldsymbol{\theta}}}{r\sin\theta}\frac{\partial \tilde{A}_e}{\partial\phi} - \frac{\hat{\boldsymbol{\phi}}}{r}\frac{\partial \tilde{A}_e}{\partial\theta} \qquad\qquad (5.134)$$

we see that $\mathbf{B} = \nabla \times \tilde{\mathbf{A}}_e$ has no r-component. Since

$$\nabla \times (\nabla \times \tilde{\mathbf{A}}_e) = -\frac{\hat{\mathbf{r}}}{r\sin\theta}\left[\frac{1}{r}\frac{\partial}{\partial\theta}\left(\sin\theta\frac{\partial \tilde{A}_e}{\partial\theta}\right) + \frac{1}{r\sin\theta}\frac{\partial^2 \tilde{A}_e}{\partial\phi^2}\right] + \frac{\hat{\boldsymbol{\theta}}}{r}\frac{\partial^2 \tilde{A}_e}{\partial r\partial\theta} + \frac{\hat{\boldsymbol{\phi}}}{r\sin\theta}\frac{\partial^2 \tilde{A}_e}{\partial r\partial\phi}$$

we see that $\tilde{\mathbf{E}} \sim \nabla \times (\nabla \times \tilde{\mathbf{A}}_e)$ has all three components. This choice of $\tilde{\mathbf{A}}_e$ produces a field TM to the r-direction. We need only choose $\nabla \cdot \tilde{\mathbf{A}}_e$ so that the resulting differential equation is convenient to solve. Substituting the above expressions into (5.133), we find that

$$-\frac{\hat{\mathbf{r}}}{r\sin\theta}\left[\frac{1}{r}\frac{\partial}{\partial\theta}\left(\sin\theta\frac{\partial\tilde{A}_e}{\partial\theta}\right) + \frac{1}{r\sin\theta}\frac{\partial^2\tilde{A}_e}{\partial\phi^2}\right] + \frac{\hat{\boldsymbol{\theta}}}{r}\frac{\partial^2\tilde{A}_e}{\partial r\partial\theta} + \frac{\hat{\boldsymbol{\phi}}}{r\sin\theta}\frac{\partial^2\tilde{A}_e}{\partial r\partial\phi} - \hat{\mathbf{r}}k^2\tilde{A}_e =$$
$$\hat{\mathbf{r}}\frac{k^2}{j\omega}\frac{\partial\tilde{\phi}_e}{\partial r} + \frac{\hat{\boldsymbol{\theta}}}{r}\frac{k^2}{j\omega}\frac{\partial\tilde{\phi}_e}{\partial\theta} + \frac{\hat{\boldsymbol{\phi}}}{r\sin\theta}\frac{k^2}{j\omega}\frac{\partial\tilde{\phi}_e}{\partial\phi}. \tag{5.135}$$

Since $\nabla \cdot \tilde{\mathbf{A}}_e$ only involves the derivatives of $\tilde{\mathbf{A}}_e$ with respect to r, we may specify $\nabla \cdot \tilde{\mathbf{A}}_e$ indirectly through

$$\tilde{\phi}_e = \frac{j\omega}{k^2}\frac{\partial\tilde{A}_e}{\partial r}.$$

With this, (5.135) becomes

$$\frac{1}{r\sin\theta}\left[\frac{1}{r}\frac{\partial}{\partial\theta}\left(\sin\theta\frac{\partial\tilde{A}_e}{\partial\theta}\right) + \frac{1}{r\sin\theta}\frac{\partial^2\tilde{A}_e}{\partial\phi^2}\right] + k^2\tilde{A}_e + \frac{\partial^2\tilde{A}_e}{\partial r^2} = 0.$$

Using

$$\frac{1}{r}\frac{\partial}{\partial r}\left[r^2\frac{\partial}{\partial r}\left(\frac{\tilde{A}_e}{r}\right)\right] = \frac{\partial^2\tilde{A}_e}{\partial r^2}$$

we can write the differential equation as

$$\frac{1}{r^2}\frac{\partial}{\partial r}\left[r^2\frac{\partial(\tilde{A}_e/r)}{\partial r}\right] + \frac{1}{r^2\sin\theta}\frac{\partial}{\partial\theta}\left[\sin\theta\frac{\partial(\tilde{A}_e/r)}{\partial\theta}\right] + \frac{1}{r^2\sin^2\theta}\frac{\partial^2(\tilde{A}_e/r)}{\partial\phi^2} + k^2\frac{\tilde{A}_e}{r} = 0.$$

The first three terms of this expression are precisely the Laplacian of $\tilde{A}_e/r$. Thus

$$(\nabla^2 + k^2)\left(\frac{\tilde{A}_e}{r}\right) = 0 \tag{5.136}$$

and the quantity $\tilde{A}_e/r$ satisfies the homogeneous Helmholtz equation.

The TM fields generated by the vector potential $\tilde{\mathbf{A}}_e = \hat{\mathbf{r}}\tilde{A}_e$ may be found by using (5.131)–(5.132). By (5.131) we have

$$\tilde{\mathbf{E}} = -j\omega\tilde{\mathbf{A}}_e - \nabla\tilde{\phi}_e = -j\omega\hat{\mathbf{r}}\tilde{A}_e - \nabla\left(\frac{j\omega}{k^2}\frac{\partial\tilde{A}_e}{\partial r}\right).$$

Expanding the gradient, we have the field components

$$\tilde{E}_r = \frac{1}{j\omega\tilde{\mu}\tilde{\epsilon}^c}\left(\frac{\partial^2}{\partial r^2} + k^2\right)\tilde{A}_e, \tag{5.137}$$

$$\tilde{E}_\theta = \frac{1}{j\omega\tilde{\mu}\tilde{\epsilon}^c}\frac{1}{r}\frac{\partial^2\tilde{A}_e}{\partial r\partial\theta}, \tag{5.138}$$

$$\tilde{E}_\phi = \frac{1}{j\omega\tilde{\mu}\tilde{\epsilon}^c}\frac{1}{r\sin\theta}\frac{\partial^2\tilde{A}_e}{\partial r\partial\phi}. \tag{5.139}$$

The magnetic field components are found using (5.132) and (5.134):

$$\tilde{H}_\theta = \frac{1}{\tilde{\mu}} \frac{1}{r \sin\theta} \frac{\partial \tilde{A}_e}{\partial \phi}, \tag{5.140}$$

$$\tilde{H}_\phi = -\frac{1}{\tilde{\mu}} \frac{1}{r} \frac{\partial \tilde{A}_e}{\partial \theta}. \tag{5.141}$$

2. TE fields. To generate fields TE to r we recall that the electromagnetic fields in a source-free region may be written in terms of magnetic vector and scalar potentials as

$$\tilde{\mathbf{H}} = -j\omega\tilde{\mathbf{A}}_h - \nabla\phi_h, \tag{5.142}$$

$$\tilde{\mathbf{D}} = -\nabla \times \tilde{\mathbf{A}}_h. \tag{5.143}$$

In a source-free region, we have from Faraday's law,

$$\tilde{\mathbf{H}} = \frac{1}{-j\omega\tilde{\mu}\tilde{\epsilon}^c} \nabla \times \tilde{\mathbf{D}} = \frac{1}{j\omega\tilde{\mu}\tilde{\epsilon}^c} \nabla \times (\nabla \times \tilde{\mathbf{A}}_h).$$

Here $\tilde{\phi}_h$ and $\tilde{\mathbf{A}}_h$ must satisfy a differential equation that may be derived by examining

$$\nabla \times (\nabla \times \tilde{\mathbf{H}}) = j\omega\nabla \times \tilde{\mathbf{D}} = j\omega\tilde{\epsilon}^c(-j\omega\tilde{\mu}\tilde{\mathbf{H}}) = k^2\tilde{\mathbf{H}},$$

where $k^2 = \omega^2\tilde{\mu}\tilde{\epsilon}^c$. Substitution from (5.142) gives

$$\nabla \times \left(\nabla \times [-j\omega\tilde{\mathbf{A}}_h - \nabla\tilde{\phi}_h]\right) = k^2[-j\omega\tilde{\mathbf{A}}_h - \nabla\tilde{\phi}_h]$$

or

$$\nabla \times (\nabla \times \tilde{\mathbf{A}}_h) - k^2\tilde{\mathbf{A}}_h = \frac{k^2}{j\omega}\nabla\tilde{\phi}_h.$$

Choosing $\tilde{\mathbf{A}}_h = \hat{\mathbf{r}}\tilde{A}_h$ and

$$\tilde{\phi}_h = \frac{j\omega}{k^2} \frac{\partial \tilde{A}_h}{\partial r}$$

we find, as with the TM fields,

$$(\nabla^2 + k^2)\left(\frac{\tilde{A}_h}{r}\right) = 0. \tag{5.144}$$

Thus the quantity $\tilde{A}_h/r$ obeys the Helmholtz equation.

We can find the TE fields using (5.142)–(5.143):

$$\tilde{H}_r = \frac{1}{j\omega\tilde{\mu}\tilde{\epsilon}^c}\left(\frac{\partial^2}{\partial r^2} + k^2\right)\tilde{A}_h, \tag{5.145}$$

$$\tilde{H}_\theta = \frac{1}{j\omega\tilde{\mu}\tilde{\epsilon}^c} \frac{1}{r} \frac{\partial^2 \tilde{A}_h}{\partial r \partial \theta}, \tag{5.146}$$

$$\tilde{H}_\phi = \frac{1}{j\omega\tilde{\mu}\tilde{\epsilon}^c} \frac{1}{r \sin\theta} \frac{\partial^2 \tilde{A}_h}{\partial r \partial \phi}, \tag{5.147}$$

$$\tilde{E}_\theta = -\frac{1}{\tilde{\epsilon}^c} \frac{1}{r \sin\theta} \frac{\partial \tilde{A}_h}{\partial \phi}, \tag{5.148}$$

$$\tilde{E}_\phi = \frac{1}{\tilde{\epsilon}^c} \frac{1}{r} \frac{\partial \tilde{A}_h}{\partial \theta}. \tag{5.149}$$

▶ **Example 5.8:** Spherical TE–TM decomposition of a plane wave

A uniform plane wave propagates along the z-direction in a homogeneous material of complex permittivity $\tilde{\epsilon}^c$ and permeability $\tilde{\mu}$. Represent its electromagnetic field

$$\tilde{\mathbf{E}}(\mathbf{r}, \omega) = \hat{\mathbf{x}}\tilde{E}_0(\omega)e^{-jkz} = \hat{\mathbf{x}}\tilde{E}_0(\omega)e^{-jkr\cos\theta},$$

$$\tilde{\mathbf{H}}(\mathbf{r}, \omega) = \hat{\mathbf{y}}\frac{\tilde{E}_0(\omega)}{\eta}e^{-jkz} = \hat{\mathbf{x}}\frac{\tilde{E}_0(\omega)}{\eta}e^{-jkr\cos\theta},$$

as a superposition of fields TE to r and TM to r.

Solution: After finding the potential functions $\tilde{\mathbf{A}}_e = \hat{\mathbf{r}}\tilde{A}_e$ and $\tilde{\mathbf{A}}_h = \hat{\mathbf{r}}\tilde{A}_h$ that represent the field, we can use (5.137)–(5.141) and (5.145)–(5.149) to get the TE and TM representations.

From (5.137) we see that $\tilde{A}_e$ is related to $\tilde{E}_r$, where

$$\tilde{E}_r = \tilde{E}_0 \sin\theta \cos\phi\, e^{-jkr\cos\theta}$$

$$= \frac{\tilde{E}_0 \cos\phi}{jkr}\frac{\partial}{\partial\theta}\left[e^{-jkr\cos\theta}\right].$$

The identity (E.103) lets us separate the r and θ dependences of the exponential function. Since

$$j_n(-z) = (-1)^n j_n(z)$$

$$= j^{-2n} j_n(z)$$

we have

$$e^{-jkr\cos\theta} = \sum_{n=0}^{\infty} j^{-n}(2n+1)j_n(kr)P_n(\cos\theta). \tag{5.150}$$

Using

$$\frac{\partial P_n(\cos\theta)}{\partial\theta} = \frac{\partial P_n^0(\cos\theta)}{\partial\theta}$$

$$= P_n^1(\cos\theta)$$

we thus have

$$\tilde{E}_r = -\frac{j\tilde{E}_0 \cos\phi}{kr}\sum_{n=1}^{\infty} j^{-n}(2n+1)j_n(kr)P_n^1(\cos\theta).$$

Here we start the sum at $n = 1$ since $P_0^1(x) = 0$. We can now identify the vector potential as

$$\frac{\tilde{A}_e}{r} = \frac{\tilde{E}_0 k}{\omega}\cos\phi\sum_{n=1}^{\infty}\frac{j^{-n}(2n+1)}{n(n+1)}j_n(kr)P_n^1(\cos\theta) \tag{5.151}$$

since by direct differentiation we have

$$\tilde{E}_r = \frac{1}{j\omega\tilde{\mu}\tilde{\epsilon}^c}\left(\frac{\partial^2}{\partial r^2} + k^2\right)\tilde{A}_e$$

$$= \frac{\tilde{E}_0 k}{j\omega^2\tilde{\mu}\tilde{\epsilon}^c}\cos\phi\sum_{n=1}^{\infty}\frac{j^{-n}(2n+1)}{n(n+1)}P_n^1(\cos\theta)\left(\frac{\partial^2}{\partial r^2} + k^2\right)[rj_n(kr)]$$

$$= -\frac{j\tilde{E}_0 \cos\phi}{kr}\sum_{n=1}^{\infty} j^{-n}(2n+1)j_n(kr)P_n^1(\cos\theta),$$

which satisfies (5.137). Here we have used the defining equation of the spherical Bessel functions (E.15) to show that

$$
\left(\frac{\partial^2}{\partial r^2} + k^2 \right) [r j_n(kr)] = r \frac{\partial^2}{\partial r^2} j_n(kr) + 2 \frac{\partial}{\partial r} j_n(kr) + k^2 r j_n(kr)
$$

$$
= k^2 r \left[\frac{\partial^2}{\partial (kr)^2} + \frac{2}{kr} \frac{\partial}{\partial (kr)} \right] j_n(kr) + k^2 r j_n(kr)
$$

$$
= -k^2 r \left[1 - \frac{n(n+1)}{(kr)^2} \right] j_n(kr) + k^2 r j_n(kr)
$$

$$
= \frac{n(n+1)}{r} j_n(kr).
$$

Observe that $\tilde{A}_e/r$ satisfies the Helmholtz equation (5.136), since it has the form of the separation of variables solution (D.113).

We may find the vector potential $\mathbf{\tilde{A}}_h = \hat{\mathbf{r}} \tilde{A}_h$ in the same manner. Noting that

$$
\tilde{H}_r = \frac{\tilde{E}_0}{\eta} \sin\theta \sin\phi e^{-jkr\cos\theta} = \frac{\tilde{E}_0 \sin\phi}{\eta j k r} \frac{\partial}{\partial\theta} \left[e^{-jkr\cos\theta} \right]
$$

$$
= \frac{1}{j\omega\tilde{\mu}\tilde{\epsilon}^c} \left(\frac{\partial^2}{\partial r^2} + k^2 \right) \tilde{A}_h,
$$

we have

$$
\frac{\tilde{A}_h}{r} = \frac{\tilde{E}_0 k}{\eta \omega} \sin\phi \sum_{n=1}^{\infty} \frac{j^{-n}(2n+1)}{n(n+1)} j_n(kr) P_n^1(\cos\theta). \tag{5.152}
$$

We may now compute the transverse components of the TM field using (5.138)–(5.141). For convenience let us define $\hat{J}_n(x) = x j_n(x)$. Then

$$
\tilde{E}_r = -\frac{j\tilde{E}_0 \cos\phi}{(kr)^2} \sum_{n=1}^{\infty} j^{-n}(2n+1) \hat{J}_n(kr) P_n^1(\cos\theta), \tag{5.153}
$$

$$
\tilde{E}_\theta = \frac{j\tilde{E}_0}{kr} \sin\theta \cos\phi \sum_{n=1}^{\infty} a_n \hat{J}_n'(kr) P_n^{1'}(\cos\theta), \tag{5.154}
$$

$$
\tilde{E}_\phi = \frac{j\tilde{E}_0}{kr \sin\theta} \sin\phi \sum_{n=1}^{\infty} a_n \hat{J}_n'(kr) P_n^1(\cos\theta), \tag{5.155}
$$

$$
\tilde{H}_\theta = -\frac{\tilde{E}_0}{kr\eta \sin\theta} \sin\phi \sum_{n=1}^{\infty} a_n \hat{J}_n(kr) P_n^1(\cos\theta), \tag{5.156}
$$

$$
\tilde{H}_\phi = \frac{\tilde{E}_0}{kr\eta} \sin\theta \cos\phi \sum_{n=1}^{\infty} a_n \hat{J}_n(kr) P_n^{1'}(\cos\theta). \tag{5.157}
$$

Here

$$
\hat{J}_n'(x) = \frac{d}{dx} \hat{J}_n(x) = \frac{d}{dx} [x j_n(x)] = x j_n'(x) + j_n(x)
$$

and

$$
a_n = \frac{j^{-n}(2n+1)}{n(n+1)}. \tag{5.158}
$$

Similarly, we have the TE fields from (5.146)–(5.149):

$$
\tilde{H}_r = -\frac{j\tilde{E}_0 \sin\phi}{\eta(kr)^2} \sum_{n=1}^{\infty} j^{-n}(2n+1) \hat{J}_n(kr) P_n^1(\cos\theta), \tag{5.159}
$$

$$\tilde{H}_\theta = j\frac{\tilde{E}_0}{\eta kr}\sin\theta\sin\phi\sum_{n=1}^{\infty}a_n\hat{J}_n'(kr)P_n^{1'}(\cos\theta), \tag{5.160}$$

$$\tilde{H}_\phi = -j\frac{\tilde{E}_0}{\eta kr\sin\theta}\cos\phi\sum_{n=1}^{\infty}a_n\hat{J}_n'(kr)P_n^{1}(\cos\theta), \tag{5.161}$$

$$\tilde{E}_\theta = -\frac{\tilde{E}_0}{kr\sin\theta}\cos\phi\sum_{n=1}^{\infty}a_n\hat{J}_n(kr)P_n^{1}(\cos\theta), \tag{5.162}$$

$$\tilde{E}_\phi = -\frac{\tilde{E}_0}{kr}\sin\theta\sin\phi\sum_{n=1}^{\infty}a_n\hat{J}_n(kr)P_n^{1'}(\cos\theta). \tag{5.163}$$

The total field is the sum of the TE and TM components. ◄

▶ **Example 5.9:** Scattering by a sphere using spherical TE–TM decomposition

Consider a PEC sphere of radius a centered at the origin and embedded in a homogeneous, isotropic material having parameters $\tilde{\mu}$ and $\tilde{\epsilon}^c$. The sphere is illuminated by a plane wave incident along the z-axis with the fields

$$\tilde{\mathbf{E}}(\mathbf{r},\omega) = \hat{\mathbf{x}}\tilde{E}_0(\omega)e^{-jkz} = \hat{\mathbf{x}}\tilde{E}_0(\omega)e^{-jkr\cos\theta},$$

$$\tilde{\mathbf{H}}(\mathbf{r},\omega) = \hat{\mathbf{y}}\frac{\tilde{E}_0(\omega)}{\eta}e^{-jkz} = \hat{\mathbf{x}}\frac{\tilde{E}_0(\omega)}{\eta}e^{-jkr\cos\theta}.$$

Find the field scattered by the sphere.

Solution: The boundary condition determining the scattered field is that the total (incident plus scattered) electric field tangential to the sphere must vanish. Example 5.8 showed that the incident electric field may be written as a sum of fields TE and TM to the r-direction. Since the region outside the sphere is source-free, we may also represent the scattered field as a sum of TE and TM fields. These may be obtained from the functions $\tilde{A}_e^s$ and $\tilde{A}_h^s$, which obey Helmholtz equations (5.136) and (5.144). Separation of variables applied to the Helmholtz equation in spherical coordinates (§ A.5.3) yields the general solution

$$\left\{\begin{matrix}\tilde{A}_e^s/r\\\tilde{A}_h^s/r\end{matrix}\right\} = \sum_{n=0}^{\infty}\sum_{m=-n}^{n}C_{nm}Y_{nm}(\theta,\phi)h_n^{(2)}(kr).$$

Here Y_{nm} is the spherical harmonic and we take the spherical Hankel function $h_n^{(2)}$ as the radial dependence, since it represents the expected outward-going wave behavior of the scattered field. Since the incident field generated by the potentials (5.151) and (5.152) exactly cancels the field generated by $\tilde{A}_e^s$ and $\tilde{A}_h^s$ on the surface of the sphere, by orthogonality the scattered potential must have ϕ and θ dependences matching those of the incident field. Thus

$$\frac{\tilde{A}_e^s}{r} = \frac{\tilde{E}_0 k}{\omega}\cos\phi\sum_{n=1}^{\infty}b_n h_n^{(2)}(kr)P_n^{1}(\cos\theta),$$

$$\frac{\tilde{A}_h^s}{r} = \frac{\tilde{E}_0 k}{\eta\omega}\sin\phi\sum_{n=1}^{\infty}c_n h_n^{(2)}(kr)P_n^{1}(\cos\theta),$$

where b_n and c_n are constants to be determined by the boundary conditions. By superposition, the total field may be computed from the total potentials, which are sums of incident and scattered potentials:

$$\frac{\tilde{A}_e^t}{r} = \frac{\tilde{E}_0 k}{\omega} \cos\phi \sum_{n=1}^{\infty} \left[a_n j_n(kr) + b_n h_n^{(2)}(kr) \right] P_n^1(\cos\theta),$$

$$\frac{\tilde{A}_h^t}{r} = \frac{\tilde{E}_0 k}{\eta\omega} \sin\phi \sum_{n=1}^{\infty} \left[a_n j_n(kr) + c_n h_n^{(2)}(kr) \right] P_n^1(\cos\theta),$$

where a_n is given by (5.158).

The total transverse electric field is found by superposing the TE and TM transverse fields found from the total potentials. We have already computed the transverse incident fields and may easily generalize these results to the total potentials. By (5.154) and (5.162) we have

$$\tilde{E}_\theta^t(a) = \frac{j\tilde{E}_0}{ka} \sin\theta \cos\phi \sum_{n=1}^{\infty} \left[a_n \hat{J}_n'(ka) + b_n \hat{H}_n^{(2)\prime}(ka) \right] P_n^{1\prime}(\cos\theta)$$

$$- \frac{\tilde{E}_0}{ka\sin\theta} \cos\phi \sum_{n=1}^{\infty} \left[a_n \hat{J}_n(ka) + c_n \hat{H}_n^{(2)}(ka) \right] P_n^1(\cos\theta) = 0,$$

where

$$\hat{H}_n^{(2)}(x) = x h_n^{(2)}(x).$$

By (5.155) and (5.163) we have

$$\tilde{E}_\phi^t(a) = \frac{j\tilde{E}_0}{ka\sin\theta} \sin\phi \sum_{n=1}^{\infty} \left[a_n \hat{J}_n'(ka) + b_n \hat{H}_n^{(2)\prime}(ka) \right] P_n^1(\cos\theta)$$

$$- \frac{\tilde{E}_0}{ka} \sin\theta \sin\phi \sum_{n=1}^{\infty} \left[a_n \hat{J}_n(ka) + c_n \hat{H}_n^{(2)}(ka) \right] P_n^{1\prime}(\cos\theta) = 0.$$

These two sets of equations are satisfied by the conditions

$$b_n = -\frac{\hat{J}_n'(ka)}{\hat{H}_n^{(2)\prime}(ka)} a_n, \qquad c_n = -\frac{\hat{J}_n(ka)}{\hat{H}_n^{(2)}(ka)} a_n.$$

Therefore

$$\tilde{\mathbf{E}}_r^s = -j\tilde{E}_0 \cos\phi \sum_{n=1}^{\infty} b_n \left[\hat{H}_n^{(2)\prime\prime}(kr) + \hat{H}_n^{(2)}(kr) \right] P_n^1(\cos\theta), \tag{5.164}$$

$$\tilde{\mathbf{E}}_\theta^s = \frac{\tilde{E}_0}{kr} \cos\phi \sum_{n=1}^{\infty} \left[j b_n \sin\theta \hat{H}_n^{(2)\prime}(kr) P_n^{1\prime}(\cos\theta) - c_n \frac{1}{\sin\theta} \hat{H}_n^{(2)}(kr) P_n^1(\cos\theta) \right], \tag{5.165}$$

$$\tilde{\mathbf{E}}_\phi^s = \frac{\tilde{E}_0}{kr} \sin\phi \sum_{n=1}^{\infty} \left[j b_n \frac{1}{\sin\theta} \hat{H}_n^{(2)\prime}(kr) P_n^1(\cos\theta) - c_n \sin\theta \hat{H}_n^{(2)}(kr) P_n^{1\prime}(\cos\theta) \right]. \blacktriangleleft \tag{5.166}$$

▶ **Example 5.10:** Radar cross-section of a sphere

Specialize the fields scattered by a sphere to the far zone and compute the radar cross-section of the sphere.

Solution: To approximate the scattered field for observation points far from the sphere, we approximate the spherical Hankel functions using (E.70) as

$$\hat{H}_n^{(2)}(z) = z h_n^{(2)}(z) \approx j^{n+1} e^{-jz}, \quad \hat{H}_n^{(2)\prime}(z) \approx j^n e^{-jz}, \quad \hat{H}_n^{(2)\prime\prime}(z) \approx -j^{n+1} e^{-jz}.$$

Substituting these, we find that $\tilde{E}_r \to 0$ as expected for the far-zone field, while

$$\tilde{E}_\theta^s \approx \tilde{E}_0 \frac{e^{-jkr}}{kr} \cos\phi \sum_{n=1}^{\infty} j^{n+1} \left[b_n \sin\theta P_n^{1\prime}(\cos\theta) - c_n \frac{1}{\sin\theta} P_n^1(\cos\theta) \right],$$

$$\tilde{E}_\phi^s \approx \tilde{E}_0 \frac{e^{-jkr}}{kr} \sin\phi \sum_{n=1}^{\infty} j^{n+1} \left[b_n \frac{1}{\sin\theta} P_n^1(\cos\theta) - c_n \sin\theta P_n^{1\prime}(\cos\theta) \right].$$

From the far-zone fields we can compute the *radar cross-section* (RCS) or *echo area* of the sphere, defined by

$$\sigma = \lim_{r\to\infty} \left(4\pi r^2 \frac{|\tilde{\mathbf{E}}^s|^2}{|\tilde{\mathbf{E}}^i|^2} \right). \tag{5.167}$$

Carrying units of m^2, this quantity describes the relative energy density of the scattered field normalized by the distance from the scattering object. Figure 5.5 shows the RCS of a conducting sphere in free space for the *monostatic* case: when the observation direction is aligned with the direction of the incident wave (i.e., $\theta = \pi$), also called the *backscatter* direction. At low frequencies the RCS is proportional to λ^{-4}; this is the range of *Rayleigh scattering*, showing that higher-frequency light scatters more strongly from microscopic particles in the atmosphere (explaining why the sky is blue) [22]. At high frequencies the result approaches that of geometrical optics, and the RCS becomes the interception area of the sphere, πa^2. This is the region of *optical scattering*. Between these two regions lies the *resonance region*, or the region of *Mie scattering*, named for G. Mie who in 1908 published the first rigorous solution for scattering by a sphere (followed soon after by Debye in 1909).

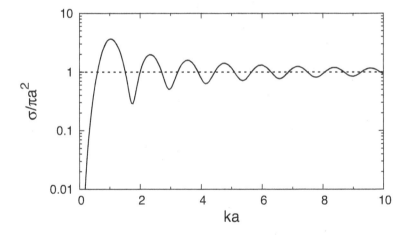

FIGURE 5.5
Monostatic radar cross-section of a conducting sphere. ◄

▶ **Example 5.11:** Time-domain scattering from a sphere

Consider the far-zone field scattered by a sphere found in Example 5.10. Transform the field into the time-domain using the inverse Fourier transform, and identify the significant scattering events.

Solution: Several interesting phenomena of sphere scattering are best examined in the time domain. We may compute the temporal scattered field numerically by performing an inverse fast Fourier transform of the frequency-domain field. Figure 5.6 shows $E_\theta(t)$ computed in the backscatter direction ($\theta = \pi$) when the incident field waveform $E_0(t)$ is a Gaussian pulse and the sphere is in free space. Two distinct features are seen in the scattered field waveform. The first is a sharp pulse almost duplicating the incident field waveform, but of opposite polarity. This is the *specular reflection* produced when the incident field first contacts the sphere and begins to induce a current on the sphere surface. The second feature, called the *creeping wave*, occurs at a time approximately $(2 + \pi)a/c$ seconds after the specular reflection. This represents the field radiated back along the incident direction by a wave of current excited by the incident field at the tangent point, which travels around the sphere at approximately the speed of light in free space. Although this wave continues to traverse the sphere, its amplitude is reduced so significantly by radiation damping that only a single feature is seen.

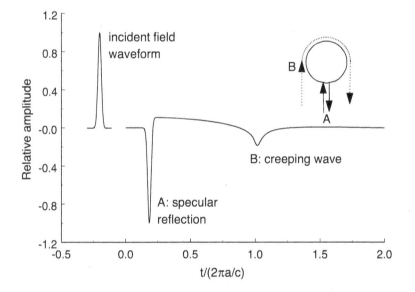

FIGURE 5.6

Time-domain field back-scattered by a conducting sphere. ◄

5.4.4   TE–TM decomposition for anisotropic media

To obtain a decomposition in terms of TE and TM fields, we set the sources in (5.116)–(5.117) to zero. It is clear that we can use superposition to find the transverse fields as a sum of those fields having $\tilde{E}_u = 0$ (TE fields) and those having $\tilde{H}_u = 0$ (TM fields). For TE fields, we have

$$\tilde{\mathbf{H}}_t^u = \bar{\boldsymbol{\Omega}}^{-1} \cdot \left\{ -jk_u \nabla_t \tilde{H}_u^u - \omega^2 \left[(\hat{\mathbf{u}} \times \tilde{\bar{\boldsymbol{\epsilon}}}) \cdot (\hat{\mathbf{u}} \times \tilde{\bar{\boldsymbol{\mu}}}) \right] \cdot \hat{\mathbf{u}} \tilde{H}_u^u \right\}, \qquad (5.168)$$

$$\tilde{\mathbf{E}}_t^u = \bar{\boldsymbol{\Phi}}^{-1} \cdot \left\{ -j\omega (\hat{\mathbf{u}} \times \tilde{\bar{\boldsymbol{\mu}}}) \cdot \left[\nabla_t \tilde{H}_u^u + \hat{\mathbf{u}} jk_u \tilde{H}_u^u \right] \right\}, \qquad (5.169)$$

while for TM fields

$$\tilde{\mathbf{H}}_t^u = \bar{\mathbf{\Omega}}^{-1} \cdot \left\{ j\omega(\hat{\mathbf{u}} \times \tilde{\bar{\boldsymbol{\epsilon}}}) \cdot [\nabla_t \tilde{E}_u^u + \hat{\mathbf{u}} j k_u \tilde{E}_u^u] \right\}, \tag{5.170}$$

$$\tilde{\mathbf{E}}_t^u = \bar{\mathbf{\Phi}}^{-1} \cdot \left\{ -j k_u \nabla_t \tilde{E}_u^u - \omega^2 \left[(\hat{\mathbf{u}} \times \tilde{\bar{\boldsymbol{\mu}}}) \cdot (\hat{\mathbf{u}} \times \tilde{\bar{\boldsymbol{\epsilon}}}) \right] \cdot \hat{\mathbf{u}} \tilde{E}_u^u \right\}. \tag{5.171}$$

To employ the TE–TM decomposition in a practical situation, it is necessary to determine differential equations for $\tilde{E}_u^u$ and $\tilde{H}_u^u$. Once these are solved, the transverse fields may be found using (5.168)–(5.171).

We first find the differential equation for $\tilde{H}_u^u$ for TE fields. From (5.99) and (5.103) we see that

$$\nabla_t \times \tilde{\mathbf{E}}_t^u = -j\omega \hat{\mathbf{u}} (\tilde{\bar{\boldsymbol{\mu}}} \cdot \tilde{\mathbf{H}}_t^u + \tilde{\mu}_{uu} \tilde{H}_u^u). \tag{5.172}$$

Substitution from (5.168)–(5.169) gives

$$\nabla_t \times \left[\bar{\mathbf{\Phi}}^{-1} \cdot \left\{ -j\omega(\hat{\mathbf{u}} \times \tilde{\bar{\boldsymbol{\mu}}}) \cdot [\nabla_t \tilde{H}_u^u + \hat{\mathbf{u}} j k_u \tilde{H}_u^u] \right\} \right] + j\omega \hat{\mathbf{u}} \tilde{\mu}_{uu} \tilde{H}_u^u$$
$$+ j\omega \hat{\mathbf{u}} \left[(\hat{\mathbf{u}} \cdot \tilde{\bar{\boldsymbol{\mu}}}) \cdot \bar{\mathbf{\Omega}}^{-1} \cdot \left\{ -j k_u \nabla_t \tilde{H}_u^u - \omega^2 \left[(\hat{\mathbf{u}} \times \tilde{\bar{\boldsymbol{\epsilon}}}) \cdot (\hat{\mathbf{u}} \times \tilde{\bar{\boldsymbol{\mu}}}) \right] \cdot \hat{\mathbf{u}} \tilde{H}_u^u \right\} \right] = 0, \tag{5.173}$$

which is a differential equation for $\tilde{H}_u^u$. For TM fields we use (5.101) and (5.105) to find

$$\nabla_t \times \tilde{\mathbf{H}}_t^u = j\omega \hat{\mathbf{u}} (\tilde{\bar{\boldsymbol{\epsilon}}} \cdot \tilde{\mathbf{E}}_t^u + \tilde{\epsilon}_{uu} \tilde{E}_u^u).$$

Substitution from (5.170)–(5.171) gives a differential equation for $\tilde{E}_u^u$:

$$\nabla_t \times \left[\bar{\mathbf{\Omega}}^{-1} \cdot \left\{ j\omega(\hat{\mathbf{u}} \times \tilde{\bar{\boldsymbol{\epsilon}}}) \cdot [\nabla_t \tilde{E}_u^u + \hat{\mathbf{u}} j k_u \tilde{E}_u^u] \right\} \right] - j\omega \hat{\mathbf{u}} \tilde{\epsilon}_{uu} \tilde{E}_u^u$$
$$- j\omega \hat{\mathbf{u}} \left[(\hat{\mathbf{u}} \cdot \tilde{\bar{\boldsymbol{\epsilon}}}) \cdot \bar{\mathbf{\Phi}}^{-1} \cdot \left\{ -j k_u \nabla_t \tilde{E}_u^u - \omega^2 \left[(\hat{\mathbf{u}} \times \tilde{\bar{\boldsymbol{\mu}}}) \cdot (\hat{\mathbf{u}} \times \tilde{\bar{\boldsymbol{\epsilon}}}) \right] \cdot \hat{\mathbf{u}} \tilde{E}_u^u \right\} \right] = 0.$$

▶ **Example 5.12:** TE fields in a lossless magnetized ferrite

Consider fields TE_z in a lossless ferrite magnetized along the y-direction. Find the wave equation for $\tilde{E}_z^z$, and find the relations between the transverse fields and $\tilde{E}_z^z$.

Solution: For a lossless ferrite magnetized along $\hat{\mathbf{y}}$ we have (§ 4.6.5) the permittivity dyadic $\tilde{\bar{\boldsymbol{\epsilon}}} = \epsilon \bar{\mathbf{I}}$ and the permeability dyadic

$$[\tilde{\bar{\boldsymbol{\mu}}}(\omega)] = \begin{bmatrix} \mu & 0 & j\kappa \\ 0 & \mu_0 & 0 \\ -j\kappa & 0 & \mu \end{bmatrix} = \hat{\mathbf{x}}\mu\hat{\mathbf{x}} + \hat{\mathbf{x}}j\kappa\hat{\mathbf{z}} - \hat{\mathbf{z}}j\kappa\hat{\mathbf{x}} + \hat{\mathbf{y}}\mu_0\hat{\mathbf{y}} + \hat{\mathbf{z}}\mu\hat{\mathbf{z}},$$

where μ and κ are given by (4.125)–(4.126). Using the relations

$$\hat{\mathbf{u}} \times \tilde{\bar{\boldsymbol{\mu}}} = \hat{\mathbf{z}} \times [\hat{\mathbf{x}}\mu\hat{\mathbf{x}} + \hat{\mathbf{x}}j\kappa\hat{\mathbf{z}} - \hat{\mathbf{z}}j\kappa\hat{\mathbf{x}} + \hat{\mathbf{y}}\mu_0\hat{\mathbf{y}} + \hat{\mathbf{z}}\mu\hat{\mathbf{z}}] = \hat{\mathbf{y}}\mu\hat{\mathbf{x}} + \hat{\mathbf{y}}j\kappa\hat{\mathbf{z}} - \hat{\mathbf{x}}\mu_0\hat{\mathbf{y}},$$

$$\hat{\mathbf{u}} \times \tilde{\bar{\boldsymbol{\epsilon}}} = \hat{\mathbf{z}} \times (\epsilon\bar{\mathbf{I}}) = \epsilon\hat{\mathbf{z}} \times (\hat{\mathbf{x}}\hat{\mathbf{x}} + \hat{\mathbf{y}}\hat{\mathbf{y}} + \hat{\mathbf{z}}\hat{\mathbf{z}}) = \hat{\mathbf{y}}\epsilon\hat{\mathbf{x}} - \hat{\mathbf{x}}\epsilon\hat{\mathbf{y}},$$

we can construct the dyadics

$$(\hat{\mathbf{u}} \times \tilde{\bar{\boldsymbol{\mu}}}) \cdot (\hat{\mathbf{u}} \times \tilde{\bar{\boldsymbol{\epsilon}}}) = [\hat{\mathbf{y}}\mu\hat{\mathbf{x}} + \hat{\mathbf{y}}j\kappa\hat{\mathbf{z}} - \hat{\mathbf{x}}\mu_0\hat{\mathbf{y}}] \cdot [\hat{\mathbf{y}}\epsilon\hat{\mathbf{x}} - \hat{\mathbf{x}}\epsilon\hat{\mathbf{y}}] = -\hat{\mathbf{x}}\epsilon\mu_0\hat{\mathbf{x}} - \hat{\mathbf{y}}\epsilon\mu\hat{\mathbf{y}},$$

$$(\hat{\mathbf{u}} \times \tilde{\bar{\boldsymbol{\epsilon}}}) \cdot (\hat{\mathbf{u}} \times \tilde{\bar{\boldsymbol{\mu}}}) = [\hat{\mathbf{y}}\epsilon\hat{\mathbf{x}} - \hat{\mathbf{x}}\epsilon\hat{\mathbf{y}}] \cdot [\hat{\mathbf{y}}\mu\hat{\mathbf{x}} + \hat{\mathbf{y}}j\kappa\hat{\mathbf{z}} - \hat{\mathbf{x}}\mu_0\hat{\mathbf{y}}] = -\hat{\mathbf{x}}\epsilon\mu\hat{\mathbf{x}} - \hat{\mathbf{y}}\epsilon\mu_0\hat{\mathbf{y}} - \hat{\mathbf{x}}j\kappa\epsilon\hat{\mathbf{z}}.$$

Thus

$$\bar{\mathbf{\Phi}} = k_u^2 \bar{\mathbf{I}} + \omega^2 (\hat{\mathbf{u}} \times \tilde{\bar{\boldsymbol{\mu}}}) \cdot (\hat{\mathbf{u}} \times \tilde{\bar{\boldsymbol{\epsilon}}}) = \hat{\mathbf{x}}(k_z^2 - \omega^2 \mu_0 \epsilon)\hat{\mathbf{x}} + \hat{\mathbf{y}}(k_z^2 - \omega^2 \mu \epsilon)\hat{\mathbf{y}} + \hat{\mathbf{z}}k_z^2\hat{\mathbf{z}}$$

$$= \begin{bmatrix} k_z^2 - \omega^2 \mu_0 \epsilon & 0 & 0 \\ 0 & k_z^2 - \omega^2 \mu \epsilon & 0 \\ 0 & 0 & k_z^2 \end{bmatrix}.$$

The inverse of $\bar{\mathbf{\Phi}}$ is trivial to compute, since $\bar{\mathbf{\Phi}}$ is diagonal:

$$\bar{\mathbf{\Phi}}^{-1} = \begin{bmatrix} \frac{1}{k_z^2 - \omega^2 \mu_0 \epsilon} & 0 & 0 \\ 0 & \frac{1}{k_z^2 - \omega^2 \mu \epsilon} & 0 \\ 0 & 0 & \frac{1}{k_z^2} \end{bmatrix}$$

$$= \hat{\mathbf{x}}(k_z^2 - \omega^2 \mu_0 \epsilon)^{-1}\hat{\mathbf{x}} + \hat{\mathbf{y}}(k_z^2 - \omega^2 \mu \epsilon)^{-1}\hat{\mathbf{y}} + \hat{\mathbf{z}}k_z^{-2}\hat{\mathbf{z}}. \qquad (5.174)$$

We also have

$$\bar{\mathbf{\Omega}} = k_u^2 \bar{\mathbf{I}} + \omega^2 (\hat{\mathbf{u}} \times \tilde{\bar{\boldsymbol{\epsilon}}}) \cdot (\hat{\mathbf{u}} \times \tilde{\bar{\boldsymbol{\mu}}})$$

$$= \hat{\mathbf{x}}(k_z^2 - \omega^2 \mu \epsilon)\hat{\mathbf{x}} + \hat{\mathbf{y}}(k_z^2 - \omega^2 \mu_0 \epsilon)\hat{\mathbf{y}} + \hat{\mathbf{z}}k_z^2\hat{\mathbf{z}} - \hat{\mathbf{x}}j\omega^2 \kappa \epsilon \hat{\mathbf{z}}$$

$$= \begin{bmatrix} k_z^2 - \omega^2 \mu \epsilon & 0 & j\omega\kappa\epsilon \\ 0 & k_z^2 - \omega^2 \mu_0 \epsilon & 0 \\ 0 & 0 & k_z^2 \end{bmatrix}.$$

The inverse of $\bar{\mathbf{\Omega}}$ is slightly more complicated. Noting that

$$\begin{bmatrix} A & 0 & D \\ 0 & B & 0 \\ 0 & 0 & C \end{bmatrix}^{-1} = \begin{bmatrix} \frac{1}{A} & 0 & -\frac{D}{AC} \\ 0 & \frac{1}{B} & 0 \\ 0 & 0 & \frac{1}{C} \end{bmatrix},$$

we can write

$$\bar{\mathbf{\Omega}}^{-1} = \hat{\mathbf{x}}(k_z^2 - \omega^2 \mu \epsilon)^{-1}\hat{\mathbf{x}} + \hat{\mathbf{y}}(k_z^2 - \omega^2 \mu_0 \epsilon)^{-1}\hat{\mathbf{y}} + \hat{\mathbf{z}}k_z^{-2}\hat{\mathbf{z}} + \hat{\mathbf{x}}j\omega^2 \kappa\epsilon(k_z^2 - \omega^2 \mu \epsilon)^{-1}k_z^{-2}\hat{\mathbf{z}}.$$

Finally, noting that

$$(\hat{\mathbf{u}} \times \tilde{\bar{\boldsymbol{\mu}}}) \cdot \hat{\mathbf{u}} = j\kappa\hat{\mathbf{y}}$$

and

$$(\hat{\mathbf{u}} \times \tilde{\bar{\boldsymbol{\mu}}}) \cdot \nabla_t \tilde{H}_u^u = [\hat{\mathbf{y}}\mu\hat{\mathbf{x}} + \hat{\mathbf{y}}j\kappa\hat{\mathbf{z}} - \hat{\mathbf{x}}\mu_0\hat{\mathbf{y}}] \cdot \left[\hat{\mathbf{x}}\frac{\partial \tilde{H}_z^z}{\partial x} + \hat{\mathbf{y}}\frac{\partial \tilde{H}_z^z}{\partial y} \right]$$

$$= \hat{\mathbf{y}}\mu\frac{\partial \tilde{H}_z^z}{\partial x} - \hat{\mathbf{x}}\mu_0\frac{\partial \tilde{H}_z^z}{\partial y},$$

we have the information necessary to compute the transverse fields.

From (5.169) we have the transverse electric field

$$\tilde{\mathbf{E}}_t^z = \bar{\mathbf{\Phi}}^{-1} \cdot \left(-j\omega\mu\hat{\mathbf{y}}\frac{\partial \tilde{H}_z^z}{\partial x} + j\omega\mu_0\hat{\mathbf{x}}\frac{\partial \tilde{H}_z^z}{\partial y} + \hat{\mathbf{z}}j\omega\kappa k_z \tilde{H}_z^z \right),$$

and thus

$$\tilde{\mathbf{E}}_t^z = \hat{\mathbf{x}}\frac{j\omega\mu_0}{k_z^2 - \omega^2 \mu_0 \epsilon}\frac{\partial \tilde{H}_z^z}{\partial y} + \hat{\mathbf{y}}\left[\frac{-j\omega\mu}{k_z^2 - \omega^2 \mu \epsilon}\frac{\partial \tilde{H}_z^z}{\partial x} + \frac{j\omega\kappa k_z}{k_z^2 - \omega^2 \mu \epsilon}\tilde{H}_z^z \right]$$

by (5.174). From (5.168) we have the transverse magnetic field

$$\tilde{\mathbf{H}}_t^z = \bar{\mathbf{\Omega}}^{-1} \cdot \left(-jk_z \hat{\mathbf{x}} \frac{\partial \tilde{H}_z^z}{\partial x} - jk_z \hat{\mathbf{y}} \frac{\partial \tilde{H}_z^z}{\partial y} + \hat{\mathbf{x}} j \omega^2 \kappa \epsilon \tilde{H}_z^z \right)$$

$$= \hat{\mathbf{x}} \left[\frac{-jk_z}{k_z^2 - \omega^2 \mu \epsilon} \frac{\partial \tilde{H}_z^z}{\partial x} + \frac{j \omega^2 \kappa \epsilon}{k_z^2 - \omega^2 \mu \epsilon} \tilde{H}_z^z \right] + \hat{\mathbf{y}} \frac{-jk_z}{k_z^2 - \omega^2 \mu_0 \epsilon} \frac{\partial \tilde{H}_z^z}{\partial y}.$$

With the transverse fields determined, it is straightforward to derive the wave equation for $\tilde{H}_z^z$ by the same procedure used to obtain (5.173). From (5.172) we have

$$\nabla_t \times \tilde{\mathbf{E}}_t^z = -j\omega \hat{\mathbf{z}} (\hat{\mathbf{z}} \times \tilde{\bar{\mu}}) \cdot \tilde{\mathbf{H}}_t^z - j\omega \hat{\mathbf{z}} \mu_{zz} \tilde{H}_z^z.$$

Substitution gives

$$\left(\hat{\mathbf{x}} \frac{\partial}{\partial x} + \hat{\mathbf{y}} \frac{\partial}{\partial y} \right) \times \left\{ \hat{\mathbf{x}} \frac{j \omega \mu_0}{k_z^2 - \omega^2 \mu_0 \epsilon} \frac{\partial \tilde{H}_z^z}{\partial y} + \hat{\mathbf{y}} \left[\frac{-j \omega \mu}{k_z^2 - \omega^2 \mu \epsilon} \frac{\partial \tilde{H}_z^z}{\partial x} + \frac{j \omega \kappa k_z}{k_z^2 - \omega^2 \mu \epsilon} \tilde{H}_z^z \right] \right\}$$

$$= -j\omega \hat{\mathbf{z}} (-j\kappa \hat{\mathbf{x}} + \mu \hat{\mathbf{z}}) \cdot \left\{ \hat{\mathbf{x}} \left[\frac{-jk_z}{k_z^2 - \omega^2 \mu \epsilon} \frac{\partial \tilde{H}_z^z}{\partial x} + \frac{j \omega^2 \kappa \epsilon}{k_z^2 - \omega^2 \mu \epsilon} \tilde{H}_z^z \right] + \hat{\mathbf{y}} \frac{-jk_z}{k_z^2 - \omega^2 \mu_0 \epsilon} \frac{\partial \tilde{H}_z^z}{\partial y} \right\}$$

$$- j\omega \mu \hat{\mathbf{z}} \tilde{H}_z^z.$$

Carrying out the vector operations, we find that only z-components survive, and

$$-\frac{j\omega \mu}{k_z^2 - \omega^2 \mu \epsilon} \frac{\partial^2 \tilde{H}_z^z}{\partial x^2} + \frac{j\omega \kappa k_z}{k_z^2 - \omega^2 \mu \epsilon} \frac{\partial \tilde{H}_z^z}{\partial x} - \frac{j\omega \mu_0}{k_z^2 - \omega^2 \mu_0 \epsilon} \frac{\partial^2 \tilde{H}_z^z}{\partial y^2}$$

$$= -j\omega \left(-\frac{k_z \kappa}{k_z^2 - \omega^2 \mu \epsilon} \frac{\partial \tilde{H}_z^z}{\partial x} + \frac{\omega^2 \kappa^2 \epsilon}{k_z^2 - \omega^2 \mu \epsilon} \tilde{H}_z^z \right) - j\omega \mu \tilde{H}_z^z.$$

Simplification yields the desired differential equation for $\tilde{H}_z^z$:

$$\left[\frac{\partial^2}{\partial x^2} + \left(\frac{\mu_0}{\mu} \frac{k_z^2 - \omega^2 \mu \epsilon}{k_z^2 - \omega^2 \mu_0 \epsilon} \right) \frac{\partial^2}{\partial y^2} + k_c^2 \right] \tilde{H}_z^z = 0,$$

where

$$k_c^2 = \omega^2 \mu \epsilon \left(1 - \frac{\kappa^2}{\mu^2} \right) - k_z^2. \quad \blacktriangleleft$$

5.5 Solenoidal–lamellar decomposition of solutions to the vector wave equation and the vector spherical wave functions

In § 5.2 we showed how potential functions may be used to represent a general vector field in terms of solenoidal and lamellar components. Here we restrict our study to solutions of the homogeneous vector Helmholtz equation. Since the electromagnetic field in a source-free homogeneous region of space satisfies this equation, we may use the solutions to construct solutions to Maxwell's equations in such regions. In the process we find a solenoidal–lamellar decomposition for the vector solution in terms of vector functions called *vector spherical wave functions* [183]. Our approach follows closely that of the book by Chen [31], which includes additional details and applications.

5.5.1   The vector spherical wave functions

Consider a vector function $\mathbf{V}(\mathbf{r})$ that satisfies the homogeneous vector Helmholtz equation,

$$\nabla^2 \mathbf{V} + k^2 \mathbf{V} = 0,$$

where k is a complex constant. Expanding the Laplacian, we can write this as

$$\nabla(\nabla \cdot \mathbf{V}) - \nabla \times (\nabla \times \mathbf{V}) + k^2 \mathbf{V} = 0. \tag{5.175}$$

We seek general representations of the vector function $\mathbf{V}$.

A first candidate for solution is simply

$$\mathbf{L}(\mathbf{r}) = \nabla f(\mathbf{r}) \tag{5.176}$$

where $f(\mathbf{r})$ is a scalar function satisfying the homogeneous scalar Helmholtz equation

$$\nabla^2 f + k^2 f = 0. \tag{5.177}$$

This is shown to be a valid solution by substitution. Putting (5.176) into (5.175), we find that

$$\nabla(\nabla \cdot \nabla f) - \nabla \times (\nabla \times \nabla f) + k^2 \nabla f = 0.$$

The middle term is zero by (B.56). Thus,

$$\nabla \left[\nabla \cdot \nabla f + k^2 f \right] = 0.$$

Since $\nabla \cdot \nabla f = \nabla^2 f$, the left side is zero by virtue of (5.177). So $\mathbf{L}$ is a solution to (5.175). Moreover, (5.176) shows that

$$\nabla \times \mathbf{L}(\mathbf{r}) = 0,$$

so $\mathbf{L}$ is irrotational and thus lamellar. Since

$$\nabla \cdot \mathbf{L}(\mathbf{r}) = \nabla \cdot \nabla f = \nabla^2 f = -k^2 f \neq 0,$$

the function $\mathbf{L}$ is not useful for representing the electromagnetic field in a source-free homogeneous region where $\nabla \cdot \mathbf{E} = \nabla \cdot \mathbf{H} = 0$.

Solenoidal solutions to (5.175) also exist. Consider the candidate solution

$$\mathbf{M}(\mathbf{r}) = \nabla \times (\mathbf{r}f),$$

where f satisfies (5.177). Substitution into (5.175) gives

$$-\nabla \times \nabla \times \mathbf{M} + k^2 \mathbf{M} = 0, \tag{5.178}$$

since $\nabla \cdot \mathbf{M} = \nabla \cdot [\nabla \times (\mathbf{r}f)] = 0$. Substituting for $\mathbf{M}$ and using $k^2 f = -\nabla^2 f$, we find that

$$\nabla \times \left[\nabla \times \nabla \times (\mathbf{r}f) + \mathbf{r}\nabla^2 f \right] = 0.$$

Let the bracketed term be a vector field $\mathbf{W}(\mathbf{r})$. If we can show that $\mathbf{W}$ is the gradient of a scalar field, then the equality is satisfied and $\mathbf{M}$ is verified as a solution. Use

$$\mathbf{M} = \nabla \times (\mathbf{r}f) = f\nabla \times \mathbf{r} - \mathbf{r} \times \nabla f = -\mathbf{r} \times \nabla f$$

since $\nabla \times \mathbf{r} = 0$. It expedient to represent $\mathbf{W}$ in spherical coordinates. Use

$$\mathbf{M} = -\mathbf{r} \times \nabla f = -r\hat{\mathbf{r}} \times \left[\hat{\mathbf{r}}\frac{\partial f}{\partial r} + \hat{\boldsymbol{\theta}}\frac{1}{r}\frac{\partial f}{\partial \theta} + \hat{\boldsymbol{\phi}}\frac{1}{r\sin\theta}\frac{\partial f}{\partial \phi}\right] = -\hat{\boldsymbol{\phi}}\frac{\partial f}{\partial \theta} + \hat{\boldsymbol{\theta}}\frac{1}{\sin\theta}\frac{\partial f}{\partial \phi}, \quad (5.179)$$

so that

$$\nabla \times (\mathbf{r} \times \nabla f)$$
$$= \frac{\hat{\mathbf{r}}}{r\sin\theta}\left[\frac{\partial}{\partial \theta}\left(\sin\theta\frac{\partial f}{\partial \theta}\right) + \frac{1}{\sin\theta}\frac{\partial^2 f}{\partial \phi^2}\right] - \frac{\hat{\boldsymbol{\theta}}}{r}\left[\frac{\partial}{\partial r}\left(r\frac{\partial f}{\partial \theta}\right)\right] - \frac{\hat{\boldsymbol{\phi}}}{r}\left[\frac{\partial}{\partial r}\left(\frac{r}{\sin\theta}\frac{\partial f}{\partial \phi}\right)\right].$$

With this we have

$$\mathbf{W} = -\nabla \times (\mathbf{r} \times \nabla f) + r\hat{\mathbf{r}}\nabla^2 f$$
$$= \hat{\mathbf{r}}\left[-\frac{1}{r\sin\theta}\frac{\partial}{\partial \theta}\left(\sin\theta\frac{\partial f}{\partial \theta}\right) - \frac{1}{r\sin^2\theta}\frac{\partial^2 f}{\partial \phi^2} + \frac{1}{r}\frac{\partial}{\partial r}\left(r^2\frac{\partial f}{\partial r}\right) + \frac{1}{r\sin\theta}\frac{\partial}{\partial \theta}\left(\sin\theta\frac{\partial f}{\partial \theta}\right)\right.$$
$$\left. + \frac{1}{r\sin^2\theta}\frac{\partial^2 f}{\partial \phi^2}\right] + \frac{\hat{\boldsymbol{\theta}}}{r}\left[\frac{\partial}{\partial r}\left(r\frac{\partial f}{\partial \theta}\right)\right] + \frac{\hat{\boldsymbol{\phi}}}{r}\left[\frac{\partial}{\partial r}\left(\frac{r}{\sin\theta}\frac{\partial f}{\partial \phi}\right)\right]$$
$$= \hat{\mathbf{r}}\frac{1}{r}\left[2r\frac{\partial f}{\partial r} + r^2\frac{\partial^2 f}{\partial r^2}\right] + \frac{\hat{\boldsymbol{\theta}}}{r}\left[\frac{\partial f}{\partial \theta} + \frac{\partial}{\partial \theta}\left(r\frac{\partial f}{\partial r}\right)\right] + \frac{\hat{\boldsymbol{\phi}}}{r\sin\theta}\left[\frac{\partial f}{\partial \phi} + \frac{\partial}{\partial \phi}\left(r\frac{\partial f}{\partial r}\right)\right].$$

Factorization yields

$$\mathbf{W} = \hat{\mathbf{r}}\left[\frac{\partial}{\partial r}\left(f + r\frac{\partial f}{\partial r}\right)\right] + \frac{\hat{\boldsymbol{\theta}}}{r}\left[\frac{\partial}{\partial \theta}\left(f + r\frac{\partial f}{\partial r}\right)\right] + \frac{\hat{\boldsymbol{\phi}}}{r\sin\theta}\left[\frac{\partial}{\partial \phi}\left(f + r\frac{\partial f}{\partial r}\right)\right]$$
$$= \nabla\left(f + r\frac{\partial f}{\partial r}\right)$$
$$= \nabla\frac{\partial}{\partial r}(rf),$$

and thus $\mathbf{M}$ is a solution to (5.175).

Note from (5.179) that $\mathbf{M}$ has no r-component, and is thus insufficient for representing general solenoidal fields. Consider a third candidate solution to (5.175)

$$\mathbf{N} = k^{-1}\nabla \times \mathbf{M},$$

which is clearly solenoidal. Substitution into (5.175) gives

$$-k^{-1}\nabla \times [\nabla \times (\nabla \times \mathbf{M})] + k\nabla \times \mathbf{M} = 0,$$

since $\nabla \cdot \mathbf{N} = 0$. Writing this as

$$\nabla \times [-\nabla \times (\nabla \times \mathbf{M}) + k^2\mathbf{M}] = 0,$$

we see that the equation is satisfied, since we have demonstrated that the term in brackets is zero in (5.178). As we shall show, the vector field $\mathbf{N}$ does have a radial component.

In summary, the three vector spherical wave functions are

$$\mathbf{L}(\mathbf{r}) = \nabla f(\mathbf{r}), \qquad \text{(lamellar)}$$
$$\mathbf{M}(\mathbf{r}) = \nabla \times [\mathbf{r}f(\mathbf{r})], \qquad \text{(solenoidal)}$$
$$\mathbf{N}(\mathbf{r}) = k^{-1}\nabla \times \mathbf{M}(\mathbf{r}). \qquad \text{(solenoidal)}$$

5.5.2 Representation of the vector spherical wave functions in spherical coordinates

The vector spherical wave functions are most often employed to solve Maxwell's equations for systems embedded in regions unbounded in the θ and ϕ directions. For this situation it is convenient to represent the function f as the solution to the scalar Helmholtz equation in spherical coordinates using the separation of variables approach described in Appendix A. Since the z-axis is included in the solution space, and since no boundaries restrict the value of ϕ, we may write

$$f_{\substack{e \\ o}mn}(\mathbf{r}) = z_n(kr)P_n^m(\cos\theta)\begin{Bmatrix} \cos(m\phi) \\ \sin(m\phi) \end{Bmatrix}.$$

Here z_n is a properly chosen spherical Bessel function, and the subscripts e and o denote functions even or odd in ϕ. Computing the gradient of f, we get

$$\mathbf{L}_{\substack{e \\ o}mn} = \hat{\mathbf{r}}\left[\frac{d}{dr}z_n(kr)\right]P_n^m(\cos\theta)\begin{Bmatrix} \cos(m\phi) \\ \sin(m\phi) \end{Bmatrix}$$

$$+ \hat{\boldsymbol{\theta}}\frac{1}{r}z_n(kr)\left[\frac{d}{d\theta}P_n^m(\cos\theta)\right]\begin{Bmatrix} \cos(m\phi) \\ \sin(m\phi) \end{Bmatrix}$$

$$\mp \hat{\boldsymbol{\phi}}\frac{m}{r\sin\theta}z_n(kr)P_n^m(\cos\theta)\begin{Bmatrix} \sin(m\phi) \\ \cos(m\phi) \end{Bmatrix}.$$

The function $\mathbf{M}$ is easily found using (5.179):

$$\mathbf{M}_{\substack{e \\ o}mn} = \mp\hat{\boldsymbol{\theta}}\frac{m}{\sin\theta}z_n(kr)P_n^m(\cos\theta)\begin{Bmatrix} \sin(m\phi) \\ \cos(m\phi) \end{Bmatrix}$$

$$- \hat{\boldsymbol{\phi}}z_n(kr)\left[\frac{d}{d\theta}P_n^m(\cos\theta)\right]\begin{Bmatrix} \cos(m\phi) \\ \sin(m\phi) \end{Bmatrix}. \tag{5.180}$$

To find $\mathbf{N}$ we note that

$$\mathbf{N} = \frac{1}{k}\left[\mathbf{W} - r\hat{\mathbf{r}}\nabla^2 f\right] = \frac{1}{k}\left[\nabla\frac{\partial}{\partial r}(rf) + \hat{\mathbf{r}}rk^2 f\right].$$

The r-component of $\mathbf{N}$ is

$$N_r = \frac{1}{k}\left[r\frac{\partial^2 f}{\partial r^2} + 2\frac{\partial f}{\partial r} + k^2 rf\right]$$

$$= P_n^m(\cos\theta)\begin{Bmatrix} \cos(m\phi) \\ \sin(m\phi) \end{Bmatrix}kr\left[z_n''(kr) + \frac{2}{kr}z_n'(kr) + z_n(kr)\right]$$

$$= P_n^m(\cos\theta)\begin{Bmatrix} \cos(m\phi) \\ \sin(m\phi) \end{Bmatrix}\frac{n(n+1)}{kr}z_n(kr)$$

by virtue of (E.15). Computing the remaining components of the gradient then gives

$$\mathbf{N}_{\substack{e \\ o}mn} = \hat{\mathbf{r}}\frac{n(n+1)}{kr}z_n(kr)P_n^m(\cos\theta)\begin{Bmatrix} \cos(m\phi) \\ \sin(m\phi) \end{Bmatrix}$$

$$+ \hat{\boldsymbol{\theta}}\frac{1}{kr}\left[\frac{d}{dr}\{rz_n(kr)\}\right]\left[\frac{d}{d\theta}P_n^m(\cos\theta)\right]\begin{Bmatrix} \cos(m\phi) \\ \sin(m\phi) \end{Bmatrix}$$

$$\mp \hat{\boldsymbol{\phi}}\frac{m}{kr\sin\theta}\left[\frac{d}{dr}\{rz_n(kr)\}\right]P_n^m(\cos\theta)\begin{Bmatrix} \sin(m\phi) \\ \cos(m\phi) \end{Bmatrix}. \tag{5.181}$$

Note that the present notation does not distinguish between, for instance, the function **M** that uses $z_n(kr) = j_n(kr)$ and the function **M** that uses $z_n(kr) = n_n(kr)$. It is useful to designate

$$z_n^{(1)}(x) = j_n(x), \qquad z_n^{(2)}(x) = n_n(x), \qquad z_n^{(3)}(x) = h_n^{(1)}(x), \qquad z_n^{(4)}(x) = h_n^{(2)}(x),$$

such that $\mathbf{M}^{(1)}$ uses $j_n(kr)$, etc.

▶ **Example 5.13:** Representation of a plane wave using vector spherical wave functions

A uniform plane wave propagates in the z-direction in a homogeneous, isotropic material of permittivity $\tilde{\epsilon}^c$ and permeability $\tilde{\mu}$ with its electric field polarized along x. Represent the electric and magnetic fields in terms of vector spherical wave functions.

Solution: We can write the electric and magnetic fields of the plane waves as

$$\tilde{\mathbf{E}}(\mathbf{r}, \omega) = \hat{\mathbf{x}}\tilde{E}_0(\omega)e^{-jkz},$$

$$\tilde{\mathbf{H}}(\mathbf{r}, \omega) = \hat{\mathbf{y}}\frac{\tilde{E}_0(\omega)}{\eta}e^{-jkz}.$$

Converting the coordinate variable z and the unit vectors to spherical coordinates, we have

$$\tilde{\mathbf{E}} = [\hat{\mathbf{r}}\sin\theta\cos\phi + \hat{\boldsymbol{\theta}}\cos\theta\cos\phi - \hat{\boldsymbol{\phi}}\sin\phi]\tilde{E}_0 e^{-jkr\cos\theta}, \tag{5.182}$$

$$\tilde{\mathbf{H}} = [\hat{\mathbf{r}}\sin\theta\sin\phi + \hat{\boldsymbol{\theta}}\cos\theta\sin\phi + \hat{\boldsymbol{\phi}}\cos\phi]\frac{\tilde{E}_0}{\eta}e^{-jkr\cos\theta}. \tag{5.183}$$

Since $\nabla \cdot \tilde{\mathbf{E}} = 0$, we do not need **L** functions in the expansion of $\tilde{\mathbf{E}}$. Comparing the ϕ dependence in (5.182) to that in (5.180) and (5.181), we see that only the $\mathbf{M}_{o1n}$ and $\mathbf{N}_{e1n}$ functions are required. We write

$$\tilde{\mathbf{E}} = \sum_{n=0}^{\infty} [a_n\mathbf{M}_{o1n} + b_n\mathbf{N}_{e1n}], \tag{5.184}$$

and seek the amplitudes a_n and b_n that make this expression the same as (5.182).

To find b_n we equate the r-component of (5.182) with that of (5.184). The fact that

$$\tilde{E}_r = \tilde{E}_0 \cos\phi \sin\theta e^{-jkr\cos\theta} = \tilde{E}_0 \cos\phi \frac{1}{jkr}\frac{\partial}{\partial\theta}\left[e^{-jkr\cos\theta}\right] \tag{5.185}$$

will facilitate comparison of like terms. Substituting the expansion (5.150) for the exponential function into (5.185), we get

$$\tilde{E}_r = \tilde{E}_0 \cos\phi \frac{1}{jkr}\sum_{n=0}^{\infty}\frac{\partial}{\partial\theta}[j^{-n}(2n+1)j_n(kr)P_n(\cos\theta)].$$

Using the derivative relationship

$$\frac{d}{d\theta}P_n(\cos\theta) = -P_n^1(\cos\theta)$$

we have

$$\tilde{E}_r = -\tilde{E}_0 \cos\phi \frac{1}{jkr}\sum_{n=0}^{\infty}j^{-n}(2n+1)j_n(kr)P_n^1(\cos\theta).$$

Now we equate this expression with the r-component of (5.184) to find

$$-\tilde{E}_0 \frac{1}{jkr} \sum_{n=1}^{\infty} j^{-n}(2n+1)j_n(kr)P_n^1(\cos\theta) = \sum_{n=1}^{\infty} b_n \frac{n(n+1)}{kr}j_n(kr)P_n^1(\cos\theta).$$

Here we have used $P_0^1(x) = 0$ and have chosen $z_n = j_n$ as the spherical Bessel functions in the spherical vector wave function expansion. By orthogonality of the spherical harmonics we obtain $j\tilde{E}_0 j^{-n}(2n+1) = n(n+1)b_n$ so that

$$b_n = \tilde{E}_0(-j)^{n-1}\frac{2n+1}{n(n+1)}.$$

To find a_n we equate the r-components of the magnetic field. By Faraday's law

$$\tilde{\mathbf{H}} = \frac{j}{\omega\mu}\nabla \times \tilde{\mathbf{E}} = \frac{j}{\omega\mu}\sum_{n=0}^{\infty}[a_n\nabla \times \mathbf{M}_{o1n} + b_n\nabla \times \mathbf{N}_{e1n}].$$

But

$$\nabla \times \mathbf{M}_{o1n} = k\mathbf{N}_{o1n},$$

and by (5.178),

$$\nabla \times \mathbf{N}_{e1n} = \nabla \times \nabla \times \mathbf{M}_{e1n} = k\mathbf{M}_{e1n}.$$

Thus,

$$\tilde{\mathbf{H}} = \frac{j}{\eta}\sum_{n=0}^{\infty}[a_n\mathbf{N}_{o1n} + b_n\mathbf{M}_{e1n}]. \tag{5.186}$$

Equating the r-component of (5.183) with the r-component of (5.186) and proceeding as before, we quickly find

$$a_n = \tilde{E}_0(-j)^n\frac{2n+1}{n(n+1)}.$$

With a_n and b_n identified, we have the final expansion for the plane wave:

$$\tilde{\mathbf{E}} = \tilde{E}_0 \sum_{n=1}^{\infty}(-j)^n\frac{2n+1}{n(n+1)}\left[\mathbf{M}_{o1n}^{(1)} + j\mathbf{N}_{e1n}^{(1)}\right], \tag{5.187}$$

$$\tilde{\mathbf{H}} = -\frac{\tilde{E}_0}{\eta} \sum_{n=1}^{\infty}(-j)^n\frac{2n+1}{n(n+1)}\left[\mathbf{M}_{e1n}^{(1)} - j\mathbf{N}_{o1n}^{(1)}\right], \tag{5.188}$$

where the superscripts on the vector spherical wave functions indicate that we are using $z_n(kr) = j_n(kr)$. Substitution of the explicit expressions for the vector spherical wave functions shows that the fields (5.187)–(5.188) are identical to those found using potential functions, i.e., (5.153)–(5.157) superposed with (5.159)–(5.163). ◀

▶ **Example 5.14:** Scattering by a conducting sphere using vector spherical wave functions

Consider a PEC sphere of radius a centered at the origin and embedded in a homogeneous, isotropic material of permittivity $\tilde{\epsilon}^c$ and permeability $\tilde{\mu}$. The sphere is illuminated by a plane wave incident along the z-direction with its electric field polarized along x. Find the scattered electric and magnetic fields in terms of vector spherical wave functions.

Solution: The fields of the incident plane wave are given by (5.187)–(5.188). We expect the symmetry of the scattered field to conform to that of the incident field, hence the former will involve only the $m = 1$ vector spherical wave functions. Expecting the scattered field to take the form of an outward traveling wave, we choose the radial functions $z_n = h_n^{(2)}$. With these observations we write the scattered fields as vector spherical wave function expansions:

$$\tilde{\mathbf{E}}^s = \tilde{E}_0 \sum_{n=1}^{\infty} (-j)^n \frac{2n+1}{n(n+1)} \left[d_n \mathbf{M}_{o1n}^{(4)} + je_n \mathbf{N}_{e1n}^{(4)} \right], \qquad (5.189)$$

$$\tilde{\mathbf{H}}^s = -\frac{\tilde{E}_0}{\eta} \sum_{n=1}^{\infty} (-j)^n \frac{2n+1}{n(n+1)} \left[e_n \mathbf{M}_{e1n}^{(4)} - jd_n \mathbf{N}_{o1n}^{(4)} \right],$$

and seek d_n and e_n so that the fields satisfy the boundary conditions on the surface of the sphere. Note that these expansions satisfy Faraday's law.

The total electric field is the sum of the incident plane-wave field (5.187) and the scattered field (5.189). Setting the θ-component of the total field to zero at $r = a$ produces the equation

$$\sum_{n=1}^{\infty} (-j)^n \frac{2n+1}{n(n+1)} \left\{ \frac{\cos\phi}{\sin\theta} j_n(ka) P_n^1(\cos\theta) \right.$$

$$\left. + j\frac{\cos\phi}{ka} \left[\frac{d}{dr}\{rj_n(kr)\} \right] \Big|_{r=a} \left[\frac{d}{d\theta} P_n^1(\cos\theta) \right] \right\}$$

$$= -\sum_{n=1}^{\infty} (-j)^n \frac{2n+1}{n(n+1)} \left\{ d_n \frac{\cos\phi}{\sin\theta} h_n^{(2)}(ka) P_n^1(\cos\theta) \right.$$

$$\left. + je_n \frac{\cos\phi}{ka} \left[\frac{d}{dr}\{rh_n^{(2)}(kr)\} \right] \Big|_{r=a} \left[\frac{d}{d\theta} P_n^1(\cos\theta) \right] \right\}.$$

Orthogonality of the spherical harmonics requires

$$j_n(ka) = -d_n h_n^{(2)}(ka),$$

$$\left[\frac{d}{dr}\{rj_n(kr)\} \right] \Big|_{r=a} = -e_n \left[\frac{d}{dr}\{rh_n^{(2)}(kr)\} \right] \Big|_{r=a},$$

and so

$$d_n = -\frac{j_n(ka)}{h_n^{(2)}(ka)}, \qquad e_n = -\frac{\left[\frac{d}{dr}\{rj_n(kr)\} \right]\Big|_{r=a}}{\left[\frac{d}{dr}\{rh_n^{(2)}(kr)\} \right]\Big|_{r=a}}.$$

Substituting these coefficients into (5.189) and using the explicit representations of the vector spherical wave functions, we may show that the electric field is identical to that found using potentials, as given by (5.164)–(5.166).

Note that using the boundary condition on the ϕ-component of the total electric field produces identical results for d_n and e_n. ◄

5.6 Application: guided waves and transmission lines

Guided wave structures use material boundaries to guide electromagnetic energy and encoded information along specified directions. The direction may be along an axis (as in hollow-pipe waveguides), radially in cylindrical coordinates (as between two parallel plates), or radially in spherical coordinates (as between coaxial cones). Guided wave structures are critically important in communications, radar, and digital electronics, and are the subject of numerous books, both general and specialized. We will primarily restrict ourselves to the fundamental electromagnetic aspects of guided waves, including modal field structure, dispersion, and power transmission. While we touch on some advanced topics such as mode excitation and discontinuities, much more detail may be found in specialized texts such as [104, 38, 39].

Guided wave structures fall into *closed boundary* and *open boundary* types. A closed boundary structure has a conducting shell that encloses the fields; as a result it can neither radiate nor interact with adjacent systems or structures. Its modal spectrum is described by discrete eigenvalues and associated modal field structures. Examples include hollow-pipe waveguides, coaxial cables, and shielded transmission lines. An open boundary structure is not enclosed by a conductor, hence can radiate into the surrounding medium and couple strongly with adjacent structures. It is characterized by both discrete and continuous spectra of eigenvalues and associated fields. Examples include fiber-optic cables, twin-wire transmission lines, and microstrip.

The modal field structure of guided waves may be categorized as TE, TM, TEM, or hybrid. In certain situations the boundary conditions may be satisfied by a field structure that is purely TE or TM to a specified direction. These modes are often convenient because of their simple field structures (possibly rendering their excitation easier) and their propagation characteristics. They generally have an associated *cutoff frequency* below which waves of a specific mode no longer propagate. TEM guided waves are only supported by structures with two or more conductors; these structures, known as *transmission lines*, have the benefit of low (potentially zero) dispersion and are therefore desirable for guiding analog-encoded information such as voice. Their absence of a lower cutoff frequency makes them useful in wideband (and baseband) applications. TEM structures can also support TE and TM fields (called *higher-order modes*) with nonzero cutoff, and these modes determine the usable bandwidth of the transmission line. Hybrid modes have field structures that cannot be categorized as purely TE or TM (or TEM) to any particular direction. Their field structure is generally complicated, but may be dominated by an electric or magnetic field in the guiding direction and thus categorized as "TE-like" or "TM-like." Hybrid modes generally occur in structures with more than one material, such as coated optical fibers. A notable example is microstrip, only supporting hybrid modes but having a "TEM-like" mode where the axial fields are much weaker than the transverse fields, and hence useful as an ersatz transmission line.

5.6.1 Hollow-pipe waveguides

A hollow-pipe waveguide consists of a conducting tube aligned with a preferred axis, often chosen as the z-axis for convenience. The pipe may contain multiple materials, both isotropic and anisotropic, but the cross-section geometry is assumed to be invariant along the axial direction. For the purpose of analyzing the modal structure of the fields, the waveguide is taken to be infinitely long and source free. The presence of sources leads to the more complicated problem of *excitation*; this problem, often handled by superposing source-free solutions using a Green's function, is not considered in its full generality here.

5.6.1.1 Hollow-pipe waveguides with homogeneous, isotropic filling

Let us begin by examining a hollow-pipe waveguide with PEC walls, completely filled with a homogeneous, isotropic material and aligned with the z-axis. The filling material has permeability $\tilde{\mu}$ and complex permittivity $\tilde{\epsilon}^c$. The guide cross-sectional shape is independent of z.

The fields in a waveguide homogeneously filled with an isotropic material may be decomposed into modes that are respectively TE to the z-direction (designated TE$_z$ modes) and TM to the z-direction (TM$_z$ modes). Each mode set independently satisfies the boundary condition that the tangential electric field must vanish on the conducting boundary. We may construct the modal fields either in terms of the longitudinal field components $\tilde{E}_z$ and $\tilde{H}_z$ or the Hertzian potentials. We adopt the Hertzian potentials here. Later we will use the longitudinal fields to determine field behavior in anisotropic materials (§ 5.6.1.2). For TM fields we choose $\tilde{\mathbf{\Pi}}_e = \hat{\mathbf{z}}\tilde{\Pi}_e$, $\tilde{\mathbf{\Pi}}_h = 0$; for TE fields we choose $\tilde{\mathbf{\Pi}}_h = \hat{\mathbf{z}}\tilde{\Pi}_h$, $\tilde{\mathbf{\Pi}}_e = 0$. Both of the potentials must obey the Helmholtz equation

$$\left(\nabla^2 + k^2\right)\tilde{\Pi}_z = 0,$$

where $\tilde{\Pi}_z$ represents either $\tilde{\Pi}_e$ or $\tilde{\Pi}_h$. This may be solved by a Fourier transform approach or by separation of variables. In either case we introduce the transverse position vector $\boldsymbol{\rho}$ by the relation $\mathbf{r} = \hat{\mathbf{z}}z + \boldsymbol{\rho}$ and write

$$\tilde{\Pi}_z(\mathbf{r}, \omega) = \tilde{\Pi}_z(z, \boldsymbol{\rho}, \omega).$$

We also split the Laplacian operator into transverse and longitudinal parts, writing the Helmholtz equation as

$$\left(\nabla_t^2 + \frac{\partial^2}{\partial z^2} + k^2\right)\tilde{\Pi}_z(z, \boldsymbol{\rho}, \omega) = 0. \tag{5.190}$$

Solution for $\tilde{\Pi}_z$ by Fourier transform approach. To solve (5.190) using Fourier transforms, we write $\tilde{\Pi}_z$ as an inverse transform

$$\tilde{\Pi}_z(\mathbf{r}, \omega) = \frac{1}{2\pi}\int_{-\infty}^{\infty}\tilde{\psi}(k_z, \boldsymbol{\rho}, \omega)e^{jk_z z}\,dk_z, \tag{5.191}$$

where $\tilde{\psi}$ is the spatial Fourier spectrum of $\tilde{\Pi}_z$:

$$\tilde{\psi}(k_z, \boldsymbol{\rho}, \omega) = \int_{-\infty}^{\infty}\tilde{\Pi}_z(\mathbf{r}, \omega)e^{-jk_z z}\,dz.$$

Substituting (5.191) into the Helmholtz equation (5.190) and taking the derivatives, we have

$$\frac{1}{2\pi}\int_{-\infty}^{\infty}\left[\left(\nabla_t^2 + k^2 - k_z^2\right)\tilde{\psi}(k_z, \boldsymbol{\rho}, \omega)\right]e^{jk_z z}\,dk_z = 0.$$

By the Fourier integral theorem, the bracketed term must be zero. Thus, we have a differential equation for $\tilde{\psi}$:

$$\nabla_t^2\tilde{\psi}(k_z, \boldsymbol{\rho}, \omega) + k_c^2\tilde{\psi}(k_z, \boldsymbol{\rho}, \omega) = 0 \tag{5.192}$$

where

$$k_c = \sqrt{k^2 - k_z^2} \tag{5.193}$$

is called the *cutoff wavenumber*. A hollow-pipe waveguide is a closed boundary structure, with k_c (and hence k_z) taking discrete values. The spectrum (5.191) reduces to a discrete sum of *modal* contributions to the potential. Each term has the form

$$\tilde{\Pi}_z(\mathbf{r}, \omega) = \tilde{\psi}(\boldsymbol{\rho}, \omega)e^{\mp jk_z z},$$

where $\tilde{\psi}$ satisfies (5.192).

Solution for $\tilde{\Pi}_z$ by separation of variables. Alternatively, we may try a product solution $\tilde{\Pi}_z(\mathbf{r}, \omega) = \tilde{Z}(z, \omega)\tilde{\psi}(\boldsymbol{\rho}, \omega)$. Substitution into (5.190) yields

$$\frac{1}{\tilde{\psi}(\boldsymbol{\rho}, \omega)}\nabla_t^2\tilde{\psi}(\boldsymbol{\rho}, \omega) + k^2 = -\frac{1}{Z(z, \omega)}\frac{\partial^2}{\partial z^2}Z(z, \omega).$$

Because the left side has positional dependence only on $\boldsymbol{\rho}$ while the right side has dependence only on z, both must equal the same constant, say k_z^2. The resulting ordinary differential equation

$$\frac{\partial^2 Z}{\partial z^2} + k_z^2 Z = 0$$

has solutions $Z = e^{\mp jk_z z}$. The second implication

$$\nabla_t^2\tilde{\psi}(\boldsymbol{\rho}, \omega) + k_c^2\tilde{\psi}(\boldsymbol{\rho}, \omega) = 0 \tag{5.194}$$

is a wave equation with k_c given by (5.193). The complete solution is a superposition of terms of the form

$$\tilde{\Pi}_z(\mathbf{r}, \omega) = \tilde{\psi}(\boldsymbol{\rho}, \omega)e^{\mp jk_z z}$$

with k_z taking discrete values.

Solution to the differential equation for $\tilde{\psi}$. To find the eigenvalues and associated field distributions for a hollow-pipe waveguide with homogeneous isotropic filling, we must solve (5.192) or (5.194). For certain geometries this can be done in closed form using separation of variables. If the boundary conforms to a coordinate level surface in a separable coordinate system, the separation of variables solutions define individual modes. Classic examples include the rectangular and circular guides considered below, along with elliptical guides [72] and variants of circular guides such as coaxial cables and sectoral guides. A few interesting geometries such as equilateral triangles and the isosceles right triangle considered below may be solved by superposing finitely many individual separation of variables solutions to construct each mode.

Numerical approaches may be required for complex geometries. Examples include the finite element method [2] and the integral equation method (§ 7.4.4).

Field representation for TE and TM modes. The fields may be computed from the Hertzian potentials using $u = z$ in (5.122)–(5.123) and (5.124)–(5.125). Because the fields all contain the term $e^{\mp jk_z z}$, we define new field quantities $\tilde{\mathbf{e}}$ and $\tilde{\mathbf{h}}$:

$$\tilde{\mathbf{E}}(\mathbf{r}, \omega) = \tilde{\mathbf{e}}(\boldsymbol{\rho}, \omega)e^{\mp jk_z z}, \qquad \tilde{\mathbf{H}}(\mathbf{r}, \omega) = \tilde{\mathbf{h}}(\boldsymbol{\rho}, \omega)e^{\mp jk_z z}.$$

Substituting $\tilde{\Pi}_e = \tilde{\psi}_e e^{\mp jk_z z}$, we have for TM fields

$$\tilde{\mathbf{e}} = \mp jk_z\nabla_t\tilde{\psi}_e + \hat{\mathbf{z}}k_c^2\tilde{\psi}_e, \qquad \tilde{\mathbf{h}} = -j\omega\tilde{\epsilon}^c\hat{\mathbf{z}} \times \nabla_t\tilde{\psi}_e.$$

The simple relationship between the transverse parts of $\tilde{\mathbf{E}}$ and $\tilde{\mathbf{H}}$ permits us to write

$$\tilde{e}_z = k_c^2\tilde{\psi}_e, \tag{5.195}$$

$$\tilde{\mathbf{e}}_t = \mp jk_z\nabla_t\tilde{\psi}_e, \tag{5.196}$$

$$\tilde{\mathbf{h}}_t = \pm Y_e(\hat{\mathbf{z}} \times \tilde{\mathbf{e}}_t). \tag{5.197}$$

Here

$$Y_e = \omega\tilde{\epsilon}^c/k_z$$

is the complex *TM wave admittance*. We can also write $\hat{\mathbf{z}} \times \tilde{\mathbf{h}}_t = \pm Y_e \hat{\mathbf{z}} \times (\hat{\mathbf{z}} \times \tilde{\mathbf{e}}_t) = \mp Y_e \tilde{\mathbf{e}}_t$ so that

$$\tilde{\mathbf{e}}_t = \mp Z_e(\hat{\mathbf{z}} \times \tilde{\mathbf{h}}_t),$$

where $Z_e = 1/Y_e$ is the *TM wave impedance*.

For TE fields, we have with $\tilde{\Pi}_h = \tilde{\psi}_h e^{\mp jk_z z}$,

$$\tilde{\mathbf{e}} = j\omega\tilde{\mu}\hat{\mathbf{z}} \times \nabla_t \tilde{\psi}_h, \qquad \tilde{\mathbf{h}} = \mp jk_z \nabla_t \tilde{\psi}_h + \hat{\mathbf{z}} k_c^2 \tilde{\psi}_h,$$

or

$$\tilde{h}_z = k_c^2 \tilde{\psi}_h, \tag{5.198}$$

$$\tilde{\mathbf{h}}_t = \mp jk_z \nabla_t \tilde{\psi}_h, \tag{5.199}$$

$$\tilde{\mathbf{e}}_t = \mp Z_h(\hat{\mathbf{z}} \times \tilde{\mathbf{h}}_t), \tag{5.200}$$

where

$$Z_h = \omega\tilde{\mu}/k_z \tag{5.201}$$

is the *TE wave impedance*.

Modal solutions for the transverse field dependence. Equation (5.194) governs the transverse behavior of the waveguide fields. When coupled with an appropriate boundary condition, this homogeneous equation has an infinite spectrum of discrete solutions called *eigenmodes* or simply *modes*. Each mode is associated with a real *eigenvalue* k_c that depends on the cross-sectional geometry but not on frequency or homogeneous material parameters. We number the modes so that $k_c = k_{cn}$ for the nth mode. The amplitude of each modal solution depends on the excitation source within the waveguide.

The appropriate boundary conditions are implied by the condition that for both TM and TE fields the tangential component of $\tilde{\mathbf{E}}$ must vanish on the waveguide walls: $\hat{\mathbf{n}} \times \tilde{\mathbf{E}} = 0$, where $\hat{\mathbf{n}}$ is the unit interior normal to the waveguide wall. For TM fields we have $\tilde{E}_z = 0$ and thus

$$\tilde{\psi}_e(\boldsymbol{\rho}, \omega) = 0 \qquad (\boldsymbol{\rho} \in \Gamma) \tag{5.202}$$

where Γ is the contour describing the waveguide boundary. For TE fields we have $\hat{\mathbf{n}} \times \tilde{\mathbf{E}}_t = 0$ or

$$\hat{\mathbf{n}} \times (\hat{\mathbf{z}} \times \nabla_t \tilde{\psi}_h) = 0.$$

Using $\hat{\mathbf{n}} \times (\hat{\mathbf{z}} \times \nabla_t \tilde{\psi}_h) = \hat{\mathbf{z}}(\hat{\mathbf{n}} \cdot \nabla_t \tilde{\psi}_h) - (\hat{\mathbf{n}} \cdot \hat{\mathbf{z}})\nabla_t \tilde{\psi}_h$ and noting that $\hat{\mathbf{n}} \cdot \hat{\mathbf{z}} = 0$, we have

$$\hat{\mathbf{n}} \cdot \nabla_t \tilde{\psi}_h(\boldsymbol{\rho}, \omega) = \frac{\partial \tilde{\psi}_h(\boldsymbol{\rho}, \omega)}{\partial n} = 0 \qquad (\boldsymbol{\rho} \in \Gamma).$$

Wave nature of the waveguide fields. We have seen that all waveguide field components, for both TE and TM modes, vary as $e^{\mp jk_{zn} z}$. Here $k_{zn}^2 = k^2 - k_{cn}^2$ is the *propagation constant* of the nth mode. Letting

$$k_z = \sqrt{k^2 - k_c^2} = \beta - j\alpha \tag{5.203}$$

we see that $\tilde{\mathbf{E}}, \tilde{\mathbf{H}} \sim e^{\mp j\beta z} e^{\mp \alpha z}$. Assuming β is nonzero, and choosing the branch of the square root function such that $\beta > 0$, the minus sign yields a wave propagating in the direction of increasing z, while the plus sign is associated with a wave propagating in the opposite direction.

When the guide is filled with a good dielectric we may assume $\tilde{\mu} = \mu$ is real and frequency independent and use (4.239) to show that

$$k_z = \beta - j\alpha = \sqrt{[\omega^2 \mu \operatorname{Re} \tilde{\epsilon}^c - k_c^2] - j\omega^2 \mu \operatorname{Re} \tilde{\epsilon}^c \tan \delta_c}$$

$$= \sqrt{\mu \operatorname{Re} \tilde{\epsilon}^c} \sqrt{\omega^2 - \omega_c^2} \sqrt{1 - j\frac{\tan \delta_c}{1 - (\omega_c/\omega)^2}}$$

where δ_c is the loss tangent (4.238) and where

$$\omega_c = k_c/\sqrt{\mu \operatorname{Re} \tilde{\epsilon}^c}$$

is the *cutoff frequency*. Under the condition

$$\frac{\tan \delta_c}{1 - (\omega_c/\omega)^2} \ll 1 \tag{5.204}$$

we may approximate the square root using the first two terms of the binomial series to show that

$$\beta - j\alpha \approx \sqrt{\mu \operatorname{Re} \tilde{\epsilon}^c} \sqrt{\omega^2 - \omega_c^2} \left[1 - j\frac{1}{2}\frac{\tan \delta_c}{1 - (\omega_c/\omega)^2} \right]. \tag{5.205}$$

Condition (5.204) requires that ω be sufficiently removed from ω_c, either by having $\omega > \omega_c$ or $\omega < \omega_c$. When the frequency is *above cutoff* ($\omega > \omega_c$) we find from (5.205) that

$$\beta \approx \omega\sqrt{\mu \operatorname{Re} \tilde{\epsilon}^c}\sqrt{1 - \omega_c^2/\omega^2}, \qquad \alpha \approx \frac{\omega^2 \mu \operatorname{Re} \tilde{\epsilon}^c}{2\beta} \tan \delta_c.$$

Here $\alpha \ll \beta$ and the wave propagates down the guide with relatively little loss. When the frequency is *below cutoff* or the *waveguide is cut off* ($\omega < \omega_c$) we find that

$$\alpha \approx \omega\sqrt{\mu \operatorname{Re} \tilde{\epsilon}^c}\sqrt{\omega_c^2/\omega^2 - 1}, \qquad \beta \approx \frac{\omega^2 \mu \operatorname{Re} \tilde{\epsilon}^c}{2\alpha} \tan \delta_c.$$

Here the phase constant is small and the attenuation rate is large. For frequencies near ω_c the transition between these two types of wave behavior is rapid but continuous.

When the filling material is lossless, having permittivity ϵ and permeability μ, the transition across the cutoff frequency is discontinuous. For $\omega > \omega_c$ we have

$$\beta = \omega\sqrt{\mu\epsilon}\sqrt{1 - \omega_c^2/\omega^2}, \qquad \alpha = 0,$$

and the wave propagates without loss. For $\omega < \omega_c$

$$\alpha = \omega\sqrt{\mu\epsilon}\sqrt{\omega_c^2/\omega^2 - 1}, \qquad \beta = 0,$$

and the wave is evanescent. The dispersion diagram in Figure 5.7 displays the abrupt cutoff phenomenon. We can compute the phase and group velocities of the wave above cutoff just as we did for plane waves:

$$v_p = \frac{\omega}{\beta} = \frac{v}{\sqrt{1 - \omega_c^2/\omega^2}},$$

$$v_g = \frac{d\omega}{d\beta} = v\sqrt{1 - \omega_c^2/\omega^2}, \tag{5.206}$$

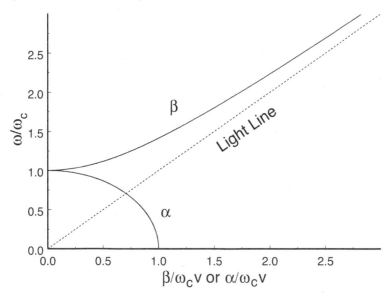

FIGURE 5.7
Dispersion plot for a hollow-pipe waveguide. Light line computed using $v = 1/\sqrt{\mu\epsilon}$.

where $v = 1/\sqrt{\mu\epsilon}$. Note that $v_g v_p = v^2$. We show later that v_g is the velocity of energy transport within a lossless guide. Note that $v_p \to v$ and $v_g \to v$ as $\omega \to \infty$. More interestingly, as $\omega \to \omega_c$ we find that $v_p \to \infty$ and $v_g \to 0$ (Figure 5.8).

We may also speak of the *guided wavelength* of a monochromatic wave propagating with frequency $\check{\omega}$ in a waveguide. We define this wavelength as $\lambda_g = 2\pi/\beta$. When the filling material is lossless the guided wavelength is

$$\lambda_g = \frac{2\pi}{\beta} = \frac{\lambda}{\sqrt{1 - \omega_c^2/\check{\omega}^2}} = \frac{\lambda}{\sqrt{1 - \lambda^2/\lambda_c^2}}.$$

Here

$$\lambda = \frac{2\pi}{\check{\omega}\sqrt{\mu\epsilon}}, \qquad \lambda_c = \frac{2\pi}{k_c}.$$

Orthogonality of waveguide modes. The modal fields in a closed-pipe waveguide obey several orthogonality relations. Let $(\check{\mathbf{E}}_n, \check{\mathbf{H}}_n)$ be the time-harmonic electric and magnetic fields of one particular waveguide mode (TE or TM), and let $(\check{\mathbf{E}}_m, \check{\mathbf{H}}_m)$ be the fields of a different mode (TE or TM). One very useful relation states that for a waveguide containing lossless materials,

$$\int_{CS} \hat{\mathbf{z}} \cdot (\check{\mathbf{e}}_n \times \check{\mathbf{h}}_m^*) \, dS = 0 \qquad (m \neq n), \tag{5.207}$$

where CS is the guide cross-section. This is used to establish that the total power carried by a wave is the sum of the powers carried by individual modes (see below).

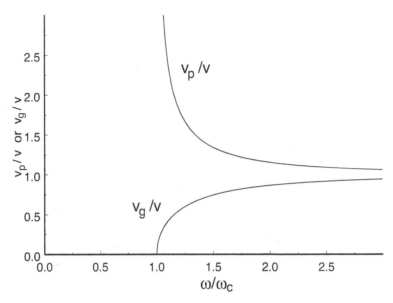

FIGURE 5.8
Phase and group velocity for a hollow-pipe waveguide.

Other important relationships include the orthogonality of the longitudinal fields,

$$\int_{CS} \check{E}_{zm}\check{E}_{zn}\, dS = 0 \qquad (m \neq n), \tag{5.208}$$

$$\int_{CS} \check{H}_{zm}\check{H}_{zn}\, dS = 0 \qquad (m \neq n), \tag{5.209}$$

and the orthogonality of transverse fields,

$$\int_{CS} \check{\mathbf{E}}_{tm} \cdot \check{\mathbf{E}}_{tn}\, dS = 0 \qquad (m \neq n), \tag{5.210}$$

$$\int_{CS} \check{\mathbf{H}}_{tm} \cdot \check{\mathbf{H}}_{tn}\, dS = 0 \qquad (m \neq n).$$

These may be combined as orthogonality properties of the total fields:

$$\int_{CS} \check{\mathbf{E}}_{m} \cdot \check{\mathbf{E}}_{n}\, dS = 0 \qquad (m \neq n), \tag{5.211}$$

$$\int_{CS} \check{\mathbf{H}}_{m} \cdot \check{\mathbf{H}}_{n}\, dS = 0 \qquad (m \neq n). \tag{5.212}$$

See Collin [38] for proofs of these relations.

Power carried by time-harmonic waves in lossless waveguides. The power carried by a time-harmonic wave propagating down a waveguide is defined as the time-average Poynting flux passing through the guide cross-section:

$$P_{av} = \frac{1}{2}\int_{CS} \operatorname{Re}(\check{\mathbf{E}} \times \check{\mathbf{H}}^{*}) \cdot \hat{\mathbf{z}}\, dS.$$

The field within the guide is assumed to be a superposition of all possible waveguide modes. For waves traveling in the $+z$-direction this implies

$$\check{\mathbf{E}} = \sum_m \left(\check{\mathbf{e}}_{tm} + \hat{\mathbf{z}}\check{e}_{zm}\right) e^{-jk_{zm}z}, \qquad \check{\mathbf{H}} = \sum_n \left(\check{\mathbf{h}}_{tn} + \hat{\mathbf{z}}\check{h}_{zn}\right) e^{-jk_{zn}z}.$$

Substitution gives

$$P_{av} = \tfrac{1}{2}\operatorname{Re}\left\{\int_{CS}\left[\sum_m \left(\check{\mathbf{e}}_{tm} + \hat{\mathbf{z}}\check{e}_{zm}\right) e^{-jk_{zm}z} \times \sum_n \left(\check{\mathbf{h}}_{tn}^* + \hat{\mathbf{z}}\check{h}_{zn}^*\right) e^{jk_{zn}^* z}\right]\cdot\hat{\mathbf{z}}\,dS\right\}$$

$$= \tfrac{1}{2}\operatorname{Re}\left\{\sum_m\sum_n e^{-j(k_{zm}-k_{zn}^*)z}\int_{CS}\hat{\mathbf{z}}\cdot\left(\check{\mathbf{e}}_{tm}\times\check{\mathbf{h}}_{tn}^*\right)\,dS\right\}$$

$$= \tfrac{1}{2}\operatorname{Re}\left\{\sum_n e^{-j(k_{zn}-k_{zn}^*)z}\int_{CS}\hat{\mathbf{z}}\cdot\left(\check{\mathbf{e}}_{tn}\times\check{\mathbf{h}}_{tn}^*\right)\,dS\right\}$$

by (5.207). For modes propagating in a lossless guide, $k_{zn} = \beta_{zn}$. For modes that are cut off, $k_{zn} = -j\alpha_{zn}$. However, we find below that terms in this series representing modes that are cut off are zero. Thus

$$P_{av} = \sum_n \tfrac{1}{2}\operatorname{Re}\left\{\int_{CS}\hat{\mathbf{z}}\cdot\left(\check{\mathbf{e}}_{tn}\times\check{\mathbf{h}}_{tn}^*\right)\,dS\right\} = \sum_n P_{n,av}.$$

Hence for waveguides filled with lossless media the total time-average power flow is given by the superposition of the individual modal powers.

Simple formulas for the individual modal powers in a lossless guide may be obtained by substituting the expressions for the fields. For TM modes we use (5.196)–(5.197) to get

$$P_{av} = \tfrac{1}{2}\operatorname{Re}\left\{|k_z|^2 Y_e^* e^{-j(k_z-k_z^*)}\int_{CS}\hat{\mathbf{z}}\cdot\left(\nabla_t\check{\psi}_e\times[\hat{\mathbf{z}}\times\nabla_t\check{\psi}_e^*]\right)\,dS\right\}$$

$$= \tfrac{1}{2}|k_z|^2\operatorname{Re}\left\{Y_e^*\right\}e^{-j(k_z-k_z^*)}\int_{CS}\nabla_t\check{\psi}_e\cdot\nabla_t\check{\psi}_e^*\,dS.$$

Here we have used (B.7) and $\hat{\mathbf{z}}\cdot\nabla_t\check{\psi}_e = 0$. This expression can be simplified via the two-dimensional version of Green's first identity (B.35):

$$\int_S \left(\nabla_t a\cdot\nabla_t b + a\nabla_t^2 b\right)\,dS = \oint_\Gamma a\frac{\partial b}{\partial n}\,dl.$$

Using $a = \check{\psi}_e$ and $b = \check{\psi}_e^*$ and integrating over the waveguide cross-section, we have

$$\int_{CS}\left(\nabla_t\check{\psi}_e\cdot\nabla_t\check{\psi}_e^* + \check{\psi}_e\nabla^2\check{\psi}_e^*\right)\,dS = \oint_\Gamma \check{\psi}_e\frac{\partial\check{\psi}_e^*}{\partial n}\,dl.$$

Substituting $\nabla_t^2\check{\psi}_e^* = -k_c^2\check{\psi}_e^*$ and remembering that $\check{\psi}_e = 0$ on Γ, we reduce this to

$$\int_{CS}\nabla_t\check{\psi}_e\cdot\nabla_t\check{\psi}_e^*\,dS = k_c^2\int_{CS}\check{\psi}_e\check{\psi}_e^*\,dS. \tag{5.213}$$

Thus,

$$P_{av} = \tfrac{1}{2}\operatorname{Re}\left\{Y_e^*\right\}|k_z|^2 k_c^2 e^{-j(k_z-k_z^*)z}\int_{CS}\check{\psi}_e\check{\psi}_e^*\,dS.$$

For modes above cutoff we have $k_z = \beta$ and $Y_e = \breve{\omega}\epsilon/k_z = \breve{\omega}\epsilon/\beta$. The power carried by these modes is therefore

$$P_{av} = \tfrac{1}{2}\breve{\omega}\epsilon\beta k_c^2 \int_{CS} \breve{\psi}_e \breve{\psi}_e^* \, dS.$$

For modes below cutoff we have $k_z = -j\alpha$ and $Y_e = j\breve{\omega}\epsilon/\alpha$. Thus, $\mathrm{Re}\{Y_e^*\} = 0$ and $P_{av} = 0$. At frequencies below cutoff, the fields are evanescent and do not carry power in the manner of propagating waves.

The corresponding result for TE modes,

$$P_{av} = \tfrac{1}{2}\breve{\omega}\mu\beta k_c^2 \int_{CS} \breve{\psi}_h \breve{\psi}_h^* \, dS, \tag{5.214}$$

is left as an exercise.

Stored energy in a waveguide and velocity of energy transport. Consider a source-free section of lossless waveguide bounded on its two ends by the cross-sectional surfaces CS_1 and CS_2. Setting $\breve{\mathbf{J}}^i = \breve{\mathbf{J}}^c = 0$ in (4.156), we have

$$\frac{1}{2} \oint_S (\breve{\mathbf{E}} \times \breve{\mathbf{H}}^*) \cdot d\mathbf{S} = 2j\breve{\omega} \int_V [\langle w_e \rangle - \langle w_m \rangle] \, dV,$$

where V is the region of the guide between CS_1 and CS_2. The right-hand side represents the difference between the total time-average stored electric and magnetic energies. Thus

$$2j\breve{\omega} [\langle W_e \rangle - \langle W_m \rangle] =$$

$$\frac{1}{2} \int_{CS_1} -\hat{\mathbf{z}} \cdot (\breve{\mathbf{E}} \times \breve{\mathbf{H}}^*) \, dS + \frac{1}{2} \int_{CS_2} \hat{\mathbf{z}} \cdot (\breve{\mathbf{E}} \times \breve{\mathbf{H}}^*) \, dS - \frac{1}{2} \int_{S_{\mathrm{cond}}} (\breve{\mathbf{E}} \times \breve{\mathbf{H}}^*) \cdot d\mathbf{S},$$

where S_{cond} consists of the conducting waveguide walls and $\hat{\mathbf{n}}$ points into the guide. For a propagating mode, the first two terms on the right cancel, since with no loss $\breve{\mathbf{E}} \times \breve{\mathbf{H}}^*$ is the same on CS_1 and CS_2. The third term is zero because $(\breve{\mathbf{E}} \times \breve{\mathbf{H}}^*) \cdot \hat{\mathbf{n}} = (\hat{\mathbf{n}} \times \breve{\mathbf{E}}) \cdot \breve{\mathbf{H}}^*$, and $\hat{\mathbf{n}} \times \breve{\mathbf{E}} = 0$ on the waveguide walls. Therefore

$$\langle W_e \rangle = \langle W_m \rangle$$

for any section of a lossless waveguide.

We may compute the time-average stored magnetic energy in a section of lossless waveguide of length l as

$$\langle W_m \rangle = \frac{\mu}{4} \int_0^l \int_{CS} \breve{\mathbf{H}} \cdot \breve{\mathbf{H}}^* \, dS \, dz.$$

For propagating TM modes, we can substitute (5.197) to find

$$\langle W_m \rangle/l = \frac{\mu}{4} (\beta Y_e)^2 \int_{CS} (\hat{\mathbf{z}} \times \nabla_t \breve{\psi}_e) \cdot (\hat{\mathbf{z}} \times \nabla_t \breve{\psi}_e^*) \, dS.$$

But $(\hat{\mathbf{z}} \times \nabla_t \breve{\psi}_e) \cdot (\hat{\mathbf{z}} \times \nabla_t \breve{\psi}_e^*) = \hat{\mathbf{z}} \cdot [\nabla_t \breve{\psi}_e^* \times (\hat{\mathbf{z}} \times \nabla_t \breve{\psi}_e)] = \nabla_t \breve{\psi}_e \cdot \nabla_t \breve{\psi}_e^*$ so

$$\langle W_m \rangle/l = \frac{\mu}{4} (\beta Y_e)^2 \int_{CS} \nabla_t \breve{\psi}_e \cdot \nabla_t \breve{\psi}_e^* \, dS.$$

Finally, by (5.213) we have the stored energy per unit length for a propagating TM mode:

$$\langle W_m \rangle / l = \langle W_e \rangle / l = \frac{\mu}{4} (\breve{\omega}\epsilon)^2 k_c^2 \int_{CS} \breve{\psi}_e \breve{\psi}_e^* \, dS.$$

The corresponding result for a TE mode,

$$\langle W_e \rangle / l = \langle W_m \rangle / l = \frac{\epsilon}{4} (\breve{\omega}\mu)^2 k_c^2 \int_{CS} \breve{\psi}_h \breve{\psi}_h^* \, dS,$$

is left as an exercise.

As with plane waves in (4.241) we may describe the velocity of energy transport as the ratio of the Poynting flux density to the total stored energy density:

$$\mathbf{S}_{av} = \langle w_T \rangle \mathbf{v}_e.$$

For TM modes this energy velocity is

$$v_e = \frac{\frac{1}{2}\breve{\omega}\epsilon\beta k_c^2 \breve{\psi}_e \breve{\psi}_e^*}{2 \frac{\mu}{4}(\breve{\omega}\epsilon)^2 k_c^2 \breve{\psi}_e \breve{\psi}_e^*} = \frac{\beta}{\breve{\omega}\mu\epsilon} = v\sqrt{1 - \omega_c^2 / \breve{\omega}^2},$$

which is identical to the group velocity (5.206). This is also the case for TE modes, for which

$$v_e = \frac{\frac{1}{2}\breve{\omega}\mu\beta k_c^2 \breve{\psi}_h \breve{\psi}_h^*}{2 \frac{\epsilon}{4}(\breve{\omega}\mu)^2 k_c^2 \breve{\psi}_h \breve{\psi}_h^*} = \frac{\beta}{\breve{\omega}\mu\epsilon} = v\sqrt{1 - \omega_c^2 / \breve{\omega}^2}.$$

Attenuation due to wall loss; perturbation approximation. The attenuation constant resulting from losses in the material filling a waveguide with PEC walls is determined using (5.203):

$$\alpha = -\,\mathrm{Im}\left\{ \sqrt{\omega^2 \tilde{\mu}\tilde{\epsilon}^c - k_c^2} \right\}.$$

There will be additional attenuation of the propagating wave if the walls are imperfectly conducting, due to power transfered into the walls by the fields in the guide. Computing this power loss requires knowledge of the fields when the walls are not PEC. These fields are difficult to determine because the boundary conditions established for PEC boundaries no longer apply, and a separation of variables solution is usually not possible. In fact, often the only reliable way to determine the fields with high accuracy is to use a purely numerical solution. However, an approximation to the power delivered to the walls may be established by assuming that when the walls are good (but not perfect) conductors, the fields inside the guide are identical to those found when the walls are PEC. Using these fields, the power delivered to the wall can be found using the diffusion equations from §3.6.2, provided the walls are locally planar.

A frequency-domain field diffuses into a conductor with conductivity σ and permeability μ such that the current induced in the conductor is given by (3.213). For a time-harmonic field the phasor current can be written as

$$\breve{J}(x) = \sigma \breve{E}_0 e^{-\frac{x}{\delta}} e^{-j\frac{x}{\delta}}$$

where $\delta = 1/\sqrt{\pi \breve{f}\mu\sigma}$ is the skin depth and $\breve{E}_0$ is the phasor amplitude of the tangential electric field at the surface. The time-average power dissipated in the conductor is, by the complex Poynting theorem,

$$P_d = \frac{1}{2} \int_V \breve{\mathbf{E}} \cdot \breve{\mathbf{J}}^* \, dV = \frac{1}{2\sigma} \int_V |\breve{\mathbf{J}}|^2 \, dV.$$

Substitution yields the power dissipated per unit length at position z:

$$\frac{P_d(z)}{l} = \frac{\sigma}{2} \int \int_0^\infty |\check{E}_0|^2 e^{-2\frac{x}{\delta}}\, dx\, dy = \frac{\sigma \delta}{4} \int |\check{E}_0|^2 dy.$$

We also know that the tangential magnetic field is given by (3.215), such that

$$|\check{E}_0|^2 = 2 \frac{|\check{\mathbf{H}}_{\text{tan}}|^2}{\sigma^2 \delta^2}$$

where $|\check{\mathbf{H}}_{\text{tan}}| = |\hat{\mathbf{n}} \times \check{\mathbf{H}}|$. Substitution gives

$$\frac{P_d(z)}{l} = \int \frac{R_s}{2} |\check{\mathbf{H}}_{\text{tan}}|^2\, dy$$

where $R_s = 1/\sigma\delta$ is the surface resistance (3.216). If the surface is non-planar with varying magnetic field, we generalize this expression to

$$\frac{P_d(z)}{l} = \int_\Gamma \frac{R_s}{2} |\check{\mathbf{H}}_{\text{tan}}|^2\, dl, \tag{5.215}$$

where Γ describes the contour of the surface transverse to z.

To use (5.215) to obtain the attenuation coefficient for a waveguide with lossy walls, we note that the power passing through the waveguide cross-section at axial position z can be written as

$$P_{av}(z) = P_0 e^{-2\alpha_c z},$$

where $P_0 = P_{av}(z = 0)$ and α_c is the attenuation constant due to conductor loss. Thus

$$\frac{dP_{av}(z)}{dz} = -2\alpha_c P_0 e^{-2\alpha_c z} = -2\alpha_c P_{av}(z).$$

But the derivative is just the power dissipated per unit length at position z. Equating this with (5.215) gives a formula for the attenuation constant

$$\alpha_c = \frac{P_d(z)/l}{2P_{av}(z)} = \frac{P_d(0)/l}{2P_0}. \tag{5.216}$$

We employ this formula as follows. First the fields are computed assuming PEC waveguide walls. The time-average power carried through the cross-section of the waveguide by the propagating wave, P_0, is then computed. Then the power dissipated in the walls per unit length is computed from (5.215) using the magnetic field found with the walls assumed to be PEC. Finally, the attenuation constant is found from (5.216). This *perturbational* approach assumes that the change in magnetic field introduced by the wall losses is negligible.

Fields of a rectangular waveguide.   Consider a rectangular waveguide with a cross-section occupying $0 \leq x \leq a$ and $0 \leq y \leq b$. The filling is assumed to be a lossless dielectric of permittivity ϵ and permeability μ. We seek the modal fields within the guide.

Both TE and TM fields can exist. In each case we must solve the differential equation

$$\nabla_t^2 \tilde{\psi} + k_c^2 \tilde{\psi} = 0,$$

where $k_z^2 + k_c^2 = k^2$. A product solution in rectangular coordinates results from the separation of variables technique (§ A.5.3):

$$\tilde{\psi}(x, y, \omega) = [A_x \sin k_x x + B_x \cos k_x x] [A_y \sin k_y y + B_y \cos k_y y] \qquad (5.217)$$

where $k_x^2 + k_y^2 = k_c^2$. This is easily checked by substitution.

For TM modes the solution is subject to the boundary condition (5.202):

$$\tilde{\psi}_e(\boldsymbol{\rho}, \omega) = 0 \qquad (\boldsymbol{\rho} \in \Gamma). \qquad (5.218)$$

Applying this at $x = 0$ and $y = 0$ we find $B_x = B_y = 0$. Applying the boundary condition at $x = a$ we then find $\sin k_x a = 0$ and thus $k_x = n\pi/a$ for $n = 1, 2, \ldots$. (Note that $n = 0$ corresponds to the trivial solution $\tilde{\psi}_e = 0$.) Similarly, from the condition at $y = b$ we find that $k_y = m\pi/b$ for $m = 1, 2, \ldots$. Thus

$$\tilde{\psi}_e(x, y, \omega) = A_{nm} \sin\left(\frac{n\pi x}{a}\right) \sin\left(\frac{m\pi y}{b}\right).$$

By (5.195)–(5.197) the fields are

$$\tilde{E}_z = k_{c_{nm}}^2 A_{nm} \left[\sin\frac{n\pi x}{a} \sin\frac{m\pi y}{b}\right] e^{\mp j k_z z}, \qquad (5.219)$$

$$\tilde{\mathbf{E}}_t = \mp j k_z A_{nm} \left[\hat{\mathbf{x}}\frac{n\pi}{a} \cos\frac{n\pi x}{a} \sin\frac{m\pi y}{b} + \hat{\mathbf{y}}\frac{m\pi}{b} \sin\frac{n\pi x}{a} \cos\frac{m\pi y}{b}\right] e^{\mp j k_z z}, \qquad (5.220)$$

$$\tilde{\mathbf{H}}_t = j k_z Y_e A_{nm} \left[\hat{\mathbf{x}}\frac{m\pi}{b} \sin\frac{n\pi x}{a} \cos\frac{m\pi y}{b} - \hat{\mathbf{y}}\frac{n\pi}{a} \cos\frac{n\pi x}{a} \sin\frac{m\pi y}{b}\right] e^{\mp j k_z z}, \qquad (5.221)$$

where

$$Y_e = \frac{1}{\eta\sqrt{1 - \omega_{c_{nm}}^2/\omega^2}}, \qquad \eta = (\mu/\epsilon)^{1/2}.$$

Each pair m, n describes a different field pattern and thus a different mode, designated TM$_{nm}$. The cutoff wavenumber of the TM$_{nm}$ mode is

$$k_{c_{nm}} = \sqrt{\left(\frac{n\pi}{a}\right)^2 + \left(\frac{m\pi}{b}\right)^2} \qquad (m, n = 1, 2, 3, \ldots)$$

and the cutoff frequency is

$$\omega_{c_{nm}} = v\sqrt{\left(\frac{n\pi}{a}\right)^2 + \left(\frac{m\pi}{b}\right)^2} \qquad (m, n = 1, 2, 3, \ldots)$$

where $v = 1/(\mu\epsilon)^{1/2}$. Thus the TM$_{11}$ mode has the lowest cutoff frequency of any TM mode. There is a range of frequencies for which this is the only propagating TM mode.

For TE modes the solution is subject to

$$\hat{\mathbf{n}} \cdot \nabla_t \tilde{\psi}_h(\boldsymbol{\rho}, \omega) = \frac{\partial \tilde{\psi}_h(\boldsymbol{\rho}, \omega)}{\partial n} = 0 \qquad (\boldsymbol{\rho} \in \Gamma).$$

At $x = 0$ we have $\partial\tilde{\psi}_h/\partial x = 0$ leading to $A_x = 0$. At $y = 0$ we have $\partial\tilde{\psi}_h/\partial y = 0$ leading to $A_y = 0$. At $x = a$ we require $\sin k_x a = 0$ and thus $k_x = n\pi/a$ for $n = 0, 1, 2, \ldots$. Similarly, from the condition at $y = b$ we find $k_y = m\pi/b$ ($m = 0, 1, 2, \ldots$). The case $n = m = 0$ produces the trivial solution. Thus

$$\tilde{\psi}_h(x, y, \omega) = B_{nm} \cos\left(\frac{n\pi x}{a}\right) \cos\left(\frac{m\pi y}{b}\right) \qquad (m, n = 0, 1, 2, \ldots, m + n > 0). \quad (5.222)$$

By (5.198)–(5.200) the fields are

$$\tilde{H}_z = k_{c_{nm}}^2 B_{nm} \left[\cos \frac{n\pi x}{a} \cos \frac{m\pi y}{b} \right] e^{\mp jk_z z}, \tag{5.223}$$

$$\tilde{\mathbf{H}}_t = \pm jk_z B_{nm} \left[\hat{\mathbf{x}} \frac{n\pi}{a} \sin \frac{n\pi x}{a} \cos \frac{m\pi y}{b} + \hat{\mathbf{y}} \frac{m\pi}{b} \cos \frac{n\pi x}{a} \sin \frac{m\pi y}{b} \right] e^{\mp jk_z z}, \tag{5.224}$$

$$\tilde{\mathbf{E}}_t = jk_z Z_h B_{nm} \left[\hat{\mathbf{x}} \frac{m\pi}{b} \cos \frac{n\pi x}{a} \sin \frac{m\pi y}{b} - \hat{\mathbf{y}} \frac{n\pi}{a} \sin \frac{n\pi x}{a} \cos \frac{m\pi y}{b} \right] e^{\mp jk_z z}, \tag{5.225}$$

where

$$Z_h = \frac{\eta}{\sqrt{1 - \omega_{c_{nm}}^2 / \omega^2}}.$$

In this case the modes are designated TE$_{nm}$. The cutoff wavenumber of the TE$_{nm}$ mode is

$$k_{c_{nm}} = \sqrt{\left(\frac{n\pi}{a} \right)^2 + \left(\frac{m\pi}{b} \right)^2} \qquad (m, n = 0, 1, 2, \ldots, \ m + n > 0)$$

and the cutoff frequency is

$$\omega_{c_{nm}} = v \sqrt{\left(\frac{n\pi}{a} \right)^2 + \left(\frac{m\pi}{b} \right)^2} \qquad (m, n = 0, 1, 2, \ldots, \ m + n > 0)$$

where $v = 1/(\mu\epsilon)^{1/2}$. Modes having the same cutoff frequency are said to be *degenerate*. This is the case with the TE and TM modes. But the field distributions differ so the modes are distinct. Note that we may also have degeneracy among the TE or TM modes. For instance, if $a = b$ then the cutoff frequency of the TE$_{nm}$ mode is identical to that of the TE$_{mn}$ mode. If $a \geq b$ then the TE$_{10}$ mode has the lowest cutoff frequency and is termed the *dominant* mode in a rectangular guide. There is a finite band of frequencies in which this is the only mode propagating (although the bandwidth is small if $a \approx b$).

▶ **Example 5.15:** Lowest-order mode in a rectangular waveguide

Find the cutoff frequency and the fields for the lowest-order mode in a rectangular waveguide filled with a lossless isotropic material of permittivity ϵ and permeability μ. Assume $a > b$.

Solution: The lowest order mode in a rectangular waveguide with $a > b$ is the TE$_{10}$ mode with cutoff wavenumber $k_{c_{10}} = \pi/a$ and cutoff frequency

$$f_{c_{10}} = \frac{v}{2a},$$

where $v = 1/\sqrt{\mu\epsilon}$. Note that $f_{c_{10}}$ is independent of b. The fields are given by (5.223)–(5.225):

$$\tilde{H}_z = k_{c_{10}}^2 B_{10} \cos \frac{\pi x}{a} e^{\mp jk_z z},$$

$$\tilde{\mathbf{H}}_t = \pm jk_z B_{10} \frac{\pi}{a} \hat{\mathbf{x}} \sin \frac{\pi x}{a} e^{\mp jk_z z},$$

$$\tilde{\mathbf{E}}_t = -jk_z Z_h B_{10} \frac{\pi}{a} \hat{\mathbf{y}} \sin \frac{\pi x}{a} e^{\mp jk_z z}.$$

Note that the lowest-order TM mode is the TM$_{11}$ mode with cutoff wavenumber

$$k_{c_{11}} = \sqrt{\left(\frac{\pi}{a} \right)^2 + \left(\frac{\pi}{b} \right)^2} = \frac{\pi}{a} \sqrt{1 + \left(\frac{a}{b} \right)^2},$$

and cutoff frequency

$$f_{c_{11}} = \frac{v}{2a}\sqrt{1 + \left(\frac{a}{b}\right)^2}.$$

The second TE mode to propagate is either the TE_{01} mode with cutoff frequency $f_{c_{01}} = v/(2b)$ or the TE_{20} mode with cutoff frequency $f_{c_{20}} = v/a$, depending on the ratio a/b. Often $a/b = 2$ is used, making the cutoff frequencies of the TE_{20} and TE_{01} modes both v/a, and the cutoff frequencies of the TE_{11} and TM_{11} modes both $\sqrt{5/4}v/a$.

For a square waveguide, the cutoff frequencies of the TE_{10} and the TE_{01} modes coincide (the modes are degenerate). ◄

▶ **Example 5.16:** Power flow of the TE_{10} mode in a rectangular waveguide

Find the time-average power flow for the lowest order mode in a rectangular waveguide filled with a lossless isotropic material of permittivity ϵ and permeability μ. Assume $a > b$.

Solution: We specialize the potential (5.222) to time harmonic form for the TE_{10} mode to get

$$\breve{\psi}_h = B_{10}\cos\frac{\pi x}{a}.$$

Substitution into (5.214) gives

$$P_{av} = \tfrac{1}{2}|B_{10}|^2\breve{\omega}\mu\beta\left(\frac{\pi}{a}\right)^2\int_0^b\int_0^a\cos^2\frac{\pi x}{a}\,dx\,dy = \frac{b}{4a}\pi^2\breve{\omega}\mu\beta|B_{10}|^2. \quad ◄ \tag{5.226}$$

▶ **Example 5.17:** Attenuation constant for the TE_{10} mode in a rectangular waveguide

Find the attenuation constant due to wall loss, α_c, for the lowest order mode in a rectangular waveguide filled with a lossless isotropic material of permittivity ϵ and permeability μ. Assume $a > b$.

Solution: The power dissipated in the waveguide walls per unit length is given by (5.215). Substituting for the magnetic field from Example 5.15 we see that on the walls at $x = 0$ and $x = a$,

$$|\hat{\mathbf{n}} \times \breve{\mathbf{H}}|^2\big|_{z=0} = k_{c_{10}}^4|B_{10}|^2.$$

Similarly, on the walls at $y = 0$ and $y = b$,

$$|\hat{\mathbf{n}} \times \breve{\mathbf{H}}|^2\big|_{z=0} = k_{c_{10}}^4|B_{10}|^2\cos^2\frac{\pi x}{a} + k_z^2|B_{10}|^2\left(\frac{\pi}{a}\right)^2\sin^2\frac{\pi x}{a}.$$

So the time-average power dissipated per unit length in the waveguide walls is

$$\frac{P_d(0)}{l} = 2\frac{R_s}{2}\int_0^b k_{c_{10}}^4|B_{10}|^2\,dy + 2\frac{R_s}{2}\int_0^a k_{c_{10}}^4|B_{10}|^2\cos^2\frac{\pi x}{a}\,dx$$
$$+ 2\frac{R_s}{2}\int_0^a k_z^2|B_{10}|^2\left(\frac{\pi}{a}\right)^2\sin^2\frac{\pi x}{a}\,dx.$$

Integration gives

$$\frac{P_d(0)}{l} = R_s|B_{10}|^2\left(\frac{\pi}{a}\right)^2\left\{b\left(\frac{\pi}{a}\right)^2 + \frac{a}{2}\left[\left(\frac{\pi}{a}\right)^2 + k_z^2\right]\right\}$$
$$= \frac{R_s|B_{10}|^2\pi^2}{2a^4}[2b\pi^2 + k^2a^3]. \tag{5.227}$$

Substituting the power dissipated (5.227) and the power flux (5.226) into the perturbational formula for the attenuation constant (5.216), we have

$$\alpha_c = \frac{P_d(0)/l}{2P_0} = \frac{R_s |B_{10}|^2 \pi^2}{2a^4} [2b\pi^2 + k^2 a^3] \frac{4a}{2b\pi^2 \tilde{\omega}\mu\beta |B_{10}|^2}$$

$$= \frac{R_s}{\tilde{\omega}\mu a^3 b\beta} [2b\pi^2 + k^2 a^3].$$

The behavior of α_c with frequency is important practically. Consider an X-band WR-90 rectangular guide ($a = 22.86$ mm, $b = 10.16$ mm) made of brass with conductivity $\sigma = 1.4 \times 10^7$ S/m, permittivity ϵ_0, and permeability μ_0. Recalling that $R_s = 1/\sigma\delta$ where $\delta = 1/\sqrt{\pi f \mu \sigma}$ is the skin depth, and plotting α_c vs. frequency, we obtain the curve shown in Figure 5.9. Two important characteristics of α_c are exhibited. First, the attenuation constant increases without bound near the 6.557 GHz cutoff frequency. This implies that one should observe an operational safety margin to avoid excessive loss. WR-90 waveguide has a suggested operational band of 8.2–12.5 GHz (the upper frequency determined by the cutoff frequency of the first higher-order mode). Second, the attenuation reaches a minimum at a certain frequency (in this case 15.4 GHz) and then increases with frequency. This is characteristic of all rectangular waveguide modes.

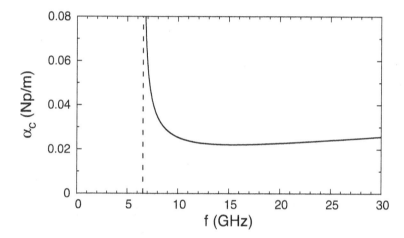

FIGURE 5.9
Attenuation constant due to wall loss for an air-filled WR-90 X-band waveguide constructed from brass. The vertical dashed line indicates the cutoff frequency. ◄

Fields of a circular waveguide. Consider a circular waveguide with a cross-section occupying $0 \leq \rho \leq a$ and $0 \leq \phi < 2\pi$. The filling is a lossless dielectric of permittivity ϵ and permeability μ. We seek the modal fields.

Both TE and TM fields can exist here as well. In each case we must solve the differential equation

$$\nabla_t^2 \tilde{\psi} + k_c^2 \tilde{\psi} = 0,$$

where $k_c^2 + k_z^2 = k^2$. Separation of variables in cylindrical coordinates (§ A.5.3) yields solutions of the form

$$\tilde{\psi}(\rho, \phi, \omega) = [A_\rho J_{k_\phi}(k_c \rho) + B_\rho N_{k_\phi}(k_c \rho)][A_\phi \sin k_\phi \phi + B_\phi \cos k_\phi \phi]. \tag{5.228}$$

Finiteness of the fields on the z-axis requires $B_\rho = 0$. Periodicity in ϕ implies that $k_\phi = n = 0, 1, 2, \ldots$. Note that $n = 0$ corresponds to the case of azimuthal symmetry (ϕ-invariance). The relationship between A_{k_ϕ} and B_{k_ϕ} cannot be determined outside the excitation problem; we therefore take the ϕ-dependence as $\cos(n\phi + \Phi_n)$ where Φ_n is an undetermined phase angle. Thus, the solution is written as

$$\tilde{\psi}(\rho, \phi, \omega) = A_n J_n(k_c \rho) \cos(n\phi + \Phi_n).$$

For TM modes the solution is subject to the boundary condition (5.202):

$$\tilde{\psi}_e(\boldsymbol{\rho}, \omega) = 0 \qquad (\boldsymbol{\rho} \in \Gamma).$$

Applying this at $\rho = a$ for $0 \leq \phi < 2\pi$ we find that $J_n(k_c a) = 0$. If p_{nm} denotes the mth zero of $J_n(x)$, the cutoff wavenumber for the TM_{nm} mode is $k_{c_{nm}} = p_{nm}/a$, the cutoff frequency is $\omega_{c_{nm}} = p_{nm} v/a$ where $v = 1/\sqrt{\mu\epsilon}$, and the propagation constant is

$$k_z = \sqrt{k^2 - \left(\frac{p_{nm}}{a}\right)^2}.$$

With these we have the potential function

$$\tilde{\psi}_e(\rho, \phi, \omega) = A_{nm} J_n\left(\rho \frac{p_{nm}}{a}\right) \cos(n\phi + \Phi_n).$$

By (5.195)–(5.197) the fields are

$$\tilde{E}_z = \left(\frac{p_{nm}}{a}\right)^2 A_{nm} J_n\left(\rho \frac{p_{nm}}{a}\right) \cos(n\phi + \Phi_n) e^{\mp j k_z z},$$

$$\tilde{E}_\rho = \mp j k_z A_{nm} \frac{p_{nm}}{a} J_n'\left(\rho \frac{p_{nm}}{a}\right) \cos(n\phi + \Phi_n) e^{\mp j k_z z},$$

$$\tilde{E}_\phi = \pm j k_z A_{nm} \frac{n}{\rho} J_n\left(\rho \frac{p_{nm}}{a}\right) \sin(n\phi + \Phi_n) e^{\mp j k_z z},$$

$$\tilde{H}_\rho = -j k_z Y_e A_{nm} \frac{n}{\rho} J_n\left(\rho \frac{p_{nm}}{a}\right) \sin(n\phi + \Phi_n) e^{\mp j k_z z},$$

$$\tilde{H}_\phi = -j k_z Y_e A_{nm} \frac{p_{nm}}{a} J_n'\left(\rho \frac{p_{nm}}{a}\right) \cos(n\phi + \Phi_n) e^{\mp j k_z z},$$

where

$$Y_e = \frac{1}{\eta \sqrt{1 - \omega_{c_{nm}}^2/\omega^2}}, \qquad \eta = (\mu/\epsilon)^{1/2}.$$

Each pair m, n generates the field pattern of a distinct TM_{nm} mode.

For TE modes the solution is subject to

$$\hat{\mathbf{n}} \cdot \nabla_t \tilde{\psi}_h(\boldsymbol{\rho}, \omega) = \frac{\partial \tilde{\psi}_h(\boldsymbol{\rho}, \omega)}{\partial \rho} = 0 \qquad (\boldsymbol{\rho} \in \Gamma).$$

Applying this at $\rho = a$ for $0 \leq \phi < 2\pi$ we find that $J_n'(k_c a) = 0$. Denoting by p_{nm}' the mth zero of $J_n'(x)$, the cutoff wavenumber for the TE_{nm} mode is

$$k_{c_{nm}} = \frac{p_{nm}'}{a}, \qquad (5.229)$$

the cutoff frequency is

$$\omega_{c_{nm}} = p_{nm}' \frac{v}{a},$$

TABLE 5.1

Zeros of $J_n(x)$

	$m = 1$	$m = 2$	$m = 3$	$m = 4$	$m = 5$
$n = 0$	2.40483	5.52008	8.65373	11.79153	14.93092
$n = 1$	3.83171	7.01559	10.17347	13.32369	16.47063
$n = 2$	5.13562	8.41724	11.61984	14.79595	17.95982
$n = 3$	6.38016	9.76102	13.01520	16.22347	19.40942
$n = 4$	7.58834	11.06471	14.37254	17.61597	20.82693
$n = 5$	8.77148	12.33860	15.70017	18.98013	22.21780

and the propagation constant is

$$k_z = \sqrt{k^2 - \left(\frac{p'_{nm}}{a}\right)^2}.$$

With these we have the potential function

$$\tilde{\psi}_h(\rho, \phi, \omega) = B_{nm} J_n\left(\rho\frac{p'_{nm}}{a}\right)\cos(n\phi + \Phi_n). \tag{5.230}$$

By (5.198)–(5.200) the fields are

$$\tilde{H}_z = \left(\frac{p'_{nm}}{a}\right)^2 B_{nm} J_n\left(\rho\frac{p'_{nm}}{a}\right)\cos(n\phi + \Phi_n)e^{\mp jk_z z}, \tag{5.231}$$

$$\tilde{H}_\rho = \mp jk_z B_{nm}\frac{p'_{nm}}{a} J_n'\left(\rho\frac{p'_{nm}}{a}\right)\cos(n\phi + \Phi_n)e^{\mp jk_z z}, \tag{5.232}$$

$$\tilde{H}_\phi = \pm jk_z B_{nm}\frac{n}{\rho} J_n\left(\rho\frac{p'_{nm}}{a}\right)\sin(n\phi + \Phi_n)e^{\mp jk_z z}, \tag{5.233}$$

$$\tilde{E}_\rho = jk_z Z_h B_{nm}\frac{n}{\rho} J_n\left(\rho\frac{p'_{nm}}{a}\right)\sin(n\phi + \Phi_n)e^{\mp jk_z z}, \tag{5.234}$$

$$\tilde{E}_\phi = jk_z Z_h B_{nm}\frac{p'_{nm}}{a} J_n'\left(\rho\frac{p'_{nm}}{a}\right)\cos(n\phi + \Phi_n)e^{\mp jk_z z}, \tag{5.235}$$

where

$$Z_h = \frac{\eta}{\sqrt{1 - \omega_{c_{nm}}^2/\omega^2}}.$$

Each pair m, n generates the field pattern of a distinct TE_{nm} mode.

The ordering of modes according to their cutoff frequencies is determined by the zeros of $J_n(x)$ and $J_n'(x)$. Some approximate values are given in Tables 5.1 and 5.2. Note that the cutoff frequencies are arranged as

$$TE_{11} \quad TM_{01} \quad TE_{21} \quad TE_{01} \quad TM_{11} \quad TE_{31} \quad TM_{21} \quad TE_{41} \quad TE_{12} \quad TM_{02}$$

The TE_{01} and TM_{11} modes are degenerate, since $p'_{01} = p_{11}$; in fact, the TE_{0m} mode is degenerate with the $TM_{1,m}$ mode because $J_0'(x) = -J_1(x)$.

Although the fundamental mode of the circular guide is the TE_{11} mode, the TE_{01} has advantages including low wall loss attenuation. Its azimuthal symmetry also makes it the mode with the simplest field structure. The examples below concern this mode.

TABLE 5.2

Zeros of $J'_n(x)$

	$m = 1$	$m = 2$	$m = 3$	$m = 4$	$m = 5$
$n = 0$	3.83171	7.01559	10.17347	13.32369	16.47063
$n = 1$	1.84118	5.33144	8.53632	11.70600	14.86359
$n = 2$	3.05424	6.70613	9.96947	13.17037	16.34752
$n = 3$	4.20119	8.01524	11.34592	14.58585	17.78875
$n = 4$	5.31755	9.28240	12.68191	15.96411	19.19603
$n = 5$	6.41562	10.51986	13.98719	17.31284	20.57551

▶ **Example 5.18:** TE_{01} mode in a circular waveguide

Find the cutoff frequency and the fields for the TE_{01} mode in a circular waveguide filled with a lossless isotropic material of permittivity ϵ and permeability μ.

Solution: The cutoff wavenumber for the TE_{01} mode is found using (5.229) and Table 5.2. Since $p'_{01} = 3.83171$,

$$k_{c_{01}} = \frac{3.83171}{a},$$

and the cutoff frequency is

$$f_{c_{10}} = \frac{k_{c_{01}} v}{2\pi} = \frac{0.609836 v}{a}, \tag{5.236}$$

where $v = 1/\sqrt{\mu\epsilon}$. The fields are given by (5.231)–(5.235):

$$\tilde{H}_z = \left(\frac{p'_{01}}{a}\right)^2 B_{01} J_0\left(\rho\frac{p'_{01}}{a}\right) e^{\mp jk_z z},$$

$$\tilde{H}_\rho = \pm jk_z B_{01}\frac{p'_{01}}{a} J_1\left(\rho\frac{p'_{01}}{a}\right) e^{\mp jk_z z},$$

$$\tilde{E}_\phi = -jk_z Z_h B_{01}\frac{p'_{01}}{a} J_1\left(\rho\frac{p'_{01}}{a}\right) e^{\mp jk_z z},$$

$$\tilde{H}_\phi = \tilde{E}_\rho = 0,$$

where we have incorporated the phase Φ into the amplitude B_{01} and have used $J'_0(x) = -J_1(x)$. ◀

▶ **Example 5.19:** Power flow of the TE_{01} mode in a circular waveguide

Find the time-average power flow for the TE_{01} mode in a circular waveguide filled with a lossless isotropic material of permittivity ϵ and permeability μ.

Solution: We specialize the potential (5.230) to time harmonic form for the TE_{01} mode to get

$$\check{\psi}_h = B_{01} J_0\left(\rho\frac{p'_{01}}{a}\right).$$

Substitution into (5.214) gives

$$P_{av} = \tfrac{1}{2}|B_{10}|^2 \check{\omega}\mu\beta \left(\frac{p'_{01}}{a}\right)^2 \int_0^{2\pi} \int_0^a J_0^2\left(\rho\frac{p'_{01}}{a}\right) \rho\, d\rho\, d\phi.$$

Use of

$$\int_0^a J_0^2\left(\rho\frac{p'_{01}}{a}\right)\rho\,d\rho = \frac{a^2}{2}J_0^2(p'_{01})$$

gives

$$P_{av} = |B_{10}|^2\check{\omega}\mu\beta\frac{\pi}{2}(p'_{01})^2 J_0^2(p'_{01}).\ \blacktriangleleft \qquad (5.237)$$

▶ **Example 5.20:** Attenuation constant for the TE$_{01}$ mode in a circular waveguide

Find the attenuation constant due to wall loss, α_c, for the TE$_{01}$ mode in a circular waveguide filled with a lossless isotropic material of permittivity ϵ and permeability μ.

Solution: The power dissipated in the waveguide walls per unit length is given by (5.215). Substituting for the magnetic field from Example 5.18, we see that on the wall at $\rho = a$,

$$|\hat{\mathbf{n}}\times\check{\mathbf{H}}|^2\big|_{z=0} = \left(\frac{p'_{01}}{a}\right)^4 |B_{01}|^2 J_0^2(p'_{01}).$$

So the time-average power dissipated per unit length in the waveguide walls is

$$\frac{P_d(0)}{\ell} = \frac{R_s}{2}\int_0^{2\pi}\left(\frac{p'_{01}}{a}\right)^4 |B_{01}|^2 J_0^2(p'_{01})a\,d\phi$$

$$= \pi a R_s\left(\frac{p'_{01}}{a}\right)^4 |B_{01}|^2 J_0^2(p'_{01}). \qquad (5.238)$$

Substituting (5.237) and (5.238) into (5.216), we find that

$$\alpha_c = \frac{P_d(0)/\ell}{2P_0}$$

$$= \frac{\pi a R_s\left(\frac{p'_{01}}{a}\right)^4 |B_{01}|^2 J_0^2(p'_{01})}{|B_{10}|^2\check{\omega}\mu\beta\pi\,(p'_{01})^2\,J_0^2(p'_{01})}$$

$$= \frac{R_s}{\check{\omega}\mu a\beta}\left(\frac{p'_{01}}{a}\right)^2. \qquad (5.239)$$

The behavior of α_c with frequency for the TE$_{01}$ mode is quite different from that of the attenuation constant for a rectangular waveguide (and from that of most modes in a circular guide). Consider a circular guide with radius $a = 27.88$ mm made from brass with conductivity $\sigma = 1.4\times10^7$ S/m, permittivity ϵ_0, and permeability μ_0. Using (5.236), we see that this value of a produces a TE$_{01}$ mode cutoff frequency of 6.557 GHz, identical to that of the fundamental (TE$_{10}$) mode in the WR-90 rectangular guide considered in Example 5.17. Recalling that

$$R_s = \frac{1}{\sigma\delta}$$

where

$$\delta = \frac{1}{\sqrt{\pi f\mu\sigma}}$$

and plotting α_c vs. frequency, we obtain the curve shown in Figure 5.10. Also shown is the attenuation constant for the WR-90 rectangular waveguide considered in Example 5.17.

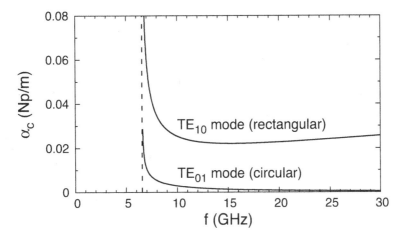

FIGURE 5.10
Attenuation constants due to wall loss for an air-filled circular waveguide and an air-filled WR-90 X-band waveguide, both constructed from brass. Vertical dashed line indicates the cutoff frequency.

Not only is the attenuation constant for the TE_{01} mode in a circular guide much smaller than for the TE_{10} mode in a rectangular guide with the same cutoff frequency, the attenuation constant of the TE_{01} circular waveguide mode decreases monotonically with frequency. This monotonic behavior is shared by all azimuthally symmetric TE modes in a circular waveguide (i.e., all TE_{0m} modes). In contrast, all other TE modes and all TM modes (even azimuthally symmetric modes) in a circular guide have attenuation constants that increase with frequency after some point, as do all modes in rectangular guides. Of course, the circular guide will have three other modes (TE_{11}, TM_{01}, and TE_{21}) that propagate along with the TE_{01} mode. It is necessary to suppress these other modes to truly take advantage of the low attenuation properties of the TE_{01} mode. This was the subject of extensive research at Bell Laboratories before the advent of low-loss fiber optic cables [133]. ◄

Waveguide excitation by current sheets. An important property of the spectrum of eigenfields for a perfectly conducting closed-pipe waveguide is that they represent a *complete* and *orthogonal* set of functions that can be used to construct an arbitrary solution to Maxwell's equations within the guide. Consider, for instance, a surface electric current in the plane $z = 0$ in a waveguide (Figure 5.11). The current will excite TE_z and TM_z modes in the regions $z > 0$ and $z < 0$. We seek the amplitudes of the modal fields.

The fields in each region may be constructed from superpositions of waveguide modes. We consider only the transverse fields, since only they are needed for application of the boundary conditions at the position of the source. For $z < 0$ we have only waves traveling (or evanescent) in the $-z$ direction, with the transverse fields given by

$$\tilde{\mathbf{E}}_t = \sum_{n=1}^{\infty} a_n \tilde{\mathbf{e}}_{tn}(\boldsymbol{\rho}, \omega) e^{+jk_{zn}z},$$

$$\tilde{\mathbf{H}}_t = \sum_{n=1}^{\infty} a_n \tilde{\mathbf{h}}_{tn}(\boldsymbol{\rho}, \omega) e^{+jk_{zn}z}.$$

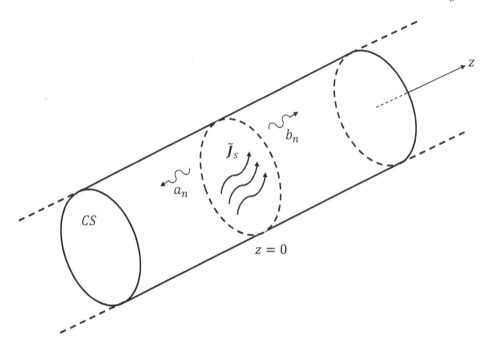

FIGURE 5.11

Surface current sheet in the cross-sectional plane of a hollow-pipe waveguide.

For $z > 0$ we have only waves traveling (or evanescent) in the $+z$ direction, with

$$\tilde{\mathbf{E}}_t = \sum_{n=1}^{\infty} b_n \tilde{\mathbf{e}}_{tn}(\boldsymbol{\rho}, \omega) e^{-jk_{zn}z},$$

$$\tilde{\mathbf{H}}_t = -\sum_{n=1}^{\infty} b_n \tilde{\mathbf{h}}_{tn}(\boldsymbol{\rho}, \omega) e^{-jk_{zn}z}.$$

Note that we have embedded the sign difference on the transverse magnetic fields traveling in the $\pm z$ directions, as described in (5.197) and (5.199). The amplitudes a_n and b_n are determined by imposing the boundary conditions on the transverse fields at $z = 0$.

Continuity of the transverse electric field at $z = 0$ requires

$$\sum_{n=1}^{\infty} a_n \tilde{\mathbf{e}}_{tn}(\boldsymbol{\rho}, \omega) = \sum_{n=1}^{\infty} b_n \tilde{\mathbf{e}}_{tn}(\boldsymbol{\rho}, \omega) \qquad (\boldsymbol{\rho} \in CS).$$

Dotting both sides by $\tilde{\mathbf{e}}_{tm}(\boldsymbol{\rho}, \omega)$ and integrating over the cross-section of the guide, we have

$$\sum_{n=1}^{\infty} a_n \int_{CS} \tilde{\mathbf{e}}_{tn} \cdot \tilde{\mathbf{e}}_{tm} \, dS = \sum_{n=1}^{\infty} b_n \int_{CS} \tilde{\mathbf{e}}_{tn} \cdot \tilde{\mathbf{e}}_{tm} \, dS.$$

The orthogonality condition (5.210) implies that only the $n = m$ term survives from each series, giving $a_n = b_n$ for all n. The transverse magnetic field is discontinuous by the surface current according to

$$\sum_{n=1}^{\infty} (-b_n - a_n) \hat{\mathbf{z}} \times \tilde{\mathbf{h}}_{tn}(\boldsymbol{\rho}, \omega) = \tilde{\mathbf{J}}_s(\boldsymbol{\rho}, \omega) \qquad (\boldsymbol{\rho} \in CS).$$

But

$$\hat{\mathbf{z}} \times \tilde{\mathbf{h}}_{tn} = \frac{\tilde{\mathbf{e}}_{tn}}{Z_n}$$

for both TE and TM modes, where Z_n is the wave impedance for the nth mode. Using this along with $a_n = b_n$, we have

$$\sum_{n=1}^{\infty} \frac{1}{Z_n} (-2a_n) \tilde{\mathbf{e}}_{tn}(\boldsymbol{\rho}, \omega) = \tilde{\mathbf{J}}_s(\boldsymbol{\rho}, \omega) \qquad (\boldsymbol{\rho} \in CS).$$

Dotting both sides with $\tilde{\mathbf{e}}_{tm}(\boldsymbol{\rho}, \omega)$ and integrating over CS, we get

$$-\sum_{n=1}^{\infty} \frac{2}{Z_n} a_n \int_{CS} \tilde{\mathbf{e}}_{tn} \cdot \tilde{\mathbf{e}}_{tm} \, dS = \int_{CS} \tilde{\mathbf{e}}_{tm} \cdot \tilde{\mathbf{J}}_s \, dS.$$

By orthogonality of the transverse fields,

$$\int_{CS} \tilde{\mathbf{e}}_{tn} \cdot \tilde{\mathbf{e}}_{tm} \, dS = C_n \delta_{mn} \tag{5.240}$$

where δ_{mn} is the Kronecker delta. The modal amplitudes are therefore

$$a_n = -\frac{1}{2} \frac{Z_n}{C_n} \int_{CS} \tilde{\mathbf{e}}_{tn} \cdot \tilde{\mathbf{J}}_s \, dS. \tag{5.241}$$

▶ **Example 5.21:** Vertical current sheet in a rectangular waveguide

A uniform, vertically directed current sheet of density $\hat{\mathbf{y}} J_{s0}$ resides in the plane $z = 0$ within a rectangular waveguide. Find the amplitudes of the modes excited in the guide.

Solution: As a y-independent current produces y-independent fields, only the TE$_{n0}$ mode is excited. From (5.225) we find that

$$\tilde{\mathbf{e}}_{tn} = \hat{\mathbf{y}} \sin\left(\frac{n\pi}{a} x\right).$$

Equations (5.240) and (5.241) give

$$C_n = \int_0^b \int_0^a \sin^2\left(\frac{n\pi}{a} x\right) = \frac{ab}{2}, \qquad a_n = -\frac{Z_n}{ab} J_{s0} \int_0^b \int_0^a \sin\left(\frac{n\pi}{a} x\right) dx.$$

Integration shows that $a_n = 0$ for even n, while for odd n

$$a_n = -\frac{2 Z_n J_{s0}}{n\pi}. \quad ◀$$

Higher-order modes in a coaxial cable. The coaxial cable is normally operated in its principal mode, which is the transmission-line (TEM) mode (§ 5.6.3). Because this mode has no cutoff, the cable may be operated at any desired frequency. But *higher-order modes* begin to propagate at higher frequencies. Hence the usable frequency band of a coaxial cable extends to the cutoff frequency of the first higher-order mode. It is clearly important to understand the behavior of these modes.

An ideal coaxial cable consists of coaxial conducting cylinders of radii a and b ($a < b$). The guiding region between the cylinders is filled with material having permittivity $\tilde{\epsilon}^c$

and permeability $\tilde{\mu}$. It is possible to satisfy the boundary conditions at both conductors with fields purely TE or purely TM to the z-direction. The solution to the wave equation for the potential function is given by (5.228), and since we must retain both the Bessel functions of the first and second kinds in the region $a \leq \rho \leq b$, we can write the potential as

$$\tilde{\psi}(\rho, \phi, \omega) = [A_n J_n(k_c \rho) + B_n N_n(k_c \rho)] \cos(n\phi + \Phi_n),$$

where k_c will be determined from the boundary conditions.

For TM modes, the boundary condition is that

$$\tilde{\psi}_e = 0$$

at $\rho = a$ and $\rho = b$, $0 \leq \phi < 2\pi$. This gives the set of homogeneous equations

$$\begin{bmatrix} J_n(k_c a) & N_n(k_c a) \\ J_n(k_c b) & N_n(k_c b) \end{bmatrix} \begin{bmatrix} A_n \\ B_n \end{bmatrix} = \begin{bmatrix} 0 \\ 0 \end{bmatrix}.$$

A nontrivial solution is only possible when the determinant of the coefficient matrix is zero:

$$J_n(k_c a)N_n(k_c b) - J_n(k_c b)N_n(k_c a) = 0.$$

This *characteristic equation* must be satisfied by the cutoff wavenumber k_c of the TM modes. For TE modes, the boundary condition requires

$$\hat{\mathbf{n}} \cdot \nabla_t \tilde{\psi}_h = \frac{\partial \tilde{\psi}_h}{\partial \rho} = 0,$$

at $\rho = a$ and $\rho = b$, $0 \leq \phi < 2\pi$. We obtain

$$\begin{bmatrix} J_n'(k_c a) & N_n'(k_c a) \\ J_n'(k_c b) & N_n'(k_c b) \end{bmatrix} \begin{bmatrix} A_n \\ B_n \end{bmatrix} = \begin{bmatrix} 0 \\ 0 \end{bmatrix}$$

and equate the determinant of the coefficient matrix to zero to get the characteristic equation for TE modes:

$$J_n'(k_c a)N_n'(k_c b) - J_n'(k_c b)N_n'(k_c a) = 0.$$

Neither characteristic equation has a closed-form solution, so a numerical root search is required.

The first higher-order mode is TE_{11}. Since its one full sinusoidal variation of the fields in $0 \leq \phi < 2\pi$ is reminiscent of the behavior of the TE_{20} mode in a rectangular guide, we might expect the fields to be similar if the rectangular guide were wrapped into a circle. The cutoff wavenumber of the rectangular mode is $k_c = 2\pi/A$ where A is the rectangular guide width. Replacing A with the average circumference of the coaxial guide, $2\pi(a+b)/2$, we get an approximate formula

$$k_{ca} = \frac{2}{a+b}$$

for the cutoff wavenumber of the TE_{11} mode of the coaxial guide. The cutoff frequency of a lossless guide with $v = 1/\sqrt{\epsilon\mu}$ should be approximately

$$f_{ca} = \frac{vk_c}{2\pi} = \frac{v}{\pi(a+b)}.$$

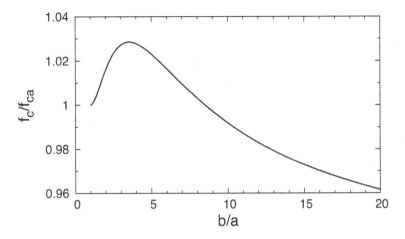

FIGURE 5.12

Ratio of exact and approximate cutoff frequencies for the TE_{11} mode of a coaxial guide.

The validity of this approximation depends on the ratio b/a. To test its accuracy, we write the characteristic equation for the TE_{11} mode as

$$J_1'(k_c a) N_1'\left(k_c a \frac{b}{a}\right) - N_1'(k_c a) J_1'\left(k_c a \frac{b}{a}\right) = 0, \tag{5.242}$$

and find $k_c a$ in terms of b/a. We then compare this to the approximate value

$$k_{ca} a = \frac{2}{1 + \frac{b}{a}}.$$

The ratio of these $k_c a$ values is also the ratio of the exact and approximate cutoff frequencies. It is plotted in Figure 5.12 vs. b/a. When $b \approx a$, the approximation for f_c is very accurate. As b/a increases, the approximation becomes too small, with a maximum error of about 3% at $b/a \approx 3.5$. As b/a increases further, the error is reduced until the approximation is exact at $b/a \approx 8.5$. For b/a above this value, the approximate cutoff frequency is too large, and the error continues to increase with b/a.

▶ **Example 5.22:** Operating band of a precision 3.5 mm coaxial connector

The inner radius of the outer conductor of a precision 3.5 mm coaxial connector is $3.5/2 = 1.75$ mm. If the connector mates with a coaxial line having characteristic resistance 50 Ω, find the radius of the inner conductor and the operating band of the coaxial line.

Solution: The characteristic resistance of a coaxial cable is given by (5.312):

$$R_c = \frac{\eta}{2\pi} \ln\left(\frac{b}{a}\right).$$

A precision connector uses an air-filled cable with $\eta = \eta_0$. Hence for $R_c = 50\,\Omega$ we have

$$b/a = e^{\frac{100\pi}{\eta_0}} = 2.302.$$

With $b = 1.75$ mm we get $a = 0.7601$ mm.

The operating band of the connector is from dc to the cutoff frequency of the TE_{11} mode. The approximate formula yields

$$f_{ca} = \frac{c}{\pi(a+b)} = 38.02 \text{ GHz.}$$

Solving (5.242) using a root search we find that $k_c a = 0.6182$ when $b/a = 2.302$. This gives an exact cutoff frequency

$$f_c = \frac{c}{2\pi a} k_c a = 38.81 \text{ GHz.}$$

Note that $f_c/f_{ca} = 1.021$, which matches the value shown in Figure 5.12. ◄

Fields of an isosceles triangle waveguide.   Rectangular and circular waveguides are examples of geometries where each separation of variables solution defines an individual mode. Certain triangular waveguides may be analyzed using separation of variables in rectangular coordinates, but the boundary conditions can only be satisfied by a finite superposition of separation of variables solutions [142]. Classic examples include the equilateral triangle [170] and the right isosceles triangle of Figure 5.13 [97]. We will consider the latter problem.

For TM modes, consider the separation of variables solution (5.217) with the boundary condition (5.218). To make $\tilde{\psi}_e = 0$ at both $x = 0$ and $y = 0$, we require

$$\tilde{\psi}_e(x, y, \omega) = A \sin \frac{n\pi x}{a} \sin k_y y \qquad (n = 1, 2, \ldots).$$

It is clear that a single term of this form cannot satisfy $\tilde{\psi}_e = 0$ on the remaining boundary. However, a superposition of two terms can.

Consider the solution

$$\tilde{\psi}_e(x, y, \omega) = A_1 \sin \frac{n\pi x}{a} \sin k_{y1} y + A_2 \sin \frac{m\pi x}{a} \sin k_{y2} y. \qquad (5.243)$$

We know that for a rectangular guide, each of these terms can satisfy the boundary conditions at $x = a$ and $y = a$ while having individual, *different* cutoff wavenumbers k_c. Our present goal is to have (5.243) represent a *single* mode of the triangular guide associated with a *single* cutoff frequency k_c. Substituting (5.243) into the transverse wave equation (5.194), we get

$$A_1 \sin \frac{n\pi x}{a} \sin k_{y1} y \left[-\left(\frac{n\pi}{a}\right)^2 - k_{y1}^2 + k_c^2 \right]$$

$$+ A_2 \sin \frac{m\pi x}{a} \sin k_{y2} y \left[-\left(\frac{m\pi}{a}\right)^2 - k_{y2}^2 + k_c^2 \right] = 0 \qquad ((x, y) \in CS)$$

where CS is the guide cross-section. This requires

$$k_c^2 = \left(\frac{n\pi}{a}\right)^2 + k_{y1}^2 = \left(\frac{m\pi}{a}\right)^2 + k_{y2}^2. \qquad (5.244)$$

The remaining boundary condition requires $\tilde{\psi}_e(x, y, \omega) = 0$ on the line $y = x$. Substitution gives

$$\tilde{\psi}_e(x, y = x, \omega) = A_1 \sin \frac{n\pi x}{a} \sin k_{y1} x + A_2 \sin \frac{m\pi x}{a} \sin k_{y2} x = 0 \quad (0 \le x \le a). \quad (5.245)$$

Clearly the choice $-A_2 = A_1 = A$ along with

$$k_{y1} = \frac{m\pi}{a}, \qquad k_{y2} = \frac{n\pi}{a},$$

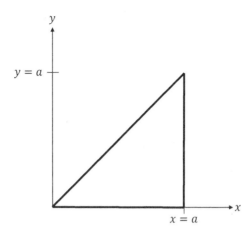

FIGURE 5.13

Right isosceles triangle waveguide.

satisfies boundary condition (5.245). But since it also satisfies wavenumber condition (5.244), the resulting potential function satisfies the transverse wave equation. The potential for TM fields is

$$\tilde{\psi}_e(x, y, \omega) = A_{nm} \left[\sin\left(\frac{n\pi x}{a}\right) \sin\left(\frac{m\pi y}{a}\right) - \sin\left(\frac{m\pi x}{a}\right) \sin\left(\frac{n\pi y}{a}\right) \right]. \qquad (5.246)$$

Each pair m, n defines an individual waveguide mode. Swapping m and n produces only a sign change, so the m, n mode and the n, m mode are the same. Since $n = m$ produces the trivial solution, as does $n = 0$, we restrict the indices according to $0 < m < n$.

By (5.246) the fields in a right isosceles triangular guide are

$$\tilde{E}_z = k_{c_{nm}}^2 A_{nm} \left\{ \left[\sin\frac{n\pi x}{a} \sin\frac{m\pi y}{a} \right] - \left[\sin\frac{m\pi x}{a} \sin\frac{n\pi y}{a} \right] \right\} e^{\mp jk_z z}, \qquad (5.247)$$

$$\tilde{E}_t = \mp jk_z A_{nm} \left\{ \left[\hat{\mathbf{x}}\frac{n\pi}{a} \cos\frac{n\pi x}{a} \sin\frac{m\pi y}{a} + \hat{\mathbf{y}}\frac{m\pi}{a} \sin\frac{n\pi x}{a} \cos\frac{m\pi y}{a} \right] \right.$$
$$\left. - \left[\hat{\mathbf{x}}\frac{m\pi}{a} \cos\frac{m\pi x}{a} \sin\frac{n\pi y}{a} + \hat{\mathbf{y}}\frac{n\pi}{a} \sin\frac{m\pi x}{a} \cos\frac{n\pi y}{a} \right] \right\} e^{\mp jk_z z}, \qquad (5.248)$$

$$\tilde{H}_t = jk_z Y_e A_{nm} \left\{ \left[\hat{\mathbf{x}}\frac{m\pi}{b} \sin\frac{n\pi x}{a} \cos\frac{m\pi y}{a} - \hat{\mathbf{y}}\frac{n\pi}{a} \cos\frac{n\pi x}{a} \sin\frac{m\pi y}{a} \right] \right.$$
$$\left. - \left[\hat{\mathbf{x}}\frac{n\pi}{a} \sin\frac{m\pi x}{a} \cos\frac{n\pi y}{a} - \hat{\mathbf{y}}\frac{m\pi}{a} \cos\frac{m\pi x}{a} \sin\frac{n\pi y}{a} \right] \right\} e^{\mp jk_z z}. \qquad (5.249)$$

Here

$$Y_e = \frac{1}{\eta\sqrt{1 - \omega_{c_{nm}}^2/\omega^2}},$$

with $\eta = (\mu/\epsilon)^{1/2}$ and

$$k_{c_{nm}} = \sqrt{\left(\frac{n\pi}{a}\right)^2 + \left(\frac{m\pi}{a}\right)^2}.$$

Comparing (5.247)–(5.249) with (5.219)–(5.221) and noting that $k_{c_{nm}} = k_{c_{mn}}$, we recognize the fields in a right isosceles triangular waveguide as superpositions of the (m, n) and (n, m) mode fields in a square waveguide. Johnson [97] gives a nice geometrical description of how this superposition satisfies the boundary condition on the slanted wall at $y = x$.

For TE modes, the potential function

$$\tilde{\psi}_h(x, y, \omega) = B_{nm} \left[\cos\left(\frac{n\pi x}{a}\right) \cos\left(\frac{m\pi y}{a}\right) + \cos\left(\frac{m\pi x}{a}\right) \cos\left(\frac{n\pi y}{a}\right) \right] \qquad (5.250)$$

satisfies (5.194). The fields are

$$\tilde{H}_z = k_{c_{nm}}^2 B_{nm} \left\{ \left[\cos\frac{n\pi x}{a} \cos\frac{m\pi y}{a} \right] + \left[\cos\frac{m\pi x}{a} \cos\frac{n\pi y}{a} \right] \right\} e^{\mp jk_z z}, \qquad (5.251)$$

$$\tilde{\mathbf{H}}_t = \pm jk_z B_{nm} \left\{ \left[\hat{\mathbf{x}}\frac{n\pi}{a} \sin\frac{n\pi x}{a} \cos\frac{m\pi y}{a} + \hat{\mathbf{y}}\frac{m\pi}{a} \cos\frac{n\pi x}{a} \sin\frac{m\pi y}{a} \right] \right.$$
$$\left. + \left[\hat{\mathbf{x}}\frac{m\pi}{a} \sin\frac{m\pi x}{a} \cos\frac{n\pi y}{a} + \hat{\mathbf{y}}\frac{n\pi}{a} \cos\frac{m\pi x}{a} \sin\frac{n\pi y}{a} \right] \right\} e^{\mp jk_z z}, \qquad (5.252)$$

$$\tilde{\mathbf{E}}_t = jk_z Z_h B_{nm} \left\{ \left[\hat{\mathbf{x}}\frac{m\pi}{a} \cos\frac{n\pi x}{a} \sin\frac{m\pi y}{a} - \hat{\mathbf{y}}\frac{n\pi}{a} \sin\frac{n\pi x}{a} \cos\frac{m\pi y}{a} \right] \right.$$
$$\left. + \left[\hat{\mathbf{x}}\frac{n\pi}{a} \cos\frac{m\pi x}{a} \sin\frac{n\pi y}{a} - \hat{\mathbf{y}}\frac{m\pi}{a} \sin\frac{m\pi x}{a} \cos\frac{n\pi y}{a} \right] \right\} e^{\mp jk_z z}, \qquad (5.253)$$

where

$$Z_h = \frac{\eta}{\sqrt{1 - \omega_{c_{nm}}^2/\omega^2}}, \qquad k_{c_{nm}} = \sqrt{\left(\frac{n\pi}{a}\right)^2 + \left(\frac{m\pi}{a}\right)^2}.$$

The cutoff wavenumber formula is the same as that for TM modes.

Note that $\tilde{E}_y = 0$ on the wall at $x = 0$ and $\tilde{E}_x = 0$ on the wall at $y = 0$. To check whether the tangential electric field is zero on the remaining wall, we compute

$$\tilde{\mathbf{E}}_{\text{tan}} = \hat{\mathbf{n}} \times \tilde{\mathbf{E}}_t,$$

where the unit normal to the wall at $x = y$ is

$$\hat{\mathbf{n}} = \frac{\hat{\mathbf{x}} - \hat{\mathbf{y}}}{\sqrt{2}}.$$

These relations do yield $\tilde{\mathbf{E}}_{\text{tan}} = 0$. So the wave equation and boundary conditions are satisfied, and the potential function (5.250) represents TE mode solutions.

Further, we find that the triangular guide fields (5.251)–(5.253) are superpositions of the (m, n) and (n, m) TE mode fields of a square guide. In the TE case, $n = m$ is allowed, as is either $m = 0$ or $n = 0$ but not both. Since swapping m and n produces the same mode, we restrict the indices to $0 \leq m \leq n$, $m + n > 0$.

▶ **Example 5.23:** Lowest order mode in a triangular waveguide

Find the cutoff frequency and the fields for the lowest order mode in a right isosceles triangular waveguide filled with a lossless isotropic material of permittivity ϵ and permeability μ.

Solution: The lowest order mode in a triangular guide is the TE$_{10}$ mode with cutoff wavenumber $k_{c_{10}} = \pi/a$ and cutoff frequency

$$f_{c_{10}} = \frac{v}{2a}, \qquad v = 1/\sqrt{\mu\epsilon}.$$

This is also the cutoff frequency for a TE$_{10}$ mode in a square guide. The fields are given by

(5.251)–(5.253):

$$\tilde{H}_z = k_{c_{10}}^2 B_{10} \left\{ \cos \frac{\pi x}{a} + \cos \frac{\pi y}{a} \right\} e^{\mp j k_z z},$$

$$\tilde{\mathbf{H}}_t = \pm j k_z B_{10} \frac{\pi}{a} \left\{ \hat{\mathbf{x}} \sin \frac{\pi x}{a} + \hat{\mathbf{y}} \sin \frac{\pi y}{a} \right\} e^{\mp j k_z z},$$

$$\tilde{\mathbf{E}}_t = j k_z Z_h B_{10} \frac{\pi}{a} \left\{ \hat{\mathbf{x}} \sin \frac{\pi y}{a} - \hat{\mathbf{y}} \sin \frac{\pi x}{a} \right\} e^{\mp j k_z z}.$$

Note that the lowest order TM mode is TM_{21} with cutoff wavenumber

$$k_{c_{21}} = \sqrt{\left(\frac{2\pi}{a}\right)^2 + \left(\frac{\pi}{a}\right)^2} = \sqrt{5}\frac{\pi}{a}$$

and cutoff frequency

$$f_{c_{10}} = \frac{\sqrt{5}v}{2a}. \quad \blacktriangleleft$$

▶ **Example 5.24:** Power flow in a triangular waveguide

Find the time-average power flow for the lowest order mode in a right isosceles triangular waveguide filled with a lossless isotropic material of permittivity ϵ and permeability μ. Compare to the power carried by the lowest order mode in a square waveguide.

Solution: We specialize the potential (5.250) to time harmonic form for the TE_{10} mode,

$$\check{\psi}_h = B_{10} \left[\cos \frac{\pi x}{a} + \cos \frac{\pi y}{a} \right],$$

and substitute into (5.214):

$$P_{av} = \tfrac{1}{2} |B_{10}|^2 \check{\omega} \mu \beta \left(\frac{\pi}{a}\right)^2 \int_{CS} \left[\cos^2 \frac{\pi x}{a} + \cos^2 \frac{\pi y}{a} + 2 \cos \frac{\pi x}{a} \cos \frac{\pi y}{a} \right] dS.$$

The third integral is

$$\int_0^a \cos \frac{\pi x}{a} \int_0^x \cos \frac{\pi y}{a} \, dy \, dx = \frac{a}{\pi} \int_0^a \cos \frac{\pi x}{a} \sin \frac{\pi x}{a} \, dx = 0.$$

The first two integrals combine to give

$$\int_0^a \int_y^a \cos^2 \frac{\pi x}{a} \, dx \, dy + \int_0^a \int_0^x \cos^2 \frac{\pi y}{a} \, dy \, dx$$

$$= \int_0^a \left[\frac{a}{2} - \frac{y}{2} - \frac{\sin \frac{2\pi y}{a}}{4\frac{\pi}{a}} \right] dy + \int_0^a \left[\frac{x}{2} + \frac{\sin \frac{2\pi x}{a}}{4\frac{\pi}{a}} \right] dx = \frac{a^2}{4} + \frac{a^2}{4} = \frac{a^2}{2}$$

and thus

$$P_{av} = \frac{\pi^2}{4} \check{\omega} \mu \beta |B_{10}|^2.$$

Comparison with (5.226) shows that the power flow for the TE_{10} mode in the isosceles right triangle waveguide is the same as that for the TE_{10} mode in a square waveguide of side length a. Although it has only half the cross-sectional area of the rectangular guide, the triangular guide carries the TE_{10} and TE_{01} modes of the square guide simultaneously. ◀

5.6.1.2 A hollow-pipe waveguide filled with a homogeneous anisotropic material

Both gyromagnetic materials (such as ferrites) and gyroelectric materials (such as plasmas) may be placed in waveguides for special purposes. Because anisotropic materials couple orthogonal field components, it is often impossible to satisfy the boundary conditions at the waveguide walls with purely TE or TM modes. But pure TE or TM modes do exist in certain situations. One of these is considered next.

Rectangular waveguide filled with a lossless magnetized ferrite. So far we have only considered waveguides filled with isotopic materials, computing the fields from a potential function. For a waveguide filled with a ferrite material, which is anisotropic, we compute the transverse fields by first obtaining the longitudinal fields. This field-based approach also applies to waveguides filled with isotropic materials as an alternative to the potential approach (e.g., [156]), but is particularly useful for waveguide problems with anisotropic materials.

Consider a rectangular waveguide with cross-section occupying $0 \leq x \leq a$ and $0 \leq y \leq b$, filled with a lossless ferrite magnetized in the y-direction. We will restrict ourselves to solutions for fields that are TE to the z-direction and, like the TE_{n0} modes in a waveguide filled with an isotropic medium, are y-invariant. Using the results of Example 5.12 with y-derivatives set to zero, we have the transverse field formulas

$$\tilde{\mathbf{E}}_t^z = \hat{\mathbf{y}} \left[\frac{-j\omega\mu}{k_z^2 - \omega^2\mu\epsilon} \frac{\partial \tilde{H}_z^z}{\partial x} + \frac{j\omega\kappa k_z}{k_z^2 - \omega^2\mu\epsilon} \tilde{H}_z^z \right], \tag{5.254}$$

$$\tilde{\mathbf{H}}_t^z = \hat{\mathbf{x}} \left[\frac{-jk_z}{k_z^2 - \omega^2\mu\epsilon} \frac{\partial \tilde{H}_z^z}{\partial x} + \frac{j\omega^2\kappa\epsilon}{k_z^2 - \omega^2\mu\epsilon} \tilde{H}_z^z \right], \tag{5.255}$$

where $\tilde{H}_z^z$ obeys the wave equation

$$\left(\frac{\partial^2}{\partial x^2} + k_c^2 \right) \tilde{H}_z^z = 0 \tag{5.256}$$

with

$$k_c^2 = \omega^2\mu\epsilon \left(1 - \frac{\kappa^2}{\mu^2} \right) - k_z^2. \tag{5.257}$$

The solution to (5.256) is

$$\tilde{H}_z^z(x, k_z, \omega) = A\sin(k_c x) + B\cos(k_c x).$$

Substitution into (5.254) gives the transverse electric field

$$\tilde{E}_y^z = \frac{j\omega}{k_z^2 - \omega^2\mu\epsilon} \left[(-A\mu k_c + B\kappa k_z)\cos(k_c x) + (B\mu k_c + A\kappa k_z)\sin(k_c x) \right].$$

Applying the boundary condition on $\tilde{E}_y^z$ at $x = 0$, we get $-A\mu k_c + B\kappa k_z = 0$. Solving for A and substituting into (5.254), we find that

$$\tilde{E}_y^z = \frac{j\omega}{k_z^2 - \omega^2\mu\epsilon} B \frac{\kappa^2}{\mu k_c} \left(\frac{k_c^2\mu^2}{\kappa^2} + k_z^2 \right) \sin(k_c x).$$

This can be simplified using (5.257):

$$\tilde{E}_{y_-}^z = -j\frac{\omega\mu}{k_c}B\left(1 - \frac{\kappa^2}{\mu^2}\right)\sin(k_c x).$$

Applying the boundary condition on $\tilde{E}_y^z$ at $x = a$, we have $\sin(k_c a) = 0$. So k_c takes the discrete values

$$k_{cn} = n\pi/a \qquad (n = 1, 2, \ldots)$$

just as in a rectangular guide filled with an isotropic material. Since k_c is discrete, so are k_z and the modal spectrum. Use of (5.255) and some simplification yield the complete field set

$$\tilde{H}_{zn}(x, z, \omega) = B_n\left[\cos\left(\frac{n\pi}{a}x\right) + \frac{\kappa k_{zn}}{\mu k_{cn}}\sin\left(\frac{n\pi}{a}x\right)\right]e^{jk_{zn}z},$$

$$\tilde{H}_{xn}(x, z, \omega) = -j\frac{k_{zn}}{k_{cn}}B_n\left[\sin\left(\frac{n\pi}{a}x\right) + \frac{\kappa k_{cn}}{\mu k_{zn}}\cos\left(\frac{n\pi}{a}x\right)\right]e^{jk_{zn}z},$$

$$\tilde{E}_{yn}(x, z, \omega) = -j\frac{\omega\mu}{k_{cn}}\left(1 - \frac{\kappa^2}{\mu^2}\right)B_n\sin\left(\frac{n\pi}{a}x\right)e^{jk_{zn}z},$$

where

$$k_{zn}^2 = \omega^2\mu\epsilon\left(1 - \frac{\kappa^2}{\mu^2}\right) - \left(\frac{n\pi}{a}\right)^2. \tag{5.258}$$

These expressions hold for waves traveling in the $-z$ direction; for waves traveling in the $+z$ direction, we replace k_z by $-k_z$. This means that both $\tilde{H}_{xn}$ and $\tilde{E}_{yn}$ have different forms for waves propagating in the $\pm z$ directions, an important observation practically.

It is clear that, unlike in the case of a waveguide filled with an isotropic material, we cannot represent the ratio of transverse electric to transverse magnetic fields as a spatially invariant wave impedance. Furthermore, since μ and κ depend on frequency, the cutoff effect is more complicated than in the case of a guide filled with a lossless dielectric. By (4.125) and (4.126) we have

$$\mu = \mu_0\frac{f^2 - f_0^2 - f_0 f_M}{f^2 - f_0^2}, \qquad \kappa = \mu_0\frac{f f_M}{f^2 - f_0^2}.$$

Substituting these into (5.258) and simplifying, we get

$$k_{zn}^2 = \left(\frac{2\pi}{v_0}\right)^2\left[f^2\left(\frac{f^2 - f_B^2}{f^2 - f_A^2}\right) - f_{c0}^2\right].$$

Here $f_A = \sqrt{f_0(f_0 + f_M)}$, $f_B = f_0 + f_M$, $v_0 = 1/\sqrt{\mu_0\epsilon}$, and $f_{c0} = nv_0/(2a)$. So v_0 is the velocity of the wave in an unbounded dielectric medium with parameters μ_0 and ϵ, and f_{c0} is the cutoff frequency of the TE_{10} mode in a rectangular guide filled with a dielectric medium with parameters μ_0 and ϵ. For $f \gg f_B$, the wavenumber becomes

$$k_{zn}^2 = \left(\frac{2\pi}{v_0}\right)^2\left[f^2 - f_{c0}^2\right],$$

which is that of the dielectric-filled guide. There is a low frequency bandstop region where the wave will not propagate. Clearly, for $f_A < f < f_B$ we have $k_{zn}^2 < 0$ and the waveguide is in cutoff. A more careful analysis shows that $k_{zn}^2 < 0$ for $f < f_1$ and $f_A < f < f_2$, where

$$f_{1,2}^2 = \frac{f_B^2 + f_{c0}^2}{2} \mp \sqrt{\left(\frac{f_B^2 + f_{c0}^2}{2}\right)^2 - f_A^2 f_{c0}^2}.$$

Since $f_1 < f_{c0}$, the lower cutoff frequency of the ferrite-filled guide is less than that for a guide filled with dielectric of permittivity ϵ and permeability μ_0.

We can compute the time average power carried by a time-harmonic wave propagating in a ferrite-loaded waveguide by first integrating the time-average Poynting flux over the cross-section:

$$P_{av} = \int_{CS} \frac{1}{2}\hat{\mathbf{z}} \cdot \mathrm{Re}\left\{\check{\mathbf{E}} \times \check{\mathbf{H}}^*\right\} dS$$

$$= \int_0^b \int_0^a \frac{1}{2}\mathrm{Re}\{-\check{E}_y \check{H}_x^*\}\, dx\, dy.$$

It is convenient to normalize the fields such that the phasor amplitude of the electric field is E_0. Then, for a wave traveling in the $+z$ direction,

$$\check{E}_{yn}(x,z) = E_0 \sin\left(\frac{n\pi}{a}x\right) e^{-jk_{zn}z},$$

$$\check{H}_{xn}(x,z) = -\frac{k_{zn}}{\check{\omega}\mu} \frac{E_0}{\left(1 - \frac{\kappa^2}{\mu^2}\right)} \left[\sin\left(\frac{n\pi}{a}x\right) - \frac{\kappa k_{cn}}{\mu k_{zn}}\cos\left(\frac{n\pi}{a}x\right)\right] e^{-jk_{zn}z},$$

and

$$P_{av} = \mathrm{Re}\left\{\frac{k_{zn}^*}{2\check{\omega}\mu} \frac{|E_0|^2}{\left(1 - \frac{\kappa^2}{\mu^2}\right)} \int_0^b \int_0^a \sin^2\left(\frac{n\pi}{a}x\right) dx dy\right\}$$

$$- \frac{\kappa k_{cn}}{2\check{\omega}\mu^2} \frac{|E_0|^2}{\left(1 - \frac{\kappa^2}{\mu^2}\right)} \int_0^b \int_0^a \sin\left(\frac{n\pi}{a}x\right)\cos\left(\frac{n\pi}{a}x\right) dx dy$$

$$= \frac{ab}{4} \frac{\mathrm{Re}\{k_{zn}\}}{\check{\omega}\mu} \frac{|E_0|^2}{\left(1 - \frac{\kappa^2}{\mu^2}\right)}. \tag{5.259}$$

▶ **Example 5.25:** Rectangular waveguide filled with ferrite

Assume the commercial ferrite G-1010 of Example 4.10 fills a WR-90 X-band rectangular waveguide of dimensions $a = 22.86$ by $b = 10.16$ mm, and is biased at $H_0 = 1500$ Oe. Determine the lower cutoff frequency and the bandstop region for the TE$_{10}$ mode. Plot the dispersion curve for $0 < f < 10$ GHz, and plot the phase and group velocities and the normalized power flow vs. frequency.

Solution: By Example 4.10 we have $f_0 = 4.2$ GHz and $f_M = 2.8$ GHz. To compute the cutoff frequencies we need the permittivity value. A typical value for G-1010 is $\epsilon_r = 14.2$; use of this value gives $f_{c0} = 1.740$ GHz. We also have $f_1 = 1.331$ GHz, $f_A = 5.422$ GHz, $f_B = 7$ GHz, and $f_2 = 7.089$ GHz. Thus, the wave is cut off for $f < 1.331$ GHz, and in the band $5.422 < f < 7.089$ GHz.

The dispersion curve is shown in Figure 5.14. Here $\beta = \sqrt{k_z^2}$ when $k_z^2 \geq 0$. The lower cutoff at 1.331 GHz and the bandstop region between 5.422 and 7.089 GHz are clearly visible. Also shown is the dispersion curve for a rectangular guide filled with a dielectric with relative permittivity $\epsilon_r = 14.2$. At high frequencies the behavior of the ferrite-filled guide approaches that of the dielectric-filled guide. The latter has a higher cutoff frequency and shows no bandstop effect.

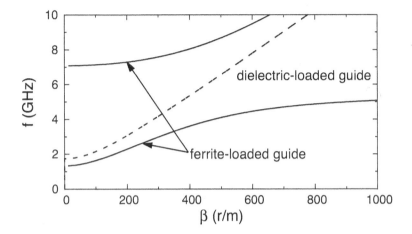

FIGURE 5.14

Dispersion plots for material-filled rectangular waveguides. Solid line corresponds to a ferrite-filled guide; dashed line corresponds to a dielectric-filled guide with $\epsilon_r = 14.2$.

The phase and group velocities may be found from the dispersion curve and the formulas $v_p = \omega/\beta$ and $v_g = d\omega/d\beta$. These velocities are plotted in Figure 5.15 for the TE_{10} mode, relative to the velocity in an unbounded medium, $v_0 = c/\sqrt{\epsilon_r} = 7.956 \times 10^7$ m/s. Note that the phase and group velocities both approach zero near the bottom of the stop band.

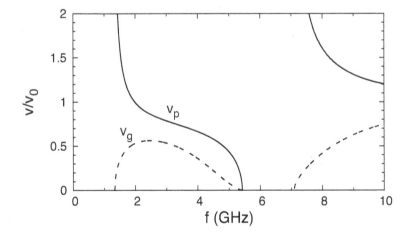

FIGURE 5.15

Phase and group velocities for the TE_{10} mode in a ferrite-filled rectangular waveguide.

To plot the power flow, it is convenient to normalize (5.259). Let us divide by the time-average power carried across an equal size aperture by a plane wave of amplitude E_0 traveling in a material with permittivity $\epsilon = \epsilon_r \epsilon_0$, where $\epsilon_r = 14.2$. This power is

$$P_0 = \frac{|E_0|^2}{2\eta} ab$$

where $\eta = \eta_0/\sqrt{\epsilon_r}$. So

$$\frac{P_{av}}{P_0} = \frac{\text{Re}\{k_{zn}\}\,\eta}{2\breve{\omega}}\,\frac{\mu}{\mu^2 - \kappa^2}.$$

The normalized power is plotted in Figure 5.16 for the TE_{10} mode. Since k_z is imaginary in the cutoff region, there is no time-average power flow there.

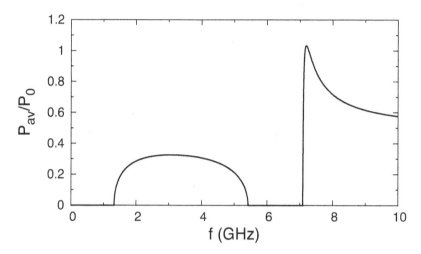

FIGURE 5.16

Normalized power carried by the TE_{10} fields in a ferrite-filled rectangular waveguide. ◄

5.6.1.3 Hollow-pipe waveguides filled with more than one material

Partially filled hollow-pipe waveguides are encountered in a variety of circumstances. Dielectrics may be added to alter the phase velocity of the guided wave, or to create a filter. Ferrite slabs are used in waveguides to create nonreciprocal devices such as circulators or isolators. The simplest examples involve isotropic materials symmetrically filling regions of circular or rectangular guides. We consider two simple examples below.

A rectangular waveguide with a centered material slab: TE_{n0} modes. Consider a rectangular waveguide with a lossless slab of material of permittivity ϵ and permeability μ centered along the x-direction (Figure 5.17). The material occupies region 2 $(-d/2 \leq x \leq d/2)$, while free space (air) occupies regions 1 and 3 $(-a/2 \leq x < -d/2$ and $d/2 < x \leq a/2$, respectively). We find that the boundary conditions cannot be satisfied by fields TE or TM to the z-direction, unless the fields are y-invariant. The boundary conditions *can* be satisfied, however, with fields TE or TM to the x-direction (also called *longitudinal section modes*). We consider only the y-invariant TE to z case here; the reader interested in constructing solutions using fields TE or TM to the x-direction may consult [82].

We proceed as in the rectangular waveguide case, seeking a separation of variables

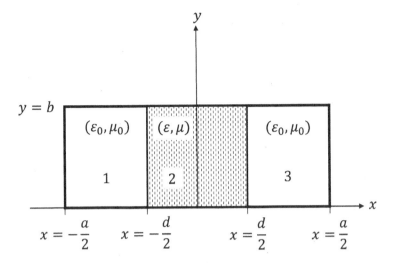

FIGURE 5.17
Partially filled rectangular waveguide.

solution for the potential function $\tilde{\psi}_h$. By (5.217) the potential function takes the form

$$\tilde{\psi}_h = C_1 \sin k_x x + C_2 \cos k_x x$$

since the fields are y-invariant. Here $k_x^2 = k^2 - k_z^2$. Since the material in region 2 differs from that in regions 1 and 3, the values of k differ as do those of k_x. However, to satisfy the boundary conditions, k_z must be the same in each region. This makes sense because the wave velocity depends on k_z, and the wave must propagate with the same velocity in each region to ensure field continuity.

We also note that the structure is symmetric about $x = 0$. We thus expect that fields with even or odd symmetry should satisfy the boundary conditions independently, and so we consider even and odd modes according to the symmetry of the potential function $\tilde{\psi}_h$. The boundary condition on the walls at $x = \pm a/2$ is

$$\frac{\partial \tilde{\psi}_h}{\partial x} = 0.$$

Thus, for even modes

$$\tilde{\psi}_{h1}(x) = A \cos k_{x0} \left(x + \frac{a}{2} \right),$$

$$\tilde{\psi}_{h2}(x) = B \cos k_x x,$$

$$\tilde{\psi}_{h3}(x) = A \cos k_{x0} \left(x - \frac{a}{2} \right).$$

Clearly these potentials are even about $x = 0$ and satisfy the boundary conditions at $x = \pm a/2$. We also have

$$k_{x0}^2 = k_0^2 - k_z^2, \qquad k_x^2 = k^2 - k_z^2, \tag{5.260}$$

where $k_0^2 = \omega^2 \mu_0 \epsilon_0$ and $k^2 = \omega^2 \mu \epsilon$. By (5.198) the axial magnetic fields are

$$\tilde{h}_{z1} = A k_{x0}^2 \cos k_{x0} \left(x + \frac{a}{2} \right),$$

$$\tilde{h}_{z2} = B k_x^2 \cos k_x x,$$

$$\tilde{h}_{z3} = A k_{x0}^2 \cos k_{x0} \left(x - \frac{a}{2} \right),$$

and by (5.200) the transverse electric fields are

$$\tilde{e}_{y1} = -A j k_{x0} \omega \mu_0 \sin k_{x0} \left(x + \frac{a}{2} \right),$$

$$\tilde{e}_{y2} = -B j k_x \omega \mu \sin k_x x,$$

$$\tilde{e}_{y3} = -A j k_{x0} \omega \mu_0 \sin k_{x0} \left(x - \frac{a}{2} \right),$$

for waves traveling in the $+z$ direction. Note that for an even potential function, $\tilde{H}_z$ is even but $\tilde{E}_y$ is odd.

The constants A and B and the axial wavenumber k_z are determined by applying the boundary conditions at $x = \pm d/2$. By symmetry, boundary conditions enforced at $x = d/2$ will also be satisfied at $x = -d/2$. Continuity of $\tilde{H}_z$ and $\tilde{E}_y$ at $x = d/2$ gives

$$B k_x^2 \cos k_x \frac{d}{2} = A k_{x0}^2 \cos k_{x0} \frac{d-a}{2},$$

$$B k_x \mu \sin k_x \frac{d}{2} = A k_{x0} \mu_0 \sin k_{x0} \frac{d-a}{2}.$$

Setting the determinant of the coefficient matrix to zero gives a characteristic equation relating k_x and k_{x0}:

$$\frac{\mu}{k_x} \tan k_x \frac{d}{2} = -\frac{\mu_0}{k_{x0}} \tan k_{x0} \frac{a-d}{2}. \tag{5.261}$$

Subtracting the equations in (5.260) gives a second relationship between k_x and k_{x0}:

$$k_x^2 - k_{x0}^2 = k^2 - k_0^2 = k_0^2 (\mu_r \epsilon_r - 1). \tag{5.262}$$

Simultaneous solution of these two equations determines k_x and k_{x0}, and thus k_z through (5.260).

For odd modes

$$\tilde{\psi}_1(x) = -A \cos k_{x0} \left(x + \frac{a}{2} \right),$$

$$\tilde{\psi}_2(x) = B \sin k_x x,$$

$$\tilde{\psi}_3(x) = A \cos k_{x0} \left(x - \frac{a}{2} \right),$$

which are odd about $x = 0$ and satisfy the boundary conditions at $x = \pm a/2$. With these we have the relations (5.260). By (5.198) the axial magnetic fields are

$$\tilde{h}_{z1} = -A k_{x0}^2 \cos k_{x0} \left(x + \frac{a}{2} \right),$$

$$\tilde{h}_{z2} = B k_x^2 \sin k_x x,$$

$$\tilde{h}_{z3} = A k_{x0}^2 \cos k_{x0} \left(x - \frac{a}{2} \right),$$

and by (5.200) the transverse electric fields are

$$\tilde{e}_{y1} = A j k_{x0} \omega \mu_0 \sin k_{x0} \left(x + \frac{a}{2} \right), \tag{5.263}$$

$$\tilde{e}_{y2} = B j k_x \omega \mu \cos k_x x, \tag{5.264}$$

$$\tilde{e}_{y3} = -A j k_{x0} \omega \mu_0 \sin k_{x0} \left(x - \frac{a}{2} \right). \tag{5.265}$$

Note that for an odd potential function, $\tilde{H}_z$ is odd but $\tilde{E}_y$ is even.

As with the even-mode case, we apply continuity of $\tilde{H}_z$ and $\tilde{E}_y$ at $x = d/2$ to get

$$B k_x^2 \sin k_x \frac{d}{2} = A k_{x0}^2 \cos k_{x0} \frac{d-a}{2}, \tag{5.266}$$

$$B k_x \mu \cos k_x \frac{d}{2} = -A k_{x0} \mu_0 \sin k_{x0} \frac{d-a}{2}. \tag{5.267}$$

Setting the determinant of the coefficient matrix to zero gives a characteristic equation relating k_x and k_{x0}:

$$\frac{\mu}{k_x} \cot k_x \frac{d}{2} = \frac{\mu_0}{k_{x0}} \tan k_{x0} \frac{a-d}{2}. \tag{5.268}$$

Simultaneous solution of this and (5.262) determines k_x and k_{x0}, and thus k_z through (5.260).

▶ **Example 5.26:** Rectangular waveguide partially filled with Teflon: first odd mode

Assume that a Teflon sample partially fills a WR-90 X-band rectangular waveguide of dimensions $a = 22.86$ by $b = 10.16$ mm, as shown in Figure 5.17. The sample has width $d = a/4$ and material properties $\epsilon_r = 2.1$, $\mu_r = 1$. Plot the dispersion curve for the first odd mode and compare it to that for the TE$_{10}$ mode when the waveguide is completely filled with air or with Teflon. Also plot $|\tilde{E}_y|$ for both a partially filled waveguide and an air-filled waveguide at 10 GHz.

Solution: The dispersion curve for the odd modes of the partially filled guide can be plotted by numerically solving the characteristic equation (5.268) along with (5.260). Since the materials are lossless, $k_z = \beta$. Typically, the frequency is set and a search is conducted for a value of β satisfying the two equations. With increasing frequency, more solutions for β will appear, corresponding to both the fundamental mode and the higher-order modes. Below the cutoff frequency of the fundamental mode, no solutions for β exist.

Figure 5.18 shows the dispersion curve of the lowest order odd mode for the guide partially filled with Teflon. Also shown are the dispersion curves for the same guide completely filled with air, and completely filled with Teflon. These curves can be generated from the usual relationships

$$\beta = \sqrt{k^2 - k_c^2},$$

$$k_c = \pi/a.$$

They can also be generated by solving (5.268) and (5.260), with either $d \approx 0$ (for the air-filled guide) or $d \approx a$ (for the Teflon-filled guide). Comparing this limiting case for the two solutions is a good way to check the solution for the partially filled guide.

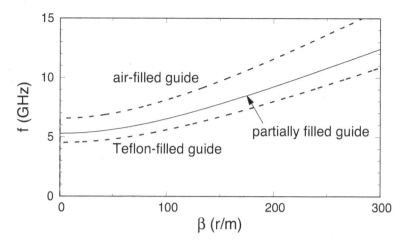

FIGURE 5.18
Dispersion plot for a waveguide partially filled with Teflon. Solid line is for the first odd mode of the partially filled guide. Dashed lines correspond to the TE_{10} mode in a fully filled guide.

Notice that the presence of Teflon lowers the cutoff frequency from 6.56 GHz for the air-filled guide to 5.28 GHz. The dispersion curve is also lowered, but follows the same trend as the curve for the air-filled guide. The cutoff frequency is above that of the Teflon-filled guide (4.52 GHz). Also, since the field maximum is in the material region, and the field minimum is in the air region, the dispersion curve is closer to that for Teflon than that for air. The opposite effect occurs for the second mode (Example 5.27). The addition of the dielectric also lowers the phase velocity (Figure 5.19). Hence dielectric-loaded waveguides act as *slow-wave structures*; such structures are revisited in Example 5.28.

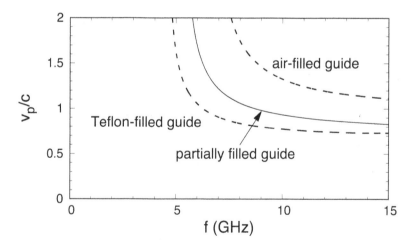

FIGURE 5.19
Phase velocity for a waveguide partially filled with Teflon. Solid line is for the first odd mode of the partially filled guide. Dashed lines correspond to the TE_{10} mode in a fully filled guide.

The magnitude of $\tilde{E}_y$ may be computed from (5.263)–(5.265). The needed relation between A and B can be obtained from either (5.266) or (5.267). Figure 5.20 shows the normalized field along with the TE_{10} field in an air-filled guide (both fields are normalized to a maximum value of unity). Note that the presence of the material concentrates the electric field within the material region at the center of the guide, and reduces it accordingly in the air region. Also note that $\tilde{E}_y$ is continuous at the air-material boundary, as required by the boundary conditions.

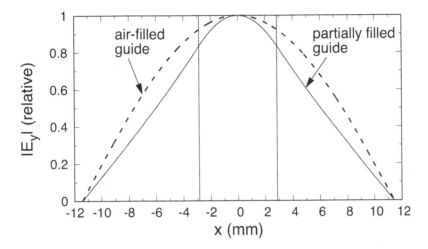

FIGURE 5.20

Normalized electric field for a waveguide partially filled with Teflon. $f = 10$ GHz. Solid line is for the first odd mode of the partially filled guide. Dashed lines correspond to the TE_{10} mode in an air-filled guide. Vertical lines indicate extent of the material. ◄

▶ **Example 5.27:** Rectangular waveguide partially filled with Teflon: first even mode

Repeat Example 5.26 to plot the dispersion curve for the first even mode of the partially filled rectangular waveguide. Compare the curve to those of the TE_{20} mode for a guide fully filled with air and with Teflon. Also, plot the fields at 15 GHz, and compare to those in a fully filled guide.

Solution: The dispersion curve for the even modes of the partially filled guide can be plotted by numerically solving the characteristic equation (5.261) along with the relationship (5.260). Figure 5.21 shows the dispersion curve of the lowest-order even mode for the guide partially filled with Teflon. Also shown are the dispersion curves for the TE_{20} mode in guides completely filled with air and Teflon, respectively. As with the first odd mode of Example 5.26, the effect of adding the Teflon is to lower the cutoff frequency from that of the air-filled guide. The resulting dispersion curve is closer to that for an air-filled guide than for a Teflon-filled guide — the opposite of the effect seen in Figure 5.18 for the first odd mode. This is because the field maxima for the first even mode occur in the air region of the partially filled guide, rather than in the material region. As seen in Figure 5.22, the electric field is concentrated somewhat by the presence of the Teflon, but the field null is maintained at the center of the guide.

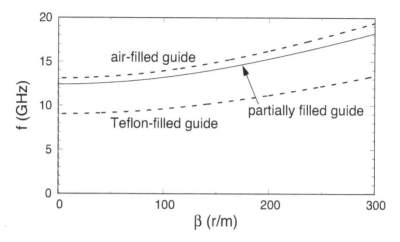

FIGURE 5.21
Dispersion plot for a waveguide partially filled with Teflon. Solid line is for the first even mode of the partially filled guide. Dashed lines correspond to the TE_{20} mode in a fully filled guide.

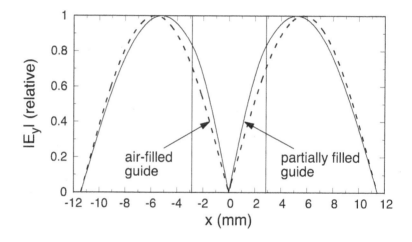

FIGURE 5.22
Normalized electric field for a waveguide partially filled with Teflon. $f = 15$ GHz. Solid line is for the first even mode of the partially filled guide. Dashed lines correspond to the TE_{20} mode in an air-filled guide. Vertical lines indicate extent of the material. ◄

A material-lined circular waveguide. Lining a circular waveguide with a material layer can provide a variety of advantages over an empty guide, including bandwidth improvement [129] and lowering of the phase velocity [51]. Consider a circular guide of radius b with a material layer occupying the region $a < \rho < b$ (region 2) adjacent to the conductor (Figure 5.23). The material has permittivity $\tilde{\epsilon}^c$ and permeability $\tilde{\mu}$, while free space (air) occupies region 1 ($0 \leq \rho \leq a$). We find that the boundary conditions cannot be satisfied using fields TE or TM to the z-direction unless they are ϕ-invariant

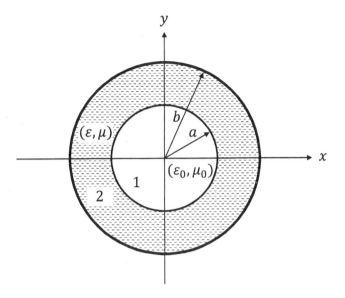

FIGURE 5.23
Material-lined waveguide.

(i.e., azimuthally symmetric). We will consider only the TE to z azimuthally symmetric modes here.

We proceed as in the circular waveguide case, seeking a separation of variables solution for the potential function $\tilde{\psi}$. By (5.228) this function takes the form

$$\tilde{\psi}_h = C_1 J_0(k_\rho \rho) + C_2 N_0(k_\rho \rho)$$

since the fields are ϕ-invariant. Here $k_\rho^2 = k^2 - k_z^2$. Since the materials of regions 2 and 1 differ, the k values also differ and so do the k_ρ. However, to satisfy the boundary conditions, k_z must be the same in each region. We must omit the term containing $N_0(k_\rho \rho)$ in region 1, as this function is unbounded near the z-axis. So we write

$$\tilde{\psi}_{h1} = A J_0(k_{\rho 0} \rho), \qquad \tilde{\psi}_{h2} = C J_0(k_\rho \rho) + D N_0(k_\rho \rho),$$

where

$$k_{\rho 0}^2 = k_0^2 - k_z^2, \qquad k_\rho^2 = k^2 - k_z^2, \tag{5.269}$$

with $k_0^2 = \omega^2 \mu_0 \epsilon_0$ and $k^2 = \omega^2 \tilde{\mu} \tilde{\epsilon}^c$. The boundary condition on the conducting wall requires

$$\left. \frac{\partial \tilde{\psi}_{h2}}{\partial \rho} \right|_{\rho=b} = 0,$$

leading to $D = -C J_0'(k_\rho b)/N_0'(k_\rho b)$, and therefore

$$\tilde{\psi}_{h2} = C \left[J_0(k_\rho \rho) - N_0(k_\rho \rho) \frac{J_0'(k_\rho b)}{N_0'(k_\rho b)} \right] = B \left[J_0(k_\rho \rho) N_0'(k_\rho b) - N_0(k_\rho \rho) J_0'(k_\rho b) \right].$$

By (5.198) the axial magnetic fields are

$$\tilde{h}_{z1} = A k_{\rho 0}^2 J_0(k_{\rho 0} \rho),$$
$$\tilde{h}_{z2} = B k_\rho^2 \left[J_0(k_\rho \rho) N_0'(k_\rho b) - N_0(k_\rho \rho) J_0'(k_\rho b) \right],$$

and by (5.200) the transverse electric fields are

$$\tilde{e}_{\phi 1} = A j k_{\rho 0} \omega \mu_0 J_0'(k_{\rho 0} \rho),$$
$$\tilde{e}_{\phi 2} = B j k_\rho \omega \mu \left[J_0'(k_\rho \rho) N_0'(k_\rho b) - N_0'(k_\rho \rho) J_0'(k_\rho b) \right].$$

Constants A and B are determined from the boundary conditions at $\rho = a$. Equating $\tilde{H}_{z1} = \tilde{H}_{z2}$ and $\tilde{E}_{\phi 1} = \tilde{E}_{\phi 2}$ gives the simultaneous equations

$$A k_{\rho 0}^2 J_0(k_{\rho 0} a) = B k_\rho^2 f_1,$$
$$-A k_{\rho 0} \mu_0 J_1(k_{\rho 0} a) = B k_\rho \tilde{\mu} f_2,$$

where

$$f_1 = N_0(k_\rho a) J_1(k_\rho b) - J_0(k_\rho a) N_1(k_\rho b),$$
$$f_2 = J_1(k_\rho a) N_1(k_\rho b) - N_1(k_\rho a) J_1(k_\rho b),$$

and where the relations $J_0'(x) = -J_1(x)$ and $N_0'(x) = -N_1(x)$ have been used. Setting the determinant of the coefficient matrix to zero gives the characteristic equation for the azimuthally symmetric TE modes,

$$\frac{\tilde{\mu}}{\mu_0} P(k_{\rho 0}) + Q(k_\rho) = 0, \tag{5.270}$$

where

$$P(k_{\rho_0}) = k_{\rho 0} \frac{J_0(k_{\rho_0} a)}{J_1(k_{\rho_0} a)}, \qquad Q(k_\rho) = k_\rho \frac{f_1}{f_2}. \tag{5.271}$$

Simultaneous solution of (5.270) and (5.269) yields the dispersion relationship between ω and k_z.

▶ **Example 5.28:** Circular waveguide partially filled with barium tetratitinate

A circular waveguide having radius $a = 27.88$ mm is lined with a layer of the ceramic barium tetratitinate (BaTi$_{04}$O$_9$), with an inner radius $b = 0.9a = 25.09$ mm. Assume the sample is lossless, with $\epsilon_r = 38$ and $\mu_r = 1$. Plot the dispersion curve for the first $n = 0$ TE mode and compare it to that for the TE$_{01}$ mode when the guide is filled with air or with barium tetratitinate. Plot the phase velocity for the partially filled and air-filled guides.

Solution: The dispersion curve for the partially filled guide can be plotted by numerically solving the characteristic equation (5.270) along with the relations (5.269). Since the materials are lossless, $k_z = \beta$. Some caution is required in the numerical solution process as one or both of $k_{\rho 0}$ and k_ρ will be imaginary. Hence we use (5.271) only when $k_0^2 \geq \beta^2$ and $k^2 \geq \beta^2$. If $k_0^2 < \beta^2$, we define $\sqrt{k_0^2 - \beta^2} = j\tau$ and use

$$J_0(jx) = I_0(x), \qquad J_1(jx) = j I_1(x), \tag{5.272}$$

to find

$$P(\tau) = \tau \frac{I_0(\tau a)}{I_1(\tau a)}.$$

If $k^2 < \beta^2$ we define $\sqrt{k^2 - \beta^2} = j\chi$ and use (5.272) along with

$$N_0(jx) = j I_0(x) - \frac{2}{\pi} K_0(x), \qquad N_1(jx) = -I_1(x) + j\frac{2}{\pi} K_1(x),$$

to find

$$Q(\chi) = \chi \frac{K_0(\chi a)I_1(\chi b) + I_0(\chi a)K_1(\chi b)}{K_1(\chi a)I_1(\chi b) - I_1(\chi a)K_1(\chi b)}.$$

With these definitions, the characteristic equation (5.270) can be used as is.

Typically, the frequency is set and one searches for β satisfying the two equations. Note that additional solutions for β will appear with increasing frequency; these correspond to both the fundamental mode and the higher-order modes. For frequencies below fundamental mode cutoff, no solutions for β exist.

Figure 5.24 shows the dispersion curve of the lowest-order mode for the circular guide lined with barium tetratitinate. Also shown are the dispersion curves for the TE_{01} mode in the same guide completely filled with air, and completely filled with barium tetratitinate. These can be generated from the usual relationships

$$\beta = \sqrt{k^2 - k_c^2},$$

$$k_c = p'_{01}/a,$$

or by solving (5.270) and (5.269) with either $a \approx b$ (for the air-filled guide) or $a \approx 0$ (for the barium-tetratitinate-filled guide). Comparing this limiting case for the two solutions is a good way to verify the solution for the partially filled guide.

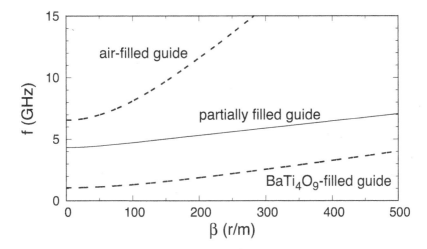

FIGURE 5.24
Dispersion plot for a circular waveguide lined with barium tetratitinate. Solid line is for the first $n = 0$ TE mode of the partially filled guide. Dashed lines correspond to the TE_{01} mode in a fully filled guide.

Figure 5.24 shows that the presence of barium tetratitinate lowers the cutoff frequency and thus significantly reduces the phase velocity in the dielectric-lined guide. Since the dielectric constant of the ceramic is relatively large, only a thin layer (one tenth the guide radius) is needed for the low phase velocity. This slow-wave effect is used in microwave amplifiers where a beam of electrons is sent down the axis of the guide with a velocity approximately equal to the phase velocity of the wave. The interaction force between the beam and the electromagnetic fields serves to transfer the kinetic energy of the beam to the wave, producing the desired amplification. Figure 5.25 shows that the dielectric lowers the phase velocity to less than 20% of the speed of light by 12 GHz.

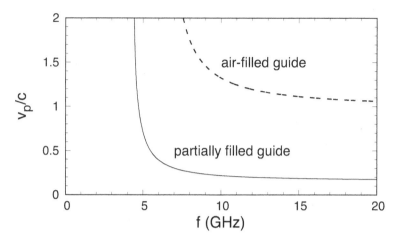

FIGURE 5.25

Phase velocity for a circular waveguide lined with barium tetratitinate. Solid line is for the first $n = 0$ TE mode of the partially filled guide. Dashed line corresponds to the TE_{01} mode in an air-filled guide. ◄

5.6.2 Cascaded hollow-pipe waveguides

Waveguide junctions arise when two dissimilar waveguides are connected together in a plane, as might happen in the case of a transition from, say, a rectangular to a circular guide [125], or when waveguides connect through an iris [101]. Cascaded connections of dissimilar guides through multiple junctions are often used as filters or transformers [145], and for the measurements of the properties of materials [49].

Analysis of multiply connected guides may be undertaken by extending the two-port T-parameter cascading technique of § 4.11.5.7 to multi-mode systems. However, this approach requires the inversion of ill-conditioned matrices. An alternative to cascading network matrices is a recursive approach as described in § 4.11.5.6 for plane waves in planar layered media. Analysis begins at one end of the system and an input is propagated through to the other end. This technique was proposed by Gessel and Ciric using a mode-matching approach [68, 69], but without much physical insight into the meaning of the equations. Franza and Chew provided a physical basis for a recursive mode-matching approach [65] by employing transmission and reflection operators as is done with layered media [35]. This approach provides some physical insight, but it is possible to provide a much more compact form with a simpler perspective on the origination of the various terms. Here we describe a simple implementation of the recursion approach in terms of interfacial transmission and reflection matrices that relate the amplitudes of the modes across the junction between two waveguide sections. These matrices may be obtained by any convenient means, such as moment method, finite element method, or mode matching. We give examples of computing the matrices for simple capacitive and inductive steps in rectangular guides using mode matching.

5.6.2.1 Recursive technique for cascaded multi-mode waveguide sections

Consider a system composed of N uniform waveguide sections connected at $N - 1$ junctions (Figure 5.26). It is assumed that the first waveguide (region 0) and the last wave-

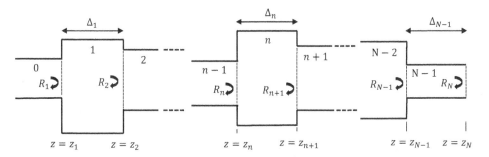

FIGURE 5.26

Cascaded system on N uniform waveguides.

guide (region $N - 1$) propagate a single dominant mode (although they need not be identical waveguides). For convenience, the input port to the cascaded system of waveguides is taken to be at $z = z_1$, even though the first section (region 0) may be of nonzero length to allow for attenuation of the higher-order modes. If desired, the port may be moved to a position $z < z_1$ by accommodating the propagation of the dominant mode. The output port is taken to be at $z = z_N$.

As shown in Figure 5.27, the modal amplitudes of the waves traveling in the $+z$ direction immediately to the left of interface n are given by the vector a_n, while those traveling in the $-z$ direction are given by the vector b_n. The global reflection matrix R_n relates b_n to a_n through $b_n = R_n a_n$, and thus describes the reflection from the entire structure to the right of interface n.

Assuming the dominant mode amplitudes are the first entries in the amplitude vectors, the global reflection and transmission coefficients of the cascaded system are

$$ S_{11} = \frac{(b_1)_1}{(a_1)_1}, \qquad S_{21} = \frac{(a_N)_1}{(a_1)_1}, $$

respectively, with the output port assumed matched, so that $b_N = 0$. Here S_{11} and S_{21} are the *scattering parameters* or S-parameters of the cascaded system [156]. All the waveguides are assumed to contain isotropic materials, so $S_{12} = S_{21}$ by reciprocity. However, $S_{11} = S_{22}$ only if the structure is symmetric.

The permittivity and permeability of the material uniformly filling section n are $\tilde{\epsilon}_n^c$ and $\tilde{\mu}_n$, respectively, and the length of section n is Δ_n. The interface between region $n - 1$ and region n is located at $z = z_n$. The interfacial reflection coefficient matrix at this junction, Γ_n, relates the modal amplitudes of the waves reflected at the interface to those incident from the left in the case that regions $n - 1$ and n are semi-infinite. The interfacial transmission coefficient matrix T_n relates the modal amplitudes of the waves transmitted through the interface to those incident at the interface from the left. Similarly, $\bar{\Gamma}_n$ is the interfacial reflection coefficient matrix at z_n for waves incident on the interface from the right, while $\bar{T}_n$ is the interfacial transmission coefficient matrix at z_n for waves incident on the interface from the right.

As the mth modal wave (propagating or evanescent) transitions from the interface at z_n to the interface at z_{n+1}, it will undergo a phase shift and attenuation given by $\exp\{-jk_{zm}^n \Delta_n\}$, where k_{zm}^n is the propagation constant of the mth mode of region n. It is thus expedient to define the propagation matrix for the modes of section n as a diagonal matrix P_n with entries $[P_n]_{i,j} = \delta_{ij} \exp\{-jk_{zi}^n \Delta_n\}$, where δ_{ij} is the Kronecker

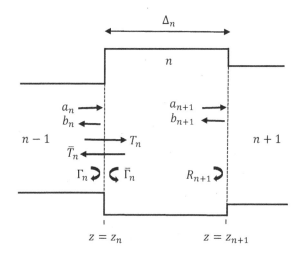

FIGURE 5.27

Region n of cascaded waveguide system.

delta.

The wave amplitudes incident on interface $n + 1$ may be viewed as consisting of two terms. The first is given by the wave amplitudes incident on interface n multiplied by the interfacial transmission matrix T_n to account for transmission through interface n, and then multiplied by the propagation matrix P_n to account for propagation across region n. Added to that term are the wave amplitudes reflected by the entire structure to the right of region n, which are propagated back to interface n, reflected at interface n, and then propagated back to interface $n + 1$. The net result is

$$a_{n+1} = P_n(T_n a_n) + P_n \bar{\Gamma}_n P_n (R_{n+1} a_{n+1}). \tag{5.273}$$

Similarly, the wave amplitudes transmitted back into region $n - 1$ from region n can be written as a sum of two terms:

$$b_n = \Gamma_n a_n + \bar{T}_n P_n (R_{n+1} a_{n+1}). \tag{5.274}$$

The reflection coefficient for the cascaded structure can be computed recursively using (5.273)–(5.274). From (5.273) it is seen that

$$a_{n+1} = \tau_n a_n \tag{5.275}$$

where

$$\tau_n = [I - \bar{\Gamma}'_n R'_{n+1}]^{-1} T'_n$$

with $\bar{\Gamma}'_n = P_n \bar{\Gamma}_n$, $T'_n = P_n T_n$, $R'_{n+1} = P_n R_{n+1}$, and I the identity matrix. Substituting (5.275) into (5.274), we obtain

$$b_n = \Gamma_n a_n + \bar{T}_n R'_{n+1} \tau_n a_n.$$

But $b_n = R_n a_n$, so

$$R_n = \Gamma_n + \bar{T}_n R'_{n+1} \tau_n. \tag{5.276}$$

Equation (5.276) is the key recursive equation. To find S_{11} for the cascaded system, the output port is matched terminated so that $R_N = 0$. This implies $R_{N-1} = \Gamma_{N-1}$. Using (5.276), R_{N-2} is computed as

$$R_{N-2} = \Gamma_{N-2} + \bar{T}_{N-2} R'_{N-1} \tau_{N-2}.$$

This process is repeated until we reach R_1. Then, $S_{11} = (R_1)_{11}$ since the $(1,1)$ entry in the reflection matrix corresponds to the dominant mode.

The transmission coefficient for the cascaded system may be computed using the values of τ generated while performing recursion on S_{11}. From (5.275),

$$a_N = \tau_{N-1} a_{N-1},$$

$$a_{N-1} = \tau_{N-2} a_{N-2},$$

$$a_{N-2} = \tau_{N-3} a_{N-3},$$

$$\vdots$$

Backsubstitution beginning with $a_2 = \tau_1 a_1$ gives

$$a_N = (\tau_{N-1} \tau_{N-2} \tau_{N-3} \cdots \tau_1) a_1 = \tau a_1.$$

Thus, $S_{21} = (\tau)_{11}$.

If S_{12} and S_{22} are sought, the process may be repeated by recursing in the opposite direction.

Note that this recursive technique may be used for any termination at the output port, such as a short or open circuit, a known impedance, or the known reflection coefficient for some other structure. This is useful when altering only a portion of a cascaded system. Since the structure to the right of the altered portion is unchanged, its known reflection may be used as the starting point in the recursion. For instance, if the unknown material properties of a sample held in a waveguide sample holder are to be found through some iterative search, then the structure to the right of the sample need not be reanalyzed, since only the properties of the sample are changed during the search.

5.6.2.2 Junction matrices found using mode matching

The transmission and reflection matrices at a waveguide junction may be found using any convenient method, including purely numerical techniques like finite element or integral equation methods. Here the use of mode matching is detailed for the case where a smaller waveguide opens completely into a larger guide (waveguide expansion), or where a larger guide connects with a smaller guide (waveguide reduction). See Figure 5.28. In each case, the smaller aperture is assumed to be completely contained within the larger. Note that the waveguides need not have the same cross-sectional shape.

Assume guide A occupies the semi-infinite region $z < 0$ while guide B occupies $z > 0$, as shown in Figure 5.29 for a simple capacitive step between two rectangular guides of the same width. The transverse fields in region A ($z < 0$) may be expanded in a modal series as

$$\tilde{\mathbf{E}}_t(\mathbf{r}) = \sum_{n=1}^{N_A} \left[a_n e^{-jk_{z,n}^A z} + b_n e^{jk_{z,n}^A z} \right] \tilde{\mathbf{e}}_n^A(\boldsymbol{\rho}),$$

$$\tilde{\mathbf{H}}_t(\mathbf{r}) = \sum_{n=1}^{N_A} \left[a_n e^{-jk_{z,n}^A z} - b_n e^{jk_{z,n}^A z} \right] \tilde{\mathbf{h}}_n^A(\boldsymbol{\rho}),$$

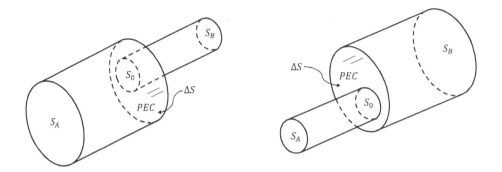

FIGURE 5.28
Junction between two waveguides. Left: waveguide reduction. Right: waveguide expansion.

while for the fields in region B,

$$\tilde{\mathbf{E}}_t(\mathbf{r}) = \sum_{n=1}^{N_B} \left[c_n e^{-jk_{z,n}^B z} + d_n e^{jk_{z,n}^B z} \right] \tilde{\mathbf{e}}_n^B(\boldsymbol{\rho}),$$

$$\tilde{\mathbf{H}}_t(\mathbf{r}) = \sum_{n=1}^{N_B} \left[c_n e^{-jk_{z,n}^B z} - d_n e^{jk_{z,n}^B z} \right] \tilde{\mathbf{h}}_n^B(\boldsymbol{\rho}).$$

For numerical expediency, the expansions have been truncated at N_A modes for the fields in region A and N_B modes for region B. The propagation constants for the nth mode in region $\{A, B\}$ are given by $k_{z,n}^{A,B}$, while $\tilde{\mathbf{e}}_n^{A,B}$ are the transverse electric field wave functions of the nth mode in region $\{A, B\}$.

Caution is warranted when combining the fields for TE and TM modes, as the modal amplitudes must have the same units. For instance, the units carried by amplitude A_{nm} in (5.220) differ from those of amplitude B_{nm} in (5.225). In the expansions above, we choose to specify $\tilde{\mathbf{e}}_n^{A,B}$ for both TE and TM modes, then find the transverse magnetic fields using

$$\tilde{\mathbf{h}}_n^{A,B} = \frac{\hat{z} \times \tilde{\mathbf{e}}_n^{A,B}}{Z_n^{A,B}}.$$

Here the wave impedances are, for modes TE or TM with respect to the z-direction,

$$Z_n^{A,B} = \begin{cases} \dfrac{\omega \tilde{\mu}_{A,B}}{k_{z,n}^{A,B}}, & \mathrm{TE}_z, \\[2mm] \dfrac{k_{z,n}^{A,B}}{\omega \tilde{\epsilon}_{A,B}^c}, & \mathrm{TM}_z. \end{cases}$$

Application of the boundary conditions on tangential electric field and tangential magnetic field, respectively, at $z = 0$ gives

$$\sum_{n=1}^{N_A} [a_n + b_n] \tilde{\mathbf{e}}_n^A(\boldsymbol{\rho}) = \begin{cases} \displaystyle\sum_{n=1}^{N_B} [c_n + d_n] \tilde{\mathbf{e}}_n^B(\boldsymbol{\rho}), & \boldsymbol{\rho} \in S_B, \\[4mm] 0, & \boldsymbol{\rho} \in \Delta S, \end{cases} \qquad (5.277)$$

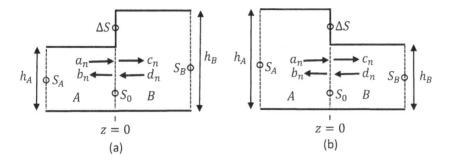

FIGURE 5.29
Step junction between two rectangular waveguides sharing a common bottom wall. (a) waveguide expansion (b) waveguide reduction.

$$\sum_{n=1}^{N_A} [a_n - b_n]\tilde{\mathbf{h}}_n^A(\boldsymbol{\rho}) = \sum_{n=1}^{N_B} [c_n - d_n]\tilde{\mathbf{h}}_n^B(\boldsymbol{\rho}), \quad \boldsymbol{\rho} \in S_B, \tag{5.278}$$

for the case of waveguide reduction, and

$$\sum_{n=1}^{N_B} [c_n + d_n]\tilde{\mathbf{e}}_n^B(\boldsymbol{\rho}) = \begin{cases} \sum_{n=1}^{N_A} [a_n + b_n]\tilde{\mathbf{e}}_n^A(\boldsymbol{\rho}), & \boldsymbol{\rho} \in S_A, \\ 0, & \boldsymbol{\rho} \in \Delta S, \end{cases} \tag{5.279}$$

$$\sum_{n=1}^{N_B} [c_n - d_n]\tilde{\mathbf{h}}_n^B(\boldsymbol{\rho}) = \sum_{n=1}^{N_A} [a_n - b_n]\tilde{\mathbf{h}}_n^A(\boldsymbol{\rho}), \quad \boldsymbol{\rho} \in S_A, \tag{5.280}$$

for the case of waveguide expansion. Here $\{S_A, S_B\}$ is the cross-sectional surface of waveguide $\{A, B\}$, and Δ_S is the PEC surface joining the two guides at the junction. See Figure 5.28.

To obtain a system of linear mode-matching equations for waveguide reduction, (5.277) is multiplied by $\tilde{\mathbf{e}}_m^A(\boldsymbol{\rho})$ ($1 \leq m \leq N_A$) and integrated over S_A, while (5.278) is multiplied by $\tilde{\mathbf{h}}_m^B(\boldsymbol{\rho})$ ($1 \leq m \leq N_B$) and integrated over S_B. Similarly, for waveguide expansion, (5.279) is multiplied by $\tilde{\mathbf{e}}_m^B(\boldsymbol{\rho})$ ($1 \leq m \leq N_B$) and integrated over S_B, while (5.280) is multiplied by $\tilde{\mathbf{h}}_m^A(\boldsymbol{\rho})$ ($1 \leq m \leq N_A$) and integrated over S_A. This gives, in both cases, two sets of simultaneous equations of the form

$$\sum_{n=1}^{N_A} a_n C_{mn} + \sum_{n=1}^{N_A} b_n C_{mn} - \sum_{n=1}^{N_B} c_n D_{mn} - \sum_{n=1}^{N_B} d_n D_{mn} = 0 \quad (1 \leq m \leq N_A), \tag{5.281}$$

$$\sum_{n=1}^{N_A} a_n E_{mn} - \sum_{n=1}^{N_A} b_n E_{mn} - \sum_{n=1}^{N_B} c_n F_{mn} + \sum_{n=1}^{N_B} d_n F_{mn} = 0 \quad (1 \leq m \leq N_B). \tag{5.282}$$

For waveguide reduction, the coefficients may be written as

$$C_{mn} = f_n^A \delta_{mn}, \tag{5.283}$$

$$D_{mn} = Q_{mn}, \tag{5.284}$$

$$E_{mn} = \frac{1}{Z_m^B Z_n^A} Q_{nm}, \tag{5.285}$$

$$F_{mn} = \left(\frac{1}{Z_n^B}\right)^2 f_n^B \delta_{mn}, \tag{5.286}$$

where

$$f_n^A = \int_{S_A} \mathbf{e}_n^A(\boldsymbol{\rho}) \cdot \mathbf{e}_n^A(\boldsymbol{\rho}) \, dS, \qquad f_n^B = \int_{S_B} \mathbf{e}_n^B(\boldsymbol{\rho}) \cdot \mathbf{e}_n^B(\boldsymbol{\rho}) \, dS, \tag{5.287}$$

and δ_{mn} is the Kronecker delta. Here mode orthogonality in each waveguide system has been used to obtain (5.283) and (5.286). For waveguide expansion, the coefficients are

$$C_{mn} = Q_{nm}, \tag{5.288}$$

$$D_{mn} = f_n^B \delta_{mn}, \tag{5.289}$$

$$E_{mn} = \left(\frac{1}{Z_n^A}\right)^2 f_n^A \delta_{mn}, \tag{5.290}$$

$$F_{mn} = \frac{1}{Z_m^A Z_n^B} Q_{mn}. \tag{5.291}$$

In these equations,

$$Q_{mn} = \int_{S_0} \tilde{\mathbf{e}}_n^B(\boldsymbol{\rho}) \cdot \tilde{\mathbf{e}}_m^A(\boldsymbol{\rho}) \, dS, \tag{5.292}$$

where S_0 is the common aperture at the waveguide junction: S_A for a waveguide expansion, and S_B for a waveguide reduction.

To find the reflection and transmission matrices, let $d_n = 0$. In the case of waveguide reduction, the matrices C and F are diagonal by mode orthogonality, and it is expedient to write (5.281)–(5.282) as the vector equations

$$-a - b + (C^{-1}D)c = 0,$$

$$-(F^{-1}E)a + (F^{-1}E)b + c = 0,$$

since the inverses of C and F are trivial to compute. Defining the interfacial reflection matrix Γ through $b = \Gamma a$ and the interfacial transmission matrix through $c = Ta$, the equations may be solved to give

$$T = 2(VU + I)^{-1}V, \qquad \Gamma = (UT - I), \qquad U = C^{-1}D, \qquad V = F^{-1}E$$

where I is the identity matrix. In the case of waveguide expansion, the matrices D and E are diagonal, and similar steps give

$$T = 2(VU + I)^{-1}V, \qquad \Gamma = -(UT - I), \qquad U = E^{-1}F, \qquad V = D^{-1}C.$$

Finally, a compact form for Γ and T may be obtained by noting that Q_{mn} for waveguide reduction is identical to Q_{nm} for waveguide expansion. Thus, substituting from (5.283)–(5.291), we have the compact form

Waveguide reduction

$$T = 2\,(BA + I)^{-1}\,B, \qquad \Gamma = (AT - I),$$

$$A_{mn} = \frac{Q_{mn}}{f_m^A} \qquad (1 \le m \le N_A,\ 1 \le n \le N_B),$$

$$B_{mn} = \frac{Z_m^B}{Z_n^A}\frac{Q_{nm}}{f_m^B} \qquad (1 \le m \le N_B,\ 1 \le n \le N_A).$$

Waveguide expansion

$$T = 2\,(BA + I)^{-1}\,B, \qquad \Gamma = -(AT - I),$$

$$A_{mn} = \frac{Z_m^A}{Z_n^B}\frac{Q_{mn}}{f_m^A} \qquad (1 \le m \le N_A,\ 1 \le n \le N_B),$$

$$B_{mn} = \frac{Q_{nm}}{f_m^B} \qquad (1 \le m \le N_B,\ 1 \le n \le N_A).$$

In these expressions, Q_{mn} is computed for a waveguide reduction.

Note that the only term used in computing Γ and T that is frequency or material-parameter dependent is the ratio of the wave impedances. This fact may be used to reduce the computation time in iterative applications or during sweeps across frequency bands.

▶ **Example 5.29:** Capacitive step in a rectangular waveguide

Consider a capacitive junction [82] between two rectangular waveguides of different heights h_A and h_B but identical widths, both sharing a common lower wall (see Figure 5.29). Find the values f_n and Q_{mn} that determine the reflection and transmission matrices.

Solution: If the waves originating from the left transition from a shorter to a taller guide, such that $h_B > h_A$, then the junction represents a waveguide expansion. If $h_B < h_A$, the transition is a waveguide reduction.

It is assumed that all junctions in the waveguide system are capacitive, and that the first junction is excited solely by a TE_{10} mode. Then, by symmetry, the modes in all the guides will be either $\mathrm{TE}_{1\nu_n}$ or $\mathrm{TM}_{1\nu_n}$, where ν_n is the index of mode n according to

$$\nu_n = \begin{cases} \dfrac{n-1}{2}, & \mathrm{TE}_{1\nu_n},\ n = 1, 3, 5, \dots \\[2mm] \dfrac{n}{2}, & \mathrm{TM}_{1\nu_n},\ n = 2, 4, 6, \dots \end{cases}$$

The transverse wave functions of the nth mode in region $\{A, B\}$ for a rectangular guide are thus given by

$$\tilde{\mathbf{e}}_n^{A,B} = \begin{cases} \hat{x}k_{y,n}^{A,B}\cos\left(\dfrac{\pi}{a}x\right)\sin(k_{y,n}^{A,B}y) \\[1mm] \quad -\hat{y}\dfrac{\pi}{a}\sin\left(\dfrac{\pi}{a}x\right)\cos(k_{y,n}^{A,B}y), & \mathrm{TE}_z \\[3mm] -\hat{x}\dfrac{\pi}{a}\cos\left(\dfrac{\pi}{a}x\right)\sin(k_{y,n}^{A,B}y) \\[1mm] \quad -\hat{y}k_{y,n}^{A,B}\sin\left(\dfrac{\pi}{a}x\right)\cos(k_{y,n}^{A,B}y), & \mathrm{TM}_z \end{cases} \qquad (5.293)$$

where the various wavenumbers are related through

$$(k_{c,n}^{A,B})^2 = (\pi/a)^2 + (k_{y,n}^{A,B})^2,$$

$$(k_{A,B})^2 = \omega^2 \tilde{\mu}_{A,B} \tilde{\epsilon}_{A,B}^c = (k_{c,n}^{A,B})^2 + (k_{z,n}^{A,B})^2,$$

$$k_{y,n}^{A,B} = \frac{\nu_n \pi}{h_{A,B}}.$$

The functions $f_n^{A,B}$ are found using (5.287). Substituting for $\tilde{\mathbf{e}}_n^{A,B}$ and integrating, we have

$$f_n^{A,B} = \frac{a}{2} \frac{h_{A,B}}{\epsilon_{\nu_n}} (k_{c,n}^{A,B})^2, \qquad \epsilon_{\nu_n} = \begin{cases} 1, & \nu_n = 0, \\ 2, & \nu_n > 0. \end{cases}$$

The reflection and transmission matrices are determined by the values of Q_{mn} found from (5.292). Substituting from (5.293), it is found that Q_{mn} depends on whether the wave functions in the product in (5.292) are TE_z or TM_z.

$\mathbf{TE}_m \cdot \mathbf{TE}_n$

$$Q_{mn} = \begin{cases} f_n^B, & k_{ym}^A = k_{yn}^B \\ \dfrac{a}{2} (-1)^{\nu_n} \dfrac{k_{ym}^A (k_{cn}^B)^2 \sin(k_{ym}^A h_B)}{(k_{ym}^A)^2 - (k_{yn}^B)^2}, & k_{ym}^A \neq k_{yn}^B \end{cases}$$

$\mathbf{TM}_m \cdot \mathbf{TM}_n$

$$Q_{mn} = \begin{cases} f_n^B, & k_{ym}^A = k_{yn}^B \\ \dfrac{a}{2} (-1)^{\nu_n} \dfrac{k_{yn}^B (k_{cm}^A)^2 \sin(k_{ym}^A h_B)}{(k_{ym}^A)^2 - (k_{yn}^B)^2}, & k_{ym}^A \neq k_{yn}^B \end{cases}$$

$\mathbf{TM}_m \cdot \mathbf{TE}_n$

$$Q_{mn} = \begin{cases} 0, & k_{ym}^A = k_{yn}^B \\ \dfrac{\pi}{2} (-1)^{\nu_n} \sin(k_{ym}^A h_B), & k_{ym}^A \neq k_{yn}^B \end{cases}$$

$\mathbf{TE}_m \cdot \mathbf{TM}_n$

$$Q_{mn} = 0. \ \blacktriangleleft$$

▶ **Example 5.30:** Capacitive waveguide iris

Consider a thick iris in an X-band rectangular waveguide (Figure 5.30), corresponding to three cascaded rectangular guides ($N = 3$). Guides 0 and 2 are standard air-filled WR-90 waveguides with dimensions $a = 0.9$ inch by $b = 0.4$ inch (22.86 by 10.16 mm). The iris is also air filled, with thickness $\Delta_1 = 0.125$ inch (3.175 mm) and step height $h_1 = 0.125$ inches (3.175 mm). If the output port is taken to be at the second interface (with $\Delta_2 = 0$), compute the S-parameters of the iris waveguide system. Explore the dependence of these parameters on the number of modes used for field expansion.

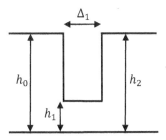

FIGURE 5.30

Side view of a three-region capacitive X-band iris. All regions are empty. $h_0 = h_2 = 10.16$ mm, $h_1 = 3.175$ mm, $\Delta_1 = 3.175$ mm.

Solution: From the given dimensions, we have $h_0 = h_2 = 0.4$ inch (10.16 mm). To ensure rapid convergence of the mode-matching technique, the number of modes in each region should be chosen carefully [68, 65, 117]. For the waveguide step problem, the ratio of the number of modes used to the height of the opening should be kept constant (i.e., N_i/h_i should be the same for each region). Since there is no TM_{10} mode, it is useful to choose the number of modes as $N_i = 2MR_i+1$. Here R_i is an integer determined from the least common multiple of the step heights, and M is an integer multiplier. For the present waveguide iris example, $h_0 = h_2 = 0.4$ and $h_1 = 0.125$. Thus, $h_1/h_0 = 16/5$, and $R_0 = R_2 = 16$, $R_1 = 5$. The table below shows the convergence of S_{11} and S_{21} at $f = 12.4$ GHz as M is increased.

M	N_0	N_1	$|S_{11}|$ (dB)	$\angle S_{11}$ (°)	$|S_{21}|$ (dB)	$\angle S_{21}$ (°)
1	33	11	-1.08495	-167.779	-6.55492	-77.7787
2	65	21	-1.08367	-167.800	-6.55941	-77.8001
5	161	51	-1.08327	-167.807	-6.56085	-77.8069
10	321	101	-1.08323	-167.808	-6.56099	-77.8076

The S-parameters for the waveguide iris were generated across X-band using mode matching with $M = 10$. Because the structure is symmetric, $S_{22} = S_{11}$. The results are shown in Figures 5.31 and 5.32.

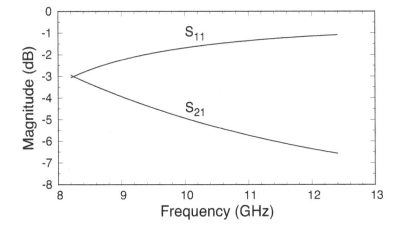

FIGURE 5.31

Magnitude of S-parameters of a capacitive waveguide iris.

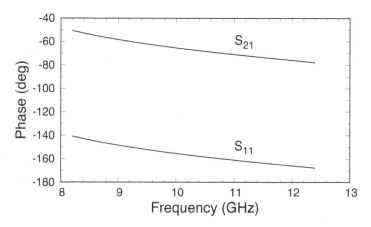

FIGURE 5.32

Phase of S-parameters of a capacitive waveguide iris. ◄

▶ **Example 5.31:** Waveguide sample holder

Consider a sample holder comprising a material sandwiched between two capacitive wave-guide irises in an X-band rectangular waveguide (Figure 5.33). This is a cascading of five rectangular waveguides, with aperture heights $h_0 = 10.16$, $h_1 = 2.54$, $h_2 = 5.08$, $h_3 = 2.54$, and $h_4 = 10.16$ mm. Guides 0 and 4 are standard air-filled WR-90 waveguides with dimensions $a = 0.9$ inch by $b = 0.4$ inch (22.86 by 10.16 mm). All regions are air-filled except for the central region (region 2), which contains a lossless dielectric material of relative permittivity $\epsilon_{r2} = 2.1$ (Teflon). Taking the output port at the fourth interface (with $\Delta_4 = 0$), compute the S-parameters of the iris waveguide system.

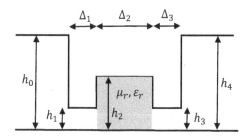

FIGURE 5.33

Side view of a five-region X-band sample holder. All regions are empty except region 2, which contains a material with $\mu_r = 1$, $\epsilon_r = 2.1$. $h_0 = h_4 = 10.16$ mm, $h_1 = h_3 = 2.54$ mm, $h_2 = 5.08$ mm, $\Delta_1 = \Delta_3 = 2.54$ mm, $\Delta_2 = 5.08$ mm.

Solution: The number of modes in each region is based on the least common multiple of the heights as $N_i = 2MR_i + 1$, with $R_0 = 4$, $R_1 = 1$, $R_2 = 2$, $R_3 = 1$, and $R_4 = 4$. The S-parameters for the sample holder as a function of frequency are shown in Figures 5.34 and 5.35, where they are computed using $M = 50$. Note the deep null in S_{11} near 10 GHz where there is near complete transmission through the system. For frequencies near the null, convergence requires a large number of modes. Computation can be accelerated by extrapolating from smaller values of M.

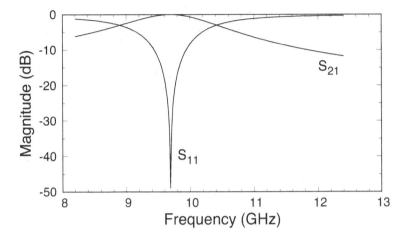

FIGURE 5.34
Magnitude of S-parameters of a waveguide sample holder.

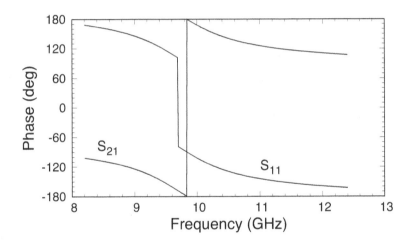

FIGURE 5.35
Phase of S-parameters of a waveguide sample holder. ◄

▶ **Example 5.32:** Inductive step in a rectangular waveguide

Consider an inductive junction [82] between two rectangular waveguides of different widths W_A and W_B but identical heights, both centered on the z-axis (Figure 5.36). Find the values f_n and Q_{mn} that determine the reflection and transmission matrices.

Solution: If the waves originating from the left transition from a narrower to a wider guide, such that $W_B > W_A$, then the junction represents a waveguide expansion. If $W_B < W_A$, the transition is a waveguide reduction.

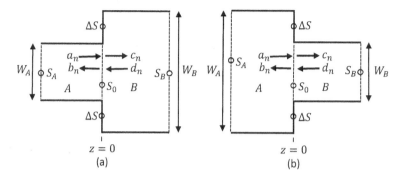

FIGURE 5.36

Top view of an inductive step junction between two centered rectangular waveguides of different widths. (a) waveguide expansion (b) waveguide reduction.

It is assumed that all junctions in the waveguide system are inductive, with each guide centered on the z-axis, and that the first junction is excited solely by a TE_{10} mode. Then, by symmetry, the modes in all the guides will be TE_{n0}, where n is an odd integer. Since all the guides are centered on the z-axis, the transverse wave functions of the nth mode in region $\{A, B\}$ for a rectangular guide are given by

$$\tilde{\mathbf{e}}_n^{A,B} = \hat{\mathbf{y}} k_{xn}^{A,B} \sin\left[k_{xn}^{A,B}\left(x - \tfrac{1}{2} W_{A,B}\right)\right] \quad (|x| \leq \tfrac{1}{2} W_{A,B}, \ n = 1, 2, \ldots). \tag{5.294}$$

Here

$$k_{xn}^{A,B} = \frac{(2n-1)\pi}{W_{A,B}},$$

and the relationship between the wavenumbers is

$$(k_{A,B})^2 = \omega^2 \tilde{\mu}_{A,B} \tilde{\epsilon}_{A,B}^c = (k_{xn}^{A,B})^2 + (k_{zn}^{A,B})^2.$$

Note that it is convenient for computational purposes to rewrite the wave functions (5.294) using a trigonometric identity as

$$\tilde{\mathbf{e}}_n^{A,B} = \hat{\mathbf{y}} (-1)^n k_{xn}^{A,B} \cos(k_{xn}^{A,B} x). \tag{5.295}$$

The functions $f_n^{A,B}$ are found using (5.287). Substitution for $\tilde{\mathbf{e}}_n^{A,B}$ from (5.295) and integration give

$$f_n^{A,B} = \frac{W_{A,B} b}{2} (k_{x,n}^{A,B})^2.$$

The reflection and transmission matrices are determined by the values of Q_{mn} found from (5.292). Substituting from (5.295) and integrating, Q_{mn} is found to be

$$Q_{mn} = \begin{cases} (-1)^{m+n} f_n^B, & k_{xm}^A = k_{xn}^B \\ (-1)^{m+1} 2b \cos\left(k_{xm}^A \dfrac{W_B}{2}\right) \dfrac{(k_{xn}^B)^2 k_{xm}^A}{(k_{xn}^B)^2 - (k_{xm}^A)^2}, & k_{xm}^A \neq k_{xn}^B \end{cases}$$

for a waveguide reduction, and

$$Q_{mn} = \begin{cases} (-1)^{m+n} f_m^A, & k_{xm}^A = k_{xn}^B \\ (-1)^{n+1} 2b \cos\left(k_{xn}^B \dfrac{W_A}{2}\right) \dfrac{(k_{xm}^A)^2 k_{xn}^B}{(k_{xm}^A)^2 - (k_{xn}^B)^2}, & k_{xm}^A \neq k_{xn}^B \end{cases}$$

for a waveguide expansion. ◄

▶ **Example 5.33:** Inductive waveguide iris

Consider a thick inductive iris in an S-band rectangular waveguide, as shown in Figure 5.37, corresponding to three cascaded rectangular guides ($N = 3$). Guides 0 and 2 are standard air-filled WR-284 waveguides with dimensions $a = 2.84$ inch by $b = 1.34$ inch (72.136 by 34.036 mm). The iris region is cubical shaped, with thickness $\Delta_1 = 34.036$ mm and width $W_1 = 34.036$ mm. The iris is filled with a dielectric material of relative permittivity $\epsilon_r = 2.1$ (Teflon). Taking the output port at the second interface (with $\Delta_2 = 0$), compute the S-parameters of the iris waveguide system.

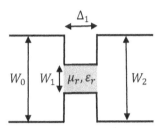

FIGURE 5.37
Top view of a three-region inductive S-band iris. $\epsilon_r = 2.1$, $\mu_r = 1$, $W_0 = W_2 = 72.136$ mm, $W_1 = \Delta_1 = 34.036$ mm.

Solution: From the given dimensions, we have $W_0 = W_2 = 2.84$ inch (72.136 mm). The ratio of W_0 to W_1 does not allow for a simple relationship between the number of modes in each waveguide region, so the field expansion is done with $N_0 = N_2 = 200$ modes, and $N_1 = 100$ modes. This provides reasonable accuracy in the magnitude and phase of the S-parameters, which are shown in Figures 5.38 and 5.39 as a function of frequency. As was noted for the sample holder of Example 5.31, there is a deep null in S_{11} where there is near complete transmission through the system, in this case near 10 GHz.

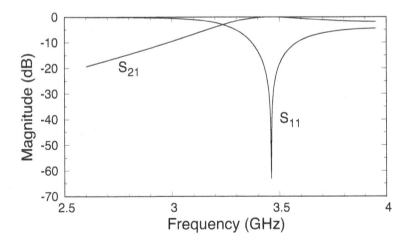

FIGURE 5.38
Magnitude of S-parameters of an inductive iris.

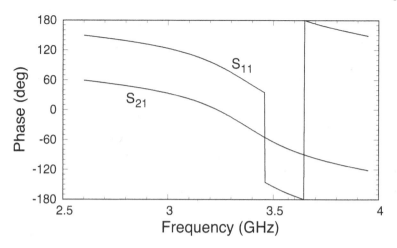

FIGURE 5.39

Phase of S-parameters of an inductive iris. ◀

5.6.3 TEM modes in axial waveguiding structures

TEM modes are the primary operating modes of transmission lines. They offer zero cutoff frequency and, under lossless conditions, zero dispersion. Transmission line theory is extensive, and here we only consider the basic concepts of the propagation of TEM modes. We assume the geometry of the guiding structure is independent of a chosen axial direction, and that the structure is composed of conductors immersed in a homogeneous isotropic material with permittivity $\tilde{\epsilon}^c(\omega)$ and permeability $\tilde{\mu}(\omega)$.

5.6.3.1 Field relationships for TEM modes

In §5.4.2.1 we found that TEM fields can be represented by either a magnetic or an electric Hertzian potential. In either case the potential satisfies

$$\nabla^2\tilde{\Pi} = 0, \qquad \left(\frac{\partial^2}{\partial u^2} + k^2\right)\tilde{\Pi} = 0.$$

With z as the guided-wave axis, this equation has solutions of the form

$$\tilde{\Pi}(\boldsymbol{\rho}, z, \omega) = \tilde{\psi}(\boldsymbol{\rho}, \omega)e^{\mp jkz},$$

where the $\mp$ signs yield waves propagating in the $\pm z$ directions, respectively, and $\tilde{\psi}$ satisfies $\nabla_t^2\tilde{\psi} = 0$. Note that the propagation constant is given by the wavenumber k; unlike TE and TM modes in hollow-pipe waveguides, guided TEM modes lack cutoff frequencies and therefore propagate at all frequencies. This behavior is similar to that of a uniform plane wave.

If we adopt the electric Hertzian potential, the electric field is given by (5.128). It is traditional to define a new potential function $\tilde{\Phi}$ as

$$\tilde{\Phi}(\boldsymbol{\rho}, \omega) = \pm jk\tilde{\psi}(\boldsymbol{\rho}, \omega),$$

such that

$$\tilde{\mathbf{E}}(\mathbf{r}, \omega) = \tilde{\mathbf{e}}_t(\boldsymbol{\rho}, \omega) e^{\mp jkz}, \tag{5.296}$$

$$\tilde{\mathbf{H}}(\mathbf{r}, \omega) = \tilde{\mathbf{h}}_t(\boldsymbol{\rho}, \omega) e^{\mp jkz},$$

where

$$\tilde{\mathbf{e}}_t = -\nabla_t \tilde{\Phi}, \qquad \tilde{\mathbf{h}}_t = \pm \frac{\omega \tilde{\epsilon}^c}{k} \hat{\mathbf{z}} \times \nabla_t \tilde{\Phi}.$$

Then

$$\tilde{\mathbf{e}}_t = -\nabla_t \tilde{\Phi}, \tag{5.297}$$

$$\tilde{\mathbf{h}}_t = \pm \frac{\hat{\mathbf{z}} \times \tilde{\mathbf{e}}_t}{Z_{TEM}}, \tag{5.298}$$

where $Z_{TEM} = \eta = \sqrt{\mu/\tilde{\epsilon}^c}$ is the TEM wave impedance. Note that since $\nabla_t^2 \tilde{\psi} = 0$,

$$\nabla_t^2 \tilde{\Phi} = 0. \tag{5.299}$$

The wave nature of the TEM fields. The field components for TEM modes vary as $e^{\mp jkz}$. Thus, the propagation constant is identical to the wavenumber in the medium supporting the waveguiding structure and the wave characteristics are the same as those for a plane wave. We can write the propagation constant as $k = \beta - j\alpha$ so that $\tilde{\mathbf{E}}, \tilde{\mathbf{H}} \sim e^{\mp j\beta z} e^{\mp \alpha z}$. If the source of the fields lies at $z = d$, then for $z > d$ we choose the minus sign to obtain a wave propagating away from the source; for $z < -d$ we choose the plus sign.

When the guide material is a good dielectric, we may assume $\tilde{\mu} = \mu$ is real and frequency independent and use (4.239) to show that

$$k = \beta - j\alpha = \omega \sqrt{\mu \operatorname{Re} \tilde{\epsilon}^c} \sqrt{1 - j \tan \delta_c}$$

where δ_c is the loss tangent (4.238). Under low-loss conditions where $\tan \delta_c \ll 1$, we may approximate the square root using the first two terms of the binomial series to get

$$\beta - j\alpha \approx \omega \sqrt{\mu \operatorname{Re} \tilde{\epsilon}^c} \left(1 - j \tfrac{1}{2} \tan \delta_c\right), \tag{5.300}$$

and therefore

$$\beta \approx \omega \sqrt{\mu \operatorname{Re} \tilde{\epsilon}^c}, \qquad \alpha = -\frac{\omega \operatorname{Im} \tilde{\epsilon}^c}{2\eta} \tag{5.301}$$

where $\eta = \sqrt{\mu / \operatorname{Re} \tilde{\epsilon}^c}$ is the lossless intrinsic impedance. Comparing (5.300) with (5.205), we see that the cutoff frequency of TEM modes is zero, hence they propagate at all frequencies.

We can compute the phase and group velocities of the wave just as we did for plane waves: $v_p = \omega/\beta$ and $v_g = d\omega/d\beta$. The guided wavelength of a TEM wave is $\lambda_g = 2\pi/\beta$. Under low-loss conditions, (5.301) yields

$$v_p = v_g = \frac{1}{\sqrt{\mu \operatorname{Re} \tilde{\epsilon}^c}},$$

so the wave propagates (approximately) without dispersion.

Attenuation due to conductor losses for TEM modes. The attenuation of TEM modes due to imperfectly conducting metals in the guiding structure may be computed using a perturbation method, exactly as with hollow-pipe guides. The attenuation constant is given by (5.216):

$$\alpha_c = \frac{P_d(0)/l}{2P_0},$$

where P_0 is the time-average power carried by the guide at $z = 0$ and $P_d(0)/\ell$ is the power dissipated per unit length in the conductors.

5.6.3.2 Voltage and current on a transmission line

We define the frequency-domain voltage between two points P_1 and P_2 *in a specific cross-sectional plane* of the TEM-mode waveguiding structure as the line integral of the electric field (5.296):

$$\tilde{V}_{21}(z,\omega) = -\int_{P_1}^{P_2} \tilde{\mathbf{E}}(\boldsymbol{\rho}, z, \omega) \cdot \mathbf{dl}.$$

Use of (5.297) gives

$$\tilde{V}_{21}(z,\omega) = e^{\mp jkz} \int_{P_1}^{P_2} \nabla_t \tilde{\Phi}(\boldsymbol{\rho}, \omega) \cdot \mathbf{dl} = \left[\tilde{\Phi}(\boldsymbol{\rho}_2, \omega) - \tilde{\Phi}(\boldsymbol{\rho}_1, \omega)\right] e^{\mp jkz} = \tilde{V}_0 e^{\mp jkz}$$

regardless of the path taken from P_1 to P_2. So the voltage $\tilde{V}_{21}$ is a traveling wave. Moreover, the spatial dependence of the potential in the transverse plane is identical to that of the static potential. Hence we can define an absolute potential in the manner of the absolute static potential, such that the potential difference is the difference in absolute potentials evaluated at $\boldsymbol{\rho}_2$ and $\boldsymbol{\rho}_1$ with a reference potential chosen at a convenient point.

The voltage difference $\tilde{V}_0$ can be maintained by using a pair of perfect electric conductors with a z-invariant geometry (Figure 5.40). Observe that we *cannot* maintain the voltage difference with a single conductor, so a hollow-pipe guide cannot support a TEM mode. Extension of the voltage concept to a set of more than two conductors is left for the interested reader. By proper choice of reference potential, we can define the potential on conductor 1 to be a voltage $-\tilde{V}_0/2$, and on conductor 2 to be a voltage $\tilde{V}_0/2$, such that the difference is $\tilde{V}_0$. Then the electric field lines will emanate from conductor 1, which carries a positive charge, and terminate on conductor 2, which carries a negative charge. Because the fields are TEM, both the electric and magnetic fields lie entirely in the transverse plane.

There is an important relationship between the transmission line voltage and the current carried by each conductor. Assume for simplicity that the conductors are embedded in a lossless material so that $\tilde{\epsilon}^c = \epsilon$ and $\tilde{\mu} = \mu$ are both real. The current can be found by first computing the surface charge density on the conductors using the boundary condition on normal $\tilde{\mathbf{D}}$:

$$\tilde{\rho}_s(\mathbf{r}, \omega) = e^{\mp jkz} \hat{\mathbf{n}} \cdot \epsilon \tilde{\mathbf{e}}_t(\boldsymbol{\rho}, \omega), \qquad (5.302)$$

where ρ describes points on the surface of the conductors. Hence the charge is a traveling wave. Similarly, the surface current density on the surface of the conductors is

$$\tilde{\mathbf{J}}_s(\mathbf{r}, \omega) = e^{\mp jkz} \hat{\mathbf{n}} \times \tilde{\mathbf{h}}_t(\boldsymbol{\rho}, \omega) = \pm e^{\mp jkz} \hat{\mathbf{n}} \times \frac{\hat{\mathbf{z}} \times \tilde{\mathbf{e}}_t(\boldsymbol{\rho}, \omega)}{Z_{TEM}},$$

or

$$\tilde{\mathbf{J}}_s = \pm \hat{\mathbf{z}} e^{\mp jkz} \frac{\hat{\mathbf{n}} \cdot \tilde{\mathbf{e}}_t}{Z_{TEM}}.$$

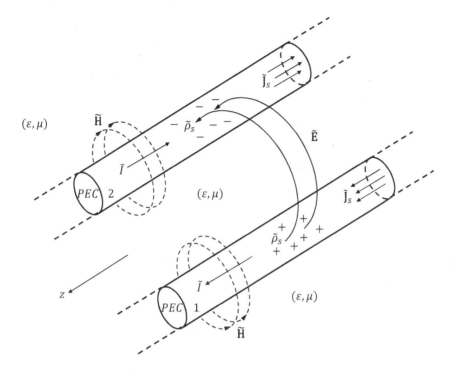

FIGURE 5.40
Two-conductor transmission line.

The current is also a traveling wave, and flows in the axial direction. The current flowing at some axial distance z can be found by integrating the surface current density. Using (5.302) and integrating around the conductor with the positive charge, we have

$$\tilde{I}(z,\omega) = \oint_\Gamma \hat{\mathbf{z}} \cdot \tilde{\mathbf{J}}_s(\mathbf{r},\omega)\, dl = \pm \frac{1}{\epsilon Z_{TEM}} \oint_\Gamma \tilde{\rho}_s(\mathbf{r},\omega)\, dl = \tilde{I}_0 e^{\mp jkz}.$$

Next examine Figure 5.41. The total charge lying on the surface of the positively charged conductor between $z_0 - \ell/2$ and $z_0 + \ell/2$ is

$$\tilde{Q} = \int_S \tilde{\rho}_s \, dS = \int_{z_0-\ell/2}^{z_0+\ell/2} \oint_\Gamma \tilde{\rho}_s \, dl \, dz.$$

Thus, for a wave traveling in the $+z$ direction,

$$\tilde{Q} = \epsilon Z_{TEM} \tilde{I}_0 \int_{z_0-\ell/2}^{z_0+\ell/2} e^{-jkz} \, dl = \epsilon Z_{TEM} \tilde{I}_0 e^{-jkz_0} \frac{2\sin\left(k\frac{\ell}{2}\right)}{k}.$$

Assuming $k\ell \ll 1$ and using $\sin x \approx x$ for small x, we have the charge per unit length at $z = z_0$:

$$\frac{\tilde{Q}}{\ell} = \epsilon Z_{TEM} \tilde{I}_0 e^{-jkz_0}.$$

With this we can compute the capacitance per unit length,

$$\mathcal{C} = \frac{C}{\ell} = \frac{\tilde{Q}/\ell}{\tilde{V}} = \epsilon Z_{TEM} \frac{\tilde{I}_0 e^{-jkz_0}}{\tilde{V}_0 e^{-jkz_0}} = \epsilon \eta \frac{\tilde{I}_0}{\tilde{V}_0} = \epsilon \frac{\eta}{R_c},$$

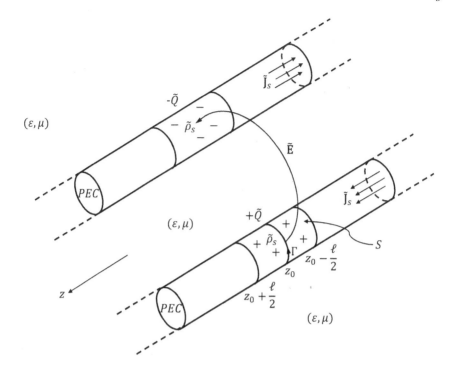

FIGURE 5.41
Calculation of the capacitance of a two-conductor transmission line.

and write the current as

$$\tilde{I} = \frac{\tilde{V}_0}{R_c} e^{\mp jkz}$$

where

$$R_c = \frac{\tilde{V}_0}{\tilde{I}_0} = \frac{\epsilon Z_{TEM}}{\mathcal{C}} = \frac{\epsilon \eta}{\mathcal{C}} \tag{5.303}$$

is the *characteristic resistance* of the transmission line.

Transmission lines have inductance as well as capacitance [39]. Consider Figure 5.42; we may think of the two conductors as forming a long loop carrying current $\tilde{I}(z) = \tilde{I}_0 e^{-jkz}$. The magnetic flux through surface S is

$$\tilde{\psi}_m = \int_{z_0-\ell/2}^{z_0+\ell/2} \int_{P_1}^{P_2} \mu \tilde{\mathbf{H}} \cdot \hat{\mathbf{n}} \, dS.$$

Use of (5.298) gives, for a wave traveling in the $+z$ direction,

$$\tilde{\psi}_m = \int_{z_0-\ell/2}^{z_0+\ell/2} \int_{P_1}^{P_2} \frac{\mu}{Z_{TEM}} \hat{\mathbf{n}} \cdot (\hat{\mathbf{z}} \times \tilde{\mathbf{e}}_t) e^{-jkz} \, dS = \frac{\mu}{Z_{TEM}} \int_{z_0-\ell/2}^{z_0+\ell/2} \left[\int_{P_1}^{P_2} -\tilde{\mathbf{e}}_t \cdot \mathbf{dl} \right] e^{-jkz} \, dz.$$

The inner integral is just the voltage, so

$$\tilde{\psi}_m = \frac{\mu \tilde{V}_0}{Z_{TEM}} \int_{z_0-\ell/2}^{z_0+\ell/2} e^{-jkz} \, dz = \frac{\mu \tilde{V}_0}{Z_{TEM}} e^{-jkz_0} \frac{2 \sin\left(k\frac{\ell}{2}\right)}{k}.$$

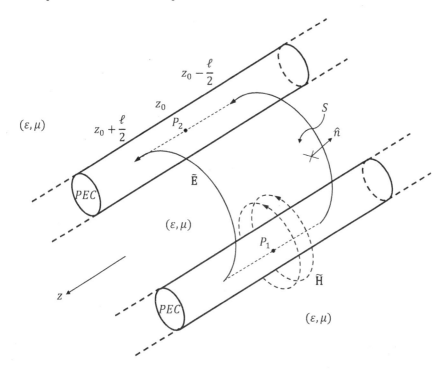

FIGURE 5.42
Calculation of the inductance of a two-conductor transmission line.

Assuming $k\ell << 1$ and using $\sin(x) \approx x$ for small x, we have the flux per unit length at $z = z_0$:

$$\frac{\tilde{\psi}_m}{\ell} = \frac{\mu}{Z_{TEM}} \tilde{V}_0 e^{-jkz_0}.$$

Using this, we compute the inductance per unit length:

$$\mathcal{L} = \frac{\tilde{\psi}_m/\ell}{\tilde{I}} = \frac{\mu}{Z_{TEM}} \frac{\tilde{V}_0 e^{-jkz_0}}{\tilde{I}_0 e^{-jkz_0}} = \frac{\mu}{\eta} R_c.$$

But, since $R_c = \epsilon\eta/\mathcal{C}$, we have

$$\mathcal{L}\mathcal{C} = \left(\frac{\mu}{\eta} R_c\right) \left(\frac{\epsilon\eta}{R_c}\right) = \mu\epsilon$$

and

$$\frac{\mathcal{L}}{\mathcal{C}} = \left(\frac{\mu}{\eta} R_c\right) \left(\frac{R_c}{\epsilon\eta}\right) = R_c^2$$

so that

$$R_c = \sqrt{\mathcal{L}/\mathcal{C}}. \tag{5.304}$$

5.6.3.3 Telegraphist's equations

With perfect conductors and lossless materials, the method used to compute capacitance and inductance for a transmission line is identical to that used in statics and quasistatics.

The formulas obtained in those sections can be used to determine the characteristic resistance of a transmission line. For lossy lines, the behaviors of the voltage and current are often described by a set of coupled differential equations called the *telegraphist's equations*. These may be derived from Maxwell's equations via some simplifying assumptions.

Suppose the conductors have conductivity $\tilde{\sigma}_c$ and are embedded in a material having frequency-independent permeability μ and complex permittivity $\tilde{\epsilon}^c$. Because the conductors are imperfect, the current will penetrate into the volume, and as a result of Ohm's law, $\tilde{\mathbf{J}} = \sigma\tilde{\mathbf{E}}$, the electric field inside the conductor will have a z-component. So the transmission line mode is not purely TEM in this circumstance and, in fact, can support radiation (see [104] for a rigorous development of transmission line theory based on Maxwell's equations). For simplicity we consider the addition of loss to only slightly perturb the ideal fields found with no loss and continue to regard the transmission-line mode as TEM.

Consider the path Γ shown in Figure 5.43, having segments along the surface of the conductors and transverse to the conductors. The large-scale form of Faraday's law reads

$$\oint_\Gamma \tilde{\mathbf{E}} \cdot \mathbf{dl} = -j\omega\tilde{\psi}_m.$$

The integrals transverse to the conductors give the voltages at z and $z + \Delta z$. For Δz small, the magnetic flux is approximately

$$\tilde{\psi}_m = \Delta z \left[\int_{P_1}^{P_2} \mu\tilde{\mathbf{H}} \cdot \mathbf{n} \right] = \Delta z \mathcal{L}_e \tilde{I}(z, \omega)$$

where $\mathcal{L}_e$ is the external inductance per unit length of the transmission line. The integrals along the surface of the conductors may be approximated using the planar surface impedance relationship (3.214). For conductor 1 we have

$$\int_{C_1} \tilde{\mathbf{E}} \cdot \mathbf{dl} = \tilde{I} \int_z^{z+\Delta z} \frac{1+j}{\delta\sigma_c w_1} \, dl = \Delta z\tilde{I}\mathcal{R}_1 + j\omega\Delta z\mathcal{L}_1\tilde{I},$$

while for conductor 2,

$$\int_{C_2} \tilde{\mathbf{E}} \cdot \mathbf{dl} = \Delta z\tilde{I}\mathcal{R}_2 + j\omega\Delta z\mathcal{L}_2\tilde{I}.$$

Here,

$$\mathcal{R}_n = \frac{1}{\sigma_c\delta w_n} \quad \text{and} \quad \mathcal{L}_n = \frac{1}{\omega\sigma_c\delta w_n}$$

are, respectively, the resistance and internal inductance per unit length of conductor n. Also, δ is the skin depth of the conductor and w_n is an appropriate width parameter depending on the cross-sectional geometry of conductor n. For circular wires of radius a, (3.225) reveals that $w = 2\pi a$ when $\delta \ll a$. As shown below, $\mathcal{R}$ may also be computed using power considerations. We now write Faraday's law as

$$\tilde{V}(z + \Delta z) - \tilde{V}(z) + \Delta z\tilde{I}(\mathcal{R}_1 + \mathcal{R}_2) = -j\omega\Delta z(\mathcal{L}_1 + \mathcal{L}_2 + \mathcal{L}_e)$$

or

$$\frac{\tilde{V}(z + \Delta z) - \tilde{V}(z)}{\Delta z} = -\tilde{I}(\mathcal{R} + j\omega\mathcal{L})$$

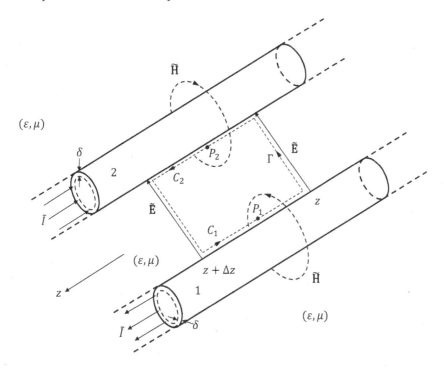

FIGURE 5.43
Diagram for derivation of the first telegraphist's equation.

where $\mathcal{R} = \mathcal{R}_1 + \mathcal{R}_2$ is the total resistance per unit length and $\mathcal{L} = \mathcal{L}_1 + \mathcal{L}_2 + \mathcal{L}_e$ is the total inductance per unit length of the transmission line. As $\Delta z \to 0$, the left side becomes a derivative so that

$$\frac{d\tilde{V}}{dz} = -\tilde{I}\mathcal{Z} \qquad (5.305)$$

where $\mathcal{Z} = \mathcal{R} + j\omega\mathcal{L}$ is the series impedance per unit length of the line. Equation (5.305) is the first telegraphist's equation.

The second telegraphist's equation follows from the large-scale form of Ampere's law. We start with

$$\nabla \times \tilde{\mathbf{H}} = \tilde{\mathbf{J}}^i + j\omega\tilde{\epsilon}^c\tilde{\mathbf{E}} = \tilde{\mathbf{J}}^i + j\omega(\mathrm{Re}\,\tilde{\epsilon}^c + j\,\mathrm{Im}\,\tilde{\epsilon}^c)\tilde{\mathbf{E}}$$

and take the divergence to get

$$\nabla \cdot \tilde{\mathbf{J}}^i = -j\omega(\mathrm{Re}\,\tilde{\epsilon}^c)\nabla \cdot \tilde{\mathbf{E}} + \omega(\mathrm{Im}\,\tilde{\epsilon}^c)\nabla \cdot \tilde{\mathbf{E}}.$$

Consider Figure 5.44. Integrating over the closed surface S consisting of the surfaces S_i for $i = 1, 2, 3$, we have

$$\oint_S \tilde{\mathbf{J}}^i \cdot \hat{\mathbf{n}}\,dS = -j\omega(\mathrm{Re}\,\tilde{\epsilon}^c)\oint_S \tilde{\mathbf{E}} \cdot \hat{\mathbf{n}}\,dS + \oint_S \omega\,\mathrm{Im}(\tilde{\epsilon}^c)\tilde{\mathbf{E}} \cdot \hat{\mathbf{n}}\,dS. \qquad (5.306)$$

The left side is the flux of the current through S, consisting of the current $\tilde{I}(z + \Delta z)$ flowing out through S_1 and the current $\tilde{I}(z)$ flowing in through S_2. Since we consider $\tilde{\mathbf{E}}$

to be predominantly transverse to z, the integrals on the right side only contribute over S_3. Noting that for the lossless transmission line

$$\oint_S \tilde{\mathbf{E}} \cdot \hat{\mathbf{n}}\, dS = \frac{1}{\epsilon} \oint_s \tilde{\mathbf{D}} \cdot \hat{\mathbf{n}}\, dS = \tilde{Q}/\epsilon,$$

and that the capacitance per unit length is

$$\mathcal{C} = \frac{\tilde{Q}/\tilde{V}}{\Delta z},$$

we have

$$\oint_S \tilde{\mathbf{E}} \cdot \hat{\mathbf{n}}\, dS = \frac{\Delta z \mathcal{C}}{\epsilon} \tilde{V}$$

and by (5.306)

$$\tilde{I}(z + \Delta z) - \tilde{I}(z) = \omega \frac{\operatorname{Im} \tilde{\epsilon}^c}{\operatorname{Re} \tilde{\epsilon}^c} \Delta z \mathcal{C} \tilde{V} - j\omega \Delta z \mathcal{C} \tilde{V}.$$

Division by Δz and the limit passage $\Delta z \to 0$ yield

$$\frac{d\tilde{I}}{dz} = -\tilde{V}(\mathcal{G} + j\omega\mathcal{C})$$

where

$$\mathcal{G} = -\omega \frac{\operatorname{Im} \tilde{\epsilon}^c}{\operatorname{Re} \tilde{\epsilon}^c} \mathcal{C} = \omega \tan \delta_c \mathcal{C} \tag{5.307}$$

is the conductance per unit length. Here $\tan \delta_c$ is the loss tangent of the material supporting the conductors. Note that $\mathcal{G}\tilde{V}$ is the shunt current per unit length flowing between the conductors through the lossy material in between. If the material is a dielectric with real permittivity ϵ and conductivity σ_d, then

$$\mathcal{G} = \frac{\sigma_d}{\epsilon}\mathcal{C}.$$

From this we have

$$\frac{\mathcal{C}}{\mathcal{G}} = \frac{\epsilon}{\sigma_d},$$

which is identical to the RC product relationship (3.181) found using the quasistatic definitions of resistance and capacitance. Finally, defining the shunt susceptance per unit length,

$$\mathcal{Y} = \mathcal{G} + j\omega\mathcal{C},$$

we have the second telegraphist's equation

$$\frac{d\tilde{I}}{dz} = -\tilde{V}\mathcal{Y}. \tag{5.308}$$

Equations (5.305) and (5.308) form a coupled system of differential equations that may be easily separated using substitution. The result is a set of wave equations,

$$\left(\frac{d^2}{dz^2} - \gamma^2\right) \left\{ \begin{array}{c} \tilde{V}(z) \\ \tilde{I}(z) \end{array} \right\} = 0,$$

where

$$\gamma = \sqrt{\mathcal{Y}\mathcal{Z}} = \sqrt{(\mathcal{R} + j\omega\mathcal{L})(\mathcal{G} + j\omega\mathcal{C})} = j\omega\sqrt{\mathcal{L}\mathcal{C}} \sqrt{1 - \frac{\mathcal{R}}{\omega\mathcal{L}}\frac{\mathcal{G}}{\omega\mathcal{C}} - j\left(\frac{\mathcal{R}}{\omega\mathcal{L}} + \frac{\mathcal{G}}{\omega\mathcal{C}}\right)} = \alpha + j\beta$$

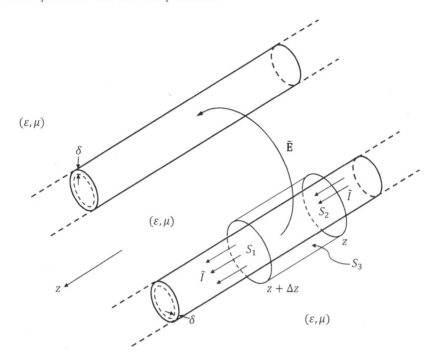

FIGURE 5.44

Diagram for derivation of the second telegraphist's equation.

is the *propagation constant* for the transmission line. Here $\alpha = \mathrm{Re}\{\gamma\}$ is the attenuation constant and $\beta = \mathrm{Im}\{\gamma\}$ is the phase constant of the traveling wave. As usual, we can use β to determine the phase and group velocities $v_p = \omega/\beta$ and $v_g = d\omega/d\beta$ along with the wavelength $\lambda = \omega/\beta$. Solution for the voltage gives

$$\tilde{V}(z) = \tilde{V}^+ e^{-\gamma z} + \tilde{V}^- e^{\gamma z},$$

a superposition of waves traveling in the $+z$ and $-z$ directions. Equation (5.305) yields the current in terms of the voltage:

$$\tilde{I}(z) = -\frac{1}{\mathcal{Z}}\frac{d\tilde{V}(z)}{dz} = \frac{1}{Z_c}[\tilde{V}^+ e^{-\gamma z} - \tilde{V}^- e^{\gamma z}]$$

where

$$Z_c = \sqrt{\frac{\mathcal{Z}}{\mathcal{Y}}} = \sqrt{\frac{\mathcal{R} + j\omega\mathcal{L}}{\mathcal{G} + j\omega\mathcal{C}}} = \sqrt{\frac{\mathcal{L}}{\mathcal{C}}}\sqrt{\frac{1 - j\frac{\mathcal{R}}{\omega\mathcal{L}}}{1 - j\frac{\mathcal{G}}{\omega\mathcal{C}}}}$$

is the *characteristic impedance* of the transmission line. Note that for a lossless line, where $\mathcal{R} = \mathcal{G} = 0$, we have

$$\gamma = j\omega\sqrt{\mathcal{L}\mathcal{C}} = j\omega\sqrt{\mu\epsilon} = jk, \qquad Z_c = \sqrt{\mathcal{L}/\mathcal{C}} = R_c.$$

These results match those obtained earlier for a lossless line.

Most transmission lines are designed with minimal losses (and hence minimal attenuation and dispersion) in mind. By definition, a *low-loss transmission line* has

$$\frac{\mathcal{R}}{\omega\mathcal{L}} \ll 1, \qquad \frac{\mathcal{G}}{\omega\mathcal{C}} \ll 1.$$

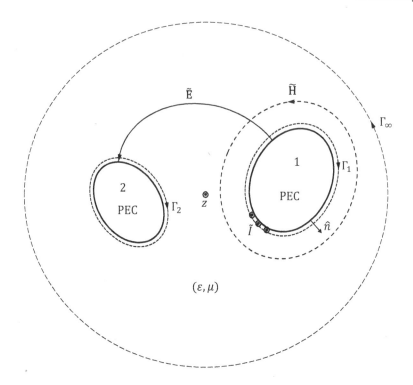

FIGURE 5.45
Cross-section of a lossless transmission line.

Under these conditions, a binomial series approximation gives

$$\gamma \approx j\omega\sqrt{\mathcal{LC}}\left[1 - j\frac{1}{2}\left(\frac{\mathcal{R}}{\omega\mathcal{L}} + \frac{\mathcal{G}}{\omega\mathcal{C}}\right)\right],$$

so that

$$\alpha \approx \frac{1}{2}\left(\frac{\mathcal{R}}{R_c} + \mathcal{G}R_c\right), \qquad \beta \approx j\omega\sqrt{\mathcal{LC}}. \tag{5.309}$$

Here $R_c = \sqrt{\mathcal{L}/\mathcal{C}}$ is the characteristic resistance of a lossless transmission line. To this order of approximation, the phase constant is identical to that of a lossless line.

5.6.3.4   Power carried by time-harmonic waves on lossless transmission lines

The power carried by a time-harmonic wave propagating on a transmission line is defined as the time-average Poynting flux passing through the cross-section:

$$P_{av} = \frac{1}{2}\int_{CS} \mathrm{Re}\left\{\check{\mathbf{E}} \times \check{\mathbf{H}}^*\right\} \cdot \hat{\mathbf{z}}\,dS.$$

For simplicity we assume the line is lossless; then

$$P_{av} = \frac{1}{2}\int_{CS} \mathrm{Re}\left\{\check{\mathbf{E}}_t \times \check{\mathbf{H}}_t^*\right\} \cdot \hat{\mathbf{z}}\,dS$$

since the fields are transverse to z. It is convenient to express this in terms of the transmission line voltage and current. Figure 5.45 shows the transverse (cross-sectional) plane at axial position z. The voltage difference between the conductors is $\check{V}_0$, with conductor 2 held at a reference zero potential. Using $\check{\mathbf{e}}_t = -\nabla\check{\Phi}$, and assuming the wave is propagating in the $+z$ direction, we have

$$P_{av} = -\frac{1}{2}\int_{CS} \mathrm{Re}\left\{e^{-jkz}\nabla\check{\Phi}\times\check{\mathbf{H}}_t^*\right\}\cdot\hat{\mathbf{z}}\,dS,$$

hence

$$P_{av} = -\frac{1}{2}\int_{CS}\mathrm{Re}\left\{e^{-jkz}\nabla\times(\check{\Phi}\check{\mathbf{H}}_t^*)\right\}\cdot\hat{\mathbf{z}}\,dS + \frac{1}{2}\int_{CS}\mathrm{Re}\left\{e^{-jkz}\check{\Phi}\nabla\times\check{\mathbf{H}}_t^*\right\}\cdot\hat{\mathbf{z}}\,dS$$

using (B.49). By Ampere's law, the curl of the magnetic field in the second integral is the volume current density flowing in the axial direction, which is zero as no axial conduction current flows in the material medium. Hence the second integral vanishes. Use of Stokes' theorem on the first integral gives

$$P_{av} = -\tfrac{1}{2}\,\mathrm{Re}\left\{e^{-jkz}\int_{\Gamma}\check{\Phi}\check{\mathbf{H}}_t^*\cdot\mathbf{dl}\right\}.$$

Because the potential decays to zero as Γ_∞ recedes to infinity, and since the potential is zero on Γ_2, only the contribution from Γ_1 survives, and on that contour the potential is $\check{V}_0$. Noting that $\mathbf{dl} = (\hat{\mathbf{n}}\times\hat{\mathbf{z}})\,dl$ we can write $\check{\mathbf{H}}_t^*\cdot\mathbf{dl} = \check{\mathbf{H}}_t^*\cdot(\hat{\mathbf{n}}\times\hat{\mathbf{z}})\,dl = -(\hat{\mathbf{n}}\times\check{\mathbf{H}}_t^*)\cdot\hat{\mathbf{z}}\,dl$:

$$P_{av} = \tfrac{1}{2}\,\mathrm{Re}\left\{\check{V}_0 e^{-jkz}\int_{\Gamma_1}\check{\mathbf{J}}^*\cdot\hat{\mathbf{z}}\,dl\right\}.$$

The integral is the complex conjugate of the axial current in the wire, so

$$P_{av} = \tfrac{1}{2}\,\mathrm{Re}\left\{\check{V}\check{I}^*\right\}$$

exactly as in circuit theory. In fact, since $\check{I} = \check{V}/R_c$ for a lossless line, we also have

$$P_{av} = \frac{|\check{V}|^2}{2R_c} = \tfrac{1}{2}|\check{I}|^2 R_c. \tag{5.310}$$

With this expression for time-average power, it is possible to find a formula for the resistance per unit length of a transmission line. The attenuation coefficient for a low-loss line with only conductor loss is, by (5.309),

$$\alpha = \frac{\mathcal{R}}{2R_c}.$$

Equating this with the perturbational formula for the attenuation constant (5.309), we have

$$\frac{\mathcal{R}}{2R_c} = \frac{P_d(0)/l}{2P_0}.$$

Then by (5.310),

$$\mathcal{R} = \frac{2P_d(0)/l}{|\check{I}|^2} = \frac{2P_d(0)/l}{|\check{V}|^2}R_c^2. \tag{5.311}$$

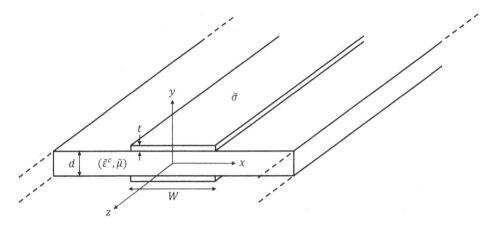

FIGURE 5.46
Strip transmission line.

5.6.3.5   Example: the strip (parallel-plate) transmission line

A simple transmission line can be made from two long parallel conducting plates of width W, thickness t, and separation d, as shown in Figure 5.46. The plates have conductivity $\tilde{\sigma}$ and are embedded in a medium with permeability $\tilde{\mu}$ and complex permittivity $\tilde{\epsilon}^c$. A voltage difference $\tilde{V}_0$ is placed across the conductors.

If $W/d \gg 1$, we can neglect fringing with little error in the resulting transmission line parameters. We first determine the fields between the plates, assuming the latter are perfectly conducting. Neglecting fringing, the potential is approximately $\tilde{\Phi}(\mathbf{r},\omega) = \tilde{\Phi}(y,\omega)$ and, hence, by (5.299),

$$\nabla_t^2 \tilde{\Phi} = \frac{\partial^2}{\partial y^2}\tilde{\Phi} = 0.$$

Two integrations yield $\tilde{\Phi} = C_1 y + C_2$. Evaluating the constants through the boundary values $\tilde{\Phi}(y=0) = \tilde{V}_0$ and $\tilde{\Phi}(y=d) = 0$, we get

$$\tilde{\Phi} = \tilde{V}_0 \left(1 - \frac{y}{d}\right).$$

By (5.297) and (5.298) the fields are

$$\tilde{\mathbf{e}}_t = \hat{\mathbf{y}}\frac{\tilde{V}_0}{d}, \qquad \tilde{\mathbf{h}}_t = -\hat{\mathbf{x}}\frac{\tilde{V}_0}{\eta d}.$$

The capacitance per unit length of the parallel plate transmission line can be found by assuming that the material in which the plates are embedded is lossless with permittivity ϵ and permeability μ, and computing the charge per unit length. For a wave traveling in the $+z$ direction, the charge density on the lower conductor is

$$\tilde{\rho}_s = \epsilon\hat{\mathbf{y}} \cdot \tilde{\mathbf{E}} = \frac{\epsilon\tilde{V}_0}{d}e^{-jkz}.$$

Integration gives

$$\frac{\tilde{Q}}{\ell} = \int_{-W/2}^{W/2} \tilde{\rho}_s \, dx = \frac{W\epsilon\tilde{V}_0}{d}e^{-jkz}.$$

Thus, the capacitance per unit length is

$$\mathcal{C} = \frac{\tilde{Q}/\ell}{\tilde{V}} = \frac{\frac{W\epsilon\tilde{V}_0}{d}e^{-jkz}}{\tilde{V}_0 e^{-jkz}} = \epsilon W/d.$$

From (5.303) we also have the characteristic resistance of the transmission line,

$$R_c = \epsilon\eta/\mathcal{C} = \eta d/W.$$

The external inductance per unit length can be found using (5.304):

$$\mathcal{L}_e = \mathcal{C}R_c^2 = \mu d/W.$$

To account for losses, we compute the resistance, conductance, and internal inductance per unit length. By (5.307), the conductance per unit length is

$$\mathcal{G} = -\omega\frac{\operatorname{Im}\tilde{\epsilon}^c}{\operatorname{Re}\tilde{\epsilon}^c}\mathcal{C} = W\omega\tan\delta/d.$$

The internal inductance per unit length for each conductor is found by assuming that $t \gg \delta$ and using (3.217):

$$\mathcal{L}_i = \frac{1}{W\omega\tilde{\sigma}\delta}.$$

Since the inductance of the upper plate is identical, the total internal inductance is

$$\mathcal{L}_i = \frac{2}{W\omega\tilde{\sigma}\delta}.$$

To find the resistance per unit length, we first use (5.215) to compute the power dissipated in the conductors. For the bottom conductor we have

$$\frac{P_d(z)}{l} = \frac{R_s}{2}\int_{-W/2}^{W/2}\frac{|\check{V}_0|^2}{\eta^2 d^2}\,dx = \frac{R_s}{2}W\frac{|\check{V}_0|^2}{\eta^2 d^2},$$

hence, for both conductors,

$$\frac{P_d(z)}{l} = R_s W\frac{|\check{V}_0|^2}{\eta^2 d^2}.$$

This and (5.311) yield the resistance per unit length:

$$\mathcal{R} = 2R_s/W.$$

The simple relationship

$$\mathcal{R}/\mathcal{L}_i = R_s\omega\tilde{\sigma}\delta = \omega$$

seen here holds for all transmission lines involving conductors much thicker than a skin depth [156].

▶ **Example 5.34**: Characteristics of a strip transmission line

Design a strip transmission line using Rogers RT/duroid® 6002 high-frequency laminate. The strip is etched onto the top and bottom of the laminate with a width chosen to produce a characteristic resistance of $50\,\Omega$. The properties of the laminate are $\epsilon_r = 2.94$, $\tan\delta = 0.0012$, and $d = 0.02$ inches (0.508 mm). The weight of the copper cladding from which the strips are etched is $1\,\mathrm{oz/ft^2}$ ($t = 0.35\,\mu\mathrm{m}$). Determine the width W of the strip that will produce a $50\,\Omega$ characteristic resistance. Determine all the relevant characteristics of this transmission line at $f = 10$ GHz.

Solution: The characteristic resistance of the line is $R_c = \eta d/W = 50\,\Omega$. Using $\eta =$

$\sqrt{\mu_0/(\epsilon_0\epsilon_r)} = 219.7\,\Omega$ we find that $W = 2.232$ mm and therefore $W/d = 4.394$. Although this ratio is not exceptionally large compared to unity, it should be large enough that fringing can be neglected without introducing significant error.

The capacitance per unit length of the lossless line is $\mathcal{C} = \epsilon W/d = 114.4$ pF/m, so the external inductance per unit length is $\mathcal{L}_e = \mathcal{C}R_c^2 = 0.286\,\mu$H/m. These quantities are all frequency independent. The remaining quantities depend on frequency. The skin depth at 10 GHz is

$$\delta = \frac{1}{\sqrt{\pi f \mu_0 \sigma}} = 0.6609 \times 10^{-6} \text{ m},$$

making $t/\delta = 52.96$. Since the conductor is many skin depths thick, the resistance and internal inductance formulas should be fairly accurate. With the surface resistance value

$$R_s = \frac{1}{\sigma\delta} = 0.0261\ \Omega/\square,$$

the resistance per unit length and internal inductance per unit length are

$$\mathcal{R} = 2R_s/W = 23.4\ \Omega/\text{m}, \qquad \mathcal{L}_i = \mathcal{R}/\omega = 3.72 \times 10^{-4}\,\mu\text{H/m}.$$

Because it is much smaller than the external inductance, the internal inductance is often ignored. The conductance per unit length is

$$\mathcal{G} = \omega\mathcal{C}\tan\delta = 8.62 \times 10^{-3} \text{ S/m}.$$

Now we can check whether the line is low loss by computing

$$\frac{\mathcal{R}}{\omega\mathcal{L}} = 1.30 \times 10^{-3}, \qquad \frac{\mathcal{G}}{\omega\mathcal{C}} = 1.20 \times 10^{-3},$$

and noting that ratios are small compared to unity. It is interesting to compute the characteristic impedance

$$Z_c = \sqrt{\frac{\mathcal{R} + j\omega\mathcal{L}}{\mathcal{G} + j\omega\mathcal{C}}} = 50.03 - j0.002482\ \Omega$$

and see that the lossless value of $R_c = 50\ \Omega$ matches the result when losses are included (to three significant figures). The small imaginary part (in this case a small fraction of an ohm) is usually ignored. The propagation constant is

$$\gamma = \sqrt{(\mathcal{R} + j\omega\mathcal{L})(\mathcal{G} + j\omega\mathcal{C})} = 0.44935 + j359.60 \text{ m}^{-1}.$$

Hence $\alpha = 0.44935$ Np/m and $\beta = 359.60$ r/m. For this low-loss line we can use (5.309) to find the attenuation and phase constants:

$$\alpha_c = \frac{1}{2}\frac{\mathcal{R}}{R_c} = 0.23375 \text{ Np/m},$$

$$\alpha_d = \frac{1}{2}\mathcal{G}R_c = 0.21562 \text{ Np/m},$$

$$\beta = \omega\sqrt{\mathcal{L}\mathcal{C}} = 359.60 \text{ r/m}.$$

Note that α_c and α_d are comparable in magnitude; this contrasts with the case of the coaxial transmission line considered in Example 5.34 where $\alpha_c \gg \alpha_d$ (albeit for a different frequency and dielectric). The sum $\alpha_d + \alpha_c = 0.44937$ matches well with the value of α found by computing γ. Also, the lossless value of β matches the value found by computing γ to more than five significant figures. Often the attenuation is given in terms of dB per 100 meters. The conversion formula

$$\text{atten} = -20\log_{10}e^{-\alpha d} = 8.686\alpha d$$

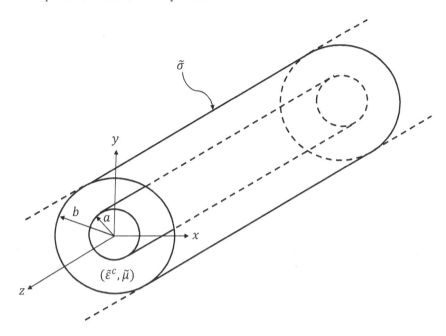

FIGURE 5.47
Coaxial cable transmission line.

with $d = 100$ m yields atten $= 868.6\alpha = 390.3$ dB.

Lastly, we can compute the velocity of the transmission line wave relative to the speed of light in vacuum: $v_p/c = (\omega/\beta)/c = 0.583$. ◄

5.6.3.6   Example: the coaxial transmission line

As a more complicated example, consider a coaxial cable consisting of a solid inner conductor and an outer conducting tube (Figure 5.47). The conductors have conductivity $\tilde{\sigma}$, and the intervening material has permeability $\tilde{\mu}$ and complex permittivity $\tilde{\epsilon}^c$. There is a voltage difference $\tilde{V}_0$ between the conductors.

We start with the fields inside the cable, assuming perfectly conducting walls. By symmetry, $\tilde{\Phi}(\boldsymbol{\rho}, \omega) = \tilde{\Phi}(\rho, \omega)$, and, hence, by (5.299),

$$\nabla_t^2 \tilde{\Phi} = \frac{1}{\rho}\frac{\partial}{\partial\rho}\left(\rho\frac{\partial\tilde{\Phi}}{\partial\rho}\right) = 0.$$

Successive integrations yield $\tilde{\Phi} = C_1 \ln\rho + C_2$; application of the boundary conditions $\tilde{\Phi}(\rho = a) = \tilde{V}_0$ and $\tilde{\Phi}(\rho = b) = 0$ yields

$$\tilde{\Phi} = -\tilde{V}_0 \frac{\ln(\rho/b)}{\ln(b/a)}.$$

By (5.297) and (5.298) the fields are

$$\tilde{\mathbf{e}}_t = \hat{\boldsymbol{\rho}}\frac{\tilde{V}_0}{\rho\ln(b/a)}, \qquad \tilde{\mathbf{h}}_t = \hat{\boldsymbol{\phi}}\frac{\tilde{V}_0}{\eta\rho\ln(b/a)}.$$

The capacitance per unit length can be found by assuming that the material filling the cable is lossless with permittivity ϵ and permeability μ, and computing the charge per unit length. For a wave traveling in the $+z$ direction, the charge density on the inner conductor is

$$\tilde{\rho}_s = \epsilon \hat{\boldsymbol{\rho}} \cdot \tilde{\mathbf{E}} = \frac{\epsilon \tilde{V}_0}{a \ln(b/a)} e^{-jkz}.$$

Integration gives

$$\frac{\tilde{Q}}{\ell} = \int_0^{2\pi} \tilde{\rho}_s a \, d\phi = \frac{2\pi \epsilon \tilde{V}_0}{\ln(b/a)} e^{-jkz}.$$

So the capacitance per unit length is

$$\mathcal{C} = \frac{\tilde{Q}/\ell}{\tilde{V}} = \frac{\frac{2\pi \epsilon \tilde{V}_0}{\ln(b/a)} e^{-jkz}}{\tilde{V}_0 e^{-jkz}} = \frac{2\pi \epsilon}{\ln(b/a)}.$$

From (5.303) we obtain the characteristic resistance of the coaxial cable:

$$R_c = \frac{\epsilon \eta}{\mathcal{C}} = \frac{\eta}{2\pi} \ln(b/a). \tag{5.312}$$

The external inductance per unit length can be found using (5.304):

$$\mathcal{L}_e = \mathcal{C} R_c^2 = \frac{\mu}{2\pi} \ln(b/a).$$

To account for losses in the cable, we compute the resistance, conductance, and internal inductance per unit length. By (5.307), the conductance per unit length is simply

$$\mathcal{G} = \omega \mathcal{C} \tan \delta = -\frac{2\pi \omega \operatorname{Im} \tilde{\epsilon}^c}{\ln(b/a)}.$$

The internal inductance per unit length for the inner conductor is found by assuming that $a \gg \delta$, and using (3.226) from Example 3.57:

$$\mathcal{L}_i = \frac{\mu}{4\pi} \frac{\delta}{a}.$$

We may use the same formula for the outer conductor by replacing a with b. Thus, the total internal inductance is

$$\mathcal{L}_i = \frac{\mu}{4\pi} \frac{\delta}{a} \left(1 + \frac{a}{b}\right).$$

To find the resistance per unit length, we first compute the power dissipated in the conductors using (5.215). For the conductor at $\rho = a$ we have

$$\frac{P_d(z)}{l} = \frac{R_s}{2} \int_0^{2\pi} \frac{|\check{V}_0|^2}{\eta^2 a^2 \ln^2(b/a)} a \, d\phi = \pi R_s \frac{|\check{V}_0|^2}{\eta^2 a \ln^2(b/a)}.$$

Repeating for $\rho = b$ and adding gives the total power dissipated in both conductors as

$$\frac{P_d(z)}{l} = \pi R_s \frac{|\check{V}_0|^2}{\eta^2 \ln^2(b/a)} \left(\frac{1}{a} + \frac{1}{b}\right).$$

Substitution into (5.311) then gives the resistance per unit length as

$$\mathcal{R} = \frac{R_s}{2\pi a} \left(1 + \frac{a}{b}\right).$$

Note that the ratio of resistance per unit length to internal inductance per unit length is

$$\frac{\mathcal{R}}{\mathcal{L}_i} = \frac{2R_s}{\delta\mu} = \omega.$$

This simple relation holds for any transmission line having conductors much thicker than a skin depth [156].

▶ **Example 5.35:** Characteristics of RG-405/U coaxial cable

RG-405/U coaxial cable has a solid inner conductor of radius $a = 0.254$ mm and a solid outer conductor of inner radius $b = 0.7874$ mm and thickness $t = 0.5$ mm. The conductors are copper with $\sigma = 5.8 \times 10^7$ S/m. The material between the conductors is Teflon with dielectric constant $\epsilon_r = 2.05$, loss tangent $\tan\delta = 2 \times 10^{-4}$, and permeability μ_0. Determine all the relevant characteristics of this transmission line at $f = 1$ GHz.

Solution: The characteristic resistance and capacitance per unit length of the lossless line are

$$R_c = \frac{\eta}{2\pi}\ln(b/a) = 47.4 \ \Omega, \qquad \mathcal{C} = \frac{2\pi\epsilon_r\epsilon_0}{\ln(b/a)} = 100.8 \text{ pF/m},$$

so the external inductance per unit length is

$$\mathcal{L}_e = \mathcal{C}R_c^2 = 0.226 \ \mu\text{H/m}.$$

These quantities are frequency independent. The remaining quantities depend on frequency. The skin depth at 1 GHz is

$$\delta = \frac{1}{\sqrt{\pi f \mu_0 \sigma}} = 2.09 \times 10^{-6} \text{ m}.$$

Hence the normalized radius of the inner conductor and thickness of the outer conductor are $a/\delta = 121.4$ and $t/\delta = 23.9$. Both conductors are many skin depths thick; the resistance and internal inductance formulas should be fairly accurate. The surface resistance is

$$R_s = \frac{1}{\sigma\delta} = 8.25 \times 10^{-3} \ \Omega/\square,$$

so the resistance and internal inductance per unit length are

$$\mathcal{R} = \frac{R_s}{2\pi a}\left(1 + \frac{a}{b}\right) = 6.84 \ \Omega/\text{m},$$

$$\mathcal{L}_i = \frac{\mathcal{R}}{\omega} = 1.09 \times 10^{-3} \mu\text{H/m}.$$

Much smaller than the external inductance, the internal inductance is often ignored. The conductance per unit length is

$$\mathcal{G} = \omega\mathcal{C}\tan\delta = 1.27 \times 10^{-4} \text{ S/m}.$$

At this point we can verify that the line is low loss by computing

$$\frac{\mathcal{R}}{\omega\mathcal{L}} = 4.79 \times 10^{-3}, \qquad \frac{\mathcal{G}}{\omega\mathcal{C}} = 2 \times 10^{-4}$$

and noting that these ratios are small compared to unity. The characteristic impedance is

$$Z_c = \sqrt{\frac{\mathcal{R} + j\omega\mathcal{L}}{\mathcal{G} + j\omega\mathcal{C}}} = 47.5 - j0.109 \ \Omega.$$

The lossless value of $R_c = 47.4 \ \Omega$ is thus very close to the result when losses are included. The small imaginary part (in this case one tenth of an ohm) is usually ignored. The propagation constant is

$$\gamma = \sqrt{(\mathcal{R} + j\omega\mathcal{L})(\mathcal{G} + j\omega\mathcal{C})} = 7.50 \times 10^{-2} + j30.1 \ \mathrm{m}^{-1}$$

so that $\alpha = 7.50 \times 10^{-2}$ Np/m and $\beta = 30.1$ r/m. For this low-loss line, we can also compute the attenuation and phase constants using (5.309):

$$\alpha_c = \frac{1}{2}\frac{\mathcal{R}}{R_c} = 7.22 \times 10^{-2} \ \mathrm{Np/m},$$

$$\alpha_d = \frac{1}{2}\mathcal{G}R_c = 3.00 \times 10^{-3} \ \mathrm{Np/m},$$

$$\beta = \omega\sqrt{\mathcal{L}\mathcal{C}} = 30.1 \ \mathrm{r/m}.$$

It is clear that conductor losses far exceed dielectric losses. The total of α_c and α_d closely matches the value of α found by computing γ. Moreover, the lossless value of β is close to the value found by computing γ. Often the attenuation is expressed in dB per 100 meters. The conversion is

$$\mathrm{atten} = -20\log_{10} e^{-\alpha d} = 8.686\alpha d$$

where $d = 100$ m; in other words, atten $= 868.6\alpha = 65.1$ dB.

Lastly, we can compute the velocity of the transmission line wave relative to the speed of light in vacuum using

$$\frac{v_p}{c} = \frac{\omega}{\beta c} = 0.697. \ \blacktriangleleft$$

5.6.4   Open-boundary axial waveguides

Forming an important class of open-boundary waveguides, *optical guides* employ total internal reflection to guide a wave in a purely dielectric structure or a combined dielectric/metallic structure with a dielectric boundary open to the surrounding region. Important examples include planar integrated optical guides, which are important components of optical communications and computation circuitry, and fiber optical cables, which are used for long-haul low-loss digital and analog communications.

Open-boundary guides have both discrete and continuous eigenvalue spectra, with the continuous spectrum describing radiation from the guiding structure. We will consider the discrete spectra describing guided-wave modes, investigating the conditions under which these modes can be supported. We will find that these modes are characterized by fields that travel axially and decay (are evanescent) in the transverse direction, and are thus concentrated within the guiding structure and externally very near its surface. This field concentration remains true even when the structures are bent or curved (although radiation can result). Because the waves remain "attached" to the surface profile, they are called *surface waves*.

Two simple examples of optical guides, the slab waveguide and the uniform optical fiber, are considered below.

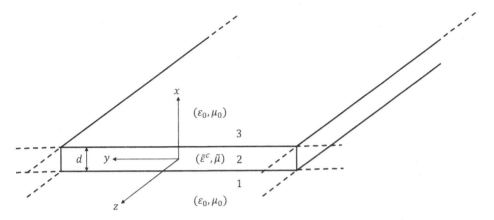

FIGURE 5.48
Slab waveguide.

5.6.4.1 Example: the symmetric slab waveguide

The slab waveguide is the simplest planar integrated optical waveguide. As shown in Figure 5.48, it consists of a material slab with permittivity $\tilde{\epsilon}^c$ and permeability $\tilde{\mu}$, surrounded by free space. Since both the region above (the *superstrate*) and the region below (the *substrate*) have the same properties, the guide is called a *symmetric* slab. The important case where the substrate and superstrate differ is more complicated to analyze, but the procedure for doing so is analogous with that used for the symmetric guide, and is left as an exercise.

The slab occupies the region $-d/2 \leq x \leq d/2$. We seek y-invariant solutions to Maxwell's equations representing waves traveling along the z direction. The boundary conditions can be satisfied by fields purely TE or TM to the z-direction, and thus we may use the potential (5.217) obtained from separating variables in rectangular coordinates. By the y-invariance we have

$$\tilde{\psi}(x, y, \omega) = C_1 \sin k_x x + C_2 \cos k_x x$$

where $k_x^2 = k^2 - k_z^2$. In regions 1 and 3, $k = k_0 = \omega\sqrt{\mu_0 \epsilon_0}$; in region 2, $k = \omega\sqrt{\tilde{\mu}\tilde{\epsilon}^c}$. Since k differs between the slab region and the surround, so does k_x. However, to satisfy the boundary conditions k_z must be the same in each region. Indeed, we compute wave velocity from k_z and expect the wave to propagate identically in each region. So we write $k_0^2 = k_{x0}^2 + k_z^2$ in regions 1 and 3, and $k^2 = k_x^2 + k_z^2$ in region 2.

Symmetry about $x = 0$ leads us to expect that fields with even or odd symmetry will satisfy the boundary conditions independently. Therefore we consider even and odd modes according to the symmetry of the potential function $\tilde{\psi}$. Also, since the waves propagate in the slab by total internal reflection, we expect standing waves in the slab along the x-direction. The fields outside the slab should be evanescent along x. The condition for evanescence in a lossless guide is $k_0 < k_z$, such that k_{x0} becomes an imaginary value $j\tau$ and in regions 1 and 3 we can write

$$k_0^2 = k_z^2 - \tau^2, \tag{5.313}$$

while in region 2

$$k^2 = k_z^2 + k_x^2. \tag{5.314}$$

In this case the trigonometric functions can be replaced by exponentials that decay away from the slab. Hence for even modes the potential functions in each region are

$$\tilde{\psi}_1^e(x) = Ae^{\tau x}, \qquad \tilde{\psi}_2^e(x) = B\cos k_x x, \qquad \tilde{\psi}_3^e(x) = Ae^{-\tau x}.$$

Clearly these are even about $x = 0$. For odd modes,

$$\tilde{\psi}_1^o(x) = -Ae^{\tau x}, \qquad \tilde{\psi}_2^o(x) = B\sin k_x x, \qquad \tilde{\psi}_3^o(x) = Ae^{-\tau x},$$

which are odd about $x = 0$.

For modes that are TE_z, relation (5.198) yields the axial magnetic fields:

$$\tilde{h}_{z1}^e = -A\tau^2 e^{\tau x}, \qquad\qquad \tilde{h}_{z1}^o = A\tau^2 e^{\tau x},$$
$$\tilde{h}_{z2}^e = Bk_x^2 \cos k_x x, \qquad\qquad \tilde{h}_{z2}^o = Bk_x^2 \sin k_x x,$$
$$\tilde{h}_{z3}^e = -A\tau^2 e^{-\tau x}, \qquad\qquad \tilde{h}_{z3}^o = -A\tau^2 e^{-\tau x}.$$

Relation (5.199) yields the transverse magnetic fields,

$$\tilde{h}_{x1}^e = -Ajk_z\tau e^{\tau x}, \qquad\qquad \tilde{h}_{x1}^o = Ajk_z\tau e^{\tau x},$$
$$\tilde{h}_{x2}^e = Bjk_x k_z \sin k_x x, \qquad\qquad \tilde{h}_{x2}^o = -Bjk_x k_z \cos k_x x,$$
$$\tilde{h}_{x3}^e = Ajk_z\tau e^{-\tau x}, \qquad\qquad \tilde{h}_{x3}^o = Ajk_z\tau e^{-\tau x},$$

and (5.200) yields the transverse electric fields:

$$\tilde{e}_{y1}^e = Aj\tau\omega\mu_0 e^{\tau x}, \qquad\qquad \tilde{e}_{y1}^o = -Aj\tau\omega\mu_0 e^{\tau x},$$
$$\tilde{e}_{y2}^e = -Bjk_x\omega\tilde{\mu}\sin k_x x, \qquad\qquad \tilde{e}_{y2}^o = Bjk_x\omega\tilde{\mu}\cos k_x x,$$
$$\tilde{e}_{y3}^e = -Aj\tau\omega\mu_0 e^{-\tau x}, \qquad\qquad \tilde{e}_{y3}^o = -Aj\tau\omega\mu_0 e^{-\tau x}.$$

Note that for an even potential function, $\tilde{H}_z$ is even but $\tilde{E}_y$ is odd. For an odd potential function, $\tilde{H}_z$ is odd but $\tilde{E}_y$ is even.

Constants A, B and the propagation constant k_z are determined from the boundary conditions at $x = \pm d/2$. By symmetry, enforcement of these conditions at $x = d/2$ will suffice. Continuity of $\tilde{H}_z$ and $\tilde{E}_y$ at $x = d/2$ shows that for even TE modes,

$$Bk_x^2 \cos k_x \frac{d}{2} = -A\tau^2 e^{-\tau d/2},$$
$$Bk_x\tilde{\mu}\sin k_x \frac{d}{2} = A\tau\mu_0 e^{-\tau d/2}, \qquad\qquad (5.315)$$

and for odd TE modes,

$$Bk_x^2 \sin k_x \frac{d}{2} = -A\tau^2 e^{-\tau d/2},$$
$$Bk_x\tilde{\mu}\cos k_x \frac{d}{2} = -A\tau\mu_0 e^{-\tau d/2}.$$

Equating the determinant of the coefficient matrix to zero, we obtain a characteristic equation relating k_x and τ. For even TE modes we have

$$\cot k_x \frac{d}{2} = -\frac{\tilde{\mu}}{\mu_0}\frac{\tau}{k_x},$$

while for odd TE modes,

$$\tan k_x \frac{d}{2} = \frac{\tilde{\mu}}{\mu_0} \frac{\tau}{k_x}.$$

Subtracting (5.313) from (5.314) gives a second relationship between k_x and τ:

$$k_x^2 + \tau^2 = k^2 - k_0^2. \tag{5.316}$$

Simultaneous solution of this and the characteristic equation determines k_x and τ, and thus k_z.

For modes that are TM$_z$, relation (5.195) yields the axial electric fields

$$\tilde{e}_{z1}^e = -A\tau^2 e^{\tau x}, \qquad\qquad \tilde{e}_{z1}^o = A\tau^2 e^{\tau x},$$
$$\tilde{e}_{z2}^e = Bk_x^2 \cos k_x x, \qquad\qquad \tilde{e}_{z2}^o = Bk_x^2 \sin k_x x,$$
$$\tilde{e}_{z3}^e = -A\tau^2 e^{-\tau x}, \qquad\qquad \tilde{e}_{z3}^o = -A\tau^2 e^{-\tau x},$$

relation (5.196) gives the transverse electric fields

$$\tilde{e}_{x1}^e = -Ajk_z\tau e^{\tau x}, \qquad\qquad \tilde{e}_{x1}^o = Ajk_z\tau e^{\tau x},$$
$$\tilde{e}_{x2}^e = Bjk_xk_z \sin k_x x, \qquad\qquad \tilde{e}_{x2}^o = -Bjk_xk_z \cos k_x x,$$
$$\tilde{e}_{x3}^e = Ajk_z\tau e^{-\tau x}, \qquad\qquad \tilde{e}_{x3}^o = Ajk_z\tau e^{-\tau x},$$

and relation (5.197) gives the transverse magnetic fields

$$\tilde{h}_{y1}^e = -Aj\tau\omega\epsilon_0 e^{\tau x}, \qquad\qquad \tilde{h}_{y1}^o = Aj\tau\omega\epsilon_0 e^{\tau x},$$
$$\tilde{h}_{y2}^e = Bjk_x\omega\tilde{\epsilon}^c \sin k_x x, \qquad\qquad \tilde{h}_{y2}^o = -Bjk_x\omega\tilde{\epsilon}^c \cos k_x x,$$
$$\tilde{h}_{y3}^e = Aj\tau\omega\epsilon_0 e^{-\tau x}, \qquad\qquad \tilde{h}_{y3}^o = Aj\tau\omega\epsilon_0 e^{-\tau x}.$$

For an even potential function, $\tilde{E}_z$ is even and $\tilde{H}_y$ is odd. For an odd potential function, $\tilde{E}_z$ is odd and $\tilde{H}_y$ is even.

Continuity of $\tilde{E}_z$ and $\tilde{H}_y$ at $x = d/2$ imposes two conditions on A and B. For even TM modes we have

$$Bk_x^2 \cos k_x \frac{d}{2} = -A\tau^2 e^{-\tau d/2}, \qquad Bk_x\tilde{\epsilon}^c \sin k_x \frac{d}{2} = A\tau\epsilon_0 e^{-\tau d/2},$$

and for odd TM modes,

$$Bk_x^2 \sin k_x \frac{d}{2} = -A\tau^2 e^{-\tau d/2}, \qquad -Bk_x\tilde{\epsilon}^c \cos k_x \frac{d}{2} = A\tau\epsilon_0 e^{-\tau d/2}.$$

Setting the determinant of the coefficient matrix to zero gives a characteristic equation relating k_x and τ. For even TM modes this reads

$$\cot k_x \frac{d}{2} = -\frac{\tilde{\epsilon}^c}{\epsilon_0} \frac{\tau}{k_x},$$

while for odd TM modes it reads

$$\tan k_x \frac{d}{2} = \frac{\tilde{\epsilon}^c}{\epsilon_0} \frac{\tau}{k_x}.$$

Solving the characteristic equation simultaneously with (5.316), we get k_x, τ, and thus k_z.

The cutoff frequency of a lossless slab waveguide is the lowest frequency at which the waves are guided by total internal reflection without radiation. Since $\tau = \sqrt{k_z^2 - k_0^2}$ must be real for the external fields to be evanescent, cutoff occurs when $\tau = 0$, or equivalently $k_z = k_0$. For even modes (both TE and TM) this requires

$$\cot k_x \frac{d}{2} = 0,$$

and for odd modes

$$\tan k_x \frac{d}{2} = 0.$$

Hence $k_x d/2 = n\pi/2$, where n is odd for even modes and even for odd modes. But when $\tau = 0$ we also have from (5.316) that

$$k_x = \sqrt{k^2 - k_0^2} = k_0 \sqrt{\epsilon_r \mu_r - 1}.$$

With $k_0 = 2\pi f_c/c$, the cutoff frequencies are given by

$$f_c = \frac{nc}{2d\sqrt{\epsilon_r \mu_r - 1}} \tag{5.317}$$

where $n = 1, 2, 3, \ldots$ for even modes and $n = 0, 2, 4, \ldots$ for odd modes.

It is clear that in a lossless guide, the dominant modes are the first odd modes, which have $n = 0$, designated TE$_0$ and TM$_0$. Both have $f_c = 0$ and therefore propagate to zero frequency.

▶ **Example 5.36:** Mode diagram for a circuit board substrate

A circuit board made with an FR-4 substrate comprises a slab of material with thickness 2.4 mm, relative permittivity $\epsilon_r = 4.8$, and relative permeability $\mu_r = 1$. Assume the substrate is lossless and resides in free space. Plot the ω-β curves for several slab modes. Also plot phase velocity as a function of frequency for the TE$_0$ and TM$_0$ modes.

Solution: The ω-β curves may be generated by solving the characteristic equations for the odd and even TE and TM modes. This can be done numerically using a root-search algorithm. It is straightforward to fix the frequency ω and search for values $k_z = \beta$ that satisfy the equations. As $k_0 \leq \beta \leq k$ for propagating modes, we can restrict our search to that regime. Figure 5.49 shows the curves for the first four sets of modes. The dashed lines correspond to the phase velocity of waves propagating in unbounded free space (where $\beta = k_0$) and in an unbounded material (where $\beta = k$). The first two odd modes ($n = 0$) clearly have no lower cutoff frequency. All higher-order modes have nonzero cutoff frequencies given by (5.317); the first three are 32.04 GHz, 64.08 GHz, and 96.12 GHz.

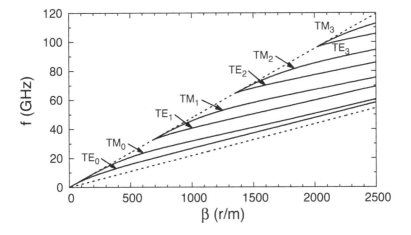

FIGURE 5.49

Dispersion plot for first four sets of modes in a slab waveguide. Dashed lines show the light lines for $\beta = k_0$ and $\beta = k$.

Near cutoff, $\beta \approx k_0$ and the decay constant τ is small. In this case the fields of the slab guide extend significantly into the surrounding regions, and the wave propagates much as if it is a plane wave in free space, with a phase velocity near the speed of light in vacuum. With increasing frequency, β becomes nearer to k, τ becomes smaller, and the fields become more confined within the slab. In this case the waves behave as if they are propagating within a material with dielectric constant ϵ_r. This is clearly seen in Figure 5.50, which shows the phase velocity of the TE_0 and TM_0 modes as functions of frequency. At low frequency the waves have phase velocities near the speed of light in vacuum. As frequency increases and the waves concentrate within the slab, the phase velocity approaches the speed of light in the material.

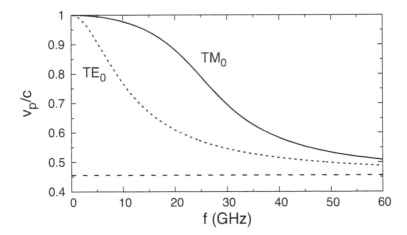

FIGURE 5.50

Phase velocity for TE_0 and TM_0 modes in a slab waveguide with $d = 2.4$ mm and $\epsilon_r = 4.8$. Horizontal dashed line indicates $c/\sqrt{\epsilon_r}$. ◄

Power flow in a lossless slab for frequencies above cutoff may be computed from Poynting's theorem. Consider the TE modes, for example; under the time-harmonic assumption, the time-average Poynting vector is

$$\mathbf{S}_{av} = \tfrac{1}{2}\operatorname{Re}\{\check{\mathbf{E}} \times \check{\mathbf{H}}^*\} = \hat{\mathbf{x}}\tfrac{1}{2}\operatorname{Re}\{\check{e}_y\check{h}_z^*\} - \hat{\mathbf{z}}\tfrac{1}{2}\operatorname{Re}\{\check{e}_y\check{h}_x^*\}.$$

Note the component in the transverse direction; upon substitution we find that $\check{e}_y\check{h}_z^*$ is imaginary, however, and thus the transverse power flow is purely reactive (with zero time-average value). Substituting the field expressions for even modes, we have

$$\mathbf{S}_{av} = \begin{cases} \hat{\mathbf{z}}\tfrac{1}{2}|B|^2k_x^2k_z\check{\omega}\mu\sin^2 k_x x, & |x| < d/2, \\ \hat{\mathbf{z}}\tfrac{1}{2}|A|^2\tau^2k_z\check{\omega}\mu_0 e^{-2\tau|x|}, & |x| > d/2. \end{cases}$$

Dotting with $\hat{\mathbf{z}}$ and integrating from $x = -\infty$ to $x = \infty$, we obtain the power per unit width carried by the traveling wave:

$$\frac{P_{av}}{w} = 2\int_0^{d/2} \tfrac{1}{2}|B|^2k_x^2k_z\check{\omega}\mu\sin^2 k_x x\, dx + 2\int_{d/2}^{\infty} \tfrac{1}{2}|A|^2\tau^2k_z\check{\omega}\mu_0 e^{-2\tau x}\, dx$$

$$= \frac{|B|^2}{4}\check{\omega}\mu_0 k_z \left\{ \mu_r k_x[k_x d - \sin k_x d] + 2\frac{|A|^2}{|B|^2}\tau e^{-\tau d} \right\}.$$

But by (5.315),

$$\frac{A}{B} = \frac{k_x}{\tau}\mu_r e^{\tau d/2}\sin k_x \frac{d}{2}.$$

Substitution and simplification yield

$$\frac{P_{av}}{w} = \frac{|B|^2}{4}\check{\omega}\mu k_x k_z \left\{ [k_x d - \sin k_x d] + \mu_r\frac{k_x}{\tau}[1 - \cos k_x d] \right\}.$$

The first term represents the power carried by the fields in the slab, while the second represents the power carried by the fields in the surrounding air. The ratio of the power carried in the air to that carried in the slab is

$$\frac{P_{\text{air}}}{P_{\text{slab}}} = \frac{\mu_r}{\tau d}\frac{1 - \cos k_x d}{1 - \frac{\sin k_x d}{k_x d}}. \tag{5.318}$$

Similar expressions may be obtained for odd TE modes and for TM modes.

▶ **Example 5.37:** Power ratio for a circuit board substrate

Consider the circuit board in Example 5.36. Plot the ratio of the power per unit width carried by the TE_1 mode in the slab to the power carried in the surrounding air.

Solution: We plot (5.318) vs. frequency for the TE_1 mode in Figure 5.51. Near cutoff, most of the power is carried in the surrounding air and the ratio is large. Far above cutoff, the field in the air region decays rapidly with distance and most of the power is carried in the slab. At 35 GHz more than ten times as much power is carried in the air than in the slab, whereas at 100 GHz more than five times as much power is carried in the slab than in air.

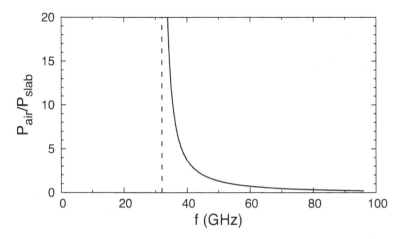

FIGURE 5.51

Ratio of average power per unit width carried by the TE_1 mode in the slab of Example 5.36 to the power carried in air. Dashed vertical line shows the cutoff frequency for the TE_1 mode. ◀

5.6.4.2 Optical fiber

The optical fiber, or *fiber optic cable*, is an important guiding structure in modern communication systems. In its simplest form it is a long tube of homogeneous dielectric, such as glass or plastic, with circular cross-section. The fiber guides electromagnetic waves via total internal reflection, and thus there is an evanescent field outside the fiber. Because the fiber contains no metal, the attenuation is often less than for a classical waveguide or transmission line. In addition, by varying the dielectric constant over the cross-section of the fiber, dispersion can be decreased and the bandwidth increased. We will consider the simplest case of a fiber with uniform dielectric constant.

Consider a circular optical fiber of radius a (Figure 5.52). The fiber has permittivity $\tilde{\epsilon}_1^c$ and permeability $\tilde{\mu}_1$, while the region outside the fiber has permittivity $\tilde{\epsilon}_2^c$ and permeability $\tilde{\mu}_2$. The boundary conditions cannot be satisfied by fields that are TE or TM to the z-direction, unless the fields are ϕ-invariant (i.e., azimuthally symmetric). So we restrict ourselves to these azimuthally symmetric modes.

We proceed as with the circular waveguide, seeking a separation of variables solution for the potential function $\tilde{\psi}$. By (5.228), this function takes the form

$$\tilde{\psi} = AJ_0(k_\rho\rho) + BN_0(k_\rho\rho)$$

by the assumed ϕ-independence. Here $k_\rho^2 = k^2 - k_z^2$. As the materials in region 1 and 2 differ, so do the respective values of k and hence of k_ρ. But satisfaction of the boundary conditions implies that k_z is the same for each region.

Since the waves are guided by total internal reflection within the fiber, we expect a standing wave inside the fiber and an evanescent wave outside. Finiteness of the fields along the z-axis requires that we omit terms of the type $N_0(k_\rho\rho)$ in region 1. We write

$$\tilde{\psi}_{h1} = AJ_0(k_{\rho 1}\rho), \qquad \tilde{\psi}_{e1} = BJ_0(k_{\rho 1}\rho),$$

where

$$k_{\rho 1}^2 = k_1^2 - k_z^2 \tag{5.319}$$

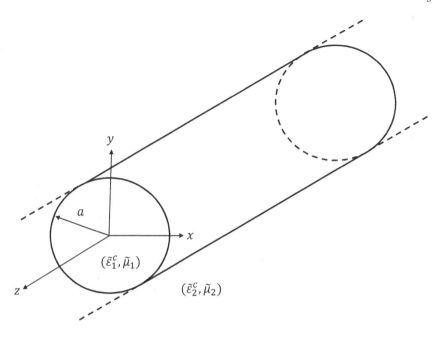

FIGURE 5.52

Optical fiber waveguide.

with $k_1^2 = \omega^2 \tilde{\mu}_1 \tilde{\epsilon}_1^c$.

The potential in region 2 requires more care. Assume there is no loss so that k_1 and k_2 are real. The Bessel functions $J_0(k_\rho \rho)$ and $N_0(k_\rho \rho)$ represent standing waves, but there is no boundary external to the fiber against which standing waves can form. However, appropriate combinations of the Bessel functions yield Hankel functions $H_0^{(1)}$ and $H_0^{(2)}$ that represent waves traveling radially inward and outward, respectively. Because inward traveling wave are nonphysical here, we write for region 2

$$\tilde{\psi} = AH_0^{(2)}(k_{\rho 2}\rho) \tag{5.320}$$

where $k_{\rho 2}^2 = k_2^2 - k_z^2$. If the fiber is lossless and k_z is real, then for $k_2 > k_z$ this solution represents radiation from the fiber since the outward traveling wave carries power away from the structure. This is not an appropriate condition for waves guided by total internal reflection, where the external waves are evanescent and thus carry no outward-directed power. However, if $k_2 < k_z$ and k_z is real, then $k_{\rho 2}$ becomes imaginary, and we can write $k_{\rho 2} = -j\tau$. Then, since

$$H_0^{(2)}(-jx) = j\frac{2}{\pi}K_0(x),$$

the potential function (5.320) can be written as

$$\tilde{\psi} = AK_0(\tau\rho),$$

where

$$\tau^2 = k_z^2 - k_2^2. \tag{5.321}$$

The modified Bessel function $K_n(x)$ decreases with increasing x, so this function represents a wave evanescent in ρ when $k_z > k_2$. We therefore write the potentials in region

2 as

$$\tilde{\psi}_{h2} = CK_0(\tau\rho), \qquad \tilde{\psi}_{e1} = DK_0(\tau\rho),$$

Finally, we note that if the fiber has losses then k_1 will be complex and k_z will take the form $k_z = \beta + j\alpha$, incorporating both phase and attenuation constants.

For modes that are TM$_z$, the relations (5.195) and (5.196) yield the transverse fields

$$\tilde{e}_{z1} = k_{\rho1}^2 BJ_0(k_{\rho1}\rho), \qquad \tilde{h}_{\phi1} = j\omega\tilde{\epsilon}_1^c Bk_{\rho1}J_1(k_{\rho1}\rho),$$

$$\tilde{e}_{z2} = \tau^2 DK_0(\tau\rho), \qquad \tilde{h}_{\phi2} = -j\omega\tilde{\epsilon}_2^c D\tau K_1(k_{\rho1}\rho),$$

since $J_0'(x) = -J_1(x)$ and $K_0'(x) = K_1(x)$. Continuity of the tangential fields at $\rho = a$ yields

$$k_{\rho1}^2 BJ_0(k_{\rho1}a) = \tau^2 DK_0(\tau a), \qquad \tilde{\epsilon}_1^c Bk_{\rho1}J_1(k_{\rho1}a) = -\tilde{\epsilon}_2^c D\tau K_1(k_{\rho1}a),$$

and hence the characteristic equation

$$\frac{k_{\rho1}}{\tilde{\epsilon}_1^c} \frac{J_0(k_{\rho1}a)}{J_1(k_{\rho1}a)} = -\frac{\tau}{\tilde{\epsilon}_2^c} \frac{K_0(\tau a)}{K_1(\tau a)}. \tag{5.322}$$

Simultaneous solution of this with the sum of (5.319) and (5.321)

$$k_{\rho1}^2 + \tau^2 = k_1^2 - k_2^2 \tag{5.323}$$

gives the k_z-ω dispersion relation for azimuthally symmetric TM modes.

For TE$_z$ modes,

$$\tilde{h}_{z1} = k_{\rho1}^2 AJ_0(k_{\rho1}\rho), \qquad \tilde{e}_{\phi1} = -j\omega\tilde{\mu}_1 Ak_{\rho1}J_1(k_{\rho1}\rho),$$

$$\tilde{h}_{z2} = \tau^2 CK_0(\tau\rho), \qquad \tilde{e}_{\phi2} = j\omega\tilde{\mu}_2 C\tau K_1(k_{\rho1}\rho).$$

Continuity of tangential fields at $\rho = a$ produces

$$k_{\rho1}^2 AJ_0(k_{\rho1}a) = \tau^2 CK_0(\tau a), \qquad -\tilde{\mu}_1 Ak_{\rho1}J_1(k_{\rho1}a) = \tilde{\mu}_2 C\tau K_1(k_{\rho1}a),$$

and in turn the characteristic equation

$$\frac{k_{\rho1}}{\tilde{\mu}_1} \frac{J_0(k_{\rho1}a)}{J_1(k_{\rho1}a)} = -\frac{\tau}{\tilde{\mu}_2} \frac{K_0(\tau a)}{K_1(\tau a)}. \tag{5.324}$$

Its solution simultaneously with (5.323) gives the k_z-ω dispersion relation for azimuthally symmetric TE modes.

The cutoff frequency of a lossless optical fiber is the lowest frequency at which the waves are guided by total internal reflection without radiation. Since $\tau = \sqrt{k_z^2 - k_2^2}$ must be real for evanescent external fields, cutoff occurs when $\tau = 0$ or equivalently $k_z = k_2$. The small argument approximation of the modified Bessel functions (E.57) and (E.58),

$$K_0(x) \sim -\ln x, \qquad K_1(x) \sim 1/x,$$

shows that the right-hand sides of (5.322) and (5.324) behave as

$$\lim_{\tau a \to 0} \frac{\tau a K_0(\tau a)}{K_1(\tau a)} = \lim_{\tau a \to 0} -(\tau a)^2 \ln(\tau a) = 0.$$

Hence cutoff occurs when $J_0(k_{\rho_1} a) = 0$, or $k_{\rho_1} = p_{0m}/a$. Using $\tau = 0$ in (5.323), we can also write the cutoff condition as

$$\sqrt{k_1^2 - k_2^2} = \frac{p_{0m}}{a}$$

or

$$f_c = \frac{1}{2\pi\sqrt{\mu_1\epsilon_1 - \mu_2\epsilon_2}}\frac{p_{0m}}{a}.$$

Note $\mu_1\epsilon_1 > \mu_2\epsilon_2$ is needed for total internal reflection.

▶ **Example 5.38:** TE_{01} and TM_{01} modes in an optical fiber

An optical fiber made from silica glass has dielectric constant $\epsilon_r = 2.15$ and diameter 0.1 mm. If the fiber is immersed in free space, find the cutoff frequencies of the TE_{01} and TM_{01} modes and plot the respective dispersion curves. Also, plot the phase velocity and compare wave behavior near cutoff to that of a hollow-pipe waveguide.

Solution: Using $p_{01} = 2.40483$ and noting that glass is nonmagnetic, we find that

$$f_c = \frac{2.40483}{2\pi a}\frac{c}{\sqrt{\epsilon_r - 1}} = 2.140 \text{ THz}$$

for both the TE_{01} and TM_{01} modes.

Figure 5.53 shows the dispersion plot for TM modes obtained from (5.322), and for TE modes obtained from (5.324). The curves differ only slightly. Near cutoff, the decay constant τ is small, and the fields extend into the surrounding free-space region. In this case the wave behaves much as if it is propagating in vacuum, with a phase velocity near the vacuum speed of light. As frequency increases, so does τ, the fields becoming more confined to the fiber. This is verified in Figure 5.54, which shows phase velocity as a function of frequency. At cutoff, the phase velocity for the optical fiber is the speed of light in free space. As the frequency increases, the phase velocity of the optical fiber approaches the velocity of light in the fiber material, $c/\sqrt{\epsilon_r}$. This is similar to the behavior of the phase velocity for the slab waveguide considered in Example 5.36.

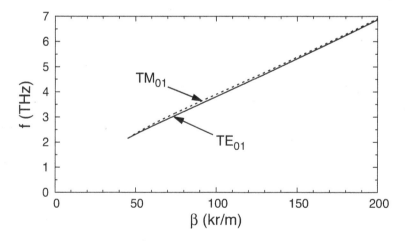

FIGURE 5.53

Dispersion plot for TE_{01} and TM_{01} modes in an optical fiber with $a = 0.05$ mm and $\epsilon_r = 2.15$.

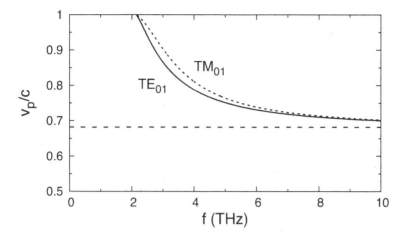

FIGURE 5.54

Phase velocity for TE_{01} and TM_{01} modes in an optical fiber with $a = 0.05$ mm and $\epsilon_r = 2.15$. Horizontal dashed line indicates $v_p = c/\sqrt{\epsilon_r}$.

The wave behavior of the optical fiber can be explored by examining the distribution of the transverse fields. Figure 5.55 shows $|\tilde{E}_\phi|$ normalized to unity maximum for two different frequencies. At $f = 2.146$ THz the fiber operates just above the 2.140 THz cutoff frequency and the field outside the fiber decays slowly with distance. With the field in the outside region dominant, the phase velocity $v_p/c = 0.9995$ is close to that for a wave traveling in air. In contrast, at $f = 6.870$ THz the fiber operates well above cutoff and the field is well contained within the fiber. So the phase velocity $v_p/c = 0.720$ is close to the value $v_p/c = 1/\sqrt{\epsilon_r} = 0.682$ for a wave traveling in the glass material composing the fiber.

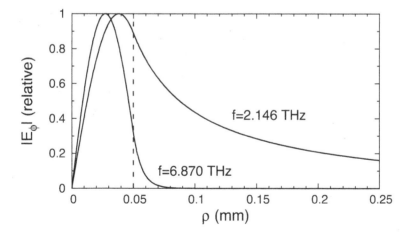

FIGURE 5.55

Normalized $|E_\phi|$ for the TE_{01} modes of an optical fiber with $a = 0.05$ mm and $\epsilon_r = 2.15$. Vertical dashed line indicates the boundary of the fiber. ◄

5.6.4.3   Accurate calculation of attenuation for a circular hollow-pipe waveguide

The circular hollow-pipe waveguide can be treated as a boundary value problem by assuming a waveguide wall many skin depths thick. Under this condition the fields decay to small values before reaching the region external to the waveguide, and the wall may be replaced by an infinite medium with material parameters equal to those of the wall without affecting the value of the propagation constant. Since we have solved this boundary value problem for the optical fiber, we need only specify that the region internal to the fiber is air and the region external is a conductor. We can then solve either (5.322) or (5.324) for k_z to approximate the attenuation constants of the TM_{0m} modes or the TE_{0m} modes in a circular hollow-pipe waveguide.

As an example, consider the characteristic equation (5.324) for the TE_{0m} modes. Let the waveguide walls have conductivity σ, permittivity ϵ_0, and permeability μ_0. Then

$$\tilde{\epsilon}_2^c = \epsilon_0 \left(1 - j \frac{\sigma}{\omega \epsilon_0} \right) \quad \text{so that} \quad k_2 = \frac{\omega}{c} \sqrt{1 - j \frac{\sigma}{\omega \epsilon_0}}.$$

Since k_2 is complex, so is $k_z = \beta - j\alpha$, with α corresponding to the attenuation constant for the waveguide. If $\sigma \gg \omega \epsilon_0$, then k_2 is large and so is $\tau = \sqrt{k_z^2 - k_2^2}$. This can cause troubles with the terms in (5.324), as the modified Bessel functions tend to overflow for large complex arguments. We can circumvent this by using the asymptotic expansion for the Bessel functions [1]

$$K_\nu(z) \sim \sqrt{\frac{\pi}{2z}} e^{-z} \sum_{n=0}^{N} \frac{a_{\nu,n}}{(8z)^n}, \quad a_{\nu,0} = 1, \quad a_{\nu,i} = a_{\nu,i-1} \frac{4\nu^2 - (2i-1)^2}{i}, \quad i = 1, 2, \ldots.$$

Here N must be chosen large enough for sufficient accuracy but not so large that the series begins to diverge. For a wall comprising typical metals, a few terms are appropriate. With this we have the ratio

$$\frac{K_0(\tau a)}{K_1(\tau a)} = \frac{\sum_{n=0}^{N} \frac{a_{0,n}}{(8\tau a)^n}}{\sum_{n=0}^{N} \frac{a_{1,n}}{(8\tau a)^n}},$$

which may be used to compute (5.324) without worry of overflow.

▶ **Example 5.39:** Accurate calculation of attenuation for a circular hollow-pipe waveguide

Solve the characteristic equation for an optical fiber to determine the attenuation constant for the TE_{01} mode of a circular hollow-pipe waveguide of radius $a = 27.88$ mm with a wall made from brass having conductivity $\sigma = 1.4 \times 10^7$ S/m. Compare to the results found in Example 5.20 using the perturbational formula.

Solution: It is important to note that phase constant β and attenuation constant α both depend on wall loss. While the perturbational approach gives an estimate for α, it does not predict the amount of error that results from assuming β to be that of the lossless waveguide.

By solving the characteristic equation (5.324) with finite conductivity in the outer region, the change in β can be accurately computed along with an accurate value of α. It is found that the error in the perturbational formula (5.239) is greatest very close to the 6.557 GHz cutoff frequency, but diminishes quickly as the frequency increases away from cutoff. For example, at 6.560 GHz the solution to the characteristic equation gives $\beta = 3.92195$ r/m as compared to the lossless value of 3.77296 r/m, a difference of about 4%. The computed value of α at that frequency is 0.143430 Np/m, compared to 0.149107 Np/m as predicted by the perturbational formula — again a difference of 4%. However, at 7 GHz the characteristic equation gives $\beta = 51.3430$ r/m and $\alpha = 0.0106063$ Np/m, compared to a value of $\beta = 51.3324$ r/m for the lossless guide and $\alpha = 0.0106094$ Np/m from the perturbational formula. This is a difference of only 0.02% for β and 0.03% for α. At 10 GHz the differences are only 0.01%. We conclude that for operating frequencies reasonably above cutoff, the perturbational formula gives excellent estimates of the attenuation coefficient for the TE$_{01}$ mode of a circular waveguide. ◀

5.6.5   Waves guided in radial directions

5.6.5.1   Cylindrically guided waves: E-plane and H-plane sectoral guides

It is possible to guide waves radially in cylindrical coordinates. Guides with rectangular cross-sections dependent on radial distance are used to form horn antennas, which are particularly important in radar and other microwave applications. Two configurations are popular: the E-plane and H-plane sectoral horns. Waveguides used to create these horns are considered below.

E-plane sectoral waveguide.   The waveguide shown in Figure 5.56 has perfectly conducting walls and is filled with a homogeneous material of permeability $\tilde{\mu}$ and permittivity $\tilde{\epsilon}^c$. The top walls are parallel perfect conductors, while the side walls are PEC and form an angle α. We seek solutions to Maxwell's equations that represent waves traveling in the ρ direction in cylindrical coordinates. As a simple case, assume the sectoral guide is connected to a rectangular waveguide at its mouth at $\rho = \rho_0$, which is operating in the TE$_{10}$ mode with the long axis aligned with z. Because the rectangular waveguide fields are y-invariant, we expect the fields in the rectangular guide to transition into ϕ-invariant sectoral waveguide fields. We also expect the field directions to be maintained across the transition such that the sectoral waveguide fields have components $\tilde{E}_\phi(\rho, z)$, $\tilde{H}_z(\rho, z)$, and $\tilde{H}_\rho(\rho, z)$. Then Faraday's law requires

$$\frac{1}{\rho}\frac{\partial}{\partial \rho}\left(\rho \tilde{E}_\phi\right) = -j\omega\tilde{\mu}\tilde{H}_z, \qquad (5.325)$$

$$-\frac{\partial \tilde{E}_\phi}{\partial z} = -j\omega\tilde{\mu}\tilde{H}_\rho, \qquad (5.326)$$

while the source-free Ampere's law requires

$$\frac{\partial \tilde{H}_\rho}{\partial z} - \frac{\partial \tilde{H}_z}{\partial \rho} = j\omega\tilde{\epsilon}^c\tilde{E}_\phi. \qquad (5.327)$$

An equation for $\tilde{E}_\phi$ may be found by substituting (5.325) and (5.326) into (5.327):

$$\frac{\partial^2 \tilde{E}_\phi}{\partial \rho^2} + \frac{1}{\rho}\frac{\partial \tilde{E}_\phi}{\partial \rho} + \frac{\partial^2 \tilde{E}_\phi}{\partial z^2} + \left(k^2 - \frac{1}{\rho^2}\right)\tilde{E}_\phi = 0. \qquad (5.328)$$

We seek a product solution to (5.328) subject to $\tilde{E}_\phi = 0$ at $z = 0$ and $z = a$. Let $\tilde{E}_\phi = P(\rho)Z(z)$. Substitution gives

$$\left[\frac{1}{P}\left(\frac{\partial^2 P}{\partial \rho^2} + \frac{1}{\rho}\frac{\partial P}{\partial \rho}\right) + k^2 - \frac{1}{\rho^2}\right] + \left[\frac{1}{Z}\frac{\partial^2 Z}{\partial z^2}\right] = 0.$$

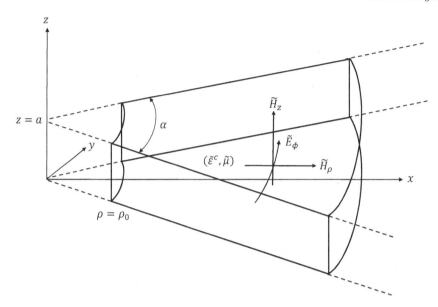

FIGURE 5.56

E-plane sectoral waveguide.

This equation can only be satisfied if the terms in brackets are equal to constants. Let the first term equal a constant k_z^2 and examine

$$\frac{1}{Z}\frac{\partial^2 Z}{\partial z^2} = -k_z^2.$$

This has solutions

$$Z(z) = C_1 \sin(k_z z) + C_2 \cos(k_z z).$$

To satisfy $\tilde{E}_\phi = 0$ at $z = 0$, we must have $Z(0) = 0$ and thus $C_2 = 0$. To satisfy $\tilde{E}_\phi = 0$ at $z = a$, we must have $\sin(k_z a) = 0$ and thus

$$k_z = n\pi/a \qquad (n = 1, 2, 3, \ldots).$$

Because the sectoral guide is fed by a rectangular waveguide operating in the TE_{10} mode, we choose $n = 1$ to provide for smooth transition of the fields. Thus we have the equation for P:

$$\frac{1}{P}\left(\frac{\partial^2 P}{\partial \rho^2} + \frac{1}{\rho}\frac{\partial P}{\partial \rho}\right) + k^2 - \frac{1}{\rho^2} = \left(\frac{\pi}{a}\right)^2,$$

or

$$\frac{\partial^2 P}{\partial \rho^2} + \frac{1}{\rho}\frac{\partial P}{\partial \rho} + \left(k_\rho^2 - \frac{1}{\rho^2}\right)P = 0,$$

where

$$k_\rho = \sqrt{k^2 - \left(\frac{\pi}{a}\right)^2}.$$

This is the ordinary Bessel equation of order one. Its solutions

$$P(\rho) = AH_1^{(2)}(k_\rho\rho) + BH_1^{(1)}(k_\rho\rho)$$

represent waves traveling in the $\pm\rho$ directions. With this we have the transverse fields in the guide,

$$\tilde{E}_\phi = [AH_1^{(2)}(k_\rho\rho) + BH_1^{(1)}(k_\rho\rho)]\sin\left(\frac{\pi z}{a}\right)$$

and

$$\tilde{H}_z = j\frac{1}{\omega\tilde{\mu}}\frac{1}{\rho}\frac{\partial}{\partial\rho}\left(\rho\tilde{E}_\phi\right) = j\frac{1}{\omega\tilde{\mu}}\frac{1}{\rho}\frac{\partial}{\partial\rho}[A\rho H_1^{(2)}(k_\rho\rho) + B\rho H_1^{(1)}(k_\rho\rho)]\sin\left(\frac{\pi z}{a}\right).$$

Using the derivative relationship

$$\frac{d}{dz}[zH_1^{(1,2)}(z)] = zH_0^{(1,2)}(z)$$

we obtain

$$\tilde{H}_z = j\frac{k_\rho}{\omega\tilde{\mu}}[AH_0^{(2)}(k_\rho\rho) + BH_0^{(1)}(k_\rho\rho)]\sin\left(\frac{\pi z}{a}\right).$$

The wave propagating in the E-plane sectoral waveguide has properties similar to a wave propagating in a rectangular guide. In a lossless guide where $\tilde{\mu} = \mu$ and $\tilde{\epsilon}^c = \epsilon$, the propagation constant k_ρ is real only when $k > \pi/a = k_c$, so there is a cutoff effect. For frequencies above cutoff we have $k_\rho = \beta$ and the outward traveling wave behaves like

$$\tilde{E}_\phi = AH_1^{(2)}(\beta\rho)\sin\left(\frac{\pi z}{a}\right) \approx jA\sqrt{\frac{2j}{\pi\beta\rho}}e^{-j\beta\rho}\sin\left(\frac{\pi z}{a}\right)$$

for large ρ. Below cutoff we have $k_\rho = j\alpha$ and thus for large ρ

$$\tilde{E}_\phi = AH_1^{(2)}(j\alpha\rho)\sin\left(\frac{\pi z}{a}\right) \approx jA\sqrt{\frac{2}{\pi\alpha\rho}}e^{-\alpha\rho}\sin\left(\frac{\pi z}{a}\right);$$

the wave is evanescent. For large ρ we also have the phase velocity

$$v_p = \frac{\omega}{\beta} = \frac{v}{\sqrt{1 - f_c^2/f^2}}$$

where $v = 1/\sqrt{\mu\epsilon}$ and where $f_c = v/(2a)$ is the cutoff frequency, exactly as in a rectangular guide.

We can also define a transverse wave impedance as the ratio of the transverse fields for an outward traveling wave. Let

$$Z_h = \frac{\tilde{E}_\phi}{\tilde{H}_z} = -j\frac{\omega\tilde{\mu}}{k_\rho}\frac{H_1^{(2)}(k_\rho\rho)}{H_0^{(2)}(k_\rho\rho)}.$$

We see that, unlike the case of the rectangular guide, this transverse wave impedance is dependent on position. But for large ρ

$$\frac{H_1^{(2)}(k_\rho\rho)}{H_0^{(2)}(k_\rho\rho)} \approx j$$

and thus

$$Z_h \approx \omega\tilde{\mu}/k_\rho,$$

which is identical to the wave impedance of the TE_{10} mode in a rectangular guide.

H-plane sectoral waveguide. The H-plane sectoral waveguide is similar to the E-plane guide, except that the rectangular waveguide feeding the sectoral guide is rotated by 90°. See Figure 5.57. The guide has perfectly conducting walls and is filled with a homogeneous material of permeability $\tilde{\mu}$ and permittivity $\tilde{\epsilon}^c$. As with the E-plane sectoral guide, the top walls are parallel perfect conductors, while the side walls are PEC and form an angle α. We again seek solutions to Maxwell's equations that represent radially traveling waves. Let the sectoral guide be connected to a rectangular waveguide at its mouth at $\rho = \rho_0$ that is operating in the TE_{10} mode. However, in contrast to the E-plane guide, the short axis of the rectangular waveguide is aligned with z. Because the rectangular waveguide fields are independent of z, we expect the fields in the sectoral waveguide to be independent of z. We also expect the field directions to be maintained across the transition such that the sectoral guide fields have components $\tilde{E}_z(\rho, \phi)$, $\tilde{H}_\phi(\rho, \phi)$, and $\tilde{H}_\rho(\rho, \phi)$. Using these, we find that Faraday's law requires

$$\frac{1}{\rho}\frac{\partial}{\partial \phi}\left(\rho \tilde{E}_z\right) = -j\omega\tilde{\mu}\tilde{H}_\rho, \tag{5.329}$$

$$-\frac{\partial \tilde{E}_z}{\partial \rho} = -j\omega\tilde{\mu}\tilde{H}_\phi, \tag{5.330}$$

while the source-free Ampere's law requires

$$\frac{1}{\rho}\frac{\partial}{\partial \rho}\left(\rho \tilde{H}_\phi\right) - \frac{1}{\rho}\frac{\partial \tilde{H}_\rho}{\partial \rho} = j\omega\tilde{\epsilon}^c\tilde{E}_z. \tag{5.331}$$

An equation for $\tilde{E}_z$ may be found by substituting (5.329) and (5.330) into (5.331):

$$\frac{\partial^2 \tilde{E}_z}{\partial \rho^2} + \frac{1}{\rho}\frac{\partial \tilde{E}_z}{\partial \rho} + \frac{1}{\rho^2}\frac{\partial^2 \tilde{E}_z}{\partial \phi^2} + k^2 \tilde{E}_z = 0. \tag{5.332}$$

We seek a product solution to (5.332) satisfying $\tilde{E}_z = 0$ at $\phi = -\alpha/2$ and $\phi = \alpha/2$. Let $\tilde{E}_z = P(\rho)\Phi(\phi)$. Substitution gives

$$\left[\frac{\rho^2}{P}\frac{\partial^2 P}{\partial \rho^2} + \frac{\rho}{P}\frac{\partial P}{\partial \rho} + k^2\rho^2\right] + \left[\frac{1}{\Phi}\frac{\partial^2 \Phi}{\partial \phi^2}\right] = 0.$$

This equation can only be satisfied if the terms in brackets are equal to constants. Set the first term equal to a constant k_ϕ^2 and examine

$$\frac{1}{\Phi}\frac{\partial^2 \Phi}{\partial \phi^2} = -k_\phi^2.$$

This has solutions

$$\Phi(\phi) = C_1 \sin(k_\phi \phi) + C_2 \cos(k_\phi \phi).$$

Setting $\tilde{E}_z = 0$ at $\phi = \pm\alpha/2$ gives the system of equations

$$\begin{bmatrix} \sin(k_\phi \alpha/2) & \cos(k_\phi \alpha/2) \\ -\sin(k_\phi \alpha/2) & \cos(k_\phi \alpha/2) \end{bmatrix} \begin{bmatrix} C_1 \\ C_2 \end{bmatrix} = \begin{bmatrix} 0 \\ 0 \end{bmatrix}.$$

Setting the determinant to zero to force a nontrivial solution gives $\sin(k_\phi \alpha) = 0$, and thus

$$k_\phi = n\pi/\alpha \qquad (n = 1, 2, 3, \ldots).$$

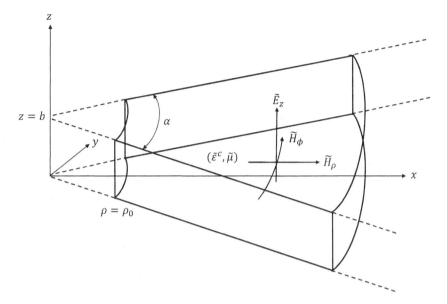

FIGURE 5.57

H-plane sectoral waveguide.

Because the sectoral guide is fed by a rectangular waveguide operating in the TE_{10} mode, we choose $n = 1$ to permit the fields to transition smoothly. Then, we see that $C_1 = 0$ and

$$\Phi(\phi) = C \cos\left(\frac{\pi}{\alpha}\phi\right).$$

Next we have the equation for P:

$$\frac{\partial^2 P}{\partial \rho^2} + \frac{1}{\rho}\frac{\partial P}{\partial \rho} + \left(k^2 - \frac{k_\phi^2}{\rho^2}\right)P = 0,$$

which is the ordinary Bessel equation of order k_ϕ. Its solutions

$$P(\rho) = AH_{k_\phi}^{(2)}(k\rho) + BH_{k_\phi}^{(1)}(k\rho)$$

represent waves traveling in the $\pm\rho$ directions. With this we have the transverse fields in the guide:

$$\tilde{E}_z = \left[AH_{k_\phi}^{(2)}(k\rho) + BH_{k_\phi}^{(1)}(k\rho)\right]\cos\left(\frac{\pi}{\alpha}\phi\right)$$

and

$$\tilde{H}_\phi = -j\frac{1}{\omega\tilde{\mu}}\frac{\partial \tilde{E}_z}{\partial \rho} = -j\frac{k}{\omega\tilde{\mu}}\left[AH_{k_\phi}^{(2)\prime}(k\rho) + BH_{k_\phi}^{(1)\prime}(k\rho)\right]\cos\left(\frac{\pi}{\alpha}\phi\right).$$

Unlike a wave propagating in the E-plane sectoral waveguide, the wave in the H-plane sectoral guide does not experience a cutoff effect and may propagate down to zero frequency. For a lossless guide we have at large ρ the phase velocity $v_p = 1/\sqrt{\mu\epsilon}$, which is the velocity of a plane wave in the medium filling the guide.

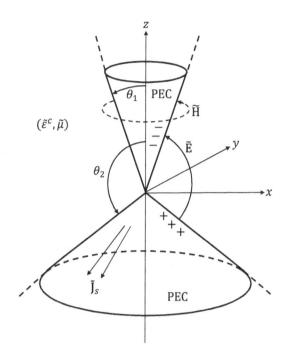

FIGURE 5.58
Conical transmission line.

As with the E-plane sectoral guide, we can define a transverse wave impedance for outward traveling waves as the ratio of transverse fields,

$$Z_h = \frac{\tilde{E}_z}{\tilde{H}_\phi} = j\frac{\omega\tilde{\mu}}{k}\frac{H^{(2)}_{k_\phi}(k\rho)}{H^{(2)'}_{k_\phi}(k\rho)},$$

which depends on position. However, for large ρ

$$\frac{H^{(2)}_{k_\phi}(k\rho)}{H^{(2)'}_{k_\phi}(k\rho)} \approx -j$$

and, thus,

$$Z_h \approx \omega\tilde{\mu}/k,$$

which is the wave impedance of a TEM wave propagating with no cutoff.

5.6.5.2   Spherically guided waves: the biconical transmission line

It is possible to guide waves radially in spherical coordinates. The biconical transmission line, consisting of nested conducting cones embedded in a medium of permittivity $\tilde{\epsilon}^c$ and permeability $\tilde{\mu}$, is shown in Figure 5.58. The structure is excited at the apex of the cones, and a radially directed TEM wave travels outward with the field confined between the cones.

The TE-TM decomposition for spherical coordinates (§ 5.4.3) can be specialized to the case where $\tilde{E}_r = \tilde{H}_r = 0$. Either the TE_r or the TM_r decomposition can be used; here we use TM. From (5.137) we see that $\tilde{E}_r = 0$ when

$$\left(\frac{\partial^2}{\partial r^2} + k^2\right)\tilde{A}_e = 0.$$

Symmetry requires $\tilde{A}_e(\mathbf{r}, \omega) = \tilde{A}_e(r, \theta, \omega)$, so we seek a product solution of the form $\tilde{A}_e = R(r)\Theta(\theta)$. Substitution gives

$$\left(\frac{\partial^2}{\partial r^2} + k^2\right)R(r) = 0,$$

hence $R(r) = e^{-jkr}$ for a wave propagating in the $+r$ direction. We must also satisfy (5.136) and so

$$\frac{\Theta}{r^2}\frac{\partial}{\partial r}\left[r^2\frac{\partial}{\partial r}\left(\frac{R}{r}\right)\right] + \frac{R}{r^2 \sin\theta}\frac{\partial}{\partial \theta}\left[\sin\theta\frac{\partial}{\partial \theta}\left(\frac{\Theta}{r}\right)\right] + k^2\Theta\frac{R}{r} = 0.$$

Substituting for $R(r)$ and performing the derivatives, we find that the first and third terms cancel, giving

$$\frac{\partial}{\partial \theta}\left[\sin\theta\frac{\partial\Theta}{\partial \theta}\right] = 0.$$

Integration gives

$$\frac{\partial\Theta}{\partial \theta} = \frac{K}{\sin\theta},$$

where K is a constant. A second integration gives

$$\Theta = K\ln\left(\cot\frac{\theta}{2}\right).$$

By (5.138) and (5.141), we obtain

$$\tilde{E}_\theta = -\frac{\omega}{k}\frac{e^{-jkr}}{r}\frac{K}{\sin\theta}, \qquad \tilde{H}_\phi = -\frac{1}{\tilde{\mu}}\frac{e^{-jkr}}{r}\frac{K}{\sin\theta}.$$

The wave impedance is

$$\frac{\tilde{E}_\theta}{\tilde{H}_\phi} = \frac{\omega\tilde{\mu}}{k} = \eta$$

as expected for a TEM wave.

Since the transmission-line wave is traveling in the radial direction, we compute the voltage difference between the two cones at a specified value of r:

$$V(r) = -\int_{\theta_2}^{\theta_1}\tilde{E}_\theta r\, d\theta = \frac{\omega}{k}e^{-jkr}\int_{\theta_2}^{\theta_1}\frac{K}{\sin\theta}\,d\theta = K\frac{\omega}{k}e^{-jkr}\ln\left[\frac{\cot(\theta_1/2)}{\cot(\theta_2/2)}\right].$$

Since K is arbitrary, we can define

$$K = \tilde{V}_0\frac{k}{\omega\ln\left[\frac{\cot(\theta_1/2)}{\cot(\theta_2/2)}\right]}$$

so that

$$V(r) = \tilde{V}_0 e^{-jkr}.$$

The magnetic field is then

$$\tilde{H}_\phi = -\frac{\tilde{V}_0}{\eta \ln\left[\frac{\cot(\theta_1/2)}{\cot(\theta_2/2)}\right]}\frac{e^{-jkr}}{r}\frac{1}{\sin\theta}$$

so that the surface current on the cone at $\theta = \theta_2$, where $\hat{\mathbf{n}} = -\hat{\boldsymbol{\theta}}$, is

$$\tilde{\mathbf{J}}_s = \hat{\mathbf{r}}\frac{\tilde{V}_0}{\eta \ln\left[\frac{\cot(\theta_1/2)}{\cot(\theta_2/2)}\right]}\frac{e^{-jkr}}{r}\frac{1}{\sin\theta_2}.$$

The current flowing on the cone is

$$\tilde{I}(r) = \int_0^{2\pi} \tilde{\mathbf{J}}_s \cdot \hat{\mathbf{r}} r \sin\theta_2 \, d\phi = 2\pi\frac{\tilde{V}_0}{\eta \ln\left[\frac{\cot(\theta_1/2)}{\cot(\theta_2/2)}\right]}e^{-jkr}.$$

We thus have the characteristic impedance of the biconical transmission line as the ratio of voltage to current at radius r:

$$Z_c = \frac{\tilde{V}(r)}{\tilde{I}(r)} = \frac{\eta}{2\pi}\ln\left[\frac{\cot(\theta_1/2)}{\cot(\theta_2/2)}\right].$$

▶ **Example 5.40:** Impedance of a bicone antenna

A classic application for a biconical transmission line is the bicone antenna. The cones are made identical with $\theta_2 = \pi - \theta_1$ and are fed at the apex by a two-wire transmission line (or a coaxial cable with a balun). Assuming the cones reside in free space, find the cone angle that makes the characteristic impedance of the biconical transmission line 300 Ω.

Solution: Setting $\theta_2 = \pi - \theta_1$ gives

$$\cot\frac{\theta_2}{2} = \cot\left(\frac{\pi}{2} - \frac{\theta_1}{2}\right) = \tan\frac{\theta_1}{2}$$

so that

$$Z_c = \frac{\eta_0}{2\pi}\ln\cot^2\frac{\theta_1}{2} = \frac{\eta_0}{\pi}\ln\cot\frac{\theta_1}{2} = 300 \ \Omega.$$

Solving for θ_1 we have

$$\theta_1 = 2\tan^{-1}\left[e^{-\frac{300\pi}{\eta_0}}\right] = 9.37°. \ \blacktriangleleft$$

▶ **Example 5.41:** Impedance of a discone antenna

Another antenna application for a biconical transmission line is the discone antenna. The angle of the top cone is chosen to be $\theta_1 = 90°$ so that the cone becomes a flat ground plane below which the second cone extends. (Since the ground plane must in practice be truncated, the top cone becomes a disc, giving the antenna its name.) The antenna is fed using a coaxial cable passed up through the bottom cone. If the cones reside in free space, find the angle of the bottom cone that makes the characteristic impedance of the conical transmission line 50 Ω.

Solution: Setting $\theta_1 = \pi/2$ gives the characteristic impedance

$$Z_c = \frac{\eta_0}{2\pi} \ln \tan \frac{\theta_2}{2} = 50 \ \Omega.$$

Solving for θ_2 we have

$$\theta_2 = 2 \tan^{-1} \left[e^{\frac{100\pi}{\eta_0}} \right] = 133.0°.$$

The bottom cone makes an angle of $47°$ with the ground plane. ◄

5.7   Problems

5.1 Verify that the fields and sources obeying even planar reflection symmetry obey the component Maxwell's equations (5.1)–(5.6). Repeat for fields and sources obeying odd planar reflection symmetry.

5.2 Consider an electric Hertzian dipole located on the z-axis at $z = h$. Show that if the dipole is parallel to the plane $z = 0$, then adding an oppositely directed dipole of the same strength at $z = -h$ produces zero electric field tangential to the plane. Also show that if the dipole is z-directed, then adding another z-directed dipole at $z = -h$ produces zero electric field tangential to the $z = 0$ plane. Since the field for $z > 0$ is unaltered in each case, if we place a PEC in the $z = 0$ plane, we establish that tangential components of electric current image in the opposite direction, while vertical components image in the same direction.

5.3 Consider a z-directed electric line source $\tilde{I}_0$ located at $y = h, x = 0$ between conducting plates at $y = \pm d$, $d > h$. The material between the plates has permeability $\tilde{\mu}(\omega)$ and complex permittivity $\tilde{\epsilon}^c(\omega)$. Write the impressed and scattered fields in terms of Fourier transforms and apply the boundary conditions at $z = \pm d$ to determine the electric field between the plates. Show that the result is identical to the expression (5.8) obtained using symmetry decomposition, which required the boundary condition to be applied only on the top plate.

5.4 Consider an unbounded, homogeneous, isotropic medium described by permeability $\tilde{\mu}(\omega)$ and complex permittivity $\tilde{\epsilon}^c(\omega)$. Assuming there are magnetic sources present, but no electric sources, show that the fields may be written as

$$\tilde{\mathbf{H}}(\mathbf{r}) = -j\omega\tilde{\epsilon}^c \int_V \bar{\mathbf{G}}_e(\mathbf{r}|\mathbf{r}';\omega) \cdot \tilde{\mathbf{J}}^i_m(\mathbf{r}',\omega) \, dV',$$

$$\tilde{\mathbf{E}}(\mathbf{r}) = \int_V \bar{\mathbf{G}}_m(\mathbf{r}|\mathbf{r}';\omega) \cdot \tilde{\mathbf{J}}^i_m(\mathbf{r}',\omega) \, dV',$$

where $\bar{\mathbf{G}}_e$ is given by (5.80) and $\bar{\mathbf{G}}_m$ is given by (5.81).

5.5 Show that for a cubical excluding volume, the depolarizing dyadic is $\bar{\mathbf{L}} = \bar{\mathbf{I}}/3$.

5.6 Consider the depolarizing dyadic for a cylindrical excluding volume with height and diameter both $2a$, and with the limit taken as $a \to 0$. Let $\tilde{\mathbf{J}} = \tilde{J}_z\hat{\mathbf{z}}$ and show that $L_{zz} = 0.293$.

5.7 Show that the spherical wave function

$$\tilde{\psi}(\mathbf{r}, \omega) = \frac{e^{-jkr}}{4\pi r}$$

obeys the radiation conditions (5.84) and (5.85).

5.8 Verify that the transverse component of the Laplacian of $\mathbf{A}$ is

$$(\nabla^2 \mathbf{A})_t = \left[\nabla_t(\nabla_t \cdot \mathbf{A}_t) + \frac{\partial^2 \mathbf{A}_t}{\partial u^2} - \nabla_t \times \nabla_t \times \mathbf{A}_t \right].$$

Verify that the longitudinal component of the Laplacian of $\mathbf{A}$ is

$$\hat{\mathbf{u}}\left(\hat{\mathbf{u}} \cdot \nabla^2 \mathbf{A}\right) = \hat{\mathbf{u}}\nabla^2 A_u.$$

5.9 Verify the identities (B.88)–(B.99).

5.10 Verify the identities (B.100)–(B.104).

5.11 Derive the formula (5.96) for the transverse component of the electric field.

5.12 The longitudinal/transverse decomposition can be performed beginning with the time-domain Maxwell's equations. Show that for a homogeneous, lossless, isotropic region described by permittivity ϵ and permeability μ, the longitudinal fields obey the wave equations

$$\left(\frac{\partial^2}{\partial u^2} - \frac{1}{v^2}\frac{\partial^2}{\partial t^2} \right) \mathbf{H}_t = \nabla_t \frac{\partial H_u}{\partial u} - \epsilon\hat{\mathbf{u}} \times \nabla_t\frac{\partial E_u}{\partial t} + \frac{\partial \mathbf{J}_{mt}}{\partial t} - \hat{\mathbf{u}} \times \frac{\partial \mathbf{J}_t}{\partial u},$$

$$\left(\frac{\partial^2}{\partial u^2} - \frac{1}{v^2}\frac{\partial^2}{\partial t^2} \right) \mathbf{E}_t = \nabla_t \frac{\partial E_u}{\partial u} + \mu\hat{\mathbf{u}} \times \nabla_t\frac{\partial H_u}{\partial t} + \hat{\mathbf{u}} \times \frac{\partial \mathbf{J}_{mt}}{\partial u} + \mu\frac{\partial \mathbf{J}_t}{\partial t}.$$

Also show that the transverse fields may be found from the longitudinal fields by solving

$$\left(\nabla^2 - \frac{1}{v^2}\frac{\partial}{\partial t^2} \right) E_u = \frac{1}{\epsilon}\frac{\partial \rho}{\partial u} + \mu\frac{\partial J_u}{\partial t} + \hat{\mathbf{u}} \cdot (\nabla_t \times \mathbf{J}_{mt}),$$

$$\left(\nabla^2 - \frac{1}{v^2}\frac{\partial}{\partial t^2} \right) H_u = \frac{1}{\mu}\frac{\partial \rho_m}{\partial u} + \epsilon\frac{\partial J_{mu}}{\partial t} - \hat{\mathbf{u}} \cdot (\nabla_t \times \mathbf{J}_t).$$

Here $v = 1/\sqrt{\mu\epsilon}$.

5.13 Consider a homogeneous, lossless, isotropic region of space described by permittivity ϵ and permeability μ. Beginning with the source-free time-domain Maxwell equations in rectangular coordinates, choose z as the longitudinal direction and show that the TE–TM decomposition is given by

$$\left(\frac{\partial^2}{\partial z^2} - \frac{1}{v^2}\frac{\partial^2}{\partial t^2}\right) E_y = \frac{\partial^2 E_z}{\partial z \partial y} + \mu \frac{\partial^2 H_z}{\partial x \partial t},$$

$$\left(\frac{\partial^2}{\partial z^2} - \frac{1}{v^2}\frac{\partial^2}{\partial t^2}\right) E_x = \frac{\partial^2 E_z}{\partial x \partial z} - \mu \frac{\partial^2 H_z}{\partial y \partial t},$$

$$\left(\frac{\partial^2}{\partial z^2} - \frac{1}{v^2}\frac{\partial^2}{\partial t^2}\right) H_y = -\epsilon \frac{\partial^2 E_z}{\partial x \partial t} + \frac{\partial^2 H_z}{\partial y \partial z},$$

$$\left(\frac{\partial^2}{\partial z^2} - \frac{1}{v^2}\frac{\partial^2}{\partial t^2}\right) H_x = \epsilon \frac{\partial^2 E_z}{\partial y \partial t} + \frac{\partial^2 H_z}{\partial x \partial z},$$

with

$$\left(\nabla^2 - \frac{1}{v^2}\frac{\partial^2}{\partial t^2}\right) E_z = 0,$$

$$\left(\nabla^2 - \frac{1}{v^2}\frac{\partial^2}{\partial t^2}\right) H_z = 0.$$

Here $v = 1/\sqrt{\mu\epsilon}$.

5.14 Consider the case of TM fields in the time domain. Show that for a homogeneous, isotropic, lossless medium with permittivity ϵ and permeability μ, the fields may be derived from a single Hertzian potential $\mathbf{\Pi}_e(\mathbf{r}, t) = \hat{\mathbf{u}}\tilde{\Pi}_e(\mathbf{r}, t)$ that satisfies the wave equation

$$\left(\nabla^2 - \frac{1}{v^2}\frac{\partial^2}{\partial t^2}\right) \Pi_e = 0$$

and that the fields are

$$\mathbf{E} = \nabla_t \frac{\partial \Pi_e}{\partial u} + \hat{\mathbf{u}}\left(\frac{\partial^2}{\partial u^2} - \frac{1}{v^2}\frac{\partial^2}{\partial t^2}\right)\Pi_e, \qquad \mathbf{H} = -\epsilon\hat{\mathbf{u}} \times \nabla_t \frac{\partial \Pi_e}{\partial t}.$$

5.15 Consider the case of TE fields in the time domain. Show that for a homogeneous, isotropic, lossless medium with permittivity ϵ and permeability μ, the fields may be derived from a single Hertzian potential $\mathbf{\Pi}_h(\mathbf{r}, t) = \hat{\mathbf{u}}\tilde{\Pi}_h(\mathbf{r}, t)$ that satisfies the wave equation

$$\left(\nabla^2 - \frac{1}{v^2}\frac{\partial^2}{\partial t^2}\right) \Pi_h = 0$$

and that the fields are

$$\mathbf{E} = \mu\hat{\mathbf{u}} \times \nabla_t \frac{\partial \Pi_h}{\partial t}, \qquad \mathbf{H} = \nabla_t \frac{\partial \Pi_h}{\partial u} + \hat{\mathbf{u}}\left(\frac{\partial^2}{\partial u^2} - \frac{1}{v^2}\frac{\partial^2}{\partial t^2}\right)\Pi_h.$$

5.16 Show that in the time domain, TEM fields may be written for a homogeneous, isotropic, lossless medium with permittivity ϵ and permeability μ in terms of a Hertzian potential $\mathbf{\Pi}_e = \hat{\mathbf{u}}\Pi_e$ that satisfies

$$\nabla_t^2 \Pi_e = 0$$

and that the fields are

$$\mathbf{E} = \nabla_t \frac{\partial \Pi_e}{\partial u}, \qquad \mathbf{H} = -\epsilon \hat{\mathbf{u}} \times \nabla_t \frac{\partial \Pi_e}{\partial t}.$$

5.17 Show that in the time domain, TEM fields may be written for a homogeneous, isotropic, lossless medium with permittivity ϵ and permeability μ in terms of a Hertzian potential $\mathbf{\Pi}_h = \hat{\mathbf{u}} \Pi_h$ that satisfies

$$\nabla_t^2 \Pi_h = 0$$

and that the fields are

$$\mathbf{E} = \mu \hat{\mathbf{u}} \times \nabla_t \frac{\partial \Pi_h}{\partial t}, \qquad \mathbf{H} = \nabla_t \frac{\partial \Pi_h}{\partial u}.$$

5.18 Consider a TEM plane-wave field of the form

$$\tilde{\mathbf{E}} = \hat{\mathbf{x}} \tilde{E}_0 e^{-jkz}, \qquad \tilde{\mathbf{H}} = \hat{\mathbf{y}} \frac{\tilde{E}_0}{\eta} e^{-jkz},$$

where $k = \omega \sqrt{\mu \epsilon}$ and $\eta = \sqrt{\mu/\epsilon}$. Show that:

(a) $\tilde{\mathbf{E}}$ may be obtained from $\tilde{\mathbf{H}}$ using the equations for a field that is TE_y;

(b) $\tilde{\mathbf{H}}$ may be obtained from $\tilde{\mathbf{E}}$ using the equations for a field that is TM_x;

(c) $\tilde{\mathbf{E}}$ and $\tilde{\mathbf{H}}$ may be obtained from the potential $\tilde{\mathbf{\Pi}}_h = \hat{\mathbf{y}}(\tilde{E}_0/k^2\eta)e^{-jkz}$;

(d) $\tilde{\mathbf{E}}$ and $\tilde{\mathbf{H}}$ may be obtained from the potential $\tilde{\mathbf{\Pi}}_e = \hat{\mathbf{x}}(\tilde{E}_0/k^2)e^{-jkz}$;

(e) $\tilde{\mathbf{E}}$ and $\tilde{\mathbf{H}}$ may be obtained from the potential $\tilde{\mathbf{\Pi}}_e = \hat{\mathbf{z}}(j\tilde{E}_0 x/k)e^{-jkz}$;

(f) $\tilde{\mathbf{E}}$ and $\tilde{\mathbf{H}}$ may be obtained from the potential $\tilde{\mathbf{\Pi}}_h = \hat{\mathbf{z}}(j\tilde{E}_0 y/k\eta)e^{-jkz}$.

5.19 Prove the orthogonality relationships (5.208) and (5.209) for the longitudinal fields in a lossless waveguide. *Hint*: Use the surface version of Green's second identity

$$\int_S (a\nabla^2 b - b\nabla^2 a)\, dS = \oint_\Gamma \left(a\frac{\partial b}{\partial n} + b\frac{\partial a}{\partial n} \right) dl$$

and let $a = \check{\psi}_m$, $b = \check{\psi}_n$ where $\check{\psi} = \check{\psi}_e$ for TM modes and $\check{\psi} = \check{\psi}_h$ for TE modes, with m and n designating different modes.

5.20 Verify the waveguide orthogonality conditions (5.211)–(5.212) by substituting the field expressions for a rectangular waveguide.

5.21 Show that the time-average power carried by a propagating TE mode in a lossless waveguide is given by

$$P_{av} = \frac{1}{2}\check{\omega}\mu\beta k_c^2 \int_{CS} \check{\psi}_h \check{\psi}_h^* \, dS.$$

5.22 Show that the time-average stored energy per unit length for a propagating TE mode in a lossless waveguide is

$$\langle W_e \rangle / l = \langle W_m \rangle / l = \frac{\epsilon}{4} (\check{\omega}\mu)^2 k_c^2 \int_{CS} \check{\psi}_h \check{\psi}_h^* \, dS.$$

5.23 Consider a propagating TM mode in a lossless rectangular waveguide. Show that the time-average power carried by the propagating wave is

$$P_{av_{nm}} = \frac{1}{2} \check{\omega}\epsilon\beta_{nm} k_{c_{nm}}^2 |A_{nm}|^2 \frac{ab}{4}.$$

5.24 Consider a propagating TE mode in a lossless rectangular waveguide. Show that the time-average power carried by the propagating wave is

$$P_{av_{nm}} = \frac{1}{2} \check{\omega}\mu\beta_{nm} k_{c_{nm}}^2 |B_{nm}|^2 \frac{ab}{4}.$$

5.25 Consider a homogeneous, lossless region of space characterized by permeability μ and permittivity ϵ. Beginning with the time-domain Maxwell equations, show that the θ and ϕ components of the electromagnetic fields can be written in terms of the radial components. From this give the TE$_r$–TM$_r$ field decomposition.

5.26 Consider the formula for the radar cross-section of a PEC sphere (5.167). Show that for the monostatic case the RCS becomes

$$\sigma = \frac{\lambda^2}{4\pi} \left| \sum_{n=1}^{\infty} \frac{(-1)^n (2n+1)}{\hat{H}_n^{(2)\prime}(ka) \hat{H}_n^{(2)}(ka)} \right|^2.$$

5.27 Beginning with the monostatic formula for the RCS of a conducting sphere given in Problem 5.26, use the small-argument approximation to the spherical Hankel functions to show that the RCS is proportional to λ^{-4} when $ka \ll 1$.

5.28 Consider an inhomogeneous region with $\mu = \mu(\mathbf{r})$ and $\epsilon = \epsilon_0$. Only electric sources exist, and the fields may be written in terms of the potentials as

$$\mathbf{E} = -\frac{\partial \mathbf{A}}{\partial t} - \nabla\phi, \qquad \mathbf{B} = \nabla \times \mathbf{A}.$$

Using the Lorenz condition, obtain a partial differential equation for $\mathbf{A}$.

5.29 The vector potential in an unbounded space region has a Fourier spectrum given by

$$\tilde{\mathbf{A}}(\mathbf{r}, \omega) = \frac{\mu_0}{4\pi} \int_V \tilde{\mathbf{J}}(\mathbf{r}', \omega) \frac{e^{-jkR}}{R} \, dV'$$

where $k = \omega/c$. Compute the inverse temporal transform of $\tilde{\mathbf{A}}(\mathbf{r}, \omega)$ and explain why the result is known as a *retarded potential*.

5.30 The two-dimensional Green's function for the Helmholtz equation is the solution of

$$\nabla^2 \tilde{G}(\boldsymbol{\rho}|\boldsymbol{\rho}') + k^2 \tilde{G}(\boldsymbol{\rho}|\boldsymbol{\rho}') = -\delta(\boldsymbol{\rho} - \boldsymbol{\rho}')$$

where $\boldsymbol{\rho} = \hat{\mathbf{x}}x + \hat{\mathbf{y}}y$ is the two-dimensional position vector. Using spatial Fourier transforms, derive the following form of the Green's function:

$$\tilde{G}(\boldsymbol{\rho}|\boldsymbol{\rho}') = \frac{1}{4j} H_0^{(2)}(k|\boldsymbol{\rho} - \boldsymbol{\rho}'|)$$

where $H_0^{(2)}(\cdot)$ is the Hankel function of the second kind. (See § 7.4.1 for another approach.)

5.31 A dielectric slab of permittivity $\tilde{\epsilon}$ and permeability $\tilde{\mu}$ occupies the region $-a/2 \leq x \leq a/2$. Perfectly conducting plates are placed parallel to the slab at $x = \pm d/2$, forming a "covered" slab waveguide. Here $d > a$. Derive the characteristic equation for modes TM_z when the field $\tilde{E}_z$ is odd about $x = 0$.

6

Integral solutions of Maxwell's equations

6.1 Vector Kirchhoff solution: method of Stratton and Chu

One of the most powerful tools in electromagnetics is the integral solution to Maxwell's equations formulated by Stratton and Chu [183, 184]. These authors used the vector Green's theorem to solve for $\tilde{\mathbf{E}}$ and $\tilde{\mathbf{H}}$ in much the same way as is done in static fields with the scalar Green's theorem. An alternative approach is to use the Lorentz reciprocity theorem of § 4.10.2, as done by Fradin [64]. The reciprocity approach allows the identification of terms arising from surface discontinuities, which must be added to the result obtained from the other approach [183].

6.1.1 The Stratton–Chu formula

Consider an isotropic, homogeneous medium occupying a bounded region V in space. The medium is described by permeability $\tilde{\mu}(\omega)$, permittivity $\tilde{\epsilon}(\omega)$, and conductivity $\tilde{\sigma}(\omega)$. The region V is bounded by a surface S, which can be multiply connected so that S is the union of several surfaces $S_1, \ldots, S_N$ as shown in Figure 6.1; these are used to exclude unknown sources and to formulate the *vector Huygens principle*. Impressed electric and magnetic sources may thus reside both inside and outside V.

We wish to solve for the electric and magnetic fields at a point $\mathbf{r}$ within V. To do this we employ the Lorentz reciprocity theorem (4.170), written here using the frequency-domain fields as an integral over primed coordinates:

$$- \oint_S [\tilde{\mathbf{E}}_a(\mathbf{r}',\omega) \times \tilde{\mathbf{H}}_b(\mathbf{r}',\omega) - \tilde{\mathbf{E}}_b(\mathbf{r}',\omega) \times \tilde{\mathbf{H}}_a(\mathbf{r}',\omega)] \cdot \hat{\mathbf{n}}' dS'$$
$$= \int_V [\tilde{\mathbf{E}}_b(\mathbf{r}',\omega) \cdot \tilde{\mathbf{J}}_a(\mathbf{r}',\omega) - \tilde{\mathbf{E}}_a(\mathbf{r}',\omega) \cdot \tilde{\mathbf{J}}_b(\mathbf{r}',\omega)$$
$$- \tilde{\mathbf{H}}_b(\mathbf{r}',\omega) \cdot \tilde{\mathbf{J}}_{ma}(\mathbf{r}',\omega) + \tilde{\mathbf{H}}_a(\mathbf{r}',\omega) \cdot \tilde{\mathbf{J}}_{mb}(\mathbf{r}',\omega)] \, dV'. \tag{6.1}$$

Note that the negative sign on the left arises from the definition of $\hat{\mathbf{n}}$ as the *inward* normal to V, as shown in Figure 6.1. We place an electric Hertzian dipole at the point $\mathbf{r} = \mathbf{r}_p$ where we wish to compute the field, and set $\tilde{\mathbf{E}}_b = \tilde{\mathbf{E}}_p$ and $\tilde{\mathbf{H}}_b = \tilde{\mathbf{H}}_p$ in the reciprocity theorem, where $\tilde{\mathbf{E}}_p$ and $\tilde{\mathbf{H}}_p$ are the fields produced by the dipole (5.74)–(5.75):

$$\tilde{\mathbf{H}}_p(\mathbf{r},\omega) = j\omega\nabla \times [\tilde{\mathbf{p}}G(\mathbf{r}|\mathbf{r}_p;\omega)], \tag{6.2}$$

$$\tilde{\mathbf{E}}_p(\mathbf{r},\omega) = \frac{1}{\tilde{\epsilon}^c}\nabla \times (\nabla \times [\tilde{\mathbf{p}}G(\mathbf{r}|\mathbf{r}_p;\omega)]). \tag{6.3}$$

We also let $\tilde{\mathbf{E}}_a = \tilde{\mathbf{E}}$ and $\tilde{\mathbf{H}}_a = \tilde{\mathbf{H}}$, where $\tilde{\mathbf{E}}$ and $\tilde{\mathbf{H}}$ are the fields produced by the impressed sources $\tilde{\mathbf{J}}_a = \tilde{\mathbf{J}}^i$ and $\tilde{\mathbf{J}}_{ma} = \tilde{\mathbf{J}}_m^i$ within V that we wish to find at $\mathbf{r} = \mathbf{r}_p$. Since the dipole fields are singular at $\mathbf{r} = \mathbf{r}_p$, we must exclude the point $\mathbf{r}_p$ with a small spherical surface

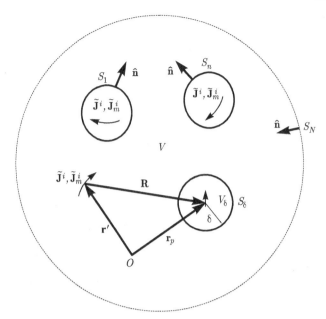

FIGURE 6.1
Geometry used to derive the Stratton–Chu formula.

S_δ surrounding the volume V_δ as shown in Figure 6.1. Substituting these fields into (6.1), we obtain

$$-\oint_{S+S_\delta}[\tilde{\mathbf{E}} \times \tilde{\mathbf{H}}_p - \tilde{\mathbf{E}}_p \times \tilde{\mathbf{H}}] \cdot \hat{\mathbf{n}}'\, dS' = \int_{V-V_\delta}[\tilde{\mathbf{E}}_p \cdot \tilde{\mathbf{J}}^i - \tilde{\mathbf{H}}_p \cdot \tilde{\mathbf{J}}_m^i]\, dV'. \qquad (6.4)$$

A useful identity involves the spatially constant vector $\tilde{\mathbf{p}}$ and the Green's function $\tilde{G}(\mathbf{r}'|\mathbf{r}_p)$:

$$\begin{aligned}
\nabla' \times [\nabla' \times (\tilde{G}\tilde{\mathbf{p}})] &= \nabla'[\nabla' \cdot (\tilde{G}\tilde{\mathbf{p}})] - \nabla'^2(\tilde{G}\tilde{\mathbf{p}}) \\
&= \nabla'[\nabla' \cdot (\tilde{G}\tilde{\mathbf{p}})] - \tilde{\mathbf{p}}\nabla'^2\tilde{G} \\
&= \nabla'(\tilde{\mathbf{p}} \cdot \nabla'\tilde{G}) + \tilde{\mathbf{p}}k^2\tilde{G}, \qquad (6.5)
\end{aligned}$$

where we have used $\nabla'^2\tilde{G} = -k^2\tilde{G}$ for $\mathbf{r}' \neq \mathbf{r}_p$.

We begin by computing the terms on the left side of (6.4). We suppress the $\mathbf{r}'$ dependence of the fields and also the dependencies of $\tilde{G}(\mathbf{r}'|\mathbf{r}_p)$. Substituting from (6.2), we have

$$\oint_{S+S_\delta}[\tilde{\mathbf{E}} \times \tilde{\mathbf{H}}_p] \cdot \hat{\mathbf{n}}'\, dS' = j\omega\oint_{S+S_\delta}[\tilde{\mathbf{E}} \times \nabla' \times (\tilde{G}\tilde{\mathbf{p}})] \cdot \hat{\mathbf{n}}'\, dS'.$$

Using $\hat{\mathbf{n}}' \cdot [\tilde{\mathbf{E}} \times \nabla' \times (\tilde{G}\tilde{\mathbf{p}})] = \hat{\mathbf{n}}' \cdot [\tilde{\mathbf{E}} \times (\nabla'\tilde{G} \times \tilde{\mathbf{p}})] = (\hat{\mathbf{n}}' \times \tilde{\mathbf{E}}) \cdot (\nabla'\tilde{G} \times \tilde{\mathbf{p}})$ we can write

$$\oint_{S+S_\delta}[\tilde{\mathbf{E}} \times \tilde{\mathbf{H}}_p] \cdot \hat{\mathbf{n}}'\, dS' = j\omega\tilde{\mathbf{p}} \cdot \oint_{S+S_\delta}[\hat{\mathbf{n}}' \times \tilde{\mathbf{E}}] \times \nabla'\tilde{G}\, dS'.$$

Next we examine

$$\oint_{S+S_\delta}[\tilde{\mathbf{E}}_p \times \tilde{\mathbf{H}}] \cdot \hat{\mathbf{n}}'\, dS' = -\frac{1}{\tilde{\epsilon}^c}\oint_{S+S_\delta}[\tilde{\mathbf{H}} \times \nabla' \times \nabla' \times (\tilde{G}\tilde{\mathbf{p}})] \cdot \hat{\mathbf{n}}'\, dS'.$$

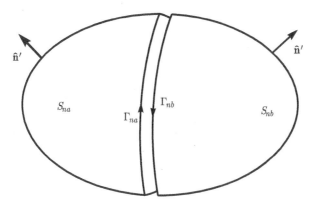

FIGURE 6.2
Decomposition of surface S_n to isolate surface field discontinuity.

Use of (6.5) along with the identity (B.49) gives

$$\oint_{S+S_\delta} [\tilde{\mathbf{E}}_p \times \tilde{\mathbf{H}}] \cdot \hat{\mathbf{n}}' \, dS' = -\frac{1}{\tilde{\epsilon}^c} \oint_{S+S_\delta} \{(\tilde{\mathbf{H}} \times \tilde{\mathbf{p}})k^2 \tilde{G}$$
$$- \nabla' \times [(\tilde{\mathbf{p}} \cdot \nabla' \tilde{G})\tilde{\mathbf{H}}] + (\tilde{\mathbf{p}} \cdot \nabla' \tilde{G})(\nabla' \times \tilde{\mathbf{H}})\} \cdot \hat{\mathbf{n}}' \, dS'.$$

We would like to use Stokes's theorem on the second term of the right-hand side. Since the theorem is not valid for surfaces on which $\tilde{\mathbf{H}}$ has discontinuities, we break the closed surfaces in Figure 6.1 into open surfaces whose boundary contours isolate the discontinuities as shown in Figure 6.2. Then we may write

$$\oint_{S_n=S_{na}+S_{nb}} \hat{\mathbf{n}}' \cdot \nabla' \times [(\tilde{\mathbf{p}} \cdot \nabla' \tilde{G})\tilde{\mathbf{H}}] \, dS' = \oint_{\Gamma_{na}+\Gamma_{nb}} \mathbf{dl}' \cdot \tilde{\mathbf{H}}(\tilde{\mathbf{p}} \cdot \nabla' \tilde{G}).$$

For surfaces not containing discontinuities of $\tilde{\mathbf{H}}$, the two contour integrals provide equal and opposite contributions and this term vanishes. Thus the left side of (6.4) is

$$-\oint_{S+S_\delta} [\tilde{\mathbf{E}} \times \tilde{\mathbf{H}}_p - \tilde{\mathbf{E}}_p \times \tilde{\mathbf{H}}] \cdot \hat{\mathbf{n}}' \, dS' =$$
$$-\frac{1}{\tilde{\epsilon}^c} \tilde{\mathbf{p}} \cdot \oint_{S+S_\delta} [j\omega \tilde{\epsilon}^c(\hat{\mathbf{n}}' \times \tilde{\mathbf{E}}) \times \nabla' \tilde{G} + k^2(\hat{\mathbf{n}}' \times \tilde{\mathbf{H}})\tilde{G} + \hat{\mathbf{n}}' \cdot (\tilde{\mathbf{J}}^i + j\omega \tilde{\epsilon}^c \tilde{\mathbf{E}})\nabla' \tilde{G}] \, dS'$$

where we have substituted $\tilde{\mathbf{J}}^i + j\omega \tilde{\epsilon}^c \tilde{\mathbf{E}}$ for $\nabla' \times \tilde{\mathbf{H}}$ and used $(\tilde{\mathbf{H}} \times \tilde{\mathbf{p}}) \cdot \hat{\mathbf{n}}' = \tilde{\mathbf{p}} \cdot (\hat{\mathbf{n}}' \times \tilde{\mathbf{H}})$.
Now consider the right side of (6.4). By (6.3) we have

$$\int_{V-V_\delta} \tilde{\mathbf{E}}_p \cdot \tilde{\mathbf{J}}^i \, dV' = \frac{1}{\tilde{\epsilon}^c} \int_{V-V_\delta} \tilde{\mathbf{J}}^i \cdot \nabla' \times \nabla' \times (\tilde{\mathbf{p}}G)] \, dV'.$$

Using (6.5) and (B.48), we have

$$\int_{V-V_\delta} \tilde{\mathbf{E}}_p \cdot \tilde{\mathbf{J}}^i \, dV' = \frac{1}{\tilde{\epsilon}^c} \int_{V-V_\delta} \{k^2 (\tilde{\mathbf{p}} \cdot \tilde{\mathbf{J}}^i)\tilde{G} + \nabla' \cdot [\tilde{\mathbf{J}}^i(\tilde{\mathbf{p}} \cdot \nabla' \tilde{G})] - (\tilde{\mathbf{p}} \cdot \nabla' \tilde{G})\nabla' \cdot \tilde{\mathbf{J}}^i\} \, dV'.$$

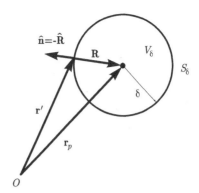

FIGURE 6.3
Geometry of surface integral used to extract $\mathbf{E}$ at $\mathbf{r}_p$.

Replacing $\nabla' \cdot \tilde{\mathbf{J}}^i$ with $-j\omega\tilde{\rho}^i$ from the continuity equation and using the divergence theorem on the second term on the right-hand side, we then have

$$\int_{V-V_\delta} \tilde{\mathbf{E}}_p \cdot \tilde{\mathbf{J}}^i \, dV' = \frac{1}{\tilde{\epsilon}^c}\tilde{\mathbf{p}} \cdot \left[\int_{V-V_\delta} (k^2\tilde{\mathbf{J}}^i\tilde{G} + j\omega\tilde{\rho}^i\nabla'\tilde{G}) \, dV' - \oint_{S+S_\delta} (\hat{\mathbf{n}}' \cdot \tilde{\mathbf{J}}^i)\nabla'\tilde{G} \, dS' \right].$$

Lastly we examine

$$\int_{V-V_\delta} \tilde{\mathbf{H}}_p \cdot \tilde{\mathbf{J}}_m^i \, dV' = j\omega \int_{V-V_\delta} \tilde{\mathbf{J}}_m^i \cdot \nabla' \times (\tilde{G}\tilde{\mathbf{p}}) \, dV'.$$

Use of $\tilde{\mathbf{J}}_m^i \cdot \nabla' \times (\tilde{G}\tilde{\mathbf{p}}) = \tilde{\mathbf{J}}_m^i \cdot (\nabla'\tilde{G} \times \tilde{\mathbf{p}}) = \tilde{\mathbf{p}} \cdot (\tilde{\mathbf{J}}_m^i \times \nabla'\tilde{G})$ gives

$$\int_{V-V_\delta} \tilde{\mathbf{H}}_p \cdot \tilde{\mathbf{J}}_m^i \, dV' = j\omega\tilde{\mathbf{p}} \cdot \int_{V-V_\delta} \tilde{\mathbf{J}}_m^i \times \nabla'\tilde{G} \, dV'.$$

We now substitute all terms into (6.4) and note that each term involves a dot product with $\tilde{\mathbf{p}}$. But $\tilde{\mathbf{p}}$ is arbitrary so

$$-\oint_{S+S_\delta} [(\hat{\mathbf{n}}' \times \tilde{\mathbf{E}}) \times \nabla'G + (\hat{\mathbf{n}}' \cdot \tilde{\mathbf{E}})\nabla'\tilde{G} - j\omega\tilde{\mu}(\hat{\mathbf{n}}' \times \tilde{\mathbf{H}}\tilde{G}] \, dS'$$
$$+ \frac{1}{j\omega\tilde{\epsilon}^c} \oint_{\Gamma_a+\Gamma_b} (\mathbf{dl}' \cdot \tilde{\mathbf{H}})\nabla'\tilde{G} = \int_{V-V_\delta} \left[-\tilde{\mathbf{J}}_m^i \times \nabla'\tilde{G} + \frac{\tilde{\rho}^i}{\tilde{\epsilon}^c}\nabla'\tilde{G} - j\omega\tilde{\mu}\tilde{\mathbf{J}}^i\tilde{G} \right] dV'.$$

The electric field may be extracted from the above expression by letting the radius of the excluding volume V_δ approach zero. We first consider the surface integral over S_δ. Figure 6.3 shows that $R = |\mathbf{r}_p - \mathbf{r}'| = \delta$, $\hat{\mathbf{n}}' = -\hat{\mathbf{R}}$, and

$$\nabla'\tilde{G}(\mathbf{r}'|\mathbf{r}_p) = \frac{d}{dR}\left(\frac{e^{-jkR}}{4\pi R}\right)\nabla'R = \hat{\mathbf{R}}\left(\frac{1+jk\delta}{4\pi\delta^2}\right)e^{-jk\delta} \approx \frac{\hat{\mathbf{R}}}{\delta^2} \quad \text{as } \delta \to 0.$$

Assuming $\tilde{\mathbf{E}}$ is continuous at $\mathbf{r}' = \mathbf{r}_p$, we can write

$$
-\lim_{\delta \to 0} \oint_{S_\delta} [(\hat{\mathbf{n}}' \times \tilde{\mathbf{E}}) \times \nabla' \tilde{G} + (\hat{\mathbf{n}}' \cdot \tilde{\mathbf{E}}) \nabla' \tilde{G} - j\omega\tilde{\mu}(\hat{\mathbf{n}}' \times \tilde{\mathbf{H}})\tilde{G}] \, dS'
$$
$$
= \lim_{\delta \to 0} \int_\Omega \frac{1}{4\pi} \left[(\hat{\mathbf{R}} \times \tilde{\mathbf{E}}) \times \frac{\hat{\mathbf{R}}}{\delta^2} + (\hat{\mathbf{R}} \cdot \tilde{\mathbf{E}})\frac{\hat{\mathbf{R}}}{\delta^2} - j\omega\tilde{\mu}(\hat{\mathbf{R}} \times \tilde{\mathbf{H}})\frac{1}{\delta} \right] \delta^2 \, d\Omega
$$
$$
= \lim_{\delta \to 0} \int_\Omega \frac{1}{4\pi} \left[-(\hat{\mathbf{R}} \cdot \tilde{\mathbf{E}})\hat{\mathbf{R}} + (\hat{\mathbf{R}} \cdot \hat{\mathbf{R}})\tilde{\mathbf{E}} + (\hat{\mathbf{R}} \cdot \tilde{\mathbf{E}})\hat{\mathbf{R}} \right] d\Omega = \tilde{\mathbf{E}}(\mathbf{r}_p).
$$

Here we have used $\int_\Omega d\Omega = 4\pi$ for the total solid angle subtending the sphere S_δ. Finally, assuming that the volume sources are continuous, the volume integral over V_δ vanishes and we have

$$
\tilde{\mathbf{E}}(\mathbf{r}, \omega) = \int_V \left(-\tilde{\mathbf{J}}_m^i \times \nabla' \tilde{G} + \frac{\tilde{\rho}^i}{\tilde{\epsilon}^c} \nabla' \tilde{G} - j\omega\tilde{\mu}\tilde{\mathbf{J}}^i \tilde{G} \right) dV'
$$
$$
+ \sum_{n=1}^N \oint_{S_n} \left[(\hat{\mathbf{n}}' \times \tilde{\mathbf{E}}) \times \nabla' \tilde{G} + (\hat{\mathbf{n}}' \cdot \tilde{\mathbf{E}}) \nabla' \tilde{G} - j\omega\tilde{\mu}(\hat{\mathbf{n}}' \times \tilde{\mathbf{H}})\tilde{G} \right] dS'
$$
$$
- \sum_{n=1}^N \frac{1}{j\omega\tilde{\epsilon}^c} \oint_{\Gamma_{na}+\Gamma_{nb}} (\mathbf{dl}' \cdot \tilde{\mathbf{H}})\nabla' \tilde{G}. \tag{6.6}
$$

A similar formula for $\tilde{\mathbf{H}}$ can be derived by placing a magnetic dipole of moment $\tilde{\mathbf{p}}_m$ at $\mathbf{r} = \mathbf{r}_p$ and proceeding as above. This leads to

$$
\tilde{\mathbf{H}}(\mathbf{r}, \omega) = \int_V \left(\tilde{\mathbf{J}}^i \times \nabla' \tilde{G} + \frac{\tilde{\rho}_m^i}{\tilde{\mu}} \nabla' \tilde{G} - j\omega\tilde{\epsilon}^c\tilde{\mathbf{J}}_m^i \tilde{G} \right) dV'
$$
$$
+ \sum_{n=1}^N \oint_{S_n} \left[(\hat{\mathbf{n}}' \times \tilde{\mathbf{H}}) \times \nabla' \tilde{G} + (\hat{\mathbf{n}}' \cdot \tilde{\mathbf{H}}) \nabla' \tilde{G} + j\omega\tilde{\epsilon}^c(\hat{\mathbf{n}}' \times \tilde{\mathbf{E}})\tilde{G} \right] dS'
$$
$$
+ \sum_{n=1}^N \frac{1}{j\omega\tilde{\mu}} \oint_{\Gamma_{na}+\Gamma_{nb}} (\mathbf{dl}' \cdot \tilde{\mathbf{E}})\nabla' \tilde{G}. \tag{6.7}
$$

We can also obtain (6.7) by substituting (6.6) into Faraday's law.

▶ **Example 6.1:** Specialization of the Stratton–Chu formulas for static fields

Specialize the Stratton–Chu formulas (6.6) and (6.7) for the case of static fields when the sources are embedded in unbounded space. Show that for the case of electric sources in unbounded space, the expected forms of Coulomb's law and the Biot-Savart law are obtained.

Solution: The surface and line integrals in the Stratton–Chu formulas are not needed for unbounded space. The specialization to statics may be obtained by setting $\omega = 0$, which also results in $k = 0$. Under these conditions (6.6) becomes

$$
\mathbf{E}(\mathbf{r}) = \int_V \left(-\mathbf{J}_m^i \times \nabla' G + \frac{\rho^i}{\epsilon} \nabla' G \right) dV', \tag{6.8}
$$

while (6.7) becomes

$$\mathbf{H}(\mathbf{r}) = \int_V \left(\mathbf{J}^i \times \nabla' G + \frac{\rho_m^i}{\mu} \nabla' G \right) dV'. \tag{6.9}$$

Here ρ^i and ρ_m^i are static charge densities, $\mathbf{J}^i$ and $\mathbf{J}_m^i$ are steady current densities, and G is the static Green's function

$$G(\mathbf{r}|\mathbf{r}') = \frac{1}{4\pi|\mathbf{r} - \mathbf{r}'|}.$$

When the magnetic sources are zero, the electric field is from (6.8)

$$\mathbf{E}(\mathbf{r}) = \int_V \frac{\rho^i}{\epsilon} \nabla' G \, dV'.$$

Using

$$\nabla' G = \frac{1}{4\pi} \frac{\mathbf{r} - \mathbf{r}'}{|\mathbf{r} - \mathbf{r}'|^3}$$

we recover Coulomb's law (3.43):

$$\mathbf{E}(\mathbf{r}) = \frac{1}{4\pi\epsilon} \int_V \rho^i(\mathbf{r}') \frac{\mathbf{r} - \mathbf{r}'}{|\mathbf{r} - \mathbf{r}'|^3} \, dV'.$$

When the magnetic sources are zero, the magnetic field is from (6.9)

$$\mathbf{H}(\mathbf{r}) = \int_V \mathbf{J}^i \times \nabla' G \, dV'.$$

Again substituting for $\nabla' G$ and using $\mathbf{B} = \mu\mathbf{H}$, we get

$$\mathbf{B}(\mathbf{r}) = \frac{\mu}{4\pi} \int_V \mathbf{J}^i(\mathbf{r}') \times \frac{\mathbf{r} - \mathbf{r}'}{|\mathbf{r} - \mathbf{r}'|^3} \, dV',$$

which is the Biot–Savart law (3.131). ◀

▶ **Example 6.2:** Static Ampere's law from the Stratton–Chu formulas

Obtain from the static specialization for the magnetic field in unbounded space (6.9) the point form of Ampere's law.

Solution: Taking the curl of (6.9), we have

$$\nabla \times \mathbf{H} = \int_V \nabla \times \left[\frac{\rho_m^i(\mathbf{r}')}{\mu} \nabla' G \right] dV' + \int_V \nabla \times [\mathbf{J}^i(\mathbf{r}') \times \nabla' G] \, dV'. \tag{6.10}$$

But

$$\nabla \times \left[\frac{\rho_m^i(\mathbf{r}')}{\mu} \nabla' G \right] = \frac{\rho_m^i(\mathbf{r}')}{\mu} \nabla \times \nabla' G - \nabla' G \times \nabla \left(\frac{\rho_m^i(\mathbf{r}')}{\mu} \right)$$

$$= \frac{\rho_m^i(\mathbf{r}')}{\mu} \nabla \times \nabla' G$$

since $\nabla \rho_m^i(\mathbf{r}') = 0$. We also have

$$\nabla \times \nabla' G = -\nabla \times \nabla G = 0,$$

so the first integral in (6.10) is zero. Next, note that

$$\nabla \times \left[\mathbf{J}^i(\mathbf{r}') \times \nabla'G\right] = -\nabla \times \left[\mathbf{J}^i(\mathbf{r}') \times \nabla G\right]$$
$$= -\nabla \times \left[G\nabla \times \mathbf{J}^i(\mathbf{r}') - \nabla \times \left(G\mathbf{J}^i(\mathbf{r}')\right)\right]$$
$$= \nabla \times \nabla \times \left[G\mathbf{J}^i(\mathbf{r}')\right]$$

since $\nabla \times \mathbf{J}^i(\mathbf{r}') = 0$. Expansion of the $\nabla \times \nabla\times$ operation gives

$$\nabla \times \mathbf{H} = \int_V \left(\nabla\left\{\nabla \cdot [G\mathbf{J}^i(\mathbf{r}')]\right\} - \nabla^2[G\mathbf{J}^i(\mathbf{r}')]\right) dV'.$$

Use of

$$\nabla \cdot [G\mathbf{J}^i(\mathbf{r}')] = G\nabla \cdot \mathbf{J}^i(\mathbf{r}') + \mathbf{J}^i(\mathbf{r}') \cdot \nabla G$$
$$= -\mathbf{J}^i(\mathbf{r}') \cdot \nabla'G$$

along with

$$\nabla^2[G\mathbf{J}^i(\mathbf{r}')] = G\nabla^2\mathbf{J}^i(\mathbf{r}') + \mathbf{J}^i(\mathbf{r}')\nabla^2G + 2(\nabla G \cdot \nabla)\mathbf{J}^i(\mathbf{r}') = \mathbf{J}^i(\mathbf{r}')\nabla^2G$$

where $\nabla^2G = -4\pi\delta(\mathbf{r} - \mathbf{r}')$, gives

$$\nabla \times \mathbf{H} = -\nabla \int_V [\mathbf{J}^i(\mathbf{r}') \cdot \nabla'G] \, dV' + \mathbf{J}^i(\mathbf{r}).$$

As a last step, we write the remaining integral as

$$\int_V [\mathbf{J}^i(\mathbf{r}') \cdot \nabla'G] \, dV' = \int_V \left(-\nabla' \cdot [\mathbf{J}^i(\mathbf{r}')G] + G\nabla' \cdot \mathbf{J}^i(\mathbf{r}')\right) dV'.$$

For steady currents, $\nabla' \cdot \mathbf{J}(\mathbf{r}') = 0$ and the divergence theorem gives

$$\int_V [\mathbf{J}^i(\mathbf{r}') \cdot \nabla'G] \, dV' = -\oint_{S\to\infty} [\mathbf{J}^i(\mathbf{r}')G] \cdot \hat{\mathbf{n}}' \, dS'.$$

But all sources are bounded so that this last integral is zero, and we have, finally, $\nabla \times \mathbf{H} = \mathbf{J}^i$ as desired. ◄

6.1.2 The Sommerfeld radiation condition

We saw (§ 5.2.9) that if the potentials are not to be influenced by effects that are infinitely removed, they must obey a radiation condition. We can make the same argument about the fields from (6.6) and (6.7). Let us allow one of the excluding surfaces, say S_N, to recede to infinity (enclosing all of the sources as it expands). Under this limit passage, any contributions from the fields on this surface to the fields at $\mathbf{r}$ should vanish.

Letting S_N be a sphere centered at the origin, we note that $\hat{\mathbf{n}}' = -\hat{\mathbf{r}}'$ and that as $r' \to \infty$

$$\tilde{G}(\mathbf{r}|\mathbf{r}';\omega) = \frac{e^{-jk|\mathbf{r}-\mathbf{r}'|}}{4\pi|\mathbf{r}-\mathbf{r}'|} \approx \frac{e^{-jkr'}}{4\pi r'},$$

$$\nabla'\tilde{G}(\mathbf{r}|\mathbf{r}';\omega) = \hat{\mathbf{R}}\left(\frac{1+jkR}{4\pi R^2}\right)e^{-jkR} \approx -\hat{\mathbf{r}}'\left(\frac{1+jkr'}{r'}\right)\frac{e^{-jkr'}}{4\pi r'}.$$

Substituting these expressions into (6.6) we find that

$$\lim_{S_N \to S_\infty} \oint_{S_N} [(\hat{\mathbf{n}}' \times \tilde{\mathbf{E}}) \times \nabla' \tilde{G} + (\hat{\mathbf{n}}' \cdot \tilde{\mathbf{E}}) \nabla' \tilde{G} - j\omega\tilde{\mu}(\hat{\mathbf{n}}' \times \tilde{\mathbf{H}})\tilde{G}] \, dS'$$

$$\approx \lim_{r' \to \infty} \int_0^{2\pi} \int_0^\pi \left\{ [(\hat{\mathbf{r}}' \times \tilde{\mathbf{E}}) \times \hat{\mathbf{r}}' + (\hat{\mathbf{r}}' \cdot \tilde{\mathbf{E}})\hat{\mathbf{r}}'] \left(\frac{1+jkr'}{r'} \right) \right.$$

$$\left. + j\omega\tilde{\mu}(\hat{\mathbf{r}}' \times \tilde{\mathbf{H}}) \right\} \frac{e^{-jkr'}}{4\pi r'} r'^2 \sin\theta' \, d\theta' \, d\phi'$$

$$\approx \lim_{r' \to \infty} \int_0^{2\pi} \int_0^\pi \{r'[jk\tilde{\mathbf{E}} + j\omega\tilde{\mu}(\hat{\mathbf{r}}' \times \tilde{\mathbf{H}})] + \tilde{\mathbf{E}}\} \frac{e^{-jkr'}}{4\pi} \sin\theta' \, d\theta' \, d\phi'.$$

Since this gives the contribution to the field in V from the fields on the surface receding to infinity, we expect that this term should be zero. If the medium has loss, the exponential term decays and drives the contribution to zero. For a lossless medium, the contributions are zero if

$$\lim_{r \to \infty} r\tilde{\mathbf{E}}(\mathbf{r}, \omega) < \infty, \tag{6.11}$$

$$\lim_{r \to \infty} r \left[\hat{\mathbf{r}} \times \eta\tilde{\mathbf{H}}(\mathbf{r}, \omega) + \tilde{\mathbf{E}}(\mathbf{r}, \omega) \right] = 0. \tag{6.12}$$

To accompany (6.7) we also have

$$\lim_{r \to \infty} r\tilde{\mathbf{H}}(\mathbf{r}, \omega) < \infty, \tag{6.13}$$

$$\lim_{r \to \infty} r \left[\eta\tilde{\mathbf{H}}(\mathbf{r}, \omega) - \hat{\mathbf{r}} \times \tilde{\mathbf{E}}(\mathbf{r}, \omega) \right] = 0. \tag{6.14}$$

We refer to (6.11) and (6.13) as the *finiteness conditions*, and to (6.12) and (6.14) as the *Sommerfeld radiation condition*, for the electromagnetic field. They show that far from the sources, the fields must behave as a wave TEM to the r-direction. We shall see in § 6.2 that the waves are in fact *spherical TEM waves*.

6.1.3   Fields in the excluded region: the extinction theorem

The Stratton–Chu formula provides a solution for the field within V, external to the excluded regions. As an interesting consequence, and one that helps us identify the equivalence principle, it gives the null result $\tilde{\mathbf{H}} = \tilde{\mathbf{E}} = 0$ when evaluated at points within the excluded regions.

Let us consider two cases. In the first case we do *not* exclude the particular region V_m, but do exclude the remaining regions V_n for $n \neq m$. The electric field everywhere outside the remaining excluded regions (including at points within V_m) is, by (6.6),

$$\tilde{\mathbf{E}}(\mathbf{r}, \omega) = \int_{V+V_m} \left(-\tilde{\mathbf{J}}_m^i \times \nabla' \tilde{G} + \frac{\tilde{\rho}^i}{\tilde{\epsilon}^c} \nabla' \tilde{G} - j\omega\tilde{\mu}\tilde{\mathbf{J}}^i \tilde{G} \right) dV'$$

$$+ \sum_{n \neq m} \oint_{S_n} [(\hat{\mathbf{n}}' \times \tilde{\mathbf{E}}) \times \nabla' \tilde{G} + (\hat{\mathbf{n}}' \cdot \tilde{\mathbf{E}}) \nabla' \tilde{G} - j\omega\tilde{\mu}(\hat{\mathbf{n}}' \times \tilde{\mathbf{H}})\tilde{G}] \, dS'$$

$$- \sum_{n \neq m} \frac{1}{j\omega\tilde{\epsilon}^c} \oint_{\Gamma_{na}+\Gamma_{nb}} (\mathbf{dl}' \cdot \tilde{\mathbf{H}})\nabla' \tilde{G} \qquad (\mathbf{r} \in V + V_m).$$

In the second case we apply the Stratton–Chu formula only to V_m, and exclude all other regions. We incur a sign change on the surface and line integrals compared to the first case because the normal is now directed oppositely. By (6.6) we have

$$\tilde{\mathbf{E}}(\mathbf{r},\omega) = \int_{V_m} \left(-\tilde{\mathbf{J}}^i_m \times \nabla'\tilde{G} + \frac{\tilde{\rho}^i}{\tilde{\epsilon}^c}\nabla'\tilde{G} - j\omega\tilde{\mu}\tilde{\mathbf{J}}^i\tilde{G} \right) dV'$$

$$- \oint_{S_m} [(\hat{\mathbf{n}}' \times \tilde{\mathbf{E}}) \times \nabla'\tilde{G} + (\hat{\mathbf{n}}' \cdot \tilde{\mathbf{E}})\nabla'\tilde{G} - j\omega\tilde{\mu}(\hat{\mathbf{n}}' \times \tilde{\mathbf{H}})\tilde{G}] \, dS'$$

$$+ \frac{1}{j\omega\tilde{\epsilon}^c} \oint_{\Gamma_{ma}+\Gamma_{mb}} (\mathbf{dl}' \cdot \tilde{\mathbf{H}})\nabla'\tilde{G} \qquad (\mathbf{r} \in V_m).$$

Both expressions for $\tilde{\mathbf{E}}$ hold at points of V_m; subtraction gives

$$0 = \int_V \left(-\tilde{\mathbf{J}}^i_m \times \nabla'\tilde{G} + \frac{\tilde{\rho}^i}{\tilde{\epsilon}^c}\nabla'\tilde{G} - j\omega\tilde{\mu}\tilde{\mathbf{J}}^i\tilde{G} \right) dV'$$

$$+ \sum_{n=1}^{N} \oint_{S_n} [(\hat{\mathbf{n}}' \times \tilde{\mathbf{E}}) \times \nabla'\tilde{G} + (\hat{\mathbf{n}}' \cdot \tilde{\mathbf{E}})\nabla'\tilde{G} - j\omega\tilde{\mu}(\hat{\mathbf{n}}' \times \tilde{\mathbf{H}})\tilde{G}] \, dS'$$

$$- \sum_{n=1}^{N} \frac{1}{j\omega\tilde{\epsilon}^c} \oint_{\Gamma_{na}+\Gamma_{nb}} (\mathbf{dl}' \cdot \tilde{\mathbf{H}})\nabla'\tilde{G} \qquad (\mathbf{r} \in V_m).$$

This expression is exactly the Stratton–Chu formula (6.6) evaluated at points within the excluded region V_m. The treatment of $\tilde{\mathbf{H}}$ is analogous and is left as an exercise. Since we may repeat this for any excluded region, we find that the Stratton–Chu formula returns the null field when evaluated at points outside V. This is sometimes called the *vector Ewald–Oseen extinction theorem* [91]. We emphasize that the fields within the excluded regions are *not* generally zero; the Stratton–Chu formula merely returns this result when evaluated there.

6.2   Fields in an unbounded medium

Two special cases of the Stratton–Chu formula are important because of their application to antenna theory. The first is that of sources radiating into an unbounded region. The second involves a bounded region with all sources excluded. We shall consider the former here and the latter in § 6.3.

If there are no bounding surfaces in (6.6) and (6.7), except for one surface that has been allowed to recede to infinity and therefore provides no surface contribution, we find that the electromagnetic fields in unbounded space are given by

$$\tilde{\mathbf{E}} = \int_V \left(-\tilde{\mathbf{J}}^i_m \times \nabla'\tilde{G} + \frac{\tilde{\rho}^i}{\tilde{\epsilon}^c}\nabla'\tilde{G} - j\omega\tilde{\mu}\tilde{\mathbf{J}}^i\tilde{G} \right) dV',$$

$$\tilde{\mathbf{H}} = \int_V \left(\tilde{\mathbf{J}}^i \times \nabla'\tilde{G} + \frac{\tilde{\rho}^i_m}{\tilde{\mu}}\nabla'\tilde{G} - j\omega\tilde{\epsilon}^c\tilde{\mathbf{J}}^i_m\tilde{G} \right) dV'.$$

We can view the right-hand sides as superpositions of the fields present in the cases where (1) electric sources are present exclusively, and (2) magnetic sources are present exclusively. With $\tilde{\rho}_m^i = 0$ and $\tilde{\mathbf{J}}_m^i = 0$ we find that

$$\tilde{\mathbf{E}} = \int_V \left(\frac{\tilde{\rho}^i}{\tilde{\epsilon}^c} \nabla' \tilde{G} - j\omega\tilde{\mu}\tilde{\mathbf{J}}^i \tilde{G} \right) dV', \tag{6.15}$$

$$\tilde{\mathbf{H}} = \int_V \tilde{\mathbf{J}}^i \times \nabla' \tilde{G} \, dV'.$$

But $\nabla' \tilde{G} = -\nabla \tilde{G}$, so

$$\tilde{\mathbf{E}}(\mathbf{r}, \omega) = -\nabla\tilde{\phi}_e(\mathbf{r}, \omega) - j\omega\tilde{\mathbf{A}}_e(\mathbf{r}, \omega)$$

where

$$\tilde{\phi}_e(\mathbf{r}, \omega) = \int_V \frac{\tilde{\rho}^i(\mathbf{r}', \omega)}{\tilde{\epsilon}^c(\omega)} \tilde{G}(\mathbf{r}|\mathbf{r}'; \omega) \, dV',$$

$$\tilde{\mathbf{A}}_e(\mathbf{r}, \omega) = \int_V \tilde{\mu}(\omega)\tilde{\mathbf{J}}^i(\mathbf{r}', \omega)\tilde{G}(\mathbf{r}|\mathbf{r}'; \omega) \, dV', \tag{6.16}$$

are the electric scalar and vector potential functions introduced in § 5.2. Using

$$\tilde{\mathbf{J}}^i \times \nabla' \tilde{G} = -\tilde{\mathbf{J}}^i \times \nabla \tilde{G} = \nabla \times (\tilde{\mathbf{J}}^i \tilde{G}),$$

we have

$$\tilde{\mathbf{H}}(\mathbf{r}, \omega) = \frac{1}{\tilde{\mu}(\omega)} \nabla \times \int_V \tilde{\mu}(\omega)\tilde{\mathbf{J}}^i(\mathbf{r}', \omega)\tilde{G}(\mathbf{r}|\mathbf{r}'; \omega) \, dV' = \frac{1}{\tilde{\mu}(\omega)} \nabla \times \tilde{\mathbf{A}}_e(\mathbf{r}, \omega). \tag{6.17}$$

These expressions for the fields are identical to those of (5.50) and (5.51), hence the integral formula for the electromagnetic fields produces a result identical to that obtained using potential relations. Similarly, with $\tilde{\rho}^i = 0$ and $\tilde{\mathbf{J}}^i = 0$, we have

$$\tilde{\mathbf{E}} = -\int_V \tilde{\mathbf{J}}_m^i \times \nabla' \tilde{G} \, dV',$$

$$\tilde{\mathbf{H}} = \int_V \left(\frac{\tilde{\rho}_m^i}{\tilde{\mu}} \nabla' \tilde{G} - j\omega\tilde{\epsilon}^c\tilde{\mathbf{J}}_m^i \tilde{G} \right) dV',$$

or

$$\tilde{\mathbf{E}}(\mathbf{r}, \omega) = -\frac{1}{\tilde{\epsilon}^c(\omega)} \nabla \times \tilde{\mathbf{A}}_h(\mathbf{r}, \omega),$$

$$\tilde{\mathbf{H}}(\mathbf{r}, \omega) = -\nabla\tilde{\phi}_h(\mathbf{r}, \omega) - j\omega\tilde{\mathbf{A}}_h(\mathbf{r}, \omega),$$

where

$$\tilde{\phi}_h(\mathbf{r}, \omega) = \int_V \frac{\tilde{\rho}_m^i(\mathbf{r}', \omega)}{\tilde{\mu}(\omega)} \tilde{G}(\mathbf{r}|\mathbf{r}'; \omega) \, dV',$$

$$\tilde{\mathbf{A}}_h(\mathbf{r}, \omega) = \int_V \tilde{\epsilon}^c(\omega)\tilde{\mathbf{J}}_m^i(\mathbf{r}', \omega)\tilde{G}(\mathbf{r}|\mathbf{r}'; \omega) \, dV',$$

are the magnetic scalar and vector potentials introduced in § 5.2.

► **Example 6.3:** Fields of a dipole antenna

A dipole antenna consists of a thin wire of length $2l$ and radius a, fed at the center by a voltage generator, as shown in Figure 6.4. The antenna is embedded in an unbounded, lossless medium with parameters μ, ϵ. The generator induces an impressed current on the surface of the wire, which in turn radiates an electromagnetic wave. For very thin wires ($a \ll \lambda$, $a \ll l$) embedded in a lossless medium, the current may be approximated to reasonable accuracy using a standing wave distribution:

$$\tilde{\mathbf{J}}^i(\mathbf{r}, \omega) = \hat{\mathbf{z}} \tilde{I}(\omega) \sin\left[k(l - |z|)\right] \delta(x) \delta(y). \tag{6.18}$$

Compute the field produced by the dipole antenna.

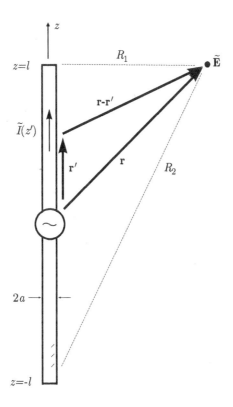

FIGURE 6.4
Dipole antenna in a lossless unbounded medium.

Solution: We use (6.16) to get the vector potential, (6.17) to get the magnetic field, and Ampere's law to get the electric field. Substituting the current expression into (6.16) and integrating over x and y, we have

$$\tilde{\mathbf{A}}_e(\mathbf{r}, \omega) = \hat{\mathbf{z}} \frac{\mu \tilde{I}}{4\pi} \int_{-l}^{l} \sin k(l - |z'|) \frac{e^{-jkR}}{R} \, dz' \tag{6.19}$$

where $R = \sqrt{(z - z')^2 + \rho^2}$ and $\rho^2 = x^2 + y^2$. By (6.17),

$$\tilde{\mathbf{H}} = \nabla \times \frac{1}{\mu} \tilde{\mathbf{A}}_e = -\hat{\phi} \frac{1}{\mu} \frac{\partial \tilde{A}_{ez}}{\partial \rho}.$$

Writing the sine function in (6.19) in terms of exponentials, we then have

$$\tilde{H}_\phi = j\frac{\tilde{I}}{8\pi}\left[e^{jkl}\int_{-l}^{0}\frac{\partial}{\partial\rho}\frac{e^{-jk(R-z')}}{R}\,dz' - e^{-jkl}\int_{-l}^{0}\frac{\partial}{\partial\rho}\frac{e^{-jk(R+z')}}{R}\,dz'\right.$$
$$\left. + e^{jkl}\int_{0}^{l}\frac{\partial}{\partial\rho}\frac{e^{-jk(R+z')}}{R}\,dz' - e^{-jkl}\int_{0}^{l}\frac{\partial}{\partial\rho}\frac{e^{-jk(R-z')}}{R}\,dz'\right].$$

Noting that

$$\frac{\partial}{\partial\rho}\frac{e^{-jk(R\pm z')}}{R} = \pm\rho\frac{\partial}{\partial z'}\frac{e^{-jk(R\pm z')}}{R\left[R\mp(z-z')\right]} = -\rho\frac{1+jkR}{R^3}e^{-jk(R\pm z')}$$

we can write

$$\tilde{H}_\phi = j\frac{\tilde{I}\rho}{8\pi}\left[-e^{jkl}\frac{e^{-jk(R-z')}}{R\left[R+(z-z')\right]}\bigg|_{-l}^{0} - e^{-jkl}\frac{e^{-jk(R+z')}}{R\left[R-(z-z')\right]}\bigg|_{-l}^{0}\right.$$
$$\left. + e^{jkl}\frac{e^{-jk(R+z')}}{R\left[R-(z-z')\right]}\bigg|_{0}^{l} + e^{-jkl}\frac{e^{-jk(R-z')}}{R\left[R+(z-z')\right]}\bigg|_{0}^{l}\right].$$

Collecting terms and simplifying, we get

$$\tilde{H}_\phi(\mathbf{r},\omega) = j\frac{\tilde{I}(\omega)}{4\pi\rho}\left[e^{-jkR_1} + e^{-jkR_2} - (2\cos kl)e^{-jkr}\right] \tag{6.20}$$

where $R_1 = \sqrt{\rho^2 + (z-l)^2}$ and $R_2 = \sqrt{\rho^2 + (z+l)^2}$. For points external to the dipole, the source current is zero, and thus

$$\tilde{\mathbf{E}}(\mathbf{r},\omega) = \frac{1}{j\omega\epsilon}\nabla\times\tilde{\mathbf{H}}(\mathbf{r},\omega) = \frac{1}{j\omega\epsilon}\left\{-\hat{\rho}\frac{\partial}{\partial z}\tilde{H}_\phi(\mathbf{r},\omega) + \hat{z}\frac{1}{\rho}\frac{\partial}{\partial\rho}[\rho\tilde{H}_\phi(\mathbf{r},\omega)]\right\}.$$

Performing the derivatives, we have

$$\tilde{E}_\rho(\mathbf{r},\omega) = j\frac{\eta\tilde{I}(\omega)}{4\pi}\left[\frac{z-l}{\rho}\frac{e^{-jkR_1}}{R_1} + \frac{z+l}{\rho}\frac{e^{-jkR_2}}{R_2} - \frac{z}{\rho}(2\cos kl)\frac{e^{-jkr}}{r}\right], \tag{6.21}$$

$$\tilde{E}_z(\mathbf{r},\omega) = -j\frac{\eta\tilde{I}(\omega)}{4\pi}\left[\frac{e^{-jkR_1}}{R_1} + \frac{e^{-jkR_2}}{R_2} - (2\cos kl)\frac{e^{-jkr}}{r}\right]. \tag{6.22}$$

These closed-form expressions for the near-zone fields will be used in § 6.4.4 to find an approximate formula for the input impedance of a dipole antenna. ◄

6.2.1   The far-zone fields produced by sources in unbounded space

Many antennas may be analyzed in terms of electric currents and charges radiating in unbounded space. Since antennas serve to transmit information over great distances, the fields far from the sources are often sought. Assume the sources are contained within a sphere of radius r_s centered at the origin. We define the *far zone* of the sources to consist of all observation points satisfying both $r \gg r_s$ (and thus $r \gg r'$) and $kr \gg 1$. At such points we may approximate the unit vector $\hat{\mathbf{R}}$ directed from the sources to the observation point by the unit vector $\hat{\mathbf{r}}$ directed from the origin to the observation point. Moreover,

$$\nabla'\tilde{G} = \frac{d}{dR}\left(\frac{e^{-jkR}}{4\pi R}\right)\nabla'R = \hat{\mathbf{R}}\left(\frac{1+jkR}{R}\right)\frac{e^{-jkR}}{4\pi R} \approx \hat{\mathbf{r}}jk\frac{e^{-jkR}}{4\pi R} = \hat{\mathbf{r}}jkG. \tag{6.23}$$

Using this, we can obtain expressions for $\tilde{\mathbf{E}}$ and $\tilde{\mathbf{H}}$ in the far zone of the sources. The approximation (6.23) leads directly to

$$\tilde{\rho}^i \nabla' \tilde{G} \approx \left[j \frac{\nabla' \cdot \tilde{\mathbf{J}}^i}{\omega} \right] \left(\hat{\mathbf{r}} j k \tilde{G} \right) = -\frac{k}{\omega} \hat{\mathbf{r}} [\nabla' \cdot (\tilde{G} \tilde{\mathbf{J}}^i) - \tilde{\mathbf{J}}^i \cdot \nabla' \tilde{G}].$$

Substituting this into (6.15), again using (6.23) and also using the divergence theorem, we have

$$\tilde{\mathbf{E}}(\mathbf{r}, \omega) \approx -\int_V j\omega\tilde{\mu}[\tilde{\mathbf{J}}^i - \hat{\mathbf{r}}(\hat{\mathbf{r}} \cdot \tilde{\mathbf{J}}^i)]\tilde{G}\, dV' + \hat{\mathbf{r}}\frac{k}{\omega\tilde{\epsilon}^c} \oint_S (\hat{\mathbf{n}}' \cdot \tilde{\mathbf{J}}^i)\tilde{G}\, dS',$$

where the surface S surrounds the volume V that contains the impressed sources. If we let this volume slightly exceed that needed to contain the sources, then we do not change the value of the volume integral above; however, the surface integral vanishes since $\hat{\mathbf{n}}' \cdot \tilde{\mathbf{J}}^i = 0$ everywhere on the surface. Using $\hat{\mathbf{r}} \times (\hat{\mathbf{r}} \times \tilde{\mathbf{J}}^i) = \hat{\mathbf{r}}(\hat{\mathbf{r}} \cdot \tilde{\mathbf{J}}^i) - \tilde{\mathbf{J}}^i$, we then obtain the far-zone expression

$$\tilde{\mathbf{E}}(\mathbf{r}, \omega) \approx j\omega\hat{\mathbf{r}} \times \left[\hat{\mathbf{r}} \times \int_V \tilde{\mu}(\omega)\tilde{\mathbf{J}}^i(\mathbf{r}', \omega)\tilde{G}(\mathbf{r}|\mathbf{r}'; \omega)\, dV' \right] = j\omega\hat{\mathbf{r}} \times [\hat{\mathbf{r}} \times \tilde{\mathbf{A}}_e(\mathbf{r}, \omega)],$$

where $\tilde{\mathbf{A}}_e$ is the electric vector potential. The far-zone electric field has no r-component, and it is often convenient to write

$$\tilde{\mathbf{E}}(\mathbf{r}, \omega) \approx -j\omega\tilde{\mathbf{A}}_{eT}(\mathbf{r}, \omega) \tag{6.24}$$

where $\tilde{\mathbf{A}}_{eT}$ is the vector component of $\tilde{\mathbf{A}}_e$ transverse to the r-direction:

$$\tilde{\mathbf{A}}_{eT} = -\hat{\mathbf{r}} \times [\hat{\mathbf{r}} \times \tilde{\mathbf{A}}_e] = \tilde{\mathbf{A}}_e - \hat{\mathbf{r}}(\hat{\mathbf{r}} \cdot \tilde{\mathbf{A}}_e) = \hat{\boldsymbol{\theta}}\tilde{A}_{e\theta} + \hat{\boldsymbol{\phi}}\tilde{A}_{e\phi}.$$

We can approximate the magnetic field similarly. Noting that $\tilde{\mathbf{J}}^i \times \nabla'\tilde{G} = \tilde{\mathbf{J}}^i \times (jk\hat{\mathbf{r}}\tilde{G})$ we have

$$\tilde{\mathbf{H}}(\mathbf{r}, \omega) \approx -j\frac{k}{\tilde{\mu}(\omega)}\hat{\mathbf{r}} \times \int_V \tilde{\mu}(\omega)\tilde{\mathbf{J}}^i(\mathbf{r}', \omega)\tilde{G}(\mathbf{r}|\mathbf{r}', \omega)\, dV' \approx -\frac{1}{\eta}j\omega\hat{\mathbf{r}} \times \tilde{\mathbf{A}}_e(\mathbf{r}, \omega).$$

Therefore

$$\tilde{\mathbf{E}}(\mathbf{r}, \omega) = -\eta\hat{\mathbf{r}} \times \tilde{\mathbf{H}}(\mathbf{r}, \omega), \qquad \tilde{\mathbf{H}}(\mathbf{r}, \omega) = \frac{\hat{\mathbf{r}} \times \tilde{\mathbf{E}}(\mathbf{r}, \omega)}{\eta},$$

in the far zone.

To simplify the computations, we often approximate the vector potential in the far zone. Noting that

$$R = \sqrt{(\mathbf{r} - \mathbf{r}') \cdot (\mathbf{r} - \mathbf{r}')} = \sqrt{r^2 + r'^2 - 2(\mathbf{r} \cdot \mathbf{r}')}$$

and remembering that $r \gg r'$ for $\mathbf{r}$ in the far zone, we can use the leading terms of a binomial expansion of the square root to get

$$R = r\sqrt{1 - \frac{2(\hat{\mathbf{r}} \cdot \mathbf{r}')}{r} + \left(\frac{r'}{r}\right)^2} \approx r\sqrt{1 - \frac{2(\hat{\mathbf{r}} \cdot \mathbf{r}')}{r}} \approx r\left[1 - \frac{\hat{\mathbf{r}} \cdot \mathbf{r}'}{r}\right] \approx r - \hat{\mathbf{r}} \cdot \mathbf{r}' \tag{6.25}$$

and hence

$$\tilde{G}(\mathbf{r}|\mathbf{r}'; \omega) \approx \frac{e^{-jkr}}{4\pi r}e^{jk\hat{\mathbf{r}} \cdot \mathbf{r}'}. \tag{6.26}$$

Here we have kept the approximation (6.25) intact in the phase of G but have used $1/R \approx 1/r$ in the amplitude of G. We must keep a more accurate approximation for the phase since $k(\hat{\mathbf{r}} \cdot \mathbf{r}')$ may be an appreciable fraction of a radian. We thus have the far-zone approximation for the vector potential

$$\tilde{\mathbf{A}}_e(\mathbf{r}, \omega) \approx \tilde{\mu}(\omega) \frac{e^{-jkr}}{4\pi r} \int_V \tilde{\mathbf{J}}^i(\mathbf{r}', \omega) e^{jk\hat{\mathbf{r}} \cdot \mathbf{r}'} \, dV', \qquad (6.27)$$

which we may use in computing (6.24).

Let us summarize the far-zone field expressions:

$$\tilde{\mathbf{E}}(\mathbf{r}, \omega) = -j\omega[\hat{\boldsymbol{\theta}}\tilde{A}_{e\theta}(\mathbf{r}, \omega) + \hat{\boldsymbol{\phi}}\tilde{A}_{e\phi}(\mathbf{r}, \omega)], \qquad (6.28)$$

$$\tilde{\mathbf{H}}(\mathbf{r}, \omega) = \frac{\hat{\mathbf{r}} \times \tilde{\mathbf{E}}(\mathbf{r}, \omega)}{\eta}, \qquad (6.29)$$

$$\tilde{\mathbf{A}}_e(\mathbf{r}, \omega) = \frac{e^{-jkr}}{4\pi r} \tilde{\mu}(\omega) \tilde{\mathbf{a}}_e(\theta, \phi, \omega), \qquad (6.30)$$

$$\tilde{\mathbf{a}}_e(\theta, \phi, \omega) = \int_V \tilde{\mathbf{J}}^i(\mathbf{r}', \omega) e^{jk\hat{\mathbf{r}} \cdot \mathbf{r}'} \, dV', \qquad (6.31)$$

where the *directional weighting function* $\tilde{\mathbf{a}}_e$ is independent of r and describes the angular variation, or *pattern*, of the fields.

In the far zone $\tilde{\mathbf{E}}, \tilde{\mathbf{H}}, \hat{\mathbf{r}}$ are mutually orthogonal. Because of this, and because the fields vary as e^{-jkr}/r, the electromagnetic field in the far zone takes the form of a spherical TEM wave, which is consistent with the Sommerfeld radiation condition.

6.2.1.1 Power radiated by time-harmonic sources in unbounded space

In § 5.2.7 we defined the power radiated by a time-harmonic source in unbounded space as the total time-average power passing through a sphere of large radius. We found that for a Hertzian dipole the radiated power could be computed from the far-zone fields through

$$P_{av} = \lim_{r \to \infty} \int_0^{2\pi} \int_0^\pi \mathbf{S}_{av} \cdot \hat{\mathbf{r}} r^2 \sin\theta \, d\theta \, d\phi \qquad (6.32)$$

where

$$\mathbf{S}_{av} = \tfrac{1}{2} \operatorname{Re}\left\{ \check{\mathbf{E}} \times \check{\mathbf{H}}^* \right\}$$

is the time-average Poynting vector. By superposition this holds for any localized source. Assuming a lossless medium and using phasor notation to describe the time-harmonic fields, we have, by (6.29),

$$\mathbf{S}_{av} = \tfrac{1}{2} \operatorname{Re}\left\{ \frac{\check{\mathbf{E}} \times (\hat{\mathbf{r}} \times \check{\mathbf{E}}^*)}{\eta} \right\} = \hat{\mathbf{r}} \frac{\check{\mathbf{E}} \cdot \check{\mathbf{E}}^*}{2\eta}.$$

Substituting from (6.28), we can also write $\mathbf{S}_{av}$ in terms of the directional weighting function as

$$\mathbf{S}_{av} = \hat{\mathbf{r}} \frac{\check{\omega}^2}{2\eta} \left(\check{A}_{e\theta} \check{A}_{e\theta}^* + \check{A}_{e\phi} \check{A}_{e\phi}^* \right) = \hat{\mathbf{r}} \frac{k^2 \eta}{(4\pi r)^2} \left(\tfrac{1}{2} \check{a}_{e\theta} \check{a}_{e\theta}^* + \tfrac{1}{2} \check{a}_{e\phi} \check{a}_{e\phi}^* \right). \qquad (6.33)$$

As $\mathbf{S}_{av}$ describes the variation of the power density with θ, ϕ, it is sometimes used as a descriptor of the *power pattern* of the sources.

6.3 Fields in a bounded, source-free region

In § 6.2 we considered the first important special case of the Stratton–Chu formula: sources in an unbounded medium. We now consider the second important special case of a bounded, source-free region. This has applications to the study of microwave antennas and, in its scalar form, to the study of the diffraction of light.

6.3.1 The vector Huygens principle

We may derive the formula for a bounded, source-free region of space by specializing the general Stratton–Chu formulas. We assume that all sources of the fields are within the excluded regions and thus set the sources to zero within V. From (6.6)–(6.7) we have

$$\tilde{\mathbf{E}}(\mathbf{r}, \omega) = \sum_{n=1}^{N} \oint_{S_n} \left[(\hat{\mathbf{n}}' \times \tilde{\mathbf{E}}) \times \nabla' \tilde{G} + (\hat{\mathbf{n}}' \cdot \tilde{\mathbf{E}}) \nabla' \tilde{G} - j\omega\tilde{\mu}(\hat{\mathbf{n}}' \times \tilde{\mathbf{H}}) \tilde{G} \right] dS'$$
$$- \sum_{n=1}^{N} \frac{1}{j\omega\tilde{\epsilon}^c} \oint_{\Gamma_{na}+\Gamma_{nb}} (\mathbf{dl}' \cdot \tilde{\mathbf{H}}) \nabla' \tilde{G}, \tag{6.34}$$

and

$$\tilde{\mathbf{H}}(\mathbf{r}, \omega) = \sum_{n=1}^{\dot{N}} \oint_{S_n} \left[(\hat{\mathbf{n}}' \times \tilde{\mathbf{H}}) \times \nabla' \tilde{G} + (\hat{\mathbf{n}}' \cdot \tilde{\mathbf{H}}) \nabla' \tilde{G} + j\omega\tilde{\epsilon}^c(\hat{\mathbf{n}}' \times \tilde{\mathbf{E}}) \tilde{G} \right] dS'$$
$$+ \sum_{n=1}^{N} \frac{1}{j\omega\tilde{\mu}} \oint_{\Gamma_{na}+\Gamma_{nb}} (\mathbf{dl}' \cdot \tilde{\mathbf{E}}) \nabla' \tilde{G}. \tag{6.35}$$

This is known as the *vector Huygens principle* after the Dutch physicist C. Huygens, who formulated his "secondary source concept" to explain the propagation of light. According to his idea, published in *Traité de la lumière* in 1690, points on a propagating wavefront are secondary sources of spherical waves that add together in just the right way to produce the field on any successive wavefront. We can interpret (6.34)–(6.35) in much the same way. The field at each point within V, where there are no sources, can be imagined to arise from spherical waves emanated from every point on the surface bounding V. The amplitudes of these waves are determined by the values of the fields on the boundaries. Hence we may consider the boundary fields to be equivalent to secondary sources of the fields within V. We will expand on this concept below by introducing the concept of equivalence and identifying the specific form of the secondary sources.

6.3.2 The Franz formula

The vector Huygens principle as derived above requires secondary sources for the fields within V that involve both the tangential and normal components of the fields on the bounding surface. Since only tangential components are required to guarantee uniqueness within V, we seek an expression involving only $\hat{\mathbf{n}} \times \tilde{\mathbf{H}}$ and $\hat{\mathbf{n}} \times \tilde{\mathbf{E}}$. Physically, the normal component of the field is equivalent to a secondary charge source on the surface, while the tangential component is equivalent to a secondary current source. Since charge and current are related by the continuity equation, specification of the normal component is superfluous.

To derive a version of the vector Huygens principle that omits the normal fields, we take the curl of (6.35) to get

$$\nabla \times \tilde{\mathbf{H}}(\mathbf{r}, \omega) = \sum_{n=1}^{N} \nabla \times \oint_{S_n} (\hat{\mathbf{n}}' \times \tilde{\mathbf{H}}) \times \nabla' \tilde{G} \, dS' + \sum_{n=1}^{N} \oint_{S_n} \nabla \times [(\hat{\mathbf{n}}' \cdot \tilde{\mathbf{H}}) \nabla' \tilde{G}] \, dS'$$

$$+ \sum_{n=1}^{N} \nabla \times \oint_{S_n} j\omega \tilde{\epsilon}^c (\hat{\mathbf{n}}' \times \tilde{\mathbf{E}}) \tilde{G} \, dS' + \sum_{n=1}^{N} \frac{1}{j\omega \tilde{\mu}} \oint_{\Gamma_{na} + \Gamma_{nb}} \nabla \times [(\mathbf{dl}' \cdot \tilde{\mathbf{E}}) \nabla' \tilde{G}] \, dS'. \quad (6.36)$$

Now, using $\nabla' G = -\nabla G$ and employing (B.49), we can show that

$$\nabla \times \left[f(\mathbf{r}') \nabla' \tilde{G}(\mathbf{r}|\mathbf{r}') \right] = -f(\mathbf{r}') \left\{ \nabla \times \left[\nabla \tilde{G}(\mathbf{r}|\mathbf{r}') \right] \right\} + \left[\nabla \tilde{G}(\mathbf{r}|\mathbf{r}') \right] \times \nabla f(\mathbf{r}') = 0,$$

since $\nabla \times \nabla \tilde{G} = 0$ and $\nabla f(\mathbf{r}') = 0$. This implies that the second and fourth terms of (6.36) are zero. The first term can be modified using

$$\nabla \times \{ [\hat{\mathbf{n}}' \times \tilde{\mathbf{H}}(\mathbf{r}')] \tilde{G}(\mathbf{r}|\mathbf{r}') \} = \tilde{G}(\mathbf{r}|\mathbf{r}') \nabla \times [\hat{\mathbf{n}}' \times \tilde{\mathbf{H}}(\mathbf{r}')] - [\hat{\mathbf{n}}' \times \tilde{\mathbf{H}}(\mathbf{r}')] \times \nabla \tilde{G}(\mathbf{r}|\mathbf{r}')$$

$$= [\hat{\mathbf{n}}' \times \tilde{\mathbf{H}}(\mathbf{r}')] \times \nabla' \tilde{G}(\mathbf{r}|\mathbf{r}'),$$

giving

$$\nabla \times \tilde{\mathbf{H}}(\mathbf{r}, \omega) = \sum_{n=1}^{N} \nabla \times \oint_{S_n} \nabla \times [(\hat{\mathbf{n}}' \times \tilde{\mathbf{H}}) \tilde{G}] \, dS' + \sum_{n=1}^{N} \nabla \times \oint_{S_n} j\omega \tilde{\epsilon}^c (\hat{\mathbf{n}}' \times \tilde{\mathbf{E}}) \tilde{G} \, dS'.$$

Finally, using Ampere's law $\nabla \times \tilde{\mathbf{H}} = j\omega \tilde{\epsilon}^c \tilde{\mathbf{E}}$ in the source free region V, and taking the curl in the first term outside the integral, we have

$$\tilde{\mathbf{E}}(\mathbf{r}, \omega) = \sum_{n=1}^{N} \nabla \times \nabla \times \oint_{S_n} \frac{1}{j\omega \tilde{\epsilon}^c} (\hat{\mathbf{n}}' \times \tilde{\mathbf{H}}) \tilde{G} \, dS' + \sum_{n=1}^{N} \nabla \times \oint_{S_n} (\hat{\mathbf{n}}' \times \tilde{\mathbf{E}}) \tilde{G} \, dS'. \quad (6.37)$$

Similarly

$$\tilde{\mathbf{H}}(\mathbf{r}, \omega) = -\sum_{n=1}^{N} \nabla \times \nabla \times \oint_{S_n} \frac{1}{j\omega \tilde{\mu}} (\hat{\mathbf{n}}' \times \tilde{\mathbf{E}}) \tilde{G} \, dS' + \sum_{n=1}^{N} \nabla \times \oint_{S_n} (\hat{\mathbf{n}}' \times \tilde{\mathbf{H}}) \tilde{G} \, dS'. \quad (6.38)$$

These expressions together constitute the *Franz formula* for the vector Huygens principle [189].

6.3.3 Love's equivalence principle

Love's equivalence principle allows us to identify the equivalent Huygens sources for the fields within a bounded, source-free region V. It then allows us to replace a problem in the bounded region with an "equivalent" problem in unbounded space where the source-excluding surfaces are replaced by equivalent sources. The field produced by both the real and the equivalent sources gives a field in V identical to that of the original problem. This is particularly useful since we know how to compute the fields within an unbounded region via potential functions.

We identify the equivalent sources by considering the electric and magnetic Hertzian potentials produced by electric and magnetic current sources. Let an impressed electric surface current $\tilde{\mathbf{J}}_s^{eq}$ and a magnetic surface current $\tilde{\mathbf{J}}_{ms}^{eq}$ flow on the closed surface S in a homogeneous, isotropic medium with permeability $\tilde{\mu}(\omega)$ and complex permittivity $\tilde{\epsilon}^c(\omega)$. These sources produce

$$\tilde{\boldsymbol{\Pi}}_e(\mathbf{r}, \omega) = \oint_S \frac{\tilde{\mathbf{J}}_s^{eq}(\mathbf{r}', \omega)}{j\omega\tilde{\epsilon}^c(\omega)} \tilde{G}(\mathbf{r}|\mathbf{r}'; \omega) \, dS', \tag{6.39}$$

$$\tilde{\boldsymbol{\Pi}}_h(\mathbf{r}, \omega) = \oint_S \frac{\tilde{\mathbf{J}}_{ms}^{eq}(\mathbf{r}', \omega)}{j\omega\tilde{\mu}(\omega)} \tilde{G}(\mathbf{r}|\mathbf{r}'; \omega) \, dS', \tag{6.40}$$

which in turn can be used to find

$$\tilde{\mathbf{E}} = \nabla \times (\nabla \times \tilde{\boldsymbol{\Pi}}_e) - j\omega\tilde{\mu}\nabla \times \tilde{\boldsymbol{\Pi}}_h,$$

$$\tilde{\mathbf{H}} = j\omega\tilde{\epsilon}^c\nabla \times \tilde{\boldsymbol{\Pi}}_e + \nabla \times (\nabla \times \tilde{\boldsymbol{\Pi}}_h).$$

Substitution yields

$$\tilde{\mathbf{E}}(\mathbf{r}, \omega) = \nabla \times \nabla \times \oint_S \frac{1}{j\omega\tilde{\epsilon}^c}[\tilde{\mathbf{J}}_s^{eq}\tilde{G}] \, dS' + \nabla \times \oint_S [-\tilde{\mathbf{J}}_{ms}^{eq}]\tilde{G} \, dS',$$

$$\tilde{\mathbf{H}}(\mathbf{r}, \omega) = -\nabla \times \nabla \times \oint_S \frac{1}{j\omega\tilde{\mu}}[-\tilde{\mathbf{J}}_{ms}^{eq}\tilde{G}] \, dS' + \nabla \times \oint_S \tilde{\mathbf{J}}_s^{eq}\tilde{G} \, dS'.$$

These are identical to the Franz equations (6.37)–(6.38) if we identify

$$\tilde{\mathbf{J}}_s^{eq} = \hat{\mathbf{n}} \times \tilde{\mathbf{H}}, \qquad \tilde{\mathbf{J}}_{ms}^{eq} = -\hat{\mathbf{n}} \times \tilde{\mathbf{E}}. \tag{6.41}$$

These are the equivalent source densities for the Huygens principle.

We now state *Love's equivalence principle* [38]. Consider the fields within a homogeneous, source-free region V with parameters $(\tilde{\epsilon}^c, \tilde{\mu})$ bounded by a surface S. We know how to compute the fields using the Franz formula and the surface fields. Now consider a second problem in which the same surface S exists in an unbounded medium with identical parameters. If the surface carries the equivalent sources (6.41), then the electromagnetic fields within V calculated using the Hertzian potentials (6.39)–(6.40) are identical to those of the first problem, while the fields calculated outside V are zero. We see that this must be true since the Franz formulas and the field/potential formulas are identical, and the Franz formula (since it was derived from the Stratton–Chu formula) gives the null field outside V. The two problems are *equivalent* in the sense that they produce identical fields within V.

The fields produced by the equivalent sources obey the appropriate boundary conditions across S. From (2.158)–(2.159) we have the boundary conditions

$$\hat{\mathbf{n}} \times (\tilde{\mathbf{H}}_1 - \tilde{\mathbf{H}}_2) = \tilde{\mathbf{J}}_s, \qquad \hat{\mathbf{n}} \times (\tilde{\mathbf{E}}_1 - \tilde{\mathbf{E}}_2) = -\tilde{\mathbf{J}}_{ms}.$$

Here $\hat{\mathbf{n}}$ points inward to V, $(\tilde{\mathbf{E}}_1, \tilde{\mathbf{H}}_1)$ are the fields within V, and $(\tilde{\mathbf{E}}_2, \tilde{\mathbf{H}}_2)$ are the fields within the excluded region. If the fields produced by the equivalent sources within the excluded region are zero, then the fields must obey

$$\hat{\mathbf{n}} \times \tilde{\mathbf{H}}_1 = \tilde{\mathbf{J}}_s^{eq}, \qquad \hat{\mathbf{n}} \times \tilde{\mathbf{E}}_1 = -\tilde{\mathbf{J}}_{ms}^{eq},$$

which is true by the definition of $(\tilde{\mathbf{J}}_s^{eq}, \tilde{\mathbf{J}}_{sm}^{eq})$.

Note that we can extend the equivalence principle to the case where the media are different internal to V than external to V. See Chen [33].

With the equivalent sources identified, we may compute the electromagnetic field in V using standard techniques. Specifically, we may use the Hertzian potentials as shown above or, since the Hertzian potentials are a simple remapping of the vector potentials, we may use (5.54)–(5.55) to write

$$\tilde{\mathbf{E}} = -j\frac{\omega}{k^2}[\nabla(\nabla \cdot \tilde{\mathbf{A}}_e) + k^2\tilde{\mathbf{A}}_e] - \frac{1}{\tilde{\epsilon}^c}\nabla \times \tilde{\mathbf{A}}_h, \tag{6.42}$$

$$\tilde{\mathbf{H}} = -j\frac{\omega}{k^2}[\nabla(\nabla \cdot \tilde{\mathbf{A}}_h) + k^2\tilde{\mathbf{A}}_h] + \frac{1}{\tilde{\mu}}\nabla \times \tilde{\mathbf{A}}_e, \tag{6.43}$$

where

$$\tilde{\mathbf{A}}_e(\mathbf{r}, \omega) = \oint_S \tilde{\mu}(\omega)\tilde{\mathbf{J}}_s^{eq}(\mathbf{r}', \omega)\tilde{G}(\mathbf{r}|\mathbf{r}'; \omega)\, dS' \tag{6.44}$$

$$= \oint_S \tilde{\mu}(\omega)[\hat{\mathbf{n}}' \times \tilde{\mathbf{H}}(\mathbf{r}', \omega)]\tilde{G}(\mathbf{r}|\mathbf{r}'; \omega)\, dS', \tag{6.45}$$

$$\tilde{\mathbf{A}}_h(\mathbf{r}, \omega) = \oint_S \tilde{\epsilon}^c(\omega)\tilde{\mathbf{J}}_{ms}^{eq}(\mathbf{r}', \omega)\tilde{G}(\mathbf{r}|\mathbf{r}'; \omega)\, dS' \tag{6.46}$$

$$= \oint_S \tilde{\epsilon}^c(\omega)[-\hat{\mathbf{n}}' \times \tilde{\mathbf{E}}(\mathbf{r}', \omega)]\tilde{G}(\mathbf{r}|\mathbf{r}'; \omega)\, dS'. \tag{6.47}$$

At points where the source is zero we can write the fields in the alternative form

$$\tilde{\mathbf{E}} = -j\frac{\omega}{k^2}\nabla \times \nabla \times \tilde{\mathbf{A}}_e - \frac{1}{\tilde{\epsilon}^c}\nabla \times \tilde{\mathbf{A}}_h, \tag{6.48}$$

$$\tilde{\mathbf{H}} = -j\frac{\omega}{k^2}\nabla \times \nabla \times \tilde{\mathbf{A}}_h + \frac{1}{\tilde{\mu}}\nabla \times \tilde{\mathbf{A}}. \tag{6.49}$$

By superposition, if there are volume sources within V, we merely add the fields due to these sources as computed from the potential functions.

6.3.4 The Schelkunoff equivalence principle

With Love's equivalence principle we create an equivalent problem by replacing an excluded region with equivalent electric and magnetic sources. These require knowledge of both the tangential electric and magnetic fields over the bounding surface. However, the uniqueness theorem says that only one of either the tangential electric or the tangential magnetic fields need be specified to make the fields within V unique. Thus we may wonder whether it is possible to formulate an equivalent problem involving only tangential $\tilde{\mathbf{E}}$ or tangential $\tilde{\mathbf{H}}$. It is indeed possible, as shown by Schelkunoff [38, 171].

When we use the equivalent sources to form the equivalent problem, we know that they produce a null field within the excluded region. Thus we may form a different equivalent problem by filling the excluded region with a perfect conductor, and keeping the same equivalent sources. The boundary conditions across S are not changed, and thus by the uniqueness theorem the fields within V are not altered. However, the manner in which we must compute the fields within V is changed. We can no longer use formulas for the fields produced by sources in free space, but must use formulas for fields produced by sources in the vicinity of a conducting body. In general this can be difficult since it

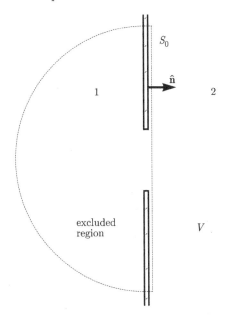

FIGURE 6.5

Geometry for problem of an aperture in a perfectly conducting ground screen illuminated by an impressed source.

requires the formation of a new Green's function that satisfies the boundary condition over the conducting body (which could possess a peculiar shape). Fortunately, we showed (§ 4.10.2.5) that an electric source adjacent and tangential to a perfect electric conductor produces no field, hence we need not consider the equivalent electric sources ($\hat{n} \times \tilde{\mathbf{H}}$) when computing the fields in V. Thus, in our new equivalent problem we need the single tangential field $-\hat{n} \times \tilde{\mathbf{E}}$. This is the *Schelkunoff equivalence principle*.

There is one situation in which it is relatively easy to use the Schelkunoff equivalence. Consider a perfectly conducting ground screen with an aperture in it, as shown in Figure 6.5. We assume that the aperture has been illuminated in some way by an electromagnetic wave produced by sources in region 1 so that there are both fields within the aperture and electric current flowing on the region-2 side of the screen due to diffraction from the edges of the aperture. We wish to compute the fields in region 2. We can create an equivalent problem by placing a planar surface S_0 adjacent to the screen, but slightly offset into region 2, and then closing the surface at infinity so that all of the screen plus region 1 is excluded. Then we replace region 1 with homogeneous space and place on S_0 the equivalent currents $\tilde{\mathbf{J}}_s^{eq} = \hat{n} \times \tilde{\mathbf{H}}$ and $\tilde{\mathbf{J}}_{ms}^{eq} = -\hat{n} \times \tilde{\mathbf{E}}$, where $\tilde{\mathbf{H}}$ and $\tilde{\mathbf{E}}$ are the fields on S_0 in the original problem. Over the portion of S_0 adjacent to the screen, $\tilde{\mathbf{J}}_{ms}^{eq} = 0$, since $\hat{n} \times \tilde{\mathbf{E}} = 0$, but $\tilde{\mathbf{J}}_s^{eq} \neq 0$. From the equivalent currents we can compute the fields in region 2 using the potential functions. However, it is often difficult to determine $\tilde{\mathbf{J}}_s^{eq}$ over the conducting surface. If we apply Schelkunoff's equivalence, we can formulate a second equivalent problem in which we place into region 1 a perfect conductor. Then we have the equivalent source currents $\tilde{\mathbf{J}}_s^{eq}$ and $\tilde{\mathbf{J}}_{ms}^{eq}$ adjacent and tangential to a perfect conductor. By the image theorem of § 5.1 we can replace this problem with yet another equivalent problem in which the conductor is replaced by the images of $\tilde{\mathbf{J}}_s^{eq}$ and $\tilde{\mathbf{J}}_{ms}^{eq}$ in homogeneous space. Since the image of the tangential electric current $\tilde{\mathbf{J}}_s^{eq}$ is oppositely directed, the fields of the electric current and its image cancel. The image of the magnetic current is co-directed with $\tilde{\mathbf{J}}_{ms}^{eq}$, so the fields produced by the magnetic current and its image add. Note that $\tilde{\mathbf{J}}_{ms}^{eq}$ is nonzero only over the aperture (since $\hat{\mathbf{n}} \times \tilde{\mathbf{E}} = 0$ on the screen), and thus the field in region 1 can be found from

$$\tilde{\mathbf{E}}(\mathbf{r}, \omega) = -\frac{1}{\tilde{\epsilon}^c(\omega)} \nabla \times \tilde{\mathbf{A}}_h(\mathbf{r}, \omega),$$

where

$$\tilde{\mathbf{A}}_h(\mathbf{r}, \omega) = \int_{S_0} \tilde{\epsilon}^c(\omega)[-2\hat{\mathbf{n}}' \times \tilde{\mathbf{E}}_{ap}(\mathbf{r}', \omega)]\tilde{G}(\mathbf{r}|\mathbf{r}'; \omega) \, dS' \tag{6.50}$$

and $\tilde{\mathbf{E}}_{ap}$ is the electric field in the aperture in the original problem. We present examples in § 6.4.5.

6.3.5 Far-zone fields produced by equivalent sources

The equivalence principle is useful for analyzing antennas with complicated source distributions. The sources may be excluded using a surface S, and then knowledge of the fields over S (found, for example, by estimation or measurement) can be used to compute the fields external to the antenna. Here we describe how to compute these fields in the far zone.

Given that $\tilde{\mathbf{J}}_s^{eq} = \hat{\mathbf{n}} \times \tilde{\mathbf{H}}$ and $\tilde{\mathbf{J}}_{ms}^{eq} = -\hat{\mathbf{n}} \times \tilde{\mathbf{E}}$ are the equivalent sources on S, we may compute the fields using the potentials (6.45) and (6.47). Using (6.26) these can be approximated in the far zone ($r \gg r'$, $kr \gg 1$) as

$$\tilde{\mathbf{A}}_e(\mathbf{r}, \omega) = \tilde{\mu}(\omega)\frac{e^{-jkr}}{4\pi r}\tilde{\mathbf{a}}_e(\theta, \phi, \omega), \tag{6.51}$$

$$\tilde{\mathbf{A}}_h(\mathbf{r}, \omega) = \tilde{\epsilon}^c(\omega)\frac{e^{-jkr}}{4\pi r}\tilde{\mathbf{a}}_h(\theta, \phi, \omega), \tag{6.52}$$

where

$$\tilde{\mathbf{a}}_e(\theta, \phi, \omega) = \oint_S \tilde{\mathbf{J}}_s^{eq}(\mathbf{r}', \omega)e^{jk\hat{\mathbf{r}}\cdot\mathbf{r}'} \, dS', \tag{6.53}$$

$$\tilde{\mathbf{a}}_h(\theta, \phi, \omega) = \oint_S \tilde{\mathbf{J}}_{ms}^{eq}(\mathbf{r}', \omega)e^{jk\hat{\mathbf{r}}\cdot\mathbf{r}'} \, dS' \tag{6.54}$$

are the directional weighting functions.

To compute the fields from the potentials we must apply the curl operator. So we must evaluate

$$\nabla \times \left[\frac{e^{-jkr}}{r}\mathbf{V}(\theta, \phi)\right] = \frac{e^{-jkr}}{r}\nabla \times \mathbf{V}(\theta, \phi) + \nabla\left(\frac{e^{-jkr}}{r}\right) \times \mathbf{V}(\theta, \phi).$$

The curl of $\mathbf{V}$ is proportional to $1/r$ in spherical coordinates, hence the first term on the right is proportional to $1/r^2$. Since we are interested in the far-zone fields, this term can be discarded in favor of $1/r$-type terms. Using

$$\nabla\left(\frac{e^{-jkr}}{r}\right) = -\hat{\mathbf{r}}\left(\frac{1 + jkr}{r}\right)\frac{e^{-jkr}}{r} \approx -\hat{\mathbf{r}}jk\frac{e^{-jkr}}{r} \qquad (kr \gg 1)$$

we have

$$\nabla \times \left[\frac{e^{-jkr}}{r}\mathbf{V}(\theta, \phi)\right] \approx -jk\hat{\mathbf{r}} \times \left[\frac{e^{-jkr}}{r}\mathbf{V}(\theta, \phi)\right].$$

Under this approximation we also establish

$$\nabla \times \nabla \times \left[\frac{e^{-jkr}}{r}\mathbf{V}(\theta,\phi)\right] \approx -k^2\hat{\mathbf{r}} \times \hat{\mathbf{r}} \times \left[\frac{e^{-jkr}}{r}\mathbf{V}(\theta,\phi)\right] = k^2\frac{e^{-jkr}}{r}\mathbf{V}_T(\theta,\phi)$$

where $\mathbf{V}_T = \mathbf{V} - \hat{\mathbf{r}}(\hat{\mathbf{r}} \cdot \mathbf{V})$ is the vector component of $\mathbf{V}$ transverse to the r-direction. With these formulas we can approximate (6.48)–(6.49) as

$$\tilde{\mathbf{E}}(\mathbf{r},\omega) = -j\omega\tilde{\mathbf{A}}_{eT}(\mathbf{r},\omega) + \frac{jk}{\tilde{\epsilon}^c(\omega)}\hat{\mathbf{r}} \times \tilde{\mathbf{A}}_h(\mathbf{r},\omega), \tag{6.55}$$

$$\tilde{\mathbf{H}}(\mathbf{r},\omega) = -j\omega\tilde{\mathbf{A}}_{hT}(\mathbf{r},\omega) - \frac{jk}{\tilde{\mu}(\omega)}\hat{\mathbf{r}} \times \tilde{\mathbf{A}}_e(\mathbf{r},\omega). \tag{6.56}$$

Note that

$$\hat{\mathbf{r}} \times \tilde{\mathbf{E}} = -j\omega\hat{\mathbf{r}} \times \tilde{\mathbf{A}}_{eT} + \frac{jk}{\tilde{\epsilon}^c}\hat{\mathbf{r}} \times \hat{\mathbf{r}} \times \tilde{\mathbf{A}}_h.$$

Since $\hat{\mathbf{r}} \times \tilde{\mathbf{A}}_{eT} = \hat{\mathbf{r}} \times \tilde{\mathbf{A}}_e$ and $\hat{\mathbf{r}} \times \hat{\mathbf{r}} \times \tilde{\mathbf{A}}_h = -\tilde{\mathbf{A}}_{hT}$, we have

$$\hat{\mathbf{r}} \times \tilde{\mathbf{E}} = \eta\left[-j\omega\tilde{\mathbf{A}}_{hT} - \frac{jk}{\tilde{\mu}}\hat{\mathbf{r}} \times \tilde{\mathbf{A}}_e\right] = \eta\tilde{\mathbf{H}}.$$

Thus

$$\tilde{\mathbf{H}} = \frac{\hat{\mathbf{r}} \times \tilde{\mathbf{E}}}{\eta}$$

and the electromagnetic field in the far zone is a TEM spherical wave as expected.

Finally, we can write the far-zone electric and magnetic fields in terms of the directional weighting functions by substituting (6.51)–(6.52) into (6.55)–(6.56):

$$\tilde{\mathbf{E}}(\mathbf{r},\omega) = -j\omega\tilde{\mu}\frac{e^{-jkr}}{4\pi r}\tilde{\mathbf{a}}_{eT}(\mathbf{r},\omega) + jk\frac{e^{-jkr}}{4\pi r}\hat{\mathbf{r}} \times \tilde{\mathbf{a}}_h(\mathbf{r},\omega), \tag{6.57}$$

$$\tilde{\mathbf{H}}(\mathbf{r},\omega) = -j\omega\tilde{\epsilon}^c\frac{e^{-jkr}}{4\pi r}\tilde{\mathbf{a}}_{hT}(\mathbf{r},\omega) - jk\frac{e^{-jkr}}{4\pi r}\hat{\mathbf{r}} \times \tilde{\mathbf{a}}_e(\mathbf{r},\omega). \tag{6.58}$$

6.4 Application: antennas

Antennas constitute a ubiquitous, but often unobserved, technology. Essential to a society predicated on wireless connectivity, they can be found not only in cell phones and computers, but in appliances, cameras, clothing, and any object that can benefit from information transfer. Wireless sensors are crucial for security applications and for monitoring the health of both people and physical structures. Automobiles may have a dozen or more antennas for AM and FM radio, satellite radio and internet, GPS, cell phone, keyless entry, Bluetooth connectivity, collision avoidance radar, and automatic parking systems; autonomous vehicles will soon require new wireless systems for communication and control.

Because the uses for wireless systems are so extensive and varied, antenna technology is diverse and complex. But all antennas serve similar purposes, and a simple definition is tempting: an antenna is a device that is designed to create or receive electromagnetic waves. This definition may have held in the early days of radio when antennas were primarily used for communicating wirelessly over long distances, which could only be done through the propagation of electromagnetic waves, but the definition is now inadequate. Classic radio and television systems, radars, and modern cellular telephone networks utilize antennas that are widely spaced and operate in the *far zone* of electromagnetic fields. These antennas are designed as if they are operating in empty space, although interaction with the ground or distant objects such as buildings may be factored in. In contrast, many modern wireless systems transmit information over very short distances. These systems interact electromagnetically through the near-zone fields and do not depend on the creation of traveling waves. Examples in the medical industry include implanted devices such as pacemakers and wireless glucose sensors, radio-frequency cancer ablation, and transcranial magnetic stimulation. Each of these applications requires a careful design of an electromagnetic sensor or probe that interacts with nearby material objects, and this interaction must be included in the design. Similarly, radio-frequency identification systems (RFID) use transmitters and receivers in close proximity, and thus use near-zone fields. The coils used to create and sense the electromagnetic fields in these applications are often referred to as antennas, thus broadening the early view of antennas as creators of electromagnetic waves.

In this section we give a brief overview of antennas in the original sense, as creators and receivers of electromagnetic waves, and consider the electromagnetic aspects of the technology. We include basic definitions and give some simple classic examples. Various technical books cover the many specialized applications of antennas, and several excellent textbooks cover the basic theory behind antennas to a much deeper extent than our basic introduction (see, e.g., [11, 110, 186]). Note that in the discussion below, all time-varying quantities are assumed time-harmonic and phasors are used throughout.

6.4.1   Types of antennas

Antennas are categorized in many ways, often based on their operational properties (high gain, wideband, circularly polarized, receiving) or their physical attributes (dish, leaky-wave, dielectric rod, patch). For emphasizing fundamental operating principles, we are attracted to the simple dyad used by Elliott [57].

- **Type I** antennas have current distributions that are known or can be approximated. These include wire antennas such as dipoles, loops, and helices.

- **Type II** antennas have field distributions that are known or can be approximated. These include *aperture antennas* such as slots, horns, and dishes, and also patches and other integrated antennas.

The approach for computing various antenna properties depends on whether the antenna is type I or type II. We will illustrate both types in later sections. The far-zone fields produced by a type I antenna are determined from the known current using (6.28)–(6.31). The far-zone fields of a type II antenna are determined from known fields over an enclosing surface using equivalent sources and (6.51)–(6.54) and (6.57)–(6.58).

6.4.2   Basic antenna properties

Antenna properties fall into two broad categories: circuit properties and radiation properties. Each antenna is connected to a system by a transmission line or waveguide, and

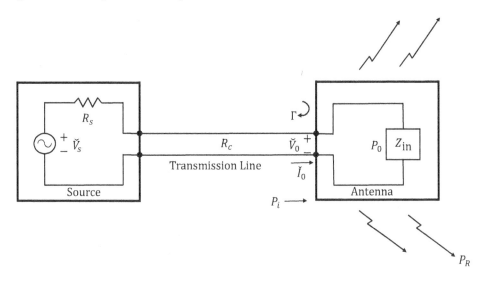

FIGURE 6.6
Transmitting antenna system, with the antenna acting as a circuit element.

thus may be viewed as a circuit element for the purpose of system design. A pair of terminals or a terminal plane is defined for the antenna, and a *terminal voltage* $\check{V}_0$ is described either in terms of the transmission line voltage or the transverse electric field amplitude. Similarly, a *terminal current* $\check{I}_0$ is defined in terms of a transmission line current or a transverse magnetic field. The frequency responses of the voltage and current determine the circuit properties of the antenna as a one-port device. For transmitter design it may only be necessary to know the impedance of the antenna, i.e., the ratio of terminal voltage to current. In this sense an antenna behaves as any other load. The circuit properties are determined by a knowledge of the fields near to the antenna. A simple diagram of the antenna/transmitter system is shown in Figure 6.6.

In contrast, for communication system design a knowledge of radiation properties — such as radiated power, pattern, and polarization — is essential. These are described using the far-zone electromagnetic fields. Other properties, such as receiving cross-section, efficiency, and gain, depend on knowledge of both the near and far-zone fields, and thus combine information about the circuit and radiation properties.

Most of the properties are defined for the antenna acting as a transmitting element. Of course, antennas can also be used to intercept electromagnetic waves, and there are important properties that describe the behavior of the antenna as a receiving element. Fortunately, a concept called *antenna reciprocity* allows us to relate some of the transmitting and receiving properties of a specific antenna.

A simple diagram of an antenna/receiver system is shown in Figure 6.7. An incident electromagnetic wave induces a current in the receiving antenna, which in turn produces a voltage drop (due to the near-zone electric field created by the induced charge) across the input to a transmission line, which in turn is connected to a receiver. The induced voltage is dropped across the impedance Z_{TL} at the input to the transmission line, and this voltage drop may be regarded as exciting the input of the antenna just as if the antenna is transmitting (but with a reverse polarity due to the voltage drop, rather than the voltage rise for a transmitting antenna). The net result is that some of the induced

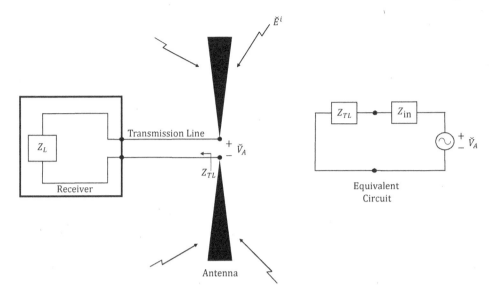

FIGURE 6.7
Receiving antenna system, and equivalent circuit.

voltage is dropped across Z_{TL} and some across Z_{in}. The simple equivalent circuit shown in Figure 6.7 is the result, where the voltage generator describes the voltage induced by the incident field. Note that maximum received power occurs when $Z_{\mathrm{TL}} = Z_{\mathrm{in}}^*$. This is usually achieved by matching the receiver to both the transmission line and to the antenna.

Several important antenna properties are described below. Definitions are paraphrased from the *IEEE Standard for Definitions of Terms for Antennas* [90]. All references to power indicate time-average power unless otherwise indicated.

6.4.2.1 Radiation properties

Polarization. The *polarization* of a transmitting antenna is the polarization of the electromagnetic wave produced by the antenna. This polarization is very similar to the plane wave polarization (§ 4.11.4.3) since the far-zone field of an antenna is TEM, as is a plane wave. So an antenna can be linearly polarized, elliptically polarized, or circularly polarized.

Radiation intensity. The *radiation intensity*, U, of a transmitting antenna is the power radiated from the antenna per unit solid angle. This can be found from the r-component of the far-zone Poynting vector by compensating for the distance traveled by the spherical wave, and is dependent on the angles θ and ϕ in spherical coordinates. From (6.33) we have

$$U(\theta, \phi) = r^2 \hat{\mathbf{r}} \cdot \mathbf{S}_{av} = \frac{k^2 \eta}{(4\pi)^2} \left(\tfrac{1}{2} \breve{a}_{e\theta} \breve{a}_{e\theta}^* + \tfrac{1}{2} \breve{a}_{e\phi} \breve{a}_{e\phi}^* \right). \tag{6.59}$$

Radiated power. The *radiated power* P_R produced by a transmitting antenna is that power that is emitted by the antenna and does not return. It is thus the time-average

Poynting flux that passes through a sphere of infinite radius. By (6.32),

$$P_R = \int_0^{2\pi} \int_0^{\pi} U(\theta, \phi) \sin\theta \, d\theta \, d\phi.$$

Antenna pattern. The *antenna pattern* or *radiation pattern* of a transmitting antenna is the spatial distribution of some property of the far-zone field. Usually either the radiation intensity or the magnitude of the electric field is used to describe the pattern, but other properties such as the phase or polarization of the field may be used. When the radiation intensity is used, the pattern is called the *power pattern*. When the magnitude of the electric field is used it is called the *field pattern*. The pattern is a three-dimensional function of the angles θ and ϕ, but often two-dimensional cuts are made in certain representative planes, which are often planes in which a certain field component is dominant. Sometimes individual components of the electric field are plotted, and a plot of the magnitude of the total electric field is then called the *total field pattern*.

A cut of a typical power pattern is shown in Figure 6.8. Several pattern characteristics may be identified. A number of pattern minima or *nulls* are present. A null that defines a direction where the radiation intensity vanishes is called a *pattern zero*. The angular region between these nulls describes a *pattern lobe*. The lobe with the largest amplitude, U_{max}, is called the *main beam* of the antenna. There may be more than one main beam, since many antennas produce patterns that are periodic in angle. Lobes with smaller amplitudes U_{SL} are called *side lobes*. The lobe opposite to the main beam with amplitude U_{BL} is called the *back lobe*. This gives rise to the *side-lobe-level* defined in terms of dB:

$$SLL = 10 \log_{10} \frac{U_{\text{SL}}}{U_{\text{max}}} \ \text{dB}.$$

There is a value of SLL associated with each side lobe, of which there may be many. The reciprocal of the side lobe level associated specifically with the back lobe is called the *front-to-back ratio*,

$$FBR = 10 \log_{10} \frac{U_{\text{max}}}{U_{\text{BL}}} \ \text{dB}.$$

Note that FBR is a nonnegative number in dB.

Beamwidth. The antenna pattern beamwidth describes the angular extent of the antenna main beam in a pattern cut that contains the maximum of the radiation intensity. Usually this is the angle between half-power points ($U_{\text{max}}/2$), called the *3-dB beamwidth* or *half-power beamwidth* (HPBW). But in complex patterns that contain many nulls (such as those produced by large aperture antennas) the beamwidth is often defined as the angular distance between the nulls defining the main beam. See Figure 6.8. In a field pattern, the 3-dB beamwidth is the angular distance between points where $|\check{\mathbf{E}}| = E_{\text{max}}/\sqrt{2}$. Here E_{max} is the amplitude of the maximum main-beam electric field.

Isotropic radiator. An isotropic radiator is a fictitious antenna that radiates equally in all directions. Its radiation intensity is constant with angle:

$$U(\theta, \phi) = \frac{P_R}{4\pi}.$$

Its three-dimensional antenna pattern is a sphere, and each pattern cut is a circle. There are no nulls in the pattern.

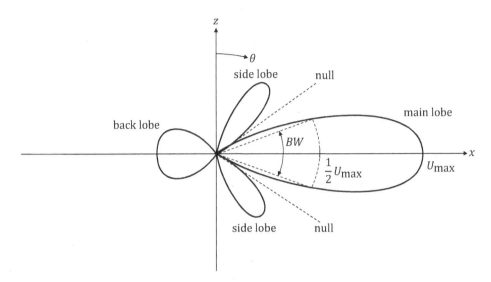

FIGURE 6.8
Typical power pattern cut.

Directivity. The *directivity* (formerly called *directive gain*) of a transmitting antenna is the ratio of the radiation intensity in a given direction to the radiation intensity averaged over all directions. Since the average radiation intensity is the radiated power divided by 4π, the directivity is

$$D(\theta, \phi) = 4\pi \frac{U(\theta, \phi)}{P_R}. \tag{6.60}$$

The directivity of an isotropic radiator is thus unity. For angles where the directivity exceeds unity, the antenna radiates with an intensity greater than that of an isotropic radiator. All physical antennas have directivities that are greater than unity for some ranges of angles and less than unity for others. Note that the maximum value of the directivity is often written merely as D, with no angle dependence indicated, or as $D_{\max}$.

6.4.2.2 Circuit properties

Input impedance. The *input impedance* of a transmitting antenna is the ratio of terminal voltage to terminal current:

$$Z_{\text{in}} = \check{V}_0/\check{I}_0 = R_{\text{in}} + jX_{\text{in}}.$$

The complex power accepted by the antenna is given by circuit theory as

$$P_{\text{acc}} = \tfrac{1}{2}\check{V}_0\check{I}_0^* = \tfrac{1}{2}|\check{I}_0|^2 R_{\text{in}} + j\tfrac{1}{2}|\check{I}_0|^2 X_{\text{in}}.$$

The real part of the input impedance, the *input resistance* R_{in}, describes the transfer of power to the antenna. Part of this power is dissipated by losses in the antenna, while the rest is radiated by the antenna. The dissipated power is determined entirely by the fields within the antenna (in the form of ohmic loss). The radiated power, although defined above in terms of the far-zone fields, may be found from the near-zone fields that determine the input current and voltage. Note, however, that the expression for these fields involves the free-space Green's function, which inherently includes the unbounded nature of the region surrounding the antenna. The imaginary part of the input impedance, the *input reactance* X_{in}, describes the transfer of reactive power to the antenna. The reactance is produced by energy storage in the near field region surrounding the antenna, just as the reactance of a circuit is produced by energy stored in circuit capacitance and inductance.

Resonance frequency. A *resonance frequency* of a transmitting antenna is any frequency at which the input reactance of the antenna vanishes. At this frequency the electric and magnetic energies stored in the near-zone fields balance, and no reactive power is accepted by the antenna.

Antenna mismatch factor. Figure 6.6 shows a circuit diagram for a transmitting antenna. The transmission line is assumed lossless with characteristic resistance R_c. The reflection coefficient for the transmission-line wave at the antenna is

$$\Gamma = \frac{Z_{\text{in}} - R_c}{Z_{\text{in}} + R_c},$$

and the time-average power accepted by the antenna is

$$P_0 = (1 - |\Gamma|^2)P_i,$$

where P_i is the time-average power incident on the antenna. The term

$$M = 1 - |\Gamma|^2$$

is the *antenna mismatch factor*. If the input impedance of the antenna equals the characteristic resistance of the transmission line, the antenna is *matched*. A matched antenna has a zero reflection coefficient and a unity antenna mismatch factor.

Two other terms are used to describe the quality of the antenna match. The *return loss RL* is defined as

$$RL = -20 \log_{10} |\Gamma| \text{ dB},$$

which has the range $0 \le RL < \infty$ dB. Note that RL is a nonnegative number. The standing wave ratio S is defined as

$$S = \frac{1 + |\Gamma|}{1 - |\Gamma|}$$

and satisfies $1 \le S < \infty$.

Impedance bandwidth. The *impedance bandwidth* (or merely *bandwidth*) of an antenna is the range of frequencies over which the impedance of the antenna satisfies some specified criterion. Two common definitions are the 10-dB bandwidth, which is the range of frequencies for which $RL \ge 10$ dB, and the $S = 2$ bandwidth, which is the range of frequencies for which $S \le 2$. As these conditions may hold over more than one band, an antenna may have *multi-band* properties. The *absolute bandwidth* is specified as the absolute frequency range, while the *fractional bandwidth* is given as a fraction of the resonance frequency. The *percentage bandwidth* is the fractional bandwidth measured in per cent.

Radiation resistance. The *radiation resistance* R_r of a transmitting antenna is that portion of the input resistance that is associated with radiated power. The time-average power accepted by the antenna is

$$P_0 = \operatorname{Re} P_{\text{acc}} = \tfrac{1}{2}|\breve{I}_0|^2 R_{\text{in}} = \tfrac{1}{2}|\breve{I}_0|^2 (R_r + R_L) = P_R + P_L,$$

where P_L is the power dissipated in the antenna due to ohmic losses as associated with a resistance R_L, and P_R is the radiated power as associated with a resistance R_r. Thus, the radiated power is

$$P_R = \tfrac{1}{2}|\breve{I}_0|^2 R_r,$$

and the radiation resistance is

$$R_r = 2P_R/|\breve{I}_0|^2. \tag{6.61}$$

Radiation efficiency. The *radiation efficiency* (or merely *efficiency*) η_R of a transmitting antenna is the ratio of the radiated to the time-average accepted power:

$$\eta_R = \frac{P_R}{P_0} = \frac{R_r}{R_r + R_L} \leq 1.$$

6.4.2.3 Properties combining both circuit and radiation effects

Gain. The *gain* (or *total gain*) $G(\theta, \phi)$ of a transmitting antenna is the ratio of the radiation intensity in a given direction to the radiation intensity of an isotropic radiator that radiates the power accepted by the antenna. Since the power radiated by the antenna is the radiation efficiency times the accepted power,

$$G(\theta, \phi) = \eta_R D(\theta, \phi).$$

The maximum value of the gain is written without angular indication:

$$G = \eta_R D.$$

Often the gain is expressed in dB. Then

$$G(\theta, \phi) = 10 \log_{10} \eta_R + 10 \log_{10} D(\theta, \phi) \text{ dB}.$$

The *realized gain*, RG, of a transmitting antenna is the gain reduced by the mismatch loss:

$$RG(\theta, \phi) = (1 - |\Gamma|^2)\eta_R D(\theta, \phi),$$

which has a maximum value of

$$RG = (1 - |\Gamma|^2)\eta_R D.$$

Realized gain is often expressed in dB as

$$RG(\theta, \phi) = 10 \log_{10} D(\theta, \phi) + 10 \log_{10} \eta_R + 10 \log_{10}(1 - |\Gamma|^2) \text{ dB}.$$

When the gain of an antenna is plotted as a function of angle, the *gain pattern* results; this pattern is often plotted in polar coordinates on a dB scale.

Effective area. The *effective area* (or *effective cross-section*) $A_e(\theta, \phi)$ of a receiving antenna is the ratio of the power available at the terminals of the receiving antenna to the power density of a plane wave incident from a certain direction on the antenna:

$$A_e(\theta, \phi) = \frac{P_A}{|\mathbf{S}_{av}(\theta, \phi)|}. \tag{6.62}$$

Here θ and ϕ are the arrival angles of the plane wave. Note that the units of A_e are m^2. Note also that using available power in this definition implies that the antenna is *polarization matched* to the plane wave and that the load and transmission line are impedance matched. That is, the orientation of the antenna with respect to the polarization of the incident wave and also the impedance of the transmission line and load are such that maximum power is transferred to the load. There are modifications to the effective area that take polarization and impedance mismatches into consideration.

Since the effective area describes the dependence of the received power on the orientation of the receiving antenna, $A_e(\theta, \phi)$ describes the *receiving pattern* of the receiving antenna. If the receiving antenna is oriented to point the main beam of its receiving pattern in the direction of the incident plane wave, the intercepted power is maximum and the effective area is written as A_e.

Antenna reciprocity. For antennas embedded in isotropic media, certain *antenna reciprocity* relationships follow from the reciprocity theorem. Chief among these is the relation between effective area and directivity:

$$A_e(\theta, \phi) = \frac{\lambda^2}{4\pi} D(\theta, \phi). \tag{6.63}$$

See [57] for a derivation.

Since the power pattern of a transmitting antenna varies with angle in the same manner as the directivity, the receiving pattern of an antenna when the antenna acts as a receiver is identical to the power pattern of the antenna when that same antenna acts as a transmitter.

Link budget equation. Consider now the transfer of power from a transmitting system to a receiving system. The transmitting antenna has directivity

$$D_T(\theta, \phi) = 4\pi \frac{U(\theta, \phi)}{P_R},$$

where the radiation intensity U is the power density of the outward propagating spherical wave per unit solid angle. Now suppose the transmitting antenna is oriented so that the receiving antenna lies along the main lobe of its pattern. In that direction, $D_T(\theta, \phi) = D_T$. The spherical wave may be considered locally planar over the aperture of the receiving antenna with power density $|\mathbf{S}_{av}| = U_{\max}/r^2$, since the antennas are assumed to be separated by a large distance r. Thus

$$D_T = 4\pi r^2 \frac{|\mathbf{S}_{av}|}{P_R}.$$

This and (6.62) give

$$A_e(\theta, \phi) = 4\pi r^2 \frac{P_A}{P_R D_T},$$

where the angular dependence describes the orientation of the receiving antenna. Using reciprocity relation (6.63), we can replace the effective area of the receiving antenna with its directivity. If the receiving antenna is oriented to point its main beam toward the transmitter, the effective area is

$$A_e = \frac{\lambda^2}{4\pi} D_R = 4\pi r^2 \frac{P_A}{P_R D_T}.$$

The resulting *link budget* or *Friis equation*

$$P_A = P_R \frac{\lambda^2 D_T D_R}{(4\pi r)^2}$$

describes the relationship between the power radiated by the transmitting antenna and the maximum power available to the receiver, and is important in wireless communication system design. Note that improper orientation of the antennas, misalignment of the polarization, mismatch of the receiver or transmitter impedances, and antenna losses all reduce the amount of power delivered to the receiver. Modified link budget equations are available to account for these effects by substituting gain or realized gain for directivity, and by including a polarization mismatch factor.

6.4.3   Characteristics of some type-I antennas

In this section we present some examples of type-I antennas, investigating such radiation properties as pattern, beamwidth, radiated power, radiation resistance, and directivity. Properties that involve circuit characteristics, such as impedance and bandwidth, are difficult to investigate without resorting to numerical techniques. We will find the impedances of dipole and loop antennas in Chapter 7 by numerically solving integral equations, but also give later in the present chapter an approximate method for finding input impedances of wire antennas.

6.4.3.1   The Hertzian dipole antenna

We presented the Hertzian dipole in Example 5.3 to demonstrate calculating fields from potentials. Here we reconsider the Hertzian dipole as an actual antenna applicable in situations where short antennas are needed. Consider the antenna shown in Figure 6.9. A wire is connected between two conducting spheres, with a break in the center for the antenna terminals. This is the structure of the transmitting antenna used by Heinrich Hertz in many of his experiments to investigate the properties of electromagnetic waves. He used two one-meter-long copper wires terminated by zinc spheres of diameter 30 cm, and when he applied a voltage to the antenna input using an induction coil the antenna resonated at approximately 100 MHz [5]. Since this dipole antenna is somewhat larger than a half-wavelength long, its inherent input reactance is inductive; the spheres add capacitance to bring the antenna into resonance.

We now know that if a dipole antenna is considerably shorter than a half-wavelength it will have a nonuniform current distribution that is triangular in shape, tapering to zero current at the ends to satisfy the continuity equation. The conducting spheres at the ends provide places for charge storage, and the current no longer must taper to zero. In fact, for very short antennas the current is very nearly uniform in phase and magnitude. This is the antenna now often referred to as a Hertzian dipole antenna.

The antenna properties of a Hertzian dipole are easily calculated from its directional weighting functions. Consider the antenna in free space (Figure 6.9). A filamentary

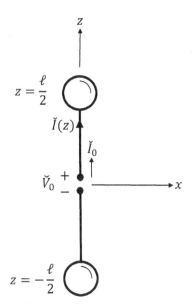

FIGURE 6.9
A Hertzian dipole antenna.

current in free space flows on the z-axis from $z = -\ell/2$ to $z = \ell/2$. If $\ell \ll \lambda$, we can take the current to be uniform with $\breve{I}(z) = \breve{I}_0$. Neglecting radiation from the spheres, the directional weighting function is by (6.31)

$$\breve{\boldsymbol{a}}_e(\theta, \phi) = \hat{\mathbf{z}}\breve{I}_0 \int_{-\ell/2}^{\ell/2} e^{jk_0 z' \cos\theta}\, dz'.$$

Integration gives

$$\breve{\boldsymbol{a}}_e(\theta, \phi) = \hat{\mathbf{z}}\breve{I}_0\ell\frac{\sin\left(k_0\frac{\ell}{2}\cos\theta\right)}{k_0\frac{\ell}{2}\cos\theta},$$

which, since $k_0\ell = 2\pi\ell/\lambda \ll 1$, can be simplified to

$$\breve{\boldsymbol{a}}_e(\theta) \approx \hat{\mathbf{z}}\breve{I}_0\ell = (\hat{\mathbf{r}}\cos\theta - \hat{\boldsymbol{\theta}}\sin\theta)\breve{I}_0\ell. \tag{6.64}$$

By (6.30) the far-zone vector potential is

$$\breve{\mathbf{A}}_e = \frac{e^{-jk_0 r}}{4\pi r}\mu_0(\hat{\mathbf{r}}\cos\theta - \hat{\boldsymbol{\theta}}\sin\theta)\breve{I}_0\ell$$

and (6.28)–(6.29) give

$$\breve{\mathbf{E}} = \hat{\boldsymbol{\theta}}j\eta_0\breve{I}_0 k_0\ell\frac{e^{-jk_0 r}}{4\pi r}\sin\theta, \tag{6.65}$$

$$\breve{\mathbf{H}} = \hat{\boldsymbol{\phi}}j\breve{I}_0 k_0\ell\frac{e^{-jk_0 r}}{4\pi r}\sin\theta.$$

Various antenna properties may be found from these expressions, as shown in the following examples.

▶ **Example 6.4:** Pattern of a Hertzian dipole antenna

Plot the field pattern of a Hertzian dipole antenna and find the 3-dB beamwidth.

Solution: The field pattern is a plot of the electric field magnitude. Field patterns are usually plotted on linear scales, whereas power and gain patterns are often plotted in dB. Both polar and rectangular plots of field pattern cuts may be used. Rectangular plots allow easier identification of the angular positions of pattern characteristics such as nulls, while polar plots give a better spatial sense of the behavior of the pattern. In addition, field patterns are usually normalized to the amplitude of the maximum main beam electric field, $E_{\max}$. Recalling (6.65)

$$\check{\mathbf{E}} = \hat{\boldsymbol{\theta}} j\eta_0 \check{I}_0 k_0 \ell \frac{e^{-jk_0 r}}{4\pi r} \sin\theta,$$

the field pattern of a Hertzian dipole antenna is a plot of

$$\frac{|\check{\mathbf{E}}(\theta)|}{E_{\max}} = |\sin\theta|.$$

Figure 6.10 shows a three-dimensional plot of the field pattern. Clearly the pattern is independent of the angle ϕ, and the main lobe lies along the direction $\theta = \pi/2$. There is a null along the z-axis. A polar plot of the field pattern cut of the xz-plane is shown in Figure 6.11. Here $\theta = 0$ is indicated at the top of the plot, and increasing θ is counter-clockwise from this point (as usual in spherical coordinates). Since $0 \leq \theta \leq 180°$, the right side of the pattern plot corresponds to $x \geq 0$, while the left half corresponds to $x \leq 0$. (Some authors extend the range of θ to $0 \leq \theta \leq 360°$, but this can be confusing.)

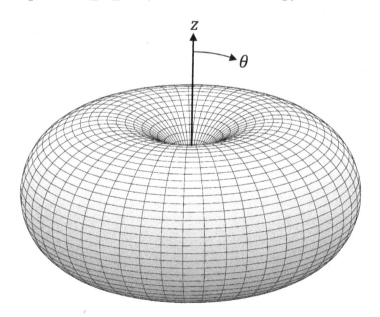

FIGURE 6.10
3-D field pattern of a Hertzian dipole antenna.

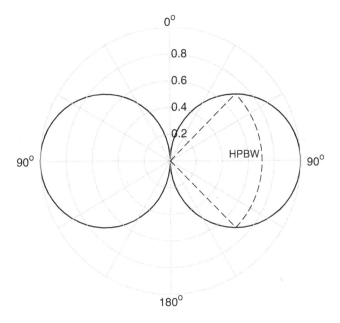

FIGURE 6.11

xz-plane field pattern cut of a Hertzian dipole antenna.

The 3-dB beamwidth is the angular extent between points where

$$\frac{|\check{\mathbf{E}}(\theta)|}{E_{\max}} = |\sin \theta| = 1/\sqrt{2}.$$

Thus, $HPBW = 2 \sin^{-1}(1/\sqrt{2}) = 90°$. The extent of the main beam is indicated in Figure 6.11. ◄

▶ **Example 6.5:** Gain of a Hertzian dipole antenna

Find the maximum directivity and radiation resistance of a Hertzian dipole antenna, and plot the gain pattern.

Solution: To compute the directivity and radiation resistance we must first find the radiated power. From (6.59) we obtain the radiation intensity

$$U(\theta, \phi) = \frac{k_0^2 \eta_0}{(4\pi)^2} \frac{1}{2} \check{a}_{e\theta} \check{a}_{e\theta}^* = \frac{k_0^2 \eta_0}{(4\pi)^2} \frac{1}{2} |\check{I}_0|^2 \ell^2 \sin^2 \theta = \frac{\eta_0}{8} |\check{I}_0|^2 (\ell/\lambda)^2 \sin^2 \theta.$$

Thus, the radiated power is

$$P_R = \int_0^{2\pi} \int_0^{\pi} \frac{\eta_0}{8} |\check{I}_0|^2 (\ell/\lambda)^2 \sin^3 \theta \, d\theta \, d\phi = \eta_0 \frac{\pi}{3} |\check{I}_0|^2 (\ell/\lambda)^2.$$

By (6.60) the directivity is

$$D(\theta, \phi) = 4\pi \frac{U(\theta, \phi)}{P_R} = 4\pi \frac{\frac{\eta_0}{8} |\check{I}_0|^2 (\ell/\lambda)^2 \sin^2 \theta}{\eta_0 \frac{\pi}{3} |\check{I}_0|^2 (\ell/\lambda)^2} = \frac{3}{2} \sin^2 \theta,$$

so the maximum directivity is $D = 1.5$. By (6.61) the radiation resistance is

$$R_r = \frac{2P_R}{|\check{I}_0|^2} = \eta_0 \frac{2\pi}{3}(\ell/\lambda)^2. \tag{6.66}$$

This important result states that for a short antenna the radiation resistance is small, hence a large input current is required to radiate significant power. For instance, when $\ell/\lambda = 1/10$, the radiation resistance is a scant $7.9\,\Omega$.

If the Hertzian dipole is assumed lossless, then $\eta_R = 1$ and the gain is the same as the directivity. We often specify the gain in dB, with the maximum gain given as $G = 10\log_{10}(3/2) = 1.76$ dB. Figure 6.12 plots $10\log_{10} G(\theta)$ vs. θ. Here we have chosen a rectangular plot. Note that the zeros of the antenna pattern at $\theta = 0$ and $\theta = 180°$ are truncated at convenient values so the dependence of the gain on angle may be clearly seen.

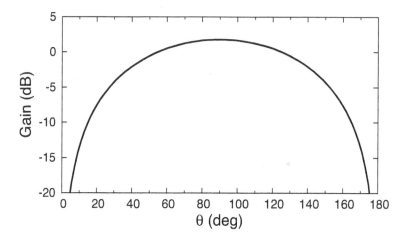

FIGURE 6.12
Gain of a lossless Hertzian dipole antenna. ◄

6.4.3.2   The dipole antenna

The dipole antenna was introduced in § 6.2 as an example of a radiating current distribution. See Figure 6.4. We found that the *near-zone* fields could be computed exactly for a filamentary current with a sinusoidal distribution. Defining the phasor terminal current as $\check{I}_0$, we can write the sinusoidal current distribution for a dipole antenna in free space as

$$\check{\mathbf{I}}(z) = \hat{\mathbf{z}}\check{I}_0 \frac{\sin[k_0(l - |z|)]}{\sin k_0 l} \qquad (-l/2 \le z \le l/2). \tag{6.67}$$

Thus we see that at the terminals, where $z = 0$, $\check{I}_z(0) = \check{I}_0$. To find the antenna parameters, we use the general far-zone expressions (6.28)–(6.31). Substituting (6.67) into (6.31) we have the directional weighting function

$$\check{a}_e(\theta, \phi, \omega) = \int_{-l}^{l} \hat{\mathbf{z}} \frac{\check{I}_0}{\sin k_0 l} \sin k_0(l - |z'|)e^{jk_0 z' \cos\theta}\, dz'.$$

Writing the sine functions in terms of exponentials, we have

$$\check{a}_e(\theta, \phi, \omega) = \frac{\hat{z}\check{I}_0}{2j \sin k_0 l} \left[e^{jk_0 l} \int_0^l e^{jk_0 z'(\cos\theta - 1)} dz' - e^{-jk_0 l} \int_0^l e^{jk_0 z'(\cos\theta + 1)} dz' \right.$$
$$\left. + e^{jk_0 l} \int_{-l}^0 e^{jk_0 z'(\cos\theta + 1)} - e^{-jk_0 l} \int_{-l}^0 e^{jk_0 z'(\cos\theta - 1)} \right].$$

Carrying out the integrals and simplifying, we obtain

$$\check{a}_e(\theta, \phi, \omega) = \hat{z} \frac{2\check{I}_0}{k_0} \frac{F(\theta, k_0 l)}{\sin\theta},$$

where

$$F(\theta, k_0 l) = \frac{\cos(k_0 l \cos\theta) - \cos k_0 l}{\sin k_0 l \sin\theta}$$

is called the *radiation function*. Using $\hat{z} = \hat{r}\cos\theta - \hat{\theta}\sin\theta$ we find that

$$\check{a}_{e\theta}(\theta, \phi, \omega) = -\frac{2\check{I}_0}{k_0} F(\theta, k_0 l), \qquad \check{a}_{e\phi}(\theta, \phi, \omega) = 0. \qquad (6.68)$$

Thus we have from (6.30) and (6.28) the electric field

$$\check{\mathbf{E}} = \hat{\theta} \frac{j\eta_0 \check{I}_0}{2\pi} \frac{e^{-jk_0 r}}{r} F(\theta, k_0 l) \qquad (6.69)$$

and from (6.29) the magnetic field

$$\check{\mathbf{H}} = \hat{\phi} \frac{j\check{I}_0}{2\pi} \frac{e^{-jk_0 r}}{r} F(\theta, k_0 l). \qquad (6.70)$$

▶ **Example 6.6:** Pattern of a dipole antenna

Plot the field pattern of a dipole antenna for $2l = \lambda/2$, $2l = \lambda$, and $2l = 3\lambda/2$. Find the 3-dB beamwidth for the first two of these cases.

Solution: The field pattern is a plot of the magnitude of the electric field, usually normalized to the maximum main beam electric field, $E_{\max}$. Thus, the field pattern of a dipole antenna is a plot of

$$\frac{|\check{\mathbf{E}}(\theta)|}{E_{\max}} = \frac{|F(\theta, k_0 l)|}{\max |F(\theta, k_0 l)|}.$$

Here $\max |F(\theta, k_0 l)|$ is the maximum over $0 \le \theta \le \pi$ for a given value of $k_0 l$. For a half-wavelength antenna with $2l = \lambda/2$, we have

$$F(\theta, \pi/2) = \frac{\cos\left(\frac{\pi}{2}\cos\theta\right)}{\sin\theta}, \qquad (6.71)$$

which attains the maximum value $F(\pi/2, \pi/2) = 1$. The half-power beamwidth is obtained by solving

$$\frac{\cos\left(\frac{\pi}{2}\cos\theta\right)}{\sin\theta} = \frac{1}{\sqrt{2}},$$

and finding $\theta = 129.04°$ and $\theta = 50.96°$. The resulting beamwidth is about $78°$, somewhat smaller than the $90°$ beamwidth of the Hertzian dipole antenna found above. This is evident from Figure 6.13, which shows a polar plot of the field pattern cut of the xz-plane for the Hertzian and half-wave dipole antennas.

For a full-wavelength antenna with $2l = \lambda$ we have the immediate concern that $\sin k_0 l = \sin \pi = 0$, which drives F to infinity. This is a result of the postulated sinusoidal current distribution for a filamentary full-wave dipole falling to zero at the input. In Chapter 7 we will find that the actual current distribution for a dipole with nonzero wire radius has a nonzero minimum at the input. Thus, for purposes of plotting the pattern, we suppress the $\sin k_0 l$ term in the denominator of F and write, for $2l = \lambda$,

$$F(\theta, \pi) = \frac{1 + \cos\left(\pi \cos\theta\right)}{\sin\theta},$$

which has the value 2 at $\theta = \pi/2$. The half-power beamwidth is found by solving

$$\frac{1 + \cos\left(\pi \cos\theta\right)}{2\sin\theta} = \frac{1}{\sqrt{2}},$$

which yields $\theta = 113.92°$ and $\theta = 66.08°$. The resulting beamwidth of about $47.8°$ is considerably narrower than the $78°$ beamwidth of the half-wave dipole antenna found above. This is evident from Figure 6.13, which shows a polar plot of the field pattern cut of the x-z plane for both the half-wave and the full-wave dipole antennas.

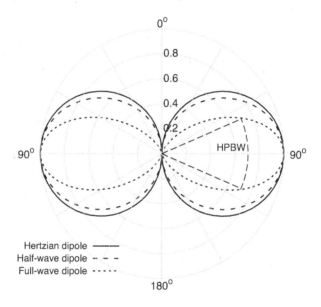

FIGURE 6.13

xz-plane field pattern cuts of half and full-wave dipole antennas, compared to a Hertzian dipole antenna.

Dipole antennas of other lengths have their own distinctive patterns. In general, longer antennas have more complex patterns since the cosine function in $F(\theta, k_0 l)$ goes through more cycles as θ varies within $[0, \pi]$. An interesting pattern occurs when $2l = 3\lambda/2$:

$$F(\theta, 3\pi/2) = -\frac{\cos\left(\frac{3\pi}{2}\cos\theta\right)}{\sin\theta}.$$

The absolute value of this function is maximum at $\theta = 42.563°$ where $|F(42.563°, 3\pi/2)| = 1.399$. In fact, the pattern repeats and an additional maximum appears at $\theta = 137.436°$. A 3D pattern is shown in Figure 6.14, and a pattern cut in the x-z plane is shown in Figure 6.15. There is also a smaller lobe at $\theta = 90°$.

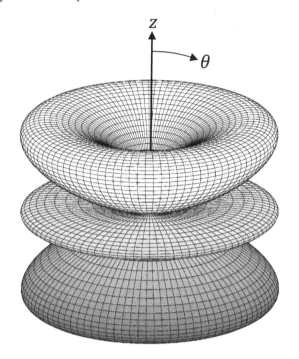

FIGURE 6.14
3-D field pattern of a 3/2-wave dipole antenna.

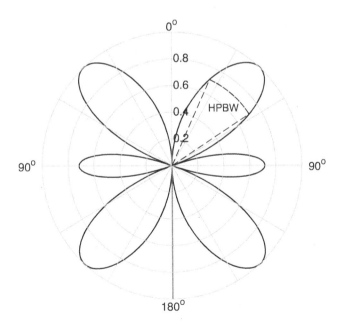

FIGURE 6.15
xz-plane field pattern cut of 3/2-wave dipole antenna.

Identifying a "main beam" for repeating patterns is problematic, but often the beamwidth of the largest repeating lobes is still identified. Solving

$$\frac{\cos\left(\frac{3\pi}{2}\cos\theta\right)}{1.399\sin\theta} = \frac{1}{\sqrt{2}},$$

we find $\theta = 24.4°$ and $\theta = 57.2°$ and hence a $32.8°$ half-power beamwidth (Figure 6.15). ◄

▶ **Example 6.7:** Gain of a half-wave dipole antenna

Find the maximum directivity and radiation resistance of a half-wave dipole antenna, and plot the gain pattern.

Solution: To compute the directivity and radiation resistance, we must first find the radiated power. Substituting (6.68) into (6.59), we have the radiation intensity

$$U(\theta,\phi) = \frac{k_0^2\eta_0}{(4\pi)^2}\frac{1}{2}\breve{a}_{e\theta}\breve{a}_{e\theta}^* = \frac{k_0^2\eta_0}{(4\pi)^2}\frac{1}{2}\frac{4|\breve{I}_0|^2}{k_0^2}F^2(\theta,k_0l) = \frac{\eta_0|\breve{I}_0|^2}{8\pi^2}F^2(\theta,k_0l).$$

Thus, the radiated power is

$$P_R = \int_0^{2\pi}\int_0^\pi \frac{\eta_0|\breve{I}_0|^2}{8\pi^2}F^2(\theta,k_0l)\sin\theta\,d\theta\,d\phi = \frac{\eta_0|\breve{I}_0|^2}{4\pi}\int_0^\pi F^2(\theta,k_0l)\sin\theta\,d\theta. \qquad (6.72)$$

Letting

$$C(k_0l) = \int_0^\pi F^2(\theta,k_0l)\sin\theta\,d\theta, \qquad (6.73)$$

we have the radiated power

$$P_R = \frac{\eta_0|\breve{I}_0|^2}{4\pi}C(k_0l),$$

and we obtain from (6.60) the directivity

$$D(\theta,\phi) = 4\pi\frac{U(\theta,\phi)}{P_R} = 4\pi\frac{\frac{\eta_0|\breve{I}_0|^2}{8\pi^2}F^2(\theta,k_0l)}{\frac{\eta_0|\breve{I}_0|^2}{4\pi}C(k_0l)} = 2\frac{F^2(\theta,k_0l)}{C(k_0l)}. \qquad (6.74)$$

By (6.61) the radiation resistance is

$$R_r = \frac{2P_R}{|\breve{I}_0|^2} = \frac{\eta_0}{2\pi}C(k_0l).$$

For the case of a half-wave dipole, we substitute (6.71) into (6.73) to find

$$C(k_0l) = \int_0^\pi \frac{\cos^2\left(\frac{\pi}{2}\cos\theta\right)}{\sin\theta}\,d\theta$$
$$= 1.22.$$

— Thus from (6.72) we have the radiated power

$$P_R = 1.22\frac{\eta_0|\breve{I}_0|^2}{4\pi} = 36.6|\breve{I}_0|^2 \text{ W}$$

and so radiation resistance

$$R_r = \frac{2P_R}{|\breve{I}_0|^2} = 73.2 \ \Omega.$$

We also have from (6.74) the directivity

$$D(\theta) = \frac{2}{1.22} \frac{\cos^2\left(\frac{\pi}{2}\cos\theta\right)}{\sin^2\theta}.$$

The maximum value of the directivity is

$$D = D(\pi/2) = 2/1.22 = 1.64.$$

Thus, the dipole antenna has a slightly larger maximum directivity than the $D = 1.5$ of the Hertzian dipole; this is reflected in the slightly narrower beamwidth of the half-wave dipole antenna.

If the half-wave dipole is assumed lossless, then $\eta_R = 1$ and the gain is the same as the directivity. We often specify the gain in dB, with the maximum gain given as

$$G = 10\log_{10}(1.64) = 2.15\,\text{dB},$$

which is slightly larger than the $G = 1.76$ dB for the Hertzian dipole. Figure 6.16 shows a plot of $10\log_{10} G(\theta)$ as a function of θ. We see that the main beam is slightly narrower than that of a Hertzian dipole seen in Figure 6.12.

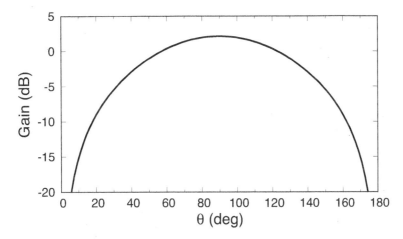

FIGURE 6.16

Gain of a lossless half-wave dipole antenna. ◄

▶ **Example 6.8:** Short dipole antenna

Consider a short dipole antenna ($l \ll \lambda$). Find simple approximations for the current distribution, directivity, and radiation resistance.

Solution: When $l \ll \lambda$, we may approximate the current using the first term in the power series for $\sin x$: $\sin x \approx x$. This gives

$$\frac{\breve{I}_z(z)}{\breve{I}_0} = \frac{\sin[k_0(l - |z|)]}{\sin k_0 l} \approx 1 - \frac{|z|}{l},$$

which is a simple triangle function. To test the accuracy of this approximation, consider a short dipole antenna with $2l = \lambda/5$. A plot of the current distribution normalized to the input current is shown in Figure 6.17 using both the sinusoidal and triangular distributions. Clearly for a dipole of this length the triangular distribution is an excellent approximation for the current. Thus we expect the corresponding approximations for the pattern and directivity to be equally accurate.

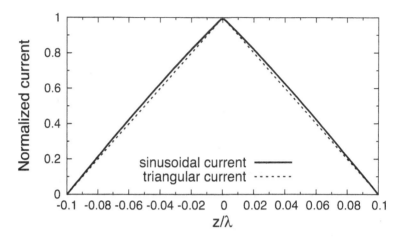

FIGURE 6.17
Normalized magnitude of the current distribution on a dipole antenna with length $2l = \lambda/5$.

When $l \ll \lambda$ we can approximate the radiation function using $\sin x \approx x$ and $\cos x \approx 1 - x^2/2$. This gives

$$F(\theta, k_0 l) \approx \frac{1 - \frac{1}{2}(k_0 l \cos\theta)^2 - 1 + \frac{1}{2}(k_0 l)^2}{(k_0 l)\sin\theta} = \frac{1}{2}k_0 l \sin\theta.$$

Since $F(\theta, k_0 l)$ determines the pattern of the dipole antenna, we see immediately that the pattern of a short dipole antenna is identical to that of a Hertzian dipole antenna, and thus so are the beamwidth, directivity and gain. To find the radiated power we compute from (6.73)

$$C(k_0 l) = \int_0^\pi \frac{1}{4}(k_0 l)^2 \sin^3\theta \, d\theta = \frac{4}{3}\pi^2 (l/\lambda)^2$$

and then

$$P_R = \frac{\eta_0 |\check{I}_0|^2}{4\pi} C(k_0 l) = \frac{\eta_0 \pi}{3}(l/\lambda)^2 |\check{I}_0|^2.$$

We also have the radiation resistance

$$R_r = \frac{2P_R}{|\check{I}_0|^2} = \eta_0 \frac{2\pi}{3}(l/\lambda)^2.$$

Comparing this to (6.66) we see that the formula for the radiation resistance of a short dipole antenna is the same as that for the Hertzian dipole antenna. However, we should be careful to note that the variable ℓ in the formula for the Hertzian dipole antenna is the *full length* of the antenna, whereas l for the short dipole antenna is the *half length*. So the radiation resistance of a short dipole antenna is only one fourth that of a Hertzian dipole of the same length. The reason for this is that the Hertzian dipole antenna has a uniform current of strength $\check{I}_0$, whereas the short dipole only achieves that value at the input. Thus, the smaller amount of current over the majority of the length of the dipole antenna produces less radiated power than the Hertzian dipole antenna, and thus a smaller radiation resistance. ◄

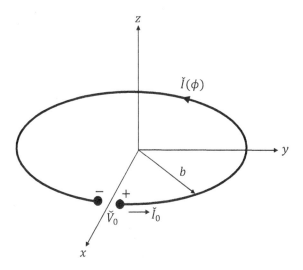

FIGURE 6.18
Circular loop antenna.

6.4.3.3 The circular loop antenna

A circular loop antenna consists of a wire formed into a circular loop, fed at a gap in the loop as shown in Figure 6.18. The loop has radius b and carries filamentary current $\check{\mathbf{I}}(\phi) = \hat{\boldsymbol{\phi}}\check{I}_0 f(\phi)$. Here the current distribution function $f(\phi)$ is normalized so that $f(0) = 1$ and thus $\check{I}_0$ is the input current at $\phi = 0$. In Chapter 7 we solve an integral equation to determine $f(\phi)$; we find that for a thin wire, the current distribution is approximately sinusoidal, especially at frequencies near the first resonance. In essence, the current acts as a traveling wave on a short-circuit terminated transmission line such that the current has a maximum at $\phi = \pi$. So a simple approximation for the current distribution function is

$$f(\phi) = \frac{\cos(k_0 b[\pi - |\phi|])}{\cos(k_0 b\pi)} \qquad (-\pi < \phi \le \pi).$$

The directional weighting function is, from (6.31),

$$\check{\mathbf{a}}_e(\theta, \phi) = \check{I}_0 \int_{-\pi}^{\pi} \hat{\boldsymbol{\phi}}' f(\phi') e^{jk_0 b\hat{\mathbf{r}}\cdot\hat{\boldsymbol{\rho}}'} b\, d\phi'.$$

Using $\hat{\mathbf{r}} \cdot \hat{\boldsymbol{\rho}}' = \sin\theta\cos(\phi - \phi')$, $\hat{\boldsymbol{\phi}} \cdot \hat{\boldsymbol{\phi}}' = \cos(\phi - \phi')$, and $\hat{\boldsymbol{\theta}} \cdot \hat{\boldsymbol{\phi}}' = \cos\theta\sin(\phi - \phi')$, we find

$$\left\{ \begin{matrix} \check{a}_{e\phi} \\ \check{a}_{e\theta} \end{matrix} \right\} = \check{I}_0 b \left\{ \begin{matrix} 1 \\ \cos\theta \end{matrix} \right\} \int_{-\pi}^{\pi} f(\phi') \left\{ \begin{matrix} \cos(\phi - \phi') \\ \sin(\phi - \phi') \end{matrix} \right\} e^{jk_0 b\sin\theta\cos(\phi-\phi')}\, d\phi'. \qquad (6.75)$$

Substituting for $f(\phi)$ and using the substitution $u = \phi - \phi'$, we obtain

$$\left\{ \begin{matrix} \check{a}_{e\phi} \\ \check{a}_{e\theta} \end{matrix} \right\} = \frac{\check{I}_0 b}{\cos(k_0 b\pi)} \left\{ \begin{matrix} 1 \\ \cos\theta \end{matrix} \right\} \int_{\phi-\pi}^{\phi+\pi} \cos(k_0 b[\pi - |\phi - u|]) \left\{ \begin{matrix} \cos u \\ \sin u \end{matrix} \right\} e^{jk_0 b\sin\theta\cos u}\, du.$$

The xz-plane is of particular interest as it contains the antenna feed. Setting $\phi = 0$ for the half-plane $x < 0$ gives

$$\begin{Bmatrix} \breve{a}_{e\phi} \\ \breve{a}_{e\theta} \end{Bmatrix} = \frac{\breve{I}_0 b}{\cos(k_0 b\pi)} \begin{Bmatrix} 1 \\ \cos\theta \end{Bmatrix} \int_{-\pi}^{\pi} \cos(k_0 b[\pi - |u|]) \begin{Bmatrix} \cos u \\ \sin u \end{Bmatrix} e^{jk_0 b \sin\theta \cos u}\, du.$$

Thus, $\breve{a}_{e\theta} = 0$ and

$$\breve{a}_{e\phi} = \frac{\breve{I}_0 b}{\cos(k_0 b\pi)} \int_{-\pi}^{\pi} \cos(k_0 b[\pi - |u|]) \cos u\, e^{jk_0 b \sin\theta \cos u}\, du$$

$$= 2\frac{\breve{I}_0 b}{\cos(k_0 b\pi)} \int_0^{\pi} \cos(k_0 b[\pi - u]) \cos u\, e^{jk_0 b \sin\theta \cos u}\, du.$$

The change of variable $v = \pi - u$ yields

$$\breve{a}_{e\phi} = -2\frac{\breve{I}_0 b}{\cos(k_0 b\pi)} \int_0^{\pi} \cos(k_0 b v) \cos v\, e^{-jk_0 b \sin\theta \cos v}\, dv.$$

Setting $\phi = \pi$ for the half-plane $x > 0$ gives

$$\begin{Bmatrix} \breve{a}_{e\phi} \\ \breve{a}_{e\theta} \end{Bmatrix} = \frac{\breve{I}_0 b}{\cos(k_0 b\pi)} \begin{Bmatrix} 1 \\ \cos\theta \end{Bmatrix} \int_0^{2\pi} \cos(k_0 b[\pi - |\pi - u|]) \begin{Bmatrix} \cos u \\ \sin u \end{Bmatrix} e^{jk_0 b \sin\theta \cos u}\, du.$$

Thus, $\breve{a}_{e\theta} = 0$ and

$$\breve{a}_{e\phi} = 2\frac{\breve{I}_0 b}{\cos(k_0 b\pi)} \int_0^{\pi} \cos(k_0 b u) \cos u\, e^{jk_0 b \sin\theta \cos u}\, du. \tag{6.76}$$

Note that

$$|\breve{a}_{e\phi}(\theta, \phi = 0)| = |\breve{a}_{e\phi}(\theta, \phi = \pi)|,$$

and thus the pattern is symmetric in the xz-plane. Similarly we can show that $\breve{a}_{e\phi} = 0$ in the yz-plane, while for $\phi = \pi/2$

$$\breve{a}_{e\theta} = -2\frac{\breve{I}_0 b}{\cos(k_0 b\pi)} \cos\theta \int_0^{\pi} \cos(k_0 b u) \cos u \cos(k_0 b \sin\theta \sin u)\, du. \tag{6.77}$$

Again, the pattern is symmetric in the plane.

▶ **Example 6.9:** Small loop antenna

Consider a loop antenna that is small compared to a wavelength ($k_0 b \ll 1$). Find simple approximations for the current distribution, directivity, beamwidth, and radiation resistance.

Solution: When $k_0 b \ll 1$, we may approximate the current distribution using the first term in the power series for $\cos x$,

$$f(\phi) = 1,$$

and thus the current is constant with respect to position. Substitution into (6.75) gives

$$\begin{Bmatrix} \breve{a}_{e\phi} \\ \breve{a}_{e\theta} \end{Bmatrix} = \breve{I}_0 b \begin{Bmatrix} 1 \\ \cos\theta \end{Bmatrix} \int_{-\pi}^{\pi} \begin{Bmatrix} \cos(\phi - \phi') \\ \sin(\phi - \phi') \end{Bmatrix} e^{jk_0 b \sin\theta \cos(\phi - \phi')}\, d\phi'. \tag{6.78}$$

Thus, $\breve{a}_{e\theta} = 0$ and

$$\breve{a}_{e\phi} = 2\breve{I}_0 b \int_0^{\pi} \cos u\, e^{jk_0 b \sin\theta \cos u}\, du.$$

Using (E.85) we can write

$$\int_0^\pi \cos u e^{jx\cos u}\, du = j\pi J_1(x)$$

so that

$$\check{a}_{e\phi} = 2j\pi \check{I}_0 b J_1(k_0 b \sin\theta).$$

But since $k_0 \ll 1$, we can use the small argument approximation (E.50) of $J_1(x) \approx x/2$ for $x \ll 1$:

$$\check{a}_{e\phi} = jk_0 \pi b^2 \check{I}_0 \sin\theta. \tag{6.79}$$

Examining (6.64) we recall that the directional weighting function for a Hertzian dipole antenna is

$$\check{a}_{e\theta}(\theta) = -\sin\theta \check{I}_0 \ell.$$

Thus, the pattern of a small loop antenna is identical to that of a Hertzian dipole antenna (although the polarizations are perpendicular). It follows that the directivity and beamwidth are identical as well: $D = 1.5$, $HPBW = 90°$. The deep null in the antenna pattern along the loop axis is useful for radio direction finding, and small loops have been used on aircraft since the early days of radio.

To find the radiation resistance for a small loop antenna, we first compute the radiated power. From (6.59) we obtain the radiation intensity

$$U(\theta,\phi) = \frac{k_0^2 \eta_0}{(4\pi)^2}\frac{1}{2}\check{a}_{e\phi}\check{a}_{e\phi}^* = \frac{k_0^2 \eta_0}{(4\pi)^2}\frac{1}{2}|\check{I}_0|^2 k_0^2 (\pi b^2)^2 \sin^2\theta = \frac{\eta_0}{32}\pi^4 |\check{I}_0|^2 (d/\lambda)^4 \sin^2\theta,$$

where d is the loop diameter. The radiated power is

$$P_R = \int_0^{2\pi}\int_0^\pi \frac{\eta_0}{32}\pi^4 |\check{I}_0|^2 (d/\lambda)^4 \sin^3\theta\, d\theta\, d\phi = \eta_0 \frac{\pi^5}{12}|\check{I}_0|^2 (d/\lambda)^4,$$

and the radiation resistance is

$$R_r = \eta_0 \frac{\pi^5}{6}(d/\lambda)^4.$$

Note that while the radiation resistance of the Hertzian dipole antenna diminishes with length as $(\ell/\lambda)^2$, that of the small loop antenna diminishes with diameter much more quickly, as $(d/\lambda)^4$. This is because the current on opposite sides of the loop flow oppositely and the radiation zone fields due to opposing currents tend to cancel. ◄

▶ **Example 6.10:** Patterns of a loop antenna

Plot the xz- and yz-plane patterns of a loop antenna carrying a sinusoidal standing wave current.

Solution: We show in Chapter 7 that a circular loop made from thin wire is resonant when the circumference of the loop is about a full wavelength, or

$$k_0 b = 2\pi b/\lambda \approx 1.$$

Figure 6.19 shows the pattern of $\check{E}_\phi$ in the xz-plane for a small loop found from (6.79), and also for loops up to resonant size found by computing (6.76) numerically. The pattern of the small loop shows a deep null along the loop axis (the z-axis). As the loop becomes electrically larger, the null becomes shallower. When the loop is about 0.6 wavelengths in circumference, the null has nearly disappeared. A resonant loop with $k_0 b = 1$ has a maximum value of $\check{E}_\phi$ in the xz-plane along the loop axis rather than off the side of the loop.

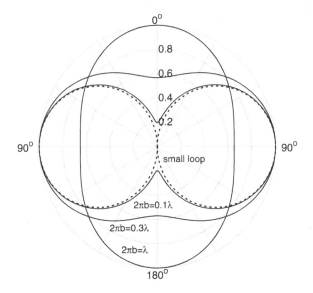

FIGURE 6.19

xz-plane cut of the $\check{E}_\phi$ field pattern of a circular loop antenna.

Figure 6.20 shows the pattern of $\check{E}_\theta$ in the yz-plane for a loop of resonant size and for a smaller loop, both found by computing (6.77) numerically. There is little difference between the patterns; both show field maxima along the loop axis. However, recall that these patterns are normalized. The field strength varies greatly with loop size, with a smaller loop having a weaker field. As the loop radius tends to zero, $\check{E}_\theta$ vanishes, as noted in Example 6.9 where it is shown that a very small loop only has a significant ϕ component of electric field.

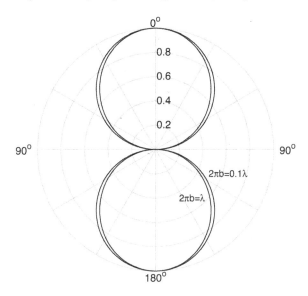

FIGURE 6.20

yz-plane cut of the $\check{E}_\theta$ field pattern of a circular loop antenna. ◄

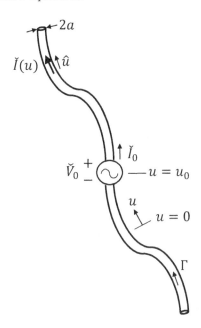

FIGURE 6.21

A generic wire antenna of circular cross-section.

6.4.4 Induced-emf formula for the input impedance of wire antennas

In the examples above we computed the radiation resistance of various antennas, but said nothing about how to find the input reactance. Typically this must be done using a numerical solution for the input current; we give examples for dipole and loop antennas in Chapter 7. However, there are ways to approximate the input impedance of antennas, including the *induced-emf method* [99], which is particularly useful for near-resonant length dipole antennas.

Consider a perfectly conducting wire antenna of circular cross-section but arbitrary shape, residing in free space (Figure 6.21). Distance along the wire axis is measured using a variable u, with L being the total axial length of the wire. The surface tangent in the axial direction is denoted by $\hat{\mathbf{u}}$. We assume that $L \gg a$ and that $a \ll \lambda$. In this case the current induced on the wire surface flows primarily axially rather than circumferentially. We further assume that any bends in the wire are sufficiently smooth that the induced current depends only on the axial position. Thus,

$$\check{\mathbf{J}}_s(\mathbf{r}) \approx \hat{\mathbf{u}} \frac{\check{I}(u)}{2\pi a},$$

where $\check{I}(u)$ is the total axially directed current carried on the wire surface. This surface current supports a scattered field $\check{\mathbf{E}}^s(\mathbf{r})$ everywhere external to the antenna. At the wire surface this field is primarily directed axially. A more detailed discussion is provided in § 7.3.

The antenna feed is located at axial position u_0 and consists of a voltage $\check{V}_0$ applied across a narrow gap in the wire. We can approximate the impressed field in the gap region as

$$\check{E}_u^i(u) = \check{V}_0 \delta(u - u_0),$$

such that when we integrate the field across the gap we get the voltage $\check{V}_0$. The boundary condition on the surface of the perfectly conducting wire requires that the total tangential electric field be zero. Thus, for all points on the wire surface

$$\check{V}_0 \delta(u - u_0) = -\hat{\mathbf{u}} \cdot \check{\mathbf{E}}^s(u).$$

Now, we choose a contour Γ along the *surface* of the wire in the axial direction, multiply the equation by the current distribution $\check{I}(u)$, and integrate:

$$\check{V}_0 \int_\Gamma \delta(u - u_0) \check{I}(u)\, du = -\int_\Gamma \hat{\mathbf{u}} \cdot \check{\mathbf{E}}^s(u) \check{I}(u)\, du.$$

The integral on the left is $\check{I}(u = u_0) = \check{I}_0$ by the sifting property. Then, replacing $\check{V}_0$ by $Z_{\text{in}} \check{I}_0$, we have a formula for the input impedance:

$$Z_{\text{in}} = -\frac{1}{\check{I}_0^2} \int_\Gamma \hat{\mathbf{u}} \cdot \check{\mathbf{E}}^s(u) \check{I}(u) du. \tag{6.80}$$

Equation (6.80) is the *induced emf* formula for the input impedance of a wire antenna. To use it we must know, or have a good approximation for, the near-zone electric field on the antenna surface. Unfortunately this field is readily estimated for only a few geometries. The good news is that (6.80) is a *stationary variational* formula [57], and only a rough approximation to the current is needed to produce useful values of the input impedance. The dipole antenna is a case where we can determine the near-zone field for an approximate current distribution, and apply the formula. See the following example.

▶ **Example 6.11:** Approximate input impedance of a dipole antenna

Consider the dipole antenna shown in Figure 6.4. Find an approximate formula for the input impedance using the induced emf method.

Solution: The standing wave approximation for the current distribution on a thin dipole antenna is from (6.18)

$$\check{I}(z) = \check{I}_m \sin\left[k_0(l - |z|)\right], \tag{6.81}$$

where $\check{I}_m$ is the maximum current on the antenna, related to the input current through

$$\check{I}_m = \frac{\check{I}_0}{\sin(k_0 l)}.$$

Points on the dipole surface are described using the axial variable $u = z$ and radial variable $\rho = a$. The axial component of the surface field at axial position z is, by (6.22),

$$\check{E}_z(z) = -j\frac{\eta_0 \check{I}_m}{4\pi} \left[\frac{e^{-jk_0 R_1}}{R_1} + \frac{e^{-jk_0 R_2}}{R_2} - (2\cos k_0 l)\frac{e^{-jk_0 r}}{r}\right]$$

where $R_1 = \sqrt{a^2 + (z - l)^2}$, $R_2 = \sqrt{a^2 + (z + l)^2}$, and $r = \sqrt{a^2 + z^2}$. Substitution into (6.80) gives a formula for the input impedance:

$$Z_{\text{in}} = \frac{j\eta_0}{4\pi \sin^2(k_0 l)} \int_{-l}^{l} \sin[k_0(l - |z|)] \left[\frac{e^{-jk_0 R_1}}{R_1} + \frac{e^{-jk_0 R_2}}{R_2} - 2\cos k_0 l \frac{e^{-jk_0 r}}{r}\right] dz.$$

It is possible to write this in terms of standard functions; Elliott [57] expresses the final result as

$$Z_{\text{in}} = j\frac{\eta_0}{2\pi \sin^2(k_0 l)} \left\{(4\cos^2 k_0 l)S_1 - (\cos 2k_0 l)S_2 - (\sin 2k_0 l)[2C_1 - C_2]\right\}.$$

Here

$$S_1 = -k_0 a - jF(2k_0 l),$$
$$S_2 = -k_0 a - jF(4k_0 l),$$
$$C_1 = \ln(2l/a) - F(2k_0 l),$$
$$C_2 = \ln(2l/a) - F(4k_0 l),$$

with

$$F(x) = \tfrac{1}{2}[\text{Cin}(x) + j\text{Si}(x)].$$

In this expression $\text{Si}(x)$ is the *sine integral*

$$\text{Si}(x) = \int_0^x \frac{\sin u}{u}\, du,$$

and $\text{Cin}(x)$ is the *modified cosine integral*

$$\text{Cin}(x) = \int_0^x \frac{1 - \cos u}{u}\, du.$$

Figure 6.22 shows a plot of $Z_{\text{in}} = R_{\text{in}} + jX_{\text{in}}$ for $l = 0.25$ m and $a = 0.005$ m. Compare this to Figure 7.16, showing the impedance for the same antenna found by solving an integral equation. The results compare reasonably well near the first and third resonances, but the induced-emf method fails near the second resonance, which is actually an antiresonance. This is because the simple formula for the current (6.81) is zero at the input when $k_0 l = \pi$, which corresponds to the length of a full-wave antenna, $2l = \lambda$. So the computed impedance diverges at $f = 600$ MHz. The current is actually small but nonzero at the antiresonance. However, the results at the first resonance are fairly consistent. The induced-emf formula returns a 283-MHz resonance frequency with a 62-Ω input resistance there, while the integral equation approach returns values of 278 MHz and 73 Ω. The third resonance also matches reasonably well, with the induced-emf method returning values of 876 MHz and 95 Ω as compared with 883 MHz and 119 Ω from the integral equation approach.

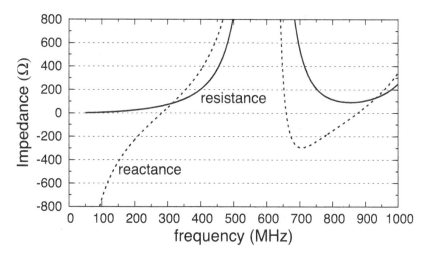

FIGURE 6.22

Input impedance of a dipole antenna found using the induced-emf method. $l = 0.25$ m, $a = 0.005$ m. ◄

▶ **Example 6.12:** Bandwidth of a dipole antenna

Consider a dipole antenna with length $l = 0.25$ m and radius $a = 0.005$ m. Find the bandwidth relative to 75 Ω for the first two true resonances.

Solution: To find the 10-dB bandwidth, we compute the reflection coefficient for the dipole antenna terminating a transmission line. Dipole antennas are often fed using coaxial cables with 75 Ω characteristic resistance. The reflection coefficient is then

$$\Gamma = \frac{Z_{in} - 75}{Z_{in} + 75}.$$

The input impedance of this antenna was computed in Example 6.11. Figure 6.23 shows $|\Gamma|$ in dB, with the $RL = 10$ dB line marked. The first resonance occurs at 283 MHz, with a return loss of 20.5 dB. At this frequency the antenna is approximately a half wavelength long. The $RL = 10$ dB points are 269 and 302 MHz, giving a bandwidth of 33 MHz, which is about 12% of the resonance frequency. The second true resonance (ignoring the first antiresonance) occurs at 876 MHz, with $RL = 10$ dB frequencies of 850 and 899 MHz. This gives a bandwidth of 49 MHz, or about 6%. Note that although the absolute bandwidth of the second resonance is greater than that of the first, the fractional bandwidth is less.

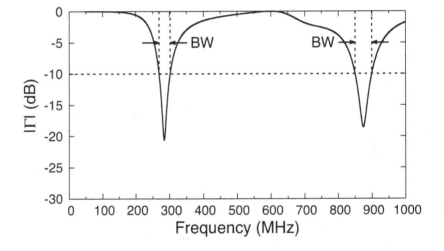

FIGURE 6.23

Reflection coefficient relative to 75 Ω for a dipole antenna of length $l = 0.25$ m and radius $a = 0.005$ m. ◄

6.4.5   Characteristics of some type-II antennas

Let us turn to some radiation properties of typical type-II antennas, also known as *aperture antennas*. We concentrate on computing the pattern and determining the beamwidth. Determination of circuit properties is complicated by the variety of methods used to feed these antennas. For instance, a large reflecting dish is fed using a primary radiator, which could be dipole, patch, horn, or other antenna type. Each of these primary antennas might have different terminal characteristics, but if they illuminate the aperture of the reflector identically, they will give rise to the same far-field pattern.

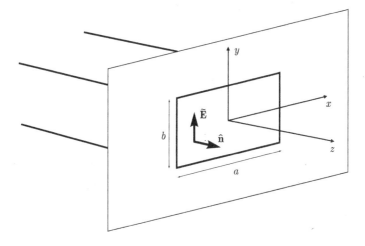

FIGURE 6.24
Aperture antenna consisting of a rectangular waveguide opening into a conducting ground screen of infinite extent.

6.4.5.1 Rectangular waveguide opening into a ground plane

A waveguide left open to free space cannot be considered an "open circuit" in the sense of circuits or transmission lines. The tangential electromagnetic fields at the aperture of the guide may be interpreted as equivalent sources radiating into the region surrounding the guide and sending power away from the structure. In addition, some current flowing on the inner surface of the guide will leak around to the outside surface and produce radiation. Thus, the wave reflected from the mouth of the guide must have a smaller amplitude than the incident wave. If the desire is to produce radiation, then an open-ended waveguide may be considered a useful antenna for microwave applications. In fact, by flaring the ends of the guide outward to form a horn, the radiation may be increased, and such *horn antennas* form an important part of the pantheon of microwave antennas.

One difficulty in calculating the fields produced by an open-ended waveguide is lack of knowledge about leakage current mentioned above. Sometimes this current is ignored in a desire to produce an approximate but useful result; see the example of a reflector antenna below. However, if the waveguide opens into a large (assumed infinite) conducting ground plane, we may use the Schelkunoff equivalence principle without need for knowledge of the currents on the screen.

Consider the case of a rectangular waveguide opening into a perfectly conducting ground screen of infinite extent (Figure 6.24). For simplicity, assume the waveguide propagates a pure TE_{10} mode, and that all higher-order modes excited by aperture reflections can be ignored (see [176] for confirmation that this assumption is reasonable). The electric field in the aperture S_0 is

$$\check{\mathbf{E}}_a(x, y) = \hat{\mathbf{y}} E_0 \cos\left(\frac{\pi}{a} x\right).$$

We may compute the far-zone field using the Schelkunoff equivalence principle (§ 6.3.4). We exclude the region $z < 0^+$ using a planar surface S, which we close at infinity, then fill the region $z < 0$ with a perfect conductor. By image theory the equivalent electric sources on S cancel while the equivalent magnetic sources double. Since the only nonzero magnetic sources are on S_0 ($\hat{\mathbf{n}} \times \tilde{\mathbf{E}} = 0$ on the screen), we have the equivalent problem of the source

$$\check{\mathbf{J}}_{ms}^{eq} = -2\hat{\mathbf{n}} \times \check{\mathbf{E}}_a = 2\hat{\mathbf{x}}E_0 \cos\left(\frac{\pi}{a}x\right)$$

on S_0 in free space, where the equivalence holds for $z > 0$.

We may find the far-zone field created by this equivalent current by first computing the directional weighting function (6.54). Since

$$\hat{\mathbf{r}} \cdot \mathbf{r}' = \hat{\mathbf{r}} \cdot (x'\hat{\mathbf{x}} + y'\hat{\mathbf{y}}) = x' \sin\theta \cos\phi + y' \sin\theta \sin\phi,$$

we find that

$$\check{\mathbf{a}}_h(\theta, \phi) = \int_{-b/2}^{b/2} \int_{-a/2}^{a/2} \hat{\mathbf{x}}2E_0 \cos\left(\frac{\pi}{a}x'\right) e^{jkx' \sin\theta \cos\phi} e^{jky' \sin\theta \sin\phi}\, dx'\, dy'$$

$$= \hat{\mathbf{x}}4\pi E_0 ab \frac{\cos\pi X}{\pi^2 - 4(\pi X)^2} \frac{\sin\pi Y}{\pi Y}$$

where

$$X = \frac{a}{\lambda} \sin\theta \cos\phi, \qquad Y = \frac{b}{\lambda} \sin\theta \sin\phi.$$

Here λ is the free-space wavelength. By (6.55) the electric field is

$$\check{\mathbf{E}} = \frac{jk_0}{\epsilon_0} \hat{\mathbf{r}} \times \check{\mathbf{A}}_h$$

where $\check{\mathbf{A}}_h$ is given in (6.52). Using $\hat{\mathbf{r}} \times \hat{\mathbf{x}} = \hat{\boldsymbol{\phi}} \cos\theta \cos\phi + \hat{\boldsymbol{\theta}} \sin\phi$, we find that

$$\check{\mathbf{E}} = jk_0 ab E_0 \frac{e^{-jkr}}{r}(\hat{\boldsymbol{\theta}} \sin\phi + \hat{\boldsymbol{\phi}} \cos\theta \cos\phi) \frac{\cos(\pi X)}{\pi^2 - 4(\pi X)^2} \frac{\sin(\pi Y)}{\pi Y}. \qquad (6.82)$$

The magnetic field is merely $\check{\mathbf{H}} = (\hat{\mathbf{r}} \times \check{\mathbf{E}})/\eta_0$.

▶ **Example 6.13:** Pattern of a rectangular waveguide opening into a ground plane

Consider a WR-90 X-band waveguide opening into an infinite ground plane. Plot the antenna pattern and find the beamwidth at center band, $f = 10$ GHz.

Solution: The WR-90 waveguide has dimensions $a = 0.9$ inches by $b = 0.4$ inches (22.86 by 10.16 mm). The field is typically plotted in the xz and yz planes. In the yz-plane we have $\phi = \pi/2$ or $\phi = 3\pi/2$, and by (6.82) the electric field has only a θ-component, while the magnetic field has only a ϕ-component. Since the electric field is tangent to the yz-plane, this plane is called the *E-plane*. The magnitude of the electric field in this plane is proportional to

$$F(\theta) = \left| \frac{\sin\left(\pi \frac{b}{\lambda} \sin\theta\right)}{\pi \frac{b}{\lambda} \sin\theta} \right|,$$

and thus a plot of this function is the E-plane cut of the antenna pattern. Note that $F(\theta)$ attains a maximum value of unity at $\theta = 0$. The pattern appears in Figure 6.25. Note that the field is zero in the half-space $z < 0$ because of the ground plane, so the field is only shown in the upper half of the figure. Note also that the field is nonzero on the surface of the ground plane, since the direction of $\check{\mathbf{E}}$ is perpendicular there. As the field is never less than $1/\sqrt{2}$ of the maximum field, the half-power beamwidth is undefined (alternatively, we may say that the beam extends over the full $180°$ range of possible angles).

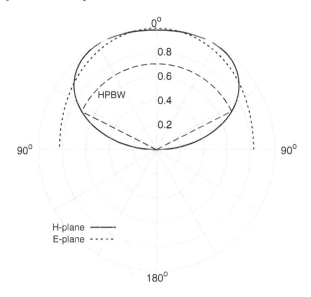

FIGURE 6.25

Pattern of an X-band rectangular waveguide opening into an infinite ground plane. $f = 10$ GHz.

In the xz-plane we have $\phi = 0$ or $\phi = \pi$, and by (6.82) the electric field has only a ϕ-component, while the magnetic field has only a θ-component. Since the magnetic field is tangent to the yz-plane, this plane is called the *H-plane*. The magnitude of the electric field in this plane is proportional to

$$F(\theta) = \left| \cos\theta \, \frac{\cos\left(\pi \frac{a}{\lambda} \cos\theta\right)}{\pi^2 - 4\left(\pi \frac{a}{\lambda} \cos\theta\right)^2} \right|,$$

so a plot of this function is the H-plane cut of the antenna pattern. Note that the maximum value of $F(\theta)$ does not occur at $\theta = 0$, but rather at $\theta_{\max} = 26.30°$, where it takes on the value $F_{\max} = 0.056987$. Thus, to plot the pattern we normalize to this value, with the result shown in Figure 6.25. Although the maximum is not along $\theta = 0$, there is still a main beam oriented in this direction, and we can find the beamwidth by solving

$$\left| \cos\theta \cos\left(\pi \frac{a}{\lambda} \cos\theta\right) \right| = \frac{0.056987}{\sqrt{2}} \left| \pi^2 - 4\left(\pi \frac{a}{\lambda} \cos\theta\right)^2 \right|.$$

The solution is $\theta = 63.69°$, so the half-power beamwidth is $127.38°$. ◄

6.4.5.2   Dish antenna with uniform plane-wave illumination

Practical aperture antennas normally do not have large flanges or ground planes for their radiating openings, but rather radiate directly from the antenna structures. An example is a *reflector* or *dish antenna* (Figure 6.26). A conducting reflector is illuminated by a primary radiator (a smaller antenna) and the reflected electromagnetic fields form an aperture field distribution over the mouth of the dish, which is a circular disk of radius a. Although the majority of the surface current on the dish flows on the illuminated side, some current leaks to the outer surface. We will ignore the radiated field due to this contribution in order to simplify the analysis. To compute the radiated field, we surround the dish with a surface S composed of S_0, which is congruent to the conducting

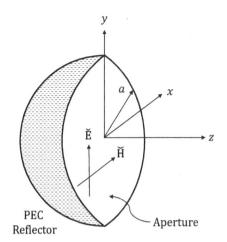
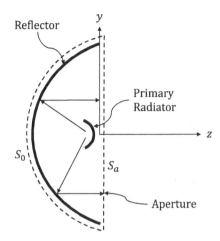

FIGURE 6.26
Dish antenna.

surface of the dish, and S_a, which describes the aperture of the dish. We can then find the radiated field by replacing the dish with equivalent sources over the closed surface S. Because the surface S_0 lies against a perfect conductor, the tangential electric field is zero and there are no equivalent magnetic currents on this surface. Also, as we have neglected current leakage to the outside of the dish, there is no tangential magnetic current on the surface S_0. So all equivalent sources lie on the aperture surface S_a.

The aperture fields are determined by the shape of the reflector and the pattern of the primary source. Often these are chosen to produce a specific aperture field distribution and consequent far-field pattern. The simplest aperture illumination is that of a plane wave, such as would be produced by a point source located at the focus of a parabolic reflector. Ignoring diffraction from the dish edge, the spherical wave emanating from the point source is collimated into a planar wave that appears at the mouth of the dish. We define the planar surface of the dish aperture as the zero-phase reference, and choose the electric field polarization along y to obtain the simple aperture field expressions

$$\check{\mathbf{E}} = \hat{\mathbf{y}} E_0, \qquad \check{\mathbf{H}} = -\hat{\mathbf{x}} E_0/\eta_0.$$

So the aperture hosts both equivalent electric and magnetic sources. The equivalent magnetic source is

$$\check{\mathbf{J}}^{eq}_{ms} = -\hat{\mathbf{n}} \times \check{\mathbf{E}}_a = \hat{\mathbf{x}} E_0,$$

which produces a directional weighting function given by (6.54). Since

$$\hat{\mathbf{r}} \cdot \mathbf{r}' = (\hat{\mathbf{x}} \sin\theta\cos\phi + \hat{\mathbf{y}}\sin\theta\sin\phi + \hat{\mathbf{z}}\cos\theta) \cdot (\hat{\mathbf{x}}\rho'\cos\phi' + \hat{\mathbf{y}}\rho'\sin\phi') = \rho'\sin\theta\cos(\phi - \phi')$$

we have

$$\check{\boldsymbol{a}}_h(\theta, \phi) = \hat{\mathbf{x}} E_0 \int_0^a \int_0^{2\pi} e^{jk_0\rho'\sin\theta\cos(\phi-\phi')} \, d\phi' \rho' \, d\rho'.$$

Using (E.85) to show that

$$\int_0^{2\pi} e^{jk_0\rho'\sin\theta\cos(\phi-\phi')} \, d\phi' = 2\int_0^{\pi} e^{jk_0\rho'\sin\theta\cos u} \, du = 2\pi J_0(k_0\rho'\sin\theta),$$

we obtain

$$\breve{\mathbf{a}}_h(\theta, \phi, \omega) = \hat{\mathbf{x}} 2\pi E_0 \int_0^a J_0(k_0\rho' \sin\theta)\rho'\, d\rho'.$$

Using the integral (E.106) with $n = 0$,

$$\int x J_0(x)\, dx = x J_1(x) + C,$$

gives directional weighting function

$$\breve{\mathbf{a}}_h(\theta, \phi) = \hat{\mathbf{x}} \frac{2\pi a E_0}{k_0 \sin\theta} J_1(k_0 a \sin\theta)$$

$$= (\hat{\mathbf{r}}\sin\theta\cos\phi + \hat{\boldsymbol{\theta}}\cos\theta\cos\phi - \hat{\boldsymbol{\phi}}\sin\phi)\frac{2\pi a E_0}{k_0 \sin\theta} J_1(k_0 a \sin\theta).$$

Similarly, the equivalent electric source is

$$\breve{\mathbf{J}}_s^{eq} = \hat{\mathbf{n}} \times \breve{\mathbf{H}}_a = -\hat{\mathbf{y}} E_0/\eta_0,$$

which by (6.53) produces a directional weighting function

$$\breve{\mathbf{a}}_e(\theta, \phi) = -\hat{\mathbf{y}}\frac{2\pi a E_0}{\eta_0 k_0 \sin\theta} J_1(k_0 a \sin\theta)$$

$$= -(\hat{\mathbf{r}}\sin\theta\sin\phi + \hat{\boldsymbol{\theta}}\cos\theta\sin\phi + \hat{\boldsymbol{\phi}}\cos\phi)\frac{2\pi a E_0}{\eta_0 k_0 \sin\theta} J_1(k_0 a \sin\theta).$$

Lastly, we substitute the directional weighting functions into (6.57) to get the expression for the far-zone electric field

$$\breve{\mathbf{E}} = (\hat{\boldsymbol{\theta}}\sin\phi + \hat{\boldsymbol{\phi}}\cos\phi)\frac{jk_0}{2}\frac{e^{-jk_0 r}}{r} E_0 a \frac{1 + \cos\theta}{\sin\theta} J_1(k_0 a \sin\theta). \tag{6.83}$$

▶ **Example 6.14:** Pattern of a circular dish antenna

Consider a circular dish antenna with plane-wave aperture illumination. If the radius of the dish is $a = 5\lambda$, plot the total field pattern.

Solution: The total field pattern is a plot of the electric field magnitude. From (6.83) we see that this quantity is proportional to

$$F(\theta) = \left| \frac{1 + \cos\theta}{\sin\theta} J_1(k_0 a \sin\theta) \right|, \tag{6.84}$$

which is independent of ϕ. When $a = 5\lambda$ we have $k_0 a = 10\pi$, and plotting (6.84) produces Figure 6.27. There is one large main beam along the dish axis (z-axis) and several sidelobes. Note that the aspect ratio of the 3D pattern has been exaggerated to better visualize the pattern shape. The main beam is actually much narrower than pictured. Figure 6.28 shows a polar plot of the xz-plane cut of total field pattern. Note that the pattern has been normalized to the maximum value of $F(\theta)$, which is

$$F_{\max} = F(0) = k_0 a = 10\pi.$$

The narrow main beam is characteristic of large dish antennas. Reflector antennas are used to create radiation patterns with narrow beamwidths and corresponding high gains.

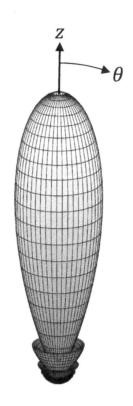

FIGURE 6.27
3D total field pattern of a dish antenna with aperture size $a = 5\lambda$.

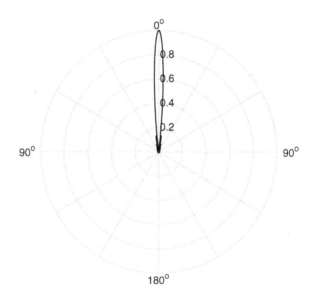

FIGURE 6.28
xz-plane cut of total field pattern of a dish antenna with aperture size $a = 5\lambda$. ◀

▶ **Example 6.15:** Beamwidth and sidelobes of a circular dish antenna

Consider a circular dish antenna with plane-wave aperture illumination. If the radius of the dish is $a = 5\lambda$, determine the beamwidth. Find the position of the pattern nulls and the sidelobe level for each of the first five sidelobes.

Solution: The main beam and sidelobes are easiest to visualize using a rectangular plot with a log scale. Figure 6.29 shows a plot of

$$F(\theta) = 20 \log_{10} \left\{ \frac{1}{k_0 a} \left| \frac{1 + \cos\theta}{\sin\theta} J_1(k_0 a \sin\theta) \right| \right\} \text{ dB.}$$

The field pattern is 3 dB below its maximum value at an angle of 2.947°, and thus the half-power beamwidth is twice this, or about 5.89°. Often with patterns that have many sidelobes the beamwidth is defined in terms of the pattern nulls. There are true zeros in the pattern when $J_1(k_0 a \sin\theta) = 0$. Since $J_1(x)$ has zeros at 3.83171, 7.01559, 10.17347, 13.32369, ..., the pattern has zeros at $\sin^{-1}(3.83171/10\pi) = 7.006°$, 12.904°, 18.895°, 25.094°, ... Thus, the main beam beamwidth as measured between the first pattern nulls is 14.01°. When a/λ is large, the first null occurs for a small θ value, and thus the beamwidth may be approximated using the simple formula

$$BW = 2 \sin^{-1} \frac{3.83171}{k_0 a}$$

$$\approx \frac{7.66342}{k_0 a} \text{ rad}$$

or

$$BW \approx \frac{69.882}{a/\lambda} \text{ deg.}$$

Using $a/\lambda = 5$ gives $BW = 13.98°$, which is quite close to the value of 14.01° found above.

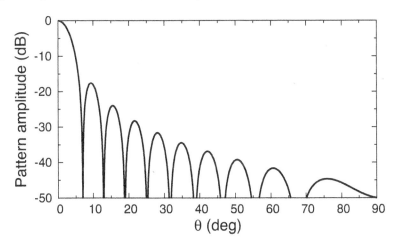

FIGURE 6.29
xz-plane cut of total field pattern of a dish antenna with aperture size $a = 5\lambda$.

Because of the additional term $(1 + \cos\theta)/\sin\theta$, the sidelobes do not appear at the maxima of $J_1(k_0 a \sin\theta)$. Instead, their positions and the values of the sidelobe levels must be found numerically. Values for the first five sidelobes are shown in the table below.

n	θ (deg)	SLL (dB)
1	9.404	-17.63
2	15.53	-23.97
3	21.70	-28.27
4	28.08	-31.61
5	34.84	-34.41

◀

6.5 Problems

6.1 Beginning with the Lorentz reciprocity theorem, derive (6.7).

6.2 Obtain (6.7) by substitution of (6.6) into Faraday's law.

6.3 Show that (6.7) returns the null result when evaluated within the excluded regions.

6.4 Consider the specialization of the Stratton-Chu formula for the static electric field given by (6.8). By taking the divergence of **E**, show that the point form of Gauss' law may be obtained.

6.5 Show that under the condition $kr \gg 1$ the formula for the magnetic field of a dipole antenna (6.20) reduces to (6.70), while the formulas for the electric fields (6.21) and (6.22) reduce to (6.69).

6.6 Consider the dipole antenna shown in Figure 6.4. Instead of a standing-wave current distribution, assume the antenna carries a *traveling-wave current distribution*

$$\tilde{\mathbf{J}}^i(\mathbf{r}, \omega) = \hat{\mathbf{z}} \tilde{I}(\omega) e^{-jk|z|} \delta(x)\delta(y), \qquad -l \leq z \leq l.$$

Find the electric and magnetic fields at all points away from the current distribution. Specialize the result for $kr \gg 1$.

6.7 A circular loop of thin wire has radius a and lies in the $z = 0$ plane in free space. The loop is fed at $\phi = 0$ and is terminated by a load at $\phi = \pi$ such that the current behaves as a traveling wave. Assume the phasor current induced on the wire has the density

$$\check{\mathbf{J}}(\mathbf{r}) = \hat{\boldsymbol{\phi}} \check{I}_0 e^{-jk_0 a|\phi|} \delta(r-a) \frac{\delta(\theta - \pi/2)}{r}, \qquad |\phi| \leq \pi.$$

Determine formulas for the directional weighting functions. Plot the radiation resistance of the loop antenna as a function of $k_0 a$.

6.8 A circular loop of thin wire has radius a and lies in the $z = 0$ plane in free space. Assume the loop is small so that $k_0 a \ll 1$. The loop is fed at $\phi = 0$ and is open-circuited at $\phi = \pi$ such that the phasor current induced on the wire has the density

$$\check{\mathbf{J}}(\mathbf{r}) = \hat{\boldsymbol{\phi}} \check{I}_0 \cos\left(\frac{\phi}{2}\right) \delta(r - a) \frac{\delta(\theta - \pi/2)}{r}, \qquad |\phi| \leq \pi.$$

Determine the radiation resistance and compare it to that for a small short-circuited loop antenna. Explain the difference.

6.9 Consider a plane wave with the fields

$$\tilde{\mathbf{E}} = \tilde{E}_0 \hat{\mathbf{x}} e^{-jkz}, \qquad \tilde{\mathbf{H}} = \frac{\tilde{E}_0}{\eta} \hat{\mathbf{y}} e^{-jkz},$$

normally incident from $z < 0$ on a square aperture of side a in a PEC ground screen at $z = 0$. Assume that the field in the aperture is identical to the field of the plane wave with the screen absent (this is called the *Kirchhoff approximation*). Compute the far-zone electromagnetic fields for $z > 0$.

6.10 Consider a coaxial cable of inner radius a and outer radius b, opening into a PEC ground plane at $z = 0$. Assume that only the TEM wave exists in the line and that no higher-order modes are created when the wave reflects from the aperture. Compute the far-zone electric and magnetic fields of this aperture antenna.

6.11 A linear dipole antenna has its ends terminated by loads in order to create the traveling-wave current

$$\tilde{I}(z) = \tilde{I}_0 e^{-jk|z|}.$$

Find the vector potential in the far zone of the antenna.

6.12 A thin, perfectly conducting sheet occupies the entire $z = 0$ plane. There is a square aperture cut in the plane of width $2a$ centered at the origin. A source of frequency ω is placed in the region $z < 0$ so that the electric field within the aperture is $\tilde{\mathbf{E}} = \hat{\mathbf{x}} \tilde{E}_0$ where $\tilde{E}_0$ is a constant. Find $\tilde{E}_\phi$ in the far zone of the aperture ($z > 0$).

7

Integral equations in electromagnetics

7.1 A brief overview of integral equations

In much of this book, we formulate our mathematical description of electromagnetics in terms of differential equations — the wave equation in particular. Sometimes, however, the quantity of interest resides within an integral, and to determine it we must solve an *integral equation*. Such equations often arise via the integral solutions to Maxwell's equations described in Chapter 6. Typically the desired current, voltage, or impedance is embedded within a spatial integral describing the interaction of fields with material bodies, either penetrable or perfectly conducting. In this chapter we examine several interesting and pertinent problems in electromagnetics, each formulated as an integral equation. We seek numerical solutions and interpret them physically. We also address some difficulties and challenges inherent in the solution process.

Since our aim is introductory, we restrict ourselves to simple (but relevant) canonical problems. Modern practical problems typically involve complex material bodies of complicated shape. Many require "tricks of the trade" that can only be learned after mastering more elementary methods. We therefore focus on one- and two-dimensional problems, leaving three-dimensional problems for later study by the reader.

A note on numerical computation. Each numerical solution shown in this book is carried out using Fortran 77 with single-precision arithmetic (unless otherwise noted). Simple routines are used for integration, matrix solution, and the generation of special functions; see, e.g., [154, 216, 139].

7.1.1 Classification of integral equations

Remember that in an integral equation, the unknown function occurs under the integral sign. We restrict ourselves to the *linear* integral equations that commonly arise in electromagnetics. These can be classified as follows [138].

1. If the integration limits are constant, the equation is of *Fredholm* type. An equation of the form

$$\alpha f(x) = F(x) + \lambda \int_a^b K(x|x')f(x')\,dx', \qquad (7.1)$$

holding for all x in the interval $[a, b]$, is a Fredholm integral equation in one space dimension. Here $f(x)$ is the unknown function and a and b are constants. We call $[a, b]$ the *domain* of the integral equation. The remaining quantities $F(x)$, $K(x|x')$, λ, and α are discussed below.

2. If the upper limit is variable, the equation is of *Volterra* type:

$$\alpha f(x) = F(x) + \lambda \int_a^x K(x|x')f(x')\,dx' \qquad (a \le x \le b). \tag{7.2}$$

In each case the function $F(x)$ is assumed given and λ is a constant parameter. If $F(x)$ happens to be zero, the equation is *homogeneous*.

The function $K(x|x')$, depending on two variables, is also known *a priori*. It is called the *kernel* of the integral equation. The kernel is termed *symmetric* if

$$K(x|x') = K(x'|x)$$

for all x, x' in $[a, b]$. It is *convolutional* if

$$K(x|x') = K(x - x').$$

A kernel that can be written in the form

$$K(x|x') = \sum_n G_n(x)H_n(x')$$

is said to be *separable*. If $K(x|x')$ has singularities or discontinuities within the range of integration, or if the integration limits are infinite, the kernel is *singular* [138]. Kernels can be real or complex. They can be *hermitian, positive definite*, etc. — all properties important in the mathematical study of integral equations.

The constant α in (7.1)–(7.2) is either zero or unity. If $\alpha = 0$, the equation is *of the first kind* and $f(x)$ appears only within the integral. If $\alpha = 1$, we have an integral equation of the *second kind*.

Integral equations can involve functions of more than one variable. For instance,

$$f(x, y) = F(x, y) + \lambda \int_a^b \int_c^d K(x, y|x', y')f(x', y')\,dx'\,dy'$$

is a Fredholm equation of the second kind in the two space dimensions x, y. Its domain is the rectangle $a \le x \le b$, $c \le y \le d$.

Linear operator notation. In a linear integral equation, we may regard the integral as a linear operator acting on the unknown function. Such an operator $\mathcal{L}$ must satisfy

$$\mathcal{L}[\alpha f(x) + \beta g(x)] = \alpha\,\mathcal{L}[f(x)] + \beta\,\mathcal{L}[g(x)]$$

for any two operand functions $f(x), g(x)$ and any two constants α, β. With this notation we may cast an integral equation in *linear operator form*. By setting

$$\mathcal{L}[f(x)] = f(x) - \lambda \int_a^x K(x|x')f(x')\,dx',$$

for example, we may write the second-kind Volterra equation (7.2) as

$$\mathcal{L}[f(x)] = F(x).$$

7.1.2   Analytic solution of integral equations

A number of techniques can be used to solve integral equations analytically. Fredholm equations with separable kernels are amenable to so-called *direct solutions* (see [87] for details). *Eigenfunction expansion* techniques result in series solutions. Consider, for example, a second-kind homogeneous Fredholm equation

$$f(x) = \lambda \int_a^b K(x|x')f(x')\,dx'$$

with a real symmetric kernel. This equation has solutions only for real discrete values of λ known as *eigenvalues*. The solutions themselves are termed *eigenfunctions* for the equation. It turns out that a solution to the corresponding *nonhomogeneous* problem may be constructed as a weighted superposition of the eigenfunctions. See [87] for further details on this method. Through the use of *transform techniques* (involving the Fourier or Laplace transform) we may be able to convert an integral equation into an algebraic equation. This can be particularly effective for equations having convolutional kernels. Finally, many second-kind equations may be solved by *successive substitution*, resulting in *Neumann series* expressions.

Unfortunately, however, most of the integral equations encountered in electromagnetics will not yield to these elegant classical techniques. We therefore turn to numerical approaches.

7.1.3   Numerical solution of integral equations

When integral equations cannot be solved analytically, either in an exact or an approximate manner, a numerical solution may be sought.

The method of successive substitution, mentioned above as an analytical approach, can also be used as a numerical approach. Let us describe the method in a bit more detail. Given a Fredholm equation of the second kind, we write it in the form

$$f(x) = F(x) + \lambda \int_a^b K(x|x')f(x')\,dx',$$

i.e., with the two occurrences of the unknown function $f(x)$ separated by the equals sign. We use this to set up a *recursion relation*:

$$f_{n+1}(x) = F(x) + \lambda \int_a^b K(x|x')f_n(x')\,dx' \qquad (n = 0, 1, 2, \ldots). \qquad (7.3)$$

We begin by selecting a function $f_0(x)$, of relatively simple form, which we expect to approximate $f(x)$ in some manner. We set $n = 0$ in (7.3), substitute $f_0(x)$ into the right-hand side, and use the equation to generate a function $f_1(x)$:

$$f_1(x) = F(x) + \lambda \int_a^b K(x|x')f_0(x')\,dx'.$$

Depending on the properties of $F(x)$, $K(x|x')$, and so on, it may happen that $f_1(x)$ is closer to $f(x)$ than $f_0(x)$ is. If so, we can set $n = 1$ in (7.3) and use $f_1(x)$ to generate an even better estimate $f_2(x)$. Such an *iteration* process may lead to a sequence of approximations $f_n(x)$ that *converges* to $f(x)$. Conditions for convergence are discussed, e.g., in [198].

A very different approach is to expand the unknown function as a linear combination of *basis functions*. The numerical expansion coefficients are determined by applying weighting functions to the resulting equation. This procedure, commonly called the *method of moments* [81], is discussed in the following section.

7.1.4 The method of moments (MoM)

Consider a linear integral equation

$$\mathcal{L}[f(x)] = F(x) \qquad (a \leq x \leq b). \tag{7.4}$$

Let us choose a set of N basis functions $\phi_n(x)$ and expand the unknown function $f(x)$ as a linear combination of these:

$$f(x) \approx \bar{f}(x) = \sum_{n=1}^{N} a_n \phi_n(x). \tag{7.5}$$

Here the a_n are expansion coefficients to be determined. If $f(x)$ is well-behaved and the $\phi_n(x)$ form a *complete set*, then the representation (7.5) becomes exact as $N \to \infty$. Substitution of (7.5) into (7.4) gives

$$r(x) = \sum_{n=1}^{N} a_n \, \mathcal{L}[\phi_n(x)] - F(x) \approx 0 \qquad (a \leq x \leq b) \tag{7.6}$$

in view of the linearity of $\mathcal{L}$. Here $r(x)$ is called the *residual*. If we can find coefficients a_n that make the residual small for all x such that $a \leq x \leq b$, then we take $\bar{f}(x)$ to be a good approximation for $f(x)$. To determine a_n, one of two routes is generally taken.

7.1.4.1 Method of collocation

Rather than seeking a_n that require the residual to be small for all x in $[a, b]$, we force the residual to vanish at N discrete *match points* x_m that lie within this interval:

$$r(x = x_m) = 0 \qquad (m = 1, \ldots, N).$$

This yields N simultaneous algebraic equations in the N unknowns $a_1, \ldots, a_N$,

$$\sum_{n=1}^{N} a_n \, \mathcal{L}[\phi_n(x)]|_{x=x_m} = F(x_m) \qquad (m = 1, \ldots, N),$$

which can be written in matrix form as

$$[A_{mn}][a_n] = [b_m]. \tag{7.7}$$

Here

$$A_{mn} = \mathcal{L}[\phi_n(x)]|_{x=x_m} \qquad (m, n = 1, \ldots, N),$$
$$b_m = F(x_m) \qquad (m = 1, \ldots, N).$$

Once the a_n have been determined, they can be used in (7.5) to obtain the approximation $\bar{f}(x)$ to $f(x)$.

7.1.4.2 Method of weighting functions

An alternative to collocation (or *point matching*) is to set N weighted averages of the residual to zero:

$$\int_a^b w_m(x) r(x) \, dx = 0 \qquad (m = 1, \ldots, N).$$

Here $w_m(x)$ are called *weighting functions*. This produces the set of equations

$$\sum_{n=1}^N a_n \int_a^b \{\mathcal{L}[\phi_n(x)]\} \, w_m(x) \, dx = \int_a^b F(x) w_m(x) \, dx \qquad (m = 1, \ldots, N), \qquad (7.8)$$

which takes the form (7.7) with

$$A_{mn} = \int_a^b \{\mathcal{L}[\phi_n(x)]\} \, w_m(x) \, dx, \qquad b_m = \int_a^b F(x) w_m(x) \, dx.$$

For instance, a Volterra equation has

$$A_{mn} = \int_a^b \left\{ \int_a^x K(x|x') \phi_n(x') \, dx' \right\} w_m(x) \, dx.$$

It is clear that collocation is a special case of the weighting function method, with

$$w_m(x) = \delta(x - x_m) \qquad (m = 1, \ldots, N).$$

By choosing

$$w_m(x) = x^{m-1} \qquad (m = 1, \ldots, N)$$

we require equality of the first $m - 1$ *moments* of Equation (7.6). This is the origin of the term *method of moments* (or *moment method*, abbreviated MoM). However, the term is now used (at least in the electromagnetics community) in the broader sense presented above. The MoM was pioneered by Harrington [81, 80], and has been extended by many researchers over the years.

7.1.4.3 Choice of basis and weighting functions

We mentioned that if the ϕ_n form a complete set, then (7.5) becomes an exact representation for $f(x)$. We will never achieve $N \to \infty$ in a numerical solution, however, and must be satisfied with an approximation in any case. Hence we may abandon the completeness requirement and choose a basis set that is numerically convenient and provides an acceptable level of accuracy.

Basis and weighting functions can be broadly classed as follows.

1. *Subdomain functions* are defined only over part of the domain of the unknown function. Included in this category are rectangular pulses, triangle functions, and piecewise sinusoidal functions.

2. *Entire domain functions* are defined on the whole domain of the unknown function. Included here are power series, orthogonal polynomials (e.g., Chebyshev polynomials), and various forms of Fourier series.

How does one select the "best" weighting functions for a given choice of basis functions? One method is to require the basis functions to be *orthogonal* to the residual in the sense [44]

$$\int_a^b \phi_m(x) r(x)\, dx = 0 \qquad (m = 1, \ldots, N).$$

This produces the set of equations

$$\sum_{n=1}^N a_n \int_a^b \{\mathcal{L}[\phi_n(x)]\}\, \phi_m(x)\, dx = \int_a^b F(x)\phi_m(x)\, dx \qquad (m = 1, \ldots, N).$$

Comparing to (7.8), we see that this is identical to choosing $w_m(x) = \phi_m(x)$. This special case of the MoM in which the basis and weighting functions are identical is called *Galerkin's method*.

Galerkin's method is very close in principle to the method of *least squares* [87]. In this approach we seek the coefficients a_n that minimize the average of the squared residual

$$\epsilon = \int_a^b r^2(x)\, dx = \int_a^b \left[\sum_{n=1}^N a_n \mathcal{L}[\phi_n(x)] - F(x) \right]^2 dx.$$

Differentiating with respect to a_m and setting the result to zero, we obtain the system of equations

$$\int_a^b \mathcal{L}[\phi_m(x)] \left[\sum_{n=1}^N a_n \mathcal{L}[\phi_n(x)] - F(x) \right] dx = 0 \qquad (m = 1, \ldots, N)$$

or

$$\sum_{n=1}^N a_n \int_a^b \{\mathcal{L}[\phi_n(x)]\, \mathcal{L}[\phi_m(x)]\}\, dx = \int_a^b F(x)\phi_m(x)\, dx \qquad (m = 1, \ldots, N).$$

This is simply (7.8) with $w_m(x) = \mathcal{L}[\phi_m(x)]$.

We should emphasize that MoM does not always yield accurate solutions to integral equations. For first-kind equations having smooth kernels, it leads to ill-conditioned systems of linear equations and associated difficulties with propagation of roundoff error [44]. Suppose, for instance, that $f(t)$ and $g(t)$ are smooth functions defined on $[0, T_f]$ and $[0, T_g]$, respectively. Their convolution is given by

$$c(t) = \int_0^t f(t')g(t - t')\, dt' \qquad (0 \le t \le T_f + T_g).$$

If $c(t)$ and $g(t)$ are known (e.g., measured), then $f(t)$ satisfies a Volterra equation of the first kind; the problem of finding $f(t)$ is referred to as *deconvolution*. In this case MoM yields a linear system of algebraic equations that is notoriously ill-conditioned [163]. This becomes apparent when we realize that convolution is a smoothing operation; since $c(t)$ carries less "information" than either $f(t)$ or $g(t)$, the recovery of $f(t)$ from $c(t)$ is highly sensitive to numerical or experimental errors in the data.

7.1.5 Writing a boundary value problem as an integral equation

A boundary value problem (BVP) consisting of a linear differential equation and a set of boundary values can often be represented as an integral equation by directly integrating

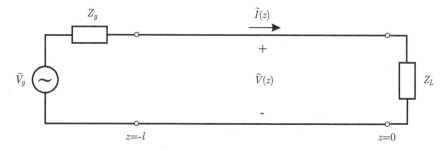

FIGURE 7.1

Transmission line with steady-state excitation.

the differential equation. (Not all integral equations have a differential equation counterpart, however.) An example is the Sturm–Liouville problem discussed in § A.5. Because the resulting integral equation incorporates the boundary conditions, the problem statement may be simpler and its solution more convenient.

As an example of this approach, let us consider the simple problem of a transmission line with AC excitation (Figure 7.1). The voltage and current waves on the line are given by the telegraphist's equations (5.305) and (5.308)

$$\frac{d\tilde{V}(z)}{dz} = -\mathcal{Z}\tilde{I}(z), \tag{7.9}$$

$$\frac{d\tilde{I}(z)}{dz} = -\mathcal{Y}\tilde{V}(z), \tag{7.10}$$

where $\mathcal{Z} = \mathcal{R} + j\omega\mathcal{L}$ is the series impedance per unit length and $\mathcal{Y} = \mathcal{G} + j\omega\mathcal{C}$ is the shunt admittance per unit length. Differentiating (7.9) and substituting into (7.10), we obtain

$$\frac{d^2\tilde{V}(z)}{dz^2} - \gamma^2\tilde{V}(z) = 0 \tag{7.11}$$

where $\gamma^2 = \mathcal{Z}\mathcal{Y}$. Reference to the figure shows that the required boundary conditions are

$$\tilde{V}_g = Z_g\tilde{I}(-l) + \tilde{V}(-l) \tag{7.12}$$

and

$$\tilde{V}(0) = Z_L\tilde{I}(0). \tag{7.13}$$

To convert (7.11) into an integral equation, we integrate twice. The first integration gives

$$\frac{d\tilde{V}(z)}{dz} = \gamma^2 \int_{-l}^{z} \tilde{V}(t)\, dt + C_1, \tag{7.14}$$

where C_1 is a constant. Note that the lower limit is arbitrary; we choose $-l$ for later convenience. A second integration gives

$$\tilde{V}(z) = \gamma^2 \int_{-l}^{z} \int_{-l}^{u} \tilde{V}(t)\, dt\, du + C_1 z + C_2. \tag{7.15}$$

The identity [6]

$$\int_{a}^{z} \int_{a}^{u} F(t)\, dt\, du = \int_{a}^{z} (z - t) F(t)\, dt \tag{7.16}$$

permits us to write (7.15) as

$$\tilde{V}(z) = \gamma^2 \int_{-l}^{z} (z-t)\tilde{V}(t)\,dt + C_1 z + C_2. \tag{7.17}$$

To determine C_1 and C_2 we apply the boundary conditions. Taking $Z_g = 0$ for simplicity, we have $\tilde{V}(-l) = \tilde{V}_g$. At $z = -l$, (7.17) reduces to

$$\tilde{V}_g = -C_1 l + C_2. \tag{7.18}$$

To enforce (7.13), we obtain $\tilde{I}(z)$ from (7.9) and apply (7.14) to get

$$-\gamma^2 \int_{-l}^{0} t\tilde{V}(t)\,dt + C_2 = -\gamma^2 \frac{Z_L}{\mathcal{Z}} \int_{-l}^{0} \tilde{V}(t)\,dt - C_1 \frac{Z_L}{\mathcal{Z}}. \tag{7.19}$$

Solving (7.18) and (7.19) for C_1 and C_2, and substituting back into (7.17), we obtain the integral equation for $\tilde{V}(z)$:

$$\tilde{V}(z) = \gamma^2 \int_{-l}^{z} (z-t)\tilde{V}(t)\,dt + \gamma^2(l+z) \int_{-l}^{0} \frac{\mathcal{Z}t - Z_L}{\mathcal{Z}l + Z_L}\tilde{V}(t)\,dt - \tilde{V}_g \frac{\mathcal{Z}z - Z_L}{\mathcal{Z}l + Z_L}, \tag{7.20}$$

valid for $-l \leq z \leq 0$. Note that $\tilde{V}(z)$ appears both inside and outside the integrals. Furthermore, both constant and variable integration limits are present. By rewriting the kernel of the first integral in terms of a step function, one can combine the integrals into a single integral over the range $[-l, 0]$. The result is a Fredholm equation of the second kind, with a discontinuous kernel.

Let us apply the MoM to (7.20). We choose pulse function expansion of $\tilde{V}(z)$ and collocation. The voltage is expressed as

$$\tilde{V}(z) = \sum_{n=1}^{N} a_n P_n(z) \tag{7.21}$$

where $P_n(z)$ is a rectangular pulse function of unit amplitude:

$$P_n(z) = \begin{cases} 1, & -l + (n-1)\Delta \leq z \leq -l + n\Delta, \\ 0, & \text{elsewhere,} \end{cases}$$

with $\Delta = l/N$. This amounts to partitioning the transmission line into N segments and treating $\tilde{V}(z)$ as a constant a_n over the nth segment. The resulting *staircase approxima-tion* to $\tilde{V}(z)$ is indicated in Figure 7.2; it is reasonable to expect this approximation to improve as N is increased.

Substituting (7.21) into (7.20) and point matching at N discrete points $\{z_m\}$, we obtain

$$\sum_{n=1}^{N} a_n P_n(z_m) = \sum_{n=1}^{N} a_n \gamma^2 \int_{-l}^{z_m} (z_m - t) P_n(t)\,dt$$

$$+ \sum_{n=1}^{N} a_n \gamma^2 (l + z_m) \int_{-l}^{0} \frac{\mathcal{Z}t - Z_L}{\mathcal{Z}l + Z_L} P_n(t)\,dt - \tilde{V}_g \frac{\mathcal{Z}z_m - Z_L}{\mathcal{Z}l + Z_L}$$

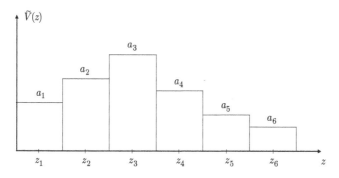

FIGURE 7.2
Representation of transmission-line voltage using rectangular-pulse basis functions.

for $m = 1, \ldots, N$. We can rewrite this as

$$\sum_{n=1}^{N} a_n \left[\delta_{mn} + U_{mn} + V_{mn} \right] = b_m, \tag{7.22}$$

where δ_{mn} is the Kronecker delta function,

$$U_{mn} = \gamma^2 \int_{-l}^{z_m} (t - z_m) P_n(t)\, dt = \begin{cases} 0, & m < n, \\ -\frac{1}{8}\gamma^2 \Delta^2, & m = n, \\ \gamma^2 \Delta (z_n - z_m), & m > n, \end{cases}$$

$$V_{mn} = -\gamma^2 (l + z_m) \frac{1}{\mathcal{Z}l + Z_L} \int_{-l}^{0} [\mathcal{Z}t - Z_L] P_n(t)\, dt = -\gamma^2 (l + z_m) \Delta \frac{\mathcal{Z}z_n - Z_L}{\mathcal{Z}l + Z_L},$$

and

$$b_m = -\tilde{V}_g \frac{\mathcal{Z}z_m - Z_L}{\mathcal{Z}l + Z_L}.$$

Here we have placed the match points at the centers of the segments so that

$$z_m = -l + (m - \tfrac{1}{2})\Delta.$$

▶ **Example 7.1:** Voltage on a lossless transmission line

Consider a lossless transmission line ($\mathcal{R} = \mathcal{G} = 0$). Assume $Z_0 = 50\,\Omega$, $Z_L = 100 - j100\,\Omega$, and $l = 2.3\lambda$. Plot the voltage on the line as a solution to the integral equation, and compare to the analytic result. Determine the standing wave ratio on the line.

Solution: Assume time-harmonic excitation at frequency $\breve{\omega}$. Since the line is lossless, $\mathcal{Z} = j\breve{\omega}\mathcal{L}$ and $\mathcal{Y} = j\breve{\omega}\mathcal{C}$, hence $\gamma = j2\pi/\lambda$ where $\lambda = 1/f\sqrt{\mathcal{L}\mathcal{C}}$. Defining the characteristic impedance as $Z_0 = \sqrt{\mathcal{Z}/\mathcal{Y}}$, we have $\mathcal{Z} = \gamma Z_0$. Figure 7.3 shows the real and imaginary parts of the transmission line voltage, computed from (7.22) using $N = 500$. Note that $\breve{V}(-l) = \tilde{V}_g$. Also shown is the voltage found analytically (see, e.g., [34]):

$$\breve{V}(z) = \breve{V}_g \frac{e^{-\gamma z} + \Gamma e^{\gamma z}}{e^{\gamma l} + \Gamma e^{-\gamma l}},$$

where

$$\Gamma = \frac{Z_L - Z_0}{Z_L + Z_0}$$

is the load reflection coefficient. The standing wave ratio S, defined as the ratio of the largest to smallest voltage magnitudes, is $S = 4.266$ from the MoM solution. This closely matches the analytic result

$$S = \frac{1 + |\Gamma|}{1 - |\Gamma|},$$

which is, to six significant digits, $S = 4.26556$.

Because the matrix equation is well conditioned, we can improve our approximation to $\tilde{V}(z)$ by increasing N. A study of the approximation error is left to the reader (a similar study will be carried out in § 7.2).

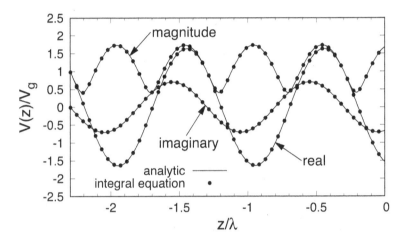

FIGURE 7.3

Voltage on a lossless transmission line. Integral equation solved numerically using 500 pulse functions. To avoid clutter, every tenth point of the integral equation solution is plotted. ◄

7.1.6   How integral equations arise in electromagnetics

Integral equations occur in areas such as antennas, radar, guided waves, and microwave systems. In a *scattering problem*, an impressed electromagnetic field arising from an immutable source interacts with some structure, inducing currents and charges on and within the structure. These induced sources create a secondary or *scattered* electromagnetic field. When a radar target is illuminated by a wave produced by a transmitter, for example, the wave "reflects" or "scatters" from the target, and this scattered field can be analyzed to determine the distance, speed, and trajectory of the target.

7.1.6.1   Integral equation for scattering from a penetrable body

Suppose an impressed electric field $\tilde{\mathbf{E}}^i$ interacts with an inhomogeneous penetrable body having medium parameters μ_0, $\epsilon(\mathbf{r})$, and $\sigma(\mathbf{r})$. The body occupies region V and is immersed in free space (Figure 7.4). The impressed field induces both polarization and conduction currents within the body, and these in turn maintain a scattered field $\tilde{\mathbf{E}}^s$. The total field $(\tilde{\mathbf{E}}, \tilde{\mathbf{H}})$, consisting of the sum of the impressed and scattered fields, must

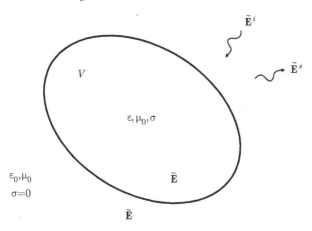

FIGURE 7.4
Scattering from a penetrable body.

obey Maxwell's equations. Ampere's law requires that

$$\nabla \times \tilde{\mathbf{H}} = \tilde{\mathbf{J}} + j\omega \tilde{\mathbf{D}},$$

where $\tilde{\mathbf{J}} = \sigma \tilde{\mathbf{E}}$ and $\tilde{\mathbf{D}} = \epsilon \tilde{\mathbf{E}}$. Hence

$$\begin{aligned}
\nabla \times \tilde{\mathbf{H}}(\mathbf{r}) &= \sigma(\mathbf{r})\tilde{\mathbf{E}}(\mathbf{r}) + j\omega\epsilon(\mathbf{r})\tilde{\mathbf{E}}(\mathbf{r}) \\
&= \{\sigma(\mathbf{r}) + j\omega\left[\epsilon(\mathbf{r}) - \epsilon_0\right]\}\,\tilde{\mathbf{E}}(\mathbf{r}) + j\omega\epsilon_0\tilde{\mathbf{E}}(\mathbf{r}) \\
&= \tilde{\mathbf{J}}^{eq}(\mathbf{r}) + j\omega\epsilon_0\tilde{\mathbf{E}}(\mathbf{r}).
\end{aligned}$$

So the total electric and magnetic fields satisfy Maxwell's equations when an *equivalent current* given by

$$\tilde{\mathbf{J}}^{eq}(\mathbf{r}) = \{\sigma(\mathbf{r}) + j\omega\left[\epsilon(\mathbf{r}) - \epsilon_0\right]\}\,\tilde{\mathbf{E}}(\mathbf{r}) = f(\mathbf{r})\tilde{\mathbf{E}}(\mathbf{r}) \tag{7.23}$$

is embedded in free space.

If we can determine $\tilde{\mathbf{J}}^{eq}(\mathbf{r})$, we can use the formulas for potentials due to sources in free space (§ 5.2) to compute the scattered field. In the absence of magnetic sources, (5.50) reads

$$\tilde{\mathbf{E}}^s = -j\omega\tilde{\mathbf{A}}^s_e - \nabla\tilde{\phi}^s_e, \tag{7.24}$$

where

$$\tilde{\mathbf{A}}^s_e(\mathbf{r}) = \mu_0 \int_V \tilde{\mathbf{J}}^{eq}(\mathbf{r}')\tilde{G}(\mathbf{r}|\mathbf{r}')\,dV', \qquad \tilde{\phi}^s_e(\mathbf{r}) = \frac{1}{\epsilon_0}\int_V \tilde{\rho}^{eq}(\mathbf{r}')\tilde{G}(\mathbf{r}|\mathbf{r}')\,dV',$$

are the "scattered" potentials produced by the equivalent current. The equivalent charge density $\tilde{\rho}^{eq}$ is related to the equivalent current through the continuity equation

$$\nabla \cdot \tilde{\mathbf{J}}^{eq}(\mathbf{r}) = -j\omega\tilde{\rho}^{eq}(\mathbf{r}),$$

and $\tilde{G}(\mathbf{r}|\mathbf{r}')$ is the free-space Green's function (5.70). Substitution into (7.24) yields

$$\tilde{\mathbf{E}}^s(\mathbf{r}) = -j\omega\mu_0 \int_V \tilde{\mathbf{J}}^{eq}(\mathbf{r}')\tilde{G}(\mathbf{r}|\mathbf{r}')\,dV' - \frac{j}{\omega\epsilon_0}\nabla\int_V \nabla' \cdot \tilde{\mathbf{J}}^{eq}(\mathbf{r}')\tilde{G}(\mathbf{r}|\mathbf{r}')\,dV' \tag{7.25}$$

$$= -\frac{j\eta_0}{k_0}\int_V [\nabla' \cdot \tilde{\mathbf{J}}^{eq}(\mathbf{r}')\nabla + k_0^2\tilde{\mathbf{J}}^{eq}(\mathbf{r}')]\frac{e^{-jk_0 R}}{4\pi R}\,dV'. \tag{7.26}$$

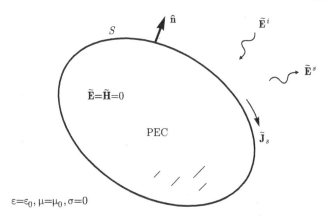

FIGURE 7.5

Scattering from a perfect electrical conductor body.

Recall that the total field is the sum of the impressed and scattered fields:

$$\tilde{\mathbf{E}} = \tilde{\mathbf{E}}^s + \tilde{\mathbf{E}}^i. \tag{7.27}$$

By (7.26) and (7.23) we have

$$\tilde{\mathbf{E}}(\mathbf{r}) + \frac{j\eta_0}{k_0} \int_V \left\{ \nabla' \cdot [f(\mathbf{r}')\tilde{\mathbf{E}}(\mathbf{r}')]\nabla + k_0^2 f(\mathbf{r}')\tilde{\mathbf{E}}(\mathbf{r}') \right\} \frac{e^{-jk_0 R}}{4\pi R} \, dV' = \tilde{\mathbf{E}}^i(\mathbf{r}). \tag{7.28}$$

This result, holding for all $\mathbf{r} \in V$, is a volume *electric field integral equation* (EFIE) for the total electric field $\tilde{\mathbf{E}}$ within V. Note that it is a second-kind Fredholm equation. If $\tilde{\mathbf{E}}$ can be determined from (7.28), then $\tilde{\mathbf{J}}^{eq}(\mathbf{r})$ can be determined from (7.23) and $\tilde{\mathbf{E}}^s(\mathbf{r})$ from (7.26).

7.1.6.2 Integral equation for scattering from a perfectly conducting body

Now suppose $\tilde{\mathbf{E}}^i$ interacts with a PEC body (Figure 7.5). The impressed field induces a conduction current $\tilde{\mathbf{J}}_s$ on the surface S, which in turn generates a scattered field $\tilde{\mathbf{E}}^s$. Equation (7.27) continues to hold in this case. The boundary condition that the total tangential electric field be zero on the surface of the body requires

$$\hat{\mathbf{n}} \times [\tilde{\mathbf{E}}^i(\mathbf{r}) + \tilde{\mathbf{E}}^s(\mathbf{r})] = 0 \qquad (\mathbf{r} \in S).$$

Substitution for the scattered field gives

$$\frac{j\eta_0}{k_0} \hat{\mathbf{n}} \times \int_S [\nabla' \cdot \tilde{\mathbf{J}}_s(\mathbf{r}')\nabla + k_0^2 \tilde{\mathbf{J}}_s(\mathbf{r}')] \frac{e^{-jk_0 R}}{4\pi R} \, dS' = \hat{\mathbf{n}} \times \tilde{\mathbf{E}}^i(\mathbf{r}) \qquad (\mathbf{r} \in S).$$

This is a surface EFIE for the unknown $\tilde{\mathbf{J}}_s$. It is a Fredholm equation of the first kind.

7.2 Plane-wave reflection from an inhomogeneous region

Integral equations arise naturally in the study of the scattering and radiation of electromagnetic waves, but can also replace more conventional formulations involving differential equations. A simple example of this is given in § 7.1.5. In this section we apply the approach to a problem in plane-wave reflection; we show how the differential equation for the field within a layered medium may be converted to an integral equation and treated by the MoM technique.

7.2.1 Reflection from a medium inhomogeneous in the z-direction

In § 4.11.5 we analyze the reflection of plane waves by planarly layered materials, with each layer assumed homogeneous. An interesting extension is to consider a planar medium with inhomogeneous material properties. If the properties vary only in the direction perpendicular to the interfaces, the problem can be formulated as a boundary-value problem and the field found as a solution to a differential equation. Here we will consider the differential equation arising from a simple inhomogeneous medium, convert to an integral equation, and solve using MoM.

Consider a y-polarized plane wave (TE polarization), incident from free space onto an interface presented by a conductor-backed material layer (Figure 7.6). The layer has thickness d, inhomogeneous conductivity $\tilde{\sigma}(z,\omega)$, and complex inhomogeneous permittivity

$$\tilde{\epsilon}^c(z,\omega) = \tilde{\epsilon}(z,\omega) - j\tilde{\sigma}(z,\omega)/\omega.$$

We assume $\tilde{\mu}$ is spatially constant. The electric field in region 0 ($z < 0$) or in region 1 ($d > z > 0$) may be written as $\tilde{\mathbf{E}}(x,z,\omega) = \hat{\mathbf{y}}\tilde{E}_y(x,z,\omega)$, while the magnetic field is by Faraday's law

$$\tilde{\mathbf{H}}(x,z,\omega) = \frac{j}{\omega\tilde{\mu}(\omega)}\left[-\hat{\mathbf{x}}\frac{\partial \tilde{E}_y(x,z,\omega)}{\partial z} + \hat{\mathbf{z}}\frac{\partial \tilde{E}_y(x,z,\omega)}{\partial x}\right]. \tag{7.29}$$

To obtain a differential equation for $\tilde{E}_y$, we substitute (7.29) into Ampere's law and get

$$\left(\frac{\partial^2}{\partial x^2} + \frac{\partial^2}{\partial z^2} + k^2(z,\omega)\right)\tilde{E}_y(x,z,\omega) = 0 \tag{7.30}$$

where $k(z,\omega) = \omega\sqrt{\tilde{\mu}(\omega)\tilde{\epsilon}^c(z,\omega)}$ is the wavenumber.

We may solve (7.30) using separation of variables (Appendix A). Writing $\tilde{E}_y(x,z,\omega) = f(x,\omega)g(z,\omega)$, we obtain from (7.30) the separated ordinary differential equations

$$\frac{d^2 f(x,\omega)}{dx^2} + k_x^2(\omega)f(x,\omega) = 0, \tag{7.31}$$

$$\frac{d^2 g(z,\omega)}{dz^2} + k_z^2(z,\omega)g(z,\omega) = 0, \tag{7.32}$$

where $k_z^2(z,\omega) = k^2(z,\omega) - k_x^2(\omega)$. Equation (7.31) has solution

$$f(x,\omega) = e^{-jk_x(\omega)x},$$

hence

$$\tilde{E}_y(x,z,\omega) = e^{-jk_x(\omega)x}g(z,\omega)$$

in both regions. In region 0 there are both incident and reflected waves. Based on our experience with reflection from a homogeneous medium, we expect these waves to share the same value of k_x, i.e., $k_x(\omega) = k_{x0}(\omega) = k_0(\omega)\sin\phi_0$, where ϕ_0 is the angle of incidence and $k_0(\omega) = \omega\sqrt{\mu_0\epsilon_0}$.

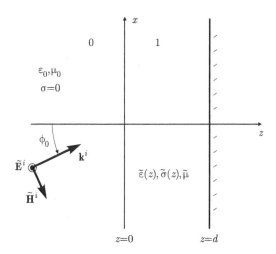

FIGURE 7.6
Reflection of a TE-polarized plane wave from a conductor-backed, inhomogeneous dielectric layer.

In region 0, where k does not depend on z, (7.32) is also the harmonic differential equation. The incident wave is given by

$$g(z,\omega) = \tilde{E}_0(\omega)e^{-jk_{z0}z}$$

where $k_{z0}(\omega) = [k_0^2(\omega) - k_x^2(\omega)]^{1/2} = k_0(\omega)\cos\phi_0$, and the reflected wave by

$$g(z,\omega) = \tilde{E}_r(\omega)e^{jk_{z0}z}.$$

Hence the total electric field in region 0 is

$$\tilde{E}_y(x,z,\omega) = \tilde{E}_0(\omega)e^{-jk_{x0}(\omega)x}e^{-jk_{z0}(\omega)z} + \tilde{E}_r(\omega)e^{-jk_{x0}(\omega)x}e^{jk_{z0}(\omega)z}.$$

To determine the magnetic field in region 0 we use (7.29):

$$\tilde{H}_x(x,z,\omega) = -\frac{\cos\phi_0}{\eta_0}\tilde{E}_0(\omega)e^{-jk_{x0}(\omega)x}e^{-jk_{z0}(\omega)z} + \frac{\cos\phi_0}{\eta_0}\tilde{E}_r(\omega)e^{-jk_{x0}(\omega)x}e^{jk_{z0}(\omega)z}.$$

To satisfy the boundary conditions at $z = 0$, the wave vector component k_x must have the same value in regions 0 and 1. Hence the fields in region 1 are

$$\tilde{E}_y(x,z,\omega) = g(z,\omega)e^{-jk_{x0}(\omega)x},$$

$$\tilde{H}_x(x,z,\omega) = \frac{-j}{\omega\tilde{\mu}(\omega)}e^{-jk_{x0}(\omega)x}\frac{\partial g(z,\omega)}{\partial z}.$$

These represent superpositions of forward and backward traveling waves.

The jump conditions on the tangential fields are required to provide a unique solution to (7.32). Setting $E_y = 0$ on the surface of the conductor, we get

$$g(d,\omega) = 0. \tag{7.33}$$

The conditions for continuity of tangential $\tilde{\mathbf{E}}$ and $\tilde{\mathbf{H}}$ at $z = 0$ are

$$\tilde{E}_0(\omega) + \tilde{E}_r(\omega) = g(0, \omega), \tag{7.34}$$

$$-\tilde{E}_0(\omega) + \tilde{E}_r(\omega) = -j \frac{\eta_0}{\cos \phi_0} \frac{1}{\omega \tilde{\mu}(\omega)} \left. \frac{\partial g(z, \omega)}{\partial z} \right|_{z=0}, \tag{7.35}$$

respectively; subtraction yields

$$2\tilde{E}_0(\omega) = g(0, \omega) + a(\omega) \left. \frac{\partial g(z, \omega)}{\partial z} \right|_{z=0} \tag{7.36}$$

where $a(\omega) = j\eta_0/[\omega \tilde{\mu}(\omega) \cos \phi_0]$.

Once $g(z)$ has been determined either analytically or numerically, the reflection coefficient may be found from (7.34) as

$$\tilde{\Gamma}(\omega) = \frac{\tilde{E}_r(\omega)}{\tilde{E}_0(\omega)} = -1 + \frac{g(0, \omega)}{\tilde{E}_0}. \tag{7.37}$$

7.2.2 Conversion to an integral equation

To determine the field inside the inhomogeneous layer and hence the reflection coefficient for this problem, we must solve (7.32). Analytic solutions are available only for certain permittivity and conductivity profiles (see, e.g., [208, 164, 102]). As an alternative to solving the differential equation numerically, we may convert it to an integral equation.

Let us suppress the dependence on ω and write (7.32) in the form

$$g''(z) + k_z^2(z)g(z) = 0 \tag{7.38}$$

and integrate twice. We get

$$g'(z) + \int_0^z k_z^2(t)g(t)\, dt = C_1 \tag{7.39}$$

and then, by (7.16),

$$g(z) = -\int_0^z (z - t)k_z^2(t)g(t)\, dt + C_1 z + C_2. \tag{7.40}$$

It is worth checking that (7.40) satisfies (7.38). See Problem 7.3.

The constants C_1 and C_2 must satisfy (7.36) and (7.33). Substitution of (7.40) and (7.39) into (7.36) gives

$$2\tilde{E}_0 = C_2 + aC_1,$$

while substitution of (7.40) into (7.33) gives

$$-\int_0^d (d - t)k_z^2(t)g(t)\, dt + C_1 d + C_2 = 0.$$

Solving simultaneously, we obtain

$$C_1 = \frac{1}{a - d}\left[2\tilde{E}_0 - \int_0^d (d - t)k_z^2(t)g(t)\, dt \right],$$

$$C_2 = 2\tilde{E}_0 \frac{d}{a - d} + \frac{a}{a - d}\int_0^d (d - t)k_z^2(t)g(t)\, dt.$$

Substitution into (7.40) yields the integral equation for $g(z)$:

$$g(z) = -\int_0^z (z-t)k_z^2(t)g(t)\,dt + \frac{a-z}{a-d}\int_0^d (d-t)k_z^2(t)g(t)\,dt + 2\tilde{E}_0\frac{z-d}{a-d}. \quad (7.41)$$

This Fredholm equation of the second kind can be rewritten in terms of the unit step function (A.6) as

$$g(z) = \int_0^d \left[\frac{a-z}{a-d}(d-t) - (z-t)U(z-t)\right]k_z^2(t)g(t)\,dt + 2\tilde{E}_0\frac{z-d}{a-d}. \quad (7.42)$$

We can solve (7.41) for arbitrary permittivity and conductivity profiles, including piecewise continuous functions describing a sequence of homogeneous layers. As a simple example, consider a single homogeneous layer with parameters $\tilde{\epsilon}(\omega)$ and $\tilde{\sigma}(\omega)$. In this case we may solve (7.32) analytically for comparison. Since k_z is independent of z, the solution of (7.32) is

$$g(z,\omega) = C_1 \sin k_z z + C_2 \cos k_z z.$$

An application of (7.34) and (7.35) will determine C_1 and C_2; the result is that

$$g(z,\omega) = 2\tilde{E}_0\frac{\sin k_z(d-z)}{\sin k_z d - ak_z\cos k_z d}. \quad (7.43)$$

By (7.37), the reflection coefficient is

$$\tilde{\Gamma}(\omega) = \frac{\sin k_z d + ak_z\cos k_z d}{\sin k_z d - ak_z\cos k_z d}. \quad (7.44)$$

With a bit of manipulation, (7.44) may be put into the form

$$\tilde{\Gamma}(\omega) = \frac{\tilde{\Gamma}_0(\omega) - \tilde{P}^2(\omega)}{1 - \tilde{\Gamma}_0(\omega)\tilde{P}^2(\omega)} \quad (7.45)$$

where $\tilde{P}(\omega) = e^{-jk_z d}$ and

$$\tilde{\Gamma}_0(\omega) = \frac{1 + ja(\omega)k_z(\omega)}{1 - ja(\omega)k_z(\omega)} = \frac{\tilde{\mu}_r(\omega)\cos\phi_0 - \sqrt{\tilde{\mu}_r(\omega)\tilde{\epsilon}_r^c(\omega) - \sin^2\phi_0}}{\tilde{\mu}_r(\omega)\cos\phi_0 + \sqrt{\tilde{\mu}_r(\omega)\tilde{\epsilon}_r^c(\omega) - \sin^2\phi_0}}.$$

See Problem 7.9. Here $\tilde{\epsilon}_r^c = \tilde{\epsilon}^c/\epsilon_0$ and $\tilde{\Gamma}_0$ is the interfacial reflection coefficient. The same result was obtained in § 4.11.5.

7.2.3 Solution to the integral equation

We may solve (7.41) using MoM. Collocation yields excellent results. We expand $g(z)$ in a set of pulse functions

$$g(z) = \sum_{n=1}^N a_n P_n(z) \quad (7.46)$$

where

$$P_n(z) = \begin{cases} 1, & (n-1)\Delta \le z \le n\Delta, \\ 0 & \text{elsewhere}, \end{cases}$$

with $\Delta = d/N$. This effectively partitions the slab into N regions with $g(z)$ a constant a_n in the nth region. Substituting (7.46) into (7.41) and matching the equation at the points $z = z_m = (m - 1/2)\Delta$, we obtain

$$\sum_{n=1}^{N} a_n P_n(z_m) - k_z^2 \sum_{n=1}^{N} a_n \int_0^d \left[\frac{a - z_m}{a - d}(d - t) \right] P_n(t)\, dt$$

$$+ k_z^2 \sum_{n=1}^{N} a_n \int_0^{z_m} (z_m - t) P_n(t)\, dt = 2\tilde{E}_0 \frac{z_m - d}{a - d}.$$

This simultaneous system can be written as

$$\sum_{n=1}^{N} a_n [\delta_{mn} + U_{mn} + V_{mn}] = b_m \qquad (m = 1, \ldots, N) \tag{7.47}$$

where δ_{mn} is the Kronecker delta,

$$V_{mn} = -k_z^2 \int_0^d \left[\frac{a - z_m}{a - d}(d - t) \right] P_n(t)\, dt$$

$$= -k_z^2 \left[\frac{a - z_m}{a - d} \right] \int_{(n-1)\Delta}^{n\Delta} (d - t)\, dt$$

$$= -k_z^2 \left[\frac{a - z_m}{a - d} \right] \Delta(d - z_n),$$

$$U_{mn} = k_z^2 \int_0^{z_m} (z_m - t) P_n(t)\, dt = \begin{cases} 0, & m < n, \\ \dfrac{k_z^2}{8}\Delta^2, & m = n, \\ k_z^2 \Delta(z_m - z_n), & m > n, \end{cases}$$

and

$$b_m = 2\tilde{E}_0 \frac{z_m - d}{a - d}.$$

▶ **Example 7.2:** Accuracy of moment-method solution for a time-harmonic plane wave incident on a conductor-backed medium

Consider a conductor-backed slab with $d = 0.025$ m, $\epsilon_r = 8$, $\mu_r = 4$, $\sigma = 0.3$ S/m, $\phi_0 = 30°$, and time-harmonic excitation at $f = 10$ GHz. Explore the dependence of the solution accuracy on the number of basis functions used.

Solution: To determine how the solution accuracy depends on N, we compute the reflection coefficient from the numerical solution to (7.37) and compare the result with the analytic solution (7.44). Note that (7.37) requires knowledge of $g(z)$ at $z = 0$. Unfortunately, a_1 itself is a rather poor approximation to this. Instead we use an extrapolation technique, passing a second-order polynomial through the first three pulse-function amplitudes and then evaluating it at $z = 0$ (see Problem 7.24). The result is

$$g(0) \approx \tfrac{1}{8}(15a_1 - 10a_2 + 3a_3). \tag{7.48}$$

Figure 7.7 shows how the fractional error in the reflection coefficient, defined as

$$\epsilon_\Gamma(N) = \frac{|\Gamma_{\text{numerical}} - \Gamma_{\text{analytic}}|}{|\Gamma_{\text{analytic}}|},$$

depends on N. As N increases beyond 50 or so, ϵ_Γ begins to decrease. In fact, a tenfold increase in N yields about two digits of improvement. For $N = 2000$, the reflection coefficient agrees with the analytic result $\Gamma = -0.1340 + j\, 0.09345$ to four significant digits.

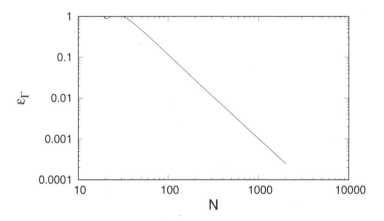

FIGURE 7.7
Fractional error in the reflection coefficient computed by solving the integral equation with N pulse functions. ◀

▶ **Example 7.3:** Field profile for a plane wave incident on a conductor-backed medium

Consider the conductor-backed slab of Example 7.2. Plot the electric field profile $g(z)$.

Solution: Figure 7.8 shows the magnitude of $g(z)$ obtained from (7.41) with $N = 2000$. The a_n values appear as circles plotted at the points z_n, and the analytic solution (7.43) is shown for comparison. Note the expected standing-wave field in the slab, and the attenuation with distance into the slab (an effect of nonzero conductivity).

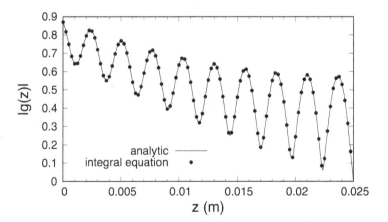

FIGURE 7.8
Magnitude of $g(z)$ within slab region. Integral equation solved numerically using 2000 pulse functions. To avoid clutter, every twentieth point of the integral equation solution is plotted. ◀

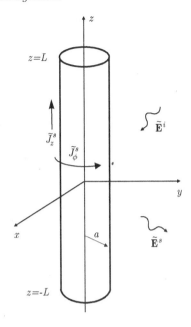

FIGURE 7.9
Excitation of a straight circular wire by an impressed electric field.

7.3 Solution to problems involving thin wires

Thin wires arise often in the study of electromagnetic fields. Many antennas are constructed from wires; these include simple dipoles and loops, and more complicated log-periodic and Yagi-Uda arrays, bowties, rhombics, bicones, and helices. Reflectors and shields often consist of wire screens, and conducting surfaces are sometimes simulated using wire meshes. In all of these cases, the electromagnetic interactions between the structures and the fields exciting them can be described using integral equations. In this section we examine some simple situations in both the time and frequency domains.

7.3.1 The straight wire

Consider a straight segment having length $2L$ and circular cross-section of radius a, immersed in free space as shown in Figure 7.9. When an impressed static electric field interacts with the wire, such as from an applied voltage, a charge is induced on its surface, and this charge creates a secondary electric field. The arrangement of the charge is such that the total tangential electric field, consisting of the superposition of the impressed and secondary fields, must be zero at the wire surface. Although the static problem provides an introduction to the use of integral equations in electromagnetics, the dynamic problem is more illuminating. In the dynamic problem, an impressed field $\tilde{\mathbf{E}}^i(\mathbf{r}, \omega)$ excites a current on the wire surface, and this current creates a secondary, or scattered, field $\tilde{\mathbf{E}}^s(\mathbf{r}, \omega)$. By applying the boundary condition on the tangential electric field on the wire surface, we can obtain an integral equation governing the behavior of the surface current.

7.3.1.1   Derivation of the electric-field integral equation

Because the structure is finite, it is actually a right circular cylinder. When the wire is thin, however, as defined by the two conditions

$$a/\lambda \ll 1, \qquad a/L \ll 1,$$

we can ignore the influence of the wire ends on the surface current. Hence we shall only enforce the boundary condition on tangential electric field over the remaining surface:

$$\hat{\mathbf{n}} \times \tilde{\mathbf{E}}(\mathbf{r}) = \hat{\boldsymbol{\rho}} \times [\tilde{\mathbf{E}}^i(\mathbf{r}) + \tilde{\mathbf{E}}^s(\mathbf{r})] = 0 \qquad (\mathbf{r} \in S),$$

where S is the surface $\rho = a$, $0 \leq \phi < 2\pi$, $-L \leq z \leq L$, and we have suppressed the dependence on ω. Thus

$$\hat{\boldsymbol{\rho}} \times [\hat{\mathbf{z}}\tilde{E}_z^s + \hat{\boldsymbol{\phi}}\tilde{E}_\phi^s + \hat{\mathbf{z}}\tilde{E}_z^i + \hat{\boldsymbol{\phi}}\tilde{E}_\phi^i]\big|_{\rho=a} = 0. \tag{7.49}$$

Now let us assume that either (a) $\tilde{E}_\phi^i = 0$ so that $\tilde{E}_\phi^s = 0$, or (b) $\tilde{E}_z^i \gg \tilde{E}_\phi^i$ so that $\tilde{E}_\phi^s$ can be ignored in favor of $\tilde{E}_z^s$. The first condition occurs when the wire is symmetrically excited as in the case of a linear antenna. The second occurs when an impressed field (such as an incident plane wave) is polarized primarily along the wire axis. In either case, (7.49) becomes simply

$$\tilde{E}_z^s(\mathbf{r}) + \tilde{E}_z^i(\mathbf{r}) = 0 \qquad (\rho = a,\ 0 \leq \phi < 2\pi,\ -L \leq z \leq L). \tag{7.50}$$

The desired integral equation is obtained when the scattered field in (7.50) is represented as a superposition integral. Since the impressed field is predominantly axial, so are the induced surface current and resulting vector potential $\tilde{\mathbf{A}}^s$. Thus, the scattered electric field is

$$\tilde{\mathbf{E}}^s(\mathbf{r}) = -j\frac{\omega}{k_0^2}[\nabla(\nabla \cdot \tilde{\mathbf{A}}^s) + k_0^2 \tilde{\mathbf{A}}^s] \tag{7.51}$$

or

$$\tilde{E}_z^s(\mathbf{r}) = -j\frac{\omega}{k_0^2}\left[\frac{\partial^2 \tilde{A}_z^s}{\partial z^2} + k_0^2 \tilde{A}_z^s\right].$$

Substitution into (7.50) gives

$$-j\frac{\omega}{k_0^2}\left(\frac{\partial^2}{\partial z^2} + k_0^2\right)\tilde{A}_z^s(a,\phi,z) = -\tilde{E}_z^i(a,\phi,z) \qquad (0 \leq \phi < 2\pi,\ -L \leq z \leq L), \tag{7.52}$$

where the vector potential on the surface of the wire is

$$\tilde{A}_z^s(a,\phi,z) = \frac{\mu_0}{4\pi}\int_{-L}^{L}\int_{0}^{2\pi}\tilde{J}_z^s(\phi',z')\frac{e^{-jk_0 R}}{R}a\,d\phi'\,dz'. \tag{7.53}$$

Here, $R = |\mathbf{r} - \mathbf{r}'|$ is the distance from a source point (a,ϕ',z') on the surface to a field point (a,ϕ,z) on the surface (Figure 7.10):

$$R = R(\phi - \phi', z - z') = \sqrt{4a^2 \sin^2\frac{1}{2}(\phi - \phi') + (z - z')^2}.$$

To proceed further, we assume that either (a) $\tilde{E}_z^i$ is independent of ϕ and hence by (7.52) so is $\tilde{A}_z^s$, or (b) the wire is thin so that $\tilde{E}_z^i$ depends only weakly on ϕ and hence

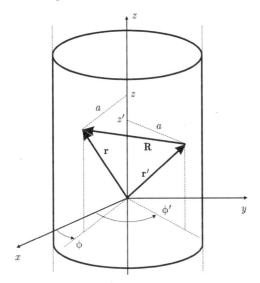

FIGURE 7.10
Geometry of a circular wire showing distance between a source point on the surface of the wire at $\mathbf{r}'$ and an observation point on the surface at $\mathbf{r}$.

so does $\tilde{A}_z^s$. Then, instead of requiring (7.52) to hold for all ϕ, we equate the *average values* of the two sides:

$$-j\frac{\omega}{k_0^2}\left(\frac{\partial^2}{\partial z^2}+k_0^2\right)\left[\frac{1}{2\pi}\int_0^{2\pi}\tilde{A}_z^s(a,\phi,z)d\phi\right] = -\frac{1}{2\pi}\int_0^{2\pi}\tilde{E}_z^i(a,\phi,z)d\phi \qquad (-L\le z\le L). \tag{7.54}$$

The term in brackets is the average value of the scattered vector potential, which we refer to as $\tilde{A}^s$. By (7.53) this is

$$\tilde{A}^s(z) = \frac{1}{2\pi}\frac{\mu_0}{4\pi}\int_{-L}^{L}\left\{\int_0^{2\pi}\tilde{J}_z^s(\phi',z')\left[\int_0^{2\pi}\frac{e^{-jk_0R(\phi-\phi',z-z')}}{R(\phi-\phi',z-z')}\,d\phi\right]a\,d\phi'\right\}dz'.$$

Setting $\xi = \phi - \phi'$ and noting that R is periodic with period 2π, we obtain

$$\tilde{A}^s(z) = \frac{1}{2\pi}\frac{\mu_0}{4\pi}\int_{-L}^{L}\left\{\int_0^{2\pi}\tilde{J}_z^s(\phi',z')\left[\int_0^{2\pi}\frac{e^{-jk_0R(\xi,z-z')}}{R(\xi,z-z')}\,d\xi\right]a\,d\phi'\right\}dz'.$$

Interchanging the order of integration, we have

$$\tilde{A}^s(z) = \frac{1}{2\pi}\frac{\mu_0}{4\pi}\int_{-L}^{L}\left\{\int_0^{2\pi}\tilde{J}_z^s(\phi',z')a\,d\phi'\int_0^{2\pi}\frac{e^{-jk_0R(\xi,z-z')}}{R(\xi,z-z')}\,d\xi\right\}dz'.$$

We recognize

$$\tilde{I}(z) = \int_0^{2\pi}\tilde{J}_z^s(\phi',z)a\,d\phi'$$

as the current on the wire surface at axial point z, and define the *average* Green's function as

$$\tilde{G}(z-z') = \int_0^{2\pi}\frac{e^{-jk_0R(\xi,z-z')}}{R(\xi,z-z')}\,d\xi.$$

The average vector potential can then be written as

$$\tilde{A}^s(z) = \frac{1}{2\pi}\frac{\mu_0}{4\pi} \int_{-L}^{L} \tilde{I}(z')\tilde{G}(z-z')\,dz'. \tag{7.55}$$

In terms of $\tilde{A}^s$, the boundary condition equation (7.54) becomes

$$\left(\frac{d^2}{dz^2} + k_0^2\right)\tilde{A}^s(z) = -j\frac{k_0^2}{\omega}\tilde{E}^i(z) \qquad (-L \le z \le L), \tag{7.56}$$

where we designate the average value of the impressed field as

$$\tilde{E}^i(z) = \frac{1}{2\pi}\int_0^{2\pi} \tilde{E}_z^i(a,\phi,z)\,d\phi. \tag{7.57}$$

Equation (7.56) is an ordinary differential equation for the average vector potential $\tilde{A}^s$. Its solution consists of a superposition of particular and complementary components:

$$\tilde{A}^s(z) = f_p(z) + f_c(z), \tag{7.58}$$

where [23]

$$f_p(z) = \frac{1}{k_0}\int_{z_0}^{z}\left[-j\frac{k_0^2}{\omega}\tilde{E}^i(u)\right]\sin k_0(z-u)\,du \tag{7.59}$$

and

$$f_c(z) = \bar{C}_1 \sin k_0 z + \bar{C}_2 \cos k_0 z.$$

Here z_0 is an arbitrary constant chosen for convenience, and constants $\bar{C}_1$ and $\bar{C}_2$ must satisfy the boundary conditions on the current at the ends of the wire. Upon substitution of (7.55) for $\tilde{A}^s$, (7.58) becomes

$$\int_{-L}^{L}\tilde{I}(z')G(z-z')\,dz' + C_1\sin k_0 z + C_2\cos k_0 z = -j\frac{8\pi^2}{\eta_0}\int_{-L}^{z}\tilde{E}^i(u)\sin k_0(z-u)\,du \tag{7.60}$$

where C_1 and C_2 are two new constants to be determined. We have chosen $z_0 = -L$ in (7.59) and have used $\omega\mu_0/k_0 = \eta_0$.

Equation (7.60) is the EFIE for $\tilde{I}(z)$, since the boundary condition employed in its derivation involves tangential electric field. It is also called the *Hallén* integral equation after Erik Hallén, the Swedish researcher who first sought solutions to this type of equation [77, 78]. We note that this is a Fredholm equation of the first kind, because the limits on the integral involving $\tilde{I}(z)$ are constant, and the unknown function only appears within the integral.

7.3.1.2 Solution to the electric-field integral equation

Equation (7.60) may be solved by the MoM. The simplest approach is to use collocation with pulse-function expansion of $\tilde{I}(z)$. Let

$$\tilde{I}(z) = \sum_{n=1}^{N} a_n P_n(z) \tag{7.61}$$

where $P_n(z)$ is the nth pulse function given by

$$P_n(z) = \begin{cases} 1, & (n-1)\Delta \le z \le n\Delta, \\ 0, & \text{elsewhere,} \end{cases}$$

with $\Delta = 2L/N$. In essence we partition the wire into N segments carrying constant current values a_n. Substituting (7.61) into (7.60) and matching the equation at the points $z = z_m = -L + (m - 1/2)\Delta$, we obtain a system of N equations in $N + 2$ unknowns:

$$\sum_{n=1}^{N} a_n A_{mn} + C_1 \sin k_0 z_m + C_2 \cos k_0 z_m = b_m, \tag{7.62}$$

where

$$A_{mn} = \int_{z_n - \Delta/2}^{z_n + \Delta/2} \tilde{G}(z_m - z') \, dz',$$

$$b_m = -j\frac{8\pi^2}{\eta_0} \int_{-L}^{z_m} \tilde{E}^i(u) \sin k_0(z_m - u) \, du. \tag{7.63}$$

Two more equations are obtained by imposing the conditions on the current at the wire ends. Since the wire is thin, there is little surface area at the end for charge to accumulate. Hence the continuity equation implies that the current must be very weak there. For simplicity we assume the current vanishes at the ends:

$$\tilde{I}(-L) = \tilde{I}(L) = 0.$$

Since the currents are assumed constant within the partitions, these conditions are most easily invoked by requiring that $a_1 = a_N = 0$. However, a_1 and a_N more accurately approximate the current at the *centers* of the partitions. With little additional effort, we can implement the quadratic extrapolation (7.48) to estimate the currents at the actual wire ends:

$$\tilde{I}(-L) \approx \frac{1}{8}[15a_1 - 10a_2 + 3a_3] = 0,$$

$$\tilde{I}(L) \approx \frac{1}{8}[3a_{N-2} - 10a_{N-1} + 15a_N] = 0.$$

Using these, we can write (7.62) in the matrix form

$$\begin{bmatrix} A_{11} & A_{12} & A_{13} & A_{14} & \cdots & A_{1,N-3} & A_{1,N-2} & A_{1,N-1} & A_{1,N} & \sin k_0 z_1 & \cos k_0 z_1 \\ A_{21} & A_{22} & A_{23} & A_{24} & \cdots & A_{2,N-3} & A_{2,N-2} & A_{2,N-1} & A_{2,N} & \sin k_0 z_2 & \cos k_0 z_2 \\ \vdots & \vdots & \vdots & \vdots & & \vdots & \vdots & \vdots & \vdots & \vdots & \vdots \\ A_{N1} & A_{N2} & A_{N3} & A_{N4} & \cdots & A_{N,N-3} & A_{N,N-2} & A_{N,N-1} & A_{N,N} & \sin k_0 z_N & \cos k_0 z_N \\ \frac{15}{8} & -\frac{10}{8} & \frac{3}{8} & 0 & \cdots & 0 & 0 & 0 & 0 & 0 & 0 \\ 0 & 0 & 0 & 0 & \cdots & 0 & \frac{3}{8} & -\frac{10}{8} & \frac{15}{8} & 0 & 0 \end{bmatrix} \begin{bmatrix} a_1 \\ a_2 \\ \vdots \\ a_N \\ C_1 \\ C_2 \end{bmatrix} = \begin{bmatrix} b_1 \\ b_2 \\ \vdots \\ b_N \\ 0 \\ 0 \end{bmatrix}. \tag{7.64}$$

We can establish two important properties of the entries A_{mn} through a simple change of variables. Let $u = z_m - z'$. Then

$$A_{mn} = \int_{(m-n-1/2)\Delta}^{(m-n+1/2)\Delta} \tilde{G}(u) \, du, \tag{7.65}$$

where

$$\tilde{G}(u) = \int_0^{2\pi} \frac{e^{-jk_0 R(\xi, u)}}{R(\xi, u)} \, d\xi \quad \text{with} \quad R(\xi, u) = \sqrt{4a^2 \sin^2 \frac{\xi}{2} + u^2}.$$

Note that A_{mn} depends on the indices m and n only through the difference $m - n$. Another change of variables $v = -u$ quickly establishes that $A_{mn} = A_{nm}$, which is a result of the reciprocity of sources and fields in free space. Together, these conditions significantly reduce the computational expense of filling the matrix. A vector may be filled with the distinct values of A_{mn} for all allowed differences $m - n$, and the matrix filled by selection from this vector.

Most of the A_{mn} may be computed easily using numerical integration. However, the $m = n$ "self" terms involve a singularity in the Green's function; the integrand of $\tilde{G}(z - z')$ becomes infinite when the source and observation points coincide. Although this singularity is integrable, it might cause difficulties with numerical integration. Let us "extract" the singularity by isolating a singular term and integrating it analytically. From (7.65) we have

$$A_{mm} = 4 \int_{-\Delta/2}^{\Delta/2} \int_0^{\pi/2} \frac{e^{-jk_0 R(\psi,u)}}{R(\psi,u)} \, d\psi \, du,$$

where we have defined $\psi = \xi/2$. Next, we split the integrand into two pieces:

$$A_{mm} = 4 \int_{-\Delta/2}^{\Delta/2} \int_0^{\pi/2} \frac{e^{-jk_0 R(\psi,u)} - 1}{R(\psi,u)} \, d\psi \, du + 8 \int_0^{\Delta/2} \int_0^{\pi/2} \frac{d\psi \, du}{R(\psi,u)}. \tag{7.66}$$

The first term has no singularity at $R = 0$ (the singularity can be removed by defining the integrand to be $-jk_0$ at $\psi = u = 0$. The second term can be computed partially in closed form. Noting that

$$\int_0^{\Delta/2} \frac{du}{\sqrt{4a^2 \sin^2 \psi + u^2}} = \ln \left(\frac{\Delta}{2a} + \sqrt{4 \sin^2 \psi + \left(\frac{\Delta}{2a} \right)^2} \right) - \ln 2 - \ln(|\sin \psi|),$$

and using [74]

$$\int_0^{\pi/2} \ln(\sin \psi) \, d\psi = -\frac{\pi}{2} \ln 2,$$

we rewrite (7.66) as

$$A_{mm} = 8 \int_0^{\Delta/2} \int_0^{\pi/2} \frac{e^{-jk_0 R(\psi,u)} - 1}{R(\psi,u)} \, d\psi \, du + 8 \int_0^{\pi/2} \ln \left(\frac{\Delta}{2a} + \sqrt{4 \sin^2 \psi + \left(\frac{\Delta}{2a} \right)^2} \right) \, d\psi.$$

Neither of the above integrals has a singularity, hence both can be done efficiently using numerical integration.

7.3.1.3   The thin-wire approximation

Since the computation of A_{mn} entails double numerical integration, suitable approximations have been sought to promote computational efficiency. One example is the *thin wire approximation*. Through numerical experimentation [57], it is found that when $\Delta/a \gtrsim 4$, the A_{mn} may be approximated using

$$\tilde{G}(z - z') \approx \tilde{G}_t(z - z') = 2\pi \frac{e^{-jk_0 R_0(z-z')}}{R_0(z - z')},$$

where

$$R_0(z - z') = \sqrt{a^2 + (z - z')^2}.$$

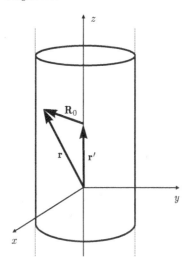

FIGURE 7.11
Geometry of a circular wire showing distance between a source point on the axis of the wire at $\mathbf{r}'$ and an observation point on the surface at $\mathbf{r}$.

Here $\tilde{G}_t(z-z')$ is called the *thin wire kernel*, and R_0 represents the distance from a source point on the axis of the wire to a field point on the surface of the wire (Figure 7.11). Intuitively, the surface current is regarded as a concentrated line current on the wire axis, and the boundary condition on the tangential field is employed on the wire surface. This approximation is often applied to thin curved wires (§ 7.3.2).

Because the distance R_0 never becomes zero, $\tilde{G}_t(z - z')$ does not have a singularity at $z = z'$. Thus, computational efficiency is improved by reducing A_{mn} to a single integral, and by eliminating the singularity. The matrix elements are now

$$A_{mn} = 2\pi \int_{z_n-\Delta/2}^{z_n+\Delta/2} \frac{e^{-jk_0\sqrt{a^2+(z_m-z')^2}}}{\sqrt{a^2 + (z_m - z')^2}} \, dz'.$$

Although there is no singularity, when $z' = z_m$ the denominator of the integrand becomes small and the integrand peaks sharply. This can cause difficulties in numerical integration. Thus, for the self term we can do an extraction much like for the singular integrand. Let

$$A_{mm} = 2\pi \int_{-\Delta/2}^{\Delta/2} \frac{e^{-jk_0\sqrt{a^2+u^2}} - e^{-jk_0a}}{\sqrt{a^2 + u^2}} \, du + 2\pi e^{-jk_0 a} \int_{-\Delta/2}^{\Delta/2} \frac{du}{\sqrt{a^2 + u^2}}.$$

The second integral is integrated in closed form to give

$$A_{mm} = 4\pi \int_{0}^{\Delta/2} \frac{e^{-jk_0\sqrt{a^2+u^2}} - e^{-jk_0a}}{\sqrt{a^2 + u^2}} + 4\pi e^{-jk_0 a} \ln\left(\frac{\Delta}{2a} + \sqrt{\left(\frac{\Delta}{2a}\right)^2 + 1}\right).$$

7.3.1.4 Impressed field models for antennas

It remains to specify the impressed field $\tilde{E}^i$ in (7.60). The form of this expression depends on what it represents. If the wire acts as a transmitting antenna, $\tilde{E}^i$ should represent the field in a source region near the attachment point of the feed cable.

Slice gap model. The simplest way to represent the impressed field for an antenna is with a *slice gap generator*. We assume the wire has a very thin slice of material removed at $z = z_0$, leaving a gap of width 2δ where a feed cable would be attached to impress a voltage between the two legs of the antenna, which is appropriately called a *dipole antenna* (Figure 7.12). Since the gap is assumed narrow compared to the radius of the wire, the field within the gap is parallel to the wire axis:

$$\tilde{\mathbf{E}}^i = \begin{cases} f(z)\hat{\mathbf{z}}, & z_0 - \delta \leq z \leq z_0 + \delta, \\ 0, & \text{elsewhere.} \end{cases}$$

Moreover, the line integral of the electric field is equal to the voltage difference $\tilde{V}_0$ between the antenna legs:

$$\int_{z_0-\delta}^{z_0+\delta} \tilde{\mathbf{E}}^i \cdot \mathbf{dl} = \int_{z_0-\delta}^{z_0+\delta} f(z)\,dz = \tilde{V}_0. \tag{7.67}$$

Since the gap is narrow, $f(z)$ may be assumed constant. Then (7.67) implies

$$f(z) = \frac{\tilde{V}_0}{2\delta}.$$

For simplicity the gap width is often taken as infinitesimal. Then

$$\lim_{\delta \to 0} \int_{z_0-\delta}^{z_0+\delta} f(z)dz = \tilde{V}_0$$

is satisfied by

$$\tilde{E}_z^i(z) = f(z) = \tilde{V}_0 \delta(z - z_0). \tag{7.68}$$

Equation (7.68) is the classic slice-gap impressed field. Using this in (7.63), we obtain

$$b_m = \begin{cases} -j\frac{8\pi^2}{\eta_0}\tilde{V}_0 \sin k_0(z_m - z_0), & z_0 < z_m, \\ 0, & z_0 \geq z_m. \end{cases}$$

In practice z_0 is taken at the junction between two segments, say p and $p+1$. The *input impedance* of the antenna, Z_{in}, is defined as the ratio of the applied voltage $\tilde{V}_0$ to the input current $\tilde{I}_0$ at the junction. This can be approximated rather inaccurately by using the current on segment p or $p+1$ as the input current. More accurate is the quadratic extrapolation

$$\tilde{I}_0 \approx \tfrac{1}{8}[15a_{p+1} - 10a_{p+2} + 3a_{p+3}].$$

Then

$$Z_{\text{in}} = \tilde{V}_0/\tilde{I}_0. \tag{7.69}$$

Often the generator is located at the center of the dipole; this leads to a balanced antenna with a symmetric current.

Magnetic frill model. It is important to note that (7.68) can lead to computational problems since the source region has infinite capacitance [105]. Even so, numerical solutions to the EFIE using the slice-gap generator model are useful approximations to the currents produced by practical generators. An alternative generator model is the *frill model*. This accurately models the case of a monopole antenna formed by extending the center conductor of a coaxial cable through a ground plane (Figure 7.13). The radial

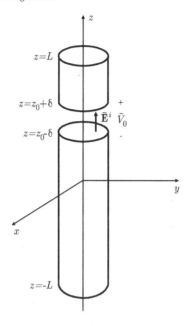

FIGURE 7.12
Dipole antenna fed by a slice-gap generator located at $z = z_0$.

electric field in the aperture at the junction of the cable and the ground plane is taken to be that of an unterminated cable, and is replaced by an equivalent magnetic current using the equivalence principle of § 6.3.4. The resulting equivalent current is imaged into the ground plane, as is the current on the wire, to form a center-fed dipole antenna. Unfortunately, the form of the impressed field $\tilde{E}_z^i$ on the wire surface produced by the aperture field is complicated [200]. Electrostatic approximations [38, 89] are available but inconvenient. A simpler approach is to use the field on the wire axis and assume this holds for points on the wire surface. The result, as given in [186], is

$$\tilde{E}_z^i(a, z) = \tilde{E}^i(z) = \frac{\tilde{V}_0}{2\ln(b/a)}\left[\frac{e^{-jk_0 R_1(z)}}{R_1(z)} - \frac{e^{-jk_0 R_2(z)}}{R_2(z)}\right] \qquad (7.70)$$

where $R_1 = \sqrt{z^2 + a^2}$, $R_2 = \sqrt{z^2 + b^2}$, and b is the coaxial cable outer conductor radius (Figure 7.13). Note that this expression has a strong peak near $z = 0$, hence the strongest impressed field is near the generator. However, unlike the available expressions for the field on the wire surface, there is no singularity. When computing b_m via (7.63), the peaking nature of the integrand can be extracted to make the integral easier to compute.

▶ **Example 7.4:** Impedance of a thin dipole antenna

Consider a very thin dipole antenna of half length $L = 0.25$ m and radius $a = 0.0001$ m. The wire is excited at its center by a time-harmonic source of frequency $f = 300$ MHz ($\lambda \approx 1$ m) with unit voltage applied using either a gap model or a frill model with $b/a = 2.3$ (this value of b/a produces a characteristic impedance of 50 Ω in an air-filled cable). Compute the input impedance of the dipole using both the full kernel and the thin-wire approximation.

Solution: We compute the input impedance by solving the integral equation (7.60). The

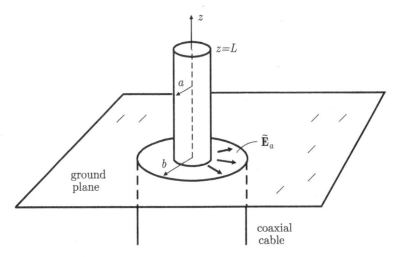

FIGURE 7.13

Monopole antenna fed by a coaxial cable through a ground plane.

matrix equation (7.64) is solved both with and without the thin-wire approximation for the matrix elements, and Z_{in} is computed using (7.69) with extrapolation to the center of the dipole. Figure 7.14 shows how the input resistance and reactance depend on the number of pulse functions used to represent the current. Here, identical results are obtained to three decimal places when using the slice-gap and frill models, and with or without the thin-wire kernel. Thus, the curves represent any of the four possible combinations. It can be seen that for this wire, which is quite thin ($L/a = 2500$), the impedance converges quickly, with stable results reached within $N = 50$ pulses. Recall that the thin-wire approximation holds for partitions satisfying $\Delta/a \gtrsim 4$. For this wire, $\Delta/a = 16.7$ when $N = 300$, so even at the finest partitioning used, the thin-wire kernel gives results comparable to using the full kernel.

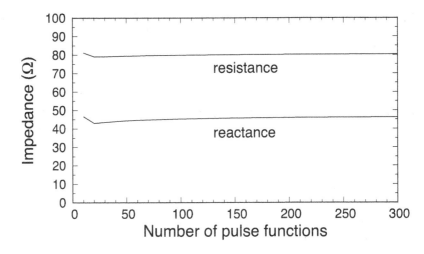

FIGURE 7.14

Input impedance of a dipole antenna with half-length $L = 0.25$ m and radius $a = 0.0001$ m. ◀

▶ **Example 7.5:** Impedance of a thick dipole antenna

Consider a thick dipole antenna of half length $L = 0.25$ m and radius $a = 0.01$ m ($L/a = 25$). As in Example 7.4, the wire is excited at its center at $f = 300$ MHz ($\lambda \approx 1$ m) with unit voltage applied using either a gap model or a frill model with $b/a = 2.3$. Compute the input impedance of the dipole using both the full kernel and the thin-wire approximation.

Solution: Once again we compute the input impedance by solving the integral equation (7.60). The matrix equation (7.64) is solved both with and without the thin-wire approximation for the matrix elements, and Z_{in} is computed using (7.69) with extrapolation to the center of the dipole. Figure 7.15 shows Z_{in} as a function of the number of pulses used to represent the current. Here the input impedance is seen to converge more slowly than for the thin dipole. For a frill generator with the full kernel, the impedance has settled down fairly well by $N = 100$. However, for a slice-gap generator the reactance is changing appreciably even when $N = 300$ partitions is reached. This is a consequence of the infinite capacitance of the gap region. Note that with this antenna, the thin wire approximation is violated with a partitioning of as little as $N = 12$. Hence we do not expect the results found using the thin-wire kernel to duplicate those found from the full kernel. This is apparent in Figure 7.15, where the thin-wire results begin to diverge from the full kernel results as N is increased. When $N \approx 100$ is reached, the thin wire approximation breaks down.

Figure 7.15 shows that when the dipole radius is appreciable, the input impedance obtained from the integral equation is highly dependent on the model used to describe the feed region. This limits the usefulness of the simple slice-gap and frill feed models for describing more complicated and realistic feed structures. The impedance results obtained in this manner must be regarded as only useful approximations. An accurate calculation of the impedance of a realistic antenna requires a highly realistic model of the feed.

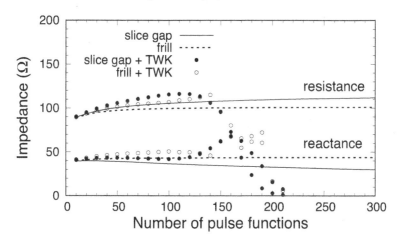

FIGURE 7.15
Input impedance of a dipole antenna with half-length $L = 0.25$ m and radius $a = 0.01$ m. TWK indicates the use of the thin-wire kernel. ◀

▶ **Example 7.6:** Resonances of a dipole antenna

Consider a dipole antenna of length $L = 0.25$ m and radius $a = 0.005$ m. Compute and plot the input impedance as a function of frequency using a slice-gap generator with the full kernel and $N = 200$. Identify the resonance frequencies of the antenna.

Solution: A dipole antenna exhibits resonance properties similar to that of an electric circuit. When viewed over a wide frequency range, the input impedance shows points of resonance where the reactance is zero; these are generally taken as the operating points of the antenna. Figure 7.16 shows the input impedance as a function of frequency. At low frequencies the gap capacitance dominates, and Z_{in} is predominantly capacitive. As the frequency increases, the resistance and reactance increase until the first resonance is reached at 278 MHz, where the input resistance is $R_{in} = 73$ Ω. This frequency corresponds to a dipole length of $2L/\lambda = 0.463$; thus the antenna is slightly less than a half-wavelength long at its first resonance. The resistance continues to increase with frequency, while the reactance peaks and then decreases rapidly back to zero at 444 MHz. This is the second resonance of the antenna, at which $R_{in} = 622$ Ω and $2L/\lambda = 0.74$. This resonance is often referred to as an "antiresonance" in analogy with electric circuits. Because the reactance shows a strong dependence on frequency at the antiresonance, and because R_{in} is so large there, the antenna is generally not operated at this frequency. The next two resonances occur at 883 MHz ($2L/\lambda = 1.47$, $R_{in} = 119$ Ω) and 963 MHz ($2L/\lambda = 1.61$, $R_{in} = 264$ Ω).

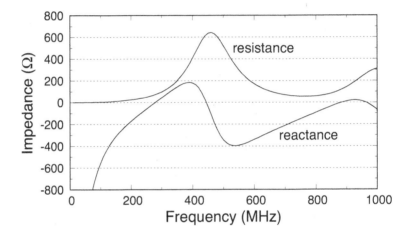

FIGURE 7.16
Input impedance of a dipole antenna with half-length $L = 0.25$ m and radius $a = 0.005$ m found using $N = 200$ pulse functions. ◄

▶ **Example 7.7:** Current distribution on a dipole antenna

For the dipole antenna of Example 7.6, plot the current distributions at the first three resonant frequencies.

Solution: We plot $\tilde{I}(z)$ from the solution of the matrix equation (7.64). Figures 7.17 and 7.18 show the magnitude and phase of the antenna current; each point on the curve is the value of the current at the center of a given partition. At the first resonance, the current distribution is nearly sinusoidal, with the phase varying only a few degrees over the length of the antenna. At the second resonance, the current magnitude shows nearly a full period of sinusoidal variation, but since the length of the antenna at resonance is significantly shorter than one full wavelength, the current does not approach zero at the input. At the third resonance, the current shows three half-periods of variation. At both the second and third resonances, the phase of the current varies significantly over the length of the wire.

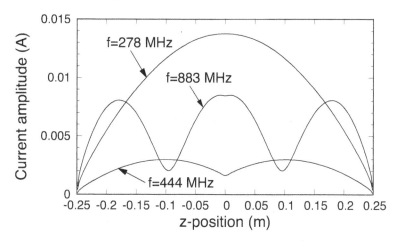

FIGURE 7.17

Magnitude of the current distribution on a dipole antenna with half-length $L = 0.25$ m and radius $a = 0.005$ m found using $N = 200$ pulse functions. Applied voltage is 1 V.

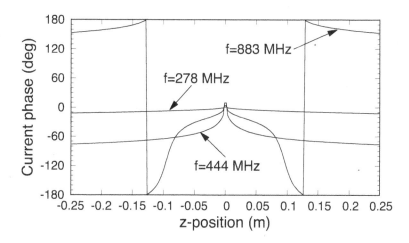

FIGURE 7.18

Phase of the current distribution on a dipole antenna with half-length $L = 0.25$ m and radius $a = 0.005$ m found using $N = 200$ pulse functions. Applied voltage is 1 V. ◀

7.3.1.5 Impressed field models for scatterers

When the scattering properties of a thin wire are sought, we model the impressed field as arising from some far source. Consider a plane wave incident obliquely upon a wire located in free space (Figure 7.19). Our assumption (§ 7.3.1.1) is that $\tilde{\mathbf{E}}^i$ is polarized so that its axial component dominates over the component perpendicular to the wire. So we assume the incident electric field is

$$\tilde{\mathbf{E}}^i(\mathbf{r}) = \tilde{\mathbf{E}}_0 e^{-j\mathbf{k}^i \cdot \mathbf{r}}$$

with $\tilde{\mathbf{E}}_0 = \tilde{E}_0 \hat{\mathbf{z}} \sin\theta_i + \tilde{E}_0 \hat{\mathbf{x}} \cos\theta_i$ and $\mathbf{k}^i = k_0 \hat{\mathbf{x}} \sin\theta_i - k_0 \hat{\mathbf{z}} \cos\theta_i$, with θ_i the incidence

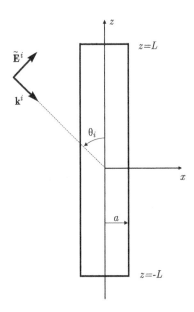

FIGURE 7.19
Circular wire excited by an incident plane wave.

angle measured from the wire axis. The average impressed field defined by (7.57) is

$$\tilde{E}^i(z) = \frac{1}{2\pi}\tilde{E}_0 e^{jk_0 z \cos\theta_i} \sin\theta_i \int_0^{2\pi} e^{-jk_0 a \cos\phi \sin\theta_i}\, d\phi$$

$$= \frac{1}{2\pi}\tilde{E}_0 e^{jk_0 z \cos\theta_i} \sin\theta_i \int_0^{2\pi} e^{jk_0 a \sin u \sin\theta_i}\, du$$

$$= \tilde{E}_0 e^{jk_0 z \cos\theta_i} \sin\theta_i J_0(k_0 a \sin\theta_i). \tag{7.71}$$

For a thin wire with $k_0 a \ll 1$, we have $J_0(k_0 a \sin\theta_i) \approx 1$.

The b_m terms in (7.64) may be found by substituting (7.71) into (7.63). This gives

$$b_m = -j\frac{8\pi^2}{\eta_0}\tilde{E}_0 \sin\theta_i J_0(k_0 a \sin\theta_i) \int_{-L}^{z_m} e^{jk_0 u \cos\theta_i} \sin k_0(z_m - u)\, du.$$

Integrating and simplifying, we obtain

$$b_m = -j\frac{8\pi^2}{\eta_0}\tilde{E}_0 \frac{J_0(k_0 a \sin\theta_i)}{k_0 \sin\theta_i}\left[e^{jk_0 z_m \cos\theta_i} - j\cos\theta_i \sin k_0(z_m + L) - \cos k_0(z_m + L)\right].$$

$$\tag{7.72}$$

The details are left as an exercise.

▶ **Example 7.8:** Current induced on a wire scatterer

A wire of half-length $L = 0.25$ m and radius $a = 0.0025$ m is illuminated by a plane wave (Figure 7.19). The incidence angle is $\theta_i = 30°$ and the amplitude of the incident electric field is $\tilde{E}_0 = 1$ V/m. Compute the current at the center of the wire ($z = 0$) as function of frequency using the thin-wire kernel.

Solution: Figure 7.20 shows the current magnitude at the center of the wire ($z = 0$) computed with $N = 200$ pulse functions by solving (7.64). Several peaks are seen in this spectrum of the current, representing resonances of the wire. These suggest that the temporal response of the antenna should consist of natural oscillations due to interfering current waves traveling along the antenna. This is considered in the next example.

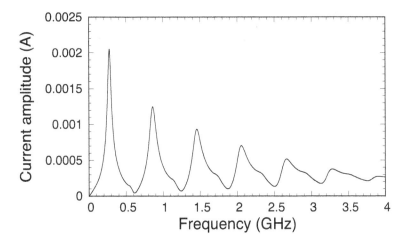

FIGURE 7.20
Magnitude of current at the center of a wire with half-length $L = 0.25$ m and radius $a = 0.0025$ m. Incident field has angle $\theta_i = 30°$ and amplitude $\tilde{E}_0 = 1$ V/m. ◀

▶ **Example 7.9:** Transient scattering by a thin wire

A transient plane wave with a Gaussian temporal waveform is incident at angle $\theta_i = 30°$ on a wire of half-length $L = 0.25$ m and radius $a = 0.0025$ m. Plot the current at the center of the wire as a function of time.

Solution: The temporal response of the wire may be found by computing the inverse FFT of the current spectrum of Figure 7.20. Figure 7.21 shows the time-dependent current found by windowing the spectrum with a Gaussian function and transforming into the time domain. Such windowing is equivalent to interrogating the wire with a transient plane wave having a Gaussian time waveform. Here, the width of the Gaussian is chosen to be 0.3 ns at 50% amplitude. The current waveform consists of several peaks representing the current waves induced on the wire. The time origin is referred to the center of the wire, so the first peak occurs at $t = 0$ ns when the incident wave first reaches the observation point at the wire center. Subsequent peaks occur at intervals of approximately 1.67 ns, which is the one-way transit time of the wire for a wave traveling at the free-space speed of light. Thus, after the transient excitation field has passed over the wire, a current wave remains on the wire, traveling back and forth between its ends. Each time the current pulse passes the observation point, another peak is seen. Interestingly, the pulse shape changes as it propagates along the wire, broadening due to radiative dispersion. That is, since different frequency components in the pulse radiate with different efficiencies, the pulse shape changes with time. After several transit times have elapsed, the pulse has broadened into a waveform that closely resembles a damped sinusoid. This is the dominant resonance of the wire, which radiates least effectively. It is important to note that this complicated temporal response may be represented precisely using a finite sum of damped sinusoidal waveforms representing the natural resonances of the wire excited by the frequency content in the incident Gaussian pulse. The technique used to describe the natural response of a body is called the "singularity expansion method" and is considered in § 7.3.3.

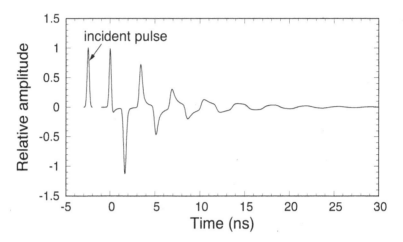

FIGURE 7.21

Temporal current at the center of a wire with half-length $L = 0.25$ m and radius $a = 0.0025$ m found by performing the inverse FFT of the current spectrum. The Gaussian window function produces the incident field waveform shown. ◄

Once the matrix equation is solved and the current amplitudes a_n determined, the field scattered by the wire may be computed in the far zone using the expressions (6.27). Thus

$$\tilde{E}_\theta(r,\theta) = j\omega \frac{\mu_0}{4\pi} \frac{e^{-jk_0 r}}{r} \sin\theta \int_{-L}^{L} \tilde{I}(z') e^{jk_0 \hat{\mathbf{r}} \cdot \mathbf{r}'} \, dz'. \tag{7.73}$$

Using (7.61), we obtain

$$\tilde{E}_\theta(r,\theta) = j\omega \frac{\mu_0}{4\pi} \frac{e^{-jk_0 r}}{r} \sin\theta \sum_{n=1}^{N} a_n \int_{z_n-\Delta/2}^{z_n+\Delta/2} e^{jk_0 z' \cos\theta} \, dz'$$

$$= j\eta_0 \frac{e^{-jk_0 r}}{2\pi r} \tan\theta \sin\left(k_0 \frac{\Delta}{2} \cos\theta\right) \sum_{n=1}^{N} a_n e^{jk_0 z_n \cos\theta}. \tag{7.74}$$

For the special case of broadside observation ($\theta = \pi/2$) this simplifies to

$$\tilde{E}_\theta(r, \theta = \pi/2) = j\frac{k_0 \eta_0}{4\pi} \frac{e^{-jk_0 r}}{r} \Delta \sum_{n=1}^{N} a_n.$$

In scattering problems we must often compute the radar cross-section (RCS) of the object. This is defined in (5.167) as

$$\sigma = \lim_{r\to\infty} \left(4\pi r^2 \frac{|\tilde{\mathbf{E}}^s|^2}{|\tilde{\mathbf{E}}^i|^2} \right). \tag{7.75}$$

Substitution of (7.74) gives

$$\sigma(\theta) = \frac{\eta_0^2}{\pi} \tan^2\theta \sin^2\left(k_0 \frac{\Delta}{2}\cos\theta\right) \left|\sum_{n=1}^{N} \frac{a_n}{\tilde{E}_0} e^{jk_0 z_n \cos\theta}\right|^2.$$

This is the *bistatic* radar cross-section, where the illumination and observation angles may differ. When observed at broadside, the RCS becomes the simple expression

$$\sigma(\pi/2) = \pi\eta_0^2 \left(\frac{\Delta}{\lambda}\right)^2 \left|\sum_{n=1}^{N} \frac{a_n}{\tilde{E}_0}\right|^2.$$

This is still a bistatic RCS, since the incidence angle may vary.

▶ **Example 7.10:** RCS of a wire scatterer

Compute the bistatic radar cross-section of a wire of radius $a/\lambda = 0.005$ as a function of wire length for various incidence angles.

Solution: Figure 7.22 shows the bistatic RCS found using $N = 200$ pulse functions with the full kernel. The angle of the incident wave is varied, while the observation is taken to be at broadside ($\theta = 90°$). As the wire is lengthened, the RCS increases until reaching a peak at $2L = 0.46\lambda$. This is the same length at which the input impedance of a center-fed dipole achieves resonance.

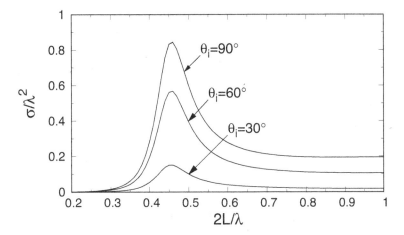

FIGURE 7.22
Bistatic RCS of a wire of radius $a/\lambda = 0.005$ for various angles of incidence. Observation angle is $\theta = 90°$ (broadside). ◀

7.3.2   Curved wires

7.3.2.1   Pocklington equation for curved wires

We can extend the analysis of the straight thin wire to obtain an integral equation for the current on a curved thin wire. Consider a curved wire of circular cross-sectional radius

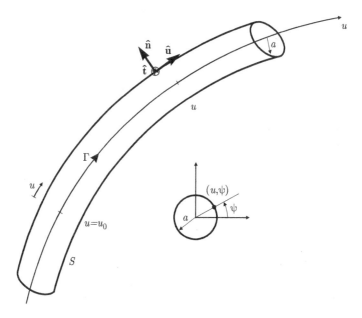

FIGURE 7.23

Geometry of a curved thin wire of circular cross-section.

a located in free space (Figure 7.23). Distance is reckoned along the antenna axis by the arc-length variable u, with coordinate origin at u_0. The position of a point on the wire periphery is described using u along with an angular variable ψ measured in the plane of the wire cross-section at u (Figure 7.23). At the point (u, ψ) on the wire surface, the tangential unit vector can be decomposed into a component $\hat{\mathbf{u}}$ parallel to the wire axis, and a component $\hat{\mathbf{t}}$ such that $\hat{\mathbf{n}} \times \hat{\mathbf{u}} = \hat{\mathbf{t}}$, where $\hat{\mathbf{n}}$ is the outward unit normal at that point.

We continue to assume that the axial component of surface current, $\tilde{J}_u = \hat{\mathbf{u}} \cdot \tilde{\mathbf{J}}_s$, is much larger than the transverse component $\tilde{J}_t = \hat{\mathbf{t}} \cdot \tilde{\mathbf{J}}_s$. Hence the axial component of the scattered field on the surface of the wire produced by the induced current dominates the transverse component. The boundary condition on the tangential total field on the wire surface is

$$\tilde{E}_u^s(\mathbf{r}) + \tilde{E}_u^i(\mathbf{r}) = 0 \qquad (\mathbf{r} \in S) \tag{7.76}$$

where $\tilde{E}_u = \hat{\mathbf{u}} \cdot \tilde{\mathbf{E}}$. This equation leads to an EFIE for the induced current, as follows.

Using (7.51), we can write the scattered electric field in terms of the scattered vector potential as

$$\tilde{E}_u^s = -j\frac{\omega}{k_0^2}\left[\frac{\partial}{\partial u}(\nabla \cdot \tilde{\mathbf{A}}^s) + k_0^2 \tilde{A}_u^s\right]$$

where $\tilde{A}_u^s = \hat{\mathbf{u}} \cdot \tilde{\mathbf{A}}^s$. The scattered vector potential, produced by the induced axially directed current, is

$$\tilde{\mathbf{A}}^s(\mathbf{r}) = \frac{\mu_0}{4\pi}\int_\Gamma\int_0^{2\pi}\hat{\mathbf{u}}'\tilde{J}_u^s(\psi', u')\frac{e^{-jk_0 R}}{R}\,dS'$$

where Γ denotes the axial path of the wire and $R = R(\psi, \psi', u, u')$ is the distance from a source point on the wire surface at (ψ', u') to an observation point at (ψ, u). The divergence of the scattered potential is

$$\nabla \cdot \tilde{\mathbf{A}}^s(\mathbf{r}) = \frac{\mu_0}{4\pi} \int_\Gamma \int_0^{2\pi} \nabla \cdot \left[\hat{\mathbf{u}}' \tilde{J}_u^s(\psi', u') \frac{e^{-jk_0 R}}{R} \right] dS'.$$

Using (B.48), we rewrite this as

$$\nabla \cdot \tilde{\mathbf{A}}^s(\mathbf{r}) = \frac{\mu_0}{4\pi} \int_\Gamma \int_0^{2\pi} \tilde{J}_u^s(\psi', u') \hat{\mathbf{u}}' \cdot \nabla \left[\frac{e^{-jk_0 R}}{R} \right] dS'.$$

Use of the identity

$$\nabla \left\{ \exp(-jk_0 R)/R \right\} = -\nabla' \left\{ \exp(-jk_0 R)/R \right\}$$

gives

$$\nabla \cdot \tilde{\mathbf{A}}^s(\mathbf{r}) = -\frac{\mu_0}{4\pi} \int_\Gamma \int_0^{2\pi} \tilde{J}_u^s(\psi', u') \frac{\partial}{\partial u'} \left[\frac{e^{-jk_0 R}}{R} \right] dS',$$

hence

$$\tilde{E}_u^s(\mathbf{r}) = j \frac{\omega}{k_0^2} \frac{\mu_0}{4\pi} \int_\Gamma \int_0^{2\pi} \tilde{J}_u^s(\psi', u') \left\{ \frac{\partial^2}{\partial u \partial u'} - k_0^2 (\hat{\mathbf{u}} \cdot \hat{\mathbf{u}}') \right\} \frac{e^{-jk_0 R}}{R} dS'.$$

Substituting this into the boundary condition (7.76), we obtain

$$\int_\Gamma \int_0^{2\pi} \tilde{J}_u^s(\psi', u') \left\{ \frac{\partial^2}{\partial u \partial u'} - k_0^2 (\hat{\mathbf{u}} \cdot \hat{\mathbf{u}}') \right\} \frac{e^{-jk_0 R}}{R} du' a \, d\psi'$$
$$= j \frac{4\pi k_0}{\eta_0} \tilde{E}_u^i(\psi, u) \qquad (0 \leq \psi < 2\pi, \ u \in \Gamma). \tag{7.77}$$

Instead of requiring (7.77) to hold for all ψ, we equate the average values of the two sides:

$$\frac{1}{2\pi} \int_0^{2\pi} \left[\int_\Gamma \int_0^{2\pi} \tilde{J}_u^s(\psi', u') \left\{ \frac{\partial^2}{\partial u \partial u'} - k_0^2 (\hat{\mathbf{u}} \cdot \hat{\mathbf{u}}') \right\} \frac{e^{-jk_0 R}}{R} du' a \, d\psi' \right] d\psi$$
$$= j \frac{4\pi k_0}{\eta_0} \frac{1}{2\pi} \int_0^{2\pi} \tilde{E}_u^i(\psi, u) \, d\psi \qquad (u \in \Gamma).$$

Rearrangement gives

$$\int_\Gamma \int_0^{2\pi} \tilde{J}_u^s(\psi', u') \left\{ \frac{\partial^2}{\partial u \partial u'} - k_0^2 (\hat{\mathbf{u}} \cdot \hat{\mathbf{u}}') \right\} \left[\int_0^{2\pi} \frac{e^{-jk_0 R}}{R} d\psi \right] du' a \, d\psi'$$
$$= j \frac{8\pi^2 k_0}{\eta_0} \frac{1}{2\pi} \int_0^{2\pi} \tilde{E}_u^i(\psi, u) \, d\psi \qquad (u \in \Gamma).$$

In general, $R(\psi, \psi', u, u') \neq R(\psi - \psi', u, u')$. However, the following hold if the radius of curvature of the wire is much less than the wire radius.

1. When $|u - u'|$ is small, $R \approx R(\psi - \psi')$, since that portion of the wire is nearly straight.

2. When $|u - u'|$ is large, R is approximately the distance between points on the axis and hence independent of ψ and ψ'.

Thus, we use the approximation $R(\psi, \psi', u, u') \approx \bar{R}(\psi - \psi', u, u')$, where $\bar{R}$ is chosen to satisfy the above criteria. Then

$$\int_\Gamma \left[\int_0^{2\pi} \tilde{J}_u^s(\psi', u') a \, d\psi' \right] \left\{ \frac{\partial^2}{\partial u \partial u'} - k_0^2 (\hat{\mathbf{u}} \cdot \hat{\mathbf{u}}') \right\} \left[\int_0^{2\pi} \frac{e^{-jk_0 \bar{R}(\xi, u, u')}}{\bar{R}(\xi, u, u')} \, d\xi \right] du'$$

$$= j \frac{8\pi^2 k_0}{\eta_0} \tilde{E}^i(u) \qquad (u \in \Gamma), \qquad (7.78)$$

where

$$\tilde{E}^i(u) = \frac{1}{2\pi} \int_0^{2\pi} \tilde{E}_u^i(\psi, u) \, d\psi$$

is the average value of the impressed field over the wire periphery. We recognize

$$\tilde{I}(u) = \int_0^{2\pi} \tilde{J}_u^s(\psi', u) a \, d\psi'$$

as the total current flowing on the wire surface at axial point u, and define the average Green's function as

$$\tilde{G}(u, u') = \int_0^{2\pi} \frac{e^{-jk_0 \bar{R}(\xi, u, u')}}{\bar{R}(\xi, u, u')} \, d\xi. \qquad (7.79)$$

Then (7.78) becomes

$$\int_\Gamma \tilde{I}(u') \left\{ \frac{\partial^2}{\partial u \partial u'} - k_0^2 (\hat{\mathbf{u}} \cdot \hat{\mathbf{u}}') \right\} \tilde{G}(u, u') \, du' = j \frac{8\pi^2 k_0}{\eta_0} \tilde{E}^i(u) \qquad (u \in \Gamma). \qquad (7.80)$$

This is the desired EFIE for the unknown current $\tilde{I}(u)$ induced on the curved wire. It is known as *Pocklington's integral equation* after Henry Cabourn Pocklington, who derived a form of the equation in 1897 [153]. Its specialization for a straight wire, and solution using MoM, is left as an exercise.

As in the case of the straight wire, the thin wire approximation may be used to approximate $\tilde{G}(u, u')$ as

$$\tilde{G}(u, u') = 2\pi \frac{e^{-jk_0 \bar{R}_0(u, u')}}{\bar{R}_0(u, u')}.$$

Here $\bar{R}_0(u, u') = \bar{R}(0, u, u')$ is the distance from a point on the axis at u' to a point on the wire surface at u.

7.3.2.2 Hallén equation for curved wires

The integral equation (7.80) is less convenient numerically than an equation of Hallén form, such as (7.60), because the two derivatives on the Green's function $\tilde{G}(u, u')$ increase its singularity at $u = u'$. K.K. Mei [128] devised a method for converting Pocklington's equation to a Hallén-type equation with fewer derivatives, at the expense of introducing an additional integral.

Define the kernel of Pocklington's equation as

$$K(u, u') = \left\{ \frac{\partial^2}{\partial u \partial u'} - k_0^2 (\hat{\mathbf{u}} \cdot \hat{\mathbf{u}}') \right\} \tilde{G}(u, u').$$

Then (7.80) becomes

$$\int_\Gamma \tilde{I}(u') K(u, u') \, du' = j \frac{8\pi^2 k_0}{\eta_0} \tilde{E}^i(u) \qquad (u \in \Gamma). \qquad (7.81)$$

We wish to transform this into an equation of the form

$$\left(\frac{d^2}{du^2} + k_0^2\right) \int_\Gamma \tilde{I}(u')\Pi(u,u')\,du' = -j\frac{8\pi^2 k_0}{\eta_0}\tilde{E}^i(u) \qquad (u \in \Gamma). \qquad (7.82)$$

Since this is an ordinary differential equation of the same form as (7.56), we have

$$\int_\Gamma \tilde{I}(u')\Pi(u,u')\,du' = A\sin k_0 u + B\cos k_0 u - j\frac{8\pi^2}{\eta_0}\int_{u_1}^u \tilde{E}^i(\zeta)\sin k_0(u-\zeta)\,d\zeta \qquad (u \in \Gamma) \qquad (7.83)$$

where A, B, and u_1 are constants. This is the desired Hallén integral equation.

It remains to identify the kernel $\Pi(u,u')$ of Hallén's equation. Comparing (7.81) with (7.82), we see that

$$\left(\frac{d^2}{du^2} + k_0^2\right)\Pi(u,u') = -K(u,u').$$

Solution of this differential equation gives

$$\Pi(u,u') = -\frac{1}{k_0}\int_{u_2}^u K(\chi,u')\sin k_0(u-\chi)\,d\chi, \qquad (7.84)$$

where u_2 is a constant. The homogeneous solution, when substituted into (7.83), produces a term that augments the homogeneous solution on the right-hand-side of (7.83). Substituting $K(u,u')$ into (7.84) we get

$$\Pi(u,u') = -\int_{u_2}^u \left\{\left[\frac{\partial^2}{\partial\chi\partial u'} - k_0^2(\hat{\boldsymbol{\chi}}\cdot\hat{\mathbf{u}}')\right]\tilde{G}(\chi,u')\right\}\frac{\sin k_0(u-\chi)}{k_0}\,d\chi.$$

One of the derivatives on G may be removed using integration by parts. Write

$$-\int_{u_2}^u \frac{\partial^2\tilde{G}(\chi,u')}{\partial\chi\partial u'}\frac{\sin k_0(u-\chi)}{k_0}\,d\chi = -\frac{\sin k_0(u-\chi)}{k_0}\frac{\partial\tilde{G}(\chi,u')}{\partial u'}\bigg|_{u_2}^u$$
$$-\int_{u_2}^u \frac{\partial\tilde{G}(\chi,u')}{\partial u'}\cos k_0(u-\chi)\,d\chi.$$

Since the first term merely contributes to the homogeneous solution in (7.83), it is not considered further. Next use integration by parts to write

$$\int_{u_2}^u k_0^2(\hat{\boldsymbol{\chi}}\cdot\hat{\mathbf{u}}')\tilde{G}(\chi,u')\frac{\sin k_0(u-\chi)}{k_0} = (\hat{\boldsymbol{\chi}}\cdot\hat{\mathbf{u}}')\tilde{G}(\chi,u')\cos k_0(u-\chi)|_{u_2}^u$$
$$-\int_{u_2}^u \frac{\partial}{\partial\chi}[(\hat{\boldsymbol{\chi}}\cdot\hat{\mathbf{u}}')\tilde{G}(\chi,u')]\cos k_0(u-\chi)\,d\chi.$$

The lower limit of the first term on the right-hand side again contributes to the homogeneous solution in (7.83). Finally then,

$$\Pi(u,u') = (\hat{\mathbf{u}}\cdot\hat{\mathbf{u}}')\tilde{G}(u,u') - \int_{u_2}^u \left\{\frac{\partial}{\partial\chi}[(\hat{\boldsymbol{\chi}}\cdot\hat{\mathbf{u}}')\tilde{G}(\chi,u')] + \frac{\partial\tilde{G}(\chi,u')}{\partial u'}\right\}\cos k_0(u-\chi)\,d\chi.$$

Depending on the geometry of the wire, it may be possible to remove the remaining derivatives by using integration by parts once more. This holds for a circular loop (Problem 7.15).

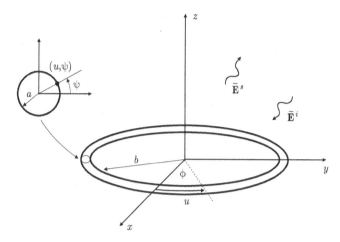

FIGURE 7.24
Circular loop.

7.3.2.3 Example: the circular loop antenna

Consider a circular wire loop of radius b (Figure 7.24). The arc length along the wire axis is given by $u = b\phi$, and the distance between points at u and u' on the axis is

$$D(\phi, \phi') = 2b \sin\left(\frac{|\phi - \phi'|}{2}\right).$$

Similarly, the distance between points (ψ, u) and (ψ', u) on the surface is

$$d(\psi, \psi') = 2a \sin\left(\frac{|\psi - \psi'|}{2}\right).$$

Thus, a good choice for $\bar{R}$ in the Green's function (7.79) is $\bar{R} = \sqrt{D^2 + d^2}$, or

$$\bar{R}(\xi, u, u') = \sqrt{4b^2 \sin^2\left(\frac{\phi - \phi'}{2}\right) + 4a^2 \sin^2\left(\frac{\xi}{2}\right)}.$$

Importantly, this choice of $\bar{R}$ produces $\tilde{G}(\phi, \phi') = \tilde{G}(\phi - \phi')$.

The kernel of Pocklington equation (7.81) can be specialized to the wire loop by using

$$u = b\phi, \quad \hat{\mathbf{u}} = \hat{\boldsymbol{\phi}}, \quad \frac{\partial}{\partial u} = \frac{1}{b}\frac{\partial}{\partial \phi}, \quad \hat{\mathbf{u}} \cdot \hat{\mathbf{u}}' = \cos(\phi - \phi').$$

Noting that

$$\frac{\partial \tilde{G}(\phi - \phi')}{\partial \phi'} = -\frac{\partial \tilde{G}(\phi - \phi')}{\partial \phi}$$

we obtain

$$K(\phi, \phi') = K(\phi - \phi') = -\left[\frac{1}{b^2}\frac{\partial^2}{\partial \phi^2} + k_0^2 \cos(\phi - \phi')\right]\tilde{G}(\phi - \phi'). \qquad (7.85)$$

Hence, (7.81) becomes

$$\int_{-\pi}^{\pi} \tilde{I}(\phi')K(\phi - \phi')b\,d\phi' = j\frac{8\pi^2 k_0}{\eta_0}\tilde{E}^i(\phi) \qquad (-\pi < \phi \le \pi). \qquad (7.86)$$

Since $\tilde{I}(\zeta)$ and $K(\zeta)$ are periodic in ζ, (7.86) can be solved using Fourier series. This approach was first taken by Hallén [77] and implemented by Storer in 1956 [181]. We begin by expanding the Green's function in a Fourier series:

$$\tilde{G}(\zeta) = \sum_{n=-\infty}^{\infty} K_n e^{jn\zeta} \qquad (-\pi < \zeta \le \pi),$$

where

$$K_n = \frac{1}{2\pi} \int_{-\pi}^{\pi} \tilde{G}(\zeta) e^{-jn\zeta} \, d\zeta. \tag{7.87}$$

Using this, we rewrite (7.85) as

$$K(\phi - \phi') = -\sum_{n=-\infty}^{\infty} K_n \left[\frac{1}{b^2} \frac{\partial^2}{\partial \phi^2} + k_0^2 \cos(\phi - \phi') \right] e^{jn(\phi - \phi')}.$$

Computing the derivatives and writing the cosine function in terms of exponentials, we get

$$K(\phi - \phi') = \sum_{n=-\infty}^{\infty} \left[\frac{n^2}{b^2} K_n - \frac{k_0^2}{2}(K_{n-1} + K_{n+1}) \right] e^{jn(\phi - \phi')}$$

$$= \sum_{n=-\infty}^{\infty} \alpha_n e^{jn(\phi - \phi')}.$$

So the α_n are the Fourier coefficients of the kernel function. Note that since $G(\xi)$ is even, $K_{-n} = K_n$ and thus $\alpha_{-n} = \alpha_n$. Putting the Fourier expansion for $K(\phi - \phi')$ back into the integral equation (7.86) and rearranging, we get

$$\sum_{n=-\infty}^{\infty} \alpha_n e^{jn\phi} \left[\int_{-\pi}^{\pi} \tilde{I}(\phi') e^{-jn\phi'} \, d\phi' \right] = j \frac{8\pi^2 k_0}{\eta_0} \tilde{E}^i(\phi).$$

The term in brackets is 2π times the Fourier coefficient for the current:

$$I_n = \frac{1}{2\pi} \int_{-\pi}^{\pi} \tilde{I}(\phi') e^{-jn\phi'} \, d\phi'.$$

So

$$\sum_{n=-\infty}^{\infty} \left[\alpha_n 2\pi I_n \right] e^{jn\phi} = j \frac{8\pi^2 k_0}{\eta_0} \tilde{E}^i(\phi). \tag{7.88}$$

This is a Fourier series for the term on the right-hand side. Letting

$$E_n = \frac{1}{2\pi} \int_{-\pi}^{\pi} \tilde{E}^i(\phi) e^{-jn\phi} \, d\phi$$

be the Fourier coefficients of the impressed field, from (7.88) we have the Fourier coefficients of the current:

$$I_n = j \frac{4\pi k_0}{\eta_0} \frac{E_n}{\alpha_n}.$$

So Pocklington's integral equation has solution

$$\tilde{I}(\phi) = \sum_{n=-\infty}^{\infty} j \frac{4\pi k_0}{\eta_0} \frac{E_n}{\alpha_n} e^{jn\phi}. \tag{7.89}$$

In computing K_n, a singularity of $G(\xi)$ at $\xi = 0$ is encountered. This may be extracted by the technique used for the straight wire. Substituting (7.79) into (7.87), we have

$$K_n = \frac{1}{2\pi} \int_{-\pi}^{\pi} e^{-jn\phi} \left[\frac{1}{2\pi} \int_{-\pi}^{\pi} \frac{e^{-jk_0\bar{R}}}{\bar{R}} \, d\xi \right] d\phi.$$

Since the singularity occurs at $\bar{R} = 0$, we add and subtract the singular term $1/R'$ within the integral:

$$K_n = \frac{1}{4\pi^2} \int_{-\pi}^{\pi} \left\{ \int_{-\pi}^{\pi} \left[e^{-jn\phi} \frac{e^{-jk_0\bar{R}}}{\bar{R}} - \frac{1}{R'} \right] d\xi \right\} d\phi + \frac{1}{4\pi^2} \int_{-\pi}^{\pi} \int_{-\pi}^{\pi} \frac{d\xi \, d\phi}{R'}. \tag{7.90}$$

Here R' may be any function having similar behavior to $\bar{R}$ as $\bar{R} \to 0$. For simplicity we choose

$$R'(\xi, \phi) = \sqrt{b^2\phi^2 + a^2\xi^2}.$$

This allows the second integral in (7.90) to be done in closed form. Employing (F.1), we get

$$\frac{1}{4\pi^2} \int_{-\pi}^{\pi} \int_{-\pi}^{\pi} \frac{d\xi \, d\phi}{R'} = \frac{1}{\pi a} \left[Q \ln \left(\frac{1}{Q} + \sqrt{1 + \frac{1}{Q^2}} \right) + \ln \left(Q + \sqrt{1 + Q^2} \right) \right]$$

where $Q = a/b$.

Let us apply the Fourier series solution to a wire loop antenna with a gap generator at $\phi = 0$. Assume the generator occupies a gap in the wire from $\phi = -\delta$ to $\phi = \delta$. The impressed field is

$$\tilde{E}_\phi^i = \begin{cases} \frac{\tilde{V}_0}{2b\delta}, & -\delta \leq \phi \leq \delta, \\ 0, & \text{elsewhere,} \end{cases}$$

and so

$$E_n = \frac{1}{2\pi} \int_{-\delta}^{\delta} \tilde{E}^i(\phi) e^{-jn\phi} \, d\phi$$

$$= j \frac{\tilde{V}_0}{2\pi b} \frac{\sin(n\delta)}{n\delta}. \tag{7.91}$$

As $\delta \to 0$ we obtain a slice-gap generator and

$$E_n = j \frac{\tilde{V}_0}{2\pi b}.$$

This generator has the same problem of infinite capacitance as with the straight wire, so $\delta > 0$ is often used to improve the behavior of the series (7.89).

The input admittance of the loop antenna is easily found from (7.89). The input current is

$$\tilde{I}_{\text{in}} = \tilde{I}(\phi = 0)$$

$$= \sum_{n=-\infty}^{\infty} j \frac{4\pi k_0}{\eta_0} \frac{E_n}{\alpha_n}, \tag{7.92}$$

and the admittance is $Y_{\text{in}} = \tilde{I}_{\text{in}}/\tilde{V}_0$.

▶ **Example 7.11:** Input admittance of a circular loop antenna

Compute the input admittance of a circular loop antenna of radius $b = 0.2$ m at frequency 300 MHz, as a function of the number of terms N used in the series (7.92).

Solution: As N increases, the input conductance converges rapidly, but the input susceptance changes in a more complicated manner. Figure 7.25 shows the input susceptance for $a = 0.002$ m ($b/a = 100$) and $a = 0.01$ m ($b/a = 20$). When the slice gap generator is used ($\delta = 0$ in (7.91)), the susceptance continues to increase as N increases, with the result for the larger wire radius varying more dramatically. When a narrow gap of $\delta = 1°$ is used for the generator region, the susceptance converges relatively quickly, with stable results by $N = 100$ for $b/a = 100$, and $N = 200$ for $b/a = 20$.

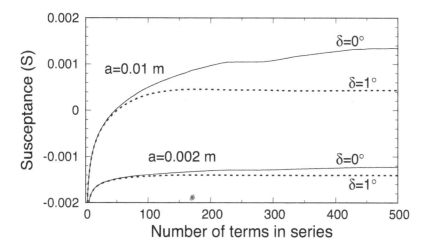

FIGURE 7.25

Input susceptance of a circular loop antenna. $f = 300$ MHz, $b = 0.2$ m. ◀

▶ **Example 7.12:** Resonances of a circular loop antenna

Compute the input admittance of a circular loop antenna of radius $b = 0.2$ m for wire radii of $a = 0.002$ m and $a = 0.01$ m. Identify the resonance frequencies of the antenna.

Solution: Figures 7.26 and 7.27 show the input conductance and susceptance, respectively. The admittances are computed using $\delta = 1°$, and $N = 100$ terms for $a = 0.002$ m and $N = 200$ terms for $a = 0.01$ m. For the loop with the thinner wire, several resonance points are seen where the input susceptance is zero. The first three are located at 111 MHz, 256 MHz, and 346 MHz. The first and third resonances are actually antiresonances with small input conductances and low input currents. This is explored further in the next example.

Interestingly, for the thicker wire, there is only one actual resonance, since, although it oscillates with frequency, the susceptance passes through zero only at one frequency.

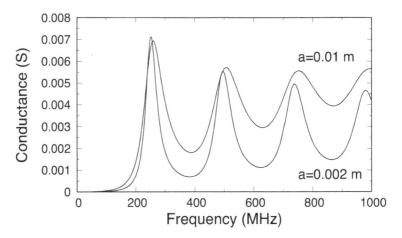

FIGURE 7.26
Input conductance of a circular loop antenna. $b = 0.2$ m, $\delta = 1°$.

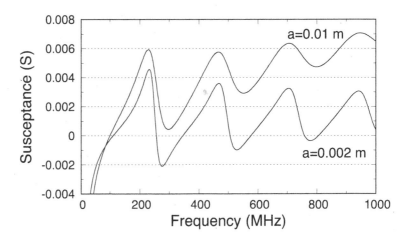

FIGURE 7.27
Input susceptance of a circular loop antenna. $b = 0.2$ m, $\delta = 1°$. ◀

▶ **Example 7.13:** Current distribution on a circular loop antenna

Consider the circular loop antenna of Example 7.12. Plot the current distributions at the first three resonant frequencies.

Solution: Figure 7.28 shows the magnitude of the current on a circular loop of radius $b = 0.2$ and wire radius $a = 0.002$ m, computed at the first three resonant frequencies. The first resonance at 111 MHz is actually an antiresonance. The conductance at this frequency is very small ($G = 0.000030$ S); thus the input current is small and the antenna is difficult to drive. Note that the current is nearly zero at the input, and that the magnitude goes through approximately one half cycle around the loop (which has a circumference of 0.46λ at 111 MHz). The next resonance, at 256 MHz, is a true resonance, with an input conductance of $G = 0.0069$ S. In this case, the current is maximum at the input and varies through one full cycle around the loop (circumference 1.07λ at 256 MHz). The third resonance, at 346 MHz, is again an antiresonance, with input conductance $G = 0.00084$ S. The current has a minimum at the input, and goes through not quite one and a half cycles around the loop (circumference 1.45λ at 346 MHz).

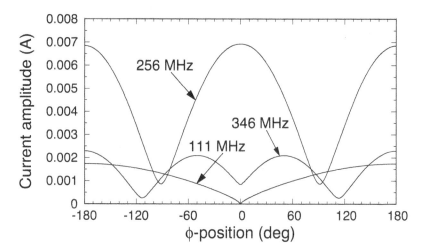

FIGURE 7.28

Magnitude of current on a circular loop antenna. $\tilde{V}_0 = 1$ V, $a = 0.002$ m, $b = 0.2$ m, $N = 100$, $\delta = 1°$.

Figure 7.29 shows the phase of the current. At the primary resonance frequency of 256 MHz, the phase is fairly constant, with a sign change near $\phi = \pm 90°$. At the two antiresonances, the phase changes rapidly near the source where the current magnitude is small.

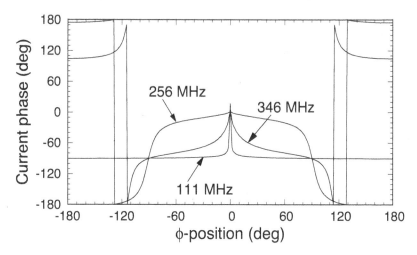

FIGURE 7.29

Phase of current on a circular loop antenna. $\tilde{V}_0 = 1$ V, $a = 0.002$ m, $b = 0.2$ m, $N = 100$, $\delta = 1°$. ◀

7.3.3 Singularity expansion method for time-domain current on a straight wire

When a mechanical object is subjected to a distributed transient force, the object oscillates, or "rings," at its natural frequencies. (Consider a bell struck by a hammer as an example.) This effect is also called a natural resonance. Due to losses, the amplitudes of the oscillations die exponentially with time, and we say that the resonances are "damped." A similar effect can be observed in scattering and radiation problems. In § 7.3.1.5, we see that the time-domain current on a wire illuminated by a plane wave demonstrates an oscillatory behavior, and that the oscillations dampen with time as the current radiates away its energy. In the 1970's, Carl Baum proposed using this effect to describe the behavior of the current in a number of transient problems [13, 14]. He hypothesized that if a conducting object is subjected to a distributed transient electromagnetic field, the induced current will oscillate in the natural modes of the object, and that this oscillation will begin after the transient excitation has passed. The ensuing period of natural oscillation is called the "late time" of the response, while the time during which the object is under excitation is called the "early time." Thus, an antenna with a transient voltage applied at its terminals will oscillate freely after the exciting field has spread across the extent of the antenna. Similarly, a radar target illuminated by a transient plane wave will oscillate freely after the excitation field has passed completely over the object.

Baum developed a fairly complete technique for analyzing the natural response of objects. The *Singularity Expansion Method* (SEM) has been applied to many different types of objects, including both conducting and material bodies. We will consider a simple application to a straight thin wire.

Consider a straight wire (Figure 7.9) excited by some transient event. The event may be local, such as a voltage applied to terminals on the wire, or distributed, as in the case of excitation by a time-domain plane wave sweeping across the wire. We assume that the temporal waveform of the excitation is time-limited, and model the current at any point on the wire as a sum of damped sinusoidal functions (natural oscillations) in the late-time period. The onset of the late-time period occurs at a time $T_L(z)$ that depends on the observation position of the current and the duration of the excitation. Prior to this time, the current behaves in a more complicated fashion, which is difficult to describe. If $T_0(z)$ is the first time at which a nonzero current is induced at position z, then we write the current as

$$I(z,t) = U\left(t - T_L(z)\right) \sum_{n=1}^{2N} a_n I_n(z) e^{s_n t} + \left[U\left(t - T_0(z)\right) - U\left(t - T_L(z)\right)\right] w(z,t)$$

$$= I_L(z,t) + I_E(z,t). \tag{7.93}$$

Here $I_L(z,t)$ is a natural-mode expansion of the late-time current, while $I_E(z,t)$ is the early-time current. Each term in the natural-mode series has a complex natural frequency $s_n = \sigma_n + j\omega_n$, a complex modal amplitude a_n, and a modal current distribution $I_n(z)$. Unless $\omega_n = 0$, the natural frequencies and modal amplitudes must occur in conjugate pairs so that $I(z,t)$ is real. Hence N represents the number of modes of the wire that are excited by the incident field. Also, we assume all modes decay with time due to radiation damping, and thus $\sigma_n < 0$.

Since our study of integral equations has concentrated on frequency-domain representations of the induced current, let us explore the behavior of the current expression (7.93) in the frequency domain. As in systems theory, it is customary to employ the Laplace frequency variable s. Substituting (7.93) into the two-sided Laplace transform integral, we get

$$I(z, s) = \int_{-\infty}^{\infty} \left[I_L(z, t) + I_E(z, t) \right] e^{-st} \, dt$$

$$= \sum_{n=1}^{2N} a_n I_n(z) \int_{T_L(z)}^{\infty} e^{(s_n - s)t} \, dt + \int_{T_0(z)}^{T_L(z)} w(z, t) e^{-st} \, dt = I_L(z, s) + I_E(z, s).$$

The first integral may be easily computed, giving

$$I_L(z, s) = \sum_{n=1}^{2N} \frac{a_n I_n(z) e^{(s_n - s)T_L(z)}}{s - s_n}. \tag{7.94}$$

Thus, the late-time natural mode series becomes a pole series in the s-domain, with all poles in the left-half plane. Since the integral converges only when $\mathrm{Re}\{s\} > \sigma_n$, any inversion contour used to obtain a time-domain expression involving the late time must be taken to the right of $\max_n \{\sigma_n\}$. Note that since the early-time current $w(z, t)$ is time-limited, its Laplace spectrum $I_E(z, s)$ is an entire function of s.

7.3.3.1 Integral equation for natural frequencies and modal current distributions

An integral equation for the modal current distribution $I_n(z)$ may be found by using the results from § 7.3.1.1. Consider a transient plane wave incident on a straight thin wire, with the same geometry as shown in Figure 7.19 for frequency-domain excitation. The s-domain vector potential produced by the current $I(z, s)$ induced on the wire may be found by substituting $\omega = s/j$ into (7.55):

$$A^s(z, s) = \frac{1}{2\pi} \frac{\mu_0}{4\pi} \int_{-L}^{L} I(z', s) \tilde{G}(z - z', s) \, dz', \tag{7.95}$$

where

$$\tilde{G}(z - z', s) = \int_0^{2\pi} \frac{e^{-(s/c)R(\xi, z - z')}}{R(\xi, z - z')} \, d\xi.$$

The equation (7.56) generated from the boundary condition on the tangential electric field at the wire surface becomes the s-domain equation

$$\left(\frac{d^2}{dz^2} - \frac{s^2}{c^2} \right) A^s(z, s) = -\frac{s}{c^2} E^i(z, s) \qquad (-L \leq z \leq L),$$

where $E^i(z, s)$ is the Laplace spectrum of the transient excitation field. Substituting (7.95) and using the plane-wave field (7.71), we get

$$\int_{-L}^{L} I(z', s) K(z - z', s) \, dz' = -8\pi^2 s \epsilon_0 E^i(z, s) \qquad (-L \leq z \leq L), \tag{7.96}$$

where

$$K(z - z', s) = \left(\frac{d^2}{dz^2} - \gamma^2 \right) \tilde{G}(z - z', s).$$

Here we use the abbreviation $\gamma = s/c$ to represent a quantity akin to a wavenumber. Since the excitation is assumed to be time-limited, $E^i(z, s)$ is an entire function. Also note that we have moved the spatial derivatives into the vector potential integral to form a new kernel function $K(z - z', s)$.

Equation (7.96) represents an s-domain integral equation for the current $I(z, s)$ on the wire. But since we already have a model for this current, we can use the integral equation to determine the modal current distribution functions $I_n(z)$. With $I(z, s) = I_L(z, s) + I_E(z, s)$ and (7.94) we get

$$\int_{-L}^{L} \sum_{n=1}^{2N} a_n \frac{I_n(z') e^{(s_n - s) T_L(z')}}{s - s_n} K(z - z', s) \, dz' + \int_{-L}^{L} I_E(z', s) K(z - z', s) \, dz'$$
$$= -8\pi^2 s \epsilon_0 E^i(z, s) \qquad (-L \leq z \leq L). \qquad (7.97)$$

Now we multiply through by $(s - s_m)$ and let $s \to s_m$. Only the $n = m$ term in the sum survives the limit; the other two terms vanish because $I_E(z, s)$ and $E^i(s, z)$ are entire functions of s. The result is

$$\int_{-L}^{L} I_n(z') K(z - z', s_n) \, dz' = 0 \qquad (-L \leq z \leq L). \qquad (7.98)$$

This is the desired integral equation for $I_n(z)$. Since it is homogeneous (independent of the excitation function), $I_n(z)$ must describe a natural mode. Furthermore, it has solutions only for certain discrete values of s_n; these are the natural frequencies of the wire.

Once the modal currents are found, the modal amplitudes a_n, which Baum [13] calls the *coupling coefficients*, are determined as follows. Multiply (7.97) by $I_m(z)$ and integrate:

$$\sum_{n=1}^{2N} \frac{a_n}{s - s_n} \int_{-L}^{L} I_m(z) \int_{-L}^{L} e^{(s_n - s) T_L(z')} I_n(z') K(z - z', s) \, dz'$$
$$+ \int_{-L}^{L} I_m(z) \int_{-L}^{L} I_E(s, z') K(z - z', s) \, dz' \, dz = -8\pi^2 s \epsilon_0 \int_{-L}^{L} I_m(z) E^i(z, s) \, dz.$$

Now use the fact that $K(z - z', s) = K(z' - z, s)$ and rearrange to get

$$\sum_{n=1}^{2N} \frac{a_n}{s - s_n} \int_{-L}^{L} e^{(s_n - s) T_L(z')} I_n(z') \left[\int_{-L}^{L} I_m(z) K(z' - z, s) \, dz \right] dz'$$
$$+ \int_{-L}^{L} I_E(s, z') \left[\int_{-L}^{L} I_m(z) K(z' - z, s) \, dz \right] dz' = -8\pi^2 s \epsilon_0 \int_{-L}^{L} I_m(z) E^i(z, s) \, dz.$$

As $s \to s_m$, the bracketed terms approach zero by (7.98). Hence the term involving $I_E(s, z)$ vanishes, as do all terms in the sum except the $n = m$ term, which produces an indeterminate form requiring

$$\lim_{s \to s_m} \frac{a_m}{s - s_m} \int_{-L}^{L} I_m(z') \left[\int_{-L}^{L} I_m(z) K(z' - z, s) \, dz \right] dz' = R_m$$

where

$$R_m = -8\pi^2 s_m \epsilon_0 \int_{-L}^{L} I_m(z) E^i(z, s_m) \, dz.$$

By L'Hopital's rule we obtain

$$a_m \lim_{s \to s_m} \frac{\frac{\partial}{\partial s} \left\{ \int_{-L}^{L} I_m(z') \int_{-L}^{L} I_m(z) K(z' - z, s) \, dz \, dz' \right\}}{\frac{\partial}{\partial s}(s - s_m)} = R_m.$$

So the modal amplitude is simply

$$a_m = R_m / C_m \tag{7.99}$$

where C_m is a normalization constant:

$$C_m = \int_{-L}^{L} I_m(z') \int_{-L}^{L} I_m(z) K'(z' - z, s_m) \, dz \, dz'.$$

Here

$$K'(z - z', s) = \frac{\partial K(z - z', s)}{\partial s}.$$

The s-domain excitation field can be found by using (7.71) with $\omega = s/j$:

$$E^i(z, s) = \sin \theta_i E_0(s) J_0(-jsa \sin \theta_i / c) e^{-(s/c)z \cos \theta_i},$$

where $E_0(s)$ is the Laplace spectrum of the time-limited excitation waveform. Thus

$$R_m = -8\pi^2 s_m \epsilon_0 \sin \theta_i E_0(s_m) J_0(-js_m a \sin \theta_i / c) \int_{-L}^{L} I_m(z) e^{-(s_m/c)z \cos \theta_i} \, dz. \tag{7.100}$$

7.3.3.2 Numerical solution for natural-mode current

The natural frequencies of the wire are the values of s_m for which (7.98) has nontrivial solutions. The current distributions $I_n(z)$ associated with the s_n are the modal currents for the wire. We can write (7.98) as

$$\left(\frac{d^2}{dz^2} - \gamma_n^2 \right) \int_{-L}^{L} I_n(z') \tilde{G}(z - z', s_n) \, dz' = 0 \qquad (-L \leq z \leq L), \tag{7.101}$$

where $\gamma_n = s_n / c$. Solution of this differential equation gives the Hallén form of the integral equation

$$\int_{-L}^{L} I_n(z') \tilde{G}(z - z', s_n) \, dz' + C_1 \sinh(\gamma_n z) + C_2 \cosh(\gamma_n z) = 0 \qquad (-L \leq z \leq L). \tag{7.102}$$

This is identical to the Hallén equation for the current on a wire scatterer, (7.60), except that (7.102) is homogeneous and uses the parameter γ in place of jk. Thus, if we solve (7.102) using pulse functions and point matching, we may use the same matrix equation (7.64) with the substitutions $k \to -j\gamma$ and $\omega \to -js$. However, we must set the right-hand side to zero to obtain the required homogeneous equation

$$[Q(s)][x] = 0,$$

where Q is an $(N + 2) \times (N + 2)$ matrix (with N the number of pulses used to represent the current). Nontrivial solutions are only possible when the determinant of the matrix is zero:

$$\det[Q(s)] = 0. \tag{7.103}$$

Since the entries in the MoM matrix $[Q]$ depend on the complex frequency s, (7.103) is a transcendental equation for the natural frequencies. For each value of s_n satisfying (7.103), there is a nontrivial solution $[x]$ in the nullspace of $[Q]$ [182]. The first N entries of $[x]$ are the amplitudes of the pulses describing the natural-mode current $I_n(z)$.

▶ **Example 7.14:** Natural resonance frequencies of a thin straight wire

Compute the first 10 natural resonance frequencies of a thin wire of half length $L = 0.25$ m and radius $a = 0.0025$ m.

Solution: The natural frequencies are found by computing the MoM matrix and searching for zeros of its determinant. The zeros can be determined by a simple secant method root search [67]. Initial guesses for the search may be obtained from the simple approximation that the natural mode current distributions are integer multiples of a half-wavelength standing wave along the wire, i.e., that $2L = n\lambda/2$. Using $c = \lambda f$, we find

$$\omega_n \approx \pm n\frac{\pi c}{2L}. \tag{7.104}$$

With this as the initial guess, the first ten natural frequencies found to satisfy (7.103) are shown in the following table. (Only the poles with positive ω_n are shown since the poles form conjugate pairs.)

n	$\frac{2L}{\pi c}\sigma_n$	$\frac{2L}{\pi c}\sigma_n$
1	-0.08176	0.9131
2	-0.1206	1.885
3	-0.1486	2.865
4	-0.1714	3.849
5	-0.1911	4.835
6	-0.2087	5.824
7	-0.2247	6.811
8	-0.2396	7.799
9	-0.2536	8.792
10	-0.2668	9.779

To obtain these results, the full kernel was used to compute the MoM matrix entries, with $N = 200$ partitions in each case. The natural frequencies are normalized such that if (7.104) was an accurate representation of ω_n, its normalized value would simply be n. Note that the normalized values of ω_n are somewhat smaller than the integer values predicted by (7.104). Also note that the damping coefficients σ_n are negative so that the current decreases exponentially with time as the traveling current wave radiates its energy. Since $|\sigma_n|$ is significantly smaller than $|\omega_n|$, we call the wire a high-Q structure. Not all structures that exhibit natural resonances are high-Q; structures that have a higher volume to surface area ratio, such as spheres, have a much lower Q [32]. Interestingly, the normalized natural frequencies of the wire depend only on the ratio $2L/a$. So the values in the table hold for any wire with ratio $2L/a = 200$.

The natural frequencies of the table above occupy a "layer" in the complex plane near the imaginary axis. This is shown more clearly in Figure 7.30. However, these poles are not the only solutions to (7.103). Infinitely many pole layers are present, each successive layer having a larger damping coefficient than the previous one [194]. It is found that for thin wires, the second layer has much larger damping coefficients than the first. Since these natural frequencies correspond to modes that decay quite rapidly, they do not manifest themselves significantly in the late-time response. This is demonstrated clearly in Example 7.16 below, where we show that the first layer of poles produces a natural response closely matching that predicted by the inverse Fourier transform of the frequency-domain response.

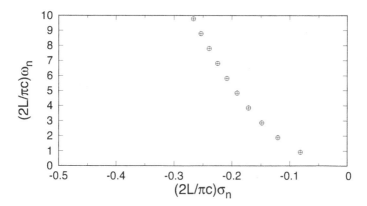

FIGURE 7.30

Normalized natural frequencies $s_n = \sigma_n + j\omega_n$ of a thin wire with $2L/a = 200$. ◀

▶ **Example 7.15:** Natural resonance currents of a thin straight wire

Plot the current distribution for the $n = 1$ and $n = 4$ natural resonance modes of a thin wire of half length $L = 0.25$ m and radius $a = 0.0025$ m. Compare to simple sinusoidal distributions.

Solution: Figure 7.31 shows the current functions found by solving for the nullspace of the MoM matrix, for $n = 1$ and $n = 4$. Here the functions have been normalized by their maximum complex values. Also plotted are the sinusoidal functions

$$I_n(z) = \sin\left(\frac{n\pi}{2L}[z - L]\right). \tag{7.105}$$

Note that the real part of the natural mode current is nearly sinusoidal, while the imaginary part is small in comparison. Hence we are justified in using (7.104) to estimate the natural frequencies. This simple behavior of the current forms a convenient method for approximating the natural-mode response of thin wires of arbitrary shape [162].

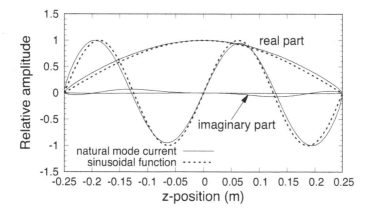

FIGURE 7.31

Natural mode current distributions on a thin wire for $n = 1$ and $n = 4$, compared to sinusoidal functions. $L/a = 200$. ◀

Once the natural frequencies and modal currents are known, the coupling coefficients can be determined from (7.99). To evaluate C_m we need the derivative of the kernel

$$K'(z - z', s) = \frac{\partial}{\partial s} \left\{ \left(\frac{\partial^2}{\partial z^2} - \gamma^2 \right) \int_0^{2\pi} \frac{e^{-\gamma R}}{R} \, d\xi \right\}.$$

Since $\gamma = s/c$, we have

$$K'(z - z', s) = -\frac{1}{c} \frac{\partial^2}{\partial z^2} g(z - z', s) - \frac{2}{c\gamma} \left[\gamma^2 \tilde{G}(z - z', s) \right] + \frac{\gamma^2}{c} g(z - z', s)$$

where

$$g(z - z', s) = \int_0^{2\pi} e^{-\gamma R} \, d\xi.$$

Thus,

$$C_m = \int_{-L}^{L} I_m(z) \left\{ \int_{-L}^{L} I_m(z') \left[-\frac{1}{c} \frac{\partial^2}{\partial z^2} g(z - z', s_m) \right] dz' \right\} dz$$

$$- \frac{2}{c\gamma_m} \int_{-L}^{L} I_m(z) \left\{ \gamma_m^2 \int_{-L}^{L} I_m(z') \tilde{G}(z - z', s_m) \, dz' \right\} dz$$

$$+ \frac{\gamma_m^2}{c} \int_{-L}^{L} I_m(z) \left\{ \int_{-L}^{L} I_m(z') g(z - z', s_m) \, dz' \right\} dz.$$

The second integral may be replaced using (7.101):

$$C_m = - \int_{-L}^{L} I_m(z) \left\{ \int_{-L}^{L} I_m(z') \frac{\partial^2}{\partial z^2} F(z - z', s) \, dz' \right\} dz$$

$$+ \frac{\gamma_m^2}{c} \int_{-L}^{L} I_m(z) \left\{ \int_{-L}^{L} I_m(z') g(z - z', s_m) \, dz' \right\} dz, \qquad (7.106)$$

where

$$F(z - z', s) = \frac{1}{c} g(z - z', s) + \frac{2}{c\gamma} \tilde{G}(z - z', s).$$

Noting that

$$\frac{\partial F(z - z', s)}{\partial z} = -\frac{\partial F(z - z', s)}{\partial z'},$$

we can write the first integral in (7.106) as

$$\int_{-L}^{L} I_m(z) \left\{ \int_{-L}^{L} I_m(z') \frac{\partial}{\partial z'} \frac{\partial F(z - z', s)}{\partial z} \, dz' \right\} dz$$

$$= - \int_{-L}^{L} I_m(z) \left[\int_{-L}^{L} \frac{\partial I_m(z')}{\partial z'} \frac{\partial F(z - z', s)}{\partial z} \, dz' \right] dz$$

$$= - \int_{-L}^{L} \frac{\partial I_m(z')}{\partial z'} \left[\int_{-L}^{L} I_m(z) \frac{\partial F(z - z', s)}{\partial z} \, dz \right] dz'.$$

Here we have used integration by parts along with the condition $I(-L) = I(L) = 0$. Integrating by parts once more, we get the final formula for C_m:

$$C_m = C_{m1} + C_{m2}$$

where

$$C_{m1} = \int_{-L}^{L} \int_{-L}^{L} \frac{\partial I_m(z)}{\partial z} \frac{\partial I_m(z')}{\partial z'} F(z - z', s_m) \, dz' \, dz, \tag{7.107}$$

$$C_{m2} = \int_{-L}^{L} \int_{-L}^{L} I_m(z) I_m(z') \frac{\gamma_m^2}{c} g(z - z', s_m) \, dz' \, dz. \tag{7.108}$$

For pulse function expansion of the current, a numerically convenient formula for C_m may be developed. Differentiating (7.61), we obtain

$$\frac{\partial I_m(z)}{\partial z} = \sum_{n=1}^{N} a_n \left[\delta \left(z - \left[z_n - \frac{\Delta}{2} \right] \right) - \delta \left(z - \left[z_n + \frac{\Delta}{2} \right] \right) \right].$$

Substituting this into (7.107) and computing the integrals using the sifting property of the impulse, we get

$$C_{m1} = \sum_{i=1}^{N} \sum_{j=1}^{N} a_i a_j \left[2F(z_{ij}) - F(z_{ij} - \Delta) - F(z_{ij} + \Delta) \right], \tag{7.109}$$

where $z_{ij} = z_i - z_j = (i - j)\Delta$. Using the pulse function representation for the current in (7.108), we obtain

$$C_{m2} = \frac{\gamma_m^2}{c} \sum_{i=1}^{N} \sum_{j=1}^{N} a_i a_j \int_{z_i - \frac{\Delta}{2}}^{z_i + \frac{\Delta}{2}} \left[\int_{z_j - \frac{\Delta}{2}}^{z_j + \frac{\Delta}{2}} g(z - z', s_m) \, dz' \right] dz.$$

Although these integrals may be computed numerically, the integrand is smooth and a simple approximation may be developed by using trapezoidal-rule integration. For the inner integral, this gives

$$\int_{z_j - \frac{\Delta}{2}}^{z_j + \frac{\Delta}{2}} g(z - z', s_m) \, dz' \approx \frac{\Delta}{2} \left[g \left(z - \left[z_j - \frac{\Delta}{2} \right] \right) + g \left(z - \left[z_j + \frac{\Delta}{2} \right] \right) \right].$$

Applying the trapezoidal rule to the outer integration, we have

$$C_{m2} = \frac{\gamma_m^2}{c} \frac{\Delta^2}{4} \sum_{i=1}^{N} \sum_{j=1}^{N} a_i a_j \left[2g(z_{ij}) + g(z_{ij} - \Delta) + g(z_{ij} + \Delta) \right]. \tag{7.110}$$

We can also evaluate R_m for the case of pulse function expansion of the current and plane-wave incidence. Substituting (7.61) into (7.100), we get

$$R_m = -16\pi^2 \frac{E_0(s_m)}{\eta_0} \tan \theta_i J_0(-js_m a \sin \theta_i / c) \sinh \left(\gamma_m \frac{\Delta}{2} \cos \theta_i \right) \sum_{n=1}^{N} a_n e^{-\gamma_m z_n \cos \theta_i}. \tag{7.111}$$

► **Example 7.16:** Temporal behavior of current at the center of a thin straight wire

Consider a plane wave incident at $\theta_i = 30°$ on a wire of half-length $L = 0.25$ m and radius $a = 0.0025$ m. Compute the time-domain current at the center of the wire using SEM and compare to the result fund using the inverse FFT. From this comparison, determine the onset of the late-time period.

Solution: In § 7.3.1.5 we determined the time-domain response of the current at the center of the wire by computing the inverse FFT of the frequency-domain current. The results are shown in Figure 7.21. To compute the equivalent current distribution using SEM, Equation (7.93) is used. Since we have not developed a method for determining the early-time current $\tilde{I}_E$, we concentrate on the late-time current

$$I_L(z,t) = U\left(t - T_L(z)\right) \sum_{n=1}^{2N} a_n I_n(z) e^{s_n t}. \tag{7.112}$$

We also do not know exactly when the late time begins, but we can explore this by comparing to results found using the inverse FFT. Hence we compute (7.112) without using the unit step function, and see how far back in time we can extend the formula and still have it match the inverse FFT of the frequency-domain result. The quantities a_n, $I_n(z)$, and s_n all occur in conjugate pairs, so

$$I_L(z,t) = \sum_{n=1}^{N} e^{\sigma_n t} 2 \operatorname{Re}\left\{a_n I_n(z) e^{j\omega_n t}\right\}. \tag{7.113}$$

The natural frequencies and modal currents are found by solving (7.103) as discussed earlier, while the amplitude constants are found from (7.99) with C_m computed using (7.109) and (7.110), and R_m computed using (7.111). To keep the computations simple, C_m is found using the thin-wire approximation, while R_m is computed with the Bessel function set to unity (again, the thin wire assumption).

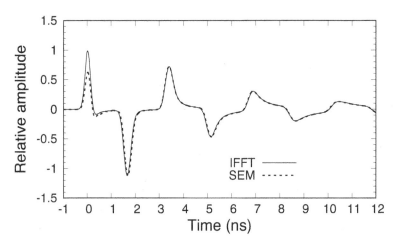

FIGURE 7.32

Temporal current at the center of a wire with half-length $L = 0.25$ m and radius $a = 0.0025$ m for a plane wave of amplitude $E_0 = 1$ V/m incident at $\theta_i = 30°$. Current found using inverse FFT of current spectrum is compared to current from SEM. The window function is equivalent to a Gaussian incident pulse of half-width 0.3 ns.

Figure 7.32 shows the result of computing the current at the center of the wire using (7.113) with the first ten natural modes whose frequencies are shown in the table from Example 7.14. To compare directly with the result found using the inverse FFT of the frequency-domain results, the time series is transformed into the frequency domain using the FFT, windowed in the same manner as the frequency-domain results, and transformed back into the time domain. This procedure produces an effect identical to having an incident plane-wave field with a Gaussian waveform. Here the temporal width of the Gaussian is chosen as 0.3 ns at half amplitude. Figure 7.32 shows that the first event in the current waveform is a pulse at $t = 0$. Since the coordinate origin is at the center of the wire, $t = 0$ is when the incident wave first arrives at the wire center. The next event occurs at $t = 1.66$ ns. Since the one-way transit time of the wire for a wave propagating at the speed of light is $t = 2L/c = 1.66$ ns, the second event occurs when the current wave excited on the wire travels to the end of the wire, reflects, and returns to the center. Note that before this time, the waveform found using SEM does not match well with the inverse FFT of the frequency-domain results. But subsequently the match is excellent. Thus, we take the onset of late-time at the center of the wire to be $T_L = 1.66$ ns. This is physically reasonable, since the natural oscillations depend on the size of the wire, and $t = 1.66$ ns is the first time that the observer at the center of the wire has information about the extent of the wire (provided by the returning current wave reflected from the end). ◄

7.3.4 Time-domain integral equations for a straight wire

Options for obtaining the time-domain current on a straight wire include inverse Fourier transformation of the frequency-domain current (§ 7.3.1.5) and the singularity expansion method (§ 7.3.3). An alternative is to solve an integral equation derived directly in the time domain. Time-domain integral equations have received less attention than frequency-domain equations, but they provide a viable technique for determining the transient response of a wire scatterer or antenna.

Consider a straight wire of length $2L$ and radius a (Figure 7.9). The wire is illuminated by a transient excitation field $\mathbf{E}^i(\mathbf{r}, t)$ that could arise from a generator located on the wire (when used as an antenna) or an external source (in a scattering problem). The excitation field induces a temporal current on the wire surface of such value that the total tangential electric field vanishes for all points on the wire surface at all times. Assuming the wire is thin, we have

$$E_z^s(\mathbf{r}, t) = -E_z^i(\mathbf{r}, t) \qquad (\mathbf{r} \in S), \tag{7.114}$$

where $E_z^s(\mathbf{r}, t)$ is the scattered field produced by the induced current.

The scattered electric field produced by the current induced on the wire surface can be written in terms of the potential functions. By (5.44) we have

$$\mathbf{E}^s = -\frac{\partial \mathbf{A}_e^s}{\partial t} - \nabla \phi_e^s,$$

where the vector potential is given in (5.66). Differentiating with respect to time and substituting (7.114), we obtain

$$-\frac{\partial^2 A_z^s(\mathbf{r}, t)}{\partial t^2} - \hat{\mathbf{z}} \cdot \nabla \frac{\partial \phi_e^s(\mathbf{r}, t)}{\partial t} = -\frac{\partial E_z^i(\mathbf{r}, t)}{\partial t} \qquad (\mathbf{r} \in S),$$

where $A_z^s = \hat{\mathbf{z}} \cdot \mathbf{A}_e^s$. By the Lorenz condition (5.47),

$$-\frac{\partial^2 A_z^s(\mathbf{r}, t)}{\partial t^2} + c^2 \frac{\partial^2 A_z^s(\mathbf{r}, t)}{\partial z^2} = -\frac{\partial E_z^i(\mathbf{r}, t)}{\partial t} \qquad (\mathbf{r} \in S). \tag{7.115}$$

If we employ a thin wire assumption as we did in the frequency domain, we can assume the current supporting A_z^s is a line current on the wire axis. Then (7.115) becomes

$$\left(\frac{\partial^2}{\partial z^2} - \frac{1}{c^2}\frac{\partial^2}{\partial t^2}\right) A_z^s(z,t) = -\frac{1}{c^2}\frac{\partial E_z^i(z,t)}{\partial t} \qquad (-L \le z \le L), \tag{7.116}$$

where

$$A_z^s(z,t) = \frac{\mu_0}{4\pi} \int_0^L \frac{I(z',t - R/c)}{R}\, dz' \tag{7.117}$$

and $R = \sqrt{(z - z')^2 + a^2}$.

Equation (7.116) is a spatio-temporal integro-differential equation for the time-domain current on the wire. It can be solved numerically using the *marching on in time* technique. This approach is fairly complicated, and we will not attempt a full description of its implementation (see, e.g., [95, 135, 195]). Alternatively, it may be viewed as a one-dimensional partial differential equation for the temporal vector potential. Upon solution, the result is a pure spatio-temporal integral equation for the current on the wire. This approach was developed by Hallén [195], and the resulting equation was solved numerically by Liu and Mei [121] using marching-on-in-time.

7.3.4.1 Time-domain Hallén equation

To develop the Hallén equation we may use the solution to the one-dimensional wave equation derived in Appendix A. Use of (A.29) allows the solution of (7.116) to be written as

$$A_z^s(z,t) = -\frac{1}{2c}\int_0^z \int_{t-\frac{z-z'}{c}}^{t+\frac{z-z'}{c}} \frac{\partial E_z^i(z',\tau)}{\partial \tau}\, d\tau\, dz'$$

$$+ f(t - z/c) + g(t + z/c).$$

Computing the temporal integral, we get

$$A_z^s(z,t) = -\frac{1}{2c}\int_0^z E_z^i\left(z', t + \frac{z - z'}{c}\right) dz'$$

$$+ \frac{1}{2c}\int_0^z E_z^i\left(z', t - \frac{z - z'}{c}\right) dz'$$

$$+ f(t - z/c) + g(t + z/c).$$

We may change the lower limits in these integrals, since this only augments the homogeneous solutions $f(t - z/c)$ and $g(t + z/c)$. Thus

$$A_z^s(z,t) = \frac{1}{2c}\int_z^L E_z^i\left(z', t - \frac{|z - z'|}{c}\right) dz'$$

$$+ \frac{1}{2c}\int_{-L}^z E_z^i\left(z', t - \frac{z - z'}{c}\right) dz'$$

$$+ f(t - z/c) + g(t + z/c),$$

or

$$A_z^s(z,t) = \frac{1}{2c}\int_{-L}^L E_z^i\left(z', t - \frac{|z - z'|}{c}\right) dz' + f(t - z/c) + g(t + z/c).$$

Substituting (7.117), we finally have

$$\int_{-L}^{L} \frac{I(z', t - R/c)}{4\pi R} \, dz' = \frac{1}{2\eta_0} \int_{-L}^{L} E_z^i \left(z', t - \frac{|z - z'|}{c} \right) dz'$$
$$+ f(t - z/c) + g(t + z/c) \qquad (-L \le z \le L), \qquad (7.118)$$

where μ_0 has been absorbed into $f(t)$ and $g(t)$.

As an example, consider a wire illuminated by a transient plane-wave field (Figure 7.19). From (7.71), we know that the axial component of the excitation field in the frequency domain is

$$\tilde{E}_z^i(z, \omega) = \tilde{E}_0(\omega) \sin \theta_0 e^{jk_0 z \cos \theta_0}.$$

Here we have assumed that the wire is electrically thin over the frequency band of interest, so that the Bessel function is unity. Computation of the inverse Fourier transform gives the time-domain field

$$E_z^i(z, t) = \sin \theta_0 \frac{1}{2\pi} \int_{-\infty}^{\infty} \tilde{E}_0(\omega) e^{j \frac{\omega}{c} z \cos \theta_0} e^{j\omega t} \, d\omega$$
$$= \sin \theta_0 \frac{1}{2\pi} \int_{-\infty}^{\infty} \tilde{E}_0(\omega) e^{j\omega \left(t + \frac{z}{c} \cos \theta_0 \right)} \, d\omega$$
$$= \sin \theta_0 E_0 \left(t + \frac{z}{c} \cos \theta_0 \right)$$

where $E_0(t)$ is the inverse transform of $\tilde{E}_0(\omega)$. Substituting this into (7.118), we have

$$\int_{-L}^{L} \frac{I(z', t - R/c)}{4\pi R} \, dz' = \frac{\sin \theta_0}{2\eta_0} \int_{-L}^{L} E_0 \left(z', t - \frac{|z - z'|}{c} + \frac{z'}{c} \cos \theta_0 \right) dz'$$
$$+ f(t - z/c) + g(t + z/c) \qquad (-L \le z \le L). \qquad (7.119)$$

7.3.4.2 Approximate solution for the early-time current

As mentioned above, (7.119) may be solved using the marching-on-in-time technique. Here we instead consider a very simple approximate solution to this equation — one that will provide some insight into the mechanism creating the early-time portion of the response. When the wire is thin, the quantity $1/4\pi R$ is highly peaked, and thus a majority of the integral on the left-hand side of (7.119) will arise from values of the current near $z = z'$. So a rough approximation for the left-hand side results in

$$I(z, t) \int_{-L}^{L} \frac{1}{4\pi R} \, dz' = \frac{\sin \theta_0}{2\eta_0} \int_{-L}^{L} E_0 \left(z', t - \frac{|z - z'|}{c} + \frac{z'}{c} \cos \theta_0 \right) dz'$$
$$+ f(t - z/c) + g(t + z/c), \qquad (7.120)$$

which provides a simple solution for the current $I(z,t)$. The remaining integral on the left-hand side may be computed in closed form as

$$
\begin{aligned}
Q(z) &= \int_{-L}^{L} \frac{1}{4\pi R} \, dz' \\
&= \frac{1}{4\pi} \int_{-L}^{L} \frac{dz'}{\sqrt{a^2 + (z - z')^2}} \\
&= \frac{1}{4\pi} \ln \left(\frac{z + L + \sqrt{(z + L)^2 + a^2}}{z - L + \sqrt{(z - L)^2 + a^2}} \right).
\end{aligned}
$$

The solution for $I(z,t)$ given by (7.120) has an interesting interpretation. The current consists of a term describing the interaction of the incident field with the wire as it propagates along the wire, augmented by the two homogeneous solutions $f(t - z/c)$ and $g(t + z/c)$. The latter terms represent waves propagating along the wire in the $\pm z$ directions, respectively. Note that these waves must be present to satisfy the boundary conditions that the current must vanish at the ends of the wire, since the term involving the impressed field does not itself vanish. Also, after the impressed field has passed completely over the wire, these terms will continue the response into the late time, and thus are a major component of the natural response of the wire. Of course, it is important to remember that this solution is only a rough approximation to the true solution of the integral equation.

To estimate the homogeneous solutions during the early time period, let us assume the dominant contribution to $g(t + z/c)$ arises from the first interaction of the impressed field with the wire end at $z = L$. The requirement that the current vanish at this end excites a downward propagating wave. There is as yet no upward propagating wave $f(t - z/c)$, because we assume this term arises from the interaction of the impressed field with the wire end at $z = -L$. Setting the current in (7.120) to zero at $z = L$ results in

$$
0 = \frac{\sin\theta_0}{2\eta_0} \int_{-L}^{L} E_0 \left(z', t - \frac{L - z'}{c} + \frac{z'}{c} \cos\theta_0 \right) dz' + g\left(t + \frac{L}{c} \right).
$$

Letting $\tau = t + L/c$, we have

$$
g(\tau) = -\frac{\sin\theta_0}{2\eta_0} \int_{-L}^{L} E_0 \left(z', \tau - \frac{2L}{c} + \frac{z'}{c} + \frac{z'}{c} \cos\theta_0 \right) dz',
$$

so that

$$
g\left(t + \frac{z}{c} \right) = -\frac{\sin\theta_0}{2\eta_0} \int_{-L}^{L} E_0 \left(z', t - \frac{2L}{c} + \frac{z + z'}{c} + \frac{z'}{c} \cos\theta_0 \right) dz'.
$$

Using this, the current becomes from (7.120)

$$
\begin{aligned}
I(z,t)Q(z) = \frac{\sin\theta_0}{2\eta_0} \int_{-L}^{L} &\left\{ E_0 \left(z', t - \frac{|z - z'|}{c} + \frac{z'}{c} \cos\theta_0 \right) \right. \\
&\left. - E_0 \left(z', t - \frac{2L}{c} + \frac{z + z'}{c} + \frac{z'}{c} \cos\theta_0 \right) \right\} dz'. \quad (7.121)
\end{aligned}
$$

This current will travel down the wire until reaching the bottom, where the upward traveling wave $f(t - z/c)$ will be excited. To determine the form of this wave, the boundary condition of zero current at $z = -L$ is enforced. Augmenting (7.121) with $f(t - z/c)$ and applying the boundary condition, we obtain

$$
f\left(t + \frac{L}{c}\right) = -\frac{\sin\theta_0}{2\eta_0} \int_{-L}^{L} \left\{ E_0\left(z', t - \frac{L + z'}{c} + \frac{z'}{c}\cos\theta_0\right) \right.
$$
$$
\left. - E_0\left(z', t - \frac{3L}{c} + \frac{z'}{c} + \frac{z'}{c}\cos\theta_0\right) \right\} dz',
$$

and so

$$
f\left(t - \frac{z}{c}\right) = -\frac{\sin\theta_0}{2\eta_0} \int_{-L}^{L} \left\{ E_0\left(z', t - \frac{2L}{c} - \frac{z + z'}{c} + \frac{z'}{c}\cos\theta_0\right) \right.
$$
$$
\left. - E_0\left(z', t - \frac{4L}{c} - \frac{z - z'}{c} + \frac{z'}{c}\cos\theta_0\right) \right\} dz'.
$$

With this, the final approximation for the current in the early time is

$$
I(z, t) = \frac{1}{Q(z)}\frac{\sin\theta_0}{2\eta_0} \int_{-L}^{L} \left\{ E_0\left(z', t - \frac{|z - z'|}{c} + \frac{z'}{c}\cos\theta_0\right) \right.
$$
$$
- E_0\left(z', t - \frac{2L}{c} + \frac{z + z'}{c} + \frac{z'}{c}\cos\theta_0\right)
$$
$$
- E_0\left(z', t - \frac{2L}{c} - \frac{z + z'}{c} + \frac{z'}{c}\cos\theta_0\right)
$$
$$
\left. + E_0\left(z', t - \frac{4L}{c} - \frac{z - z'}{c} + \frac{z'}{c}\cos\theta_0\right) \right\} dz'. \tag{7.122}
$$

▶ **Example 7.17:** Approximation of the early-time current on a wire

A transient plane wave with a Gaussian waveform is incident at an angle $\theta_i = 30°$ on a wire of half-length $L = 0.25$ m and radius $a = 0.0025$ m. Compute the early-time current at the center of the wire as a function of time. Compare the approximation (7.122) to the current obtained via the inverse FFT.

Solution: To compute the early-time current using the inverse FFT of $\tilde{I}(z, \omega)$, as found in § 7.3.1.5, it is necessary to window the frequency domain data generated by evaluating $\tilde{I}(z, \omega)$ before performing the inverse FFT. The inverse transform of the window function is thus equivalent to the time-domain excitation waveform of the incident field in the time-domain solution. As in § 7.3.1.5 we use a Gaussian window, which inverse transforms to a Gaussian excitation function of the form $E_0(t) = E_0 e^{-(\ln 16)(t/T)^2}$ where T is the temporal width of the Gaussian at half height. Figure 7.33 shows the approximate early-time current computed using a Gaussian incident field with a pulse width $T = 0.3$ ns. Plotted are both (7.121), which includes just the downward propagating homogeneous solution, and (7.122), which includes both the downward and upward homogeneous solutions. Also plotted is the inverse FFT of the frequency domain result obtained in § 7.3.1.5. It can be seen that the initial pulse of current is directly induced by the incident field passing over the wire, augmented by the downward traveling homogeneous solution. When the upward-traveling homogeneous solution is added in, the second pulse is produced, and thus we attribute this pulse to the wave reflected from the bottom end of the wire. At this point, the incident field has passed, and the natural response of the current may begin because the extent of the wire has been determined. Of course, due to the approximate nature of the solution, the pulses are not reproduced exactly.

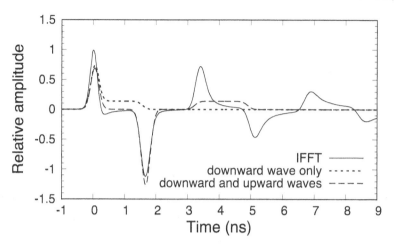

FIGURE 7.33

Temporal current at the center of a wire with half-length $L = 0.25$ m and radius $a = 0.0025$ m for a plane wave of amplitude $E_0 = 1$ V/m incident at $\theta_i = 30°$. Current found using inverse FFT of current spectrum is compared to approximate early-time current. The Gaussian incident pulse has a half-width of 0.3 ns. ◄

7.4 Solution to problems involving two-dimensional conductors

A two-dimensional electromagnetics problem is one in which the physical structure is invariant along a given direction, usually taken as a coordinate axis. An infinitely long cylinder of circular or square cross-section is an example of a two-dimensional problem. If the structure is invariant with respect to one of the remaining coordinate variables (e.g., with ϕ in the case of a circular cylinder), then the problem may yield to separation of variables. For instance, we address scattering from a circular dielectric cylinder as a boundary-value problem in § 4.11.8.4. In contrast, scattering from a cylinder of rectangular cross-section cannot be treated using separation of variables, and we must resort to numerical techniques. One approach is to write an integral equation for the induced current and solve it by MoM.

The form of the excitation also influences the complexity of an electromagnetics problem. Although the geometry of a structure may be two-dimensional, the excitation may be such that the induced current depends on all three coordinate variables. For instance, a finite-length dipole antenna adjacent to an infinitely long dielectric circular cylinder aligned along the z-axis induces a polarization current dependent on ρ, ϕ, and z. If the cylinder is PEC, the current depends on the two variables ϕ and z. If the excitation is a normally incident plane wave, then the dependence of the current is determined by the incident wave polarization. If the electric field is TE to z, the current depends only on ϕ; if TM to z, then only on z. Thus, the simplest problems involve PEC bodies, with the excitation polarized either along the direction of invariance or perpendicular to it. In this section we restrict ourselves to these simple problems, decomposing the excitation field into components TE and TM to the direction of invariance.

7.4.1 The two-dimensional Green's function

Integral equations for two-dimensional problems involve superposition of a current that is independent of one variable. Thus, the integral over this variable only involves the Green's function, which may be computed analytically to produce a *two-dimensional Green's function*. Consider a current $\tilde{\mathbf{J}}(\mathbf{r})$ that is independent of some direction, say z, and immersed in free space. It produces an electric vector potential given by

$$\tilde{\mathbf{A}}_e(\mathbf{r}) = \mu_0 \int_V \tilde{\mathbf{J}}(\boldsymbol{\rho}')\frac{e^{-jk_0R}}{4\pi R}\, dV' = \mu_0 \int_V \tilde{\mathbf{J}}(\boldsymbol{\rho}')\tilde{G}(\mathbf{r}|\mathbf{r}')\, dV',$$

where $\tilde{G}(\mathbf{r}|\mathbf{r}')$ is the three-dimensional Green's function and $\boldsymbol{\rho}$ is the two-dimensional position vector. The distance between the source and observation points is given in terms of the position vectors $\mathbf{r} = z\hat{\mathbf{z}} + \rho\hat{\boldsymbol{\rho}}$ and $\mathbf{r}' = z'\hat{\mathbf{z}} + \rho'\hat{\boldsymbol{\rho}}'$:

$$R^2 = (\mathbf{r}-\mathbf{r}')\cdot(\mathbf{r}-\mathbf{r}') = (z-z')^2 + (\boldsymbol{\rho}-\boldsymbol{\rho}')\cdot(\boldsymbol{\rho}-\boldsymbol{\rho}') = (z-z')^2 + P^2$$

where

$$P = |\boldsymbol{\rho}-\boldsymbol{\rho}'| = \sqrt{\rho^2 + \rho'^2 - 2\rho\rho'\cos(\phi-\phi')}.$$

Writing the volume integral as an integral along z and an integral over the cross-section of the current, we express the vector potential as

$$\tilde{\mathbf{A}}_e(\boldsymbol{\rho}) = \mu_0 \int_{-\infty}^{\infty}\int_{CS} \tilde{\mathbf{J}}(\boldsymbol{\rho}')\frac{e^{-jk_0\sqrt{(z-z')^2+P^2}}}{4\pi\sqrt{(z-z')^2+P^2}}\, dS'\, dz'$$

$$= \mu_0 \int_{CS} \tilde{\mathbf{J}}(\boldsymbol{\rho}')\left[\int_{-\infty}^{\infty}\frac{e^{-jk_0\sqrt{u^2+P^2}}}{4\pi\sqrt{u^2+P^2}}\, du\right] dS'$$

$$= \mu_0 \int_{CS} \tilde{\mathbf{J}}(\boldsymbol{\rho}')\tilde{G}_{2D}(\boldsymbol{\rho}|\boldsymbol{\rho}')\, dS', \tag{7.123}$$

where $\tilde{G}_{2D}$ is the two-dimensional Green's function.

To obtain a formula for $\tilde{G}_{2D}$, we write

$$\tilde{G}_{2D}(\boldsymbol{\rho}|\boldsymbol{\rho}') = 2\int_0^{\infty}\frac{e^{-jk_0P\sqrt{(u/P)^2+1}}}{4\pi P\sqrt{(u/P)^2+1}}\, du$$

and apply the change of variables $w = \sqrt{(u/P)^2+1}$ to get

$$\tilde{G}_{2D}(\boldsymbol{\rho}|\boldsymbol{\rho}') = 2\int_1^{\infty}\frac{e^{-j(k_0P)w}}{4\pi\sqrt{w^2-1}}\, dw.$$

Finally, using the integral representation of the Hankel function [74]

$$H_0^{(2)}(x) = \frac{2j}{\pi}\int_1^{\infty}\frac{e^{-jxt}}{\sqrt{t^2-1}}\, dt \qquad (x>0)$$

we arrive at

$$\tilde{G}_{2D}(\boldsymbol{\rho}|\boldsymbol{\rho}') = \frac{1}{4j}H_0^{(2)}(k_0|\boldsymbol{\rho}-\boldsymbol{\rho}'|). \tag{7.124}$$

Thus we have the following formula for the vector potential produced by a z-invariant current:

$$\tilde{\mathbf{A}}_e(\boldsymbol{\rho}) = \mu_0 \int_{CS} \tilde{\mathbf{J}}(\boldsymbol{\rho}')\frac{1}{4j}H_0^{(2)}(k_0|\boldsymbol{\rho}-\boldsymbol{\rho}'|)\, dS'.$$

If the current flows on the surface of a perfect conductor, then

$$\tilde{\mathbf{A}}_e(\boldsymbol{\rho}) = \mu_0 \int_\Gamma \tilde{\mathbf{J}}_s(\boldsymbol{\rho}') \frac{1}{4j} H_0^{(2)}(k_0 |\boldsymbol{\rho} - \boldsymbol{\rho}'|) \, dl', \tag{7.125}$$

where Γ describes the contour of the perfect conductor in the cross-sectional plane.

As a simple example, consider the electric field at a position $\boldsymbol{\rho}$ produced by an electric line source of amplitude $\tilde{I}_0$ located at $\boldsymbol{\rho}_0$. From (7.123), we see that $\tilde{\mathbf{A}}_e(\boldsymbol{\rho})$ has only a z-component. Hence

$$\tilde{\mathbf{E}} = -\frac{j\omega}{k_0^2} \nabla(\nabla \cdot \tilde{\mathbf{A}}_e) - j\omega \tilde{\mathbf{A}}_e.$$

Since

$$\nabla \cdot \tilde{\mathbf{A}}_e = \frac{\partial \tilde{A}_{ez}}{\partial z} = 0,$$

we have

$$\tilde{\mathbf{E}}(\boldsymbol{\rho}) = -j\omega\mu_0 \int_{CS} \tilde{I}_0 \delta(\boldsymbol{\rho}' - \boldsymbol{\rho}_0) \tilde{G}_{2D}(\boldsymbol{\rho}|\boldsymbol{\rho}') \, dS' = -j\omega\mu_0 \tilde{I}_0 \tilde{G}_{2D}(\boldsymbol{\rho}|\boldsymbol{\rho}_0),$$

which matches (4.343).

7.4.2 Scattering by a conducting strip

Consider an infinitely long, perfectly conducting strip of width $2w$, immersed in free space (Figure 7.34). A plane wave incident on the strip is polarized either with its electric field along the z-axis (TM$_z$ polarization) or with its magnetic field along the z-axis (TE$_z$ polarization). The incident field induces a surface current on the strip, either along the z-direction (for TM$_z$ polarization) or transverse to the z-direction (for TE$_z$ polarization). Current is induced on both the top and bottom surfaces of the strip, but we assume the strip is infinitesimally thin, and that the currents on the top and bottom surfaces are nearly adjacent so that their sum may be treated as the source of the scattered field. Using (7.125), we can write the scattered vector potential in terms of the two-dimensional Green's function:

$$\tilde{\mathbf{A}}_e^s(x, y) = \frac{\mu_0}{4j} \int_{-w}^{w} \tilde{\mathbf{J}}_s(x') H_0^{(2)}\left(k_0 \sqrt{(x - x')^2 + y^2}\right) dx'. \tag{7.126}$$

The scattered electric field is then

$$\tilde{\mathbf{E}}^s = -\frac{j\omega}{k_0^2} \nabla(\nabla \cdot \tilde{\mathbf{A}}_e^s) - j\omega \tilde{\mathbf{A}}_e^s. \tag{7.127}$$

The direction of the induced current depends on the polarization of the incident field, so each case will be considered separately.

7.4.2.1 TM polarization

A TM$_z$-polarized incident plane wave is indicated in Figure 7.34. Its electric field,

$$\tilde{\mathbf{E}}^i = \hat{\mathbf{z}} \tilde{E}_0(\omega) e^{jk_0(x\cos\phi_0 + y\sin\phi_0)}, \tag{7.128}$$

induces a z-directed current on the strip, which in turn produces a z-directed scattered vector potential through (7.126). The scattered electric field is determined from (7.127). Since the vector potential is independent of z, its divergence is zero and thus

$$\tilde{\mathbf{E}}^s = -\hat{\mathbf{z}} \frac{\omega\mu_0}{4} \int_{-w}^{w} \tilde{J}_z(x') H_0^{(2)}\left(k_0 \sqrt{(x - x')^2 + y^2}\right) dx'. \tag{7.129}$$

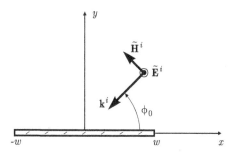

FIGURE 7.34
Scattering of a TM-polarized plane wave from a perfectly conducting strip.

Here $\tilde{J}_z$ is the sum of the z-directed surface currents on the top and bottom surfaces of the strip. The boundary condition on the surface of the perfect conductor at $y = 0$ requires the total tangential electric field there to be zero. Since the field has only a z-component, substitution from (7.128) and (7.129) produces the equation

$$\frac{4}{\omega\mu_0}\tilde{E}_0(\omega)e^{jk_0 x \cos\phi_0} = \int_{-w}^{w} \tilde{J}_z(x')H_0^{(2)}(k_0|x - x'|)\,dx' \qquad (-w \le x \le w). \qquad (7.130)$$

This is an EFIE for the current $\tilde{J}_z(x)$ on the strip.

The integral equation can be solved using MoM. We expand $\tilde{J}_z(x)$ in terms of pulse functions:

$$\tilde{J}_z(x) = \sum_{n=1}^{N} a_n P_n(x), \qquad (7.131)$$

where

$$P_n(x) = \begin{cases} 1, & x_n - \Delta/2 \le x \le x_n + \Delta/2, \\ 0, & \text{elsewhere,} \end{cases}$$

with $x_n = -w + (n - 1/2)\Delta$ and $\Delta = 2w/N$. This is equivalent to partitioning the strip into narrower strips, each with constant current. Substituting the expansion into (7.130) and point matching at $x = x_m$, we obtain the set of linear equations

$$\sum_{n=1}^{N} a_n \int_{x_n-\Delta/2}^{x_n+\Delta/2} H_0^{(2)}(k_0|x_m - x'|)\,dx' = \frac{4}{\omega\mu_0}\tilde{E}_0(\omega)e^{jk_0 x_m \cos\phi_0} \qquad (m = 1,\dots,N),$$

or

$$\sum_{n=1}^{N} a_n A_{mn} = b_m.$$

To better understand the matrix entries, we use the change of variables $u = k_0(x' - x_m)$ and get

$$A_{mn} = \frac{1}{k_0}\int_{k_0\Delta(m-n-1/2)}^{k_0\Delta(m-n+1/2)} H_0^{(2)}(|u|)\,du. \qquad (7.132)$$

A second change of variables of $v = -u$ shows that $A_{mn} = A_{nm}$. We also see that A_{mn} depends on m and n only through the difference $m - n$. As with the wire, these two conditions significantly reduce the computational costs of filling the MoM matrix.

Most of the A_{mn} can be computed easily using numerical integration. However, the Hankel function has a singularity for small argument, so the "self term" A_{mm} requires caution. As with the wire, we may extract the singularity by subtracting a function whose limiting behavior matches that of the integrand as the integration variable approaches zero. We integrate the extracted term in closed form. Using the small argument approximation of the Bessel functions (E.50) and (E.52), we find

$$H_0^{(2)}(x) = J_0(x) - jN_0(x) \approx 1 - j\frac{2}{\pi}(\ln x + \gamma - \ln 2)$$

where $\gamma = 0.5772157\ldots$. We therefore write

$$A_{mm} = \frac{1}{k_0}\int_{-k_0\Delta/2}^{k_0\Delta/2} H_0^{(2)}(|u|)\,du$$

$$= \frac{2}{k_0}\int_0^{k_0\Delta/2} H_0^{(2)}(u)\,du$$

$$= \frac{2}{k_0}\int_0^{k_0\Delta/2} [H_0^{(2)}(u) - f_0(u)]\,du + \frac{2}{k_0}\int_0^{k_0\Delta/2} f_0(u)\,du,$$

where

$$f_0(u) = 1 - j\frac{2}{\pi}(\ln u + \gamma - \ln 2). \tag{7.133}$$

Integrating $f_0(u)$ in closed form, we obtain

$$A_{mm} = \frac{2}{k_0}\int_0^{k_0\Delta/2} [H_0^{(2)}(u) - f_0(u)]\,du + \Delta - j\frac{2\Delta}{\pi}\left[\gamma - \ln 2 + \ln\left(k_0\frac{\Delta}{2}\right) - 1\right].$$

Since the integrand no longer has a singularity at $u = 0$ (it may be removed by defining the integrand to be zero at $u = 0$), the integral may be computed efficiently using numerical integration. Note that we can use this same trick when the source and observation partitions are adjacent. Although there is no singularity in this case, the integrand may still vary rapidly. See Problem 7.22.

Calculation of far-zone scattered field and RCS.   Having determined the current, we may find the electric field using (7.129):

$$\tilde{\mathbf{E}}^s(\rho, \phi) = -\hat{\mathbf{z}}\frac{\omega\mu_0}{4}\int_{-w}^{w} \tilde{J}_z(x')H_0^{(2)}(k_0 R)\,dx',$$

where $R = \sqrt{\rho^2 - 2\rho x'\cos\phi}$. When $\rho \gg w$, the binomial approximation yields

$$H_0^{(2)}(k_0 R) \approx H_0^{(2)}\left(k_0\rho\left[1 - \frac{x'}{\rho}\cos\phi\right]\right).$$

In the far zone where both $\rho \gg w$ and $k\rho \gg 1$, we may use the large argument approximation for the Hankel function (E.64) to show that

$$H_0^{(2)}(k_0 R) \approx \sqrt{\frac{2j}{\pi k_0\rho}}e^{-jk_0\rho}e^{jk_0 x'\cos\phi}.$$

Thus

$$\tilde{E}_z^s(\rho, \phi) = -\frac{\omega\mu_0}{4}\sqrt{\frac{2j}{\pi k_0}}\frac{e^{-jk_0\rho}}{\sqrt{\rho}}\int_{-w}^{w}\tilde{J}_z(x')e^{jk_0 x'\cos\phi}\,dx'. \qquad (7.134)$$

Substitution of (7.131) gives

$$\tilde{E}_z^s(\rho, \phi) = -\frac{\omega\mu_0}{4}\sqrt{\frac{2j}{\pi k_0}}\frac{e^{-jk_0\rho}}{\sqrt{\rho}}\sum_{n=1}^{N}a_n\int_{x_n-\Delta/2}^{x_n+\Delta/2}e^{jk_0 x'\cos\phi}\,dx'. \qquad (7.135)$$

Carrying out the integral and simplifying, we obtain the desired approximation for the scattered field:

$$\tilde{E}_z^s(\rho, \phi) = -\frac{\eta_0}{2}\sqrt{\frac{2j}{\pi k_0}}\frac{e^{-jk_0\rho}}{\sqrt{\rho}}\frac{k_0\Delta}{2}\frac{\sin\left(k_0\frac{\Delta}{2}\cos\phi\right)}{k_0\frac{\Delta}{2}\cos\phi}\sum_{n=1}^{N}a_n e^{jk_0 x_n\cos\phi}. \qquad (7.136)$$

Since the strip is infinitely long, the RCS defined by (7.75) is infinite. But we can compute an RCS per unit length, called the *radar cross-sectional width* or *scattering width*:

$$\sigma_{2D} = \lim_{\rho\to\infty}\left[2\pi\rho\frac{|\tilde{\mathbf{E}}^s(\rho, \phi)|^2}{|\tilde{\mathbf{E}}^i(\phi_0)|^2}\right]. \qquad (7.137)$$

This is a bistatic quantity since it depends on both the incidence and observation angles ϕ_0 and ϕ; a monostatic quantity may be obtained by setting $\phi = \phi_0$. We may compute σ_{2D} for the strip by substituting (7.136) into (7.137). This gives the normalized width

$$\frac{\sigma_{2D}}{\lambda} = \eta_0^2\frac{\pi}{2}\left(\frac{\Delta}{\lambda}\right)^2\left|\frac{\sin\left(k_0\frac{\Delta}{2}\cos\phi\right)}{k_0\frac{\Delta}{2}\cos\phi}\right|^2\left|\sum_{n=1}^{N}\frac{a_n}{\tilde{E}_0}e^{jk_0 x_n\cos\phi}\right|^2. \qquad (7.138)$$

Note that the ϕ_0 dependence of σ_{2D} is carried by the amplitudes a_n.

Physical optics approximation for the current and RCS of a strip — TM case. The solution to the integral equation gives an accurate value for the current induced on a conducting strip by an incident plane wave. Although straightforward, the numerical solution does require significant computational effort. We may obtain a simple approximation to the current from the principle of physical optics (PO). Since the upper surface of the strip is completely illuminated by the incident plane wave, while the lower surface is completely in shadow, the total PO current in the $y = 0$ plane is given by

$$\tilde{\mathbf{J}}_s^{PO}(x) = 2\hat{\mathbf{n}} \times \tilde{\mathbf{H}}^i|_{y=0} \qquad (-w \le x \le w).$$

Here $\hat{\mathbf{n}} = \hat{\mathbf{y}}$ and

$$\tilde{\mathbf{H}}^i(x, y) = \frac{\hat{\mathbf{k}}^i \times \tilde{\mathbf{E}}^i(x, y)}{\eta_0} = \frac{\tilde{E}_0}{\eta_0}\left(\hat{\mathbf{y}}\cos\phi_0 - \hat{\mathbf{x}}\sin\phi_0\right)e^{jk_0(x\cos\phi_0+y\sin\phi_0)}$$

is the magnetic field of the incident plane wave. Thus

$$\tilde{\mathbf{J}}_s^{PO}(x) = \hat{\mathbf{z}}\frac{2\tilde{E}_0}{\eta_0}\sin\phi_0 e^{jk_0 x\cos\phi_0}. \qquad (7.139)$$

The far-zone field produced by the PO current may be computed by substituting (7.139) into (7.134):

$$\tilde{E}_z^{PO}(\rho,\phi) = -\frac{\omega\mu_0}{4}\sqrt{\frac{2j}{\pi k_0}}\frac{e^{-jk_0\rho}}{\sqrt{\rho}}\int_{-w}^{w}\frac{2\tilde{E}_0}{\eta_0}\sin\phi_0 e^{jk_0 x'\cos\phi_0}e^{jk_0 x'\cos\phi}\,dx'.$$

Integration gives

$$\tilde{E}_z^{PO}(\rho,\phi) = -(k_0 w)\sqrt{\frac{2j}{\pi k_0}}\frac{e^{-jk_0\rho}}{\sqrt{\rho}}\tilde{E}_0\sin\phi_0\frac{\sin\left[k_0 w(\cos\phi+\cos\phi_0)\right]}{k_0 w(\cos\phi+\cos\phi_0)}. \tag{7.140}$$

The PO scattering width can be computed by substituting (7.140) into (7.137). This results in the normalized scattering width

$$\frac{\sigma_{2D}^{PO}}{\lambda} = \left(\frac{w}{\lambda}\right)^2 8\pi\sin^2\phi_0\left[\frac{\sin\left[k_0 w(\cos\phi+\cos\phi_0)\right]}{k_0 w(\cos\phi+\cos\phi_0)}\right]^2. \tag{7.141}$$

▶ **Example 7.18:** Current induced on a conducting strip by a plane wave — TM case

Consider a 300 MHz TM-polarized plane wave of amplitude $\tilde{E}_0 = 1$ V/m that illuminates a strip of width $2w = 2$ m (electrical width 2λ at 300 MHz). Plot the magnitude and phase of the current induced on the strip.

Solution: Figure 7.35 shows the magnitude of the current induced on the strip for two different incidence angles, found using both the MoM and the PO approximation (7.139). Note that the current diverges near the strip edge as expected (§ 4.11.8.6). Except near the edge, the current converges quickly as the number of partitions is increased. Shown are results for both $N = 201$ and $N = 11$ partitions. Near the center of the strip, accurate results are obtained even for small N. However, a small number of partitions cannot accurately describe the rapid variation of current near the edge. Note that near the center, the current predicted by PO is very close to that determined from the MoM. However, the PO approximation does not predict the edge singularity. Figure 7.36 shows the phase of the current on the strip. Again, good results may be obtained with as few as 11 basis functions.

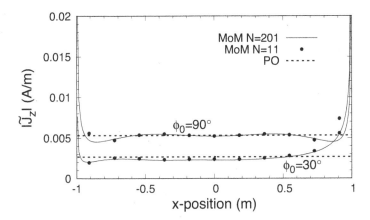

FIGURE 7.35
Magnitude of current induced on a conducting strip of width $2w = 2$ m at 300 MHz. TM polarization. $\tilde{E}_0 = 1$ V/m.

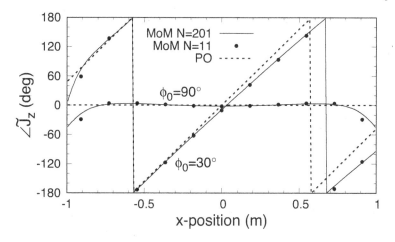

FIGURE 7.36
Phase of current induced on a conducting strip of width $2w = 2$ m at 300 MHz. TM polarization. $\tilde{E}_0 = 1$ V/m. ◄

▶ **Example 7.19:** Scattering width of a conducting strip — TM case

Plot the scattering width of a strip of width $2w = 2$ m at 300 MHz with TM illumination.

Solution: The monostatic ($\phi = \phi_0$) radar scattering width of the strip, calculated using (7.138), is shown in Figure 7.37. The width is greatest near broadside ($\phi = 90°$) and decreases rapidly away from broadside. The PO result obtained using (7.141) compares well to the MoM result near broadside, but decreases in accuracy away from broadside, with several non-physical nulls seen.

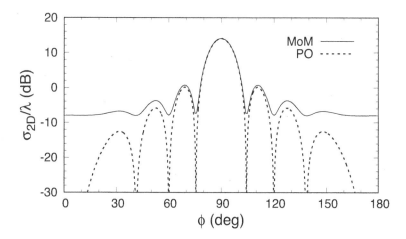

FIGURE 7.37
Normalized monostatic scattering width for a conducting strip of width $2w = 2$ m at 300 MHz. TM polarization. $N = 201$ partitions used for MoM solution. ◄

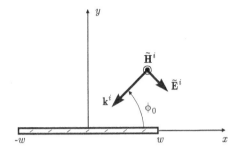

FIGURE 7.38
Scattering of a TE-polarized plane wave from a perfectly conducting strip.

7.4.2.2 TE polarization

Obtaining the field scattered by a strip conductor under TE_z polarization is somewhat more involved than for the TM case. A TE-polarized incident plane wave is indicated in Figure 7.38. Its fields are given by

$$\tilde{\mathbf{H}}^i = \hat{\mathbf{z}} \frac{\tilde{E}_0(\omega)}{\eta_0} e^{jk_0(x\cos\phi_0 + y\sin\phi_0)}, \tag{7.142}$$

$$\tilde{\mathbf{E}}^i = -\eta_0 \hat{\mathbf{k}}^i \times \tilde{\mathbf{H}}^i = \tilde{E}_0(\omega)\left[-\hat{\mathbf{y}}\cos\phi_0 + \hat{\mathbf{x}}\sin\phi_0\right]e^{jk_0(x\cos\phi_0 + y\sin\phi_0)}. \tag{7.143}$$

The incident electric field induces an x-directed current $\tilde{\mathbf{J}}_s(x) = \hat{\mathbf{x}}\tilde{J}_x(x)$ on the strip, which produces an x-directed scattered vector potential through (7.126). The scattered electric field is found using (7.127):

$$\tilde{\mathbf{E}}^s = -\frac{j\omega}{k_0^2}\nabla\left(\frac{\partial \tilde{A}_{ex}^s}{\partial x}\right) - j\omega\hat{\mathbf{x}}\tilde{A}_{ex}^s.$$

Thus, the x-component of the scattered electric field is

$$\tilde{E}_x^s = -\frac{j\omega}{k_0^2}\frac{\partial^2 \tilde{A}_{ex}^s}{\partial x^2} - j\omega\tilde{A}_{ex}^s.$$

The boundary condition on the surface of the strip (i.e., that the total tangential electric field must vanish) produces

$$\left(\frac{\partial^2}{\partial x^2} + k_0^2\right)\tilde{A}_{ex}^s(x) = -j\frac{k_0^2}{\omega}\sin\phi_0\tilde{E}_0 e^{jk_0 x\cos\phi_0} \qquad (-w \le x \le w). \tag{7.144}$$

Here the potential on the strip is

$$\tilde{A}_{ex}^s(x) = \frac{\mu_0}{4j}\int_{-w}^{w}\tilde{J}_x(x')H_0^{(2)}(k_0|x - x'|)\,dx'. \tag{7.145}$$

As in the TM case, $\tilde{\mathbf{J}}_s$ is the sum of the currents on the top and bottom strip surfaces.

The ordinary differential equation (7.144) has the same form as (7.56) for the thin wire, hence has solution

$$\tilde{A}_{ex}^s(x) = f_p(x) + f_c(x) \tag{7.146}$$

where the particular and complementary solutions are, respectively,

$$f_p(x) = \frac{1}{k_0} \int_{x_0}^{z} \left[-j \frac{k_0^2}{\omega} \sin \phi_0 \tilde{E}_0 e^{jk_0 u \cos \phi_0} \sin k_0(x - u) \right] du$$

and

$$f_c(x) = C_1 \sin k_0 x + C_2 \cos k_0 x.$$

The choice of x_0 is arbitrary and affects only the constants in the complementary solution. Choosing $x_0 = 0$ and computing the integral, we get the particular solution

$$f_p(x) = -j \frac{\tilde{E}_0}{\omega} \frac{1}{\sin \phi_0} (e^{jk_0 x \cos \phi_0} - \cos k_0 x - j \cos \phi_0 \sin k_0 x).$$

The last two bracketed terms merely contribute to the complementary solution, hence we may take

$$f_p(x) = -j \frac{\tilde{E}_0}{\omega} \frac{1}{\sin \phi_0} e^{jk_0 x \cos \phi_0}.$$

Equating (7.146) and (7.145), we get

$$\int_{-w}^{w} \tilde{J}_x(x') H_0^{(2)}(k_0|x - x'|) \, dx' + C_1 \sin k_0 x + C_2 \cos k_0 x = \frac{4\tilde{E}_0}{k_0 \eta_0} \frac{1}{\sin \phi_0} e^{jk_0 x \cos \phi_0}. \quad (7.147)$$

Equation (7.147) is a Hallén-type EFIE for the current on the strip. The constants C_1 and C_2 may be found by applying the edge condition on $\tilde{J}_s$ at $x = \pm w$. In § 4.11.8.6 we show that the current normal to a sharp edge vanishes at the edge. Hence we have the additional conditions

$$\tilde{J}_x(-w) = \tilde{J}_x(w) = 0.$$

Expanding the current $\tilde{J}_x(x)$ in pulse functions as in (7.131) and point matching at x_m, we get a matrix equation identical in form to (7.64) for a thin wire. In the present expression, the A_{mn} are given by (7.132), which are the entries for the TM case, while the b_m are those of the TM case divided by $\sin \phi_0$:

$$b_m = \frac{4\tilde{E}_0}{k_0 \eta_0} \frac{1}{\sin \phi_0} e^{jk_0 x_m \cos \phi_0}.$$

Calculation of far-zone scattered field and RCS. To compute the radar scattering width, we need the scattered field in the far zone. It is easiest to find the scattered magnetic field, given by

$$\tilde{\mathbf{H}}^s = \frac{1}{\mu_0} \nabla \times \tilde{\mathbf{A}}_e^s = -\hat{\mathbf{z}} \frac{1}{\mu_0} \frac{\partial \tilde{A}_{ex}^s}{\partial y}.$$

The derivative of the vector potential (7.145) is

$$\frac{\partial \tilde{A}_{ex}^s}{\partial y} = \frac{\mu_0}{4j} \int_{-w}^{w} \tilde{J}_x(x') \frac{\partial}{\partial u} [H_0^{(2)}(u)] \frac{\partial u}{\partial y} \, dx'$$

where $u = k_0 R = k_0 \sqrt{(x - x')^2 + y^2}$. Computing the derivatives and remembering that

$$\frac{\partial}{\partial u} H_0^{(2)}(u) = -H_1^{(2)}(u),$$

we obtain

$$\tilde{\mathbf{H}}^s(x,y) = \hat{\mathbf{z}} y \frac{k_0^2}{4j} \int_{-w}^{w} \tilde{J}_x(x') \frac{H_1^{(2)}\left(k_0\sqrt{(x-x')^2+y^2}\right)}{k_0\sqrt{(x-x')^2+y^2}} \, dx'. \tag{7.148}$$

A far-zone expression for the scattered magnetic field can be found using $R \approx \rho - x'\cos\phi$ along with (E.64). We get

$$H_1^{(2)}(k_0 R) \approx j\sqrt{\frac{2j}{\pi k_0}} \frac{e^{-jk_0\rho}}{\sqrt{\rho}} e^{jk_0 x'\cos\phi}.$$

Substituting this into (7.148) and using $y = \rho\sin\phi$, we arrive at the far-zone field

$$\tilde{\mathbf{H}}^s(\rho,\phi) = \hat{\mathbf{z}} \frac{k_0}{4}\sin\phi\sqrt{\frac{2j}{\pi k_0}} \frac{e^{-jk_0\rho}}{\sqrt{\rho}} \int_{-w}^{w} \tilde{J}_x(x') e^{jk_0 x'\cos\phi} \, dx'. \tag{7.149}$$

The pulse function expansion (7.131) yields

$$\tilde{\mathbf{H}}^s(\rho,\phi) = \hat{\mathbf{z}} \frac{k_0}{4}\sin\phi\sqrt{\frac{2j}{\pi k_0}} \frac{e^{-jk_0\rho}}{\sqrt{\rho}} \sum_{n=1}^{N} a_n \int_{x_n-\Delta/2}^{x_n+\Delta/2} e^{jk_0 x'\cos\phi} \, dx'$$

$$= \hat{\mathbf{z}} \frac{k_0\Delta}{4}\sin\phi\sqrt{\frac{2j}{\pi k_0}} \frac{e^{-jk_0\rho}}{\sqrt{\rho}} \frac{\sin\left(k_0\frac{\Delta}{2}\cos\phi\right)}{k_0\frac{\Delta}{2}\cos\phi} \sum_{n=1}^{N} a_n e^{jk_0 x_n\cos\phi}. \tag{7.150}$$

With the scattered field determined, the radar cross-sectional width may be found from

$$\sigma_{2D} = \lim_{\rho\to\infty}\left[2\pi\rho\frac{|\tilde{\mathbf{H}}^s(\rho,\phi)|^2}{|\tilde{\mathbf{H}}^i(\phi_0)|^2}\right]. \tag{7.151}$$

Use of (7.142) and (7.150) gives the normalized result

$$\frac{\sigma_{2D}}{\lambda} = \eta_0^2 \frac{\pi}{2}\left(\frac{\Delta}{\lambda}\right)^2\sin^2\phi\left|\frac{\sin\left(k_0\frac{\Delta}{2}\cos\phi\right)}{k_0\frac{\Delta}{2}\cos\phi}\right|^2\left|\sum_{n=1}^{N}\frac{a_n}{\tilde{E}_0}e^{jk_0 x_n\cos\phi}\right|^2. \tag{7.152}$$

This is the bistatic width. The monostatic width may be obtained by setting $\phi = \phi_0$.

Physical optics approximation for the current and RCS of a strip — TE case.
The PO approximation for the strip current under TE illumination is, from (7.142),

$$\tilde{\mathbf{J}}_s^{PO}(x) = 2\hat{\mathbf{n}} \times \tilde{\mathbf{H}}^i|_{y=0} = \hat{\mathbf{x}}\frac{2\tilde{E}_0}{\eta_0}e^{jk_0 x\cos\phi_0}. \tag{7.153}$$

Inserting this into (7.149) and integrating, we obtain

$$\tilde{H}_z^{PO}(\rho,\phi) = (k_0 w)\sqrt{\frac{2j}{\pi k_0}} \frac{e^{-jk_0\rho}}{\sqrt{\rho}} \frac{\tilde{E}_0}{\eta_0}\sin\phi\frac{\sin[k_0 w(\cos\phi+\cos\phi_0)]}{k_0 w(\cos\phi+\cos\phi_0)}. \tag{7.154}$$

The PO scattering width can be computed by substituting (7.154) into (7.151). The result is the normalized scattering width

$$\frac{\sigma_{2D}^{PO}}{\lambda} = \left(\frac{w}{\lambda}\right)^2 8\pi\sin^2\phi\left[\frac{\sin[k_0 w(\cos\phi+\cos\phi_0)]}{k_0 w(\cos\phi+\cos\phi_0)}\right]^2. \tag{7.155}$$

This is identical to (7.141) for the TM case, except that $\sin^2\phi_0$ is replaced by $\sin^2\phi$. Under monostatic conditions when $\phi = \phi_0$, the results coincide.

▶ **Example 7.20:** Current induced on a conducting strip by a plane wave — TE case

Repeat Example 7.18 for the case of TE polarization.

Solution: Figure 7.39 shows the magnitude of the current induced on the strip for two different incidence angles, found using both the MoM and the PO approximation (7.153). The current converges very quickly as the number of partitions is increased, with results for $N = 11$ matching fairly closely those for $N = 201$. Figure 7.40 shows the phase of the current on the strip. Again, good results may be obtained with as few as 11 basis functions.

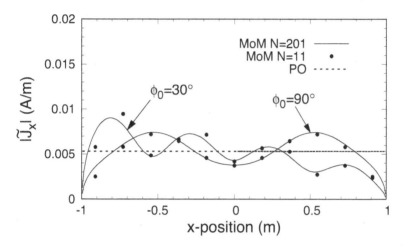

FIGURE 7.39
Magnitude of current induced on a conducting strip of width $2w = 2$ m at 300 MHz. TE polarization. $\tilde{E}_0 = 1$ V/m.

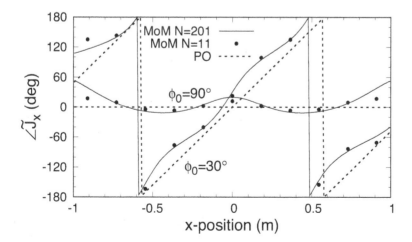

FIGURE 7.40
Phase of current induced on a conducting strip of width $2w = 2$ m at 300 MHz. TE polarization. $\tilde{E}_0 = 1$ V/m. ◀

▶ **Example 7.21:** Scattering width of a conducting strip — TE case

Repeat Example 7.19 for TE polarization.

Solution: The monostatic radar scattering width of the strip calculated using (7.152) is shown in Figure 7.41. As in the TM case, the width is greatest near broadside and decreases rapidly away from broadside. Moreover, the PO result obtained using (7.155) compares well to the MoM result near broadside, but decreases in accuracy away from broadside. ◀

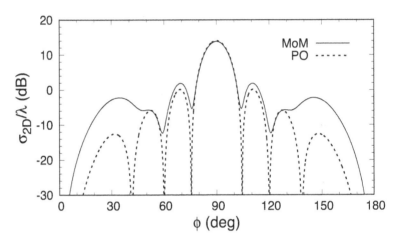

FIGURE 7.41
Normalized monostatic scattering width for a conducting strip of width $2w = 2$ m at 300 MHz. TE polarization. $N = 201$ partitions used for MoM solution. ◀

7.4.3   Scattering by a resistive strip

The solutions described above are easily modified to determine the current induced in a resistive strip (a conducting strip with finite conductivity). For example, consider a TM-polarized plane wave incident on a conducting strip as shown in Figure 7.34. Instead of taking the strip to be infinitesimally thin as in § 7.4.2, we assume a finite thickness t such that $t \ll \lambda$ and $t \ll w$. Since the strip is penetrable, the EFIE (7.28) describes the total field within the strip. However, it is more convenient to deal with the z-directed equivalent current:

$$\tilde{J}_z^{eq}(\boldsymbol{\rho}) = f(\boldsymbol{\rho})\tilde{E}_z(\boldsymbol{\rho}), \tag{7.156}$$

where $f(\boldsymbol{\rho})$ is given in (7.23). Since the strip is very thin, the scattered field is computed by replacing the volume current with a surface current obeying

$$\tilde{J}_z(x) \approx [\tilde{J}_z^{eq}(x)]t.$$

With this, (7.156) becomes

$$\frac{\tilde{J}_z(x)}{t} = f(x)[\tilde{E}_z^i(x) + \tilde{E}_z^s(x)],$$

or

$$\tilde{E}_z^s(x) - \tilde{J}_z(x)Z^i(x) = -\tilde{E}_z^i(x).$$

Here

$$Z^i(x) = \frac{1}{[\sigma(x) + j\omega (\epsilon(x) - \epsilon_0)]\, t}$$

is called the *internal impedance* of the strip. Substituting for the scattered field from (7.129), we get

$$-\frac{\omega\mu_0}{4} \int_{-w}^{w} \tilde{J}_z(x') H_0^{(2)} (k_0|x - x'|) \, dx' - \tilde{J}_z(x) Z^i(x) = -\tilde{E}_z^i(x) \qquad (-w \le x \le w),$$

(7.157)

which is an integral equation for the equivalent surface current on the strip. Once the current has been obtained, the scattered field and radar scattering width may be found as with the PEC strip.

To solve (7.157), we use (7.131) and point match at z_m:

$$\sum_{n=1}^{N} a_n \int_{x_n-\Delta/2}^{x_n+\Delta/2} H_0^{(2)} (k_0|x_m - x'|) \, dx' + \frac{4}{\omega\mu_0} Z^i(x_m) \sum_{n=1}^{N} a_n P_n(x_m)$$

$$= \frac{4}{\omega\mu_0} \tilde{E}_0(\omega) e^{jk_0 x_m \cos\phi_0} \qquad (m = 1, \ldots, N).$$

But $P_n(x_m)$ is nonzero only when $n = m$, so

$$\sum_{n=1}^{N} a_n \left[A_{mn} + \delta_{mn} \frac{4}{k_0\eta_0} Z^i(x_m) \right] = b_m$$

where A_{mn} is given in (7.132) and δ_{mn} is the Kronecker delta. Thus we have a matrix equation nearly identical to that for the PEC strip. The only difference is that the diagonal matrix entries have an additional term that depends on the material properties of the strip. Note that if $\sigma \to \infty$, these additional terms reduce to zero, and the result for the PEC strip is recovered.

A similar approach can be used to find the current and field under TE illumination (Problem 7.23).

▶ **Example 7.22:** Current induced on a resistive strip by a plane wave — TM case

Consider a 300 MHz TM-polarized plane wave of amplitude $\tilde{E}_0 = 1$ V/m that illuminates a resistive strip of width $2w = 2$ m and surface resistance 1000 ohms per square. Plot the magnitude and phase of the current induced on the strip.

Solution: The strip loss is described in terms of the *surface resistance*

$$R_s = 1/(\sigma t),$$

where σ is the conductivity and t is the strip thickness. For this example we compare the case of a perfectly conducting strip ($R_s = 0$) to that of a resistive strip with uniform surface resistance $R_s = 1000$ ohms per square. We also take the permittivity of the strip to be that of free space so that $Z^i = R_s$.

Figure 7.42 shows the magnitude of the current induced on the strip for the two cases of surface resistance. Two things are evident. First, the amplitude of the current is smaller when the strip is resistive. Second, there is no edge singularity when the strip is resistive. Also plotted is the current obtained using a PO approximation. For a resistive strip, the

PO current can be approximated using the simple formula [53]

$$\tilde{\mathbf{J}}_s^{PO}(x) = 2\hat{\mathbf{n}} \times \tilde{\mathbf{H}}^i P(\phi_0) \tag{7.158}$$

where

$$P(\phi_0) = \frac{1}{1 + \frac{2R_s}{Z_0}}$$

with $Z_0 = \eta_0 / \sin\phi_0$ for TM polarization and $Z_0 = \eta_0 \sin\phi_0$ for TE polarization. We see that the PO current is very close to that found using MoM.

Figure 7.43 shows the phase of the current on the strip. The lossy strip displays less phase variation than the PEC strip.

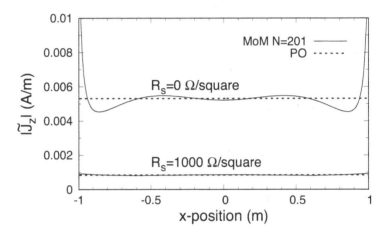

FIGURE 7.42
Magnitude of current induced on a resistive strip of width $2w = 2$ m at 300 MHz with $\phi_0 = 90°$. TM polarization. $\tilde{E}_0 = 1$ V/m.

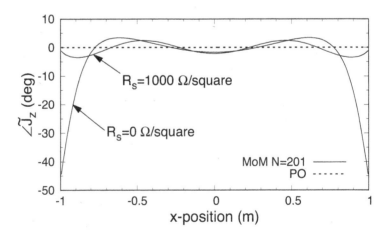

FIGURE 7.43
Phase of current induced on a resistive strip of width $2w = 2$ m at 300 MHz with $\phi_0 = 90°$. TM polarization. $\tilde{E}_0 = 1$ V/m. ◄

▶ **Example 7.23:** Scattering width of a resistive strip — TM case

Consider a resistive strip of width $2w = 2$ m and a surface resistance 1000 ohms per square. Plot the scattering width at 300 MHz with TM illumination.

Solution: The monostatic radar scattering width of the strip calculated using (7.152) is shown in Figure 7.44. The scattering width of the resistive strip is significantly smaller than that of the PEC strip, due to the smaller induced current. In addition, the scattering width of the resistive strip shows stronger nulls. Also shown is the PO scattering width computed using the PO current (7.158). The PO approximation of the scattering width matches the result obtained using MoM quite accurately, except as $\phi \to 0$ or $\phi \to 180°$.

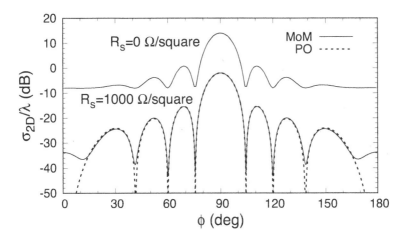

FIGURE 7.44
Normalized monostatic scattering width for a resistive strip of width $2w = 2$ m at 300 MHz. TM polarization. $N = 201$ partitions used for MoM solution. ◀

7.4.4 Cutoff wavenumbers of hollow-pipe waveguides

In § 5.6 we compute the cutoff wavenumbers k_c of certain hollow-pipe waveguides with PEC walls. Once the cutoff frequencies are known, the propagation constants can be found for the various modes through the expression

$$k_z = \sqrt{k^2 - k_c^2}.$$

For guides with simple contours, the cutoff frequencies arise from the separation-of-variables solution to a boundary-value problem. Otherwise a numerical technique is needed to find the cutoff wavenumber.

Consider an empty waveguide with perfectly conducting walls (Figure 7.45). The cross-sectional shape is z-invariant, and its boundary is described by the contour Γ. A primary source exists somewhere within the guide, creating an impressed field $\tilde{\mathbf{E}}^i$. This field induces a current on the guide's inner wall, which produces a scattered field $\tilde{\mathbf{E}}^s$. The boundary condition on the electric field at the waveguide wall is

$$\hat{\mathbf{n}} \times (\tilde{\mathbf{E}}^i + \tilde{\mathbf{E}}^s) = 0 \qquad (\mathbf{r} \in S),$$

where $\hat{\mathbf{n}}$ is the interior unit normal to the wall. When the source is extinguished, the

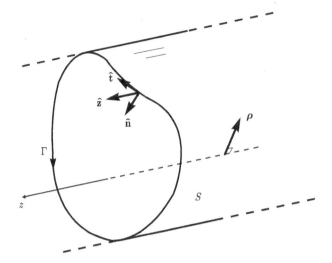

FIGURE 7.45
Conducting, hollow-pipe waveguide.

solution to the resulting homogeneous equation gives the normal mode fields of the guide. This equation,

$$\hat{\mathbf{n}} \times \tilde{\mathbf{E}}^s = 0, \tag{7.159}$$

has nontrivial solutions only for certain values of the wavenumber k. If the current induced on the guide wall is assumed z-invariant, then these eigenvalues are the cutoff wavenumbers of the guide, since only at cutoff is the field in the guide z-invariant.

7.4.4.1 Solution for TM$_z$ modes

As described in § 5.6, a TM$_z$ waveguide mode has only a z-component of electric field, hence is associated with a z-directed current $\tilde{J}^s_z$ on the guide wall. Assuming this current is z-invariant, the associated z-directed vector potential is

$$\tilde{A}^s_{ez}(\boldsymbol{\rho}) = \frac{\mu_0}{4j} \oint_\Gamma \tilde{J}_z(\boldsymbol{\rho}') H_0^{(2)}(k_0|\boldsymbol{\rho} - \boldsymbol{\rho}'|) \, dl'. \tag{7.160}$$

Since the electric field is $\tilde{E}_z = -j\omega\tilde{A}^s_{ez}$, the equation describing the normal-mode current, which also determines the allowed values of the cutoff wavenumber, is obtained from (7.159):

$$\oint_\Gamma \tilde{J}_z(\boldsymbol{\rho}') H_0^{(2)}(k_0|\boldsymbol{\rho} - \boldsymbol{\rho}'|) \, dl' = 0 \qquad (\boldsymbol{\rho} \in \Gamma). \tag{7.161}$$

We may solve (7.161) using MoM. For a waveguide of arbitrary shape, the usual approach is to approximate the boundary using flat strips (see, e.g., [187]). As a simpler example, consider a circular waveguide of radius a. Then

$$\boldsymbol{\rho} = a(\hat{\mathbf{x}}\cos\phi + \hat{\mathbf{y}}\sin\phi), \qquad \boldsymbol{\rho}' = a(\hat{\mathbf{x}}\cos\phi' + \hat{\mathbf{y}}\sin\phi'),$$

so that

$$|\boldsymbol{\rho} - \boldsymbol{\rho}'| = a\sqrt{2}\sqrt{1 - \cos(\phi - \phi')} = 2a\sin\left(\frac{|\phi - \phi'|}{2}\right).$$

Relation (7.161) becomes

$$\int_0^{2\pi} \tilde{J}_z(\phi') H_0^{(2)} \left(2k_0 a \sin \frac{|\phi - \phi'|}{2} \right) d\phi' = 0 \qquad (0 \le \phi \le 2\pi). \tag{7.162}$$

To solve (7.162), we expand the current as

$$\tilde{J}_z(\phi) = \sum_{n=1}^N a_n P_n(\phi) \tag{7.163}$$

where

$$P_n(\phi) = \begin{cases} 1, & \phi_n - \delta/2 \le \phi \le \phi_n + \delta/2, \\ 0, & \text{elsewhere}, \end{cases}$$

with $\phi_n = (n - 1/2)\delta$ and $\delta = 2\pi/N$. Point matching at the angles ϕ_m ($1 \le m \le N$), we get a matrix equation

$$\sum_{n=1}^N A_{mn} a_n = 0 \qquad (m = 1, \ldots, N). \tag{7.164}$$

This homogeneous equation has nontrivial solutions for the eigenvalues $k_0 a$ under the condition

$$\det \{A_{mn}(k_0 a)\} = 0. \tag{7.165}$$

The matrix entries in (7.164) are given by

$$A_{mn} = \int_{\phi_n - \delta/2}^{\phi_n + \delta/2} H_0^{(2)} \left(2k_0 a \sin \frac{|\phi_m - \phi'|}{2} \right) d\phi'$$

or, after a simple change of variables,

$$A_{mn} = \int_{\phi_m - \phi_n - \delta/2}^{\phi_m - \phi_n + \delta/2} H_0^{(2)} \left(2k_0 a \sin \frac{|u|}{2} \right) du. \tag{7.166}$$

Note that the matrix is symmetric and its elements depend on m and n only through the difference $m - n$.

The A_{mn} may be computed using numerical integration. The self ($m = n$) terms are singular at $u = 0$. We can extract the singularity using the technique of § 7.3.1.2. Write

$$A_{mm} = 2 \int_0^{\delta/2} \left[H_0^{(2)} \left(2k_0 a \sin \frac{u}{2} \right) - f_0(u) \right] du + 2 \int_0^{\delta/2} f_0(u) \, du.$$

Here $f_0(u)$ must share the singular behavior of the original integrand as $u \to 0$. Based on the small argument approximation for the Hankel function, we choose

$$f_0(u) = 1 - j\frac{2}{\pi} \left[\ln(k_0 a u) + \gamma \right].$$

The integral of $f_0(u)$ may be done in closed form, giving

$$2 \int_0^{\delta/2} f_0(u) \, du = \delta - j\frac{2\delta}{\pi} \left[\gamma - 1 + \ln \left(\frac{k_0 a \delta}{2} \right) \right].$$

After the cutoff wavenumbers have been found from (7.165), the amplitudes a_n may be determined by finding the nullspace of the matrix A_{mn} [182]. Then the guide field $\tilde{E}_z$ may be determined from the potential function (7.160). Substitution of (7.163) gives

$$\tilde{E}_z^s(\rho, \phi) = -\eta_0 \frac{k_0 a}{4} \sum_{n=1}^{N} a_n \int_{\phi-\phi_n-\delta/2}^{\phi-\phi_n+\delta/2} H_0^{(2)} \left(k_0 a \sqrt{1 + \bar{\rho}^2 - 2\bar{\rho} \cos u} \right) du \qquad (7.167)$$

where $\bar{\rho} = \rho/a$ is the normalized radial distance. Note that since the a_n were found as the solution to a homogeneous equation, only their relative values are meaningful.

▶ **Example 7.24:** Cutoff wavenumbers of a circular waveguide — TM modes

Determine the cutoff wavenumbers of several TM_z modes in a circular waveguide. How does the error depend on the number of basis functions used?

Solution: The following table gives results for the normalized cutoff wavenumbers $k_c a$ of several TM modes found by solving the eigenvalue equation (7.165) using 50 pulse functions in the expansion (7.163).

mode	numerical result	theoretical value
TM_{01}	2.404905	2.404826
TM_{02}	5.520261	5.520078
TM_{03}	8.654015	8.653728
TM_{11}	3.831776	3.831706
TM_{21}	5.135492	5.135623
TM_{22}	8.416518	8.417244

The secant method [67] produced complex values for $k_c a$, and the real parts of those values are given in the table. The imaginary parts of $k_c a$ were six to seven orders of magnitude smaller than the real parts. The theoretical values of $k_c a$ for mode TM_{nm} are given by the mth root of the transcendental equation (§ 5.6)

$$J_n(k_c a) = 0,$$

where $J_n(x)$ is the ordinary Bessel function of order n. The table reveals that for the modes shown, the computed values of the wavenumber are accurate to four digits in comparison with the theoretical values. Figure 7.46 shows the error between the computed and theoretical values of the cutoff wavenumbers, as defined by

$$\epsilon_{k_c}(N) = \frac{\left| k_{c_{\text{numerical}}} - k_{c_{\text{analytic}}} \right|}{\left| k_{c_{\text{analytic}}} \right|}. \qquad (7.168)$$

For the TM_{01} mode, good results are obtained even for $N = 5$ partitions ($k_c a = 2.4056$ versus a theoretical result of $k_c a = 2.4048$). This result improves slowly as N is increased. For the TM_{22} mode, low values of N result in a poor solution for $k_c a$, but increasing N increases the accuracy of the result. For the TM_{11} mode, low values of N again produce poor results for $k_c a$, but increasing N rapidly improves the accuracy of the result. However, in this case the error reaches a minimum at $N = 33$ and then increases somewhat. Note that for larger values of N, any meaningful attempt at estimating the error is hampered by the use of single-precision arithmetic (with 7 significant digits) in all of the numerical calculations.

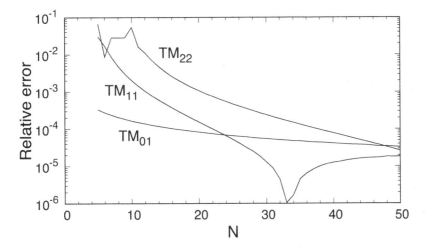

FIGURE 7.46
Error between computed and theoretical cutoff wavenumbers for a circular waveguide. ◀

▶ **Example 7.25:** Axial electric field in a circular waveguide — TM modes

Determine the dependence of the axial electric field on ρ for the TM_{01} and the TM_{22} modes in a circular waveguide. Compare the fields found using the moment method with the analytic values.

Solution: Because the theoretical values of the cutoff wavenumbers for a circular guide are known a priori, it is easy to associate a particular solution for a_n with a corresponding mode. This is not the case for an irregularly shaped guide that may have no analytic solution. In that case, to understand the type of mode considered, it is useful to examine the spatial variation of the field in the guide. For example, a mode showing one cycle of the field along a cut line may be termed mode TM_1, while another mode showing two variations might be termed TM_2. In our present example, we can compute the z-component of the electric field within the guide via (7.167) and compare it with the theoretical field. At a particular value of ϕ, the radial dependence of the theoretical field is (§ 5.6)

$$\tilde{E}_z(\rho) = \tilde{E}_0 J_n \left(k_c a \frac{\rho}{a} \right), \tag{7.169}$$

where $\tilde{E}_0$ is a complex constant. Figure 7.47 shows the numerical electric field computed from (7.167) using $N = 50$ partitions, compared to the theoretical formula (7.169) for two modes. Because the fields satisfy a homogeneous equation, their amplitudes are indeterminate; hence the maximum values of the fields are set to unity for comparison. Note that although (7.167) returns a complex value, when the maximum value of the complex field is normalized to unity, the imaginary part is six to seven orders of magnitude below the real part. Figure 7.47 compares the real part of the normalized field with (7.169), and excellent agreement is seen. The clear difference in the number of oscillations between the two modes is a distinguishing characteristic.

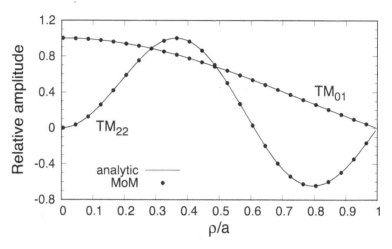

FIGURE 7.47
Relative electric field $\tilde{E}_z$ in an empty circular guide computed using $N = 50$ partitions, and
compared to theory. ◀

7.4.4.2 Solution for TE$_z$ modes

Waveguide modes that are TE$_z$ have no z-component of electric field, and thus are
associated with a transverse current $\tilde{\mathbf{J}}^s$ on the guide wall. This current produces both a
scattered vector potential

$$\tilde{\mathbf{A}}^s_e(\boldsymbol{\rho}) = \mu_0 \oint_\Gamma \tilde{\mathbf{J}}_s(\boldsymbol{\rho}')\tilde{G}_{2D}(\boldsymbol{\rho}|\boldsymbol{\rho}')\,dl' \tag{7.170}$$

and a scattered scalar potential

$$\tilde{\Phi}^s_e(\boldsymbol{\rho}) = \frac{1}{4j\epsilon_0} \oint_\Gamma \tilde{\rho}_s(\boldsymbol{\rho}')\tilde{G}_{2D}(\boldsymbol{\rho}|\boldsymbol{\rho}')\,dl'. \tag{7.171}$$

Here $\tilde{G}_{2D}(\boldsymbol{\rho}|\boldsymbol{\rho}')$ is the Green's function (7.124). The surface charge density is related to
the surface current density by the continuity equation

$$\nabla \cdot \tilde{\mathbf{J}}_s = -j\omega\tilde{\rho}_s.$$

The scattered electric field is given in terms of the potentials as

$$\tilde{\mathbf{E}}^s = -j\omega\tilde{\mathbf{A}}^s_e - \nabla\tilde{\Phi}^s_e.$$

Substituting this into (7.159), and using (7.170) and (7.171), we obtain the characteristic
equation for the eigenvalues k_c for TE$_z$ modes:

$$k_0^2 \oint_\Gamma \hat{\mathbf{t}} \cdot \tilde{\mathbf{J}}_s(\boldsymbol{\rho}')\tilde{G}_{2D}(\boldsymbol{\rho}|\boldsymbol{\rho}')\,dl' + \hat{\mathbf{t}} \cdot \nabla \oint_\Gamma \nabla' \cdot \tilde{\mathbf{J}}_s(\boldsymbol{\rho}')\tilde{G}_{2D}(\boldsymbol{\rho}|\boldsymbol{\rho}')\,dl' = 0 \quad (\boldsymbol{\rho} \in \Gamma) \tag{7.172}$$

where $\hat{\mathbf{t}} = \hat{\mathbf{n}} \times \hat{\mathbf{z}}$ is a unit vector tangent to the surface of the guide in the cross-sectional
plane (Figure 7.45).

For the circular guide we have $\tilde{\mathbf{J}}_s(\boldsymbol{\rho}) = \hat{\boldsymbol{\phi}}\tilde{J}_\phi(\phi)$ and $\hat{\mathbf{t}} = \hat{\boldsymbol{\phi}}$, and use

$$\hat{\boldsymbol{\phi}} \cdot \hat{\boldsymbol{\phi}}' = \cos(\phi - \phi'),$$

$$\hat{\mathbf{t}} \cdot \nabla = \frac{1}{a}\frac{\partial}{\partial\phi},$$

$$\nabla \cdot \tilde{\mathbf{J}}_s = \frac{1}{a}\frac{\partial\tilde{J}_\phi}{\partial\phi}$$

in (7.172) to get

$$k_0^2 \int_0^{2\pi} \cos(\phi - \phi')\tilde{J}_\phi(\phi')\tilde{G}_{2D}(\phi - \phi')a\,d\phi'$$
$$+ \frac{1}{a}\frac{\partial}{\partial\phi}\int_0^{2\pi} \frac{\partial\tilde{J}_\phi(\phi')}{\partial\phi'}\tilde{G}_{2D}(\phi - \phi')\,d\phi' = 0 \qquad (0 \le \phi \le 2\pi). \tag{7.173}$$

Here

$$\tilde{G}_{2D}(\phi - \phi') = \frac{1}{4j}H_0^{(2)}(k_0 R), \qquad R = 2a\sin\left(\frac{|\phi - \phi'|}{2}\right).$$

Equation (7.173) can be solved for the cutoff wavenumbers using MoM. Substituting the pulse expansion (7.163) and noting that

$$\frac{\partial\tilde{J}_\phi(\phi)}{\partial\phi} = \sum_{n=1}^N a_n \frac{\partial P_n(\phi)}{\partial\phi} = \sum_{n=1}^N a_n \left[\delta\left(\phi - \left[\phi_n - \frac{\delta}{2}\right]\right) - \delta\left(\phi - \left[\phi_n + \frac{\delta}{2}\right]\right)\right] \tag{7.174}$$

we obtain

$$(k_0 a)^2 \sum_{n=1}^N a_n \int_{\phi_n - \delta/2}^{\phi_n + \delta/2} \cos(\phi - \phi')\tilde{G}_{2D}(\phi - \phi')\,d\phi'$$
$$+ \sum_{n=1}^N a_n \frac{\partial}{\partial\phi}\left[\tilde{G}_{2D}\left(\phi - \phi_n + \frac{\delta}{2}\right) - \tilde{G}_{2D}\left(\phi - \phi_n - \frac{\delta}{2}\right)\right] = 0 \qquad (0 \le \phi \le 2\pi),$$
$$\tag{7.175}$$

where

$$\tilde{G}_{2D}(u) = \frac{1}{4j}H_0^{(2)}(k_0 R),$$

$$R(u) = 2a\sin\frac{|u|}{2}.$$

Differentiating, we have

$$\frac{\partial\tilde{G}_{2D}(\phi - \phi')}{\partial\phi} = (k_0 a)G_1(\phi - \phi')$$

where

$$G_1(u) = -\frac{1}{4j}H_1^{(2)}(k_0 R)\,\mathrm{sgn}(u)\cos(u/2).$$

The signum function $\operatorname{sgn} x$ arises from the derivative of $|u|$. With this, (7.175) becomes

$$(k_0 a)^2 \sum_{n=1}^{N} a_n \int_{\phi_n - \delta/2}^{\phi_n + \delta/2} \cos(\phi - \phi') \tilde{G}_{2D}(\phi - \phi') \, d\phi'$$

$$+ k_0 a \sum_{n=1}^{N} a_n \left[G_1 \left(\phi - \phi_n + \frac{\delta}{2} \right) - G_1 \left(\phi - \phi_n - \frac{\delta}{2} \right) \right] = 0 \qquad (0 \le \phi \le 2\pi).$$

Finally, point matching at $\phi = \phi_m$ and using a change of variables, we can cast the resulting matrix equation into the form (7.164), where

$$A_{mn} = (k_0 a)^2 \int_{\phi_m - \phi_n - \delta/2}^{\phi_m - \phi_n + \delta/2} \cos u \, \tilde{G}_{2D}(u) \, du$$

$$+ k_0 a \left[G_1 \left(\phi_m - \phi_n + \frac{\delta}{2} \right) - G_1 \left(\phi_m - \phi_n - \frac{\delta}{2} \right) \right]. \qquad (7.176)$$

Note that these matrix entries depend on m and n only through the difference $m - n$, hence A_{mn} may be computed efficiently. Also, the matrix is symmetric: $A_{mn} = A_{nm}$.

When $m = n$, the integrand in (7.176) has a singularity, which can be extracted using the same approach used with the TM modes.

Once the cutoff wavenumbers have been found using (7.165), the amplitudes a_n may be determined by finding the nullspace of the matrix A_{mn} [182]. With these, the fields within the guide may be found using the potential functions (7.170) and (7.171). As an example, the ϕ component of the electric field is

$$\tilde{E}_\phi^s = -j\omega \hat{\phi} \cdot \tilde{\mathbf{A}}_e^s - \hat{\phi} \cdot \nabla \tilde{\Phi}_e^s,$$

or

$$\tilde{E}_\phi^s = -j\omega\mu_0 \int_0^{2\pi} (\hat{\phi} \cdot \hat{\phi}') \tilde{J}_\phi(\phi') \tilde{G}_{2D}(\rho, \phi - \phi') a \, d\phi'$$

$$- \frac{1}{\rho} \frac{\partial}{\partial \phi} \frac{1}{\epsilon_0} \int_0^{2\pi} \tilde{\rho}_s(\phi') \tilde{G}_{2D}(\rho, \phi - \phi') a \, d\phi', \qquad (7.177)$$

where

$$\tilde{G}_{2D}(\rho, \phi - \phi') = \frac{1}{4j} H_0^{(2)}(k_0 R)$$

with

$$k_0 R = k_0 a \sqrt{1 + \bar{\rho}^2 - 2\bar{\rho} \cos(\phi - \phi')}.$$

Here $\bar{\rho} = \rho/a$. Noting that

$$\tilde{\rho}_s(\phi') = \frac{j}{\omega} \frac{1}{a} \frac{\partial \tilde{J}_\phi(\phi')}{\partial \phi'}$$

we can write (7.177) as

$$\frac{\tilde{E}_\phi^s}{-j\eta_0(k_0 a)} = \int_0^{2\pi} \cos(\phi - \phi') \tilde{J}_\phi(\phi') \tilde{G}_{2D}(\rho, \phi - \phi') \, d\phi'$$

$$+ \frac{1}{\bar{\rho}} \frac{1}{(k_0 a)^2} \int_0^{2\pi} \frac{\partial \tilde{J}_\phi(\phi')}{\partial \phi'} \frac{\partial \tilde{G}_{2D}(\rho, \phi - \phi')}{\partial \phi} \, d\phi'. \qquad (7.178)$$

Computing the derivative, we get

$$\frac{\partial \tilde{G}_{2D}(\rho, \phi - \phi')}{\partial \phi} = (ka) G_1(\rho, \phi - \phi'),$$

where

$$G_1(\rho, \phi - \phi') = -\frac{1}{4j} H_1^{(2)}(k_0 R) \frac{\bar{\rho} \sin(\phi - \phi')}{R}.$$

Relation (7.178) becomes

$$\frac{\tilde{E}_\phi^s}{-j\eta_0(k_0 a)} = \int_0^{2\pi} \cos(\phi - \phi') \tilde{J}_\phi(\phi') \tilde{G}_{2D}(\rho, \phi - \phi') \, d\phi'$$

$$+ \frac{1}{\bar{\rho}} \frac{1}{(ka)} \int_0^{2\pi} \frac{\partial \tilde{J}_\phi(\phi')}{\partial \phi'} G_1(\rho, \phi - \phi') \, d\phi'.$$

Finally, by (7.163) and (7.174) we obtain

$$\frac{\tilde{E}_\phi^s}{-j\eta_0(k_0 a)} = \sum_{n=1}^{N} a_n \int_{\phi - \phi_n - \delta/2}^{\phi - \phi_n + \delta/2} \cos(u) \tilde{G}_{2D}(\rho, u) \, du$$

$$+ \frac{1}{\bar{\rho}} \frac{1}{(k_0 a)} \sum_{n=1}^{N} a_n \left[G_1\left(\rho, \phi - \phi_n + \frac{\delta}{2}\right) - G_1\left(\rho, \phi - \phi_n - \frac{\delta}{2}\right) \right].$$

$$(7.179)$$

▶ **Example 7.26:** Cutoff wavenumbers of a circular waveguide – TE modes

Determine the cutoff wavenumbers of several TE$_z$ modes in a circular waveguide. Explore the dependence of the error on the number of basis functions used.

Solution: The following table gives results for the normalized cutoff wavenumbers $k_c a$ of several TE modes found by solving the eigenvalue equation (7.165) with 50 pulse functions.

mode	numerical result	theoretical value
TE$_{01}$	3.831833	3.831706
TE$_{02}$	7.015819	7.015587
TE$_{03}$	10.17380	10.17347
TE$_{11}$	1.841250	1.841184
TE$_{21}$	3.054258	3.054237
TE$_{22}$	6.706162	6.706133

The theoretical values of $k_c a$ for mode TE$_{nm}$ are given by the mth root of the transcendental equation (§ 5.6)

$$J_n'(k_c a) = 0,$$

where $J_n'(x)$ is the derivative of the ordinary Bessel function with respect to its argument. As with the results for the TM modes, the computed values of the wavenumber are accurate to about four digits when compared with the theoretical values. Figure 7.48 shows the error between the computed and theoretical values of the cutoff wavenumbers, as defined by (7.168). Convergence is similar to the TM modes.

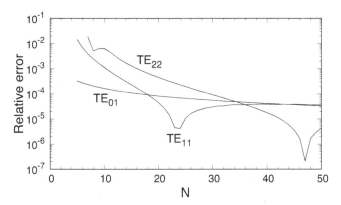

FIGURE 7.48
Error between computed and theoretical cutoff wavenumbers for a circular waveguide. ◄

► **Example 7.27:** Azimuthal electric field in a circular waveguide — TE modes

Determine the dependence of the azimuthal electric field on ρ for the TE_{01} and the TE_{22} modes in a circular waveguide. Compare the fields found using the moment method with the analytic values.

Solution: We can compute the ϕ-component of the electric field within the guide using (7.179) and compare it with the theoretical field. At a particular value of ϕ the radial dependence of the theoretical field is (§ 5.6)

$$\tilde{E}_\phi(\rho) = \tilde{E}_0 J_n' \left(k_c a \frac{\rho}{a} \right), \tag{7.180}$$

where $\tilde{E}_0$ is a complex constant. Figure 7.49 shows the numerical electric field computed from (7.179) using $N = 50$ partitions, compared to the theoretical formula (7.180) for two modes. As with the field plots for the TM modes, the field amplitudes are normalized to a maximum value of unity for comparison. Note that, as expected, $\tilde{E}_\phi$ is zero at $\rho = a$ since it is tangential to the waveguide wall there.

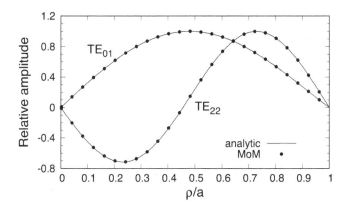

FIGURE 7.49
Relative electric field $\tilde{E}_\phi$ in an empty circular guide computed using $N = 50$ partitions, and compared to theory . ◄

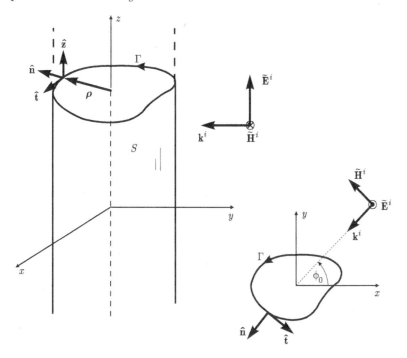

FIGURE 7.50

Conducting cylinder illuminated by normally incident plane wave.

7.4.5 Scattering by a conducting cylinder

Somewhat more involved than the problem of scattering by a strip is scattering by a conducting cylinder. This problem is interesting because it is possible to form an integral equation for the current induced on the surface of the cylinder by applying a boundary condition on the magnetic field. The result is a *magnetic field integral equation* or MFIE. This is in contrast to the preceding sections, in which only EFIEs were considered. We shall compare the results for the MFIE with those for the EFIE, and discuss the benefits and limitations of each equation.

Consider an infinitely long, perfectly conducting cylinder, of arbitrary cross-section and immersed in free space (Figure 7.50). A normally incident plane wave is polarized either with its electric field along the z-axis (TM_z polarization), or with its magnetic field along the z-axis (TE_z polarization). A surface current is induced on the cylinder, flowing either in the z-direction (for TM_z polarization) or transverse to the z-direction (for TE_z polarization). The cylinder is assumed solid with no internal field. The induced current maintains a scattered vector potential

$$\tilde{\mathbf{A}}_e^s(\boldsymbol{\rho}) = \mu_0 \oint_\Gamma \tilde{\mathbf{J}}_s(\boldsymbol{\rho}')\tilde{G}_{2D}(\boldsymbol{\rho}|\boldsymbol{\rho}')\,dl', \tag{7.181}$$

where $\tilde{G}_{2D}(\boldsymbol{\rho}|\boldsymbol{\rho}')$ is the two-dimensional Green's function (7.124) and Γ describes the contour of the cylinder in the cross-sectional plane. In the case of TE polarization, there is also a scattered scalar potential

$$\tilde{\Phi}_e^s(\boldsymbol{\rho}) = \frac{1}{4j\epsilon_0} \oint_\Gamma \tilde{\rho}_s(\boldsymbol{\rho}')\tilde{G}_{2D}(\boldsymbol{\rho}|\boldsymbol{\rho}')\,dl', \tag{7.182}$$

where the surface charge density is related to the surface current density by the continuity equation

$$\nabla \cdot \tilde{\mathbf{J}}_s = -j\omega\tilde{\rho}_s.$$

In terms of the potentials, the scattered electric field is

$$\tilde{\mathbf{E}}^s = -j\omega\tilde{\mathbf{A}}_e^s - \nabla\tilde{\Phi}_e^s. \tag{7.183}$$

Alternatively, the Lorenz condition can be used:

$$\tilde{\mathbf{E}}^s = -\frac{j\omega}{k_0^2}\nabla(\nabla \cdot \tilde{\mathbf{A}}_e^s) - j\omega\tilde{\mathbf{A}}_e^s. \tag{7.184}$$

The boundary condition for the electric field on the cylinder surface requires the total field to vanish:

$$\hat{\mathbf{n}} \times [\tilde{\mathbf{E}}^i(\boldsymbol{\rho}) + \tilde{\mathbf{E}}^s(\boldsymbol{\rho})] = 0 \qquad (\boldsymbol{\rho} \in \Gamma). \tag{7.185}$$

This leads to the EFIE for the surface current $\tilde{\mathbf{J}}_s(\boldsymbol{\rho})$. As noted above, the direction of the induced current depends on the polarization of the incident field, so each case is considered separately.

7.4.5.1   TM polarization — EFIE

The electric field of the TM-polarized incident plane wave indicated in Figure 7.50 is given by

$$\tilde{\mathbf{E}}^i = \hat{\mathbf{z}}\tilde{E}_0(\omega)e^{jk_0(x\cos\phi_0 + y\sin\phi_0)} = \hat{\mathbf{z}}\tilde{E}_0(\omega)e^{jk_0\rho\cos(\phi-\phi_0)}.$$

This field induces a z-directed surface current $\tilde{\mathbf{J}}_s = \hat{\mathbf{z}}\tilde{J}_z$ on the cylinder, which in turn produces a z-directed scattered vector potential through (7.181). The scattered electric field is found using (7.184). Since the vector potential is independent of z, its divergence is zero and

$$\tilde{E}_z^s(\boldsymbol{\rho}) = -\frac{\omega\mu_0}{4}\oint_\Gamma \tilde{J}_z(\boldsymbol{\rho}')H_0^{(2)}\left(k_0|\boldsymbol{\rho}-\boldsymbol{\rho}'|\right)dl'. \tag{7.186}$$

The boundary condition on the tangential electric field at the surface of the cylinder yields the desired integral equation. Equating the total z-directed electric field (incident plus scattered) to zero, we obtain

$$\oint_\Gamma \tilde{J}_z(\boldsymbol{\rho}')H_0^{(2)}\left(k_0|\boldsymbol{\rho}-\boldsymbol{\rho}'|\right)dl' = \frac{4\tilde{E}_0}{k_0\eta_0}e^{jk_0\rho\cos(\phi-\phi_0)} \qquad (\boldsymbol{\rho} \in \Gamma).$$

This is the EFIE for the current $\tilde{J}_z$.

As an example, consider a cylinder with a circular cross-section. The EFIE can be solved for the current using MoM, exactly as was done in § 7.4.4 for the circular waveguide. Expanding the current in pulse functions as in (7.163), and point matching at angles ϕ_m, we obtain the matrix equation

$$\sum_{n=1}^{N} a_n A_{mn} = b_m, \tag{7.187}$$

where A_{mn} is given in (7.166), and

$$b_m = \frac{4\tilde{E}_0}{\eta_0 k_0 a}e^{jk_0 a\cos(\phi_m-\phi_0)}.$$

This equation may be solved using any values of ka except those that cause the matrix to be singular.

Once the current has been found, the scattered electric field may be computed from (7.186). When the observation point is far from the cylinder, we can use the approximation

$$R = |\boldsymbol{\rho} - \boldsymbol{\rho}'| = \sqrt{\rho^2 + a^2 - 2a\rho\cos(\phi - \phi')} \approx \rho - a\cos(\phi - \phi')$$

along with (E.64) to get

$$\tilde{E}_z^s = -\eta_0 k_0 a 4 \sqrt{\frac{2j}{\pi k_0 a}} \frac{e^{-jk_0 a(\rho/a)}}{\sqrt{\rho/a}} \int_0^{2\pi} e^{jk_0 a\cos(\phi - \phi')} \tilde{J}_z(\phi')\, d\phi'.$$

The current expansion (7.163) then gives

$$\tilde{E}_z^s = -\frac{\eta_0}{4} k_0 a \sqrt{\frac{2j}{\pi k_0 a}} \frac{e^{-jk_0 a(\rho/a)}}{\sqrt{\rho/a}} \sum_{n=1}^N a_n \int_{\phi - \phi_n - \delta/2}^{\phi - \phi_n + \delta/2} e^{jk_0 a\cos u}\, du. \tag{7.188}$$

This can be used with (7.137) to obtain the bistatic scattering width

$$\frac{\sigma_{2D}(\phi)}{\lambda} = \eta_0^2 \frac{\pi}{2} \left(\frac{a}{\lambda}\right)^2 \left| \sum_{n=1}^N \frac{a_n}{\tilde{E}_0} \int_{\phi - \phi_n - \delta/2}^{\phi - \phi_n + \delta/2} e^{jk_0 a\cos u}\, du \right|^2. \tag{7.189}$$

PO approximations for the current and scattering width may also be obtained. The current is simply

$$\tilde{\mathbf{J}}_s^{PO} = 2\hat{\mathbf{n}} \times \tilde{\mathbf{H}}^i,$$

where $\hat{\mathbf{n}} = \hat{\boldsymbol{\rho}}$ and $\tilde{\mathbf{H}}^i = \hat{\mathbf{k}}^i \times \tilde{\mathbf{E}}^i / \eta_0$. This gives

$$\tilde{\mathbf{J}}_s^{PO}(\phi) = \begin{cases} 2\hat{\mathbf{z}}\cos(\phi - \phi_0)\dfrac{\tilde{E}_0}{\eta_0} e^{jk_0 a\cos(\phi - \phi_0)}, & -\dfrac{\pi}{2} + \phi_0 \leq \phi \leq \dfrac{\pi}{2} + \phi_0, \\ 0, & \text{elsewhere.} \end{cases} \tag{7.190}$$

Note that the PO current is nonzero only in the region illuminated by the incident wave. Substituting (7.190) into (7.188), we obtain the far-zone PO scattered field

$$\tilde{E}_z^s = \tilde{E}_0 \frac{k_0 a}{2} \sqrt{\frac{2j}{\pi k_0 a}} \frac{e^{-jk_0 a(\rho/a)}}{\sqrt{\rho/a}} \int_{-\pi/2}^{\pi/2} \cos u \, e^{jk_0 a[\cos u + \cos(u + \phi_0 - \phi)]}\, du.$$

Because of the circular symmetry of the cylinder, $\tilde{E}_z^s$ depends only on the difference of the angles ϕ and ϕ_0. The expression for the far-zone scattered field can be combined with (7.137) to produce the bistatic radar scattering width

$$\frac{\sigma_{2D}^{PO}(\phi)}{\lambda} = 2\pi \left(\frac{a}{\lambda}\right)^2 \left| \int_{-\pi/2}^{\pi/2} \cos u \, e^{jk_0 a[\cos u + \cos(u + \phi_0 - \phi)]}\, du \right|^2. \tag{7.191}$$

▶ **Example 7.28:** Current induced on a conducting circular cylinder by a plane wave — TM case

Consider a conducting circular cylinder of radius $a/\lambda = 2$ and a TM plane wave incident at $\phi_0 = 180°$ with amplitude $\tilde{E}_0 = 1$ V/m. Plot the magnitude and phase of the current induced on the cylinder. Compare them to those obtained using the series solution, and to those obtained using PO.

Solution: Figure 7.51 compares the amplitude of the current found by solving the MoM matrix equation (7.187) with the series solution from § 4.11.8, found by summing (4.363). Approximately 100 terms of the series are required to give 5 digits of accuracy in the sum, with $N = 200$ pulse functions in the MoM giving about 3 digits of accuracy. Similar agreement between the MoM and series solution is seen in Figure 7.52, which shows the phase of the current. Also shown is the PO current (7.190). The amplitude and phase of the PO current agree well with the MoM solution near the specular reflection point ($\phi = 180°$), but PO is incapable of computing the current that extends into the shadow zone on the cylinder ($-90° < \phi < 90°$).

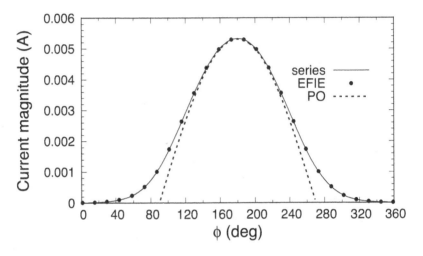

FIGURE 7.51

Magnitude of current induced on a circular cylinder of radius $a/\lambda = 2$ by a TM plane wave incident at $\phi_0 = 180°$. $\tilde{E}_0 = 1$ V/m. $N = 200$ partitions used for MoM solution.

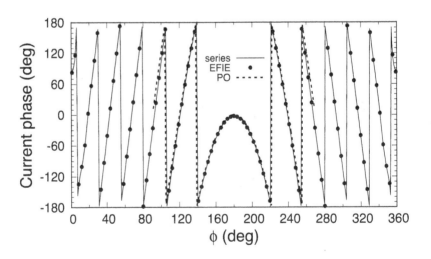

FIGURE 7.52

Phase of current induced on a circular cylinder of radius $a/\lambda = 2$ by a TM plane wave incident at $\phi_0 = 180°$. $\tilde{E}_0 = 1$ V/m. $N = 200$ partitions used for MoM solution. ◄

▶ **Example 7.29:** Scattering width of a conducting circular cylinder — TM case

Plot the scattering width of a conducting circular cylinder of radius $a/\lambda = 2$ with TM illumination.

Solution: Figure 7.53 shows the bistatic radar scattering width of the cylinder found by computing (7.189), and also by using the PO result (7.191). The PO result compares well in the forward scatter ($\phi = 0°$) and backscatter ($\phi = 180°$) directions, but fails to accurately estimate the scattering width away from these points.

Also shown is the scattering width obtained using the series solution for the scattered field. From (4.361) we have

$$\tilde{E}_z^s(\rho, \phi) = -\tilde{E}_0 \sum_{n=0}^{\infty} \epsilon_n j^{-n} \frac{J_n(ka)}{H_n^{(2)}(ka)} H_n^{(2)}(k\rho) \cos n\phi.$$

Using the large argument approximation for the Hankel function (E.64), we obtain in the far zone

$$\tilde{E}_z^s(\rho, \phi) \approx -\tilde{E}_0 \sqrt{\frac{2j}{\pi k}} \frac{e^{-jk\rho}}{\sqrt{\rho}} \sum_{n=0}^{\infty} \epsilon_n \frac{J_n(ka)}{H_n^{(2)}(ka)} \cos n\phi \qquad (k\rho \gg 1).$$

Substituting this into (7.137), we get the scattering width

$$\frac{\sigma_{2D}(\phi)}{\lambda} = \frac{2}{\pi} \left| \sum_{n=0}^{\infty} \epsilon_n \frac{J_n(ka)}{H_n^{(2)}(ka)} \cos n\phi \right|^2.$$

The results for σ_{2D} found using the series and the MoM solutions are close enough to appear as a single line in Figure 7.53.

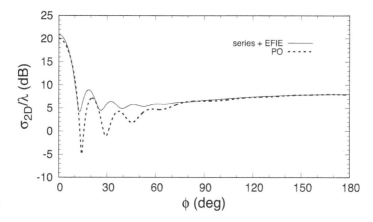

FIGURE 7.53
Bistatic scattering width of a circular cylinder of radius $a/\lambda = 2$ for a TM plane wave incident at $\phi_0 = 180°$. $N = 200$ partitions used for MoM solution. ◀

7.4.5.2   TE polarization — EFIE

A TE-polarized incident plane wave has a z-directed magnetic field, and an electric field given by (7.143), which can also be written as

$$\tilde{\mathbf{E}}^i = \tilde{E}_0 \left[-\hat{\mathbf{y}} \cos \phi_0 + \hat{\mathbf{x}} \sin \phi_0 \right] e^{jk\rho \cos(\phi - \phi_0)}.$$

This field induces a current on the cylinder transverse to the z-direction, which produces both scattered vector and scalar potentials (7.181) and (7.182). The scattered electric field is found using (7.183). Substituting this expression into the boundary condition (7.185) we find that

$$j\omega\mu_0 \oint_\Gamma \hat{\mathbf{t}}\cdot\tilde{\mathbf{J}}_s(\boldsymbol{\rho}')\tilde{G}_{2D}(\boldsymbol{\rho}|\boldsymbol{\rho}')\,dl' + \hat{\mathbf{t}}\cdot\nabla\frac{j}{\omega\epsilon_0}\oint_\Gamma \nabla'\cdot\tilde{\mathbf{J}}_s(\boldsymbol{\rho}')\tilde{G}_{2D}(\boldsymbol{\rho}|\boldsymbol{\rho}')\,dl' = \hat{\mathbf{t}}\cdot\tilde{\mathbf{E}}^i(\boldsymbol{\rho}) \qquad (\boldsymbol{\rho}\in\Gamma)$$

where $\hat{\mathbf{t}} = \hat{\mathbf{z}}\times\hat{\mathbf{n}}$ is the unit tangent to the cylinder in the cross-sectional plane (Figure 7.50). This is the EFIE for the current induced on the cylinder.

As a specific example, let the cylinder have a circular cross-section. The EFIE can be solved for the current using MoM just as in § 7.4.4 for the circular waveguide. Expanding the current in pulse functions as in (7.163), and point matching at angles ϕ_m, we obtain the matrix equation

$$\sum_{n=1}^{N} a_n A_{mn} = b_m, \tag{7.192}$$

where A_{mn} is given in (7.176), and

$$b_m = j\tilde{E}_0\frac{k_0 a}{\eta_0}\cos(\phi_m - \phi_0)e^{jk_0 a\cos(\phi_m - \phi_0)}.$$

This may be solved using any value of $k_0 a$ that does not render the matrix singular.

To compute the radar scattering width of the cylinder, we must find the far-zone scattered field. It is easier to compute the scattered magnetic field than the scattered electric field. We begin with

$$\tilde{\mathbf{H}}^s = \frac{1}{\mu_0}\nabla\times\tilde{\mathbf{A}}_e^s = \nabla\times\oint_\Gamma \tilde{\mathbf{J}}_s(\boldsymbol{\rho}')\tilde{G}_{2D}(\boldsymbol{\rho}|\boldsymbol{\rho}')\,dl'.$$

Moving the curl operation inside the integral and employing (B.49), we get

$$\tilde{\mathbf{H}}^s = -\oint_\Gamma \tilde{\mathbf{J}}_s(\boldsymbol{\rho}')\times\nabla\tilde{G}_{2D}(\boldsymbol{\rho}|\boldsymbol{\rho}')\,dl'$$

since $\nabla\times\tilde{\mathbf{J}}_s(\boldsymbol{\rho}') = 0$. Recalling that $\tilde{G}_{2D}(\boldsymbol{\rho}|\boldsymbol{\rho}') = H_0^{(2)}(k_0 R)/(4j)$ where $R = |\boldsymbol{\rho} - \boldsymbol{\rho}'|$, we find that

$$\nabla\tilde{G}_{2D} = -\frac{k_0}{4j}H_1^{(2)}(k_0 R)\nabla R$$

where

$$\nabla R = \frac{\boldsymbol{\rho} - \boldsymbol{\rho}'}{|\boldsymbol{\rho} - \boldsymbol{\rho}'|} = \hat{\mathbf{R}}.$$

So

$$\tilde{\mathbf{H}}^s = \frac{k_0}{4j}\oint_\Gamma H_1^{(2)}(k_0 R)\tilde{\mathbf{J}}_s(\boldsymbol{\rho}')\times\hat{\mathbf{R}}\,dl'. \tag{7.193}$$

In the far zone of the circular cylinder where $\rho\gg a$, we may write $\hat{\mathbf{R}}\approx\hat{\boldsymbol{\rho}}$ and $R\approx\rho - a\cos(\phi - \phi')$. Then, using (E.64) and noting that $\hat{\boldsymbol{\phi}}'\times\hat{\boldsymbol{\rho}} = -\hat{\mathbf{z}}\cos(\phi - \phi')$, we have

$$\tilde{\mathbf{H}}^s \approx -\hat{\mathbf{z}}\frac{k_0 a}{4}\sqrt{\frac{2j}{\pi k_0}}\frac{e^{-jk_0\rho}}{\sqrt{\rho}}\int_0^{2\pi}\tilde{J}_\phi(\phi')e^{jk_0 a\cos(\phi - \phi')}\cos(\phi - \phi')\,d\phi'.$$

This is the desired far-zone expression. By (7.163) we obtain

$$\tilde{H}_z^s = -\frac{k_0 a}{4}\sqrt{\frac{2j}{\pi k_0 a}}\frac{e^{-jk_0 a(\rho/a)}}{\sqrt{\rho/a}}\sum_{n=1}^{N} a_n \int_{\phi-\phi_n-\delta/2}^{\phi-\phi_n+\delta/2} e^{jk_0 a \cos u}\cos u\,du. \qquad (7.194)$$

This and (7.151) yield

$$\frac{\sigma_{2D}(\phi)}{\lambda} = \eta_0^2 \frac{\pi}{2}\left(\frac{a}{\lambda}\right)^2 \left|\sum_{n=1}^{N}\frac{a_n}{\tilde{E}_0}\int_{\phi-\phi_n-\delta/2}^{\phi-\phi_n+\delta/2} e^{jk_0 a \cos u}\cos u\,du\right|^2, \qquad (7.195)$$

where we have used

$$\tilde{\mathbf{H}}^i = \hat{\mathbf{z}}\frac{\tilde{E}_0}{\eta_0}e^{jk_0 a \cos(\phi-\phi_0)}. \qquad (7.196)$$

Note that (7.195) is the bistatic scattering width, since it depends on the observation angle ϕ.

A PO approximation for the current and scattering width may also be obtained. The current is simply

$$\tilde{\mathbf{J}}_s = 2\hat{\mathbf{n}}\times\tilde{\mathbf{H}}^i,$$

where $\hat{\mathbf{n}} = \hat{\rho}$ and $\tilde{\mathbf{H}}^i$ is given in (7.196). Thus,

$$\tilde{\mathbf{J}}_s(\phi) = \begin{cases} -2\hat{\phi}\frac{\tilde{E}_0}{\eta_0}e^{jk_0 a \cos(\phi-\phi_0)}, & -\frac{\pi}{2}+\phi_0 \le \phi \le \frac{\pi}{2}+\phi_0, \\ 0, & \text{elsewhere.} \end{cases} \qquad (7.197)$$

The PO current is nonzero only in the region illuminated by the incident wave: $-\pi/2 + \phi_0 \le \phi \le \pi/2 + \phi_0$. Substitution of (7.197) into (7.194) gives the far-zone PO scattered field

$$\tilde{H}_z^s = \frac{\tilde{E}_0}{\eta_0}\frac{k_0 a}{2}\sqrt{\frac{2j}{\pi k_0 a}}\frac{e^{-jk_0 a(\rho/a)}}{\sqrt{\rho/a}}\int_{-\pi/2}^{\pi/2}\cos(u+\phi_0-\phi)e^{jk_0 a[\cos u + \cos(u+\phi_0-\phi)]}\,du.$$

Because of the circular symmetry of the cylinder, $\tilde{H}_z^s$ depends only on $\phi - \phi_0$. The expression for the far-zone scattered field can be substituted into the expression (7.151) to yield the bistatic radar scattering width

$$\frac{\sigma_{2D}^{PO}(\phi)}{\lambda} = 2\pi\left(\frac{a}{\lambda}\right)^2\left|\int_{-\pi/2}^{\pi/2}\cos(u+\phi-\phi_0)e^{jk_0 a[\cos u + \cos(u+\phi_0-\phi)]}\,du\right|^2. \qquad (7.198)$$

▶ **Example 7.30:** Current induced on a conducting circular cylinder by a plane wave — TE case

Consider a conducting circular cylinder of radius $a/\lambda = 2$ and a TE plane wave incident at $\phi_0 = 180°$ with amplitude $\tilde{E}_0 = 1$ V/m. This case was considered in Example 7.28 for TM polarization. Plot the magnitude and phase of the current induced on the cylinder. Compare to the series solution and to PO.

Solution: Figure 7.54 compares the amplitude of the current found by solving the MoM matrix equation (7.192) with the series solution. Using $\tilde{\mathbf{J}}_s = \hat{\rho}\times\hat{\mathbf{z}}\tilde{H}_z = -\hat{\phi}\tilde{H}_z$ and substituting the total magnetic field (4.365), we have the series expression for the surface current

$$\tilde{J}_\phi(\phi) = -\frac{\tilde{E}_0}{\eta_0}\sum_{n=0}^{\infty}\frac{\epsilon_n j^{-n}}{H_n^{(2)\prime}(k_0 a)}\left[J_n(k_0 a)H_n^{(2)\prime}(k_0 a) - J_n'(k_0 a)H_n^{(2)}(k_0 a)\right]\cos n\phi.$$

In view of (E.95),

$$\tilde{J}_\phi(\phi) = j\frac{\tilde{E}_0}{\eta_0}\frac{2}{\pi k_0 a}\sum_{n=0}^{\infty}\frac{\epsilon_n j^{-n}}{H_n^{(2)\prime}(k_0 a)}\cos n\phi.$$

Approximately 100 terms in the series are needed for 5-digit accuracy in the sum, with $N = 200$ pulse functions in the MoM giving about 3-digit accuracy. Similar agreement between the MoM and series solution is seen in Figure 7.55, which shows the phase of the current. Also shown is the PO current (7.197). Note that the amplitude of the PO current is constant in the illuminated zone, and matches the MoM solution at the specular point. However, PO does not predict how the current rolls off as the observation point is moved away from the specular point. In contrast, the phase of the PO current agrees well with the MoM solution throughout the illuminated zone.

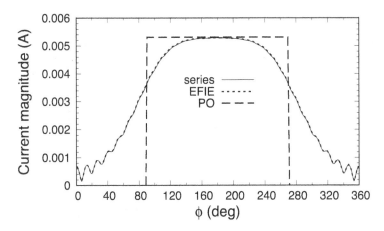

FIGURE 7.54

Magnitude of current induced on a circular cylinder of radius $a/\lambda = 2$ by a TE plane wave incident at $\phi_0 = 180°$. $\tilde{E}_0 = 1$ V/m. $N = 200$ partitions used for MoM solution.

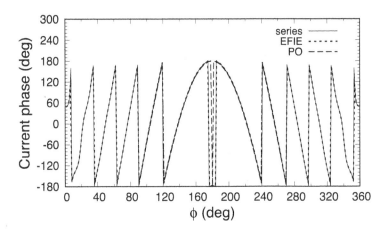

FIGURE 7.55

Phase of current induced on a circular cylinder of radius $a/\lambda = 2$ by a TE plane wave incident at $\phi_0 = 180°$. $\tilde{E}_0 = 1$ V/m. $N = 200$ partitions used for MoM solution. ◄

▶ **Example 7.31:** Scattering width of a conducting circular cylinder — TE case

Plot the scattering width of a conducting circular cylinder of radius $a/\lambda = 2$ with TE illumination.

Solution: Figure 7.56 shows the bistatic radar scattering width of the cylinder obtained from both (7.195) and (7.198). The PO result compares well in the forward-scatter ($\phi = 0°$) and backscatter ($\phi = 180°$) directions, but becomes less accurate away from these points. However, the PO approximation does a good job predicting the strong null in the scattering width near $\phi = 30°$. Note that this null is absent from the TM scattering width (Figure 7.53).

Also shown is the scattering width obtained from the series solution for the scattered field. From (4.365) we find the scattered field

$$\tilde{H}_z^s(\rho, \phi) = -\frac{\tilde{E}_0}{\eta_0} \sum_{n=0}^{\infty} \epsilon_n j^{-n} \frac{J_n'(k_0 a)}{H_n^{(2)'}(k_0 a)} H_n^{(2)}(k_0 \rho) \cos n\phi.$$

Using (E.64) we obtain in the far zone

$$\tilde{H}_z^s(\rho, \phi) \approx -\frac{\tilde{E}_0}{\eta_0} \sqrt{\frac{2j}{\pi k_0}} \frac{e^{-jk_0\rho}}{\sqrt{\rho}} \sum_{n=0}^{\infty} \epsilon_n \frac{J_n'(k_0 a)}{H_n^{(2)'}(k_0 a)} \cos n\phi \qquad (k_0 \rho \gg 1).$$

This and (7.151) give

$$\frac{\sigma_{2D}(\phi)}{\lambda} = \lim_{\rho \to \infty} \left[2\pi\rho \frac{|\tilde{\mathbf{H}}^s(\rho, \phi)|^2}{|\tilde{\mathbf{H}}^i(\phi_0)|^2} \right]$$

$$= \frac{2}{\pi} \left| \sum_{n=0}^{\infty} \epsilon_n \frac{J_n'(k_0 a)}{H_n^{(2)'}(k_0 a)} \cos n\phi \right|^2.$$

The results for σ_{2D} found using the series and the MoM solutions appear as a single line in Figure 7.56.

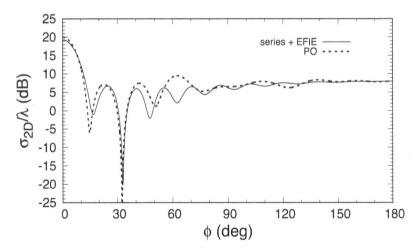

FIGURE 7.56
Bistatic scattering width of a circular cylinder having radius $a/\lambda = 2$ for a TE plane wave incident at $\phi_0 = 180°$. $N = 200$ partitions used for MoM solution. ◀

▶ **Example 7.32:** Dependence of cylinder scattering width on cylinder radius

Compare the dependence of the radar scattering width of a cylinder as a function of radius for TE and TM incident waves.

Solution: Figure 7.57 shows the radar scattering width as a function of cylinder radius for both TE and TM incident waves. The results are obtained in the backscatter direction ($\phi = \phi_0 = 180°$) using the MoM with $N = 200$ partitions, and compared to the PO scattering width (which, in the backscatter direction, is identical for TE and TM polarizations). For small cylinder radii, the results for TE and TM polarization are quite different, and the PO approximation provides a poor estimate of either scattering width. But with increasing cylinder radius the TE and TM results merge and the PO result becomes quite accurate. In fact it is true in general that the PO approximation provides useful results only when the radii of curvature of all surfaces are reasonable fractions of a wavelength.

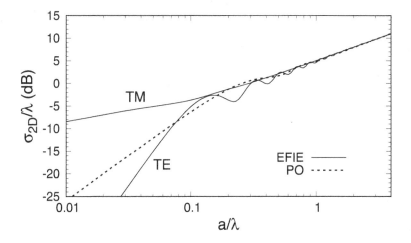

FIGURE 7.57
Scattering width of a circular cylinder vs. cylinder radius, in the backscatter direction: $\phi = \phi_0$. MoM solution with $N = 200$ partitions. ◀

7.4.5.3   TE polarization — MFIE

The integral equations examined in the previous two sections are derived from the boundary condition on the electric field at the surface of the PEC cylinder, and are thus EFIEs. We can obtain another integral equation for the current by imposing the boundary condition on the magnetic field; the result is a *magnetic field integral equation* (MFIE).

An important structural difference between the MFIE and EFIE arises from the fact that the boundary condition on the magnetic field involves a discontinuous function — the magnetic field has a discontinuity at the surface of the PEC cylinder equal to the surface current induced by the impressed field. We must be careful to properly treat this discontinuity by approaching the surface from the outside of the cylinder.

Although the MFIE can be derived for both TM and TE polarizations, it is typically applied to the TE case. The MFIE is more convenient than the EFIE for this case, while the EFIE is more convenient for the TM case.

Consider a PEC cylinder illuminated by a TE-polarized plane wave. The impressed field induces a surface current transverse to the z-direction, and this current in turn creates a z-directed scattered magnetic field. The boundary condition at the cylinder surface is that the total magnetic field (incident plus scattered) has a discontinuity equal to the current induced on the cylinder. Consider Figure 7.58. The observation point for the field outside the cylinder is at $\boldsymbol{\rho}$. This point is a distance h above the surface, and we apply the boundary condition by bringing the observation point down to the point $\boldsymbol{\rho}_0$ on the surface by moving along the outward normal $\hat{\mathbf{n}}$. Since the total magnetic field inside the cylinder is zero, the boundary condition is

$$\lim_{h \to 0} \hat{\mathbf{n}} \times [\tilde{\mathbf{H}}^i(\boldsymbol{\rho}) + \tilde{\mathbf{H}}^s(\boldsymbol{\rho})] = \tilde{\mathbf{J}}_s(\boldsymbol{\rho}_0).$$

Here $\tilde{\mathbf{H}}^i$ is the incident field and $\tilde{\mathbf{H}}^s$ is the scattered field produced by the induced current. The limit of the incident field may be evaluated by allowing $\boldsymbol{\rho} \to \boldsymbol{\rho}_0$, but we must be careful with the scattered field because of the singularity in the Green's function used for its computation. We write

$$\lim_{h \to 0} \hat{\mathbf{n}} \times \tilde{\mathbf{H}}^s(\boldsymbol{\rho}) + \hat{\mathbf{n}} \times \tilde{\mathbf{H}}^i(\boldsymbol{\rho}_0) = \tilde{\mathbf{J}}_s(\boldsymbol{\rho}_0). \tag{7.199}$$

Now define the unit tangent to the cylinder in the cross-sectional plane as

$$\hat{\mathbf{t}} = \hat{\mathbf{z}} \times \hat{\mathbf{n}}, \tag{7.200}$$

which coincides with $\hat{\boldsymbol{\phi}}$ in cylindrical coordinates. We write (7.199) as

$$\lim_{h \to 0} \hat{\mathbf{n}} \times [\hat{\mathbf{n}} \times \tilde{\mathbf{H}}^s(\boldsymbol{\rho})] + \hat{\mathbf{n}} \times [\hat{\mathbf{n}} \times \tilde{\mathbf{H}}^i(\boldsymbol{\rho}_0)] = \hat{\mathbf{n}} \times [\hat{\mathbf{t}} \tilde{J}_s(\boldsymbol{\rho}_0)].$$

Using (7.200), employing (B.7), and noting that $\hat{\mathbf{n}} \cdot \hat{\mathbf{z}} = 0$, we obtain

$$-\hat{\mathbf{z}} \lim_{h \to 0} \tilde{H}_z^s(\boldsymbol{\rho}) - \hat{\mathbf{z}} \tilde{H}_z^i(\boldsymbol{\rho}_0) = \hat{\mathbf{z}} \tilde{J}_s(\boldsymbol{\rho}_0),$$

or

$$\tilde{J}_s(\boldsymbol{\rho}_0) + \lim_{h \to 0} \tilde{H}_z^s(\boldsymbol{\rho}) = -\tilde{H}_z^i(\boldsymbol{\rho}_0) \qquad (\boldsymbol{\rho}_0 \in \Gamma). \tag{7.201}$$

Note that the boundary condition must hold at each point $\boldsymbol{\rho}_0$ on the cylinder surface.

To transform (7.201) into an MFIE, we use (7.193):

$$\tilde{\mathbf{H}}^s = \frac{k_0}{4j} \oint_{\Gamma} H_1^{(2)}(k_0 R) \tilde{\mathbf{J}}_s(\boldsymbol{\rho}') \times \hat{\mathbf{R}} \, dl'. \tag{7.202}$$

Here

$$\tilde{\mathbf{J}}_s(\boldsymbol{\rho}') \times \hat{\mathbf{R}} = \tilde{J}_s(\boldsymbol{\rho}') \hat{\mathbf{t}}' \times \hat{\mathbf{R}}.$$

Substituting $\hat{\mathbf{t}}' = \hat{\mathbf{z}} \times \hat{\mathbf{n}}'$ and employing (B.7) we see that

$$\tilde{\mathbf{J}}_s(\boldsymbol{\rho}') \times \hat{\mathbf{R}} = -\hat{\mathbf{z}} \tilde{J}_s(\boldsymbol{\rho}') \hat{\mathbf{R}} \cdot \hat{\mathbf{n}}'. \tag{7.203}$$

Substituting (7.202) into (7.201) and using (7.203), we obtain

$$\tilde{J}_s(\boldsymbol{\rho}_0) + \lim_{h \to 0} j \frac{k_0}{4} \oint_{\Gamma} (\hat{\mathbf{n}}' \cdot \hat{\mathbf{R}}) \tilde{J}_s(\boldsymbol{\rho}') H_1^{(2)}(k_0 R) \, dl' = -\tilde{H}_z^i(\boldsymbol{\rho}_0) \qquad (\boldsymbol{\rho}_0 \in \Gamma). \tag{7.204}$$

This is the MFIE for the induced current $\tilde{J}_s$.

The MFIE (7.204) was derived using the boundary condition involving the discontinuity of $\tilde{\mathbf{H}}$ across the interface between the region external to the cylinder, and the region internal, where the field is zero. Hence it does *not* apply to an open surface such as the strip studied in § 7.4.2, as this structure has no internal region where the field is zero. Also note that while the equation *does* hold for structures having thin but nonzero cross-sectional areas, the numerical solution to the MFIE tends to be inaccurate.

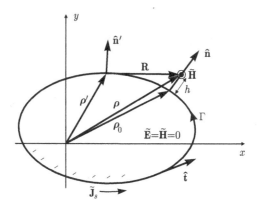

FIGURE 7.58
Geometry for obtaining MFIE from boundary condition for magnetic field.

Alternative form of the MFIE. The limit in (7.204) may be evaluated as needed in each particular application. Typically, the Green's function singularity is extracted as discussed elsewhere in this chapter. Alternatively, we may attempt to extract the singularity in a more general manner, obtaining a form of the MFIE particularly amenable to an MoM solution.

Let us split the contour Γ into a small segment Γ_0 immediately below the observation point (Figure 7.59) and its complement $\Gamma - \Gamma_0$. We choose Γ_0 so that the limiting point $\boldsymbol{\rho}_0$ resides at the "center" of the segment (in the sense that Γ_0 always contains $\boldsymbol{\rho}_0$ as the length of the segment decreases). The integral in (7.204) becomes

$$\lim_{h \to 0} j \frac{k_0}{4} \oint_\Gamma (\hat{\mathbf{n}}' \cdot \hat{\mathbf{R}}) \tilde{J}_s(\boldsymbol{\rho}') H_1^{(2)}(k_0 R) \, dl' = \lim_{h \to 0} j \frac{k_0}{4} \int_{\Gamma - \Gamma_0} (\hat{\mathbf{n}}' \cdot \hat{\mathbf{R}}) \tilde{J}_s(\boldsymbol{\rho}') H_1^{(2)}(k_0 R) \, dl'$$

$$+ \lim_{h \to 0} j \frac{k_0}{4} \int_{\Gamma_0} (\hat{\mathbf{n}}' \cdot \hat{\mathbf{R}}) \tilde{J}_s(\boldsymbol{\rho}') H_1^{(2)}(k_0 R) \, dl'. \qquad (7.205)$$

In the first integral on the right, we may substitute $\boldsymbol{\rho}$ for $\boldsymbol{\rho}_0$ since the singular point is excluded from the domain of integration. To compute the second integral we take Γ_0 so small that the segment is approximately flat. Then the integral represents the scattered magnetic field just above the center of a flat strip, and we are interested in the value of the field as the observation point approaches the strip.

To compute the magnetic field just above the center of the strip, consider Figure 7.60. We seek $\tilde{\mathbf{H}}^s$ as $h \to 0$. From (7.148) we see that

$$\tilde{H}_z^s = -h \frac{k_0^2}{4j} \int_{-w}^{w} \tilde{J}_s(x') \frac{H_1^{(2)}\left(k_0 \sqrt{(x')^2 + h^2}\right)}{k_0 \sqrt{(x')^2 + h^2}} \, dx'.$$

To compute the integral we extract the singularity at $x' = 0$ that occurs when $h \to 0$. We use the small argument approximation

$$H_1^{(2)}(z) \approx \frac{z}{2} + j \frac{1}{\pi} \frac{2}{z} \quad \text{to write} \quad \frac{H_1^{(2)}(z)}{z} \approx \frac{1}{2} + j \frac{1}{\pi} \frac{2}{z^2}.$$

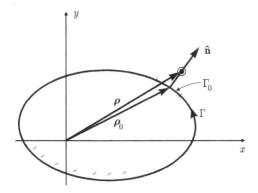

FIGURE 7.59
Geometry for deriving the alternative form of the MFIE. A small segment of the surface contour is removed beneath the observation point.

Then we form

$$\tilde{H}_z^s = -h\frac{k_0^2}{4j} \int_{-w}^{w} \tilde{J}_s(x') \left[\frac{H_1^{(2)}\left(k_0\sqrt{(x')^2+h^2}\right)}{k_0\sqrt{(x')^2+h^2}} - \frac{1}{2} - j\frac{1}{k_0^2\pi}\frac{2}{(x')^2+h^2} \right] dx'$$
$$- h\frac{k_0^2}{4j} \int_{-w}^{w} \frac{1}{2}\tilde{J}_s(x')\, dx' - h\frac{k_0^2}{4j} \int_{-w}^{w} \tilde{J}_s(x')j\frac{1}{k_0^2\pi}\frac{2}{(x')^2+h^2}\, dx'.$$

The first two of these integrals are well-behaved and make no contributions to $\tilde{\mathbf{H}}_z^s$ as $h \to 0$. Without the expression for $\tilde{J}_s$ we cannot directly compute the third integral. However, we can invoke the mean value theorem to write

$$\tilde{H}_z^s = -h\frac{1}{2\pi}\tilde{J}_s(x_0) \int_{-w}^{w} \frac{dx'}{(x')^2+h^2},$$

where x_0 lies somewhere in the interval $[-w, w]$. Computing the integral, we obtain

$$\tilde{H}_z^s = -\frac{\tilde{J}_s(x_0)}{\pi} \tan^{-1}\frac{w}{h}$$

and the limit passage is now possible:

$$\lim_{h\to 0} \tilde{H}_z^s = -\frac{\tilde{J}_s(x_0)}{\pi} \lim_{h\to 0} \tan^{-1}\frac{w}{h} = -\frac{\tilde{J}_s(x_0)}{\pi}\left(\frac{\pi}{2}\right) = -\frac{\tilde{J}_s(x_0)}{2}. \tag{7.206}$$

Although x_0 is unknown, we know that $x_0 \to 0$ as $w \to 0$. Thus, when the segment length becomes small, the field immediately above and at the center of the segment is one-half the negative of the current at the center of the segment.

Using this in (7.205) we find that

$$\lim_{h\to 0} j\frac{k_0}{4} \oint_{\Gamma} (\hat{\mathbf{n}}' \cdot \hat{\mathbf{R}})\tilde{J}_s(\rho')H_1^{(2)}(k_0 R)\, dl'$$
$$= \lim_{\Gamma_0\to 0} j\frac{k_0}{4} \int_{\Gamma-\Gamma_0} (\hat{\mathbf{n}}' \cdot \hat{\mathbf{R}})\tilde{J}_s(\rho')H_1^{(2)}(k_0 R)\, dl' - \frac{\tilde{J}_s(\rho_0)}{2}.$$

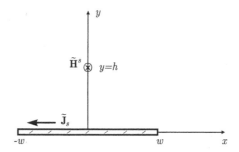

FIGURE 7.60
Geometry for calculating the magnetic field produced by the current flowing on the removed surface segment, which is assumed flat.

Finally, substitution into (7.204) gives the alternative form of the MFIE. For notational convenience we let $\boldsymbol{\rho}_0$ become $\boldsymbol{\rho}$, where $\boldsymbol{\rho}$ now lies on Γ. This gives

$$\frac{\tilde{J}_s(\boldsymbol{\rho})}{2} + \lim_{\Gamma_0 \to 0} j\frac{k_0}{4} \oint_{\Gamma-\Gamma_0} (\hat{\mathbf{n}}' \cdot \hat{\mathbf{R}}) \tilde{J}_s(\boldsymbol{\rho}') H_1^{(2)}(k_0 R)\, dl' = -\tilde{H}_z^i(\boldsymbol{\rho}) \qquad (\boldsymbol{\rho} \in \Gamma). \quad (7.207)$$

Note that we have replaced the limit on h by a limit on the contour segment Γ_0; we will find that the result is particularly amenable to MoM solution.

Example: circular cylinder. Let us specialize the alternative form of the MFIE (7.207) for a circular cylinder, so that we may compare the results to those obtained from the EFIE. Noting that $\hat{\mathbf{n}} = \hat{\boldsymbol{\rho}}$ and

$$\boldsymbol{\rho} = a(\hat{\mathbf{x}}\cos\phi + \hat{\mathbf{y}}\sin\phi), \qquad \boldsymbol{\rho}' = a(\hat{\mathbf{x}}\cos\phi' + \hat{\mathbf{y}}\sin\phi'),$$

we find

$$R = |\boldsymbol{\rho} - \boldsymbol{\rho}'| = \sqrt{2}a\sqrt{1 - \cos(\phi - \phi')} = 2a\sin\frac{|\phi - \phi'|}{2}$$

and

$$\hat{\mathbf{n}}' \cdot \mathbf{R} = a(\hat{\boldsymbol{\rho}}' \cdot \hat{\boldsymbol{\rho}} - 1) = -a[1 - \cos(\phi - \phi')]$$

so that

$$\hat{\mathbf{n}}' \cdot \hat{\mathbf{R}} = -\frac{1}{\sqrt{2}}\sqrt{1 - \cos(\phi - \phi')} = -\sin\frac{|\phi - \phi'|}{2}.$$

Substituting these into (7.207), we have

$$\frac{\tilde{J}_s(\phi)}{2} - j\frac{k_0}{4}\int_0^{2\pi} \sin\frac{|\phi - \phi'|}{2}\tilde{J}_s(\phi')H_1^{(2)}\left(2k_0 a \sin\frac{|\phi - \phi'|}{2}\right)a\,d\phi'$$

$$= -\frac{\tilde{E}_0}{\eta_0}e^{jk_0 a \cos(\phi - \phi_0)} \qquad (0 \leq \phi \leq 2\pi). \quad (7.208)$$

The dash on the integral sign denotes a *principal-value integral* in which the interval $[\phi - \delta/2, \phi + \delta/2]$ is excluded from the domain of integration and the limit as $\delta \to 0$ is taken.

The MFIE (7.208) may be solved using MoM. We expand the current in the pulse basis set (7.163) and point match at ϕ_m. To implement the principal-value integral, we merely exclude the $n = m$ term from the sum and implement the limit by increasing the value of N. Since the current in each partition is constant, the question of where to evaluate x_0 in (7.206) is moot. The resulting system of equations is

$$\frac{1}{2} \sum_{n=1}^{N} a_n P_n(\phi_m) - \sum_{\substack{n=1 \\ n \neq m}}^{N} \int_{\phi_n - \delta/2}^{\phi_n + \delta/2} \sin \frac{|\phi_m - \phi'|}{2} H_1^{(2)} \left(2 k_0 a \sin \frac{|\phi_m - \phi'|}{2} \right) a \, d\phi'$$

$$= -\frac{\tilde{E}_0}{\eta_0} e^{j k_0 a \cos(\phi_m - \phi_0)} \qquad (m = 1, \ldots, N).$$

Using the change of variables $u = \phi_m - \phi'$, we can put this into the form of a matrix equation (7.187), where $b_m = -(\tilde{E}_0/\eta_0) e^{j k_0 z \cos(\phi_m - \phi_0)}$ and

$$A_{mn} = \begin{cases} \frac{1}{2}, & m = n, \\ -j \frac{k_0 a}{4} \int_{\phi_m - \phi_n - \delta/2}^{\phi_m - \phi_n + \delta/2} \sin \frac{|u|}{2} H_1^{(2)} \left(2 k_0 a \sin \frac{|u|}{2} \right) du, & m \neq n. \end{cases}$$

Note that the matrix entries depend on m and n only through the difference $m - n$, allowing for efficient computation.

▶ **Example 7.33:** Current induced on a conducting circular cylinder by a plane wave found using the MFIE

Consider a conducting circular cylinder of radius $a/\lambda = 2$ and a TE plane wave incident at $\phi_0 = 180°$ with amplitude $\tilde{E}_0 = 1$ V/m. Compute the magnitude and phase of the current induced on the cylinder by solving the MFIE. Compare to the results from the EFIE.

Solution: Figure 7.61 compares the amplitude of the current found by solving the EFIE and the MFIE, both with $N = 200$ partitions. The methods are in excellent agreement. However, the matrix entries for the MFIE solution are considerably simpler. Similar agreement is seen in Figure 7.62 for the phase of the current.

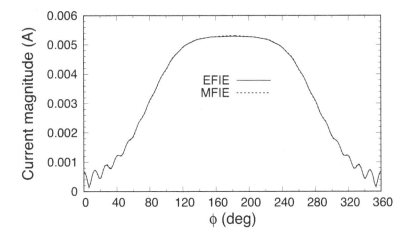

FIGURE 7.61
Magnitude of current induced on a circular cylinder of radius $a/\lambda = 2$ by a TE plane wave incident at $\phi_0 = 180°$. $\tilde{E}_0 = 1$ V/m. $N = 200$ partitions used.

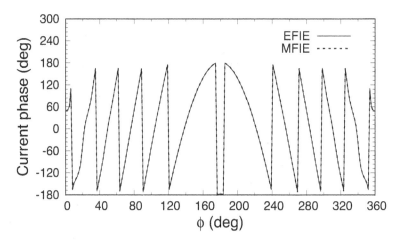

FIGURE 7.62

Phase of current induced on a circular cylinder of radius $a/\lambda = 2$ by a TE plane wave incident at $\phi_0 = 180°$. $\tilde{E}_0 = 1$ V/m. $N = 200$ partitions used. ◄

7.5   Scattering by a penetrable cylinder

As an example of solving an integral equation for a penetrable object, consider a material cylinder of arbitrary cross-sectional shape, excited by a TM-polarized incident plane wave (Figure 7.63). The cylinder has uniform permeability μ_0, nonuniform conductivity $\sigma(\mathbf{r})$, and nonuniform permittivity $\epsilon(\mathbf{r}) = \epsilon_0 \epsilon_r(\mathbf{r})$. The incident field excites polarization and conduction currents within the cylinder. To determine the resulting scattered field, we use the volume EFIE derived in § 7.1.6.

We can adapt (7.25) to the two-dimensional geometry of Figure 7.63 by using the two-dimensional Green's function (7.124). The z-directed incident electric field excites a z-directed equivalent current $\tilde{\mathbf{J}}^{eq}$ which, being z-invariant, has zero divergence. Hence

$$\tilde{E}_z(\boldsymbol{\rho}) = -j\omega\mu_0 \int_S \tilde{J}_z^{eq}(\boldsymbol{\rho}')\tilde{G}_{2D}(\boldsymbol{\rho}|\boldsymbol{\rho}') \, dS' + \tilde{E}_z^i(\boldsymbol{\rho}) \qquad (\boldsymbol{\rho} \in S).$$

By (7.23) we obtain

$$\tilde{E}_z(\boldsymbol{\rho}) = -j\omega\mu_0 \int_S f(\boldsymbol{\rho}')\tilde{E}_z(\boldsymbol{\rho}')\tilde{G}_{2D}(\boldsymbol{\rho}|\boldsymbol{\rho}') \, dS' + \tilde{E}_z^i(\boldsymbol{\rho}) \qquad (\boldsymbol{\rho} \in S), \qquad (7.209)$$

where

$$f(\boldsymbol{\rho}) = \sigma(\boldsymbol{\rho}) + j\omega[\epsilon(\boldsymbol{\rho}) - \epsilon_0].$$

The integral equations describing scattering from a PEC object, considered earlier in this chapter, are surface integral equations. In a two-dimensional situation they require the current to be found on a contour that bounds the cross-section of the object. Since (7.209) arises from a volume integral equation, in this two-dimensional situation we must determine the equivalent current over the cross-sectional surface of the cylinder. A simple approach is to partition the cross-section into small subregions, such as rectangles or

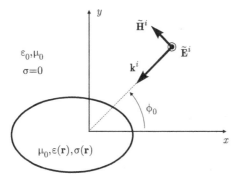

FIGURE 7.63
TM plane wave incident on a penetrable cylinder.

triangles, within which both the material properties and total field are assumed constant. By point-matching the integral equation at some point within each subregion, we create a system of linear equations

$$\tilde{E}_z(\boldsymbol{\rho}_m) = -j\omega\mu_0 \sum_{n=1}^{N} f(\boldsymbol{\rho}_n)\tilde{E}_z(\boldsymbol{\rho}_n) \int_{S_n} \tilde{G}_{2D}(\boldsymbol{\rho}_m|\boldsymbol{\rho}') \, dS' + \tilde{E}_z^i(\boldsymbol{\rho}_m) \qquad (m = 1, \ldots, N).$$

(7.210)

Here $\boldsymbol{\rho}_n$ is the transverse position vector locating a point of S_n, the domain of the nth partition. Write $a_n = \tilde{E}_z(\boldsymbol{\rho}_n)$ and $f_n = f(\boldsymbol{\rho}_n)$. Then (7.210) can be written as a matrix equation

$$\sum_{n=1}^{N} A_{mn} a_n = b_m,$$

(7.211)

where

$$b_m = \tilde{E}_0 e^{jk_0(x_m \cos\phi_0 + y_m \sin\phi_0)}$$

by (7.128), and where

$$A_{mn} = \delta_{mn} + j\omega\mu_0 f_n W_{mn}, \qquad W_{mn} = \int_{S_n} \tilde{G}_{2D}(\boldsymbol{\rho}_m|\boldsymbol{\rho}') \, dS'.$$

For simplicity, let us consider the case where each subregion is a square of side Δ. An arbitrary shape may be approximately decomposed into many squares as shown in Figure 7.64. It is clear that squares do not form a smooth boundary when representing cylinders with curved bounding contours (*staircasing effect*). Other subregion shapes, such as triangles, usually conform to the boundary better. But squares are easier to use, and as their number increases we get a reasonably accurate representation of the boundary.

With squares we have

$$W_{mn} = \int_{y_n-\Delta/2}^{y_n+\Delta/2} \int_{x_n-\Delta/2}^{x_n+\Delta/2} \frac{1}{4j} H_0^{(2)}\left(k_0\sqrt{(x_m-x')^2 + (y_m-y')^2}\right) dx' \, dy', \qquad (7.212)$$

where (x_m, y_m) is the center of the mth partition. Note the use of the wavenumber k_0 in the Green's function, since the equivalent current is assumed to be embedded in

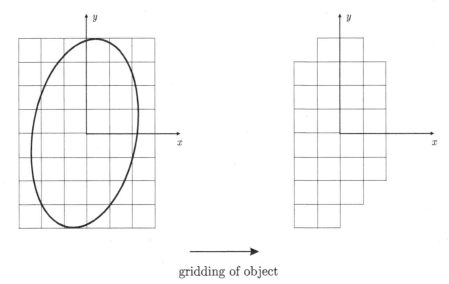

gridding of object

FIGURE 7.64
Gridding a cylindrical object into squares.

free space. The integrals in (7.212) may be computed numerically, but Richmond [159] suggests a method for approximating them in closed form. If the partitions are circular instead of square, the integrals in (7.212) may be computed analytically. If we take the area of the circle to be the same as that of the square, the analytic result for the circle may be used to approximate the integral over the square. To this end the radius a_Δ of the circle must be $a_\Delta = \Delta/\sqrt{\pi}$.

For a circular partition centered at (x_n, y_n) we can write

$$x' = x_n + p' \cos \xi', \qquad y' = y_n + p' \cos \xi',$$

where (p, ξ) describes a point within the circle in a polar coordinate system with its origin at the center of the circle. Then

$$x_m - x' = x_{mn} - p' \cos \xi', \qquad y_m - x' = y_{mn} - p' \sin \xi',$$

where $x_{mn} = x_m - x_n$ and $y_{mn} = y_m - y_n$. Thus

$$(x_m - x')^2 + (y_m - y')^2 = \rho_{mn}^2 + p'^2 - 2p' \rho_{mn} \cos(\phi_{mn} - \xi'),$$

where $x_{mn} = \rho_{mn} \cos \phi_{mn}$ and $y_{mn} = \rho_{mn} \sin \phi_{mn}$, such that

$$\rho_{mn} = \sqrt{(x_m - x_n)^2 + (y_m - y_n)^2}.$$

With this we have

$$W_{mn} = \frac{1}{4j} \int_0^{2\pi} \int_0^{a_\Delta} H_0^{(2)}\left(k_0 \sqrt{\rho_{mn}^2 + p'^2 - 2p' \rho_{mn} \cos(\phi_{mn} - \xi')}\right) p' \, dp' \, d\xi',$$

or

$$W_{mn} = \frac{1}{4j} \int_0^{2\pi} \int_0^{a_\Delta} H_0^{(2)}\left(k_0 \sqrt{\rho_{mn}^2 + p'^2 - 2p' \rho_{mn} \cos \xi'}\right) p' \, dp' \, d\xi', \qquad (7.213)$$

since the integrand is periodic in ξ' with period 2π.

To compute W_{mn}, we start with (E.101) and write

$$H_0^{(2)}\left(k_0\sqrt{\rho_{mn}^2 + p'^2 - 2p'\rho_{mn}\cos\xi'}\right) = J_0(k_0 p')H_0^{(2)}(k_0\rho_{mn})$$

$$+ 2\sum_{\ell=1}^{\infty} J_\ell(k_0 p')H_\ell^{(2)}(k_0\rho_{mn})\cos\ell\xi'.$$

Substituting this into (7.213), we find that the integral over ξ' for each term $\ell \geq 1$ in the series is zero, leaving just a contribution from the leading term. When $m \neq n$ the result is

$$W_{mn} = \frac{1}{4j}(2\pi)H_0^{(2)}(k_0\rho_{mn})\int_0^{a_\Delta} J_0(k_0 p')p'\,dp' = \frac{\pi a_\Delta}{2jk_0}H_0^{(2)}(k_0\rho_{mn})J_1(k_0 a_\Delta)$$

where (E.106) was used to compute the integral. When $m = n$ we employ (7.213) directly to obtain

$$W_{mm} = \frac{1}{4j}(2\pi)\int_0^a H_0^{(2)}(k_0 p')p'\,dp' = \frac{\pi a_\Delta}{2jk_0}H_1^{(2)}(k_0 a_\Delta) - \frac{1}{k_0^2}.$$

Here we have used the fact that

$$\lim_{u\to 0} uH_1^{(2)}(u) = j2/\pi.$$

For a nonconducting, homogeneous body of permittivity $\epsilon = \epsilon_r\epsilon_0$, the matrix elements reduce to

$$A_{mn} = \begin{cases} \left(\frac{j}{2}\right)\pi k_0 a_\Delta(\epsilon_r - 1)H_0^{(2)}(k_0\rho_{mn})J_1(k_0 a_\Delta), & m \neq n, \\ 1 + (\epsilon_r - 1)\frac{j}{2}[\pi k_0 a_\Delta H_1^{(2)}(k_0 a_\Delta) - 2j], & m = n. \end{cases} \tag{7.214}$$

Once the equivalent current has been determined from (7.211), the scattered electric field may be obtained from

$$\tilde{E}_z^s(\boldsymbol{\rho}) = -j\omega\mu_0 \int_S f(\boldsymbol{\rho}')\tilde{G}_{2D}(\boldsymbol{\rho}|\boldsymbol{\rho}')\tilde{E}_z(\boldsymbol{\rho}')\,dS',$$

which becomes

$$\tilde{E}_z^s(\rho,\phi) = -j\omega\mu_0 \sum_{n=1}^N a_n f_n \int_{S_n} \frac{1}{4j}H_0^{(2)}(k_0 R)\,dx'\,dy' \tag{7.215}$$

where

$$R^2 = (x - x')^2 + (y - y')^2 = \rho^2 + \rho'^2 - 2x'\rho\cos\phi - 2y'\rho\sin\phi.$$

In the far zone where $\rho \gg \rho'$, we have $R \approx \rho - x'\cos\phi - y'\sin\phi$. Substituting this into (7.215) and using (E.64), we get

$$\tilde{E}_z^s(\rho,\phi) = -\frac{k_0\eta_0}{4}\sqrt{\frac{2j}{\pi k_0}}\frac{e^{-jk_0\rho}}{\sqrt{\rho}}\sum_{n=1}^N a_n f_n \int_{x_n-\Delta/2}^{x_n+\Delta/2} e^{jk_0 x'\cos\phi}\,dx' \int_{y_n-\Delta/2}^{y_n+\Delta/2} e^{jk_0 y'\sin\phi}\,dy'.$$

Integration gives

$$\tilde{E}_z^s(\rho,\phi) = -\frac{k_0\eta_0}{4}\sqrt{\frac{2j}{\pi k_0}}\frac{e^{-jk_0\rho}}{\sqrt{\rho}}\Delta^2\left[\frac{\sin\left(\frac{1}{2}k_0\Delta\cos\phi\right)}{\frac{1}{2}k_0\Delta\cos\phi}\right]\left[\frac{\sin\left(\frac{1}{2}k_0\Delta\sin\phi\right)}{\frac{1}{2}k_0\Delta\sin\phi}\right].$$

$$\cdot \sum_{n=1}^N a_n f_n e^{jk_0(x_n\cos\phi + y_n\sin\phi)}. \tag{7.216}$$

To find the radar cross-sectional width of the penetrable cylinder, we merely substitute (7.216) into (7.137):

$$\frac{\sigma_{2D}}{\lambda_0} = 2\pi^3 \left(\frac{\Delta}{\lambda_0}\right)^4 \left[\frac{\sin\left(\frac{1}{2}k_0\Delta\cos\phi\right)}{\frac{1}{2}k_0\Delta\cos\phi}\right]^2 \left[\frac{\sin\left(\frac{1}{2}k_0\Delta\sin\phi\right)}{\frac{1}{2}k_0\Delta\sin\phi}\right]^2 \cdot$$

$$\cdot \left|\sum_{n=1}^{N} \frac{a_n}{\tilde{E}_0}\left(f_n\frac{\eta_0}{k_0}\right)e^{jk_0(x_n\cos\phi+y_n\sin\phi)}\right|^2 . \tag{7.217}$$

▶ **Example 7.34:** Equivalent current induced within a circular dielectric cylinder by a plane wave — TM case

Consider a dielectric cylinder having $\epsilon_r = 4$ and $a = 0.25$ m, illuminated from $\phi_0 = 180°$ at 300 MHz with $\tilde{E}_0 = 1$ V/m. The electrical diameter of the cylinder is $2a/\lambda_0 = 0.5$, or $2a/\lambda = 1$, where λ is the wavelength in the material medium. Compute the magnitude of the equivalent current induced in the cylinder.

Solution: Figure 4.37 shows a lossless, homogeneous circular material cylinder with radius a and material properties $\mu = \mu_0$, $\epsilon = \epsilon_r\epsilon_0$, and $\sigma = 0$, illuminated by a TM-polarized plane wave. The cylinder is inscribed in a square of side $2a$ as shown in Figure 7.65, and the square is partitioned into N_e^2 smaller squares of side $\Delta = 2a/N_e$. The center of each partition is computed; if it lies within the circle, the partition is taken as part of the material body and included in the computation. Otherwise it is not included. Hence we have $N \le N_e^2$ where N is the number of partitions used to represent the cylinder.

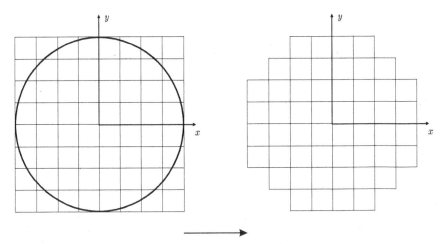

gridding of circular cylinder

FIGURE 7.65
Gridding of a circular cylinder into squares.

Since the cylinder is lossless and homogeneous, the MoM matrix entries are given by (7.214). Solution of (7.211) yields values for the total field at the partition center. From (7.23), the equivalent current is easily computed. Figure 7.66 shows the equivalent current within the cylinder found using $N_e = 50$.

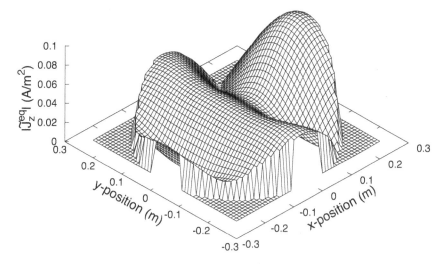

FIGURE 7.66

Magnitude of equivalent current induced within a circular dielectric cylinder of permittivity $\epsilon_r = 4$ and radius $a = 0.25$ m at 300 MHz by a TM plane wave incident at $\phi_0 = 180°$. $\tilde{E}_0 = 1$ V/m. A grid size of $N_e = 50$ partitions is used for the MoM calculation. ◄

▶ **Example 7.35**: Scattering width of a circular dielectric cylinder — TM case

Consider a dielectric cylinder having $\epsilon_r = 4$ and $a = 0.25$ m, illuminated from $\phi_0 = 180°$ at 300 MHz with $\tilde{E}_0 = 1$ V/m. The electrical diameter of the cylinder is $2a/\lambda_0 = 0.5$, or $2a/\lambda = 1$, where λ is the wavelength in the material medium. Compute the scattering width of the cylinder using MoM and compare to the result from the series solution for the scattered field.

Solution: The scattering width of the cylinder is obtained from (7.217). Figure 7.67 shows the bistatic scattering width found by illuminating the cylinder from $\phi_0 = 180°$ and varying the scatter angle ϕ. A partitioning of $N_e = 40$ is used. Also shown is the series solution. By (4.353) we have the scattered field external to the cylinder:

$$\tilde{E}_z^s = -\tilde{E}_0 \sum_{n=0}^{\infty} \epsilon_n j^{-n} \bar{D}_n \cos n\phi H_n^{(2)}(k_0\rho),$$

where

$$\bar{D}_n = \frac{\sqrt{\epsilon_r} J_n'(ka) J_n(k_0 a) - J_n'(k_0 a) J_n(ka)}{\sqrt{\epsilon_r} J_n'(ka) H_n^{(2)}(k_0 a) - H_n^{(2)'}(k_0 a) J_n(ka)}.$$

Note the use of the wavenumber for the cylinder medium:

$$k = k_0 \sqrt{\epsilon_r}.$$

In the far zone ($\rho \gg a$) we have by (E.64)

$$\tilde{E}_z^s \approx -\tilde{E}_0 \sqrt{\frac{2j}{\pi k_0}} \frac{e^{-jk_0\rho}}{\sqrt{\rho}} \sum_{n=0}^{\infty} \epsilon_n \bar{D}_n \cos n\phi.$$

Substitution into (7.137) gives the scattering width

$$\frac{\sigma_{2D}}{\lambda_0} = \frac{2}{\pi} \left| \sum_{n=0}^{\infty} \epsilon_n \bar{D}_n \cos n\phi \right|^2 .$$

Figure 7.67 shows excellent agreement between the scattering width computed using the MoM and the series, except at the bottom of the deep null near $\phi = 115°$.

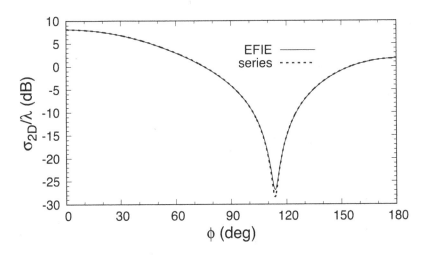

FIGURE 7.67

Bistatic scattering width of a circular dielectric cylinder of permittivity $\epsilon_r = 4$ and radius $a = 0.25$ m at 300 MHz illuminated by a TM plane wave incident at $\phi_0 = 180°$. A grid size of $N_e = 40$ partitions was used for the MoM calculation, while $N = 60$ terms were used in the series computation. ◄

▶ **Example 7.36:** Convergence of the scattering width of a circular dielectric cylinder — TM case

Consider a dielectric cylinder having $\epsilon_r = 4$ and $a = 0.25$ m, illuminated from $\phi_0 = 180°$ at 300 MHz with $\tilde{E}_0 = 1$ V/m. Explore how the scattering width of the cylinder computed using MoM depends on the partition size.

Solution: To study how the partition density affects the MoM solution, we compute the backscattering radar width ($\phi = \phi_0$) for various values of N_e (Figure 7.68). As the grid density increases, the scattering width approaches the series solution. However, the values of σ_{2D} show considerable oscillation as N_e is changed incrementally. This is due to the staircasing effect. When N_e is changed by just one, the geometry of the partitioning changes significantly. This effect can be compensated somewhat by computing the total area of the partitions compared to the area of the circular cylinder, and adjusting σ_{2D} appropriately. However, this does not eliminate the oscillations.

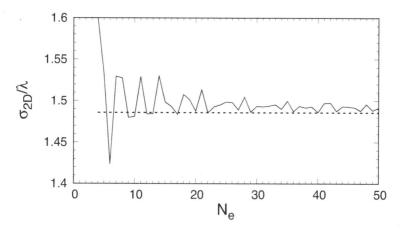

FIGURE 7.68

Backscattering width of a circular dielectric cylinder of permittivity $\epsilon_r = 4$ and radius $a = 0.25$ m at 300 MHz computed using various grid densities. ◄

▶ **Example 7.37:** Equivalent current induced within a square dielectric cylinder by a plane wave — TM case

Consider a square dielectric cylinder with the material properties of the circular dielectric cylinder in Example 7.35. The side length is chosen to be $2a$ so that the circular cylinder of that example may be inscribed within it. Compute the magnitude of the equivalent current induced in the cylinder.

Solution: Figure 7.69 shows the equivalent current induced within the cylinder. Due to the similarity in size, the current distribution within the square cylinder is similar to that seen in the circular cylinder.

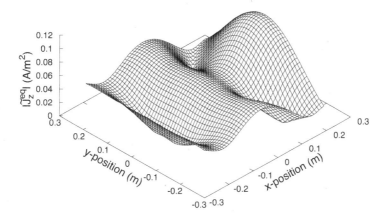

FIGURE 7.69

Magnitude of equivalent current induced within a square dielectric cylinder of permittivity $\epsilon_r = 4$ and side length $2a = 0.5$ m at 300 MHz by a TM plane wave incident at $\phi_0 = 180°$. $\tilde{E}_0 = 1$ V/m. $N_e = 50$ was used for the MoM calculation. ◄

▶ **Example 7.38:** Scattering width of a square dielectric cylinder — TM case

Compute the scattering width of the square dielectric cylinder considered in Example 7.37. Compare to the scattering width of the circular cylinder.

Solution: Figure 7.70 shows the bistatic scattering width of the square cylinder vs. observation angle. With no series solution available, comparison to an analytic scattering width is not possible. Instead, comparison is made to the circular cylinder from Example 7.37. The scattering widths are similar except for the depth and position of the null. Note that the square cylinder has a larger cross-sectional area but a smaller backscattering width ($\phi = 180°$) than the circular cylinder. In contrast, its forward scattering width ($\phi = 0°$) is larger.

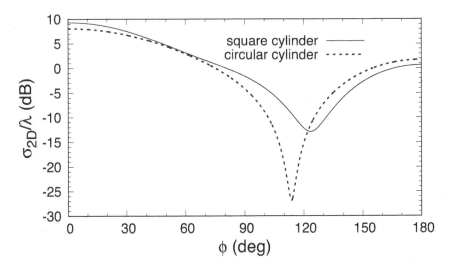

FIGURE 7.70
Bistatic scattering width of a square dielectric cylinder of permittivity $\epsilon_r = 4$ and side length $2a = 0.5$ m at 300 MHz illuminated by a TM plane wave incident at $\phi_0 = 180°$. A grid size of $N_e = 40$ partitions was used. Comparison is to a circular cylinder of identical material parameters inscribed within the square cylinder. ◀

▶ **Example 7.39:** Convergence of the scattering width of a square dielectric cylinder — TM case

Consider the square dielectric cylinder of Example 7.37. Explore how the scattering width of the cylinder computed using MoM depends on the partition size.

Solution: To study the effect of the partitioning density, the backscattering width at $\phi_0 = 180°$ was computed using various values of N_e (Figure 7.71). For a square cylinder there is no staircasing effect; the scattering width converges rapidly and monotonically as N_e is increased.

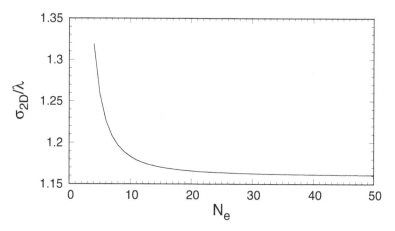

FIGURE 7.71
Backscattering width of a square dielectric cylinder of permittivity $\epsilon_r = 4$ and side length $2a = 0.5$ m at 300 MHz computed at angle $\phi = 180°$ using various grid densities. ◀

▶ **Example 7.40:** Monostatic scattering width of a square dielectric cylinder — TM case

Consider the square dielectric cylinder of Example 7.37. Plot the monostatic scattering width as a function of incidence angle.

Solution: The monostatic scattering width is the same as the backscattering width computed using $\phi = \phi_0$, and is the quantity measured by rotating the cylinder with the measurement antennas coincident and fixed. For a circular cylinder, this is a single number by azimuthal symmetry. For a square cylinder, Figure 7.72 is obtained. A variation of about 5 dB is seen, with the expected periodicity of 90° due to the square symmetry of the cylinder.

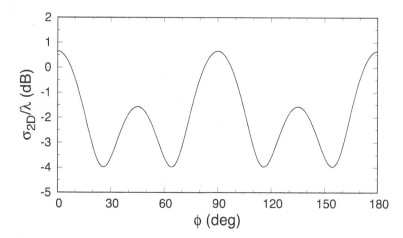

FIGURE 7.72
Monostatic scattering width of a square dielectric cylinder of permittivity $\epsilon_r = 4$ and side length $2a = 0.5$ m at 300 MHz illuminated by a TM plane wave. $N_e = 40$ was used. ◀

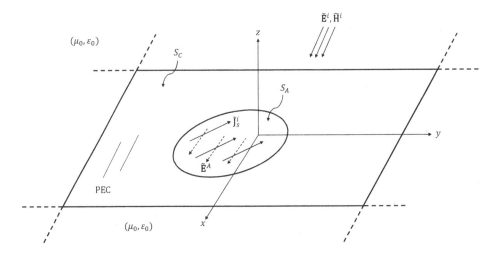

FIGURE 7.73
Aperture in a conducting ground plane.

7.6   Apertures in ground planes

Computing the interaction of electromagnetic waves with apertures in conducting screens is necessary in the solution to a number of radiation and scattering problems. The penetration of plane waves through an aperture in a screen is important in scattering theory as a fundamental problem of diffraction. It is also an important concept in electromagnetic shielding, and is addressed later in this chapter. Radiation from apertures in ground planes is important to antenna theory, as discussed in § 6.4.5. A particularly interesting problem is that of a slot antenna, where the aperture is excited using a current source placed directly in the aperture. When the aperture is present in a thin, infinite, perfectly conducting ground plane, both the scattering and radiation problems may be solved using a magnetic-field integral equation formulated in terms of the unknown aperture electric field.

7.6.1   MFIE for the unknown aperture electric field

Consider an aperture of general shape cut into a thin, infinite, perfectly conducting ground plane (Figure 7.73). The conductor occupies the $z = 0$ plane and is immersed in free space. An impressed surface current $\tilde{\mathbf{J}}_s^i(x, y, \omega)$ lies on the aperture surface S_A, and an impressed electromagnetic field $\tilde{\mathbf{E}}^i(\mathbf{r}, \omega), \tilde{\mathbf{H}}^i(\mathbf{r}, \omega)$ exists in the region $z > 0$. The radiation problem is formulated assuming $\tilde{\mathbf{E}}^i = 0$ and $\tilde{\mathbf{H}}^i = 0$, while the scattering problem assumes $\tilde{\mathbf{J}}_s^i = 0$. It is assumed that the impressed field satisfies the boundary conditions on the perfectly conducting screen everywhere in the $z = 0$ plane, and thus consists of an incident and a reflected field. (Alternatively, the total impressed field may be viewed as arising from sources above the ground plane, and by images of sources below the ground plane).

The presence of the impressed field or the aperture current will induce currents on the conducting surface S_C, which will in turn produce a scattered field $\tilde{\mathbf{E}}^s(\mathbf{r}, \omega), \tilde{\mathbf{H}}^s(\mathbf{r}, \omega)$.

(For expediency, we suppress the functional dependence on ω in the expressions below.) By symmetry, the horizontal (x, y) components of the scattered electric fields are even about $z = 0$, while the horizontal components of the scattered magnetic fields are odd. The boundary condition on S_C thus requires $\tilde{\mathbf{E}}(x, y, z = 0^+) = \tilde{\mathbf{E}}(x, y, z = 0^-) = 0$, while on S_A the electric field must be continuous. The tangential electric field is unknown in the aperture; designating it as $\tilde{\mathbf{E}}^A(x, y)$, we have

$$\tilde{\mathbf{E}}(x, y, 0^+) = \tilde{\mathbf{E}}(x, y, 0^-) = \tilde{\mathbf{E}}^A(x, y) \qquad ((x, y) \in S_A).$$

To complete the specification of the field over S_A we apply the boundary condition on tangential magnetic field:

$$\hat{\mathbf{z}} \times [\tilde{\mathbf{H}}^i(x, y, 0) + \tilde{\mathbf{H}}^s(x, y, 0^+) - \tilde{\mathbf{H}}^s(x, y, 0^-)] = \tilde{\mathbf{J}}_s^i(x, y) \qquad ((x, y) \in S_A).$$

Since $\hat{\mathbf{z}} \times \tilde{\mathbf{H}}^s$ is odd about $z = 0$, this reduces to

$$\hat{\mathbf{z}} \times [\tilde{\mathbf{H}}^i(x, y, 0) + 2\tilde{\mathbf{H}}^s(x, y, 0^+)] = \tilde{\mathbf{J}}_s^i(x, y) \qquad ((x, y) \in S_A). \tag{7.218}$$

To find the scattered magnetic field, we use Schelkunoff's equivalence principle (Section 6.3.4). The vector potential produced by the unknown aperture field is by (6.50)

$$\tilde{\mathbf{A}}_h(\mathbf{r}) = \int_{S_A} \epsilon_0 [-2\hat{\mathbf{z}} \times \tilde{\mathbf{E}}^A(x', y')] \tilde{G}(\mathbf{r}|x', y', 0) \, dS',$$

and, by (6.43), the scattered magnetic field is

$$\tilde{\mathbf{H}}^s(\mathbf{r}) = -j\frac{\omega}{k_0^2} [\nabla(\nabla \cdot \tilde{\mathbf{A}}_h(\mathbf{r})) + k_0^2 \tilde{\mathbf{A}}_h(\mathbf{r})].$$

The divergence of the vector potential may be written as

$$\nabla \cdot \tilde{\mathbf{A}}_h(\mathbf{r}) = \epsilon_0 \int_{S_A} [-2\hat{\mathbf{z}} \times \tilde{\mathbf{E}}^A(x', y')] \cdot \nabla \tilde{G}(\mathbf{r}|x', y', 0) \, dS',$$

where we have used (B.48). Thus, the scattered field is

$$\tilde{\mathbf{H}}^s(\mathbf{r}) = j\frac{2}{\omega\mu_0} \int_{S_A} \left\{ \nabla \left([\hat{\mathbf{z}} \times \tilde{\mathbf{E}}^A(x', y')] \cdot \nabla \tilde{G}(\mathbf{r}|x', y', 0) \right) \right.$$
$$\left. + k_0^2 [\hat{\mathbf{z}} \times \tilde{\mathbf{E}}^A(x', y')] \tilde{G}(\mathbf{r}|x', y', 0) \right\} dS'. \tag{7.219}$$

Finally, substitution of (7.219) into (7.218) gives

$$\frac{4j}{\omega\mu_0} \int_{S_A} \left\{ \hat{\mathbf{z}} \times \nabla \left([\hat{\mathbf{z}} \times \tilde{\mathbf{E}}^A(x', y')] \cdot \nabla \tilde{G}(\mathbf{r}|x', y', 0) \right) - k_0^2 \tilde{\mathbf{E}}^A(x', y') \tilde{G}(\mathbf{r}|x', y', 0) \right\} dS'$$
$$= \tilde{\mathbf{J}}_s^i(x, y) - \hat{\mathbf{z}} \times \tilde{\mathbf{H}}^i(x, y, 0) \qquad ((x, y) \in S_A). \tag{7.220}$$

This is the MFIE for the aperture electric field $\tilde{\mathbf{E}}^A$.

7.6.2   MFIE for a narrow slot

A classic problem is scattering from a narrow slot in a ground plane, or radiation by a narrow slot in a ground plane. Consider the slot shown in Figure 7.74. When the

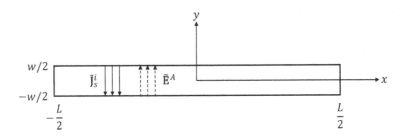

FIGURE 7.74
Narrow slot aperture.

slot is narrow ($w \ll L$), the aperture electric field is primarily y-directed. Assuming the impressed aperture current is y-directed and the impressed magnetic field is x-directed, the aperture magnetic field will be predominantly x-directed, and each of the terms in the MFIE will be y-directed. Writing $\tilde{\mathbf{E}}^A = \hat{\mathbf{y}}\tilde{E}_y^A$ and dotting (7.220) with $\hat{\mathbf{y}}$, we get

$$\frac{4j}{\omega\mu_0}\int_{S_A}\hat{\mathbf{y}}\cdot\left\{\hat{\mathbf{z}}\times\nabla\left([-\hat{\mathbf{x}}\tilde{E}_y^A(x',y')]\cdot\nabla\tilde{G}(\mathbf{r}|x',y',0)\right)-k_0^2\hat{\mathbf{y}}\tilde{E}_y^A(x',y')\tilde{G}(\mathbf{r}|x',y',0)\right\}dS'$$

$$= \tilde{J}_{sy}^i(x,y) - \hat{\mathbf{y}}\cdot[\hat{\mathbf{z}}\times\hat{\mathbf{x}}\tilde{H}_x^i(x,y,0)] \qquad ((x,y)\in S_A).$$

Use of

$$\hat{\mathbf{y}}\cdot[\hat{\mathbf{z}}\times\nabla f] = (\hat{\mathbf{y}}\times\hat{\mathbf{z}})\cdot\nabla f = \hat{\mathbf{x}}\cdot\nabla f = \frac{\partial f}{\partial x}$$

gives

$$\frac{4j}{\omega\mu_0}\int_{S_A}\left[-\tilde{E}_y^A(x',y')\frac{\partial^2\tilde{G}(x,y,0|x',y',0)}{\partial x^2}-k_0^2\tilde{E}_y^A(x',y')\tilde{G}(x,y,0|x',y',0)\right]dS'$$

$$= \tilde{J}_{sy}^i(x,y) - \tilde{H}_x^i(x,y,0) \qquad ((x,y)\in S_A)$$

or

$$\frac{4j}{\omega\mu_0}\left(\frac{\partial^2}{\partial x^2}+k_0^2\right)\int_{S_A}\tilde{E}_y^A(x',y')\tilde{G}(x,y,0|x',y',0)\,dS'$$

$$= -\tilde{J}_{sy}^i(x,y) + \tilde{H}_x^i(x,y,0) \qquad ((x,y)\in S_A). \tag{7.221}$$

For a narrow slot, the aperture field dependence may be separated into the product of two functions,

$$\tilde{E}_y^A(x,y) = \tilde{V}(x)f(y), \tag{7.222}$$

where $\tilde{V}(x)$ is the slot voltage defined by

$$\tilde{V}(x) = \int_{-w/2}^{w/2}\tilde{E}_y^A(x,y)\,dy.$$

This implies

$$\tilde{V}(x) = \tilde{V}(x)\int_{-w/2}^{w/2}f(y)\,dy.$$

or

$$\int_{-w/2}^{w/2} f(y)\,dy = 1. \tag{7.223}$$

Substituting (7.222) into (7.221), multiplying both sides by a weighting function $g(y)$, and integrating across the slot then gives

$$\frac{4j}{\omega\mu_0}\left(\frac{\partial^2}{\partial x^2} + k_0^2\right)\int_{-L/2}^{L/2}\tilde{V}(x')\left[\int_{-w/2}^{w/2}\int_{-w/2}^{w/2} f(y')g(y)\tilde{G}(x,y,0|x',y',0)\,dy\,dy'\right]dx'$$

$$= -\int_{-w/2}^{w/2} g(y)\tilde{J}_{sy}^i(x,y)\,dy + \int_{-w/2}^{w/2} g(y)\tilde{H}_x^i(x,y,0)\,dy \qquad (-L/2 \le x \le L/2). \tag{7.224}$$

The choice of $g(y)$ will be discussed later. Now, define

$$K(x - x') = \int_{-w/2}^{w/2}\int_{-w/2}^{w/2} f(y')g(y)\tilde{G}(x,y,0|x',y',0)\,dy\,dy' \tag{7.225}$$

and

$$\tilde{H}(x) = \frac{j\omega\mu_0}{4}\int_{-w/2}^{w/2} g(y)\tilde{J}_{sy}^i(x,y)\,dy - \frac{j\omega\mu_0}{4}\int_{-w/2}^{w/2} g(y)\tilde{H}_x^i(x,y,0)\,dy. \tag{7.226}$$

With these, (7.224) becomes

$$\left(\frac{\partial^2}{\partial x^2} + k_0^2\right)\int_{-L/2}^{L/2}\tilde{V}(x')K(x - x')\,dx' = \tilde{H}(x) \qquad (-L/2 \le x \le L/2).$$

Finally, solving the differential equation (as was done in Section 7.3.1.1 for the dipole antenna) gives

$$\int_{-L/2}^{L/2}\tilde{V}(x')K(x - x')\,dx' = C_1 \sin k_0 z + C_2 \cos k_0 z$$

$$+ \frac{1}{k_0}\int_{-L/2}^{x}\tilde{H}(u)\sin k_0(x - u)\,du \qquad (-L/2 \le x \le L/2), \tag{7.227}$$

which is a Hallén-type integral equation for the slot voltage.

7.6.2.1 Computing the kernel $K(x - x')$

To compute the kernel of the integral equation (7.227) we must specify the aperture field function $f(y)$ and the weighting function $g(y)$. Because the slot is narrow, it is sufficient to point match at the center of the slot by choosing $g(y) = \delta(y)$. The kernel (7.225) becomes

$$K(x - x') = \int_{-w/2}^{w/2} f(y')\tilde{G}(x,0,0|x',y',0)\,dy'.$$

The choice of $f(y)$ must reflect the physical behavior of the aperture field. The current on the ground plane near the edge of the slot obeys a square-root-type edge singularity (§ 4.11.8.6), and thus the electric field in the slot also obeys this condition near the edge. A typical choice to model this behavior is

$$f(y) = \frac{2}{\pi w\sqrt{1 - \left(\frac{y}{w/2}\right)^2}},$$

which obeys (7.223). Using this, the kernel becomes

$$K(x - x') = \frac{2}{\pi w} \int_{-w/2}^{w/2} \frac{1}{\sqrt{1 - \left(\frac{y'}{w/2}\right)^2}} \tilde{G}(x, 0, 0 | x', y', 0) \, dy'. \qquad (7.228)$$

If the free-space Green's function (5.70) is used to evaluate the kernel, both the spatial integral over x' in (7.228) and the integral over x that arises when applying the method of moments must be computed numerically. This can be troublesome because of the inherent singularities in the Green's function and in $f(y)$. Through the spectral representation of $\tilde{G}$ given by (A.56), all of the spatial integrals may be computed in closed form, leaving a single spectral integral to calculate.

Evaluation of the Green's function (A.56) at $z = z' = y = 0$ gives

$$\tilde{G}(x, 0, 0 | x', y', 0) = \frac{1}{(2\pi)^2} \int_{-\infty}^{\infty} \int_{-\infty}^{\infty} \frac{1}{2p} e^{jk_x(x-x')} e^{jk_y y'} \, dk_x \, dk_y,$$

where $p^2 = k_x^2 + k_y^2 - k_0^2$. Substitution into (7.228) then gives

$$K(x - x') = \frac{1}{(2\pi)^2} \int_{-\infty}^{\infty} \int_{-\infty}^{\infty} e^{jk_x(x-x')} \left[\frac{4}{\pi w} \int_0^{w/2} \frac{\cos(k_y y')}{\sqrt{1 - \left(\frac{y'}{w/2}\right)^2}} \, dy' \right] \frac{dk_x \, dk_y}{2p}.$$

With the substitution $u = y'/(w/2)$, the inner integral becomes

$$\frac{4}{\pi w} \int_0^{w/2} \frac{\cos(k_y y')}{\sqrt{1 - \left(\frac{y'}{w/2}\right)^2}} \, dy' = \frac{2}{\pi} \int_0^1 \frac{\cos\left(k_y \frac{w}{2} u\right)}{\sqrt{1 - u^2}} \, du.$$

But [74]

$$\int_0^1 \frac{\cos(ax)}{\sqrt{1 - x^2}} \, dx = \frac{\pi}{2} J_0(a)$$

where $J_0(x)$ is the ordinary Bessel function of the first kind and order zero. Thus, the kernel of the integral equation is

$$K(x - x') = \frac{1}{(2\pi)^2} \int_{-\infty}^{\infty} e^{jk_x(x-x')} \left[\int_0^{\infty} \frac{J_0\left(k_y \frac{w}{2}\right)}{\sqrt{k_y^2 + (k_x^2 - k_0^2)}} \, dk_y \right] dk_x.$$

Finally, use of [74]

$$\int_0^{\infty} \frac{J_0(xy)}{\sqrt{x^2 + a^2}} \, dx = I_0\left(\frac{ay}{2}\right) K_0\left(\frac{ay}{2}\right),$$

where $I_0(x)$ is the modified Bessel function of the first kind and $K_0(x)$ is the modified Bessel function of the second kind, gives

$$K(x - x') = \frac{1}{2\pi^2} \int_0^{\infty} \cos[k_x(x - x')] I(k_x) \, dk_x \qquad (7.229)$$

where

$$I(k_x) = I_0\left(\frac{w}{4}\sqrt{k_x^2 - k_0^2}\right) K_0\left(\frac{w}{4}\sqrt{k_x^2 - k_0^2}\right). \qquad (7.230)$$

7.6.3 Solution for the slot voltage using the method of moments

To solve (7.227) using the method of moments, we expand the slot voltage using pulse basis functions and point match at the center of the partitions, just as we did to solve Hallén's equation for a dipole antenna in Section 7.3.1.2. Let

$$\tilde{V}(x) = \sum_{n=1}^{N} a_n P_n(x) \tag{7.231}$$

where $P_n(x)$ is the nth pulse function given by

$$P_n(x) = \begin{cases} 1, & (n-1)\Delta \leq x \leq n\Delta, \\ 0, & \text{elsewhere,} \end{cases}$$

with $\Delta = L/N$. That is, we partition the slot into N regions, each with constant voltage a_n. Substituting (7.231) into (7.227) and matching the equation at the points $x_m = -L/2 + (m-1/2)\Delta$, we obtain a system of N equations in $N+2$ unknowns:

$$\sum_{n=1}^{N} a_n A_{mn} + C_1 \sin k_0 z_m + C_2 \cos k_0 z_m = b_m, \tag{7.232}$$

where

$$A_{mn} = \int_{x_n - \Delta/2}^{x_n + \Delta/2} K(x_m - x') \, dx', \tag{7.233}$$

$$b_m = \frac{1}{k_0} \int_{-L/2}^{x_m} \tilde{H}(u) \sin k_0(x_m - u) \, du. \tag{7.234}$$

Two more equations are obtained by imposing the conditions on the voltage at the slot ends. Since the electric field is tangential to the metal edge, the voltage must vanish at the ends:

$$\tilde{V}(-L/2) = \tilde{V}(L/2) = 0.$$

Since the voltages are assumed constant within the partitions, these conditions are most easily invoked by requiring that $a_1 = a_N = 0$. However, a_1 and a_N more accurately approximate the voltages at the *centers* of the partitions. With little additional effort, we can implement the quadratic extrapolation (7.48) to estimate the voltage at the slot ends:

$$\tilde{V}(-L/2) \approx \tfrac{1}{8}[15a_1 - 10a_2 + 3a_3] = 0,$$
$$\tilde{V}(L/2) \approx \tfrac{1}{8}[3a_{N-2} - 10a_{N-1} + 15a_N] = 0.$$

Using these, we obtain a matrix equation of a form identical to (7.64).

7.6.3.1 MoM matrix entries

Substituting the kernel expression (7.229) into (7.233), we find

$$A_{mn} = \frac{1}{2\pi^2} \int_0^\infty \left[\int_{x_n - \Delta/2}^{z_n + \Delta/2} \cos[k_x(x_m - x')] \, dx' \right] I(k_x) \, dk_x.$$

Computing the inner integral then gives

$$A_{mn} = \frac{1}{2\pi^2} F(n - m),$$

where

$$F(q) = \int_0^\infty S_q(k_x) I(k_x)\, dk_x \tag{7.235}$$

and

$$S_q(k_x) = \frac{\sin[k_x(q + \frac{1}{2})\Delta]}{k_x} - \frac{\sin[k_x(q - \frac{1}{2})\Delta]}{k_x}.$$

It is convenient to consider the two cases $k_x^2 < k_0^2$ and $k_x^2 > k_0^2$ when evaluating the expression (7.230) for $I(k_x)$. Using

$$I_0(jx) = J_0(x), \qquad K_0(jx) = -j\frac{\pi}{2} H_0^{(2)}(x)$$

in (7.230), and substituting into (7.235), we obtain

$$
\begin{aligned}
F(q) = -j\frac{\pi}{2} \int_0^{k_0} S_q(k_x) J_0\left(\frac{w}{4}\sqrt{k_0^2 - k_x^2}\right) H_0^{(2)}\left(\frac{w}{4}\sqrt{k_0^2 - k_x^2}\right) dk_x \\
+ \int_{k_0}^\infty S_q(k_x) I_0\left(\frac{w}{4}\sqrt{k_x^2 - k_0^2}\right) K_0\left(\frac{w}{4}\sqrt{k_x^2 - k_0^2}\right) dk_x.
\end{aligned}
\tag{7.236}
$$

These integrals may be calculated numerically, but the latter integral can be slow to converge. Its computation may be accelerated by adding and subtracting an asymptotic form that can be integrated analytically. Note that [1]

$$I_0(x)K_0(x) \sim \frac{1}{2x} \qquad (x \gg 1)$$

and so

$$S_q(k_x) I_0\left(\frac{w}{4}\sqrt{k_x^2 - k_0^2}\right) K_0\left(\frac{w}{4}\sqrt{k_x^2 - k_0^2}\right) \sim \frac{2}{w} \frac{\sin[k_x(q + \frac{1}{2})\Delta] - \sin[k_x(q - \frac{1}{2})\Delta]}{k_x^2}.$$

Using

$$\int_1^\infty \frac{\sin(ax)}{x^2} = \sin(a) - a\,\text{Ci}(a),$$

where

$$\text{Ci}(x) = -\int_x^\infty \frac{\cos t}{t}\, dt$$

is the cosine integral, we get

$$
\begin{aligned}
\frac{2}{w} \int_{k_0}^\infty & \frac{\sin[k_x(q + \frac{1}{2})\Delta] - \sin[k_x(q - \frac{1}{2})\Delta]}{k_x^2} \\
& = \frac{2}{w}\left\{ S_q(k_0) - (q + \tfrac{1}{2})\Delta\,\text{Ci}\left[k_0(q + \tfrac{1}{2})\Delta\right] + (q - \tfrac{1}{2})\Delta\,\text{Ci}\left[k_0(q - \tfrac{1}{2})\Delta\right]\right\}.
\end{aligned}
$$

So (7.236) can be written as

$$F(q) = -j\frac{\pi}{2} \int_0^{k_0} S_q(k_x) J_0 \left(\frac{w}{4}\sqrt{k_0^2 - k_x^2}\right) H_0^{(2)} \left(\frac{w}{4}\sqrt{k_0^2 - k_x^2}\right)$$
$$+ \int_{k_0}^{\infty} \left[S_q(k_x) I_0 \left(\frac{w}{4}\sqrt{k_x^2 - k_0^2}\right) K_0 \left(\frac{w}{4}\sqrt{k_x^2 - k_0^2}\right) - \frac{2}{wk_x}\right]$$
$$+ \frac{2}{w} \left\{ S_q(k_0) - (q + \tfrac{1}{2})\Delta \text{Ci}\left[k_0(q + \tfrac{1}{2})\Delta\right] + (q - \tfrac{1}{2})\Delta \text{Ci}\left[k_0(q - \tfrac{1}{2})\Delta\right] \right\}.$$

It is useful when computing the integrals to note that the product $I_0(x)K_0(x)$ has an asymptotic form [1]

$$I_0(x)K_0(x) \sim \frac{1}{2x} \left\{ 1 - \frac{1}{2}\frac{(-1)}{(2x)^2} + \left[\frac{1}{2}\frac{(-1)}{(2x)^2}\right]\left[\frac{3}{4}\frac{(-9)}{(2x)^2}\right] - \cdots \right\}.$$

Its implementation prevents overflow or underflow when evaluating the Bessel functions with large arguments.

Note that A_{mn} depends on m and n only through the difference $n - m$. This fact significantly reduces the computational expense of filling the matrix. A vector may be filled with the distinct values of A_{mn} for all allowed differences $n - m$, and the matrix filled by selection from this vector.

7.6.3.2 Fields produced by slot voltage — far zone

Once the slot voltage has been found using the MoM, the fields external to the slot may be computed. If the observation point is in the far-zone of the slot, the fields may be computed from the directional weighting function (6.54). We use the equivalent current

$$\tilde{\mathbf{J}}_{ms}^{eq}(x, y) = -2\hat{\mathbf{n}} \times \tilde{\mathbf{E}}^A(x, y) = -2\hat{\mathbf{z}} \times \hat{\mathbf{y}}\tilde{V}(x)f(y) = 2\hat{\mathbf{x}}\tilde{V}(x)f(y)$$

which is valid for observation points with $z > 0$. We also use

$$\hat{\mathbf{r}} \cdot \mathbf{r}' = \hat{\mathbf{r}} \cdot (x'\hat{\mathbf{x}} + y'\hat{\mathbf{y}}) = x'\sin\theta\cos\phi + y'\sin\theta\sin\phi$$

to get

$$\tilde{\mathbf{a}}_h(\theta, \phi) = 2\hat{\mathbf{x}} \int_{-L/2}^{L/2} \tilde{V}(x')e^{jk_0 x'\sin\theta\cos\phi}\,dx' \int_{-w/2}^{w/2} f(y')e^{jk_0 y'\sin\theta\sin\phi}\,dy'.$$

Since $k_0 w \ll 1$, the exponential in the second integral can be approximated by unity. The integral of $f(y)$ is also unity, giving

$$\tilde{\mathbf{a}}_h(\theta, \phi) = 2\hat{\mathbf{x}} \int_{-L/2}^{L/2} \tilde{V}(x')e^{jk_0 x'\sin\theta\cos\phi}\,dx'.$$

Next, substituting (7.231) for $\tilde{V}(x)$ and integrating we have

$$\tilde{\mathbf{a}}_h(\theta, \phi) = 4\hat{\mathbf{x}}\frac{\sin\left(k_0\frac{\Delta}{2}\sin\theta\cos\phi\right)}{k_0\sin\theta\cos\phi} \sum_{n=1}^{N} a_n e^{jk_0 x_n \sin\theta\cos\phi}.$$

This can be used to find the potential from (6.52). Finally, using

$$\hat{\mathbf{r}} \times \hat{\mathbf{x}} = \hat{\boldsymbol{\phi}}\cos\theta\cos\phi + \hat{\boldsymbol{\theta}}\sin\phi$$

in (6.55), we have the far-zone electric field

$$\tilde{\mathbf{E}} = jk_0 \frac{e^{-jk_0 r}}{r} \left(\hat{\boldsymbol{\phi}} \cos\theta \cos\phi + \hat{\boldsymbol{\theta}} \sin\phi \right) \frac{\sin\left(k_0 \frac{\Delta}{2} \sin\theta \cos\phi\right)}{k_0 \sin\theta \cos\phi} \sum_{n=1}^{N} a_n e^{jk_0 x_n \sin\theta \cos\phi}.$$

(7.237)

7.6.3.3 Fields produced by slot voltage — near zone

For points in the near zone we can substitute (6.47) into (6.48) to get

$$\tilde{\mathbf{E}}(\mathbf{r}) = - \int_{S_A} \nabla \times [-2\hat{\mathbf{n}}' \times \tilde{\mathbf{E}}_A(\mathbf{r}')] \tilde{G}(\mathbf{r}|\mathbf{r}') \, dS'.$$

Expanding the curl and using $\nabla \tilde{G} = -\nabla' \tilde{G}$, we have

$$\tilde{\mathbf{E}}(\mathbf{r}) = \int_{S_A} [2\hat{\mathbf{n}}' \times \tilde{\mathbf{E}}^A(\mathbf{r}')] \times \nabla' \tilde{G}(\mathbf{r}|\mathbf{r}') \, dS'.$$

Next we substitute $\tilde{\mathbf{E}}^A(x,y) = \hat{\mathbf{y}} \tilde{V}(x) f(y)$ and use

$$\nabla' \tilde{G} = \hat{\mathbf{R}} \frac{1 + jk_0 R}{4\pi R^2} e^{-jk_0 R}$$

to get

$$\tilde{\mathbf{E}} = \frac{1}{2\pi} \int_{-w/2}^{w/2} \int_{-L/2}^{L/2} [\pm\hat{\mathbf{z}} \times \hat{\mathbf{y}} \tilde{V}(x') f(y')] \times \hat{\mathbf{R}} \frac{1 + jk_0 R}{4\pi R^2} e^{-jk_0 R} \, dx' \, dy',$$

where $+\hat{\mathbf{z}}$ is used for $z > 0$ and $-\hat{\mathbf{z}}$ for $z < 0$. Now, if $k_0 w \ll 1$ and $R \gg w$ we can approximate

$$\mathbf{R} \approx \hat{\mathbf{x}}(x - x') + \hat{\mathbf{y}} y + \hat{\mathbf{z}} z, \qquad R \approx \sqrt{(x - x')^2 + y^2 + z^2}$$

so that

$$\tilde{\mathbf{E}} = \pm \frac{1}{2\pi} \int_{-w/2}^{w/2} f(y') \, dy' \int_{-L/2}^{L/2} [-\hat{\mathbf{x}} \times (\hat{\mathbf{y}} y + \hat{\mathbf{z}} z)] \tilde{V}(x') \frac{1 + jk_0 R}{4\pi R^3} e^{-jk_0 R} \, dx'.$$

Lastly, substitution of (7.231) gives

$$\tilde{\mathbf{E}} = \pm \frac{1}{2\pi} (\hat{\mathbf{y}} z - \hat{\mathbf{z}} y) \sum_{n=1}^{N} a_n \int_{x_n - \Delta/2}^{x_n + \Delta/2} \frac{1 + jk_0 R}{4\pi R^3} e^{-jk_0 R} \, dx',$$

(7.238)

which can be computed by numerical integration.

7.6.4 Radiation by a slot antenna

Radiation from a slot antenna may be computed by setting the incident field to zero in (7.226) and adopting an appropriate model for the impressed aperture current $\tilde{J}_{sy}^i(x,y)$. A simple method for feeding a slot antenna involves a coaxial cable with its center conductor extended across the slot. The current on the center conductor is easily modeled as a filamentary (line) current:

$$\tilde{J}_{sy}^i(x) = -\tilde{I}_0 \delta(x - x_0),$$

where x_0 is the position of the current filament. Then

$$\tilde{H}(x) = -\frac{j\omega\mu_0}{4}\tilde{I}_0\delta(x - x_0)$$

and by (7.234),

$$b_m = -\frac{j\omega\mu_0}{4k_0}\tilde{I}_0\int_{-L/2}^{x_m}\delta(u - x_0)\sin k_0(x_m - u)\,du$$

$$= \begin{cases} -j\frac{\eta_0}{4}\tilde{I}_0\sin k_0(x_m - x_0), & x_m > x_0, \\ 0, & \text{otherwise.} \end{cases}$$

For a center-fed slot with $x_0 = 0$, this is most easily implemented by choosing N even. Then

$$b_m = \begin{cases} -j\frac{\eta_0}{4}\tilde{I}_0\sin k_0(x_m), & x_m > 0, \\ 0, & \text{otherwise.} \end{cases}$$

A filamentary current source is a crude model for the actual feed of a realistic slot antenna. An alternative is to use a distributed current source, spreading out the excitation similar to the manner in which a frill voltage source spreads out the impressed electric field on a dipole surface. One approach is to model the feed as a conducting strip, with an edge singular current distribution akin to the field distribution in the slot. In the case of a center-fed slot we use

$$\tilde{J}_{sy}^i(x, y) = \begin{cases} \dfrac{2\tilde{I}_0}{\pi A\sqrt{1 - \left(\frac{x}{A/2}\right)^2}}, & |x| < A/2, \\ 0, & |x| > A/2, \end{cases}$$

where A is the width of the strip. Then $b_m = 0$ when $x_m < -A/2$. Otherwise

$$b_m = -j\frac{\tilde{I}_0\eta_0}{2\pi A}\left[\sin(k_0 x_m)\int_{-A/2}^{B}\frac{\cos u}{\sqrt{1 - \left(\frac{u}{A/2}\right)^2}}\,du\right.$$

$$\left. - \cos(k_0 x_m)\int_{-A/2}^{B}\frac{\sin u}{\sqrt{1 - \left(\frac{u}{A/2}\right)^2}}\,du\right], \tag{7.239}$$

where $B = x_m$ when $x_m < A/2$, and $B = A/2$ when $x_m > A/2$. These integrals are easily computed numerically.

With all MoM elements determined, we can solve the matrix equation and determine the voltage distribution. From this we can find the slot antenna input impedance

$$Z_{\text{in}} = \tilde{V}(x_0)/\tilde{I}_0.$$

Let us treat some examples.

▶ **Example 7.41:** Impedance of a slot antenna — effect of feed model

Consider a slot antenna of length $L = 50$ mm and width $w = 1$ mm. The slot is a half wavelength long at $f = 3$ GHz, so there should be a resonance near this frequency. Set $f = 3$ GHz and solve for the slot antenna input impedance when the slot is center-fed. Explore the effect of the number of pulse functions used, N, on the input resistance and reactance. Compare the results for the filamentary current feed and for the strip feed with a width $A = w$.

Solution: Figure 7.75 shows the input impedance of the slot antenna computed using the MoM as a function of the number of pulses used to represent the slot voltage. The input resistance of the slot found with the filamentary feed is nearly the same as that found using the strip feed; both results converge to around 305 Ω by $N = 200$ partitions. However, the input reactance differs for the two feeds. The reactance with the filamentary feed converges quickly to about -150 Ω by $N = 200$. The reactance for the strip feed varies more significantly for smaller values of N, but settles in to about -164 Ω by $N = 200$. The strip feed results require more partitions to converge due to the narrow width of the strip, since the integrals in (7.239) are dependent on the position of the match point x_m within the domain of the strip. In any case, it is apparent that the input reactance of the slot antenna depends on the feed model used, much as was observed for the dipole antenna. Thus, we again emphasize that an accurate calculation of the input impedance of a realistic antenna requires a highly realistic model of the feed.

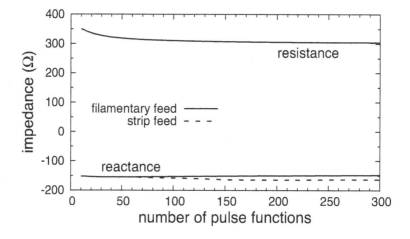

FIGURE 7.75
Impedance of a slot antenna of length $L = 50$ mm and width $w = 1$ mm. $f = 3$ GHz. ◄

▶ **Example 7.42:** Impedance of a slot antenna as a function of frequency

Consider the slot antenna of Example 7.41. Choose $N = 200$ and plot the input resistance and reactance as a function of frequency using the strip feed with $A = w$. Locate the first three resonance frequencies and plot the slot voltages as a function of position at these frequencies.

Solution: Figure 7.76 shows the impedance of the slot antenna as a function of frequency. The first resonance occurs at 2.81 GHz, with an input resistance of 494 Ω. This frequency corresponds to $L/\lambda = 0.469$, and thus the antenna is slightly less than a half-wavelength long at its first resonance, similar to the dipole antenna of Section 7.3.1. The second resonance occurs at 5.04 GHz, where the input resistance is 35 Ω. Finally, the third resonance is observed at 8.75 GHz, with an input resistance of 336 Ω. It is interesting to note that the first and third resonances are of the antiresonance type due to the rapid change in reactance and the highly peaked resistance at the resonance frequency. In contrast, the second resonance has a slowly varying reactance and relatively low resistance. This behavior is directly opposite that of the dipole antenna impedance shown in Figure 7.16, where it is the second resonance that has antiresonance characteristics, while the first and third show a more slow variation of the reactance.

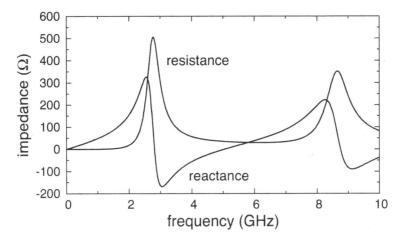

FIGURE 7.76
Impedance of a slot antenna of length $L = 50$ mm and width $w = 1$ mm found using $N = 200$ pulse functions.

The magnitude and phase of the slot voltage distribution are plotted in Figures 7.77 and 7.78, respectively, for each of the resonance frequencies identified from the impedance plot. At the first resonance, the voltage is nearly sinusoidal, with just a few degrees of phase variation across the slot. At higher resonances, both the amplitude and phase variation increase in complexity. Compare these plots with Figures 7.17 and 7.18, which show the current on a dipole antenna at the first three resonances. The striking similarity is an example of the complementarity principle [57].

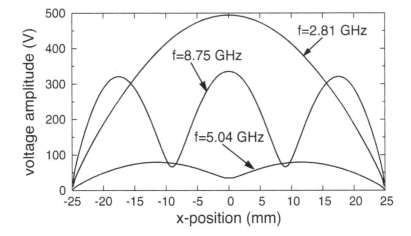

FIGURE 7.77
Magnitude of the voltage of a slot antenna of length $L = 50$ mm and width $w = 1$ mm found using $N = 200$ pulse functions. Applied current is 1 A.

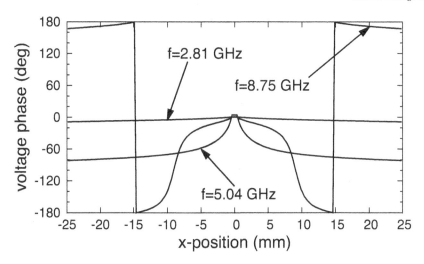

FIGURE 7.78
Phase of the voltage of a slot antenna of length $L = 50$ mm and width $w = 1$ mm found using $N = 200$ pulse functions. Applied current is 1 A. ◄

▶ **Example 7.43:** Pattern of a slot antenna

Consider the slot antenna of Example 7.42. Plot the pattern of the antenna in the xz plane as a function of θ at each of the first three resonance frequencies.

Solution: The xz-plane is identified by $\phi = 0$, $0 \leq \theta \leq \pi$ for $x \geq 0$, and $\phi = \pi$, $0 \leq \theta \leq \pi$ for $x \geq 0$. Since these conditions are cumbersome to plot, we allow θ to take on negative values to indicate the half plane $\phi = \pi$. This is only for plotting convenience. From (7.237) the field when $\phi = 0$ or $\phi = \pi$ is given by

$$\tilde{\mathbf{E}} = jk_0 \frac{e^{-jk_0 r}}{r} \left(\hat{\phi} \cos\theta \cos\phi \right) \frac{\sin\left(k_0 \frac{\Delta}{2} \sin\theta\right)}{k_0 \sin\theta} \sum_{n=1}^{N} a_n e^{jk_0 x_n \sin\theta \cos\phi}.$$

Thus, in either half plane, the magnitude of the far-zone electric field has a proportionality given by

$$|\tilde{\mathbf{E}}(\theta)| \sim \left| \cos\theta \frac{\sin\left(k_0 \frac{\Delta}{2} \sin\theta\right)}{k_0 \sin\theta} \sum_{n=1}^{N} a_n e^{jk_0 x_n \sin\theta} \right|.$$

This function describes the θ-dependence of the far-zone field, and is thus the antenna field pattern for the xz-plane. Figure 7.79 shows a plot of the antenna pattern for each of the first three resonance frequencies, with the maximum value normalized to unity. For higher resonance frequencies the slot is electrically larger, and the voltage has more variation across the slot (as seen in Figures 7.77 and 7.78.) This produces a more complex antenna pattern at higher resonance frequencies. Note that the pattern is zero at $\theta = \pm\pi/2$, since the presence of the conducting ground plane compels the tangential electric field to be zero there. Also note that we restrict our plot to the half plane $z > 0$; the pattern is symmetric about this plane since the antenna radiates identically into the lower half space.

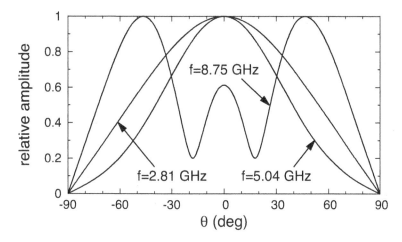

FIGURE 7.79

Field pattern in xz-plane of a slot antenna of length $L = 50$ mm and width $w = 1$ mm found using $N = 200$ pulse functions. ◀

7.7 Application: electromagnetic shielding revisited

Shielding against electromagnetic fields has become increasingly important as electronic devices have become more sophisticated and more compact. Many consumer electronics products incorporate radio-frequency devices in close proximity to digital circuitry, and the probability of interference is high. Interference can be ameliorated by surrounding sensitive circuitry with metallic shields, but gaps in butted surfaces, cracks, and ventilation holes present opportunities for interfering signals to penetrate into the shielded region. An important canonical problem for understanding leakage into shielded regions is a narrow rectangular slot in a ground plane.

In § 3.7.2.2 we studied the penetration of an electrostatic field through a circular hole in a conducting screen. The problem of penetration through an aperture is much harder for electromagnetic fields. Although approximate solutions using scalar diffraction theory are available (see [73]), accurate results require a numerical solution. Fortunately, in § 7.6.2 we put in place all the tools needed to do the calculation.

7.7.1 Penetration of a narrow slot in a ground plane

Consider a plane wave incident from the upper half space onto a narrow slot in a ground plane (Figure 7.80). We can find the field penetrating the slot by solving the matrix equation (7.232), and then computing the field in the lower half space using (7.238). To solve the matrix equation, we set the impressed aperture current $\tilde{J}^i_{sy}$ to zero, and specialize the impressed field $\tilde{H}^i_x$ in (7.226) to that of a plane wave. The matrix entries are identical to those found for the radiation case (i.e, for the slot antenna). We need only specialize the right-hand-side elements, b_m, for an incident plane wave.

Calculation of right-hand-side elements. Since the development of the MFIE for a slot in § 7.6.2 assumes that the impressed magnetic field in the aperture is x-directed, we consider the case of a TE-polarized plane wave (Figure 7.80). The wave is assumed incident with its wave vector in the xz-plane, and its electric field polarized in the y-direction. The incident wave vector is $\mathbf{k}^i = -k_0\hat{\mathbf{x}}\sin\theta_0 - k_0\hat{\mathbf{z}}\cos\theta_0$, so the fields are given by

$$\tilde{\mathbf{E}}^i = \tilde{E}_0\hat{\mathbf{y}}e^{jk_0 x\sin\theta_0}e^{jk_0 z\cos\theta_0}$$

and

$$\tilde{\mathbf{H}}^i = \frac{\mathbf{k}^i\times\tilde{\mathbf{E}}^i}{\eta_0} = -\hat{\mathbf{z}}\frac{\tilde{E}_0}{\eta_0}e^{jk_0 x\sin\theta_0}e^{jk_0 z\cos\theta_0}\sin\theta_0 + \hat{\mathbf{x}}\frac{\tilde{E}_0}{\eta_0}e^{jk_0 x\sin\theta_0}e^{jk_0 z\cos\theta_0}\cos\theta_0$$

where θ_0 is the incidence angle measured from the z-axis. In the development of the MFIE we assume that the impressed field is the sum of the incident field and the reflected field (or, equivalently, the field produced by the images of the impressed sources), and thus satisfies the boundary condition on the ground plane. For the case of a TE-polarized plane wave incident on a PEC ground plane, the reflected field will also be a plane wave, with a wave vector given by $\mathbf{k}^r = -k_0\hat{\mathbf{x}}\sin\theta_0 + k_0\hat{\mathbf{z}}\cos\theta_0$. Thus, the reflected fields are

$$\tilde{\mathbf{E}}^r = -\tilde{E}_0\hat{\mathbf{y}}e^{jk_0 x\sin\theta_0}e^{-jk_0 z\cos\theta_0}$$

and

$$\tilde{\mathbf{H}}^r = \frac{\mathbf{k}^r\times\tilde{\mathbf{E}}^r}{\eta_0} = \hat{\mathbf{z}}\frac{\tilde{E}_0}{\eta_0}e^{jk_0 x\sin\theta_0}e^{-jk_0 z\cos\theta_0}\sin\theta_0 + \hat{\mathbf{x}}\frac{\tilde{E}_0}{\eta_0}e^{jk_0 x\sin\theta_0}e^{-jk_0 z\cos\theta_0}\cos\theta_0$$

such that the total electric field tangential to the ground plane is zero. The impressed magnetic field is thus

$$\tilde{\mathbf{H}} = \tilde{\mathbf{H}}^i + \tilde{\mathbf{H}}^r$$

$$= -\hat{\mathbf{z}}2j\frac{\tilde{E}_0}{\eta_0}e^{jk_0 x\sin\theta_0}\sin(k_0 z\cos\theta_0)\sin\theta_0 + \hat{\mathbf{x}}2\frac{\tilde{E}_0}{\eta_0}e^{jk_0 x\sin\theta_0}\cos(k_0 z\cos\theta_0)\cos\theta_0.$$

Thus we have the aperture impressed field

$$\tilde{H}_x^i(x) = 2\frac{\tilde{E}_0}{\eta_0}\cos\theta_0 e^{jk_0 x\sin\theta_0}. \tag{7.240}$$

Substitution of (7.240) into (7.226) gives the aperture function

$$\tilde{H}(x) = -\frac{j\omega\mu_0}{2}\frac{\tilde{E}_0}{\eta_0}\cos\theta_0 e^{jk_0 x\sin\theta_0}.$$

Then from (7.234) we have the right-hand side elements

$$b_m = -\frac{j\omega\mu_0}{2k_0}\frac{\tilde{E}_0}{\eta_0}\cos\theta_0\int_{-L/2}^{x_m}e^{jk_0 u\sin\theta_0}\sin k_0(x_m - u)\,du$$

$$= -j\frac{\tilde{E}_0}{2}\cos\theta_0\int_{-L/2}^{x_m}e^{jk_0 u\sin\theta_0}\sin k_0(x_m - u)\,du.$$

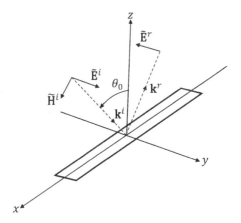

FIGURE 7.80
TE-polarized plane wave incident on a narrow rectangular slot.

Integration gives

$$b_m = j\frac{\tilde{E}_0}{2k_0\cos\theta_0}e^{-jk_0\frac{L}{2}\sin\theta_0}\left\{j\sin\theta_0\sin\left[k_0(m-\tfrac{1}{2})\Delta\right]\right.$$
$$\left. + \cos\left[k_0(m-\tfrac{1}{2})\Delta\right] - e^{jk_0\left(m-\frac{1}{2}\right)\Delta\sin\theta_0}\right\}.$$

At normal incidence ($\theta_0 = 0$) this simplifies to

$$b_m = j\frac{\tilde{E}_0}{2k_0}\left\{\cos[k_0(m-\tfrac{1}{2})\Delta] - 1\right\}. \tag{7.241}$$

▶ **Example 7.44:** Shielding effectiveness of a slot in a ground plane

A narrow slot in a perfectly conducting ground plane is illuminated by a plane electromagnetic wave of amplitude E_0, as shown in Figure 7.80. Let the slot length be $L = 50$ mm and the slot width be $w = 1$ mm. Compute the shielding effectiveness on the z-axis below the slot as a function of frequency and position using $N = 200$ partitions. Assume normal incidence ($\theta_0 = 0$).

Solution: We solve the matrix equation for the slot voltage using (7.241) for the right-hand-side elements. Then we compute the near-zone field below the slot using (7.238). For points on the z-axis, the field reduces to

$$\tilde{\mathbf{E}} = \mp\frac{1}{2\pi}\hat{\mathbf{y}}z\sum_{n=1}^{N}a_n\int_{x_n-\Delta/2}^{x_n+\Delta/2}\frac{1+jk_0R}{4\pi R^3}e^{-jk_0R}\,dx',$$

with $R = \sqrt{x'^2 + z^2}$. Using this, the shielding effectiveness is $SE = 20\log_{10}|E_0/\tilde{E}_y|$.

Figure 7.81 shows the shielding effectiveness as a function of frequency, for various observation positions beneath the ground plane. At low frequencies, where the slot is short compared to a wavelength, little field penetrates and the shielding effectiveness is high. As the frequency increases, a greater amount of field penetrates and the shield is less effective. Greatest penetration occurs at the first resonance (2.81 GHz), when the induced aperture field is strongest, producing the greatest radiation into the lower half space. Shielding effectiveness is also low near the third resonance (8.75 GHz).

Note from Figure 7.81 that as the observation point moves farther from the aperture the field decreases and SE becomes larger. Interestingly, at points near to the aperture SE becomes negative, indicating that the strength of the field penetrating the aperture is greater than that of the incident plane wave. This can be thought of as a sort of focusing effect in the near field region. Figure 7.82 shows more clearly how the shielding effectiveness increases with depth.

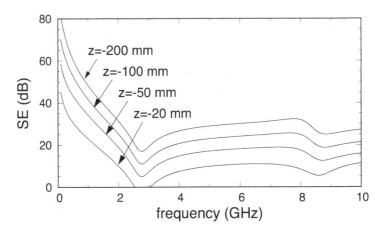

FIGURE 7.81
Shielding effectiveness on the z-axis for a plane wave normally incident on a slot in a perfectly conducting ground plane. Slot dimensions are $L = 50$ mm, $w = 1$ mm. Field found using $N = 200$ pulse functions.

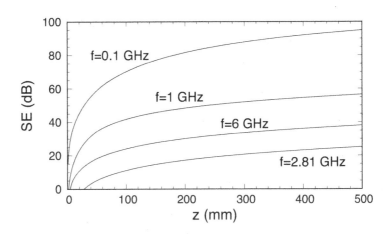

FIGURE 7.82
Shielding effectiveness on the z-axis for a plane wave normally incident on a slot in a perfectly conducting ground plane. Slot dimensions are $L = 50$ mm, $w = 1$ mm. Field found using $N = 200$ pulse functions.

It is important to realize that the shielding effectiveness of an aperture in a conducting enclosure is dependent on a variety of factors. The examples above show that SE depends strongly on frequency and on observation point. It is also dependent on the arrival angle and polarization of the incident wave. Figure 7.83 shows the dependence of SE on the incidence angle for an observation position $z = -100$ mm, showing that greatest penetration occurs at normal incidence. As the angle is increased, there is little change for lower frequencies until grazing incidence is approached (near 90°.) At 8.75 GHz there is sufficient variation in the phase of the aperture field that an increase in shielding effectiveness occurs near 25°.

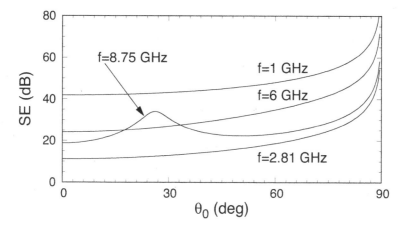

FIGURE 7.83

Shielding effectiveness on the z-axis for a plane wave incident at angle θ_0 on a slot in a perfectly conducting ground plane, observed at $z = -100$ mm. Slot dimensions are $L = 50$ mm, $w = 1$ mm. Field found using $N = 200$ pulse functions. ◀

7.8 Problems

7.1 Consider the steady-state transmission line shown in Figure 7.3. (a) Derive the integral equation for the voltage on the transmission line, assuming $Z_g \neq 0$. (b) Obtain a numerical solution to the integral equation derived in part (a) using pulse function expansion and point matching. Assume the transmission line is lossless and that $Z_0 = Z_g = 50\Omega$, $Z_L = 100 - j100\Omega$, and $\ell = 2.3\lambda$, and compare the numerical result to the analytic solution.

7.2 Verify the identity (7.16) using integration by parts.

7.3 Verify by substitution that (7.40) satisfies the differential equation (7.38).

7.4 Verify by substitution that (7.43) satisfies the integral equation (7.41).

7.5 Show that when $\tilde{\mu}$ is z-dependent, the differential equation for $g(z)$ changes from

(7.32) to

$$\frac{\partial^2 g(z,\omega)}{\partial z^2} - \frac{1}{\tilde{\mu}(z,\omega)} \frac{\partial \tilde{\mu}(z,\omega)}{\partial z} \frac{\partial g(z,\omega)}{\partial z} + k_z^2(z,\omega) g(z,\omega) = 0$$

and that the integral equation for $g(z)$ becomes

$$g(z) = - \int_0^z \{h(t) + (z-t)[f(t) - h'(t)]\} g(t)\, dt +$$

$$+ \frac{a - zh_0}{a - dh_0} \int_0^d \{h(t) + (d-t)[f(t) - h'(t)]\} g(t)\, dt + 2\tilde{E}_0 \frac{z-d}{a - dh_0}.$$

Here $a(\omega) = j\eta_0/[\omega\tilde{\mu}(0,\omega)\cos\phi_0]$, $h(z) = -\tilde{\mu}'(z)/\tilde{\mu}(z)$, and $h_0 = 1 - ah(0)$. See [164].

7.6 Since the ordinary differential equation (7.32) is of second order, it will have two independent solutions, say $g_1(z,\omega)$ and $g_2(z,\omega)$. (a) Using the boundary conditions (7.33) and (7.36), show that $g(z,\omega)$ for the conductor-backed slab is given by

$$g(z,\omega) = \frac{2\tilde{E}_0(\omega)}{F(\omega) + a(\omega)G(\omega)} \left[g_2(d,\omega)g_1(z,\omega) - g_1(d,\omega)g_2(z,\omega) \right],$$

where

$$F(\omega) = g_1(0,\omega)g_2(d,\omega) - g_1(d,\omega)g_2(0,\omega),$$
$$G(\omega) = g_1'(0,\omega)g_2(d,\omega) - g_1(d,\omega)g_2'(0,\omega),$$

with $g_1'(z,\omega) = \partial g_1(z,\omega)/\partial z$, etc. (b) Using (7.37) and the results from part (a), show that the reflection coefficient is

$$\tilde{\Gamma}(\omega) = \frac{F(\omega) - a(\omega)G(\omega)}{F(\omega) + a(\omega)G(\omega)}.$$

7.7 Consider a TE plane wave obliquely incident on a conductor-backed slab as described in § 7.2. Assume the slab has unit relative permeability, zero conductivity, and permittivity profile

$$\tilde{\epsilon}(z,\omega) = \tilde{\epsilon}_{r0}(\omega)\epsilon_0 e^{\kappa(\omega)z}.$$

Show by substitution that the solutions of the differential equation (7.32) are given by

$$g_1(z,\omega) = J_\nu(\lambda e^{\kappa z/2}), \quad g_2(z,\omega) = N_\nu(\lambda e^{\kappa z/2}),$$

where $\lambda = 2k_0\sqrt{\tilde{\mu}_r\tilde{\epsilon}_{r0}}/\kappa$ and $\nu = 2k_x/\kappa$.

7.8 Consider a TE plane wave obliquely incident on a conductor-backed slab as described in § 7.2. Assume the slab has unit relative permeability, zero conductivity, and permittivity profile

$$\tilde{\epsilon}(z,\omega) = \tilde{\epsilon}_{r0}(\omega)\epsilon_0 e^{\kappa(\omega)z}.$$

(a) Expanding the field in the slab using pulse functions as in (7.46), and using point matching, derive the expressions for the entries in the matrix equation (7.47). (b) Solve the matrix equation for $N = 2000$, $f = 10$ GHz, $\tilde{\epsilon}_{r0} = 4$, $\kappa = 20$ m^{-1}, $d = 0.03$ m, and $\phi_0 = 30°$, and determine $g(z)$ and $\tilde{\Gamma}$. Compare the results with the analytic results obtained in Problem 7.7. See [164].

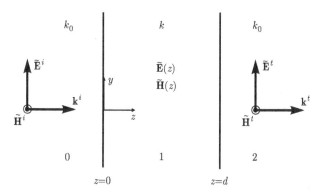

FIGURE 7.84
Problem 7.10.

7.9 Verify that (7.44) may be written in the form (7.45).

7.10 Consider a plane wave normally incident on a homogeneous air-backed slab with parameters $\tilde{\mu}$ and $\tilde{\epsilon}^c$ as shown in Figure 7.84. Show that within the slab ($0 \leq z \leq d$) the z-dependence of the electric field is given by the equation

$$g(z) = -k^2 \int_0^z (z-t)g(t)\,dt + 2\tilde{E}_0 + \left(z - j\frac{\eta_0}{\omega\tilde{\mu}}\right)\frac{k^2}{d}\int_0^d \left[(z-t) - j\frac{\eta_0}{\omega\tilde{\mu}}\right]g(t)\,dt,$$

where $\tilde{E}_0$ is the amplitude of the incident electric field.

7.11 Consider a TM plane wave obliquely incident on an inhomogeneous conductor-backed slab as shown in Figure 7.85. (a) Beginning with Maxwell's equations, derive a differential equation for the magnetic field $\tilde{H}_y(x, z)$ within the slab. Let $\tilde{H}_y(x,z) = f(x)g(z)$, and show that $g(z)$ obeys the differential equation

$$g''(z) - \frac{\tilde{\epsilon}^{c\prime}(z)}{\tilde{\epsilon}^c(z)}g'(z) + k_z^2(z)g(z) = 0,$$

where $\tilde{\epsilon}^c(z) = \tilde{\epsilon}(z) - j\tilde{\sigma}(z)/\omega$ and $k_z^2(z) = k^2(z) - k_x^2$. (b) Apply the appropriate boundary conditions and show that $g(z)$ satisfies

$$g(z) = \int_0^d \left[(z-a) - (z-t)U(z-t)\right]\left[k_z^2(t)g(t) + h(t)g'(t)\right]dt + 2\tilde{H}_0,$$

for $0 \leq z \leq d$. Here $\tilde{H}_0$ is the amplitude of the incident magnetic field, $U(z)$ is the unit step function (A.6), $a = j/[\eta_0\omega\tilde{\epsilon}^c(0)\cos\phi_0]$, and $h(z) = -\tilde{\epsilon}^{c\prime}(z)/\tilde{\epsilon}^c(z)$.

7.12 Show that the moment-method matrix entries (7.65) are symmetric: $A_{mn} = A_{nm}$.

7.13 Consider the frill generator model for a monopole antenna as shown in Figure 7.13.

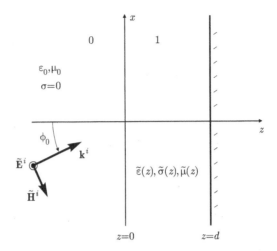

FIGURE 7.85
Problem 7.11.

The electric field in the ground-plane aperture is assumed to be the coaxial-cable field

$$\tilde{\mathbf{E}}(\mathbf{r}) = \hat{\rho}\frac{\tilde{V}_0/2}{\rho\ln(b/a)}.$$

Using the equivalence and image theorems, derive the electric field on the axis of the monopole antenna, and thus verify (7.70).

7.14 Consider Pocklington's integral equation, as given in (7.80). (a) Specialize the equation to a straight wire extending from $z = -L$ to $z = L$. Assuming that the thin-wire approximation is valid, compute the necessary derivatives of the kernel. (b) Develop an MoM solution to the equation derived in part (a), assuming that the thin-wire approximation is valid. Implement a computer solution for the current and the input impedance using pulse function expansion and point matching, with a frill generator. For the case f=300 MHz, $L = 0.25$ m, and $a = 0.0001$ m, compare the input impedance as a function of the number of pulse functions used to the results from Hallén's equation shown in Figure 7.14. Repeat for $a = 0.01$ m, and compare to Figure 7.15. Repeat for $a = 0.001$ m and compare to both figures.

7.15 Using the technique of § 7.3.2.2, convert the Pocklington integral equation for a loop antenna, (7.86), to Hallén form. Use integration by parts to remove the derivatives in the kernel.

7.16 Consider a thin wire as shown in Figure 7.19, carrying a current $\tilde{I}(z,\omega)$. The far-zone electric field produced by the current is found from (7.73). Using the inverse Fourier transform integral, show that the time-domain field in the far zone is

$$E_\theta(r,t) = \frac{\mu_0}{4\pi r}\sin\theta\int_{-L}^{L} I'\left(z', t - \frac{r}{c} + \frac{z'\cos\theta}{c}\right)dz'.$$

7.17 The coupling coefficients for the current on a thin wire excited by an incident plane wave may be approximated using the simple sinusoidal current formula (7.105). Substitute this expression and show that $a_m = R_m/C_m$, where

$$R_m \approx -8\pi^2 s_m \epsilon_0 \sin\theta_i E_0(s_m) J_0(-js_m a \sin\theta_i/c)$$

$$\times \frac{\left(\frac{m\pi}{2L}\right)}{\left(\frac{m\pi}{2L}\right)^2 + (\gamma_m \cos\theta_i)^2} \left[(-1)^m e^{\gamma_m L \cos\theta_i} - e^{-\gamma_m L \cos\theta_i}\right]$$

and

$$C_m \approx C_m^- + (-1)^m C_m^+$$

where

$$C_m^\pm = \frac{1}{2} \int_{-L}^{L} \int_{-L}^{L} \left[\left(\frac{m\pi}{2L}\right)^2 F(z-z', s_m) \mp \frac{\gamma_m^2}{c} g(z-z', s_m)\right] \cos\left(\frac{m\pi}{2L}[z\pm z']\right) dz' dz.$$

See [162].

7.18 Consider a transient plane wave incident on a circular loop of wire. The late-time current induced in the wire may be found using the singularity expansion method. Using the Fourier series representation of the current, show that the natural frequencies of the loop may be found by solving

$$\frac{n^2}{b^2} K_n(s) + \frac{\gamma^2}{2} [K_{n-1}(s) + K_{n+1}(s)] = 0,$$

where

$$K_n(s) = \frac{1}{2\pi} \int_{-\pi}^{\pi} G(\zeta, s) e^{-jn\zeta} d\zeta$$

and

$$G(\zeta, s) = \int_0^{2\pi} \frac{e^{-\gamma \bar{R}(\xi, \zeta)}}{\bar{R}(\xi, \zeta)} d\xi.$$

Note that the natural frequencies are indexed by n, which is determined by the periodicity of the current. See [17], [161].

7.19 Carry out the details of the integral to verify (7.72).

7.20 Show by substitution that the 2-D Green's function (7.124) satisfies

$$\nabla^2 G_{2D}(\boldsymbol{\rho}|\boldsymbol{\rho}') + k_0^2 G_{2D}(\boldsymbol{\rho}|\boldsymbol{\rho}') = -\delta(\boldsymbol{\rho} - \boldsymbol{\rho}').$$

7.21 Develop a solution for the current on a conducting strip when the excitation is by an electric line source centered a distance d above the strip. See Figure 7.86. Let $f = 300$ MHz, $w = 1$ m, $\tilde{I} = 1$ A, and $N = 201$ pulses. Plot the magnitude and phase of the current density as a function of position for the following cases: (a) $d = w/4$; (b) $d = w$; (c) $d = 4w$. Compare to the results for plane wave excitation shown in Figure 7.35.

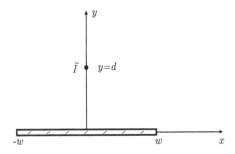

FIGURE 7.86
Problem 7.21.

7.22 Consider the computation of the moment method matrix elements for a conducting strip under TM plane-wave illumination (7.132). Although a singularity occurs only when the source and observation partitions are the same $(m = n)$, when the partitions are adjacent the integrand may still vary rapidly when the source point is close to the observation point, and thus the integral may take an undesirable amount of computational resources. Show that for adjacent partitions,

$$A_{n-1,n} = A_{n+1,n} = \frac{1}{k_0} \int_{k_0\Delta/2}^{3k_0\Delta/2} \left[H_0^{(2)}(u) - f_0(u) \right] du$$

$$+ \Delta - j\frac{2\Delta}{\pi} \left[\gamma - \ln 2 + \frac{3}{2}\ln 3 + \ln\left(k_0\frac{\Delta}{2}\right) - 1 \right],$$

where $f_0(u)$ is defined in (7.133). This may be extended to other partitions as well.

7.23 Consider the resistive strip examined in § 7.4.3. A TE-polarized plane wave is incident on the strip at an angle ϕ_0. (a) Derive a Hallén integral equation for the equivalent current on the strip. (b) Use pulse-function expansion and point matching to obtain a matrix equation for the current. (c) Using the parameters from Example 7.22, compute and plot the equivalent current induced in the strip.

7.24 A function $f(z)$, $0 \leq z \leq d$, is represented using a pulse function expansion:

$$f(z) = \sum_{n=1}^{N} a_n P_n(z)$$

where

$$P_n(z) = \begin{cases} 1, & (n-1)\Delta \leq z \leq n\Delta, \\ 0 & \text{elsewhere}, \end{cases}$$

with $\Delta = d/N$. We wish to estimate the value of $f(z)$ at $z = 0$ using the values of the first three pulse amplitudes. Let $f_a(z)$ be a quadratic function that passes through the centers of the first three pulses, and show that

$$f_a(0) = \tfrac{1}{8}\left[15a_1 - 10a_2 + 3a_3\right].$$

A

Mathematical appendix

A.1 Conservative vector fields

For a sufficiently smooth vector field $\mathbf{A} = \mathbf{A}(\mathbf{r})$, the following conditions are equivalent:

1. The net circulation of $\mathbf{A}$ around any closed path C vanishes: $\oint_C \mathbf{A} \cdot \mathbf{dl} = 0$.

2. The line integral of $\mathbf{A}$ between two locations a and b is independent of the integration path: $\int_{C_1} \mathbf{A} \cdot \mathbf{dl} = \int_{C_2} \mathbf{A} \cdot \mathbf{dl}$, where C_1 and C_2 are any two curves running from a to b.

3. There is a scalar field $\phi = \phi(\mathbf{r})$ such that $\mathbf{A} = \nabla\phi$. It is common to refer to ϕ as a *potential* function for $\mathbf{A}$.

4. $\mathbf{A}$ is irrotational: $\nabla \times \mathbf{A} = 0$.

The interested reader may see Marsden and Tromba [126] for a proof. A vector field $\mathbf{A}$ satisfying any of the conditions 1–4 is said to be *conservative*.

A.2 The Fourier transform

The Fourier transform permits us to decompose a complicated field structure into elemental components. This can simplify the computation of fields and provide physical insight into their spatiotemporal behavior. In this section we review the properties of the transform and demonstrate its usefulness in solving field equations.

A.2.1 One-dimensional Fourier transform

Let f be a function of a single variable x. The Fourier transform of $f(x)$ is the function $F(k)$ defined by the integral

$$\mathcal{F}\{f(x)\} = F(k) = \int_{-\infty}^{\infty} f(x)e^{-jkx}\, dx. \tag{A.1}$$

Note that x and the corresponding transform variable k must have reciprocal units: if x is time in seconds, then k is a *temporal frequency* in radians per second; if x is a length in meters, then k is a *spatial frequency* in radians per meter. We sometimes refer to $F(k)$ as the *frequency spectrum* of $f(x)$.

Not every function has a Fourier transform. The existence of (A.1) can be guaranteed by a set of sufficient conditions such as the following:

1. f is absolutely integrable: $\int_{-\infty}^{\infty} |f(x)|\, dx < \infty$;

2. f has no infinite discontinuities;

3. f has at most finitely many discontinuities and finitely many extrema in any finite interval (a, b).

While such rigor is certainly of mathematical value, it may be of less ultimate use to the engineer than the following heuristic observation offered by Bracewell [24]: *a good mathematical model of a physical process should be Fourier transformable.* That is, if the Fourier transform of a mathematical model does not exist, the model cannot precisely describe a physical process.

The usefulness of the transform hinges on our ability to recover f through the inverse transform:

$$\mathcal{F}^{-1}\{F(k)\} = f(x) = \frac{1}{2\pi} \int_{-\infty}^{\infty} F(k)\, e^{jkx}\, dk. \tag{A.2}$$

When this is possible we write

$$f(x) \leftrightarrow F(k)$$

and say that $f(x)$ and $F(k)$ form a Fourier transform pair. The *Fourier integral theorem* states that

$$\mathcal{F}\mathcal{F}^{-1}\{f(x)\} = \mathcal{F}^{-1}\mathcal{F}\{f(x)\} = f(x),$$

except at points of discontinuity of f. At a jump discontinuity the inversion formula returns the average value of the one-sided limits $f(x^+)$ and $f(x^-)$ of $f(x)$. At points of continuity the forward and inverse transforms are unique.

A.2.1.1   Transform theorems and properties

We now review some basic facts pertaining to the Fourier transform. Let $f(x) \leftrightarrow F(k) = R(k) + jX(k)$, and $g(x) \leftrightarrow G(k)$.

1. *Linearity.* $\alpha f(x) + \beta g(x) \leftrightarrow \alpha F(k) + \beta G(k)$ if α and β are arbitrary constants. This follows directly from the linearity of the transform integral, and makes the transform useful for solving linear differential equations (e.g., Maxwell's equations).

2. *Symmetry.* The property $F(x) \leftrightarrow 2\pi f(-k)$ is helpful when interpreting transform tables in which transforms are listed only in the forward direction.

3. *Conjugate function.* We have $f^*(x) \leftrightarrow F^*(-k)$.

4. *Real function.* If f is real, then $F(-k) = F^*(k)$. Also,

$$R(k) = \int_{-\infty}^{\infty} f(x) \cos kx\, dx, \qquad X(k) = -\int_{-\infty}^{\infty} f(x) \sin kx\, dx,$$

and

$$f(x) = \frac{1}{\pi} \operatorname{Re} \int_{0}^{\infty} F(k) e^{jkx}\, dk.$$

A real function is completely determined by its positive frequency spectrum. It is obviously advantageous to know this when planning to collect spectral data.

5. *Real function with reflection symmetry.* If f is real and even, then $X(k) \equiv 0$ and

$$R(k) = 2 \int_0^\infty f(x) \cos kx \, dx, \qquad f(x) = \frac{1}{\pi} \int_0^\infty R(k) \cos kx \, dk.$$

If f is real and odd, then $R(k) \equiv 0$ and

$$X(k) = -2 \int_0^\infty f(x) \sin kx \, dx, \qquad f(x) = -\frac{1}{\pi} \int_0^\infty X(k) \sin kx \, dk.$$

(Recall that f is even if $f(-x) = f(x)$ for all x. Similarly f is odd if $f(-x) = -f(x)$ for all x.)

6. *Causal function.* Recall that f is causal if $f(x) = 0$ for $x < 0$.

(a) If f is real and causal, then

$$X(k) = -\frac{2}{\pi} \int_0^\infty \int_0^\infty R(k') \cos k'x \sin kx \, dk' \, dx,$$

$$R(k) = -\frac{2}{\pi} \int_0^\infty \int_0^\infty X(k') \sin k'x \cos kx \, dk' \, dx.$$

(b) If f is real and causal, and $f(0)$ is finite, then $R(k)$ and $X(k)$ are related by the *Hilbert transforms*

$$X(k) = -\frac{1}{\pi} \text{P.V.} \int_{-\infty}^\infty \frac{R(k)}{k - k'} \, dk', \qquad R(k) = \frac{1}{\pi} \text{P.V.} \int_{-\infty}^\infty \frac{X(k)}{k - k'} \, dk'.$$

(c) If f is causal and has finite energy, it is not possible to have $F(k) = 0$ for $k_1 < k < k_2$. That is, the transform of a causal function cannot vanish over an interval.

A causal function is completely determined by the real or imaginary part of its spectrum. As with item 4, this is helpful when performing calculations or measurements in the frequency domain. If the function is not band-limited, however, truncation of integrals will give erroneous results.

7. *Time-limited vs. band-limited functions.* Assume $t_2 > t_1$. If $f(t) = 0$ for both $t < t_1$ and $t > t_2$, then it is not possible to have $F(k) = 0$ for both $k < k_1$ and $k > k_2$ where $k_2 > k_1$. That is, a time-limited signal cannot be band-limited. Similarly, a band-limited signal cannot be time-limited.

8. *Null function.* If the forward or inverse transform of a function is identically zero, then the function is identically zero. This important consequence of the Fourier integral theorem is useful when solving homogeneous partial differential equations in the frequency domain.

9. *Space or time shift.* For any fixed x_0,

$$f(x - x_0) \leftrightarrow F(k)e^{-jkx_0}. \tag{A.3}$$

A temporal or spatial shift affects only the phase of the transform, not the magnitude.

10. *Frequency shift.* For any fixed k_0,

$$f(x)e^{jk_0 x} \leftrightarrow F(k - k_0).$$

Note that if $f \leftrightarrow F$ where f is real, then frequency-shifting F causes f to become complex — again, this is important if F has been obtained experimentally or through computation in the frequency domain.

11. *Similarity.* We have

$$f(\alpha x) \leftrightarrow \frac{1}{|\alpha|} F\left(\frac{k}{\alpha}\right),$$

where α is any real constant. "Reciprocal spreading" is exhibited by the Fourier transform pair; dilation in space or time results in compression in frequency, and vice versa.

12. *Convolution.* We have

$$\int_{-\infty}^{\infty} f_1(x')f_2(x - x')\, dx' \leftrightarrow F_1(k)F_2(k), \tag{A.4}$$

$$f_1(x)f_2(x) \leftrightarrow \frac{1}{2\pi} \int_{-\infty}^{\infty} F_1(k')F_2(k - k')\, dk'.$$

The first of these is particularly useful when a problem has been solved in the frequency domain and the solution is found to be a product of two or more functions of k.

13. *Parseval's identity.* We have

$$\int_{-\infty}^{\infty} |f(x)|^2\, dx = \frac{1}{2\pi} \int_{-\infty}^{\infty} |F(k)|^2\, dk.$$

Computations of energy in the time and frequency domains always give the same result.

14. *Differentiation.* We have

$$\frac{d^n f(x)}{dx^n} \leftrightarrow (jk)^n F(k) \quad \text{and} \quad (-jx)^n f(x) \leftrightarrow \frac{d^n F(k)}{dk^n}.$$

The Fourier transform can convert a differential equation in the x domain into an algebraic equation in the k domain, and vice versa.

15. *Integration.* We have

$$\int_{-\infty}^{x} f(u)\, du \leftrightarrow \pi F(k)\delta(k) + \frac{F(k)}{jk}$$

where $\delta(k)$ is the Dirac delta or unit impulse.

A.2.1.2 Generalized Fourier transforms and distributions

It is worth noting that many useful functions are not Fourier transformable in the sense given above. An example is the signum function

$$\operatorname{sgn}(x) = \begin{cases} -1, & x < 0, \\ 1, & x > 0. \end{cases}$$

Although this function lacks a Fourier transform in the usual sense, for practical purposes it may still be safely associated with what is known as a *generalized Fourier transform*. A treatment of this notion would be out of place here; however, the reader should certainly be prepared to encounter an entry such as

$$\operatorname{sgn}(x) \leftrightarrow 2/jk$$

in a standard Fourier transform table. Other functions can be regarded as possessing transforms when *generalized functions* are permitted into the discussion. An important example of a generalized function is the Dirac delta $\delta(x)$, which has enormous value in describing distributions that are very thin, such as the charge layers often found on conductor surfaces. We shall not delve into the intricacies of distribution theory. However, we can hardly avoid dealing with generalized functions; to see this we need look no further than the simple function $\cos k_0 x$ with its transform pair

$$\cos k_0 x \leftrightarrow \pi[\delta(k + k_0) + \delta(k - k_0)].$$

The reader of this book must therefore know the standard facts about $\delta(x)$: that it acquires meaning only as part of an integrand, and that it satisfies the *sifting property*

$$\int_{-\infty}^{\infty} \delta(x - x_0) f(x)\, dx = f(x_0)$$

for any continuous function f. With $f(x) = 1$ we obtain the familiar relation

$$\int_{-\infty}^{\infty} \delta(x)\, dx = 1.$$

With $f(x) = e^{-jkx}$ we obtain

$$\int_{-\infty}^{\infty} \delta(x) e^{-jkx}\, dx = 1,$$

thus

$$\delta(x) \leftrightarrow 1.$$

It follows that

$$\frac{1}{2\pi} \int_{-\infty}^{\infty} e^{jkx}\, dk = \delta(x). \tag{A.5}$$

A.2.1.3 Useful transform pairs

Some of the more common Fourier transforms that arise in the study of electromagnetics are given in Appendix C. These often involve the simple functions defined here:

1. *Unit step function*

$$U(x) = \begin{cases} 1, & x > 0, \\ 0, & x < 0. \end{cases} \tag{A.6}$$

2. *Signum function*

$$\text{sgn}(x) = \begin{cases} -1, & x < 0, \\ 1, & x > 0. \end{cases} \tag{A.7}$$

3. *Rectangular pulse function*

$$\text{rect}(x) = \begin{cases} 1, & |x| < 1, \\ 0, & |x| > 1. \end{cases} \tag{A.8}$$

4. *Triangular pulse function*

$$\Lambda(x) = \begin{cases} 1 - |x|, & |x| < 1, \\ 0, & |x| > 1. \end{cases} \tag{A.9}$$

5. *Sinc function*

$$\text{sinc}(x) = \frac{\sin x}{x}. \tag{A.10}$$

A.2.2 Transforms of multi-variable functions

Fourier transformations can be performed over multiple variables by successive applications of (A.1). For example, the two-dimensional Fourier transform over x_1 and x_2 of the function $f(x_1, x_2, x_3, \ldots, x_N)$ is the quantity $F(k_{x_1}, k_{x_2}, x_3, \ldots, x_N)$ given by

$$\int_{-\infty}^{\infty} \left[\int_{-\infty}^{\infty} f(x_1, x_2, x_3, \ldots, x_N) e^{-jk_{x_1} x_1} \, dx_1 \right] e^{-jk_{x_2} x_2} \, dx_2$$

$$= \int_{-\infty}^{\infty} \int_{-\infty}^{\infty} f(x_1, x_2, x_3, \ldots, x_N) e^{-jk_{x_1} x_1} e^{-jk_{x_2} x_2} \, dx_1 \, dx_2.$$

The two-dimensional inverse transform is computed by multiple application of (A.2), recovering $f(x_1, x_2, x_3, \ldots, x_N)$ through the operation

$$\frac{1}{(2\pi)^2} \int_{-\infty}^{\infty} \int_{-\infty}^{\infty} F(k_{x_1}, k_{x_2}, x_3, \ldots, x_N) e^{jk_{x_1} x_1} e^{jk_{x_2} x_2} \, dk_{x_1} \, dk_{x_2}.$$

Higher-dimensional transforms and inversions are done analogously.

A.2.2.1 Transforms of separable functions

If we are able to write

$$f(x_1, x_2, x_3, \ldots, x_N) = f_1(x_1, x_3, \ldots, x_N) f_2(x_2, x_3, \ldots, x_N),$$

then successive transforms on the variables x_1 and x_2 result in

$$f(x_1, x_2, x_3, \ldots, x_N) \leftrightarrow F_1(k_{x_1}, x_3, \ldots, x_N) F_2(k_{x_2}, x_3, \ldots, x_N).$$

In this case a multi-variable transform can be obtained with the help of a table of one-dimensional transforms. If, for instance,

$$f(x, y, z) = \delta(x - x') \delta(y - y') \delta(z - z'),$$

then we obtain

$$F(k_x, k_y, k_z) = e^{-jk_x x'} e^{-jk_y y'} e^{-jk_z z'}$$

by three applications of (A.1).

A more compact notation for multi-dimensional functions and transforms makes use of the vector notation $\mathbf{k} = \hat{\mathbf{x}}k_x + \hat{\mathbf{y}}k_y + \hat{\mathbf{z}}k_z$ and $\mathbf{r} = \hat{\mathbf{x}}x + \hat{\mathbf{y}}y + \hat{\mathbf{z}}z$ where $\mathbf{r}$ is the position vector. In the example above, for instance, we could have written

$$\delta(x - x')\delta(y - y')\delta(z - z') = \delta(\mathbf{r} - \mathbf{r}'),$$

and

$$F(\mathbf{k}) = \int_{-\infty}^{\infty} \int_{-\infty}^{\infty} \int_{-\infty}^{\infty} \delta(\mathbf{r} - \mathbf{r}') e^{-j\mathbf{k} \cdot \mathbf{r}} \, dx \, dy \, dz = e^{-j\mathbf{k} \cdot \mathbf{r}'}.$$

A.2.2.2 Fourier–Bessel transform

If x_1 and x_2 have the same dimensions, it may be convenient to recast the two-dimensional Fourier transform in polar coordinates. Let $x_1 = \rho \cos\phi$, $k_{x_1} = p \cos\theta$, $x_2 = \rho \sin\phi$, and $k_{x_2} = p \sin\theta$, where p and ρ are defined on $(0, \infty)$ and ϕ and θ are defined on $(-\pi, \pi)$. Then

$$F(p, \theta, x_3, \ldots, x_N) = \int_{-\pi}^{\pi} \int_{0}^{\infty} f(\rho, \phi, x_3, \ldots, x_N) \, e^{-jp\rho \cos(\phi - \theta)} \rho \, d\rho \, d\phi. \tag{A.11}$$

If f is independent of ϕ (due to rotational symmetry about an axis transverse to x_1 and x_2), then the ϕ integral can be computed using the identity

$$J_0(x) = \frac{1}{2\pi} \int_{-\pi}^{\pi} e^{-jx \cos(\phi - \theta)} \, d\phi.$$

Thus (A.11) becomes

$$F(p, x_3, \ldots, x_N) = 2\pi \int_{0}^{\infty} f(\rho, x_3, \ldots, x_N) J_0(\rho p) \, \rho \, d\rho, \tag{A.12}$$

showing that F is independent of the angular variable θ. Expression (A.12) is termed the *Fourier–Bessel transform* of f. The reader can easily verify that f can be recovered from F through

$$f(\rho, x_3, \ldots, x_N) = \int_{0}^{\infty} F(p, x_3, \ldots, x_N) J_0(\rho p) \, p \, dp,$$

the inverse Fourier–Bessel transform.

A.2.3 A review of complex contour integration

Some powerful techniques for the evaluation of integrals rest on complex variable theory. In particular, the computation of the Fourier inversion integral is often aided by these techniques. We therefore provide a brief review of this material. For a fuller discussion the reader may refer to one of many widely available textbooks on complex analysis.

We shall denote by $f(z)$ a complex valued function of a complex variable z. That is,

$$f(z) = u(x, y) + jv(x, y),$$

where the real and imaginary parts $u(x, y)$ and $v(x, y)$ of f are each functions of the real and imaginary parts x and y of z:

$$z = x + jy = \operatorname{Re} z + j \operatorname{Im} z.$$

Here $j = \sqrt{-1}$, as is mostly standard in the electrical engineering literature.

A.2.3.1   Limits, differentiation, and analyticity

Let $w = f(z)$, and let $z_0 = x_0 + jy_0$ and $w_0 = u_0 + jv_0$ be points in the complex z and w planes, respectively. We say that w_0 is the limit of $f(z)$ as z approaches z_0, and write

$$\lim_{z \to z_0} f(z) = w_0,$$

if and only if both $u(x, y) \to u_0$ and $v(x, y) \to v_0$ as $x \to x_0$ and $y \to y_0$ independently. The derivative of $f(z)$ at a point $z = z_0$ is defined by the limit

$$f'(z_0) = \lim_{z \to z_0} \frac{f(z) - f(z_0)}{z - z_0},$$

if it exists. Existence requires that the derivative be independent of direction of approach; that is, $f'(z_0)$ cannot depend on the manner in which $z \to z_0$ in the complex plane. (This turns out to be a much stronger condition than simply requiring that the functions u and v be differentiable with respect to the variables x and y.) We say that $f(z)$ is *analytic* at z_0 if it is differentiable at z_0 and at all points in some neighborhood of z_0.

If $f(z)$ is not analytic at z_0 but every neighborhood of z_0 contains a point at which $f(z)$ is analytic, then z_0 is called a *singular point of* $f(z)$.

A.2.3.2   Laurent expansions and residues

Although Taylor series can be used to expand complex functions around points of analyticity, we must often expand functions around points z_0 at or near which the functions fail to be analytic. For this we use the *Laurent expansion*, a generalization of the Taylor expansion involving both positive and negative powers of $z - z_0$:

$$f(z) = \sum_{n=-\infty}^{\infty} a_n(z - z_0)^n = \sum_{n=1}^{\infty} \frac{a_{-n}}{(z - z_0)^n} + \sum_{n=0}^{\infty} a_n(z - z_0)^n.$$

The numbers a_n are the coefficients of the Laurent expansion of $f(z)$ at point $z = z_0$. The first series on the right is the *principal part* of the Laurent expansion, and the second series is the *regular part*. The regular part is an ordinary power series, hence it converges in some disk $|z - z_0| < R$ where $R \geq 0$. Putting $\zeta = 1/(z - z_0)$, the principal part becomes $\sum_{n=1}^{\infty} a_{-n}\zeta^n$. This power series converges for $|\zeta| < \rho$ where $\rho \geq 0$, hence the principal part converges for $|z - z_0| > 1/\rho \triangleq r$. When $r < R$, the Laurent expansion converges in the annulus $r < |z - z_0| < R$; when $r > R$, it diverges everywhere in the complex plane.

The function $f(z)$ has an *isolated singularity* at point z_0 if $f(z)$ is not analytic at z_0 but is analytic in the "punctured disk" $0 < |z - z_0| < R$ for some $R > 0$. Isolated singularities are classified by reference to the Laurent expansion. Three types can arise:

1. *Removable singularity.* The point z_0 is a removable singularity of $f(z)$ if the principal part of the Laurent expansion of $f(z)$ about z_0 is identically zero (i.e., if $a_n = 0$ for $n = -1, -2, -3, \ldots$).

2. *Pole of order k.* The point z_0 is a pole of order k if the principal part of the Laurent expansion about z_0 contains only finitely many terms that form a polynomial of degree k in $(z - z_0)^{-1}$. A pole of order 1 is called a *simple pole*.

3. *Essential singularity.* The point z_0 is an essential singularity of $f(z)$ if the principal part of the Laurent expansion of $f(z)$ about z_0 contains infinitely many terms (i.e., if $a_{-n} \neq 0$ for infinitely many n).

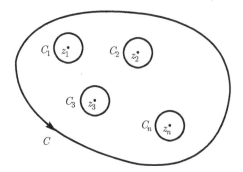

FIGURE A.1

Derivation of the residue theorem.

The coefficient a_{-1} in the Laurent expansion of $f(z)$ about an isolated singular point z_0 is the *residue of $f(z)$ at z_0*. It can be shown that

$$a_{-1} = \frac{1}{2\pi j} \oint_{\Gamma} f(z)\,dz \tag{A.13}$$

where Γ is any simple closed curve oriented counterclockwise and containing in its interior z_0 and no other singularity of $f(z)$. Particularly useful to us is the formula for evaluation of residues at pole singularities. If $f(z)$ has a pole of order k at $z = z_0$, then the residue of $f(z)$ at z_0 is given by

$$a_{-1} = \frac{1}{(k-1)!} \lim_{z \to z_0} \frac{d^{k-1}}{dz^{k-1}} [(z - z_0)^k f(z)]. \tag{A.14}$$

A.2.3.3 Cauchy–Goursat and residue theorems

It can be shown that if $f(z)$ is analytic at all points on and within a simple closed contour C, then

$$\oint_C f(z)\,dz = 0.$$

This central result is known as the *Cauchy–Goursat theorem*. We shall not offer a proof, but shall proceed instead to derive a useful consequence known as the *residue theorem*. Figure A.1 depicts a simple closed curve C enclosing n isolated singularities of a function $f(z)$. We assume that $f(z)$ is analytic on and elsewhere within C. Around each singular point z_k we have drawn a circle C_k so small that it encloses no singular point other than z_k; taken together, the C_k $(k = 1, \ldots, n)$ and C form the boundary of a region in which $f(z)$ is everywhere analytic. By the Cauchy–Goursat theorem,

$$\int_C f(z)\,dz + \sum_{k=1}^{n} \int_{C_k} f(z)\,dz = 0.$$

Hence,

$$\frac{1}{2\pi j} \int_C f(z)\,dz = \sum_{k=1}^{n} \frac{1}{2\pi j} \int_{C_k} f(z)\,dz,$$

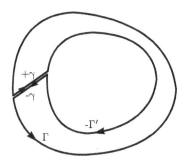

FIGURE A.2
Derivation of the contour deformation principle.

where now the integrations are all performed in a counterclockwise sense. By (A.13)

$$\int_C f(z)\,dz = 2\pi j \sum_{k=1}^{n} r_k \tag{A.15}$$

where $r_1, \ldots, r_n$ are the residues of $f(z)$ at the singularities within C.

A.2.3.4   Contour deformation

Suppose f is analytic in a region D and Γ is a simple closed curve in D. If Γ can be continuously deformed to another simple closed curve Γ' without passing out of D, then

$$\int_{\Gamma'} f(z)\,dz = \int_{\Gamma} f(z)\,dz. \tag{A.16}$$

To see this, consider Figure A.2 where we have introduced another set of curves $\pm\gamma$; these new curves are assumed parallel and infinitesimally close to each other. Let C be the composite curve consisting of Γ, $+\gamma$, $-\Gamma'$, and $-\gamma$, in that order. Since f is analytic on and within C, we have

$$\int_C f(z)\,dz = \int_{\Gamma} f(z)\,dz + \int_{+\gamma} f(z)\,dz + \int_{-\Gamma'} f(z)\,dz + \int_{-\gamma} f(z)\,dz = 0.$$

But $\int_{-\Gamma'} f(z)\,dz = -\int_{\Gamma'} f(z)\,dz$ and $\int_{-\gamma} f(z)\,dz = -\int_{+\gamma} f(z)\,dz$, hence (A.16) follows. The contour deformation principle often permits us to replace an integration contour by one that is more convenient.

A.2.3.5   Principal value integrals

We must occasionally carry out integrations of the form

$$I = \int_{-\infty}^{\infty} f(x)\,dx$$

where $f(x)$ has a finite number of singularities x_k $(k = 1, \ldots, n)$ along the real axis. Such singularities in the integrand force us to interpret I as an improper integral. With just one singularity present at point x_1, for instance, we define

$$\int_{-\infty}^{\infty} f(x)\,dx = \lim_{\varepsilon \to 0} \int_{-\infty}^{x_1-\varepsilon} f(x)\,dx + \lim_{\eta \to 0} \int_{x_1+\eta}^{\infty} f(x)\,dx$$

provided that both limits exist. When both limits do not exist, we may still be able to obtain a well-defined result by computing

$$\lim_{\varepsilon \to 0} \left(\int_{-\infty}^{x_1 - \varepsilon} f(x)\,dx + \int_{x_1 + \varepsilon}^{\infty} f(x)\,dx \right)$$

(i.e., by taking $\eta = \varepsilon$ so that the limits are "symmetric"). This quantity is called the *Cauchy principal value* of I and is denoted

$$\text{P.V.} \int_{-\infty}^{\infty} f(x)\,dx.$$

More generally, we have

$$\text{P.V.} \int_{-\infty}^{\infty} f(x)\,dx = \lim_{\varepsilon \to 0} \left(\int_{-\infty}^{x_1 - \varepsilon} f(x)\,dx + \int_{x_1 + \varepsilon}^{x_2 - \varepsilon} f(x)\,dx \right.$$

$$\left. + \cdots + \int_{x_{n-1} + \varepsilon}^{x_n - \varepsilon} f(x)\,dx + \int_{x_n + \varepsilon}^{\infty} f(x)\,dx \right)$$

for n singularities $x_1 < \cdots < x_n$.

In a large class of problems, $f(z)$ (i.e., $f(x)$ with x replaced by the complex variable z) is analytic everywhere except for the presence of finitely many simple poles. Some of these may lie on the real axis (at points $x_1 < \cdots < x_n$, say), and some may not. Consider now the integration contour C shown in Figure A.3. We choose R so large and ε so small that C encloses all the poles of f that lie in the upper half of the complex plane. In many problems of interest, the integral of f around the large semicircle tends to zero as $R \to \infty$, and the integrals around the small semicircles are well-behaved as $\varepsilon \to 0$. It may then be shown that

$$\text{P.V.} \int_{-\infty}^{\infty} f(x)\,dx = \pi j \sum_{k=1}^{n} r_k + 2\pi j \sum_{\text{UHP}} r_k$$

where r_k is the residue at the kth simple pole. The first sum on the right accounts for the contributions of those poles that lie on the real axis; note that it is associated with a factor πj instead of $2\pi j$, since these terms arose from integrals over semicircles rather than over full circles. The second sum, of course, is extended only over those poles that reside in the upper half-plane.

A.2.4 Fourier transform solution of the 1-D wave equation

Successive applications of the Fourier transform can reduce a partial differential equation to an ordinary differential equation, and finally to an algebraic equation. After the algebraic equation is solved by standard techniques, Fourier inversion can yield a solution to the original partial differential equation. We illustrate this by solving the one-dimensional inhomogeneous wave equation

$$\left(\frac{\partial^2}{\partial z^2} - \frac{1}{c^2} \frac{\partial^2}{\partial t^2} \right) \psi(x, y, z, t) = S(x, y, z, t), \tag{A.17}$$

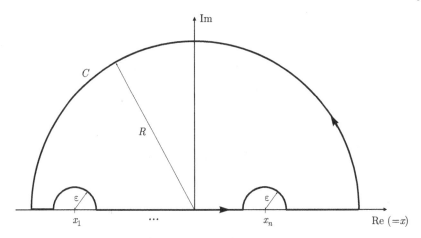

FIGURE A.3
Complex plane technique for evaluating a principal value integral.

where the field ψ is the desired unknown and S is the known source term. For uniqueness of solution we must specify ψ and $\partial\psi/\partial z$ over some $z = $ constant plane. Assume that

$$\psi(x,y,z,t)\Big|_{z=0} = f(x,y,t), \tag{A.18}$$

$$\frac{\partial}{\partial z}\psi(x,y,z,t)\Big|_{z=0} = g(x,y,t). \tag{A.19}$$

We begin by positing inverse temporal Fourier transform relationships for ψ and S:

$$\psi(x,y,z,t) = \frac{1}{2\pi}\int_{-\infty}^{\infty} \tilde{\psi}(x,y,z,\omega)e^{j\omega t}\,d\omega,$$

$$S(x,y,z,t) = \frac{1}{2\pi}\int_{-\infty}^{\infty} \tilde{S}(x,y,z,\omega)e^{j\omega t}\,d\omega.$$

Substituting into (A.17), passing the derivatives through the integral, calculating the derivatives, and combining the inverse transforms, we obtain

$$\frac{1}{2\pi}\int_{-\infty}^{\infty} \left[\left(\frac{\partial^2}{\partial z^2} + k^2\right)\tilde{\psi}(x,y,z,\omega) - \tilde{S}(x,y,z,\omega)\right]e^{j\omega t}\,d\omega = 0$$

where $k = \omega/c$. By the Fourier integral theorem

$$\left(\frac{\partial^2}{\partial z^2} + k^2\right)\tilde{\psi}(x,y,z,\omega) - \tilde{S}(x,y,z,\omega) = 0. \tag{A.20}$$

We have thus converted a partial differential equation into an ordinary differential equation. A spatial transform on z will now convert the ordinary differential equation into an algebraic equation. We write

$$\tilde{\psi}(x,y,z,\omega) = \frac{1}{2\pi}\int_{-\infty}^{\infty} \tilde{\psi}^z(x,y,k_z,\omega)e^{jk_z z}\,dk_z,$$

$$\tilde{S}(x,y,z,\omega) = \frac{1}{2\pi}\int_{-\infty}^{\infty} \tilde{S}^z(x,y,k_z,\omega)e^{jk_z z}\,dk_z,$$

in (A.20), pass the derivatives through the integral sign, compute the derivatives, and set the integrand to zero to get

$$(k^2 - k_z^2)\tilde{\psi}^z(x, y, k_z, \omega) - \tilde{S}^z(x, y, k_z, \omega) = 0;$$

hence

$$\tilde{\psi}^z(x, y, k_z, \omega) = -\frac{\tilde{S}^z(x, y, k_z, \omega)}{(k_z - k)(k_z + k)}. \tag{A.21}$$

The price we pay for such an easy solution is that we must now perform a two-dimensional Fourier inversion to obtain $\psi(x, y, z, t)$ from $\tilde{\psi}^z(x, y, k_z, \omega)$. It turns out to be easiest to perform the spatial inverse transform first, so let us examine

$$\tilde{\psi}(x, y, z, \omega) = \frac{1}{2\pi} \int_{-\infty}^{\infty} \tilde{\psi}^z(x, y, k_z, \omega) e^{jk_z z} \, dk_z.$$

By (A.21) we have

$$\tilde{\psi}(x, y, z, \omega) = \frac{1}{2\pi} \int_{-\infty}^{\infty} [\tilde{S}^z(x, y, k_z, \omega)] \left[\frac{-1}{(k_z - k)(k_z + k)} \right] e^{jk_z z} \, dk_z,$$

where the integrand involves a product of two functions. With

$$\tilde{g}^z(k_z, \omega) = \frac{-1}{(k_z - k)(k_z + k)},$$

the convolution theorem gives

$$\tilde{\psi}(x, y, z, \omega) = \int_{-\infty}^{\infty} \tilde{S}(x, y, \zeta, \omega)\tilde{g}(z - \zeta, \omega) \, d\zeta \tag{A.22}$$

where

$$\tilde{g}(z, \omega) = \frac{1}{2\pi} \int_{-\infty}^{\infty} \tilde{g}^z(k_z, \omega) e^{jk_z z} \, dk_z = \frac{1}{2\pi} \int_{-\infty}^{\infty} \frac{-1}{(k_z - k)(k_z + k)} e^{jk_z z} \, dk_z.$$

To compute this integral we use complex plane techniques. The domain of integration extends along the real k_z-axis in the complex k_z-plane; because of the poles at $k_z = \pm k$, we must treat the integral as a principal value integral. Denoting

$$I(k_z) = \frac{-e^{jk_z z}}{2\pi(k_z - k)(k_z + k)},$$

we have

$$\int_{-\infty}^{\infty} I(k_z) \, dk_z = \lim \int_{\Gamma_r} I(k_z) \, dk_z$$

$$= \lim \int_{-\Delta}^{-k-\delta} I(k_z) \, dk_z + \lim \int_{-k+\delta}^{k-\delta} I(k_z) \, dk_z + \lim \int_{k+\delta}^{\Delta} I(k_z) \, dk_z$$

where the limits take $\delta \to 0$ and $\Delta \to \infty$. Our k_z-plane contour takes detours around the poles using semicircles of radius δ, and is closed using a semicircle of radius Δ (Figure A.4). Note that if $z > 0$, we must close the contour in the upper half-plane.

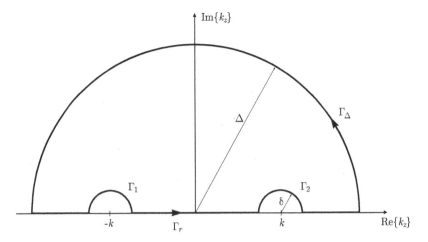

FIGURE A.4

Contour used to compute inverse transform in solution of the 1-D wave equation.

By Cauchy's integral theorem,

$$\int_{\Gamma_r} I(k_z)\,dk_z + \int_{\Gamma_1} I(k_z)\,dk_z + \int_{\Gamma_2} I(k_z)\,dk_z + \int_{\Gamma_\Delta} I(k_z)\,dk_z = 0.$$

Thus

$$\int_{-\infty}^{\infty} I(k_z)\,dk_z = -\lim_{\delta\to 0}\int_{\Gamma_1} I(k_z)\,dk_z - \lim_{\delta\to 0}\int_{\Gamma_2} I(k_z)\,dk_z - \lim_{\Delta\to\infty}\int_{\Gamma_\Delta} I(k_z)\,dk_z.$$

The contribution from the semicircle of radius Δ can be computed by writing k_z in polar coordinates as $k_z = \Delta e^{j\theta}$:

$$\lim_{\Delta\to\infty}\int_{\Gamma_\Delta} I(k_z)\,dk_z = \frac{1}{2\pi}\lim_{\Delta\to\infty}\int_0^\pi \frac{-e^{jz\Delta e^{j\theta}}}{(\Delta e^{j\theta}-k)(\Delta e^{j\theta}+k)} j\Delta e^{j\theta}\,d\theta.$$

Using Euler's identity, we can write

$$\lim_{\Delta\to\infty}\int_{\Gamma_\Delta} I(k_z)\,dk_z = \frac{1}{2\pi}\lim_{\Delta\to\infty}\int_0^\pi \frac{-e^{-\Delta z\sin\theta}e^{j\Delta z\cos\theta}}{\Delta^2 e^{2j\theta}} j\Delta e^{j\theta}\,d\theta.$$

Thus, as long as $z > 0$ the integrand will decay exponentially as $\Delta \to \infty$, and

$$\lim_{\Delta\to\infty}\int_{\Gamma_\Delta} I(k_z)\,dk_z \to 0.$$

Similarly, $\int_{\Gamma_\Delta} I(k_z)\,dk_z \to 0$ when $z < 0$ if we close the semicircle in the lower half-plane. Thus,

$$\int_{-\infty}^{\infty} I(k_z)\,dk_z = -\lim_{\delta\to 0}\int_{\Gamma_1} I(k_z)\,dk_z - \lim_{\delta\to 0}\int_{\Gamma_2} I(k_z)\,dk_z. \qquad (A.23)$$

The integrals around the poles can also be computed by writing k_z in polar coordinates. Writing $k_z = -k + \delta e^{j\theta}$, we find

$$\lim_{\delta \to 0} \int_{\Gamma_1} I(k_z)\, dk_z = \frac{1}{2\pi} \lim_{\delta \to 0} \int_{\pi}^{0} \frac{-e^{jz(-k+\delta e^{j\theta})} j\delta e^{j\theta}}{(-k + \delta e^{j\theta} - k)(-k + \delta e^{j\theta} + k)}\, d\theta$$

$$= \frac{1}{2\pi} \int_{0}^{\pi} \frac{e^{-jkz}}{-2k} j\, d\theta = -\frac{j}{4k} e^{-jkz}.$$

Similarly, using $k_z = k + \delta e^{j\theta}$, we obtain

$$\lim_{\delta \to 0} \int_{\Gamma_2} I(k_z)\, dk_z = \frac{j}{4k} e^{jkz}.$$

Substituting these into (A.23), we have

$$\tilde{g}(z,\omega) = \frac{j}{4k} e^{-jkz} - \frac{j}{4k} e^{jkz} = \frac{1}{2k} \sin kz, \tag{A.24}$$

valid for $z > 0$. For $z < 0$, we close in the lower half-plane instead and get

$$\tilde{g}(z,\omega) = -\frac{1}{2k} \sin kz. \tag{A.25}$$

Substituting (A.24) and (A.25) into (A.22), we obtain

$$\tilde{\psi}(x,y,z,\omega) = \int_{-\infty}^{z} \tilde{S}(x,y,\zeta,\omega) \frac{\sin k(z-\zeta)}{2k}\, d\zeta - \frac{1}{2k} \int_{z}^{\infty} \tilde{S}(x,y,\zeta,\omega) \frac{\sin k(z-\zeta)}{2k}\, d\zeta$$

where we have been careful to separate the two cases considered above. To make things a bit easier when we apply the boundary conditions, let us rewrite the above expression. Splitting the domain of integration, we write

$$\tilde{\psi}(x,y,z,\omega) = \int_{-\infty}^{0} \tilde{S}(x,y,\zeta,\omega) \frac{\sin k(z-\zeta)}{2k}\, d\zeta + \int_{0}^{z} \tilde{S}(x,y,\zeta,\omega) \frac{\sin k(z-\zeta)}{k}\, d\zeta$$

$$- \int_{0}^{\infty} \tilde{S}(x,y,\zeta,\omega) \frac{\sin k(z-\zeta)}{2k}\, d\zeta.$$

Expansion of the trigonometric functions then gives

$$\tilde{\psi}(x,y,z,\omega) = \int_{0}^{z} \tilde{S}(x,y,\zeta,\omega) \frac{\sin k(z-\zeta)}{k}\, d\zeta$$

$$+ \frac{\sin kz}{2k} \int_{-\infty}^{0} \tilde{S}(x,y,\zeta,\omega) \cos k\zeta\, d\zeta - \frac{\cos kz}{2k} \int_{-\infty}^{0} \tilde{S}(x,y,\zeta,\omega) \sin k\zeta\, d\zeta$$

$$- \frac{\sin kz}{2k} \int_{0}^{\infty} \tilde{S}(x,y,\zeta,\omega) \cos k\zeta\, d\zeta + \frac{\cos kz}{2k} \int_{0}^{\infty} \tilde{S}(x,y,\zeta,\omega) \sin k\zeta\, d\zeta.$$

The last four integrals are independent of z, so we can represent them with functions constant in z. Finally, rewriting the trigonometric functions as exponentials, we have

$$\tilde{\psi}(x,y,z,\omega) = \int_{0}^{z} \tilde{S}(x,y,\zeta,\omega) \frac{\sin k(z-\zeta)}{k}\, d\zeta + \tilde{A}(x,y,\omega) e^{-jkz} + \tilde{B}(x,y,\omega) e^{jkz}.$$

$$\tag{A.26}$$

This formula for $\tilde{\psi}$ was found as a solution to the inhomogeneous ordinary differential equation (A.20). Hence, to obtain the complete solution, we should add any possible solutions of the homogeneous differential equation. Since these are exponentials, (A.26) in fact represents the complete solution, where $\tilde{A}$ and $\tilde{B}$ are considered unknown and can be found using the boundary conditions.

If we are interested in the frequency-domain solution to the wave equation, then we are done. However, since our boundary conditions (A.18) and (A.19) pertain to the time domain, we must temporally inverse transform before we can apply them. Writing the sine function in (A.26) in terms of exponentials, we can express the time-domain solution as

$$\tilde{\psi}(x,y,z,t) = \int_0^z \mathcal{F}^{-1}\left\{ \frac{c}{2}\frac{\tilde{S}(x,y,\zeta,\omega)}{j\omega}e^{j\frac{\omega}{c}(z-\zeta)} - \frac{c}{2}\frac{\tilde{S}(x,y,\zeta,\omega)}{j\omega}e^{-j\frac{\omega}{c}(z-\zeta)} \right\} d\zeta$$

$$+ \mathcal{F}^{-1}\left\{\tilde{A}(x,y,\omega)e^{-j\frac{\omega}{c}z}\right\} + \mathcal{F}^{-1}\left\{\tilde{B}(x,y,\omega)e^{j\frac{\omega}{c}z}\right\}. \tag{A.27}$$

A combination of the Fourier integration and time-shifting theorems gives the general identity

$$\mathcal{F}^{-1}\left\{ \frac{\tilde{S}(x,y,\zeta,\omega)}{j\omega}e^{-j\omega t_0} \right\} = \int_{-\infty}^{t-t_0} S(x,y,\zeta,\tau)\,d\tau, \tag{A.28}$$

where we have assumed that $\tilde{S}(x,y,\zeta,0) = 0$. Using this in (A.27) along with the time-shifting theorem, we obtain

$$\psi(x,y,z,t) = \frac{c}{2}\int_0^z \left\{ \int_{-\infty}^{t-\frac{\zeta-z}{c}} S(x,y,\zeta,\tau)\,d\tau - \int_{-\infty}^{t-\frac{z-\zeta}{c}} S(x,y,\zeta,\tau)\,d\tau \right\} d\zeta$$

$$+ a\left(x,y,t-\frac{z}{c}\right) + b\left(x,y,t+\frac{z}{c}\right),$$

or

$$\psi(x,y,z,t) = \frac{c}{2}\int_0^z \int_{t-\frac{z-\zeta}{c}}^{t+\frac{z-\zeta}{c}} S(x,y,\zeta,\tau)\,d\tau\,d\zeta + a\left(x,y,t-\frac{z}{c}\right) + b\left(x,y,t+\frac{z}{c}\right) \tag{A.29}$$

where

$$a(x,y,t) = \mathcal{F}^{-1}[\tilde{A}(x,y,\omega)], \qquad b(x,y,t) = \mathcal{F}^{-1}[\tilde{B}(x,y,\omega)].$$

To calculate $a(x,y,t)$ and $b(x,y,t)$, we must use the boundary conditions (A.18) and (A.19). To apply (A.18), we put $z=0$ into (A.29) to give

$$a(x,y,t) + b(x,y,t) = f(x,y,t). \tag{A.30}$$

Using (A.19) is a bit more complicated, since we must compute $\partial\psi/\partial z$, and z is a parameter in the limits of the integral describing ψ. To compute the derivative, we apply Leibnitz' rule for differentiation:

$$\frac{d}{d\alpha}\int_{\phi(\alpha)}^{\theta(\alpha)} f(x,\alpha)\,dx = \left(\frac{d\theta}{d\alpha}\right)f\left(\theta(\alpha),\alpha\right) - \left(\frac{d\phi}{d\alpha}\right)f\left(\phi(\alpha),\alpha\right) + \int_{\phi(\alpha)}^{\theta(\alpha)}\frac{\partial f}{\partial\alpha}\,dx. \tag{A.31}$$

Using this on the integral term in (A.29), we have

$$\frac{\partial}{\partial z}\left[\frac{c}{2}\int_0^z \left(\int_{t-\frac{z-\zeta}{c}}^{t+\frac{z-\zeta}{c}} S(x,y,\zeta,\tau)\,d\tau \right) d\zeta \right] = \frac{c}{2}\int_0^z \frac{\partial}{\partial z}\left(\int_{t-\frac{z-\zeta}{c}}^{t+\frac{z-\zeta}{c}} S(x,y,\zeta,\tau)\,d\tau \right) d\zeta,$$

which is zero at $z = 0$. Thus,

$$\frac{\partial \psi}{\partial z}\bigg|_{z=0} = g(x, y, t) = -\frac{1}{c}a'(x, y, t) + \frac{1}{c}b'(x, y, t)$$

where $a' = \partial a/\partial t$ and $b' = \partial b/\partial t$. Integration gives

$$-a(x, y, t) + b(x, y, t) = c \int_{-\infty}^{t} g(x, y, \tau) \, d\tau. \qquad (A.32)$$

Equations (A.30) and (A.32) represent two algebraic equations in the two unknown functions a and b. The solutions are

$$2a(x, y, t) = f(x, y, t) - c \int_{-\infty}^{t} g(x, y, \tau) \, d\tau,$$

$$2b(x, y, t) = f(x, y, t) + c \int_{-\infty}^{t} g(x, y, \tau) \, d\tau.$$

Finally, substitution of these into (A.29) gives us the solution to the inhomogeneous wave equation

$$\psi(x, y, z, t) = \frac{c}{2} \int_{0}^{z} \int_{t-\frac{z-\zeta}{c}}^{t+\frac{z-\zeta}{c}} S(x, y, \zeta, \tau) \, d\tau \, d\zeta + \frac{1}{2}\left[f\left(x, y, t - \frac{z}{c}\right) + f\left(x, y, t + \frac{z}{c}\right) \right]$$

$$+ \frac{c}{2} \int_{t-\frac{z}{c}}^{t+\frac{z}{c}} g(x, y, \tau) \, d\tau. \qquad (A.33)$$

This is known as the *D'Alembert solution*. The terms $f(x, y, t \mp z/c)$ contribute to ψ as waves propagating away from the plane $z = 0$ in the $\pm z$-directions, respectively. The integral over the forcing term S is seen to accumulate values of S over a time interval determined by $z - \zeta$.

The boundary conditions could have been applied while still in the temporal frequency domain (but not the spatial frequency domain, since the spatial position z is lost). But to do this, we would need the boundary conditions to be in the temporal frequency domain. This is easily accomplished by transforming them to give

$$\tilde{\psi}(x, y, z, \omega)\bigg|_{z=0} = \tilde{f}(x, y, \omega),$$

$$\frac{\partial}{\partial z}\tilde{\psi}(x, y, z, \omega)\bigg|_{z=0} = \tilde{g}(x, y, \omega).$$

Applying these to (A.26) (and again using Leibnitz's rule), we have

$$\tilde{A}(x, y, \omega) + \tilde{B}(x, y, \omega) = \tilde{f}(x, y, \omega),$$

$$- jk\tilde{A}(x, y, \omega) + jk\tilde{B}(x, y, \omega) = \tilde{g}(x, y, \omega),$$

hence,

$$2\tilde{A}(x, y, \omega) = \tilde{f}(x, y, \omega) - c\frac{\tilde{g}(x, y, \omega)}{j\omega},$$

$$2\tilde{B}(x, y, \omega) = \tilde{f}(x, y, \omega) + c\frac{\tilde{g}(x, y, \omega)}{j\omega}.$$

Finally, substituting these back into (A.26) and expanding the sine function, we obtain the frequency-domain solution that obeys the given boundary conditions:

$$\tilde{\psi}(x,y,z,\omega) = \frac{c}{2} \int_0^z \left[\frac{\tilde{S}(x,y,\zeta,\omega)e^{j\frac{\omega}{c}(z-\zeta)}}{j\omega} - \frac{\tilde{S}(x,y,\zeta,\omega)e^{-j\frac{\omega}{c}(z-\zeta)}}{j\omega} \right] d\zeta$$

$$+ \frac{1}{2} \left[\tilde{f}(x,y,\omega)e^{j\frac{\omega}{c}z} + \tilde{f}(x,y,\omega)e^{-j\frac{\omega}{c}z} \right]$$

$$+ \frac{c}{2} \left[\frac{\tilde{g}(x,y,\omega)e^{j\frac{\omega}{c}z}}{j\omega} - \frac{\tilde{g}(x,y,\omega)e^{-j\frac{\omega}{c}z}}{j\omega} \right].$$

This is easily inverted using (A.28) to give (A.33).

A.2.5 Fourier transform solution of the 1-D homogeneous wave equation for dissipative media

Wave propagation in dissipative media can be studied using the one-dimensional wave equation

$$\left(\frac{\partial^2}{\partial z^2} - \frac{2\Omega}{v^2}\frac{\partial}{\partial t} - \frac{1}{v^2}\frac{\partial^2}{\partial t^2} \right) \psi(x,y,z,t) = S(x,y,z,t). \tag{A.34}$$

This equation is nearly identical to the wave equation for lossless media studied in the previous section, except for the addition of the $\partial\psi/\partial t$ term. This extra term will lead to important physical consequences regarding the behavior of the wave solutions.

We shall solve (A.34) using the Fourier transform approach of the previous section, but to keep the solution simple we shall only consider the homogeneous problem. We begin by writing ψ in terms of its inverse temporal Fourier transform:

$$\psi(x,y,z,t) = \frac{1}{2\pi} \int_{-\infty}^{\infty} \tilde{\psi}(x,y,z,\omega)e^{j\omega t}\, d\omega.$$

Substituting this into the homogeneous version of (A.34) and taking the time derivatives, we obtain

$$\frac{1}{2\pi} \int_{-\infty}^{\infty} \left[(j\omega)^2 + 2\Omega(j\omega) - v^2\frac{\partial^2}{\partial z^2} \right] \tilde{\psi}(x,y,z,\omega)e^{j\omega t}\, d\omega = 0.$$

The Fourier integral theorem leads to

$$\frac{\partial^2 \tilde{\psi}(x,y,z,\omega)}{\partial z^2} - \kappa^2 \tilde{\psi}(x,y,z,\omega) = 0 \tag{A.35}$$

where

$$\kappa = \frac{1}{v}\sqrt{p^2 + 2\Omega p}$$

with $p = j\omega$.

We can solve the homogeneous ordinary differential equation (A.35) by inspection:

$$\tilde{\psi}(x,y,z,\omega) = \tilde{A}(x,y,\omega)e^{-\kappa z} + \tilde{B}(x,y,\omega)e^{\kappa z}. \tag{A.36}$$

Here $\tilde{A}$ and $\tilde{B}$ are frequency-domain coefficients to be determined. We can either specify these coefficients directly, or solve for them by applying specific boundary conditions. We examine each possibility below.

A.2.5.1 Solution to the wave equation by direct application of boundary conditions

The solution to the wave equation (A.34) will be unique if we specify functions $f(x, y, t)$ and $g(x, y, t)$ such that

$$\psi(x, y, z, t)\Big|_{z=0} = f(x, y, t),$$

$$\frac{\partial}{\partial z}\psi(x, y, z, t)\Big|_{z=0} = g(x, y, t). \qquad (A.37)$$

Assuming the Fourier transform pairs $f(x, y, t) \leftrightarrow \tilde{f}(x, y, \omega)$ and $g(x, y, t) \leftrightarrow \tilde{g}(x, y, \omega)$, we can apply the boundary conditions (A.37) in the frequency domain:

$$\tilde{\psi}(x, y, z, \omega)\Big|_{z=0} = \tilde{f}(x, y, \omega),$$

$$\frac{\partial}{\partial z}\tilde{\psi}(x, y, z, \omega)\Big|_{z=0} = \tilde{g}(x, y, \omega).$$

From these we find

$$\tilde{A} + \tilde{B} = \tilde{f}, \qquad -\kappa\tilde{A} + \kappa\tilde{B} = \tilde{g},$$

or

$$\tilde{A} = \frac{1}{2}\left(\tilde{f} - \frac{\tilde{g}}{\kappa}\right), \qquad \tilde{B} = \frac{1}{2}\left(\tilde{f} + \frac{\tilde{g}}{\kappa}\right).$$

Substitution into (A.36) gives

$$\tilde{\psi}(x, y, z, \omega) = \tilde{f}(x, y, \omega)\cosh\kappa z + \tilde{g}(x, y, \omega)\frac{\sinh\kappa z}{\kappa}$$

$$= \tilde{f}(x, y, \omega)\frac{\partial}{\partial z}\tilde{Q}(x, y, z, \omega) + \tilde{g}(x, y, \omega)\tilde{Q}(x, y, z, \omega)$$

$$= \tilde{\psi}_1(x, y, z, \omega) + \tilde{\psi}_2(x, y, z, \omega)$$

where $\tilde{Q} = \sinh\kappa z/\kappa$. Assuming that $Q(x, y, z, t) \leftrightarrow \tilde{Q}(x, y, z, \omega)$, we can employ the convolution theorem to immediately write down $\psi(x, y, z, t)$:

$$\psi(x, y, z, t) = f(x, y, t) * \frac{\partial}{\partial z}Q(x, y, z, t) + g(x, y, z, t) * Q(x, y, z, t)$$

$$= \psi_1(x, y, z, t) + \psi_2(x, y, z, t). \qquad (A.38)$$

To find ψ we must first compute the inverse transform of $\tilde{Q}$. We resort to a tabulated result [28]:

$$\frac{\sinh\left[a\sqrt{p+\lambda}\sqrt{p+\mu}\right]}{\sqrt{p+\lambda}\sqrt{p+\mu}} \leftrightarrow \tfrac{1}{2}e^{-\frac{1}{2}(\mu+\lambda)t}J_0\left(\tfrac{1}{2}(\lambda-\mu)\sqrt{a^2-t^2}\right) \qquad (|t| < a).$$

Here a is a positive, finite real quantity, and λ and μ are finite complex quantities. Outside the range $|t| < a$ the time-domain function is zero.

Letting $a = z/v$, $\mu = 0$, and $\lambda = 2\Omega$ in the above expression, we find

$$Q(x, y, z, t) = \frac{v}{2}e^{-\Omega t}J_0\left(\frac{\Omega}{v}\sqrt{z^2 - v^2 t^2}\right)[U(t + z/v) - U(t - z/v)] \qquad (A.39)$$

where $U(x)$ is the unit step function (A.6). From (A.38) we see that

$$\psi_2(x,y,z,t) = \int_{-\infty}^{\infty} g(x,y,t-\tau)Q(x,y,z,\tau)\,d\tau = \int_{-z/v}^{z/v} g(x,y,t-\tau)Q(x,y,z,\tau)\,d\tau.$$

Using the change of variables $u = t - \tau$ and substituting (A.39), we then have

$$\psi_2(x,y,z,t) = \frac{v}{2}e^{-\Omega t}\int_{t-\frac{z}{v}}^{t+\frac{z}{v}} g(x,y,u)e^{\Omega u}J_0\left(\frac{\Omega}{v}\sqrt{z^2-(t-u)^2v^2}\right)du. \qquad \text{(A.40)}$$

To find ψ_1 we must compute $\partial Q/\partial z$. Using the product rule we have

$$\frac{\partial Q(x,y,z,t)}{\partial z} = \frac{v}{2}e^{-\Omega t}J_0\left(\frac{\Omega}{v}\sqrt{z^2-v^2t^2}\right)\frac{\partial}{\partial z}[U(t+z/v)-U(t-z/v)]$$

$$+ \frac{v}{2}e^{-\Omega t}[U(t+z/v)-U(t-z/v)]\frac{\partial}{\partial z}J_0\left(\frac{\Omega}{v}\sqrt{z^2-v^2t^2}\right).$$

Next, using $dU(x)/dx = \delta(x)$ and remembering that $J_0'(x) = -J_1(x)$ and $J_0(0) = 1$, we can write

$$\frac{\partial Q(x,y,z,t)}{\partial z} = \tfrac{1}{2}e^{-\Omega t}[\delta(t+z/v)+\delta(t-z/v)]$$

$$- \frac{z\Omega^2}{2v}e^{-\Omega t}\frac{J_1\left(\frac{\Omega}{v}\sqrt{z^2-v^2t^2}\right)}{\frac{\Omega}{v}\sqrt{z^2-v^2t^2}}[U(t+z/v)-U(t-z/v)].$$

Convolving this expression with $f(x,y,t)$ we obtain

$$\psi_1(x,y,z,t) = \tfrac{1}{2}e^{-\frac{\Omega}{v}z}f\left(x,y,t-\frac{z}{v}\right)+\tfrac{1}{2}e^{\frac{\Omega}{v}z}f\left(x,y,t+\frac{z}{v}\right)$$

$$- \frac{z\Omega^2}{2v}e^{-\Omega t}\int_{t-\frac{z}{v}}^{t+\frac{z}{v}} f(x,y,u)e^{\Omega u}\frac{J_1\left(\frac{\Omega}{v}\sqrt{z^2-(t-u)^2v^2}\right)}{\frac{\Omega}{v}\sqrt{z^2-(t-u)^2v^2}}du. \qquad \text{(A.41)}$$

Finally, adding (A.41) and (A.40), we obtain

$$\psi(x,y,z,t) = \tfrac{1}{2}e^{-\frac{\Omega}{v}z}f\left(x,y,t-\frac{z}{v}\right)+\tfrac{1}{2}e^{\frac{\Omega}{v}z}f\left(x,y,t+\frac{z}{v}\right)$$

$$- \frac{z\Omega^2}{2v}e^{-\Omega t}\int_{t-\frac{z}{v}}^{t+\frac{z}{v}} f(x,y,u)e^{\Omega u}\frac{J_1\left(\frac{\Omega}{v}\sqrt{z^2-(t-u)^2v^2}\right)}{\frac{\Omega}{v}\sqrt{z^2-(t-u)^2v^2}}du$$

$$+ \frac{v}{2}e^{-\Omega t}\int_{t-\frac{z}{v}}^{t+\frac{z}{v}} g(x,y,u)e^{\Omega u}J_0\left(\frac{\Omega}{v}\sqrt{z^2-(t-u)^2v^2}\right)du. \qquad \text{(A.42)}$$

Note that when $\Omega = 0$ this reduces to

$$\psi(x,y,z,t) = \tfrac{1}{2}f\left(x,y,t-\frac{z}{v}\right)+\tfrac{1}{2}f\left(x,y,t+\frac{z}{v}\right)+\frac{v}{2}\int_{t-\frac{z}{v}}^{t+\frac{z}{v}} g(x,y,u)\,du,$$

which matches (A.33) for the homogeneous case where $S = 0$.

A.2.5.2 Solution to the wave equation by specification of wave amplitudes

An alternative to direct specification of boundary conditions is specification of the amplitude functions $\tilde{A}(x, y, \omega)$ and $\tilde{B}(x, y, \omega)$ or their inverse transforms $A(x, y, t)$ and $B(x, y, t)$. If we specify the time-domain functions, we can write $\psi(x, y, z, t)$ as the inverse transform of (A.36). For example, a wave traveling in the $+z$-direction behaves as

$$\psi(x, y, z, t) = A(x, y, t) * F^+(x, y, z, t) \tag{A.43}$$

where

$$F^+(x, y, z, t) \leftrightarrow e^{-\kappa z} = e^{-\frac{z}{v}\sqrt{p^2 + 2\Omega p}}.$$

We can find F^+ using the Fourier transform pair [28]

$$e^{-\frac{x}{v}\sqrt{(p+\rho)^2 - \sigma^2}} \leftrightarrow e^{-\frac{\rho}{v}x}\delta(t - x/v) + \frac{\sigma x}{v}e^{-\rho t}\frac{I_1\left(\sigma\sqrt{t^2 - (x/v)^2}\right)}{\sqrt{t^2 - (x/v)^2}} \tag{A.44}$$

valid for $x/v < t$. Here x is real and positive and $I_1(x)$ is the modified Bessel function of the first kind and order 1. Outside the range $x/v < t$ the time-domain function is zero. Letting $\rho = \Omega$ and $\sigma = \Omega$, we find

$$F^+(x, y, z, t) = \frac{\Omega^2 z}{v}e^{-\Omega t}\frac{I_1(\Omega\sqrt{t^2 - (z/v)^2})}{\Omega\sqrt{t^2 - (z/v)^2}}U(t - z/v) + e^{-\frac{\Omega}{v}z}\delta(t - z/v). \tag{A.45}$$

Note that F^+ is a real functions of time, as expected.

Substituting (A.45) into (A.43) and writing the convolution in integral form, we have

$$\psi(x, y, z, t) = \int_{z/v}^{\infty} A(x, y, t - \tau)\left[\frac{\Omega^2 z}{v}e^{-\Omega\tau}\frac{I_1(\Omega\sqrt{\tau^2 - (z/v)^2})}{\Omega\sqrt{\tau^2 - (z/v)^2}}\right]d\tau$$

$$+ e^{-\frac{\Omega}{v}z}A\left(x, y, t - \frac{z}{v}\right) \qquad (z > 0). \tag{A.46}$$

A.2.6 The 3-D Green's function for waves in dissipative media

To understand the fields produced by bounded sources within a dissipative medium, we may wish to investigate solutions to the wave equation in three dimensions. The Green's function approach requires the solution to

$$\left(\nabla^2 - \frac{2\Omega}{v^2}\frac{\partial}{\partial t} - \frac{1}{v^2}\frac{\partial^2}{\partial t^2}\right)G(\mathbf{r}|\mathbf{r}'; t) = -\delta(t)\delta(\mathbf{r} - \mathbf{r}')$$

$$= -\delta(t)\delta(x - x')\delta(y - y')\delta(z - z').$$

That is, we are interested in the impulse response of a point source located at $\mathbf{r} = \mathbf{r}'$. We begin by substituting the inverse temporal Fourier transform relations

$$G(\mathbf{r}|\mathbf{r}'; t) = \frac{1}{2\pi}\int_{-\infty}^{\infty}\tilde{G}(\mathbf{r}|\mathbf{r}'; \omega)e^{j\omega t}\,d\omega, \qquad \delta(t) = \frac{1}{2\pi}\int_{-\infty}^{\infty}e^{j\omega t}\,d\omega,$$

obtaining

$$\frac{1}{2\pi}\int_{-\infty}^{\infty}\left[\left(\nabla^2 - j\omega\frac{2\Omega}{v^2} - \frac{1}{v^2}(j\omega)^2\right)\tilde{G}(\mathbf{r}|\mathbf{r}'; \omega) + \delta(\mathbf{r} - \mathbf{r}')\right]e^{j\omega t}\,d\omega = 0.$$

By the Fourier integral theorem, we have

$$(\nabla^2 + k^2)\tilde{G}(\mathbf{r}|\mathbf{r}';\omega) = -\delta(\mathbf{r} - \mathbf{r}'). \tag{A.47}$$

This is known as the *Helmholtz equation.* Here

$$k = \frac{1}{v}\sqrt{\omega^2 - j2\omega\Omega} \tag{A.48}$$

is called the *wavenumber.*

To solve the Helmholtz equation we write $\tilde{G}$ in terms of a 3-dimensional inverse Fourier transform. Substitution of

$$\tilde{G}(\mathbf{r}|\mathbf{r}';\omega) = \frac{1}{(2\pi)^3} \int_{-\infty}^{\infty} \tilde{G}^r(\mathbf{k}|\mathbf{r}';\omega)e^{j\mathbf{k}\cdot\mathbf{r}}\,d^3k,$$

$$\delta(\mathbf{r} - \mathbf{r}') = \frac{1}{(2\pi)^3} \int_{-\infty}^{\infty} e^{j\mathbf{k}\cdot(\mathbf{r}-\mathbf{r}')}\,d^3k,$$

into (A.47) gives

$$\frac{1}{(2\pi)^3} \int_{-\infty}^{\infty} \left[\nabla^2\left(\tilde{G}^r(\mathbf{k}|\mathbf{r}';\omega)e^{j\mathbf{k}\cdot\mathbf{r}}\right) + k^2\tilde{G}^r(\mathbf{k}|\mathbf{r}';\omega)e^{j\mathbf{k}\cdot\mathbf{r}} + e^{j\mathbf{k}\cdot(\mathbf{r}-\mathbf{r}')} \right] d^3k = 0.$$

Here

$$\mathbf{k} = \hat{\mathbf{x}}k_x + \hat{\mathbf{y}}k_y + \hat{\mathbf{z}}k_z$$

with $|\mathbf{k}|^2 = k_x^2 + k_y^2 + k_z^2 = K^2$. Carrying out the derivatives and invoking the Fourier integral theorem, we have

$$(K^2 - k^2)\tilde{G}^r(\mathbf{k}|\mathbf{r}';\omega) = e^{-j\mathbf{k}\cdot\mathbf{r}'}.$$

Solving for $\tilde{G}$ and substituting it into the inverse transform relation, we have

$$\tilde{G}(\mathbf{r}|\mathbf{r}';\omega) = \frac{1}{(2\pi)^3} \int_{-\infty}^{\infty} \frac{e^{j\mathbf{k}\cdot(\mathbf{r}-\mathbf{r}')}}{(K-k)(K+k)}\,d^3k. \tag{A.49}$$

To compute the inverse transform integral in (A.49), we write the 3-D transform variable in spherical coordinates:

$$\mathbf{k} \cdot (\mathbf{r} - \mathbf{r}') = KR\cos\theta, \qquad d^3k = K^2\sin\theta\,dK\,d\theta\,d\phi,$$

where $R = |\mathbf{r} - \mathbf{r}'|$ and θ is the angle between $\mathbf{k}$ and $\mathbf{r} - \mathbf{r}'$. Hence (A.49) becomes

$$\tilde{G}(\mathbf{r}|\mathbf{r}';\omega) = \frac{1}{(2\pi)^3} \int_0^{\infty} \frac{K^2\,dK}{(K-k)(K+k)} \int_0^{2\pi} d\phi \int_0^{\pi} e^{jKR\cos\theta}\sin\theta\,d\theta$$

$$= \frac{2}{(2\pi)^2 R} \int_0^{\infty} \frac{K\sin(KR)}{(K-k)(K+k)}\,dK,$$

or, equivalently,

$$\tilde{G}(\mathbf{r}|\mathbf{r}';\omega) = \frac{1}{2jR(2\pi)^2} \int_{-\infty}^{\infty} \frac{e^{jKR}}{(K-k)(K+k)}K\,dK$$

$$- \frac{1}{2jR(2\pi)^2} \int_{-\infty}^{\infty} \frac{e^{-jkR}}{(K-k)(K+k)}K\,dK.$$

We can compute the integrals over K using the complex plane technique. We consider K to be a complex variable, and note that for dissipative media we have $k = k_r + jk_i$, where $k_r > 0$ and $k_i < 0$. Thus the integrand has poles at $K = \pm k$. For the integral involving e^{+jKR} we close the contour in the upper half-plane using a semicircle of radius Δ and use Cauchy's residue theorem. Then at all points on the semicircle the integrand decays exponentially as $\Delta \to \infty$, and there is no contribution to the integral from this part of the contour. The real-line integral is thus equal to $2\pi j$ times the residue at $K = -k$:

$$\int_{-\infty}^{\infty} \frac{e^{jKR}}{(K-k)(K+k)} K \, dK = 2\pi j \frac{e^{-jkR}}{-2k}(-k).$$

For the term involving e^{-jKR} we close in the lower half-plane and again the contribution from the infinite semicircle vanishes. In this case our contour is clockwise and so the real line integral is $-2\pi j$ times the residue at $K = k$:

$$\int_{-\infty}^{\infty} \frac{e^{-jKR}}{(K-k)(K+k)} K \, dK = -2\pi j \frac{e^{-jkR}}{2k} k.$$

Thus

$$\tilde{G}(\mathbf{r}|\mathbf{r}';\omega) = \frac{e^{-jkR}}{4\pi R}. \tag{A.50}$$

Note that if $\Omega = 0$ then this reduces to

$$\tilde{G}(\mathbf{r}|\mathbf{r}';\omega) = \frac{e^{-j\omega R/v}}{4\pi R}. \tag{A.51}$$

Our last step is to find the temporal Green's function. Let $p = j\omega$. Then we can write

$$\tilde{G}(\mathbf{r}|\mathbf{r}';\omega) = \frac{e^{\kappa R}}{4\pi R} \quad \text{where} \quad \kappa = -jk = \frac{1}{v}\sqrt{p^2 + 2\Omega p}.$$

We may find the inverse transform using (A.44). Letting $x = R$, $\rho = \Omega$, and $\sigma = \Omega$, we find

$$G(\mathbf{r}|\mathbf{r}';t) = e^{-\frac{\Omega}{v}R} \frac{\delta(t - R/v)}{4\pi R} + \frac{\Omega^2}{4\pi v} e^{-\Omega t} \frac{I_1\left(\Omega\sqrt{t^2 - (R/v)^2}\right)}{\Omega\sqrt{t^2 - (R/v)^2}} U\left(t - \frac{R}{v}\right).$$

We note that in the case of no dissipation where $\Omega = 0$, this reduces to

$$G(\mathbf{r}|\mathbf{r}';t) = \frac{\delta(t - R/v)}{4\pi R}$$

which is the inverse transform of (A.51).

A.2.7 Fourier transform representation of the 3-D Green's function: the Weyl identity

Consider the solution to the Helmholtz equation with a point source

$$\nabla^2 G(\mathbf{r}|\mathbf{r}') + k^2 G(\mathbf{r}|\mathbf{r}') = -\delta(\mathbf{r} - \mathbf{r}') = -\delta(x - x')\delta(y - y')\delta(z - z'), \tag{A.52}$$

where k is a complex constant with $\text{Re}\{k\} \geq 0$ and $\text{Im}\{k\} \leq 0$. Here $G(\mathbf{r}|\mathbf{r}')$ is the three-dimensional Green's function for the Helmholtz equation, representing the wave function at point $\mathbf{r}$ produced by a unit point source at point $\mathbf{r}'$.

In Chapter 5 we find that $G(\mathbf{r}|\mathbf{r}') = e^{-jkR}/4\pi R$ where $R = |\mathbf{r} - \mathbf{r}'|$. In a variety of problems it is also useful to express G as an inverse Fourier transform over the variables x and y. Letting G^r form a three-dimensional Fourier transform pair with G, we can write

$$G(\mathbf{r}|\mathbf{r}') = \frac{1}{(2\pi)^3} \int_{-\infty}^{\infty} G^r(k_x, k_y, k_z|\mathbf{r}') e^{jk_x x} e^{jk_y y} e^{jk_z z} \, dk_x \, dk_y \, dk_z.$$

Substitution into (A.52) along with the inverse transform representation for the delta function (A.5) gives

$$\frac{1}{(2\pi)^3} \left(\nabla^2 + k^2\right) \int_{-\infty}^{\infty} G^r(k_x, k_y, k_z|\mathbf{r}') e^{jk_x x} e^{jk_y y} e^{jk_z z} \, dk_x \, dk_y \, dk_z$$

$$= -\frac{1}{(2\pi)^3} \int_{-\infty}^{\infty} e^{jk_x (x-x')} e^{jk_y (y-y')} e^{jk_z (z-z')} \, dk_x \, dk_y \, dk_z.$$

We then combine the integrands and move the Laplacian operator through the integral to obtain

$$\frac{1}{(2\pi)^3} \int_{-\infty}^{\infty} \left[\left(\nabla^2 + k^2\right) \left(G^r(\mathbf{k}|\mathbf{r}') e^{j\mathbf{k}\cdot\mathbf{r}}\right) + e^{j\mathbf{k}\cdot(\mathbf{r}-\mathbf{r}')}\right] d^3k = 0,$$

where $\mathbf{k} = \hat{\mathbf{x}} k_x + \hat{\mathbf{y}} k_y + \hat{\mathbf{z}} k_z$. Carrying out the derivatives, we get

$$\frac{1}{(2\pi)^3} \int_{-\infty}^{\infty} \left[\left(-k_x^2 - k_y^2 - k_z^2 + k^2\right) G^r(\mathbf{k}|\mathbf{r}') + e^{-j\mathbf{k}\cdot\mathbf{r}'}\right] e^{j\mathbf{k}\cdot\mathbf{r}} \, d^3k = 0.$$

Letting $k_x^2 + k_y^2 = k_\rho^2$, and invoking the Fourier integral theorem, we get the algebraic equation

$$\left(k^2 - k_\rho^2 - k_z^2\right) G^r(\mathbf{k}|\mathbf{r}') + e^{-j\mathbf{k}\cdot\mathbf{r}'} = 0,$$

which we can easily solve for G^r:

$$G^r(\mathbf{k}|\mathbf{r}') = \frac{e^{-j\mathbf{k}\cdot\mathbf{r}'}}{k_\rho^2 + k_z^2 - k^2}. \tag{A.53}$$

Equation (A.53) yields a 3-D transform representation for the Green's function. To obtain the 2-D representation, we must carry out the inverse transform over k_z. Writing

$$G^{xy}(k_x, k_y, z|\mathbf{r}') = \frac{1}{2\pi} \int_{-\infty}^{\infty} G^r(k_x, k_y, k_z|\mathbf{r}') e^{jk_z z} \, dk_z$$

we have

$$G^{xy}(k_x, k_y, z|\mathbf{r}') = \frac{1}{2\pi} \int_{-\infty}^{\infty} \frac{e^{-jk_x x'} e^{-jk_y y'} e^{jk_z (z-z')}}{k_z^2 + (k_\rho^2 - k^2)} \, dk_z.$$

Factorization of the denominator term gives

$$G^{xy}(k_x, k_y, z|\mathbf{r}') = \frac{1}{2\pi} \int_{-\infty}^{\infty} \frac{e^{-jk_x x'} e^{-jk_y y'} e^{jk_z (z-z')}}{\left(k_z - j\sqrt{k_\rho^2 - k^2}\right) \left(k_z + j\sqrt{k_\rho^2 - k^2}\right)} \, dk_z. \tag{A.54}$$

Recall that k is in the fourth quadrant of the complex plane. Hence k^2 is in the lower half-plane, $k_\rho^2 - k^2$ is in the upper half-plane, and $\sqrt{k_\rho^2 - k^2}$ is in the first quadrant. So $j\sqrt{k_\rho^2 - k^2}$ and $-j\sqrt{k_\rho^2 - k^2}$ are in the second and fourth quadrants, respectively.

To compute this integral, we let k_z be a complex variable and consider a closed contour in the complex plane, consisting of a semicircle and the real axis. As previously discussed, we compute the principal value integral as the semicircle radius $\Delta \to \infty$, and find that the contribution along the semicircle reduces to zero. Hence we can use Cauchy's residue theorem (A.15) to obtain the real-line integral:

$$G^{xy}(k_x, k_y, z|\mathbf{r}') = 2\pi j \operatorname{res}\left\{\frac{1}{2\pi} \frac{e^{-jk_x x'} e^{-jk_y y'} e^{jk_z(z-z')}}{\left(k_z - j\sqrt{k_\rho^2 - k^2}\right)\left(k_z + j\sqrt{k_\rho^2 - k^2}\right)}\right\}.$$

Here $\operatorname{res}\{f(k_z)\}$ denotes the residues of the function $f(k_z)$. The integrand in (A.54) has poles of order 1 at $k_z = j\sqrt{k^2 - k_\rho^2}$ and $k_z = -j\sqrt{k^2 - k_\rho^2}$, which are in the second and fourth quadrants, respectively. If $z - z' > 0$ we close in the upper half-plane and enclose only the pole at $k_z = j\sqrt{k_\rho^2 - k^2}$. Computing the residue using (A.14), we obtain

$$G^{xy}(k_x, k_y, z|\mathbf{r}') = \frac{e^{-jk_x x'} e^{-jk_y y'} e^{-\sqrt{k_\rho^2 - k^2}(z-z')}}{2\sqrt{k_\rho^2 - k^2}} \qquad (z > z').$$

Since $z > z'$, this function represents a wave that propagates in the $+z$ direction and decays for increasing z, as expected physically. For $z - z' < 0$ we close in the lower half-plane, enclosing the pole at $k_z = -j\sqrt{k_\rho^2 - k^2}$ and incurring an additional negative sign since our contour is now clockwise. The residue evaluation gives

$$G^{xy}(k_x, k_y, z|\mathbf{r}') = \frac{e^{-jk_x x'} e^{-jk_y y'} e^{\sqrt{k_\rho^2 - k^2}(z-z')}}{2\sqrt{k_\rho^2 - k^2}} \qquad (z < z').$$

We can combine both cases $z > z'$ and $z < z'$ by using the absolute value function:

$$G^{xy}(k_x, k_y, z|\mathbf{r}') = \frac{e^{-jk_x x'} e^{-jk_y y'} e^{-p|z-z'|}}{2p}, \qquad (A.55)$$

where $p = \sqrt{k_\rho^2 - k^2}$.

Finally, we substitute (A.55) into the inverse transform formula. This gives the Green's function representation

$$G(\mathbf{r}|\mathbf{r}') = \frac{e^{-jk|\mathbf{r}-\mathbf{r}'|}}{4\pi|\mathbf{r} - \mathbf{r}'|} = \frac{1}{(2\pi)^2} \int_{-\infty}^{\infty} \frac{e^{-p|z-z'|}}{2p} e^{j\mathbf{k}_\rho \cdot (\mathbf{r}-\mathbf{r}')} d^2 k_\rho, \qquad (A.56)$$

where $\mathbf{k}_\rho = \hat{\mathbf{x}} k_x + \hat{\mathbf{y}} k_y$, $k_\rho = |\mathbf{k}_\rho|$, and $d^2 k_\rho = dk_x\, dk_y$. Equation (A.56) has been called the *Weyl identity* [22].

A.2.8 Fourier transform representation of the static Green's function

In the study of static fields, we are interested in solving the partial differential equation

$$\nabla^2 G(\mathbf{r}|\mathbf{r}') = -\delta(\mathbf{r} - \mathbf{r}') = -\delta(x - x')\delta(y - y')\delta(z - z').$$

Here $G(\mathbf{r}|\mathbf{r}')$, called the "static Green's function," represents the potential at location $\mathbf{r}$ produced by a unit point source at location $\mathbf{r}'$.

In Chapter 3, we found that $G(\mathbf{r}|\mathbf{r}') = 1/4\pi|\mathbf{r} - \mathbf{r}'|$, but it is also useful to express G as an inverse Fourier transform over the variables x and y. We can easily obtain the desired relation by letting $k \to 0$ in (A.56). Then $p = k_\rho$ and we have the Green's function representation

$$G(\mathbf{r}|\mathbf{r}') = \frac{1}{4\pi|\mathbf{r} - \mathbf{r}'|} = \frac{1}{(2\pi)^2} \int_{-\infty}^{\infty} \frac{e^{-k_\rho|z-z'|}}{2k_\rho} e^{j\mathbf{k}_\rho \cdot (\mathbf{r}-\mathbf{r}')} \, d^2 k_\rho,$$

where $\mathbf{k}_\rho = \hat{x}k_x + \hat{y}k_y$, $k_\rho = |\mathbf{k}_\rho|$, and $d^2 k_\rho = dk_x \, dk_y$.

A.2.9 Fourier transform solution to Laplace's equation

On occasion we may wish to represent the solution of the homogeneous (Laplace) equation

$$\nabla^2 \psi(\mathbf{r}) = 0$$

in terms of a 2-D Fourier transform. In this case we represent ψ as a 2-D inverse transform as in § A.2.7 and substitute the inverse transform to obtain

$$\frac{1}{(2\pi)^2} \int_{-\infty}^{\infty} \nabla^2 \left(\psi^{xy}(k_x, k_y, z) e^{jk_x x} e^{jk_y y} \right) dk_x \, dk_y = 0.$$

Carrying out the derivatives and invoking the Fourier integral theorem, we find that

$$\left(\frac{\partial^2}{\partial z^2} - k_\rho^2 \right) \psi^{xy}(k_x, k_y, z) = 0.$$

Hence

$$\psi^{xy}(k_x, k_y, z) = A e^{k_\rho z} + B e^{-k_\rho z}$$

where A and B are constants with respect to z. Inverse transformation gives

$$\psi(\mathbf{r}) = \frac{1}{(2\pi)^2} \int_{-\infty}^{\infty} \left[A(\mathbf{k}_\rho) e^{k_\rho z} + B(\mathbf{k}_\rho) e^{-k_\rho z} \right] e^{j\mathbf{k}_\rho \cdot \mathbf{r}} \, d^2 k_\rho. \tag{A.57}$$

It is convenient to write this expression in polar coordinates for problems with azimuthal symmetry. Let $x = \rho\cos\phi$, $y = \rho\sin\phi$, $k_x = k_\rho\cos\alpha$, and $k_y = k_\rho\sin\alpha$. Then $\mathbf{k}_\rho \cdot \mathbf{r} = k_\rho\rho\cos(\alpha - \phi)$ and

$$\psi(\mathbf{r}) = \frac{1}{(2\pi)^2} \int_0^{2\pi} \int_0^{\infty} \left[A(\mathbf{k}_\rho) e^{k_\rho z} + B(\mathbf{k}_\rho) e^{-k_\rho z} \right] e^{jk_\rho\rho\cos(\alpha - \phi)} \, k_\rho \, dk_\rho \, d\alpha.$$

If the problem has azimuthal symmetry such that ψ is independent of ϕ, then both $A(\mathbf{k}_\rho)$ and $B(\mathbf{k}_\rho)$ are independent of α. (This can be shown by taking the forward transform of $\psi(\rho, z)$.) In this case

$$\psi(\rho, z) = \frac{1}{(2\pi)^2} \int_0^{\infty} \left[A(k_\rho) e^{k_\rho z} + B(k_\rho) e^{-k_\rho z} \right] \int_0^{2\pi} e^{jk_\rho\rho\cos(\alpha - \phi)} \, d\alpha \, k_\rho \, dk_\rho.$$

Using the change of variables $\xi = \alpha - \phi$ then gives

$$\psi(\rho, z) = \frac{1}{(2\pi)^2} \int_0^{\infty} \left[A(k_\rho) e^{k_\rho z} + B(k_\rho) e^{-k_\rho z} \right] \int_{-\pi}^{\pi} e^{jk_\rho\rho\cos\xi} \, d\xi \, k_\rho \, dk_\rho.$$

The integral over ξ is just $2\pi J_0(k_\rho\rho)$. Absorbing the factors 2π and k_ρ into the terms $A(k_\rho)$ and $B(k_\rho)$ gives the solution to Laplace's equation as

$$\psi(\rho, z) = \int_0^{\infty} \left[A(k_\rho) e^{k_\rho z} + B(k_\rho) e^{-k_\rho z} \right] J_0(k_\rho\rho) \, dk_\rho. \tag{A.58}$$

A.3 Vector transport theorems

We are often interested in the time rate of change of some field integrated over a moving volume or surface. Such a derivative may be used to describe the transport of a physical quantity (e.g., charge, momentum, energy) through space. Many of the relevant theorems are derived in this section. The results find application in the development of the large-scale forms of Maxwell equations, the continuity equation, and the Poynting theorem.

A.3.1 Partial, total, and material derivatives

The key to understanding transport theorems lies in the difference between the various means of time-differentiating a field. Consider a scalar field $T(\mathbf{r}, t)$ (which could represent one component of a vector or dyadic field). If we fix our position within the field and examine how the field varies with time, we describe the *partial derivative* of T. However, this may not be the most useful means of measuring the time rate of change of a field. For instance, in mechanics we might be interested in the rate at which water cools as it sinks to the bottom of a container. In this case, T could represent temperature. We could create a "depth profile" at any given time (i.e., measure $T(\mathbf{r}, t_0)$ for some fixed t_0) by taking simultaneous data from a series of temperature probes at varying depths. We could also create a temporal profile at any given depth (i.e., measure $T(\mathbf{r}_0, t)$ for some fixed $\mathbf{r}_0$) by taking continuous data from a probe fixed at that depth. But neither of these would describe how an individual sinking water particle "experiences" a change in temperature over time.

Instead, we could use a probe that descends along with a particular water packet (i.e., volume element), measuring the time rate of temperature change of that element. This rate of change is called the *convective* or *material derivative*, since it corresponds to a situation in which a physical material quantity is followed as the derivative is calculated. We anticipate that this quantity will depend on (1) the time rate of change of T at each fixed point that the particle passes, and (2) the spatial rate of change of T as well as the rapidity with which the packet of interest is swept through that space gradient. The faster the packet descends, or the faster the temperature cools with depth, the larger the material derivative should be.

To compute the material derivative, we describe the position of a water packet by the vector

$$\mathbf{r}(t) = \hat{\mathbf{x}} x(t) + \hat{\mathbf{y}} y(t) + \hat{\mathbf{z}} z(t).$$

Because no two packets can occupy the same place at the same time, the specification of $\mathbf{r}(0) = \mathbf{r}_0$ uniquely describes (or "tags") a particular packet. The time rate of change of $\mathbf{r}$ with $\mathbf{r}_0$ held constant (the material derivative of the position vector) is thus the velocity field $\mathbf{u}(\mathbf{r}, t)$ of the fluid:

$$\left(\frac{d\mathbf{r}}{dt} \right)_{\mathbf{r}_0} = \frac{D\mathbf{r}}{Dt} = \mathbf{u}. \tag{A.59}$$

Here we use the "big D" notation to denote the material derivative, thereby avoiding confusion with the partial and total derivatives described below.

To describe the time rate of change of the temperature of a particular water packet, we only need to hold $\mathbf{r}_0$ constant while we examine the change. If we write the temperature as

$$T(\mathbf{r}, t) = T(\mathbf{r}(\mathbf{r}_0, t), t) = T[x(\mathbf{r}_0, t), y(\mathbf{r}_0, t), z(\mathbf{r}_0, t), t],$$

then we can use the chain rule to find the time rate of change of T with $\mathbf{r}_0$ held constant:

$$\frac{DT}{Dt} = \left(\frac{dT}{dt}\right)_{\mathbf{r}_0} = \left(\frac{\partial T}{\partial x}\right)\left(\frac{dx}{dt}\right)_{\mathbf{r}_0} + \left(\frac{\partial T}{\partial y}\right)\left(\frac{dy}{dt}\right)_{\mathbf{r}_0} + \left(\frac{\partial T}{\partial z}\right)\left(\frac{dz}{dt}\right)_{\mathbf{r}_0} + \frac{\partial T}{\partial t}.$$

We recognize the partial derivatives of the coordinates as the components of the material velocity (A.59), and thus can write

$$\frac{DT}{Dt} = \frac{\partial T}{\partial t} + u_x\frac{\partial T}{\partial x} + u_y\frac{\partial T}{\partial y} + u_z\frac{\partial T}{\partial z} = \frac{\partial T}{\partial t} + \mathbf{u}\cdot\nabla T.$$

As expected, the material derivative depends on both the local time rate of change and the spatial rate of change of temperature.

Suppose next that our probe is motorized and can travel about in the sinking water. If the probe sinks faster than the surrounding water, the time rate of change (measured by the probe) should exceed the material derivative. Let the probe position and velocity be

$$\mathbf{r}(t) = \hat{\mathbf{x}}x(t) + \hat{\mathbf{y}}y(t) + \hat{\mathbf{z}}z(t), \qquad \mathbf{v}(\mathbf{r},t) = \hat{\mathbf{x}}\frac{dx(t)}{dt} + \hat{\mathbf{y}}\frac{dy(t)}{dt} + \hat{\mathbf{z}}\frac{dz(t)}{dt}.$$

We can use the chain rule to determine the time rate of change of the temperature observed by the probe, but in this case we do *not* constrain the velocity components to represent the moving fluid. Thus, we merely obtain

$$\frac{dT}{dt} = \frac{\partial T}{\partial x}\frac{dx}{dt} + \frac{\partial T}{\partial y}\frac{dy}{dt} + \frac{\partial T}{\partial z}\frac{dz}{dt} + \frac{\partial T}{\partial t} = \frac{\partial T}{\partial t} + \mathbf{v}\cdot\nabla T.$$

This is called the *total derivative* of the temperature field.

In summary, the time rate of change of a scalar field T seen by an observer moving with arbitrary velocity $\mathbf{v}$ is given by the total derivative

$$\frac{dT}{dt} = \frac{\partial T}{\partial t} + \mathbf{v}\cdot\nabla T. \tag{A.60}$$

If the velocity of the observer happens to match the velocity $\mathbf{u}$ of a moving substance, the time rate of change is the material derivative

$$\frac{DT}{Dt} = \frac{\partial T}{\partial t} + \mathbf{u}\cdot\nabla T. \tag{A.61}$$

We can obtain the material derivative of a vector field $\mathbf{F}$ by component-wise application of (A.61):

$$\frac{D\mathbf{F}}{Dt} = \frac{D}{Dt}\left[\hat{\mathbf{x}}F_x + \hat{\mathbf{y}}F_y + \hat{\mathbf{z}}F_z\right]$$

$$= \hat{\mathbf{x}}\frac{\partial F_x}{\partial t} + \hat{\mathbf{y}}\frac{\partial F_y}{\partial t} + \hat{\mathbf{z}}\frac{\partial F_z}{\partial t} + \hat{\mathbf{x}}\left[\mathbf{u}\cdot(\nabla F_x)\right] + \hat{\mathbf{y}}\left[\mathbf{u}\cdot(\nabla F_y)\right] + \hat{\mathbf{z}}\left[\mathbf{u}\cdot(\nabla F_z)\right].$$

Using the notation

$$\mathbf{u}\cdot\nabla = u_x\frac{\partial}{\partial x} + u_y\frac{\partial}{\partial y} + u_z\frac{\partial}{\partial z}$$

we can write

$$\frac{D\mathbf{F}}{Dt} = \frac{\partial\mathbf{F}}{\partial t} + (\mathbf{u}\cdot\nabla)\mathbf{F}. \tag{A.62}$$

This is the material derivative of a vector field $\mathbf{F}$ when $\mathbf{u}$ describes the motion of a physical material. Similarly, the total derivative of a vector field is

$$\frac{d\mathbf{F}}{dt} = \frac{\partial\mathbf{F}}{\partial t} + (\mathbf{v}\cdot\nabla)\mathbf{F}$$

where $\mathbf{v}$ is arbitrary.

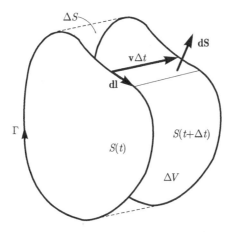

FIGURE A.5

Derivation of the Helmholtz transport theorem.

A.3.2 The Helmholtz and Reynolds transport theorems

We choose the intuitive approach taken by Tai [188] and Whitaker [214]. Consider an open surface $S(t)$ moving through space and possibly deforming as it moves. The velocity of the points composing the surface is given by the vector field $\mathbf{v}(\mathbf{r}, t)$. We are interested in computing the time derivative of the flux of a vector field $\mathbf{F}(\mathbf{r}, t)$ through $S(t)$:

$$\psi(t) = \frac{d}{dt} \int_{S(t)} \mathbf{F}(\mathbf{r}, t) \cdot d\mathbf{S} = \lim_{\Delta t \to 0} \frac{\int_{S(t+\Delta t)} \mathbf{F}(\mathbf{r}, t + \Delta t) \cdot d\mathbf{S} - \int_{S(t)} \mathbf{F}(\mathbf{r}, t) \cdot d\mathbf{S}}{\Delta t}. \quad (A.63)$$

Here $S(t+\Delta t) = S_2$ is found by extending each point on $S(t) = S_1$ through a displacement $\mathbf{v}\Delta t$, as shown in Figure A.5. Substituting the Taylor expansion

$$\mathbf{F}(\mathbf{r}, t + \Delta t) = \mathbf{F}(\mathbf{r}, t) + \frac{\partial \mathbf{F}(\mathbf{r}, t)}{\partial t} \Delta t + \cdots$$

into (A.63), we find that only the first two terms give nonzero contributions to the integral, and

$$\psi(t) = \int_{S(t)} \frac{\partial \mathbf{F}(\mathbf{r}, t)}{\partial t} \cdot d\mathbf{S} + \lim_{\Delta t \to 0} \frac{\int_{S_2} \mathbf{F}(\mathbf{r}, t) \cdot d\mathbf{S} - \int_{S_1} \mathbf{F}(\mathbf{r}, t) \cdot d\mathbf{S}}{\Delta t}. \quad (A.64)$$

The second term on the right can be evaluated with the help of Figure A.5. As the surface moves through a displacement $\mathbf{v}\Delta t$ it sweeps out a volume region ΔV that is bounded on the back by S_1, on the front by S_2, and on the side by a surface $S_3 = \Delta S$. We can thus compute the two surface integrals in (A.64) as the difference between contributions from the surface enclosing ΔV and the side surface ΔS (remembering that the normal to S_1 in (A.64) points *into* ΔV). Thus

$$\psi(t) = \int_{S(t)} \frac{\partial \mathbf{F}(\mathbf{r}, t)}{\partial t} \cdot d\mathbf{S} + \lim_{\Delta t \to 0} \frac{\oint_{S_1 + S_2 + \Delta S} \mathbf{F}(\mathbf{r}, t) \cdot d\mathbf{S} - \int_{\Delta S} \mathbf{F}(\mathbf{r}, t) \cdot d\mathbf{S}_3}{\Delta t}$$

$$= \int_{S(t)} \frac{\partial \mathbf{F}(\mathbf{r}, t)}{\partial t} \cdot d\mathbf{S} + \lim_{\Delta t \to 0} \frac{\int_{\Delta V} \nabla \cdot \mathbf{F}(\mathbf{r}, t) \, dV_3 - \int_{\Delta S} \mathbf{F}(\mathbf{r}, t) \cdot d\mathbf{S}_3}{\Delta t}$$

by the divergence theorem. To compute the integrals over ΔS and ΔV, we note from Figure A.5 that the incremental surface and volume elements are just

$$\mathbf{dS}_3 = \mathbf{dl} \times (\mathbf{v}\Delta t), \qquad dV_3 = (\mathbf{v}\Delta t) \cdot \mathbf{dS}.$$

Then, since $\mathbf{F} \cdot [\mathbf{dl} \times (\mathbf{v}\Delta t)] = \Delta t(\mathbf{v} \times \mathbf{F}) \cdot \mathbf{dl}$, we have

$$\psi(t) = \int_{S(t)} \frac{\partial \mathbf{F}(\mathbf{r},t)}{\partial t} \cdot \mathbf{dS} + \lim_{\Delta t \to 0} \frac{\Delta t \int_{S(t)} [\mathbf{v}\nabla \cdot \mathbf{F}(\mathbf{r},t)] \cdot \mathbf{dS}}{\Delta t} - \lim_{\Delta t \to 0} \frac{\Delta t \oint_{\Gamma} [\mathbf{v} \times \mathbf{F}(\mathbf{r},t)] \cdot \mathbf{dl}}{\Delta t}.$$

Taking the limit and using Stokes's theorem on the last integral, we finally have

$$\frac{d}{dt} \int_{S(t)} \mathbf{F} \cdot \mathbf{dS} = \int_{S(t)} \left[\frac{\partial \mathbf{F}}{\partial t} + \mathbf{v}\nabla \cdot \mathbf{F} - \nabla \times (\mathbf{v} \times \mathbf{F}) \right] \cdot \mathbf{dS}, \qquad (A.65)$$

which is the *Helmholtz transport theorem* [188, 40].

In case the surface corresponds to a moving physical material, we may wish to write the Helmholtz transport theorem in terms of the material derivative. We can set $\mathbf{v} = \mathbf{u}$ and use

$$\nabla \times (\mathbf{u} \times \mathbf{F}) = \mathbf{u}(\nabla \cdot \mathbf{F}) - \mathbf{F}(\nabla \cdot \mathbf{u}) + (\mathbf{F} \cdot \nabla)\mathbf{u} - (\mathbf{u} \cdot \nabla)\mathbf{F}$$

and (A.62) to obtain

$$\frac{d}{dt} \int_{S(t)} \mathbf{F} \cdot \mathbf{dS} = \int_{S(t)} \left[\frac{D\mathbf{F}}{Dt} + \mathbf{F}(\nabla \cdot \mathbf{u}) - (\mathbf{F} \cdot \nabla)\mathbf{u} \right] \cdot \mathbf{dS}.$$

If $S(t)$ in (A.65) is closed, enclosing a volume region $V(t)$, then

$$\oint_{S(t)} [\nabla \times (\mathbf{v} \times \mathbf{F})] \cdot \mathbf{dS} = \int_{V(t)} \nabla \cdot [\nabla \times (\mathbf{v} \times \mathbf{F})] \, dV = 0$$

by the divergence theorem and (B.55). In this case the Helmholtz transport theorem becomes

$$\frac{d}{dt} \oint_{S(t)} \mathbf{F} \cdot \mathbf{dS} = \oint_{S(t)} \left[\frac{\partial \mathbf{F}}{\partial t} + \mathbf{v}\nabla \cdot \mathbf{F} \right] \cdot \mathbf{dS}. \qquad (A.66)$$

We now come to an essential tool that we employ throughout the book. Using the divergence theorem, we can rewrite (A.66) as

$$\frac{d}{dt} \int_{V(t)} \nabla \cdot \mathbf{F} \, dV = \int_{V(t)} \nabla \cdot \frac{\partial \mathbf{F}}{\partial t} \, dV + \oint_{S(t)} (\nabla \cdot \mathbf{F})\mathbf{v} \cdot \mathbf{dS}.$$

Replacing $\nabla \cdot \mathbf{F}$ by the scalar field ρ, we have

$$\frac{d}{dt} \int_{V(t)} \rho \, dV = \int_{V(t)} \frac{\partial \rho}{\partial t} \, dV + \oint_{S(t)} \rho \mathbf{v} \cdot \mathbf{dS}. \qquad (A.67)$$

In this *general form* of the transport theorem, $\mathbf{v}$ is an arbitrary velocity. In most applications $\mathbf{v} = \mathbf{u}$ describes the motion of a material substance; then

$$\frac{D}{Dt} \int_{V(t)} \rho \, dV = \int_{V(t)} \frac{\partial \rho}{\partial t} \, dV + \oint_{S(t)} \rho \mathbf{u} \cdot \mathbf{dS}, \qquad (A.68)$$

which is the *Reynolds transport theorem* [214]. The D/Dt notation implies that $V(t)$ retains exactly the same material elements as it moves and deforms to follow the material substance.

We may rewrite the Reynolds transport theorem in various forms. By the divergence theorem, we have

$$\frac{d}{dt} \int_{V(t)} \rho \, dV = \int_{V(t)} \left[\frac{\partial \rho}{\partial t} + \nabla \cdot (\rho \mathbf{v}) \right] dV.$$

Setting $\mathbf{v} = \mathbf{u}$, using (B.48), and using (A.61) for the material derivative of ρ, we obtain

$$\frac{D}{Dt} \int_{V(t)} \rho \, dV = \int_{V(t)} \left[\frac{D\rho}{Dt} + \rho \nabla \cdot \mathbf{u} \right] dV. \tag{A.69}$$

We may also generate a vector form of the general transport theorem by taking ρ in (A.67) to be a component of a vector. Assembling all of the components, we have

$$\frac{d}{dt} \int_{V(t)} \mathbf{A} \, dV = \int_{V(t)} \frac{\partial \mathbf{A}}{\partial t} \, dV + \oint_{S(t)} \mathbf{A}(\mathbf{v} \cdot \hat{\mathbf{n}}) \, dS. \tag{A.70}$$

A.4 Dyadic analysis

Dyadic analysis was introduced in the late nineteenth century by Gibbs to generalize vector analysis to problems in which the components of vectors are related in a linear manner. It has now been widely supplanted by tensor theory, but maintains a foothold in engineering, where the transformation properties of tensors are not paramount (except, of course, in considerations such as those involving special relativity). Terms such as "tensor permittivity" and "dyadic permittivity" are often used interchangeably.

A.4.1 Component form representation

We wish to write one vector field $\mathbf{A}(\mathbf{r}, t)$ as a linear function of another vector field $\mathbf{B}(\mathbf{r}, t)$:

$$\mathbf{A} = f(\mathbf{B}).$$

By this we mean that each component of $\mathbf{A}$ is a linear combination of the components of $\mathbf{B}$:

$$A_1(\mathbf{r}, t) = a_{11'} B_{1'}(\mathbf{r}, t) + a_{12'} B_{2'}(\mathbf{r}, t) + a_{13'} B_{3'}(\mathbf{r}, t),$$
$$A_2(\mathbf{r}, t) = a_{21'} B_{1'}(\mathbf{r}, t) + a_{22'} B_{2'}(\mathbf{r}, t) + a_{23'} B_{3'}(\mathbf{r}, t),$$
$$A_3(\mathbf{r}, t) = a_{31'} B_{1'}(\mathbf{r}, t) + a_{32'} B_{2'}(\mathbf{r}, t) + a_{33'} B_{3'}(\mathbf{r}, t).$$

Here the $a_{ij'}$ may depend on space and time (or frequency). The prime on the second index indicates that $\mathbf{A}$ and $\mathbf{B}$ may be expressed in distinct coordinate frames $(\hat{\mathbf{i}}_1, \hat{\mathbf{i}}_2, \hat{\mathbf{i}}_3)$ and $(\hat{\mathbf{i}}_{1'}, \hat{\mathbf{i}}_{2'}, \hat{\mathbf{i}}_{3'})$, respectively. We have

$$A_1 = (a_{11'}\hat{\mathbf{i}}_{1'} + a_{12'}\hat{\mathbf{i}}_{2'} + a_{13'}\hat{\mathbf{i}}_{3'}) \cdot (\hat{\mathbf{i}}_{1'} B_{1'} + \hat{\mathbf{i}}_{2'} B_{2'} + \hat{\mathbf{i}}_{3'} B_{3'}),$$
$$A_2 = (a_{21'}\hat{\mathbf{i}}_{1'} + a_{22'}\hat{\mathbf{i}}_{2'} + a_{23'}\hat{\mathbf{i}}_{3'}) \cdot (\hat{\mathbf{i}}_{1'} B_{1'} + \hat{\mathbf{i}}_{2'} B_{2'} + \hat{\mathbf{i}}_{3'} B_{3'}),$$
$$A_3 = (a_{31'}\hat{\mathbf{i}}_{1'} + a_{32'}\hat{\mathbf{i}}_{2'} + a_{33'}\hat{\mathbf{i}}_{3'}) \cdot (\hat{\mathbf{i}}_{1'} B_{1'} + \hat{\mathbf{i}}_{2'} B_{2'} + \hat{\mathbf{i}}_{3'} B_{3'}),$$

and since $\mathbf{B} = \hat{\mathbf{i}}_{1'} B_{1'} + \hat{\mathbf{i}}_{2'} B_{2'} + \hat{\mathbf{i}}_{3'} B_{3'}$ we can write

$$\mathbf{A} = \hat{\mathbf{i}}_1(\mathbf{a}_1' \cdot \mathbf{B}) + \hat{\mathbf{i}}_2(\mathbf{a}_2' \cdot \mathbf{B}) + \hat{\mathbf{i}}_3(\mathbf{a}_3' \cdot \mathbf{B})$$

where

$$\mathbf{a}_1' = a_{11'}\hat{\mathbf{i}}_{1'} + a_{12'}\hat{\mathbf{i}}_{2'} + a_{13'}\hat{\mathbf{i}}_{3'},$$

$$\mathbf{a}_2' = a_{21'}\hat{\mathbf{i}}_{1'} + a_{22'}\hat{\mathbf{i}}_{2'} + a_{23'}\hat{\mathbf{i}}_{3'},$$

$$\mathbf{a}_3' = a_{31'}\hat{\mathbf{i}}_{1'} + a_{32'}\hat{\mathbf{i}}_{2'} + a_{33'}\hat{\mathbf{i}}_{3'}.$$

In shorthand notation

$$\mathbf{A} = \bar{\mathbf{a}} \cdot \mathbf{B} \qquad (A.71)$$

where

$$\bar{\mathbf{a}} = \hat{\mathbf{i}}_1 \mathbf{a}_1' + \hat{\mathbf{i}}_2 \mathbf{a}_2' + \hat{\mathbf{i}}_3 \mathbf{a}_3'. \qquad (A.72)$$

Written out, the quantity $\bar{\mathbf{a}}$ looks like

$$\bar{\mathbf{a}} = a_{11'}(\hat{\mathbf{i}}_1\hat{\mathbf{i}}_{1'}) + a_{12'}(\hat{\mathbf{i}}_1\hat{\mathbf{i}}_{2'}) + a_{13'}(\hat{\mathbf{i}}_1\hat{\mathbf{i}}_{3'})$$

$$+ a_{21'}(\hat{\mathbf{i}}_2\hat{\mathbf{i}}_{1'}) + a_{22'}(\hat{\mathbf{i}}_2\hat{\mathbf{i}}_{2'}) + a_{23'}(\hat{\mathbf{i}}_2\hat{\mathbf{i}}_{3'})$$

$$+ a_{31'}(\hat{\mathbf{i}}_3\hat{\mathbf{i}}_{1'}) + a_{32'}(\hat{\mathbf{i}}_3\hat{\mathbf{i}}_{2'}) + a_{33'}(\hat{\mathbf{i}}_3\hat{\mathbf{i}}_{3'}).$$

Terms such as $\hat{\mathbf{i}}_1\hat{\mathbf{i}}_{1'}$ are called *dyads*, while sums of dyads such as $\bar{\mathbf{a}}$ are called *dyadics*. The components $a_{ij'}$ of $\bar{\mathbf{a}}$ may be conveniently placed into an array:

$$[\bar{\mathbf{a}}] = \begin{bmatrix} a_{11'} & a_{12'} & a_{13'} \\ a_{21'} & a_{22'} & a_{23'} \\ a_{31'} & a_{32'} & a_{33'} \end{bmatrix}.$$

Writing

$$[\mathbf{A}] = \begin{bmatrix} A_1 \\ A_2 \\ A_3 \end{bmatrix}, \qquad [\mathbf{B}] = \begin{bmatrix} B_{1'} \\ B_{2'} \\ B_{3'} \end{bmatrix},$$

we see that $\mathbf{A} = \bar{\mathbf{a}} \cdot \mathbf{B}$ can be written as

$$[\mathbf{A}] = [\bar{\mathbf{a}}]\,[\mathbf{B}] = \begin{bmatrix} a_{11'} & a_{12'} & a_{13'} \\ a_{21'} & a_{22'} & a_{23'} \\ a_{31'} & a_{32'} & a_{33'} \end{bmatrix} \begin{bmatrix} B_{1'} \\ B_{2'} \\ B_{3'} \end{bmatrix}.$$

Note carefully that in (A.71) $\bar{\mathbf{a}}$ operates on $\mathbf{B}$ from the left. A reorganization of the components of $\bar{\mathbf{a}}$ allows us to write

$$\bar{\mathbf{a}} = \mathbf{a}_1\hat{\mathbf{i}}_{1'} + \mathbf{a}_2\hat{\mathbf{i}}_{2'} + \mathbf{a}_3\hat{\mathbf{i}}_{3'} \qquad (A.73)$$

where

$$\mathbf{a}_1 = a_{11'}\hat{\mathbf{i}}_1 + a_{21'}\hat{\mathbf{i}}_2 + a_{31'}\hat{\mathbf{i}}_3,$$

$$\mathbf{a}_2 = a_{12'}\hat{\mathbf{i}}_1 + a_{22'}\hat{\mathbf{i}}_2 + a_{32'}\hat{\mathbf{i}}_3,$$

$$\mathbf{a}_3 = a_{13'}\hat{\mathbf{i}}_1 + a_{23'}\hat{\mathbf{i}}_2 + a_{33'}\hat{\mathbf{i}}_3.$$

We may now consider using $\bar{\mathbf{a}}$ to operate on a vector $\mathbf{C} = \hat{\mathbf{i}}_1 C_1 + \hat{\mathbf{i}}_2 C_2 + \hat{\mathbf{i}}_3 C_3$ from the right:

$$\mathbf{C} \cdot \bar{\mathbf{a}} = (\mathbf{C} \cdot \mathbf{a}_1)\hat{\mathbf{i}}_{1'} + (\mathbf{C} \cdot \mathbf{a}_2)\hat{\mathbf{i}}_{2'} + (\mathbf{C} \cdot \mathbf{a}_3)\hat{\mathbf{i}}_{3'}.$$

In matrix form $\mathbf{C} \cdot \bar{\mathbf{a}}$ is

$$[\bar{\mathbf{a}}]^T [\mathbf{C}] = \begin{bmatrix} a_{11'} & a_{21'} & a_{31'} \\ a_{12'} & a_{22'} & a_{32'} \\ a_{13'} & a_{23'} & a_{33'} \end{bmatrix} \begin{bmatrix} C_1 \\ C_2 \\ C_3 \end{bmatrix}$$

where the superscript "T" denotes the matrix transpose operation. That is,

$$\mathbf{C} \cdot \bar{\mathbf{a}} = \bar{\mathbf{a}}^T \cdot \mathbf{C}$$

where $\bar{\mathbf{a}}^T$ is the transpose of $\bar{\mathbf{a}}$.

If the primed and unprimed frames coincide, then

$$\begin{aligned} \bar{\mathbf{a}} = {}& a_{11}(\hat{\mathbf{i}}_1\hat{\mathbf{i}}_1) + a_{12}(\hat{\mathbf{i}}_1\hat{\mathbf{i}}_2) + a_{13}(\hat{\mathbf{i}}_1\hat{\mathbf{i}}_3) \\ & + a_{21}(\hat{\mathbf{i}}_2\hat{\mathbf{i}}_1) + a_{22}(\hat{\mathbf{i}}_2\hat{\mathbf{i}}_2) + a_{23}(\hat{\mathbf{i}}_2\hat{\mathbf{i}}_3) \\ & + a_{31}(\hat{\mathbf{i}}_3\hat{\mathbf{i}}_1) + a_{32}(\hat{\mathbf{i}}_3\hat{\mathbf{i}}_2) + a_{33}(\hat{\mathbf{i}}_3\hat{\mathbf{i}}_3). \end{aligned}$$

In this case we may compare the results of $\bar{\mathbf{a}} \cdot \mathbf{B}$ and $\mathbf{B} \cdot \bar{\mathbf{a}}$ for a given vector $\mathbf{B} = \hat{\mathbf{i}}_1 B_1 + \hat{\mathbf{i}}_2 B_2 + \hat{\mathbf{i}}_3 B_3$. We leave it to the reader to verify that in general, $\mathbf{B} \cdot \bar{\mathbf{a}} \neq \bar{\mathbf{a}} \cdot \mathbf{B}$.

A.4.2 Vector form representation

We can express dyadics in coordinate-free fashion if we expand the concept of a dyad to permit entities such as $\mathbf{AB}$. Here $\mathbf{A}$ and $\mathbf{B}$ are called the *antecedent* and *consequent*, respectively. The operation rules

$$(\mathbf{AB}) \cdot \mathbf{C} = \mathbf{A}(\mathbf{B} \cdot \mathbf{C}), \qquad \mathbf{C} \cdot (\mathbf{AB}) = (\mathbf{C} \cdot \mathbf{A})\mathbf{B},$$

define the anterior and posterior products of $\mathbf{AB}$ with a vector $\mathbf{C}$, and give results consistent with our prior component notation. Sums of dyads such as $\mathbf{AB} + \mathbf{CD}$ are called *dyadic polynomials*, or dyadics. The simple dyadic

$$\mathbf{AB} = (A_1\hat{\mathbf{i}}_1 + A_2\hat{\mathbf{i}}_2 + A_3\hat{\mathbf{i}}_3)(B_{1'}\hat{\mathbf{i}}_{1'} + B_{2'}\hat{\mathbf{i}}_{2'} + B_{3'}\hat{\mathbf{i}}_{3'})$$

can be represented in component form using

$$\mathbf{AB} = \hat{\mathbf{i}}_1\mathbf{a}_1' + \hat{\mathbf{i}}_2\mathbf{a}_2' + \hat{\mathbf{i}}_3\mathbf{a}_3'$$

where

$$\begin{aligned} \mathbf{a}_1' &= A_1 B_{1'} \hat{\mathbf{i}}_{1'} + A_1 B_{2'} \hat{\mathbf{i}}_{2'} + A_1 B_{3'} \hat{\mathbf{i}}_{3'}, \\ \mathbf{a}_2' &= A_2 B_{1'} \hat{\mathbf{i}}_{1'} + A_2 B_{2'} \hat{\mathbf{i}}_{2'} + A_2 B_{3'} \hat{\mathbf{i}}_{3'}, \\ \mathbf{a}_3' &= A_3 B_{1'} \hat{\mathbf{i}}_{1'} + A_3 B_{2'} \hat{\mathbf{i}}_{2'} + A_3 B_{3'} \hat{\mathbf{i}}_{3'}, \end{aligned}$$

or using

$$\mathbf{AB} = \mathbf{a}_1\hat{\mathbf{i}}_{1'} + \mathbf{a}_2\hat{\mathbf{i}}_{2'} + \mathbf{a}_3\hat{\mathbf{i}}_{3'}$$

where

$$\begin{aligned} \mathbf{a}_1 &= \hat{\mathbf{i}}_1 A_1 B_{1'} + \hat{\mathbf{i}}_2 A_2 B_{1'} + \hat{\mathbf{i}}_3 A_3 B_{1'}, \\ \mathbf{a}_2 &= \hat{\mathbf{i}}_1 A_1 B_{2'} + \hat{\mathbf{i}}_2 A_2 B_{2'} + \hat{\mathbf{i}}_3 A_3 B_{2'}, \\ \mathbf{a}_3 &= \hat{\mathbf{i}}_1 A_1 B_{3'} + \hat{\mathbf{i}}_2 A_2 B_{3'} + \hat{\mathbf{i}}_3 A_3 B_{3'}. \end{aligned}$$

Note that if we write $\bar{\mathbf{a}} = \mathbf{AB}$, then $a_{ij} = A_i B_{j'}$.

A simple dyad $\mathbf{AB}$ by itself cannot represent a general dyadic $\bar{\mathbf{a}}$; only six independent quantities are available in $\mathbf{AB}$ (the three components of $\mathbf{A}$ and the three components of $\mathbf{B}$), while an arbitrary dyadic has nine independent components. However, it can be shown that any dyadic can be written as a sum of three dyads:

$$\bar{\mathbf{a}} = \mathbf{AB} + \mathbf{CD} + \mathbf{EF}.$$

This is called a *vector representation* of $\bar{\mathbf{a}}$. If $\mathbf{V}$ is a vector, the distributive laws

$$\bar{\mathbf{a}} \cdot \mathbf{V} = (\mathbf{AB} + \mathbf{CD} + \mathbf{EF}) \cdot \mathbf{V} = \mathbf{A}(\mathbf{B} \cdot \mathbf{V}) + \mathbf{C}(\mathbf{D} \cdot \mathbf{V}) + \mathbf{E}(\mathbf{F} \cdot \mathbf{V}),$$
$$\mathbf{V} \cdot \bar{\mathbf{a}} = \mathbf{V} \cdot (\mathbf{AB} + \mathbf{CD} + \mathbf{EF}) = (\mathbf{V} \cdot \mathbf{A})\mathbf{B} + (\mathbf{V} \cdot \mathbf{C})\mathbf{D} + (\mathbf{V} \cdot \mathbf{E})\mathbf{F},$$

apply.

A.4.3 Dyadic algebra and calculus

The cross product of a vector with a dyadic produces another dyadic. If $\bar{\mathbf{a}} = \mathbf{AB} + \mathbf{CD} + \mathbf{EF}$, then by definition

$$\bar{\mathbf{a}} \times \mathbf{V} = \mathbf{A}(\mathbf{B} \times \mathbf{V}) + \mathbf{C}(\mathbf{D} \times \mathbf{V}) + \mathbf{E}(\mathbf{F} \times \mathbf{V}),$$
$$\mathbf{V} \times \bar{\mathbf{a}} = (\mathbf{V} \times \mathbf{A})\mathbf{B} + (\mathbf{V} \times \mathbf{C})\mathbf{D} + (\mathbf{V} \times \mathbf{E})\mathbf{F}.$$

The corresponding component forms are

$$\bar{\mathbf{a}} \times \mathbf{V} = \hat{\mathbf{i}}_1 (\mathbf{a}_1' \times \mathbf{V}) + \hat{\mathbf{i}}_2 (\mathbf{a}_2' \times \mathbf{V}) + \hat{\mathbf{i}}_3 (\mathbf{a}_3' \times \mathbf{V}),$$
$$\mathbf{V} \times \bar{\mathbf{a}} = (\mathbf{V} \times \mathbf{a}_1)\hat{\mathbf{i}}_{1'} + (\mathbf{V} \times \mathbf{a}_2)\hat{\mathbf{i}}_{2'} + (\mathbf{V} \times \mathbf{a}_3)\hat{\mathbf{i}}_{3'},$$

where we have used (A.72) and (A.73), respectively. Interactions between dyads or dyadics may also be defined. The dot product of two dyads $\mathbf{AB}$ and $\mathbf{CD}$ is a dyad given by

$$(\mathbf{AB}) \cdot (\mathbf{CD}) = \mathbf{A}(\mathbf{B} \cdot \mathbf{C})\mathbf{D} = (\mathbf{B} \cdot \mathbf{C})(\mathbf{AD}).$$

The dot product of two dyadics can be found by applying the distributive property.

If α is a scalar, then the product $\alpha\bar{\mathbf{a}}$ is a dyadic with components equal to α times the components of $\bar{\mathbf{a}}$. Dyadic addition may be accomplished by adding individual dyadic components as long as the dyadics are expressed in the same coordinate system. Subtraction is accomplished by adding the negative of a dyadic, which is defined through scalar multiplication by -1.

Some useful dyadic identities appear in Appendix B. Many more can be found in Van Bladel [203].

The various vector derivatives may also be extended to dyadics. Computations are easiest in rectangular coordinates, since $\hat{\mathbf{i}}_1 = \hat{\mathbf{x}}$, $\hat{\mathbf{i}}_2 = \hat{\mathbf{y}}$, and $\hat{\mathbf{i}}_3 = \hat{\mathbf{z}}$ are constant with position. The dyadic

$$\bar{\mathbf{a}} = \mathbf{a}_x \hat{\mathbf{x}} + \mathbf{a}_y \hat{\mathbf{y}} + \mathbf{a}_z \hat{\mathbf{z}}$$

has divergence and curl

$$\nabla \cdot \bar{\mathbf{a}} = (\nabla \cdot \mathbf{a}_x)\hat{\mathbf{x}} + (\nabla \cdot \mathbf{a}_y)\hat{\mathbf{y}} + (\nabla \cdot \mathbf{a}_z)\hat{\mathbf{z}},$$
$$\nabla \times \bar{\mathbf{a}} = (\nabla \times \mathbf{a}_x)\hat{\mathbf{x}} + (\nabla \times \mathbf{a}_y)\hat{\mathbf{y}} + (\nabla \times \mathbf{a}_z)\hat{\mathbf{z}}.$$

Note that the divergence of a dyadic is a vector while the curl of a dyadic is a dyadic. The gradient of a vector $\mathbf{a} = a_x\hat{\mathbf{x}} + a_y\hat{\mathbf{y}} + a_z\hat{\mathbf{z}}$ is the dyadic quantity

$$\nabla\mathbf{a} = (\nabla a_x)\hat{\mathbf{x}} + (\nabla a_y)\hat{\mathbf{y}} + (\nabla a_z)\hat{\mathbf{z}}.$$

The dyadic derivatives may be expressed in coordinate-free notation by using the vector representation. The dyad $\mathbf{AB}$ has divergence and curl

$$\nabla \cdot (\mathbf{AB}) = (\nabla \cdot \mathbf{A})\mathbf{B} + \mathbf{A} \cdot (\nabla\mathbf{B}),$$
$$\nabla \times (\mathbf{AB}) = (\nabla \times \mathbf{A})\mathbf{B} - \mathbf{A} \times (\nabla\mathbf{B}).$$

The Laplacian of a dyadic is a dyadic given by

$$\nabla^2\bar{\mathbf{a}} = \nabla(\nabla \cdot \bar{\mathbf{a}}) - \nabla \times (\nabla \times \bar{\mathbf{a}}).$$

The divergence theorem for dyadics is

$$\int_V \nabla \cdot \bar{\mathbf{a}}\, dV = \oint_S \hat{\mathbf{n}} \cdot \bar{\mathbf{a}}\, dS.$$

Some of the other common differential and integral identities for dyadics can be found in Van Bladel [203] and Tai [189].

A.4.4 Special dyadics

We say that $\bar{\mathbf{a}}$ is *symmetric* if

$$\mathbf{B} \cdot \bar{\mathbf{a}} = \bar{\mathbf{a}} \cdot \mathbf{B}$$

for any vector $\mathbf{B}$. This requires $\bar{\mathbf{a}}^T = \bar{\mathbf{a}}$, i.e., $a_{ij'} = a_{ji'}$. We say that $\bar{\mathbf{a}}$ is *antisymmetric* if

$$\mathbf{B} \cdot \bar{\mathbf{a}} = -\bar{\mathbf{a}} \cdot \mathbf{B}$$

for any $\mathbf{B}$. In this case $\bar{\mathbf{a}}^T = -\bar{\mathbf{a}}$. That is, $a_{ij'} = -a_{ji'}$ and $a_{ii'} = 0$. A symmetric dyadic has only six independent components while an antisymmetric dyadic has only three. The reader can verify that any dyadic can be decomposed into symmetric and antisymmetric parts as

$$\bar{\mathbf{a}} = \tfrac{1}{2}(\bar{\mathbf{a}} + \bar{\mathbf{a}}^T) + \tfrac{1}{2}(\bar{\mathbf{a}} - \bar{\mathbf{a}}^T).$$

A simple example of a symmetric dyadic is the *unit dyadic* $\bar{\mathbf{I}}$ defined by

$$\bar{\mathbf{I}} = \hat{\mathbf{i}}_1\hat{\mathbf{i}}_1 + \hat{\mathbf{i}}_2\hat{\mathbf{i}}_2 + \hat{\mathbf{i}}_3\hat{\mathbf{i}}_3.$$

This quantity often arises in the manipulation of dyadic equations, and satisfies

$$\mathbf{A} \cdot \bar{\mathbf{I}} = \bar{\mathbf{I}} \cdot \mathbf{A} = \mathbf{A}$$

for any vector $\mathbf{A}$. In matrix form, $\bar{\mathbf{I}}$ is the identity matrix:

$$[\bar{\mathbf{I}}] = \begin{bmatrix} 1 & 0 & 0 \\ 0 & 1 & 0 \\ 0 & 0 & 1 \end{bmatrix}.$$

The components of a dyadic may be complex. We say that $\bar{\mathbf{a}}$ is *hermitian* if

$$\mathbf{B} \cdot \bar{\mathbf{a}} = \bar{\mathbf{a}}^* \cdot \mathbf{B} \tag{A.74}$$

holds for any $\mathbf{B}$. This requires that $\bar{\mathbf{a}}^* = \bar{\mathbf{a}}^T$. Taking the transpose, we can write

$$\bar{\mathbf{a}} = (\bar{\mathbf{a}}^*)^T = \bar{\mathbf{a}}^\dagger$$

where "$\dagger$" stands for the conjugate-transpose operation. We say that $\bar{\mathbf{a}}$ is *anti-hermitian* if

$$\mathbf{B} \cdot \bar{\mathbf{a}} = -\bar{\mathbf{a}}^* \cdot \mathbf{B} \qquad (A.75)$$

for arbitrary $\mathbf{B}$. In this case $\bar{\mathbf{a}}^* = -\bar{\mathbf{a}}^T$. Any complex dyadic can be decomposed into hermitian and anti-hermitian parts:

$$\bar{\mathbf{a}} = \tfrac{1}{2}(\bar{\mathbf{a}}^H + \bar{\mathbf{a}}^A) \qquad (A.76)$$

where

$$\bar{\mathbf{a}}^H = \bar{\mathbf{a}} + \bar{\mathbf{a}}^\dagger, \qquad \bar{\mathbf{a}}^A = \bar{\mathbf{a}} - \bar{\mathbf{a}}^\dagger. \qquad (A.77)$$

A dyadic identity important in the study of material parameters is

$$\mathbf{B} \cdot \bar{\mathbf{a}}^* \cdot \mathbf{B}^* = \mathbf{B}^* \cdot \bar{\mathbf{a}}^\dagger \cdot \mathbf{B}. \qquad (A.78)$$

We show this by decomposing $\bar{\mathbf{a}}$ according to (A.76), giving

$$\mathbf{B} \cdot \bar{\mathbf{a}}^* \cdot \mathbf{B}^* = \tfrac{1}{2} \left([\mathbf{B}^* \cdot \bar{\mathbf{a}}^H]^* + \mathbf{B}^* \cdot \bar{\mathbf{a}}^A]^* \right) \cdot \mathbf{B}^*$$

where we have used $(\mathbf{B} \cdot \bar{\mathbf{a}})^* = (\mathbf{B}^* \cdot \bar{\mathbf{a}}^*)$. Applying (A.74) and (A.75), we obtain

$$\mathbf{B} \cdot \bar{\mathbf{a}}^* \cdot \mathbf{B}^* = \tfrac{1}{2} \left([\bar{\mathbf{a}}^{H*} \cdot \mathbf{B}^*]^* - [\bar{\mathbf{a}}^{A*} \cdot \mathbf{B}^*]^* \right) \cdot \mathbf{B}^*$$
$$= \mathbf{B}^* \cdot \tfrac{1}{2} \left([\bar{\mathbf{a}}^H \cdot \mathbf{B}] - [\bar{\mathbf{a}}^A \cdot \mathbf{B}] \right)$$
$$= \mathbf{B}^* \cdot \left(\tfrac{1}{2}[\bar{\mathbf{a}}^H - \bar{\mathbf{a}}^A] \cdot \mathbf{B} \right).$$

Since the bracketed term is $\bar{\mathbf{a}}^H - \bar{\mathbf{a}}^A = 2\bar{\mathbf{a}}^\dagger$ by (A.77), the identity is proved.

A.5 Boundary value problems

Many physical phenomena may be described mathematically as the solutions to *boundary value problems*. The desired physical quantity (usually called a "field") in a certain region of space is found by solving one or more partial differential equations subject to certain conditions over the boundary surface. The boundary conditions may specify the values of the field, some manipulated version of the field (such as the normal derivative), or a relationship between fields in adjoining regions. If the field varies with time as well as space, initial or final values of the field must also be specified. Particularly important is whether a boundary value problem is *well-posed* and therefore has a unique solution that depends continuously on the data supplied. This depends on the forms of the differential equation and boundary conditions. The well-posedness of Maxwell's equations is discussed in § 2.2.

The importance of boundary value problems has led to an array of techniques, both analytical and numerical, for solving them. Many problems (such as boundary value problems involving Laplace's equation) may be solved in several different ways. Uniqueness permits an engineer to focus attention on which technique will yield the most efficient solution. In this section we concentrate on the separation of variables technique, which is widely applied in the solution of Maxwell's equations. We first discuss eigenvalue problems and then give an overview of separation of variables. Finally, we consider a number of example problems in each of the three common coordinate systems.

A.5.1 Sturm–Liouville problems and eigenvalues

The partial differential equations of electromagnetics can often be reduced to ordinary differential equations. In some cases, symmetry permits us to reduce the number of dimensions by inspection; in other cases, we may employ an integral transform (e.g., the Fourier transform) or separation of variables. The resulting ordinary differential equations may be viewed as particular cases of the *Sturm–Liouville differential equation*

$$\frac{d}{dx}\left[p(x)\frac{d\psi(x)}{dx}\right] + q(x)\psi(x) + \lambda\sigma(x)\psi(x) = 0 \qquad (a \leq x \leq b). \qquad (A.79)$$

In linear operator notation

$$\mathcal{L}\left[\psi(x)\right] = -\lambda\sigma(x)\psi(x), \qquad (A.80)$$

where $\mathcal{L}$ is the linear Sturm–Liouville operator

$$\mathcal{L} = \left(\frac{d}{dx}\left[p(x)\frac{d}{dx}\right] + q(x)\right).$$

Obviously $\psi(x) = 0$ satisfies (A.80). However, for certain values of λ dependent on p, q, σ, and the boundary conditions we impose, (A.80) has non-trivial solutions. Each λ that satisfies (A.80) is an *eigenvalue* of $\mathcal{L}$, and any non-trivial solution associated with that eigenvalue is an *eigenfunction*. Taken together, the eigenvalues of an operator form its *eigenvalue spectrum*.

We shall restrict ourselves to the case in which $\mathcal{L}$ is *self-adjoint*. Assume p, q, and σ are real and continuous on $[a, b]$. It is straightforward to show that for any two functions $u(x)$ and $v(x)$, Lagrange's identity

$$u\,\mathcal{L}[v] - v\,\mathcal{L}[u] = \frac{d}{dx}\left[p\left(u\frac{dv}{dx} - v\frac{du}{dx}\right)\right] \qquad (A.81)$$

holds. Integration gives Green's formula

$$\int_a^b (u\,\mathcal{L}[v] - v\,\mathcal{L}[u])\,dx = p\left(u\frac{dv}{dx} - v\frac{du}{dx}\right)\Big|_a^b.$$

The operator $\mathcal{L}$ is self-adjoint if its associated boundary conditions are such that

$$p\left(u\frac{dv}{dx} - v\frac{du}{dx}\right)\Big|_a^b = 0. \qquad (A.82)$$

Possible sets of conditions include the *homogeneous* boundary conditions

$$\alpha_1\psi(a) + \beta_1\psi'(a) = 0, \qquad \alpha_2\psi(b) + \beta_2\psi'(b) = 0, \qquad (A.83)$$

and the *periodic* boundary conditions

$$\psi(a) = \psi(b), \qquad p(a)\psi'(a) = p(b)\psi'(b). \qquad (A.84)$$

By imposing one of these sets on (A.80) we obtain a *Sturm–Liouville problem*.

The self-adjoint Sturm–Liouville operator has some nice properties. Each eigenvalue is real, and the eigenvalues form a denumerable set with no cluster point. Moreover, eigenfunctions corresponding to distinct eigenvalues are orthogonal, and the eigenfunctions form a complete set. Hence we can expand any sufficiently smooth function in terms of the eigenfunctions of a problem. We discuss this further below.

A *regular* Sturm–Liouville problem involves a self-adjoint operator $\mathcal{L}$ with $p(x) > 0$ and $\sigma(x) > 0$ everywhere, and the homogeneous boundary conditions (A.83). If p or σ vanishes at an endpoint of $[a, b]$, or an endpoint is at infinity, the problem is *singular*. The harmonic differential equation can form the basis of regular problems, while problems involving Bessel's and Legendre's equations are singular. Regular Sturm–Liouville problems have additional properties. There are infinitely many eigenvalues. There is a smallest eigenvalue but no largest eigenvalue, and the eigenvalues can be ordered as $\lambda_0 < \lambda_1 < \cdots < \lambda_n \cdots$. Associated with each λ_n is a unique (to an arbitrary multiplicative constant) eigenfunction ψ_n that has exactly n zeros in (a, b).

If a problem is singular because $p = 0$ at an endpoint, we can also satisfy (A.82) by demanding that ψ be bounded at that endpoint (a *singularity condition*) and that any regular Sturm–Liouville boundary condition hold at the other endpoint. This is the case for Bessel's and Legendre's equations discussed below.

A.5.1.1   Orthogonality of the eigenfunctions

Let $\mathcal{L}$ be self-adjoint, and let ψ_m and ψ_n be eigenfunctions associated with λ_m and λ_n, respectively. Then by (A.82) we have

$$\int_a^b \left(\psi_m(x)\, \mathcal{L}[\psi_n(x)] - \psi_n(x)\, \mathcal{L}[\psi_m(x)] \right) dx = 0.$$

But $\mathcal{L}[\psi_n(x)] = -\lambda_n \sigma(x)\psi_n(x)$ and $\mathcal{L}[\psi_m(x)] = -\lambda_m \sigma(x)\psi_m(x)$. Hence

$$(\lambda_m - \lambda_n) \int_a^b \psi_m(x)\psi_n(x)\sigma(x)\, dx = 0,$$

and $\lambda_m \neq \lambda_n$ implies that

$$\int_a^b \psi_m(x)\psi_n(x)\sigma(x)\, dx = 0. \tag{A.85}$$

We say that ψ_m and ψ_n are orthogonal with respect to the weight function $\sigma(x)$.

A.5.1.2   Eigenfunction expansion of an arbitrary function

If $\mathcal{L}$ is self-adjoint, then its eigenfunctions form a *complete set*. This means that any piecewise smooth function may be represented as a weighted series of eigenfunctions. Specifically, if f and f' are piecewise continuous on $[a, b]$, then f may be represented as the *generalized Fourier series*

$$f(x) = \sum_{n=0}^{\infty} c_n \psi_n(x). \tag{A.86}$$

Convergence of the series is uniform and gives, at any point of (a, b), the average value $[f(x^+) + f(x^-)]/2$ of the one-sided limits $f(x^+)$ and $f(x^-)$ of $f(x)$. The c_n can be found using orthogonality condition (A.85): multiply (A.86) by $\psi_m \sigma$ and integrate to obtain

$$\int_a^b f(x)\psi_m(x)\sigma(x)\,dx = \sum_{n=0}^{\infty} c_n \int_a^b \psi_n(x)\psi_m(x)\sigma(x)\,dx,$$

hence

$$c_n = \frac{\int_a^b f(x)\psi_n(x)\sigma(x)\,dx}{\int_a^b \psi_n^2(x)\sigma(x)\,dx}. \tag{A.87}$$

These coefficients ensure that the series *converges in mean* to f; i.e., the mean-square error

$$\int_a^b \left| f(x) - \sum_{n=0}^{\infty} c_n \psi_n(x) \right|^2 \sigma(x)\,dx$$

is minimized. Truncation to finitely many terms generally results in oscillations (*Gibbs phenomenon*) near points of discontinuity of f. The c_n are easier to compute if the ψ_n are orthonormal with

$$\int_a^b \psi_n^2(x)\sigma(x)\,dx = 1$$

for each n.

A.5.1.3 Uniqueness of the eigenfunctions

If both ψ_1 and ψ_2 are associated with the same eigenvalue λ, then

$$\mathcal{L}[\psi_1(x)] + \lambda\sigma(x)\psi_1(x) = 0, \qquad \mathcal{L}[\psi_2(x)] + \lambda\sigma(x)\psi_2(x) = 0,$$

hence

$$\psi_1(x)\,\mathcal{L}[\psi_2(x)] - \psi_2(x)\,\mathcal{L}[\psi_1(x)] = 0.$$

By (A.81) we have

$$\frac{d}{dx}\left[p(x)\left(\psi_1(x)\frac{d\psi_2(x)}{dx} - \psi_2(x)\frac{d\psi_1(x)}{dx} \right) \right] = 0$$

or

$$p(x)\left(\psi_1(x)\frac{d\psi_2(x)}{dx} - \psi_2(x)\frac{d\psi_1(x)}{dx} \right) = C$$

where C is constant. Either of (A.83) implies $C = 0$, hence

$$\frac{d}{dx}\left(\frac{\psi_2(x)}{\psi_1(x)} \right) = 0$$

so that $\psi_1(x) = K\psi_2(x)$ for some constant K. So under homogeneous boundary conditions, every eigenvalue is associated with a unique eigenfunction.

This is false for the periodic boundary conditions (A.84). Eigenfunction expansion then becomes difficult, as we can no longer assume eigenfunction orthogonality. However, the Gram–Schmidt algorithm may be used to construct orthogonal eigenfunctions. We refer the interested reader to Haberman [75].

A.5.1.4   The harmonic differential equation

The ordinary differential equation

$$\frac{d^2\psi(x)}{dx^2} = -k^2\psi(x) \tag{A.88}$$

is Sturm–Liouville with $p \equiv 1$, $q \equiv 0$, $\sigma \equiv 1$, and $\lambda = k^2$. Suppose we take $[a, b] = [0, L]$ and adopt the homogeneous boundary conditions

$$\psi(0) = 0 \quad \text{and} \quad \psi(L) = 0. \tag{A.89}$$

Since $p(x) > 0$ and $\sigma(x) > 0$ on $[0, L]$, equations (A.88) and (A.89) form a regular Sturm–Liouville problem. Thus we should have an infinite number of discrete eigenvalues. A power series technique yields the two independent solutions

$$\psi_a(x) = A_a \sin kx, \qquad \psi_b(x) = A_b \cos kx,$$

to (A.88); hence by linearity the most general solution is

$$\psi(x) = A_a \sin kx + A_b \cos kx. \tag{A.90}$$

The condition at $x = 0$ gives $A_a \sin 0 + A_b \cos 0 = 0$, hence $A_b = 0$. The other condition then requires

$$A_a \sin kL = 0. \tag{A.91}$$

Since $A_a = 0$ would give $\psi \equiv 0$, we satisfy (A.91) by choosing $k = k_n = n\pi/L$ for $n = 1, 2, \ldots$. Because $\lambda = k^2$, the eigenvalues are $\lambda_n = (n\pi/L)^2$ with corresponding eigenfunctions

$$\psi_n(x) = \sin k_n x.$$

Note that $\lambda = 0$ is not an eigenvalue; eigenfunctions are nontrivial by definition, and $\sin(0\pi x/L) \equiv 0$. Likewise, the differential equation associated with $\lambda = 0$ can be solved easily, but only its trivial solution can fit homogeneous boundary conditions: with $k = 0$, (A.88) becomes $d^2\psi(x)/dx^2 = 0$, giving $\psi(x) = ax + b$; this can satisfy (A.89) only with $a = b = 0$.

These "eigensolutions" obey the properties outlined earlier. In particular the ψ_n are orthogonal,

$$\int_0^L \sin\left(\frac{n\pi x}{L}\right) \sin\left(\frac{m\pi x}{L}\right) dx = \frac{L}{2}\delta_{mn},$$

and the eigenfunction expansion of a piecewise continuous function f is given by

$$f(x) = \sum_{n=1}^{\infty} c_n \sin\left(\frac{n\pi x}{L}\right)$$

where, with $\sigma(x) = 1$ in (A.87), we have

$$c_n = \frac{\int_0^L f(x) \sin\left(\frac{n\pi x}{L}\right) dx}{\int_0^L \sin^2\left(\frac{n\pi x}{L}\right) dx} = \frac{2}{L} \int_0^L f(x) \sin\left(\frac{n\pi x}{L}\right) dx.$$

Hence we recover the standard Fourier sine series for $f(x)$.

With little extra effort we can examine the eigenfunctions resulting from enforcement of the periodic boundary conditions

$$\psi(0) = \psi(L) \quad \text{and} \quad \psi'(0) = \psi'(L).$$

The general solution (A.90) still holds, so we have the choices $\psi(x) = \sin kx$ and $\psi(x) = \cos kx$. Evidently both

$$\psi(x) = \sin\left(\frac{2n\pi x}{L}\right) \quad \text{and} \quad \psi(x) = \cos\left(\frac{2n\pi x}{L}\right)$$

satisfy the boundary conditions for $n = 1, 2, \ldots$. Thus each eigenvalue $(2n\pi/L)^2$ is associated with two eigenfunctions.

A.5.1.5 Bessel's differential equation

Bessel's equation

$$\frac{d}{dx}\left(x\frac{d\psi(x)}{dx}\right) + \left(k^2 x - \frac{\nu^2}{x}\right)\psi(x) = 0 \tag{A.92}$$

occurs when problems are solved in circular-cylindrical coordinates. Comparison with (A.79) shows that $\lambda = k^2$, $p(x) = x$, $q(x) = -\nu^2/x$, and $\sigma(x) = x$. We take $[a, b] = [0, L]$ along with the boundary conditions

$$\psi(L) = 0 \quad \text{and} \quad |\psi(0)| < \infty. \tag{A.93}$$

Although the resulting Sturm–Liouville problem is singular, the specified conditions (A.93) maintain satisfaction of (A.82). The eigenfunctions are orthogonal because (A.82) is satisfied by having $\psi(L) = 0$ and $p(x)\,d\psi(x)/dx \to 0$ as $x \to 0$.

As a second-order ordinary differential equation, (A.92) has two solutions denoted by

$$J_\nu(kx) \quad \text{and} \quad N_\nu(kx),$$

and termed *Bessel functions*. Their properties are summarized in Appendix E.1. The function $J_\nu(x)$, the Bessel function of the first kind and order ν, is well-behaved in $[0, L]$. The function $N_\nu(x)$, the Bessel function of the second kind and order ν, is unbounded at $x = 0$; hence it is excluded as an eigenfunction of the Sturm–Liouville problem.

The condition at $x = L$ shows that the eigenvalues are defined by

$$J_\nu(kL) = 0.$$

We denote the mth root of $J_\nu(x) = 0$ by $p_{\nu m}$. Then

$$k_{\nu m} = \sqrt{\lambda_{\nu m}} = p_{\nu m}/L.$$

The infinitely many eigenvalues are ordered as $\lambda_{\nu 1} < \lambda_{\nu 2} < \ldots$. Associated with eigenvalue $\lambda_{\nu m}$ is a single eigenfunction $J_\nu(\sqrt{\lambda_{\nu m}}x)$. The orthogonality relation is

$$\int_0^L J_\nu\left(\frac{p_{\nu m}}{L}x\right) J_\nu\left(\frac{p_{\nu n}}{L}x\right) x\, dx = 0 \quad (m \neq n).$$

Since the eigenfunctions are also complete, we can expand any piecewise continuous function f in a *Fourier–Bessel series*

$$f(x) = \sum_{m=1}^{\infty} c_m J_\nu\left(p_{\nu m}\frac{x}{L}\right) \quad (0 \leq x \leq L, \, \nu > -1).$$

By (A.87) and (E.22) we have

$$c_m = \frac{2}{L^2 J_{\nu+1}^2(p_{\nu m})} \int_0^L f(x) J_\nu\left(p_{\nu m}\frac{x}{L}\right) x\, dx.$$

A.5.1.6 The associated Legendre equation

Legendre's equation occurs when problems are solved in spherical coordinates. It is often written in one of two forms. Letting θ be the polar angle of spherical coordinates ($0 \leq \theta \leq \pi$), the equation is

$$\frac{d}{d\theta}\left(\sin\theta\frac{d\psi(\theta)}{d\theta}\right) + \left(\lambda\sin\theta - \frac{m^2}{\sin\theta}\right)\psi(\theta) = 0.$$

This is Sturm–Liouville with $p(\theta) = \sin\theta$, $\sigma(\theta) = \sin\theta$, and $q(\theta) = -m^2/\sin\theta$. The boundary conditions

$$|\psi(0)| < \infty \quad \text{and} \quad |\psi(\pi)| < \infty$$

define a singular problem: the conditions are not homogeneous, $p(\theta) = 0$ at both endpoints, and $q(\theta) < 0$. Despite this, the Legendre problem does share properties of a regular Sturm–Liouville problem — including eigenfunction orthogonality and completeness.

Using $x = \cos\theta$, we can put Legendre's equation into its other common form

$$\frac{d}{dx}\left([1-x^2]\frac{d\psi(x)}{dx}\right) + \left(\lambda - \frac{m^2}{1-x^2}\right)\psi(x) = 0, \tag{A.94}$$

where $-1 \leq x \leq 1$. It is found that ψ is bounded at $x = \pm 1$ only if

$$\lambda = n(n+1)$$

where $n \geq m$ is an integer. These λ are the eigenvalues of the Sturm–Liouville problem, and the corresponding $\psi_n(x)$ are the eigenfunctions.

As a second-order partial differential equation, (A.94) has two solutions known as *associated Legendre functions*. The solution bounded at both $x = \pm 1$ is the associated Legendre function of the first kind, denoted $P_n^m(x)$. The second solution, unbounded at $x = \pm 1$, is the associated Legendre function of the second kind $Q_n^m(x)$. Appendix E.2 tabulates some properties of these functions.

For fixed m, each λ_{mn} is associated with a single eigenfunction $P_n^m(x)$. Since $P_n^m(x)$ is bounded at $x = \pm 1$, and since $p(\pm 1) = 0$, the eigenfunctions obey Lagrange's identity (A.81), hence are orthogonal on $[-1, 1]$ with respect to the weight function $\sigma(x) = 1$. Evaluation of the orthogonality integral leads to

$$\int_{-1}^1 P_l^m(x)P_n^m(x)\, dx = \delta_{ln}\frac{2}{2n+1}\frac{(n+m)!}{(n-m)!} \tag{A.95}$$

or equivalently

$$\int_0^\pi P_l^m(\cos\theta)P_n^m(\cos\theta)\sin\theta\, d\theta = \delta_{ln}\frac{2}{2n+1}\frac{(n+m)!}{(n-m)!}.$$

For $m = 0$, $P_n^m(x)$ is a polynomial of degree n. Each such *Legendre polynomial*, denoted $P_n(x)$, is given by

$$P_n(x) = \frac{1}{2^n n!}\frac{d^n(x^2-1)^n}{dx^n}.$$

It turns out that

$$P_n^m(x) = (-1)^m (1 - x^2)^{m/2} \frac{d^m P_n(x)}{dx^m},$$

giving $P_n^m(x) = 0$ for $m > n$.

Because the Legendre polynomials form a complete set in the interval $[-1, 1]$, we may expand any sufficiently smooth function in a *Fourier–Legendre series*

$$f(x) = \sum_{n=0}^{\infty} c_n P_n(x).$$

Convergence in mean is guaranteed if

$$c_n = \frac{2n+1}{2} \int_{-1}^{1} f(x) P_n(x) \, dx,$$

found using (A.87) along with (A.95).

In practice, the associated Legendre functions appear along with exponential functions in the solutions to spherical boundary value problems. The combined functions are known as *spherical harmonics*, and form solutions to two-dimensional Sturm–Liouville problems. We consider these next.

A.5.2   Higher-dimensional SL problems: Helmholtz's equation

Replacing d/dx by ∇, we generalize the Sturm–Liouville equation to higher dimensions:

$$\nabla \cdot [p(\mathbf{r}) \nabla \psi(\mathbf{r})] + q(\mathbf{r}) \psi(\mathbf{r}) + \lambda \sigma(\mathbf{r}) \psi(\mathbf{r}) = 0,$$

where q, p, σ, ψ are real functions. Of particular interest is the case $q(\mathbf{r}) = 0$, $p(\mathbf{r}) = \sigma(\mathbf{r}) = 1$, giving the Helmholtz equation

$$\nabla^2 \psi(\mathbf{r}) + \lambda \psi(\mathbf{r}) = 0. \tag{A.96}$$

In most boundary value problems, ψ or its normal derivative is specified on the surface of a bounded region. We obtain a three-dimensional analogue to the regular Sturm–Liouville problem by assuming the homogeneous boundary conditions

$$\alpha \psi(\mathbf{r}) + \beta \hat{\mathbf{n}} \cdot \nabla \psi(\mathbf{r}) = 0 \tag{A.97}$$

on the closed surface, where $\hat{\mathbf{n}}$ is the outward unit normal.

The problem consisting of (A.96) and (A.97) has properties analogous to those of the regular one-dimensional Sturm–Liouville problem. All eigenvalues are real. There are infinitely many eigenvalues. There is a smallest eigenvalue but no largest eigenvalue. However, associated with an eigenvalue there may be many eigenfunctions $\psi_\lambda(\mathbf{r})$. The eigenfunctions are orthogonal with

$$\int_V \psi_{\lambda_1}(\mathbf{r}) \psi_{\lambda_2}(\mathbf{r}) \, dV = 0 \qquad (\lambda_1 \neq \lambda_2).$$

They are also complete and can be used to represent any piecewise smooth function $f(\mathbf{r})$ according to

$$f(\mathbf{r}) = \sum_\lambda a_\lambda \psi_\lambda(\mathbf{r}),$$

which converges in mean when

$$a_{\lambda_m} = \frac{\int_V f(\mathbf{r}) \psi_{\lambda_m}(\mathbf{r}) \, dV}{\int_V \psi_{\lambda_m}^2(\mathbf{r}) \, dV}.$$

These properties are shared by the two-dimensional eigenvalue problem involving an open surface S with boundary contour Γ.

A.5.2.1 Spherical harmonics

We now inspect solutions to the two-dimensional eigenvalue problem

$$\nabla^2 Y(\theta, \phi) + \frac{\lambda}{a^2} Y(\theta, \phi) = 0$$

over the surface of a sphere of radius a. Since the sphere has no boundary contour, we demand that $Y(\theta, \phi)$ be bounded in θ and periodic in ϕ. In the next section we shall apply separation of variables and show that

$$Y_{nm}(\theta, \phi) = \sqrt{\frac{2n+1}{4\pi} \frac{(n-m)!}{(n+m)!}} \, P_n^m(\cos\theta) e^{jm\phi}$$

where $\lambda = n(n+1)$. Note that Q_n^m does not appear as it is not bounded at $\theta = 0, \pi$. The functions Y_{nm} are called *spherical harmonics* (sometimes *zonal* or *tesseral* harmonics, depending on the values of n and m). As expressed above they are in orthonormal form, because the orthogonality relationships for the exponential and associated Legendre functions yield

$$\int_{-\pi}^{\pi} \int_0^{\pi} Y_{n'm'}^*(\theta, \phi) Y_{nm}(\theta, \phi) \sin\theta \, d\theta \, d\phi = \delta_{n'n} \delta_{m'm}. \tag{A.98}$$

As solutions to the Sturm–Liouville problem, these functions form a complete set on the surface of a sphere. Hence they can be used to represent any piecewise smooth function $f(\theta, \phi)$ as

$$f(\theta, \phi) = \sum_{n=0}^{\infty} \sum_{m=-n}^{n} a_{nm} Y_{nm}(\theta, \phi),$$

where

$$a_{nm} = \int_{-\pi}^{\pi} \int_0^{\pi} f(\theta, \phi) Y_{nm}^*(\theta, \phi) \sin\theta \, d\theta \, d\phi$$

by (A.98). The summation index m ranges from $-n$ to n because $P_n^m = 0$ for $m > n$. For negative index we can use

$$Y_{n,-m}(\theta, \phi) = (-1)^m Y_{nm}^*(\theta, \phi).$$

Some properties of the spherical harmonics are tabulated in Appendix E.3.

A.5.3 Separation of variables

We now consider a technique that finds widespread application in solving boundary value problems, applying as it does to many important partial differential equations such as

Laplace's equation, the diffusion equation, and the scalar and vector wave equations. These equations are related to the scalar Helmholtz equation

$$\nabla^2 \psi(\mathbf{r}) + k^2 \psi(\mathbf{r}) = 0 \qquad (A.99)$$

where k is a complex constant. If k is real and we supply the appropriate boundary conditions, we have the higher-dimensional Sturm–Liouville problem with $\lambda = k^2$. We shall not pursue the extension of Sturm–Liouville theory to complex values of k.

Laplace's equation is Helmholtz's equation with $k = 0$. With $\lambda = k^2 = 0$ it might appear that Laplace's equation does not involve eigenvalues; however, separation of variables does lead us to lower-dimensional eigenvalue problems to which our previous methods apply. Solutions to the scalar or vector wave equations usually begin with Fourier transformation on the time variable, or with an initial separation of the time variable to reach a Helmholtz form.

The separation of variables idea is simple. We seek a solution to (A.99) in the form of a product of functions, each of a single variable. If ψ depends on all three spatial dimensions, then we seek a solution of the type

$$\psi(u, v, w) = U(u)V(v)W(w),$$

where u, v, and w are the coordinate variables used to describe the problem. If ψ depends on only two coordinates, we may seek a product solution involving two functions, each dependent on a single coordinate; alternatively, we may use the three-variable solution and choose constants so that the result shows no variation with one coordinate. The Helmholtz equation is considered *separable* if it can be reduced to a set of independent ordinary differential equations, each involving a single coordinate variable. The ordinary differential equations, generally of second order, can be solved by conventional techniques resulting in solutions of the form

$$U(u) = A_u U_A(u, k_u, k_v, k_w) + B_u U_B(u, k_u, k_v, k_w),$$
$$V(v) = A_v V_A(v, k_u, k_v, k_w) + B_v V_B(v, k_u, k_v, k_w),$$
$$W(w) = A_w W_A(w, k_u, k_v, k_w) + B_w W_B(w, k_u, k_v, k_w).$$

The constants k_u, k_v, k_w are called *separation constants* and are found, along with the amplitude constants A, B, by applying boundary conditions appropriate for a given problem. At least one separation constant depends on (or equals) k, so only two are independent. In many cases k_u, k_v, and k_w become the discrete eigenvalues of the respective differential equations, and correspond to eigenfunctions $U(u, k_u, k_v, k_w)$, $V(v, k_u, k_v, k_w)$, and $W(w, k_u, k_v, k_w)$. In other cases the separation constants form a continuous spectrum of values, often when a Fourier transform solution is employed.

The Helmholtz equation can be separated in eleven different orthogonal coordinate systems [137]. Undoubtedly the most important of these are the rectangular, circular-cylindrical, and spherical systems, and we shall consider each in detail. We do note, however, that separability in a certain coordinate system does not imply that all problems expressed in that coordinate system can be easily handled using the resulting solutions. Only when the geometry and boundary conditions are simple do the solutions lend themselves to easy application; often other solution techniques are more appropriate.

Although rigorous conditions can be set forth to guarantee solvability by separation of variables [124], we prefer the following, more heuristic list:

1. Use a coordinate system that allows the given partial differential equation to separate into ordinary differential equations.

2. The problem's boundaries must be such that those boundaries not at infinity coincide with a single level surface of the coordinate system.

3. Use superposition to reduce the problem to one involving a single nonhomogeneous boundary condition. Then:

   (a) Solve the resulting Sturm–Liouville problem in one or two dimensions, with homogeneous boundary conditions on all boundaries. Then use a discrete eigenvalue expansion (Fourier series) and eigenfunction orthogonality to satisfy the remaining nonhomogeneous condition.

   (b) If a Sturm–Liouville problem cannot be formulated with the homogeneous boundary conditions (because, for instance, one boundary is at infinity), use a Fourier integral (continuous expansion) to satisfy the remaining nonhomogeneous condition.

If a Sturm–Liouville problem cannot be formulated, discovering the form of the integral transform to use can be difficult. In these cases other approaches, such as conformal mapping, may prove easier.

A.5.3.1 Solutions in rectangular coordinates

In rectangular coordinates the Helmholtz equation is

$$\frac{\partial^2 \psi(x,y,z)}{\partial x^2} + \frac{\partial^2 \psi(x,y,z)}{\partial y^2} + \frac{\partial^2 \psi(x,y,z)}{\partial z^2} + k^2 \psi(x,y,z) = 0. \tag{A.100}$$

We seek a solution of the form $\psi(x,y,z) = X(x)Y(y)Z(z)$; substitution into (A.100) followed by division through by $X(x)Y(y)Z(z)$ gives

$$\frac{1}{X(x)}\frac{d^2 X(x)}{dx^2} + \frac{1}{Y(y)}\frac{d^2 Y(y)}{dy^2} + \frac{1}{Z(z)}\frac{d^2 Z(z)}{dz^2} = -k^2. \tag{A.101}$$

At this point we require the *separation argument*. The left-hand side of (A.101) is a sum of three functions, each involving a single independent variable, whereas the right-hand side is constant. But the only functions of independent variables that always sum to a constant are themselves constants. Thus we may equate each term on the left to a different constant:

$$\frac{1}{X(x)}\frac{d^2 X(x)}{dx^2} = -k_x^2,$$

$$\frac{1}{Y(y)}\frac{d^2 Y(y)}{dy^2} = -k_y^2, \tag{A.102}$$

$$\frac{1}{Z(z)}\frac{d^2 Z(z)}{dz^2} = -k_z^2,$$

provided that

$$k_x^2 + k_y^2 + k_z^2 = k^2.$$

The negative signs in (A.102) have been introduced for convenience.

Let us discuss the general solutions of equations (A.102). If $k_x = 0$, the two independent solutions for $X(x)$ are

$$X(x) = a_x x \quad \text{and} \quad X(x) = b_x$$

where a_x and b_x are constants. If $k_x \neq 0$, solutions may be chosen from the list of functions

$$e^{-jk_x x}, \qquad e^{jk_x x}, \qquad \sin k_x x, \qquad \cos k_x x,$$

any two of which are independent. Because

$$\sin x = (e^{jx} - e^{-jx})/2j \quad \text{and} \quad \cos x = (e^{jx} + e^{-jx})/2, \tag{A.103}$$

the six possible solutions for $k_x \neq 0$ are

$$X(x) = \begin{cases} A_x e^{jk_x x} + B_x e^{-jk_x x}, \\ A_x \sin k_x x + B_x \cos k_x x, \\ A_x \sin k_x x + B_x e^{-jk_x x}, \\ A_x e^{jk_x x} + B_x \sin k_x x, \\ A_x e^{jk_x x} + B_x \cos k_x x, \\ A_x e^{-jk_x x} + B_x \cos k_x x. \end{cases} \tag{A.104}$$

We may base our choice on convenience (e.g., the boundary conditions may be amenable to one particular form) or on the desired behavior of the solution (e.g., standing waves vs. traveling waves). If k is complex, then so may be k_x, k_y, or k_z; observe that with imaginary arguments the complex exponentials are actually real exponentials, and the trigonometric functions are actually hyperbolic functions.

The solutions for $Y(y)$ and $Z(z)$ are identical to those for $X(x)$. We can write, for instance,

$$X(x) = \begin{cases} A_x e^{jk_x x} + B_x e^{-jk_x x}, & k_x \neq 0, \\ a_x x + b_x, & k_x = 0, \end{cases} \tag{A.105}$$

$$Y(y) = \begin{cases} A_y e^{jk_y y} + B_y e^{-jk_y y}, & k_y \neq 0, \\ a_y y + b_y, & k_y = 0, \end{cases} \tag{A.106}$$

$$Z(z) = \begin{cases} A_z e^{jk_z z} + B_z e^{-jk_z z}, & k_z \neq 0, \\ a_z z + b_z, & k_z = 0. \end{cases} \tag{A.107}$$

▶ **Example A.1:** Solution to Laplace's equation in rectangular coordinates with one dimension

Solve $\nabla^2 V(x) = 0$.

Solution: Since V depends only on x we can use (A.105)–(A.107) with $k_y = k_z = 0$ and $a_y = a_z = 0$. Moreover, $k_x = 0$, because $k_x^2 + k_y^2 + k_z^2 = k^2 = 0$ for Laplace's equation. The general solution is therefore $V(x) = a_x x + b_x$. Boundary conditions must be specified to determine a_x and b_x; for instance, the conditions $V(0) = 0$ and $V(L) = V_0$ yield $V(x) = V_0 x/L$. ◀

▶ **Example A.2:** Solution to Laplace's equation in rectangular coordinates with two dimensions

Solve $\nabla^2 \psi(x, y) = 0$.

Solution: We produce a lack of z-dependence in ψ by letting $k_z = 0$ and choosing $a_z = 0$. Moreover, $k_x^2 = -k_y^2$ since Laplace's equation requires $k = 0$. This leads to three possibilities.

If $k_x = k_y = 0$, we have the product solution

$$\psi(x,y) = (a_x x + b_x)(a_y y + b_y). \tag{A.108}$$

If k_y is real and nonzero, then

$$\psi(x,y) = (A_x e^{-k_y x} + B_x e^{k_y x})(A_y e^{jk_y y} + B_y e^{-jk_y y}). \tag{A.109}$$

Using the relations

$$\sinh u = (e^u - e^{-u})/2 \quad \text{and} \quad \cosh u = (e^u + e^{-u})/2 \tag{A.110}$$

along with (A.103), we can rewrite (A.109) as

$$\psi(x,y) = (A_x \sinh k_y x + B_x \cosh k_y x)(A_y \sin k_y y + B_y \cos k_y y). \tag{A.111}$$

(We can reuse the constant names A_x, B_x, A_y, B_y, since the constants are unknown at this point.) If k_x is real and nonzero, we have

$$\psi(x,y) = (A_x \sin k_x x + B_x \cos k_x x)(A_y \sinh k_x y + B_y \cosh k_x y). \ \blacktriangleleft \tag{A.112}$$

▶ **Example A.3:** Solution to Laplace's equation in rectangular coordinates with two dimensions and boundary values — example 1

Consider the problem consisting of Laplace's equation

$$\nabla^2 V(x,y) = 0 \tag{A.113}$$

holding in the region $0 < x < L_1$, $0 < y < L_2$, $-\infty < z < \infty$, together with the boundary conditions

$$V(0,y) = V_1, \qquad V(L_1,y) = V_2, \qquad V(x,0) = V_3, \qquad V(x,L_2) = V_4.$$

Find $V(x,y)$.

Solution: The solution $V(x,y)$ represents the potential within a conducting tube with each wall held at a different potential. Superposition applies; since Laplace's equation is linear, we can write the solution as the sum of solutions to four different subproblems. Each subproblem has homogeneous boundary conditions on one independent variable and inhomogeneous conditions on the other, giving a Sturm–Liouville problem in one of the variables. For instance, let us examine the solutions found above in relation to the subproblem consisting of Laplace's equation (A.113) in the region $0 < x < L_1$, $0 < y < L_2$, $-\infty < z < \infty$, subject to the conditions

$$V(0,y) = V(L_1,y) = V(x,0) = 0, \quad V(x,L_2) = V_4 \neq 0.$$

First we try (A.108). The boundary condition at $x = 0$ gives

$$V(0,y) = (a_x(0) + b_x)(a_y y + b_y) = 0,$$

which holds for all $y \in (0, L_2)$ only if $b_x = 0$. The condition at $x = L_1$,

$$V(L_1,y) = a_x L_1 (a_y y + b_y) = 0,$$

then requires $a_x = 0$. But $a_x = b_x = 0$ gives $V(x,y) = 0$, and the condition at $y = L_2$ cannot be satisfied; clearly (A.108) was inappropriate. Next we examine (A.111). The condition at

$x = 0$ gives

$$V(0, y) = (A_x \sinh 0 + B_x \cosh 0)(A_y \sin k_y y + B_y \cos k_y y) = 0,$$

hence $B_x = 0$. The condition at $x = L_1$ implies

$$V(L_1, y) = [A_x \sinh(k_y L_1)](A_y \sin k_y y + B_y \cos k_y y) = 0.$$

This can hold if either $A_x = 0$ or $k_y = 0$, but the case $k_y = 0 \, (= k_x)$ was already considered. Thus $A_x = 0$ and the trivial solution reappears. Our last candidate is (A.112). The condition at $x = 0$ requires

$$V(0, y) = (A_x \sin 0 + B_x \cos 0)(A_y \sinh k_x y + B_y \cosh k_x y) = 0,$$

which implies $B_x = 0$. Next we have

$$V(L_1, y) = [A_x \sin(k_x L_1)](A_y \sinh k_y y + B_y \cosh k_y y) = 0.$$

We avoid $A_x = 0$ by setting $\sin(k_x L_1) = 0$ so that $k_{x_n} = n\pi/L_1$ for $n = 1, 2, \ldots$. (Here $n = 0$ is omitted because it would produce a trivial solution.) These are eigenvalues corresponding to the eigenfunctions $X_n(x) = \sin(k_{x_n} x)$, and were found in § A.5.1 for the harmonic equation. At this point, we have a family of solutions:

$$V_n(x, y) = \sin(k_{x_n} x)[A_{y_n} \sinh(k_{x_n} y) + B_{y_n} \cosh(k_{x_n} y)] \qquad (n = 1, 2, \ldots).$$

The subscript n on the left identifies V_n as the eigensolution associated with eigenvalue k_{x_n}. It remains to satisfy boundary conditions at $y = 0, L_2$. At $y = 0$ we have

$$V_n(x, 0) = \sin(k_{x_n} x)[A_{y_n} \sinh 0 + B_{y_n} \cosh 0] = 0,$$

hence $B_{y_n} = 0$ and

$$V_n(x, y) = A_{y_n} \sin(k_{x_n} x) \sinh(k_{x_n} y) \qquad (n = 1, 2, \ldots). \tag{A.114}$$

It is clear that no single eigensolution (A.114) can satisfy the one remaining boundary condition. However, we are guaranteed that a series of solutions can represent the constant potential on $y = L_2$; recall that as a solution to a regular Sturm–Liouville problem, the trigonometric functions are complete (hence they could represent any well-behaved function on the interval $0 \leq x \leq L_1$). In fact, the resulting series is a Fourier sine series for the constant potential at $y = L_2$. So let

$$V(x, y) = \sum_{n=1}^{\infty} V_n(x, y) = \sum_{n=1}^{\infty} A_{y_n} \sin(k_{x_n} x) \sinh(k_{x_n} y).$$

The remaining boundary condition requires

$$V(x, L_2) = \sum_{n=1}^{\infty} A_{y_n} \sin(k_{x_n} x) \sinh(k_{x_n} L_2) = V_4.$$

The constants A_{y_n} can be found using orthogonality; multiplying through by $\sin(k_{x_m} x)$ and integrating, we have

$$\sum_{n=1}^{\infty} A_{y_n} \sinh(k_{x_n} L_2) \int_0^{L_1} \sin\left(\frac{m\pi x}{L_1}\right) \sin\left(\frac{n\pi x}{L_1}\right) dx = V_4 \int_0^{L_1} \sin\left(\frac{m\pi x}{L_1}\right) dx.$$

The integral on the left equals $\delta_{mn} L_1/2$ where δ_{mn} is the Kronecker delta given by

$$\delta_{mn} = \begin{cases} 1, & m = n, \\ 0, & n \neq m. \end{cases}$$

After evaluating the integral on the right we obtain

$$\sum_{n=1}^{\infty} A_{y_n} \delta_{mn} \sinh(k_{x_n} L_2) = \frac{2V_4(1 - \cos m\pi)}{m\pi},$$

hence,

$$A_{y_m} = \frac{2V_4(1 - \cos m\pi)}{m\pi \sinh(k_{x_m} L_2)}.$$

The final solution for this subproblem is therefore

$$V(x, y) = \sum_{n=1}^{\infty} \frac{2V_4(1 - \cos n\pi)}{n\pi \sinh\left(\frac{n\pi L_2}{L_1}\right)} \sin\left(\frac{n\pi x}{L_1}\right) \sinh\left(\frac{n\pi y}{L_1}\right).$$

The remaining three subproblems are left for the reader. ◄

▶ **Example A.4:** Solution to Laplace's equation in rectangular coordinates with two dimensions and boundary values — example 2

Consider the problem consisting of Laplace's equation

$$\nabla^2 V(x, y) = 0$$

holding in the region

$$0 \leq x \leq L_1, \qquad 0 \leq y < \infty, \qquad -\infty < z < \infty,$$

together with the boundary conditions

$$V(0, y) = V(L_1, y) = 0, \qquad V(x, 0) = V_0.$$

Solve for $V(x, y)$.

Solution: Let us try the solution form that worked in the previous example:

$$V(x, y) = [A_x \sin(k_x x) + B_x \cos(k_x x)][A_y \sinh(k_x y) + B_y \cosh(k_x y)].$$

The boundary conditions at $x = 0, L_1$ are the same as before, so we have

$$V_n(x, y) = \sin(k_{x_n} x)[A_{y_n} \sinh(k_{x_n} y) + B_{y_n} \cosh(k_{x_n} y)] \qquad (n = 1, 2, \ldots).$$

To find A_{y_n} and B_{y_n} we note that V cannot grow without bound as $y \to \infty$. Individually the hyperbolic functions grow exponentially. However, using (A.110) we see that $B_{y_n} = -A_{y_n}$ gives

$$V_n(x, y) = A_{y_n} \sin(k_{x_n} x) e^{-k_{x_n} y}$$

where A_{y_n} is a new unknown constant. (Of course, we could have chosen this exponential dependence at the beginning.) Lastly, we can impose the boundary condition at $y = 0$ on the infinite series of eigenfunctions

$$V(x,y) = \sum_{n=1}^{\infty} A_{y_n} \sin(k_{x_n} x) e^{-k_{x_n} y}$$

to find A_{y_n}. The result is

$$V(x,y) = \sum_{n=1}^{\infty} \frac{2V_0}{\pi n} (1 - \cos n\pi) \sin(k_{x_n} x) e^{-k_{x_n} y}.$$

As in the previous example, the solution is a discrete superposition of eigenfunctions. ◄

▶ **Example A.5:** Solution to Laplace's equation in rectangular coordinates with two dimensions and boundary values — example 3

Consider the problem consisting of Laplace's equation

$$\nabla^2 V(x,y) = 0$$

holding in the region

$$0 \leq x \leq L_1, \quad 0 \leq y < \infty, \quad -\infty < z < \infty,$$

together with the boundary conditions

$$V(0,y) = 0, \quad V(L_1, y) = V_0 e^{-ay}, \quad V(x,0) = 0.$$

Solve for $V(x,y)$.

Solution: The solution requires a continuous superposition of eigenfunctions to satisfy the boundary conditions. Let us try

$$V(x,y) = [A_x \sinh k_y x + B_x \cosh k_y x][A_y \sin k_y y + B_y \cos k_y y].$$

The conditions at $x = 0$ and $y = 0$ require that $B_x = B_y = 0$. Thus

$$V_{k_y}(x,y) = A \sinh k_y x \sin k_y y.$$

A single function of this form cannot satisfy the remaining condition at $x = L_1$. So we form a continuous superposition

$$V(x,y) = \int_0^{\infty} A(k_y) \sinh k_y x \sin k_y y \, dk_y. \tag{A.115}$$

By the condition at $x = L_1$

$$\int_0^{\infty} A(k_y) \sinh(k_y L_1) \sin k_y y \, dk_y = V_0 e^{-ay}. \tag{A.116}$$

We can find the amplitude function $A(k_y)$ by using the orthogonality property

$$\delta(y - y') = \frac{2}{\pi} \int_0^{\infty} \sin xy \sin xy' \, dx. \tag{A.117}$$

Multiplying both sides of (A.116) by $\sin k_y' y$ and integrating, we have

$$\int_0^\infty A(k_y) \sinh(k_y L_1) \left[\int_0^\infty \sin k_y y \sin k_y' y \, dy \right] dk_y = \int_0^\infty V_0 e^{-ay} \sin k_y' y \, dy.$$

We can evaluate the term in brackets using (A.117) to obtain

$$\int_0^\infty A(k_y) \sinh(k_y L_1) \frac{\pi}{2} \delta(k_y - k_y') \, dk_y = \int_0^\infty V_0 e^{-ay} \sin k_y' y \, dy,$$

hence,

$$\frac{\pi}{2} A(k_y') \sinh(k_y' L_1) = V_0 \int_0^\infty e^{-ay} \sin k_y' y \, dy.$$

We then evaluate the integral on the right, solve for $A(k_y)$, and substitute into (A.115) to obtain

$$V(x,y) = \frac{2V_0}{\pi} \int_0^\infty \frac{k_y}{a^2 + k_y^2} \frac{\sinh(k_y x)}{\sinh(k_y L_1)} \sin k_y y \, dk_y.$$

Note that our application of the orthogonality property is merely a calculation of the inverse Fourier sine transform. Thus we could have found the amplitude coefficient by reference to a table of transforms. ◄

▶ **Example A.6:** Solution to Laplace's equation in rectangular coordinates with two dimensions and boundary values — example 4

We can use the Fourier transform solution even when the domain is infinite in more than one dimension. Consider the problem consisting of Laplace's equation

$$\nabla^2 V(x,y) = 0$$

holding in the region

$$0 \le x < \infty, \qquad 0 \le y < \infty, \qquad -\infty < z < \infty,$$

together with the boundary conditions

$$V(0,y) = V_0 e^{-ay}, \qquad V(x,0) = 0.$$

Solve for $V(x,y)$.

Solution: Because of the condition at $y = 0$, let us use

$$V(x,y) = (A_x e^{-k_y x} + B_x e^{k_y x})(A_y \sin k_y y + B_y \cos k_y y).$$

The solution form

$$V_{k_y}(x,y) = B(k_y) e^{-k_y x} \sin k_y y$$

satisfies the finiteness condition and the homogeneous condition at $y = 0$. The remaining condition can be satisfied by a continuous superposition of solutions:

$$V(x,y) = \int_0^\infty B(k_y) e^{-k_y x} \sin k_y y \, dk_y.$$

We must have

$$V_0 e^{-ay} = \int_0^\infty B(k_y) \sin k_y y \, dk_y.$$

Use of the orthogonality relationship (A.117) yields the amplitude spectrum $B(k_y)$, and we find that

$$V(x,y) = \frac{2}{\pi} \int_0^\infty e^{-k_y x} \frac{k_y}{a^2 + k_y^2} \sin k_y y \, dk_y. \quad \blacktriangleleft \qquad (A.118)$$

▶ **Example A.7:** Solution to the Helmholtz equation in rectangular coordinates with three dimensions and boundary values

Solve

$$\nabla^2 \psi(x,y,z) + k^2 \psi(x,y,z) = 0$$

for

$$0 \le x \le L_1, \qquad 0 \le y \le L_2, \qquad 0 \le z \le L_3,$$

subject to

$$\psi(0,y,z) = \psi(L_1,y,z) = 0,$$
$$\psi(x,0,z) = \psi(x,L_2,z) = 0,$$
$$\psi(x,y,0) = \psi(x,y,L_3) = 0.$$

Here $k \ne 0$ is a constant.

Solution: This is a three-dimensional eigenvalue problem as described in § A.5.1, where $\lambda = k^2$ are the eigenvalues and the closed surface is a rectangular box. Physically, the wave function ψ represents the so-called *eigenvalue* or *normal mode* solutions for the "TM modes" of a rectangular cavity. Since $k_x^2 + k_y^2 + k_z^2 = k^2$, we might have one or two separation constants equal to zero, but not all three. We find, however, that the only solution with a zero separation constant that can fit the boundary conditions is the trivial solution. In light of the boundary conditions and because we expect standing waves in the box, we take

$$\psi(x,y,z) = [A_x \sin(k_x x) + B_x \cos(k_x x)]$$
$$\times [A_y \sin(k_y y) + B_y \cos(k_y y)]$$
$$\times [A_z \sin(k_z z) + B_z \cos(k_z z)].$$

The conditions $\psi(0,y,z) = \psi(x,0,z) = \psi(x,y,0) = 0$ give $B_x = B_y = B_z = 0$. The conditions at $x = L_1$, $y = L_2$, and $z = L_3$ require the separation constants to assume the discrete values $k_x = k_{x_m} = m\pi/L_1$, $k_y = k_{y_n} = n\pi/L_2$, and $k_z = k_{z_p} = p\pi/L_3$, where $k_{x_m}^2 + k_{y_n}^2 + k_{z_p}^2 = k_{mnp}^2$ and $m,n,p = 1,2,\ldots$. Associated with each of these eigenvalues is an eigenfunction of a one-dimensional Sturm–Liouville problem. For the three-dimensional problem, an eigenfunction

$$\psi_{mnp}(x,y,z) = A_{mnp} \sin(k_{x_m} x) \sin(k_{y_n} y) \sin(k_{z_p} z)$$

is associated with each three-dimensional eigenvalue k_{mnp}. Each choice of m,n,p produces a discrete cavity resonance frequency at which the boundary conditions can be satisfied. Depending on the values of $L_{1,2,3}$, we may have more than one eigenfunction associated with an eigenvalue. For example, if $L_1 = L_2 = L_3 = L$ then $k_{121} = k_{211} = k_{112} = \sqrt{6}\pi/L$. However, the eigenfunctions associated with this single eigenvalue are all different:

$$\psi_{121} = \sin(k_{x_1} x) \sin(k_{y_2} y) \sin(k_{z_1} z),$$
$$\psi_{211} = \sin(k_{x_2} x) \sin(k_{y_1} y) \sin(k_{z_1} z),$$
$$\psi_{112} = \sin(k_{x_1} x) \sin(k_{y_1} y) \sin(k_{z_2} z).$$

When more than one cavity mode corresponds to a given resonant frequency, we call the modes *degenerate*. By completeness we can represent any well-behaved function as

$$f(x, y, z) = \sum_{m,n,p} A_{mnp} \sin(k_{x_m} x) \sin(k_{y_n} y) \sin(k_{z_p} z).$$

The A_{mnp} are found using orthogonality. When such expansions are used to solve problems involving objects (such as excitation probes) inside the cavity, they are termed *normal mode expansions* of the cavity field. ◀

A.5.3.2 Solutions in cylindrical coordinates

In cylindrical coordinates the Helmholtz equation is

$$\frac{1}{\rho}\frac{\partial}{\partial \rho}\left(\rho \frac{\partial \psi(\rho,\phi,z)}{\partial \rho}\right) + \frac{1}{\rho^2}\frac{\partial^2 \psi(\rho,\phi,z)}{\partial \phi^2} + \frac{\partial^2 \psi(\rho,\phi,z)}{\partial z^2} + k^2 \psi(\rho,\phi,z) = 0. \quad \text{(A.119)}$$

With $\psi(\rho, \phi, z) = P(\rho)\Phi(\phi)Z(z)$ we obtain

$$\frac{1}{\rho}\frac{\partial}{\partial \rho}\left(\rho \frac{\partial (P\Phi Z)}{\partial \rho}\right) + \frac{1}{\rho^2}\frac{\partial^2 (P\Phi Z)}{\partial \phi^2} + \frac{\partial^2 (P\Phi Z)}{\partial z^2} + k^2 (P\Phi Z) = 0;$$

carrying out the ρ derivatives and dividing through by $P\Phi Z$, we have

$$-\frac{1}{Z}\frac{d^2 Z}{dz^2} = k^2 + \frac{1}{\rho^2 \Phi}\frac{d^2 \Phi}{d\phi^2} + \frac{1}{\rho P}\frac{dP}{d\rho} + \frac{1}{P}\frac{d^2 P}{d\rho^2}.$$

The left side depends on z while the right side depends on ρ and ϕ, hence both must equal the same constant k_z^2:

$$-\frac{1}{Z}\frac{d^2 Z}{dz^2} = k_z^2, \quad \text{(A.120)}$$

$$k^2 + \frac{1}{\rho^2 \Phi}\frac{d^2 \Phi}{d\phi^2} + \frac{1}{\rho P}\frac{dP}{d\rho} + \frac{1}{P}\frac{d^2 P}{d\rho^2} = k_z^2. \quad \text{(A.121)}$$

We have separated the z-dependence from the dependence on the other variables. For the harmonic equation (A.120),

$$Z(z) = \begin{cases} A_z \sin k_z z + B_z \cos k_z z, & k_z \neq 0, \\ a_z z + b_z, & k_z = 0. \end{cases} \quad \text{(A.122)}$$

Of course we could use exponentials or a combination of exponentials and trigonometric functions instead. Rearranging (A.121) and multiplying through by ρ^2, we obtain

$$-\frac{1}{\Phi}\frac{d^2 \Phi}{d\phi^2} = \left(k^2 - k_z^2\right)\rho^2 + \frac{\rho}{P}\frac{dP}{d\rho} + \frac{\rho^2}{P}\frac{d^2 P}{d\rho^2}.$$

The left and right sides depend only on ϕ and ρ, respectively; both must equal some constant k_ϕ^2:

$$-\frac{1}{\Phi}\frac{d^2 \Phi}{d\phi^2} = k_\phi^2, \quad \text{(A.123)}$$

$$\left(k^2 - k_z^2\right)\rho^2 + \frac{\rho}{P}\frac{dP}{d\rho} + \frac{\rho^2}{P}\frac{d^2 P}{d\rho^2} = k_\phi^2. \quad \text{(A.124)}$$

The variables ρ and ϕ are thus separated, and harmonic equation (A.123) has solutions

$$\Phi(\phi) = \begin{cases} A_\phi \sin k_\phi \phi + B_\phi \cos k_\phi \phi, & k_\phi \neq 0, \\ a_\phi \phi + b_\phi, & k_\phi = 0. \end{cases} \qquad \text{(A.125)}$$

Equation (A.124) is a bit more involved. In rearranged form it is

$$\frac{d^2 P}{d\rho^2} + \frac{1}{\rho}\frac{dP}{d\rho} + \left(k_c^2 - \frac{k_\phi^2}{\rho^2} \right) P = 0 \qquad \text{(A.126)}$$

where

$$k_c^2 = k^2 - k_z^2.$$

The solution depends on whether any of k_z, k_ϕ, or k_c are zero. If $k_c = k_\phi = 0$, then

$$\frac{d^2 P}{d\rho^2} + \frac{1}{\rho}\frac{dP}{d\rho} = 0$$

so that

$$P(\rho) = a_\rho \ln \rho + b_\rho.$$

If $k_c = 0$ but $k_\phi \neq 0$, we have

$$\frac{d^2 P}{d\rho^2} + \frac{1}{\rho}\frac{dP}{d\rho} - \frac{k_\phi^2}{\rho^2} P = 0$$

so that

$$P(\rho) = a_\rho \rho^{-k_\phi} + b_\rho \rho^{k_\phi}. \qquad \text{(A.127)}$$

This includes the case $k = k_z = 0$ (Laplace's equation). If $k_c \neq 0$ then (A.126) is Bessel's differential equation. For noninteger k_ϕ the two independent solutions are denoted $J_{k_\phi}(z)$ and $J_{-k_\phi}(z)$, where $J_\nu(z)$ is the ordinary Bessel function of the first kind of order ν. For k_ϕ an integer n, $J_n(z)$ and $J_{-n}(z)$ are not independent and a second independent solution denoted $N_n(z)$ must be introduced. This is the ordinary Bessel function of the second kind, order n. As it is also independent when the order is noninteger, $J_\nu(z)$ and $N_\nu(z)$ are often chosen as solutions whether ν is integer or not. Linear combinations of these independent solutions may be used to produce new independent solutions. The functions $H_\nu^{(1)}(z)$ and $H_\nu^{(2)}(z)$ are the *Hankel functions* of the first and second kind of order ν, and are related to the Bessel functions by

$$H_\nu^{(1)}(z) = J_\nu(z) + j N_\nu(z),$$
$$H_\nu^{(2)}(z) = J_\nu(z) - j N_\nu(z).$$

The argument z can be complex (as can ν, but this shall not concern us). When z is imaginary we introduce two new functions $I_\nu(z)$ and $K_\nu(z)$, defined for integer order by

$$I_n(z) = j^{-n} J_n(jz),$$
$$K_n(z) = \frac{\pi}{2} j^{n+1} H_n^{(1)}(jz).$$

Expressions for noninteger order are given in Appendix E.1.

Bessel functions cannot be expressed in terms of simple, standard functions. However, a series solution to (A.126) produces many useful relationships between Bessel functions of differing order and argument. The *recursion relations* for Bessel functions serve to connect functions of various orders and their derivatives. See Appendix E.1.

Of the six possible solutions to (A.126),

$$R(\rho) = \begin{cases} A_\rho J_\nu(k_c\rho) + B_\rho N_\nu(k_c\rho), \\ A_\rho J_\nu(k_c\rho) + B_\rho H_\nu^{(1)}(k_c\rho), \\ A_\rho J_\nu(k_c\rho) + B_\rho H_\nu^{(2)}(k_c\rho), \\ A_\rho N_\nu(k_c\rho) + B_\rho H_\nu^{(1)}(k_c\rho), \\ A_\rho N_\nu(k_c\rho) + B_\rho H_\nu^{(2)}(k_c\rho), \\ A_\rho H_\nu^{(1)}(k_c\rho) + B_\rho H_\nu^{(2)}(k_c\rho), \end{cases}$$

which do we choose? Again, we are motivated by convenience and the physical nature of the problem. If the argument is real or imaginary, we often consider large or small argument behavior. For x real and large,

$$J_\nu(x) \to \sqrt{\frac{2}{\pi x}} \cos\left(x - \frac{\pi}{4} - \nu\frac{\pi}{2}\right), \qquad N_\nu(x) \to \sqrt{\frac{2}{\pi x}} \sin\left(x - \frac{\pi}{4} - \nu\frac{\pi}{2}\right),$$

$$H_\nu^{(1)}(x) \to \sqrt{\frac{2}{\pi x}} e^{j\left(x - \frac{\pi}{4} - \nu\frac{\pi}{2}\right)}, \qquad H_\nu^{(2)}(x) \to \sqrt{\frac{2}{\pi x}} e^{-j\left(x - \frac{\pi}{4} - \nu\frac{\pi}{2}\right)},$$

$$I_\nu(x) \to \sqrt{\frac{1}{2\pi x}} e^x, \qquad K_\nu(x) \to \sqrt{\frac{\pi}{2x}} e^{-x},$$

while for x real and small,

$$J_0(x) \to 1, \qquad\qquad\qquad J_\nu(x) \to \frac{1}{\nu!}\left(\frac{x}{2}\right)^\nu,$$

$$N_0(x) \to \frac{2}{\pi}(\ln x + 0.5772157 - \ln 2), \qquad N_\nu(x) \to -\frac{(\nu-1)!}{\pi}\left(\frac{2}{x}\right)^\nu.$$

Because $J_\nu(x)$ and $N_\nu(x)$ oscillate for large argument, they can represent standing waves along the radial direction. However, $N_\nu(x)$ is unbounded for small x and is inappropriate for regions containing the origin. The Hankel functions become complex exponentials for large argument, hence, they represent traveling waves. Finally, $K_\nu(x)$ is unbounded for small x and cannot be used for regions containing the origin, while $I_\nu(x)$ increases exponentially for large x and cannot be used for unbounded regions.

▶ **Example A.8:** Solution to Laplace's equation in cylindrical coordinates with two dimensions and boundary values — example 1

Solve

$$\nabla^2 V(\rho, \phi) = 0$$

in the region

$$0 \le \rho \le \infty, \qquad 0 \le \phi \le \phi_0, \qquad -\infty < z < \infty,$$

with the boundary conditions

$$V(\rho, 0) = 0, \qquad V(\rho, \phi_0) = V_0.$$

Solution: Since there is no z-dependence we let $k_z = 0$ in (A.122) and choose $a_z = 0$. Then

$k_c^2 = k^2 - k_z^2 = 0$ since $k = 0$. There are two possible solutions, depending on whether k_ϕ is zero. First let us try $k_\phi \neq 0$. Using (A.125) and (A.127), we have

$$V(\rho, \phi) = [A_\phi \sin(k_\phi \phi) + B_\phi \cos(k_\phi \phi)][a_\rho \rho^{-k_\phi} + b_\rho \rho^{k_\phi}]. \tag{A.128}$$

Assuming $k_\phi > 0$ we must have $b_\rho = 0$ to keep the solution finite. The condition $V(\rho, 0) = 0$ requires $B_\phi = 0$. Thus

$$V(\rho, \phi) = A_\phi \sin(k_\phi \phi) \rho^{-k_\phi}.$$

Our final boundary condition requires

$$V(\rho, \phi_0) = V_0 = A_\phi \sin(k_\phi \phi_0) \rho^{-k_\phi}.$$

Because this cannot hold for all ρ, we must resort to $k_\phi = 0$ and

$$V(\rho, \phi) = (a_\phi \phi + b_\phi)(a_\rho \ln \rho + b_\rho). \tag{A.129}$$

Proper behavior as $\rho \to \infty$ dictates that $a_\rho = 0$. $V(\rho, 0) = 0$ requires $b_\phi = 0$. Thus $V(\rho, \phi) = V(\phi) = b_\phi \phi$. The constant b_ϕ is found from the remaining boundary condition: $V(\phi_0) = V_0 = b_\phi \phi_0$ so that $b_\phi = V_0/\phi_0$. The final solution is

$$V(\phi) = V_0 \phi/\phi_0.$$

It is worthwhile to specialize this to $\phi_0 = \pi/2$ and compare with the solution to the same problem found earlier using rectangular coordinates. With $a = 0$ in (A.118) we have

$$V(x, y) = \frac{2}{\pi} \int_0^\infty e^{-k_y x} \frac{\sin k_y y}{k_y} \, dk_y.$$

Despite its much more complicated form, this must be the same solution by uniqueness. ◄

▶ **Example A.9**: Solution to Laplace's equation in cylindrical coordinates with two dimensions and boundary values — example 2

Solve

$$\nabla^2 V(\rho, \phi) = 0$$

in the region

$$0 \leq \rho \leq \infty, \qquad 0 \leq \phi \leq 2\pi, \qquad -\infty < z < \infty,$$

where the boundary conditions are those of a "split cylinder":

$$V(a, \phi) = \begin{cases} V_0, & 0 < \phi < \pi, \\ 0, & -\pi < \phi < 0. \end{cases}$$

Solution: Because there is no z-dependence, we choose $k_z = a_z = 0$ and have $k_c^2 = k^2 - k_z^2 = 0$. Since $k_\phi = 0$ would violate the boundary conditions at $\rho = a$, we use

$$V(\rho, \phi) = (a_\rho \rho^{-k_\phi} + b_\rho \rho^{k_\phi})(A_\phi \sin k_\phi \phi + B_\phi \cos k_\phi \phi).$$

The potential must be single-valued in ϕ: $V(\rho, \phi + 2n\pi) = V(\rho, \phi)$. This is only possible if k_ϕ is an integer, say $k_\phi = m$. Then

$$V_m(\rho, \phi) = \begin{cases} (A_m \sin m\phi + B_m \cos m\phi)\rho^m, & \rho < a, \\ (C_m \sin m\phi + D_m \cos m\phi)\rho^{-m}, & \rho > a. \end{cases}$$

On physical grounds we have discarded ρ^{-m} for $\rho < a$ and ρ^m for $\rho > a$. To satisfy the boundary conditions at $\rho = a$ we must use an infinite series of the complete set of eigensolutions. For $\rho < a$, the boundary condition requires

$$B_0 + \sum_{m=1}^{\infty} (A_m \sin m\phi + B_m \cos m\phi)a^m = \begin{cases} V_0, & 0 < \phi < \pi, \\ 0, & -\pi < \phi < 0. \end{cases}$$

Application of the orthogonality relations

$$\int_{-\pi}^{\pi} \cos m\phi \cos n\phi \, d\phi = \frac{2\pi}{\epsilon_n} \delta_{mn} \quad (m, n = 0, 1, 2, \ldots), \tag{A.130}$$

$$\int_{-\pi}^{\pi} \sin m\phi \sin n\phi \, d\phi = \pi \delta_{mn} \quad (m, n = 1, 2, \ldots), \tag{A.131}$$

$$\int_{-\pi}^{\pi} \cos m\phi \sin n\phi \, d\phi = 0 \quad (m, n = 0, 1, 2, \ldots), \tag{A.132}$$

where

$$\epsilon_n = \begin{cases} 1, & n = 0, \\ 2, & n > 0, \end{cases} \tag{A.133}$$

is Neumann's number, produces appropriate values for the constants A_m and B_m. The full solution is

$$V(\rho, \phi) = \begin{cases} \dfrac{V_0}{2} + \dfrac{V_0}{\pi} \displaystyle\sum_{n=1}^{\infty} \dfrac{[1 - (-1)^n]}{n} \left(\dfrac{\rho}{a}\right)^n \sin n\phi, & \rho < a, \\ \dfrac{V_0}{2} + \dfrac{V_0}{\pi} \displaystyle\sum_{n=1}^{\infty} \dfrac{[1 - (-1)^n]}{n} \left(\dfrac{a}{\rho}\right)^n \sin n\phi, & \rho > a. \end{cases} \blacktriangleleft$$

▶ **Example A.10:** Solution to Laplace's equation in cylindrical coordinates with three dimensions and boundary values

Solve

$$\nabla^2 V(\rho, \phi, z) = 0$$

in the region

$$0 \leq \rho \leq a, \qquad -\pi \leq \phi \leq \pi, \qquad 0 \leq z \leq h,$$

where the boundary conditions are those of a "grounded cannister" with top at potential V_0:

$$V(\rho, \phi, 0) = 0 \quad (0 \leq \rho \leq a, -\pi \leq \phi \leq \pi),$$
$$V(a, \phi, z) = 0 \quad (-\pi \leq \phi \leq \pi, 0 \leq z \leq h),$$
$$V(\rho, \phi, h) = V_0 \quad (0 \leq \rho \leq a, -\pi \leq \phi \leq \pi).$$

Solution: Symmetry precludes ϕ-dependence, hence $k_\phi = a_\phi = 0$. Since $k = 0$ (Laplace's equation) we also have $k_c^2 = k^2 - k_z^2 = -k_z^2$. Thus we have either k_z real and $k_c = jk_z$, or k_c real and $k_z = jk_c$. With k_z real we have

$$V(\rho, z) = [A_z \sin k_z z + B_z \cos k_z z][A_\rho K_0(k_z \rho) + B_\rho I_0(k_z \rho)];$$

with k_c real we have

$$V(\rho, z) = [A_z \sinh k_c z + B_z \cosh k_c z][A_\rho J_0(k_c \rho) + B_\rho N_0(k_c \rho)]. \qquad (A.134)$$

The functions K_0 and I_0 are inappropriate for use in this problem, and we proceed to (A.134). Since N_0 is unbounded for small argument, we need $B_\rho = 0$. The condition $V(\rho, 0) = 0$ gives $B_z = 0$, thus

$$V(\rho, z) = A_z \sinh(k_c z) J_0(k_c \rho).$$

The oscillatory nature of J_0 means that we can satisfy the condition at $\rho = a$:

$$V(a, z) = A_z \sinh(k_c z) J_0(k_c a) = 0 \quad \text{for} \quad 0 \le z < h$$

if $J_0(k_c a) = 0$. Letting p_{0m} denote the mth root of $J_0(x) = 0$ for $m = 1, 2, \ldots$, we have $k_{c_m} = p_{0m}/a$. Because we cannot satisfy the boundary condition at $z = h$ with a single eigensolution, we use the superposition

$$V(\rho, z) = \sum_{m=1}^{\infty} A_m \sinh\left(\frac{p_{0m} z}{a}\right) J_0\left(\frac{p_{0m} \rho}{a}\right).$$

We require

$$V(\rho, h) = \sum_{m=1}^{\infty} A_m \sinh\left(\frac{p_{0m} h}{a}\right) J_0\left(\frac{p_{0m} \rho}{a}\right) = V_0, \qquad (A.135)$$

where the A_m can be evaluated by orthogonality of the functions $J_0(p_{0m}\rho/a)$. If $p_{\nu m}$ is the mth root of $J_\nu(x) = 0$, then

$$\int_0^a J_\nu\left(\frac{p_{\nu m}\rho}{a}\right) J_\nu\left(\frac{p_{\nu n}\rho}{a}\right) \rho\, d\rho = \delta_{mn} \frac{a^2}{2} J_\nu'^2(p_{\nu n}) = \delta_{mn} \frac{a^2}{2} J_{\nu+1}^2(p_{\nu n}) \qquad (A.136)$$

where $J_\nu'(x) = dJ_\nu(x)/dx$. Multiplying (A.135) by $\rho J_0(p_{0n}\rho/a)$ and integrating, we have

$$A_n \sinh\left(\frac{p_{0n} h}{a}\right) \frac{a^2}{2} J_0'^2(p_{0n}a) = \int_0^a V_0 J_0\left(\frac{p_{0n}\rho}{a}\right) \rho\, d\rho.$$

Use of (E.107),

$$\int x^{n+1} J_n(x)\, dx = x^{n+1} J_{n+1}(x) + C,$$

allows us to evaluate

$$\int_0^a J_0\left(\frac{p_{0n}\rho}{a}\right) \rho\, d\rho = \frac{a^2}{p_{0n}} J_1(p_{0n}).$$

With this we finish calculating A_m and have

$$V(\rho, z) = 2V_0 \sum_{m=1}^{\infty} \frac{\sinh(\frac{p_{0m}}{a} z) J_0(\frac{p_{0m}}{a}\rho)}{p_{0m} \sinh(\frac{p_{0m}}{a} h) J_1(p_{0m})}$$

as the desired solution. ◄

▶ **Example A.11:** Solution to the Helmholtz equation in cylindrical coordinates with three dimensions and boundary values

Solve

$$\nabla^2 \psi(\rho, \phi, z) + k^2 \psi(\rho, \phi, z) = 0$$

in the region

$$0 \le \rho \le a, \qquad -\pi \le \phi \le \pi, \qquad -\infty < z < \infty$$

with the boundary conditions

$$\hat{\mathbf{n}} \cdot \nabla \psi(\rho, \phi, z)\Big|_{\rho=a} = \frac{\partial \psi(\rho, \phi, z)}{\partial \rho}\Big|_{\rho=a} = 0$$

for $-\pi \le \phi \le \pi$ and $-\infty < z < \infty$.

Solution: The solution to this problem leads to the transverse-electric (TE_z) fields in a lossless circular waveguide, where ψ represents the z-component of the magnetic field. Although there is symmetry with respect to ϕ, we seek ϕ-dependent solutions; the resulting complete eigenmode solution will permit us to expand any well-behaved function within the waveguide in terms of a normal mode (eigenfunction) series. In this problem none of the constants k, k_z, or k_ϕ equal zero, except as a special case. However, the field must be single-valued in ϕ and thus k_ϕ must be an integer m. We consider our possible choices for $P(\rho)$, $Z(z)$, and $\Phi(\phi)$. Since $k_c^2 = k^2 - k_z^2$ and $k^2 > 0$ is arbitrary, we must consider various possibilities for the signs of k_c^2 and k_z^2. We can rule out $k_c^2 < 0$ based on consideration of the behavior of the functions I_m and K_m. We also need not consider $k_c < 0$, since this gives the same solution as $k_c > 0$. We are then left with two possible cases. Writing $k_z^2 = k^2 - k_c^2$, we see that either $k > k_c$ and $k_z^2 > 0$, or $k < k_c$ and $k_z^2 < 0$. For $k_z^2 > 0$ we write

$$\psi(\rho, \phi, z) = [A_z e^{-jk_z z} + B_z e^{jk_z z}][A_\phi \sin m\phi + B_\phi \cos m\phi] J_m(k_c \rho).$$

Here the terms involving $e^{\mp jk_z z}$ represent waves propagating in the $\pm z$ directions. The boundary condition at $\rho = a$ requires

$$J_m'(k_c a) = 0$$

where $J_m'(x) = dJ_m(x)/dx$. Denoting the nth zero of $J_m'(x)$ by p_{mn}' we have $k_c = k_{cm} = p_{mn}'/a$. This gives the eigensolutions

$$\psi_m = [A_{zm} e^{-jk_z z} + B_{zm} e^{jk_z z}][A_{\phi m} \sin m\phi + B_{\phi m} \cos m\phi] k_c J_m\left(\frac{p_{mn}' \rho}{a}\right).$$

The undetermined constants $A_{zm}, B_{zm}, A_{\rho m}, B_{\rho m}$ could be evaluated when the individual eigensolutions are used to represent a function in terms of a modal expansion. For the case $k_z^2 < 0$ we again choose complex exponentials in z; however, $k_z = -j\alpha$ gives $e^{\mp jk_z z} = e^{\mp \alpha z}$ and attenuation along z. The reader can verify that the eigensolutions are

$$\psi_m = [A_{zm} e^{-\alpha z} + B_{zm} e^{\alpha z}][A_{\phi m} \sin m\phi + B_{\phi m} \cos m\phi] k_c J_m\left(\frac{p_{mn}' \rho}{a}\right)$$

where now $k_c^2 = k^2 + \alpha^2$. ◄

We have used Bessel function completeness in the examples above. This property is a consequence of the Sturm–Liouville problem first studied in § A.5.1. We often use Fourier–Bessel series to express functions over finite intervals. Over infinite intervals we use the Fourier–Bessel transform.

The Fourier–Bessel series can be generalized to Bessel functions of noninteger order, and to the derivatives of Bessel functions. Let $f(\rho)$ be well-behaved over the interval $[0, a]$. Then the series

$$f(\rho) = \sum_{m=1}^{\infty} C_m J_\nu\left(p_{\nu m} \frac{\rho}{a}\right) \qquad (0 \le \rho \le a, \ \nu > -1)$$

converges, and the constants are

$$C_m = \frac{2}{a^2 J_{\nu+1}^2(p_{\nu m})} \int_0^a f(\rho) J_\nu \left(p_{\nu m} \frac{\rho}{a} \right) \rho \, d\rho$$

by (A.136). Here $p_{\nu m}$ is the mth root of $J_\nu(x)$. An alternative form of the series uses $p'_{\nu m}$, the roots of $J'_\nu(x)$, and is given by

$$f(\rho) = \sum_{m=1}^{\infty} D_m J_\nu \left(p'_{\nu m} \frac{\rho}{a} \right) \qquad (0 \le \rho \le a, \ \nu > -1).$$

In this case the expansion coefficients are found using the orthogonality relationship

$$\int_0^a J_\nu \left(\frac{p'_{\nu m}}{a} \rho \right) J_\nu \left(\frac{p'_{\nu n}}{a} \rho \right) \rho \, d\rho = \delta_{mn} \frac{a^2}{2} \left(1 - \frac{\nu^2}{p'^2_{\nu m}} \right) J_\nu^2(p'_{\nu m}),$$

and are

$$D_m = \frac{2}{a^2 \left(1 - \frac{\nu^2}{p'^2_{\nu m}} J_\nu^2(p'_{\nu m}) \right)} \int_0^a f(\rho) J_\nu \left(\frac{p'_{\nu m}}{a} \rho \right) \rho \, d\rho.$$

A.5.3.3   Solutions in spherical coordinates

If into Helmholtz's equation

$$\frac{1}{r^2} \frac{\partial}{\partial r} \left(r^2 \frac{\partial \psi(r, \theta, \phi)}{\partial r} \right) + \frac{1}{r^2 \sin \theta} \frac{\partial}{\partial \theta} \left(\sin \theta \frac{\partial \psi(r, \theta, \phi)}{\partial \theta} \right)$$

$$+ \frac{1}{r^2 \sin^2 \theta} \frac{\partial^2 \psi(r, \theta, \phi)}{\partial \phi^2} + k^2 \psi(r, \theta, \phi) = 0$$

we put $\psi(r, \theta, \phi) = R(r)\Theta(\theta)\Phi(\phi)$ and multiply through by $r^2 \sin^2 \theta / \psi(r, \theta, \phi)$, we obtain

$$\frac{\sin^2 \theta}{R(r)} \frac{d}{dr} \left(r^2 \frac{dR(r)}{dr} \right) + \frac{\sin \theta}{\Theta(\theta)} \frac{d}{d\theta} \left(\sin \theta \frac{d\Theta(\theta)}{d\theta} \right) + k^2 r^2 \sin^2 \theta = -\frac{1}{\Phi(\phi)} \frac{d^2 \Phi(\phi)}{d\phi^2}.$$

Since the right side depends only on ϕ while the left side depends only on r and θ, both sides must equal some constant μ^2:

$$\frac{\sin^2 \theta}{R(r)} \frac{d}{dr} \left(r^2 \frac{dR(r)}{dr} \right) + \frac{\sin \theta}{\Theta(\theta)} \frac{d}{d\theta} \left(\sin \theta \frac{d\Theta(\theta)}{d\theta} \right) + k^2 r^2 \sin^2 \theta = \mu^2, \qquad (A.137)$$

$$\frac{d^2 \Phi(\phi)}{d\phi^2} + \mu^2 \Phi(\phi) = 0. \qquad (A.138)$$

We have thus separated off the ϕ-dependence. Harmonic ordinary differential equation (A.138) has solutions

$$\Phi(\phi) = \begin{cases} A_\phi \sin \mu\phi + B_\phi \cos \mu\phi, & \mu \neq 0, \\ a_\phi \phi + b_\phi, & \mu = 0. \end{cases}$$

(We could have used complex exponentials to describe $\Phi(\phi)$, or some combination of exponentials and trigonometric functions, but it is conventional to use only trigonometric functions.) Rearranging (A.137) and dividing through by $\sin^2 \theta$, we have

$$\frac{1}{R(r)} \frac{d}{dr} \left(r^2 \frac{dR(r)}{dr} \right) + k^2 r^2 = -\frac{1}{\sin \theta \, \Theta(\theta)} \frac{d}{d\theta} \left(\sin \theta \frac{d\Theta(\theta)}{d\theta} \right) + \frac{\mu^2}{\sin^2 \theta}.$$

We introduce a new constant k_θ^2 to separate r from θ:

$$\frac{1}{R(r)}\frac{d}{dr}\left(r^2\frac{dR(r)}{dr}\right) + k^2 r^2 = k_\theta^2, \tag{A.139}$$

$$-\frac{1}{\sin\theta\,\Theta(\theta)}\frac{d}{d\theta}\left(\sin\theta\frac{d\Theta(\theta)}{d\theta}\right) + \frac{\mu^2}{\sin^2\theta} = k_\theta^2. \tag{A.140}$$

Equation (A.140),

$$\frac{1}{\sin\theta}\frac{d}{d\theta}\left(\sin\theta\frac{d\Theta(\theta)}{d\theta}\right) + \left(k_\theta^2 - \frac{\mu^2}{\sin^2\theta}\right)\Theta(\theta) = 0,$$

can be put into a standard form by letting

$$\eta = \cos\theta \tag{A.141}$$

and $k_\theta^2 = \nu(\nu+1)$ where ν is a parameter:

$$(1-\eta^2)\frac{d^2\Theta(\eta)}{d\eta^2} - 2\eta\frac{d\Theta(\eta)}{d\eta} + \left[\nu(\nu+1) - \frac{\mu^2}{1-\eta^2}\right]\Theta(\eta) = 0 \qquad (|\eta| \le 1).$$

This is the *associated Legendre equation*. It has two independent solutions called *associated Legendre functions of the first* and *second kinds*, denoted $P_\nu^\mu(\eta)$ and $Q_\nu^\mu(\eta)$, respectively. In these functions, all three quantities μ, ν, η may be arbitrary complex constants as long as $\nu + \mu \ne -1, -2, \ldots$. But (A.141) shows that η is real in our discussion; μ will generally be real also, and will be an integer whenever $\Phi(\phi)$ is single-valued. The choice of ν is somewhat more complicated. The function $P_\nu^\mu(\eta)$ diverges at $\eta = \pm 1$ unless ν is an integer, while $Q_\nu^\mu(\eta)$ diverges at $\eta = \pm 1$ regardless of whether ν is an integer. In § A.5.1 we required that $P_\nu^\mu(\eta)$ be bounded on $[-1,1]$ to have a Sturm–Liouville problem with suitable orthogonality properties. By (A.141) we must exclude $Q_\nu^\mu(\eta)$ for problems containing the z-axis, and restrict ν to be an integer n in $P_\nu^\mu(\eta)$ for such problems. In case the z-axis is excluded, we choose $\nu = n$ whenever possible, because the finite sums $P_n^m(\eta)$ and $Q_n^m(\eta)$ are much easier to manipulate than $P_\nu^\mu(\eta)$ and $Q_\nu^\mu(\eta)$. In many problems we must count on completeness of the Legendre polynomials $P_n(\eta) = P_n^0(\eta)$ or spherical harmonics $Y_{mn}(\theta,\phi)$ in order to satisfy the boundary conditions. In this book we shall consider only those boundary value problems that can be solved using integer values of ν and μ, hence choose

$$\Theta(\theta) = A_\theta P_n^m(\cos\theta) + B_\theta Q_n^m(\cos\theta). \tag{A.142}$$

Single-valuedness in $\Phi(\phi)$ is a consequence of having $\mu = m$, and $\phi = $ constant boundary surfaces are thereby disallowed.

The associated Legendre functions have many important properties. For instance,

$$P_n^m(\eta) = \begin{cases} 0, & m > n, \\ (-1)^m \dfrac{(1-\eta^2)^{m/2}}{2^n n!}\dfrac{d^{n+m}(\eta^2-1)^n}{d\eta^{n+m}}, & m \le n. \end{cases} \tag{A.143}$$

The case $m = 0$ receives particular attention because it corresponds to azimuthal invariance (ϕ-independence). We define $P_n^0(\eta) = P_n(\eta)$ where $P_n(\eta)$ is the Legendre polynomial of order n. From (A.143), we see that*

$$P_n(\eta) = \frac{1}{2^n n!} \frac{d^n (\eta^2 - 1)^n}{d\eta^n}$$

is a polynomial of degree n, and that

$$P_n^m(\eta) = (-1)^m (1 - \eta^2)^{m/2} \frac{d^m}{d\eta^m} P_n(\eta).$$

Both the associated Legendre functions and the Legendre polynomials obey orthogonality relations and many recursion formulas.

In problems where the z-axis is included, the product $\Theta(\theta)\Phi(\phi)$ is sometimes defined as the spherical harmonic

$$Y_{nm}(\theta, \phi) = \sqrt{\frac{2n + 1}{4\pi} \frac{(n - m)!}{(n + m)!}} P_n^m(\cos\theta) e^{jm\phi}.$$

These functions, which are complete over the surface of a sphere, were treated earlier in this section.

Remembering that $k_r^2 = \nu(\nu + 1)$, the r-dependent equation (A.139) becomes

$$\frac{1}{r^2} \frac{d}{dr} \left(r^2 \frac{dR(r)}{dr} \right) + \left(k^2 + \frac{n(n + 1)}{r^2} \right) R(r) = 0. \qquad \text{(A.144)}$$

When $k = 0$ we have

$$\frac{d^2 R(r)}{dr^2} + \frac{2}{r} \frac{dR(r)}{dr} - \frac{n(n + 1)}{r^2} R(r) = 0$$

so that

$$R(r) = A_r r^n + B_r r^{-(n+1)}.$$

When $k \neq 0$, the substitution $\bar{R}(r) = \sqrt{kr} R(r)$ puts (A.144) into the form

$$r^2 \frac{d^2 \bar{R}(r)}{dr^2} + r \frac{d\bar{R}(r)}{dr} + \left[k^2 r^2 - (n + \tfrac{1}{2})^2 \right] \bar{R}(r) = 0,$$

which we recognize as Bessel's equation of half-integer order. Thus

$$R(r) = \frac{\bar{R}(r)}{\sqrt{kr}} = \frac{Z_{n+\frac{1}{2}}(kr)}{\sqrt{kr}}.$$

For convenience, we define the *spherical Bessel functions*

$$j_n(z) = \sqrt{\frac{\pi}{2z}} J_{n+\frac{1}{2}}(z),$$

$$n_n(z) = \sqrt{\frac{\pi}{2z}} N_{n+\frac{1}{2}}(z) = (-1)^{n+1} \sqrt{\frac{\pi}{2z}} J_{-(n+\frac{1}{2})}(z),$$

$$h_n^{(1)}(z) = \sqrt{\frac{\pi}{2z}} H_{n+\frac{1}{2}}^{(1)}(z) = j_n(z) + jn_n(z),$$

$$h_n^{(2)}(z) = \sqrt{\frac{\pi}{2z}} H_{n+\frac{1}{2}}^{(2)}(z) = j_n(z) - jn_n(z).$$

---

*Care must be taken when consulting tables of Legendre functions and their properties. In particular, one must be on the lookout for possible disparities regarding the factor $(-1)^m$ (cf., [74, 1, 115, 12] vs. [6, 183]). Similar care is needed with $Q_n^m(x)$.

These can be written as finite sums involving trigonometric functions and inverse powers of z. We have, for instance,

$$j_0(z) = \frac{\sin z}{z}, \qquad\qquad j_1(z) = \frac{\sin z}{z^2} - \frac{\cos z}{z},$$

$$n_0(z) = -\frac{\cos z}{z}, \qquad\qquad n_1(z) = -\frac{\cos z}{z^2} - \frac{\sin z}{z}.$$

We can now write $R(r)$ as a linear combination of any two of the spherical Bessel functions $j_n, n_n, h_n^{(1)}, h_n^{(2)}$:

$$R(r) = \begin{cases} A_r j_n(kr) + B_r n_n(kr), \\ A_r j_n(kr) + B_r h_n^{(1)}(kr), \\ A_r j_n(kr) + B_r h_n^{(2)}(kr), \\ A_r n_n(kr) + B_r h_n^{(1)}(kr), \\ A_r n_n(kr) + B_r h_n^{(2)}(kr), \\ A_r h_n^{(1)}(kr) + B_r h_n^{(2)}(kr). \end{cases} \tag{A.145}$$

Imaginary arguments produce *modified spherical Bessel functions*; the interested reader is referred to Gradshteyn [74] or Abramowitz [1].

▶ **Example A.12:** Solution to Laplace's equation in spherical coordinates with three dimensions and boundary values — example 1

Solve

$$\nabla^2 V(r, \theta, \phi) = 0$$

in the region

$$\theta_0 \le \theta \le \pi/2, \qquad 0 \le r < \infty, \qquad -\pi \le \phi \le \pi$$

where the boundary conditions are those of a cone held at a potential V_0 with respect to the $z = 0$ plane:

$$V(r, \theta_0, \phi) = V_0 \qquad (-\pi \le \phi \le \pi, \ 0 \le r < \infty),$$
$$V(r, \pi/2, \phi) = 0 \qquad (-\pi \le \phi \le \pi, \ 0 \le r < \infty).$$

Solution: Azimuthal symmetry prompts us to choose $\mu = a_\phi = 0$. Since $k = 0$ we have

$$R(r) = A_r r^n + B_r r^{-(n+1)}. \tag{A.146}$$

Noting that positive and negative powers of r are unbounded for large and small r, respectively, we take $n = B_r = 0$. Hence the solution depends only on θ:

$$V(r, \theta, \phi) = V(\theta) = A_\theta P_0^0(\cos\theta) + B_\theta Q_0^0(\cos\theta).$$

We must retain Q_0^0 since the solution region does not contain the z-axis. Using

$$P_0^0(\cos\theta) = 1 \quad \text{and} \quad Q_0^0(\cos\theta) = \ln\cot(\theta/2)$$

(cf., Appendix E.2), we have

$$V(\theta) = A_\theta + B_\theta \ln\cot(\theta/2).$$

A straightforward application of the boundary conditions gives $A_\theta = 0$ and $B_\theta = V_0 / \ln \cot(\theta_0/2)$, hence

$$V(\theta) = V_0 \frac{\ln \cot(\theta/2)}{\ln \cot(\theta_0/2)}. \quad \blacktriangleleft$$

▶ **Example A.13:** Solution to Laplace's equation in spherical coordinates with three dimensions and boundary values — example 2

Solve

$$\nabla^2 V(r, \theta, \phi) = 0$$

in the region

$$\theta_0 \leq \theta \leq \pi, \qquad 0 \leq r < \infty, \qquad -\pi \leq \phi \leq \pi$$

where the boundary conditions are those of a conducting sphere split into top and bottom hemispheres and held at a potential difference of $2V_0$:

$$V(a, \theta, \phi) = -V_0, \qquad (\pi/2 \leq \theta < \pi, \ -\pi \leq \phi \leq \pi),$$
$$V(a, \theta, \phi) = +V_0, \qquad (0 < \theta \leq \pi/2, \ -\pi \leq \phi \leq \pi).$$

Find $V(r, \theta, \phi)$ for $r < a$.

Solution: Azimuthal symmetry gives $\mu = 0$. The two possible solutions for $\Theta(\theta)$ are

$$\Theta(\theta) = \begin{cases} A_\theta + B_\theta \ln \cot(\theta/2), & n = 0, \\ A_\theta P_n(\cos\theta), & n \neq 0, \end{cases}$$

where we have discarded $Q_0^0(\cos\theta)$ because the region of interest contains the z-axis. The $n = 0$ solution cannot match the boundary conditions; neither can a single term of the type $A_\theta P_n(\cos\theta)$, but a series of these latter terms can. We use

$$V(r, \theta) = \sum_{n=0}^{\infty} V_n(r, \theta) = \sum_{n=0}^{\infty} [A_r r^n + B_r r^{-(n+1)}] P_n(\cos\theta). \qquad (A.147)$$

The terms $r^{-(n+1)}$ and r^n are not allowed, respectively, for $r < a$ and $r > a$. For $r < a$ then,

$$V(r, \theta) = \sum_{n=0}^{\infty} A_n r^n P_n(\cos\theta).$$

Letting $V_0(\theta)$ be the potential on the surface of the split sphere, we impose the boundary condition:

$$V(a, \theta) = V_0(\theta) = \sum_{n=0}^{\infty} A_n a^n P_n(\cos\theta) \qquad (0 \leq \theta \leq \pi).$$

This is a Fourier–Legendre expansion of $V_0(\theta)$. The A_n are evaluated by orthogonality. Multiplying by $P_m(\cos\theta) \sin\theta$ and integrating from $\theta = 0$ to π, we obtain

$$\sum_{n=0}^{\infty} A_n a^n \int_0^\pi P_n(\cos\theta) P_m(\cos\theta) \sin\theta \, d\theta = \int_0^\pi V_0(\theta) P_m(\cos\theta) \sin\theta \, d\theta.$$

Using orthogonality relationship (A.95) and the given $V_0(\theta)$, we have

$$A_m a^m \frac{2}{2m+1} = V_0 \int_0^{\pi/2} P_m(\cos\theta) \sin\theta \, d\theta - V_0 \int_{\pi/2}^\pi P_m(\cos\theta) \sin\theta \, d\theta.$$

The substitution $\eta = \cos\theta$ gives

$$A_m a^m \frac{2}{2m+1} = V_0 \int_0^1 P_m(\eta)\, d\eta - V_0 \int_{-1}^0 P_m(\eta)\, d\eta$$

$$= V_0 \int_0^1 P_m(\eta)\, d\eta - V_0 \int_0^1 P_m(-\eta)\, d\eta;$$

then $P_m(-\eta) = (-1)^m P_m(\eta)$ gives

$$A_m = a^{-m} \frac{2m+1}{2} V_0 [1 - (-1)^m] \int_0^1 P_m(\eta)\, d\eta.$$

Because $A_m = 0$ for m even, we can put $m = 2n+1$ $(n = 0, 1, 2, \ldots)$ and have

$$A_{2n+1} = \frac{(4n+3)V_0}{a^{2n+1}} \int_0^1 P_{2n+1}(\eta)\, d\eta = \frac{V_0(-1)^n}{a^{2n+1}} \frac{4n+3}{2n+2} \frac{(2n)!}{(2^n n!)^2}$$

by (E.178). Hence

$$V(r,\theta) = \sum_{n=0}^{\infty} V_0 (-1)^n \frac{4n+3}{2n+2} \frac{(2n)!}{(2^n n!)^2} \left(\frac{r}{a}\right)^{2n+1} P_{2n+1}(\cos\theta)$$

for $r < a$. The case $r > a$ is left to the reader. ◄

▶ **Example A.14:** Solution to the Helmholtz equation in spherical coordinates with three dimensions and boundary values

Solve

$$\nabla^2 \psi(x,y,z) + k^2 \psi(x,y,z) = 0$$

in the region

$$0 \le r \le a, \qquad 0 \le \theta \le \pi, \qquad -\pi \le \phi \le \pi$$

with the boundary condition

$$\psi(a,\theta,\phi) = 0, \qquad 0 \le \theta \le \pi, \qquad -\pi \le \phi \le \pi.$$

Solution: The wave function ψ represents the solutions for the electromagnetic field within a spherical cavity for modes TE to r. Despite the prevailing symmetry, we choose solutions that vary with both θ and ϕ. We are motivated by a desire to solve problems involving cavity excitation, and eigenmode completeness will enable us to represent any piecewise continuous function within the cavity. We employ spherical harmonics because the boundary surface is a sphere. These exclude $Q_m^n(\cos\theta)$, which is appropriate since our problem contains the z-axis. Since $k \neq 0$ we must choose a radial dependence from (A.145). Small-argument behavior rules out n_n, $h_n^{(1)}$, and $h_n^{(2)}$, leaving us with

$$\psi(r,\theta,\phi) = A_{mn} j_n(kr) Y_{nm}(\theta,\phi)$$

or, equivalently,

$$\psi(r,\theta,\phi) = A_{mn} j_n(kr) P_n^m(\cos\theta) e^{jm\phi}.$$

The eigenvalues $\lambda = k^2$ are found by applying the condition at $r = a$:

$$\psi(a,\theta,\phi) = A_{mn} j_n(ka) Y_{nm}(\theta,\phi) = 0,$$

requiring $j_n(ka) = 0$. Denoting the qth root of $j_n(x) = 0$ by α_{nq}, we have $k_{nq} = \alpha_{nq}/a$ and corresponding eigenfunctions

$$\psi_{mnq}(r, \theta, \phi) = A_{mnq} j_n(k_{nq} r) Y_{nm}(\theta, \phi).$$

The eigenvalues are proportional to the resonant frequencies of the cavity, and the eigenfunctions can be used to find the modal field distributions. Since the eigenvalues are independent of m, we may have several eigenfunctions ψ_{mnq} associated with each k_{mnq}. The only limitation is that we must keep $m \leq n$ to have $P_m^n(\cos \theta)$ nonzero. This is another instance of mode degeneracy. There are $2n$ degenerate modes associated with each resonant frequency (one for each of $e^{\pm jn\phi}$). By completeness we can expand any piecewise continuous function within or on the sphere as a series

$$f(r, \theta, \phi) = \sum_{m,n,q} A_{mnq} j_n(k_{nq} r) Y_{nm}(\theta, \phi). \quad \blacktriangleleft$$

B

Useful identities

Algebraic identities for vectors and dyadics

$$\mathbf{A} + \mathbf{B} = \mathbf{B} + \mathbf{A} \tag{B.1}$$

$$\mathbf{A} \cdot \mathbf{B} = \mathbf{B} \cdot \mathbf{A} \tag{B.2}$$

$$\mathbf{A} \times \mathbf{B} = -\mathbf{B} \times \mathbf{A} \tag{B.3}$$

$$\mathbf{A} \cdot (\mathbf{B} + \mathbf{C}) = \mathbf{A} \cdot \mathbf{B} + \mathbf{A} \cdot \mathbf{C} \tag{B.4}$$

$$\mathbf{A} \times (\mathbf{B} + \mathbf{C}) = \mathbf{A} \times \mathbf{B} + \mathbf{A} \times \mathbf{C} \tag{B.5}$$

$$\mathbf{A} \cdot (\mathbf{B} \times \mathbf{C}) = \mathbf{B} \cdot (\mathbf{C} \times \mathbf{A}) = \mathbf{C} \cdot (\mathbf{A} \times \mathbf{B}) \tag{B.6}$$

$$\mathbf{A} \times (\mathbf{B} \times \mathbf{C}) = \mathbf{B}(\mathbf{A} \cdot \mathbf{C}) - \mathbf{C}(\mathbf{A} \cdot \mathbf{B}) = \mathbf{B} \times (\mathbf{A} \times \mathbf{C}) + \mathbf{C} \times (\mathbf{B} \times \mathbf{A}) \tag{B.7}$$

$$(\mathbf{A} \times \mathbf{B}) \cdot (\mathbf{C} \times \mathbf{D}) = \mathbf{A} \cdot [\mathbf{B} \times (\mathbf{C} \times \mathbf{D})] = (\mathbf{B} \cdot \mathbf{D})(\mathbf{A} \cdot \mathbf{C}) - (\mathbf{B} \cdot \mathbf{C})(\mathbf{A} \cdot \mathbf{D}) \tag{B.8}$$

$$(\mathbf{A} \times \mathbf{B}) \times (\mathbf{C} \times \mathbf{D}) = \mathbf{C}[\mathbf{A} \cdot (\mathbf{B} \times \mathbf{D})] - \mathbf{D}[\mathbf{A} \cdot (\mathbf{B} \times \mathbf{C})] \tag{B.9}$$

$$\mathbf{A} \times [\mathbf{B} \times (\mathbf{C} \times \mathbf{D})] = (\mathbf{B} \cdot \mathbf{D})(\mathbf{A} \times \mathbf{C}) - (\mathbf{B} \cdot \mathbf{C})(\mathbf{A} \times \mathbf{D}) \tag{B.10}$$

$$\mathbf{A} \cdot (\bar{\mathbf{c}} \cdot \mathbf{B}) = (\mathbf{A} \cdot \bar{\mathbf{c}}) \cdot \mathbf{B} \tag{B.11}$$

$$\mathbf{A} \times (\bar{\mathbf{c}} \cdot \mathbf{B}) = (\mathbf{A} \times \bar{\mathbf{c}}) \cdot \mathbf{B} \tag{B.12}$$

$$\mathbf{A} \times (\bar{\mathbf{c}} \times \mathbf{B}) = (\mathbf{A} \times \bar{\mathbf{c}}) \times \mathbf{B} \tag{B.13}$$

$$\mathbf{C} \cdot (\bar{\mathbf{a}} \cdot \bar{\mathbf{b}}) = (\mathbf{C} \cdot \bar{\mathbf{a}}) \cdot \bar{\mathbf{b}} \tag{B.14}$$

$$(\bar{\mathbf{a}} \cdot \bar{\mathbf{b}}) \cdot \mathbf{C} = \bar{\mathbf{a}} \cdot (\bar{\mathbf{b}} \cdot \mathbf{C}) \tag{B.15}$$

$$\mathbf{A} \cdot (\mathbf{B} \times \bar{\mathbf{c}}) = -\mathbf{B} \cdot (\mathbf{A} \times \bar{\mathbf{c}}) = (\mathbf{A} \times \mathbf{B}) \cdot \bar{\mathbf{c}} \tag{B.16}$$

$$\mathbf{A} \times (\mathbf{B} \times \bar{\mathbf{c}}) = \mathbf{B} \cdot (\mathbf{A} \times \bar{\mathbf{c}}) - \bar{\mathbf{c}}(\mathbf{A} \cdot \mathbf{B}) \tag{B.17}$$

$$\mathbf{A} \cdot \bar{\mathbf{I}} = \bar{\mathbf{I}} \cdot \mathbf{A} = \mathbf{A} \tag{B.18}$$

$$\mathbf{A} \times \bar{\mathbf{I}} = \bar{\mathbf{I}} \times \mathbf{A} \tag{B.19}$$

$$(\bar{\mathbf{I}} \times \mathbf{A}) \cdot \mathbf{B} = \mathbf{A} \times \mathbf{B} \tag{B.20}$$

$$\mathbf{A} \cdot (\bar{\mathbf{I}} \times \mathbf{B}) = \mathbf{A} \times \mathbf{B} \tag{B.21}$$

$$(\bar{\mathbf{I}} \times \mathbf{A}) \cdot \bar{\mathbf{c}} = \mathbf{A} \times \bar{\mathbf{c}} \tag{B.22}$$

$$\bar{\mathbf{c}} \cdot (\bar{\mathbf{I}} \times \mathbf{A}) = \bar{\mathbf{c}} \times \mathbf{A} \tag{B.23}$$

Integral theorems

In the relations below, surface S bounds volume V, curve Γ bounds surface S, unit vector $\hat{\mathbf{n}}$ is normal to surface S at position $\mathbf{r}$, unit vectors $\hat{\mathbf{l}}$ and $\hat{\mathbf{m}}$ are tangential to surface S at position $\mathbf{r}$, unit vector $\hat{\mathbf{l}}$ is tangential to contour Γ, and $\hat{\mathbf{m}} \times \hat{\mathbf{l}} = \hat{\mathbf{n}}$, $\mathbf{dl} = \hat{\mathbf{l}} \, dl$, while $\mathbf{dS} = \hat{\mathbf{n}} \, dS$.

Divergence theorem

$$\int_V \nabla \cdot \mathbf{A} \, dV = \oint_S \mathbf{A} \cdot \mathbf{dS} \tag{B.24}$$

$$\int_V \nabla \cdot \bar{\mathbf{a}} \, dV = \oint_S \hat{\mathbf{n}} \cdot \bar{\mathbf{a}} \, dS \tag{B.25}$$

$$\int_S \nabla_s \cdot \mathbf{A} \, dS = \oint_\Gamma \hat{\mathbf{m}} \cdot \mathbf{A} \, dl \tag{B.26}$$

Gradient theorem

$$\int_V \nabla a \, dV = \oint_S a \mathbf{dS} \tag{B.27}$$

$$\int_V \nabla \mathbf{A} \, dV = \oint_S \hat{\mathbf{n}} \mathbf{A} \, dS \tag{B.28}$$

$$\int_V \nabla_s a \, dS = \oint_\Gamma \hat{\mathbf{m}} a \, dl \tag{B.29}$$

Curl theorem

$$\int_V (\nabla \times \mathbf{A}) \, dV = - \oint_S \mathbf{A} \times \mathbf{dS} \tag{B.30}$$

$$\int_V (\nabla \times \bar{\mathbf{a}}) \, dV = \oint_S \hat{\mathbf{n}} \times \bar{\mathbf{a}} \, dS \tag{B.31}$$

$$\int_S \nabla_s \times \mathbf{A} \, dS = \oint_\Gamma \hat{\mathbf{m}} \times \mathbf{A} \, dl \tag{B.32}$$

Stokes's theorem

$$\int_S (\nabla \times \mathbf{A}) \cdot \mathbf{dS} = \oint_\Gamma \mathbf{A} \cdot \mathbf{dl} \tag{B.33}$$

$$\int_S \hat{\mathbf{n}} \cdot (\nabla \times \bar{\mathbf{a}}) \, dS = \oint_\Gamma \mathbf{dl} \cdot \bar{\mathbf{a}} \tag{B.34}$$

Green's first identity for scalar fields

$$\int_V (\nabla a \cdot \nabla b + a \nabla^2 b)\, dV = \oint_S a \frac{\partial b}{\partial n}\, dS \tag{B.35}$$

Green's second identity for scalar fields (Green's theorem)

$$\int_V (a \nabla^2 b - b \nabla^2 a)\, dV = \oint_S \left(a \frac{\partial b}{\partial n} - b \frac{\partial a}{\partial n} \right) dS \tag{B.36}$$

Green's first identity for vector fields

$$\int_V \left\{ (\nabla \times \mathbf{A}) \cdot (\nabla \times \mathbf{B}) - \mathbf{A} \cdot [\nabla \times (\nabla \times \mathbf{B})] \right\} dV$$

$$= \int_V \nabla \cdot [\mathbf{A} \times (\nabla \times \mathbf{B})]\, dV$$

$$= \oint_S [\mathbf{A} \times (\nabla \times \mathbf{B})] \cdot \mathbf{dS} \tag{B.37}$$

Green's second identity for vector fields

$$\int_V \left\{ \mathbf{B} \cdot [\nabla \times (\nabla \times \mathbf{A})] - \mathbf{A} \cdot [\nabla \times (\nabla \times \mathbf{B})] \right\} dV$$

$$= \oint_S [\mathbf{A} \times (\nabla \times \mathbf{B}) - \mathbf{B} \times (\nabla \times \mathbf{A})] \cdot \mathbf{dS} \tag{B.38}$$

Helmholtz theorem

$$\mathbf{A}(\mathbf{r}) = -\nabla \left[\int_V \frac{\nabla' \cdot \mathbf{A}(\mathbf{r}')}{4\pi |\mathbf{r} - \mathbf{r}'|}\, dV' - \oint_S \frac{\mathbf{A}(\mathbf{r}') \cdot \hat{\mathbf{n}}'}{4\pi |\mathbf{r} - \mathbf{r}'|}\, dS' \right]$$

$$+ \nabla \times \left[\int_V \frac{\nabla' \times \mathbf{A}(\mathbf{r}')}{4\pi |\mathbf{r} - \mathbf{r}'|}\, dV' + \oint_S \frac{\mathbf{A}(\mathbf{r}') \times \hat{\mathbf{n}}'}{4\pi |\mathbf{r} - \mathbf{r}'|}\, dS' \right] \tag{B.39}$$

Miscellaneous identities

$$\oint_S \mathbf{dS} = 0 \tag{B.40}$$

$$\int_S \hat{\mathbf{n}} \times (\nabla a)\, dS = \oint_\Gamma a\, \mathbf{dl} \tag{B.41}$$

$$\int_S (\nabla a \times \nabla b) \cdot \mathbf{dS} = \int_\Gamma a \nabla b \cdot \mathbf{dl} = -\int_\Gamma b \nabla a \cdot \mathbf{dl} \tag{B.42}$$

$$\oint \mathbf{dl}\, \mathbf{A} = \int_S \hat{\mathbf{n}} \times (\nabla \mathbf{A})\, dS \tag{B.43}$$

Derivative identities

$$\nabla (a + b) = \nabla a + \nabla b \tag{B.44}$$

$$\nabla \cdot (\mathbf{A} + \mathbf{B}) = \nabla \cdot \mathbf{A} + \nabla \cdot \mathbf{B} \tag{B.45}$$

$$\nabla \times (\mathbf{A} + \mathbf{B}) = \nabla \times \mathbf{A} + \nabla \times \mathbf{B} \tag{B.46}$$

$$\nabla(ab) = a\nabla b + b\nabla a \tag{B.47}$$

$$\nabla \cdot (a\mathbf{B}) = a\nabla \cdot \mathbf{B} + \mathbf{B} \cdot \nabla a \tag{B.48}$$

$$\nabla \times (a\mathbf{B}) = a\nabla \times \mathbf{B} - \mathbf{B} \times \nabla a \tag{B.49}$$

$$\nabla \cdot (\mathbf{A} \times \mathbf{B}) = \mathbf{B} \cdot \nabla \times \mathbf{A} - \mathbf{A} \cdot \nabla \times \mathbf{B} \tag{B.50}$$

$$\nabla \times (\mathbf{A} \times \mathbf{B}) = \mathbf{A}(\nabla \cdot \mathbf{B}) - \mathbf{B}(\nabla \cdot \mathbf{A}) + (\mathbf{B} \cdot \nabla)\mathbf{A} - (\mathbf{A} \cdot \nabla)\mathbf{B} \tag{B.51}$$

$$\nabla(\mathbf{A} \cdot \mathbf{B}) = \mathbf{A} \times (\nabla \times \mathbf{B}) + \mathbf{B} \times (\nabla \times \mathbf{A}) + (\mathbf{A} \cdot \nabla)\mathbf{B} + (\mathbf{B} \cdot \nabla)\mathbf{A} \tag{B.52}$$

$$\nabla \times (\nabla \times \mathbf{A}) = \nabla(\nabla \cdot \mathbf{A}) - \nabla^2 \mathbf{A} \tag{B.53}$$

$$\nabla \cdot (\nabla a) = \nabla^2 a \tag{B.54}$$

$$\nabla \cdot (\nabla \times \mathbf{A}) = 0 \tag{B.55}$$

$$\nabla \times (\nabla a) = 0 \tag{B.56}$$

$$\nabla \times (a\nabla b) = \nabla a \times \nabla b \tag{B.57}$$

$$\nabla^2(ab) = a\nabla^2 b + 2(\nabla a) \cdot (\nabla b) + b\nabla^2 a \tag{B.58}$$

$$\nabla^2(a\mathbf{B}) = a\nabla^2 \mathbf{B} + \mathbf{B}\nabla^2 a + 2(\nabla a \cdot \nabla)\mathbf{B} \tag{B.59}$$

$$\nabla^2\bar{\mathbf{a}} = \nabla(\nabla \cdot \bar{\mathbf{a}}) - \nabla \times (\nabla \times \bar{\mathbf{a}}) \tag{B.60}$$

$$\nabla \cdot (\mathbf{A}\mathbf{B}) = (\nabla \cdot \mathbf{A})\mathbf{B} + \mathbf{A} \cdot (\nabla\mathbf{B}) = (\nabla \cdot \mathbf{A})\mathbf{B} + (\mathbf{A} \cdot \nabla)\mathbf{B} \tag{B.61}$$

$$\nabla \times (\mathbf{A}\mathbf{B}) = (\nabla \times \mathbf{A})\mathbf{B} - \mathbf{A} \times (\nabla\mathbf{B}) \tag{B.62}$$

$$\nabla \cdot (\nabla \times \bar{\mathbf{a}}) = 0 \tag{B.63}$$

$$\nabla \times (\nabla\mathbf{A}) = 0 \tag{B.64}$$

$$\nabla(\mathbf{A} \times \mathbf{B}) = (\nabla\mathbf{A}) \times \mathbf{B} - (\nabla\mathbf{B}) \times \mathbf{A} \tag{B.65}$$

$$\nabla(a\mathbf{B}) = (\nabla a)\mathbf{B} + a(\nabla\mathbf{B}) \tag{B.66}$$

$$\nabla \cdot (a\bar{\mathbf{b}}) = (\nabla a) \cdot \bar{\mathbf{b}} + a(\nabla \cdot \bar{\mathbf{b}}) \tag{B.67}$$

$$\nabla \times (a\bar{\mathbf{b}}) = (\nabla a) \times \bar{\mathbf{b}} + a(\nabla \times \bar{\mathbf{b}}) \tag{B.68}$$

$$\nabla \cdot (a\bar{\mathbf{I}}) = \nabla a \tag{B.69}$$

$$\nabla \times (a\bar{\mathbf{I}}) = \nabla a \times \bar{\mathbf{I}} \tag{B.70}$$

Identities involving the displacement vector

Here $\mathbf{R} = \mathbf{r} - \mathbf{r}'$, $R = |\mathbf{R}|$, $\hat{\mathbf{R}} = \mathbf{R}/R$, and $f'(x) = df(x)/dx$.

$$\nabla f(R) = -\nabla' f(R) = \hat{\mathbf{R}} f'(R) \tag{B.71}$$

$$\nabla R = \hat{\mathbf{R}} \tag{B.72}$$

$$\nabla\left(\frac{1}{R}\right) = -\frac{\hat{\mathbf{R}}}{R^2} \tag{B.73}$$

$$\nabla\left(\frac{e^{-jkR}}{R}\right) = -\hat{\mathbf{R}}\left(\frac{1}{R} + jk\right)\frac{e^{-jkR}}{R} \tag{B.74}$$

$$\nabla \cdot [f(R)\hat{\mathbf{R}}] = -\nabla' \cdot [f(R)\hat{\mathbf{R}}] = 2\frac{f(R)}{R} + f'(R) \tag{B.75}$$

$$\nabla \cdot \mathbf{R} = 3 \tag{B.76}$$

$$\nabla \cdot \hat{\mathbf{R}} = \frac{2}{R} \tag{B.77}$$

$$\nabla \cdot \left(\hat{\mathbf{R}}\frac{e^{-jkR}}{R}\right) = \left(\frac{1}{R} - jk\right)\frac{e^{-jkR}}{R} \tag{B.78}$$

$$\nabla \times [f(R)\hat{\mathbf{R}}] = 0 \tag{B.79}$$

$$\nabla^2\left(\frac{1}{R}\right) = -4\pi\delta(\mathbf{R}) \tag{B.80}$$

$$(\nabla^2 + k^2)\frac{e^{-jkR}}{R} = -4\pi\delta(\mathbf{R}) \tag{B.81}$$

Identities involving the plane-wave function

Here $\mathbf{E}$ is a constant vector and $k = |\mathbf{k}|$.

$$\nabla\left(e^{-j\mathbf{k}\cdot\mathbf{r}}\right) = -j\mathbf{k}e^{-j\mathbf{k}\cdot\mathbf{r}} \tag{B.82}$$

$$\nabla \cdot \left(\mathbf{E}e^{-j\mathbf{k}\cdot\mathbf{r}}\right) = -j\mathbf{k} \cdot \mathbf{E}e^{-j\mathbf{k}\cdot\mathbf{r}} \tag{B.83}$$

$$\nabla \times \left(\mathbf{E}e^{-j\mathbf{k}\cdot\mathbf{r}}\right) = -j\mathbf{k} \times \mathbf{E}e^{-j\mathbf{k}\cdot\mathbf{r}} \tag{B.84}$$

$$\nabla^2\left(\mathbf{E}e^{-j\mathbf{k}\cdot\mathbf{r}}\right) = -k^2\mathbf{E}e^{-j\mathbf{k}\cdot\mathbf{r}} \tag{B.85}$$

Identities involving the transverse/longitudinal decomposition

In the relations below, $\hat{\mathbf{u}}$ is a constant unit vector, $A_u \equiv \hat{\mathbf{u}} \cdot \mathbf{A}$, $\partial/\partial u \equiv \hat{\mathbf{u}} \cdot \nabla$, $\mathbf{A}_t \equiv \mathbf{A} - \hat{\mathbf{u}}A_u$, and $\nabla_t \equiv \nabla - \hat{\mathbf{u}}\partial/\partial u$.

$$\mathbf{A} = \mathbf{A}_t + \hat{\mathbf{u}}A_u \tag{B.86}$$

$$\nabla = \nabla_t + \hat{\mathbf{u}}\frac{\partial}{\partial u} \tag{B.87}$$

$$\hat{\mathbf{u}} \cdot \mathbf{A}_t = 0 \tag{B.88}$$

$$(\hat{\mathbf{u}} \cdot \nabla_t)\,\phi = 0 \tag{B.89}$$

$$\nabla_t\phi = \nabla\phi - \hat{\mathbf{u}}\frac{\partial\phi}{\partial u} \tag{B.90}$$

$$\hat{\mathbf{u}} \cdot (\nabla\phi) = (\hat{\mathbf{u}} \cdot \nabla)\phi = \frac{\partial\phi}{\partial u} \tag{B.91}$$

$$\hat{\mathbf{u}} \cdot (\nabla_t\phi) = 0 \tag{B.92}$$

$$\nabla_t \cdot (\hat{\mathbf{u}}\phi) = 0 \tag{B.93}$$

$$\nabla_t \times (\hat{\mathbf{u}}\phi) = -\hat{\mathbf{u}} \times \nabla_t\phi \tag{B.94}$$

$$\nabla_t \times (\hat{\mathbf{u}} \times \mathbf{A}) = \hat{\mathbf{u}}\nabla_t \cdot \mathbf{A}_t \tag{B.95}$$

$$\hat{\mathbf{u}} \times (\nabla_t \times \mathbf{A}) = \nabla_t A_u \tag{B.96}$$

$$\hat{\mathbf{u}} \times (\nabla_t \times \mathbf{A}_t) = 0 \tag{B.97}$$

$$\hat{\mathbf{u}} \cdot (\hat{\mathbf{u}} \times \mathbf{A}) = 0 \tag{B.98}$$

$$\hat{\mathbf{u}} \times (\hat{\mathbf{u}} \times \mathbf{A}) = -\mathbf{A}_t \tag{B.99}$$

$$\nabla\phi = \nabla_t\phi + \hat{\mathbf{u}}\frac{\partial\phi}{\partial u} \tag{B.100}$$

$$\nabla \cdot \mathbf{A} = \nabla_t \cdot \mathbf{A}_t + \frac{\partial A_u}{\partial u} \tag{B.101}$$

$$\nabla \times \mathbf{A} = \nabla_t \times \mathbf{A}_t + \hat{\mathbf{u}} \times \left[\frac{\partial \mathbf{A}_t}{\partial u} - \nabla_t A_u\right] \tag{B.102}$$

$$\nabla^2\phi = \nabla_t^2\phi + \frac{\partial^2\phi}{\partial u^2} \tag{B.103}$$

$$\nabla \times \nabla \times \mathbf{A} = \left[\nabla_t \times \nabla_t \times \mathbf{A}_t - \frac{\partial^2 \mathbf{A}_t}{\partial u^2} + \nabla_t\frac{\partial A_u}{\partial u}\right] + \hat{\mathbf{u}}\left[\frac{\partial}{\partial u}(\nabla_t \cdot \mathbf{A}_t) - \nabla_t^2 A_u\right] \tag{B.104}$$

$$\nabla^2\mathbf{A} = \left[\nabla_t(\nabla_t \cdot \mathbf{A}_t) + \frac{\partial^2 \mathbf{A}_t}{\partial u^2} - \nabla_t \times \nabla_t \times \mathbf{A}_t\right] + \hat{\mathbf{u}}\nabla^2 A_u \tag{B.105}$$

C

Fourier transform pairs

Listed below are some pairs of the form $g(x) \leftrightarrow G(k)$ where

$$G(k) = \int_{-\infty}^{\infty} g(x)e^{-jkx}\, dx \qquad \text{and} \qquad g(x) = \frac{1}{2\pi}\int_{-\infty}^{\infty} G(k)e^{jkx}\, dk.$$

$$\mathrm{rect}(x) \leftrightarrow 2\,\mathrm{sinc}\,k \tag{C.1}$$

$$\Lambda(x) \leftrightarrow \mathrm{sinc}^2 \frac{k}{2} \tag{C.2}$$

$$\mathrm{sgn}(x) \leftrightarrow \frac{2}{jk} \tag{C.3}$$

$$e^{jk_0 x} \leftrightarrow 2\pi\delta(k - k_0) \tag{C.4}$$

$$\delta(x) \leftrightarrow 1 \tag{C.5}$$

$$1 \leftrightarrow 2\pi\delta(k) \tag{C.6}$$

$$\frac{d^n\delta(x)}{dx^n} \leftrightarrow (jk)^n \tag{C.7}$$

$$x^n \leftrightarrow 2\pi j^n \frac{d^n\delta(k)}{dk^n} \tag{C.8}$$

$$U(x) \leftrightarrow \pi\delta(k) + \frac{1}{jk} \tag{C.9}$$

$$\sum_{n=-\infty}^{\infty} \delta\left(t - n\frac{2\pi}{k_0}\right) \leftrightarrow k_0 \sum_{n=-\infty}^{\infty} \delta(k - nk_0) \tag{C.10}$$

$$e^{-ax^2} \leftrightarrow \sqrt{\frac{\pi}{a}}\,e^{-\frac{k^2}{4a}} \tag{C.11}$$

$$e^{-ax}U(x) \leftrightarrow \frac{1}{a + jk} \tag{C.12}$$

$$e^{-a|x|} \leftrightarrow \frac{2a}{a^2 + k^2} \tag{C.13}$$

$$e^{-ax} \cos bx \, U(x) \leftrightarrow \frac{a + jk}{(a + jk)^2 + b^2} \qquad (C.14)$$

$$e^{-ax} \sin bx \, U(x) \leftrightarrow \frac{b}{(a + jk)^2 + b^2} \qquad (C.15)$$

$$\cos k_0 x \leftrightarrow \pi[\delta(k + k_0) + \delta(k - k_0)] \qquad (C.16)$$

$$\sin k_0 x \leftrightarrow j\pi[\delta(k + k_0) - \delta(k - k_0)] \qquad (C.17)$$

$$\tfrac{1}{2} b e^{-\frac{1}{2}bx} [I_0(\tfrac{1}{2}bx) + I_1(\tfrac{1}{2}bx) U(x) \leftrightarrow \sqrt{\frac{jk + b}{jk}} - 1 \qquad (C.18)$$

$$g(x) - a e^{-ax} \int_{-\infty}^{x} e^{au} g(u) \, du \leftrightarrow \frac{jk}{jk + a} G(k) \qquad (C.19)$$

D

Coordinate systems

Rectangular coordinate system

Coordinate variables

$$u = x \qquad (-\infty < x < \infty) \tag{D.1}$$

$$v = y \qquad (-\infty < y < \infty) \tag{D.2}$$

$$w = z \qquad (-\infty < z < \infty) \tag{D.3}$$

Vector algebra

$$\mathbf{A} = \hat{\mathbf{x}}A_x + \hat{\mathbf{y}}A_y + \hat{\mathbf{z}}A_z \tag{D.4}$$

$$\mathbf{A} \cdot \mathbf{B} = A_x B_x + A_y B_y + A_z B_z \tag{D.5}$$

$$\mathbf{A} \times \mathbf{B} = \begin{vmatrix} \hat{\mathbf{x}} & \hat{\mathbf{y}} & \hat{\mathbf{z}} \\ A_x & A_y & A_z \\ B_x & B_y & B_z \end{vmatrix}$$

$$= \hat{\mathbf{x}}(A_y B_z - A_z B_y) + \hat{\mathbf{y}}(A_z B_x - a_x B_z) + \hat{\mathbf{z}}(A_x B_y - A_y B_x) \tag{D.6}$$

Dyadic representation

$$\begin{aligned} \bar{\mathbf{a}} &= \hat{\mathbf{x}}a_{xx}\hat{\mathbf{x}} + \hat{\mathbf{x}}a_{xy}\hat{\mathbf{y}} + \hat{\mathbf{x}}a_{xz}\hat{\mathbf{z}} \\ &+ \hat{\mathbf{y}}a_{yx}\hat{\mathbf{x}} + \hat{\mathbf{y}}a_{yy}\hat{\mathbf{y}} + \hat{\mathbf{y}}a_{yz}\hat{\mathbf{z}} \\ &+ \hat{\mathbf{z}}a_{zx}\hat{\mathbf{x}} + \hat{\mathbf{z}}a_{zy}\hat{\mathbf{y}} + \hat{\mathbf{z}}a_{zz}\hat{\mathbf{z}} \end{aligned} \tag{D.7}$$

$$\bar{\mathbf{a}} = \hat{\mathbf{x}}\mathbf{a}'_x + \hat{\mathbf{y}}\mathbf{a}'_y + \hat{\mathbf{z}}\mathbf{a}'_z = \mathbf{a}_x\hat{\mathbf{x}} + \mathbf{a}_y\hat{\mathbf{y}} + \mathbf{a}_z\hat{\mathbf{z}} \tag{D.8}$$

$$\mathbf{a}'_x = a_{xx}\hat{\mathbf{x}} + a_{xy}\hat{\mathbf{y}} + a_{xz}\hat{\mathbf{z}} \tag{D.9}$$

$$\mathbf{a}'_y = a_{yx}\hat{\mathbf{x}} + a_{yy}\hat{\mathbf{y}} + a_{yz}\hat{\mathbf{z}} \tag{D.10}$$

$$\mathbf{a}'_z = a_{zx}\hat{\mathbf{x}} + a_{zy}\hat{\mathbf{y}} + a_{zz}\hat{\mathbf{z}} \tag{D.11}$$

$$\mathbf{a}_x = a_{xx}\hat{\mathbf{x}} + a_{yx}\hat{\mathbf{y}} + a_{zx}\hat{\mathbf{z}} \tag{D.12}$$

$$\mathbf{a}_y = a_{xy}\hat{\mathbf{x}} + a_{yy}\hat{\mathbf{y}} + a_{zy}\hat{\mathbf{z}} \tag{D.13}$$

$$\mathbf{a}_z = a_{xz}\hat{\mathbf{x}} + a_{yz}\hat{\mathbf{y}} + a_{zz}\hat{\mathbf{z}} \tag{D.14}$$

Differential operations

$$\mathbf{dl} = \hat{\mathbf{x}}\, dx + \hat{\mathbf{y}}\, dy + \hat{\mathbf{z}}\, dz \tag{D.15}$$

$$dV = dx\, dy\, dz \tag{D.16}$$

$$dS_x = dy\, dz \tag{D.17}$$
$$dS_y = dx\, dz \tag{D.18}$$
$$dS_z = dx\, dy \tag{D.19}$$

$$\nabla f = \hat{\mathbf{x}}\frac{\partial f}{\partial x} + \hat{\mathbf{y}}\frac{\partial f}{\partial y} + \hat{\mathbf{z}}\frac{\partial f}{\partial z} \tag{D.20}$$

$$\nabla \cdot \mathbf{F} = \frac{\partial F_x}{\partial x} + \frac{\partial F_y}{\partial y} + \frac{\partial F_z}{\partial z} \tag{D.21}$$

$$\nabla \times \mathbf{F} = \begin{vmatrix} \hat{\mathbf{x}} & \hat{\mathbf{y}} & \hat{\mathbf{z}} \\ \frac{\partial}{\partial x} & \frac{\partial}{\partial y} & \frac{\partial}{\partial z} \\ F_x & F_y & F_z \end{vmatrix}$$

$$= \hat{\mathbf{x}}\left(\frac{\partial F_z}{\partial y} - \frac{\partial F_y}{\partial z}\right) + \hat{\mathbf{y}}\left(\frac{\partial F_x}{\partial z} - \frac{\partial F_z}{\partial x}\right) + \hat{\mathbf{z}}\left(\frac{\partial F_y}{\partial x} - \frac{\partial F_x}{\partial y}\right) \tag{D.22}$$

$$\nabla^2 f = \frac{\partial^2 f}{\partial x^2} + \frac{\partial^2 f}{\partial y^2} + \frac{\partial^2 f}{\partial z^2} \tag{D.23}$$

$$\nabla^2 \mathbf{F} = \hat{\mathbf{x}}\nabla^2 F_x + \hat{\mathbf{y}}\nabla^2 F_y + \hat{\mathbf{z}}\nabla^2 F_z \tag{D.24}$$

Separation of the Helmholtz equation

$$\frac{\partial^2 \psi(x,y,z)}{\partial x^2} + \frac{\partial^2 \psi(x,y,z)}{\partial y^2} + \frac{\partial^2 \psi(x,y,z)}{\partial z^2} + k^2 \psi(x,y,z) = 0 \tag{D.25}$$

$$\psi(x,y,z) = X(x)Y(y)Z(z) \tag{D.26}$$

$$k_x^2 + k_y^2 + k_z^2 = k^2 \tag{D.27}$$

$$\frac{d^2 X(x)}{dx^2} + k_x^2 X(x) = 0 \tag{D.28}$$

$$\frac{d^2 Y(y)}{dy^2} + k_y^2 Y(y) = 0 \tag{D.29}$$

$$\frac{d^2 Z(z)}{dz^2} + k_z^2 Z(z) = 0 \tag{D.30}$$

$$X(x) = \begin{cases} A_x F_1(k_x x) + B_x F_2(k_x x), & k_x \neq 0, \\ a_x x + b_x, & k_x = 0. \end{cases} \tag{D.31}$$

$$Y(y) = \begin{cases} A_y F_1(k_y y) + B_y F_2(k_y y), & k_y \neq 0, \\ a_y y + b_y, & k_y = 0. \end{cases} \tag{D.32}$$

$$Z(z) = \begin{cases} A_z F_1(k_z z) + B_z F_2(k_z z), & k_z \neq 0, \\ a_z z + b_z, & k_z = 0. \end{cases} \tag{D.33}$$

$$F_1(\xi), F_2(\xi) = \begin{cases} e^{j\xi} \\ e^{-j\xi} \\ \sin(\xi) \\ \cos(\xi) \end{cases} \tag{D.34}$$

Cylindrical coordinate system

Coordinate variables

$$u = \rho \quad (0 \leq \rho < \infty) \tag{D.35}$$
$$v = \phi \quad (-\pi \leq \phi \leq \pi) \tag{D.36}$$
$$w = z \quad (-\infty < z < \infty) \tag{D.37}$$

$$x = \rho \cos \phi \tag{D.38}$$
$$y = \rho \sin \phi \tag{D.39}$$
$$z = z \tag{D.40}$$

$$\rho = \sqrt{x^2 + y^2} \tag{D.41}$$
$$\phi = \tan^{-1} \frac{y}{x} \tag{D.42}$$
$$z = z \tag{D.43}$$

Vector algebra

$$\hat{\boldsymbol{\rho}} = \hat{\mathbf{x}} \cos \phi + \hat{\mathbf{y}} \sin \phi \tag{D.44}$$
$$\hat{\boldsymbol{\phi}} = -\hat{\mathbf{x}} \sin \phi + \hat{\mathbf{y}} \cos \phi \tag{D.45}$$
$$\hat{\mathbf{z}} = \hat{\mathbf{z}} \tag{D.46}$$

$$\mathbf{A} = \hat{\boldsymbol{\rho}} A_\rho + \hat{\boldsymbol{\phi}} A_\phi + \hat{\mathbf{z}} A_z \tag{D.47}$$

$$\mathbf{A} \cdot \mathbf{B} = A_\rho B_\rho + A_\phi B_\phi + A_z B_z \tag{D.48}$$

$$\mathbf{A} \times \mathbf{B} = \begin{vmatrix} \hat{\boldsymbol{\rho}} & \hat{\boldsymbol{\phi}} & \hat{\mathbf{z}} \\ A_\rho & A_\phi & A_z \\ B_\rho & B_\phi & B_z \end{vmatrix}$$

$$= \hat{\boldsymbol{\rho}}(A_\phi B_z - A_z B_\phi) + \hat{\boldsymbol{\phi}}(A_z B_\rho - A_\rho B_z) + \hat{\mathbf{z}}(A_\rho B_\phi - A_\phi B_\rho) \tag{D.49}$$

Dyadic representation

$$\bar{\mathbf{a}} = \hat{\boldsymbol{\rho}} a_{\rho\rho} \hat{\boldsymbol{\rho}} + \hat{\boldsymbol{\rho}} a_{\rho\phi} \hat{\boldsymbol{\phi}} + \hat{\boldsymbol{\rho}} a_{\rho z} \hat{\mathbf{z}}$$
$$+ \hat{\boldsymbol{\phi}} a_{\phi\rho} \hat{\boldsymbol{\rho}} + \hat{\boldsymbol{\phi}} a_{\phi\phi} \hat{\boldsymbol{\phi}} + \hat{\boldsymbol{\phi}} a_{\phi z} \hat{\mathbf{z}}$$
$$+ \hat{\mathbf{z}} a_{z\rho} \hat{\boldsymbol{\rho}} + \hat{\mathbf{z}} a_{z\phi} \hat{\boldsymbol{\phi}} + \hat{\mathbf{z}} a_{zz} \hat{\mathbf{z}} \tag{D.50}$$

$$\bar{\mathbf{a}} = \hat{\boldsymbol{\rho}} \mathbf{a}'_\rho + \hat{\boldsymbol{\phi}} \mathbf{a}'_\phi + \hat{\mathbf{z}} \mathbf{a}'_z = \mathbf{a}_\rho \hat{\boldsymbol{\rho}} + \mathbf{a}_\phi \hat{\boldsymbol{\phi}} + \mathbf{a}_z \hat{\mathbf{z}} \tag{D.51}$$

$$\mathbf{a}'_\rho = a_{\rho\rho} \hat{\boldsymbol{\rho}} + a_{\rho\phi} \hat{\boldsymbol{\phi}} + a_{\rho z} \hat{\mathbf{z}} \tag{D.52}$$
$$\mathbf{a}'_\phi = a_{\phi\rho} \hat{\boldsymbol{\rho}} + a_{\phi\phi} \hat{\boldsymbol{\phi}} + a_{\phi z} \hat{\mathbf{z}} \tag{D.53}$$
$$\mathbf{a}'_z = a_{z\rho} \hat{\boldsymbol{\rho}} + a_{z\phi} \hat{\boldsymbol{\phi}} + a_{zz} \hat{\mathbf{z}} \tag{D.54}$$

$$\mathbf{a}_\rho = a_{\rho\rho} \hat{\boldsymbol{\rho}} + a_{\phi\rho} \hat{\boldsymbol{\phi}} + a_{z\rho} \hat{\mathbf{z}} \tag{D.55}$$
$$\mathbf{a}_\phi = a_{\rho\phi} \hat{\boldsymbol{\rho}} + a_{\phi\phi} \hat{\boldsymbol{\phi}} + a_{z\phi} \hat{\mathbf{z}} \tag{D.56}$$
$$\mathbf{a}_z = a_{\rho z} \hat{\boldsymbol{\rho}} + a_{\phi z} \hat{\boldsymbol{\phi}} + a_{zz} \hat{\mathbf{z}} \tag{D.57}$$

Differential operations

$$\mathbf{dl} = \hat{\boldsymbol{\rho}} \, d\rho + \hat{\boldsymbol{\phi}} \rho \, d\phi + \hat{\mathbf{z}} \, dz \tag{D.58}$$

$$dV = \rho \, d\rho \, d\phi \, dz \tag{D.59}$$

$$dS_\rho = \rho \, d\phi \, dz, \tag{D.60}$$
$$dS_\phi = d\rho \, dz, \tag{D.61}$$
$$dS_z = \rho \, d\rho \, d\phi \tag{D.62}$$

$$\nabla f = \hat{\boldsymbol{\rho}} \frac{\partial f}{\partial \rho} + \hat{\boldsymbol{\phi}} \frac{1}{\rho} \frac{\partial f}{\partial \phi} + \hat{\mathbf{z}} \frac{\partial f}{\partial z} \tag{D.63}$$

$$\nabla \cdot \mathbf{F} = \frac{1}{\rho} \frac{\partial}{\partial \rho} (\rho F_\rho) + \frac{1}{\rho} \frac{\partial F_\phi}{\partial \phi} + \frac{\partial F_z}{\partial z} \tag{D.64}$$

$$\nabla \times \mathbf{F} = \frac{1}{\rho} \begin{vmatrix} \hat{\boldsymbol{\rho}} & \rho\hat{\boldsymbol{\phi}} & \hat{\mathbf{z}} \\ \frac{\partial}{\partial\rho} & \frac{\partial}{\partial\phi} & \frac{\partial}{\partial z} \\ F_\rho & \rho F_\phi & F_z \end{vmatrix}$$

$$= \hat{\boldsymbol{\rho}} \left(\frac{1}{\rho} \frac{\partial F_z}{\partial\phi} - \frac{\partial F_\phi}{\partial z} \right) + \hat{\boldsymbol{\phi}} \left(\frac{\partial F_\rho}{\partial z} - \frac{\partial F_z}{\partial\rho} \right) + \hat{\mathbf{z}} \left(\frac{1}{\rho} \frac{\partial[\rho F_\phi]}{\partial\rho} - \frac{1}{\rho} \frac{\partial F_\rho}{\partial\phi} \right) \tag{D.65}$$

$$\nabla^2 f = \frac{1}{\rho} \frac{\partial}{\partial\rho} \left(\rho \frac{\partial f}{\partial\rho} \right) + \frac{1}{\rho^2} \frac{\partial^2 f}{\partial\phi^2} + \frac{\partial^2 f}{\partial z^2} \tag{D.66}$$

$$\nabla^2 \mathbf{F} = \hat{\boldsymbol{\rho}} \left(\nabla^2 F_\rho - \frac{2}{\rho^2} \frac{\partial F_\phi}{\partial\phi} - \frac{F_\rho}{\rho^2} \right) + \hat{\boldsymbol{\phi}} \left(\nabla^2 F_\phi + \frac{2}{\rho^2} \frac{\partial F_\rho}{\partial\phi} - \frac{F_\phi}{\rho^2} \right) + \hat{\mathbf{z}} \nabla^2 F_z \tag{D.67}$$

Separation of the Helmholtz equation

$$\frac{1}{\rho} \frac{\partial}{\partial\rho} \left(\rho \frac{\partial\psi(\rho,\phi,z)}{\partial\rho} \right) + \frac{1}{\rho^2} \frac{\partial^2\psi(\rho,\phi,z)}{\partial\phi^2} + \frac{\partial^2\psi(\rho,\phi,z)}{\partial z^2} + k^2\psi(\rho,\phi,z) = 0 \tag{D.68}$$

$$\psi(\rho,\phi,z) = P(\rho)\Phi(\phi)Z(z) \tag{D.69}$$

$$k_c^2 = k^2 - k_z^2 \tag{D.70}$$

$$\frac{d^2 P(\rho)}{d\rho^2} + \frac{1}{\rho} \frac{dP(\rho)}{d\rho} + \left(k_c^2 - \frac{k_\phi^2}{\rho^2} \right) P(\rho) = 0 \tag{D.71}$$

$$\frac{\partial^2\Phi(\phi)}{\partial\phi^2} + k_\phi^2\Phi(\phi) = 0 \tag{D.72}$$

$$\frac{d^2 Z(z)}{dz^2} + k_z^2 Z(z) = 0 \tag{D.73}$$

$$Z(z) = \begin{cases} A_z F_1(k_z z) + B_z F_2(k_z z), & k_z \neq 0, \\ a_z z + b_z, & k_z = 0. \end{cases} \tag{D.74}$$

$$\Phi(\phi) = \begin{cases} A_\phi F_1(k_\phi \phi) + B_\phi F_2(k_\phi \phi), & k_\phi \neq 0, \\ a_\phi \phi + b_\phi, & k_\phi = 0. \end{cases} \tag{D.75}$$

$$P(\rho) = \begin{cases} a_\rho \ln\rho + b_\rho, & k_c = k_\phi = 0, \\ a_\rho \rho^{-k_\phi} + b_\rho \rho^{k_\phi}, & k_c = 0 \text{ and } k_\phi \neq 0, \\ A_\rho G_1(k_c\rho) + B_\rho G_2(k_c\rho), & \text{otherwise.} \end{cases} \tag{D.76}$$

$$F_1(\xi), F_2(\xi) = \begin{cases} e^{j\xi} \\ e^{-j\xi} \\ \sin(\xi) \\ \cos(\xi) \end{cases} \tag{D.77}$$

$$G_1(\xi), G_2(\xi) = \begin{cases} J_{k_\phi}(\xi) \\ N_{k_\phi}(\xi) \\ H^{(1)}_{k_\phi}(\xi) \\ H^{(2)}_{k_\phi}(\xi) \end{cases} \tag{D.78}$$

Spherical coordinate system

Coordinate variables

$$u = r \qquad (0 \le r < \infty) \tag{D.79}$$

$$v = \theta \qquad (0 \le \theta \le \pi) \tag{D.80}$$

$$w = \phi \qquad (-\pi \le \phi \le \pi) \tag{D.81}$$

$$x = r \sin\theta \cos\phi \tag{D.82}$$

$$y = r \sin\theta \sin\phi \tag{D.83}$$

$$z = r \cos\theta \tag{D.84}$$

$$r = \sqrt{x^2 + y^2 + z^2} \tag{D.85}$$

$$\theta = \tan^{-1} \frac{\sqrt{x^2 + y^2}}{z} \tag{D.86}$$

$$\phi = \tan^{-1} \frac{y}{x} \tag{D.87}$$

Vector algebra

$$\hat{\mathbf{r}} = \hat{\mathbf{x}} \sin\theta \cos\phi + \hat{\mathbf{y}} \sin\theta \sin\phi + \hat{\mathbf{z}} \cos\theta \tag{D.88}$$

$$\hat{\boldsymbol{\theta}} = \hat{\mathbf{x}} \cos\theta \cos\phi + \hat{\mathbf{y}} \cos\theta \sin\phi - \hat{\mathbf{z}} \sin\theta \tag{D.89}$$

$$\hat{\boldsymbol{\phi}} = -\hat{\mathbf{x}} \sin\phi + \hat{\mathbf{y}} \cos\phi \tag{D.90}$$

$$\mathbf{A} = \hat{\mathbf{r}} A_r + \hat{\boldsymbol{\theta}} A_\theta + \hat{\boldsymbol{\phi}} A_\phi \tag{D.91}$$

$$\mathbf{A} \cdot \mathbf{B} = A_r B_r + A_\theta B_\theta + A_\phi B_\phi \tag{D.92}$$

$$\mathbf{A} \times \mathbf{B} = \begin{vmatrix} \hat{\mathbf{r}} & \hat{\boldsymbol{\theta}} & \hat{\boldsymbol{\phi}} \\ A_r & A_\theta & A_\phi \\ B_r & B_\theta & B_\phi \end{vmatrix}$$

$$= \hat{\mathbf{r}}(A_\theta B_\phi - A_\phi B_\theta) + \hat{\boldsymbol{\theta}}(A_\phi B_r - A_r B_\phi) + \hat{\boldsymbol{\phi}}(A_r B_\theta - A_\theta B_r) \tag{D.93}$$

Dyadic representation

$$\bar{a} = \hat{\mathbf{r}} a_{rr} \hat{\mathbf{r}} + \hat{\mathbf{r}} a_{r\theta} \hat{\boldsymbol{\theta}} + \hat{\mathbf{r}} a_{r\phi} \hat{\boldsymbol{\phi}}$$
$$+ \hat{\boldsymbol{\theta}} a_{\theta r} \hat{\mathbf{r}} + \hat{\boldsymbol{\theta}} a_{\theta\theta} \hat{\boldsymbol{\theta}} + \hat{\boldsymbol{\theta}} a_{\theta\phi} \hat{\boldsymbol{\phi}}$$
$$+ \hat{\boldsymbol{\phi}} a_{\phi r} \hat{\mathbf{r}} + \hat{\boldsymbol{\phi}} a_{\phi\theta} \hat{\boldsymbol{\theta}} + \hat{\boldsymbol{\phi}} a_{\phi\phi} \hat{\boldsymbol{\phi}} \tag{D.94}$$

$$\bar{a} = \hat{\mathbf{r}} \mathbf{a}'_r + \hat{\boldsymbol{\theta}} \mathbf{a}'_\theta + \hat{\boldsymbol{\phi}} \mathbf{a}'_\phi = \mathbf{a}_r \hat{\mathbf{r}} + \mathbf{a}_\theta \hat{\boldsymbol{\theta}} + \mathbf{a}_\phi \hat{\boldsymbol{\phi}} \tag{D.95}$$

$$\mathbf{a}'_r = a_{rr} \hat{\mathbf{r}} + a_{r\theta} \hat{\boldsymbol{\theta}} + a_{r\phi} \hat{\boldsymbol{\phi}} \tag{D.96}$$
$$\mathbf{a}'_\theta = a_{\theta r} \hat{\mathbf{r}} + a_{\theta\theta} \hat{\boldsymbol{\theta}} + a_{\theta\phi} \hat{\boldsymbol{\phi}} \tag{D.97}$$
$$\mathbf{a}'_\phi = a_{\phi r} \hat{\mathbf{r}} + a_{\phi\theta} \hat{\boldsymbol{\theta}} + a_{\phi\phi} \hat{\boldsymbol{\phi}} \tag{D.98}$$

$$\mathbf{a}_r = a_{rr} \hat{\mathbf{r}} + a_{\theta r} \hat{\boldsymbol{\theta}} + a_{\phi r} \hat{\boldsymbol{\phi}} \tag{D.99}$$
$$\mathbf{a}_\theta = a_{r\theta} \hat{\mathbf{r}} + a_{\theta\theta} \hat{\boldsymbol{\theta}} + a_{\phi\theta} \hat{\boldsymbol{\phi}} \tag{D.100}$$
$$\mathbf{a}_\phi = a_{r\phi} \hat{\mathbf{r}} + a_{\theta\phi} \hat{\boldsymbol{\theta}} + a_{\phi\phi} \hat{\boldsymbol{\phi}} \tag{D.101}$$

Differential operations

$$\mathbf{dl} = \hat{\mathbf{r}} \, dr + \hat{\boldsymbol{\theta}} r \, d\theta + \hat{\boldsymbol{\phi}} r \sin\theta \, d\phi \tag{D.102}$$

$$dV = r^2 \sin\theta \, dr \, d\theta \, d\phi \tag{D.103}$$

$$dS_r = r^2 \sin\theta \, d\theta \, d\phi \tag{D.104}$$
$$dS_\theta = r \sin\theta \, dr \, d\phi \tag{D.105}$$
$$dS_\phi = r \, dr \, d\theta \tag{D.106}$$

$$\nabla f = \hat{\mathbf{r}} \frac{\partial f}{\partial r} + \hat{\boldsymbol{\theta}} \frac{1}{r} \frac{\partial f}{\partial \theta} + \hat{\boldsymbol{\phi}} \frac{1}{r \sin\theta} \frac{\partial f}{\partial \phi} \tag{D.107}$$

$$\nabla \cdot \mathbf{F} = \frac{1}{r^2} \frac{\partial}{\partial r} \left(r^2 F_r \right) + \frac{1}{r \sin\theta} \frac{\partial}{\partial \theta} \left(\sin\theta F_\theta \right) + \frac{1}{r \sin\theta} \frac{\partial F_\phi}{\partial \phi} \tag{D.108}$$

$$\nabla \times \mathbf{F} = \frac{1}{r^2 \sin\theta} \begin{vmatrix} \hat{\mathbf{r}} & r\hat{\boldsymbol{\theta}} & r\sin\theta\hat{\boldsymbol{\phi}} \\ \frac{\partial}{\partial r} & \frac{\partial}{\partial \theta} & \frac{\partial}{\partial \phi} \\ F_r & rF_\theta & r\sin\theta F_\phi \end{vmatrix}$$

$$= \frac{\hat{\mathbf{r}}}{r \sin\theta} \left(\frac{\partial [F_\phi \sin\theta]}{\partial \theta} - \frac{\partial F_\theta}{\partial \phi} \right) + \frac{\hat{\boldsymbol{\theta}}}{r} \left(\frac{1}{\sin\theta} \frac{\partial F_r}{\partial \phi} - \frac{\partial [rF_\phi]}{\partial r} \right) + \frac{\hat{\boldsymbol{\phi}}}{r} \left(\frac{\partial [rF_\theta]}{\partial r} - \frac{\partial F_r}{\partial \theta} \right) \tag{D.109}$$

$$\nabla^2 f = \frac{1}{r^2}\frac{\partial}{\partial r}\left(r^2\frac{\partial f}{\partial r}\right) + \frac{1}{r^2\sin\theta}\frac{\partial}{\partial\theta}\left(\sin\theta\frac{\partial f}{\partial\theta}\right) + \frac{1}{r^2\sin^2\theta}\frac{\partial^2 f}{\partial\phi^2} \tag{D.110}$$

$$\nabla^2\mathbf{F} = \hat{\mathbf{r}}\left[\nabla^2 F_r - \frac{2}{r^2}\left(F_r + \frac{\cos\theta}{\sin\theta}F_\theta + \frac{1}{\sin\theta}\frac{\partial F_\phi}{\partial\phi} + \frac{\partial F_\theta}{\partial\theta}\right)\right]$$
$$+ \hat{\boldsymbol{\theta}}\left[\nabla^2 F_\theta - \frac{1}{r^2}\left(\frac{1}{\sin^2\theta}F_\theta - 2\frac{\partial F_r}{\partial\theta} + 2\frac{\cos\theta}{\sin^2\theta}\frac{\partial F_\phi}{\partial\phi}\right)\right]$$
$$+ \hat{\boldsymbol{\phi}}\left[\nabla^2 F_\phi - \frac{1}{r^2}\left(\frac{1}{\sin^2\theta}F_\phi - 2\frac{1}{\sin\theta}\frac{\partial F_r}{\partial\phi} - 2\frac{\cos\theta}{\sin^2\theta}\frac{\partial F_\theta}{\partial\phi}\right)\right] \tag{D.111}$$

Separation of the Helmholtz equation

$$\frac{1}{r^2}\frac{\partial}{\partial r}\left(r^2\frac{\partial\psi(r,\theta,\phi)}{\partial r}\right) + \frac{1}{r^2\sin\theta}\frac{\partial}{\partial\theta}\left(\sin\theta\frac{\partial\psi(r,\theta,\phi)}{\partial\theta}\right)$$
$$+ \frac{1}{r^2\sin^2\theta}\frac{\partial^2\psi(r,\theta,\phi)}{\partial\phi^2} + k^2\psi(r,\theta,\phi) = 0 \tag{D.112}$$

$$\psi(r,\theta,\phi) = R(r)\Theta(\theta)\Phi(\phi) \tag{D.113}$$

$$\eta = \cos\theta \tag{D.114}$$

$$\frac{1}{R(r)}\frac{d}{dr}\left(r^2\frac{dR(r)}{dr}\right) + k^2 r^2 = n(n+1) \tag{D.115}$$

$$(1-\eta^2)\frac{d^2\Theta(\eta)}{d\eta^2} - 2\eta\frac{d\Theta(\eta)}{d\eta} + \left[n(n+1) - \frac{\mu^2}{1-\eta^2}\right]\Theta(\eta) = 0 \quad (|\eta|\leq 1) \tag{D.116}$$

$$\frac{d^2\Phi(\phi)}{d\phi^2} + \mu^2\Phi(\phi) = 0 \tag{D.117}$$

$$\Phi(\phi) = \begin{cases} A_\phi\sin(\mu\phi) + B_\phi\cos(\mu\phi), & \mu\neq 0, \\ a_\phi\phi + b_\phi, & \mu = 0. \end{cases} \tag{D.118}$$

$$\Theta(\theta) = A_\theta P_n^\mu(\cos\theta) + B_\theta Q_n^\mu(\cos\theta) \tag{D.119}$$

$$R(r) = \begin{cases} R(r) = A_r r^n + B_r r^{-(n+1)}, & k = 0, \\ A_r F_1(kr) + B_r F_2(kr), & \text{otherwise.} \end{cases} \tag{D.120}$$

$$F_1(\xi), F_2(\xi) = \begin{cases} j_n(\xi) \\ n_n(\xi) \\ h_n^{(1)}(\xi) \\ h_n^{(2)}(\xi) \end{cases} \tag{D.121}$$

E

Properties of special functions

E.1 Bessel functions

Notation

z = complex number; ν, x = real numbers; n = integer

$J_\nu(z)$ = ordinary Bessel function of the first kind

$N_\nu(z)$ = ordinary Bessel function of the second kind

$I_\nu(z)$ = modified Bessel function of the first kind

$K_\nu(z)$ = modified Bessel function of the second kind

$H_\nu^{(1)}$ = Hankel function of the first kind

$H_\nu^{(2)}$ = Hankel function of the second kind

$j_n(z)$ = ordinary spherical Bessel function of the first kind

$n_n(z)$ = ordinary spherical Bessel function of the second kind

$h_n^{(1)}(z)$ = spherical Hankel function of the first kind

$h_n^{(2)}(z)$ = spherical Hankel function of the second kind

$f'(z) = df(z)/dz$ = derivative with respect to argument

Differential equations

$$\frac{d^2 Z_\nu(z)}{dz^2} + \frac{1}{z}\frac{dZ_\nu(z)}{dz} + \left(1 - \frac{\nu^2}{z^2}\right)Z_\nu(z) = 0 \tag{E.1}$$

$$Z_\nu(z) = \begin{cases} J_\nu(z) \\ N_\nu(z) \\ H_\nu^{(1)}(z) \\ H_\nu^{(2)}(z) \end{cases} \tag{E.2}$$

$$N_\nu(z) = \frac{\cos(\nu\pi)J_\nu(z) - J_{-\nu}(z)}{\sin(\nu\pi)} \qquad (\nu \neq n,\ |\arg z| < \pi) \tag{E.3}$$

$$H_\nu^{(1)}(z) = J_\nu(z) + jN_\nu(z) \tag{E.4}$$

$$H_\nu^{(2)}(z) = J_\nu(z) - jN_\nu(z) \tag{E.5}$$

$$\frac{d^2 \bar{Z}_\nu(x)}{dz^2} + \frac{1}{z}\frac{d\bar{Z}_\nu(z)}{dz} - \left(1 + \frac{\nu^2}{z^2}\right)\bar{Z}_\nu = 0 \tag{E.6}$$

$$\bar{Z}_\nu(z) = \begin{cases} I_\nu(z) \\ K_\nu(z) \end{cases} \tag{E.7}$$

$$L(z) = \begin{cases} I_\nu(z) \\ e^{j\nu\pi}K_\nu(z) \end{cases} \tag{E.8}$$

$$I_\nu(z) = e^{-j\nu\pi/2}J_\nu(ze^{j\pi/2}) \qquad\qquad (-\pi < \arg z \le \pi/2) \tag{E.9}$$

$$I_\nu(z) = e^{j3\nu\pi/2}J_\nu(ze^{-j3\pi/2}) \qquad\qquad (\pi/2 < \arg z \le \pi) \tag{E.10}$$

$$K_\nu(z) = \frac{j\pi}{2}e^{j\nu\pi/2}H_\nu^{(1)}(ze^{j\pi/2}) \qquad\qquad (-\pi < \arg z \le \pi/2) \tag{E.11}$$

$$K_\nu(z) = -\frac{j\pi}{2}e^{-j\nu\pi/2}H_\nu^{(2)}(ze^{-j\pi/2}) \qquad\qquad (-\pi/2 < \arg z \le \pi) \tag{E.12}$$

$$I_n(x) = j^{-n}J_n(jx) \tag{E.13}$$

$$K_n(x) = \frac{\pi}{2}j^{n+1}H_n^{(1)}(jx) \tag{E.14}$$

$$\frac{d^2 z_n(z)}{dz^2} + \frac{2}{z}\frac{dz_n(z)}{dz} + \left[1 - \frac{n(n+1)}{z^2}\right]z_n(z) = 0 \qquad (n = 0, \pm 1, \pm 2, \ldots) \tag{E.15}$$

$$z_n(z) = \begin{cases} j_n(z) \\ n_n(z) \\ h_n^{(1)}(z) \\ h_n^{(2)}(z) \end{cases} \tag{E.16}$$

$$j_n(z) = \sqrt{\frac{\pi}{2z}}J_{n+\frac{1}{2}}(z) \tag{E.17}$$

$$n_n(z) = \sqrt{\frac{\pi}{2z}}N_{n+\frac{1}{2}}(z) \tag{E.18}$$

$$h_n^{(1)}(z) = \sqrt{\frac{\pi}{2z}}H_{n+\frac{1}{2}}^{(1)}(z) = j_n(z) + jn_n(z) \tag{E.19}$$

$$h_n^{(2)}(z) = \sqrt{\frac{\pi}{2z}}H_{n+\frac{1}{2}}^{(2)}(z) = j_n(z) - jn_n(z) \tag{E.20}$$

$$n_n(z) = (-1)^{n+1}j_{-(n+1)}(z) \tag{E.21}$$

Orthogonality relationships

$$\int_0^a J_\nu\left(\frac{p_{\nu m}}{a}\rho\right) J_\nu\left(\frac{p_{\nu n}}{a}\rho\right)\rho\,d\rho = \delta_{mn}\frac{a^2}{2}J_{\nu+1}^2(p_{\nu n}) = \delta_{mn}\frac{a^2}{2}\left[J_\nu'(p_{\nu n})\right]^2 \quad (\nu > -1)$$
$$(E.22)$$

$$\int_0^a J_\nu\left(\frac{p_{\nu m}'}{a}\rho\right) J_\nu\left(\frac{p_{\nu n}'}{a}\rho\right)\rho\,d\rho = \delta_{mn}\frac{a^2}{2}\left(1 - \frac{\nu^2}{p_{\nu m}'^2}\right)J_\nu^2(p_{\nu m}') \quad (\nu > -1) \qquad (E.23)$$

$$\int_0^\infty J_\nu(\alpha x)J_\nu(\beta x)x\,dx = \frac{1}{\alpha}\delta(\alpha - \beta) \tag{E.24}$$

$$\int_0^a j_l\left(\frac{\alpha_{lm}}{a}r\right) j_l\left(\frac{\alpha_{ln}}{a}r\right) r^2\,dr = \delta_{mn}\frac{a^3}{2}j_{n+1}^2(\alpha_{ln}a) \tag{E.25}$$

$$\int_{-\infty}^\infty j_m(x)j_n(x)\,dx = \delta_{mn}\frac{\pi}{2n+1} \qquad (m, n \geq 0) \tag{E.26}$$

$$J_m(p_{mn}) = 0 \tag{E.27}$$
$$J_m'(p_{mn}') = 0 \tag{E.28}$$
$$j_m(\alpha_{mn}) = 0 \tag{E.29}$$
$$j_m'(\alpha_{mn}') = 0 \tag{E.30}$$

Specific examples

$$j_0(z) = \frac{\sin z}{z} \tag{E.31}$$

$$n_0(z) = -\frac{\cos z}{z} \tag{E.32}$$

$$h_0^{(1)}(z) = -\frac{j}{z}e^{jz} \tag{E.33}$$

$$h_0^{(2)}(z) = \frac{j}{z}e^{-jz} \tag{E.34}$$

$$j_1(z) = \frac{\sin z}{z^2} - \frac{\cos z}{z} \tag{E.35}$$

$$n_1(z) = -\frac{\cos z}{z^2} - \frac{\sin z}{z} \tag{E.36}$$

$$j_2(z) = \left(\frac{3}{z^3} - \frac{1}{z}\right)\sin z - \frac{3}{z^2}\cos z \tag{E.37}$$

$$n_2(z) = \left(-\frac{3}{z^3} + \frac{1}{z}\right)\cos z - \frac{3}{z^2}\sin z \tag{E.38}$$

Functional relationships

$$J_n(-z) = (-1)^n J_n(z) \tag{E.39}$$

$$I_n(-z) = (-1)^n I_n(z) \tag{E.40}$$

$$j_n(-z) = (-1)^n j_n(z) \tag{E.41}$$

$$n_n(-z) = (-1)^{n+1} n_n(z) \tag{E.42}$$

$$J_{-n}(z) = (-1)^n J_n(z) \tag{E.43}$$

$$N_{-n}(z) = (-1)^n N_n(z) \tag{E.44}$$

$$I_{-n}(z) = I_n(z) \tag{E.45}$$

$$K_{-n}(z) = K_n(z) \tag{E.46}$$

$$j_{-n}(z) = (-1)^n n_{n-1}(z) \quad (n > 0) \tag{E.47}$$

Power series

$$J_n(z) = \sum_{k=0}^{\infty} (-1)^k \frac{(z/2)^{n+2k}}{k!(n+k)!} \tag{E.48}$$

$$I_n(z) = \sum_{k=0}^{\infty} \frac{(z/2)^{n+2k}}{k!(n+k)!} \tag{E.49}$$

Small argument approximations $|z| \ll 1$

$$J_n(z) \approx \frac{1}{n!} \left(\frac{z}{2}\right)^n \tag{E.50}$$

$$J_\nu(z) \approx \frac{1}{\Gamma(\nu+1)} \left(\frac{z}{2}\right)^\nu \tag{E.51}$$

$$N_0(z) \approx \frac{2}{\pi} (\ln z + 0.5772157 - \ln 2) \tag{E.52}$$

$$N_n(z) \approx -\frac{(n-1)!}{\pi} \left(\frac{2}{z}\right)^n \quad (n > 0) \tag{E.53}$$

$$N_\nu(z) \approx -\frac{\Gamma(\nu)}{\pi} \left(\frac{2}{z}\right)^\nu \quad (\nu > 0) \tag{E.54}$$

$$I_n(z) \approx \frac{1}{n!} \left(\frac{z}{2}\right)^n \tag{E.55}$$

$$I_\nu(z) \approx \frac{1}{\Gamma(\nu+1)} \left(\frac{z}{2}\right)^\nu \tag{E.56}$$

$$K_0(z) \approx -\ln z \tag{E.57}$$

$$K_\nu(z) \approx \frac{\Gamma(\nu)}{2} \left(\frac{z}{2}\right)^{-\nu} \quad (\operatorname{Re}\nu > 0) \tag{E.58}$$

$$j_n(z) \approx \frac{2^n n!}{(2n+1)!} z^n \tag{E.59}$$

$$n_n(z) \approx -\frac{(2n)!}{2^n n!} z^{-(n+1)} \tag{E.60}$$

Large argument approximations $|z| \gg 1$

$$J_\nu(z) \approx \sqrt{\frac{2}{\pi z}} \cos\left(z - \frac{\pi}{4} - \frac{\nu\pi}{2}\right) \quad (|\arg z| < \pi) \tag{E.61}$$

$$N_\nu(z) \approx \sqrt{\frac{2}{\pi z}} \sin\left(z - \frac{\pi}{4} - \frac{\nu\pi}{2}\right) \quad (|\arg z| < \pi) \tag{E.62}$$

$$H_\nu^{(1)}(z) \approx \sqrt{\frac{2}{\pi z}} e^{j\left(z - \frac{\pi}{4} - \frac{\nu\pi}{2}\right)} \quad (-\pi < \arg z < 2\pi) \tag{E.63}$$

$$H_\nu^{(2)}(z) \approx \sqrt{\frac{2}{\pi z}} e^{-j\left(z - \frac{\pi}{4} - \frac{\nu\pi}{2}\right)} \quad (-2\pi < \arg z < \pi) \tag{E.64}$$

$$I_\nu(z) \approx \sqrt{\frac{1}{2\pi z}} e^z \quad (|\arg z| < \pi/2) \tag{E.65}$$

$$K_\nu(z) \approx \sqrt{\frac{\pi}{2z}} e^{-z} \quad (|\arg z| < 3\pi/2) \tag{E.66}$$

$$j_n(z) \approx \frac{1}{z} \sin\left(z - \frac{n\pi}{2}\right) \quad (|\arg z| < \pi) \tag{E.67}$$

$$n_n(z) \approx -\frac{1}{z} \cos\left(z - \frac{n\pi}{2}\right) \quad (|\arg z| < \pi) \tag{E.68}$$

$$h_n^{(1)}(z) \approx (-j)^{n+1} \frac{e^{jz}}{z} \quad (-\pi < \arg z < 2\pi) \tag{E.69}$$

$$h_n^{(2)}(z) \approx j^{n+1} \frac{e^{-jz}}{z} \quad (-2\pi < \arg z < \pi) \tag{E.70}$$

Recursion relationships

$$zZ_{\nu-1}(z) + zZ_{\nu+1}(z) = 2\nu Z_\nu(z) \tag{E.71}$$

$$Z_{\nu-1}(z) - Z_{\nu+1}(z) = 2Z_\nu'(z) \tag{E.72}$$

$$zZ_\nu'(z) + \nu Z_\nu(z) = zZ_{\nu-1}(z) \tag{E.73}$$

$$zZ_\nu'(z) - \nu Z_\nu(z) = -zZ_{\nu+1}(z) \tag{E.74}$$

$$zL_{\nu-1}(z) - zL_{\nu+1}(z) = 2\nu L_\nu(z) \tag{E.75}$$

$$L_{\nu-1}(z) + L_{\nu+1}(z) = 2L_\nu'(z) \tag{E.76}$$

$$zL_\nu'(z) + \nu L_\nu(z) = zL_{\nu-1}(z) \tag{E.77}$$

$$zL_\nu'(z) - \nu L_\nu(z) = zL_{\nu+1}(z) \tag{E.78}$$

$$zz_{n-1}(z) + zz_{n+1}(z) = (2n+1)z_n(z) \tag{E.79}$$

$$nz_{n-1}(z) - (n+1)z_{n+1}(z) = (2n+1)z_n'(z) \tag{E.80}$$

$$zz_n'(z) + (n+1)z_n(z) = zz_{n-1}(z) \tag{E.81}$$

$$-zz_n'(z) + nz_n(z) = zz_{n+1}(z) \tag{E.82}$$

Integral representations

$$J_n(z) = \frac{1}{2\pi} \int_{-\pi}^{\pi} e^{-jn\theta + jz\sin\theta} \, d\theta \tag{E.83}$$

$$J_n(z) = \frac{1}{\pi} \int_0^{\pi} \cos(n\theta - z\sin\theta) \, d\theta \tag{E.84}$$

$$J_n(z) = \frac{1}{2\pi} j^{-n} \int_{-\pi}^{\pi} e^{jz\cos\theta} \cos(n\theta) \, d\theta \tag{E.85}$$

$$I_n(z) = \frac{1}{\pi} \int_0^{\pi} e^{z\cos\theta} \cos(n\theta) \, d\theta \tag{E.86}$$

$$K_n(z) = \int_0^{\infty} e^{-z\cosh(t)} \cosh(nt) \, dt \quad (|\arg z| < \pi/2) \tag{E.87}$$

$$j_n(z) = \frac{z^n}{2^{n+1} n!} \int_0^{\pi} \cos(z\cos\theta) \sin^{2n+1}\theta \, d\theta \tag{E.88}$$

$$j_n(z) = \frac{(-j)^n}{2} \int_0^{\pi} e^{jz\cos\theta} P_n(\cos\theta) \sin\theta \, d\theta \tag{E.89}$$

Wronskians and cross products

$$J_\nu(z) N_{\nu+1}(z) - J_{\nu+1}(z) N_\nu(z) = -\frac{2}{\pi z} \tag{E.90}$$

$$H_\nu^{(2)}(z) H_{\nu+1}^{(1)}(z) - H_\nu^{(1)}(z) H_{\nu+1}^{(2)}(z) = \frac{4}{j\pi z} \tag{E.91}$$

$$I_\nu(z) K_{\nu+1}(z) + I_{\nu+1}(z) K_\nu(z) = \frac{1}{z} \tag{E.92}$$

$$I_\nu(z) K_\nu'(z) - I_\nu'(z) K_\nu(z) = -\frac{1}{z} \tag{E.93}$$

$$J_\nu(z) H_\nu^{(1)'}(z) - J_\nu'(z) H_\nu^{(1)}(z) = \frac{2j}{\pi z} \tag{E.94}$$

$$J_\nu(z) H_\nu^{(2)'}(z) - J_\nu'(z) H_\nu^{(2)}(z) = -\frac{2j}{\pi z} \tag{E.95}$$

$$H_\nu^{(1)}(z) H_\nu^{(2)'}(z) - H_\nu^{(1)'}(z) H_\nu^{(2)}(z) = -\frac{4j}{\pi z} \tag{E.96}$$

$$j_n(z) n_{n-1}(z) - j_{n-1}(z) n_n(z) = \frac{1}{z^2} \tag{E.97}$$

$$j_{n+1}(z) n_{n-1}(z) - j_{n-1}(z) n_{n+1}(z) = \frac{2n+1}{z^3} \tag{E.98}$$

$$j_n(z) n_n'(z) - j_n'(z) n_n(z) = \frac{1}{z^2} \tag{E.99}$$

$$h_n^{(1)}(z) h_n^{(2)'}(z) - h_n^{(1)'}(z) h_n^{(2)}(z) = -\frac{2j}{z^2} \tag{E.100}$$

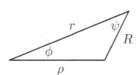

Summation formulas

R, r, ρ, ϕ, ψ as shown.

$R = \sqrt{r^2 + \rho^2 - 2r\rho\cos\phi}.$

$$e^{j\nu\psi} Z_\nu(zR) = \sum_{k=-\infty}^{\infty} J_k(z\rho) Z_{\nu+k}(zr) e^{jk\phi} \quad (\rho < r,\ 0 < \psi < \pi/2) \tag{E.101}$$

$$e^{jn\psi} J_n(zR) = \sum_{k=-\infty}^{\infty} J_k(z\rho) J_{n+k}(zr) e^{jk\phi} \tag{E.102}$$

$$e^{jz\rho\cos\phi} = \sum_{k=0}^{\infty} j^k (2k+1) j_k(z\rho) P_k(\cos\phi) \tag{E.103}$$

For $\rho < r$ and $0 < \psi < \pi/2$,

$$\frac{e^{jzR}}{R} = \frac{j\pi}{2\sqrt{r\rho}} \sum_{k=0}^{\infty} (2k+1) J_{k+\frac{1}{2}}(z\rho) H_{k+\frac{1}{2}}^{(1)}(zr) P_k(\cos\phi) \tag{E.104}$$

$$\frac{e^{-jzR}}{R} = -\frac{j\pi}{2\sqrt{r\rho}} \sum_{k=0}^{\infty} (2k+1) J_{k+\frac{1}{2}}(z\rho) H_{k+\frac{1}{2}}^{(2)}(zr) P_k(\cos\phi) \tag{E.105}$$

Integrals

$$\int x^{\nu+1} J_\nu(x)\, dx = x^{\nu+1} J_{\nu+1}(x) + C \tag{E.106}$$

$$\int Z_\nu(ax) Z_\nu(bx) x\, dx = x\frac{[b Z_\nu(ax) Z_{\nu-1}(bx) - a Z_{\nu-1}(ax) Z_\nu(bx)]}{a^2 - b^2} + C \quad (a \neq b) \tag{E.107}$$

$$\int x Z_\nu^2(ax)\, dx = \frac{x^2}{2} \left[Z_\nu^2(ax) - Z_{\nu-1}(ax) Z_{\nu+1}(ax) \right] + C \tag{E.108}$$

$$\int_0^\infty J_\nu(ax)\, dx = \frac{1}{a} \quad (\nu > -1,\ a > 0) \tag{E.109}$$

Fourier–Bessel expansion of a function

$$f(\rho) = \sum_{m=1}^{\infty} a_m J_\nu \left(p_{\nu m} \frac{\rho}{a} \right) \quad (0 \le \rho \le a, \ \nu > -1) \tag{E.110}$$

$$a_m = \frac{2}{a^2 J_{\nu+1}^2(p_{\nu m})} \int_0^a f(\rho) J_\nu \left(p_{\nu m} \frac{\rho}{a} \right) \rho \, d\rho \tag{E.111}$$

$$f(\rho) = \sum_{m=1}^{\infty} b_m J_\nu \left(p'_{\nu m} \frac{\rho}{a} \right) \quad (0 \le \rho \le a, \ \nu > -1) \cdot \tag{E.112}$$

$$b_m = \frac{2}{a^2 \left(1 - \frac{\nu^2}{p'^2_{\nu m}} J_\nu^2(p'_{\nu m}) \right)} \int_0^a f(\rho) J_\nu \left(\frac{p'_{\nu m}}{a} \rho \right) \rho \, d\rho \tag{E.113}$$

Series of Bessel functions

$$e^{jz\cos\phi} = \sum_{k=-\infty}^{\infty} j^k J_k(z) e^{jk\phi} \tag{E.114}$$

$$e^{jz\cos\phi} = J_0(z) + 2 \sum_{k=1}^{\infty} j^k J_k(z) \cos\phi \tag{E.115}$$

$$\sin z = 2 \sum_{k=0}^{\infty} (-1)^k J_{2k+1}(z) \tag{E.116}$$

$$\cos z = J_0(z) + 2 \sum_{k=1}^{\infty} (-1)^k J_{2k}(z) \tag{E.117}$$

E.2   Legendre functions

Notation

x, y, θ = real numbers; l, m, n = integers;

$P_n^m(\cos\theta)$ = associated Legendre function of the first kind

$Q_n^m(\cos\theta)$ = associated Legendre function of the second kind

$P_n(\cos\theta) = P_n^0(\cos\theta)$ = Legendre polynomial

$Q_n(\cos\theta) = Q_n^0(\cos\theta)$ = Legendre function of the second kind

Differential equation $x = \cos\theta$.

$$(1-x^2)\frac{d^2 R_n^m(x)}{dx^2} - 2x\frac{dR_n^m(x)}{dx} + \left[n(n+1) - \frac{m^2}{1-x^2} \right] R_n^m(x) = 0 \quad (|x| \le 1) \tag{E.118}$$

$$R_n^m(x) = \begin{cases} P_n^m(x) \\ Q_n^m(x) \end{cases} \tag{E.119}$$

Orthogonality relationships

$$\int_{-1}^{1} P_l^m(x)P_n^m(x)\,dx = \delta_{ln}\frac{2}{2n+1}\frac{(n+m)!}{(n-m)!} \tag{E.120}$$

$$\int_0^\pi P_l^m(\cos\theta)P_n^m(\cos\theta)\sin\theta\,d\theta = \delta_{ln}\frac{2}{2n+1}\frac{(n+m)!}{(n-m)!} \tag{E.121}$$

$$\int_{-1}^{1} \frac{P_n^m(x)P_n^k(x)}{1-x^2}\,dx = \delta_{mk}\frac{1}{m}\frac{(n+m)!}{(n-m)!} \tag{E.122}$$

$$\int_0^\pi \frac{P_n^m(\cos\theta)P_n^k(\cos\theta)}{\sin\theta}\,d\theta = \delta_{mk}\frac{1}{m}\frac{(n+m)!}{(n-m)!} \tag{E.123}$$

$$\int_{-1}^{1} P_l(x)P_n(x)\,dx = \delta_{ln}\frac{2}{2n+1} \tag{E.124}$$

$$\int_0^\pi P_l(\cos\theta)P_n(\cos\theta)\sin\theta\,d\theta = \delta_{ln}\frac{2}{2n+1} \tag{E.125}$$

Specific examples

$$P_0(x) = 1 \tag{E.126}$$

$$P_1(x) = x = \cos(\theta) \tag{E.127}$$

$$P_2(x) = \tfrac{1}{2}(3x^2 - 1) = \tfrac{1}{4}(3\cos 2\theta + 1) \tag{E.128}$$

$$P_3(x) = \tfrac{1}{2}(5x^3 - 3x) = \tfrac{1}{8}(5\cos 3\theta + 3\cos\theta) \tag{E.129}$$

$$P_4(x) = \tfrac{1}{8}(35x^4 - 30x^2 + 3) = \tfrac{1}{64}(35\cos 4\theta + 20\cos 2\theta + 9) \tag{E.130}$$

$$P_5(x) = \tfrac{1}{8}(63x^5 - 70x^3 + 15x) = \tfrac{1}{128}(63\cos 5\theta + 35\cos 3\theta + 30\cos\theta) \tag{E.131}$$

$$Q_0(x) = \frac{1}{2}\ln\left(\frac{1+x}{1-x}\right) = \ln\left(\cot\frac{\theta}{2}\right) \tag{E.132}$$

$$Q_1(x) = \frac{x}{2}\ln\left(\frac{1+x}{1-x}\right) - 1 = \cos\theta\ln\left(\cot\frac{\theta}{2}\right) - 1 \tag{E.133}$$

$$Q_2(x) = \frac{1}{4}(3x^2 - 1)\ln\left(\frac{1+x}{1-x}\right) - \frac{3}{2}x \tag{E.134}$$

$$Q_3(x) = \frac{1}{4}(5x^3 - 3x)\ln\left(\frac{1+x}{1-x}\right) - \frac{5}{2}x^2 + \frac{2}{3} \tag{E.135}$$

$$Q_4(x) = \frac{1}{16}(35x^4 - 30x^2 + 3)\ln\left(\frac{1+x}{1-x}\right) - \frac{35}{8}x^3 + \frac{55}{24}x \tag{E.136}$$

$$P_1^1(x) = -(1 - x^2)^{1/2} = -\sin\theta \tag{E.137}$$

$$P_2^1(x) = -3x(1 - x^2)^{1/2} = -3\cos\theta\sin\theta \tag{E.138}$$

$$P_2^2(x) = 3(1 - x^2) = 3\sin^2\theta \tag{E.139}$$

$$P_3^1(x) = -\tfrac{3}{2}(5x^2 - 1)(1 - x^2)^{1/2} = -\tfrac{3}{2}(5\cos^2\theta - 1)\sin\theta \tag{E.140}$$

$$P_3^2(x) = 15x(1 - x^2) = 15\cos\theta\sin^2\theta \tag{E.141}$$

$$P_3^3(x) = -15(1 - x^2)^{3/2} = -15\sin^3\theta \tag{E.142}$$

$$P_4^1(x) = -\tfrac{5}{2}(7x^3 - 3x)(1 - x^2)^{1/2} = -\tfrac{5}{2}(7\cos^3\theta - 3\cos\theta)\sin\theta \tag{E.143}$$

$$P_4^2(x) = \tfrac{15}{2}(7x^2 - 1)(1 - x^2) = \tfrac{15}{2}(7\cos^2\theta - 1)\sin^2\theta \tag{E.144}$$

$$P_4^3(x) = -105x(1 - x^2)^{3/2} = -105\cos\theta\sin^3\theta \tag{E.145}$$

$$P_4^4(x) = 105(1 - x^2)^2 = 105\sin^4\theta \tag{E.146}$$

Functional relationships

$$P_n^m(x) = \begin{cases} 0, & m > n, \\ (-1)^m \dfrac{(1-x^2)^{m/2}}{2^n n!} \dfrac{d^{n+m}(x^2-1)^n}{dx^{n+m}}, & m \le n. \end{cases} \tag{E.147}$$

$$P_n(x) = \frac{1}{2^n n!} \frac{d^n (x^2 - 1)^n}{dx^n} \tag{E.148}$$

$$R_n^m(x) = (-1)^m (1 - x^2)^{m/2} \frac{d^m R_n(x)}{dx^m} \tag{E.149}$$

$$P_n^{-m}(x) = (-1)^m \frac{(n - m)!}{(n + m)!} P_n^m(x) \tag{E.150}$$

$$P_n(-x) = (-1)^n P_n(x) \tag{E.151}$$

$$Q_n(-x) = (-1)^{n+1} Q_n(x) \tag{E.152}$$

$$P_n^m(-x) = (-1)^{n+m} P_n^m(x) \tag{E.153}$$

$$Q_n^m(-x) = (-1)^{n+m+1} Q_n^m(x) \tag{E.154}$$

$$P_n^m(1) = \begin{cases} 1, & m = 0, \\ 0, & m > 0. \end{cases} \tag{E.155}$$

$$|P_n(x)| \le P_n(1) = 1 \tag{E.156}$$

$$P_n(0) = \frac{\Gamma\left(\frac{n}{2} + \frac{1}{2}\right)}{\sqrt{\pi}\,\Gamma\left(\frac{n}{2} + 1\right)} \cos\frac{n\pi}{2} \tag{E.157}$$

$$P_n^{-m}(x) = (-1)^m \frac{(n - m)!}{(n + m)!} P_n^m(x) \tag{E.158}$$

Power series

$$P_n(x) = \sum_{k=0}^{n} \frac{(-1)^k (n+k)!}{(n-k)!(k!)^2 2^{k+1}} \left[(1-x)^k + (-1)^n (1+x)^k \right] \qquad \text{(E.159)}$$

Recursion relationships

$$(n+1-m)R_{n+1}^m(x) + (n+m)R_{n-1}^m(x) = (2n+1)xR_n^m(x) \qquad \text{(E.160)}$$

$$(1-x^2)R_n^{m'}(x) = (n+1)xR_n^m(x) - (n-m+1)R_{n+1}^m(x) \qquad \text{(E.161)}$$

$$(2n+1)xR_n(x) = (n+1)R_{n+1}(x) + nR_{n-1}(x) \qquad \text{(E.162)}$$

$$(x^2-1)R_n'(x) = (n+1)[R_{n+1}(x) - xR_n(x)] \qquad \text{(E.163)}$$

$$R_{n+1}'(x) - R_{n-1}'(x) = (2n+1)R_n(x) \qquad \text{(E.164)}$$

Integral representations

$$P_n(\cos\theta) = \frac{\sqrt{2}}{\pi} \int_0^\pi \frac{\sin\left(n+\tfrac{1}{2}\right)u}{\sqrt{\cos\theta - \cos u}}\, du \qquad \text{(E.165)}$$

$$P_n(x) = \frac{1}{\pi} \int_0^\pi [x + (x^2-1)^{1/2}\cos\theta]^n\, d\theta \qquad \text{(E.166)}$$

Addition formula

$$P_n(\cos\gamma) = P_n(\cos\theta)P_n(\cos\theta')$$
$$+ 2\sum_{m=1}^n \frac{(n-m)!}{(n+m)!} P_n^m(\cos\theta)P_n^m(\cos\theta')\cos m(\phi-\phi') \qquad \text{(E.167)}$$

$$\cos\gamma = \cos\theta\cos\theta' + \sin\theta\sin\theta'\cos(\phi-\phi') \qquad \text{(E.168)}$$

Summations

$$\frac{1}{|\mathbf{r}-\mathbf{r}'|} = \frac{1}{\sqrt{r^2 + r'^2 - 2rr'\cos\gamma}} = \sum_{n=0}^\infty \frac{r_<^n}{r_>^{n+1}} P_n(\cos\gamma) \qquad \text{(E.169)}$$

$$\cos\gamma = \cos\theta\cos\theta' + \sin\theta\sin\theta'\cos(\phi-\phi') \qquad \text{(E.170)}$$

$$r_< = \min\{|\mathbf{r}|, |\mathbf{r}'|\}, \quad r_> = \max\{|\mathbf{r}|, |\mathbf{r}'|\} \qquad \text{(E.171)}$$

Integrals

$$\int P_n(x)\,dx = \frac{P_{n+1}(x) - P_{n-1}(x)}{2n+1} + C \tag{E.172}$$

$$\int_{-1}^{1} x^m P_n(x)\,dx = 0, \quad m < n \tag{E.173}$$

$$\int_{-1}^{1} x^n P_n(x)\,dx = \frac{2^{n+1}(n!)^2}{(2n+1)!} \tag{E.174}$$

$$\int_{-1}^{1} x^{2k} P_{2n}(x)\,dx = \frac{2^{2n+1}(2k)!(k+n)!}{(2k+2n+1)!(k-n)!} \tag{E.175}$$

$$\int_{-1}^{1} \frac{P_n(x)}{\sqrt{1-x}}\,dx = \frac{2\sqrt{2}}{2n+1} \tag{E.176}$$

$$\int_{-1}^{1} \frac{P_{2n}(x)}{\sqrt{1-x^2}}\,dx = \left[\frac{\Gamma\left(n+\frac{1}{2}\right)}{n!} \right]^2 \tag{E.177}$$

$$\int_{0}^{1} P_{2n+1}(x)\,dx = (-1)^n \frac{(2n)!}{2n+2} \frac{1}{(2^n n!)^2} \tag{E.178}$$

Fourier–Legendre series expansion of a function

$$f(x) = \sum_{n=0}^{\infty} a_n P_n(x) \quad (|x| \le 1) \tag{E.179}$$

$$a_n = \frac{2n+1}{2} \int_{-1}^{1} f(x) P_n(x)\,dx \tag{E.180}$$

Limits

$$\lim_{\theta \to \pi} \frac{P_n^1(\cos\theta)}{\sin\theta} = -\frac{(-1)^n}{2} n(n+1) \tag{E.181}$$

$$\lim_{\theta \to \pi} \frac{d}{d\theta} P_n^1(\cos\theta) = \frac{(-1)^n}{2} n(n+1) \tag{E.182}$$

E.3   Spherical harmonics

Notation

θ, ϕ = real numbers; m, n = integers

$Y_{nm}(\theta, \phi)$ = spherical harmonic function

Differential equation

$$\frac{1}{\sin\theta} \frac{\partial}{\partial\theta}\left(\sin\theta \frac{\partial Y(\theta, \phi)}{\partial\theta}\right) + \frac{1}{\sin^2\theta} \frac{\partial^2 Y(\theta, \phi)}{\partial\phi^2} + \frac{1}{a^2}\lambda Y(\theta, \phi) = 0 \tag{E.183}$$

$$\lambda = a^2 n(n+1) \tag{E.184}$$

$$Y_{nm}(\theta, \phi) = \sqrt{\frac{2n+1}{4\pi} \frac{(n-m)!}{(n+m)!}} P_n^m(\cos\theta) e^{jm\phi} \tag{E.185}$$

Orthogonality relationships

$$\int_{-\pi}^{\pi} \int_0^{\pi} Y_{n'm'}^*(\theta, \phi) Y_{nm}(\theta, \phi) \sin\theta \, d\theta \, d\phi = \delta_{n'n} \delta_{m'm} \tag{E.186}$$

$$\sum_{n=0}^{\infty} \sum_{m=-n}^{n} Y_{nm}^*(\theta', \phi') Y_{nm}(\theta, \phi) = \delta(\phi - \phi') \delta(\cos\theta - \cos\theta') \tag{E.187}$$

Specific examples

$$Y_{00}(\theta, \phi) = \sqrt{\frac{1}{4\pi}} \tag{E.188}$$

$$Y_{10}(\theta, \phi) = \sqrt{\frac{3}{4\pi}} \cos\theta \tag{E.189}$$

$$Y_{11}(\theta, \phi) = -\sqrt{\frac{3}{8\pi}} \sin\theta e^{j\phi} \tag{E.190}$$

$$Y_{20}(\theta, \phi) = \sqrt{\frac{5}{4\pi}} \left(\frac{3}{2}\cos^2\theta - \frac{1}{2}\right) \tag{E.191}$$

$$Y_{21}(\theta, \phi) = -\sqrt{\frac{15}{8\pi}} \sin\theta \cos\theta e^{j\phi} \tag{E.192}$$

$$Y_{22}(\theta, \phi) = \sqrt{\frac{15}{32\pi}} \sin^2\theta e^{2j\phi} \tag{E.193}$$

$$Y_{30}(\theta, \phi) = \sqrt{\frac{7}{4\pi}} \left(\frac{5}{2}\cos^3\theta - \frac{3}{2}\cos\theta\right) \tag{E.194}$$

$$Y_{31}(\theta, \phi) = -\sqrt{\frac{21}{64\pi}} \sin\theta \left(5\cos^2\theta - 1\right) e^{j\phi} \tag{E.195}$$

$$Y_{32}(\theta, \phi) = \sqrt{\frac{105}{32\pi}} \sin^2\theta \cos\theta e^{2j\phi} \tag{E.196}$$

$$Y_{33}(\theta, \phi) = -\sqrt{\frac{35}{64\pi}} \sin^3\theta e^{3j\phi} \tag{E.197}$$

Functional relationships

$$Y_{n0}(\theta, \phi) = \sqrt{\frac{2n+1}{4\pi}} P_n(\cos\theta) \tag{E.198}$$

$$Y_{n,-m}(\theta, \phi) = (-1)^m Y_{nm}^*(\theta, \phi) \tag{E.199}$$

Addition formulas

$$P_n(\cos\gamma) = \frac{4\pi}{2n+1} \sum_{m=-n}^{n} Y_{nm}(\theta, \phi) Y_{nm}^*(\theta', \phi') \tag{E.200}$$

$$P_n(\cos\gamma) = P_n(\cos\theta)P_n(\cos\theta')$$

$$+ \sum_{m=-n}^{n} \frac{(n-m)!}{(n+m)!} P_n^m(\cos\theta)P_n^m(\cos\theta')\cos[m(\phi-\phi')] \tag{E.201}$$

$$\cos\gamma = \cos\theta\cos\theta' + \sin\theta\sin\theta'\cos(\phi-\phi') \tag{E.202}$$

Series

$$\sum_{m=-n}^{n} |Y_{nm}(\theta,\phi)|^2 = \frac{2n+1}{4\pi} \tag{E.203}$$

$$\frac{1}{|\mathbf{r}-\mathbf{r}'|} = \frac{1}{\sqrt{r^2 + r'^2 - 2rr'\cos\gamma}}$$

$$= 4\pi \sum_{n=0}^{\infty} \sum_{m=-n}^{n} \frac{1}{2n+1} \frac{r_<^n}{r_>^{n+1}} Y_{nm}^*(\theta',\phi')Y_{nm}(\theta,\phi) \tag{E.204}$$

$$r_< = \min\{|\mathbf{r}|, |\mathbf{r}'|\}, \qquad r_> = \max\{|\mathbf{r}|, |\mathbf{r}'|\} \tag{E.205}$$

Series expansion of a function

$$f(\theta,\phi) = \sum_{n=0}^{\infty} \sum_{m=-n}^{n} a_{nm}Y_{nm}(\theta,\phi) \tag{E.206}$$

$$a_{nm} = \int_{-\pi}^{\pi} \int_0^{\pi} f(\theta,\phi)Y_{nm}^*(\theta,\phi)\sin\theta\,d\theta\,d\phi \tag{E.207}$$

F

Derivation of an integral identity

We wish to evaluate the integral

$$I = \int_{y=-b}^{b} \int_{x=-a}^{a} \frac{dx\,dy}{\sqrt{x^2 + Q^2 y^2}}.$$

Because the integrand is even in both x and y we can write

$$I = 4 \int_{0}^{b} \left(\int_{0}^{a} \frac{dx}{\sqrt{x^2 + Q^2 y^2}} \right) dy.$$

The inner integral may be computed using a tabulated formula from [41]:

$$I = 4 \int_{0}^{b} \ln\left(x + \sqrt{x^2 + Q^2 y^2}\right) \Big|_{0}^{a} dy = 4 \int_{0}^{b} \ln\left(\frac{a + \sqrt{a^2 + Q^2 y^2}}{Qy}\right) dy.$$

Setting $u = Qy/a$, we obtain

$$I = 4\frac{a}{Q} \int_{0}^{Qb/a} \ln\left(\frac{1 + \sqrt{1 + u^2}}{u}\right) du$$

and can employ the identity [41]

$$\ln\left(\frac{1 + \sqrt{1 + x^2}}{x}\right) = \operatorname{csch}^{-1} x$$

to write

$$I = 4\frac{a}{Q} \int_{0}^{Qb/a} \operatorname{csch}^{-1} u \, du.$$

This integral is tabulated in Dwight [52],

$$I = 4\frac{a}{Q} \left[u \operatorname{csch}^{-1} u + \sinh^{-1} u \right] \Big|_{0}^{Qb/a},$$

and the identity $\sinh^{-1} u = \ln(u + \sqrt{u^2 + 1})$ gives

$$I = 4\frac{a}{Q} \left[u \ln\left(\frac{1 + \sqrt{1 + u^2}}{u}\right) + \ln\left(u + \sqrt{u^2 + 1}\right) \right] \Big|_{0}^{Qb/a}.$$

Evaluating and noting that $u \ln u \to 0$ as $u \to 0$, we finally obtain

$$I = 4\frac{a}{Q} \left[Q\frac{b}{a} \ln\left(\frac{a}{bQ} + \sqrt{1 + \left(\frac{a}{bQ}\right)^2}\right) + \ln\left(\frac{bQ}{a} + \sqrt{1 + \left(\frac{bQ}{a}\right)^2}\right) \right].$$

Two special cases are of interest. When $a = b$ we get

$$I = 4\frac{a}{Q}\left[Q\ln\left(\frac{1}{Q} + \sqrt{1 + \frac{1}{Q^2}}\right) + \ln\left(Q + \sqrt{1 + Q^2}\right)\right]. \tag{F.1}$$

When $Q = 1$ this reduces to

$$I = 8a\ln(1 + \sqrt{2}). \tag{F.2}$$

References

[1] Abramowitz, M., and Stegun, I., *Handbook of Mathematical Functions*, Dover Publications, New York, 1965.

[2] Ahmed, S., and Daly, P., *Waveguide Solution by the Finite-element Method*, The Radio and Electronic Engineer, vol. 38, no. 4, pp. 217–223, 1969.

[3] Anderson, J., and Ryon, J., *Electromagnetic Radiation in Accelerated Systems*, Physical Review, vol. 181, no. 5, pp. 1765–1775, May 1969.

[4] Anderson, J. C., *Dielectrics*, Reinhold Publishing Co., New York, 1964.

[5] Appleyard, R., *Pioneers of Electrical Communication — Heinrich Rudolph Hertz*, Electrical Communication, vol. 6, no. 2, pp. 63–77, October 1927.

[6] Arfken, G., *Mathematical Methods for Physicists*, Academic Press, New York, 1970.

[7] Arslanagic, S., Hansen, T. V., Mortensen, N. A., Gregersen, A. H., Sigmund, O., Ziolkowski, R. W., and Breinbjerg, O., *A Review of the Scattering-Parameter Extraction Method with Clarification of Ambiguity Issues in Relation to Metamaterial Homogenization*, IEEE Antennas and Propagation Magazine, vol. 55, no. 2, pp. 91–106, April 2013.

[8] Ayvazyan, V., et al., *Generation of GW Radiation Pulses from a VUV Free-Electron Laser Operating in the Femtosecond Regime*, Physical Review Letters, vol. 88, no. 10, pp. 1–4, 11 March 2002.

[9] Baker-Jarvis, J., Vanzura, E. J., and Kissick, W. A., *Improved Technique for Determining Complex Permittivity With the Transmission/Reflection Method*, IEEE Transactions on Microwave Theory and Techniques, vol. 38, no. 8, pp. 1096–1103, August 1990.

[10] Baker-Jarvis, J., Janezic, M. D., Domich, P. D., and Geyer, R. G., *Analysis of an Open-Ended Coaxial Probe with Lift-off for Nondestructive Testing*, IEEE Transactions on Instrumentation and Measurement, vol. 43, no. 5, pp. 711–718, October 1994.

[11] Balanis, C., *Antenna Theory: Analysis and Design*, 3rd ed., Wiley, Hoboken, NJ, 2005.

[12] Balanis, C., *Advanced Engineering Electromagnetics*, Wiley, New York, 1989.

[13] Baum, C. E., *The Singularity Expansion Method* in *Transient Electromagnetic Fields*, L. B. Felsen, Ed., Springer–Verlag, Berlin, 1976.

[14] Baum, C. E., *Emerging Technology for Transient and Broad-band Analysis and Synthesis of Antennas and Scatterers*, Proceedings of the IEEE, vol. 64, no. 11, pp. 1598–1616, 1976.

[15] Birnbaum, G., and Franeau, J., *Measurement of the Dielectric Constant and Loss of Solids and Liquids by a Cavity Perturbation Method*, Journal of Applied Physics, vol. 20, pp. 817–818, August 1949.

[16] Bittencourt, J. A., *Fundamentals of Plasma Physics*, 3rd ed., Springer–Verlag, New York, 2004.

[17] Blackburn, R., and Wilton, D., *Analysis and Synthesis of an Impedance-Loaded Loop Antenna Using the Singularity Expansion Method*, IEEE Transactions on Antennas and Propagation, vol. 26, no. 1, pp. 136–140, 1978.

[18] Boffi, L., *Electrodynamics of Moving Media*, Sc.D. Thesis, Massachusetts Institute of Technology, Cambridge, 1957.

[19] Bohm, D., *The Special Theory of Relativity*, Routledge, London, 1996.

[20] Boithias, L., *Radio Wave Propagation*, McGraw–Hill, New York, 1987.

[21] Born, M., *Einstein's Theory of Relativity*, Dover Publications, New York, 1965.

[22] Born, M., and Wolf, E., *Principles of Optics*, Pergamon Press, Oxford, 1980.

[23] Boyce, W. E., and DiPrima, R. C., *Elementary Differential Equations*, Wiley, New York, 1977.

[24] Bracewell, R., *The Fourier Transform and Its Applications*, McGraw–Hill, New York, 1986.

[25] Brillouin, L., *Wave Propagation in Periodic Structures; Electric Filters and Crystal Lattices*, Dover Publications, New York, 1953.

[26] Busse, G., Reinert, J., and Jacob, A. F., *Waveguide Characterization of Chiral Material: Experiments*, IEEE Transactions on Microwave Theory and Techniques, vol. 47, no. 3, pp. 297–301, March 1999.

[27] Cabeceira, A. C. L., Barba, I., Grande, A., and Represa, J., *A 2D-TLM Model for Electromagnetic Wave Propagation in Tellegen Media*, Microwave and Optical Technology Letters, vol. 40, no. 5, pp. 438–441, March 2004.

[28] Campbell, G. A., and Foster, R. M., *Fourier Integrals for Practical Applications*, D. Van Nostrand Co., New York, 1948.

[29] Celozzi, S., Araneo, R., and Lovat, G., *Electromagnetic Shielding*, Wiley, New York, 2008.

[30] Chen, L. F., Ong, C. K., Neo, C. P., Varadan, V. V., and Varadan, V. K., *Microwave Electronics: Measurement and Materials Characterization*, John Wiley & Sons, London, 2004.

[31] Chen, K.-M., *Special Topics in Electromagnetics*, National Taiwan University Press, Taipei, 2008.

[32] Chen, K.-M., and Westmoreland, D., *Impulse Response of a Conducting Sphere Based on Singularity Expansion Method*, Proceedings of the IEEE, vol. 69, no. 6, pp. 747–750, 1981.

[33] Chen, K.-M., *A Mathematical Formulation of the Equivalence Principle*, IEEE Transactions on Microwave Theory and Techniques, vol. 37, no. 10, pp. 1576–1580, 1989.

[34] Cheng, D. K., *Field and Wave Electromagnetics*, Addison Wesley, Reading, MA, 1983.

[35] Chew, W. C., *Waves and Fields in Inhomogeneous Media*, Van Nostrand Reinhold, New York, 1990.

[36] Churchill, R. V., Brown, J. W., and Verhey, R. F., *Complex Variables and Applications*, McGraw–Hill, New York, 1974.

[37] Cole, K. S., and Cole, R. H., *Dispersion and Absorption in Dielectrics*, Journal of Chemical Physics, no. 9, vol. 341, 1941.

[38] Collin, R., *Field Theory of Guided Waves*, IEEE Press, New York, 1991.

[39] Collin, R., *Foundations for Microwave Engineering*, 2nd ed., McGraw–Hill, New York, 1992.

[40] Costen, R., and Adamson, D., *Three-dimensional Derivation of the Electrodynamic Jump Conditions and Momentum-Energy Laws at a Moving Boundary*, Proceedings of the IEEE, vol. 53, no. 5, pp. 1181–1196, September 1965.

[41] *CRC Standard Mathematical Tables*, 24th Edition, Beyer, W.H., and Selby, S.M., eds., CRC Press, Cleveland, Ohio, 1976.

[42] Crowgey, B. R., *Rectangular Waveguide Material Characterization: Anisotropic Property Extraction and Measurement Validation*, Ph.D. Dissertation, Michigan State University, 2013.

[43] Cullwick, E., *Electromagnetism and Relativity*, Longmans, Green, and Co., London, 1957.

[44] Dahlquist, G., and Björck, A., *Numerical Methods*, Prentice Hall, Englewood Cliffs, NJ, 1974.

[45] Daniel, V., *Dielectric Relaxation*, Academic Press, London, 1967.

[46] Debye, P., *Polar Molecules*, Dover Publications, New York, 1945.

[47] Dester, G. D., Rothwell, E. J., Havrilla, M. J., and Hyde IV, M. W., *Error Analysis of a Two-layer Method for the Electromagnetic Characterization of Conductor-backed Absorbing Material Using an Open-ended Waveguide Probe*, Progress in Electromagnetics Research B, vol. 26, pp. 1–21, 2010.

[48] Devaney, A. J., and Wolf, E., *Radiating and Nonradiating Classical Current Distributions and the Fields They Generate*, Physical Review D, vol. 8, no. 4, pp. 1044–1047, August 15, 1973.

[49] Dorey, S. P., Havrilla, M. J., Frasch, L. L., Choi, C., and Rothwell, E. J., *Stepped-Waveguide Material-Characterization Technique*, IEEE Antennas and Propagation Magazine, vol. 46, no. 1, pp. 170–175, 2004.

[50] Doughty, N., *Lagrangian Interaction*, Addison Wesley, Reading, MA, 1990.

[51] Du, C.-H., and Liu, P.-K., *Millimeter-Wave Gyrotron Traveling-Wave Tube Amplifiers*, Springer–Verlag, Heidelberg, Germany, 2014.

[52] Dwight, H. B., *Tables of Integrals and Other Mathematical Data*, Macmillan, New York, 1947.

[53] Dwyer, D., Havrilla, M., Hastriter, M., Terzouli, M., and Rothwell E., *Bistatic Scattering from a Resistive Sheet Using a Modified PO Current*, URSI National Radio Science Meeting Digest, Boulder, CO, January 5–8, 2005.

[54] Einstein, A., *Relativity*, Prometheus Books, Amherst, NJ, 1995.

[55] Einstein, A., and Infeld, L., *The Evolution of Physics*, Simon and Schuster, New York, 1938.

[56] Elliott, R., *Electromagnetics*, IEEE Press, New York, 1993.

[57] Elliott, R., *Antenna Theory and Design*, Prentice Hall, Englewood Cliffs, NJ, 1981.

[58] Elliott, R. S., *An Introduction to Guided Waves and Microwave Circuits*, Prentice Hall, Englewood Cliffs, NJ, 1993.

[59] Eyges, L., *The Classical Electromagnetic Field*, Dover Publications, New York, 1980.

[60] Fan, Z., Luo, G., Zhang, Z., Zhou, L., and Wei, F., *Electromagnetic and Microwave Absorbing Properties of Multi-Walled Carbon Nanotubes/Polymer Composites*, Materials Science and Engineering: B, vol. 132, no. 1-2, pp. 85-89, 2006.

[61] Fano, R., Chu, L., and Adler, R., *Electromagnetic Fields, Energy, and Forces*, Wiley, New York, 1960.

[62] Fenner, R. A., Rothwell, E. J. and Frasch, L. L., *A Comprehensive Analysis of Free-Space and Guided-Wave Techniques for Extracting the Permeability and Permittivity of Materials Using Reflection-Only Measurements*, Radio Science, vol. 47, pp. 1004–1016, January 2012.

[63] Feynman, R., Leighton, R., and Sands, M., *The Feynman Lectures on Physics, Vol. 2*, Addison Wesley, Reading, MA, 1964.

[64] Fradin, A. Z., *Microwave Antennas*, Pergamon Press, Oxford, UK, 1961.

[65] Franza, O. P., and Chew, W. C., *Recursive Mode Matching Method for Multiple Waveguide Junction Modeling*, IEEE Transactions on Microwave Theory and Techniques, vol. 44, no. 1, pp. 87–92, 1996.

[66] Freidberg, J., *Plasma Physics and Fusion Energy*, Cambridge University Press, Cambridge, UK, 2007.

[67] Gerald, C. F., *Applied Numerical Analysis*, Addison Wesley, Reading, MA, 1978.

[68] Gessel, G. A., and Ciric, I. R., *Recurrence Modal Analysis for Multiple Waveguide Discontinuities and Its Application to Circular Structures*, IEEE Transactions on Microwave Theory and Techniques, vol. 41, no. 3, pp. 484–490, 1993.

[69] Gessel, G. A., and Ciric, I. R., *Multiple Waveguide Discontinuity Modelling with Restricted Mode Interaction*, IEEE Transactions on Microwave Theory and Techniques, vol. 42, no. 2, pp. 351–353, 1994.

[70] Ghodgaonkar, D. K., Varadan, V. V., and Varadan, V. K., *Free-Space Measurement of Complex Permittivity and Complex Permeability of Magnetic Materials at Microwave Frequencies*, IEEE Transactions on Instrumentation and Measurement, vol. 39, no. 2, pp. 387–394, April 1990.

[71] Gibson, A. A. P., and Dillon, B. M. *The variational solution of electric and magnetic circuits*, Engineering Science and Education Journal, pp. 5–10, February 1995.

[72] Goldberg, D. A., Laslett, L. J., and Rimmer, R. A., *Modes of Elliptical Waveguides: A Correction*, IEEE Transactions on Microwave Theory and Techniques, vol. 38, no. 11, pp. 1603–1608, 1990.

[73] Goodman, J., *Introduction to Fourier Optics*, McGraw–Hill, New York, 1968.

[74] Gradshteyn, I., and Ryzhik, I., *Table of Integrals, Series, and Products*, Academic Press, Boston, 1994.

[75] Haberman, R., *Elementary Applied Partial Differential Equations*, Prentice Hall, Upper Saddle River, NJ, 1998.

[76] Hakki, B. W., and Coleman, P. D., *A Dielectric Resonator Method of Measuring Inductive Capacities in the Millimeter Range*, IRE Transactions on Microwave Theory and Techniques, vol. 8, no. 4, pp. 402–410, July 1960.

[77] Hallén, E., *Theoretical Investigations Into the Transmitting and Receiving Qualities of Antennae*, Acta Universitatis Upsaliensis: Nova acta Regiae Societatis Scientiarum Upsaliensis, Almqvist & Wiksells, 1938.

[78] Hallén, E., *Exact Treatment of Antenna Current Wave Reflection at the End of a Tube-Shaped Cylindrical Antenna*, IRE Transactions on Antennas and Propagation, vol. 4, no. 3, pp. 479–491, 1956.

[79] Hansen, T. B., and Yaghjian, A. D., *Plane-Wave Theory of Time-Domain Fields*, IEEE Press, New York, 1999.

[80] Harrington, R. F., *Matrix Methods for Field Problems*, Proceedings of the IEEE, vol. 55, no. 2, pp. 136–149, 1967.

[81] Harrington, R. F., *Field Computation by Moment Methods*, IEEE Press, New York, 1993.

[82] Harrington, R., *Time Harmonic Electromagnetic Fields*, McGraw–Hill, New York, 1961.

[83] Haus, H. A., and Melcher, J. R., *Electromagnetic Fields and Energy*, Prentice Hall, Englewood Cliffs, NJ, 1989.

[84] Havrilla, M., Bogle, A., Hyde, M., and Rothwell, E., *EM Material Characterization of Conductor Backed Media using a NDE Microstrip Probe*, in Studies in Applied Electromagnetics and Mechanics: Electromagnetic Nondestructive Evaluation (XVI), vol. 38, pp. 210–218, January 2014.

[85] Hellier, C., *Handbook of Nondestructive Evaluation*, McGraw–Hill, New York, 2013.

[86] Hertz, H., *Electric Waves*, MacMillan, New York, 1900.

[87] Hildebrand, F. B., *Methods of Applied Mathematics*, Prentice Hall, Englewood Cliffs, NJ, 1952.

[88] Hoburg, J. F., *Principles of Quasistatic Magnetic Shielding with Cylindrical and Spherical Shields*, IEEE Transactions on Electromagnetic Compatibility, vol. 37, no. 4, pp. 574–579, November 1995.

[89] Hui, H. T., Yung, E. K. N., and Chan, K. Y., *On the Numerical Calculation of the Magnetic Current Frill Model*, Radio Science, vol. 36, no. 6, pp. 1683–1686, November/December 2001.

[90] *IEEE Standard for Definitions of Terms for Antennas; IEEE Std 145-2013*, IEEE Standards Association, 2013.

[91] Ishimaru, A., *Electromagnetic Wave Propagation, Radiation, and Scattering*, Prentice Hall, Englewood Cliffs, NJ, 1991.

[92] Jackson, J., *Classical Electrodynamics*, Wiley, New York, 1975.

[93] Jackson, J. D., and Okun, L. B., *Historical Roots of Gauge Invariance*, Reviews of Modern Physics, vol. 73, no. 3, pp. 663–680, July 2001.

[94] James, G. L., *Geometrical Theory of Diffraction for Electromagnetic Waves*, Peter Peregrinus Ltd., London, 1986.

[95] Ji, Z., Sarkar, T. K., Jung, B. H., and Salazar-Palma, M., *Solving Time Domain Electric Field Integral Equation for Thin-Wire Antennas Using the Laguerre Polynomials* in *Ultra-Wideband, Short-Pulse Electromagnetics 7*, F. Sabath, E. L. Mokole, U. Schenk and D. Nitsch, Ed., Springer, New York, pp. 159–171, 2007.

[96] Jiang, B., Wu, J., and Povinelli, L. A., *The Origin of Spurious Solutions in Computational Electromagnetics*, Journal of Computational Physics, vol. 125, pp. 104–123, 1996.

[97] Johnson, C. C., *Field and Wave Electrodynamics*, McGraw–Hill, New York, 1965.

[98] Jones, D., *The Theory of Electromagnetism*, Pergamon Press, New York, 1964.

[99] Jordan, E. C., and Balmain, K. G., *Electromagnetic Waves and Radiating Systems*, 2nd ed., Prentice Hall, Englewood Cliffs, NJ, 1968.

[100] Kalachev, A. A., Matitsin, S. M., Novogrudskiy, L. N., Rozanov, K. N., Sarychev, A. K., Seleznev, A. V., and Kukolev, I. V., *The Methods of Investigation of Complex Dielectric Permittivity of Layer Polymers Containing Conductive Inclusions* in *Optical and Electrical Properties of Polymers, Materials Research Society Symposia Proceedings*, Emerson, J. A., and Torkelson, J. M., Ed., vol. 214, pp. 119–124, 1991.

[101] Keller, R., and Arndt, F., *Rigorous Modal Analysis of the Asymmetric Rectangular Iris in Circular Waveguides*, IEEE Microwave and Guided Wave Letters, vol. 3, no. 6, pp. 185–187, 1993.

[102] Khalaj-Amirhosseini, M., *Analysis of Lossy Inhomogeneous Planar Layers Using Taylor's Series Expansion*, IEEE Transactions on Antennas and Propagation, vol. 54, no. 1, pp. 130–135, 2006.

[103] King, R. W. P., *Electromagnetic Engineering*, vol. I, McGraw–Hill, New York, 1945.

[104] King, R. W. P., *Transmission-Line Theory*, Dover, New York, 1965.

[105] King, R. W. P., and Harrison, C. W., *Antennas and Waves, a Modern Approach*, MIT Press, Cambridge, MA, 1969.

[106] Klemperer, O., and Barnett, M. E., *Electron Optics*, Cambridge University Press, 3rd ed., London, 1971.

[107] Koledintseva, M. Y., Pommerenke, D. J., and Drewniak, J. L., *FDTD Analysis of Printed Circuit Boards Containing Wideband Lorentzian Dielectric Dispersive Media*, Proceedings of the 2002 IEEE International Symposium on Electromagnetic Compatibility, vol. 2, pp. 830–833, August 2002.

[108] Kong, J., *Electromagnetic Wave Theory*, Wiley, New York, 1990.

[109] Kong, J., *Theorems of Bianisotropic Media*, Proceedings of the IEEE, vol. 60, no. 9, pp. 1036–1046, 1972.

[110] Kraus, J. D., and Marhefka, R. J., *Antennas for All Applications*, McGraw–Hill, New York, 2001.

[111] Kraus, J. D., and Fleisch, D. A., *Electromagnetics with Applications*, WCB McGraw–Hill, Boston, 1999.

[112] Krinsky, S., *The Physics and Properties of Free-Electron Lasers*, Brookhaven National Laboratory Report BNL–69298, June 2002.

[113] Landau, L. D., Lifshitz, E. M., and Pitaevskii, L. P., *Electrodynamics of Theoretical Physics*, 2nd Ed., Butterworth–Heinemann, Oxford, UK, 1984.

[114] Lebedev, I., *Microwave Electronics*, Mir Publishers, Moscow, English translation, 1974.

[115] Lebedev, N., *Special Functions and Their Applications*, Dover Publications, New York, 1972.

[116] LePage, W., *Complex Variables and the Laplace Transform for Engineers*, Dover Publications, New York, 1961.

[117] Leroy, M., *On the Convergence of Numerical Results in Modal Analysis*, IEEE Transactions on Antennas and Propagation, vol. 31, no. 4, pp. 655–659, 1983.

[118] Liao, S. Y., *Microwave Devices and Circuits*, 3rd ed., Prentice Hall, Englewood Cliffs, NJ, 1990.

[119] Lindell, I., Sihvola, A., Tretyakov, S., and Vitanen, A., *Electromagnetic Waves in Chiral and Bi-isotropic Media*, Artech House, Boston, 1994.

[120] Lindell, I., *Methods for Electromagnetic Field Analysis*, IEEE Press, New York, 1992.

[121] Liu, T. K., and Mei, K. K., *A Time-Domain Integral-Equation Solution for Linear Antennas and Scatterers*, Radio Science, vol. 8, nos. 8, 9, pp. 797–804, 1973.

[122] Lorentz, H. A., *Theory of Electrons*, Dover Publications, New York, 1952.

[123] Lucas, J., and Hodgson, P., *Spacetime and Electromagnetism*, Clarendon Press, Oxford, 1990.

[124] MacCluer, C., *Boundary Value Problems and Orthogonal Expansions, Physical Problems from a Sobolev Viewpoint*, IEEE Press, New York, 1994.

[125] Marcuvitz, N., *Waveguide Handbook*, Peter Peregrinus, Ltd., London, 1986.

[126] Marsden, J., and Tromba, A., *Vector Calculus*, W.H. Freeman and Company, San Francisco, 1976.

[127] Maxwell, J., *A Dynamical Theory of the Electromagnetic Field*, Royal Society Transactions, vol. CLV, reprinted in Simpson, T., *Maxwell on the Electromagnetic Field*, Rutgers University Press, New Brunswick, NJ, 1997.

[128] Mei, K., *On the Integral Equations of Thin Wire Antennas*, IEEE Transactions on Antennas and Propagation, vol. 13, no. 3, pp. 374–378, 1965.

[129] Meier, P. J., and Wheeler, H. A., *Dielectric-Lined Circular Waveguide with Increased Usable Bandwidth*, IEEE Transactions on Microwave Theory and Techniques, vol. 12, no. 2, pp. 171–175, March 1964.

[130] Meriam, J. L., *Dynamics*, Wiley, New York, 1978.

[131] Myers, L. M., *Electron Optics, Theoretical and Practical*, D. Van Nostrand Co., New York, 1939.

[132] Miller, A., *Albert Einstein's Special Theory of Relativity*, Springer–Verlag, New York, 1998.

[133] Miller, S. E., *Waveguide as a Communication Medium*, Bell System Technical Journal, vol. 33, no. 6, pp. 1209–1265, November 1954.

[134] Mills, D., *Nonlinear Optics*, Springer–Verlag, Berlin, 1998.

[135] Mittra, R., *Integral Equation Methods for Transient Scattering* in *Transient Electromagnetic Fields*, L. B. Felsen, Ed., Springer–Verlag, Berlin, 1976.

[136] Mo, C., *Theory of Electrodynamics in Media in Noninertial Frames and Applications*, Journal of Mathematical Physics, vol. 11, no. 8, pp. 2589–2610, August 1970.

[137] Moon, P., and Spencer, D., *Field Theory Handbook*, Springer–Verlag, New York, 1971.

[138] Morse, P., and Feshbach, H., *Methods of Theoretical Physics*, McGraw–Hill, New York, 1953.

[139] *Netlib Repository* at http://www.netlib.org/, July 27, 2017.

[140] Nicolson, A. M. and Ross, G. F, *Measurement of the Intrinsic Properties of Materials by Time-domain Techniques*, IEEE Transactions on Instrumentation and Measurement, vol. 19, no. 4, pp. 377–382, November 1970.

[141] O'Dell, T., *The Electrodynamics of Magneto-Electric Media*, North Holland Publishing, Amsterdam, 1970.

[142] Overfelt, P. L., and White, D. J., *TE and TM Modes of Some Triangular Cross-Section Waveguides Using Superposition of Plane Waves*, IEEE Transactions on Microwave Theory and Techniques, vol. 34, no. 1, pp. 161–167, 1986.

[143] Papas, C. H., *Theory of Electromagnetic Wave Propagation*, Dover Publications, New York, 1988.

[144] Papoulis, A., *The Fourier Integral and Its Applications*, McGraw–Hill, New York, 1962.

[145] Patzelt, H., and Arndt, F., *Double-Plane Steps in Rectangular Waveguides and their Application for Transformers, Irises, and Filters*, IEEE Transactions on Microwave Theory and Techniques, vol. 30, no. 5, pp. 771–776, 1982.

[146] Paul, C. R., *Introduction to Electromagnetic Compatibility*, Wiley, New York, 2006.

[147] Pauli, W., *Pauli Lectures on Physics: Volume I. Electrodynamics*, MIT Press, Cambridge, MA, 1973.

[148] Pauli, W., *Theory of Relativity*, Dover Publications, New York, 1981.

[149] Penfield, P., and Haus, H., *Electrodynamics of Moving Media*, MIT Press, Cambridge, MA, 1967.

[150] Peterson, A. F., Ray, S. L., and Mittra, R., *Computational Methods for Electromagnetics*, IEEE Press, New York, 1998.

[151] Phillips, R. M., *History of the Ubitron*, Nuclear Instruments and Methods in Physics Research, vol. A272, pp. 1–9, November 1995.

[152] Pierce, J. R., *Theory and Design of Electron Beams*, Van Nostrand, New York, 1954.

[153] Pocklington, H. C., *Electrical Oscillations in Wires*, Proceedings of the Cambridge Philosophical Society, vol. 9, pp. 324–332, 1897.

[154] Press, W. H., Flannery, B. P., Teukolsky, A. A., and Vetterling, W. T., *Numerical Recipes in FORTRAN 77: the Art of Scientific Computing*, 2nd ed., Cambridge University Press, Cambridge, UK, 1992.

[155] Purcell, E., *Electricity and Magnetism (Berkeley Physics Course, Vol. 2)*, McGraw–Hill, New York, 1985.

[156] Ramo, S., Whinnery, J., and Van Duzer, T., *Fields and Waves in Communication Electronics*, Wiley, New York, 1965.

[157] Ratcliffe, J. A., *An Introduction to the Ionosphere and Magnetosphere*, Cambridge University Press, London, 1972.

[158] Reid, C., *Hilbert*, Springer–Verlag, New York, 1996.

[159] Richmond, J. H., *Scattering by a Dielectric Cylinder of Arbitrary Cross Section Shape*, IEEE Transactions on Antennas and Propagation, vol. 13, no. 3, pp. 334–341, 1965.

[160] Rosser, W. V. G., *Classical Electromagnetism via Relativity*, Plenum Press, New York, 1968.

[161] Rothwell, E., and Gharsallah, N., *Determination of the Natural Frequencies of a Thin Wire Elliptical Loop*, IEEE Transactions on Antennas and Propagation, vol. 35, no. 11, pp. 1319–1324, 1987.

[162] Rothwell, E. J., Baker, J., Chen, K. M., and Nyquist, D. P., *Approximate Natural Response of an Arbitrarily Shaped Thin Wire Scatterer*, IEEE Transactions on Antennas and Propagation, vol. 39, no. 10, pp. 1457–1462, 1991.

[163] Rothwell, E. J., and Sun, W. M., *Time Domain Deconvolution of Transient Radar Data*, IEEE Transactions on Antennas and Propagation, vol. 38, no. 4, pp. 470–475, 1992.

[164] Rothwell, E. J., *Natural-Mode Representation for the Field Reflected by an Inhomogeneous Conductor-Backed Material Layer — TE Case*, Progress in Electromagnetic Research, PIER 63, pp. 1–20, 2006.

[165] Rothwell, E. J., *Exponential Approximations of the Bessel Functions $I_{0,1}(x)$, $J_{0,1}(x)$, $Y_0(x)$, and $H_0^{(1,2)}(x)$, with Applications to Electromagnetic Scattering, Radiation, and Diffraction*, IEEE Antennas and Propagation Magazine, vol. 51, no. 3, pp. 138–147, 2009.

[166] Rumsey, V. H., *Reaction Concept in Electromagnetic Theory*, Physical Review, vol. 94, pp. 1483–1491, June 1954.

[167] Sadiku, M. N. O., *Numerical Techniques in Electromagnetics*, CRC Press, Boca Raton, FL, 1992.

[168] Scanlon, P., Henriksen, R., and Allen, J., *Approaches to Electromagnetic Induction*, American Journal of Physics, vol. 37, no. 7, pp. 698–708, July 1969.

[169] Schelkunoff, S., *On Teaching the Undergraduate Electromagnetic Theory*, IEEE Transactions on Education, vol. E-15, no. 1, pp. 15–25, February 1972.

[170] Schelkunoff, S. A., *Electromagnetic Waves*, D. Van Nostrand, Princeton, NJ, 1943.

[171] Schelkunoff, S. A., *Some Equivalence Theorems of Electromagnetics and Their Applications to Radiation Problems*, Bell System Technical Journal, vol. 15, pp. 92–112, 1936.

[172] Schwartz, M., *Principles of Electrodynamics*, Dover Publications, New York, 1972.

[173] Schwinger, J., DeRaad, L., Milton, K., and Tsai, W., *Classical Electrodynamics*, Perseus Books, Reading, MA, 1998.

[174] Shen, Y., *The Principles of Nonlinear Optics*, Wiley, New York, 1984.

[175] Sihvola, A., *Electromagnetic Mixing Formulas and Applications*, Institution of Electrical Engineers, London, UK, 1999.

[176] Silver, S., ed., *Microwave Antenna Theory and Design*, Dover Publications, New York, 1965.

[177] Simpson, T., *Maxwell on the Electromagnetic Field*, Rutgers University Press, New Brunswick, NJ, 1997.

[178] Smith, D. R., Vier, D. C., Koschny, Th., Soukoulis, C. M., *Electromagnetic Parameter Retrieval from Inhomogeneous Metamaterials*, Physical Review E, vol. 71, pp. 036617-1–11, 2005.

[179] Sommerfeld, A., *Partial Differential Equations in Physics*, Academic Press, New York, 1949.

[180] Sommerfeld, A., *Electrodynamics*, Academic Press, New York, 1952.

[181] Storer, J. E., *Impedance of Thin-Wire Loop Antennas*, Transactions of the AIEE, Part I, vol. 75, pp. 600–619, 1956.

[182] Strang, G., *Linear Algebra and Its Applications*, Academic Press, New York, 1980.

[183] Stratton, J., *Electromagnetic Theory*, McGraw–Hill, New York, 1941.

[184] Stratton, J. A., and Chu, L. J., *Diffraction Theory of Electromagnetic Waves*, Physical Review, vol. 56, pp. 99–107, 1939.

[185] Stutzman, W. L., *Polarization in Electromagnetic Systems*, Artech House, Boston, 1993.

[186] Stutzman, W. L., and Thiele, G. A., *Antenna Theory and Design*, Wiley, New York, 1981.

[187] Swaminathan, M., Arvas, E., Sarkar, T. K., and Djordjevic, A. R., *Computation of Cutoff Wavenumbers of TE and TM Modes in Waveguides of Arbitrary Cross Sections Using a Surface Integral Formulation*, IEEE Transactions on Microwave Theory and Techniques, vol. 38, no. 2, pp. 154–159, 1990.

[188] Tai, C.-T., *Generalized Vector and Dyadic Analysis*, IEEE Press, New York, 1997.

[189] Tai, C.-T., *Dyadic Green's Functions in Electromagnetic Theory*, 2nd Ed., IEEE Press, New York, 1994.

[190] Tai, C.-T., *A Study of the Electrodynamics of Moving Media*, Proceedings of the IEEE, pp. 685–689, June 1964.

[191] Tai, C.-T., *On the Presentation of Maxwell's Theory*, Proceedings of the IEEE, vol. 60, no. 8, pp. 936–945, August 1972.

[192] Taylor, J. R., *An Introduction to Error Analysis*, University Science Books, Mill Valley, CA, 1982.

[193] Tellegen, B., *The Gyrator, a New Electric Network Element*, Philips Research Reports, vol. 3, pp. 81–101, April 1948.

[194] Tesche, F. M., *The Far-Field Response of a Step-Excited Linear Antenna Using SEM*, IEEE Transactions on Antennas and Propagation, vol. 23, no. 6, pp. 834–838, 1975.

[195] Tijhuis, A. G., Zhongqiu, P., and Rubio Bretones, A., *Transient Excitation of a Straight Thin-Wire Segment: A New Look at an Old Problem*, IEEE Transactions on Antennas and Propagation, vol. 40, no. 10, pp. 1132–1146, 1992.

[196] Tolstoy, I., *Wave Propagation*, McGraw–Hill, New York, 1973.

[197] Trans-Tech, http://www.skyworksinc.com/Products_TechnicalCeramics.aspx, July 27, 2017.

[198] Tricomi, F. G., *Integral Equations*, Dover Publications, New York, 1985.

[199] Truesdell, C., and Toupin, R., *The Classical Field Theories* in *Encyclopedia of Physics*, S. Flugge, Ed., vol. 3, Part 1, Springer–Verlag, Berlin, 1960.

[200] Tsai, L. L., *A Numerical Solution for the Near and Far Fields of an Annular Ring of Magnetic Current*, IEEE Transactions on Antennas and Propagation, vol. 20, no. 5, pp. 569–576, 1972.

[201] U.S. Army Corps of Engineers, *Electromagnetic Pulse (EMP) and TEMPEST Protection for Facilities*, Pamphlet No. 1110-3-2, December 1990.

[202] Van Bladel, J., *Electromagnetic Fields in the Presence of Rotating Bodies*, Proceedings of the IEEE, vol. 64, no. 3, pp. 301–318, March 1976.

[203] Van Bladel, J., *Electromagnetic Fields*, Hemisphere Publishing, New York, 1985.

[204] Van Bladel, J., *Lorenz or Lorentz*, IEEE Antennas and Propagation Magazine, vol. 33, no. 2, p. 69, April 1991.

[205] Veysoglu, M. E., Ha, Y., Shin, R. T., and Kong, J. A., *Polarimetric Passive Remote Sensing of Periodic Surfaces*, Journal of Electromagnetic Waves and Applications, vol. 5, no. 3, pp. 267–280, 1991.

[206] Visser, H. J., *Antenna Theory and Applications*, John Wiley & Sons, New York, 2012.

[207] von Aulock, W. H., Ed., *Handbook of Microwave Ferrite Materials*, Academic Press, New York, 1965.

[208] Wait, J. R., *Electromagnetic Waves in Stratified Media*, Pergamon Press, Oxford, UK, 1970.

[209] Wait, J. R., and Hill, D. A., *Electromagnetic Shielding of Sources Within a Metal-Cased Bore Hole*, IEEE Transactions on Geoscience Electronics, vol. 15, no. 2, pp. 108–112, April 1977.

[210] Wang, C.-C., *Mathematical Principles of Mechanics and Electromagnetism; Part A: Analytical and Continuum Mechanics*, Plenum Press, New York, 1979.

[211] Weigelhofer, W. S., and Hansen, S. O., *Faraday and Chiral Media Revisited–I: Fields and Sources*, IEEE Transactions on Antennas and Propagation, vol. 47, no. 5, May 1999.

[212] Weir, W. B., *Automatic Measurement of Complex Dielectric Constant and Permeability at Microwave Frequencies*, Proceedings of the IEEE, vol. 62, no. 1, pp. 33–36, January 1974.

[213] Weyl, H., *Space–Time–Matter*, Dover Publications, New York, 1952.

[214] Whitaker, S., *Intoduction to Fluid Mechanics*, Prentice Hall, Englewood Cliffs, NJ, 1968.

[215] Yaghjian, A. D., *Electric Dyadic Green's Functions in the Source Region*, Proceedings of the IEEE, vol. 68, pp. 248–263, 1980.

[216] Zhang, S., and Jin, J., *Computation of Special Functions*, Wiley, New York, 1996.

[217] Zhdanov, M. S., *Geophysical Electromagnetic Theory and Methods.* Elsevier, Amsterdam, 2009.

[218] Zierep, J., *Similarity Laws and Modeling*, Marcel Dekker, New York, 1971.

Index